GEOTECHNICAL SPECIAL PUBLICATION NO. 233

FROM SOIL BEHAVIOR FUNDAMENTALS TO INNOVATIONS IN GEOTECHNICAL ENGINEERING

HONORING ROY E. OLSON

SPONSORED BY

The Geo-Institute of the American Society of Civil Engineers

EDITED BY

Magued Iskander, Ph.D., P.E., F.ASCE
John E. Garlanger, Ph.D., P.E., M.ASCE
Mohamad H. Hussein, P.E., M.ASCE

Published by the American Society of Civil Engineers

Library of Congress Cataloging-in-Publication Data

From soil behavior fundamentals to innovations in geotechnical engineering : honoring Roy E. Olson / sponsored by the Geo-Institute of the American Society of Civil Engineers ; edited by Magued Iskander, John E. Garlanger, Mohamad H. Hussein.
pages cm. -- (Geotechnical special publication; no. 233)
Includes bibliographical references.
ISBN 978-0-7844-1326-5
1. Soil mechanics. 2. Geotechnical engineering. I. Olson, Roy E. (Roy Edwin) II. Iskander, Magued. III. Garlanger, John Ernest, 1943- IV. Hussein, Mohamad H. V. American Society of Civil Engineers. Geo-Institute.
TA710.F76 2014
624.1'5136--dc23
2014000231

Published by American Society of Civil Engineers
1801 Alexander Bell Drive
Reston, Virginia, 20191-4382
www.asce.org/bookstore | ascelibrary.org

Errata: Errata, if any, can be found at http://dx.doi.org/10.1061/9780784413265

ISBN 978-0-7844-1326-5
Manufactured in the United States of America.

20 19 18 17 16 15 14 1 2 3 4 5

Contents

Selected Papers Authored and Co-Authored by Roy Olson

All reproduced with permission

Contributed Papers

Forensics

Soil Properties

Foundations

Preface

Sponsored by the Geo-Institute of the American Society of Civil Engineers (ASCE), this Special Geotechnical Publication (GSP) honors Professor Roy E. Olson, Ph.D., P.E., NAE, Distinguished Member ASCE, for his distinguished career as a master of the art of teaching the principles and advancing the practices of geotechnical engineering. Through more than 50 years of excellence in the fields of teaching, research and consulting, initially at the University of Illinois at Urbana-Champaign and later at the University of Texas at Austin, Professor Olson has mentored hundreds of young civil engineers who have become leaders in the geotechnical community. His many awards, including the Huber Research Prize, the Croes Medal, the Norman Medal, and the ASTM Hogentogler Award, recognize his many contributions to the understanding of soil behavior and to geotechnical analysis and foundation design. The list of his published papers, which are included herein, give some evidence of the scope of his research and consulting projects.

Professor Olson is a recognized leader within the Geo-Institute and has served as Chair of the Soil Properties Committee, the Research Committee, and the Awards Committee. He also served on the Executive Committee for 10 years, the Board of Trustees of the Deep Foundations Institute, and the UK's Institution of Structural Engineers. He is a unique individual with a wry sense of humor and keen wit. His knowledge and his dedication to the profession has been an inspiration to all who know him.

Professor Olson is proud of the accomplishments and contributions made to the civil engineering profession by many of his former students, many of whom have contributed to this GSP. A list of his Masters and Ph.D. students are included herein.

This Geotechnical Special Publication contains 30 technical papers reflecting the theme of "From Soil Behavior Fundamentals to Innovations in Geotechnical Engineering". The papers contained herein are divided between contributions by invited authors and those accepted following a general call for abstracts. It also includes 21 of Professor Olson's previously published papers selected by the editors for inclusion in this volume. The papers in this GSP cover a wide range of topics including soil properties, geotechnical analysis, risk analysis, foundation design, field measurements, and case histories.

The editors thank the authors for their contributions to the civil engineering technical literature. Following ASCE-GI publishing standards and requirements, the included papers were accepted following a rigorous review process. Each paper was independently and expertly reviewed by at least two anonymous reviewers. Discussions of the papers included in this GSP can be submitted to the ASCE Journal of Geotechnical and Geoenvironmental Engineering. All papers are eligible for ASCE-GI awards.

The editors and the authors thank the reviewers of the submitted manuscripts in acknowledgement of their constructive comments and valuable suggestions. The following are the names of the reviewers who participated in this endeavor by offering their expertise and volunteering their time, often under demanding deadlines, with apologies to any whose name may have been inadvertently overlooked when compiling the list below:

LIST OF REVIEWERS

The editors are grateful to the following 88 individuals for working tirelessly under a tight schedule to peer review the papers included for publication in this Geotechnical Special Publications.

Mustapha Abboud
Cem Akguner
Joram Amir
Roger Archabal
Dipanjan Basu
Robert Bea
Craig Benson
Dale C. Biggers
Brucc Bodncr
Dan Brown
Aaron Budge
Paul Bullock
Allen Cadden
William (Billy) Camp
Michael Carter
David Chapman
Les Chernauskas
Manoj Chopra
José LM Clemente
Paul J. Cosentino
David Crapps
Alan R. Crumley
Don Degroot
Bengt Fellenius
Sean Ferguson
Luis O. García
John E. Garlanger
Robert Gilbert
Russell Green
Edward Hajduk
Pat Hannigan
Rodrigo Herrera
Wing Heung
Anwar Hirany
David Horhota
Jim Horton
Sandra Houston
James Hussin
Said Iravani
Bill Isenhower
Larry Jones
Mike Justason
Edward Kalajian
Van E. Komurka
Gary L. Kuhns
Jim Laier
Chris Lewis
Garland Likins
William Likos
Erik Loehr
Barbara Luke
Larry Madrid
Ray Mankabadi
Antonio Marinucci
Andrew Martin
Rainer Massarsch
Paul Mayne
Bill McCarron
John McCartney
Ross McGillivray
Greg McNulty
Gholamreza Mesri
Steve Meunier
Robert Mokwa
Shadi Najjar
KK Phoon
Mark Randolph
Alan Rauch
Frank Rausche
Fred Rhyner
Tom Richards
Lance Roberts
Brent Robinson
Kyle Rollins
Amr M. Sallam
Vernon R. Schaefer
John Schmertmann
Clayton Signor
John Small
Kenneth Sorensen
Robert F. Stevens
Muhannad Suleiman
Hisham Sunna
Kamal Tawfiq
Robert W. Thompson III
Frank Townsend
Sherif Wissa
Michael Woodward

This Geotechnical Special Publication is the product of a collective effort of a number of individuals within the Geo-Institute to honor one of its eminent members in appreciation for his role as teacher and mentor, researcher and practitioner.

The editors have been careful in compiling the lists of students who did Masters and Ph.D. dissertations under the direction of Professor Olson. We apologize for any names inadvertently omitted from this list.

The Editors,

Magued Iskander, Ph.D., P.E.
New York University
Brooklyn, New York
Iskander@poly.edu

John E. Garlanger, Ph.D., P.E.
Ardaman & Associates, Inc.
Orlando, Florida
jgarlanger@ardaman.com

Mohamad Hussein, P.E.
GRL Engineers, Inc.
Orlando, Florida
MHGRL@pile.com

Abridged Autobiography
Roy E. Olson

EARLY INFLUENCES

Sir Isaac Newton wrote “if I have seen further, it is by standing on the shoulders of giants”. I don’t recall standing on the shoulders of giants but, like us all, my path in life was influenced by those who came before me.

My parents grew up poor. My father left high school after his junior year, joined the Navy, and became a medical corpsman with an understanding that the Navy would help him finish high school and then send him through medical school to become a navy doctor. My mother finished nurse’s training and then joined the Navy. They married in about 1929 and were immediately discharged from the Navy because my mother was an officer, my father was enlisted, and fraternizing was prohibited. The collapse of the economy in 1929 left them with no assets, few job opportunities, and no government help. They survived by being tough and expecting nothing from anyone. I came along in 1931 and my sister followed in 1933. My mother gave up her job as a nurse and my father started a company that sold construction equipment. A run on the banks in the 1930s caused President Roosevelt to close the banks and my father lost the few assets he had. We endured and the experience taught me that you have to be strong to survive.

EDUCATION

Push to Succeed

I had an undistinguished record in grade school and middle school. After I transferred to a high school in a different neighborhood and started putting forth some effort, my grades improved. I was the star of some classes, like drafting, geometry, business law and English and survived algebra and physics. I remember having an overpowering desire to understand the sciences, especially geology and chemistry. However, I decided to attend the University of Minnesota in the “Institute of Technology” and study civil engineering.

As a freshman, I was required to attend a seminar where a faculty member informed us that only 20% of the incoming freshman would ever get a degree in engineering and, indeed, everyone I knew as a freshman disappeared.

I put forth minimal efforts academically and, instead, went out for the boxing team (lost all of my fights), took piano and ballroom dancing lessons, and started squeezing in a diversity of courses that were not part of a civil engineering curriculum. Grading was considerably tougher then and I didn’t get an A in any class until my third year. In one senior class on geometric highway design, with about 35 students, I ranked number one in the class and got a B!

With that particular professor, I decided that the only way to get an A was to learn more about the topic than he did and sit in the front row and correct him when he made errors (I got the only A in the class). I took more than the required number of courses in civil engineering, and substituted other courses to satisfy the accreditation board and inserted numerous courses in other areas.

My advisor assured me that I would never get a degree in civil engineering but I told him that I was interested in an education, not a piece of paper, and I had no intention of ever practicing civil engineering anyway. Nevertheless, after four years, I ranked 3rd in my class with a GPA slightly less than B. A friend, who was 1st in our class, had just barely above a B average. My, how times have changed!

I remained unimpressed with most of the faculty and with the whole curriculum (why would every student in the Institute have to take Semi Micro Qualitative Analysis?). I think it was in my third year that I got so displeased that I tried to lead a student strike in the civil engineering department. The strike fizzled but I noticed that the thing I was most unhappy about was changed before the next year started.

Master's Degree

We had a five-year BS degree in civil engineering with a requirement of 175 credit-hours (most programs are now around 125 credit-hours). Late in my fourth year, my advisor, Prof. Miles Kersten, informed me that the *Minnesota Surveyors and Engineers Society* had just started a scholarship program and he was concerned that if the university didn't come up with outstanding candidates, then there might not be a second year of scholarships. He asked me to apply for a fellowship and pointed out that I could graduate in four years instead of five, receiving a BS degree not designated by department, if I agreed to take the senior courses as a graduate student while also taking graduate courses and qualifying for an accredited M.S. degree in Civil Engineering. I won the graduate fellowship and took 21 semester hours of courses in my first semester (36 hours per week in class), receiving one B and six A's. I continued taking a diversity of courses and loved the relative freedom of being a graduate student and exploring things that seemed interesting. In retrospect, I regret not taking more classes in mathematics and mechanics but it was clear to me that there wasn't much you could do with knowledge in either area at that time, i.e., before digital computers appeared.

Ph.D. Studies

After excelling in the Master's program at the University of Minnesota, I wondered if I was just a "big frog in a little pond" or someone who would accomplish something in life. I decided to seek a Ph.D. degree from the best and most competitive program I could find. I literally wanted a program where the faculty would blow me away with their brilliance and I would have to struggle to survive. After months of reading college catalogs and interviewing people, it was clear that the University of Illinois best fit my interests. It was ranked number one in the U.S. in Civil Engineering but I was more interested in their geology and agronomy departments.

My time as a Ph.D. student at Illinois was a mixture of good and bad. The dichotomy between a sophisticated university faculty and elite student body on the one hand, and a rural local population on the other, meant that there was little to do except study; although perhaps this was an advantage. Besides studying soils courses in the civil engineering department, I minored in agronomy (soil physics and colloid chemistry), geology (finished my 23rd and final course in geology), and chemistry (physical chemistry, radiology, quantum mechanics).

The soils faculty in the civil engineering department at Illinois consisted of a structural engineer (Ralph Peck), a geologist (Don Deere), a forester (Tom Thornburn), and a "general" engineer (Herb Ireland). It was still the early days of what we now call geotechnical engineering and most of the programs around the country had self-taught geotechnical faculty, mostly with no more than an M.S. degree. Terzaghi taught what he termed "engineering geology" in the spring each year but otherwise was out of town and was not interested in advising a Ph.D. student.

During my first semester at Illinois, I decided to transfer to MIT and study under Donald Taylor because we had used his book on soil mechanics at Minnesota and I thought it was a marvelous text, but he died in December of my first year at Illinois (1955). I couldn't think of any of the other programs I had examined that would have offered the challenging program I was seeking.

To my surprise, the most positive aspect of the graduate program at Illinois was not the faculty but was rather my fellow graduate students. I grew up in a community dominated by descendants of Northern Europeans and, after WW II, the wartime propaganda created the impression that Americans were innately superior to everyone else. I expected to compete with other U.S. students but not with students from other countries. To my surprise, there were many students from around the world who were friendly, very bright, and highly competitive, even though they were struggling with a new culture and usually were not comfortable with the English language. I developed friendships with students from Canada, Philippines, France, Norway, Sweden, Egypt, China, Poland, Russia, and a variety of other countries and developed an enriched understanding and respect for other cultures.

Professor Peck had a wide circle of international friends and consulting clients, and many of them visited regularly. I was introduced to essentially all of them. I had extended conversations (up to a day) with Karl Terzaghi, Alex Skempton, Alan Bishop, David Henkel, Stan Wilson, Laurits Bjerrum, M. Juul Hvorslev, Ken Roscoe, Peter Wroth, and too many others to mention. Those visits gave me a much better appreciation for the world-wide community in geotechnical engineering and I owe Ralph Peck a lot for exposing me to such a varied group of great people.

Summer Jobs

I was the only student I knew at the time who never stayed in town during the summer. One summer I worked in the chemistry lab, the materials testing lab, and did field work for a testing company in St. Paul, Minnesota. I spent another summer working in northern Greenland (Thule) as a civilian scientist attached to the First Engineers Arctic Task Force and working at the edge of 700,000 square miles of ice with a maximum thickness then estimated to be five miles, and stand on top of 2000 feet of permafrost. I also worked summers for the geotechnical engineering firms of Dames & Moore (in Los Angeles, San Francisco, Portland, and Seattle) and Woodward, Clyde, Sherard & Associates (in Denver).

The summer work experiences were incredible. I got to meet a diversified group of important people out in the "real world" and everyday seemed to bring new challenges. Geotechnical problems differed significantly from one region to another and the cultures of different offices of the same firm varied widely. Most of my assignments required a level of technical knowledge that neither I nor anyone else in the office had. I learned that what was important was to have a top notch, and diversified, educational background and the ability to pull out of it the things that could be assembled to produce a useful result. I learned never to pass up an opportunity to learn new things, no matter how alien the topic was. Above all else, I learned to think "out of the box" and be enthusiastic about attacking problems for which I had no previous experience. I had to be self-confident and be able to defend my work. When I did well in dealing with the challenges, I was given more challenges and soon found myself serving as an advisor to senior engineers on even more difficult projects. I was young, enthusiastic, single, and ready to go anywhere and take on any problem whether I was qualified or not, and usually I wasn't. It was, perhaps the greatest period of my life. Beyond technical work, I got to climb mountains from southern California to Washington State, and throughout Colorado.

Most of the engineers I worked with were self-taught. Many lacked understanding of even rudimentary aspects of geotechnical engineering, like the principle of effective stress. The situation was understandable because universities were not turning out a significant number of qualified geotechnical engineers. In a way these observations were exciting because they meant that life in engineering practice held unlimited opportunities for me and others like me in the new field of geotechnical engineering.

One of my summer employers offered to have me start up a new office for the firm if I stayed with them for another six months. I was sorely tempted. However, during the week when I was made that offer, I received a research grant from the National Science Foundation. The grant would pay me half time for two years and allow me to hire another student to work for me. In addition, the University of Illinois offered me a half-time appointment as an "instructor".

NSF had morphed from the office of Naval Research only a few years before and was just beginning to develop their programs. The result was the overhead rate was only 5% and there were no restrictions whatsoever as to what I could do with the money. It was a marvelous opportunity and I reluctantly returned to academia.

Grammar

In public school I thought spelling was boring and grammar irrelevant. Because the technical content of my reports put me at the top of my classes, I "blew off" the "technical" aspects of the English language. The result was a lot of critical comments on my term papers about mysterious things like *dangling participles*, *gerunds*, and *split infinitives*. My problem with grammar was amplified when I had to pass exams in German and French to finish my Ph.D. degree. I never learned either language properly, although I read much of Terzaghi's first book, *Erdbaumekanic auf bodenphysikalischer Grundlage*, translated his entire third book, *Theorie der Setzung von Tonschichten*, and read numerous papers in German and French. I regretted my decision not to start taking a foreign language in seventh grade - it would have made my life a lot easier.

PERSONAL LIFE

Sports

I engaged in competitive sports activities essentially every day starting in grade school and extending into my early faculty years. There was touch football in the fall; basketball, hockey, skiing, boxing, and bowling in the winter; softball in the spring, and swimming and mountain climbing in the summer. However, some unpleasant experiences in the boxing ring convinced me that my future was likely to be more intellectual than athletic, and certainly less painful.

I did enjoy the challenges of climbing and climbed many trees as a child, roamed the steep roof of our house to my parent's horror, and spent years hiking and climbing mountains. I remember numerous occasions in which I found myself in a place where I could climb up no further and had no apparent way of getting down. I wondered, at least momentarily, how anyone could be so stupid as to get into such a position. I make these comments only because I suspect that there will be readers who have had similar experiences.

First Marriage

The last week that I worked in Denver, I met a girl who was enjoying life on her own but thought it was time for her to get married and settle down. She came from a highly social group in suburban New York and I thought perhaps she could round off some of my rough edges. She, in turn, detested the artificial life in "high society" and wanted someone who would be successful but not so politically correct. We were married a year later. She was a strong-willed person with a personality to match. She decreed that she was in charge of food, clothing, decorating, children, etc., and I was in charge of the structure of any house, the yard, cars, and, most importantly, my job. She stated that job selection was for me alone and she would go anywhere I needed to go (even Champaign-Urbana). After having three children and a pleasant and, doubtless, permanent marriage, she died suddenly as a result of a transfusion with blood containing the hepatitis virus. I had no clue as to how to do anything inside the house and could write a lot about mishaps that were hilarious to neighbors who were trying to help. I wasn't cut out to cook, raise children, or do other domestic things so the loss of my wife was a disaster for me and for my children.

Second Marriage

Several years later, I was introduced to the daughter of a civil engineer. She was a remarkably talented person who played violin in the Denver Symphony orchestra and sang opera in the Boston Opera company. Sadly, I only heard her sing once, at a football game. As a coloratura soprano, she hit the high note in the Star Spangled Banner like no one I ever heard before and drowned out everyone else for substantial distances. We seemed to think that our diverse backgrounds would broaden both of us, so we got married. After sixteen years of conflict, we gave up and divorced. During our marriage, my three children reportedly held a private meeting in which they decided that if ten years of college (mine) meant that you had to work seven days per week and every evening, and if getting B.S. and M.S. degrees in the "arts" meant that you might behave like their current step mother, then they were better off skipping college.

Third Marriage

After being single for a few years, I started taking ballroom dancing lessons again. I allegedly had natural talent in dancing (what else can an instructor say when you are a source of income). I met my third wife in a dancing class and we regularly learned steps faster than anyone else in the class and certainly moved around the floor well. In spite of her being a proud Texan and me a "Yankee" from Minnesota, we get along well (she tolerates my views). Our dancing days are now over as a result of pain in our feet, shoulders, and almost everywhere else as we passed into "old age" but we remember those days fondly.

FACULTY APPOINTMENT AT ILLINOIS

I spent five years in the Ph.D. program at Illinois. In the fall of 1960, I informed my advisor (Ralph Peck) that I was about to start a series of courses in nuclear physics. In turn, he informed me that I would write a dissertation about whatever I was doing at the time and then continue my research as a faculty member. I followed his dictum only to discover that I was quickly buried in work as a faculty member and my days of exploring the unknown were over. I was paid half time for several years out of university academic funds. The rest came from my research projects where I supported up to fourteen research assistants. My teaching load was four courses per semester and as a half time employee I taught "only" two courses per semester. I was an assistant professor for two years and an associate professor for three and became a full Professor in 1965.

My problems with English grammar returned to haunt me as a young faculty member. Eventually I bought the Federal book on technical writing and studied it. I also learned that Dr. Peck was a grammarian and had numerous discussions with him about English grammar. I mention technical writing in case this essay is read by a student who thinks his/her advisor is too tough on writing style as well as errors in using nouns as adjectives, splitting infinitives, and dangling participles. One study published decades ago surveyed engineering graduates and found that their main weaknesses and need for better education were writing and speaking, not technical design.

I spent ten productive years on the faculty at Illinois. It was an exhilarating, if competitive, even combative, and strenuous time and I learned a lot. However, I had grown up in a city and I disliked life in central Illinois. My second wife had conned me into believing that she liked Champaign-Urbana. However, after we were married, she was soon telling me that she was going to move somewhere else whether I did or not. I was long since ready to move so I started looking for other opportunities.

FACULTY APPOINTMENT AT TEXAS

In the spring of 1970, I went for an interview at the University of Texas and, in spite of a salary offer less than I had at Illinois, I accepted a Professorship and moved to Austin. It was difficult to leave behind a fabulous laboratory that I had equipped using research funds, and which I would never be able to duplicate. The University of Texas was in a period of rapid development that was so substantial that it was written up in Time magazine. However, I found the university to be well behind both of the previous universities in my life and there was little possibility of rebuilding the research program I had at Illinois. Nevertheless, we had some outstanding students and we recruited new and talented faculty throughout the Department.

It was a joy to participate in a period where the civil engineering graduate program at Texas moved up in national rankings from somewhere lower than 10th, to fourth. The probabilities of passing Illinois and Berkeley were not very high because of their long traditions of excellence so we probably did as well as possible.

Moving to Texas also gave me an opportunity to move into more practical aspects of geotechnical engineering and to start living in a more vibrant community. I had hoped to get involved in some of the kinds of design work that I had encountered during my summer jobs. However, local enthusiasm seemed to wane when I became an expert witness in five lawsuits against a local geotechnical engineer in a period of about two years. All of the suits were either won or settled to my client's satisfaction. There were a number of failures of slopes, retaining walls, and foundations, and that led to interesting and educational investigations, but my testimony in jury trials didn't help my popularity with local engineers. The lawsuits did give me the opportunity to work with the nemeses of engineers, *lawyers*, and develop an appreciation for how they work, and how their code of ethics differs from those of engineers, but to find that both engineers and lawyers are ethical in their own ways.

WORK WEEK

I learned to log my time when I worked for geotechnical firms, and that led me to log my time up to this very day. At the end of each year, I wrote a review of the previous year and decided whether I had invested my time wisely or not. For decades, I averaged at least 60 hours per week and one year averaged 68 hours/week. That didn't leave a lot of time for non-technical activities and that was a problem for my family.

AMERICAN SOCIETY OF CIVIL ENGINEERS

During my tenure at Illinois, technical activities in ASCE were carried out in technical divisions, ours being the *Soil Mechanics and Foundations Division* (SMFE). Divisions were administered by so-called executive committees. My interest was in research, teaching, and limited consulting activities, not society activities, and I declined invitations to join committees in ASCE, HRB, and ASTM. I was finally pressed into joining the ASCE *Soil Properties Committee* but at our first committee meeting I discovered I was the only member present and that I was the chair of the committee and, in addition, the committee was responsible for two sessions at the ASCE annual convention in a few months. The first session had been developed and I thought it was unimpressive. I was generously given a week or two to pick a topic for the second session and find speakers. I organized what I thought was a much better second session but then the executive committee cancelled the session I had organized due to lack of a time slot. I ended up chairing a session with speakers who often left me unimpressed. One speaker droned on well past his assigned time, covering trivia of no interest and refusing to leave so I finally got him in a headlock and dragged him from the stage, to cheers from the 200+ members of the audience. My impression was that the division needed some new blood.

Executive Committee

The Executive Committee apparently decided the same thing so within a few months I was moved out of my committee to become the secretary of the Executive Committee. As secretary, I wrote essentially all of the letters going out from the Executive Committee. We fired some chairs who hadn't done anything (I was told) in many years and, as the person who wrote the letters, I was the "hatchet man". In the days of manual typewriters and snail mail, I took on more than I wanted and perhaps got a reputation in some circles that I never lived down.

Division Name

At one of our first meetings, Jim Mitchell stated that the faculty at Berkeley was not pleased with the name of our division. I shared that view and thought the name was too narrow, apparently being a translation of the title of Terzaghi's 1925 book, *Erdbaumechanik auf bodenphysikalischer Grundlage* (loosely translated as *Fundamentals of Earth Construction Mechanics and Soil Physics*). The English name appeared at the time of the first international conference for the *International Society of Soil Mechanics and Foundation Engineering* (ISSMFE). The name seemed to leave out the possibility of including other areas, e.g. engineering geology, rock mechanics, geosynthetics, geo-environmental engineering, and soil dynamics. Jim and I worked as a subcommittee of two and came up with two names, *Geotechnic* and *Geotechnical Engineering*. The names did not exist in the unabridged Oxford Dictionary of the English Language so we had the right to define the names as we pleased. After two votes, ASCE members agreed on the name *Geotechnical Engineering*. We defined it to mean everything that geotechnical engineers did, and ignored the circular definition. Some friends have chosen to simply say they practice *Geotechnic*.

Management Group

In those days, members of the *Technical Activities Committee* (TAC) of the ASCE *Board of Direction*, served as liaisons with the technical divisions and that required individual members of TAC to attend a number of executive committee meetings in the divisions they monitored. That task was onerous so the decision was made to collect divisions into groups called *Management Groups* and the TAC member then met with the management group. I had served six years on our Executive Committee and retired but was elected to be our Manager so I served another four years as manager. The resulting ten years that I spent as secretary, member, and manager of our division were certainly the most interesting years of my involvement in societal activities. I met large numbers of wonderful people and developed many friendships.

As a member of a Management Group, I also met with TAC and served on their committees on *Technical Sessions* and on *Publications*. It was illuminating to find out that nine of the top ten best-setting ASCE publications were in geotechnical engineering and the only "outsider" was a pamphlet costing $0.25. We surveyed attendance at our technical sessions at several ASCE annual meetings where all divisions participated, and discovered that about 90% of the total ASCE attendance was in our sessions. In the first technical session that I chaired, we had more than 200 people in our session and the next session in that room had six people, all of them speakers.

At one meeting of the committee on technical sessions, there were some rather bitter complaints from divisions which were assigned sessions at times like Monday morning and Friday afternoon. The staff person who arranged the times was surprisingly blunt. He announced that he gave Geotech whatever we wanted and then fit other divisions in where there was space, and informed them that when they had attendance like we had, they would get better times.

EVOLUTION

I first became interested in geotechnical engineering because it was a link between geology and engineering. However, both faculty and fellow students assured me that the field had no future and geotechnical work would continue to be performed by structural engineers or general civil engineers.

When I became a faculty member, my major interest was research. At that time, it seemed like you could choose to work in any area of geotechnical engineering and do something worthwhile. I was wildly enthusiastic and energetic so I wrote multiple proposals, all of which were funded. The result was that I had more money than I could spend, but not nearly enough time to supervise the work properly and thus some of research was not as fruitful as it could have been. As a result, I cut back on research to try to balance it with the availability of qualified students.

During my time as a student, I was highly critical of almost every faculty member I encountered, and was perhaps even more critical of the degree programs. I wanted students to have lots of options as to what they studied. The result was that I served (usually chaired) curriculum committees for decades and constantly had conflicts with faculty who thought that civil engineering was really just their area and they protected their required courses strongly. At Illinois, I pushed through a curriculum in civil engineering that was objected to by most of the faculty and students. The faculty were protecting their "turf" (my view), and the students wanted the faculty to tell them what courses to take.

I also chaired a new committee at Illinois that had the students evaluating the faculty. That was an educational experience for me. I discovered, and have since verified, that student evaluations are really about whether or not they like the professor, not whether or not they got a good education. A faculty member who regularly missed more than 50% of his classes and never (my experience) came to class prepared, got perfect scores in both of his classes on every question. In the next semester, when his students came to me to complain about him, I asked why they rated him so highly in the previous semester. The universal answer was "he was such a nice guy that we didn't have the heart to tell him the truth". Experiences of that kind led me to spend more time trying to figure out students and less time doing research.

When I started teaching, we used only lectures and the blackboard to communicate. In my first class, first exam, most of the class did fine but there was a significant group of students who did miserably. I gave them a hard time in our next class but, at the end of the class, suddenly had an epiphany and told everyone to hand in all of their notes. To my surprise, the "poor" students had studied their notes but their notes were wrong. Accordingly, I typed up and reproduced notes on the primitive equipment we had then. Now, almost all of the students did fine but there would be a few who had major problems. I asked some of the students who had low scores, to come to my office so I could find out what the problem was. One student complained bitterly that he had studied my notes and gave correct answers but I still graded him down. He found my specific statements that were the problem and I had to admit that his interpretation of the notes was acceptable, even if he was the only one in class who had that problem. That meant that the problem was more mine than his, and that led to a lot of time spent trying to figure out how to communicate, and how students learned. That time also took me away from research but my class evaluations improved greatly.

I look back on my career in education as an evolutionary process that was constantly a challenge.

FINI

I think I am fortunate in that geotechnical engineering grew from a very young group when I was young (say 1950's) to a mature group now that I'm old. It was a marvelous run with many fascinating and productive people and I've always taken great pride in calling myself a *geotechnical engineer*.

CURRICULUM VITAE

Current Position

Professor (*Emeritus*)
University of Texas at Austin

Education

B.S., University of Minnesota, Engineering, 1953
M.S., University of Minnesota, Civil Engineering, 1955
Ph.D., University of Illinois, Civil Engineering, 1960

Honors & Awards

- First Minnesota Surveyors and Engineers Society Scholarship, University of Minnesota, Academic Year 1953-54
- First American Society for Testing and Materials Graduate Fellowship, January 1, 1960 to October 1, 1960
- Walter L. Huber Research Prize, American Society of Civil Engineers, October 1971
- C. A. Hogentogler Award, American Society for Testing and Materials, June 1973
- Norman Medal, American Society of Civil Engineers, November 1975 (with D. E. Daniel and T. K. Liu)
- Croes Medal, American Society of Civil Engineers, October 1984 (with D. E. Daniel)
- L. P. Gilvin Professorship, University of Texas, September 1984
- C. A. Hogentogler Award, American Society for Testing and Materials, 1987
- Terzaghi Lecture, ASCE, San Diego, October 24, 1995
- Distinguished Alumnus Award, Department of Civil Engineering, University of Illinois, April 1997
- National Academy of Engineering (2003)

Theses and Dissertations Supervised

DOCTORAL STUDENTS

Bucknam, R.E. (1964)
"A Rheological Investigation of Kaolinite Slurries in Terms of Effective Stresses," University of Illinois, 82 pages

Langfelder, L.J. (1964)
"An Investigation of Initial Negative Pore Water Pressure in Statically Compacted Cohesive Soils," University of Illinois, 123 pages

Mesri, G. (1969)
"Engineering Properties of Montmorillonite," University of Illinois, 90 pages

Painter, W.T. (1970)
"Some Engineering Properties of Kaolinite Slurries," University of Illinois, 75 pages

Garlanger, J.E. (1970)
"Pore Pressures in Partially Saturated Soils," University of Illinois, 132 pages

Toan, Nguyen Thanh (1972)
"Finite Difference Solutions of Consolidation Problems," University of Illinois, 196 pages

Campbell, Joe Davis (1973)
"Pore Pressures and Volume Changes in Unsaturated Soils," University of Illinois, 104 pages, with R. B. Peck

Daniel, David Edwin (1980)
"Moisture Movement in Soils in the Vicinity of Waste Disposal Sites," University of Texas, 235 pages

Dennis, Norman Dale, Jr. (1982)
"Development of Correlations to Improve the Prediction of Axial Pile Capacity," University of Texas, 315 pages

McNulty, Edmund Gregory (1982)
"Consolidation with Axisymetric Flow in Soils Having Radial and Vertical Variation of Properties," University of Texas, 298 pages

Iskander, Magued (1995)
"An Experimental Facility to Model the Behavior of Steel Pipe Piles in Sand", University of Texas, 272 pages

Lai, Jiunnren (1997)
"Mechanical Properties of Eagle Ford Shale", University of Texas

El Gharbawy, Sherif (1998)
''The Pullout Capacity of Suction Caisson Foundations", University of Texas, 812 pages

Alansari, Omar (1999)
"Capacity and Behavior of Steel Pipe Piles in Dry Sand", University of Texas, 550 pages

Brown, Rollins (2001)
"Predicting the Ultimate Axial Resistance of Single Driven Piles", University of Texas, 160 pages

El-Sherbiny, R. M. (2005)
"Performance of Suction Caisson Anchors in Normally Consolidated Clay", University of Texas, 332 pages, with R. B. Gilbert

Akguner, Cem (2007)
"Elastic Analysis of Axial Load-Displacement Behavior of Single Driven Piles", University of Texas, 245 pages

MASTERS THESES

Hardin, Jack (1963), "Release of Exchangeable Sodium from Purified Fithian Illite upon Dilution with Distilled Water," University of Illinois, 44 pages

Campbell, L.M. (1963), "Effect of End Confinement on the Triaxial Compression Characteristics of Sodium Kaolinite," University of Illinois, 66 pages

Gau, F.L. (1967), "Consolidation of Stratified Soil Systems," University of Illinois, 66 pages

Conlin, Miller F., Jr. (1972), "Shearing Properties of Compacted Specimens of Taylor Marl Clay as Determined in an Unconsolidated-Undrained Triaxial Tests," University of Texas, 106 pages

Daniel, David E., Jr. (1973), "Combined Vertical and Radial Flow in Consolidation Problems," University of Texas, 160 pages

Yates, Lewis B. (1975), "Finite Difference Solution for Consolidation Analyses Involving Sand Drains," University of Texas, 260 pages

Shen, Ching-Peng (1977), "A Study of the Radial Drainage Consolidation Test," University of Texas, 135 pages

Kinard, Dan T. (1977), "A Case History Study of the Time Effects in One-Dimensional Consolidation," University of Texas, 103 pages

Pelletier, John H. (1977), "Determination of Soil Properties from Analyses of a Full Scale Highway Embankment," University of Texas, 125 pages

Thomas, Roger T. (1977), "A New Radial Drainage Consolidation Cell," University of Texas, 77 pages

Fox, Timothy J. (1979), "A Study of Slope Stability in Stiff-Fissured Clays in the Austin Area, Texas," University of Texas, 51 pages

Collins, James T. (1979), "Laboratory Measurement of Soil Response During Consolidation with Sand Drains," University of Texas, 236 pages

Hamilton, Jed M. (1979), "Measurement of Permeability of Partially Saturated Soils," University of Texas, 488 pages

Trautwein, Stephen J. (1980), "Practical Measurement of Radial Flow Consolidation Properties of Clays," University of Texas, 270 pages

Pertusa, Martha (1980), "Evaluation of Caps and Liners for Low-Level Radioactive Waste Disposal Pits," University of Texas, 190 pages

Winter, David G. (1981), "Compilation and Analysis of Axial Load Tests on Friction Piles," University of Texas, 274 pages

Kleppe, J.H. (1981), "Desiccation Cracking of Compacted Soil," University of Texas, 368 pages

Aschenbrener, T.B. (1984), "The Settlement Behavior of Single Piles in Clay," University of Texas, 258 pages

McCelvey, D.P. (1984), "Hydraulic Conductivity Determination from Pressure Plate Outflow Data," University of Texas, 258 pages

Schieve, E.W. (1984), "Stability and Settlement of an Embankment on Soft Clay," University of Texas, 388 pages

Al Shafei, K. A. R. (1987), "Axial Load Capacities of Steel Pipe Piles in Sand," University of Texas, 330 pages

Alyahyai, K.S. (1987), "Use of Case Histories in Prediction of Load-Settlement Behavior of Steel Pipe Piles in Sand," University of Texas, 322 pages

Palmer, Marilyn (1988), "Evaluation of a One-Dimensional Linear Theory of Consolidation for Soils Exhibiting Secondary Consolidation", University of Texas

Van, Go (1990), "Axial Static Load Capacity of Untapered Piles in Sand", University of Texas, 146 pages

Pena, Samuel Jose (1990), "Alternate Sites for a Connecting Passage between the Atlantic and Pacific Oceans, Univ. of Texas, 109 pages

Lai, Jiunnren (1991), "The Effects of Deformation-Rate on the Drained Strength of Clays", Univ. of Texas

Alem, Ziad (1990), "A Program to Analyze Secondary Consolidation", Univ. of Texas

Malhotra, Sanjeev (1991), "Equipment for Axial Load Testing of Model Open-Ended Pipe Piles in Sand", Univ. of Texas

Pavlicek, Robert (1993), "Investigation of Axial Tensile Load Capacity of Suction Piles", Univ. of Texas, 330 pages

Chiu, Kuan (1993), "A Study of Axial Load Capacity of Piles from Case Histories", Univ. of Texas, 162 pages

Goldberg, Aaron D. (1994), "Design and Construction of a Model Suction Pile", Univ. of Texas, 140 pages

Jones, W. Casey (1994), "Response of Model Suction Piles in Sand to Cyclic Tensile Loading" University of Texas

Pedersen, Robert C. (2001), "Model Offshore Soil Deposit: Design, Preparation, and Characterization" University of Texas

Publications

REFEREED ARCHIVED JOURNAL PAPERS

Olson, R. E. (1962), "Shear Strength Properties of Calcium Illite", *Geotechnique*, Vol. 12, pp. 23-43.

Olson, R. E. (1963), "Effective Stress Theory of Soil Compaction", *Proceedings*, ASCE, Vol. 89, No. SM2, pp. 27-46

Olson, R. E. (1963), "Shear Strength Properties of a Sodium Illite," *Proceedings*, ASCE, Vol. 89, No. SM1, pp. 183-208

Olson, R. E. and L.J. Langfelder (1965), "Pore Water Pressures in Unsaturated Soils," *Proceedings* ASCE, Vol. 91, No. SM4, pp. 127-150

Olson, R. E. and K. S. Flaate (1967), "Pile-Driving Formulas for Friction Piles in Sand," *Proceedings* ASCE, Vol. 93, No. SM6, pp. 279-296

Olson, R. E. and L.M. Campbell (1967), "Bushing Friction in Triaxial Shear Testing," *Materials Research and Standards*, Vol. 7, No. 2, pp. 45-52

Mesri, G. and R. E. Olson (1970), "Shear Strength of Montmorillonite," *Geotechnique*, Vol. 20, No. 3, pp. 261-270

Olson, R. E. and G. Mesri (1970), "Mechanism Controlling Compressibility of Clays," *Proceedings,* ASCE, Vol. 96, No. SM6, pp. 1863-1878

Olson, R. E., V.J. McDonald, and J.N. Sterner (1971), "Control of Deformation Rates in Soil Testing," *Journal of the Soil Mech. and Found. Div.* ASCE, Vol. 97, No. SM1, pp. 283-286

Mesri, G. and R. E. Olson (1971), "Mechanisms Controlling the Permeability of Clays," *Clays and Clay Minerals*, Vol. 19, No. 3, pp. 151-158

Gau, F. L. and R. E. Olson (1971), "Uniformity of Specimens of a Compacted Clay," *Journal of Materials*, Vol. 6, No. 4, pp. 874-888, ASTM

Mesri, G. and R. E. Olson (1971), "Consolidation Characteristics of Montmorillonite," *Geotechnique,* Vol. 21, No. 4, pp. 341-352

Daniel, D. E. and R. E. Olson (1974), "Stress-Strain Properties of Compacted Clays," *Journal of the Geotechnical Engineering Division*, ASCE, Vol. 100, No. GT10, pp. 1123-1136

Olson, R. E. (1974), "Shearing Strengths of Kaolinite Illite, and Montmorillonite," *Journal of the Geotechnical Engineering Division*, ASCE, Vol. 100, No. GT 11, pp. 1215-1229

Olson, R. E. (1977), "Consolidation Under Time Dependent Loading," *Journal, Geotechnical Engineering Division,* Am. Soc. of Civil Engrs., Vol. 103, No. 1, pp. 55-60

Olson, R. E. and C. C. Ladd (1979), "One-Dimensional Consolidation Problems," *Jour.*, Geot Engr. Div., Amer. Soc. of Civil Engrs., Vol. 105, No. 1, pp. 11-30

Pelletier, J. H., R. E. Olson, and J. J. Rixner (1979), "Estimation of Consolidation Properties of Clay from Field Observations," *Geotechnical Testing Journal,* Am. Soc. for Testing and Materials, Vol. 2, No. 1, pp. 34-43

Olson, R. E. (1980), "Prediction of Success of Graduate Students in Civil Engineering," *Civil Engineering Education*, American Society for Engineering Education, Vol. 2, No. 2, pp. 7-12

Daniel, D. E. and R. E. Olson (1982), "Failure of an Anchored Bulkhead," *Jour.*, Geotechnical Engr. Div., ASCE, Vol. 108, No. GT10, pp. 1318-1327

Olson, R. E. (1998) "Settlement of Embankments on Soft Clays", *Jour. of Geotechnical and Geoenvironmental Engineering*, Vol. 124, No. 4, pp. 278-288

Iskander, M. G., R. E. Olson, and J. A. Bay (2001), "Design and Performance of an Electro-Pneumatic Pile Hammer for Laboratory Applications", ASTM, Geotechnical Testing Journal, Vol. 24, No. 1, pp. 72-82

Iskander, M., S. El-Gharbawy, and R. Olson (2002), "Performance of Suction Caissons in Sand and Clay", Canadian Geotechnical Journal, Vol. 39, pp. 576-584

Pedersen, R. C., R. E. Olson, and A. F. Rauch (2003), "Shear and Interface Strength of Clay at Very Low Effective Stress", ASTM *Geotechnical Testing Journal*, Vol. 26, No. 1

Olson, R. E., A. F. Rauch, A. M. Luke, D. R. Maniar, J. T. Tassoulas, and E. C. Mecham (2003), "Soil Reconsolidation Following Installation of Suction Caissons", Offshore Technology Conference, Houston, paper 15263

Luke, A. M., A. F. Rauch, R. E. Olson, and E. C. Mecham (2005), "Components of Suction Caisson Capacity Measured in Axial Pullout Tests", *Ocean Engineering*, Vol. 32, No.7, pp. 878-891

REFEREED CONFERENCE PAPERS

Olson, R. E. (1960), "Migration of Pore Fluid during Shear Consolidated-Undrained (R) Triaxial Tests," Proceedings, ASCE *Research Conference on the Shear Strength of Cohesive Soils,* Boulder, Colorado, pp. 1043-1049

Olson, R. E. and F. Mitronovas (1962), "Shear Strength and Consolidation Characteristics of Calcium and Magnesium Illite," *Clays and Clay Minerals, Proceedings, Ninth National Conference*, pp. 185-209, Pergamon Press, New York City

Olson, R. E. and J. Hardin (1963), "Shearing Properties of Remolded Sodium Illite," *Proceedings, Second Pan American Conference on Soil Mechanics and Foundation Engineering*, Vol. 1, pp. 203-218

Olson, R. E. and M. L. Kiefer (1964), "Effect of Lateral Filter Drains on the Triaxial Shear Characteristics of Soils," ASTM STP 361, pp. 482-491

Olson, R. E. and H. Kane (1965), "Dynamic Shearing Properties of Compacted Clay at High Pressures," *Proceedings, Sixth International Conference on Soil Mechanics and Foundation Engineering*, Montreal, Canada, Vol. 1, pp. 328-332

Olson, R. E. and J. F. Parola (1968), "Dynamic Shearing Properties of Compacted Clay," *Proceedings of the International Symposium on Wave Propagation and Dynamic Properties of Earth Materials*, Albuquerque, New Mexico, pp. 173-182

Richards, A. F., V.J. McDonald, R. E. Olson, and G.H. Keller (1972), "In-Place Measurement of Deep Sea Soil Shear Strength," *Underwater Soil Sampling, Testing, and Construction Control,* ASTM STP 501, Am. Soc. for Testing and Materials, pp. 55-68

McDonald, V. J., R. E. Olson, A. F. Richards, and G. H. Keller (1972), "Measurement and Control System for Deep Sea Vane Shear Device," *Ocean '72*, IEEE International Conference on Engineering in the Ocean Environment, Inst. of Electrical and Electronic Engineers, Inc., New York City, pp. 474-479

Olson, R. E., D. E. Daniel, and T. K. Liu (1974), "Finite Difference Analyses for Sand Drain Problems," *Analysis and Design in Geotechnical Engineering,* Vol. 1, pp. 85-110, Austin, Texas

Keshian, B., Jr., C.C. Ladd, and R.E. Olson (1977), Sedimentation-Consolidation Behavior of Phosphatic Clays," *Geotechnical Practice for Disposal of Solid Waste Materials,* ASCE, Geotechnical Engineering Division Specialty Conference, pp. 188-209

Fox, T.J. and R. E. Olson (1979), "A Study of Slope Stability in Stiff-Fissured Clays in the Austin Area, Texas," *Texas Section* ASCE Meeting, March 30

Pelletier, J. H. and R. E. Olson (1979), "Evaluation of Consolidation Properties of Clay from Field Observations," *Texas Section,* ASCE, March 30

Daniel, D. E. and R. E. Olson (1980), "Geotechnical Aspects in Design of Disposal Sites for Low-Level Radioactive Wastes," *Texas Section* ASCE Meeting, San Antonio, April

Trautwein, S. J., R. E. Olson, and R. L. Thomas (1981), "Radial Flow Consolidation Testing," *Proceedings*, Tenth Intern. Conf. on Soil Mech. and Foundation Engineering, Stockholm, Vol. 1, pp. 811-814

Hamilton, J. J., D. E. Daniel, and R. E. Olson (1981), "Measurement of Hydraulic Conductivity of Partially Saturated Soils," *Permeability and Groundwater Contaminant Transport,* ASTM STP 746, pp. 182-196

Olson, R. E. and D. E. Daniel (1981), "Measurement of the Hydraulic Conductivity of Fine-Grained Soils," *Permeability and Groundwater Contaminant Transport*, ASTM STP 746, pp. 18-64

Daniel, D. E., J. M. Hamilton, and R. E. Olson (1981), "Suitability of Thermocouple Psychrometers for Studying Moisture Movement in Unsaturated Soils," *Permeability and Groundwater Contaminant Transport*, ASTM STP 746, pp. 84-100

Dennis, N. D., Jr., and R. E. Olson (1983), "Axial Capacity of Steel Pipe Piles in Clay," *Geotechnical Practice in Offshore Engineering*, Amer. Soc. of Civil Engrs., Geotechnical Engr. Div. Specialty Conf., University of Texas at Austin, pp. 370-388

Dennis, N. D., Jr., and R. E. Olson (1983), "Axial Capacity of Steel Pipe Piles in Sand," *Geotechnical Practice in Offshore Engr.*, Amer. Soc. of Civil Engrs., Geotechnical Engr. Div. Specialty Conf., Univ. of Texas at Austin, pp. 389-402

Olson, R. E. and N. D. Dennis (1983), "Reliability of Pile Foundations," Preprint SC-3, *Structures Congress,* American Society of Civil Engineers, Houston, Texas

Olson, R. E., N. D. Dennis, and D. G. Winter (1984), "Prediction of Axial Pile Capacity Based on Case Histories," *Intern. Conf. on Case Histories in Geotechnical Engineering,* Univ. of Missouri-Rolla, S. Prakash, ed., Vol. 1, pp. 191-200

Olson, R. E. (1984), "Performance of an Embankment on Peat," *Intern. Conf. on Case Histories in Geot Engr.,* Univ. of Missouri-Rolla, S. Prakash, ed., Vol. 2, pp. 625-630

Aschenbrener, T. B., and R. E. Olson (1984), "Prediction of Settlement of Single Piles in Clay," *Analysis and Design of Pile Foundations,* ASCE Special Technical Publication, pp. 41-58

Kleppe, J. H., and R. E. Olson (1985), "Desiccation Cracking of Soil Barriers," *Hydraulic Barriers in Soil and Rock,* Amer. Soc. for Testing and Materials, STP 874, pp. 263-275

Olson, R. E. (1986), "State of the Art: Consolidation Testing," Consolidation Testing of Soils: Testing and Evaluation, ASTM STP 892, R. N. Yong and F. C. Townsend, eds., pp. 7-70

Olson, R. E., and W. A. Potter (1986), "Embankment Design and Construction for SH87 Bridge Over the Neches River," Texas Section, ASCE, March

Olson, R. E., and K. Al-Shafei (1988), "Axial Load Capacity of Steel Pipe Piles in Sand," Second International Conf. on Case Histories in Geotechnical Engr., Univ. of Missouri, Rolla, Vol. 3, pp. 1731-1738

Olson, R. E. (1990), "Axial Load Capacity of Steel Pipe Piles in Sand", Offshore Technology Conference, Houston, Paper OTC 6419, pp. 17-24

Olson, R. E. (1993), "Failure of a Twenty-Foot High Retaining Wall", *Proc.,* Third Intern. Conf. on Case Histories in Geotechnical Engineering, Vol. 2, pp. 791-797

Olson, R. E., and R. E. Heuer (1993), "Failure of a Large Circular Excavation", *Proc.,* Third International Conf. on Case Histories in Geotechnical Engineering, Vol. 2, pp. 799-805

Iskander, M., and R. E. Olson (1992), "Review of API Guidelines for Pipe Piles in Sand", Civil Engineering in the Oceans V, ASCE Specialty Conference, pp. 798-812

Brouillette, R. P., R. E. Olson, and J. R. Lai (1993), "Stress-Strain Characteristics of Eagle Ford Shale", Proc., International Symposium on Hard Soils-Soft Rocks, Athens, Greece, Vol. 1, pp. 397-404

Go, Van, and R. E. Olson (1993), "Axial Load Capacity of Untapered Piles in Sand", Southeast Asian Geotechnical Conference, Bangkok, Vol. 1, pp. 517-522

Iskander, M. G., R. E. Olson, and R. W. Pavlicek (1993), "Behavior of Suction Piles in Sand", *Design and Performance of Deep Foundations,* ASCE Special Publication No. 38, pp. 157-171

Iskander, M. and R. Olson (1997), "Behavior of Open Ended Pipe Piles in Sand During Installation and Loading," Proceedings XIV International Conference on Soil Mechanics and Foundation Engineering, Hamburg Germany, pp. 991-995

Olson, R. E., and M. G. Iskander (1998), "Axial Load Capacity of Piles in Sand", Proc., Fourth Intern Conf. on Case Histories in Geot Engr., Vol. 1, pp. 293-300

El-Gharbawy, S., and R. E. Olson (1998), "Behavior of Suction Caissons in the Marine Environment", Proc., Offshore Technology Conference, Houston

El-Gharbawy, S., and R. E. Olson (1998), "Laboratory Modeling of Suction Caisson Foundations", Proc., ISOPE Conference, Montreal, Vol. 1, pp. 537-542

El-Gharbawy, S., and R. E. Olson (1998), "The Pullout Capacity of Suction Caisson Foundations for Tension Leg Platforms", Proc., ISOPE Conference, Montreal, Vol. 1, pp. 531-536

El-Gharbawy, S., and R. E. Olson (1999), "Suction Anchor Installations for Deep Gulf of Mexico Applications", Proc., Offshore Technology Conference, Paper OTC 10992

El-Gharbawy, S., and R. E. Olson (1999), "Suction Caisson Foundations in the Gulf of Mexico", *Analysis, Design, Construction, and Testing of Deep Foundations*, J. M. Roesset ed., pp. 281-295

El-Gharbawy, S., and R. E. Olson (1999), "The Cyclic Pullout Capacity of Suction Caisson Foundations", ISOPE Conference, Brest, France, Vol. 1

Olson, R. E., A. F. Rauch, A. F. Tassoulas, C. P. Aubeny, and W. R. Murff (2001), "Toward the Design of New Technologics for Deep-Water Anchorages", *International Symposium on Offshore and Polar Engineering*, Scavenger, Norway, June

Rauch, A. F., Olson, R.E., Mecham, E. C., Pedersen, R.C. (2001), "A Laboratory Facility for Testing Model Suction Caissons" *Proc.*, OTRC 2001 Intl Conf. on *Geotechnical, Geological, Geophysical Properties Deepwater Sediments*, Houston, Texas, April. pp. 198-216

Rauch, A. F., R. E. Olson, A. M. Luke, and E. C. Mecham (2003), "Measured Response During Laboratory Installation of Suction Caissons", International Symposium on Offshore and Polar Engineering, Honolulu, Hawaii, p. 780-787

Luke, A. M., A. F. Rauch, R. E. Olson, and E. C. Mecham (2003), "Behavior of Suction Caissons Measured in Laboratory Pullout Tests", *Proc. OMAE03, 22nd International Conference on Offshore Mechanics and Arctic Engineering*, Cancun, Mexico, Paper 37023

Luke, A. M., A. F. Rauch, R. E. Olson, and E. C. Mecham (2003), "Components of Suction Caisson Capacity Measured in Axial Pullout Tests", Deepwater Mooring Systems: Concepts, Design, Analysis and Materials, Offshore Technology Research Center Specialty Conference, Houston, TX

Olson, R. E., A. F. Rauch, E. C. Mecham, and A. M. Luke (2003), "Self-Weight Consolidation of Large Laboratory Deposits of Clay", Proc., 12th Pan-American Conf. on Soil Mech. and Geot. Engr., MIT, Vol. 1, pp. 703-708

Coffman, R. A., R. M. El-Sherbiny, A. F. Rauch, and R. E. Olson (2004), "Measured Horizontal Capacity of Suction Caissons", *Proc., Offshore Tech. Conf.*, Houston, Paper 16161

Olson, R. E., and T. J. Shantz (2004), "Axial Load Capacity of Piles in California in Cohesionless Soils", *Current Practices and Future Trends in Deep Foundations*, ASCE Geotechnical Special Publication No. 125, pp. 1-15

El-Sherbiny, R. M., R. E. Olson, R. B. Gilbert, and S. K. Vanka (2005), "Capacity of Suction Caissons under Inclined Loading in Normally Consolidated Clay", *Intern. Symposium on Frontiers in Offshore Geotechnics*, pp. 281-287

El-Sherbiny, R. M., R. E. Olson, and R. B. Gilbert (2006), "Comparison of Different In-Situ Shear-Strength Tests in Normally Consolidated Clay", International Symposium on Offshore and Polar Engineering, Paper 2006-JSC-295, San Francisco

Olson, R. and M. Iskander (2009) "Axial Load Capacity of Un-Tapered Piles In Cohesionless Soils," Contemporary Topics in Deep Foundations, ASCE Geotechnical Special Publication No. 185, pp. 231–238

Park, K., K. H. Stokoe, R. E. Olson, and W. Seo (2009), "Settlement of Footings in Sand using Dynamic Soil Properties", Selected Papers From The International Workshop On Soil-Foundation-Structure Interaction (SFSI 09), Auckland, New Zealand

Sheehan, A., R. Olson, K. Park, and K. H. Stokoe (2010), "Estimation of Settlement of Footings under Working Loads using Equivalent-Linear Elasticity", GeoFlorida 2010: Advances in Analysis, Modeling & Design, ASCE Geotechnical Special Publication 199, pp. 1708-1717

INVITED (NON-REFEREED) CONFERENCE PAPERS

Olson, R. E. (1983), "Impact of Testing Errors on Engineering Design," *Proceedings*, 31st Annual Soil Mechanics and Foundation Engineering Conference, Minnesota Geotechnical Society, pp. 13-43

Olson, R. E., and J. H. Long (1988), "Axial Load Capacity of Tapered Piles," *The Art and Science of Geotechnical Engineering*, Prentice Hall, NYC, pp. 417-436

Olson, R. E. (1990), "Analyses of the Failure of a Twenty-Foot High Retaining Wall", Fifteenth Southwest Geotechnical Engineering Conference, Austin

Olson, R. E. (1990), "Analyses of the Failure of a Twenty-Foot High Retaining Wall", Texas Section, ASCE, Austin

Olson, R. E., and M. G. Iskander (1994), "Axial Load Capacity of Piles in Sand", Advanced Technology in Civil Engineering, ASCE-Saudi Arabian Section, Manama, Bahrain, pp. 383-394

Olson, R. E. (2002), "Axial Load Capacity of Driven Piles", 50th Annual Geotechnical Engineering Conference, University of Minnesota, pp. 161-178

TECHNICAL DISCUSSIONS AND CLOSURES

Olson, R. E. and J. D. Scott (1961), discussion of "Structure and Strength Characteristics of Compacted Clays," by H.B. Seed and C.H. Chan, *Transactions*, ASCE, Vol. 126, pp. 1398-1403

Olson, R. E. (1962), discussion, *Geotechnique*, Vol. 12, No. 4, pp. 355-358

Olson, R. E. (1963), discussion of "Shear Strength Properties of a Sample of Volcanic Material of the Valley of Mexico," by K. Y. Lo, *Geotechnique*, Vol. 13, pp. 164-166

Olson, R. E. (1964), closure to discussions of "Shear Strength Properties of a Sodium Illite," by R. E. Olson, *Proceedings*, ASCE, Vol. 90, No. SM2, pp. 167. Also a digest in Transactions, ASCE, Vol. 129, pp. 293-395

Olson, R. E. (1964), closure to "Effective Stress Theory of Soil Compaction," by R. E. Olson, *Proceedings* ASCE, Vol. 90, No. SM2, pp. 171-184. Also digest in Transactions, ASCE, Vol. 129, pp. 293-295

Olson, R. E. (1964), discussion of "The Effect of Test Techniques on the Shear Strength of Western Canadian Clays," by L.G. Chan and P.J. Rivard, ASTM STP 361, pp. 500-503

Olson, R. E. (1964), discussion of "Effect of Non-Uniform Pore Pressures on Laboratory Measurements of the Shear Strength of Soils," by G.E. Blight, ASTM STP 361, pp. 185-189

Olson, R. E. (1964), discussion of "Influence of Stress History on the Stress Paths Followed in Undrained Triaxial Tests, by D.J. Henkel and V.A. Sowa, ASTM STP 361, pp. 292-294

Olson, R. E. and L.M. Campbell (1964), discussion of "Importance of Free Ends in Triaxial Testing," by P.W. Rowe and L. Barden, Proceedings ASCE, Vol. 90, No. SM6, pp. 167-173

Olson, R. E. and L.J. Langfelder (1966), errata for "Pore Water Pressures in Unsaturated Soils," *Proceedings* ASCE, Vol. 92, No. SM3, pp. 106, digest in Transactions ASCE, Vol. 131, pp. 365

Olson, R. E. and Flaate (1969), closure to "Pile-Driving Formulas for Friction Piles in Sand," *Proceedings* ASCE, Vol. 95, SM5, pp. 1264. Digest in Transactions ASCE, Vol. 135 (1970)

Lai, J., and R. E. Olson (1998), discussion of "Consolidation Behavior of Clayey Soils Under Radial Drainage" by A. Sridharan, K. Prakash, and S. R. Asha, *Geotechnical Testing Journal*, Vol. 21, No. 1, pp. 66-67.

REVIEWS

Olson, R. E. (1970), Review of Pavements on Expansive Clays by G. Kassiff, M. Livneh, and G. Wiseman, *Materials Research and Standards,* Vol. 10, No. 7, pp. 42

Olson, R. E., J. H. Long, R. Jardine, and Yudhbir (1993), "Case Histories of Foundations", *Third Intern. Conf. on Case Histories in Geotechnical Engineering*, Vol. 3, pp. 1447-1454

Introduction to Selected Olson Papers

It was not an easy task to select which of Professor Olson's technical papers to reprint in this book. He began his career at the University of Illinois studying the effect of physico-chemical properties of clays on their engineering properties and ended his career working on the analysis and design of foundations for off-shore structures. In the 40-year period in between, he performed research on the engineering properties of partially saturated soil, conducted investigations of failed retaining structures, developed programs for predicting the consolidation of clays, conducted full-scale performance monitoring of foundations and embankments on soft clay, and performed research on the axial load capacity and settlement of pile foundations. The selected papers include most of these topics. Papers published in non-ASCE journals and conferences were given additional consideration in the selection process. Two of Professor Olson's papers, one on the state of the art of consolidation testing and one on the measurement of hydraulic conductivity of fine grained soils, could were not included because of space limitations. The following are comments provided by Professor Olson on each of the selected papers.

Olson, R. E. (1963), "Effective Stress Theory of Soil Compaction"

The moisture-density test has been a standard in geotechnical engineering since the 1930s but at the time this paper was published there were only two papers that offered explanations for the shape of the curve and neither could be verified. In this paper, data showed that there are actually two peaks. Rational physico-chemical explanations were offered to explain both peaks. Data to refute the existing theories could not be published due to length limitations on papers.

Olson, R. E. and L.J. Langfelder (1965), "Pore Water Pressures in Unsaturated Soils"

In the early 1960's, a major government agency, involved with designs of dams and other structures using compacted clays, believed that pore water could never have negative pressure and they presented experimental data supporting their view. However, agronomists had been measuring negative pore water pressures in natural soils for decades. It seemed obvious that negative pore water pressures would exert a major influence on physical properties of soils so our first step in research in this area was to make direct measurements of pore water pressures down to about -250 psi and to offer scientific explanations for how this was possible.

Olson, R. E. and K. S. Flaate (1967), "Pile-Driving Formulas for Friction Piles in Sand"

Although pile driving formulas have been almost universally replaced with wave equation analyses and dynamic measurements of stress waves at the top of the pile to determine pile capacity, this paper is of interest because it is an example of how statistical analyses of a large data base (93 load tests) can be used to determine the usefulness of an analytical equation and its relative accuracy. One of the conclusions of the paper was that there was no correlation between the measured pile capacity and the capacity computed by the Engineering News formula. Although several of the other equations provided reasonable correlations between measured and predicted capacities, the use of these formulas is not recommended in current practice.

Olson, R. E. and J. F. Parola (1968), "Dynamic Shearing Properties of Compacted Clay"

The two most widely referenced papers on the effect of strain rate on the undrained strength of clays were written by Arthur Casagrande with W. L. Shannon (high speed loading) and S. D. Wilson (creep), with the emphasis on testing of undisturbed samples of natural clays. We were concerned about the behavior of compacted clays caused by high pressures and high loading rates from nuclear detonations. To match conditions in the field, we used compacted clays at a range in water contents, with confining pressures up to 1000 psi (68 atmospheres) and times to failure as small as three milliseconds and showed that viscous effects extended to high pressures and high strain rates.

Olson, R. E. and G. Mesri (1970), "Mechanism Controlling Compressibility of Clays,"

At the time this paper was published, there was an interest in understanding where the physical properties of soils, particularly clays, came from. The usual assumption was that some properties had a physical base but there was a growing feeling that there were physico-chemical controls, e.g., from adsorbed water and from formation of diffuse double layers. We performed an extensive series of one-dimensional consolidation tests using sand, mica, kaolinite, illite, and smectite, with samples dry (pore fluid was air), or saturated with fluids of varying properties, in particular different dielectric constants and polarities. The abbreviated final conclusion was that properties of sand, mica, and kaolinite, were physically controlled once sediment had formed. Physico-chemical effects were significant for kaolinite in suspensions but the effect influenced particle arrangements and floc formation during deposition but not very significantly for effective stresses of interest in geotechnical engineering. For smectite, Gouy-Chapman double layer theory was shown to control compressibilities with sodium as the exchangeable cations but it was not very relevant for smectite with calcium cations.

Mesri, G. and R. E. Olson (1971), "Mechanisms Controlling the Permeability of Clays"

This paper represents part of an effort to explain the engineering properties of soils from a physico-chemical point of view. We investigated the effects of clay mineralogy, adsorbed cations, pore water electrolyte concentration, and dielectric constant and polarity of the pore fluid, on hydraulic conductivity and provided physico-chemical explanations for the results. The tests had possible application in the then developing Geoenvironmental area where compacted clay barriers could be used to contain wastes containing non-polar industrial wastes.

Olson, R. E. (1974), "Shearing Strengths of Kaolinite Illite, and Montmorillonite"

I was told, by a famous geotechnical engineer, that no clay could ever have an effective friction angle less than 21 deg. I argued that effective friction angles could be zero. Over a period of years, effective friction angles of clays composed of only one mineral type showed that failure envelopes were unique to a mineral type and that angles could be at least as low as two degrees. A physico-chemical explanation was provided to explain the data.

Olson, R. E., D. E. Daniel, and T. K. Liu (1974), "Finite Difference Analyses for Sand Drain Problems"

Terzaghi's theory of one-dimensional consolidation has been used by geotechnical engineers since the mid 1930's, even for cases where the assumptions do not apply. Unfortunately, there are few, if any, cases in which the theory is actually applicable and there are even fewer well documented case histories to show how well it works. When sand drains or wicks are used, an equivalent theory can be used but there are even fewer comparisons of analyses and field measurements.

We developed a computer code based on the finite difference method of analysis, that made it possible to include the following conditions in the analysis: (1) multiple soil strata, (2) presence of included sand layers that can drain, or be sealed, laterally, (3) time-dependent loading, (4) coefficients of consolidation and hydraulic conductivities that were non-linear functions of effective stress, (5) settlement-dependent submergence corrections, (6) possibility of negative pore water pressures above the water table, (7) vertical and radial consolidation, (8) insertion of either fully or partially-penetrating drains at any time, (9) anisotropy, (10) vertical strains based on either void ratios or engineering strains that can be either linear or non-linear functions of vertical effective stress, (11) the soil can be unloaded and rebound on a different stress-strain curve than for loading, and with different coefficients of consolidation and hydraulic conductivities, and (12) changes in soil properties resulting from disturbance during field installation of sand drains.

The Maine highway department used a test embankment at the I-295 crossing of the Fore River. The test embankment had four sections using different drain installation methods. The site was instrumented for settlements. Field sampling and laboratory testing were state-of-the art. Analyses involved time dependent loading, combined vertical and radial flow, effective-stress-dependent soil properties, and the effect of disturbance when drains were installed after a significant part of consolidation had occurred.

The initial analyses fit the field observations as well as possible considering the usual variation in soil properties across a site. The analyses were fit to the field observations to show the sudden change in stress-strain curves due to remolding from using driven drains. Fitted stress-dependent coefficients of consolidation agreed with laboratory measurements. At the time this paper was written, this case history coupled with an appropriate analysis, was probably the best case history yet published on this topic and the paper won the Normal Model from the American Society of Civil Engineers.

Pelletier, J. H., R. E. Olson, and J. J. Rixner (1979), "Estimation of Consolidation Properties of Clay from Field Observations"

A survey of decades of publications in geotechnical engineering show the existence of few, if any, papers where there were valid comparisons of time rates of consolidation of stratified soil deposits coupled with theoretical predictions. This paper deals with a highway embankment constructed on a deep deposit of stratified clay. Site investigation and laboratory testing were of highest quality. Field measurements involved excess pore water pressures and layer compressions. Numerical analyses were used to back out soil properties. Comparisons of measured and backed out soil properties made it possible to show the effects of sampling disturbance and to recommend methods of adjusting for such disturbance.

Trautwein, S. J., R. E. Olson, and R. L. Thomas (1981), "Radial Flow Consolidation Testing"

Radial coefficients of consolidation are needed when the engineer is predicting time rates of consolidation for cases where there is radial drainage. Devices in use at the time this paper was written used inflow to a porous stone representing the well, or outflow to a porous outer boundary. Our research showed that these devices worked badly because smear zones formed in the soil adjacent to the drainage boundary and retarded water flow significantly. The paper contains details on a device to avoid the then existing experimental problems. It also provides a method to back out the radial coefficient of consolidation that is simple, quick, and accurate.

Hamilton, J. J., D. E. Daniel, and R. E. Olson (1981), "Measurement of Hydraulic Conductivity of Partially Saturated Soils"

The Government had a goal to develop methods of storage of low-level radionuclides that would ensure that they not move more than one meter in a million years. As part of the resulting effort, this project was concerned with measurements of highly negative pore water pressures and hydraulic conductivities as low as 10^{-10} cm/sec.

Daniel, D. E. and R. E. Olson (1982), "Failure of an Anchored Bulkhead"

An anchored bulkhead, on the Houston ship channel, failed and five major geotechnical engineering firms were hired to participate in the resulting lawsuit. None of the firms could explain the failure in spite of each making dozens of soil borings and doing extensive testing. We explained the failure without soil borings or laboratory testing and showed that there was an important issue ignored in the most renowned text in the area. We were later told that the case was resolved based on our recommendations and design of anchored bulkheads was changed as a result. The paper received the ASCE Croes medal.

Dennis, N. D., Jr., and R. E. Olson (1983), "Axial Capacity of Steel Pipe Piles in Clay"

Full scale load tests on steel pipe piles demonstrated that the two methods recommended by the American Petroleum Institute in 1983 for design of offshore piles in clays, worked poorly and the axial load capacity was over-predicted for piles penetrating more than about 50 feet, with the error increasing rapidly as the penetration increased. A suggested method was considerably more accurate for piles penetrating less than 70 feet and a method proposed by Mark Randolph using our data base was better for longer piles.

Olson, R. E., and R. E. Heuer (1993), "Failure of a Large Circular Excavation"

In the 1980s, a city outgrew its ability to treat its waste products adequately and began discharging partially treated sewage into the adjacent river. Threats by a government agency forced the city to build a new sewage treatment facility quickly. The design incorporated a 117-foot diameter circular excavation, 90 feet deep, supported with ring beams. The excavation was close to the design depth when it collapsed. In the resulting litigation, one of the coauthors was retained by the plaintiff (contractor) and the other by the defendants (city and designer). Mediation was unsuccessful so the co-authors obtained permission from their clients to meet separately and work out a harmonious settlement. They did so and the settlement was accepted by all parties, eliminating an expensive trial and allowing remedial work to commence without further delay. The paper is concerned partially with design procedures for such excavations but also with the benefits of remediation to avoid expensive trials.

Olson, R. E. (1993), "Failure of a Twenty-Foot High Retaining Wall"

A twenty-foot high retaining structure consisting of a lower cantilever wall, a set-back, and then an upper cantilever wall, failed soon after construction. In the resulting litigation, the designer presented calculations showing factors of safety from two to infinity. The author performed a separate analysis and generally got factors of safety less than 0.5. There are relatively few documented case histories of the failure of such walls and even fewer where it was possible to perform analyses to explain the failure mechanism.

Olson, R. E., and M. G. Iskander (1994), "Axial Load Capacity of Piles in Sand"

For decades, it was common practice to estimate the axial load capacity of piles driven into cohesionless soils using "experience" and to verify that experience using pile driving formulas in the field. That approach was inapplicable for offshore piles and we set out to set up a load-test data base that would make it possible to compare predicted and measured capacities of piles. Part of the problem was in measuring some useful parameter relating to soil strength in the field for cohesionless soils and then finding suitable empirical parameters to be used in estimating capacities.

Iskander, M. and R. Olson (1997) "Behavior of Pipe Piles in Sand during Installation and Loading"

At the time this paper was written, the American Petroleum Institute (API) considered the design of open-ended pipe piles in sand as the most challenging foundation engineering problem. Extensive experimental facilities were developed including a laboratory pile hammer, a pressure chamber and unique double wall pipe piles capable of delineating frictional stresses acting against the internal and external surfaces of the pile. This paper presents unique measurements of the buildup of pore water pressure during pile driving in sand as well as plugging of pipe piles during installation and loading.

Iskander, M., S. El-Gharbawy, and R. Olson (2002), "Performance of Suction Caissons in Sand and Clay"

Several years before this paper was written, several experts in offshore geotechnical engineering assured me that suction caissons would never be used offshore. This paper is one in a series beginning in 1994 that described research at the University of Texas that demonstrated that suction caissons would provide a safe and economical foundation for off-shore structures.

Pedersen, R. C., R. E. Olson, and A. F. Rauch (2003), "Shear and Interface Strength of Clay at Very Low Effective Stress"

As far back as the 1930's, people ran direct shear tests on sands and obtained linear failure envelopes with significant cohesion intercepts. They were unable to find a rational explanation of a cohesion intercept for a "cohesionless" soil so they finally concluded that there was an unknown experimental error and Casagrande recommended switching from direct shear to triaxial compression tests. We can shear tests at normal stresses approaching zero and showed that the failure envelope curves down to the origin. Such data are relevant for shallow landslides.

Olson, R. E., and T. J. Shantz (2004), "Axial Load Capacity of Piles in California in Cohesionless Soils"

The California Department of Transportation (CalTrans) performed load tests on piles for decades but had not assembled the data nor used it in future designs. We re-drilled soils at numerous sites, obtained and tested soils near the piles, and performed field measurements using the cone and standard penetration tests. A database was developed with data from pile load tests conducted in California. A set of trial analyses were performed to obtain information on a variety of sometimes controversial questions, e.g., tensile versus compression side shears, effects of using a closed end versus an open end for steel pipe piles, effect of surface roughness on side shear, and a range of other questions that have existed for decades for pile foundations.

Luke, A. M., A. F. Rauch, R. E. Olson, and E. C. Mecham (2005), "Components of Suction Caisson Capacity Measured in Axial Pullout Tests

This paper is the latest in a series of papers that described the monitoring that that was performed on prototype suction caissons and the analyses that were completed to demonstrate how suction caissons were likely to perform in normally-consolidated clay during axial pullout.

Author List

Journal of the

SOIL MECHANICS AND FOUNDATIONS DIVISION

Proceedings of the American Society of Civil Engineers

EFFECTIVE STRESS THEORY OF SOIL COMPACTION

By Roy E. Olson,[1] A.M. ASCE

SYNOPSIS

The shape of the moisture-density curve obtained from compaction tests has been explained in terms of capillarity and lubrication, viscous water, and physico-chemical interactions. None of these theories has been substantiated by diagnostic laboratory experiments and none has been extended to explain the effect, on the dry density, of foot pressure or number of stress applications.

Probably the most important "tool" that has been used in the contemporary development of soil mechanics is the principle of effective stress. This principle has been used in analysis of the shearing properties of compacted soils and has been used, to a lesser extent, in analyzing the compaction process itself. However, it was not until recently (1960) that sufficient knowledge became available about the engineering properties of unsaturated soils to permit attempting a more definitive explanation of the shape of the moisture-density curve using the principle of effective stress. The effective stress theory developed herein is qualitative, as it has not been possible to determine the effective stresses and strains in the soil during compaction. Nevertheless, the theory provides an insight into the compaction process that was not available previously.

Note.—Discussion open until August 1, 1963. To extend the closing date one month, a written request must be filed with the Executive Secretary, ASCE. This paper is part of the copyrighted Journal of the Soil Mechanics and Foundations Division, Proceedings of the American Society of Civil Engineers, Vol. 89, No. SM2, March, 1963.

1 Assoc. Prof. of Civ. Engrg., Univ. of Illinois, Urbana, Ill.

INTRODUCTION

In the performance of engineering works, it is often necessary to excavate, transport, and place soil. The properties of the reworked soil depend on the placement procedure. In most cases, loose dumped soil is too weak and compressible to give satisfactory performance, so the soil is compacted. Soils are compacted to increase their shearing strength, decrease their compressibility, and, in some cases, to decrease their permeability. The effect of compaction on these properties has been studied extensively. However, the compaction process itself is only poorly understood. Considering the large quantities of soils compacted annually, it seems desirable to attain a more thorough understanding of the mechanics of the compaction process.

Soil samples used in a preliminary laboratory investigation were prepared by a kneading compaction procedure. The theory of compaction that was developed applies directly to kneading compaction but, with minor modifications, it can be applied to other compaction procedures. Furthermore, the preliminary experiments were performed using soils that contained relatively small quantities of sand and no coarser material. Modifications to the shape of the compaction curve occur when the percentages of sand and gravel become sufficiently large that the coarser particles come into physical contact during compaction. In that case, the shape and grading of these particles control whether the compaction curve is sharp or flat and whether it is symmetrical or skewed. No tests were performed with such soils; hence, the theory, as developed in this investigation, pertains only to fine grained soils.

Notation.—Letter symbols adopted for use in this paper are defined where they first appear and are arranged alphabetically in the Appendix.

REVIEW OF SHEAR STRENGTH THEORY

Introduction.—The effective stress theory of soil compaction is based on the assumption that soil resists the compaction pressure by the development of shearing stresses on the contact surfaces between the particles. Sufficient shearing stresses must be developed to resist the compaction pressure; otherwise, a complete shearing failure results. Modern soil mechanics research has shown conclusively that the shearing resistance of a soil is a function of the state of effective stress, not total stress. It follows, therefore, that this theory must utilize effective stresses.

The effective stress approach to compaction makes use of certain principles that are familiar to those engaged in research, but may not be familiar to those engaged in engineering practice. Hence, this section is a brief review of some of the pertinent knowledge relating to the shearing characteristics of cohesive soils. This section also serves to introduce the terminology and symbolism to be followed.

Coulomb Equation.—It is common practice to express the shear strength of soil with the Coulomb equation:

$$\tau = c + \sigma_n \tan \phi \quad \text{.................. (1)}$$

in which τ is the shear strength, c is the cohesion, σ_n is the normal stress on the failure plane, and ϕ is the angle of internal friction. It is convenient to use certain special symbols to represent particular types of shearing tests. Most shearing tests are performed in the triaxial device where it is convenient to define three types of tests and to represent each type with a symbol:[2] Q designates an unconsolidated-undrained test; R designates a consolidated-undrained test; and S designates a consolidated-drained test.

The parameters in the Coulomb equation are given subscripts to designate the type of triaxial test used in defining the parameter, for example ϕ_R. If the test results are expressed in terms of effective stresses, then a bar is placed over the appropriate symbol, for example, $\bar{c}_R$.

If the results of a large number of triaxial tests are plotted on a single Mohr-Coulomb diagram, the failure envelope is obscured by the presence of so many circles. Therefore, it has become customary[3] to plot data in modified Mohr-Coulomb diagrams in which $1/2(\sigma_1 - \sigma_3)$ is plotted against $1/2\ (\sigma_1 + \sigma_3)$, in terms of either total or effective stresses. The main advantage of this plot is that each test is represented by a single point rather than a circle. The slope of the resulting failure envelope is designated as ψ and the intercept as d. The same subscripts are used as with ϕ and c. The modified shearing parameters can be converted to the standard parameters using the equations:

$$\sin \phi = \tan \psi \quad \text{.................. (2)}$$

and

$$c = \frac{d}{\cos \phi} \quad \text{.................. (3)}$$

Effective Stress.—Numerous investigations have shown that the physical properties of a particular saturated soil are functions of the effective stresses in the soil, not the total stresses. Unfortunately, effective stresses are not measured directly. For saturated soils, the effective stress is calculated, with reasonable accuracy, using the equation proposed[4] by K. T. Terzaghi,

---

2 "Moderator's Report, Session 2, Testing Equipment, Techniques, and Errors," by A. Casagrande and S. D. Wilson, Proceedings, ASCE Research Conf. on Shear Strength of Cohesive Soils, Boulder, Colo., 1960, pp. 1123-1130.

3 "Factors Controlling the Strength of Partly Saturated Cohesive Soils," by A. W. Bishop, I. Alpan, G. E. Blight, and I. B. Donald, Proceedings, ASCE Research Conf. on Shear Strength of Cohesive Soils, Boulder, Colo., 1960, pp. 503-532.

4 "The Shearing Resistance of Saturated Soils," by K. T. Terzaghi, Proceedings, Internatl. Conf. on Soil Mechanics and Foundation Engrg., Harvard Univ., Cambridge, Mass. Vol. 1, 1936, pp. 54-56.

Hon. M. ASCE:

$$\bar{\sigma} = \sigma - u_w \quad \text{.................. (4)}$$

in which $\bar{\sigma}$ is the effective stress, σ is the total stress, and u_w is the pore-water pressure.

Unsaturated soils contain a third phase, air, with a pressure u_a. A. W. Bishop[5] has suggested that the effective stress in unsaturated soils could be defined using the equation:

$$\bar{\sigma} = \sigma - u_w (\chi) - u_a (1 - \chi) \quad \text{...............(5)}$$

in which χ is an empirical coefficient. The χ-coefficient varies with the degree of saturation, being unity for a saturated soil, zero for a dry soil, and having intermediate values for partially saturated soils. Bishop, et al,[3] have demonstrated the usefulness of Eq. 5 in analysis of the shear strength of soils with high degrees of saturation (usually over 60%). J. E. Jennings and J. B. Burland[6] have pointed out certain anomalies in the application of Eq. 5 to the consolidation of metastable soils with low initial degrees of saturation.

Pore Pressures.—Considerable clarity in the analysis of pore pressures has been achieved through use of the A and B parameters proposed by A. W. Skempton.[7] Skempton proposed that the pore pressure (Δu) resulting from a change in ambient stress ($\Delta\sigma_3$) and a change in stress difference ($\Delta\sigma_1 - \Delta\sigma_3$) could be expressed by the equation:

$$\Delta u = B\left[\Delta\sigma_3 + A\left(\Delta\sigma_1 - \Delta\sigma_3\right)\right] \quad \text{...............(6)}$$

in which A and B are empirical parameters. Skempton[7] did not distinguish between pore-air pressure and pore-water pressure although these pressures are different. Bishop[8] suggested use of two equations:

$$\Delta u_w = B_w\left[\Delta\sigma_3 + A_w\left(\Delta\sigma_1 - \Delta\sigma_3\right)\right] \quad \text{.............. (7a)}$$

and

$$\Delta u_a = B_a\left[\Delta\sigma_3 + A_a\left(\Delta\sigma_1 - \Delta\sigma_3\right)\right] \quad \text{.............. (7b)}$$

to express the pore-water pressure and the pore-air pressure separately. Presumably, these two equations can then be inserted into Eq. 5 to obtain the effective stress.

There is a negligible amount of data available regarding the A and B coefficients for unsaturated soils; hence, suitable values must be inferred from data obtained using saturated soils. For saturated soils, the problem is sim-

5 "The Measurement of Pore Pressure in the Triaxial Test," by A. W. Bishop, Proceedings, Conf. on Pore Pressure and Suction in Soils, 1961, pp. 38-46.

6 "Limitations to the Use of Effective Stresses in Partly Saturated Soils," by J. E. Jennings and J. B. Burland, Géotechnique, Vol. 12, pp. 125-144.

7 "The Pore-Pressure Coefficients A and B," by A. W. Skempton, Géotechnique, Vol. 4, 1954, pp. 143-147.

8 "Discussion," by A. W. Bishop, Proceedings, Conf. on Pore Pressure and Suction in Soils, 1961, pp. 38-46.

plified in that B can be taken as 1.0 within the range in stress usually encountered (typically less than 500 psi). The A-coefficient varies with strain [Fig. 1(a)]. For normally consolidated samples, the A-coefficient at failure (A_f) varies with the consolidation pressure,[9] as shown in Fig. 1(b), and with overconsolidated samples it varies with the overconsolidation ratio,[9,10,11] as shown in Fig. 1(c).

There are essentially no data available concerning the A-coefficients of unsaturated soils. It will be assumed that A_w and A_a vary with consolidation pressure and overconsolidation ratio as shown in Figs. 1(b) and 1(c) though the numerical values may be different from the values indicated in the figures.

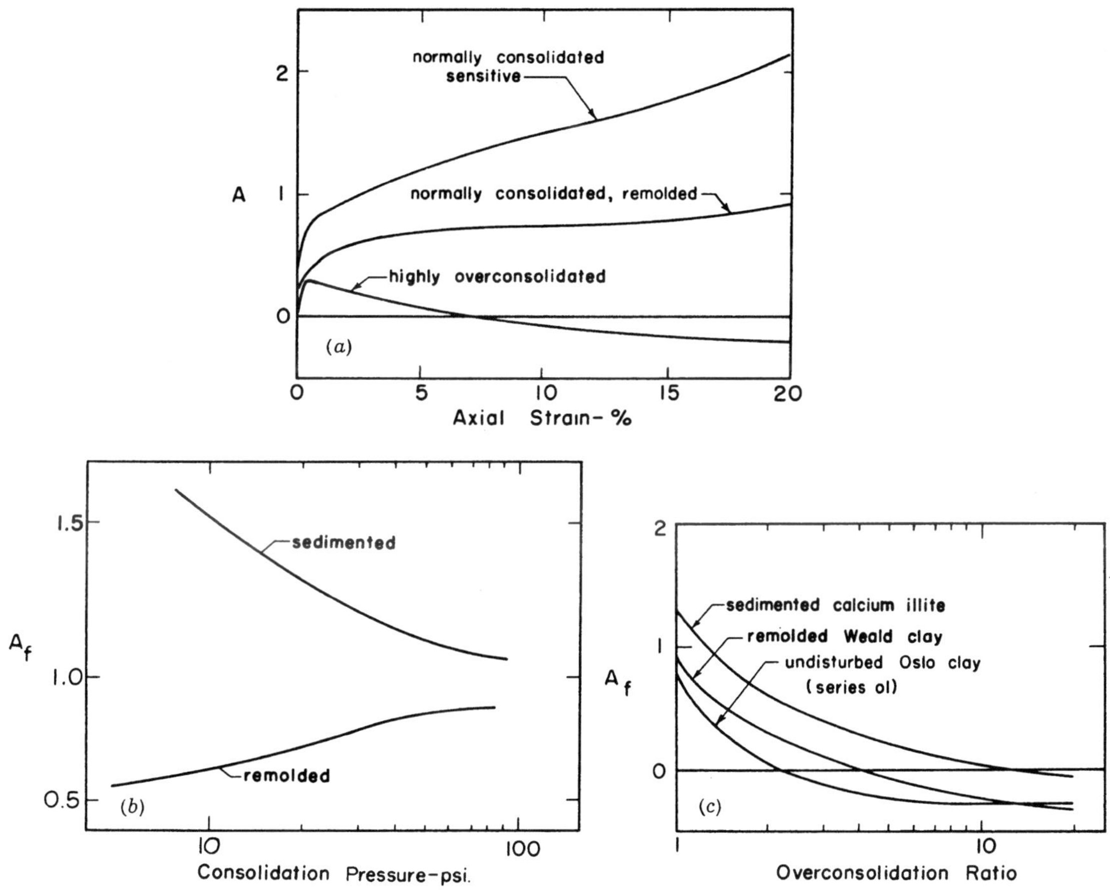

FIG. 1.—A-COEFFICIENTS OF SATURATED CLAYS

For soil samples subjected to a given ambient total stress, the A-coefficients will be inversely related to the compacted dry density; that is, dense samples will be dilatant and negative pore pressures will develop during shear,

---

9 "The Shear Strength Properties of Calcium Illite," by Roy E. Olson, Geotechnique, Vol. 12, 1962, pp. 23-43.

10 "The Effect of Overconsolidation on the Behavior of Clays During Shear," by D. J. Henkel, Géotechnique, Vol. 6, 1956, pp. 139-150.

11 "The Effect of Overconsolidation on the Shear Strength Characteristics of an Undisturbed Oslo Clay," by N. E. Simons, Proceedings, ASCE Research Conf. on Shear Strength of Cohesive Soils, Boulder, Colo., 1960, pp. 747-763.

whereas loose samples of the same soil will tend to collapse during shear and positive pore pressures will result. The A-coefficients may also vary with the initial degree of saturation, but the B-coefficients will account for much of this effect, as the shear induced pore pressures are $AB(\Delta\sigma_1 - \Delta\sigma_3)$.

Available data on B_a and B_w are also of negligible quantity although undifferentiated B-coefficients have been measured by several investigators.[7,12,13] However, on the basis of theoretical work of J. H. A. Brahtz, C. N. Zanger, and J. R. Bruggeman[14] and Jack W. Hilf,[15] F. ASCE, and the available experimental data, it seems apparent that B_a will be much smaller than B_w for most soils compacted on the dry side of the optimum moisture content. This follows from the fact that air is far more compressible than water. The coefficient B_w will vary with the initial degree of saturation and with the increment of applied ambient stress. The general form of the relationship between B_w $\Delta\sigma_3$, and S_r is indicated in Figs. 2(a) and 2(b). The

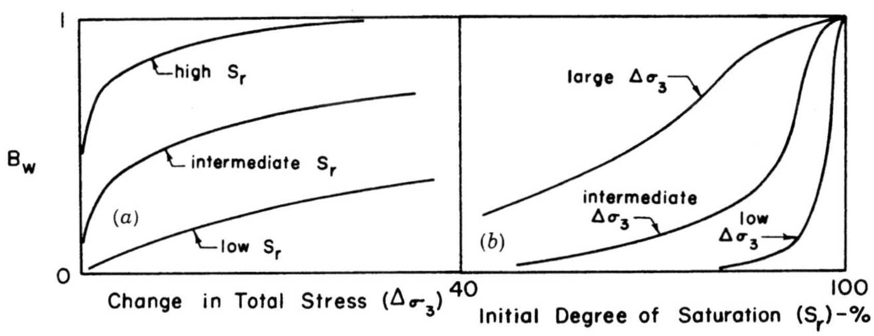

FIG. 2.—EFFECT OF INITIAL DEGREE OF SATURATION (S_r) AND STRESS INCREMENT ($\Delta\sigma_3$) ON B_w

curves in Fig. 2 are crude estimations of the type of relationships that will probably be obtained when suitable experiments can be performed; they do not represent either theoretical calculations or experimental observations.

EXISTING THEORIES OF COMPACTION

Introduction.—The main purpose of a theory of compaction is to explain the effect of moisture on the compacted dry density. However, if the theory is sufficiently general, it should also explain the effect of both moisture con-

---

12 "Investigation of Stress-Deformation and Strength Characteristics of Compacted Clays," by A. Casagrande and R. C. Hirschfeld, Harvard Soil Mechanics Series No. 61, Cambridge, Mass., 1960.

13 "Stress-Deformation and Strength Characteristics of a Clay Compacted to a Constant Dry Unit Weight," by A. Casagrande and R. C. Hirschfeld, Proceedings, ASCE Research Conf. on Shear Strength of Cohesive Soils, Boulder, Colo., 1960, pp. 359-418.

14 "Notes on Analytical Soil Mechanics," by J. H. A. Brahtz, C. N. Zangar, and J. R. Bruggeman, Technical Memorandum No. 592, Bur. of Reclamation, U. S. Dept. of the Interior, Denver, Colo., 1939.

15 "An Investigation of Pore-Water Pressure in Compacted Cohesive Soils," by Jack W. Hilf, Technical Memorandum No. 654, Bur. of Reclamation, U. S. Dept. of the Interior, Denver, Colo., 1956.

tent and density on the shear strength, consolidation characteristics, and permeability of compacted soils. The existing compaction theories have fulfilled these aims with only limited success.

Proctor's Theory.—Proctor, F. ASCE, concluded that moisture exerted a dual effect on the soil.[16] The first effect was capillarity:

> "The moisture contained in a very dry soil surrounds each particle as a thin film held in place by the force of surface tension. Where the films come in contact, the capillary force caused by the surface tension of the joined films draws the particles firmly together, causing a high frictional resistance between them."

The second effect was one of lubrication. Proctor concluded that the water lubricated the particles, reducing the interparticle friction, and therefore reducing the shearing strength. The reduction in shearing strength caused the compacted dry density to increase because the particles could slide over one another more easily during compaction. Most of Proctor's examination of compaction was based on the lubrication principle. Proctor's analysis of soil compaction predated the general understanding of the principles of effective stress. Nevertheless, his concept that capillary-induced stresses give the soil a shearing strength that resists compaction is in accordance with more recent observations. Unfortunately, he put more emphasis on the concept of lubrication.

Observations by H. M. Horn[17] have shown that water does act as a lubricant for the sheet silicates, reducing the friction by approximately two times. However, Horn's tests also showed that water is an antilubricant for the "three-dimensional network" silicates (for example, quartz and feldspar), the submerged coefficient being approximately five times the oven dry value. If the lubrication theory is valid, then a well graded quartz sand should have its maximum dry density at zero water content where the friction is a minimum. Actually, well graded quartz sands have moisture-density curves of the same general shape as those of other soils. In any case, Horn's tests indicate that nearly complete lubrication is achieved by equilibrating the mineral surfaces in an atmosphere of high relative humidity. The water contents of soils equilibrated in highly humid atmospheres are usually much lower than the optimum water content. The mineral surfaces are likely to be fully "lubricated" throughout the range in water content used in most compaction tests.

Hogentogler's Theory.—C. A. Hogentogler,[18] M. ASCE, advanced the "viscous water theory" of soil compaction. Hogentogler recognized that water is adsorbed onto the surface of soil particles. He postulated that the first layer of the adsorbed water was highly cohesive and that the cohesiveness of the water decreased with distance from the soil surface, the adsorbed water grading into the free pore water. The soil particles were assumed to be sep-

---

16 "Fundamental Principles of Soil Compaction," by R. R. Proctor, Engineering News-Record, McGraw-Hill Publ. Co., Inc., New York, N. Y., Vol. 111, 1933, p. 286.

17 "An Investigation of the Frictional Characteristics of Minerals," by H. M. Horn, thesis presented to the University of Illinois, in Urbana, Ill., in 1960, in partial fulfilment of the requirements for the degree of Doctor of Philosophy.

18 "Essentials of Soil Compaction," by C. A. Hogentogler, Proceedings, Highway Research Bd., Natl. Research Council, Washington, D. C., 1936, pp. 309-316.

arated at the contact surfaces by this viscous water. According to this theory, at low water contents, the high viscosity of the water between the contact points gives the soil a high shearing strength and results in low dry densities. As more water is added to the soil, the adsorbed water layers become thicker and less cohesive, resulting in a lower shearing strength for the soil and a higher dry density. The reduction in viscosity occurs only to a certain water content, after which the addition of more water results in "lubrication." The maximum amount of lubrication was postulated to occur at the optimum water content. The addition of water beyond the optimum water content results in displacement of the solid material by the water with a consequent reduction in the compacted dry density.

Hogentogler's theory was developed at a time when the thickness of adsorbed water films was considered to be great and when little experimental information was available concerning the properties of adsorbed water. Observations since 1955[19,20,21,22] seem to indicate that the thickness of the adsorbed water on the surface of most clays does not exceed a few molecular layers. Data on the viscosity of the adsorbed water are somewhat contradictory but none of the data seems to indicate high viscosities beyond the first few layers. The water contents at which most soils are compacted appear to be far in excess of the water contents where high adsorbed water viscosities could be important.

Lambe's Theory.—T. W. Lambe, F. ASCE, attempted to explain the shape of the compaction curve in terms of surface chemical theories.[23] At low water contents, he concluded, the electrolyte concentration in the pore water is high. The high electrolyte concentration results in reduced osmotic repulsion between soil particles, which, in turn, allows flocculation to occur. The flocculent particle arrangement was assumed to produce the low dry densities found for compaction at low water contents. Lambe concluded that, as water is added to the soil, the pore-water electrolyte concentration is reduced, allowing double layers to develop more fully and producing a more dispersed soil structure. The increased dry density was assumed to result from such dispersion that ". . . permits the particles to slide past each other into a more oriented and denser bed."[24] Lambe considered the term "lubrication" to describe properly the effect of adding more water to the soil.

Lambe's approach to compaction seems predicated on the belief that the forces producing flocculation are strong and that relatively minor changes in the pore-water electrolyte concentration result in large changes in these

---

[19] "Lattice Expansion and Rheological Behavior Relationships in Water-Montmorillonite Systems," by William R. Foster, J. G. Savins and J. M. Waite, Clays and Clay Minerals, Proceedings, 3rd Natl. Conf., Natl. Acad. of Science - Nat. Research Council Publ. No. 395, 1955, pp. 296-316.

[20] "Dehydration of Monoionic Montmorillonites," by R. A. Rowland, E. J. Weiss and W. F. Bradley, Clays and Clay Minerals, Proceedings, 4th Natl. Conf., Natl. Acad. of Science - Natl. Research Council Publ. No. 456, 1956, pp. 85-95.

[21] "The Mechanism of Dehydration of Mg-Vermiculite," by G. F. Walker, Clays and Clay Minerals, Proceedings, 4th Natl. Conf., Natl. Acad. of Science - Natl. Research Council. Publ. No. 456, 1959, pp. 101-115.

[22] "Water-Vapor Sorption on Kaolinite, Hysteresis," by R. T. Martin, Clays and Clay Minerals, Proceedings, 6th Natl. Conf., 1959, pp. 259-278.

[23] "Structure of Compacted Clay," by T. W. Lambe, Transactions, ASCE, Vol. 125, 1960, pp. 682-705.

[24] Ibid., p. 697.

forces. Lambe's theory is based on observations[25,26,27] that the Gouy-Chapman double layer theory could be used to predict the swelling curve of sodium montmorillonite and to predict the shape, but not the position, of the swelling curve of sodium illite. However, it has been shown[28] that the existing double layer theory does not apply to calcium illite and yet calcium illite has a moisture-density curve of the same shape as those of other soils.[29]

On the basis of these and other data, it is concluded that variations in the pore-water electrolyte concentration exert an influence on the shape and position of the moisture-density curve for some, but not all, soils and for some, but not all, adsorbed cations, and that the existing physico-chemical theory of compaction does not have general applicability. It should be emphasized, however, that any attempt to understand thoroughly the engineering properties of clays must ultimately resolve into a study of the physico-chemical interaction between particles. Hence, the objection to the existing theory is that it is wrong in theory, not that it is basically wrong in concept or method of approach.

Conclusions.—There have been three significant attempts to explain the effect of moisture on the compacted dry density of cohesive soils. These include Proctor's capillarity and lubrication theory, Hogentogler's viscous water theory, and Lambe's physico-chemical theory. Each theory was reasonably consistent with existing knowledge at the time the theory was developed. However, advances of knowledge in supporting fields of science have applied limitations to these theories that were not apparent when the theories were being developed.

Furthermore, none of the theories seems to have been subjected to diagnostic laboratory experiments. For example, if the function of the water is that of a lubricant, then a moisture-density curve of the usual shape should not be obtained if a non-lubricating fluid, for example, carbon tetrachloride, is used in place of water. If Lambe's application of double layer theory is correct, then a typical moisture-density curve should not result if a fluid with a low dielectric constant, for example, benzene, is used in place of water. As a result of the lack of diagnostic laboratory experiments, the existing theories are largely speculative.

The existing theories have not been extended, if in fact they could be extended, to explain the shearing strengths or stress-strain properties of compacted soils. The vague statement that the addition of moisture reduces the strength of the soil and results in densification is rather unsatisfactory, as there is no proof that such a reduction actually occurs. Certainly, for com-

---

25 "Compression Studies of Illite Suspensions," by G. H. Bolt and R. D. Miller, Proceedings, Soil Science Soc. of Amer., Vol. 19, 1955, p. 285.

26 "Physico-Chemical Analysis of the Compressibility of Pure Clays," by G. H. Bolt, Géotechnique, Vol. 6, 1956, pp. 86-93.

27 "Swelling Pressure of Montmorillonite," by B. P. Warkentin, G. H. Bolt and R. D. Miller, Proceedings, Soil Science Soc. of Amer., Vol. 21, 1957, pp. 495-497.

28 "Shear Strength and Consolidation Characteristics of Calcium and Magnesium Illite," by Roy E. Olson and F. M. Mitronovas, Clays and Clay Minerals, Proceedings, 9th Natl. Conf., 1962, pp. 185-209.

29 Discussion by R. E. Olson and J. D. Scott of "Structure and Strength Characteristics of Compacted Clays," by H. B. Seed and C. K. Chan, Transactions, ASCE, Vol. 126, Part I, 1961, pp. 1398-1404.

paction on the dry side of optimum, the unconfined compression strength is increased when the water content is increased.

EFFECTIVE STRESS THEORY OF SOIL COMPACTION

Introduction of General Concept.—Prior to compaction, the soil is in a loose condition and possesses a low shearing strength. When the compaction pressure is first applied, the shearing stresses developed between the particles reach the shearing strength of the contact surfaces, the contacts yield, the particles slide over each other, and the density of the soil increases. Simultaneously, both the total stresses and the pore pressures increase, but the pore pressures increase by less than the total stresses because the B-coefficients are less than one. Hence, the effective stresses between the particles increase as the "blow" is being applied. Deformation of the soil continues until the increasing effective stresses give the soil sufficient shearing strength to resist the foot pressure. The foot pressure is then released and the soil has been subjected to one "blow."

When the foot pressure is released, the vertical total stress becomes zero. The soil expands slightly in the vertical direction with a simultaneous reduction in the lateral total stresses. Expansion of the soil is resisted by the development of negative pore-water pressures. After a relatively small expansion, the negative pore-water pressures and the residual lateral total stresses produce sufficient compressive effective stresses in the soil so that the densified condition is maintained.

When the compaction pressure is re-applied, shearing deformations again occur, there are increases in the total stresses and pore pressures, and the effective stresses develop sufficiently for the soil to resist the compaction pressure. The yielding of some of the interparticle contacts during the stress application will result in higher densities than were developed when the first blow was applied. Again, when the foot pressure is removed, expansion of the soil is resisted by negative pore-water pressures and by total lateral stresses.

The shearing strength of the soil increases as more and more "blows" are applied. Eventually, the soil may become sufficiently strong so that no interparticle contacts yield when a blow is applied. In this case, the soil has reached its maximum density for the particular compaction procedure.

If the moisture content of the soil is increased, the higher degrees of saturation will produce higher B-coefficients and higher pore-water pressures. Because the same foot pressure must be resisted, it is apparent that approximately the same interparticle effective stresses must be developed during compaction. An important aspect of the effective stress theory of compaction is the explanation of why greater dry densities are developed as the water content is increased. The explanation is in terms of the lateral total stresses, the residual negative pore-water pressures, and the pore-pressure changes during the stress application. These pore-pressure changes may involve both the air and the water and they result from changes in both the ambient stress (B-coefficients) and the shearing stress (A-coefficients). In addition, it must be recognized that the effective stresses resulting from the changed total stresses and pore pressures are influenced by the χ-coefficient. The interplay of these various parameters as functions of foot pressure, water content, and deformation conditions controls the compacted dry density.

For compaction on the dry side of the optimum water content, the effective stress theory of compaction will demonstrate why increases in compaction water content necessitate increases in the dry density. However, for compaction at water contents slightly greater than the optimum point, an increase in water content results in a decrease in the dry density, yet the foot pressure is still fully resisted by the soil. It is apparent that additional, and perhaps different, variables must be taken into account.

Kneading compaction tests on fine-grained soils often yield moisture-density curves with two distinct peaks (Fig. 3). The effective stress theory of compaction applies only to the peak at higher water contents, that is, to the moisture-density curve corresponding to field compaction. A different theory will be examined which may explain the existence of a second peak in the moisture-density curve at low moisture contents.

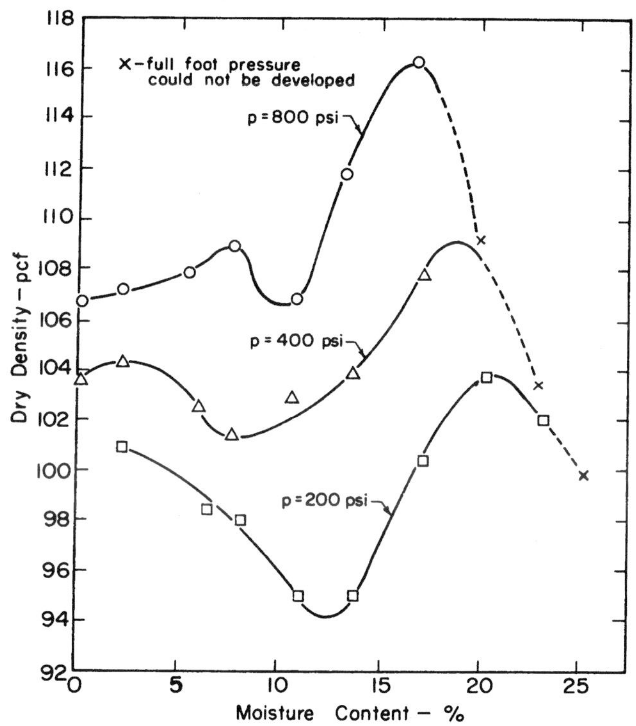

FIG. 3.—TYPICAL MOISTURE-DENSITY CURVES DERIVED FROM KNEADING COMPACTION OF AN ILLITIC CLAY

Effective Stress Theory of Compaction on Dry Side of Optimum.—The soil is assumed to be in a loose condition prior to the start of compaction. The effective stresses in the soil are small because of the low confining pressure, the low B-coefficients, and the low χ-coefficient. The low value of the χ-coefficient prevents the pore-water pressure, which may be very negative, from exerting much influence on the effective stresses. As a result of the small effective stress, the soil is weak.

When the first "blow" is applied, the element of soil receiving the compaction pressure is nearly unconfined. The compaction foot must penetrate nearly to the surface of the layer previously compacted before sufficient lateral shearing stresses can be developed to prevent complete extrusion of the soil. These horizontal shearing stresses provide the required confinement; the effective stresses increase sufficiently to resist the applied stress. During the application of the blow, the change in the pore-air pressure is insignificant (for kneading compaction) because of the great compressibility of the air and because the loose soil has a high air permeability. Assuming that the loose soil approximates a normally consolidated sediment, A_w is probably greater than 1 but the low value of B_w will prevent the development of significant pore-water pressures. The low χ-coefficient means that the pore-water pressures actually developed during compaction will exert a relatively small influence on the effective stresses.

The element of soil is in a much denser condition at the start of the second blow than at the beginning of compaction. Essentially, the same remarks apply for the application of the second blow as for the first with the exception that the denser surrounding soil provides more lateral confinement. The application of the foot pressure still causes shearing deformations, which break down some of the interparticle contacts, expel pore air, and increase the dry density. However, as a result of the greater lateral confinement, the penetration of the foot is less for the second blow than for the first, and the increase in dry density is smaller.

The same process is repeated cycle after cycle except that each time a new cycle of blows is started, the soil is stronger and undergoes smaller deformations and less densification than under the previous cycle of blows. For compaction at low water contents, the dry density increases rapidly during the first few stress applications but further density increases are small (Fig. 4).

A second batch of soil is prepared at a higher water content and compacted in the same manner. The essential difference between the first and second batches is that the higher water content of the second batch results in higher degrees of saturation for the same dry densities. Hence, B_w is larger. Furthermore, the pore-water pressure will exert a greater influence on the effective stresses because the χ-coefficient is also larger.

The effective normal stresses between particles, multiplied by the coefficient of surface friction, give the soil the strength to resist compaction. The effective normal stresses are increased when the total stresses are increased and are also increased if negative pore-water pressures are developed. The effective normal stresses are decreased by the development of positive pore-water or pore-air pressures. Decreases in the effective stresses reduce the shearing strength of the interparticle contacts and allow the particles to slide over one another during compaction, thus increasing the dry density.

When the first blow is applied to the second batch of soil, the foot again penetrates nearly to the surface of the layer previously compacted before sufficient lateral shearing stresses are developed to prevent complete extrusion of the soil. The A-coefficients are probably nearly the same as for the first batch of soil, but B_w is larger; hence, the pore-water pressure changes in a positive direction during compaction and weakens the soil. The actual pore-water pressure may remain negative during the application of the blow, however, because u_w was probably quite negative prior to compaction. The higher degree of saturation results in a higher χ-coefficient; hence, the pore-water pressure exerts a greater influence on the effective stress than it did for compaction at the lower water content. The reduced effective stresses resulting from the larger B and χ-coefficients allow the particles to slide over each other with less resistance and the full foot pressure can be developed only if the soil undergoes larger deformations that permit the development of larger lateral total stresses.

When the foot pressure is released, the total axial stress is reduced to zero. The densified condition of the soil is preserved by the residual lateral stresses and by the development of negative pore-water pressure. Naturally, the higher dry density of the second sample necessitates that the residual effective stresses be larger than they were for the first batch.

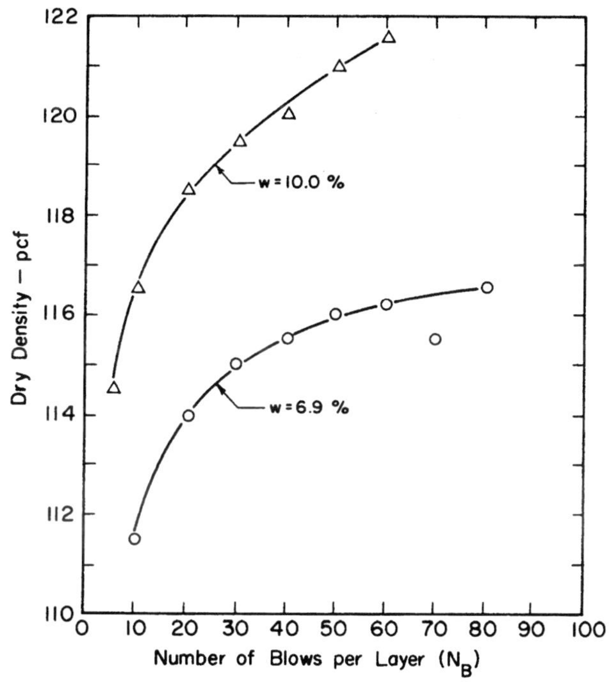

FIG. 4.—EFFECT OF NUMBER OF STRESS APPLICATIONS ON COMPACTED DRY DENSITY OF A SILTY CLAY; OPTIMUM WATER CONTENT 12%

When the second blow is applied, the increased axial total stress produces a vertical deformation that, in turn, causes an increase in the lateral confining pressure. The increased confinement increases the effective stresses in the soil; it also increases the pore pressures. The shearing stresses also induce positive changes in the pore pressure. Some of the interparticle contacts will yield and further densification occurs. This process is repeated for cycle after cycle of blows. The dry density increases slightly if any interparticle contacts are ruptured. After each blow, the increased dry density is preserved by the development of negative pore-water pressures and by the residual lateral total pressures. The increased dry density has the same general effect as overconsolidation; that is, it causes a reduction in the A_W-coefficient. Hence, the shear-induced increase in pore-water pressure, which occurs each time a blow is applied, will become less and less as the number of blows is increased. The combination of increased lateral pressure, increasingly negative residual pore-water pressure, and smaller increases in pore-water pressure when the foot pressure is applied, all lead to higher effective stresses and greater shearing strength. If enough blows are applied, the soil may become sufficiently strong so that no further densification occurs.

Deformation must also be considered. The A-coefficients used in the preceding considerations are not those corresponding to failure. When the first blow is applied, the soil is strained considerably beyond "failure" and large A-coefficients should be expected [Fig. 1(a)]. However, as the soil becomes stronger, the deformation becomes progressively smaller and the A-coefficients may become much less than the failure values. Hence, there are two reasons why the increase in pore-water pressure during the application of the foot pressure becomes progressively smaller as the number of blows is increased, namely, the increased dry density, which has the general effect of overconsolidation, and the decreased strains.

In review, for compaction at relatively low water contents, increases in the compaction water content increase the degree of saturation, which increases the B-coefficients and χ. The higher induced pore-water pressure, during the first blow, weakens the soil by reducing the effective stresses between the particles. The particles slide over one another until sufficiently large lateral stresses, and horizontal shearing stresses with the layer previously compacted, have developed to give the soil the requisite effective stress. As the soil is subjected to further blows, the effective stresses increase because of increases in the residual lateral total stresses, increasingly negative residual pore-water pressures, and because the shear-induced increases in pore-water pressure become progressively smaller. Small increases in dry density continue to occur as more and more blows are applied since local concentrations of shear stress will cause localized densification.

The same line of reasoning applies as the water content is increased further, except that the decreasing air permeability may result in the development of significant pore-air pressures. Eventually, enough water may be added to the soil so that the air channels become discontinuous and the air is trapped. When the air voids become completely discontinuous, the air permeability of the soil drops to zero (Fig. 5), and no further densification is possible. The soil has reached the so-called "optimum water content."

Cohesional Strength.—The previous analysis of the compaction process has tacitly assumed that the shearing strength of the interparticle contact surfaces is a purely frictional phenomenon. This seems to be a reasonable assumption for the average compacted soil, which has a relatively low plasticity. The development of relatively small amounts of cohesion do not modify the effective stress approach significantly.

For clays of very high plasticity, for example, sodium montmorillonite, the strength of the interparticle contacts may be purely cohesional even though the effective stress Mohr-Coulomb diagrams have an appreciable slope. In such soils, the effective stress approach to compaction may not apply. However, such soils are seldom used in the field and no attempt has been made to extend the effective stress theory of compaction to include these soils.

Compaction at Water Contents Above Optimum.—The problem remains to explain how the soil can develop sufficient strength to resist the foot pressure when the water content is increased above the optimum point because, at these water contents, increases in water content cause decreases in the compacted dry density? The solution to this problem may consist of two parts, both related to the large foot penetrations encountered on the wet side of optimum.

First, for a dense soil, A_W peaks at small strains and then decreases continuously with strain [Fig. 1 (a)] becoming negative for highly overconsolidated soils at high strains. It is possible that the pore-water pressure increases slightly as the foot pressure is applied and then decreases again, becoming sufficiently negative at large strains to allow the soil to resist the foot pressure.

Second, for a slight penetration, the compaction foot is somewhat similar to a shallow footing. The bearing capacity of a shallow footing is approximate-

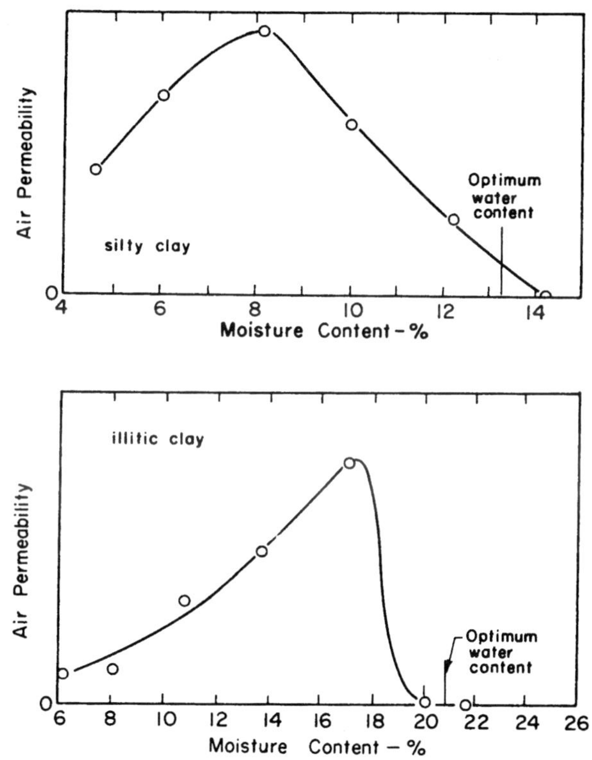

FIG 5.—AIR PERMEABILITY CURVES OF COMPACTED SOILS

ly six times the shearing strength of the soil (assuming a shear type failure). When the foot penetrates deeply into the soil, it simulates the action of a deep footing where the bearing capacity is perhaps ten times the shearing strength of the soil. Hence, the soil can resist the same foot pressure with a smaller shearing strength, provided the foot penetration increases considerably. This is precisely what happens when the water content is increased beyond that at the optimum point. If the water content is increased too much, the foot penetrates all the way to the base of the mold and the test is concluded. The same process occurs in the field when sheepsfoot rollers sink into the soil until the weight is carried by the drum.

It must be emphasized that this presentation applies to kneading-type compaction. It is possible to compact to water contents well beyond the optimum point, if dynamic compaction procedures are used, since the soil need have only sufficient shearing strength to carry the weight of the drop hammer plus enough viscosity to stop the falling hammer.

Physico-Chemical Variables.—The physico-chemical variables certainly control the shape of the moisture-density curve, because they control the electrical forces of interaction between the particles that, in turn, control the effective stress-strain properties of the soil. However, the existing physico-chemical theories pertaining to the interaction of real soil particles are in such a primitive state of development that they cannot be applied to such a complex problem as the compaction of clays. It seems preferable to use the admittedly empirical and qualitative effective stress approach to soil compaction and to take into account the effect of the physico-chemical variables through suitable adjustments to the effective stress-strain properties of the soil.

LOW WATER CONTENT MOISTURE-DENSITY CURVE

The main purpose of this study is to consider the effect of the placement water content on the compaction characteristics and engineering properties of clays within the water content range usually encountered in field compaction. Nevertheless, the peak in the moisture-density curve at low water contents in some soils (Fig. 3) is of interest. Only a few laboratory tests have been performed in this range in water content and the results of these tests have been inconclusive. Hence, the explanation of this moisture-density curve must be considered as tentative.

The surface friction between sheet silicates is high when the surfaces are dry.[17] This high friction gives the oven dry soil a high shearing strength, for a given effective stress, and results in a low dry density. The first small amount of water added to the soil hydrates the adsorbed cations and the clay surface. This water literally acts as a lubricant, because it reduces the friction between the sheet silicate particles. The soil must densify until the combination of increased confining pressure and increased particle interference gives the soil sufficient strength to resist the compaction pressure. The lubrication theory of Proctor[16] and the viscous water theory of Hogentogler[18] appear to apply to compaction in this range of water content.

The electrical forces attracting the first layer of water to the adsorbed cations and to the mineral surfaces are stronger than the forces between adjacent water molecules. Hence, no water menisci are formed at very low water contents. The lack of water menisci seems demonstrated by the fact that the soil behaves as a cohesionless dusty material at very low water contents. The forces holding the water to the adsorbed cations and to the mineral surfaces decrease as the thickness of adsorbed water increases. Hence, after several molecular layers of water have formed on all surfaces, the forces between water molecules will be of the same order of magnitude as the adsorptive forces and water menisci will begin to form. The water menisci induce a pressure difference between the air (which is at atmospheric pressure) and the pore water (which must, therefore, be negative). The development of negative pore-water pressure results in an effective stress in addition to the effective stress caused by the confining action of the mold. Hence, the foot pressure can be resisted with a slightly smaller amount of particle interference and with a smaller lateral pressure and, as a result, a smaller dry density is developed. The balance point between the decreasing soil strength due to lubrication, and the increased soil strength due to particle interference, build-up of lateral pressures, and formation of water menisci occurs at the peak point of the low water content moisture-density curve.

At the peak point of the low water content moisture-density curve, there is probably enough water to hydrate the adsorbed cations and the surfaces of the clay, and to form a limited number of water menisci, but not menisci between all particles at all points of contact. Furthermore, the first tiny menisci probably do not obey the usual equation for the pressure difference across a curved surface, since the water molecules are large compared to the radius of curvature of each meniscus. As water is added beyond the low water content optimum point, menisci form between more and more particles and undistorted menisci form. The resulting increases in effective stress cause the compacted dry density to decrease because less lateral pressure and less particle interference are required to resist the foot pressure. When all the menisci in the soil are fully developed, the minimum point between the two parts of the moisture-density curve is reached. The addition of more water flattens the menisci and reduces the pressure differential across the curved menisci; whereupon the considerations developed previously should begin to apply.

It appears that the low water content moisture-density curve can form only in soils containing a dominant percentage of plate-shaped colloidal particles. The curve has not been observed with soils that contained only small amounts of clay.

The foregoing is based on limited information and will probably need revision as further data become available. The low water content curve was included in this study only because some of the early compaction theories seem to have application to this curve rather than to the curve usually encountered in the field.

SUMMARY

The three theories that have been proposed to explain the shape of the moisture-density curve were examined in the light of knowledge of contemporary soil mechanics and found deficient.

A qualitative theory was developed in terms of effective stresses. This theory is predicated on the belief that the resistance of soil to compaction is the integrated sum of the shearing stresses developed between the particles at their points of contact and that the macroscopic manifestation of these stresses can be examined in terms of the principle of effective stress. Considerable insight into the compaction process can be obtained through this qualitative approach.

ACKNOWLEDGMENT

Many of the laboratory experiments performed during the development of this theory were performed by Brijkrishen K. Kaul, A.M. ASCE, with the assistance of Bijan Mohraz, A.M. ASCE. Rough drafts of the paper were reviewed by Ralph B. Peck, F. ASCE. The contributions of each are appreciated.

APPENDIX.—NOTATION

The following symbols have been adopted for use in this paper:

A = Skempton's[7] pore-pressure coefficient;

B = Skempton's[7] pore-pressure coefficient;

c = cohesion;

N_B = number of blows per layer;

N_L = number of layers;

O_r = overconsolidation ratio;

p = foot pressure for a kneading compaction device;

S_r = degree of saturation, in percentage;

u_a = pore-air pressure;

u_w = pore-water pressure;

w = water content;

γ_d = dry density;

ϵ = strain;

σ = normal stress;

σ_a = ambient stress;

σ_1 = maximum principal stress;

σ_3 = minimum principal stress;

τ = shearing stress;

ϕ = angle of internal friction; and

χ = Bishop's[5] coefficient.

Note.—The following subscripts may be used with various of the symbols:

a = air;

f = parameter defined at failure;

Q = an unconsolidated-undrained test;

R = a consolidated-undrained test;

S = a consolidated-drained test; and

w = water

Note.—Effective stresses are designated by placing a bar over the appropreiate symbol.

Journal of the
SOIL MECHANICS AND FOUNDATIONS DIVISION
Proceedings of the American Society of Civil Engineers

PORE WATER PRESSURES IN UNSATURATED SOILS

By Roy E. Olson[1] and Leonard J. Langfelder,[2] Associate Members, ASCE

INTRODUCTION

Unsaturated soils are three-phase systems composed of solid mineral matter, water, and air. Such soils may be formed by either natural or artifical means. Except in moist environments, surficial soils are unsaturated much of the year. Deeper deposits of unsaturated soils are found in arid regions throughout the world and are the source of severe engineering problems. Compacted soils, which are almost always unsaturated, are used extensively in engineering work. Numerous field and laboratory investigations have been performed using both natural and compacted unsaturated soils; nevertheless, most experts agree that the properties of unsaturated soils are not well understood.

To attain a better understanding of the properties of unsaturated soils, a basic investigation was begun at the University of Illinois, in 1963. The first phase of this investigation, which is reported herein, was performed to provide solutions to the following problems:

1. Pore water pressures less than approximately -1 atmosphere (gage) have rarely been measured and reported in engineering literature, however, a wealth of indirect evidence shows that pore water pressures less than -100 psi are certain in some soils in the range of moisture content encountered in engineering work, and that pore water pressures lower than -100,000 psi are probable for soils with very low moisture contents. The first problem was to design and construct a device to measure pore water pressures significantly lower than -1 atmosphere.

2. The second problem was to determine experimentally if pore water pressures significantly lower than -1 atmosphere actually exist within compacted soils in the range in moisture content commonly encountered in

Note.—Discussion open until December 1, 1965. To extend the closing date one month, a written request must be filed with the Executive Secretary, ASCE. This paper is part of the copyrighted Journal of the Soil Mechanics and Foundations Division, Proceedings of the American Society of Civil Engineers, Vol. 91, No. SM4, July, 1965.

[1] Assoc. Prof. of Civ. Engrg., Univ. of Illinois, Urbana, Ill.
[2] Asst. Prof. of Civ. Engrg., North Carolina State Univ., Raleigh, N. C.

engineering work and to determine if these negative pore water pressures are of the magnitude expected from simple capillary considerations.

3. The third problem was to provide theoretical explanations for several anomolies that arise in the interpretation of negative pore water pressures in unsaturated soils. The first anomoly is:

> If water cavitates at a pressure of approximately -1 atmosphere (gage), then how can pore water pressures of -100 psi and lower be possible in unsaturated soils where the water films have ready access to air?
>
> The second anomoly is:
>
> If the tensile strength of perfectly deaired water is approximately 2,800 psi,[3] then how is it possible to have tensile stresses greater than 100,000 psi as suggested by theoretical calculations based on measurements of vapor pressures?

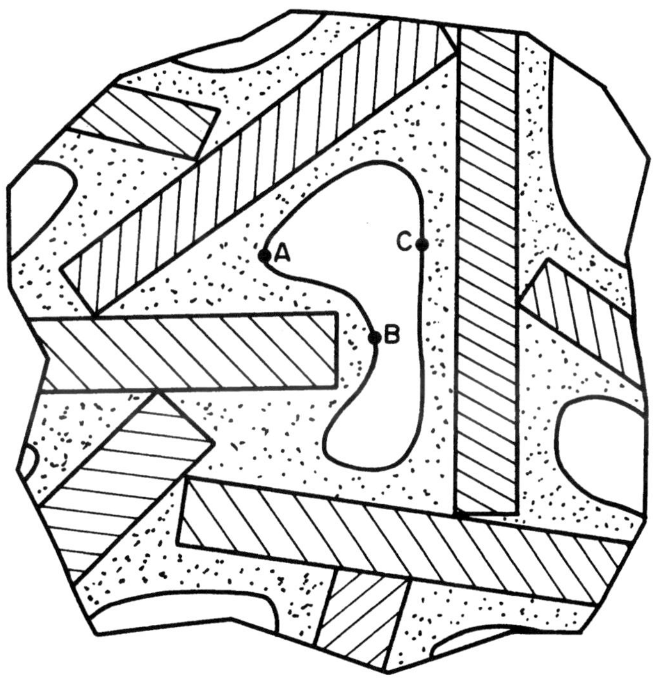

FIG. 1.—IDEALIZED SECTION THROUGH AN UNSATURATED COHESIVE SOIL

The third anomoly is suggested by the simple model shown in Fig. 1. In this model, it is assumed that the entire surface of each particle is hydrated. Capillary theory indicates that the pressure is always greater in the fluid (gaseous or liquid) on the concave side of an interface between

[3] Green, R. B., "Ordinary Liquid Water-Substance, Its Thermodynamic Properties, Dynamic Behavior, and Tensile Strength," thesis presented to the Massachusetts Institute of Technology, at Cambridge, Mass., in 1951, in partial fulfilment of the requirements for the degree of Doctor of Science.

two immiscible fluids. Referring to Fig. 1 and taking the ambient air pressure as zero (gage), the pore water pressure is negative at A, positive at B, and zero at C. The obvious question is: how is it possible for a continuous water film to have different pressures at different locations without undergoing flow? Or, more importantly, What pore water pressure is actually measured with the usual pore water pressure measuring equipment? The solutions obtained for both the experimental and theoretical problems are presented herein together with experimental data showing the magnitudes of the negative pore water pressures developed in typical compacted soils.

Notation.—The symbols used in this paper are defined where they first appear and are arranged alphabetically, in Appendix III.

DEVELOPMENT OF THE APPARATUS

Selection of the Pore Water Pressure Measuring Technique.—A survey of relevant literature was performed for the purpose of evaluating the various methods available for the measurement of negative pore water pressures in unsaturated soils in order to determine which method(s) would be suitable for use in this investigation. All of the existing techniques were either developed by soil physicists or represent modifications of techniques originated by them. Excellent reviews of the techniques used in soil physics have been presented by L. A. Richards,[4] and D. Croney and J. D. Coleman.[5] Of the various methods in use, only the pressure plate technique was found suitable for this investigation. This method has been widely used for investigations in soil mechanics.[6,7,8,9]

Pressure Plate Technique.—The main difficulty involved in the measurement of negative pore water pressures is that the water in the measuring system tends to cavitate at a pressure of approximately -1 atmosphere. Thus, only transient direct measurement of lower (more negative) pore water pressures is possible; then the water in the measuring system cavitates. In the pressure plate technique, as in the simple tensiometer technique, the pore water pressure is measured through a fine porous stone. However, in the pressure plate technique, the problem of cavitation of the water in the measuring system is avoided by increasing the pore air pressure until the pore water pressure is raised to a level that can be measured directly. The radius of curvature of the air-water menisci does not appear to

[4] Richards, L. A., "Methods of Measuring Soil Moisture Tension," Soil Science, The Williams and Wilkins Co., Baltimore, Md., Vol. 68, 1949, pp. 95-112.

[5] Croney, D., and Coleman, J. D., "Pore Pressure and Suction in Soils," Butterworths, London, 1961, pp. 31-37.

[6] Hilf, J. W., "An Investigation of Pore-Water Pressure in Compacted Cohesive Soils," Technical Memorandum No. 654, Bur. of Reclam., U.S. Dept. of the Interior, Denver, Colo., 1956.

[7] Bishop, A. W., "The Measurement of Pore Pressure in the Triaxial Test," Pore Pressure and Suction in Soils, Butterworths, London, 1961, pp. 38-46.

[8] Bishop, A. W., and Donald, I. B. "The Experimental Study of Partly Saturated Soil in the Triaxial Apparatus," Proceedings, 5th Internatl. Conf. on Soil Mechanics and Foundations Engrg., Vol. 1, 1961, pp. 13-22.

[9] Gibbs, H. J., and Coffey, C. T., "Measurement of Initial Negative Pore Pressure in Unsaturated Soil," Progress Report on Shear and Pore Pressure Research, Earth Research Program, Laboratory Report EM-665, Bur. of Reclam., U. S. Dept. of the Interior, Denver, Colo., 1963.

change significantly when the air pressure is increased, thus the difference between the final air and water pressures equals the initial negative pore water pressure. The technique of superimposing air pressures on the soil until the pore water pressure can be measured is termed "axis translation." The accuracy of this technique is examined in Appendix I.

Pressure Plate Cell.—The pressure plate cell developed for this investigation is shown in cross section in Fig. 2. The cell was designed to allow measurements of various soil parameters in addition to the negative pore water pressure. For clarity, the parts of the cell not required for the measurement of negative pore water pressures have been excluded from Fig. 2.

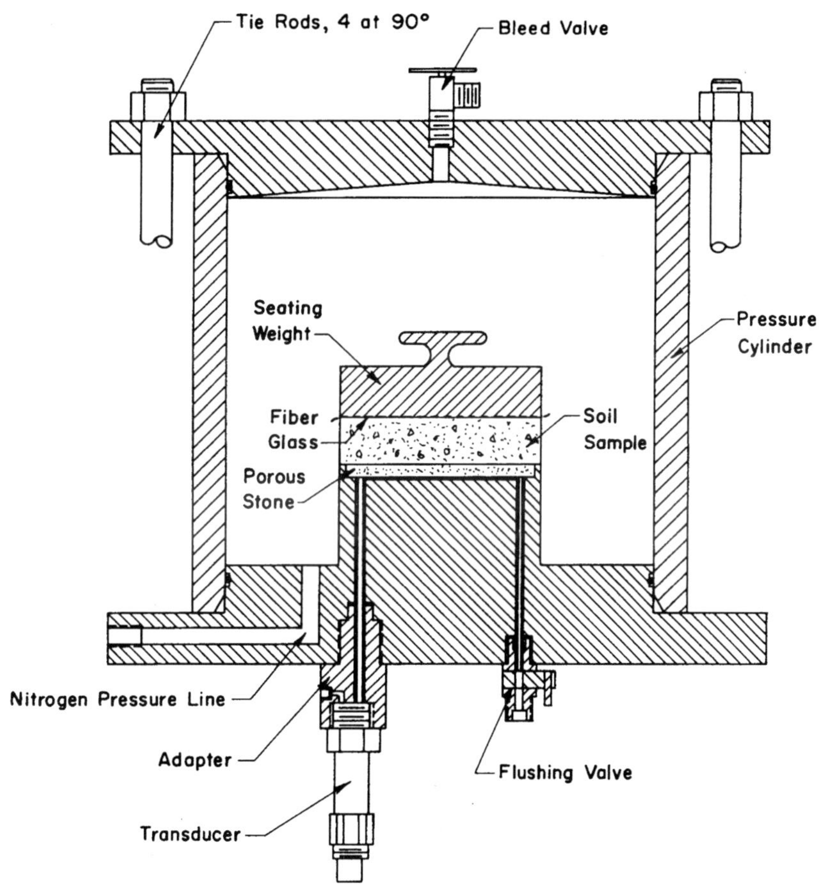

FIG. 2.—PRESSURE PLATE CELL

The base of the cell was machined from type 316 stainless steel. A porous stone with an air entry value of approximately 250 psi was cemented into the base pedestal using an epoxy resin. Two connections were made to the porous stone through the base. One connection was equipped with a minature plug valve and used for flushing the system. The second connection was equipped with an adapter and a pressure transducer (capacity 1,000 psi) and was used for the measurement of the pore water pressure. To remove possible points of nucleation and to reduce the volume of these lines to a minimum, the lines were first drilled 1/8 in. in diameter. The valves and adapter block were connected: the lines, adapter, and valves were filled with epoxy resin: and a 1/16-in. diameter greased piano wire was inserted into the resin in each connection. When the resin had partially hardened, the piano wires were withdrawn leaving a continuous hole, with smooth walls and small volume, in each connection. This technique was quite successful in eliminating points of nucleation and in reducing the volume of the pore pressure measuring system to a minimum.

The remainder of the cell consisted of a cadmium-plated steel pressure cylinder, with a design pressure capacity of 1,000 psi, and a cell top equipped with a bleed valve. Commercial compressed nitrogen was used to apply pressure in the cell. Nitrogen pressures were measured with either a 300 psi or a 1,500-psi capacity Bourdon gage.

EXPERIMENTAL PROCEDURE

Prior to performing a test, the base of the cell was thoroughly deaired by forcing deaired water, at high pressure, through the measuring system to dissolve and remove any entrapped air bubbles. When the system had been properly deaired, the water above the porous stone was drained, the top of the cell was removed, excess water was wiped from the porous stone, and a soil sample with nominal dimensions of 3-in. diameter by 3/4-in. thick was placed in contact with the stone. Close contact between the sample and the stone, which was required for rapid response of the equipment, was achieved by using a flat porous stone, a flat surface on the soil sample, and by placing a 1 kg weight on top of the soil as a seating load.

As soon as the soil sample was placed in contact with the porous stone, the negative pore water pressure in the soil tended to imbibe water from the pore water pressure lines in the base. Thus, the transducer indicated negative pore water pressures at once. If the pore water pressure was between 0 psi and -12 psi, the transducer could be although rarely was used to measure the negative pore water pressure directly.

When used as a pressure plate, the soil sample would be placed on the base as before and then the cell would be assembled as rapidly as possible in approximately 30 sec and a gas pressure applied directly to the soil sample. A sufficient gas pressure was applied to achieve a zero pore water pressure in the soil sample; the pressure transducer was as a null indicator. For most tests, the 30 sec required to assemble the cell and apply the gas pressure was a short enough period of time to avoid cavitation of the system. However, in some tests, where the stone was wiped drier than usual, where the soil sample made better than usual contact with the stone, and where the pore water pressure in the soil was quite negative, the pressure transducer would indicate a pressure of -12 psi almost instantly, and it was necessary to open the plug valve in the flushing line for a fraction of a second to increase the system pressure back to atmospheric. It was never necessary to open the valve more than once. Calculations[10] indicated that the amount of water let into the system by this procedure was much too small to make a detectible

[10] Langfelder, L. J., "An Investigation of Initial Negative Pore Water Pressure in Statically Compacted Cohesive Soils," thesis presented to the University of Illinois, at Urbana, Ill., in 1964, in partial fulfilment of the requirements for the degree of Doctor of Philosophy.

TABLE 1.—INDEX PROP

Soil (1)	L_w, in percentage (2)	P_w, in percentage (3)	Percentage of clay less than 0.002 mm (4)	Activity (5)	G (6)	S, in Square meters per gram (7)
Grundite clay	54	33	46	0.41	2.79	119
Goose Lake clay	32	16	32	0.59	2.72	111
Champaign till	25	17	16	0.73	2.73	22
Fayette-C clay	32	22	18	0.77	2.75	53
Peorian loess	29	26	12	0.43	2.75	38

[a] Mineral abbreviations are quartz, 0; illite, I: chlorite, C; Kaolinite, K; and mont

change in the pore pressure in the sample but it did delay response sufficiently to avoid cavitation. This valve opening procedure was also used by Gibbs and Coffey.[9]

The nitrogen pressure in the cell was continuously adjusted to maintain a zero reading on the transducer. Equilibrium was generally established within 30 min. An examination of the reasons for the delay in the establishment of equilibrium is presented in Appendix II.

After equilibrium was achieved, the nitrogen pressure was quickly reduced to zero, the cell disassembled and the moisture content of the sample determined. It is assumed that the initial negative pore water pressure in the soil was exactly equal to, but opposite in sign from, the applied nitrogen pressure (see Appendix I).

SOIL PROPERTIES

The five soils selected for this investigation included Grundite clay, Goose Lake clay, Champaign till, Fayette-C clay, and Peorian loess. The Grundite clay and Goose Lake clay were supplied by the Illinois Clay Products Company. The other three soils were obtained from the Central Illinois area.[11]

The index properties of the five soils are presented in Table 1. The Atterberg limits, grain size analyses, and specific gravities were determined

[11] Thompson, M. R., "Lime-Reactivity of Illinois Soils as it Relates to Compressive Strength," thesis presented to the University of Illinois, at Urbana, Ill., in 1964, in partial fulfilment of the requirements for the degree of Doctor of Philosophy.

ERTIES AND MINERALOGY

Moisture-Density Data				Mineralogy[a]				
Standard		Modified						
γ_{dm}, in pounds per cubic foot	w_{opt}, in percentage	γ_{dm}, in pounds per cubic foot	w_{opt}, in percentage	0, in percentage	I, in percentage	C, in percentage	K, in percentage	M, in percentage
(8)	(9)	(10)	(11)	(12)	(13)	(14)	(15)	(16)
102	23	115	16	5	95	--	--	--
113	15	125	11	20	15	--	65	--
122	12	136	8	20	60	20	--	--
114	16	120	13	--	30	10	--	60
107	18	114	14	--	15	10		75

morillonite, M.

according to ASTM designations[12] D-423-424, D-422, and D-854, respectively. The standard and modified moisture-density tests were performed according to ASTM designations D-698 and D-1557, respectively. The activity[13] was calculated using the modified equation suggested by H. B. Seed, R. J. Woodward, and R. Lundgren.[14] The specific surface was determined for the finer-than- No. 200 sieve fraction of the soil using the glycerol adsorption method of Diamond and Kinter.[15]

The estimated mineralogical compositions of the fraction of the soils finer than 0.002 mm (Table 1) were obtained from X-ray diffraction patterns of oriented aggregates of the soils using the methods recommended by W. D. Johns, R. E. Grim, and W. F., Bradley.[16] Mixed layer minerals with expanding lattices were included as montmorillinite. For the Fayette-C and Peorian loess soils, it was not possible to determine whether a 7Å reflection

12 "ASTM Standards, Part II, Bituminous Materials: Soils," Amer. Soc. for Testing and Materials, Philadelphia, Pa., 1964.

13 Skempton, A. W., "The Colloidal Activity of Clays," Proceedings, 3rd Internatl. Conf. on Soil Mechanics and Foundation Engrg., Zunich, Switzerland, Vol. 1, 1957, pp. 57-61.

14 Seed, H. B., Woodward, R. J., and Lundgren, R., "Prediction of Swelling Potential for Compacted Clays," Journal of the Soil Mechanics and Foundations Division, ASCE, Vol. 88, No. SM3, Proc. Paper 3169, June, 1962, pp. 53-88.

15 Diamond, S., and Kinter, E. B., "Surface Areas of Clay Minerals as Derived from Measurements of Glycerol Retention," Clays and Clay Minerals, Proceedings, 5th Natl. Conf., Univ. of Illinois, Urbana, Ill., 1958, pp. 334-347.

16 Johns, W. D., Grim, R. E., and Bradley, W. F., "Quantitative Estimations of Clay Minerals by Diffraction Methods," Journal of Sedimentary Petrology, Vol. 24, 1954, pp. 242-251.

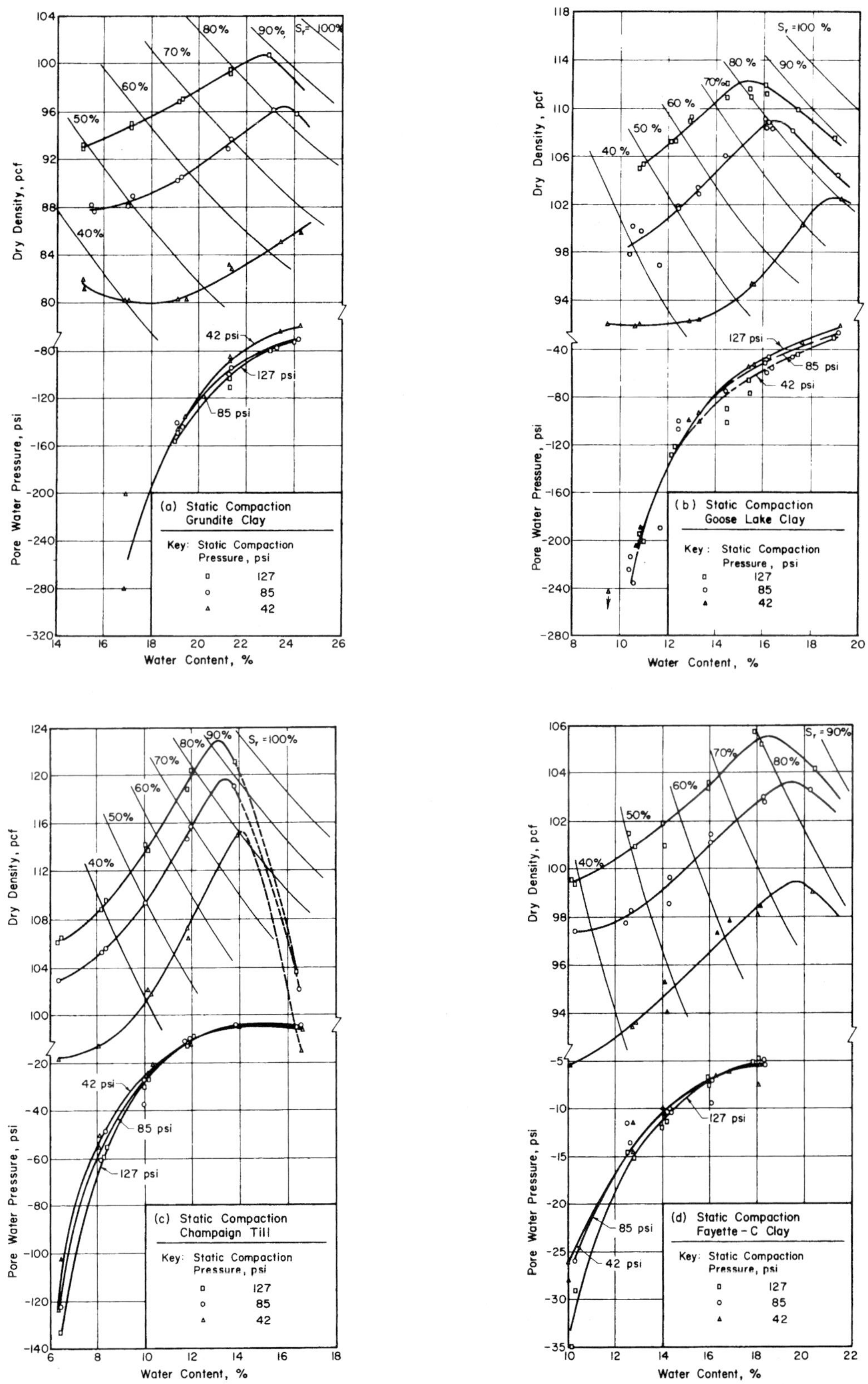

FIG. 3.—EXPERIMENTAL DATA

resulted from chlorite or kaolinite; in both cases, the peak was assumed to result from small quantities of chlorite.

SAMPLE PREPARATION

For all samples, the soil was air dried, sieved through a No. 10 sieve, mixed with the desired quantity of distilled water in a mechanical mixer, and stored inside double plastic bags for one week prior to compaction. To obtain homogeneous samples, it was necessary to resieve the soil through a No. 10 sieve just prior to compaction to break up any lumps that had formed during mixing.

For the statically compacted samples, the required quantity of soil was pressed between two polished stainless stell pistons inside a highly polished and greased stainless steel cylinder of 3.00-in. inside diameter. The compaction force was increased uniformly to its final magnitude during a 2-min period and then held constant for another 3 min. The statically compacted samples had nominal dimensions of 3 in. in diameter by 3/4-in. high. Samples were prepared in duplicate, wrapped in aluminum foil, and stored for one week in a humid atmosphere prior to testing.

The samples subjected to kneading compaction were compacted into a 3-in. diameter by 3/4-in. high ring using six evenly spaced "blows" on the upper surface, six more blows on the lower surface using a wedge shaped compaction foot having an apex angle of 60° and an area of 2 sq in. Sample storage procedures were the same as for the samples subjected to static compaction.

EXPERIMENTAL DATA

Static Compaction.—Static compaction was performed at pressures of 42 psi, 85 psi, and 127 psi using all five soils. All negative pore water pressures were measured using the axis translation technique. Both the dry densities and the negative pore water pressures are plotted against the moisture content in Figs. 3 and 4.

The most striking aspect of the pore water pressure versus moisture content curves is the magnitude of the negative pore water pressures; they are much more negative than any reported previously in engineering literature.[6,7,8,9,17,18,19,20,21] For the more plastic clays, e.g., Grundite clay

---

[17] Croney, D., and Coleman, J. D., "Soil Moisture Suction Properties and their Bearing on the Moisture Distribution in Soils," Proceedings, 3rd Internalt. Conf. on Soil Mechanics and Foundations Engrg., Zunich, Switzerland, Vol. 1, 1953, pp. 13-18.

[18] Gibbs, H. J., Hilf, J. W., Holtz, W. G., and Walker, F. C., "Shear Strength of Cohesive Soils," ASCE Research Conf. on Shear Strength of Cohesive Soils, Boulder, Colo., 1960, pp. 33-162.

[19] Bishop, A. W., Alpan, I., Blight, G. E., and Donald, I. B., "Factors Controlling the Strength of Partly Saturated Cohesive Soils," ASCE Research Conf. on Shear Strength of Cohesive Soils, Boulder, Colo., 1960, pp. 503-532.

[20] Lambe, T. W., "Residual Pore Pressure in Compacted Clay," Proceedings, 5th Internatl. Conf. on Soil Mechanics and Foundations Engrg., Paris, Vol. 1, 1961, pp. 207-212.

[21] Blight, G. E., "The Utilization of Soil Suction in the Design of Earth Dam Embankments," Proceedings, 3rd Regional Conf. for Africa on Soil Mechanics and Foundations Engrg., 1963, pp. 141-144.

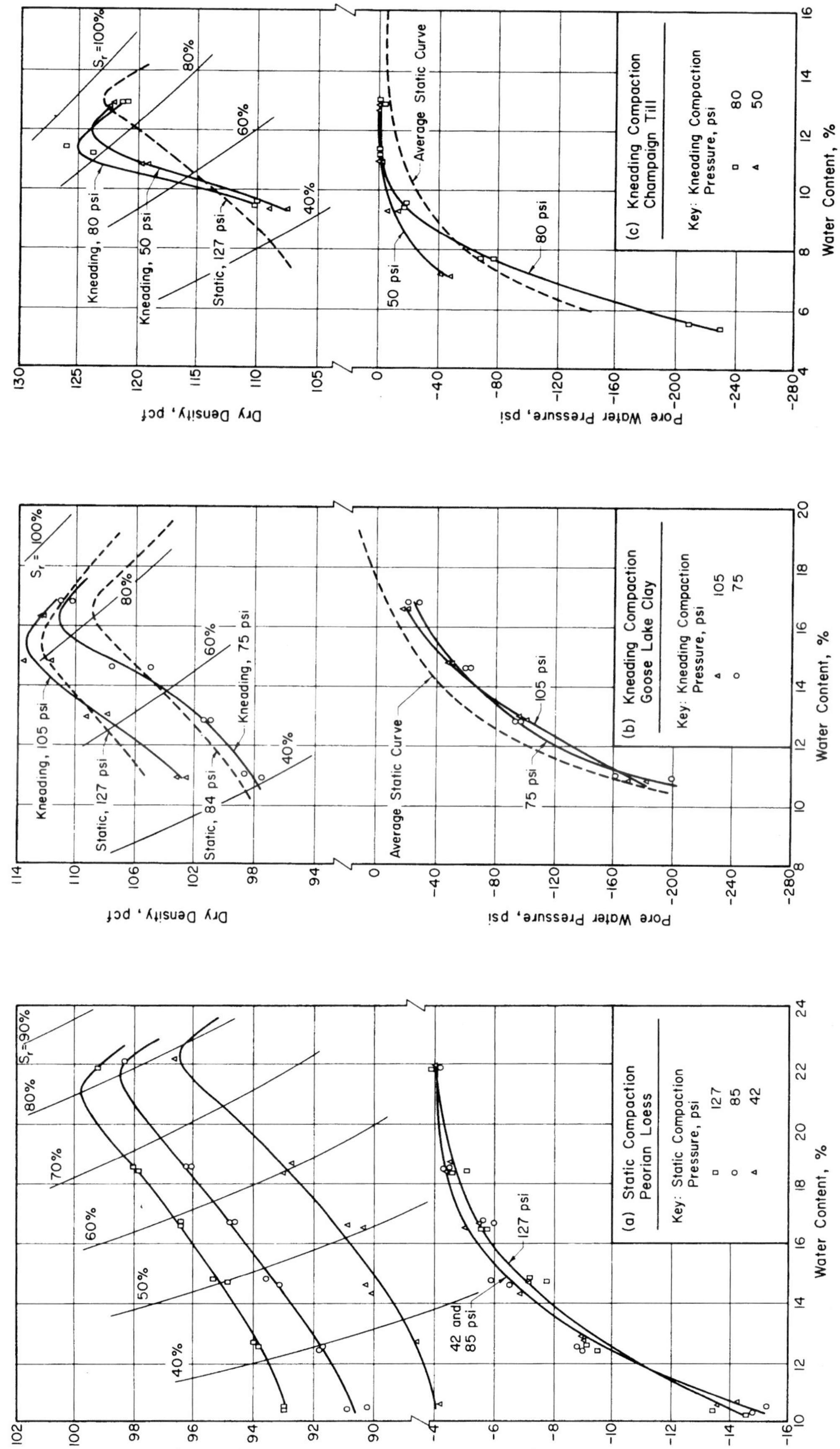

FIG. 4.—EXPERIMENTAL DATA

and Goose Lake clay, pore water pressures as low as -100 psi exist in samples compacted only 2% dry of the optimum water content. Samples of Grundite clay compacted at 127 psi had pore water pressures lower than -250 psi (the actual pressures could not be measured with the existing apparatus) when the moisture contents were only 5% less than the optimum water content.

A second striking aspect of these negative pore water pressures is that, for a given soil, they are almost independent of dry density within the range of dry density used in this investigation. For a given soil subjected to static

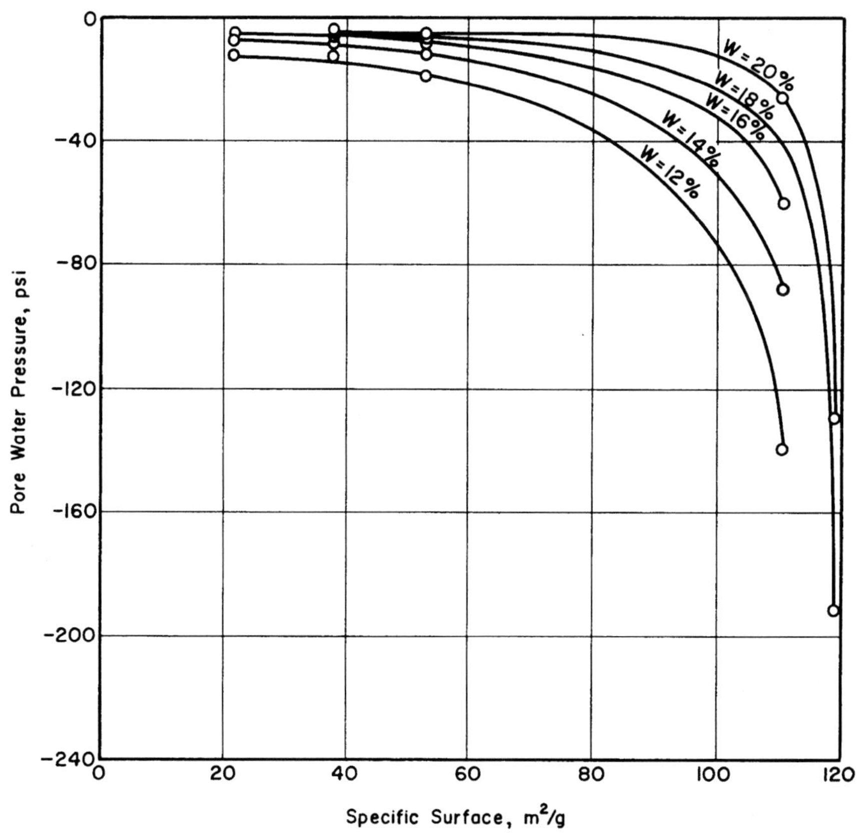

FIG. 5.—INFLUENCE OF SPECIFIC SURFACE ON THE NEGATIVE PORE WATER PRESSURE IN SAMPLES SUBJECTED TO STATIC COMPACTION AT 127 PSI

compaction, the residual negative pore water pressures are primarily a function of the moisture content.

Examination of the data in Figs. 3 and 4 indicates that the pore water pressures tend to be the most negative in the highly plastic clays, e.g., Grundite clay. It was discovered that the negative pore water pressures at a particular water content could be related to the specific surface. Typical relationships between the pore water pressure, water content, and specific surface for static compaction at 127 psi are shown in Fig. 5. Almost identical curves were obtained for static compaction at 85 psi and 42 psi, indicating that dry density had a negligible effect on these relationships.

Kneading Compaction.—Only the Goose Lake clay and Champaign till were used in the study of pore water pressures in samples prepared by kneading compaction. Compacted dry densities and pore water pressures are plotted against moisture content in Fig. 4, together with the relationships obtained for the samples prepared by static compaction (shown by dashed lines).

For the Goose Lake clay, the pore water pressures were essentially independent of the dry density, within the range of dry density used for these tests, and a function only of the moisture content. However, for the range in moisture content from 3% dry of optimum to 3% wet of optimum (the highest water content at which samples were prepared), the pore water pressures were significantly less negative for the samples subjected to kneading compaction than for samples subjected to static compaction. For example, the optimum water contents and maximum dry densities were essentially the same for the samples subjected to a static compaction pressure of 127 psi and to a nominal kneading pressure of 105 psi; however, the pore water pressure in the statically compacted sample, at the optimum water content, was -68 psi, whereas it was only -36 psi in the sample, subjected to kneading compaction at the same mositure content. In agreement with Lambe,[20] who observed similar differences the less negative pore water pressures in the samples subjected to kneading compaction probably resulted from a more dispersed soil structure. The similarity between the negative pore water pressures for samples of Goose Lake clay subjected to static and to kneading compaction at low moisture contents may indicate a similarity in soil structures.

For the Champaign till, the pore water pressures are significantly more negative for the samples subjected to kneading compaction at 80 psi than at 50 psi, i.e., the denser samples had more negative pore water pressures. This was the only series of tests performed in this investigation in which dry densities appeared to influence singificantly the pore water pressures. At moisture contents in the vicinity of optimum, the pore water pressures were much less negative for samples subjected to kneading compaction than for samples subjected to static compaction.

Capillary Models.—The curves of pore water pressure versus moisture content have the shape that would be predicted from a capillary model. At high moisture contents, the voids are nearly filled with water, the air-water interfaces are relatively flat, and the pore water pressure is only slightly negative. As the compaction water content is decreased, the menisci retreat into the interstices of the soil, the radius of curvature of the menisci decreases, and the pore water pressure becomes increasingly more negative.

Using several simple mechanical models of a soil composed of plate shaped particles, Langfelder[10] derived equations to determine the shape of the negative pore water pressure versus moisture content curves. It was demonstrated that such simple equations could be fitted almost exactly to the experimental curves (Figs. 3 and 4). However, Langfelder also pointed out that calculations of the actual moisture contents and actual negative pore water pressures would require detailed knowledge of the size and shape of the individual particles, the number of inter-particle contacts per particle, the quantity of adsorbed water, and the orientation of the particles relative to each other. Because these factors cannot be evaluated for a real soil, it is not possible to calculate absolute values of negative pore water pressures and moisture contents. The excellent fit that can be obtained between the experimental pore water pressure versus moisture content curve and the fitted theoretical curve does not prove the validity of the capillary model. Nevertheless, the capillary concepts and models have proved useful in examining pore water pressures in unsaturated soils.

Unfortunately, the simple capillary models do not provide answers for the anomolies cited in the introduction. If the capillary model is valid, then the water should cavitate when the pore water pressure approaches -1 atmosphere, and certainly the water must fail in tension when the pore water pressure drops to approximately -2,800 psi (the tensile strength of deaired water). Furthermore, there is no obvious solution for the third anomoly—the problem of having the pressure in thin water films varying from negative values at contact points between particles to zero on flat surfaces to positive values at an exposed end of a particle (Fig. 1). To investigate such problems requires a different approach, the approach taken subsequently.

EQUILIBRIUM CONDITIONS FOR PORE WATER

Introduction.—The problem of equilibrium of pore water in unsaturated soils, and the associated problem of water flow in unsaturated soils, have been studied in detail by soil physicists. The approach taken to this problem herein is essentially that of the soil physicists,[22] with certain modifications to definitions and terminology to put the concept into terms familiar to soils engineers.

Components of the Total Head.—In analyzing the flow of water through saturated soils, the soils engineer recognizes that water flows in response to a gradient in total head, not a gradient in pressure head. For flow through channels of macroscopic size, the total head is given by Bernoulli's equation

$$h = \frac{u}{\gamma_w} + z + \frac{v^2}{2g} \quad \text{. . . . . (1)}$$

in which h is the total head, u represents the pressure in the water, γ_w is the unit weight of water, z denotes the position head, v is the actual velocity of flow, and g refers to the acceleration due to gravity. For flow through most soils, the velocity of flow is so small that the term $v^2/2g$ can be ignored. Thus, for most soils, and for all soils considered herein, the flow can be analyzed in terms of position head and pressure head, provided that only flow through macroscopic channels is considered.

Eq. 1 is widely used to analyze flow of water through soils. Under the usual circumstances of application, and provided the gradients are not too small, Eq. 1 can be shown to be quite accurate. However, on a microscopic level, Eq. 1 is invalid because it does not consider the heads resulting from osmotic pressures and from adsorptive force fields. It will now be demonstrated that the anomolies examined previously can be explained provided that all components of the total head are taken into account.

For analysis of flow on a micrsocopic basis, it is convenient to postulate that the proper equation for total head is

---

[22] Bolt, G. H., and Miller, R. D., "Calculation of Component and Total Potentials of Water in Soils," Transactions, Amer. Geophysical Union, Vol. 30, 1958, pp. 917-928.

$$h = \frac{u}{\gamma_w} + \omega + \alpha + z \quad \ldots\ldots\ldots\ldots\ldots\ldots (2)$$

in which u/γ_w is the pressure head, ω denotes the osmotic head, α represents the adsorptive head, and z refers to the position head.

A brief consideration of the osmotic and adsorptive heads will be presented, and then Eq. 2 will be applied to unsaturated soils to obtain solutions to the anomolies mentioned previously.

Osmotic Head.—A schematic section through a simple osmotic cell is shown in Fig. 6. In this simple cell, distilled water is separated from a dilute aqueous salt solution by a membrane that is permeable to water but not to the salt ions. The pressure in the distilled water is maintained at atmospheric pressure. If a gage is connected to the side of the cell contain-

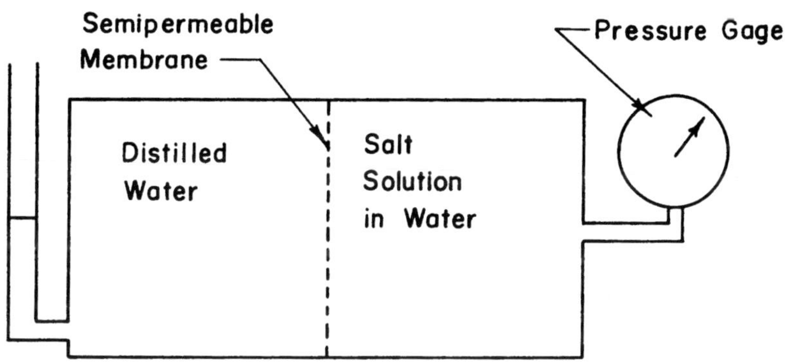

Heads

	left	right
position	0	0
pressure	0	$+\pi$
osmotic	0	$-\pi$
total	0	0

FIG. 6.—OSMOTIC CELL

ing the salt solution, it is found that the pressure is not atmospheric when equilibrium is established, but rather is approximately given by the van't Hoff equation,

$$\pi = \Delta C \, R \, T \quad \ldots\ldots\ldots\ldots\ldots\ldots (3)$$

n which π is the osmotic pressure, in atmospheres, ΔC refers to the difference in salt concentration across the membrane, in moles per liter, R represents the gas constant, in 0.08204 liter atmospheres per degree mole, and T is the absolute temperature, 273 plus the temperature in degrees Centigrade. In Eq. 3, the concentrations represent the sum of all the undissociated solute molecules and the dissociated solute ions. Thus, dissolution of 0.001 moles of NaCl in water yields 0.001 moles of Na^+ and 0.001 moles of Cl^- for a total concentration of 0.002 moles. Using Eq. 3, the osmotic pressure of a 0.1 molar solution of NaCl at 300° K.—assuming complete dissociation and no ionic interaction—is 72 psi.

To analyze the equilibrium in the osmotic cell in Fig. 6, it is necessary to define the conditions under which the total head is zero, i.e., a reference state must be defined and all measurements must then be taken relative to this state. The reference for pressure head is the ambient atmospheric pressure, i.e., all pressures are gage pressures. The osmotic and adsorptive heads (Eq. 2) will be defined as zero for distilled water that is free from all force fields except gravity; the gravitational effects are considered with the position head that will, for convenientce, be defined as zero at the elevation of the center of the cell. Thus, as indicated in Fig. 6, all components of head are zero on the distilled-water side of the membrane and, therefore, the total head is zero.

In the salt solution (Fig. 6), the pore water pressure is given by Eq. 3, the osmotic head is ω, the adsorptive head is zero, and the position head is zero. At equilibrium, there is no net flow across the membrane. Because the membrane is pervious to water, it is necessary for the total heads to be equal across the membrane. Equating the total heads and using Eq. 2,

$$0 = \frac{C\,R\,T}{\gamma_w} + \omega \qquad (4)$$

in which C is the concentration in moles per liter, as defined previously, Therefore,

$$\omega = -\frac{C\,R\,T}{\gamma_w} \qquad (5)$$

Thus, for concentration greater than zero, the osmotic head, ω, is negative and equal numerically to the pressure head. Note that ω exerts the same influence on the pressure head, when the pressure head results from osmosis, that z exerts on the pressure head when this head results from gravitational effects.

Clay particles possess a net negative charge. It is usually convenient to consider this charge as a surface charge, but it is, of course, actually a body charge. For many clays, the surface area is mainly the exposed basal surfaces of the plates. Consider these basal surfaces to possess a negative surface charge that must be counterbalanced by an excess of positive charge in the adjacent fluid. The attractive force of the surface for the excess cations in the fluid tends to pull the cations down into a monolayer, but the dispersion forces between the cations tend to spread them uniformly through the pore fluid. The balance of these opposing effects leads to a concentration of cations that decreases from a maximum near the surface to a minimum, equal to the concentration of cations in the free pore fluid, at some distance out from the surface. The distribution of ions adjacent to the surface can be estimated using the Gouy-Chapman theory.[23,24,25]

---

23 Bolt, G. H., "Analysis of the Validity of the Gouy-Champman Theory of the Electric Double Layer," Journal of Colloid Science, Vol. 10, 1955, p. 206.

24 Bolt, G. H., "Physico-Chemical Analysis of the Compressibility of Pure Clays," Geotechnique, Vol. 6, 1956, pp. 86-93.

25 Devereux, C. F., de Bruyn, P. L., "Interaction of Plane-Parallel Double Layers," Massachusetts Inst. of Tech., Press, Cambridge, Mass., 1963.

If two clay plates are arranged parallel to each other, the fluid between the plates contains a sufficient excess of cations to balance both surface charges. The concentration of cations is a maximum near each surface and a minimum at the midplane. In analyzing the problem, it is convenient to use the midplane because here the adsorptive effects (see next section) of the two plates are in exact balance. The concentration of ions at the midplane exceeds the concentration of ions in the free pore fluid. These excess cations cannot diffuse out into the pore fluid because of the attractive forces of the opposing surfaces. Thus, these attractive forces serve the same function as a semipermeable membrane in an osmotic cell; they allow water free access, but they retain the excess ions. From previous examination, it is apparent that an osmotic pressure will be developed between the particles; the magnitude of this pressure can be estimated from Eq. 3. Just as in the simple osmotic cell, the osmotic pressure between the plates is a fluid pressure; it is a fluid pressure in excess of the pressure in the free pore water.

Adsorptive Head.—The electric charge of the clay particles produces an electric field, the intensity of which decreases with distance from the surface of the particles. As a result of the gradient in field intensity, dispolar water molecules tend to become alined and are subjected to an attractive force by the particles. The water molecules are subjected to this attractive force, regardless of whether the surface charge on the particles is positive or negative. Water molecules are also subject to van der Waals attractive forces and to short range electronic forces that lead to adsorption through the formation of hydrogen bonds. As a result of the combined action of all these forces, water molecules are subjected to an attractive force by the charged surfaces of the clay particles. The component of the total head that results from the adsorptive force field is denoted by α. The sign of α can be determined by imagining what would happen if water at the reference state (zero total head) were allowed to come into contact with the adsorptive force field of the surface. Obviously, the water would flow from the reference state onto the surface of the clay. Because water flows down a gradient in total head, the adsorptive component of the ttal head must be negative.

The adsorptive head near the surface of a clay particle exerts an influence on the pore water that is similar to the effect of gravity on water in a macroscopic system. Thus, for a static body of water in the uniform gravitational field, and in the absence of other force fields, the sum of the position head (representing the gravitational force field) and the pressure head is a constant (Eq. 1). Thus, the pressure head is zero at the surface of the macroscopic body of water and increases linearly with depth because the position head decreases linearly with depth. Similarly, in the vicinity of a charged surface, where the gravitational head is sensibly constant, assume for the moment that only the pressure head and adsorptive head vary; it is necessary that the decrease in adsorptive head is balanced by an increase in pressure head as the surface of the clay is approached. The pore water pressure, therefore, increases from zero at a flat air-water interface to positive values as the surface of the particle is approached.

Unfortunately, the adsorptive head cannot be defined as zero at the air-water interface in order to simplify the problem. The adsorptive head is defined as zero where the adsorptive force field is zero. The adsorptive head at an air-water interface will be nearly zero if the water film is thick but may be quite negative if the water film is thin.

At any particular point in an adsorbed water layer, the adsorptive head will depend on the force field of the adjacent particle and on the force fields of other near-by particles and ions and will not be subject to precise deternation analytically.

Idealized Unsaturated Clay.—A section through a part of an idealized unsaturated clay is shown in Fig. 7. The air voids are continuous with each other and with the surface of the sample so that the pressure in the pore air is uniform. The pore air pressure will be used as the reference level for all pressure measurements; thus the value of the pore air pressure becomes zero (gage).

The section shown in Fig. 7 indicates that the surfaces of the clay are completely coated with water. This appears to be a reasonable assumption because

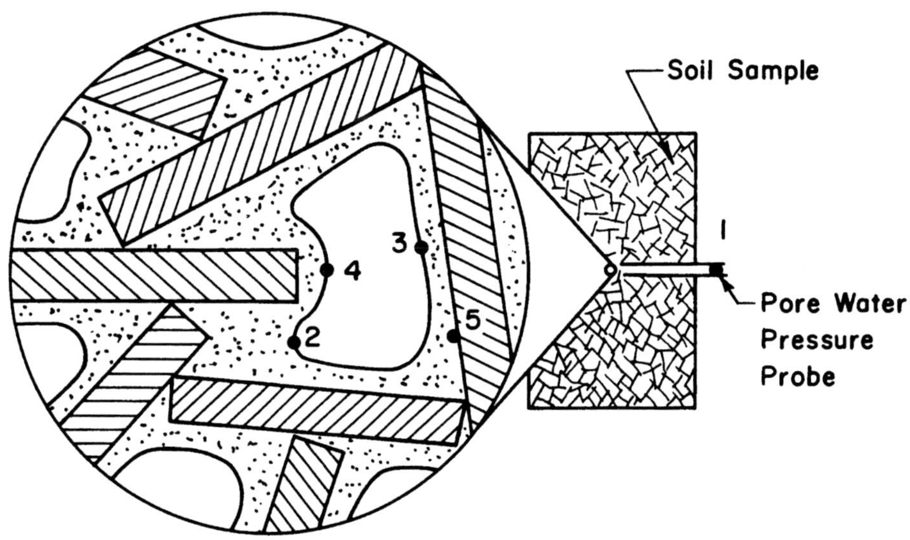

FIG. 7.—SECTION THROUGH AN IDEALIZED FLOCCULATED CLAY

adsorptive forces are exerted by both positively and negatively charged surfaces on water molecules and because even a neutral surface exerta a van der Waals attractive force. Thus, the water films are also interconnected. The foregoing assumption does not require that the water films have the same thickness; the requirement for equilibrium is that the total head of the water be the same throughout the system. Analysis of the state of stress in the water will be performed by determining values for the component heads at various points and then adding them to obtain the total heads which must all be the same.

Point 1 is located within a probe that is inserted into the unsaturates soil (Fig. 7) to measure the pore water pressure. At equilibrium there can be no diffusion of salts into or out of the probe. The probe, therefore, contains electrolyte of concentration C_1 not distilled water. The pore water pressure in the probe is u_1, and the adsorptive head is zero; there are no adsorptive force fields in the probe. Thus, the total head of the water in the probe is

$$h_1 = \frac{u_1}{\gamma_w} - \frac{C_1 R T}{\gamma_w} \quad \text{. . . . . (6)}$$

Because u_1 is negative and C_1 is positive, it follows that the water in the probe has a negative total head.

Assume that the pore water pressure at an air-water interface can be calculated using the capillary equation,

$$u = \frac{2 T_s}{r} \quad \text{. (7)}$$

in which u is the pore water pressure (gage), T_s denotes the surface tension, and r refers to the harmonic mean radius of curvature of the interface. The radius is negative if air is on the concave side of the interface (point 2, Fig. 7) and positive if water is on the concave side (point 4). The smallest value of r for which Eq. 7 is valid is not known. The writers assume that all values of r presented subsequently are larger than this unknown lower limit.

It is assumed that the element of soil under consideration is a differential element and that the probe is at the same elevation as the element; thus, gravitational heads are constant throughout the system and need not be considered further.

At point 2, the pre water pressure is $2\,T_s/r_2$ and is negative. If point 2 is within the range of influence of the force fields from adjacent particles, the osmotic head will be $-C_2\,R\,T/\gamma_w$ and the adsorptive head will be $\alpha 2$. Comparing the heads between points 1 and 2, and solving for u_1, yields

$$u_1 = \frac{2 T_s}{r_2} - \left(C_2 - C_1\right) R\, T - \alpha_2\, \gamma_w \quad \text{. . . . . (8)}$$

If point 2 is outside the range of influence of the force fields of adjacent particles, then $C_2 = C_1$ and $\alpha_2 = 0$, and the measured pore water pressure, u_1, and the pore water pressure at point 2 are equal. However, if point 2 is within the range of influence of the force fields of the adjacent particles, then C_2 exceeds C_1, and α_2 is negative. Thus, from Eq. 8, the measured pore water pressure, u_1, is lower (more negative) than the actual pore water pressure at point 2.

At point 3, the air-water interface is flat; thus, the pore water pressure at the interface must be zero. A comparison of the total heads at points 1 and 3 then leads to

$$u_1 = -\left(C_3 - C_1\right) R\, T + \alpha_3 \quad \text{. . . . . (9)}$$

For the thin water film at point 3, all the water is within the range of influence of the force fields of the adjacent particle, C_3 exceeds C_1, and α_3 is negative. Thus, the measured pore water pressure, u_1, is negative, but the actual pore water pressure is zero.

A similar line of argument of point 4 leads to

$$u_1 = \frac{2 T_s}{r_4} - \left(C_4 - C_1\right) R\, T + \alpha_4 \quad \text{. . . . . (10)}$$

The actual pore water pressure at point 4 is positive, but the water film is so thin that the force field of the adjacent particle is strong at the interface; thus C_4 substantially exceeds C_1, and α_4 is negative.

At point 5, the element of water under consideration is close to the surface of the adjacent particle. The pore water pressure equation is

$$u_1 = u_5 - \left(C_5 - C_1\right) R\ T + \alpha_5 \ldots\ldots\ldots\ldots (11)$$

Because point 5 is close to the surface of the clay, C_5 is much greater than C_1 and is substantially greater than C_4. Furthermore, α_5 should be much more negative than any α previously considered. It follows, therefore, that the actual pore water pressure at point 5 is greatly positive although the measured pore water pressure is negative.

The foregoing considerations make it clear that the actual pore water pressures in an unsaturated soil at equilibrium vary from perhaps -1 atmosphere to large positive pressures. This conclusion is not particularly surprising; the requirement for static equilibrium is a constancy of the total head, not a constancy of the pressure head.

Nature of Pore Water Pressures in Unsaturated Soils.--Based on the previous discussion, it is apparent that the simple capillary model often used in analyzing the pore water pressures in unsaturated soils will not suffice except for relatively crude analyses. To explain the anomolies that were listed in the introduction, it is necessary to use the more basic analysis examined herein. The explanations of the anomolies are as follows:

Observations that water in free contact with air caritates at a pressure of approximately -1 atmosphere, probably apply to the pore water in an unsaturated soil. The actual limiting negative pore water pressure in the soil, not probe, is approximately -1 atmosphere. If the probe records lower pressures (using the axis translation technique), the water films are so thin that all the water is within the range of influence of osmotic and adsorptive fields. Thus the pore water at some point in the soil may have a pressure of -1 atmosphere, but the total head at that point is further lowered by the presence of osmotic and adsorptive effects. A pressure lower (more negative) than -1 atmosphere must be applied to the water in the probe to reduce the total head of the water in the probe to the total head of the water at the point under consideration in the soil.

Similarly, calculated negative pore water pressures of lower than -100,000 psi for nearly dry soils mean that such a negative pore water pressure would have to be applied to the water in a probe in order to reduce the total head of the water in the probe to the total head of the water in the soil. The actual pressure in the water tightly adsorbed against the surface of the clay is likely to be positive, not negative, and therefore the anomoly of water having a tensile strength of only approximately 2,800 psi, whereas vapor pressure calculations suggest much more negative pore water pressures, is also resolved.

The third anomoly is also easily resolved. In an unsaturated soil where the probe records a negative pore water pressure, the actual pore water pressures in the soil may be negative, zero, or positive, depending on location. It does not make any difference where the probe is located in the soil because the probe does not measure the pore water pressure, but records the water pressure in the probe, which is the pressure that must be applied to that water to reduce its total head to the total head of the water in the soil. The axis translation technique operates by raising the total head of the water in the soil rather than by lowering the total head of the water in the probe, but the end result is the same.

Because the pore water pressure measured with a probe is not equal to the actual pore water pressure in the soil, except at certain locations, it would be more expressive to refer to the pore water pressure in the probe as the "equivalent pore water pressure."

The nonuniformity of pore water pressure within unsaturated soils has further ramifications in terms of defining the effective stress in unsaturated soils, in which only a small fraction of the pore water exists at the measured pore water pressure. Research into the principle of effective stress, along the lines suggested herein, is in progress; it is hoped that a more basic understanding of effective stresses will result.

CONCLUSIONS

Pore water pressures significantly more negative than approximately -1 atmosphere have not been well documented in engineering literature because of the experimental problems associated with such measurements. The writers have reported measured pore water pressures as low as -70 psi at the optimum water content and as low as -250 psi just 5% dry of optimum.

Theoretical considerations indicate that the actual pore water pressures in unsaturated soils vary from approximately -1 atmosphere to high positive values, although the measured pore water pressures may be hundreds of psi negative. The application of these principles to further clarification of effective stress in unsaturated soils requires further research.

ACKNOWLEDGMENTS

The investigation reported herein was performed in the Soil Mechanics Laboratories, Department of Civil Engineering, Univ. of Illinois, Urbana, Ill. It was sponsored by the National Science Foundation (Grant GP 2353). Leonard J. Langfelder received financial support from the Ford Foundation during his graduate studies at the University of Illinois. Several of the experiments were performed by J. D. Campbell.

APPENDIX I.—VALIDITY OF THE AXIS TRANSLATION TECHNIQUE

It can easily be demonstrated that the change in bulk volume of the pore water and solid mineral matter on the application of an ambient air pressure is so small that the curvature of the air-water interfaces cannot change significantly for the usual range of applied air pressure (e.g., less than 500 psi). Dissolution of gas into the water, according to Henry's law, would increase the volume of the water and decrease the curvature of the menisci, thus introducing an error, but the error appears to be small. If the curvature of the menisci do not change significantly, then the difference between the ambient gas pressure and the pore water pressure cannot change significantly, and the pore water pressure must increase by the same amount as the gas pressure. Thus, the original negative pore water pressure, expressed as a gage pressure, is equal to the difference between the applied gas pressure and the associated measured pore water pressure. The technique is valid provided that all the gas voids are interconnected.

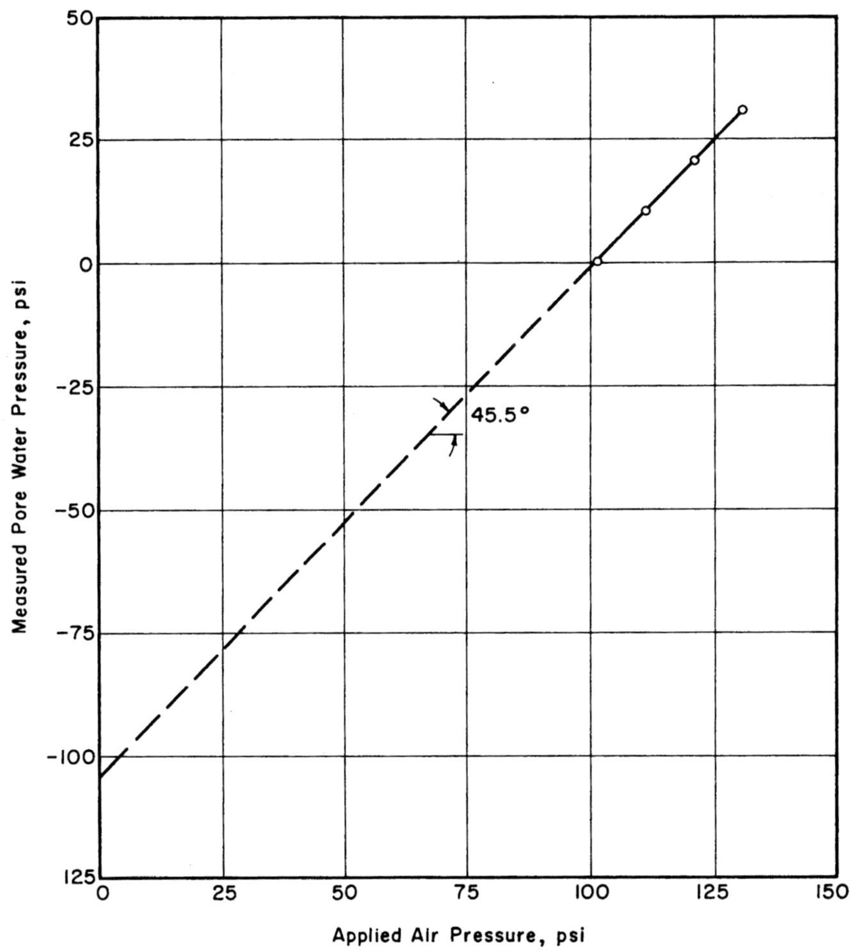

FIG. 8.—INFLUENCE OF AIR PRESSURE ON THE MEASURED PORE WATER PRESSURE

Experimental validation of the axis translation technique can be obtained by allowing a sample to come to equilibrium under some amblent air pressure and then changing the air pressure and measuring the change in pore water pressure. In soil mechanics, this method of validating the axis translation technique was used by Hilf[6] and by Bishop and Donald,[8] and the technique was found to be valid within the range of experimental error. The same technique was used several times during this investigation. Typical experimental data are shown in Fig. 8. If the axis translation technique is exact, the slope of the relationship in Fig. 8 should be 45.0°. A linear regression analysis of these data yielded a slope of 45.5°. The deviation of 0.5° is within the range of experimental error of the existing equipment.

A more precise validation of the axis translation technique would be obtained if the original negative pore water pressure were measured directly and then the axis translation technique were applied to determine if the difference between the pore water and pore air pressures really remains constant. For the range of negative pore water pressures under consideration, this technique has never been successfully applied (to the writers' knowledge) because the measuring system always cavitates. During this investigation, the axis translation technique was applied to several samples and then the ambient gas pressure was quickly reduced to zero and actual negative pore water pressures were measured. For samples with pore water pressures lower than -100 psi, the system always cavitated before a reliable estimate of the negative pore water pressure could be obtained. The technique was successfully applied to one sample of Goose Lake clay. The pore water pressure determined by the axis translation technique was -57.0 psi. With direct measurement, the pore water pressure in the measuring system had dropped to -56.0 psi when cavitation occurred. By fitting parabolas to the final points on the negative pore water pressure versus time curve it was possible to extend the curve slightly and to estimate the negative pore water pressure that would have been measured had cavitation not occurred. This estimated negative pore water pressure varied between -56.1 psi and -58.0 psi, depending on which points on the experimental curve were chosen. The difference between these pressures and -57.0 psi is within the range of accuracy of the 300-psi capacity Bourdon gage and the 1,000-psi capacity transducer used to measure the pressures. Thus, within the range of experimental error, the validity of the axis translation technique for pore water pressures significantly more negative than -1 atmosphere has been demonstrated.

APPENDIX II.—TIME LAGS IN THE ESTABLISHMENT OF EQUILIBRIUM

When the axis translation technique was used, the pore water pressure transducer was used as a null indicator and the water pressure in the lines and in the porous stone was zero throughout the test (except for minor momentary fluctuations). As the nitrogen pressure was increased, the zero pore water pressure in the stone can only be maintained if nitrogen-water interfaces of the requisite curvature develop at the surface of the stone. Such curvatures can only develop if the moisture in the pores at the surface of the stone can migrate. Because the pressure in the pore pressure connections in the base does not change, the water can migrate only into the soil sample. For the stone used in this investigation, the volume of water displaced from the stone is approximately 7.2×10^{-7} cu cm for an air pressure of 10 psi,[10] and approximately 5.8×10^{-6} cu cm for an air pressure of 100 psi. These volumes of water would have to dissipate uniformly throughout the soil before equilibrium could be established. The coefficients of permeability of the unsaturated soils could not be measured with the existing apparatus but, even assuming low coefficients of permeability, the time lag resulting from the development of capillary menisci in the stone is unlikely to exceed a few minutes. Thus this source of time lag appears to be unimportant.

A considerably larger volume of flow would result if the surface of the stone were not wiped properly free of moisture before the sample was put in place. Observations showed that time lags of more than 30 min. occurred if the stone was too moist at the beginning of the test. If the stone was wiped too dry, then time lags resulted from inadequate contact areas between the moisture of the soil and the moisture of the stone. In intermediate cases, placing the sample on the stone resulted in the immediate development of very negative pressures in the measuring system. In such cases, it was necessary to apply the nitrogen pressure to the exposed faces of the soil samples so rapidly that gas pressure gradients developed as a result of the finite gas permeability of the soil. Thus, an excessive nitrogen pressure was required at the surface of the soil to translate the pore water pressure

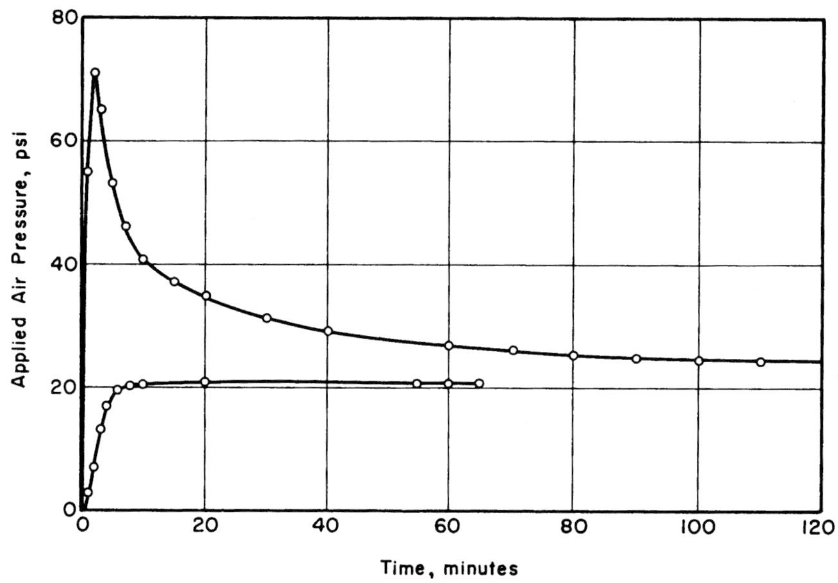

FIG. 9.—TYPICAL CURVES SHOWING THE RESPONSE TIME OF THE APPARATUS

to zero at the base of the sample where the pore water pressure was actually being measured. The nitrogen pressure gradients resulted in "overshooting." Applied nitrogen pressure versus time curves for two tests on duplicate samples are presented in Fig. 9 to demonstrate these effects.

APPENDIX III.—NOTATION

The following symbols have been adopted for use in this paper:

C = concentration of electrolyte;
G = specific gravity;
g = acceleration due to gravity;

h = total head;
L_w = liquid limit;
P_w = plastic limit;
R = gas constant;
r = radius of curvature;
S = specific surface;
T = absolute temperature;
T_s = surface tension;
u = pore pressure;
v = velocity;
w_{opt} = optimum water content;
z = depth;
α = adsorptive head;
γ_{dm} = maximum dry density;
γ_w = unit weight of water;
π = osmotic pressure; and
ω = osmotic head.

Journal of the

SOIL MECHANICS AND FOUNDATIONS DIVISION

Proceedings of the American Society of Civil Engineers

PILE-DRIVING FORMULAS FOR FRICTION PILES IN SAND

By Roy E. Olson,[1] M. ASCE, and Kaare S. Flaate,[2] A. M. ASCE

INTRODUCTION

Dynamic pile-driving formulas are widely used in predicting the load-carrying capacity of friction piles and in writing pile-driving specifications. Further, when load tests are available, dynamic pile-driving formulas are used to interpolate between, or extrapolate beyond, the load test results.

The accuracy of a pile-driving formula can be checked by comparing calculated pile capacities with capacities measured in the field. Such comparisons have demonstrated that the formulas do not generally apply to cohesive soils, especially soft cohesive soils,[3] and do not apply to piles acting as groups. They apply most accurately to individual piles driven into cohesionless soils. It has also been demonstrated that few, if any, of the existing dynamic pile-driving formulas are theoretically valid.[4] Most of the formulas were derived either using oversimplified assumptions or using empirical parameters that could be adjusted to bring the predicted capacities approximately into conformance with field measurements.

In addition to the theoretical errors, there are errors in many of the field measurements because of friction in the rams of hydraulic jacks or improper calibration of equipment. In addition, random variations in the measured capacity occur because of the problem of uniquely defining failure in the field. Thus, correlations between predicted pile capacities and capacities measured in the field are likely to involve considerable scatter.

Note.—Discussion open until April 1, 1968. To extend the closing date one month, a written request must be filed with the Executive Secretary, ASCE. This paper is part of the copyrighted Journal of the Soil Mechanics and Foundations Division, Proceedings of the American Society of Civil Engineers, Vol. 93, No. SM6, November, 1967. Manuscript was submitted for review for possible publication on January 27, 1967.

[1] Prof. of Civil Engrg., University of Illinois, Urbana, Ill.

[2] Chf. Engr., Norwegian Road Research Lab., Oslo, Norway.

[3] Terzaghi, K. T., discussion of "Pile-Driving Formulas: Progress Report of the Committee on the Bearing Value of Pile Foundations," Proceedings, ASCE, Vol. 68, No. 2, Feb., 1942, pp. 311-323.

[4] Cummings, A. E., "Dynamic Pile Driving Formulas," Contributions to Soil Mechanics, 1925-40, Boston Society of Civil Engrs., Boston, Mass., pp. 392-413.

In the investigation reported herein, seven different dynamic pile-driving formulas (Table 1) were used to predict the capacities of 93 timber, precast concrete, or steel friction piles driven into sandy soils, and the predicted capacities were compared statistically with the measured values. A linear

TABLE 1.—PILE-DRIVING FORMULAS USED IN THIS INVESTIGATION[a]

Engineering News	$Q_c = \frac{e_h E_n}{s + e}$
Gow	$Q_c = \frac{e_h E_n}{s + e\,(W_p/W_h)}$
Hiley	$Q_c = \frac{e_h E_n}{s + 1/2\,(C_1 + C_2 + C_3)} \quad \frac{W_h + n^2 W_p}{W_h + W_p}$
Pacific Coast Uniform Building Code	$Q_c = \frac{A\ E}{2\ L}\left[-s + \sqrt{s^2\ \frac{4\ e_h E_n}{A\ E}\ \frac{W_h + n^2 W_p}{W_h + W_p}}\right]$
Janbu	$Q_c = E_n/k_u s$ $k_u = C_d\left(1 + \sqrt{1 + \frac{\lambda_e}{C_d}}\right)$ $\lambda_e = E_n L/A\ E\ s^2,$ $C_d = 0.75 + 0.15\ (W_p/W_h)$
Danish	$Q_c = \frac{e_h E_n}{s + \sqrt{\frac{e_h E_n L}{2AE}}}$
Gates[b]	$Q_c = 5.6\ \sqrt{e_h E_n}\ \log\ (10/s)$

[a] As presented in this table, none of the formulas contains a factor of safety. A suitable factor of safety should be applied to obtain the design loads.

[b] In all but the Gates formula, any consistent set of units may be used. We have chosen to use inches and tons throughout. In the Gates formula, the constants contain parameters needed to make the formula dimensionally correct. Thus, any change of units from the ton-inch system necessitates changes in the constants. In the metric system of tons (1000 kg) and centimeters, the formula becomes:

$$Q_c = 4.0\ \sqrt{e_h E_n}\ \log\left(\frac{25}{s}\right)$$

relationship was assumed between the measured and computed pile capacities, and the reduced major-axis type of linear regression analysis was used to determine the slope and intercept of the regression line. Correlation coefficients were used as measures of the scatter about the regression lines. The statistical data were then used to adjust the formulas to improve their accuracy. Conclusions are drawn regarding the accuracy of the various formulas.

PILE-DRIVING FORMULAS USED IN THIS INVESTIGATION

The seven pile-driving formulas used in this investigation are presented in Table 1. In these formulas, $e_h E_n$ is the energy delivered to the cushion blocks. These energies should be multiplied by e_i, the efficiency of impact, to obtain the energy delivered to the pile. However, values of e_i were unknown. Eliminating e_i from the formulas has doubtless introduced a certain amount of scatter into the calculated capacities, but unknown types and condition of the cushion blocks preclude further refinement. In the special case of Janbu's formula, e_h and e_i were both set equal to 0.7 when the formula was originally developed[5] and were incorporated into the driving coefficient.

The Engineering News formula was developed by Wellington,[6] who deduced its general form by setting the applied energy equal to the energy obtained by graphically integrating the area under what he considered to be typical load-settlement curves for timber piles driven with drop hammers. He subsequently modified the formula for use with steam hammers. He stated that his formula "was first deduced as the correct form for a theoretically perfect equation of the bearing power of piles, barring some trifling and negligible elements to be noted; and I claim in regard to that general form that it includes in proper relation to each other every constant which ought to enter into such a theoretically perfect practical formula, and that it cannot be modified by making it more complex. . . ." It appears that the Engineering News formula achieved wide acceptance but has been less used in recent years as new formulas were introduced and the inability of the Engineering News formula to predict pile capacities with reasonable accuracy became better documented.[7] The formula was restricted to use with piles for which the average penetration per blow for the last few blows was not less than 0.25 in. and preferably not less than 0.5 in.

In the Gow formula, the denominator of the Engineering News formula was adjusted, based on intuition and experience, to account for the extra energy-absorbing characteristics of precast concrete piles.[8]

In an attempt to eliminate some of the errors associated with the theoretical evaluation of energy absorption by a pile-soil system during driving, Hiley[9] developed a formula in which the recoverable deformations of the pile cap and driving head, pile, and soil, were measured during driving and inserted into the formula as three constants, C_1, C_2, and C_3 respectively. The Hiley formula has been used extensively in the British Commonwealth and in Europe.

The Pacific Coast Uniform Building Code formula,[10] hereafter referred to as the PC formula, is typical of a number of rather cumbersome formulas in which attempts were made to include a variety of sources of energy loss. Cummings[4] objected to most of these formulas because of the inclusion of both the coefficient of restitution and a separate term for the energy loss caused by elastic compression of the pile, which is redundant.

Janbu[11] factored out of the conservation-of-energy equation a series of

---

[5] Johannsen, N., "Beregning av frittstaende pelers bruddlast og tillatte belastning," Teknisk Ukeblad, Oslo, Norway, No. 26, June 28, 1951, pp. 505-515.

[6] Wellington, A. M., discussion of "The Iron Wharf at Fort Monroe, Va.," by J. B. Duncklee, Transactions, ASCE, Vol. 27, Paper No. 543, Aug., 1892, pp. 129-137.

[7] Agerschou, H. A., "Analysis of the Engineering News Pile Formula," Journal of the Soil Mechanics and Foundations Division, ASCE, Vol. 88, No. SM5, Proc. Paper 3298, Oct., 1962, pp. 1-11.

[8] Isaacs, D. V., "Reinforced Concrete Pile Formulae," Transactions, Inst. of Engrs., Sydney, Australia, Vol. 12, 1931, pp. 305-323.

[9] Hiley, A., "The Efficiency of the Hammer Blow, and its Effects with Reference to Piling," Engineering, London, June 2, 1922, p. 673.

[10] "Uniform Building Code," Pacific Coast Building Officials Conf., Los Angeles, Calif., Vol. 1, 1955, pp. 207-208.

[11] Janbu, N., "Une Analyse Energetique du Battage des Pieux a l'aide de Paramètres sans Dimension," Publication No. 3, Norwegian Geotechnical Inst., Oslo, Norway, 1953.

variables that could not usually be evaluated and associated them together as his "driving coefficient," C_d. The driving coefficient included terms representing the efficiency of the pile hammer, the difference between the dynamic and static pile capacities, and the rate of transferral of pile load into the soil with respect to depth. It also included the length and cross-sectional area of the pile, Young's modulus for the pile, and both the pile capacity and the set. Janbu correlated his driving coefficient with the ratio of the weight of the pile to the weight of the falling parts of the hammer, W_p/W_h (Table 1).

The Danish formula[12] was developed using dimensional analysis and by simplifying some of the more complicated formulas. A total of 78 load tests was used in correlating the predicted capacities with capacities measured in the field.

The Gates formula[13] was developed by greatly simplifying the form of existing formulas and then applying a statistical adjustment which was based on approximately 100 pile load tests. The data on which the study was based were not presented and there was no indication of the amount of scatter. Apparently all types of soil were included in the study.

These formulas were selected to be representative of various types of formulas and to include most of the formulas in common use at the present time.

FIELD DATA

The field test results used in this study have been reported by Flaate.[14] He collected the results of 116 load tests on timber, precast concrete, and steel piles driven into sandy soils. In this analysis, only the timber piles with measured capacities less than 100 tons, and steel and concrete piles with measured capacities less than 250 tons were used. It seems unlikely that piles would be designed for higher loads without requiring load tests. Flaate[14] reported the following data: (1) The type of pile, (2) the approximate subsoil conditions, (3) the length, cross-sectional area, and weight of the pile, (4) the type of hammer, (5) either the weight and height of fall of the hammer or its reported energy, (6) the average penetration of the pile under the final few blows, (7) the reported capacity, and (8) the source of the information.

Attempts to apply dynamic pile-driving formulas to the piles used in these tests involve a number of uncertainties, including the following:

1. The efficiencies of the various pile hammers were not reported. The values tabulated by Chellis[15] were used in this study. The actual field values of e_h depend greatly on the condition of the hammer at the time of driving and may differ significantly from the values used in this analysis.

2. The cushion blocks used on top of the piles were not usually specified.

[12] Sorensen, T., and Hansen, B., "Pile Driving Formula—An Investigation Based on Dimensional Considerations and a Statistical Analysis," Proceedings, 4th Internatl. Conf. on Soil Mechanics and Foundations, held in London in 1956, Vol. 2, 1957, pp. 61-65.

[13] Gates, M., "Empirical Formula for Predicting Pile Bearing Capacity," Civil Engineering, Vol. 27, No. 3, Mar., 1957, pp. 65-66.

[14] Flaate, K. S., "An Investigation of the Validity of Three Pile Driving Formulae in Cohesionless Material," Publication No. 56, Norwegian Geotechnical Inst., Oslo, Norway, 1964, pp. 11-22.

[15] Chellis, R. D., Pile Foundations, 2nd ed., McGraw-Hill Book Co., Inc., New York, N. Y., 1961, pp. 28-33.

They doubtless varied widely in type and condition. These blocks exert great influence on the shape of the load pulse applied to the pile[16] and thus influence the energy actually delivered to the pile.

3. The coefficients of restitution were not known. The PC Code was followed by using $n^2 = 0.25$ for steel piles and 0.10 for concrete and timber piles.

4. The dynamic compression and recovery of the piles during driving were not generally reported. Thus, values for the constants C_1, C_2, and C_3 in Hiley's formula were taken from the tabulations in Chellis.[17]

5. The elastic modulus of the material in each pile was not reported. Average values for the static elastic modulus[14] were used in this study. The dynamic elastic modulus may exceed the static modulus, especially for timber piles. In addition, the moduli of the timber piles vary with the type of wood, storage conditions, and driving conditions.

6. Only about ten of the piles were driven entirely through cohesionless soils. At many of the sites the piles were driven through soft cohesive soils into underlying sands. In other cases the sand was interstratified with clay, silt, and sometimes organic soil, or was described as silty sand or clayey sand.

7. The capacities of most of the piles were reported without presenting the actual load-settlement diagrams. It is believed that a scatter of perhaps 15% has resulted from the use of different failure criteria.

8. Based on recent field studies[18] it is believed that many of the measured pile capacities are in error by 10% or more because of friction in hydraulic loading jacks, and improper calibration of equipment.

Attempts to account for the various sources of error by adjusting the field data were not considered desirable because: (1) The adjustment procedures would be too complex for normal field use; (2) data were not available for making most of the corrections; and (3) arbitrary choices involved in making such adjustments would introduce bias.

STATISTICAL METHODS

As expected, the measured and computed pile capacities did not correlate perfectly. It was convenient, therefore, to apply certain simple forms of statistics to the interpretation of the data. The equations used in the statistical study are presented here for convenience of reference.

The mean ($\overline{X}$) and estimated standard deviation (S_X) of N observations (X_i) of the variable X are given by

$$\overline{X} = \frac{1}{N} \sum X_i \qquad (1)$$

[16] Housel, W. S., "Michigan Study of Pile Driving Hammers," Journal of the Soil Mechanics and Foundations Division, ASCE, Vol. 91, No. SM5, Proc. Paper 4483, Sept., 1965, pp. 37-64.

[17] Chellis, R. D., op. cit., pp. 505-506.

[18] Davisson, M. T., "Summary of Knowledge Gained from Tests on Instrumented Piles," presented at ASCE Metropolitan Section, Seminar on Pile Foundations, New York, N. Y., 1966.

and $$S_X = \sqrt{\frac{\sum (X_i^2) - \frac{1}{N} (\Sigma X_i)^2}{N - 1}} \quad (2)$$

If there are two variables, X and Y, and Y is assumed to be linearly related to X according to

$$Y = AX + B \quad (3)$$

then linear regression analysis can be used to estimate the most probable values of A and B. In comparing the measured and computed pile capacities, it was assumed that significant errors existed in both, and thus it was necessary to minimize the square of the deviations measured perpendicular to the regression line, i.e., the reduced major-axis technique of linear regression analysis applied. The pertinent equations are

$$A = \frac{S_Y}{S_X} \quad (4)$$

and $$B = \overline{Y} - A\overline{X} \quad (5)$$

In addition to the parameters A and B, it would be desirable to have one or more parameters to describe the scatter of the observations relative to the regression line. Of the various available parameters, it appears that the correlation coefficient, r, is the most satisfactory because it is single-valued and varies between fixed limits of plus one for a perfect positive correlation through zero for no correlation to minus one for a perfect negative correlation (standard errors of slope and intercept, and standard deviations could also be used). The correlation coefficient is defined as

$$r = \frac{\sqrt{\sum X_i Y_i - \frac{1}{N} (\Sigma X_i)(\Sigma Y_i)}}{\sqrt{\sum (X_i^2) - \frac{1}{N} (\Sigma X_i)^2} \sqrt{\sum (Y_i^2) - \frac{1}{N} (\Sigma Y_i)^2}} \quad (6)$$

MODIFICATION OF JANBU'S FORMULA

There is no apparent reason why Janbu's driving coefficient should correlate with the ratio of the weight of the pile to the weight of the hammer. To check this correlation, the load tests reported by Flaate[14] were used to back-calculate the driving coefficients. These driving coefficients are plotted in Fig. 1. It is apparent that there is no significant relationship between C_d and W_p/W_h for the tests used in this study. The average values of C_d for timber, precast concrete, and steel piles, were 0.92, 1.06, and 1.07, respectively, with an average for all piles except S49, C15, and T20, of 1.02. Therefore, Janbu's formula was simplified by substituting $C_d = 1$.

Separate statistical studies of the pile data were performed using Janbu's original formula, using the modification with a driving coefficient of one, and using another modification in which the efficiency terms were not incorporated into the driving coefficient. The studies showed that the modification with $C_d = 1$ was both the most accurate and the simplest. Thus, in all subsequent

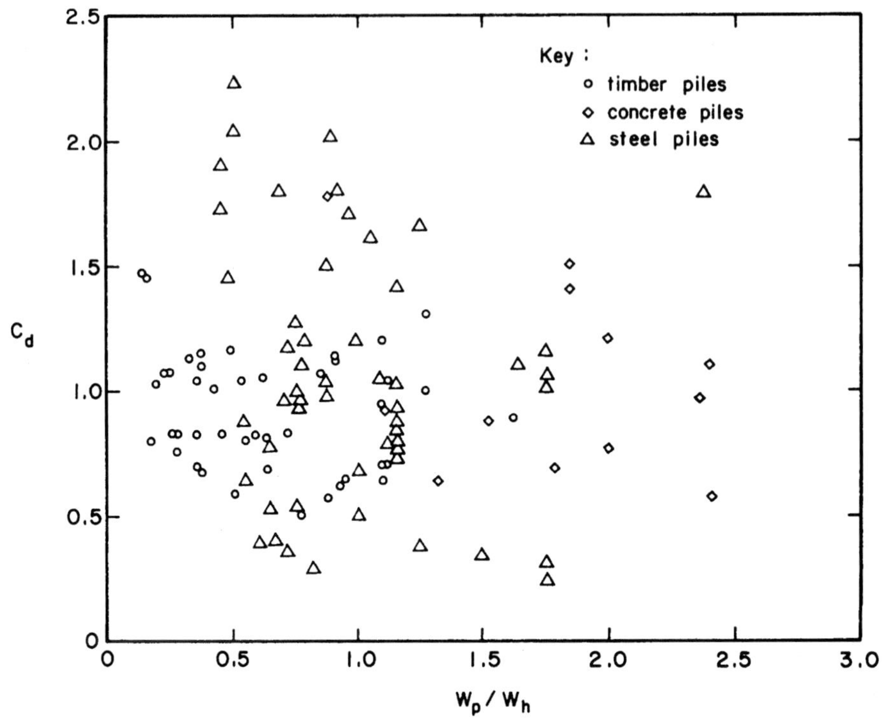

FIG. 1.—INFLUENCE OF THE RATIO W_p/W_h ON JANBU'S DRIVING COEFFICIENT

sections, any statement regarding Janbu's formula will refer to the modification in which the driving coefficient is one.

COMPARISON OF MEASURED AND COMPUTED PILE CAPACITIES

Presentation of Data.—The capacities of the piles studied by Flaate[14] were calculated using the seven formulas presented in Table 1, except that the modified form of Janbu's formula was used. The measured pile capacities and the values calculated using the seven formulas are reported in Table 2 for timber, precast concrete, and steel piles. These data were used to prepare diagrams in which the measured pile capacity, Q_m, was plotted as the ordinate and the calculated capacity, Q_c, as the abscissa. The reduced major-axis type of linear regression analysis was then used to find the best linear relationship between Q_m and Q_c according to

$$Q_m = A\,Q_c + B \quad \text{(7)}$$

For a perfect pile-driving formula and uniquely defined values of Q_m, the measured and computed capacities would be identical for each pile and the statistical parameters would be $A = 1$, $B = 0$, and $r = 1$.

Regression analyses were performed for each type of pile separately using each pile-driving formula. Representative Q_m - Q_c diagrams for timber, precast concrete, and steel piles, are presented in Figs. 2 through 4, respectively. The statistical parameters are tabulated in Table 3. Cumulative frequency curves are presented in Figs. 5 through 7.

Evaluation.—For timber piles, Janbu's formula is clearly superior to the others (Table 3). The regression line has a slope of nearly 45°, a small inter-

TABLE 2.—COMPARISON OF MEASURED AND COMPUTED CAPACITIES OF PILES

Test Number	Q_m, in tons	Q_c, in tons						
		EN	G	Ga	Da	PC	Hi	J
(1)	(2)	(3)	(4)	(5)	(6)	(7)	(8)	(9)
				(*a*) Timber				
1	38	30	31	46	50	35	29	34
2	38	33	35	51	60	42	35	42
3	35	26	32	41	37	29	26	25
4	34	28	35	43	41	32	29	28
5	23	21	28	33	25	22	22	17
6	22	25	38	40	33	28	30	23
7	107	90	134	84	117	88	108	75
8	117	90	209	110	167	113	205	123
9	40	77	76	46	40	26	26	33
10	40	62	61	42	35	22	23	28
11	40	68	67	44	37	24	24	31
12	40	152	147	60	55	33	48	44
13	43	86	87	48	40	26	27	34
14	25	40	41	31	25	20	20	20
15	26	43	44	32	27	23	25	22
16	71	155	177	60	64	53	68	58
17	37	85	103	43	43	37	39	39
18	45	113	131	49	51	41	41	47
19	187	200	183	134	258	154	184	192
20	—	—	—	—	—	—	—	—
21	121	126	141	102	170	69	124	127
22	162	138	175	114	191	127	159	144
23	98	38	48	78	91	55	64	67
24	96	38	43	78	96	55	57	71
25	50	49	37	55	68	37	30	45
26	53	63	55	65	72	45	47	53
27	45	12	12	15	8	4	4	6
28	28	39	39	29	23	15	14	19
29	44	82	85	46	42	33	33	37
30	40	92	96	48	46	36	35	41
31	68	74	123	70	104	94	105	72
32	93	90	165	84	138	116	144	102
33	68	63	104	61	84	82	88	56
34	105	84	160	79	125	112	131	90
35	68	55	97	62	84	77	83	58
36	74	85	120	74	85	66	94	63
37	53	64	60	65	69	44	46	50
38	43	101	115	47	53	41	41	48
39	59	73	81	41	39	34	34	35
40	37	91	104	45	46	37	38	41
41	47	101	113	47	54	42	40	49
42	54	161	189	61	80	75	83	73
43	68	161	188	61	76	69	78	70
44	66	188	228	65	100	93	101	90
				(*b*) Concrete				
1	77	53	63	66	103	74	73	74
2	77	39	48	64	101	70	67	75
3	99	41	37	68	125	80	64	93
4	132	42	33	71	129	75	60	97
5	121	42	29	71	148	87	60	110
6	138	42	25	71	142	77	55	106
7	70	195	201	64	135	96	68	118
8	34	29	16	45	50	23	18	33
9	35	26	17	41	42	21	17	27
10	80	37	21	73	133	68	55	100
11	73	37	21	71	128	66	54	96
12	121	75	39	113	183	86	85	134
13	132	90	42	96	188	91	66	141

TABLE 2.—*CONTINUED*

Test Number	Q_m, in tons	Q_c, in tons						
		EN	G	Ga	Da	PC	Hi	J
(1)	(2)	(3)	(4)	(5)	(6)	(7)	(8)	(9)
				(*b*) Concrete (continued)				
14	296	455	452	116	235	110	120	210
15	95	692	398	115	259	83	158	221
16	290	316	316	154	370	133	130	312
17	138	245	245	98	227	88	93	207
				(*c*) Steel				
1	56	37	34	50	65	46	46	41
2	44	37	34	50	65	46	46	41
3	45	40	36	53	72	52	51	46
4	55	40	37	54	75	55	53	48
5	67	42	38	56	82	60	59	54
6	45	45	40	59	91	66	66	61
7	79	45	40	59	91	66	66	62
8	45	45	49	60	98	73	76	65
9	67	45	49	60	98	73	76	65
10	89	49	43	65	109	79	80	77
11	79	51	61	69	131	103	107	93
12	110	51	45	69	121	87	90	87
13	155	174	212	116	236	178	213	175
14	124	119	146	97	182	139	150	133
15	220	134	202	110	215	163	210	161
16	118	63	70	78	170	131	110	122
17	107	61	75	75	149	118	107	106
18	203	114	105	106	286	205	165	210
19	244	140	127	129	365	234	241	275
20	337	115	167	108	238	179	209	178
21	233	74	70	104	209	162	145	342
22	297	127	254	180	282	199	307	456
23	305	129	262	136	291	202	327	494
24	296	129	239	136	308	210	317	515
25	310	131	246	141	320	215	327	559
26	140	600	423	111	179	99	129	145
27	150	705	474	121	189	101	135	151
28	140	725	482	122	190	102	135	152
29	90	450	344	99	161	93	113	49
30	80	470	355	100	163	94	115	38
31	150	468	574	93	121	123	119	106
32	120	450	546	91	120	78	119	105
33	99	197	220	61	102	75	55	99
34	24	26	19	41	41	26	19	26
36	290	378	406	92	166	128	100	152
37	285	454	482	98	202	151	118	186
38	135	259	275	79	136	113	115	122
39	350	790	913	122	245	181	126	461
40	315	625	684	111	229	193	137	385
41	300	467	468	118	306	229	124	500
42	360	1395	1395	163	301	174	143	263
43	170	643	740	109	107	64	52	90
44	143	693	882	115	102	61	55	85
45	258	500	499	115	176	96	59	148
46	307	1395	1384	163	300	173	143	252
47	173	302	296	84	159	115	108	144
48	462	1170	1169	163	333	186	132	280
49	44	257	319	73	142	131	143	147
50	63	265	321	74	137	116	135	125
51	55	150	164	59	87	76	74	77
52	42	128	129	56	83	64	38	70
53	27	100	101	50	63	49	41	52
54	54	180	183	63	99	75	68	89
55	60	186	185	66	98	68	60	90

cept, and a high correlation coefficient. The data are shown graphically in Fig. 2(*d*). The Danish and Gates formulas have equally high values of the correlation coefficient but less satisfactory values of A and B. The PC and Hiley formulas are somewhat less accurate, and the Engineering News and Gow formulas are considerably less accurate.

Consideration of the data for precast concrete piles is made difficult by the fact that a total of only fifteen tests was available. For the available data, Hiley's formula yields the value of A closest to one but the low correlation

TABLE 3.—COMPILATION OF STATISTICAL PARAMETERS

Pile (1)	Formula (2)	N (3)	A (4)	B, in tons (5)	r (6)
Timber	Engineering News	37	0.45	16	0.28
	Gow	37	0.37	18	0.43
	Hiley	37	0.64	19	0.77
	Pacific Coast	37	0.80	14	0.74
	Janbu (C_d = 1)	37	0.98	9	0.86
	Danish	37	0.71	9	0.86
	Gates	37	1.30	-17	0.86
Concrete	Engineering News	15	0.20	72	0.11
	Gow	15	0.32	69	0.12
	Hiley	15	1.08	24	0.43
	Pacific Coast	15	1.57	-19	0.75
	Janbu (C_d = 1)	15	0.66	23	0.64
	Danish	15	0.60	11	0.69
	Gates	15	1.62	-27	0.65
Steel	Engineering News	41	0.28	43	0.37
	Gow	41	0.28	42	0.38
	Hiley	41	1.14	-10	0.76
	Pacific Coast	41	1.07	0	0.79
	Janbu (C_d = 1)	41	0.91	7	0.83
	Danish	41	0.89	-16	0.82
	Gates	41	2.34	-83	0.84
All	Engineering News	93	0.33	37	0.29
	Gow	93	0.32	37	0.36
	Hiley	93	0.92	7	0.72
	Pacific Coast	93	1.04	2	0.76
	Janbu (C_d = 1)	93	0.87	10	0.81
	Danish	93	0.77	-2	0.81
	Gates	93	1.81	-48	0.81

coefficient indicates considerable scatter [Fig. 3(*b*)]. This scatter is also evident in the cumulative frequency curve (Fig. 6). The A-values are farther from one for the PC, Danish, Janbu, and Gates formulas, but the correlation coefficients are higher. The cumulative frequency curves in Fig. 6 suggest that the formula yielding values of Q_m/Q_c closest to one is Janbu's formula. The low correlation coefficients obtained when the Engineering News and Gow formulas were used indicate that there is essentially no correlation between the measured and computed pile capacities [Fig. 3(*a*)].

For steel piles, the PC formula yields excellent values of A and B but the correlation coefficients are slightly higher for the Danish, Janbu and Gates formulas. The Hiley formula is slightly less accurate than these and, again, the Engineering News and Gow formulas are considerably less accurate.

To obtain a measure of the over-all average accuracy of the formulas, the weighted average values of A, B, and r were calculated (Table 3). The corre-

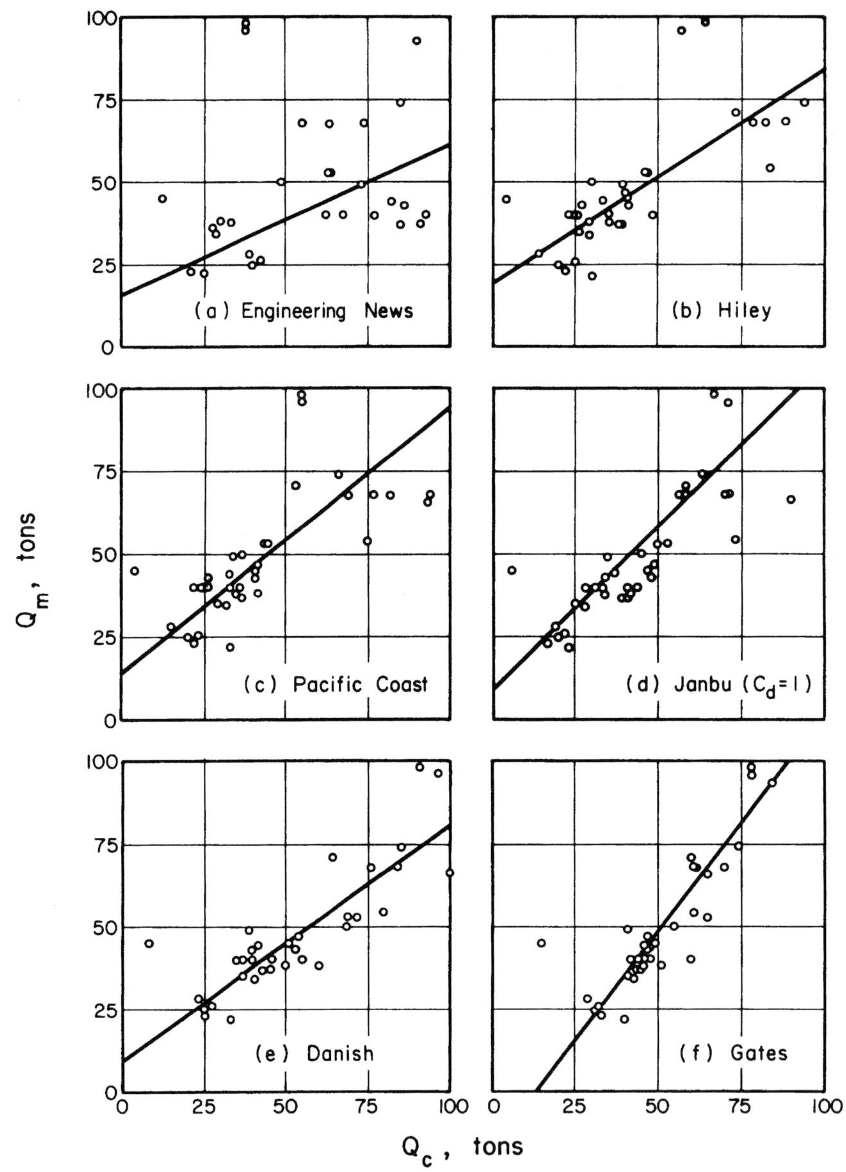

FIG. 2.—RELATIONSHIPS BETWEEN MEASURED AND COMPUTED CAPACITIES FOR TIMBER PILES

lation coefficients are highest, and equal, for the Gates, Danish, and Janbu formulas, and are slightly less for the PC and Hiley formulas. They are so low for the Engineering News and Gow formulas that these formulas can be eliminated from further consideration. When the values of A and B are taken into account, the Gates formula can be eliminated. The PC formula yields the best average values of A and B but it does so, in part, by underestimating the capacities of timber piles and overestimating the capacity of most concrete piles. It appears that valid arguments can be used in support of the Danish, PC, Hiley, or Janbu formulas; no one of the four is clearly superior to the others.

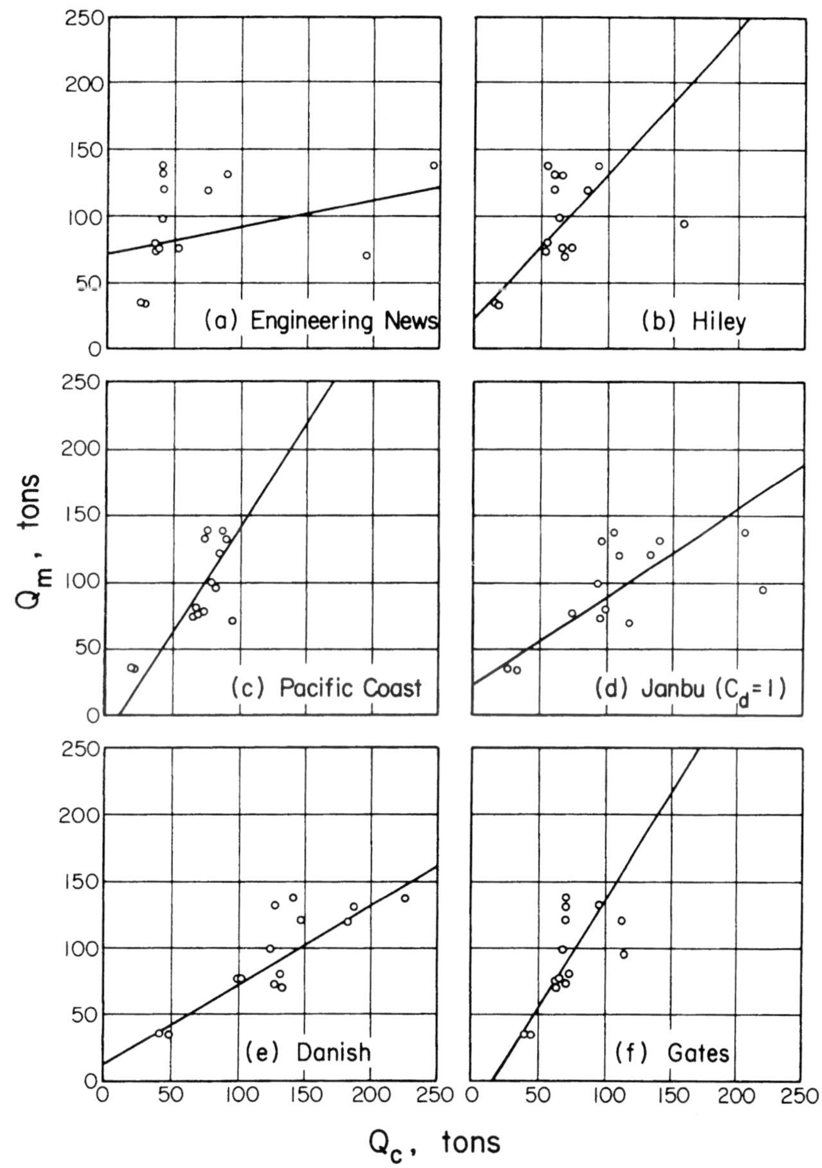

FIG. 3.—RELATIONSHIPS BETWEEN MEASURED AND COMPUTED CAPACITIES FOR CONCRETE PILES

Adjustment of the Formulas.—None of the formulas used in this study has any claim to theoretical rigor. Their usefulness is determined solely by the degree to which they successfully predict pile capacities in the field. Any adjustment to the formulas that improves their correlation with measured capacities is justified on pragmatic grounds.

All of the formulas can be improved if a new calculated capacity, Q_c^*, is defined using the equation

$$Q_c^* = A\ Q_c + B \quad (8)$$

in which the parameters A and B are tabulated in Table 3. If the values of A and B are taken separately for each type of pile, and the new calculated capacities are compared with the measured capacities, the regression line must yield $A = 1$ and $B = 0$. Thus, the accuracy of the adjusted formula depends on the correlation coefficient.

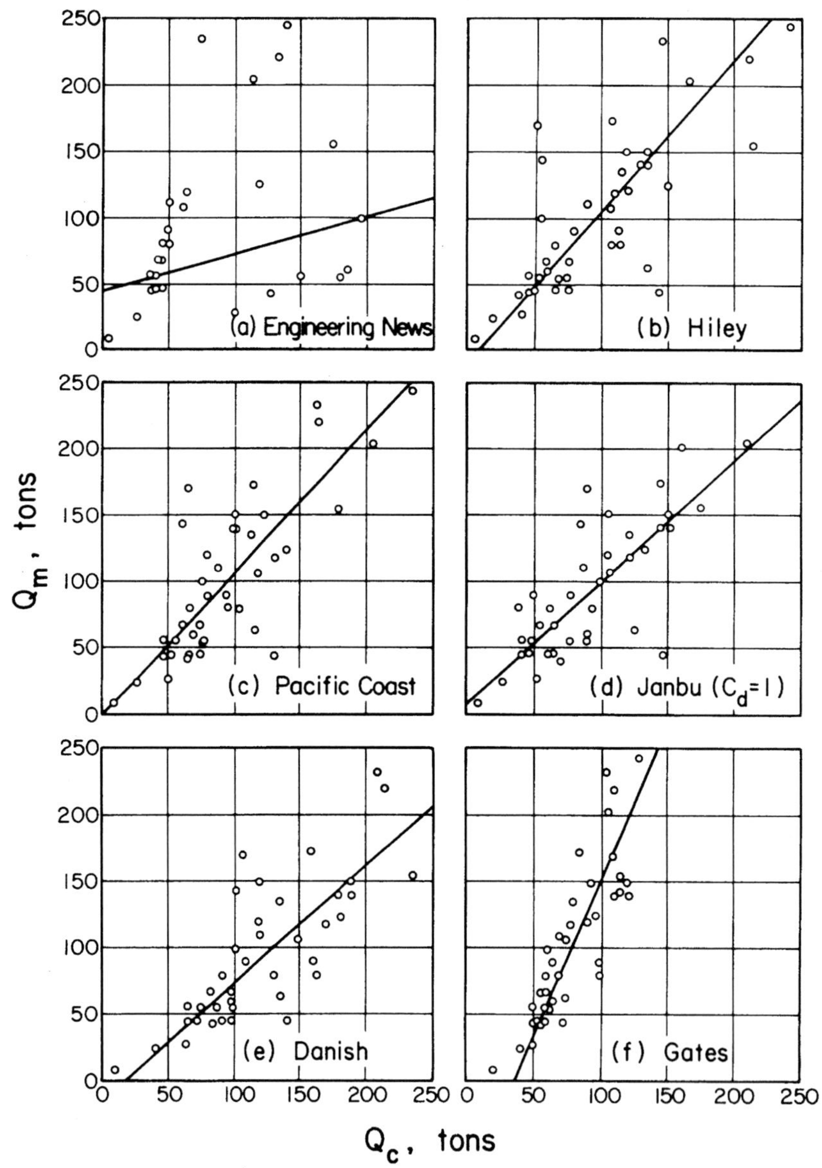

FIG. 4.—RELATIONSHIPS BETWEEN MEASURED AND COMPUTED CAPACITIES FOR STEEL PILES

In terms of high average correlation coefficients and simplicity of use, the Gates formula must rank first, followed closely by the Danish and Janbu formulas. The PC formula is less accurate and considerably more cumbersome to apply. The Hiley formula might be more accurate if the values of C_1, C_2, and C_3 were measured in the field, as Hiley recommended, but the formula

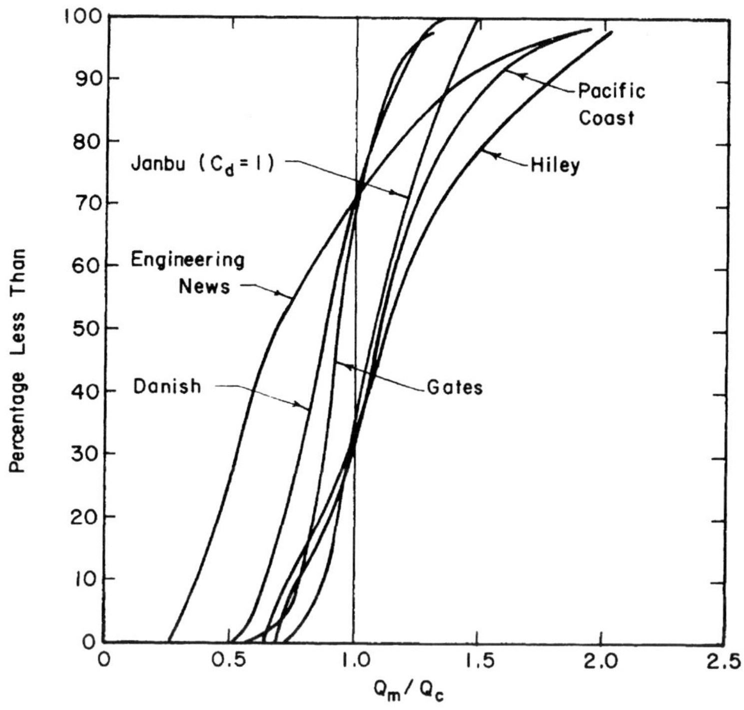

FIG. 5.—CUMULATIVE FREQUENCY CURVES FOR TIMBER PILES

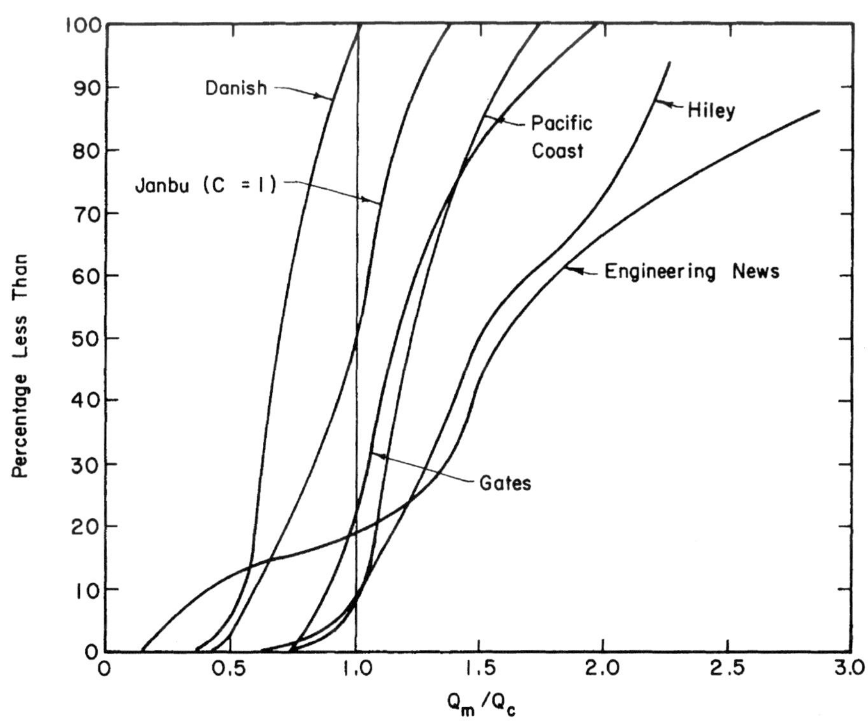

FIG. 6.—CUMULATIVE FREQUENCY CURVES FOR PRECAST CONCRETE PILES

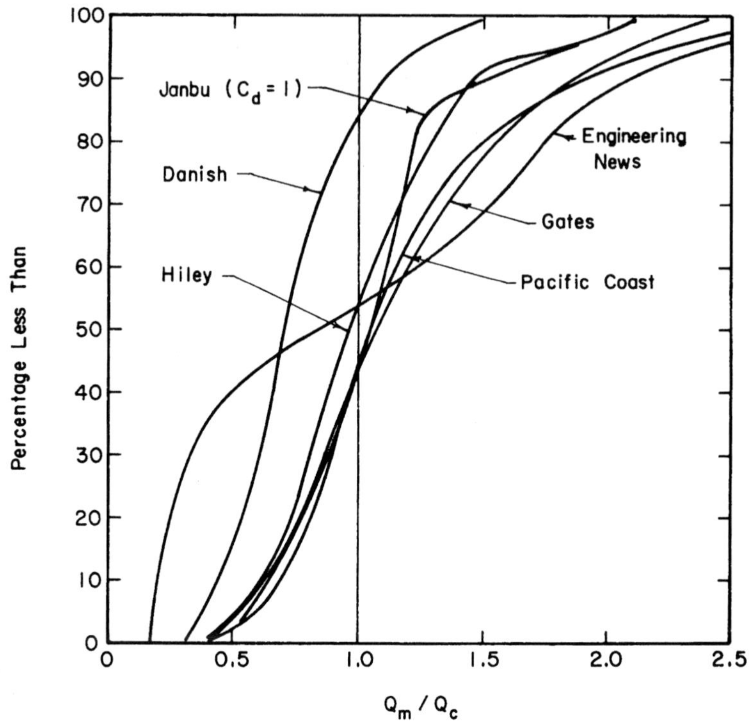

FIG. 7.—CUMULATIVE FREQUENCY CURVES FOR STEEL PILES

ranks fifth among the seven studied, when the average values of these constants are used.

If separate formulas are used for timber, precast concrete, and steel piles, the adjusted forms of the Gates formula become

$$\left.\begin{array}{ll} \text{Timber} & Q_c = 7.2\sqrt{e_h E_n}\,\log(10/s) - 17 \\ \text{Precast Concrete} & Q_c = 9.0\sqrt{e_h E_n}\,\log(10/s) - 27 \\ \text{Steel} & Q_c = 13.0\sqrt{e_h E_n}\,\log(10/s) - 83 \end{array}\right\} \quad \ldots\ldots\ldots (9)$$

in which Q_c is in tons, E_n is in in.-tons, and s is in inches.

The pile capacities were all recalculated using these three formulas. The measured and recomputed capacities are compared in Fig. 8. The equation of the regression line was $Q_m = Q_c$ and the correlation coefficient was 0.83, indicating a good correlation. The ratio Q_m/Q_c was calculated for each pile. The average value of the ratio was one and the standard deviation was estimated (Eq. 2) to be 0.28 (pile S35, which was driven with a small hammer, yielded a negative capacity and was not included in the calculation). Thus, approximately two-thirds of the piles had measured capacities within 28% of the calculated value.

If a single pile formula is desired for use with all three types of piles, then the average A and B factors (Table 3) can be used to adjust any of the formulas. It seems apparent from their high correlation coefficients and simplicity of use, that the Gates, Danish, or Janbu formulas should be recommended. As an example of the adjustment, a complete set of calculations was performed using the modified form of Janbu's formula with $A = 0.87$ and $B = 10$ tons. For all 93 piles, the regression line relating the measured capacities to the adjusted, computed capacities yielded $A = 1$, $B = 0$, and $r = 0.85$. The mean value of Q_m/Q_c was one and the coefficient of variation was 35%, i.e., about two-thirds of the piles had calculated capacities within 35% of the measured values. Similar correlations would be obtained if a single adjusted Danish or Gates formula were used.

When applying the revised pile-driving formulas in practice, use of a factor of safety of about three is recommended. There are several reasons for

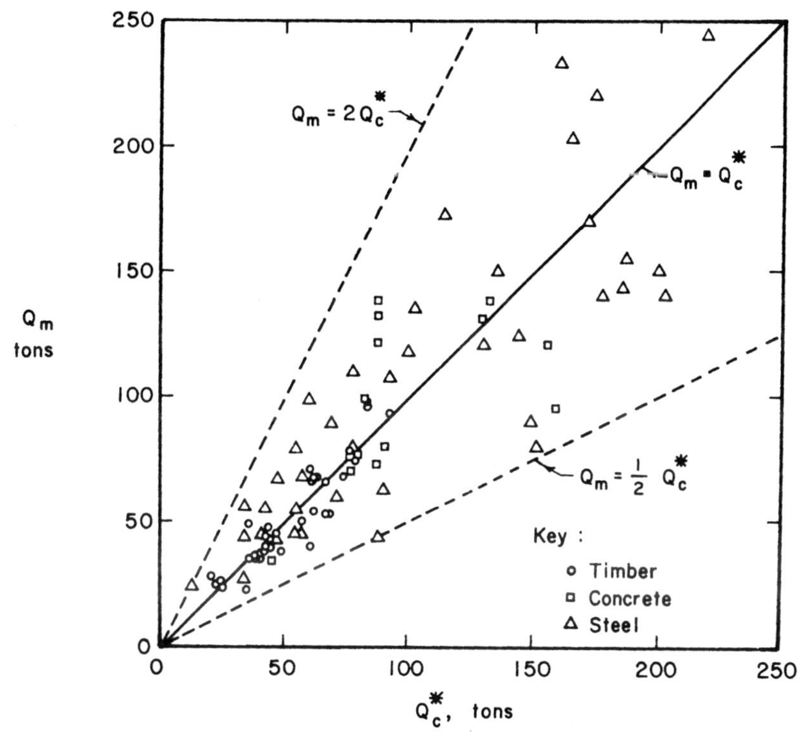

FIG. 8.—COMPARISON OF MEASURED PILE CAPACITIES AND THE CAPACITIES CALCULATED USING THE STATISTICALLY ADJUSTED FORMS OF GATES' FORMULA (EQ. 9)

choosing three. First, a factor of two is needed to account for inaccuracy in calculations (Fig. 8). Second, additional capacity is needed to account for normal uncertainties in loading. Third, the factor of safety should be increased further to ensure relatively small settlements under the design load. The factor of safety actually used in design may be adjusted to larger or smaller values depending on the consequences of excessive settlement.

Validity of the Adjustment Procedure.—A question may be raised regarding the validity of the adjusted forms of the pile-driving formulas because of the fact that they were adjusted, statistically, to fit certain data and then were compared with the same data. Can the adjusted formulas be applied to other piles with reasonable confidence? The answer would appear to be yes, provided that the data used in this study are representative of data generally obtained in the field. Whether the data are, in fact, representative can be established only by the collection of a large number of additional load-test results which would then be used to refine the values of A and B. It appears that the number of tests used in this study for timber and steel piles was sufficient to allow moderate confidence to be placed in the values of A and B presented in Table 3. However, the small number of tests using precast concrete piles, and the large amount of scatter, would suggest that more data are needed to establish values for A and B for these piles. No data are included for cast-in-place concrete piles where the casing is driven with a steel mandrel, but it appears probable that the formulas used for steel piles would apply.

In any case, it seems certain that the adjusted formulas are more accurate than the original ones. If the original formulas were good enough to use, then the adjusted formulas are likely to be better.

CONCLUSIONS

The measured capacities of 93 piles driven into sandy soils were compared with the capacities predicted using the Engineering News, Gow, Hiley, PC Code, Janbu (C_d = 1), Danish, and Gates formulas. The accuracy of the formulas varied with the type of pile. Janbu's formula was the most accurate for timber and steel piles. None of the formulas was clearly best for the precast concrete piles. In all cases, however, the Engineering News and Gow formulas were clearly inferior to the others and were eliminated from further consideration.

The three formulas that yielded the highest average correlation coefficients were the Danish, Janbu (C_d = 1), and Gates formulas. These formulas were adjusted statistically to fit the observed capacities. The adjusted formulas are believed to be more accurate in predicting the field capacities of timber and steel piles than any previous formulas. For greatest accuracy and simplicity of use, it is recommended that the adjusted forms of Gates' formula, Eq. 9, be used.

Adjusted forms of the Janbu (C_d = 1) or Danish formulas are also recommended. They yield essentially the same results as the Gates formula but are slightly more difficult to use.

Single adjusted formulas may be used for all types of piles but the predicted capacities are likely to be slightly less accurate than when a different adjusted form of the formula is used for each type of pile.

APPENDIX.—NOTATION

The following symbols are used in this paper:

A = slope of a regression line (dimensionless); cross-sectional area of a pile, in square inches;
B = intercept of a regression line, in tons;
C_1 = recoverable deformation of the pile cap and head, in inches per blow;
C_2 = recoverable deformation of the pile, in inches per blow;
C_3 = recoverable deformation of the soil, in inches per blow;
E = Young's modulus for the pile material, in tons per square inch;

E_n = nominal energy of the pile hammer, in inch-tons per blow;
e_h = efficiency of the hammer (dimensionless ratio);
e_i = efficiency of impact between the hammer and the cushion block (dimensionless ratio);
H = height of fall of the ram, in inches;
L = length of the pile, in inches;
N = number of observations;
Q_c = pile capacity calculated using a dynamic pile-driving formula, in tons;
Q_c^* = calculated pile capacity after the application of a statistical adjustment, in tons;
Q_m = measured pile capacity, in tons;
r = correlation coefficient (dimensionless number ranging from 1 to -1);
S_X = estimated standard deviation of X;
S_Y = estimated standard deviation of Y;
s = average penetration of the pile per hammer blow for the final ten blows, also known as the set, in inches per blow;
W_h = weight of the falling part of the hammer, in tons;
W_p = weight of the pile, in tons;
X = a variable;
$\overline{X}$ = mean value of X;
X_i = a particular value of X;
Y = a variable;
$\overline{Y}$ = mean value of Y; and
Y_i = a particular value of Y.

DYNAMIC SHEARING PROPERTIES OF COMPACTED CLAY

by

R. E. Olson[1] and J. F. Parola[2]

Abstract

A series of 116 Q-type triaxial compression tests was performed using specimens of Goose Lake clay compacted at water contents ranging from 9 percent dry of optimum to 3 percent on the wet side. The specimens were subjected to confining pressures ranging from 10 psi to 1000 psi and were loaded to failure in times ranging from 2 ms to an hour.

The average increase in compressive strength per decade reduction in time to failure, for specimens compacted at water contents near optimum, increased from about 2 percent for the range 100 min-10 min to 18 percent for the range 60 ms-6 ms. Specimens compacted at lower water contents underwent smaller strength increases in dynamic tests.

The secant modulus, defined at 1 percent axial strain, increased at a rate of about 15 percent per decade reduction in time to failure.

The compressive strength and stress-strain properties of the Goose Lake clay are shown to vary between wide limits depending on the compaction water content, the confining pressure, and the time to failure.

Introduction

Although the literature contains a number of papers dealing with the dynamic shearing properties of clays (Casagrande and Shannon, 1949; Richardson and Whitman, 1963; Olson and Kane, 1965; Schimming, Haas and Saxe, 1966; and others) there have been no reported studies of the dynamic shearing properties of compacted clays in which the compaction water content, the confining pressure, and the loading rate were all varied through a sufficiently wide range so that their relative influences could be determined.

In a previous study (Olson and Kane, 1965), the compressive properties of compacted specimens of Goose Lake clay were investigated using the triaxial shear apparatus, with confining pressures up to 1000 psi and times to failure as short as 3 milliseconds (3 ms), but all specimens were compacted at the same water content and density. Problems were encountered with the response time of the recording system for the 3-ms tests.

In this investigation, the compressive properties of compacted specimens of Goose Lake clay were investigated for a range of compaction water content from 9 percent on the dry side of optimum to 3 percent on the wet side, for a range of times to failure ranging from 3 ms to about an hour, and for a range in confining pressure from 10 psi to 1000 psi. The compressional properties of interest were the shearing strength, axial strain at failure, and the secant modulus defined at 1 percent strain.

General Soil Properties

The soil used in this investigation was the Goose Lake clay, supplied by the Illinois Clay Products Company, Joliet, Illinois. The clay has a liquid limit of 31 percent, a plastic limit of 17 percent, a shrinkage limit of 15 percent, and a specific gravity of 2.72. All of the soil passes the No. 30 sieve and more than 95 percent passes the No. 40. The soil is 9 percent sand (coarser than 0.06 mm), 57 percent silt, and 34 percent clay (finer than 2 microns). The activity (Skempton, 1948; Seed, Woodward, and Lundgren, 1962), defined as the ratio of the plasticity index to the quantity, the percentage

[1]Professor of Civil Engineering, University of Illinois, Urbana, Illinois.

[2]Graduate Research Assistant, Department of Civil Engineering, University of Illinois, Urbana, Illinois.

DYNAMICS OF EARTH MATERIALS

finer than 2 microns less 10 percent, is 0.58. Standard (ASTM D698-64T) and modified (ASTM D1557-64T) moisture-density tests yielded optimum points of 112.7 pcf at 14.5 percent, and 124.5 pcf at 11.3 percent, respectively. An X-ray diffraction study of the finer-than-2-micron fraction (Langfelder, 1964) indicated the presence of 20-25 percent quartz, 10-15 percent illite, and about 65 percent kaolinite.

A series of static triaxial compression tests performed using saturated specimens remolded at a water content of 28 percent and consolidated to pressures ranging from 5 psi to 120 psi, yielded $\bar{c}$ = 1.5 psi, and $\bar{\phi}$ = 25-1/2 deg., with a tendency for $\bar{\phi}$ to decrease slightly with increasing pressure.

Specimen Preparation

The specimens were compacted into a 1.5-inch diameter by 3-inch high split compaction mold by kneading compaction. A wedge-shaped foot with a 30-degree apex angle was used. The soil was compacted in ten layers using eight strokes per layer and a nominal foot pressure of 100 psi. The specimens were stored at constant water content for at least four days prior to shear.

Apparatus

The tests were performed by using essentially the same apparatus described by Olson and Kane (1965). The steel triaxial cell is designed for confining pressures up to about 1500 psi. The loading piston enters the cell through two ball bushings and an O-ring seal. The axial force applied to the soil is measured using a load cell equipped with electric-resistance-type strain gages and mounted directly beneath the specimen. Axial deformation was measured using a linear variable differential transformer (LVDT). Gas leakage through the membrane was prevented by using mercury as a cell fluid for tests with times to failure greater than 3 seconds. For faster tests the mercury was drained from the cell just prior to firing the loading machine.

The testing machine was described previously (Olson and Kane, 1965). It provides loading rates ranging from about 200 in./sec down to about 5 x 10^{-3} in./sec, corresponding to time to failure of a 3-inch-tall specimen ranging from 3 ms up to about 2 minutes. For the 3-ms tests the press is operated pneumatically. It is operated hydraulically for tests lasting over about 30 ms and up to about 2 minutes. Tests performed with times to failure longer than about 2 minutes were performed using standard mechanically driven presses.

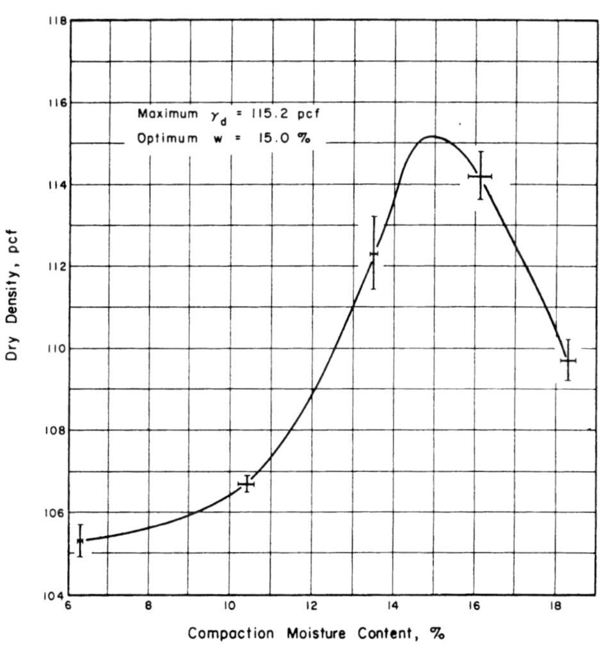

Figure 1. Kneading Compaction Curve of Goose Lake Clay

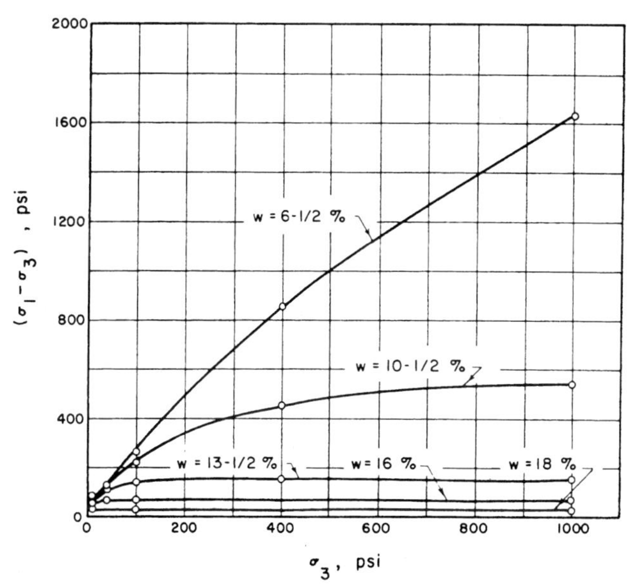

Figure 2. Static Q-Type Failure Envelopes

MATERIAL PROPERTIES

Data were hand recorded for tests lasting longer than about 2 minutes. For the faster tests, the data were recorded on an oscillograph and an FM tape system and the press and recording system were programed for automatic operation.

Moisture-Density Data

A complete tabulation of the moisture-density data has been presented previously (Olson and Parola, 1967). The average water contents and dry densities, and the estimated standard deviations are shown in Table 1. The scatter of both moisture contents and dry densities is too small to exert an important influence on any series of tests. The moisture-density curve is presented in Figure 1. The data apparently cover as wide a range in water content as is likely to be encountered in the field for such soils.

TABLE 1. MOISTURE-DENSITY DATA

Nominal water content (%) (1)	Actual water content Mean (%) (2)	Actual water content S (%) (3)	Dry density Mean (pcf) (4)	Dry density S (pcf) (5)	Number of observations (6)	Degree of saturation (7)
6-1/2	6.3	0.1	105.3	0.4	23	28
10-1/2	10.4	0.2	106.7	0.2	25	48
13-1/2	13.5	0.1	112.3	0.9	25	72
16	16.1	0.3	114.2	0.6	20	90
18	18.3	0.2	109.7	0.5	23	91

Static Shear Diagrams

The strengths of specimens in which the time-to-failure exceeded about 2 minutes, hereafter referred to as static tests, are plotted versus confining pressure in Figure 2. The points shown represent one to three tests. When more than one static test was performed at a given confining pressure, the strengths were so nearly the same that it was not possible to distinguish the separate points on the diagram; thus, only one point is shown for each water content and pressure.

The $(\sigma_1 - \sigma_3)_f$ vs σ_3 plot was chosen over the various other types of diagrams because it shows the shear data more clearly. The standard shearing parameters, ϕ and c, can be obtained from the slopes, ψ, and intercepts, d, of the failure envelopes shown in Figure 2 using the equations:

$$\phi = \sin^{-1} \frac{\tan \psi}{2 + \tan \psi}$$

$$c = d \frac{1 - \sin \phi}{2 \cos \phi} \tag{1}$$

The failure envelopes are of the shapes usually obtained. The curvature results from increasing values of B_a and B_w (Skempton, 1954; Bishop, 1961) with gradually increased degrees of saturation, whereas the variable cohesion intercepts depend largely on relative values of the residual negative pore water pressure (values for this soil given by Olson and Langfelder, 1965) and the X factors (Bishop, 1961).

It is interesting to note that the value of ϕ_Q for the specimens of clay prepared at a water content of 6-1/2 percent and subjected to confining pressures less than 200 psi is 32-1/2 deg, which is 7 deg higher than $\bar{\phi}$ for remolded specimens. The difference probably results from the differences in the soil structure (Lambe, 1958; Seed and Chan, 1959).

DYNAMICS OF EARTH MATERIALS

Influence of Deformation Rate on Compressive Strength

The compressive strengths have been plotted versus the time to failure in Figure 3. The scatter is significantly less than usually obtained from such tests, making it clear that the strength increase associated with decreasing times to failure is real. There is a marked increase in strength, in every series of tests, as the time to failure decreased.

To compare the various strength-time diagrams, the strengths were all normalized by dividing by the strength corresponding to a time to failure of 1 minute. The normalized strengths were then plotted versus time to failure. These diagrams indicated that the specimens compacted at the lowest water content underwent smaller strength increases than the other specimens, but that the specimens compacted at water contents ranging from 5 percent dry side of optimum to 3 percent on the wet side all exhibited about the same strength increases. The normalized strengths of the specimens compacted at water contents from 5 percent dry of optimum to 3 percent on the wet side have been plotted versus time to failure in Figure 4. Only specimens subjected to confining pressures of either 100 psi or 1000 psi were used in this figure. An average curve has been drawn through the points. The average normalized strengths for various times to failure are tabulated in Table 2. The strength increases per decade reduction in the time to failure were calculated as $(\tau_t - \tau_{10t})/\tau_{10t}$ and are also presented in Table 2. For the range in time to failure such as used in some Q-type laboratory triaxial tests, 10 min to 100 min, there is only about a 2-percent increase in strength for a decade decrease in the time to failure. However, the dynamic effect increases for faster tests until there is an 18-percent increase in strength for the decade from 60 ms to 6 ms. On the average, the specimens with a time to failure of 6 ms were 44 percent stronger than specimens loaded in 1 minute. For the specimens compacted at a water content 9 percent less than optimum, the strength increases per decade reduction in the time to failure were about half the values shown in Table 2 for specimens compacted nearer the optimum water content.

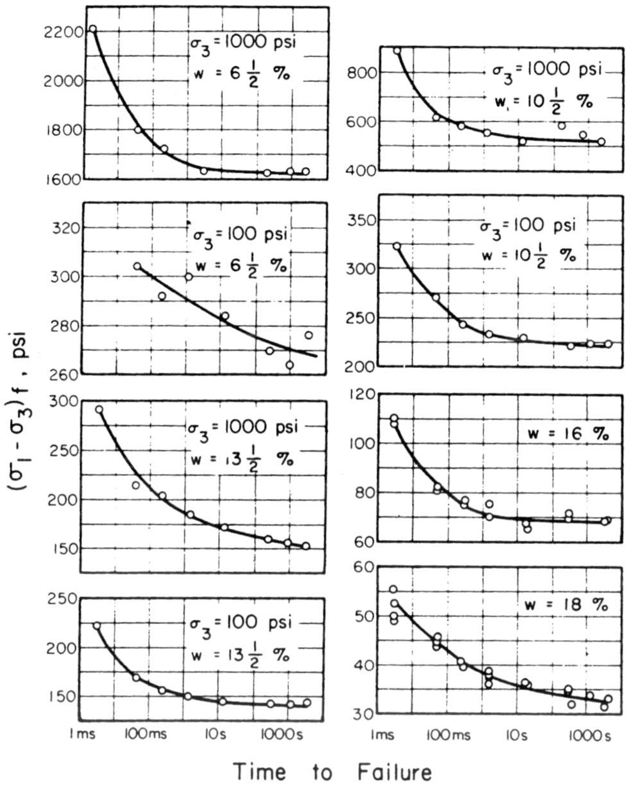

Figure 3. Influence of the Time to Failure on the Undrained Compressive Strength

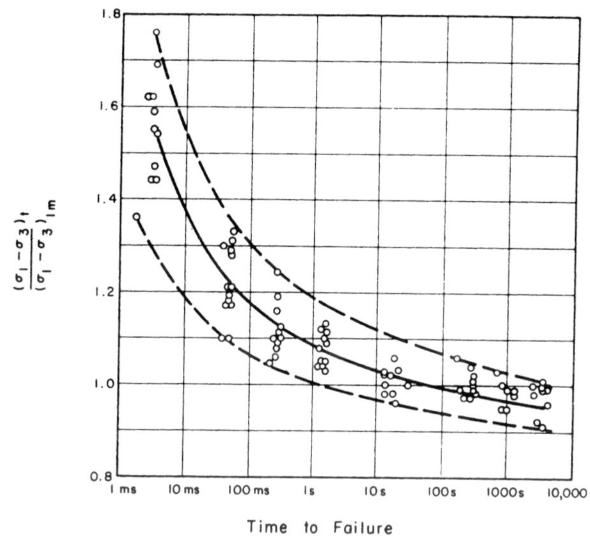

Figure 4. Normalized Strength Variation with the Time to Failure

Strains at Failure

The axial strains at failure are plotted versus the confining pressure in Figure 5. These strains were independent of the rate of loading.

TABLE 2. AVERAGED INFLUENCES OF DEFORMATION RATE ON THE STRENGTHS OF SPECIMENS COMPACTED AT WATER CONTENTS RANGING FROM 5 PERCENT DRY OF OPTIMUM TO 3 PERCENT ON THE WET SIDE

Time to failure (1)	Normalized strength (2)	Increase in strength per decade reduction in time to failure (%) (3)
100 min	0.95	
		2
10 min	0.97	
		3
1 min	1.00	
		4
6 sec	1.04	
		6
600 ms	1.10	
		11
60 ms	1.22	
		18
6 ms	1.44	

For the specimens compacted at a water content of 6-1/2 percent, the kneading action of the compaction foot caused a densification equivalent to that produced by a hydrostatic stress several times the nominal foot pressure of 100 psi. Because of the extremely low, i.e., nearly zero, value of X (Bishop, 1961), the removal of the specimen from mold resulted in a reduction of the effective stress to nearly zero. Thus, under a confining pressure of only 10 psi, the soil reacted to shear in the same manner as a highly overconsolidated clay subject to drained shear, i.e., the soil was a strain weakening material and failed at a low strain on a single shear plane. Increasing the total confining pressure increased the effective stress by nearly the same amount. Thus the soil became more nearly normally consolidated and became, therefore, a strain hardening material with an increasing strain at failure. At a confining pressure somewhere in the range of 200 to 400 psi, the specimens became essentially normally consolidated. Thereafter, the strains at failure decreased slightly because of slight dispersion of the soil structure induced by the high confining pressure.

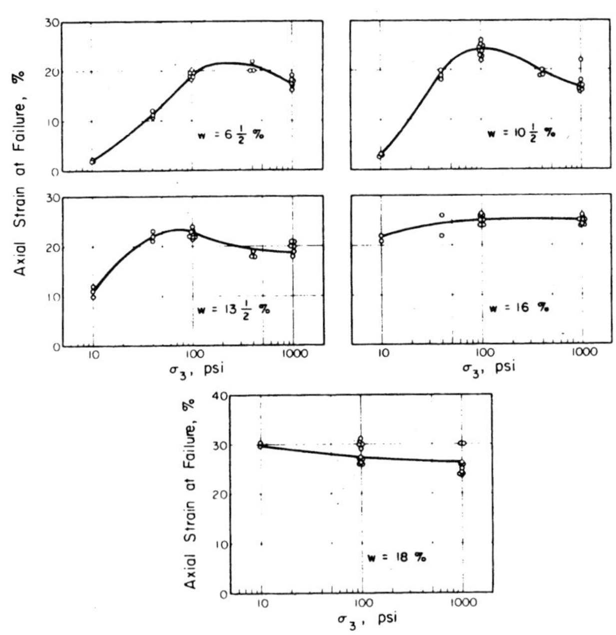

Figure 5. Axial Strains at Failure for Q-Type Compression Tests

As the compaction water content was increased, the initial effective stress also increased, and thus the specimens had less tendency to dilate when sheared at low confining pressures. The strain at failure, at low confining pressures, therefore increased as the compaction water content increased. At high confining pressures the soil reacted as for specimens compacted at low water contents, and for the same reasons.

For the specimens compacted on the wet side of optimum it was impossible to develop the full 100-psi foot pressure. The compaction foot penetrated the soil until a collar above the foot contacted the surface of the soil and stopped penetration. In the field, an analogous situation occurs when the feet of a sheepsfoot roller penetrate the soil until the drum presses directly on the soil surface. For specimens compacted at such high water contents, the effective stress at the end of compaction is quite small because the pore water pressure is only slightly negative (Olson and Langfelder, 1965). The soil has a high degree of saturation and the soil structure is dispersed. Thus, the applied total stresses produce only small changes in effective stress, and the strains at failure are large and relatively independent of the total confining pressure.

The strain at failure and the failure mode are, thus, strongly dependent on the compaction water content and the confining pressure, but independent of loading rate.

Secant Moduli

A convenient parameter to use in studying the stress-strain properties of soils at low strains is the secant modulus. In this study, the secant modulus was defined as the ratio of the stress difference to the axial strain, defined at an axial strain of 1 percent.

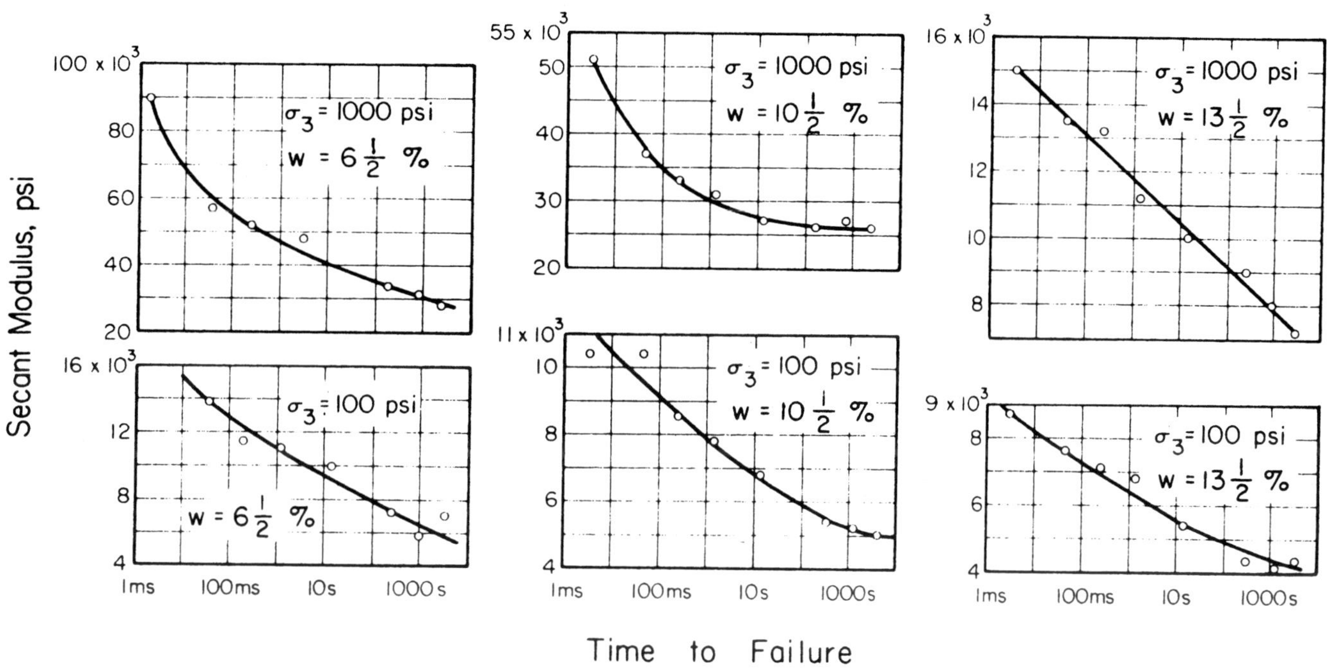

Figure 6. Effect of Loading Rate on the Secant Modulus Defined at an Axial Strain of 1 Percent

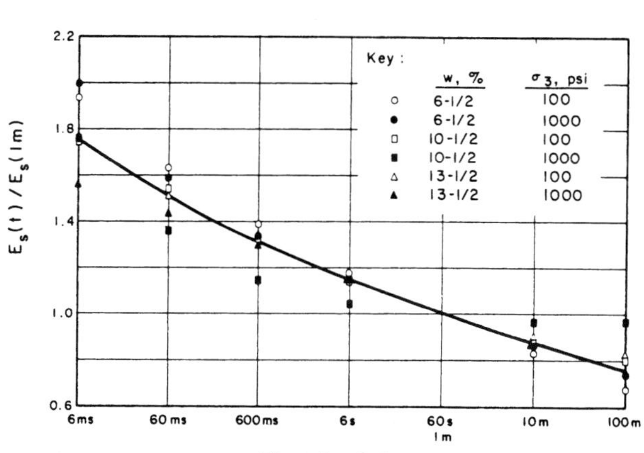

Figure 7. Influence of Time to Failure on the Normalized Secant Moduli at an Axial Strain of 1 Percent

Secant moduli for confining pressures of 100 psi and 1000 psi are plotted versus the time to failure in Figure 6 for specimens compacted at water contents of 6-1/2, 10-1/2, and 13-1/2 percent. The stresses at 1 percent axial strain were so small for specimens compacted on the wet side of optimum (water contents of 16 and 18 percent) that normal electronic recording errors, plus small variations in the seating loads, resulted in excessive scatter in the data and a strain-rate effect was not well defined. For specimens compacted at moisture contents of 16 and 18 percent, the range of secant moduli was 2300-3100, and 900-1700 psi, respectively.

The average curves shown in Figure 6 were normalized by dividing by the moduli at a time to failure of 1 minute. The normalized curves are shown in Figure 7. For practical purposes, a single line can be drawn through these points and a common set of normalized moduli can be determined. The normalized moduli and the percent increase in the normalized modulus per decade reduction in the time to failure, calculated as $(E_t - E_{10t})/E_{10t}$, are tabulated in Table 3. For this soil, compacted on the dry side of optimum, the secant modulus for confining pressures ranging from 100 psi to 1000 psi increases about 15 percent for each decade reduction in the time to failure. The same rate of increase of modulus fits the data for specimens compacted on the wet side of optimum.

The secant modulus, therefore, undergoes considerably more change as a function of deformation rate than does the compressive strength.

TABLE 3. AVERAGED INFLUENCES OF DEFORMATION RATE ON THE SECANT MODULI OF SPECIMENS OF GOOSE LAKE CLAY COMPACTED AT WATER CONTENTS LESS THAN OPTIMUM

Time to failure (1)	Normalized secant modulus (2)	Increase in secant modulus per decade decrease in the time to failure (%) (3)
100 m	0.75	
10 m	0.88	17
1 m	1.00	14
6 s	1.16	16
600 ms	1.32	14
60 ms	1.51	14
6 ms	1.76	17

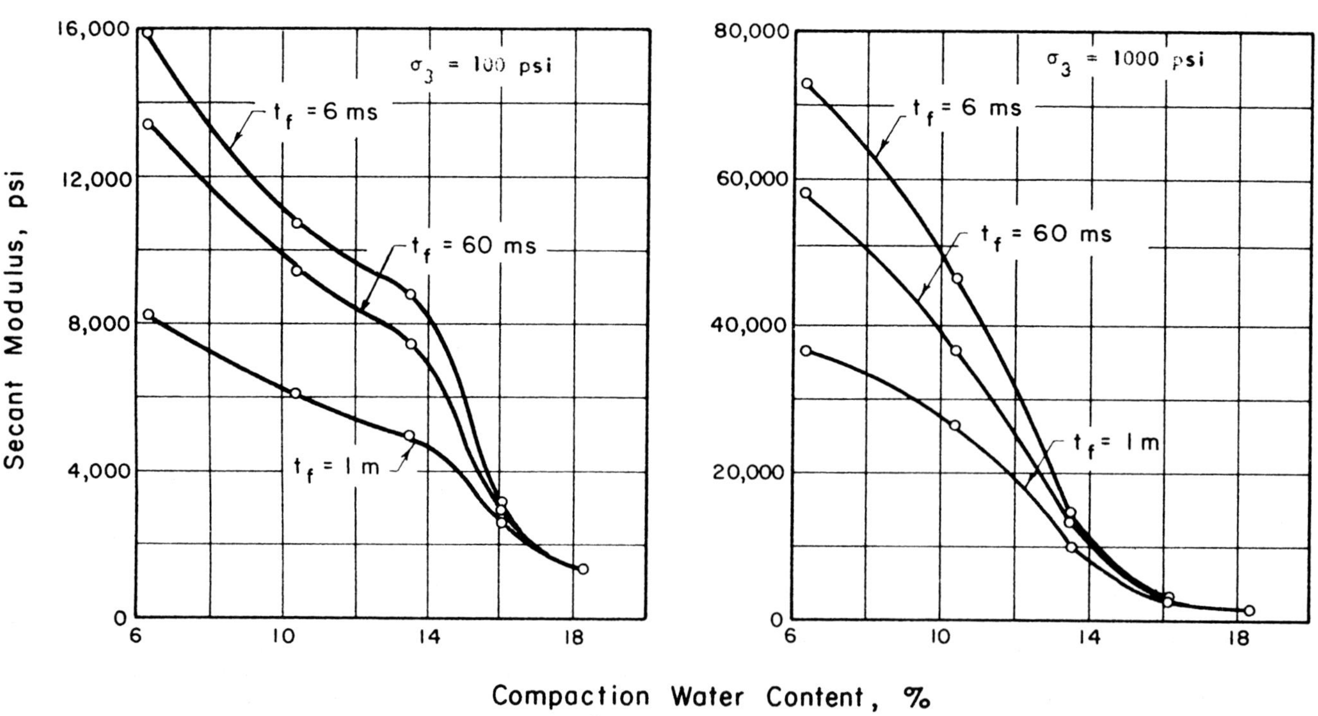

Figure 8. Variation of the Secant Modulus at 1 Percent Axial Strain with the Compaction Water Content

DYNAMICS OF EARTH MATERIALS

The influence of compaction water content on the secant modulus is shown in Figure 8 for confining pressures of 100 and 1000 psi and times to failure of 1 min, 60 ms, and 6 ms. For the confining pressure of 100 psi, which is considerably less than the equivalent hydrostatic pressure applied during compaction, the secant modulus undergoes a rapid decrease in the vicinity of the optimum water content. At a pressure of 1000 psi, any prestress induced by the compaction operation has been overcome and the secant modulus decreases more uniformly as the water content is increased. Apparently a very wide range in secant moduli can be obtained just by varying the compaction water content.

Comparison with Previous Study

A somewhat similar series of triaxial compression tests was performed previously (Olson and Kane, 1965). In that study, specimens of Goose Lake clay were compacted at a water content of 12.2 percent to an average dry density of 119.3 pcf, using a 200-psi nominal foot pressure. Pertinent aspects of the previous series of tests are summarized in Table 4.

TABLE 4. SUMMARY OF RESULTS OF EARLIER TESTS ON DENSE SPECIMENS OF GOOSE LAKE CLAY

Confining pressure (psi) (1)	Time to failure (2)	$(\sigma_1 - \sigma_3)_f$ (psi) (3)	Normalized $(\sigma_1 - \sigma_3)_f$ (psi) (4)	Increase in $(\sigma_1 - \sigma_3)_f$ per decade (%) (5)	E_s (psi) (6)	Normalized E_s (psi) (7)	Increase in E_s per decade (%) (8)
114	1 m	228	1.00		10,000	1.00	
	6 s	244	1.07	7	10,000	1.00	0
	600 ms	260	1.14	7	10,000	1.00	0
	60 ms	280	1.23	8	10,000	1.00	0
	6 ms	308	1.35	10	10,000	1.00	0
1110	1 m	427	1.00		19,400	1.00	
	6 s	438	1.03	3	22,100	1.14	14
	600 ms	452	1.06	3	24,600	1.27	11
	60 ms	484	1.13	7	27,100	1.40	10
	6 ms	560	1.31	16	29,500	1.52	9

In the previous study, the compressive strengths of specimens subjected to 1010 psi of confining pressure varied with the time to failure in essentially the same manner as did the specimens used in this study. However, specimens subjected to only 114 psi of confining pressure in the previous study underwent more uniform strength increases per decade reduction in the time to failure. The rapid acceleration of the dynamic effect for the most rapid tests may be characteristic of normally consolidated or lightly overconsolidated specimens. In the previous study, the secant moduli increased at a rate of 9 to 14 percent per decade decrease in time to failure, for specimens subjected to 1010 psi of confining pressure. In this study, the rate was 14 to 17 percent/decade. Thus, the results are comparable. However, in the previous study, the secant modulus was independent of deformation rate when the confining pressure was 114 psi. It appears possible that the application of a confining pressure much smaller than the pressure applied during compaction will produce specimens that are elastic for strains up to at least 1 percent, and which therefore yield a secant modulus that is not dependent on the rate of deformation.

MATERIAL PROPERTIES

Summary and Conclusions

Based on this investigation, it appears that triaxial compression tests can be performed on a routine basis using confining pressures as high as 1000 psi and times to failure as short as 3 milliseconds, simultaneously. Further, the shearing strengths can be determined with such a precision that dynamic effects are clearly defined. The available apparatus even allows reproducible determination of the secant modulus, defined at 1 percent axial strain, except for soft specimens. A redesigned load cell would allow these measurements as well.

One hundred and sixteen Q-type triaxial compression tests were performed using compacted specimens of Goose Lake clay. The specimens were compacted at water contents ranging from 9 percent below optimum to 3 percent above. Times to failure ranged from an hour down to 1.7 milliseconds. Based on these tests, and a set of tests performed previously, the following conclusions are drawn regarding the properties of compacted Goose Lake clay.

The Mohr-Coulomb failure envelopes were concave towards the normal-stress axis and became horizontal when the degree of saturation reached 100 percent. The envelopes had the shapes and relative positions usually anticipated for such a cohesive soil.

For specimens compacted in the range of water content from 5 percent dry side of optimum to 3 percent on the wet side, the average increase in compressive strength per decade reduction in the time to failure increased from only 2 percent for the range from 100 to 10 minutes to 18 percent for the range from 60 to 6 milliseconds. This viscous effect was slightly larger for tests performed under a confining pressure of 1000 psi than for specimens subjected to 100 psi. Specimens compacted at a water content 9 percent dry side of optimum underwent a viscous strength increase at about half the above rates.

The axial strains at failure, which were quite well defined, were independent of the rate of deformation. Specimens compacted at a water content 9 percent dry side of optimum were strain weakening when subjected to confining pressures as low as 10 psi, and failed at axial strains as low as 2 percent. As the confining pressure increased to 100 psi and higher, the soil became a strain-hardening material and the specimens failed at axial strains near 20 percent. As the compaction water content was increased, the axial strains at failure became less dependent on the confining pressure. Specimens compacted at water contents above optimum failed at axial strains between 20 and 30 percent.

The secant moduli, defined at 1 percent axial strain, were well defined for the specimens compacted dry side of optimum, but were so low for specimens compacted on the wet side of optimum that excessive scatter occurred. For the specimens compacted on the dry side of optimum, the secant modulus increased at a rate of about 15 percent per decade reduction in the time to failure. For the specimens compacted at the lowest water contents, the secant modulus increased linearly with the confining pressure. For specimens compacted on the wet side of optimum, the secant modulus was, of course, essentially independent of the confining pressure. For any given time to failure and confining pressure, the secant modulus decreased as the compaction moisture content was increased. For example, for specimens subjected to a confining pressure of 1000 psi and loaded to failure in 6 ms, the secant modulus decreased from 74,000 psi to 1500 psi as the compaction water content was increased from 6-1/2 percent to 18 percent.

Based on these tests, it appears that the compressive strength and stress-strain characteristics of compacted clays, like the Goose Lake clay, vary between wide limits depending on the compaction water content, the loading rate, and the confining pressure.

Acknowledgment

The main part of this investigation was performed as Phase I of the study "An Experimental Investigation of the Dynamic Response of Model Silo-Type Structures in Cohesive Soils," project AF 29-(601)-6811, in the soil mechanics laboratories of the University of Illinois, Urbana, Illinois. The project was sponsored by the Air Force Weapons Laboratory, Kirtland Air Force Base, New Mexico. Mr. R. H. Atkinson was the Project Officer. The main project was under the direction of Professors G. K. Sinnamon and S. J. Paul, University of Illinois, Urbana, Illinois.

A part of this investigation was also supported by the National Science Foundation as part of grant GK1284X, Professor R. E. Olson, principal investigator.

DYNAMICS OF EARTH MATERIALS

References

Bishop, A. W., 1961. "The Measurement of Pore Pressure in the Triaxial Test," Proceedings, Conf. on Pore Pressure and Suction in Soils, London, Butterworths, London.

Bishop, A. W., 1961. "Discussion," Proceedings, Conf. on Pore Pressure and Suction in Soils, London, Butterworths, London.

Casagrande, A., and Shannon, W. L., 1949. "Strength of Soils Under Dynamic Loads," Transactions, ASCE, Vol. 114.

Lambe, T. W., 1958. "The Structure of Compacted Clay," Proceedings, ASCE, Vol. 84, SM2, Paper 1654.

Langfelder, L. J., 1964. "An Investigation of Initial Negative Pore Water Pressure in Statically Compacted Cohesive Soils," Ph.D. thesis, University of Illinois, Urbana, Illinois.

Olson, R. E., and Kane, H., 1965. "Dynamic Shearing Properties of Compacted Clay at High Pressures," Proceedings, Sixth Intern. Conf. on Soil Mech. and Found. Engr., Vol. 1.

Olson, R. E., and Langfelder, L. J., 1965. "Pore Water Pressures in Unsaturated Soils," Proceedings, ASCE, Vol. 91, SM4.

Olson, R. E., and Parola, J. F., 1967. "An Experimental Investigation of the Dynamic Response of Model Silo-Type Structures in Cohesive Soils, Phase I Report, Soil Properties," report submitted to the Air Force Weapons Laboratory, Kirtland Air Force Base, New Mexico.

Richardson, A. M., and Whitman, R. V., 1963. "Effect of Strain-Rate upon Undrained Shear Resistance of a Saturated Remoulded Fat Clay," Geotechnique, Vol. 13.

Schimming, B. B., Haas, H. J., and Saxe, H. C., 1966. "Study of Dynamic and Static Failure Envelopes," Proceedings, ASCE, Vol. 92, SM2.

Seed, H. B., and Chan, C. K., 1959. "Structure and Strength Characteristics of Compacted Clays," Proceedings, ASCE, Vol. 85, SM5, Paper 2216.

Seed, H. B., Woodward, R. J., and Lundgren, R., 1962. "Prediction of the Swelling Potential for Compacted Clays," Proceedings, ASCE, Vol. 88, SM3.

Skempton, A. W., 1948. "The Colloidal 'Activity' of Clays," Proceedings, Third Intern. Conf. on Soil Mech. and Found. Engr., Vol. 1.

Skempton, A. W., 1954. "The Pore-Pressure Coefficients A and B," Geotechnique, Vol. 4.

Journal of the
SOIL MECHANICS AND FOUNDATIONS DIVISION
Proceedings of the American Society of Civil Engineers

MECHANISMS CONTROLLING COMPRESSIBILITY OF CLAYS

By Roy E. Olson,[1] M. ASCE and Gholamreza Mesri,[2] A. M. ASCE

INTRODUCTION

Karl Terzaghi seems to have been one of the first engineers to make a serious attempt to understand the mechanisms controlling the compressibility of clays. Based largely on the studies of Atterberg (1) and Gilboy (8), Terzaghi (22, p. 343) concluded that the large void ratios and high compressibilities of clays were caused by the presence of "scale shaped" mineral particles. He described swelling as "nothing but the purely elastic extension of the deformable fabric. . ."

Terzaghi had previously postulated the existence of semisolid layers of adsorbed water on clay surfaces but he used this hypothesis only to explain the low permeability of clays (21) and secondary consolidation (23).

For purposes of analysis herein, models of the type used by Terzaghi (22) will be termed mechanical in that the deformations involve a mechanical bending, compressing, and rupture of the particles coupled with shearing displacements at the contacts. Swelling is assumed to result from elastic rebound of bent particles. The interaction between adjacent particles is physical, i.e., separations at the contacts do not exceed about 10 Angstrom units. The particles need not literally touch one another but they can be separated by no more than a few layers of absorbed material. Surface chemical effects exist, even in the mechanical model, because adsorption of the pore fluid alters the coefficient of surface friction (10).

Simultaneously with the early studies in soil mechanics, investigators in chemistry and related fields became interested in the properties of suspensions. Long range electrical forces were needed to explain the various physical and chemical properties of these suspensions. These long range forces were assumed to be osmotic (double layer) repulsive forces and van der Waals at-

Note.—Discussion open until April 1, 1971. To extend the closing date one month, a written request must be filed with the Executive Director, ASCE. This paper is part of the copyrighted Journal of the Soil Mechanics and Foundations Division, Proceedings of the American Society of Civil Engineers, Vol. 96, No. SM6, November, 1970. Manuscript was submitted for review for possible publication on February 20, 1970.

[1] Prof. of Civ. Engrg., University of Texas at Austin, Austin, Tex.

[2] Asst. Prof. of Civ. Engrg., University of Illinois at Urbana, Urbana, Ill.

tractive forces (11,17,24,25). Subsequently, investigators in agronomy, and soil mechanics applied these theories to explain the swelling characteristics of clays (2,3,4,5,6,12,13,18,19,20,27,28). For purposes of analysis, the term physico-chemical will be used to describe the model in which the interaction of particles is long range and involves electrical forces of the double layer and van der Waals type.

It should be possible to perform experiments to determine which of the two models better describes the consolidation characteristics of clays. For example, the physico-chemical model predicts that large variations in pore water electrolyte concentration will cause great changes in the position of the swelling curves whereas the mechanical model predicts either no change, or a very small change, depending on whether or not the electrolyte concentration is assumed to influence surface friction. Laboratory consolidation tests on

TABLE 1.—PROPERTIES OF CLAYS

Mineral (1)	Supplier, source, trade name (2)	Liquid limit, as a percentage (3)	Plastic limit, as a percentage (4)	Specific gravity (5)	Surface area, in square meters per gram (6)	Cation exchange capacity, in milli-equivalents per 100 g (7)	Fraction finer than 0.002 mm (8)	Estimated ratio of diameter to height (9)
Kaolinite	Minerals and Chemicals Philipp Corp., Klondyke clay, McIntyre, Georgia	40 to 50	27 to 31	2.65	14	2.2	47	2-5
Illite	J. L. Eades, Dept. of Geology, Univ. of Illinois, Marblehead, Wisconsin	83 to 104	31 to 32	2.80		28	100	10-50
Smectite	American Colloid Company, Wyoming, Volclay	190 to 1,160	31 to 47	2.65 to 2.80	500 to 700	100	97	150-500

clays with wide variations in electrolyte concentration should then show which model is more nearly applicable. Similarly, replacement of the pore water with a fluid in which double layers cannot develop, e.g., carbon tetrachloride, would result in substantial reductions in void ratio, at any given effective stress, if the physico-chemical model is applicable. On the other hand, such a substitution will usually increase surface friction, and, according to the mechanical model, cause the soil to have a higher void ratio. Thus, the two models predict quite the opposite effect.

The purpose of this paper is to establish, experimentally, the relative influences of the mechanical and physico-chemical mechanisms for a variety of clays. Detailed theoretical analyses together with additional experimental data, will be presented in subsequent papers. In this paper, the term compressibility is used to indicate one-dimensional volume change resulting from a change in the applied effective stress. The time rates of volume change, including secondary effects, will not be considered. All relationships between void ratio and effective stress are defined at the end of primary consolidation (7).

Experiments were performed with specimens of kaolinite, illite, and smectite in a variety of pore fluids. When water was the fluid, both the valency and concentration of the cations in the pore water were varied. The sources of the clays, and various physical and chemical properties, are shown in Table 1.

A limited number of experiments were performed using a ground muscovite to demonstrate the type of data that are obtained for a soil composed of platey particles when mechanical effects dominate. Data from the literature are used for sand where interaction will be mechanical but the particles are equidimensional.

Experimental procedures are presented in Appendix I.

PUBLISHED COMPARISONS OF MECHANICAL AND PHYSICO-CHEMICAL MODELS

Salas and Serratose (19) reported the results of one-dimensional consolidation tests on homoionized bentonite in water and natural bentonite in methyl and ethyl alcohol, and carbon tetrachloride. For the bentonite in water they found that the compression indices, swelling indices, and void ratios at any given effective stress, were inverse functions of the valency of the adsorbed cations, in qualitative conformance with double layer theory. The indices and void ratios were lower yet when the pore fluid was an alcohol, and were a minimum for carbon tetrachloride, also in conformance with the physico-chemical model.

Waidelich (26) studied the one-dimensional consolidation characteristics of specimens of kaolin, Goose Lake clay (described as halloysite but more likely a mixed layer material), and Wyoming bentonite, in water, aniline, nitrobenzene, and four alcohols. No nonpolar fluids were used. For kaolinite and Goose Lake clay, neither the compression nor the swelling indices were greatly effected by variations in the properties of the pore fluid, thus suggesting the qualitative validity of the mechanical model. For bentonite, the swelling index was as much as 120 times higher for the specimen in water than for the specimens in other fluids, and the void ratios were also substantially higher. The inference is that the physico-chemical model would be more applicable for the bentonite.

Other studies of the influence of physico-chemical variables on the engineering properties of clays were referenced previously. They will not be presented in detail because they did not specifically compare the mechanical and physico-chemical models for compressibility.

EXPERIMENTAL OBSERVATIONS

Ground Muscovite.—The one-dimensional consolidation curves for ground muscovite in water, dry air, and carbon tetrachloride are shown in Fig. 1. Tests performed by Horn (10) on a Brazilian muscovite, yielded a coefficient of surface friction of 0.41 for a dry surface, 0.37 for a surface in carbon tetrachloride, and 0.22 for a surface in water. The virgin compression curves in Fig. 1 are in the position that would be predicted using the mechanical model, i.e., the lower the surface friction the lower the void ratio. The surface frictional effect is a maximum at low pressures on the virgin compression curves because there the angles between the particles are a maximum and surface friction is important in stabilizing the loose soil structure. At high pressures the flakey particles are more nearly parallel and compression, and swelling, are controlled mostly by flexure of the plates. The pore fluid seems to have little influence on the slope of the swelling curves, perhaps because the expansion involves a minimal amount of shearing displacement at the contacts.

Kaolinite.—One-dimensional compression curves for kaolinite in water, ethyl alcohol, carbon tetrachloride, and dry air are shown in Fig. 2(*a*). The relative positions of the curves are the same as for ground muscovite, suggesting that the mechanisms controlling compression may be the same.

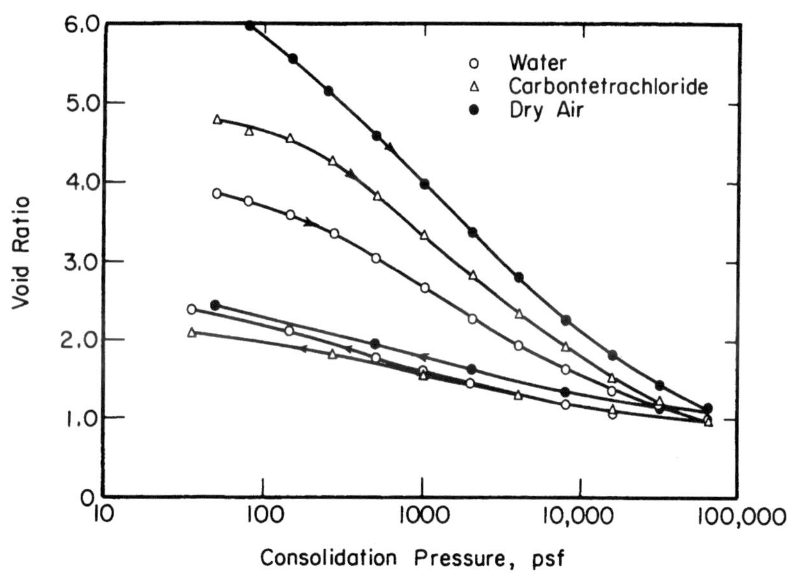

FIG. 1.—ONE-DIMENSIONAL CONSOLIDATION CURVES FOR GROUND MUSCONITE

The swelling curves for kaolinite are shown in Fig. 2(*b*). The fact that the swelling curves have approximately the same slope suggests that the main cause of swelling is again the rebound of bent particles rather than swelling of double layers.

An independent check on the applicability of the physico-chemical model was obtained by performing one-dimensional consolidation tests with wide variations in the electrolyte concentration in the pore water. Representative curves of this type are shown in Figs. 3(*a*) and 3(*b*) for calcium and sodium kaolinite, respectively. Variations in the electrolyte concentration had a negligible effect on the slopes of the swelling curves, indicating again the applicability of the mechanical model. The substantial variations in the positions of the virgin compression curves, particularly for sodium kaolinite, shows that double layers probably exerted an influence on the properties of the original suspensions (15) and affected the original geometric arrangement of the particles.

Illite.—The one-dimensional compression curves for illite are shown in Fig. 4(*a*). At the higher values of effective stress the dominating mechanical effects are again evident with the lowest void ratios for specimens in water and the highest for the specimen in dry air. At the lowest effective stresses the effect is reversed, again suggesting that the physico-chemical mechanisms exert a significant influence on the properties of the slurries from which these specimens were consolidated.

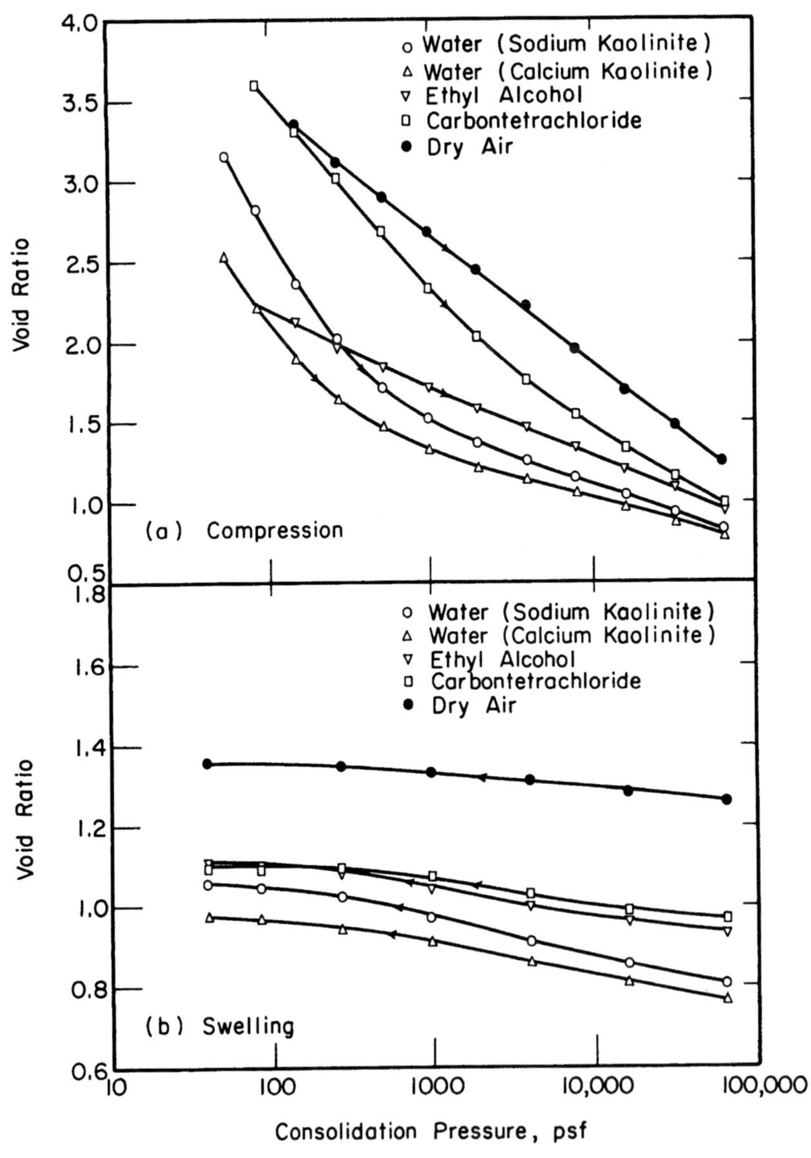

FIG. 2.—ONE-DIMENSIONAL CONSOLIDATION CURVES OF KAOLINITE IN VARIOUS PORE FLUIDS

The swelling curves for illite [Fig. 4(*b*)] in dry air and in carbon tetrachloride, where mechanical effects must dominate, are relatively flat. A slightly larger slope occurs with ethyl alcohol, perhaps as a result of adsorption (double layers are unlikely to develop in alcohol because of its low dielectric constant). The much steeper swelling curves for illite in water, particularly for the sodium illite, suggests an important physico-chemical effect (see later analysis).

Consolidation curves for calcium and sodium illite with a wide variation in pore water electrolyte concentration are shown in Figs. 5(*a*) and 5(*b*), respectively. For calcium illite the consolidation curves appear to be essentially independent of electrolyte concentration [in agreement with earlier tests on another illite reported by Olson and Mitronovas (16) suggesting either major errors in existing double layer theories or a dominating mechanical

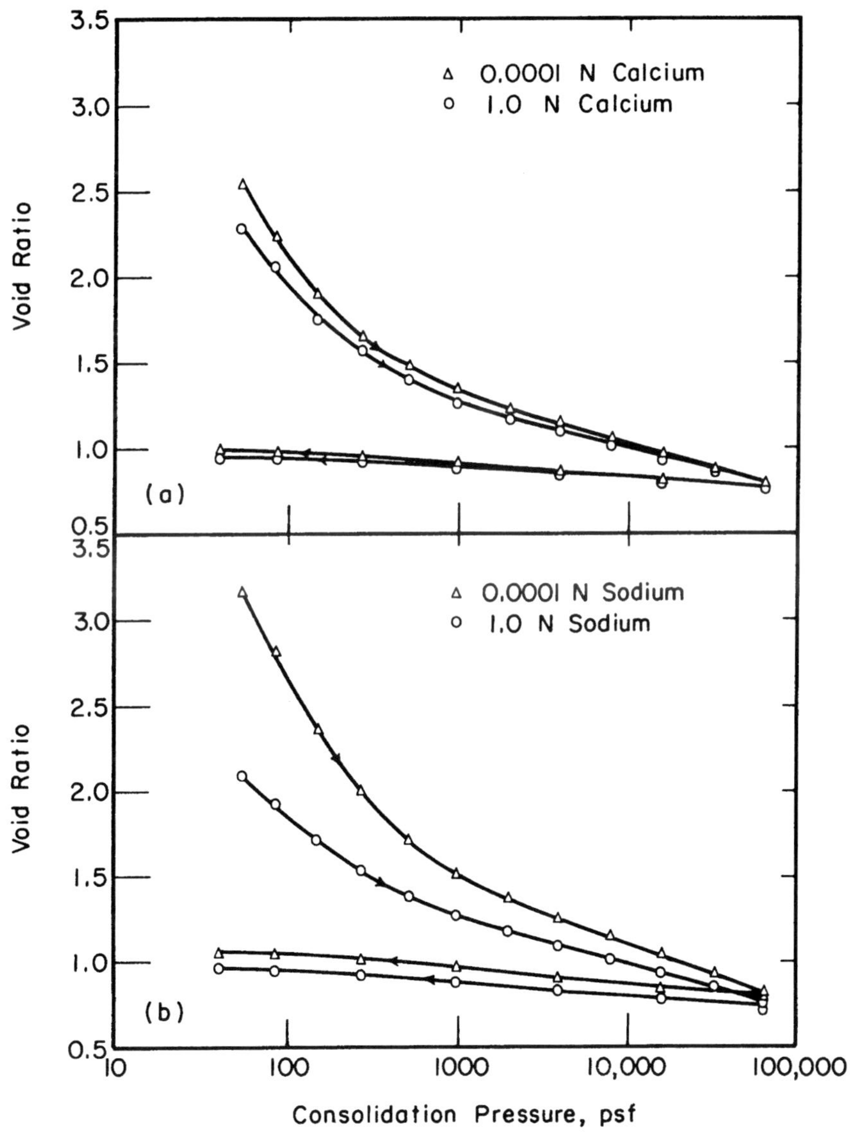

FIG. 3.—ONE-DIMENSIONAL CONSOLIDATION CURVES OF KAOLINITE IN WATER

control of swelling]. For the sodium illite the electrolyte concentration exerts a minor effect on the position of the compression curve but the position of the swelling curve, particularly at low values of effective stress, is distinctly influenced by the electrolyte concentration. A significant amount of double layer swelling is suggested for the specimen with the lowest concentration of electrolyte.

Smectite.—The one-dimensional compression and swelling curves for smectite in various pore fluids are shown in Figs. 6(*a*) and 6(*b*), respectively. The amount of swelling of the smectite in carbon tetrachloride [Fig. 6(*b*)], which is presumably purely mechanical, is negligible compared to the swelling of the sodium smectite in water. Further, the positions of both the compression and swelling curves are greatly influenced by variations in the valency of the adsorbed cations. Thus, physico-chemical effects seem to dominate.

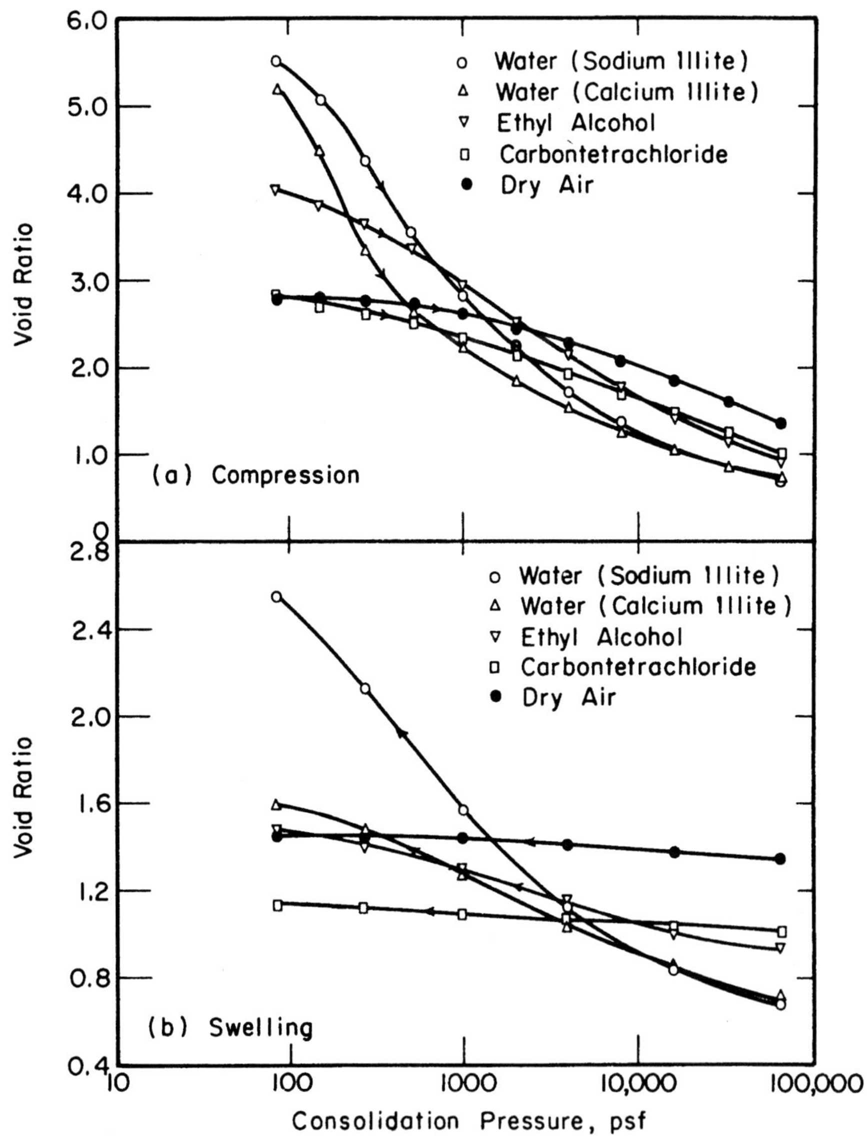

FIG. 4.—ONE-DIMENSIONAL CONSOLIDATION CURVES OF ILLITE IN VARIOUS PORE FLUIDS

Perhaps the most dramatic evidence of the long range physico-chemical effects is the fact that the sodium smectite withstood a pressure of 100 psf upon first loading with the void volume 90 times the volume of solids. Only long range (double layer) forces could give such stability.

Representative consolidation curves for calcium and sodium smectite, with a wide range of electrolyte concentration, are shown in Figs. 7(*a*) and 7(*b*), respectively. As for illite, double layer effects seem to be much more significant when the ions are monovalent.

ANALYSIS

It seems desirable to examine the variables that control the properties of the two models in more detail before attempting to draw general conclusions from the data previously presented.

Mechanical Model.

The main variables that control the mechanical model are particle shape, geometric arrangement of particles, and surface friction.

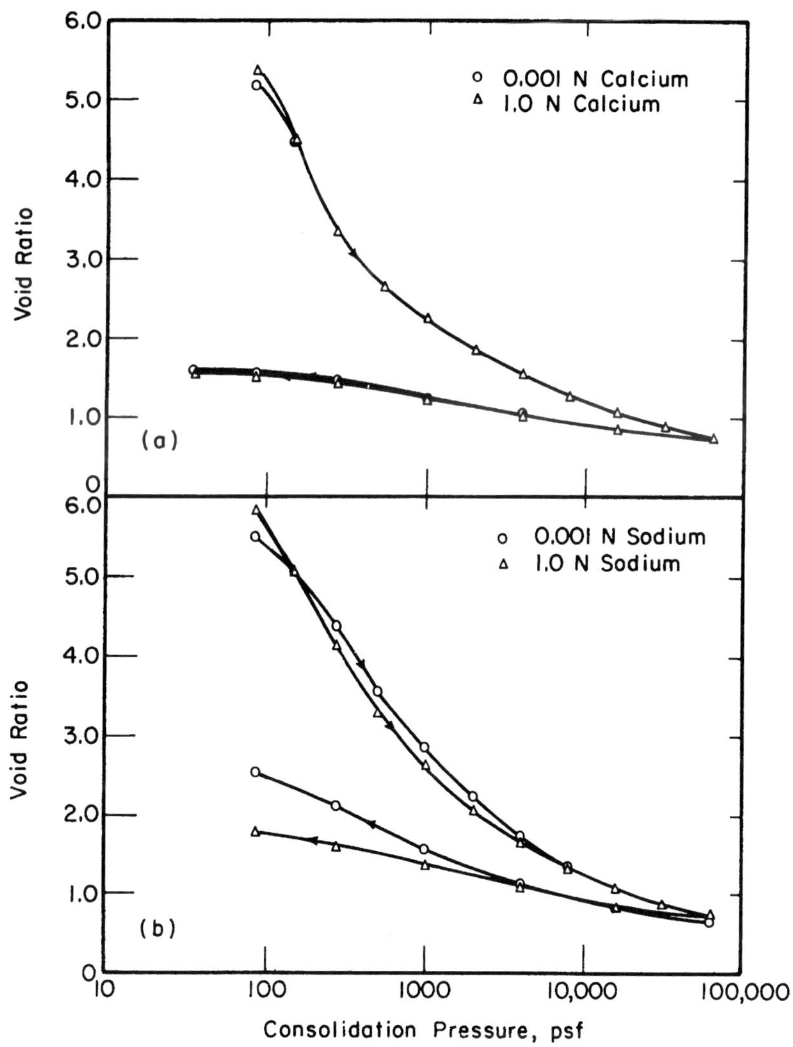

FIG. 5.—ONE-DIMENSIONAL CONSOLIDATION CURVES OF ILLITE IN WATER

Particle Shape.—For essentially equidimensional particles volume change results from distortion and rupture of the particles, and sliding at the contacts. For platey particles, as the diameter-to-height ratio increases, more energy can be stored in the soil by particle bending and the slope of the swelling curve increases (compare the swelling curves for the ground muscovite and the kaolinite). Estimates of the diameter-to-height ratio for the clays tested are indicated in Table 1. The diameter-to-height ratio of the muscovite is estimated to be twenty to fifty. Based on these ratios and the measured consolidation curves of the different clays, it appears that the particles of kaolinite, illite, and muscovite may be thought of as stiff, elastic,

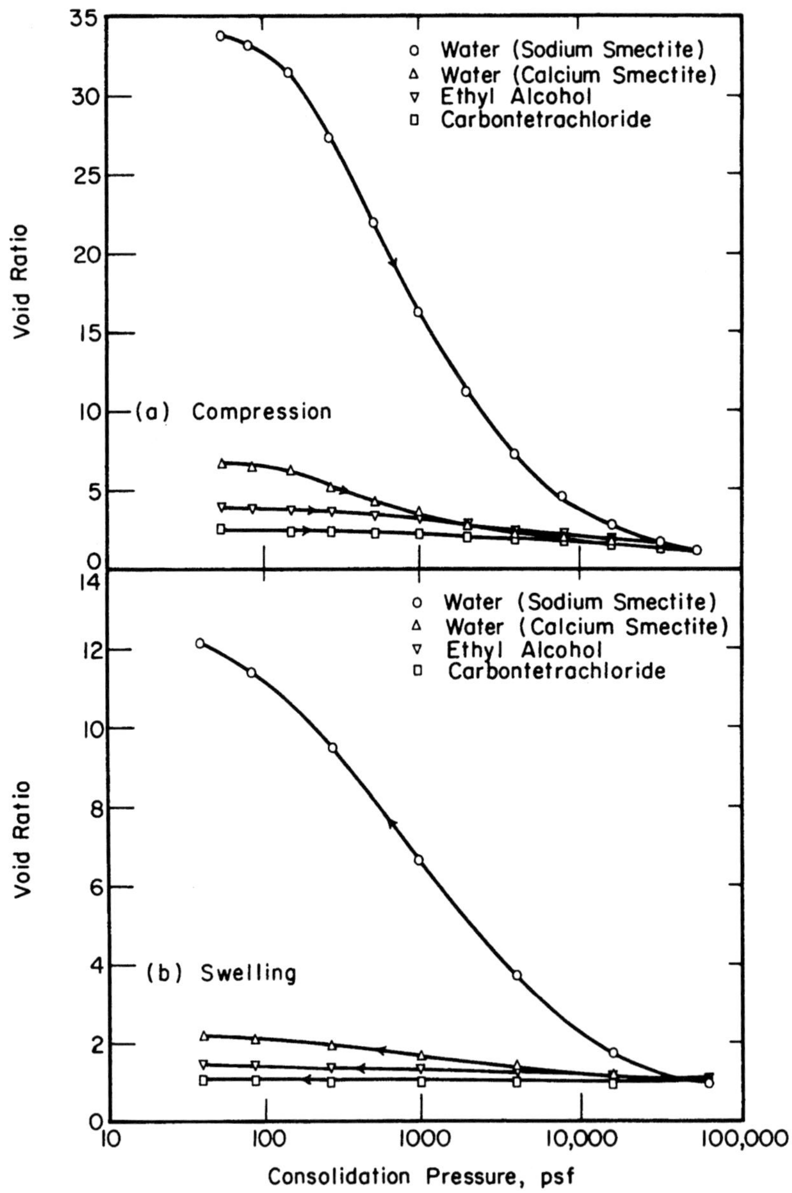

FIG. 6.—ONE-DIMENSIONAL CONSOLIDATION CURVES OF SMECTITE IN VARIOUS PORE FLUIDS

plates, whereas the particles of smectite might better be thought of as highly flexible films.

Geometric Arrangement of Particles.—If the specimens of soil are prepared in such a way that the average angle of contact between adjacent particles is large, then the mechanical properties of the particles will tend to exert a major influence on the virgin compression curve.

Surface Friction.—A high coefficient of surface friction hinders the displacement of the particles relative to each other and stabilizes the virgin compression curves at high void ratios. The virgin compression curves for muscovite, kaolinite, and illite, clearly show this effect. The influence of fluid adsorption on the coefficient of friction seems to decrease greatly if the surface becomes rough. This effect was clearly shown for quartz surfaces by Horn and Deere (10).

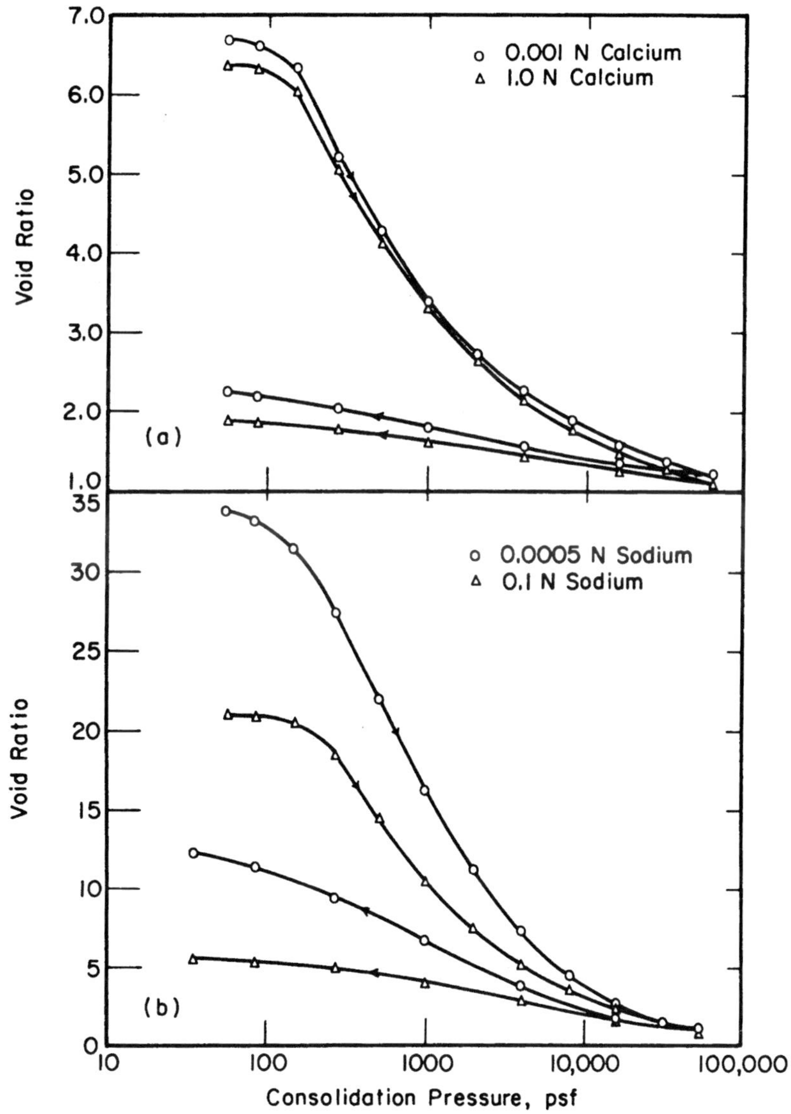

FIG. 7.—ONE-DIMENSIONAL CONSOLIDATION CURVES OF SMECTITE IN WATER

Physico-Chemical Model.

The variables that control the physico-chemical model are particle shape, geometric arrangement of particles, and the chemical variables.

Particle Shape.—Particle shape is important because wide, flat, smooth plates allow a maximum of interaction of double layers and thus promote physico-chemical effects at the same time that they tend to reduce mechanical effects.

Geometric Arrangement of Particles.—If the particles are mutually perpendicular, double layer repulsive forces are greatly diminished, thus reducing physico-chemical effects. Alternatively, a parallel arrangement of particles, as during swelling from a high pressure, maximizes the adsorptive and double layer type of swelling.

Chemical Variables.—The final variables that control the properties of the physico-chemical model are those that are dominantly chemical in nature, e.g., surface charge density, valency of the adsorbed cations, dielectric constant and dipole moment of the pore fluid, and the concentration of electrolyte in the free pore fluid.

TABLE 2.—SUMMARY OF SWELLING DATA

Mineral (1)	Pore fluid, adsorbed cations, electrolyte concentration, in gram equivalent weights per liter (2)	Void ratio at effective consolidation pressure of 100 psf (3)	Swelling index (4)
Kaolinite	water, sodium, 1	0.95	0.08
	water, sodium, 1 × 10^{-4}	1.05	0.08
	water, calcium, 1	0.94	0.07
	water, calcium, 1 × 10^{-4}	0.98	0.07
	ethyl alcohol	1.10	0.06
	carbon tetrachloride	1.10	0.05
	dry air	1.36	0.04
Illite	water, sodium, 1	1.77	0.37
	water, sodium, 1 × 10^{-3}	2.50	0.65
	water, calcium, 1	1.51	0.28
	water, calcium, 1 × 10^{-3}	1.59	0.31
	ethyl alcohol	1.48	0.19
	carbon tetrachloride	1.14	0.04
	dry air	1.46	0.04
Smectite	water, sodium, 1 × 10^{-1}	5.40	1.53
	water, sodium, 5 × 10^{-4}	11.15	3.60
	water, calcium, 1	1.84	0.26
	water, calcium, 1 × 10^{-3}	2.18	0.34
	ethyl alcohol	1.49	0.10
	carbon tetrachloride	1.21	0.03
Muscovite	water	2.19	0.42
	carbon tetrachloride	1.98	0.35
	dry air	2.29	0.41
Sand			0.01 to 0.03

It is apparent that a surface charge is required before a double layer can develop and that the higher is the charge the "stronger" is the double layer effect. The distribution of charge is important too. For example, a negative face charge and a positive edge charge lead to flocculation rather than repulsion and dispersion.

The valency of the cations is important because double layers are thicker, and thus exert more effect, for monovalent cations than for divalent cations. The concentration of electrolyte also influences the thickness of the double layers, the higher the concentration the thinner the double layers.

The dielectric constant of the fluid is important since double layers of consequence cannot form unless the dielectric constant is high, as for water. Other properties of the fluid, such as dipole moment and the details of bonding within the fluid, influence the degree to which adsorbed layers form on the surfaces.

Virgin Compression Curves.

The relative positions of the virgin compression curves of muscovite, kaolinite, and illite are in conformance with predictions based on the mechanical model. The actual positions of the curves, however, are greatly influenced by the original soil structure which was, in turn, influenced by physico-chemical effects in the original suspensions and slurries.

The dominance of mechanical effects during virgin consolidation of muscovite, kaolinite, and illite, result from a combination of effects revolving mainly around the facts that these particles tend to be large (thus leading to small surface areas and reduced surface chemical effects) and stiff, and are at large angles to each other, especially at low effective stresses.

Conversely, in the case of smectite, the particles are very thin (of the order of 10 Angstrom units), and highly flexible, with a high surface area, so that double layer effects dominate, and mechanical effects become negligible. Thus, replacement of water by carbon tetrachloride results in a collapse of double layers and large reductions in void ratio for smectite whereas it leads to larger void ratios for the muscovite, kaolinite, and illite.

Swelling Curves.

For purposes of comparison, the void ratios at an effective stress of 100 psf and the swelling indexes, defined as the average slopes of the e-log $\bar{\sigma}$ curves for the range of stress from 64,000 psf back to 100 psf, are shown in Table 2 for all reported tests. The range of values for the slopes of the swelling curves for sand were taken from Hendron (9) and are shown to indicate the extreme case of mechanical swelling of equidimensional particles.

The slopes of the swelling curves of kaolinite in air and carbon tetrachloride are somewhat greater than typical slopes for sand, as would be predicted by the mechanical model considering that the kaolinite particles are not equidimensional. The larger slopes for the swelling curves in alcohol and water may result from the lower coefficients of surface friction which allow the particles to slide over each other more easily and promotes more swelling. There is no evidence of significant double layer effects.

The slopes of the swelling curves for illite seem, at first, to be anomolous. The slopes for illite in dry air and carbon tetrachloride are only slightly greater than those of sand, and very much less than those of muscovite even though the particles of illite and muscovite seem to have about the same values of *D/H*. If the slopes of the swelling curves of illite in dry air and carbon tetrachloride are accepted as representative of purely mechanical swelling, then the much higher slopes for illite in water must indicate a dominant physico-chemical control, a conclusion in direct conflict with the data for the virgin compression curve. Further, the slopes of the swelling curves in concentrated sodium or calcium chloride, where the double layers are so severely compressed that significant double layer swelling is unlikely, are much greater than the slopes for dry air and carbon tetrachloride. The most likely explanation is that the individual particles of illite in dry air and carbon tetrachloride associated into more or less equidimensional grains that were so strong that they did not break down significantly even under an effective stress of 64,000 psf. Thus, the slopes of the swelling curves for illite in dry air and carbon tetrachloride probably represent mechanical swelling of a soil composed of particles of small D/H. The slopes are similar to those of kaolinite and not much greater than the slopes for sand. It is possible that the slopes of the swelling curves for illite in alcohol and calcium illite in water represent mechanical swelling. Based on that assumption, it follows that an important part of the swelling of sodium illite in water is controlled by physico-chemical variables, the actual degree of control apparently varying with the concentration of electrolyte in the pore water.

The analysis of the swelling of illite applies also to smectite. The small slope for the smectite in carbon tetrachloride probably indicates the presence of coherent grains, each containing many layers of montmorillonite. The low slope of the swelling curve for smectite in alcohol fits in with the concept of filmy particles (layers) of negligible mechanical strength. The great effect of cation valency and concentration on the slope of the swelling curve in water indicates that double layer forces are exerting a controlling influence on the swelling of smectite, particularly sodium smectite.

Natural Soils.

This study was concerned mainly with purified systems in the laboratory because only in that case is it possible to isolate the different variables. The conclusions should apply to natural soils as well. Certainly it is field experience that volume changes in soils rich in smectite are far in excess of those of soils containing the nonexpanding lattice minerals and that the amount of swelling of the smectite depends on the valency of the adsorbed ions. The main qualification involved in extending the results of these laboratory studies to natural soils is that most natural soils are cemented, at least to a minor degree, so that the facility with which particles can slide over one another is controlled by the cement more than by friction.

SUMMARY AND CONCLUSIONS

The compressibility characteristics of clays, in general, are influenced by both mechanical and physico-chemical effects. The term mechanical is used to denote short range particle interaction in which compressibility is controlled by the physical properties of the mineral particles, i.e., by their strength and flexibility, and by surface friction. Mechanical effects control the compressibility of such soils as sands and micaceous silts. The term physico-chemical signifies comparatively long range interaction between particles, especially through double layers; the mechanical properties of individual particles are of relatively minor importance. Variations in the dielectric constant of the pore fluid, in the concentration of dissolved electrolyte, and in the valency of adsorbed cations, cause major changes in the compressibility of soils according to the physico-chemical model, but minor changes according to the mechanical model.

One-dimensional consolidation tests were performed using specimens of ground muscovite, kaolinite, illite, and smectite. Variables included the valency of the adsorbed cations, the pore water electrolyte concentration, and the nature of the pore fluid (dry air, water, carbon tetrachloride, and alcohol).

Although mechanical and physico-chemical effects must occur simultaneously in all soils, the experiments indicate that one or the other usually dominates. In suspensions (in cases where suspensions are stable), physico-chemical effects dominate because of the large average spacing between the particles. As these physico-chemical effects influence the original soil structure, they exert an indirect influence on virgin compressibility for all soils. However, once the original soil structure is established, the virgin compression curves of kaolinite and illite seem to be controlled by mechanical effects whereas the virgin compression curves of smectite are mechanically controlled whereas those of smectite are physico-chemically controlled. Illite occupies an intermediate position with both mechanical and physico-chemical effects evident during swelling.

ACKNOWLEDGMENTS

The experiments reported herein were performed in the soil mechanics laboratories of the University of Illinois at Urbana, during the period from 1963 through 1969. They represent part of the research performed under Grants GP325 and GK531 from the National Science Foundation.

APPENDIX I.—EXPERIMENTAL PROCEDURES

Preparation of Homoionic Clays.—The homoionic clays were prepared by repeated washings with a large excess of the chloride of the appropriate cation. The exchangeable cations of the purified material were then analyzed and in all cases at least 96% of the exchangeable cations were the desired cation. Values of pH were adjusted using either hydrochloric acid or the hydroxide of the appropriate cation.

Preparation of Sedimented Specimens.—Sedimentation of all specimens was started from a slurry. The slurry was de-aired under vacuum before being poured into sedimentation tubes. The specimens for one-dimensional consolidation tests were consolidated under pressures in the range of 90 psf to 125 psf before being extruded into the consolidation rings.

Preparation of Clays in Organic Pore Fluids and "Dry Air".—Air dry clay was washed in ethyl alcohol, centrifuged to remove the excess alcohol, and dried at 110° C for 24 hr. The dry clay was cooled under vacuum in a desiccator and mixed directly with the appropriate organic fluid. The clay-organic fluid slurry was poured directly into the consolidation rings.

The specimens tested in "dry air" were tested immediately after both the soil and the consolidation cell were removed from an over at 110° C. The apparatus was packed in dry calcium chloride to minimize adsorption of water vapor and the entire experiment was performed before the specimen could cool significantly.

Consolidation Apparatus.—All one-dimensional consolidation tests were performed in 2.5-in. diam stainless steel consolidation rings. Both the inside of the rings and the outside of the loading cap were encased in tetrafluoethylene to reduce friction. The ease with which specimens could be extruded from the rings after the tests indicates that friction between the soil and the rings was negligible. The cells were designed to prevent evaporation of fluid during testing (14). Consolidation pressures ranged from 35 psf to 64,000 psf.

APPENDIX II.—REFERENCES

1. Atterberg, A., "Die Plastizität und Bindigkeit liefernden Bestandteile der Tone," *Intern. Mitt. fur Bodenkunde,* 1919.
2. Aylmore, L. A. G., and Quirk, J. P., "Swelling of Clay-Water Systems," *Nature,* Vol. 183, 1959, pp. 1752–1753.
3. Aylmore, L. A. G., and Quirk, J. P., "The Structural Status of Clay Systems," *Clays and Clay Minerals,* Vol. 9, Pergamon Press, New York, 1962, pp. 104–130.
4. Blackmore, A. W., and Miller, R. D., "Tactoid Size and Osmotic Swelling in Calcium Montmorillonite," *Proceedings,* Soil Science Society of America, Vol. 25, 1961, pp. 169–173.
5. Bolt, G. H.., "Physico-Chemical Analysis of the Compressibility of Pure Clays," *Geotechnique,* London, Vol. 6, No. 1, Mar., 1956, pp. 86–93.
6. Bolt, G. H., and Miller, R. D., "Compression Studies of Illite Suspensions," *Proceedings,* Soil Science Society of America, Vol. 19, 1955, pp. 285–288.
7. Casagrande, A., and Fadum, R. E., "Notes on Soil Testing for Engineering Purposes," Harvard Graduate School of Engineering *Soil Mechanics Series No. 8,* 1940.
8. Gilboy, G., "The Compressibility of Sand-Mica Mixtures," *Proceedings,* ASCE, 1928, pp. 555–568.
9. Hendron, A. J., "The Behavior of Sand in One-Dimensional Compression," thesis presented to the University of Illinois, at Urbana, Ill., in 1963, in partial fulfillment of the requirements for the degree of Doctor of Philosophy.
10. Horn, H. M., and Deere, D. U., "Frictional Characteristics of Minerals," *Geotechnique,* London, Vol. 12, No. 4, Dec., 1962, pp. 319–335.
11. Kruyt, H. R., "General Introduction, Chapter 1," *Colloid Science, Vol. 1, Irreversible Systems,* H. R. Kruyt, ed., Elsevier Publ. Co., New York, 1952, pp. 1–57.
12. Ladd, C. C., "Mechanisms of Swelling by Compacted Clay," Highway Research Board, *Bulletin 245,* 1959, pp. 10–26.
13. Lambe, T. W., "The Structure of Compacted Clay," *Journal of the Soil Mechanics and Foundations Division,* ASCE, Vol. 84, No. SM2, Proc. Paper 1654, 1958.
14. Mesri, G., "Engineering Properties of Montmorillonite," thesis presented to the University of Illinois, at Urbana, Ill., in 1969, in partial fulfillment of the requirements for the degree of Doctor of Philosophy.
15. Michaels, A. S., and Bolger, J. C., "Settling Rates and Sediment Volumes of Flocculated Kaolin Suspensions," *Industrial and Engineering Chemistry,* Vol. 1, 1962, pp. 24–33.
16. Olson, R. E., and Mitronovas, F., "Shear Strength and Consolidation Characteristics of Calcium and Magnesium Illite," *Clays and Clay Minerals, Proceedings,* Ninth National Conference, Pergamon Press, New York, 1962, pp. 185–209.
17. Overbeek, J. Th. G., "Electrochemistry of the Double Layer," Chapter 4, *Colloid Science, Vol. 1, Irreversible Systems,* H. R. Kruyt, ed., Elsevier Publ. Co., New York, 1952, pp. 115–190.
18. Quirk, J. P., "Particle Interaction and Soil Swelling," *Israel Journal of Chemistry,* Vol. 6, 1968, pp. 213–234.
19. Salas, J. A., and Serratosa, J. M., "Compressibility of Clays," *Proceedings, Third International Conference on Soil Mechanics and Foundation Engineering,* Zurich, Vol. 1, 1953, pp. 192–198.

20. Samuels, S. G., "The Effect of Base Exchange on the Engineering Properties of Clays," Building Research Station, Department of Science and Industrial Research, Garston, England, *Note C-186*, 1950.
21. Terzaghi, K., "The Mechanics of Adsorption and of the Swelling of Gels," Colloid Symposium Monograph, Chemical Catalog Co., Inc., New York, 1926, pp. 58–78.
22. Terzaghi, K., "Technisch-geologische Beschreibung der Bodenbeschaffenheit für bautechnische Zwecke," Chapter IX, Part A, *Ingenieurgeologie*, by K. A. Redlich, K. Terzaghi, and R. Kampe, Julis Springer, Wien und Berlin, 1929.
23. Terzaghi, K., and Frohlich, O. I., *Theorie der Setzung von Tonschichten*, Franz Deuticke, Leipzig, 1936.
24. van Olphen, H., *An Introduction to Clay Colloid Chemistry*, Interscience Publishers, New York, 1963.
25. Verwey, E. J. W., and Overbeek, J. Th. G., *Theory of the Stability of Lyophobic Colloids*, Elsevier Publishing Company, Amsterdam, 1948.
26. Waidelich, W. C., "Physico-Chemical Factors Influencing the Consolidation of Soils," thesis presented to Princeton University, at Princeton, New Jersey, in 1957, in partial fulfillment of the requirements for the degree of Master of Science.
27. Warkentin, B. P., Bolt, G. H., and Miller, R. D., "Swelling Pressures of Montmorillonite," *Proceedings*, Soil Science Society of America, Vol. 21, 1957, pp. 495–497.
28. Warkentin, B. P., and Schofield, R. K., "Swelling Pressure of Sodium Montmorillonite in Sodium Chloride Solutions," *Journal of Soil Science*, Vol. 13, 1962, pp. 98–105.

Clays and Clay Minerals, 1971, Vol. 19, pp. 151–158. Pergamon Press. Printed in Great Britain

MECHANISMS CONTROLLING THE PERMEABILITY OF CLAYS

GHOLAMREZA MESRI

Department of Civil Engineering, University of Illinois, Urbana, Ill. 61801, U.S.A.

and

ROY E. OLSON

Department of Civil Engineering, The University of Texas at Austin, Austin, Texas 78712, U.S.A.

(*Received* 9 *October* 1970)

Abstract—Coefficients of permeability, calculated using Terzaghi's theory of one-dimensional consolidation, are reported for smectite, illite, and kaolinite, in water, methyl and ethyl alcohol, benzene, and carbontetrachloride. When the pore fluid was water the clays were homoionized to either the sodium or calcium form and the pore water electrolyte concentration was varied. The coefficients of permeability are evaluated in terms of both mechanical and physico-chemical variables. It appears that the coefficients of permeability are mainly influenced by mechanical effects, particularly the distribution of void sizes and the tortuosity of the channels. The coefficient of permeability is maximized if the flow channels consist of many small channels and a relatively few large ones, through which the main flow occurs. Physico-chemical variables exert great influence on the coefficient of permeability through their influence on dispersion or aggregation of the clay particles.

INTRODUCTION

On a microscopic scale the flow of fluids in a porous medium is governed by three equations, viz. the equation of continuity, the equation of state, and the dynamical equation of motion (Lamb, 1932). Unfortunately, the mathematical difficulties involved in solving these equations, and the difficulties involved in prescribing the nature of the flow channels and the interactions between the fluid and the porous media, are such that the classical approach to solving problems involving flow in porous media has not proved useful.

Instead, flow is considered on a macroscopic scale. Based on experiments, Darcy (1856) formulated a macroscopic flow equation which may be written:

$$v = -k\frac{dh}{ds} \qquad (1)$$

where v is the discharge velocity, k is the coefficient of permeability (often termed the conductivity in the scientific literature), h is the total head, and s is the distance of macroscopic flow measured in the direction of flow.

In consideration of its widespread use there have been numerous studies of Darcy's equation. These studies have been of two types: (1) studies of the general form of the equation, and (2) studies of the fundamental factors controlling the constant. This paper is concerned with the constant.

Numerous attempts have been made to study the coefficient of permeability of granular media through the use of capillary (Scheidigger, 1953) or hydraulic radius (Kozeny, 1927; Carman, 1941, 1948; Leibenzon, 1947) models. These analyses were simplified by the assumption of no chemical interaction between the fluid and the porous media. In contrast, there have been few attempts to develop suitable models for clays although the permeability of clays has been discussed extensively.

Terzaghi (1925) emphasized the importance of the nonuniformity of the voids on clay permeability and the substantial dependence of the permeability on the void ratio. He considered that the physical properties of the pore water changed in the immediate vicinity of the clay surfaces.

Macey (1942) measured the coefficient of permeability of clays in water and in nonpolar fluids. He reported rates of flow for benzene 100,000 to 1,000,000 times greater than for water in the same clay. He considered that particle spacing, particle size as influenced by aggregation or dispersion, particle arrangement, adsorbed layers, and interlamellar swelling all influenced the permeability but he considered the most important cause for the lower permeability in water to be the anomolous viscosity of the water near the clay surfaces.

Grace (1953) demonstrated that improved dispersion is the main reason for the marked reduction of permeability caused by use of certain electrolytes and he pointed out the importance of particle reorientation as consolidation progressed.

Michaels and Lin (1954) showed experimentally that the permeability of kaolinite decreased markedly as the polarity of the permeating fluid increased and that the most important factor controlling the permeability of kaolinite was the degree of dispersion of the kaolinite in the original suspending fluid. They also concluded (see also Michaels and Lin, 1955) that electroosmotic counterflow might be of some importance when aqueous solutions were used as the permeants but that such colloidal effects as adsorbed liquid surface films appeared to have little effect on the permeability.

Olsen (1960) measured hydraulic flow rates, electrical conductivities, and streaming potentials for flow through kaolinite, illite, and Boston blue clay, and concluded that electro-osmotic counterflow, high viscosity, and tortuous flow paths failed to account for the permeability characteristics he measured. He concluded that unequal pore sizes was the most important variable.

The foregoing studies and others as well, indicate that the most important variable influencing the permeability of clays is flocculation of the clay particles, which influences the distribution of void sizes and shapes.

In the discussion to follow, values for the coefficients of permeability of kaolinite, illite, and smectite are reported as functions of void ratio and the type of pore fluid, and mechanisms controlling the permeability are discussed.

EXPERIMENTAL AND ANALYTICAL PROCEDURES

The experimental and analytical procedures have been presented in detail by Mesri (1969) and reviewed by Olson and Mesri (1970). In brief, homoionic slurries of kaolinite, illite, and smectite (Table 1) in water were prepared by repeated washing with concentrated solutions of NaCl or $CaCl_2$. Slurries in other fluids were prepared by washing air dried clay in ethyl alcohol, centrifuging, drying at 110°C, and then mixing with the desired fluid. The slurries were consolidated to pressures ranging from 90 to 125 *psf* in special sedimentation tubes (Mesri, 1969) and then transferred to 2·5-in. dia. consolidation rings where they were consolidated in increments to a pressure of 64,000 *psf* using a pressure ratio of two.

The coefficients of permeability were calculated by fitting Terzaghi's theory of consolidation (Terzaghi, 1943) to the observed laboratory time-settlement observations and extracting the coefficient of permeability from the calculated coefficient of consolidation. The fitting operation utilized the logarithmic method (Casagrande and Fadum, 1940) for all tests in which the pore fluid was water or alcohol. However, specimens in which the pore

Table 1. Properties of clays

Mineral (1)	Supplier source trade name (2)	Liquid limit in per cent (3)	Plastic limit in per cent (4)	Specific gravity (5)	Surface area in m² per g (6)	Cation exchange capacity m-equil. per 100 g (7)	Fraction finer than 0·002 (8)	Estimated ratio of dia. to thickness (9)
Kaolinite	Minerals and Chemicals Philipp Corp., Klondyke clay, McIntyre, Georgia	40 to 50	27 to 31	2·65	14	2·2	47	2–5
Illite	J. L. Eades, Dept. of Geology Univ. of Illinois, Marblehead, Wisconsin	83 to 104	31 to 32	2·80		28	100	10–50
Smectite	American Colloid Company, Wyoming, Volclay	190 to 1160	31 to 47	2·65 to 2·80	500 to 700	100	97	150–500

fluid was non-polar often underwent 50 per cent or more of their settlement during the first 6 sec after loading, thus precluding use of the logarithmic method. For these specimens the square root fitting method (Taylor, 1948) was used. In our experience, the standard square root method, in which fitting is performed at an average degree of consolidation of 90 per cent, usually yields coefficients of permeability that are higher than those calculated by either method when fitting is performed at 50 per cent consolidation. However, the coefficients of permeability in non-polar fluids were so much higher than in other fluids that a slight downwards adjustment in the data would have no effect on the interpretation.

A question may be raised as to how the coefficients of permeability calculated using Terzaghi's theory compare with values measured directly. Terzaghi (1923) made such comparisons when he first developed the theory; he found satisfactory agreement. Casagrande and Fadum (1944) reported that they always found satisfactory agreement provided the logarithmic fitting method was used and provided that there was a distinct change in curvature when the primary settlement curve merged with the secondary settlement curve. Taylor (1942) presented comparisons for remolded specimens of Boston Blue clay, based on the square root fitting method, and showed that the measured coefficients of permeability generally exceeded the calculated values. He attributed this difference in permeabilities to Terzaghi's assumption that the sole cause of delay in compression is the time required for the water to be squeezed out, i.e. to the permeability of the clay. Taylor concluded that

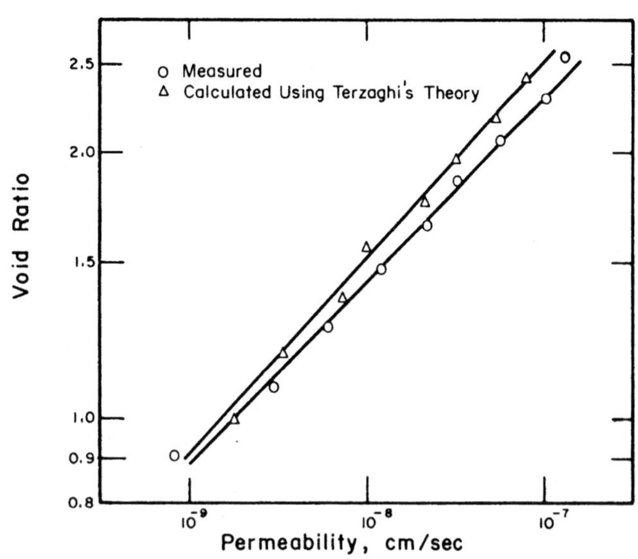

Fig. 1. Coefficients of permeability of a 0·01 N calcium illite.

the structure of clay itself possessed a time dependent resistance to compression so that the total resistance to volume change came partly from permeability and partly from the structural resistance of the clay itself. By attributing all of the resistance to low permeability, Terzaghi's theory must inevitably lead to an under-estimate of the permeability. It has been our experience, based on a number of comparisons between measured and computed coefficients of permeability on both remolded and undisturbed clays, that the calculated coefficients of permeability are low by only about 5–20 per cent provided that the clay is normally consolidated at the time of determination. All clays used in this investigation were normally consolidated. A comparison of measured and computed coefficients of permeability, which is typical of numerous other such comparisons, is shown in Fig. 1.*

EXPERIMENTAL RESULTS AND DISCUSSION

The coefficients of permeability of kaolinite, illite, and smectite, are shown as functions of void ratio in Figs. 2–4, respectively. For the cases where water was the pore fluid, all data are indicated without regard to the electrolyte concentration. The scatter in data for the sodium forms of the illite and smectite is mainly caused by the variations in electrolyte concentration, a variable that will be discussed subsequently.

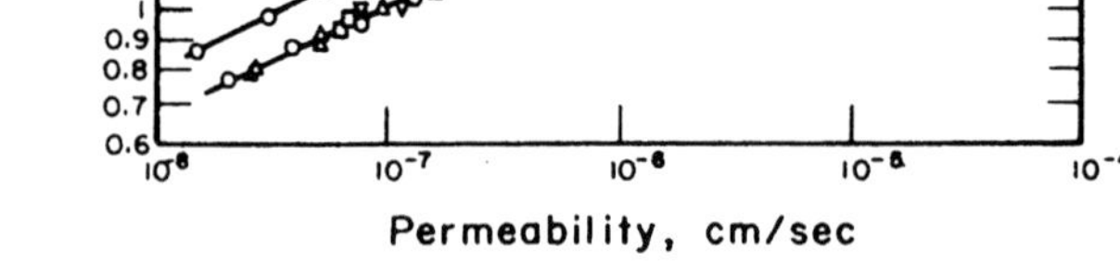

Fig. 2. Coefficients of permeability of kaolinite.

In discussing the data in Figs. 2–4, it is convenient to separate the variables that influence the coefficients of permeability into two broad and overlapping groups, viz. mechanical variables and physico-chemical variables.

*The illite used in this comparison came from Fithian, Illinois, and has slightly different properties from the illite used in the main series of experiments to be reported in this paper.

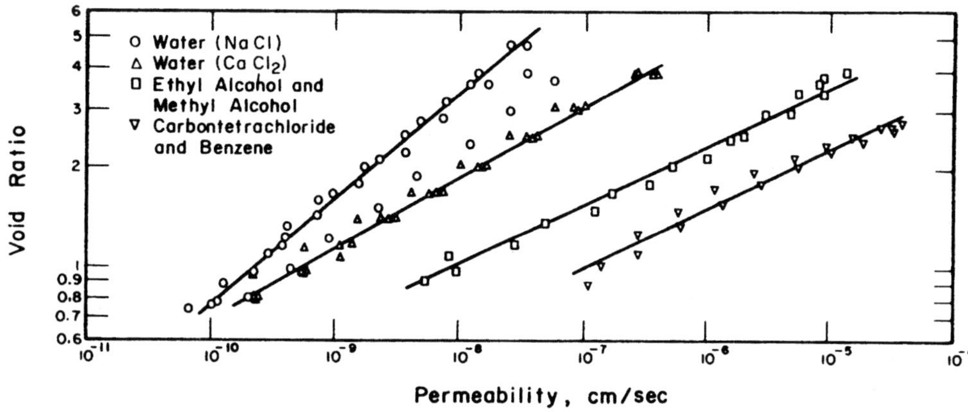

Fig. 3. Coefficients of permeability of illite.

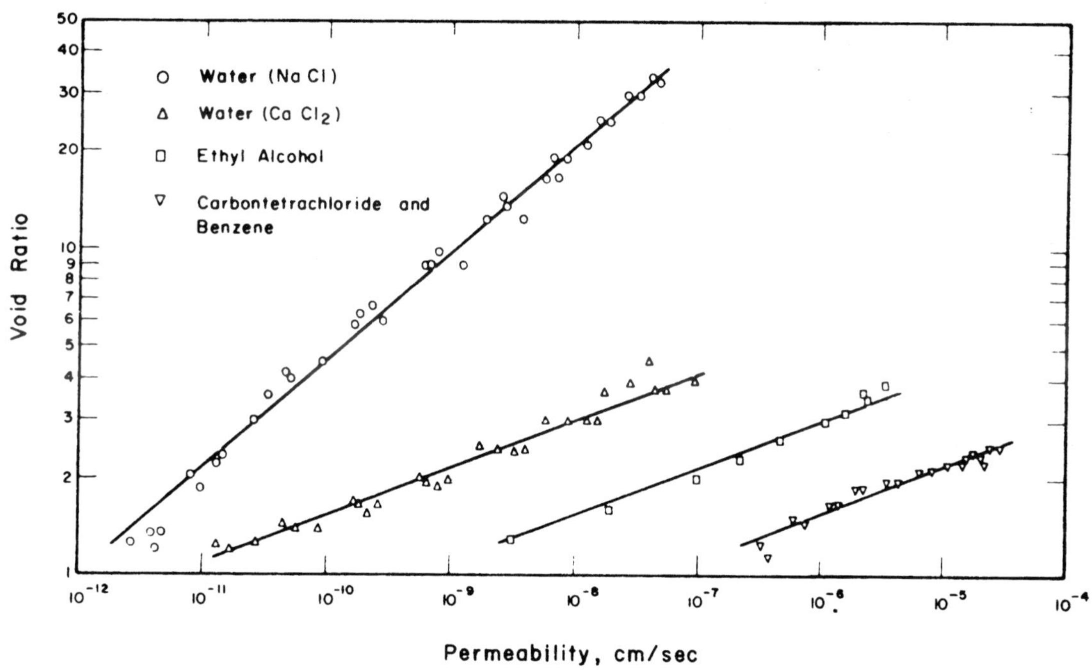

Fig. 4. Coefficients of permeability of smectite.

Mechanical variables

Various mechanical variables may be used but perhaps the most convenient are the size, shape, and geometrical arrangement of the clay particles. These variables in turn determine the geometrical properties of the pore system.

The smaller the particle size the smaller the size of the individual flow channels, all other variables being constant. The simplest theoretical models of porous media (Lamb, 1932) show that the coefficient of permeability is directly proportional to the second power of the diameter of the flow channel. The great range in size of the flow channels is one reason for the large differences in the coefficient of permeability of the three clay minerals. The smectite has the smallest particle size and the kaolinite the largest. The coefficients of permeability of the three clays, in the sodium form and in water are shown in Fig. 5. The illite is about 200 times more pervious than the smectite, and the kaolinite is about 200,000 times more pervious, at the same void ratio. Variables other than channel size also contribute to the differences in Fig. 5 as will be discussed subsequently.

The geometry of the flow channel is a function of the shape and orientation of the clay particles. For the platey clay particles, application of pressure, under fully drained conditions and without lateral strain, results in an orientation of the plates normal to the direction of maximum principal stress (Lambe, 1953; Norrish, 1954; Emerson, 1956; Mitchell, 1956; Quigley and Thompson, 1966)

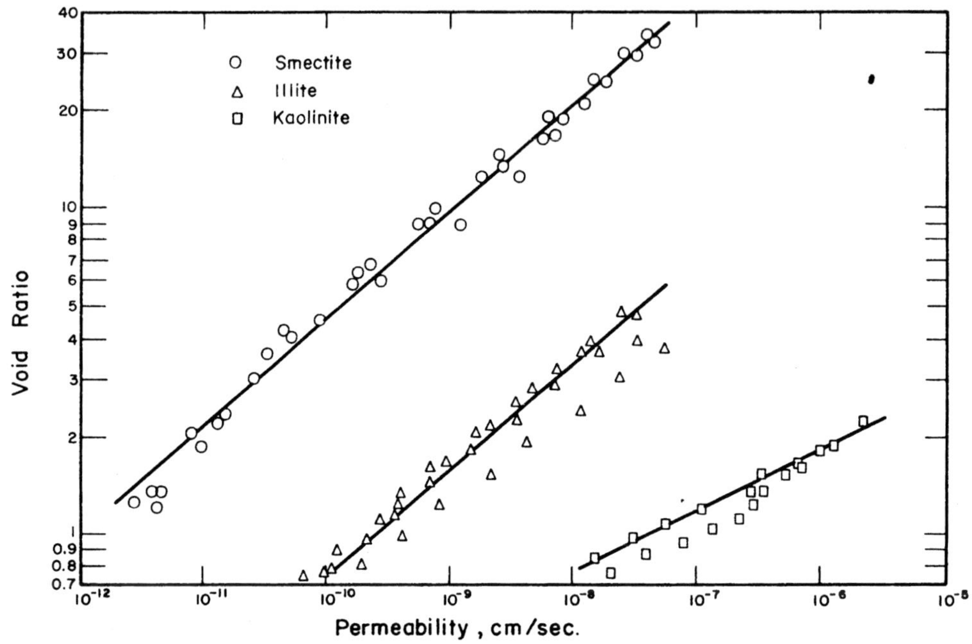

Fig. 5. Coefficient of permeability of all three sodium clays in water.

and leads to an increasingly tortuous flow path in the direction of the applied pressure. This particle orientation, or tortuosity effect, is larger for particles with larger diameter to thickness ratios. The diameter to thickness ratio (Table 1) increases from kaolinite to illite to smectite and is another source of the difference in the coefficients of permeability of the three clays (Fig. 5). Also, particle orientation increases with increasing consolidation pressure (decreasing void ratio) causing a reduction in permeability with reduction in void ratio. Of course, the coefficient of permeability also decreases with void ratio simply because of the reduction in total void space.

Physico-chemical variables

The data in Figs. 2–4 indicate that the coefficients of permeability are largest for nonpolar fluids, smaller for polar fluids of low dielectric constant, and lowest for water, which is polar and has a high dielectric constant. Further, the coefficient of permeability is generally lower when the adsorbed cations are monovalent rather than divalent. Generally, a reduction in electrolyte concentration (Fig. 6) tends to reduce the coefficient of permeability but the effect of electrolyte concentration diminishes as the valency of the cations increases, and is smaller in sequence from smectite to illite to kaolinite. It is convenient to consider these effects as physico-chemical.

The physico-chemical variables may be taken as the surface charge density and distribution, valency of the adsorbed cations, and such properties of the fluid as dielectric constant, dipole moment, and viscosity. Other variables could be chosen and some of the above variables, e.g. viscosity of the fluid, could just as well be considered mechanical variables.

The existence of a net negative charge within the clay particles leads to the adsorption of cations. For pore fluids of low dielectric constant, such as the alcohols, benzene, and carbon tetrachloride used in this study, the cations are held tightly against the surface of the particles and have a minimal tendency to block the flow channels. Further, when nonpolar fluids are used, there is little tendency for the fluid to be adsorbed by either the cations or the surface and the flow channels are almost completely open. Thus, the coefficient of permeability should be largest for benzene and carbon tetrachloride (Figs. 2–4). When a fluid such as an alcohol is used, the fluid is adsorbed either by the cations or, more likely, by the formation of hydrogen bonds with the surface, and a part of the flow channels is likely to be blocked, thus reducing the permeability (Figs. 2–4). However, these effects seem clearly inadequate to explain the very large differences between the coefficients of permeability of the smectite and illite in nonpolar pore fluids, in alcohols, and in water. Further, these effects do not explain the fact (Fig. 7) that all of the clays yielded the same void ratio vs. permeability relationship when nonpolar fluids were used.

Since there is a negligible adsorption of the nonpolar fluids and no formation of diffuse double

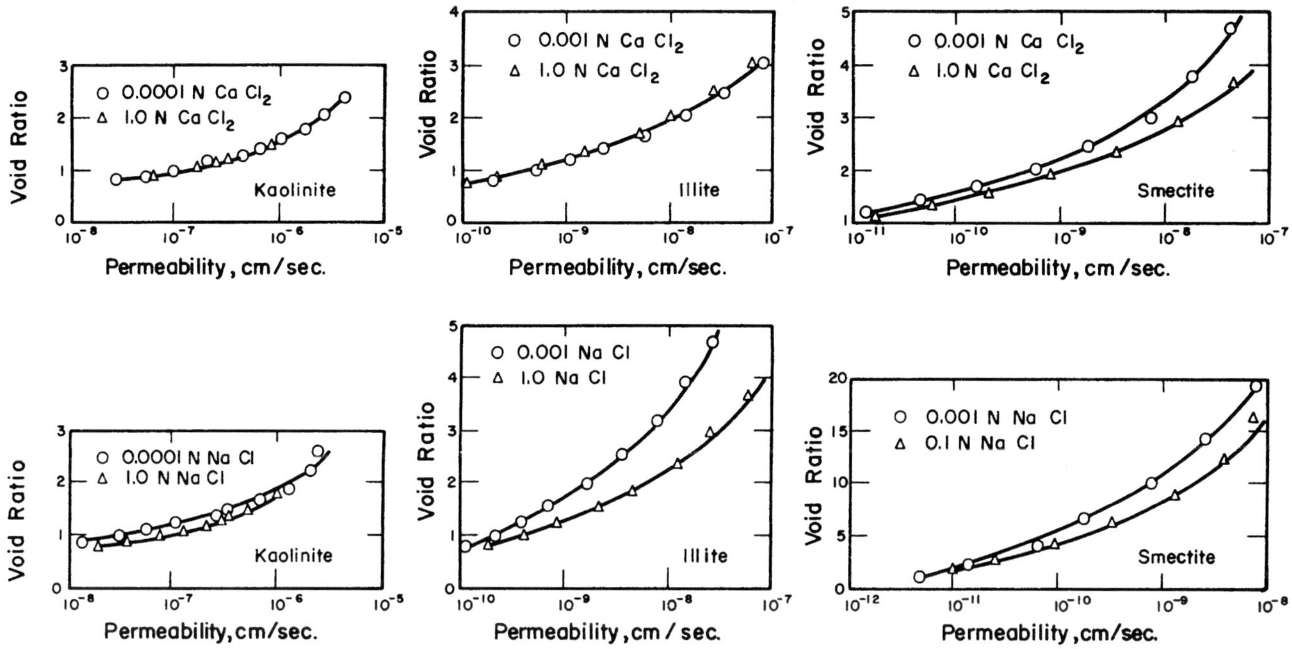

Fig. 6. Coefficients of permeability of all three clay minerals in water.

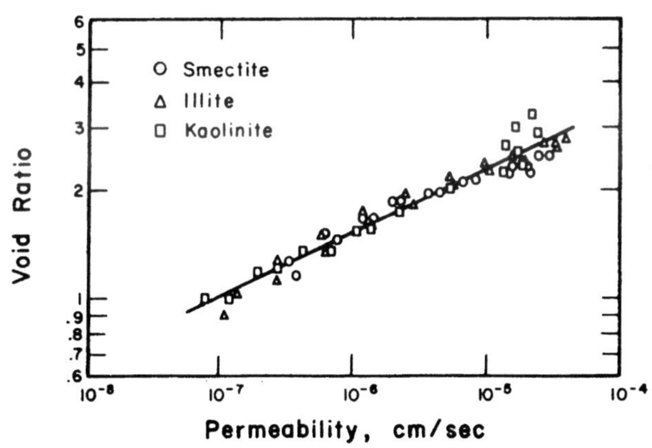

Fig. 7. Coefficients of permeability of all three clay minerals in nonpolar fluids.

layers, the mechanisms controlling the coefficients of permeability of the clays in nonpolar fluids should be mechanical. The existence of essentially the same permeability for the three minerals (Fig. 7) then implies the existence of voids of similar size and shape. Apparently the particles of illite and smectite formed stable aggregates, and the main fluid flow took place in the large voids between the aggregates. Thus, for a constant void ratio, the channels between these aggregates would be considerably larger in size than channels between evenly spaced particles and the rate of flow would be expected to increase with some power of the channel diameters. The main effect of using alcohol or water as a permeant is then to break down the aggregates, thus reducing the size of the largest flow channels at a constant void ratio, and reducing the coefficient of permeability. The existence of large aggregates was previously used to explain the swelling characteristics of the same clays (Olson and Mesri, 1970).

The clay particles in aggregates in nonpolar fluids are believed to be randomly arranged and the aggregates are probably more-or-less spherical in shape. When the pore fluid is water, it is believed that aggregates of parallel particles, termed domains (Emerson, 1959; Aylmore and Quirk, 1959), are formed for smectite and possibly for illite as well. In the case of calcium smectite the water is adsorbed between the layers until the basal spacing is about 19 Å (Norrish, 1954; Senich, Demirel and Handy, 1967) and then adsorption stops. Since the layers cannot swell sufficiently to form diffuse double layers, it is believed that domains of perhaps eight to ten layers form (Blackmore and Miller, 1961) with larger voids between the domains. The fluid trapped within the domains is probably essentially immobile until high pressures are applied. Thus, the fluid flows mainly through the larger interdomainal voids. The calcium smectite in water is more dispersed than the smectite in alcohol, but because of the presence of domains is less dispersed than the sodium smectite in water, and thus has intermediate coefficients of permeability. A similar argument apparently may be used with the illite. The kaolinite particles are too large and too nearly equidimensional to form domains that influence the permeabilities significantly.

For calcium clays a diffuse double layer probably

forms on the exposed surfaces of clay particles (Norrish, 1954) but in the case of kaolinite and illite the void spaces are so large that the diffuse double layers do not block the main flow channels significantly. Thus, changes in double layer thickness associated with variations in pore water electrolyte concentration do not influence the permeability by measurable amounts (Fig. 6). However, in the case of calcium smectite with its much smaller void spaces, the apparent double layer thickness influences permeability significantly (Fig. 6). Further, in the case of calcium smectite, it is probable that thicker domains form when the electrolyte concentration reaches about 1 N so, for a given void ratio, the interdomainal voids are larger and a higher permeability results.

For a sodium clay in water, diffuse double layers form and the clay disperses. The lower the electrolyte concentration the greater is the tendency of the clay to disperse uniformly through the available space without the occurrence of any larger voids. Thus, electrolyte concentration, through its influence on dispersion, influences the coefficient of permeability.

SUMMARY AND CONCLUSIONS

The coefficients of permeability of clays are controlled by variables that may be classified as mechanical and physico-chemical. The mechanical variables of main interest are the size, shape, and the geometrical arrangement of the clay particles. The reduction in the coefficient of permeability, at constant void ratio, from kaolinite to illite to smectite is largely the result of a reduction in the size of individual flow channels and an increase in the tortuosity of the flow paths.

Physico-chemical variables exert great influence on the coefficient of permeability by controlling the tendency of the clay to disperse or to form aggregates. Aggregation leads to the existence of many tiny flow channels through which there is likely to be little flow, and a smaller number of relatively large channels through which the main flow occurs. Dispersion leads to channels that are all of nearly the same size, and tend to be nonequidimensional, and thus reduces fluid flow.

Acknowledgments—The experiments reported in this paper were performed in the soil mechanics laboratories of the University of Illinois at Urbana, during the period from 1963 through 1969. They represent part of the research performed under grants GP325 and GK531 from the National Science Foundation.

REFERENCES

Aylmore, L. A. G. and Quirk, J. P. (1959), Swelling of clay–water systems: *Nature, Lond.* **183**, 1752–1753.

Blackmore, A. V. and Miller, R. D. (1961) Tactoid size and osmotic swelling in calcium monomorillonite: *Soil Sci. Soc. Am. Proc.* **25**, 169–173.

Carman, P. D. (1941) Capillary size and capillary movement of moisture in fine sands: *Soil Sci.* **52**, 1–14.

Casagrande, A. and Fadum, R. E. (1940) *Notes on Soil Testing for Engineering Purposes*: Soil Mech. Series No. 8, Harvard Graduate School of Engineering.

Darcy, H. (1856) *Les Fontaines Publiques de la Ville de Dijon.* Dalmont, Paris.

Emerson, W. W. (1956) Liquid crystals of sodium monomorillonite: *Nature, Lond.* **178**, 1248.

Emerson, W. W. (1959) The structure of soil crumbs: *J. Soil Sci.* **10**, 234.

Grace, H. P. (1953) Resistance and compressibility of filter cakes: *Chem. Engng Prog.* **49**, 303–318, 367–377.

Kozeny, J. (1927) Uber Kapillare Leitung des Wassers im Boden: *Akad. Wiss. Wiener* **Abt. IIa, 136**, 271–306.

Lamb, H. (1932) *Hydrodynamics* 6th Edn. Cambridge University Press, London.

Lambe, T. W. (1953) The structure of inorganic soil: *Proc. ASCE* Separate No. 315.

Leibenzon, L. S. (1947) Dvizhenie Prirodnykh Zhidkostei i Gazov v Poristoi Srede: *Gosudarstv. Izdat. Tekh.-Teoret. Lit.* Moscow.

Macey, H. H. (1942) Clay–water relationship and the internal mechanisms of drying: *Trans. Br. Ceram. Soc.* **41**, 73–121.

Mesri, G. (1969) Engineering Properties of Montmorillonite. Ph.D. Thesis, University of Illinois at Urbana.

Michaels, A. S. and Lin, C. S. (1954) The permeability of kaolinite: *Ind. Engng Chem.* **46**, 1239–1246.

Michaels, A. S. and Lin, C. S. (1955) Effects of counterelectroosmosis and sodium ion exchange on permeability of kaolinite: *Ind. Engng. Chem.* **47**, 1249–1253.

Mitchell, J. K. (1956) The fabric of natural clays and its relation to engineering properties: *Proc. High. Res. Bd.* **35**, 693–713.

Norrish, K. (1954) The swelling of montmorillonite: *Discuss Faraday Soc.* **18**, 120–134.

Olsen, H. W. (1960) Hydraulic flow through saturated clays: *Clays and Clay Minerals* **11**, 131–161.

Olson, R. E. and Mesri, G. (1970) Mechanisms controlling the compressibility of clays: *J. Am. Soc. Civ. Engrs.* **96**, SM6, 1853–1878.

Quigley, R. M. and Thompson C. D. (1966) The fabric of Anisotropically consolidated sensitive marine clay: *Can Geotech. Jl.* **3**, 61–73.

Scheidigger, A. E. (1960) *The Physics of Flow Through Porous Media.* Macmillan, New York.

Senich, D., Demirel, T. and Handy, R. L. (1967) X ray diffraction and adsorption isotherm studies of the calcium montmorillonite–H_2O system: *Highway Res. Rec.* **209**, 23–54.

Taylor, D. W. (1942) *Research on Consolidation of Clays*, Serial 82, Massachusetts Institute of Technology, Department of Civil Engineering, Cambridge.

Taylor, D. W. (1948) *Fundamentals of Soil Mechanics.* Wiley, New York.

Terzaghi, K. T. (1923), *Die Berechnung der Durchlassigkeitsziffer des Tons aus dem Verlauf der hydrodynamischen Spannungserscheinungen*, Akademie der Wissenschaften in Wien. Sitzungsberichte, Mathematisch-naturwissenschaftliche Klasse – IIa, **132**, pp. 125–138.

Terzaghi, C. (1925) Determination of the permeability of clay: *Engng News Rec.* **95**, 832–836.

Terzaghi, K. T. (1943) *Theoretical Soil Mechanics*, pp. 510. Wiley, New York.

Résumé—Les coefficients de permeabilité calculés à l'aide de la théorie de la consolidation unidimensionelle de Terzaghi sont donnés pour une smectite, une illite et une kaolinite en présence d'eau, d'alcool méthylique, d'alcool éthylique, de benzène et de tétrachlorure de carbone. Lorsque le fluide remplissant les pores est l'eau, les argiles ont été rendues homoioniques, soit sous forme sodium, soit sous forme calcium et l'on a fait varier les concentrations en électrolyte de la solution des pores. Les coefficients de perméabilité ont été évalués à la fois en terme de variables mécaniques et physico-chimiques. Il apparaît que les coefficients de perméabilité sont essentiellement influencés par les effets mécaniques, en particulier, la distribution des dimensions des vides et la tortuosité des pores. Le coefficient de perméabilité passe par un maximum lorsque les pores assurant l'écoulement sont constitués par de nombreux petits canaux et relativement peu de grands, à travers lesquels la plus grande part du débit s'établit. Les variables physico-chimiques exercent une grande influence sur le coefficient de perméabilité par le biais du rôle qu'elles jouent dans la dispersion et l'agrégation des particules d'argile.

Kurzreferat—Es wird über Permeabilitätskoeffizienten, berechnet unter Verwendung der Theorie von Terzaghi über eindimensionale Verdichtung, für Smectit, Illit und Kaolinit in Wasser, Methyl- und Äthylalkohol, Benzol und Tetrachlorkohlenstoff berichtet. Wenn die Porenflüssigkeit Wasser war wurden die Tone entweder in die Natrium-oder die Calciumform homoionisiert und die Elektrolytkonzentration des Porenwassers wurde variiert. Die Permeabilitätskoeffizienten wurden mit Hilfe mechanischer sowie physikalisch-chemischer Variabler eingeschätzt. Es scheint, dass die Permeabilitätskoeffizienten in erster Linie durch mechanische Wirkungen beeinflusst werden, besonders die Verteilung von Hohlraumgrössen und die Gewundenheit der Kanäle. Der Permeabilitätskoeffizient wird maximiert wenn die Strömungskanäle aus vielen kleinen Kanälen und verhältnismässig wenigen weiten bestehen, durch welche die Hauptströmung stattfindet. Physikalisch-chemische Variable üben infolge ihrer Wirkung auf die Dispersion oder Zusammenballung der Tonteilchen einen beträchtlichen Einfluss auf den Permeabilitätskoeffizienten aus.

Резюме — Приведены коэффициенты проницаемости для смектита, иллита и каолинита (по отношению к воде, метиловому и этиловому спиртам, бензолу и четыреххлористому углероду), подсчитанные в соответствии с теорией одномерной консолидации Терцаги. В тех случаях, когда наполнителем пор служила вода, глины гомоионизировались до натриевой или кальциевой форм, и концентрация водного электролита пор изменялась. Коэффициент проницаемости оценивался как в отношении механических, так и в отношении физико-химических переменных. Как оказалось, на коэффициент проницаемости влияют, главным образом, механические эффекты, и в особенности — характер распределения пор по размерам и степень извилистости каналов. Коэффициент проницаемости достигает максимума в тех случаях, когда каналы, по которым проходит поток жидкости, состоят из множества мелких каналов и сравнительно небольшого числа крупных. Физико-химические переменные оказывают влияние на коэффициент проницаемости лишь в той степени, в которой они влияют на дисперсию или агрегацию глинистых частиц.

JOURNAL OF THE
GEOTECHNICAL
ENGINEERING DIVISION

Shearing Strengths of Kaolinite, Illite, and Montmorillonite

By Roy E. Olson,[1] F. ASCE

Introduction

During the years 1957–1970, the writer was principal investigator on a series of projects in which attempts were made to determine the effects of physicochemical variables on the engineering properties of kaolinite, illite, and montmorillonite. From this research, and that of others, there emerged a qualitative physicochemical model that was previously applied to one-dimensional compressibility (19) and permeability (16) and is now applied to the peak shearing strengths.

Physicochemical Effects

The terms, "physical" and "chemical," may be applied rather loosely to describe the interaction of particles. The term, physical, is used to denote interactions that are controlled largely by the size, shape, packing, and physical properties of the individual grains, and to the friction between them, whereas the term, chemical, is applied to interactions through diffuse double layers, van der Waals forces, and ionic forces.

Physical Interaction.—In the case of large particles, e.g., in the sand range, the dimensions of the particles are much greater than the distance through which chemical forces act and the magnitudes of gravitational forces on individual particles vastly exceed the magnitudes of chemical forces until the distance between particles become of the order of one-millionth of the particle diameter. The comparatively large contact forces, caused by high gravitational force per particle, apparently overwhelm any diffuse double layer forces that are present and the particles either engage in direct physical contact, or may be separated only by a molecular layer or two of adsorbed fluid.

The compressibility and shearing strength of a soil composed of such particles

Note.—Discussion open until April 1, 1975. To extend the closing date one month, a written request must be filed with the Editor of Technical Publications, ASCE. This paper is part of the copyrighted Journal of the Geotechnical Engineering Division, Proceedings of the American Society of Civil Engineers, Vol. 100, No. GT11, November, 1974. Manuscript was submitted for review for possible publication on August 7, 1973.

[1] Prof. of Civ. Engrg., Univ. of Texas, Austin, Tex.

is controlled by the rigidity and strength of individual particles, their packing, and the coefficient of friction between the particles. For loose packing and low stresses, the compressibility results from particles sliding or rolling over one another, and the tendency to compress during shear leads to low strengths. As the density of packing increases, the compressibility increasingly results from elastic compression of the particles and is much higher for platey particles (7) than for spherical particles (9). The strength increases because of increased interference between grains. At high stresses the particles begin to crush at their contacts and the compressibility (9,24) and strength (30) are influenced by the physical strength of the particles. Such interactions are typically characterized as physical even though the coefficient of friction between particles depends on interaction with the ambient fluid (10).

Chemical Interaction.—From a soil mechanics point of view it is convenient to use the term, chemical, to refer to adsorption of polar fluids as well as diffuse double-layer interactions. Since chemical adsorption of fluids influences the coefficient of friction between particles, it is not possible to separate the physical and chemical effects completely.

Chemical forces of interaction have been reviewed in reasonable detail in the engineering literature by Ladd (13), Lambe (14,15), Bolt (5), and Rosenqvist (25), and covered in considerable detail in the nonengineering literature (3,4,6,26, 27,28,29,31). For our qualitative purposes it suffices to say that "long-range" chemical interactions (distances over approx 20 Å) consist mainly of diffuse double-layer repulsions, van der Waals attractions, and ionic attraction or repulsion, whereas short-range forces include also repulsions from adsorbed fluid layers, and, of course, high energy repulsion caused by electron cloud interactions associated with actual contact.

Soil Structure.—Recent studies of soil fabric (21,22,32) indicate that the comparatively simple models in use during the last decade do not describe fabric adequately. The wide range in particle size should be taken into account (8) and the fabric that was often viewed in terms of individual particles (15) and then domains (1) must now include even larger units called peds (22). Further, in natural clays the comparatively simple interaction forces of recent years must be modified to include effects of organic matter (21), amorphous coatings (32), and short-range diagenetic bonding (2,17). Such complications may be required to explain such phenomena as sensitivity but they do not seem required to develop qualitative explanations fo the peak shearing strengths measured herein and will not be used.

Most of the samples used in this investigation were prepared by sedimentation and their initial fabrics should have been controlled by particle interaction in suspension. It is generally observed that if the electrolyte concentration in a suspension is low, and particularly if the adsorbed cations are monovalent, the clay tends to disperse and when it settles out the sediment is dense and anisotropic. The simplest model used to explain such behavior postulates the development of long-range, diffuse, double-layer repulsions that cause particles to repel each other and sediment individually with their flat surfaces horizontal. Such a model seems to explain the engineering properties to be reported herein but appears at variance with the loose-ped structures reported by Pusch (21).

When the pore-water electrolyte concentration is high, the soil flocculates quickly into aggregates that are often visible to the naked eye and resulting sediments are loose. The simplest model postulates that diffuse double-layer forces are compressed in this case, allowing the particles to approach closely before significant repulsive forces develop. Pusch (21) finds that such soils have tight flocs but large voids.

Flocculation is also influenced by pH. Low pH leads to rapid flocculation and loose sediments whereas high pH leads to dispersion and dense sediments. The simplest model assumes positive edge-to-face flocculation at low pH and dispersion at high pH brought about by repulsion between negative faces and negative edges.

Although chemical effects influence the initial soil structure of clays formed by sedimentation, subsequent interaction of the particles may be controlled by physical effects, chemical effects, or both.

Application of Physicochemical Model to Mohr-Coulomb Diagrams

Soils with more or less equidimensional particles and wide distribution of particle sizes yield the highest values of the angle of internal friction, $\bar{\phi}$, apparently because of a maximum amount of particle interaction during shear. Thus, well-graded dense sands yield higher angles of internal friction than loose poorly graded sands. From such grounds, and based purely on the mechanical model, one would expect that kaolinite, with its more equidimensional grains would have a higher angle of internal friction than illite and considerably higher than for montmorillonite.

If diffuse double layers develop from the surfaces of the clay particles, the interaction of the layers from opposing surfaces should develop a repulsion that would resist part of the normal effective stress without producing any shearing resistance. Thus, the effective angle of internal friction must decrease as double-layer repulsion increases, i.e., for decreases in pore-water electrolyte concentration, decreases in valency of adsorbed cations, and for more nearly parallel particles.

General Characteristics of Clays Studied

The three groups of clay minerals that are most commonly encountered in geotechnical engineering are kaolinite, illite, and montmorillonite. Experiments were performed using one soil of each mineralogy. The sources of the minerals used in the strength investigation and their index properties are presented in Table 1. The ranges in properties result from variations in the chemical environment used in the experiments. The kaolinite and montmorillonite were not fractionated; the illite was fractionated and only the finer-than-2-μ equivalent-spherical size was used. All properties shown in Table 1 were measured except the diameter-to-thickness ratios that were based on measurements reported in the literature.

Experimental Procedures

The experimental procedures have been detailed in previous publications (16,18,19) and need only be reviewed briefly.

The illite was dried and pulverized, with large impurities removed by hand.

TABLE 1.—

Mineral (1)	Supplier, source, and trade name (2)	w_L, as a percentage (3)	w_P, as a percentage (4)
Kaolinite	Minerals and Chemicals Phillip Corp., McIntyre, Georgia, Klondyke Clay	40-50	27-31
Illite	Fithian, Illinois	65-90	35-40
Montmorillonite	American Colloids Co., Wyoming, Volclay	190-1,160	31-47

It was then neutralized with NaOH and alternately washed with NaCl and distilled water, and then fractionated by repeated decantation.

Homoionization of all clays was accomplished by repeated washing with appropriate chlorides and the pH was adjusted with the clay in suspension. Salt concentrations were also adjusted in suspension. The kaolinite reacted rapidly to changes in chemistry and came to apparent equilibrium within a few hours. Complete sample preparation and shear testing per sample was accomplished in as little as 3 days. Illite samples reacted more slowly to changed chemistry and preparation of some of the montmorillonite samples, particularly the sodium montmorillonite with a low pore-water electrolyte concentration, required more than 1 yr, but suspensions all appeared to be stable when sedimentation began.

Most specimens were prepared by sedimentation of suspensions or pouring of slurries of clay into 1.5 in. (38 mm) diam plastic tubes and then one-dimensionally consolidating the soil under pressures ranging from 5 psi-20 psi (35 kN/m^2-140 kN/m^2) to lengths from 2.8 in.-3.5 in. (70 mm-90 mm). Samples were extruded from the one-dimensional tubes, consolidated isotropically in the triaxial cells, and then subjected to compression to failure at constant rates of deformation that were low enough to ensure adequate equilibrium of pore pressures.

The triaxial tests were performed over a period of 13 yr, with the illite tests performed first and montmorillonite tests last. Experimental procedures in both chemical preparation and shear testing improved markedly during this period.

Experimental Data and Interpretation

During this study 204 triaxial compression tests were performed using kaolinite, 181 using illite, and 63 with montmorillonite. The chemistry of the specimens used for this publication are summarized in Table 2. Tests not reported in Table 2 were parts of special studies, e.g., effect of deformation rates on results of $\bar{R}$ and S tests, tests on the finer-than-5-μ fraction of illite, and measurements of sensitivity, or tests deleted because of uncertain chemical changes (high and low pH tests on illite).

No measurements of residual values of $\bar{\phi}$ were attempted. Kenney (11) has reported on residual envelopes of clays.

Kaolinite.—Failure envelopes for calcium and sodium kaolinite, at pH = 7

Properties of Clays

G_s (5)	S, square meters per gram (6)	CEC, in milliequivalents per 100 g (7)	Finer than 2μ, as a percentage (8)	Estimated diameter height ratio (9)
2.65	14	1.5-4.5	47	2-5
2.85	90-100	25	100	10-50
2.65-2.80	500-700	100	97	150-500

TABLE 2.—Summary of Triaxial Compression Tests

Mineral (1)	Cation (2)	pH (3)	Electrolyte concentration, N (4)	Test type (5)	Number of tests (6)
Kaolinite	Ca^{++}	7	0.01	$\bar{R}$	16
				S	19
	Na^{+}	5	0.001	$\bar{R}$	16
				S	17
		7	0.01	$\bar{R}$	101
		9	0.001	$\bar{R}$	14
				S	21
Illite	Ca^{++}	7	1	$\bar{R}$	3
			0.1	$\bar{R}$	3
			0.01	$\bar{R}$	4
			0.001	$\bar{R}$	3
			0.0003	$\bar{R}$	3
			0.01	$\bar{R}$	33
	Na^{+}	7	0.1	$\bar{R}$	29
			0.01	S	12
			0.007	$\bar{R}$	17
Montmorillonite	Ca^{++}	7	0.001	$\bar{R}$	4
			0.01	$\bar{R}$	13
			0.1	$\bar{R}$	4
			0.01	S	4
	Na^{+}	7	0.001	$\bar{R}$	3
			0.01	$\bar{R}$	4
			0.1	$\bar{R}$	5

and $C = 0.01$ N, are compared in Fig. 1. The negligible effect of cation valency and the relatively high angle of internal friction ($\bar{\phi}_R \sim 25°$) suggest that diffuse, double-layer repulsive forces have had little influence on the failure envelope and that physical mechanisms dominate. Remarkably, even the soil structure seems to have had a negligible effect on the envelope because overconsolidated specimens yielded the same failure envelope as did the normally consolidated specimens. Further, the envelope was identical for drained and undrained tests, as shown in Fig. 2, again with data plotted for both normally consolidated and overconsolidated specimens.

The lack of any apparent double-layer effects could have resulted from a predominant edge-to-face or edge-to-edge particle arrangement that would minimize double-layer repulsions. To investigate the effect of soil structure further,

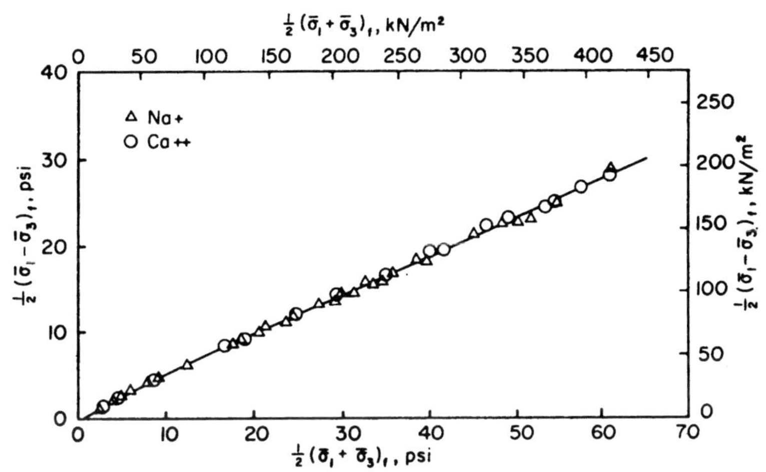

FIG. 1.—Failure Envelope for Sedimented Specimens of Homoionic Kaolinite—All Tests Are of $\bar{R}$ Type with Failure Defined at Point of Stress-Path Tangency (Both Normally Consolidated and Overconsolidated Specimens Are Included; pH = 7 and C = 0.01 N)

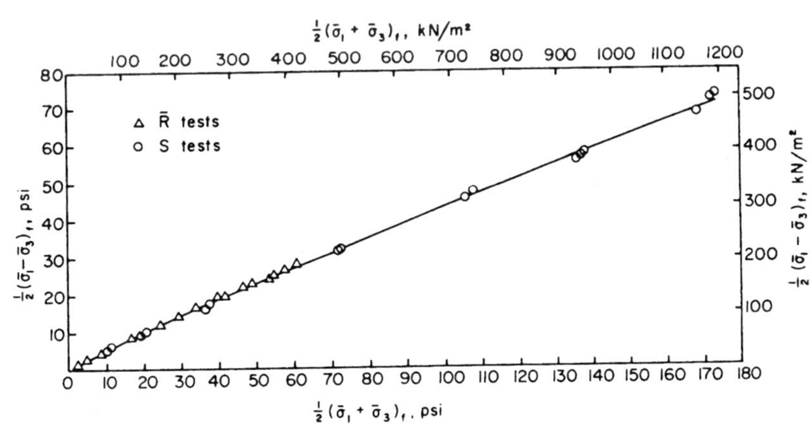

FIG. 2.—Failure Envelope for Sedimented Specimens of Calcium Kaolinite at pH = 7 and C = 0.01 N, in Drained (S) and Undrained ($\bar{R}$) Shear (In Undrained Tests Failure Was Defined as Point of Stress-Path Tangency—No Energy Correction Was Applied in Drained Tests But Correction Would Have Been Negligible in Any Case)

a series of tests was performed using sodium kaolinite at pH = 9 and C = 0.001 N, a chemistry that should have induced a dispersed soil structure with maximum opportunity for diffuse double-layer forces to influence the failure envelope, and another series at pH = 5 and C = 0.001 N that should represent an extreme case of edge-to-face flocculation. The kaolinites at values of pH = 5 and 9 must have had much different chemical interactions because at some water contents the kaolinite at pH = 9 was a slurry that could be poured easily whereas the kaolinite at pH = 5 was a paste that would retain its shape when molded. The difference in soil structure showed clearly in the consolidation

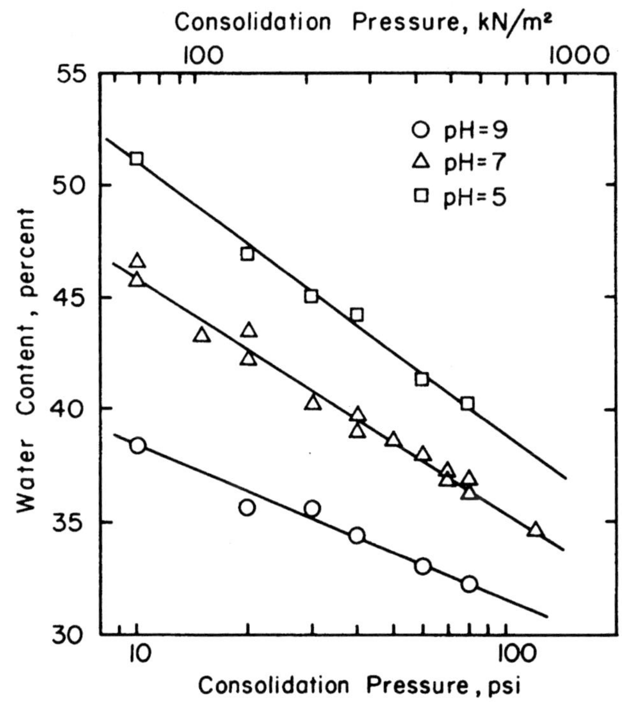

FIG. 3.—Virgin Consolidation Curves for Sedimented Specimens of Sodium Kaolinite at Values of pH = 5, 7, and 9

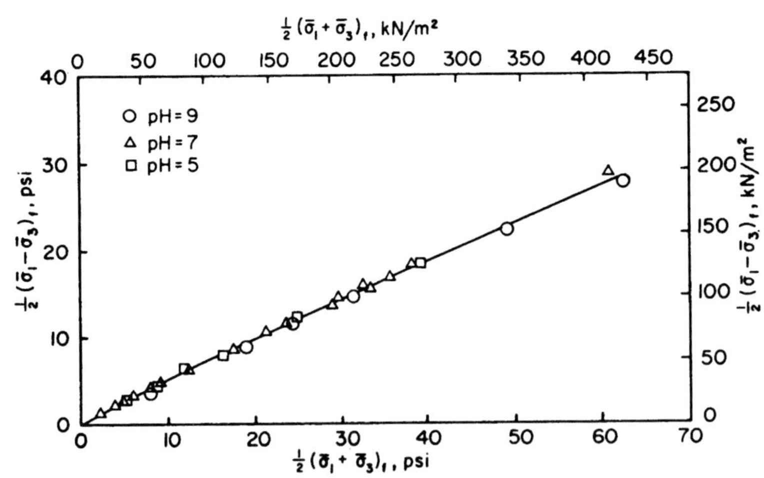

FIG. 4—Failure Envelope for Normally Consolidated, Sedimented Specimens of Sodium Kaolinite at Values of pH = 5, 7, and 9 (All Tests Are $\bar{R}$ Compression Tests with Failure Defined at Point of Stress-Path Tangency)

curves (Fig. 3) but the effective-stress failure envelopes for the specimens at values of pH = 5, 7, and 9 were identical (Fig. 4). Thus, there is no evidence of a significant chemical effect once the kaolinite has been sedimented and consolidation begins.

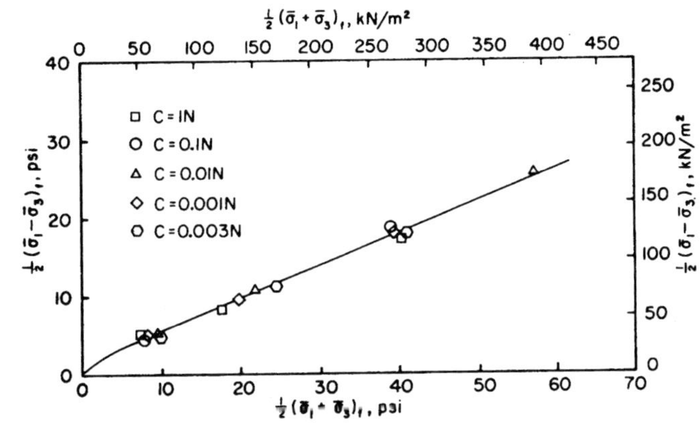

FIG. 5.—Failure Envelope for $\bar{R}$ Triaxial Compression Tests on Calcium Illite at pH = 7 and Range in Pore-Water Electrolyte Concentration (Failure Was Defined As Point of Stress Path Tangency)

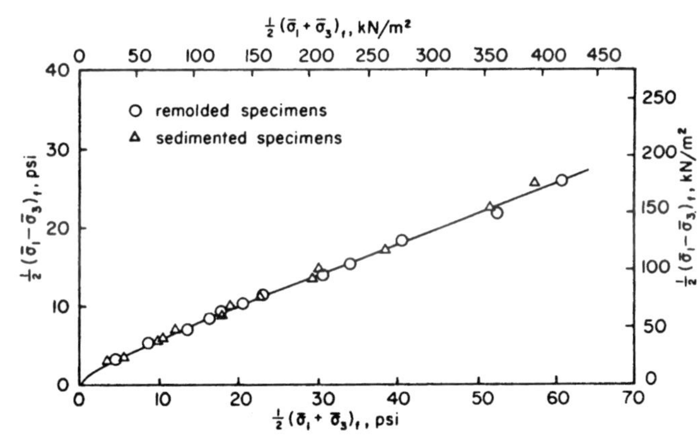

FIG. 6.—Failure Envelope for Remolded and Sedimented Specimens of 0.01 *N* Calcium Illite (Failure Was Defined as Point of Stress-Path Tangency)

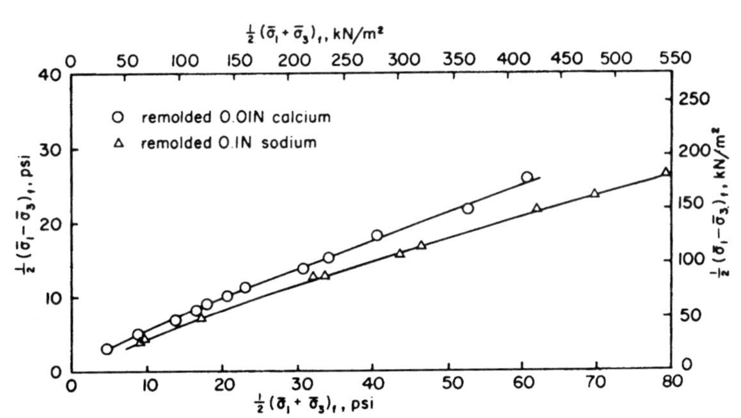

Fig. 7.—Failure Envelopes for Remolded Specimens of Calcium and Sodium Illite (Failure Was Defined at Point of Stress-Path Tangency)

The conclusion seems to be that the shearing strength of kaolinite is controlled by physical effects. Chemical variables influence the original structure of specimens prepared by sedimentation but not the effective-stress shearing strength. The lack of double-layer effects seems in keeping with the low degree of isomorphous substitution in the kaolinite lattice and the comparatively large size of the particles.

Illite.—Failure envelopes for sedimented specimens of calcium illite at pH = 7 and a wide range in pore-water electrolyte concentration are shown in Fig. 5. The immeasurable effect of electrolyte concentration on the effective-stress failure envelope suggests that no significant diffuse double layers were developed and that particle interaction is mainly physical. The same conclusion applied to the one-dimensional consolidation curves (19,20). Tests were performed on both sedimented and remolded (at liquidity index = 0.5) specimens of 0.01 *N* calcium illite. The effective-stress failure envelopes (Fig. 6) were identical, suggesting either that the arrangement of the particles had no measurable effect on the failure envelopes, expressed in effective stresses, or that the particle arrangements at failure were the same. For a soil composed of such flakey particles, the relatively high friction angle ($\bar{\phi}_R \sim 23°$) seems to imply little chemical interaction between particles.

The strengths of sodium illite specimens were lower than those of calcium illite at the same effective normal stresses, e.g., the failure envelopes for remolded specimens of 0.1 *N* sodium illite and 0.01 *N* calcium illite are compared in Fig. 7. The lower strengths of the sodium illite are ascribed to the presence of chemical repulsive forces, presumably of the diffuse double-layer type.

If the lower strengths of the sodium illite resulted from diffuse double-layer repulsions, then reductions in the pore-water electrolyte concentration should reduce the strength in an effective stress diagram even further. A series of tests was then performed on specimens of sodium illite at a pore-water electrolyte concentration of 0.007 *N* (attempts to reach lower concentrations were unsuccessful, apparently because Fe^{+++} was released from traces of pyrite and entered into exchange positions releasing sodium until the pore-water electrolyte concentration increased to about 0.007 *N*). Serious difficulties arose with the experimental procedures then in use, particularly with the manual measurement of pore pressures and with fixed-cell bushings, because the permeability of the specimens dropped markedly and consolidation and shearing times became very long. Regardless of the problems, the effective-stress failure envelope was defined with minimal scatter and the strengths were about 15% lower for the 0.007 *N* samples than for those at 0.1 *N*.

Taken collectively, the tests with illite suggested that no diffuse double layers developed for calcium illite, thus leading to particle interactions that were mainly physical and values of $\bar{\phi}_R$ of the order of 23°–26°, depending on the failure criterion being used, but not depending on soil structure or pore-water electrolyte concentration. Sodium illite had lower strengths that depended on pore-water electrolyte concentration and (assumed) particle arrangements, in qualitative conformance with predictions based on the assumption of development of diffuse double layer forces. The maximum value of $\bar{\phi}_R$ for sodium illite was about 21° for specimens at a pore-water electrolyte concentration of 0.1 *N*, and the minimum value was about 16° for the specimens with a pore-water electrolyte concentration of 0.007 *N*. The low value of 16° is in agreement with the value of $\bar{\phi}_S$, from drained direct shear tests, reported by Quigley (23). Note that Quigley's $\bar{\phi}$ for potassium illite was 23° (see Ref. 23), a value comparable to the $\bar{\phi}$ of calcium illite.

Montmorillonite.—A series of $\bar{R}$ triaxial compression tests was performed using specimens of calcium montmorillonite consolidated from dilute slurries at pH 7 and a range in pore-water electrolyte concentration from 0.001 *N* to 0.1 *N*. The failure envelopes (Fig. 8) were independent of the electrolyte concentration, indicating again that diffuse double-layer forces do not develop in the calcium clay with sufficient force to influence the shearing strength measurably.

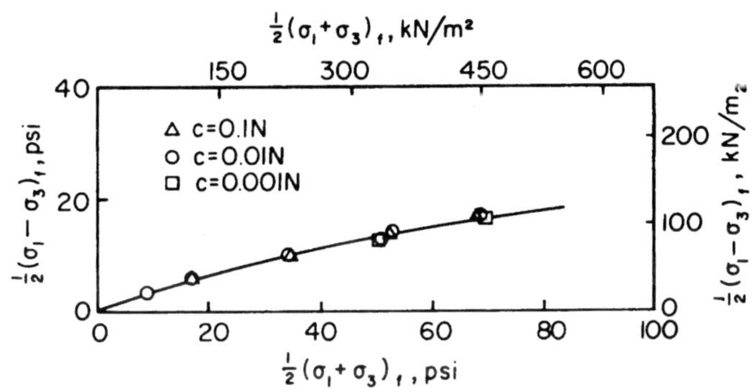

FIG. 8.—Failure Envelope for Specimens of Calcium Montmorillonite at Pore-Water Electrolyte Concentrations of 0.1 *N*, 0.01 *N*, and 0.001 *N*

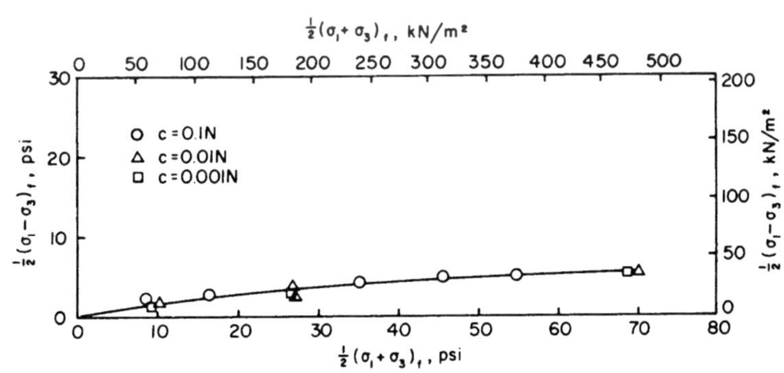

FIG. 9.—Failure Envelope for Sodium Montmorillonite at pH = 7 and Range in Pore-Water Electrolyte Concentration from 0.001 *N* to 0.1 *N*

In the case of sodium montmorillonite, the coefficients of permeability were so low that consolidation and testing times per specimen sometimes extended over several years. It was necessary to restrict the experimental program to $\bar{R}$ compression tests on normally consolidated specimens. Specimens were prepared with electrolyte concentrations ranging from 0.1 *N* down to 0.001 *N*. Although pore-water electrolyte concentrations influenced both the one-dimensional compression (19) and permeability (16), there was no apparent effect of electrolyte concentration on the failure envelope for the stresses used in this study (Fig. 9). The main reason for this lack of effect is believed to be that the double-layer forces influence the normal stress not the shearing stress; for an envelope as flat as the one for sodium montmorillonite, a significant reduction in the effective normal stress would simply move the point to the left on essentially the same envelope. Further, for the stresses used in this study, the double layers tend to be compressed so the large effects found in suspensions are depressed.

The shearing envelopes for montmorillonite suggest that double-layer effects are insignificant for calcium montmorillonite and that the strength of sodium

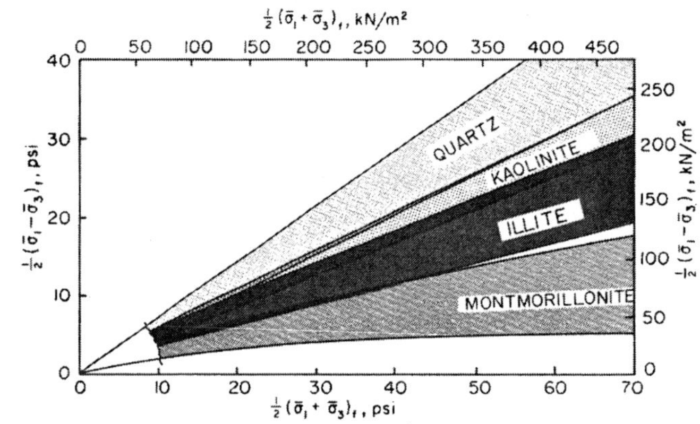

FIG. 10.—Ranges in Failure Envelopes for Soils Composed of Pure Clay Minerals or Quartz

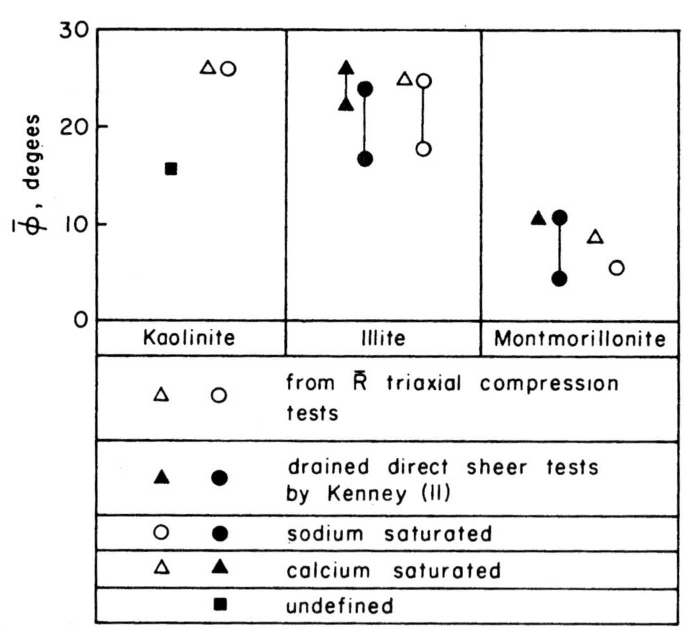

FIG. 11.—Comparison of Friction Angles Measured Using $\bar{R}$ Triaxial Compression Tests with Residual Angles Measured Using *S* Direct Shear Tests on Thin Samples

montmorillonite is so low that the effects of even substantial changes in diffuse double-layer forces would be masked. It remains to be explained why the sodium montmorillonite is so much weaker than the calcium montmorillonite. It is believed that the sodium montmorillonite breaks down to flakes only one unit cell thick (10 Å) and these flakes have little physical strength. The particles are believed to be so flexible that essentially all particle interaction must be of the face to face type, thus leading to low intergranular stresses and a strong probability that the particles are always separated by at least one layer of adsorbed water if not a full diffuse double layer. Calcium montmorillonite is believed to aggregate into particles a number of unit cells thick, which have significant strengths. Thus the higher strength of the calcium montmorillonite is believed to be caused by the higher particle strengths, i.e., by physical effects.

Comparison of Strengths of Kaolinite, Illite, and Montmorillonite

The separate examinations of the failure envelopes for each of the minerals has made it clear that the dominant interaction of particles for the effective stresses used in this study is physical, as the term was defined earlier in this paper. The strengths depended mainly on the strengths of individual particles as determined, qualitatively, by the particle sizes and their diameter to thickness ratios. Since the particles of kaolinite, illite, and montmorillonite are materially different in size and shape (Table 1) their failure envelopes should differ. The failure envelopes are compared in Fig. 10 where the hatched zones indicate the range in positions of the failure envelopes examined earlier. A typical range in location of failure envelopes for quartz sand is also shown for comparison. The relative positions of the envelopes are believed to support the contention that the primary interaction of the particles is physical, provided that adsorbed pore water is included in the physical model as influencing the interparticle friction.

Comparison with Residual $\bar{\phi}$ Values of Kenney

A comprehensive series of drained direct shear tests on paste samples about 1 mm thick was reported by Kenney (11). Kenney varied mineralogy, adsorbed cations, and pore-water electrolyte concentration. His values of $\bar{\phi}$ (defined at $\bar{\sigma}_n$ = 1 kg/cm^2) are compared with the values of $\bar{\phi}$ reported in this paper [defined as secant values at $(\bar{\sigma}_1 + \bar{\sigma}_3)_f = 50$ psi] in Fig. 11. The two sets of observations compare remarkably well for illite and montmorillonite considering that: (1) Kenney's tests are for residual conditions; (2) undrained ($\bar{R}$) triaxial data are compared with drained (S) direct shear data; and (3) different soils were used. Kenney's value of $\bar{\phi}$ for kaolinite is remarkably low considering the blocky nature of kaolinite (Table 1). The liquid limit, plastic limit, and cation exchange capacity are all rather high for Kenney's kaolinite, suggesting that it may be an unusual material.

None of the triaxial tests could be carried to high enough strains to approximate residual conditions, and no tests were performed using precut shear planes. However, in the few tests carried beyond failure there was little evidence of any tendency toward a major reduction in the developed value of $\bar{\phi}$ with further strain. The small differences between the $\bar{\phi}$ values of the samples prepared by sedimentation or remolding in the laboratory, with Kenney's residual $\bar{\phi}$ may indicate that much of the reduction from $\bar{\phi}_f$ to $\bar{\phi}_r$ in the case of natural soils, results from rupture of natural cement (12) more than a simple change in structure.

Summary and Conclusions

The engineering properties of clays may be ascribed to either physical or chemical interactions between particles. For the three clay minerals studied, the shearing strength appeared to be controlled mainly by physical effects. No diffuse double layer effects were apparent for any of the clays in the calcium form. For clays in the sodium form, no diffuse double-layer effects were found for kaolinite, possibly because of the low level of isomorphous substitution in the kaolinite lattice. A distinct diffuse double-layer effect was noted for sodium illite. For sodium montmorillonite, the strengths were so low that reductions in normal effective stresses of 50% or more would probably go undetected.

Taken collectively, the data suggest that location of the effective-stress failure envelope is controlled mainly by physical effects controlled by the size and shape of the individual particles. Particles that are large and more or less equidimensional lead to high strengths regardless of chemical effects whereas particles that are very thin with high diameter-to-thickness ratios, as in sodium montmorillonite, have very low strengths.

The values of $\bar{\phi}_R$ were similar to the residual values of $\bar{\phi}$ reported by Kenney (11).

Acknowledgment

The tests reported here were performed as parts of the work on a series of projects sponsored by the National Science Foundation, including the following: 5073, 13362, GP325, and GK531. Numerous graduate students worked on the various phases of these projects. Special acknowledgments are given to J. Hardin, M. L. Kiefer, and G. Mesri, who were responsible for most of the work on illite, kaolinite, and montmorillonite, respectively.

Appendix I.—References

1. Aylmore, L. A. G., and Quirk, J. P., "The Structural Status of Clay Systems," *Clays and Clay Minerals, Proceedings of Ninth National Conference,* Pergamon Press, Inc., New York, N.Y., 1960, pp. 104-130.
2. Bjerrum, L., "The Engineering Geology of Norwegian Normally Consolidated Clays as Related to Settlements of Buildings." *Geotechnique,* London, England, Vol. 17, 1967, pp. 81-118.
3. Blackmore, A. V., and Miller, R. D., "Tactoid Size and Osmotic Swelling in Calcium Montmorillonite," *Proceedings,* Soil Science Society of America, Vol. 25, 1961, pp. 169-173.
4. Bolt, G. H., "Analysis of the Validity of the Gouy-Chapman Theory of the Electric Double Layer," *Journal of Colloid Science,* Vol. 10, 1955, pp. 206-218.
5. Bolt, G. H., "Physico-Chemical Analysis of the Compressibility of Pure Clays," *Geotechnique,* London, England, Vol. 6, 1956, pp. 86-93.
6. Collis-George, N., and Bozeman, J. M., "A Double Layer Theory for Mixed Ion Systems as Applied to the Moisture Content of Clays under Constraint," *Australian Journal of Soil Research,* Australia, Vol. 8, 1970, pp. 249-258.
7. Gilboy, G., "The Compressibility of Sand-Mica Mixtures," *Proceedings,* ASCE, Vol. 54, No. 2, 1928, pp. 555-568.
8. Hansbo, S., "Influence of Mobile Particles in Soft Clay on Permeability," *Proceedings,* International Symposium on Soil Structure, Gothenburg, Sweden, Swedish Geotechnical Society, Stockholm, Sweden, 1973, pp. 132-135.
9. Hendron, A. J., "The Behavior of Sand in One-Dimensional Compression," thesis presented to the University of Illinois, at Urbana, Ill., in 1963, in partial fulfillment of the requirements for the degree of Doctor of Philosophy.
10. Horn, H. M., and Deere, D. U., "Frictional Characteristics of Minerals," *Geotechnique,* London, England, Vol. 12, No. 4, 1962, pp. 319-335.

11. Kenney, T. C., "The Influence of Mineral Composition on the Residual Strength of Natural Soils," *Proceedings*, Geotechnical Conference, Vol. 1, Norwegian Geotechnical Institute, Oslo, Norway, 1967, pp. 123-219.
12. Kenney, T. C., Moum, J., and Berre, T., "An Experimental Study of Bonds in a Natural Clay," *Proceedings*, Geotechnical Conference, Vol. 1, Norwegian Geotechnical Institute, Oslo, Norway, 1967, pp. 65-69.
13. Ladd, C. C., "Mechanisms of Swelling of Compacted Clay," *Highway Research Bulletin No. 245*, Highway Research Board, Washington, D.C., 1959, pp. 10-26.
14. Lambe, T. W., "The Structure of Inorganic Soil," *Proceedings*, ASCE, Vol. 79 Separate No. 315, Oct., 1953, pp. 1-49.
15. Lambe, T. W., "The Structure of Compacted Clay," *Journal of the Soil Mechanics and Foundations Division*, ASCE, Vol. 84, No. SM2, Proc. Paper 1654, May, 1958, pp. 1-34.
16. Mesri, G., and Olson, R. E., "Mechanisms Controlling the Permeability of Clays," *Clays and Clay Minerals*, Vol. 19, No. 3, 1970, pp. 151-158.
17. Nelson, J. D., "Influence of Clay Fabric on Bonds and Dilation," *Proceedings*, International Symposium on Soil Structure, Gothenburg, Sweden, Swedish Geotechnical Society, Stockholm, Sweden, 1973, pp. 153-159.
18. Olson, R. E., "The Shear Strength Properties of Calcium Illite," *Geotechnique*, London, England, Vol. 12, 1963, pp. 23-43.
19. Olson, R. E., and Mesri, G., "Mechanisms Controlling Compressibility of Clays," *Journal of the Soul Mechanics and Foundations Division*, ASCE, Vol. 96, No. SM6, Proc. Paper 7649, Nov., 1970, pp. 1863-1878.
20. Olson, R. E., and Mitronovas, F., "Shear Strength and Consolidation Characteristics of Calcium and Magnesium Illite," *Clays and Clay Minerals, Proceedings*, Ninth National Conference, Pergamon Press Inc., New York, N.Y., 1962, pp. 185-209.
21. Pusch, R., "Influence of Salinity and Organic Matter on the Formation of Clay Microstructure," *Proceedings*, International Symposium on Soil Structure, Gothenburg, Sweden, Swedish Geotechnical Society, Stockholm, Sweden, 1973, pp. 161-173.
22. Pusch, R., "Physico-Chemical Processes which Affect Soil Structure and Vice Versa," *General Report on Theme 3*, Appendix, *Proceedings* of the International Symposium on Soil Structure, Gothenburg, Sweden, Swedish Geotechnical Society, Stockholm, Sweden, pp. 27-35.
23. Quigley, R. M., "The Engineering Properties of Illite Related to Fabric and Pore Water Composition," *Proceedings*, Sixteenth Canadian Soil Mechanics Conference, National Research Council of Canada, Ottawa, Canada, 1962, pp. 21-42.
24. Roberts, J. E., and de Souza, J. M., "The Compressibility of Sands," *Proceedings*, American Society for Testing and Materials, Vol. 58, 1958, pp. 1269-1277.
25. Rosenqvist, I. T., "The Influence of Physico-Chemical Factors Upon the Mechanical Properties of Clays," *Clays and Clay Minerals, Proceedings*, Ninth National Conference, Pergamon Press Inc., New York, N.Y., 1960, pp. 12-27.
26. Stocker, P. T., "The Structure and Swelling of Divalent Montmorillonites—A Review," *Australian Road Research, Journal of the Australian Road Research Board*, Australia, Vol. 3, No. 9, 1969, pp. 49-71.
27. van Olphen, H., "Forces Between Suspended Bentonite Particles," *Clay and Clay Minerals, Proceedings, National Academy of Science-National Research Council Publication 456*, Fourth National Conference, 1956, pp. 204-224.
28. van Olphen, H., "Compaction of Clay Sediments in the Range of Molecular Particle Distances," *Clays and Clay Minerals, Proceedings*, Eleventh National Conference, Pergamon Press Inc., New York, N.Y., 1963, pp. 178-187.
29. Verwey, E. J. W., and Overbeek, J. T., G., *Theory of the Stability of Lyophobic Colloids*, Elsevier Publishing Co., New York, N.Y., 1948, 205 pp.
30. Vesic, A. S., and Clough, G. W., "Behavior of Granular Materials under High Pressures," *Journal of the Soil Mechanics and Foundations Division*, ASCE, Vol. 94, No. SM3, Proc. Paper 5954, May, 1968, pp. 661-688.
31. Warkentin, B. P., and Schofield, R. K., "Swelling Pressure of Sodium Montomorillonite in NaCl Solutions," *Journal of Soil Science*, Vol. 13, 1962, pp. 98-105.
32. Yong, R. N., and Sheeran, D. E., "Fabric Unit Interaction and Soil Behavior," *Proceedings*, International Symposium on Soil Structure, Gothenburg, Sweden, Swedish Geotechnical Society, Stockholm, Sweden, 1973, pp. 176-183.

Appendix II.—Notation

The following symbols are used in this paper:

C = concentration of electrolyte in pore water, in gram equivalent weights per liter;
CEC = cation exchange capacity, in milliequivalents per 100 g;
G_s = specific gravity of solids;
$\bar{R}$ = consolidated-undrained test;
S = consolidated-drained test, surface area, in square meters per gram;
w_L = liquid limit, as a percentage;
w_P = plastic limit, as a percentage;
$\bar{\sigma}_n$ = normal effective stress on failure plane (1 psi = 6.89 kN/m^2);
$\bar{\sigma}_1$ = maximum principal effective stress (1 psi = 6.89 kN/m^2);
$\bar{\sigma}_3$ = minor principal effective stress (1 psi = 6.89 kN/m^2);
$\bar{\phi}$ = effective angle of internal friction, in degrees; and
$\bar{\phi}_R$ = residual angle of internal friction, in degrees.

FINITE DIFFERENCE ANALYSES FOR SAND DRAIN PROBLEMS

By Roy E. Olson,[1] F. ASCE, David E. Daniel,[2] A.M. ASCE, and Thomas K. Liu,[3] F. ASCE

INTRODUCTION

The geotechnical literature contains a number of papers in which sand drain installations are described and field performance is compared with theoretical predictions. Some of the field experiences suggest that the design theories in general use correlate rather well with field measurements (19, 20, 22, 23, 27, 28, 30, 31, 49, 55, 58, 59) whereas others indicate major discrepancies (24, 29, 32, 33, 36, 50, 57, 60). Unfortunately, the theories in general use (5, 6, 9, 35) contain simplifying assumptions not fully satisfied by any of the case histories published so far. In addition, many of the case histories contain important uncertainties regarding actual field conditions and nearly all are deficient in terms of well-defined soil properties. In a few cases where excellent correlations were obtained between field measurements and theoretical predictions, attempts by the authors to re-analyze the results led to considerably poorer correlations than those reported in the original publication. The result of the uncertainties in theory and field case histories is that a designer cannot be certain that his embankment will perform in the manner predicted by his analyses.

Studies of the available case histories indicate clearly that improvements are needed in both the methods of analysis and in the techniques used for measurement of soil properties. In particular, the analysis should account for layered systems, two-dimensional water flow, time-dependent loading with the load dependent on settlement, partially saturated soils, soil properties that are functions of effective stress, nonlinear stress-settlement curves, and secondary effects. The major defect in measurement of soil properties is clearly the lack of tests in which the soil is allowed to drain radially and in which the samples are large enough to include representative stratification (17, 37, 38, 39, 40, 48).

The authors have developed a finite difference solution that accounts for some of the deficiencies in theories currently in use. The method has been applied to one case history and found to be successful in predicting the rates of settlement in the field. The data for the case history will be presented first in order to clarify the variables that must be included in the analyses. Then the analytical technique will be explained and applied to the case history.

---

[1] Professor of Civil Engineering, Univ. of Texas, Austin, Texas.
[2] Staff Engineer, Woodward-Lundgren, Oakland, California.
[3] Partner, Haley and Aldrich, Cambridge, Massachusetts.

FORE RIVER TEST EMBANKMENT

The uncertainties in analysis of sand drain installations faced the engineers of the Maine State Highway Commission during the mid-1960's when they were required to design embankments for Highway I-295 where it crossed the Fore River in Portland, Maine. Settlements up to about 6 feet were predicted with one-dimensional consolidation times ranging from a few months to a century. The decision was made to install sand drains but the field evidence in support of sand drain theories then in use were so contradictory that it seemed unwise to design such a major embankment without field tests. Accordingly, a test embankment was constructed and instrumentation was installed to measure settlements and pore water pressure, and subsequently to measure lateral movement. The conditions at the Fore River test (FRT) embankment were sufficiently simple, and the available information on soil properties and field performance sufficiently complete, so that a rational analysis of the results may be attempted. Various aspects of this case history have been discussed in the literature by Aldrich and Johnson (2). The analytical procedure used by Aldrich and Johnson has been discussed by Ladd, Rixner, and Gifford (23). The field data are included in detail in two reports by Haley and Aldrich (15, 16). The data used in this paper came from the Haley and Aldrich reports supplemented by additional data from their files.

Field Conditions at the Fore River Test Embankment

The site of the test embankment consisted of a mud flat that was barely exposed at low tide and covered by an average of about 4 feet (1.2 m) of water. The compressible soil was a stratum of slightly organic clay, about 25 feet (7.6 m) thick, containing frequent sand pockets and occasional peaty zones. Deeper soils were comparatively incompressible silty sands and gravels. Average properties of the clay were: liquid limit = 65%, natural water content = 65%, plastic limit = 35%, submerged unit weight = 38 pcf (4.6 N/m^3), degree of saturation = 100%, and sensitivity = 6. Data from consolidation tests will be presented subsequently.

Test Embankment

The test embankment was 350 feet (107 m) square at the original ground elevation and 120 feet (37 m) square at the crown elevation (Fig. 1). The first 6 feet (1.8 m) of fill was an uncompacted silty sand. Settlement platforms were installed near the top of this layer on about the sixtieth construction day. Four feet (1.2 m) of clean sand was then placed to serve as a drainage layer for the sand drains, and then compacted granular soils were used to complete the embankment. After placement of the drainage layer construction continued rapidly (Fig. 2) until the 120th construction day, when the total thickness of embankment was 14 feet (4.3 m). At this point a slide developed on the north side, with a displacement of up to 3.5 feet (1.1 m). Cracks developed along the east and south slopes as well. The embankment was cut back to a thickness of 12 feet (3.7 m) and the comparatively steep side slopes then existing were flattened. No further distress was noted.

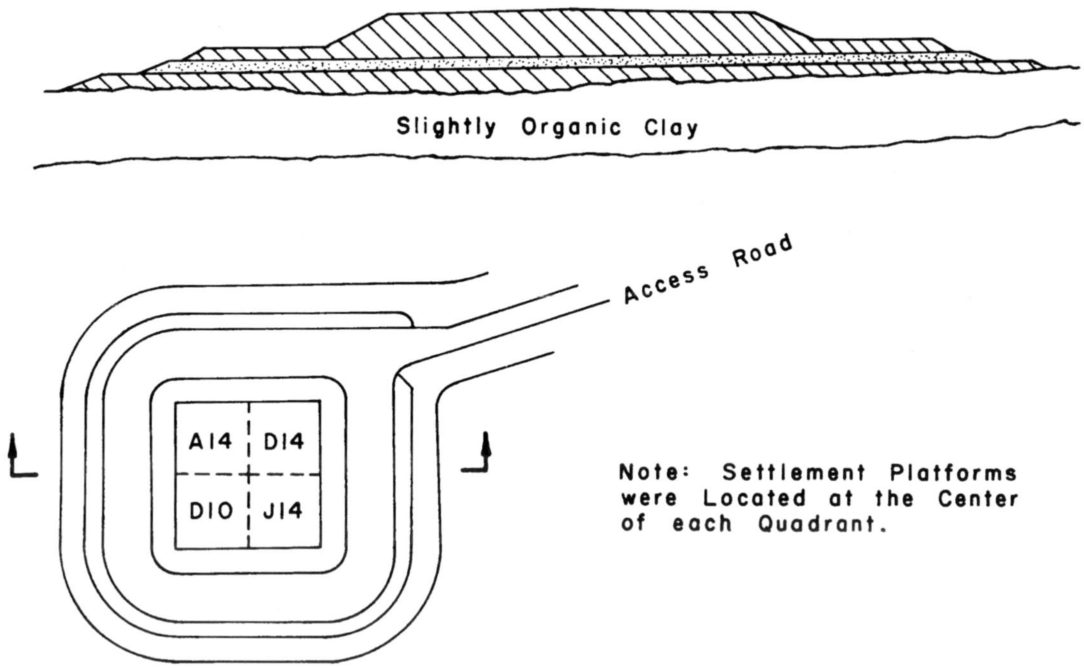

Fig. 1—Plan and Profile View of Fore River Test Embankment

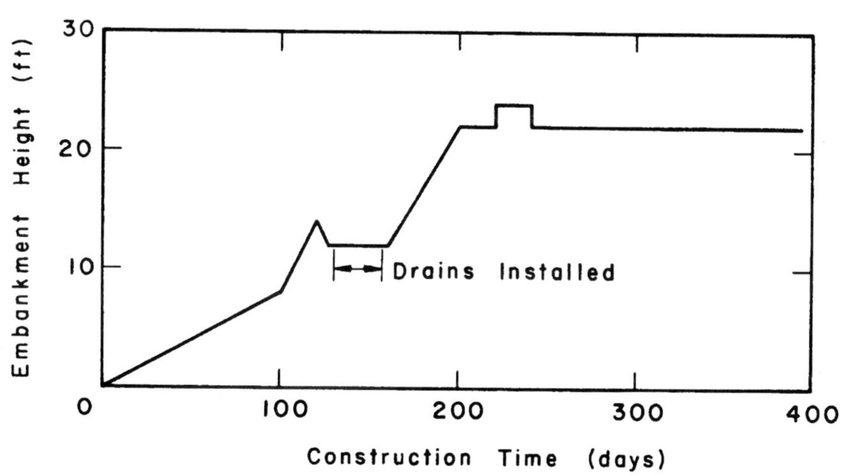

Fig. 2—Load History

The surface of the embankment was divided into four quadrants as shown in Fig. 1 and 18-inch (46-cm) diameter sand drains were installed in triangular patterns and with their bases in the silty sand and gravel layer beneath the clay layer. The method of construction and drain spacing for the quadrants were as follows: augering at 14-foot (4.3-m) centers (A14), non-displacement jetting at 14-foot (4.3-m) centers (J14), and displacement-type driving at 10 and 14-foot (3.1 and 4.3-m) centers (D10 and D14, respectively). The time from the beginning of construction to the day of installation of the drains was approximately 135 days for A14, 130 days for J14, and 155 days for D10 and D14. Placement of fill began again on day 160 and the ultimate fill height of 22 feet (6.7 m) was achieved on about construction day 210.

Field Settlement Observations

Settlement observations were made for a period of about 1000 days at which point construction of the main embankment began and the measuring equipment was destroyed. The observed time-settlement curves are presented in Fig. 3a. The shapes of the curves suggest that primary consolidation was essentially complete but no data are available on rates of secondary settlement. The measured time-settlement curves were extended slightly by curve-fitting techniques to estimate the probable primary settlement.

In order to compare the time-settlement curves, it is convenient to divide all measured settlements by the estimated final primary settlement and report this ratio as the average degree of consolidation, U. The resulting curves of average degree of consolidation versus time are presented in Fig. 3b. The fact that the three quadrants having drains at 14-foot (4.3-m) centers have more-or-less identical settlement curves is approximately in accord with theory, but the fact that the drains driven on 10-foot (3.1-m) centers yield the same time-settlement curve is not in accord with theory and requires analysis.

Laboratory Tests

About 20 one-dimensional consolidation tests were performed on samples taken using 3.5-inch (8.9-cm) diameter fixed-piston samplers. The tests were performed on samples trimmed down to 2.5 inches (6.4-cm) in diameter by 7/8-inch (2.2-cm) thick. For all tests the water flow was vertical and the soil was oriented just as in the field. No experiments were performed with radial flow of water.

Void ratio-effective stress relationships.-The relationships between void ratio and effective stress varied considerably because of natural scatter in water content, variations in field overburden effective stress ($\bar{\sigma}_o$), and variations in the maximum previous consolidation pressure ($\bar{\sigma}_{o\,max}$) brought about by desiccation.

In the test area, the maximum previous consolidation pressure was approximately 700 psf (35 kN/m^2), independent of depth (Fig. 4).

The slopes of the virgin curves of compressive strain versus log of effective stress (strain index = C_ϵ) varied from 0.15 to 0.28 with a

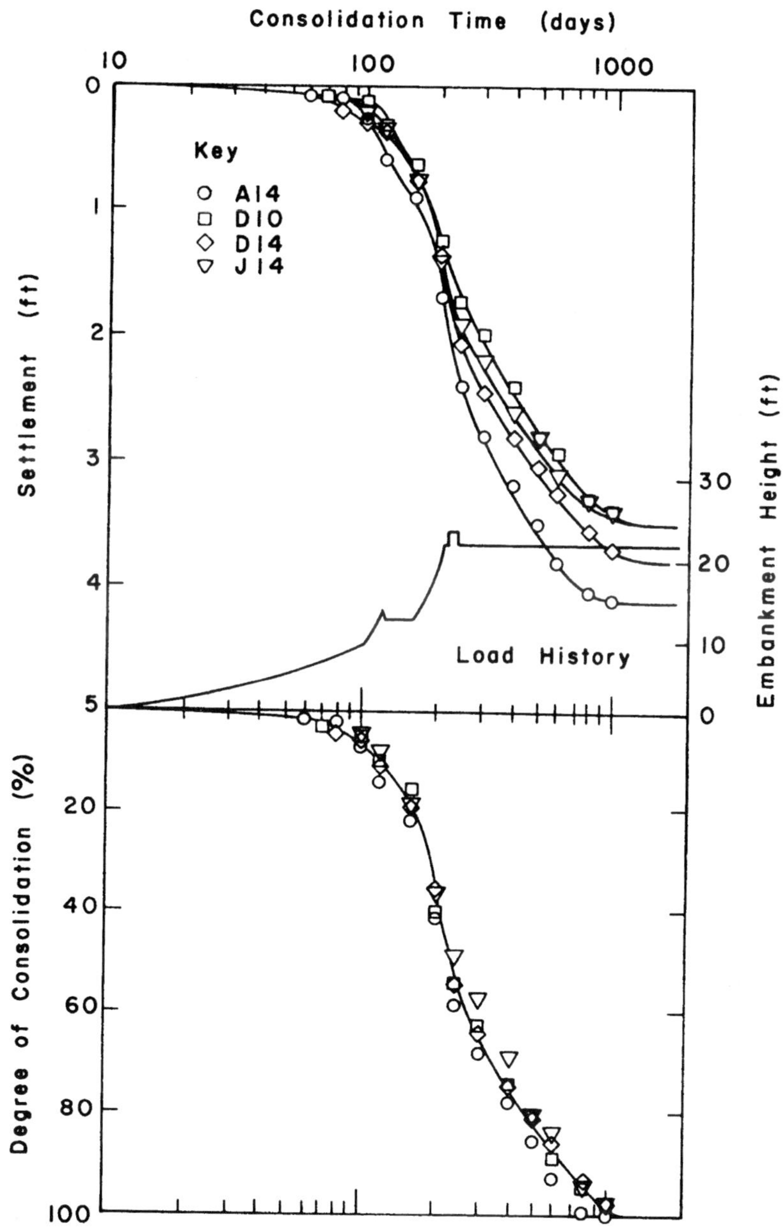

Fig. 3 Settlement Records

mean value of 0.22 and a standard deviation of 0.036. The slope of the recompression curve varied with effective stress. For analytical purposes, the slopes of the virgin and recompression curves were taken as 0.22 and 0.02, respectively.

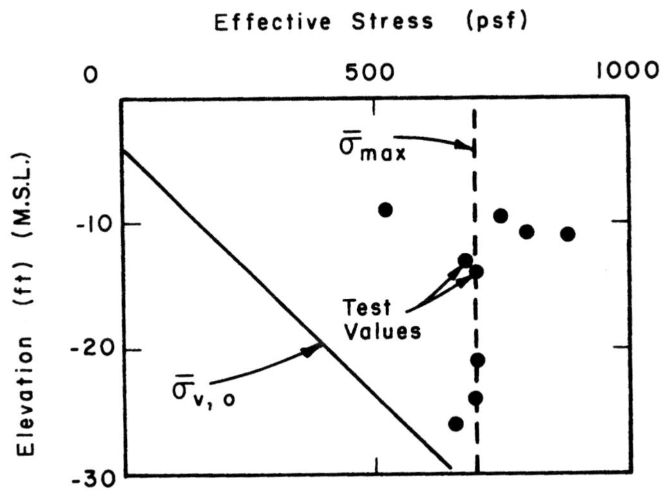

Fig. 4 Relationship Between Maximum Previous Consolidation Pressure and Depth

Coefficients of Consolidation.-Data from consolidation tests were reduced using the logarithmic and square root methods (51). The logarithmic method gave lower coefficients of consolidation and the recalculated theoretical curves compared less well with the laboratory measurements than when the square root method was used. Based on this observation, and the fact that the square root method has been reported to give better coefficients for field use (20), the square root method was used in this investigation. The coefficients of consolidation are shown in Fig. 5. As usual, the coefficients of consolidation for "undisturbed" samples were high for the reloading range in effective stress, minimized soon after the pressures reached the virgin curve, and then slowly increased again. The field coefficient of consolidation for vertical flow would be expected to be about 0.3 ft^2/day (0.03 m^2/day) for initial loading and to decrease to a low value of about 0.05 ft^2/day (0.005 m^2/day) for the highest effective stresses reached during field consolidation (about 3500 psf or 170 kN/m^2).

Analysis of Total Settlement

Method of analysis.-The settlement platforms were installed near the center of each quadrant where the deformation could be expected to be one-dimensional. In spite of the slide near the north edge of the test embankment and cracks forming at other locations, the field time-settlement curves gave no evidence of mass flow from beneath the center of the quadrants and none was assumed in the analyses.

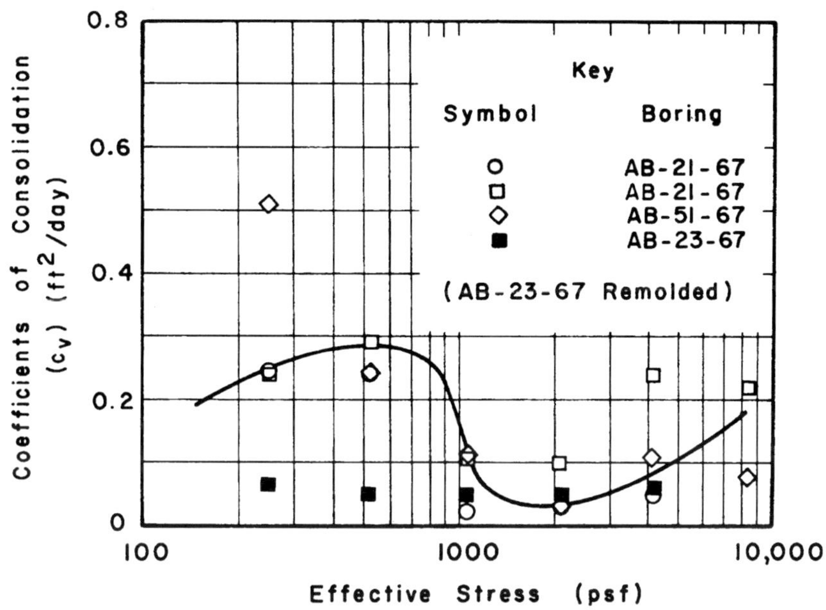

Fig. 5 Coefficients of Consolidation Measured in the Laboratory

The compressible layer was divided into N sublayers, each denoted by a value of j, and a settlement in each, S_j, was calculated using the equation:

$$S_j = \left[C_{\varepsilon r} H \log (\bar{\sigma}_{v,max}/\bar{\sigma}_{v,o}) + C_{\varepsilon} H \log (\bar{\sigma}_{v,f}/\bar{\sigma}_{v,max}) \right]_j \quad . . . . . (1)$$

and the total settlement from:

$$S = \sum_{j=1,2..}^{N} S_j \quad . . . . (2)$$

where $C_{\varepsilon r}$ and C_{ε} are slopes of the reloading and virgin consolidation curves on strain type consolidation curves (0.02 and 0.22, respectively), H is the thickness of the sublayer, and $\bar{\sigma}_{v,o}$, $\bar{\sigma}_{v,max}$, and $\bar{\sigma}_{v,f}$ are the initial, maximum previous, and final vertical effective stresses, respectively. The number of sublayers was made large enough so no change in settlement resulted from further increase in N. The applied stresses were adjusted for submergence of the embankment as settlement occurred. Field measurements of pore pressure indicated the presence of a five-foot artesian head in the lower sand. In calculating settlements it was assumed that the installation of the sand drains destroyed the artesian condition.

Comparison of measured and computed settlements.-The calculated and measured settlements are compared in Table 1. It is not possible to

Table 1

Comparison of Computed and Measured Final Settlements

Method of Installation	Sand Drain Spacing (ft)	H (ft)	Computed Ult. Sett. (ft)	Measured Ult. Sett. (ft)
Augered	14	27	3.9	4.1
Driven	10	25	3.7	3.5
Driven	14	25	3.7	3.8
Jetted	14	24	3.5	3.5

determine the extent to which counterbalancing errors have affected the comparisons. It is possible that the effects of sampling disturbance on the laboratory tests, and sand-drain-installation disturbance on the field settlements, counterbalanced each other. No such evidence comes from the field observations, however, considering that the disturbance should have been more severe in section D10 than A14, for example. It is possible that the effect of field disturbance was damped out by the large amount of applied stress as observed by Schmidt and Gould (47). It should also be noted that the calculated settlements would be changed by about 10% for a 150 psf (7 kN/m^2) change in the maximum previous consolidation pressure (Fig. 4) or a 0.02 change in the strain index (C_ε), errors that are easily possible considering the scatter in measured properties and the small number of tests.

Time rate of settlement analyses.-In order to analyze this case history properly, one needs a theory that will take into account two-dimensional flow, time-dependent loading, substantial variations in the coefficient of consolidation, and a non-linear load settlement curve. The method of analysis will be presented in the next section.

THEORETICAL ANALYSES OF TIME-SETTLEMENT CURVES

Closed-Form Analyses

The differential equation for one-dimensional radial dissipation of excess pore water pressure was derived by Rendulic (34) and the first solutions (in error) were published by Terzaghi (52). Carrillo (9) demonstrated that the two-dimensional flow problem could be uncoupled and the solutions for the vertical and radial flow problems combined to give the solution for the two-dimensional case. Independently of Carrillo, Barron (5, 6) analyzed both the problems of purely radial and combined vertical and radial flow. Based on the solution of the heat flow equation by Glover (14), Barron presented his so-called "free strain theory." Unfortunately, the free strain theory required the solution of an infinite series of Bessels functions of first and second kind and first and second order. Although Barron presented numerical solutions, the free strain theory does not seem to be in general use.

Barron (5, 6) presented a mathematically simpler solution, termed the "equal strain theory," which is based on the assumptions that the surface settlement is uniform at all times and that horizontal strains are everywhere zero. This theory is used extensively in spite of the fact that it involves an unlikely distribution of initial excess pore water pressure.

These theories and their application in practice have been reviewed by several investigators (18, 32, 35, 41). The theories cannot conveniently be extended to account for layered systems, time-dependent loading, variable coefficients of consolidation, nor inelastic stress-strain curves.

Finite Difference Analyses for Sand Drains

Finite difference equations for the solution of sand drain problems were first presented in the report by Moran, Proctor, Meuser, and Rutledge (32). The method has been applied to one-dimensional consolidation problems by a number of authors (1, 3, 4, 12, 13, 42, 44, 45, 46, 54).

Finite Difference Formulation

Carrillo (9) published the following differential equation for combined vertical and radial flow:

$$\frac{\partial \bar{u}}{\partial t} = c_v \frac{\partial^2 \bar{u}}{\partial z^2} + c_r \left(\frac{1}{r}\frac{\partial \bar{u}}{\partial r} + \frac{\partial^2 \bar{u}}{\partial r^2}\right) \quad \ldots\ldots\ldots\ldots\ldots \quad (3)$$

where $\bar{u}$ is the excess pore water pressure as a function of time (t), depth (z), and radius (r), and the relevant soil properties are the coefficients of consolidation for vertical and radial flow, c_v and c_r, respectively. Equation 3 is approximate in the sense that more rigorous constitutive relations were not employed (7, 8, 43) but there is some evidence (56) that Eq. 3 may be sufficiently accurate for most engineering purposes. The implicit, or Crank-Nicholson (10) method of analysis was chosen for this investigation because of its speed and versatility. The computer storage requirements were not found to be so large as to require use of the ADI method (21).

The zone of influence of a single sand drain and the subscript notation for individual nodes are depicted in Fig. 6. The implicit finite difference approximations for each term in Eq. 3 are as follows:

$$\frac{\partial \bar{u}}{\partial t} = \frac{\bar{u}_{z,r,t+\Delta t} - \bar{u}_{z,r,t}}{\Delta t} \quad \ldots\ldots\ldots\ldots\ldots \quad (4a)$$

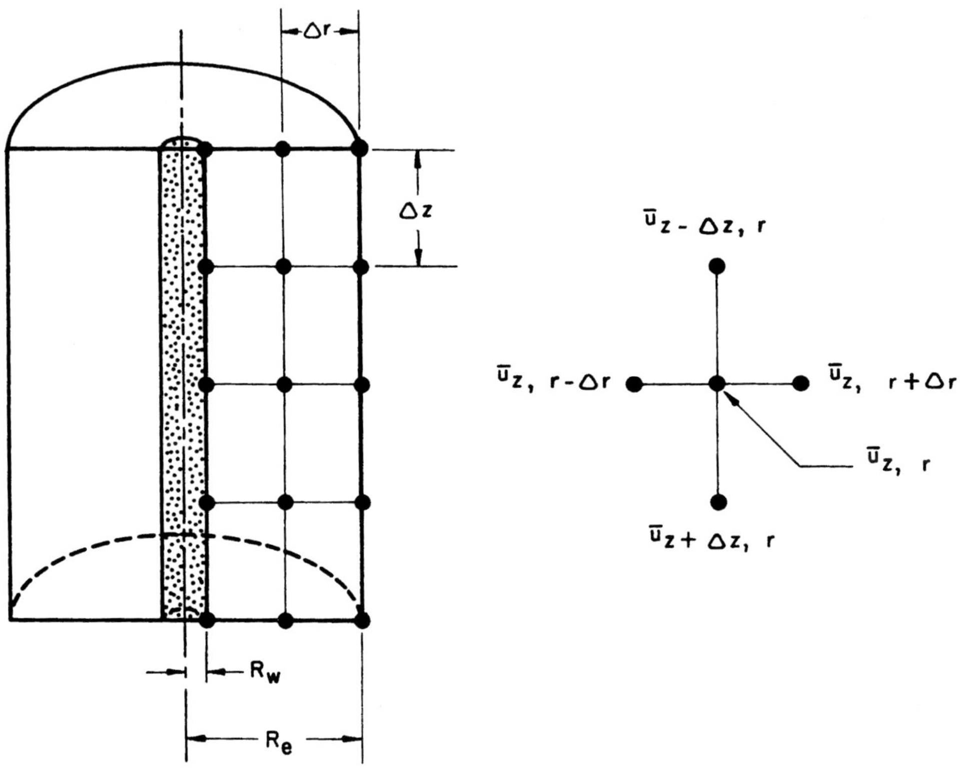

Fig. 6 Zone of Influence and Node Notation

$$\frac{\partial^2 \bar{u}}{\partial z^2} = \frac{1}{2(\Delta z)^2}\Big[\bar{u}_{z-\Delta z,r,t} - 2\bar{u}_{z,r,t} + \bar{u}_{z+\Delta z,r,t} + \bar{u}_{z-\Delta z,r,t+\Delta t} - 2\bar{u}_{z,r,t+\Delta t} + \bar{u}_{z+\Delta z,r,t+\Delta t}\Big] \quad \text{(4b)}$$

$$\frac{1}{r}\frac{\partial \bar{u}}{\partial r} = \frac{1}{4r_r\Delta r}\Big[\bar{u}_{z,r+\Delta r,t} - \bar{u}_{z,r-\Delta r,t} + u_{z,r+\Delta r,t+\Delta t} - u_{z,r-\Delta r,t+\Delta t}\Big] \quad \text{(4c)}$$

$$\frac{\partial^2 \bar{u}}{\partial r^2} = \frac{1}{2(\Delta r)^2}\Big[\bar{u}_{z,r-\Delta r,t} - 2\bar{u}_{z,r,t} + \bar{u}_{z,r+\Delta r,t} + \bar{u}_{z,r-\Delta r,t+\Delta t} - 2\bar{u}_{z,r,t+\Delta t} + u_{z,r+\Delta r,t+\Delta t}\Big] \quad \text{(4d)}$$

where Δt is the time step, Δr the radial distance between nodes, Δz the vertical distance between nodes, and r_r the radial distance from the vertical axis of symmetry to the rth node.

Equations 4a-d can be combined into the finite difference approximation of Eq. 1:

$$\bar{u}_{z-\Delta z,r,t+\Delta t} - \frac{\alpha_r}{\alpha_z}\left(\frac{\Delta r}{2r_r} - 1\right)\bar{u}_{z,r-\Delta r,t+\Delta t} - \frac{2}{\alpha_z}(1 + \alpha_r + \alpha_z)$$

$$\bar{u}_{z,r,t+\Delta t} + \frac{\alpha_r}{\alpha_z}\left(\frac{\Delta r}{2r_r} + 1\right)\bar{u}_{z,r+\Delta r,t+\Delta t} + \bar{u}_{z+\Delta z,r,t+\Delta t} =$$

$$-\bar{u}_{z-\Delta z,r,t} + \frac{\alpha_r}{\alpha_z}\left(\frac{\Delta r}{2r_r} - 1\right)\bar{u}_{z,r-\Delta r,t} - \frac{2}{\alpha_z}(1 - \alpha_r - \alpha_z)\,\bar{u}_{z,r,t} -$$

$$\frac{\alpha_r}{\alpha_z}\left(\frac{\Delta r}{2r_r} + 1\right)\bar{u}_{z,r+\Delta r,t} - \bar{u}_{z+\Delta z,r,t}. \quad \ldots \ldots \ldots \ldots (5)$$

where $\alpha_r = c_r \, \Delta t / (\Delta r)^2$ and $\alpha_z = c_v \, \Delta t / (\Delta z)^2$.

Imaginary nodes.-Some of the excess pore water pressures in Eq. 5 may be considered "imaginary." For example, if $\bar{u}_{z,r}$ is on the upper horizontal boundary of the system then $\bar{u}_{z-\Delta z,r}$ does not exist within the compressible layer under consideration and thus may be considered an imaginary node. Special equations must be applied at all boundaries and these equations are then the "boundary conditions" of the problem.

For a freely draining boundary at a point (z,r,t) the pore pressure $\bar{u}_{z,r,t}$ is equal to 0 for all values of t. Alternatively, the programmer could specify that a pore pressure exists at an imaginary node across the boundary at the same node spacing as the first node inside the boundary and with a pore pressure of opposite sign.

At an impervious horizontal or vertical boundary the rate of water flow is 0 and thus from Darcy's law, the rate of change of pore pressure with respect to the axis perpendicular to the boundary is 0:

$$\frac{\partial \bar{u}}{\partial z} = 0 \quad \text{or} \quad \frac{\partial \bar{u}}{\partial r} = 0 \quad \ldots \ldots \ldots \ldots \ldots \ldots (6)$$

For the sand drain problem, the cylindrical boundary at the external radius of influence of the drain (R_e) is an impervious boundary. The boundary condition is satisfied by setting up an imaginary node at one node spacing outside the boundary and specifying that the pore pressure at the imaginary node at any time be the same as the pore pressure at the first node inside of the boundary:

$$\bar{u}_{z-\Delta z,r} = \bar{u}_{z+\Delta z,r} \quad \text{or} \quad \bar{u}_{z,r-\Delta r} = \bar{u}_{z,r+\Delta r} \quad \ldots \ldots \ldots \ldots (7)$$

The case of an ith node located at the interface between two soil layers is illustrated in Fig. 7. Let $\bar{v}$ represent excess pore pressures in the layer above the ith node and let $\bar{w}$ represent excess pore

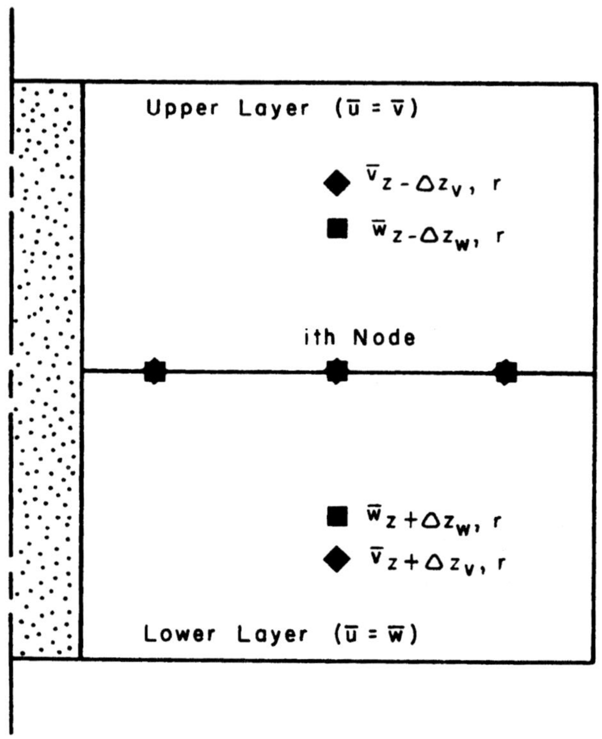

Fig. 7 Nodes at Layer Boundary

pressures in the layer below the ith node, as shown in Fig. 7. The subscript v indicates the upper layer and w indicates the lower layer. Equation 5 can be applied at the ith node for $\bar{v}_{z,r}$ and $\bar{w}_{z,r}$. The resulting equations are:

$$\bar{v}_{z-\Delta z_v,r,t+\Delta t} + A_v \bar{v}_{z,r-\Delta r,t+\Delta t} + B_{L,v} \bar{v}_{z,r,t+\Delta t} + C_v \bar{v}_{z,r+\Delta r,t+\Delta t} + \bar{v}_{z+\Delta z_v,r,t+\Delta t} = -\bar{v}_{z-\Delta z_v,r,t} - A_v \bar{v}_{z,r-\Delta r,t} + B_{R,v} \bar{v}_{z,r,t} - C_v \bar{v}_{z,r+\Delta r,t} - \bar{v}_{z+\Delta z_v,r,t} \quad \text{. . . (8a)}$$

$$\bar{w}_{z-\Delta z_w,r,t+\Delta t} + A_w \bar{w}_{z,r-\Delta r,t+\Delta t} + B_{L,w} \bar{w}_{z,r,t+\Delta t} + C_w \bar{w}_{z,r+\Delta r,t+\Delta t} + \bar{w}_{z+\Delta z_w,r,t+\Delta t} = -\bar{w}_{z-\Delta z_w,r,t} - A_w \bar{w}_{z,r-\Delta r,t} + B_{R,w} \bar{w}_{z,r,t} - C_w \bar{w}_{z,r+\Delta r,t} - \bar{w}_{z+\Delta z_w,r,t} \quad \text{. . . . . (8b)}$$

where $A = -\frac{\alpha_r}{\alpha_z}\left(\frac{\Delta r}{2r_r} - 1\right)$

$$B_L = -\frac{2}{\alpha_z}\left(1 + \alpha_r + \alpha_z\right)$$

$$B_R = -\frac{2}{\alpha_z}\left(1 - \alpha_r - \alpha_z\right)$$

$$C = \frac{\alpha_r}{\alpha_z}\left(\frac{\Delta r}{2r_r} + 1\right)$$

Equations 8a and 8b involve the four imaginary pore pressures

$\bar{v}_{z+\Delta z_v,r,t+\Delta t}$, $\bar{v}_{z+\Delta z_v,r,t}$, $\bar{w}_{z-\Delta z_w,r,t+\Delta t}$, and $\bar{w}_{z-\Delta z_w,r,t}$.

The imaginary pore pressures can be eliminated by writing continuity equations at the boundary for times t and $t+\Delta t$. The differential equation at the boundary is:

$$k_{z,v}\left(\frac{\partial \bar{u}}{\partial z}\right)_v = k_{z,w}\left(\frac{\partial \bar{u}}{\partial z}\right)_w \quad (9)$$

The average central difference approximations of Eq. 9 for times t and $t+\Delta t$ are:

$$k_{z,v}\frac{\bar{v}_{z+\Delta z_v,r,t} - \bar{v}_{z-\Delta z_v,r,t}}{2\,(\Delta z_v)} = k_{z,w}\frac{\bar{w}_{z+\Delta z_w,r,t} - \bar{w}_{z-\Delta z_w,r,t}}{2\,(\Delta z_w)} \quad (10a)$$

$$k_{z,v}\frac{\bar{v}_{z+\Delta z_v,r,t+\Delta t} - \bar{v}_{z-\Delta z_v,r,t+\Delta t}}{2\,(\Delta z_v)} =$$

$$k_{z,w}\frac{\bar{w}_{z+\Delta z_w,r,t+\Delta t} - \bar{w}_{z-\Delta z_w,r,t+\Delta t}}{2\,(\Delta z_w)} \quad (10b)$$

Equations 8a, 8b, 10a, 10b may all be combined to eliminate the imaginary pore pressures. The resulting difference equation at the layer boundary is:

$$\bar{u}_{z-\Delta z_v,r,t+\Delta t} + \rho_A\,\bar{u}_{z,r-\Delta r,t+\Delta t} + \rho_{BL}\,\bar{u}_{z,r,t+\Delta t} +$$

$$\rho_C\,\bar{u}_{z,r+\Delta r,t+\Delta t} + \gamma\,\bar{u}_{z+\Delta z_w,r,t+\Delta t} = -\bar{u}_{z-\Delta z_v,r,t} -$$

$$\rho_A\,\bar{u}_{z,r-\Delta r,t} + \rho_{BR}\,\bar{u}_{z,r,t} - \rho_C\,\bar{u}_{z,r+\Delta r,t} - \gamma\,\bar{u}_{z+\Delta z_w,r,t} \quad (11)$$

where $\gamma = \frac{k_{z,w}}{k_{z,v}} \frac{\Delta z_v}{\Delta z_w}$

$$\rho_A = \frac{A_v + \gamma A_w}{2}$$

$$\rho_{BL} = \frac{B_{L,v} + \gamma B_{L,w}}{2}$$

$$\rho_{BR} = \frac{B_{R,v} + \gamma B_{R,w}}{2}$$

$$\rho_C = \frac{C_v + \gamma C_w}{2}.$$

Solution of difference equations.-An equation can be written at every node in the following form:

$$a_1 \bar{u}_{z-\Delta z,r,t+\Delta t} + a_2 \bar{u}_{z,r-\Delta r,t+\Delta t} + a_3 \bar{u}_{z,r,t+\Delta t} +$$

$$a_4 \bar{u}_{z,r+\Delta r,t+\Delta t} + a_5 \bar{u}_{z+\Delta z,r,t+\Delta t} = b \quad \ldots\ldots\ldots\ldots \quad (12)$$

where a and b are constants. Of course, at system boundaries, imaginary pore pressures are substituted as indicated previously.

The same number of equations can be written as there are node points so that a system of linear algebraic simultaneous equations may be developed. The equations may be solved to obtain all the values of unknown pore pressures at time $t+\Delta t$.

Time-Dependent Loading

For a time-dependent loading, Eq. 3 is modified by adding to the right side $\partial q/\partial t$, where q is the applied surface load (53). The finite difference approximation is then:

$$\frac{\partial q}{\partial t} = \frac{q_{t+\Delta t} - q_t}{\Delta t} \quad \ldots\ldots\ldots\ldots \quad (13)$$

The terms on the right side of Eq. 13 are then added to the terms on the right side of Eq. 5 in order to effect a solution.

It is convenient to specify the surface load at specific points in time such that a linear variation of q with respect to t can be assumed between the points of definition. The program can then easily calculate values of q at any time t.

Use of Nonlinear Stress-Strain Curves

The actual relationship between stress and strain (or void ratio and effective stress) is not linear for real soils. When the settlements are large the error introduced into the time-settlement curve by assuming the soil is linearly elastic is a significant error. In a finite difference solution, it is possible to specify values of void ratio and effective stress at selected points and thus define a field consolidation curve of an appropriate shape. In effecting a solution, the excess pore water pressure at any time t may be used to calculate effective stresses in the soil deposit and in turn these effective stresses can be used with the input void ratio-effective stress relationship to determine void ratios as functions of depth and time. The settlement for the Lth layer is then given by:

$$\Delta S_L = H_{SL} \Delta e_L \quad \text{(14)}$$

where ΔS = settlement, H_S = height of solids (initial thickness of layer divided by the quantity one plus the initial void ratio), Δe = change in void ratio during a single time step, and the subscript L indicates the Lth compressible layer. In principle, the compressible layer could be subdivided so that each node would be the only node in a sublayer and Eq. 14 would then be applied to each node separately. It is, however, more efficient to follow this computational procedure using the average pore pressure and effective stresses for an entire soil layer. If such a procedure introduces a significant error, then the compressible layer in the field may be subdivided into a suitable number of sublayers. The equation used for computing the average effective stress within a single compressible layer is:

$$\bar{\sigma}_L = \bar{\sigma}_{o,L} + q - \bar{u}_{avg,L} \quad \text{(15)}$$

where $\bar{\sigma}_L$ = average effective stress in the Lth layer, $\bar{\sigma}_{o,L}$ = average effective overburden pressure in L^{th} layer when q = 0, and $\bar{u}_{avg,L}$ = average excess pore pressure.

Large Deformations and Submergence Effect

In problems where the change in thickness of the compressible layer is large, it may be worthwhile to take into account the changing drainage distance and also the submergence of parts of the embankment as settlement occurs. Such adjustments are relatively easy in the finite difference analysis since the settlement during any time period Δt can be subtracted from the thickness of the layer at the time of the last computation to give the new thickness. It is usually convenient to assume that the nodes remain evenly spaced though such is not really necessary. Similarly, the effect of submergence of overburden soil or embankment fill is taken into account by recalculating the applied load, taking submergence into account, after each time step.

Variable Soil Properties

Values of c_v, c_r, and k_z control the time rate of dissipation of excess pore water pressure. The coefficient of consolidation is likely to change moderately during consolidation of any soil and by more than 100 times for a few sensitive clays. Such large variations in soil properties should be taken into account in any time settlement analysis.

Variable soil properties are taken into account by defining each as a function of effective stress and storing these in the computer. At the end of each time iteration, the computer determines the average effective stress in each layer and goes to the relationship between the soil property and effective stress to determine the appropriate value of the soil property for the next iteration. Soil properties are thus held constant only for a single time step.

Computer Program

The finite difference method described in previous sections requires too much computational effort to make hand solutions practical. A computer program was developed to take into account all of the aforementioned variables by Daniel (11). This program makes it possible to take into account layered systems, time dependent loading, variable soil properties, large strains, two-dimensional flow, nonlinear load settlement curves, settlement dependent submergence of the embankment, and an initial one-dimensional consolidation stage followed by construction of the sand drains with two-dimensional consolidation of disturbed soil thereafter. The central processor times for a single run on The University of Texas CDC 6600/6400 computer system were almost always less than 150 seconds. Comparisons of the numerical solutions with closed-form solutions obtained for simple cases indicate that the numerical method is as accurate as desired for these cases.

APPLICATION TO CASE HISTORY

Analyses of the four time settlement curves for the Fore River test embankment indicated that there was a degree of disturbance associated with the installation of all sand drains with the disturbance a minimum for the quadrant where the drains were augered (25, 26) and a maximum where the drains were driven on 10-foot (3.1-m) centers. It is convenient to restrict consideration to these two quadrants.

Analysis for Quadrant A14

The first analysis (curve 1, Fig. 8) was performed using the assumptions of the free strain consolidation theory of Barron (5, 6) but accounting for time dependent loading. The assumptions that should be emphasized are: (1) the soil is linearly elastic, (2) strains are small, and (3) soil properties are constant. The first assumption makes it possible to express the settlement at any time as the ultimate settlement times the average degree of consolidation. The second assumption makes it possible to hold the vertical drainage distance constant. The coefficients of consolidation were held constant at $c_v = 0.11\ ft^2/day$ ($0.01\ m^2/day$) and $c_R = c_v$.

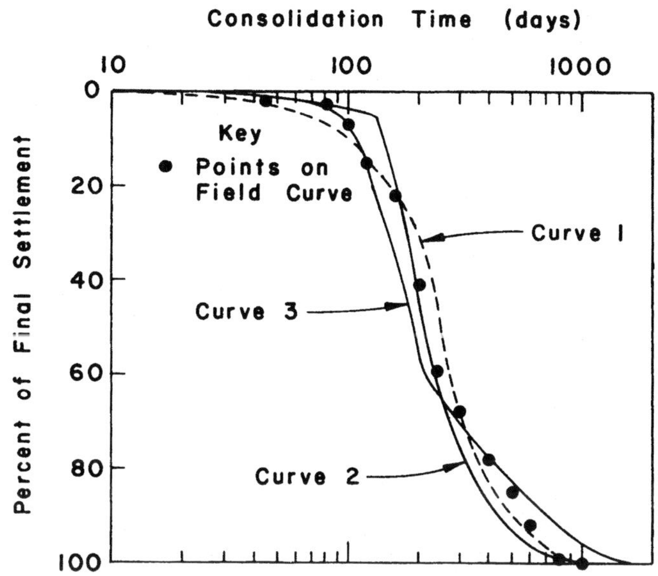

Fig. 8 Comparison of Computed and Measured Settlements at Quadrant A14.

Curve 2 (Fig. 8) was computed using the same constant coefficients of consolidation but with the soil constrained to follow the field void-ratio-versus-effective-stress curve. The influence of the non-linear stress-strain relationship is clear, especially at low stress levels.

Curve 3 (Fig. 8) was computed using the field void-ratio-versus-effective-stress curve, with the vertical coefficient of consolidation defined by the solid line in Fig. 5, and with $c_R = 1.2\ c_v$.

Further refinement of the analysis is not possible because radial coefficients of consolidation were not measured and the effects of disturbance during installation of the drains is unknown. However, trial analyses can be performed using the field void-ratio-versus-effective stress relationship and $c_R = 1.2\ c_v$ but with the vertical coefficients of consolidation adjusted to produce the best correlation between theoretical prediction and field measurement. The theoretical curve is compared with the field curve in Fig. 9 and the vertical coefficients of consolidation assumed for the analysis are compared with the coefficients measured in the laboratory in Fig. 10. The recalculated field coefficients are generally right in the middle of the laboratory values, suggesting that the choice of $c_R = 1.2\ c_v$ is about right and that the degree of disturbance associated with installation of augered drains is no more than that associated with obtaining the best undisturbed samples.

Analysis for Quadrant D10

Similar analyses were used for the field data from quadrant D10 and are shown in Fig. 11. Clearly, use of the laboratory coefficients of consolidation leads to an overestimate of the speed of consolidation

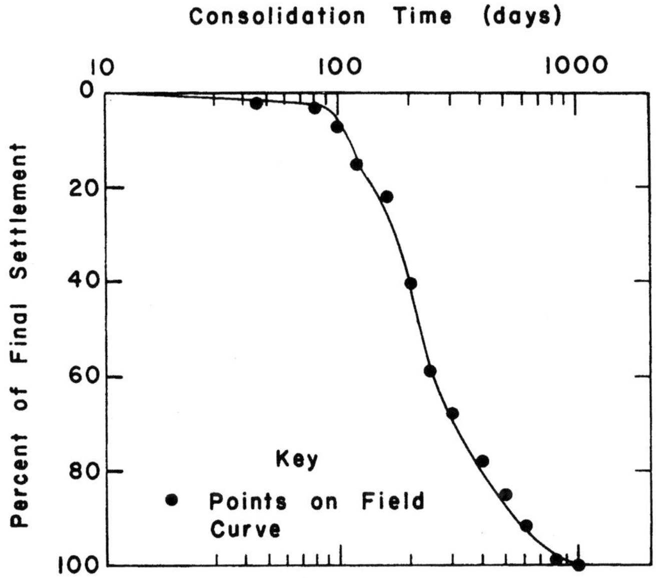

Fig. 9 Comparison of Field Measurements and the Theoretical Curve Calculated using Values of c_v from Fig. 10.

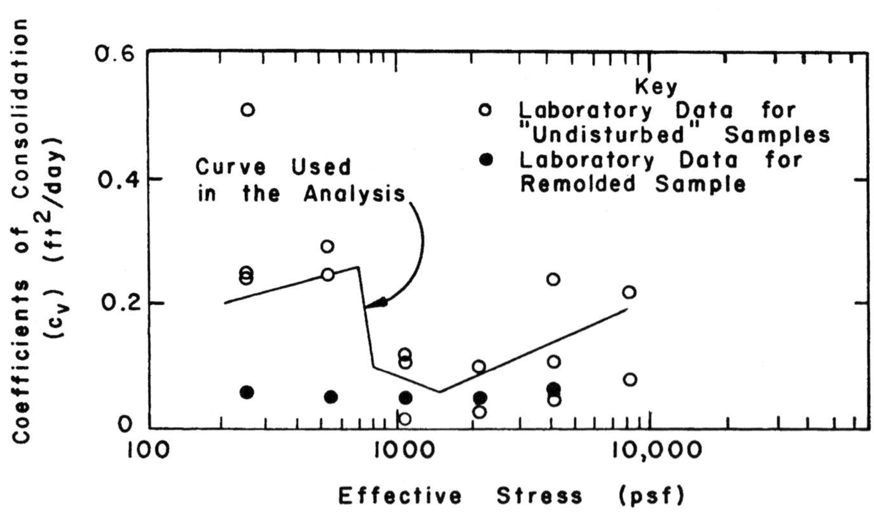

Fig. 10 $c - \bar{\sigma}$ Relationship for Theoretical Time-Settlement Curve of Fig. 9.

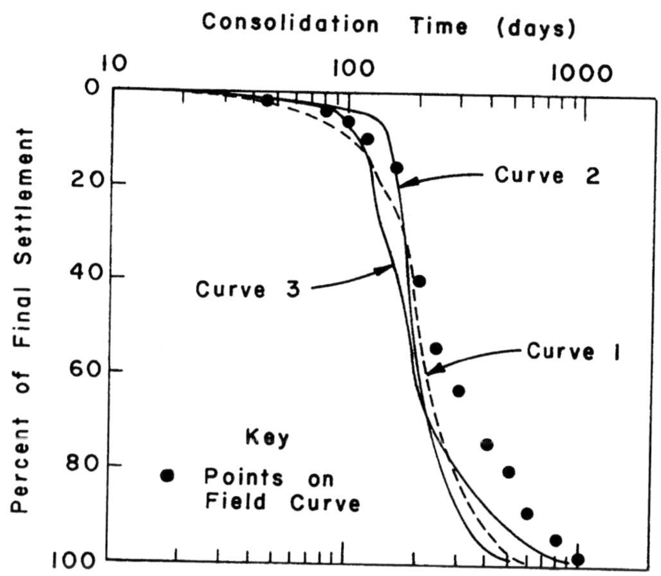

Fig. 11 Comparison of Computed and Measured Settlements at Quadrant D10

in the field. Since disturbance is known to reduce the coefficient of consolidation (Fig. 5) the most likely source of error lies in the use of the laboratory coefficients of consolidation. Of course, the disturbance involved in installing the drains would also influence the early part of the void ratio-effective stress curve but not necessarily the final settlement (47). Accordingly, both the c_v-$\bar{\sigma}$ and e-$\bar{\sigma}$ curves were adjusted until the computed time-settlement curve matched the measured curve as nearly as possible. The comparison between the theoretical and measured time-settlement curves is shown in Fig. 12, the c_v-$\bar{\sigma}$ curve

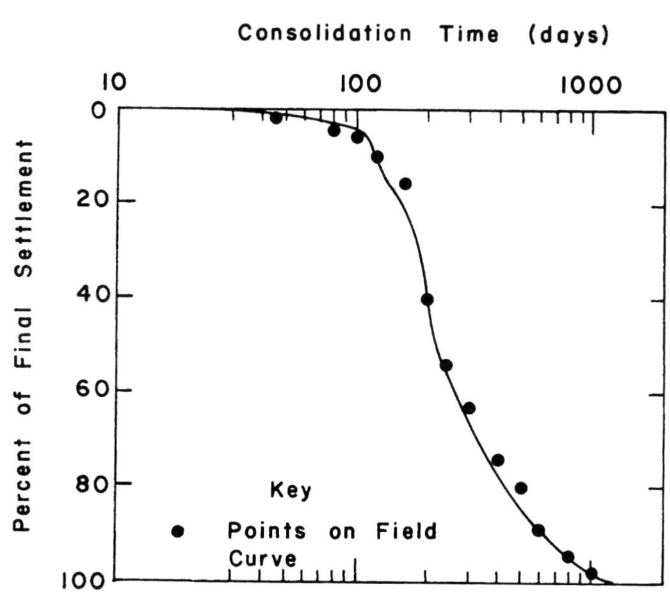

Fig. 12 Comparison of Field Measurements and the Theoretical Curve Calculated using Values of c_v from Fig. 13.

specified in the analysis ($c_r = 1.2\ c_v$) is shown in Fig. 13, and

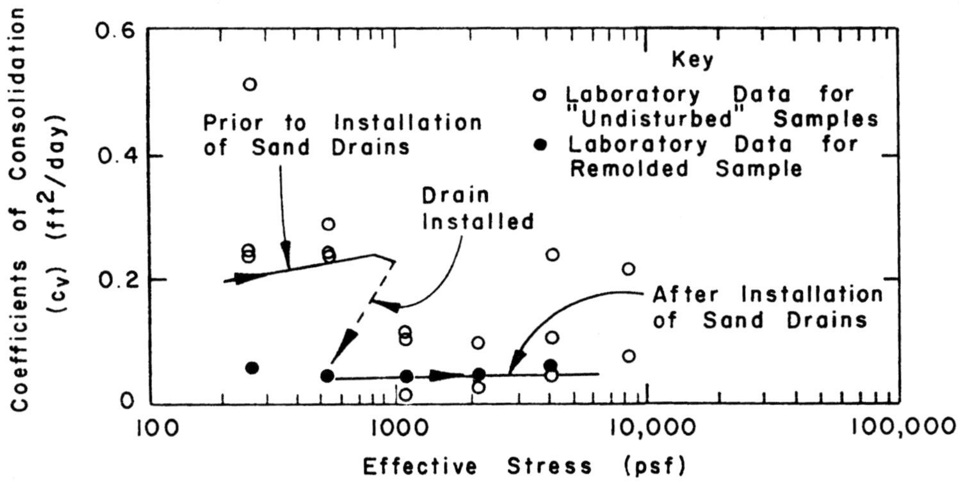

Fig. 13 c_v-$\bar{\sigma}$ Relationship for Theoretical Time-Settlement Curve of Fig. 12.

the e-$\bar{\sigma}$ curve followed is shown in Fig. 14. For the coefficients of

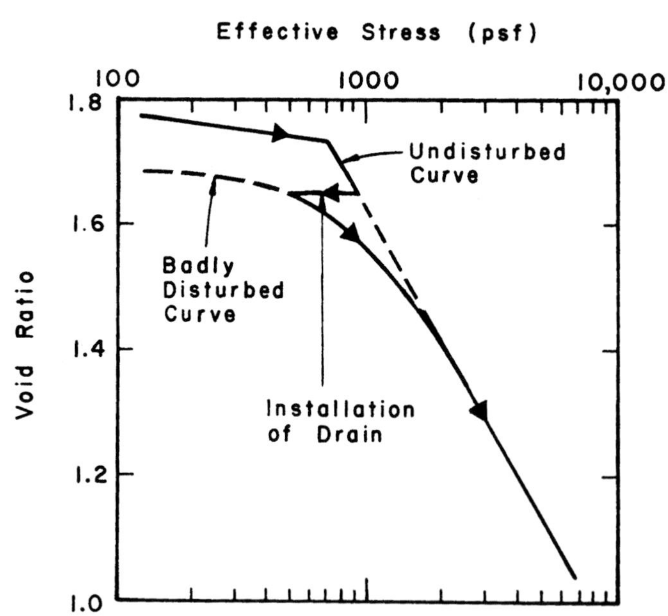

Fig. 14 e-$\bar{\sigma}$ Relationship for Theoretical Time-Settlement Curve of Fig. 12.

consolidation (Fig. 13) two curves were required, one for undisturbed soil prior to installation of sand drains and one for disturbed soil after drain installation. The dashed line of Fig. 13 indicates the instantaneous change from one curve to the other when the drains were driven. The reduction in effective stress is caused by disturbance during drain installation and is also shown in Fig. 14. The coefficients of consolidation that yielded the best fit matched precisely the coefficients from the laboratory tests on what was interpreted to be a badly disturbed sample. The comparison between the theoretical curve and the field measurements (Fig. 12) is probably the best that can be hoped for considering the uncertainties in the field measurements themselves.

Of course, the number of assumptions made is so large that a definitive conclusion cannot be reached but the data do suggest that driving of drains on 10-foot (3.1-m) centers caused nearly complete disturbance of the soil.

SUMMARY AND CONCLUSIONS

The conditions at the Fore River test embankment were nearly ideal for allowing a comparison to be made between theoretical predictions of the time rate of settlement and field measurements. In particular, the compressible soil was a single layer with two freely draining boundaries and the degree of saturation was apparently 100%. One-dimensional consolidation tests on "undisturbed" fixed-piston samples yielded data with minimal scatter and thus made it possible to define the consolidation characteristics for vertical flow with more than usual precision. No consolidation tests were performed with radial flow and thus the coefficient of consolidation for radial flow, c_r, had to be inferred from visual descriptions of the samples.

Field conditions seemed to be typical for embankments constructed on soft clays in that the loading was time dependent, the sand drains were installed after one-dimensional consolidation had begun, the coefficient of vertical consolidation was a function of the effective stress, and a degree of disturbance was associated with the installation of the drains.

The finite difference method of analysis made it possible to take into account time-dependent loading, variable coefficients of consolidation, two-dimensional flow, non-linear e-log $\bar{\sigma}$ relationships settlement dependent submergence of the embankment, and installation of the sand drains after one-dimensional consolidation had begun. Perhaps more importantly, the finite difference method of analysis made it possible to assign the soil undisturbed properties prior to installation of the drains and disturbed properties after the drains were installed.

The fitted theoretical time settlement curves fit the field observations with a high level of precision and the associated coefficients of consolidation were in excellent agreement with laboratory tests provided the degree of disturbance in the laboratory and in the field were comparable. Furthermore, the analysis made it possible for the soil to consolidate along the proper void ratio-effective stress curve, even when the undisturbed curve was used prior to drain installation and a curve for a badly disturbed soil was used after drain installation.

The observations suggest that the simple Terzaghi type theory of consolidation applied in two dimensions can be successful in predicting the proper time rate of settlement in the field provided the soil is initially saturated. The computer program is tolerably simple and the running times for the single layer system used at the Fore River site were usually less than 5 seconds.

Based on this case history and a study of the literature on other case histories it seems apparent that the most important need at the present time is for a better definition of soil properties, particularly the coefficient of consolidation for vertical deformation but radial flow.

ACKNOWLEDGMENTS

Analyses of the data for the Fore River site were supported by the National Science Foundation under grant GK-35467. Field data were provided by Haley and Aldrich, consulting geotechnical engineers, Cambridge, Massachusetts. Field data were collected by personnel of the Maine State Highway Commission, and Haley and Aldrich. Appreciation is expressed to the Maine State Highway Commission for releasing the data.

APPENDIX.—REFERENCES

1. Abbott, M. B., "One-Dimensional Consolidation of Multi-Layered Soils," Geotechnique, Vol. 10, 1960, pp. 151-165.

2. Aldrich, H. P., and Johnson, E. G., "Embankment Test Sections to Evaluate Field Performance of Vertical Sand Drains for Interstate 295," Highway Research Record, No. 405, 1972, pp. 60-71.

3. Barden, L., "Consolidation of Compacted and Unsaturated Clays," Geotechnique, Vol. 15, No. 3, 1965, pp. 267-286.

4. Barden, L., and Berry, P. L., "Consolidation of Normally Consolidated Clay," Journal of the Soil Mechanics and Foundations Division, ASCE, Vol. 91, SM5, 1965, pp. 15-35.

5. Barron, R. A., "The Influence of Drain Wells on the Consolidation of Fine-Grained Soils," U.S. Army Engineer District, Providence, 1944.

6. Barron, R. A., "Consolidation of Fine-Grained Soils by Drain Wells," Transactions, ASCE, Vol. 113, 1948, pp. 718-754.

7. Biot, M. A., "La Probleme de la Consolidation des Matieres Argileuses sous une Charge," Ann. Soc. Scient. Bruxelles, Vol. 55, 1935.

8. Biot, M. A., "General Theory of Three-Dimensional Consolidation," Journal of Applied Physics, Vol. 12, 1941, pp. 155-164.

9. Carrillo, N., "Simple Two- and Three-Dimensional Cases in the Theory of Consolidation of Soils," Journal of Mathematics and Physics, Vol. 21, No. 1, Mar., 1942, pp. 1-5.

10. Crank, J., and Nicholson, P., "A Practical Method for Numerical Evaluation of Solutions of Partial Differential Equations of the Heat Conduction Type," Proceedings, Cambridge Philosophical Society, Cambridge, Vol. 43, 1947, pp. 50-67.

11. Daniel, D. E., "Combined Vertical and Radial Flow in Consolidation Problems," thesis presented to The University of Texas, at Austin, Texas, in 1973, in partial fulfillment of the requirements for the degree of Master of Science.

12. Davis, E. H., and Lee, I. K., "One-Dimensional Consolidation of Layered Soils," Proceedings, Seventh International Conference on Soil Mechanics and Foundation Engineering, Vol. 2, 1969, pp. 65-72.

13. Gibson, R. E., and Lumb, P., "Numerical Solutions of Some Problems in the Consolidation of Clay," Proceedings, Institute of Civil Engineers (Pt I), Vol. 2, Part 2, 1953, p. 182.

14. Glover, R. E., Technical Memorandum No. 158, Bureau of Reclamation, Denver, Colorado, 1930.

15. Haley and Aldrich, "Report No. 1 - Engineering Properties of Foundation Soils at Long Creek-Fore River Areas and Back Cove," Report submitted to the Maine State Highway Commission, Oct., 1969.

16. Haley and Aldrich, "Report No. 2 - Test Sections to Evaluate Field Performance of Vertical Sand Drains," Report submitted to the Maine State Highway Commission, Jan., 1970.

17. Horne, M. R., "The Consolidation of a Stratified Soil with Vertical and Horizontal Drainage," International Journal of Mechanical Sciences, Vol. 6, 1974, pp. 187-197.

18. Johnson, S. J., "Precompression with Vertical Sand Drains," Journal of the Soil Mechanics and Foundations Division, ASCE, Vol. 96, SM1, Jan., 1970, pp. 145-175.

19. Jonas, E., "Subsurface Stabilization of Organic Clay by Precompression," Journal of the Soil Mechanics and Foundations Division, ASCE, Vol. 90, SM5, Sept., 1964, pp. 363-376.

20. Kapp, M. S., York, D. L., Aronowitz, A., and Sitomer, H., "Construction on Marshland Deposits: Treatments and Results," Highway Research Record, No. 133, 1966, pp. 1-22.

21. Koppula, S. D., and Morgenstern, N. R., "Consolidation of a Clay Layer in Two Dimensions," Journal of the Soil Mechanics and Foundations Division, ASCE, Vol. 98, SM1, 1972, pp. 79-94.

22. Kuesel, T. R., Schmidt, B., and Rafaeli, D., "Settlements and Strengthening of Soft Clay Accelerated by Sand Drains," Highway Research Record, No. 457, 1973, pp. 18-26.

23. Ladd, C. C., Rixner, J. J., and Gifford, D. G., "Performance of Embankments with Sand Drains on Sensitive Clay," ASCE Specialty Conference on Performance of Earth and Earth-Supported Structures, Purdue Univ., Vol. 1, Part 1, 1972, pp. 211-242.

24. Lake, J. R., "Pore-Pressure and Settlement Measurements During Small-Scale and Laboratory Experiments to Determine the Effectiveness of Vertical Sand Drains in Peat," Conference on Pore Pressure and Suction in Soils, London, 1961, pp. 103-107.

25. Landau, R. E., "The Post Hole Digger Comes of Age in Sand Drain Work," Construction Methods and Equipment, Apr., 1958, pp. 154-158.

26. Landau, R. E., Method of Installation as a Factor in Sand Drain Stabilization Design," Highway Research Record, No. 133, 1966, pp. 75-97.

27. Margason, E., and Arango, I., "Sand Drain Performance on a San Francisco Bay Mud Site," Proceedings, ASCE Specialty Conference on the Performance of Earth and Earth-Supported Structures, Purdue Univ., Vol. I, Part 1, 1972, pp. 181-210.

28. McAlpin, G. W., and Sinacori, M. N., "Sand Drains for Embankment on Marl Foundation," Highway Research Board, Bulletin 115, 1955, pp. 15-30.

29. Meardi, G., "Consolidation with Sand Piles of Soft Clays in Levee Foundations," Proceedings, Fourth International Conference on Soil Mechanics and Foundation Engineering, Vol. II, 1957, pp. 334-337.

30. Moore, L. H., "Summary of Treatments for Highway Embankments on Soft Foundations," Highway Research Record, No. 133, 1966, pp. 45-59.

31. Moore, L. H., and Grosert, T., "An Appraisal of Sand Drain Projects Designed and Constructed by the New York State Department of Transportation," Physical Research Report 68-1, Albany, N. Y., 1968.

32. Moran, Proctor, Meuser and Rutledge, "Study of Deep Soil Stabilization by Vertical Sand Drains," Report to Navy Bureau of Yards and Docks, 1958.

33. Murray, R. T., "Embankments Constructed on Soft Foundations: Settlement Study at Avonmouth," Road Research Laboratory Report LR419, 1971.

34. Rendulic, L., "Der Hydrodynamische Spannungsausgleich in Zentral Entwasserten Tonzylindern," Wasserwirtschaft und Technik, Wien, Vol. 2, 1935, pp. 250-253.

35. Richart, F. E., Jr., "Review of the Theories for Sand Drains," Transactions, ASCE, Vol. 124, 1959, pp. 709-739.

36. Root, A. W., "California Experience in Construction of Highways Across Marsh Deposits," Highway Research Board, Bulletin 173, 1958, pp. 46-64.

37. Rowe, P. W., "Measurement of the Coefficient of Consolidation of a Lacustrine Clay," Geotechnique, Vol. 9, No. 3, 1959, pp. 107-118.

38. Rowe, P. W., "The Calculation of the Consolidation Rates of Laminated, Varved or Layered Clays, with Particular Reference to Sand Drains," Geotechnique, Vol. 14, No. 4, 1965, pp. 321-340.

39. Rowe, P. W., and Barden, L., "A New Consolidation Cell," Geotechnique, Vol. 16, No. 2, 1966, pp. 162-170.

40. Rowe, P. W., and Shields, D. H., "The Measured Horizontal Coefficient of Consolidation of Laminated, Layered or Varved Clays," Proceedings, Sixth International Conference on Soil Mechanics and Foundation Engineering, Vol. 1, 1965, pp. 342-344.

41. Rutledge, P. C., and Johnson, S. G., "Review of Uses of Vertical Sand Drains," Highway Research Board, Bulletin 173, 1957.

42. Schiffman, R. L., "Consolidation of Soil Under Time-Dependent Loading and Varying Permeability," Proceedings, Highway Research Board, Vol. 37, 1958, pp. 584-617.

43. Schiffman, R. L., Chen, A. T-F., and Jordan, J. C., "An Analysis of Consolidation Theories," Journal of the Soil Mechanics and Foundations Division, ASCE, Vol. 95, SM1, 1969, pp. 285-312.

44. Schiffman, Robert L., and Gibson, R. E., "Consolidation of Non-Homogeneous Clay Layers," Journal of the Soil Mechanics and Foundations Division, ASCE, Vol. 90, SM5, pp. 1-30.

45. Schiffman, R. L., and Stein, J. R., "One-Dimensional Consolidation of Layered Systems," Journal of the Soil Mechanics and Foundations Division, ASCE, Vol. 96, SM4, 1970, pp. 1499-1504.

46. Schiffman, R. L., Whitman, R. V., and Jordan, J. C., "Settlement Problem Oriented Computer Language," Journal of the Soil Mechanics and Foundations Division, ASCE, Vol. 96, SM2, pp. 649-669.

47. Schmidt, T. J. and Gould, J. P., "Consolidation Properties of an Organic Clay Determined from Field Observations," Paper presented at the ASCE Session In Situ Properties of Organic Soils, Oct. 20, 1967, New York City.

48. Shields, D. H. and Rowe, P. W., "Radial Drainage Oedometer for Laminated Clays," Journal of the Soil Mechanics and Foundations Division, ASCE, Vol. 91, No. SM1, 1965, pp. 15-23.

49. Sowers, G. F., "Fill Settlement Despite Vertical Sand Drains," Journal of the Soil Mechanics and Foundations Division, ASCE, Vol. 10, SM5, Sept., 1964, pp. 289-302.

50. Takagi, S., "Determination of the Process of Consolidation by Sand Drains under Variable Loading," Proceedings, Fourth International Conference on Soil Mechanics and Foundation Engineering, Vol. I., 1957, pp. 444-448.

51. Taylor, D. W., Research on the Consolidation of Clays, Publication of the Department of Civil and Sanitary Engineering, Massachusetts Institute of Technology, Cambridge, Serial 82, 1942, p. 17.

52. Terzaghi, K., Theoretical Soil Mechanics, John Wiley and Sons, New York, 1943, 510 pp.

53. Terzaghi, K., and Frohlich, O. K., Theorie der Setzung von Tonschichten, F. Deuticke, Leipzig, 1936.

54. Toan, N. T., "Finite Difference Solutions of Consolidation Problems," thesis presented to the University of Illinois, at Champaign-Urbana, Ill., in 1972 in partial fulfillment of the requirements for the degree of Doctor of Philosophy.

55. Tsien, S. I., "Stabilization of Marsh Deposit," Highway Research Board, Bulletin 115, 1955, pp. 15-43.

56. Viggiani, C., "Discussion of 'An Analysis of Consolidation Theories'," Journal of the Soil Mechanics and Foundations Division, ASCE, Vol. 96, SM1, 1970, pp. 331-334.

57. Weber, W. G., "Experimental Sand Drain Fill at Napa River," Highway Research Record, No. 133, 1966, pp. 23-44.

58. Weber, W. G., "Performance of Embankments Constructed over Peat," Journal of the Soil Mechanics and Foundations Division, ASCE, Vol. 95, SM1, Jan., 1969, pp. 53-76.

59. Weber, R., and Hill, W. C., "Modifications of Sand Drain Principle for Pressure Relief in Stabilizing Embankment Foundation," Highway Research Board, Bulletin 115, 1955, pp. 1-14.

60. Wu, T. H., and Peck, R. B., "Field Observations on Sand Drain Construction on Two Highway Projects in Illinois," Proceedings, Highway Research Board, Vol. 35, 1956, pp. 747-753.

Authorized Reprint from
Geotechnical Testing Journal, Vol. 2, No. 1
Copyright
American Society for Testing and Materials
1916 Race Street, Phila., Pa. 19103
1979

J. H. Pelletier, [1] *R. E. Olson,* [2] *and J. J. Rixner* [3]

Estimation of Consolidation Properties of Clay from Field Observations

REFERENCE: Pelletier, J. H., Olson, R. E., and Rixner, J. J., **"Estimation of Consolidation Properties of Clay from Field Observations,"** *Geotechnical Testing Journal,* GTJODJ, Vol. 2, No. 1, Mar. 1979, pp. 34-43.

ABSTRACT: Analyses to obtain theoretical time-settlement curves for embankments typically make use of soil properties measured in the laboratory on small disturbed samples subject to strain rates and hydraulic gradients that are several orders of magnitude greater than those encountered in the field. Only analyses of well-documented case histories can demonstrate the validity of such extrapolations. Refined analyses based on detailed field observations of settlements and pore water pressures at various depths beneath the main I-295 highway embankment adjacent to the Fore River in Portland, Me., have demonstrated that: (1) the field e versus log $\bar{\sigma}$ curves are typically above the laboratory curves and have a sharper break at $\bar{\sigma}_{max}$; (2) c_v is higher in the field than in the laboratory; (3) at this site c_r/c_v ranged from 1 to 2; and (4) there was a generally good correlation between field properties and reconstructed laboratory properties. At this site the large differences between strain rates and hydraulic gradients in the laboratory and the field apparently did not introduce important errors.

KEY WORDS: soil physical properties, consolidation, settlement, sand drains, highway embankments, pore pressures, radial flow, finite differences

To perform a settlement analysis an engineer must have, first, a suitable stress-strain/time theory for the soil; second, an adequate definition of the soil properties; and third, a suitably general method of incorporating the properties into the soil model and achieving a solution under realistic field conditions.

Stress-Strain/Time Theory

The analysis for total settlement typically involves use of a nonlinear large-strain model, but the time rates of settlement are usually analyzed using Terzaghi's theory in its simplest form [*1*], in which the soil model is of the small-strain, linearly elastic type with constant properties and the soil profile and load-time relationships are similarly simplified. The potential for major differences between predicted and observed time-settlement behavior in the field is substantial. For example, Terzaghi's theory involves the assumption of the validity of Darcy's law. Mitchell and Younger [*2*] have reviewed the literature and reported numerous examples of studies in which the measured relationship between discharge velocity and hydraulic gradient was not linear, as is required by Darcy's law. Nevertheless, in typical laboratory consolidation tests the average hydraulic gradients are several orders of magnitude higher than in the field and substantial differences in the average coefficient of permeability k in the laboratory and in the field are conceivable.

In using Terzaghi's theory it is assumed that the only source of delay in consolidation is hydrodynamic, that is, is due to low permeability. In more general theories [*3-17*] it is assumed that the soil structure itself has a time-dependent resistance to volume change. If so, strain rates used in the laboratory should approximate those expected in the field. However, in typical cases a given strain is developed in the laboratory in a time ranging from a few minutes to a few hours, whereas field times are likely to range between months and decades, thus yielding average strain rates that are on the order of 10^4 times higher in the laboratory than in the field and leading to a significant potential source of error.

Definition of Soil Properties

Consolidation properties are measured in the laboratory using samples that are miniscule compared to the size of the soil deposit in the field. For example, if 20 laboratory tests are performed using samples 63.5 (2.5 in.) in diameter and 19 mm (0.75 in.) thick from a soil deposit 15 m deep beneath a highway embankment 100 m wide by 300 m long, the ratio of volume of soil in the field to volume tested is nearly 10^{10}.

In addition, the extent to which samples tested in the laboratory are disturbed is unknown. Experience shows that disturbance generally rounds off the stress-strain curve in the vicinity of the maximum past pressure [*18-20*] and reduces the coefficient of consolidation [*21,22*], thus generally leading to overestimates of both the amount of settlement and the consolidation time.

Incorporation of Appropriate Conditions in Analyses

In typical analyses, approximations must be made to account for the effects of soil stratification, time-dependent loading, large and nonuniform strains, hysteresis during unloading, effective stress-dependent strains and coefficients of consolidation, settlement-dependent submergence corrections, and other effects. These effects can be taken into account collectively only through use of numerical methods [*4,15,21-39*], although in-

[1] Research engineer, Shell Development Co., Houston, Tex. 77001.
[2] Professor of civil engineering, University of Texas, Austin, Tex. 78712. Member of ASTM.
[3] Vice-president, Haley and Aldrich, Inc., 238 Main St., Cambridge, Mass. 02142.

0149-6115/79/0003-0034$00.40

dividual effects can often be included in a generalized closed form theory [*39-51*].

The above-mentioned sources of error should lead to substantial differences between predicted and actual settlement behavior in the field. Attempts to adjust the analyses typically involve application of "engineering judgement," but the only definitive basis for such judgement is a collection of comparisons between predicted and measured field performances. The literature contains relatively few such comparisons and for a substantial fraction of these the analysis and the stress-strain/time model for the soil were so oversimplified that it is difficult to draw reliable conclusions.

One approach to resolving the problems enumerated above is to analyze data from field projects in such a way as to obtain reasonably accurate estimates of bulk soil properties in the field and to compare these field properties with values measured in the laboratory. Because relatively complex properties are to be evaluated it is necessary to obtain detailed field observations, which should include, as a minimum, measurements of settlements at various depths. The accumulation of such data for a variety of field conditions will at the very least provide a more firm basis for "engineering judgement."

In this paper the measured performance of a section of highway embankment, founded on a layered system of sands and clays containing sand drains, is analyzed to estimate the actual field dependance of void ratio *e* and the coefficients of vertical and horizontal consolidation on effective stress. These relationships are compared with values measured in the laboratory to determine whether laboratory values and an extended classical theory of consolidation can be reliably used to predict field performance for sites such as the one under analysis.

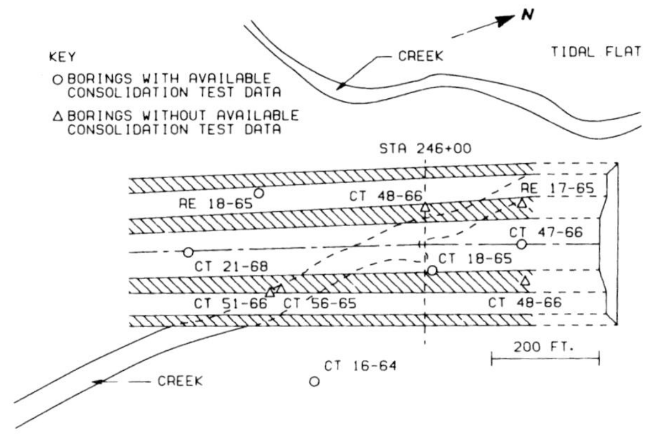

FIG. 1—*Plan view of embankment in the vicinity of Station 246+00.*

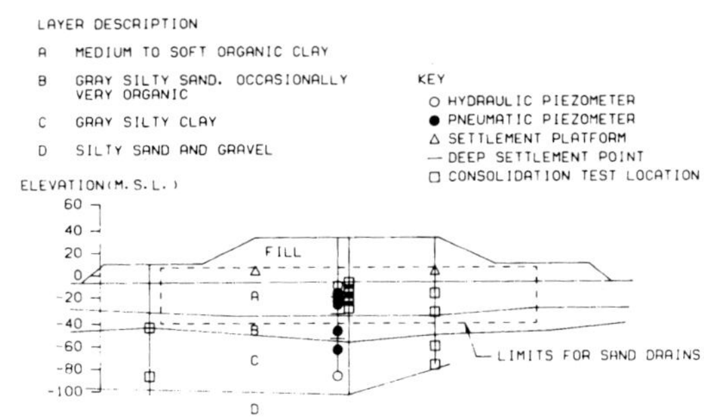

FIG. 2—*Cross-section of embankment at Station 246+00, showing instrumentation.*

Field Conditions

Between mid-1969 and late 1970 the Maine State Highway Commission (MSHC) constructed embankments for the highway I-295 on the north and south sides of the new Fore River bridge south of the city of Portland, Me. Haley and Aldrich, Inc., consulting engineers, recommended that the embankments be instrumented [*52*] and together with personnel from the MSHC monitored and interpreted the field performance of the embankments until June 1972. Monitoring equipment included surface and deep settlement markers, pneumatic and hydraulic piezometers, and slope indicators.

Data collected for a test embankment on the south side of the bridge were analyzed previously [*22*]. Analyses presented in this paper apply to the production embankment, and particularly to conditions in the vicinity of Station 246 on the north approach fill (Fig. 1). Continued developments in analytical technique make it possible to perform a more sophisticated analysis than in 1974 and thus to represent field conditions more closely.

Soil Profile

The compressible soils underlying the embankment include two layers of clay and one of silty sand (Fig. 2). The top layer, termed Layer A by Haley and Aldrich [*52*], consists of about 9.8 m (32 ft) of soft to medium slightly organic gray silty clay containing broken shells, pieces of wood, frequent sandy zones, and occasional peaty zones. The Atterberg limits plot within about 3% of the A-line, with liquid limits ranging from 30% to 90% and natural water contents near the liquid limit. The average shear strength measured using the field vane was 25 kPa (520 psf) with a coefficient of variation of about 23%.

The second layer (Layer B) is about 3.7 m (12 ft) thick and consists of loose to dense gray silty sand with occasional lenses or layers of highly organic soils.

The lowest compressible layer (Layer C) consists of about 15 m (48 ft) of medium to stiff gray silty clay. The liquid limits range from 30 to 55% and the plasticity indices plot 0 to 5% above the A-line. The natural water contents were generally a few percent less than the liquid limit. The average field vane shear strength was 66 kPa (1370 psf) with a coefficient of variation of 23%.

Sampling and Consolidation Testing

Soil samples were obtained by the MSHC using 89-mm (3.5-in.) diameter thin-walled fixed-piston samplers. All consolidation tests were performed by personnel of the MSHC using floating ring consolidation cells 64 mm (2.5 in.) in diameter, and 22 mm (0.87 in.) high. Loads were applied using a dead-weight hanger system. For most of the tests a load increment ratio of 1 was used and each load was left in place for 24 h. A few tests were performed using reduced load increments and a few with longer loading periods. Data from 16 tests in the vicinity of Station 246 were analyzed for this study.

Typical settlement versus log time curves for samples from Layer A are shown in Fig. 3. There was little evidence of primary consolidation until the applied stress $\bar{\sigma}$ was in excess of the apparent value of the maximum previous consolidation pressure

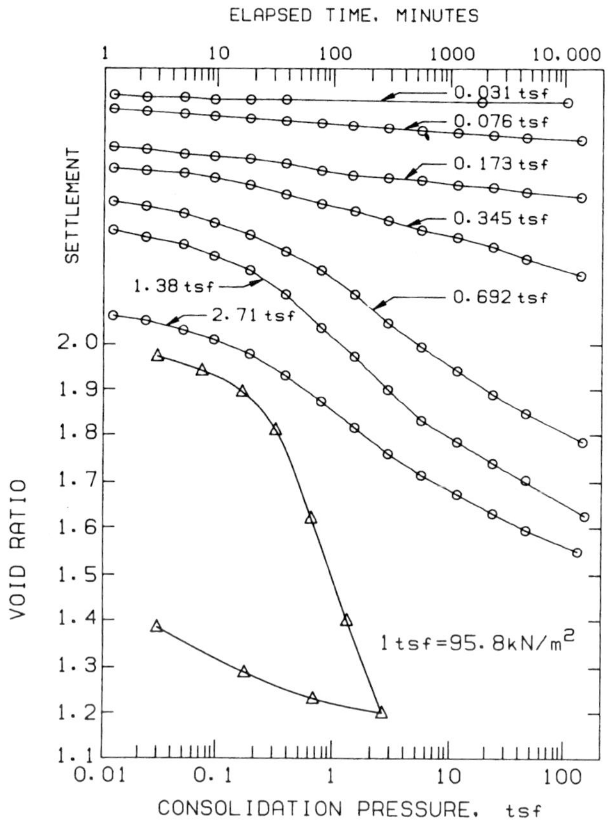

FIG. 3—*Typical laboratory consolidation data (Layer A).*

$\bar{\sigma}_{max}$. None of the secondary slopes were linear on the settlement versus log time plots and the break between primary and secondary consolidation was often indistinct. For values of $\bar{\sigma}$ just greater than $\bar{\sigma}_{max}$, the slope of the secondary curve was so great that the primary and secondary curves merged without apparent flexure and the settlement at the end of primary consolidation S_{100} could not be defined. For stresses greater than about two times $\bar{\sigma}_{max}$, secondary settlement generally represented about 20 to 40% of the total settlement. Regardless of the problems, Terzaghi's theory was fitted to the curves using standard methods and the coefficients of consolidation were evaluated.

All settlements S were also plotted against the square root of time $\sqrt{t}$; the coefficients of consolidation were evaluated using the standard procedure [*1*, p. 239] except that fitting was accomplished at $U = 45\%$ where U is the degree of consolidation. There were generally four to ten points on the linear part of the S versus $\sqrt{t}$ curve, thus allowing the S_0 and S_{90} points to be defined uniquely.

The ratio of the value of the coefficient of consolidation for vertical compression and vertical water flow c_v obtained from the $\sqrt{t}$ method to that obtained from the log t method ranged from 1 to 4 and averaged about 1.8. The ratio did not depend on the value of $\bar{\sigma}$ relative to $\bar{\sigma}_{max}$. In the analyses to follow, all laboratory values of c_v were determined using the square root method.

The values of $\bar{\sigma}_{max}$ were estimated from the e versus log $\bar{\sigma}$ curves, with e defined at the end of primary consolidation by the square root method. Values of $\bar{\sigma}_{max}$ are compared with calculated effective overburden pressure $\bar{\sigma}_o$ for the two compressible layers, A and C, in Fig. 4. In layer A the overconsolidation ratio ranged from 1.4 to 2.6 and $\bar{\sigma}_{max} - \bar{\sigma}_o$ from 9.6 to 31.1 kPa (200 to 650 psf). In Layer C, the upper part of the layer appears to be more highly overconsolidated than the lower part, perhaps because of desiccation. Here the overconsolidation ratio ranged from 1.5 to 3.0 and $\bar{\sigma}_{max} - \bar{\sigma}_o$ from 115 to 254 kPa (2400 to 5300 psf).

Data on the shapes of the e versus log $\bar{\sigma}$ and c_v versus log $\bar{\sigma}$ curves will be presented below.

Field Activities

The filling schedule at Station 246 is shown in Fig. 5. Construction began on construction day (CD) 400 on 28 August 1969. The shallow settlement markers were installed after 1.1 m

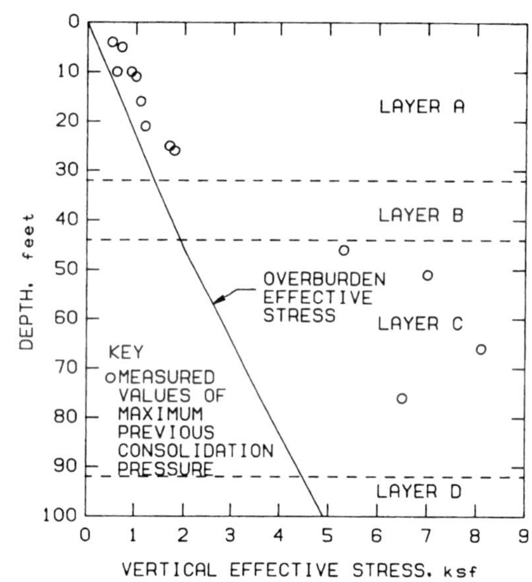

FIG. 4—*Measured values of $\bar{\sigma}_{max}$ for borings in the vicinity of Station 246+00.*

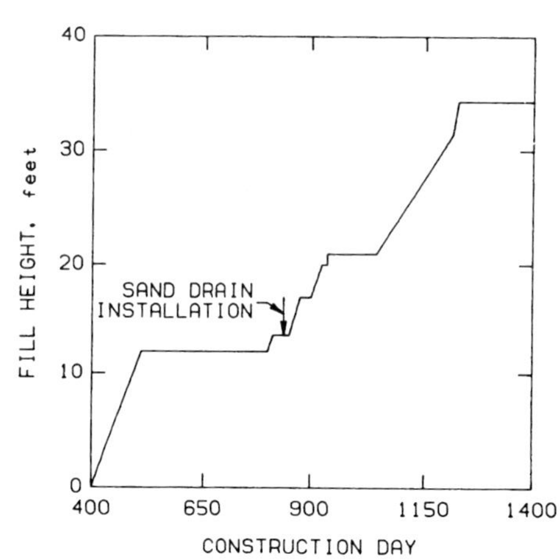

FIG. 5—*Loading history near Station 246+00.*

(3.5 ft) of fill had been placed to bring the surface up to mean sea level (*MSL*). The first stage fill was 72 m (236 ft) wide at the base, had 2:1 side slopes, and brought the surface up to elevation (El) 2.7 m (9 ft) *MSL*. Deep settlement markers were installed from 7 to 11 August 1970 on about CD 746 at the location indicated in Fig. 2. Hydraulic piezometers were installed at essentially the same time and pneumatic piezometers were installed between 24 November and 16 December 1970 (average CD 860). Slope indicators were installed on 2 to 4 September 1970 on about CD 770. On CD 797, a slide occurred between Stations 246 and 250, with a vertical displacement of about 0.5 m (1.5 ft) at the outer edge of the embankment, but no movement occurred at the centerline. Between CD 795 and 810, 0.46-m (18-in.) diameter sand drains were installed through Layer A by jetting in a triangular pattern at 2.6 m (8.5 ft) on center (N = 6)[4] to an average depth of 11 m (36 ft). Between about CD 850 and 1217, the central part of the fill was raised to El 9.5 m (31 ft) *MSL* (Fig. 5). The final width of the crown was 44 m (144 ft).

Methods of Analysis

All theoretical analyses were performed using a computer program, SD5, that is based on the Crank-Nicholson [*53*] method of solution of the finite difference approximation for the following differential equation:

$$\partial\bar{u}/\partial t = c_r/r\cdot\partial/\partial r\,(\partial\bar{u}/\partial r) + c_v\cdot\partial^2\bar{u}/\partial z \quad (1)$$

where $\bar{u}$ is the excess pore water pressure, r is radius, z is depth, and c_r is the coefficient of consolidation for vertical compression for radial water flow. The program was written to account for the following conditions:

1. The soil may be stratified with each layer assigned a separate set of soil properties.
2. For each layer, the consolidation curve is input as a series of points of e (or strain) versus effective stress and the average value of e and average effective stresses in each layer track through these points during consolidation.
3. The effects of changing layer thickness on drainage distance are taken into account but each layer is assumed to remain homogeneous.
4. Vertical coefficients of permeability and c_v and c_r are also input as functions of vertical effective stress.
5. Applied load may be any desired function of time.
6. If unloading occurs, each layer rebounds from the value of e and effective stress in existence at the time of unloading. It is assumed that the rebound e versus log $\bar{\sigma}$ curve is linear and that c_v and c_r remain constant during swelling.
7. During construction of fill, the program automatically compensates for settlement so the elevation of the top of the fill is the specified value at the end of each construction stage.
8. Submergence corrections are made continuously.
9. The elevation of the water table may change during consolidation.
10. Pore water pressures above the water table may be set at zero or the negative values corresponding to the no-flow condition.
11. Sand drains may be partially penetrating.
12. Effects of disturbance during installation of sand drains may be taken into account by specifying new curves of e versus log $\bar{\sigma}$, c_v versus log $\bar{\sigma}$ and c_r versus log $\bar{\sigma}$, to be used after drain installation. Each layer is still treated as being homogeneous.
13. Effects of stress distribution are taken into account through use of depth-dependent influence factors.
14. An initial artesian condition may be specified. The artesian pore pressures begin to dissipate as soon as sand drains penetrate the aquifer.
15. Sand layers may be included above or within the system of compressible layers.

[4] N is the dimensionless sand drain spacing factor defined as R_E/R_W where R_E is the radius of influence of the drain and R_W is the radius of the sand drain.

Three different sets of analyses were performed. First, an analysis was performed using soil properties obtained directly from laboratory consolidation tests without corrections for effects of disturbance. Second, an analysis was performed using soil properties determined from e versus log $\bar{\sigma}$ curves that were reconstructed by a method suggested by Schmertmann [*20*] using a special technique for reconstructing the c_v versus log $\bar{\sigma}$ relationship (see later discussion). Finally, a trial-and-error analysis was performed to try to ascertain the actual in-situ soil properties at the site.

For all analyses, the soil system was divided into seven sublayers. These divisions were made after visual inspection of the laboratory consolidation data. Layer A (Fig. 2) was divided into four sublayers designated A1, A2, A3, and A4 with interfaces at initial depths of 1.8, 4.6, and 6.4 m (6, 15, and 21 ft). Layer B was considered as one sublayer and Layer C was divided into two sublayers designated C1 and C2. The division between sublayers C1 and C2 was made at a depth of 18.3 m (60 ft). Several assumptions were made concerning the interpretation of the settlement data. First, the settlement platforms were not installed until 1.1 m (3.5 ft) of fill had been placed. It was assumed that this 1.1 m of fill was incompressible and that the settlement of the ground surface was equal to the settlement of the platforms. Second, the deep settlement points were not installed until 3.8 m (12.5 ft) of fill had been placed. Layer C was sufficiently precompressed that negligible settlement of this layer should have occurred during the application of 3.8 m of fill. Therefore all the surface settlement measured under the first 3.8 m of fill was assumed to be due to compression of Layer A.

Unfortunately, field settlement records were kept for only 17 months after construction operations were completed, so no data on the ultimate settlement at the site were available. The lack of a value of ultimate settlement has little effect on the accuracy of the analysis of primary consolidation but does preclude analyses of secondary compression following the primary stage.

The amount of vertical deformation caused by undrained mass movement is difficult to determine. Measured horizontal movements at the edge of the embankment, in areas where no failure occurred, reach 16 in. but these observations were made more than 100 ft away from the location of the settlement gages, which were near the centerline of the embankment (Fig. 2). Further, a shear failure that occurred in the side slope of the embankment did not influence the measured centerline settlements detectably. Consequently, undrained centerline deformations were assumed to be negligible.

Analysis Using Uncorrected Laboratory Soil Properties

All consolidation test data from Haley and Aldrich's files were re-analyzed. The square root fitting method was used and curves of e_{100} versus log $\bar{\sigma}$ were prepared for each test. Test data for samples from Layers A and C were excluded if the shapes of the curves suggested that the sample had been disturbed significantly. After visual inspection of the remaining laboratory data, one typical e versus log $\bar{\sigma}$ and c_v versus log $\bar{\sigma}$ curve was chosen for each sublayer to model that sublayer's compressive properties.

Analysis Using Reconstructed Soil Properties

The laboratory curves assigned to each of the sublayers described above were reconstructed to approximate in-situ conditions. The procedure was as follows: First, the e versus log $\bar{\sigma}$ curve assigned to each sublayer was reconstructed using a method suggested by Schmertmann [*20*]. Second, values of k were adjusted to approximate in-situ values, as discussed below.

In addition to hydrodynamic effects (delayed settlement caused by the finite k of the soil), some of the delay in consolidation may be due to soil structural effects, that is, the soil structure itself may have a time-dependent resistance to volume change. These soil structural effects are commonly termed secondary or creep effects and are not taken into account in Terzaghi's theory. Note that the secondary effects of interest occur here during the apparent primary stage.

If the secondary effects are of a viscous nature they may be more significant in the laboratory consolidation test than during an actual field loading because of the high strain rate in the laboratory test. The resulting increase in laboratory consolidation time causes k as calculated using Terzaghi's theory to be lower than the actual value of k.

It would seem that the measured k would be the value to use in field analyses where soil viscous effects may be minimal [*54*]. However, field values of k were not measured in this study; therefore, the actual values had to be estimated from the calculated values. Research has shown that for consolidation tests performed on sedimented samples the measured and calculated values of k are often in close agreement at lower values of e, that is, on the virgin compression curve, as shown in Fig. 6. Based on such research, the calculated values of k for points on the virgin consolidation curve were simply extrapolated linearly on e versus log k plots back into the range of void ratio encountered during reloading. Corrected values of c_v were then calculated using the extrapolated values of k and values of compressibility taken from the reconstructed e versus $\bar{\sigma}$ curve.

No measurements were made of c_r. In keeping with a previous analysis in the area [*22*] the value of c_r/c_v was assumed to be 1.2.

Analysis Using Fitted Soil Properties

In order to estimate the actual soil properties at the site, e versus log $\bar{\sigma}$, c_r versus log $\bar{\sigma}$, and c_v versus log $\bar{\sigma}$ curves were modified by trial and error until the computed and measured time-settlement curves and excess pore pressure curves were essentially identical.

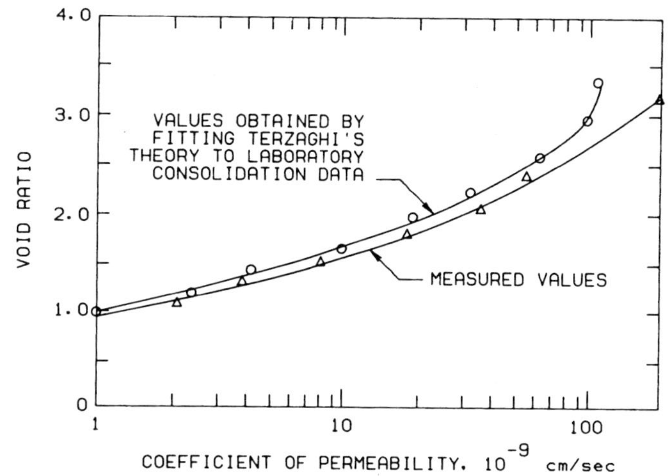

FIG. 6—*Comparison of measured and computed coefficients of permeability for a sample of clay that was prepared by sedimentation from dilute suspension in the laboratory.*

Analysis Using Soil Properties Obtained from Uncorrected Laboratory Consolidation Tests

The time-settlement curves calculated using uncorrected laboratory data are compared with the actual time-settlement curves for the surface settlement in Fig. 7a. The computed surface settlement exceeds the measured surface settlement by 410 mm (16 in.) in the early portion of the curve and by 200 mm (8 in.) near the end.

The computed and actual time-compression curves for the various layers are compared in Figs. 7b to 7d. In Layers A and C the computed compression exceeds the measured values by 330 and 80 mm (13 and 3 in.), respectively. These errors are consistent with the usual experience that moderate amounts of disturbance lead to a rounding of the e versus log $\bar{\sigma}$ curve and an increase in computed settlement. In Layer B the computed settlements are much less than the measured values. It is possible that the one consolidation test for this layer was performed on a nonrepresentative sample. A more likely explanation is that the sample was very badly disturbed. Since the program causes the soil to follow the input e versus log $\bar{\sigma}$ curve precisely, the initial e used by the program is not the real one but rather the e on the laboratory curve at the initial effective stress. Thus in the case of a severely disturbed sample the initial e used in the analysis would be much lower than the actual initial e in the field and consequently the calculated settlement would be too low. This explanation will be examined more closely when the fitted field e versus log $\bar{\sigma}$ curve is discussed.

Analysis Using Reconstructed Soil Properties

The time-settlement curves calculated using reconstructed soil properties are compared with the measured time-settlement curves for the surface, and for all three layers, in Fig. 8. Although the computed and measured surface settlements are in fair agreement, it is apparent that settlements were overestimated in Layer A and underestimated in Layers B and C. The main source of error is believed to be the estimate of $\bar{\sigma}_{max}$.

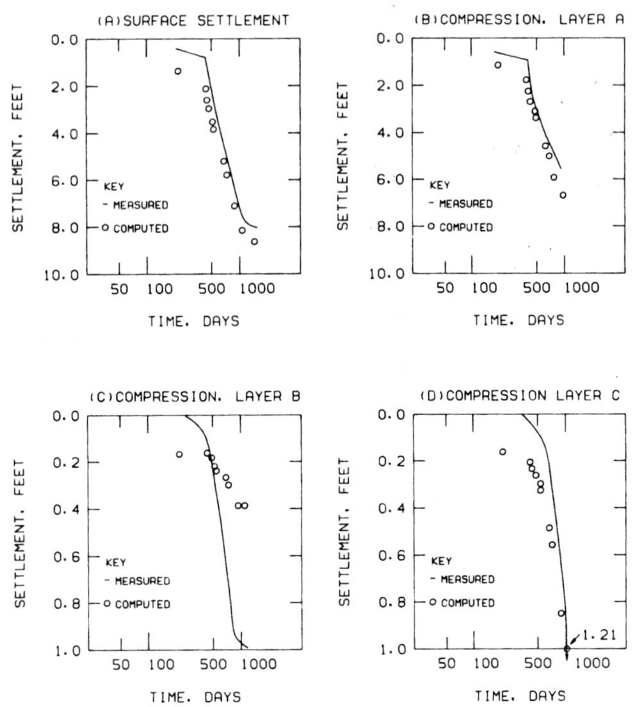

FIG. 7—*Comparison of measured and computed curves for surface settlement and compression of each of the three layers. The computed curves were based on uncorrected laboratory data.*

It should be noted that the reconstructed *e* versus log $\bar{\sigma}$ curves passed through the correct initial *e* and effective stress. The resulting great improvement in correlation between measured and computed settlements in Layer B supports the explanation offered above for the poor correlation using the uncorrected laboratory *e* versus log $\bar{\sigma}$ curve.

Analysis Using Fitted Soil Properties

A trial-and-error procedure was used to fit the theory to the measured field compression curves. The fitted theoretical time-settlement curves for the surface and each layer are shown in Fig. 9.

Layer A

The *e* versus log $\bar{\sigma}$ curves obtained by the trial-and-error fitting process are compared with the laboratory and reconstructed *e* versus log $\bar{\sigma}$ curves used previously for Sublayers A1 through A4 in Fig. 10. The reconstructed field *e* versus log $\bar{\sigma}$ curves lie above the uncorrected laboratory curves in all cases, but the field curves and reconstructed laboratory curves are close except for Sublayer A1.

The fitted values of c_v versus effective stress are shown together with all laboratory and reconstructed c_v values in Fig. 11. Although the scatter in the laboratory data was considerable, the fitted c_v values are consistently larger than the laboratory values.

Generally, the c_v values obtained using the reconstructed *e* versus log $\bar{\sigma}$ and *e* versus log *k* curves were lower than those obtained by the trial-and-error solution, although excellent agreement was achieved in a few cases. A better agreement between the reconstructed and fitted values of c_v would be achieved if the estimated *e* versus log *k* curve was displaced to higher *k* values.

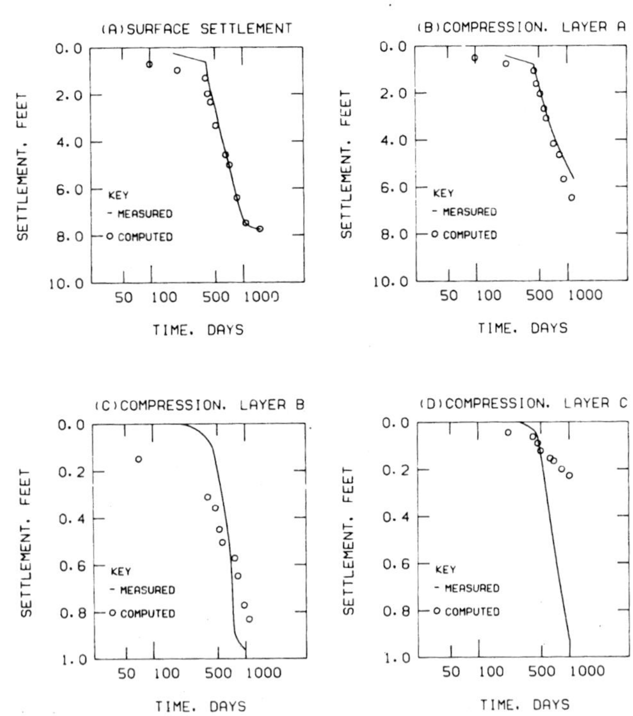

FIG. 8—*Comparison of measured and computed curves for surface settlement and compression of the three layers. The computed settlements were based on laboratory data corrected for the effects of sampling disturbance.*

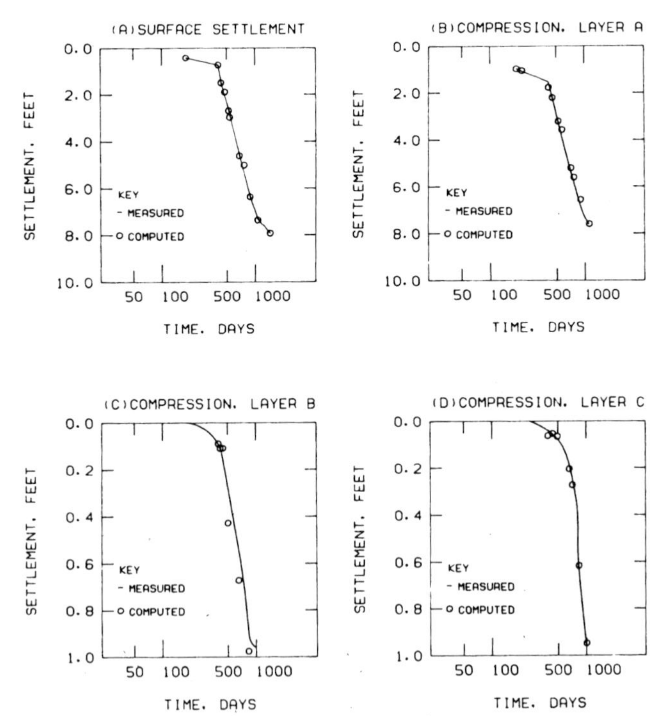

FIG. 9—*Comparison of measured curves of surface settlement and compression of individual layers with the curves fitted to the field data by trial.*

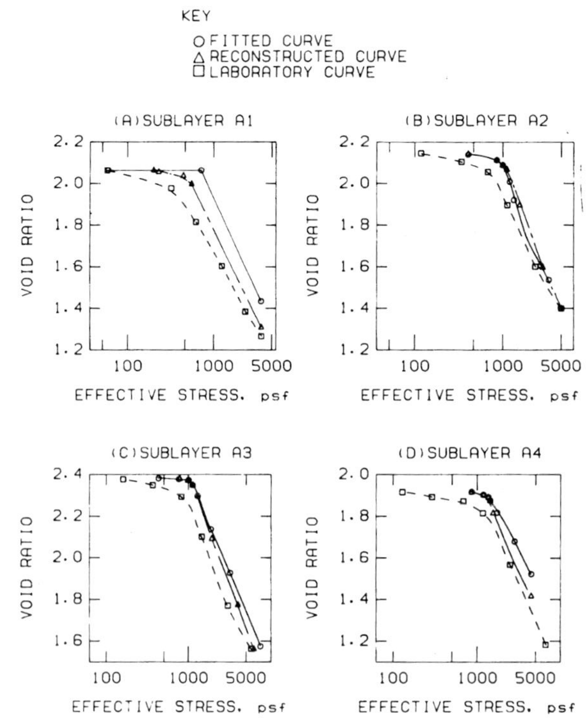

FIG. 10—*Comparison of compression curves for Layer A.*

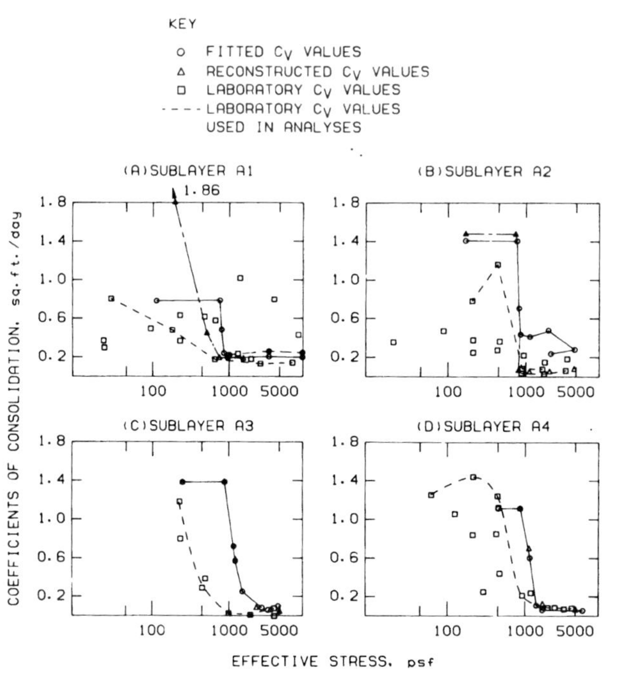

FIG. 11—*Comparison of values of* c_v *for vertical drainage for Layer A.*

Values of c_r varied in the same manner as c_v values (Fig. 12) but were higher in most cases. The ratio c_r/c_v ranged from 1.0 to 2.0 for the backfitted analysis.

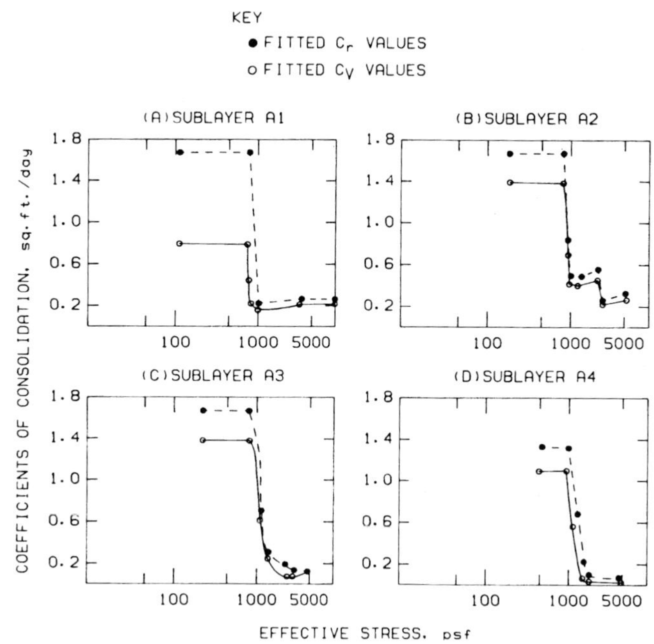

FIG. 12—*Comparison of the fitted values of* c_r *and* c_v *for Layer A.*

Layer B

The fitted e versus log $\bar{\sigma}$ curve is compared with the corrected and original laboratory e versus log $\bar{\sigma}$ curves in Fig. 13; the disturbance of the sample used for the laboratory test seems evident. The close correlation between the reconstructed laboratory e versus log $\bar{\sigma}$ curve and the fitted field curve is fortuitous; the original laboratory curve is so straight that there is little basis for the reconstruction.

Both laboratory data and field observations indicated that Layer B was freely draining.

Layer C

The e versus log $\bar{\sigma}$ curves obtained by the trial-and-error fitting process are shown together with the original laboratory and re-

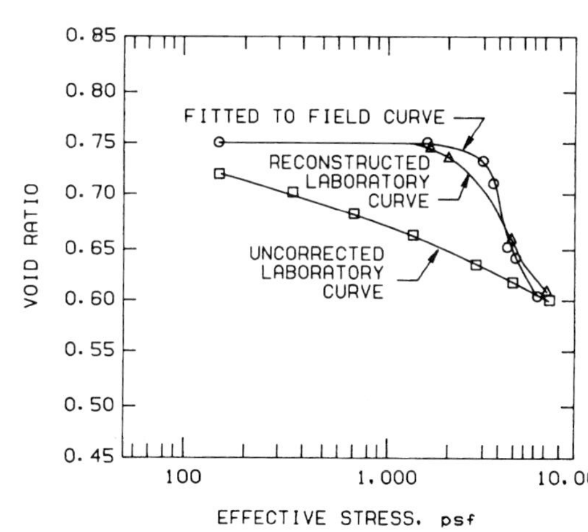

FIG. 13—*Relationship of* e *to log* $\bar{\sigma}$ *for Layer B.*

constructed laboratory e versus log $\bar{\sigma}$ curves for Sublayers C1 and C2 in Fig. 14. The laboratory curves are rounded, apparently due to sampling disturbance, with samples from Sublayer C2 showing more disturbance than those from Sublayer C1. The reconstructed laboratory and fitted field e versus log $\bar{\sigma}$ curves are similar for Sublayers C1 and C2.

The fitted field values of c_v versus effective stress are shown along with all laboratory and reconstructed values in Fig. 15. For both of the Sublayers C1 and C2, the fitted values of c_v are higher than the values determined in the laboratory, presumably because of disturbance of the samples. As might be expected, a constant value of c_v was adequate to model Sublayer C1 as the vertical effective stress in this layer never reached $\bar{\sigma}_{max}$. Sublayer C2 was loaded to an effective stress greater than $\bar{\sigma}_{max}$; c_v values at stresses beyond $\bar{\sigma}_{max}$ drop into the range of values obtained in the latter stages of the laboratory consolidation tests.

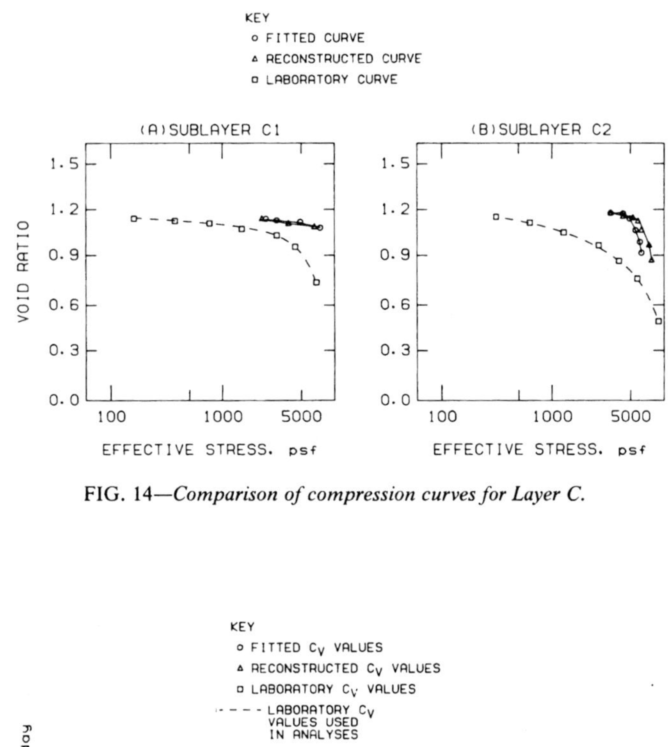

FIG. 14—*Comparison of compression curves for Layer C.*

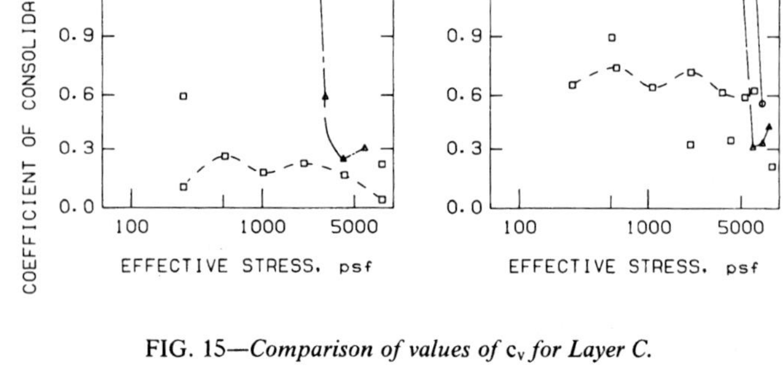

FIG. 15—*Comparison of values of* c_v *for Layer C.*

Discussion

The problems encountered in analyzing consolidation of soils when sand drains are used are significant. The samples tested represent a minute fraction of the soil deposit and may not be representative. Extrapolation from the strain rates and gradients used in the laboratory to the values encountered in the field applies a severe test for the theory. The effects of disturbance when sand drains are installed are not clear. The accuracy of the field data themselves are often suspect. The method of analysis typically used involves application of grossly simplifying assumptions. The analytical method used here is more general than the method typically used in engineering practice but is still greatly simplified. For example, it is a pseudo-two-dimensional analysis in which no attempt is made to account for constant volume shear deformation nor for such effects as formation of zones of reduced permeability near drainage boundaries. As a result of these circumstances it is not possible to prove the validity of either the analysis or the interpretation.

On the other hand it can be concluded that the field data show no clear evidence of errors in the extended version of Terzaghi's theory. Corrections to the laboratory data of a type, direction, and magnitude that are consistent with known effects of sample disturbance produce excellent correlations between predicted and measured time-compression curves for each of the seven layers used in the analysis. It appears that additional work is needed in developing sampling and testing methods that will further reduce the effects of disturbance. Further development of theory also seems justified but the lack of detailed field data on such factors as spatial variation of pore pressures and deformations and relevant soil properties will make it difficult to estimate the field applicability of an improved analysis.

Summary and Conclusions

The two-dimensional finite difference analysis for sand drains seems to account for most of the variables that are important when the analysis applies to a central region of the embankment. The analysis specifically accounts for effective stress-dependent values of e and coefficients of consolidation, stratified soils, time-dependent loading, partially penetrating drains, stress distribution with depth, artesian condition, and a number of other effects.

The use of uncorrected laboratory data coupled with the finite difference analysis led to an overprediction of surface settlements at all times, with the percentage of error largest during the early stages. Correction of the e versus log $\bar{\sigma}$ and c_v versus log $\bar{\sigma}$ relationships led to a significant improvement in predicted field performance, except for Layer C where the estimated value of $\bar{\sigma}_{max}$ was too large.

A trial-and-error procedure was used to fit the numerical analysis to the field data and curves of e versus log $\bar{\sigma}$, c_v versus log $\bar{\sigma}$, and c_r versus log $\bar{\sigma}$ were obtained. These curves differed from laboratory curves in the way one would expect if the laboratory samples were disturbed. Improved techniques for sampling and testing and improved techniques for correcting laboratory data to account for the effects of disturbance are needed to allow full use of numerical methods of analysis.

Acknowledgments

Analyses of the data for the Fore River site were supported by the National Science Foundation under grant ENG75-09609. Field data were provided by Haley and Aldrich, Inc., consulting geotechnical engineers, Cambridge, Mass. Field data were collected by personnel of the Maine State Highway Commission. Appreciation is expressed to the Maine State Highway Commission for releasing the data and to NSF for financial support.

References

[1] Taylor, D. W., *Fundamentals of Soil Mechanics,* John Wiley and Sons, New York, 1948.

[2] Mitchell, J. K. and Younger, J. S., "Abnormalities in Hydraulic Flow Through Fine Grained Soils," in *Permeability and Capillarity in Soils,* STP 417, American Society for Testing and Materials, Philadelphia, 1966, pp. 106-139.

[3] Barden, L., "Consolidation with Non-Linear Viscosity," *Geotechnique,* Vol. 15, 1965, pp. 267-286.

[4] Barden, L., "Primary and Secondary Consolidation of Clay and Peat," *Geotechnique,* Vol. 18, No. 1-24, 1968.

[5] Barden, L. and Berry, P. L., "Consolidation of Normally Consolidated Clay," *Journal of the Soil Mechanics and Foundations Division of the American Society of Civil Engineers,* Vol. 91, No. SM5, 1965, pp. 15-36.

[6] Berry, P. L. and Poskitt, T. J., "The Consolidation of Peat," *Geotechnique,* Vol. 22, 1972, pp. 27-52.

[7] Berry, P. L. and Vickers, B., "Consolidation of Fibrous Peat," *Journal of the Geotechnical Engineering Division of the American Society of Civil Engineers,* Vol. 101, No. GT8, 1975, pp. 741-753.

[8] Davis, E. H. and Raymond, G. P., "A Nonlinear Theory of Consolidation," *Geotechnique,* Vol. 15, No. 2, 1965, pp. 161-173.

[9] Lo, K. Y., "Secondary Compression of Clays," *Journal of the Soil Mechanics and Foundations Division of the American Society of Civil Engineers,* Vol. 87, No. SM4, 1961, pp. 61-88.

[10] Merchant, W., "Some Theoretical Considerations on the One-Dimensional Consolidation of Clay," M.S. thesis, Massachusetts Institute of Technology, Cambridge, Mass., 1939.

[11] Murayama, S. and Shabata, T., "On the Secondary Consolidation of Clay," *Proceedings of the Second Japan Congress on Testing Materials,* 1959, pp. 178-181.

[12] Poskitt, T. J., "A Note on the Consolidation of Clay with Nonlinear Viscosity," *Geotechnique,* Vol. 17, 1967, pp. 284-289.

[13] Poskitt, T. J., "Consolidation of Clay and Peat with Variable Properties," *Journal of the Soil Mechanics and Foundations Division of the American Society of Civil Engineers,* Vol. 97, No. SM6, 1971, pp. 841-888.

[14] Schiffman, R. L., "Secondary Consolidation of Clays: A Viscoelastic Analysis," *Advances in Consolidation Theories for Clays,* H. B. Poorooschasb, Ed., *Proceedings of Specialty Session No. 12,* Seventh International Conference on Soil Mechanics and Foundation Engineering, 1969.

[15] Schiffman, R. L., Chen, A., and Jordan, J. C., "An Analysis of Consolidation Theories," *Journal of the Soil Mechanics and Foundations Division of the American Society of Civil Engineers,* Vol. 95, No. SM1, 1969, pp. 285-312.

[16] Schiffman, R. L., Ladd, C., and Chen, A., "The Secondary Consolidation of Clay," in *Rheology and Soil Mechanics,* International Union of Theoretical and Applied Mechanics, Ed. by J. Kravtchenko and P. M. Sirieys, Springer-Verlag, West Berlin, 1964.

[17] Taylor, D. W. and Merchant, W., "A Theory of Clay Consolidation Accounting for Secondary Compressions," *Journal of Mathematics and Physics,* Vol. 19, 1940, pp. 167-185.

[18] Casagrande, A., "The Structure of Clay and its Importance in Foundation Engineering," *Contributions to Soil Mechanics,* Boston Society of Civil Engineers, Boston, 1932, pp. 72-113.

[19] Casagrande, A., "The Determination of the Preconsolidation Load and its Practical Significance," *Proceedings of the International Conference in Soil Mechanics and Foundation Engineering,* Vol. 3, 1936, pp. 60-64.

[20] J. H. Schmertmann, "The Undisturbed Consolidation Behavior of Clay," *Transactions of the American Society of Civil Engineers,* Vol. 120, 1955, pp. 1201-1233.

[21] Moran, Proctor, Meuser, and Rutledge, "Study of Deep Soil Stabilization by Vertical Sand Drains," Publication PB151692, U.S. Dept. of Commerce, Washington, D.C., 1958.

[22] Olson, R. E., Daniel, D. E., and Liu, T. K., "Finite Difference Analysis for Sand Drain Problems," *Analysis and Design in Geotechnical Engineering,* Vol. 1, American Society of Civil Engineers, New York, 1974, pp. 85-110.

[23] Abbott, M. B., "One-Dimensional Consolidation of Multi-Layered Soils," *Geotechnique,* Vol. 10, 1060, pp. 151-165.

[24] Schiffman, R. L. and Stein, J. R., "Progrs-I, A Computer Program to Calculate the Progress of Ground Settlement," *International Union of Theoretical and Applied Mechanics,* Ed. By J. Kravtchenko and P. M. Sirieys, Springer-Verlag, West Berlin, 1964.

[25] Schiffman, R. L. and Stein, J. R., "One-Dimensional Consolidation of Layered Systems, *Journal of the Soil Mechanics and Foundations Division of the American Society of Civil Engineers,* Vol. 96, No. SM4, 1970, pp. 1499-1504.

[26] Scott, R. F., "Numerical Analysis of Consolidation Problems," M.S. thesis, Dept. of Civil Engineering, Massachusetts Institute of Technology, Cambridge, Mass., 1953.

[27] Christian, J. T. and Boehmer, J. W., "Plane Strain Consolidation by Finite Elements," *Journal of the Soil Mechanics and Foundations Division of the American Society of Civil Engineers,* Vol. 96, No. SM4, 1970, pp. 1435-1457.

[28] Christian, J. T., Boehmer, J. W., and Martin, P. P., "Consolidation of a Layer Under a Strip Load," *Journal of the Soil Mechanics and Foundations Division of the American Society of Civil Engineers,* Vol. 98, No. SM7, 1972, pp. 693-707.

[29] Davis, E. H. and Lee, I. K., "One-Dimensional Consolidation of Layered Soils," *Proceedings of the Seventh International Conference on Soil Mechanics and Foundation Engineering,* Vol. 2, 1969, pp. 65-72.

[30] Desai, C. S. and Johnson, L. D., "Evaluation of Two Finite Element Formulations for One-Dimensional Consolidations," *Computers and Structures,* Vol. 2, 1972, pp. 469-486.

[31] Desai, C. S. and Johnson, L. D., "Evaluation of Some Numerical Schemes for Consolidation," *International Journal For Numerical Methods in Engineering,* Vol. 7, 1973, pp. 243-254.

[32] Gibson, R. E. and Lumb, P., "Numerical Solution of Some Problems in the Consolidation of Clay, *Proceedings of the Institution of Civil Engineers,* Pt. 1, Vol. 2, Part 2, London, p. 182.

[33] Hart, E. G., Kondner, R. L., and Boyer, W. C., "Analysis for Partially Penetrating Sand Drains," *Journal of the Soil Mechanics and Foundations Division of the American Society of Civil Engineers,* Vol. 84, No. SM4, Paper 1812, 1958.

[34] Kinard, D. T., "A Case Study of the Time Effects in One-Dimensional Consolidation, M.S. thesis, University of Texas, Austin, Tex., 1977.

[35] Krizek, R. J. and Krugmann, P. K., "Precompression Analysis for Highway Embankments," *Analysis and Design in Geotechnical Engineering,* Vol. 1, American Society of Civil Engineers, New York, 1974, pp. 111-141.

[36] Mesri, G. and Rokhsar, A., "Theory of Consolidation for Clays," *Journal of the Geotechnical Engineering Division of the American Society of Civil Engineers,* Vol. 100, No. GT8, 1974, pp. 889-904.

[37] Pelletier, J. H., "Determination of Soil Properties From Analyses of A Full Scale Highway Embankment," M.S. thesis, University of Texas, Austin, Tex., 1977.

[38] Schiffman, R. L. and Arya, S. K., "One-Dimensional Consolidation," in *Numerical Methods in Geotechnical Engineering,* C. S. Desai and J. T. Christian, Eds., McGraw-Hill Book Co., New York, 1977.

[39] Schiffman, R. L. and Gibson, R. E., "Consolidation of Nonhomogeneous Clay Layers," *Journal of the Soil Mechanics and Foundations Division of the American Society of Civil Engineers,* Vol. 90, No. SM5, 1964, pp. 1-30.

[40] Biot, M. A., "General Solutions of the Equations of Elasticity and Consolidation for a Porous Material," *Journal of Applied Mechanics,* Vol. 23, 1956, pp. 91-96.

[41] Bishop, A. W. and Gibson, R. E., "The Influence of the Provisions for Boundary Drainage on Strength and Consolidation Characteristics of Soil Measured in the Triaxial Apparatus," in *Laboratory Shear Testing of Soils,* STP 361, American Society for Testing and Materials, Philadelphia, 1964, pp. 435-451.

[42] Carrillo, N., "Simple Two- and Three-Dimensional Cases in the Theory of Consolidation of Soils, *Journal of Mathematics and Physics,* Vol. 21, No. 1, 1942, pp. 1-5.

[*43*] Christian, J. T., "Consolidation with Internal Pressure Generation," *Journal of the Geotechnical Engineering Division of the American Society of Civil Engineers,* Vol. 102, No. GT10, 1976, pp. 1111-1115.

[*44*] Christian, J. T., "Consolidation with Triangular Pressure Generation," *Journal of the Geotechnical Engineering Division of the American Society of Civil Engineering,* Vol. 102, No. GT2, 1977, pp. 133-136.

[*45*] De Wet, J. A., "Three-Dimensional Consolidation," *Highway Research Bulletin,* No. 342, 1962, pp. 342.

[*46*] Gibson, R. E., England, G. L., and Hussey, M. J. L., "The Theory of One-Dimensional Consolidation of Saturated Clays," *Geotechnique,* Vol. 17, No. 3, 1967, pp. 261-273.

[*47*] De Josselin De Jong, G., "Application of Stress Functions to Consolidation Problems," *Proceedings of the Fourth International Conference on Soil Mechanics and Foundation Engineering,* Vol. 1, 1957, pp. 320-323.

[*48*] De Josselin De Jong, G., "Consolidatie In Drie Dimensies," *L. G. M. Medelingen,* Delft, Vol. 7, No. 3, 1963, pp. 57-73.

[*49*] McNamee, J. and Gibson, R. E., "Plane Strain and Axially Symmetric Problems of the Consolidation of a Semi-Infinite Clay Stratum," *Quarterly Journal of Mechanics and Applied Mathematics,* Vol. 13, 1960, pp. 210-227.

[*50*] Rowe, P. W., "The Calculation of the Consolidation Rates of Laminated Varved or Layered Clays, with Particular Reference to Sand Drains," *Geotechnique,* Vol. 14, No. 4, 1964, pp. 321-340.

[*51*] Zaretskii, Y. K., "Theory of Soil Consolidation," (translated from the Russian,) U.S. Dept. of Commerce Publication TT70-50131, National Technical Information Services, Springfield, Va., 1972.

[*52*] Haley and Aldrich, "*Engineering Properties of Foundation Soils at Long Creek-Fore River Areas and Back Cove,*" Report No. 1 to the Maine State Highway Commission, Cambridge, Mass., 1969.

[*53*] Crank, J., and Nicholson, P., "A Practical Method for Numerical Evaluation of Solutions of Partial Differential Equations of the Heat Conduction Type," *Proceedings of the Cambridge Philosophical Society,* Cambridge, England, 1947, pp. 50-67.

[*54*] Weber, W. G., Jr., "In-Situ Permeabilities for Determining Rates of Consolidation," *Highway Research Record,* No. 243, 1968, pp. 49-61.

Radial Flow Consolidation Testing

Etude Expérimentale de la Consolidation à Drainage Radial

S.J. TRAUTWEIN Graduate Research Assistant, Civil Engineering Department, The University of Texas at Austin, Austin, Texas
R.E. OLSON Professor, Civil Engineering Department, The University of Texas at Austin, Austin, Texas
R.L. THOMAS Engineer, Phillips Petroleum Company, Bartlesville, Oklahoma.

SYNOPSIS A radial flow consolidation cell that can be used with commercially available loading frames was developed and a series of tests were performed on samples of sedimented kaolin. Simple curve fitting methods are discussed for the calculation of the radial coefficient of consolidation for both radially inwards and outwards drainage using either free or equal strain theory. Test results indicated that consolidation is retarded due to a smear zone when rigid drains are used and that the free strain theory works better in reducing the data.

INTRODUCTION

The design of sand drain projects in the United States has been based on either experience or simple closed form analyses. Even though the economic justification of using sand drains depends largely on the radial coefficient of consolidation (CR), it is seldom measured directly. When design is based on experience, soil properties are usually defined using index tests or one-dimensional consolidation tests. Even when closed form analyses are used, the simplifying assumptions are so severe that most designers are satisfied with properties obtained from normal one-dimensional consolidation tests. The lack of confidence in these types of analyses has been demonstrated by the use of full-scale instrumented test embankments during the design phase.

The recent advent of more sophisticated design techniques (Olson et al., 1974; Pelletier et al., 1979) has brought with it a need for much better definition of the relevant soil properties. As a minimum, it is necessary to be able to perform radial-flow consolidation tests routinely with samples of reasonable sizes.

Attempts to use apparatus and data-reduction techniques reported in the literature demonstrated that both needed improvement. This paper is concerned with the interpretation of laboratory tests with one-dimensional compression and radial drainage. Simple methods are discussed for the calculation of CR for both radially inwards and outwards drainage using both the equal strain and free strain theories. Experimental studies in which the measured values of CR were higher for inflow than for outflow tests are discussed and a rational explanation of the observations is advanced. This explanation indicates that apparatus used by several investigators (McKinlay, 1961; Aboshi and Monden, 1963; Rowe and Barden, 1966) most likely yielded values of CR which were too low.

EQUIPMENT AND TESTING PROGRAM

A radial-flow consolidation cell was developed that can be used with commercially available loading frames and can accommodate samples 63 mm in diameter by 50 mm high. Water can either drain into a central sand drain (an "inflow" test) or out to a freely draining outer boundary (an "outflow" test). Radial permeability tests are also possible.

Samples used for inflow tests were provided with a 13-mm diameter sand drain, surrounded by a layer of filter paper to prevent soil intrusion. The drain was cut with a thin walled tube. At the outer boundary, side friction was minimized using a tetrafluoethylene liner. For outflow tests this liner was replaced with a 3-mm thick precast porous plastic sleeve with a permeability of about 0.0001 cm/sec.

The radial permeability of the soil was measured by flowing water from a central sand drain to the outer porous liner, after the soil had consolidated to various stresses.

Outflow tests using rigid drains are generally to be preferred over inflow tests because of the simpler test set up and faster consolidation.

The soil used in this study was kaolin "Hydrite R," a commercially prepared clay with a specific gravity of 2.58, and liquid limit and plasticity index of 57% and 17%, respectively. Samples were prepared by mixing powdered kaolin with water to form a slurry with a water content of 300%, and then consolidating the slurry under 69 kPa in tubes with inside diameters of 63 mm. Both radial flow and standard one-dimensional consolidation tests were performed.

4/59

THEORY

The time rate of consolidation of a cylindrical soil sample undergoing radial consolidation may be described using either the free strain (FS) or equal strain (ES) theories. The FS theory applies to the case of a uniform surface pressure and non-uniform deformation, whereas the ES theory applies to uniform deformation and non-uniform pressure.

Free strain solutions were presented by Barron (1948) for inflow, and by Silviera (1953) and McKinlay (1961) for outflow. Barron (1948) also presented an ES solution for inflow, and Escario and Uriel (1961) obtained an outflow solution. Theoretical curves of degree of consolidation (UR) versus radial time factor (TR) are presented in Fig. 1 for ES and FS with:

$$TR = (CR)(T)/(RE)^2 \quad (1)$$

where CR is the radial coefficient of consolidation, T is time, and RE is the external radius of the sample. In Fig. 1 N is the ratio RE/RW, where RW is the radius of the drain.

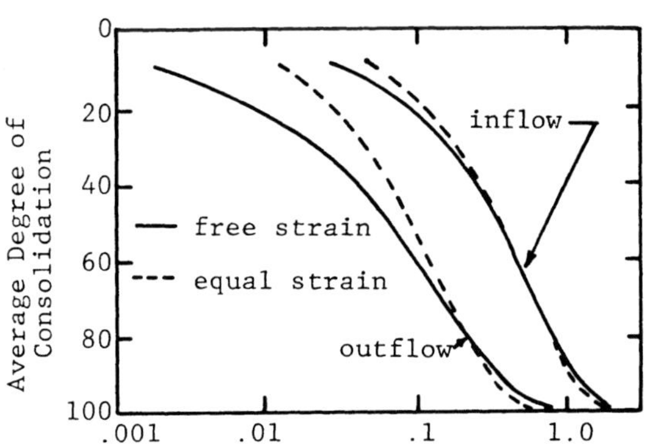

Fig. 1. Theoretical UR-TR Relationships for Inflow (N=5) and Outflow Radial Consolidation Tests

Although rigid loading caps were used in this study it is not clear which theory would be better for reducing the data. The rigid caps forced an ES condition during consolidation but the uniform initial excess pore water pressure (Collins, 1979) corresponds to the FS theory. The difference between ES and FS UR-TR curves is greater for outflow than for inflow (Fig. 1) tests, thus suggesting that outflow tests should be used in determining the relative validity of the two theories.

DATA REDUCTION

Several methods of data reduction have been suggested (Escario and Uriel, 1961; Aboshi and Monden, 1963; Shields and Rowe, 1965). However, these methods are either tedious due to the need for repetitive curve plotting or they do not yield the values of settlement corresponding to the beginning and end of primary consolidation. McKinlay (1961) presented the most convenient method of data reduction in which settlement (S) is plotted versus time (T) raised to a power n. His method, which is only good for outflow tests using free strain theory, is similar to the standard square root method of Taylor (1942) except n and the curve fitting factor (CFF) differ from 0.5 and 1.15, respectively. The approach taken by McKinlay was used to develop new methods for the remaining cases. Values for n, CFF, and TR at 45% consolidation for all the test types are summarized in Table I (Trautwein, 1980).

Table 1 - Summary of Factors Used in the Recommended Methods of Data Reduction

	Outflow		Inflow	
	ES	FS	ES	FS
n	0.83	0.47	0.83	0.67
CFF	1.61	1.22	1.61	1.31
TR(45%)	0.075	0.049	0.280	0.264

TEST RESULTS

The void ratio-effective stress curves for vertical flow tests, and both inflow and outflow radial flow tests were similar, indicating that neither ring friction nor sand drain stiffness were serious factors in the radial flow tests.

Average values of CV and CR are plotted versus effective stress in Fig. 2. Scatter was large for stresses smaller than that used in the sedimentation tubes but was minimal for higher stresses. An unexpected result was that CR was consistently larger for inflow tests than for outflow. Measured values of permeability were compared to those back-calculated from consolidation theory using:

$$k = (CR)(AV)(\gamma)/(1 + e) \quad (2)$$

where AV is the coefficient of compressibility, γ is the unit weight of water and e is the void ratio. The comparison (Fig. 3) indicates that the inflow tests were more likely to have yielded correct results.

PROBLEMS WITH THE OUTFLOW TEST

Values of CR for outflow tests could have been reduced by the presence of a non-freely draining porous sleeve but theoretical studies using a finite difference computer program indicated that the porous plastic was freely draining.

Another possibility was that a smear zone had formed in the soil next to the drain in the outflow tests. Berry and Wilkinson (1969) presented numerical solutions for the outflow case with an outer smear zone. If the inflow

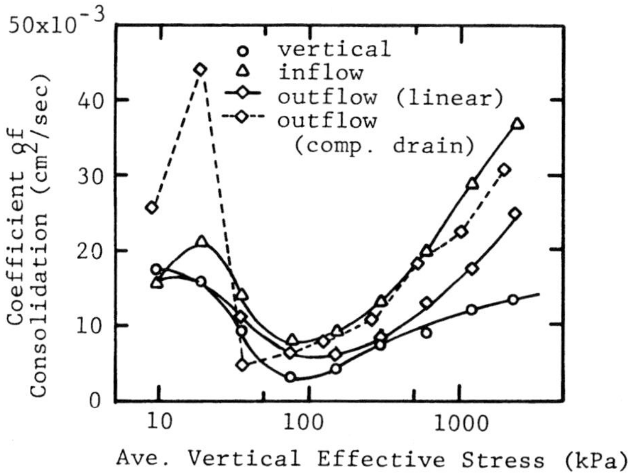

Fig. 2. Vertical and Radial Coefficients of Consolidation

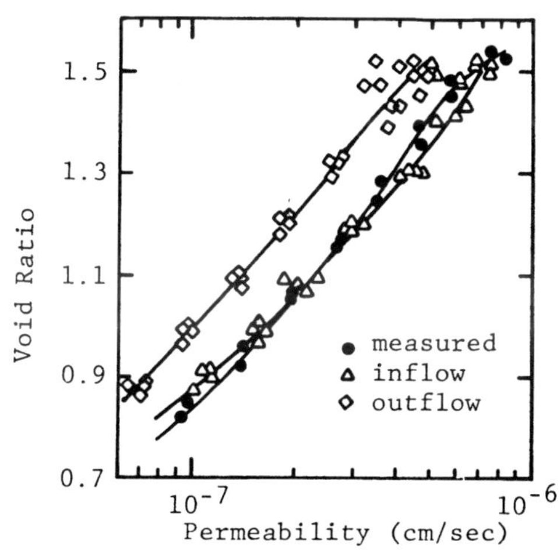

Fig. 3. Comparison of Measured and Backcalculated Permeability Using Free Strain Theory

value of CR is assumed to be correct and the radial permeability of the smear zone is taken as the vertical permeability measured in the standard one-dimensional tests, then Berry and Wilkinson's curves can be used to show that the thickness of the smear zone was about 2 mm, a value that seems reasonable.

To eliminate the smear zone caused by the rigid porous sleeve, a radial flow consolidation test was performed with a drainage zone of mixed sand and sawdust of about the same compressibility as the kaolin. The CR values obtained from this test agree reasonably well with those obtained from the inflow tests (Fig. 2), thus supporting the hypothesis that a smear zone forms as the soil is dragged down adjacent to a rigid boundary. Based on these results, the values of CR obtained from radial consolidation tests in cells employing rigid porous elements are probably less than the actual values.

Outflow tests employing compressible drains are not recommended for routine use however because of the increased side friction and the large percentage of the total area occupied by the drain. In addition, lateral deformation at the sample will occur if the compressibility of the drain is different from that of the soil being tested.

CHOICE OF THEORY

Experimental time-settlement curves for outflow tests matched theoretical curves fitted according to FS theory better than when ES was assumed. Measured and backcalculated permeabilities for the outflow tests with the compressible drain are compared in Fig. 4. Backcalculated values of permeability from the FS theory are closer to the measured values than when ES theory is used. It appears that accounting for the proper initial pore pressure distribution is more important than accounting for the proper strain conditions.

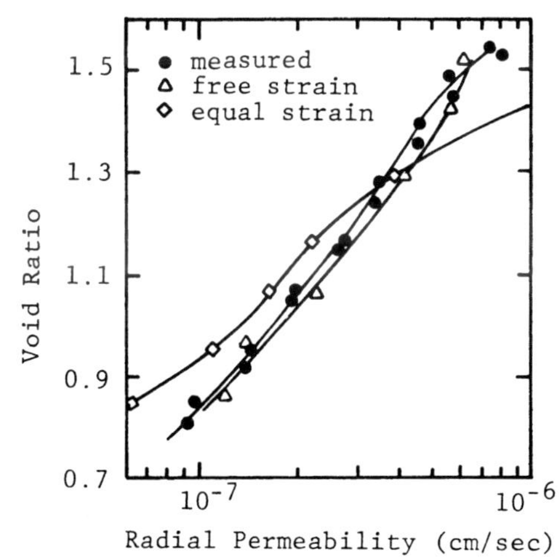

Fig. 4. Comparison of Measured and Backcalculated Permeability Using Equal Strain and Free Strain Theory for the Outflow Test with a Compressible Drain

SUMMARY AND CONCLUSIONS

The authors have presented practical methods for measuring the radial coefficient of consolidation (CR). Comparisons of test results indicated that the use of rigid drains may lead to continuous smearing of soil near the drain during consolidation, and to erroneously low values of CR. Compressible drains are recommended to avoid this problem. Test results also indicate that the free strain theory should be used in reducing test data.

4/59

ACKNOWLEDGMENTS

This research was sponsored by the National Science Foundation under Grants ENG 75-09609 and ENG-7918195.

REFERENCES

Aboshi and Monden (1963). Determination of the Horizontal Coefficient of Consolidation of an Alluvial Clay. Proc. 4th Australian-New Zealand CSMFE, 159-164, Tokyo.

Barron, R. A. (1948). Consolidation of Fine-Grained Soils by Drain Wells. Transactions (113), 718-754, ASCE.

Berry, P. L. and Wilkinson, W. B. (1969). The Radial Consolidation of Clay Soils. Geotechnique (19), 2, 253-284.

Collins, J. T. (1979). Laboratory Measurements of Soil Response During Consolidation with Sand Drains. Master's Thesis, The University of Texas at Austin.

Escario, V. and Uriel, S. (1961). Determining the Coefficient of Consolidation and Horizontal Permeability by Radial Drainage. Proc 5th ICSMFE, (1), 83-87.

McKinlay, D. G. (1961). A Laboratory Study of Consolidation in Clays with Particular Reference to Conditions of Radial Porewater Drainage. Proc 5th ICSMFE, (1), 225-228.

Olson, R.E., Daniel, D.E. and Liu, T.K. (1974) Finite Difference Analyses for Sand Drain Problems. Proc Conference on Analysis and Design in Geotechnical Engineering, ASCE, (1), 85-110.

Pelletier, J.H., Olson, R.E. and Rixner, J.J. (1979). Estimation of Consolidation Properties of Clay from Field Observations. Geotechnical Testing Journal, American Society for Testing and Materials, Vol. 2, No. 1, 34-43.

Rowe, P. W. and Barden, L. (1966). A New Consolidation Cell. Geotechnique (16), 2, 162-170.

Silviera, I. da (1953). Consolidation of a Cylindrical Clay Sample with External Radial Flow of Water. Proc 3rd ICSMFE, (1), 55-56.

Taylor, D.W. (1942). Research on Consolidation of Clays. Serial 82, Publication from the Department of Civil and Sanitary Engineering, Massachusetts Institute of Technology, Cambridge.

Trautwein, S. T. (1980). Practical Measurements of the Radial Flow Consolidation Properties of Clays. Master's Thesis, The University of Texas at Austin.

J. M. Hamilton,[1] *D. E. Daniel,*[2] *and R. E. Olson*[2]

Measurement of Hydraulic Conductivity of Partially Saturated Soils

REFERENCE: Hamilton, J. M., Daniel, D. E., and Olson, R. E., "**Measurement of Hydraulic Conductivity of Partially Saturated Soils,**" *Permeability and Groundwater Contaminant Transport, ASTM STP 746,* T. F. Zimmie and C. O. Riggs, Eds., American Society for Testing and Materials, 1981, pp. 182–196.

ABSTRACT: Increased interest in protection of the environment has led to a need to be able to predict long-term movement of moisture, and contaminants, in the vicinity of shallow land disposal sites for various toxic and radioactive wastes. The advantages of locating potential sites in arid regions has led to a need to develop means for predicting moisture movements in fairly dry, partially saturated soils and for measurements of the hydraulic conductivities of such soils.

A technique was developed in which thermocouple psychrometers and the instantaneous profile method were used to measure conductivities of soils with suctions as high as 80 atm. For one clay of low plasticity, the conductivity dropped from about 10^{-7} cm/s when saturated to almost 10^{-12} cm/s for degrees of saturation of the order of 30 percent. Measured conductivities were used in a finite element solution to Darcy's equation to predict final water content profiles. The predicted and measured profiles compared well, which suggests that the measured conductivities were tolerably accurate. Based on data presented, the proposed technique seems promising for conductivity measurements in clays with degrees of saturation between about 30 and 90 percent and in sands with 5 to 50 percent saturation.

KEY WORDS: laboratory testing, soils, partially saturated soils, unsaturated soils, psychrometers, finite element method, hydraulic conductivity, permeability, groundwater

In geology and civil engineering, problems with hydraulic conductivity have mainly involved cohesionless, highly pervious, saturated soils. Usually, efforts are made either to cut off large flows, as in the case of leakage beneath dams, or to develop large flows, as for water supply. However, hydrologists have been interested in analysis of flow in partially saturated soils, as one

[1]Geotechnical engineer, Law Engineering Testing Co., Houston, Tex. 77024.

[2]Assistant professor and professor of civil engineering, respectively, University of Texas at Austin, Department of Civil Engineering, Austin, Tex. 78712.

component in the hydrologic cycle, and agricultural engineers and agronomists have been interested in flow in partially saturated soils because of the design of irrigation systems and studies of availability of moisture to plants, respectively. The bulk of the research into permeability of partially saturated soils has been performed in agriculture. In recent years, with increased interest in protection of the environment, a need has developed to be able to predict moisture movement near disposal sites for toxic industrial wastes and radioactive wastes. There has been a tendency to locate such sites in arid regions to minimize flow into the groundwater and, hence, an increased interest in flow in fairly dry, partially saturated soils.

There are two aspects to the field problem. One is concerned with theoretical analyses of flow under realistic conditions, and the second is concerned with measurement of the relevant soil properties. In the case of toxic waste disposal sites, the interest is in movement of the contaminants rather than the water. However, the contaminants will generally move more slowly than the water, so prediction of water movement can be used as an upper bound for movement of the contaminants.

In this paper we are concerned with the problem of measuring the hydraulic conductivity of partially saturated soils in the laboratory. We will briefly consider the general problems involved, define relevant terms, discuss the method we used, and present and discuss experimental data for one soil. Measurements are reported for degrees of saturation ranging from about 90 percent to about 20 percent, that is, in samples that might be considered fairly dry.

Water Flow in Partially Saturated Soils

By analogy with saturated soils, water flow in partially saturated soils is assumed to occur in accordance with Darcy's law which, for one-dimensional flow, may be written as

$$q = -k \frac{dh}{dx} A \tag{1}$$

where q is the volume of flow per unit time [measured in terms of volume (L^3) per time (T)]; k is the constant of proportionality, usually called the hydraulic conductivity [measured in terms of length (L)/T]; h is the total head (L); x is the position coordinate (L); and A is the total cross-sectional area (L^2). In agronomy the equation is generally written using the total potential, Ψ, in place of the total head. In the remainder of this paper the term hydraulic conductivity will be shortened to conductivity.

On a macroscopic scale, the total head in a saturated soil is assumed to be composed of the pressure head, h_p, and the elevation head, h_e, with

$$h_p = u/\gamma_w \tag{2}$$

where u and γ_w denote the total pore water pressure and unit weight of water, respectively, and h_e is the elevation of the point in relation to an arbitrary datum. In unsaturated soils, other components of head must exist in order to explain the existence of total heads much more negative than the cavitation stress of water [*1*].[3] The components of the total head typically include, besides pressure and elevation, those components caused by dissolved salts in the pore water (called osmotic head) and those which arise from electrochemical interactions between pore water and soil particles (usually called the adsorptive head). In flow problems involving partially saturated soils, questions arise as to which components of the total head should be included in the calculation of the hydraulic gradient. For example, studies by Letey et al [*2*] suggest that gradients in the osmotic head do not contribute to flow of water in the same quantitative way as other components. Thus, flow due to gradients in salt concentration as well as flow due to temperature gradients are problems that must be treated separately. In this paper, we assume that gradients in the osmotic head are negligible so that gradients in the total head may be taken, approximately, as the driving force for water flow. In small laboratory samples, subject to horizontal flow, the variation in elevation head is negligible and flow is conveniently described in terms of pressure heads. As a matter of convenience, the pressure head is converted to pore pressure using Eq 2, and the use of minus signs in front of the pore water pressure is eliminated by defining suction, p, as the negative of the pore water pressure and writing Darcy's law as

$$q = \frac{k}{\gamma_w} \frac{dp}{dx} A \tag{3}$$

For suctions between 0 and 1 atm it is possible to take direct measurements provided a suitably fine porous stone is used [*3*]. The measuring apparatus is usually termed a tensiometer [*4*]. In the laboratory, higher suctions can be measured by superimposing elevated pore air pressure so the measured pore water pressure is zero or positive, and using a probe or porous end plate with pores so fine that the pressurized air cannot blow through into the measuring system [*1,5,6*]. However, the air does pass through the porous sensing element in solution and forms again as air bubbles inside the measuring system, thus causing the system to malfunction. The severe problem with measurement of suction has resulted in most experimental studies of water flow in partially saturated soils being limited to ranges in water content where tensiometers can be used. For most applications, the water flow occurs at small suctions so that deficiencies in the operation of tensiometers are not a serious problem. However, the range of tensiometers is not sufficient to permit their use with soils at most arid sites. The recent development of ther-

[3]The italic numbers in brackets refer to the list of references appended to this paper.

mocouple psychrometers has made it possible to measure total suctions equivalent to about 80 atm of negative pore pressure [*7*], which is adequate for studies of moisture movement in most relatively dry soils. Such equipment was used for the studies reported here.

Measurement of Conductivity of Partially Saturated Soil

Klute [*8*] has summarized the state of the art for measuring the conductivity of partially saturated soils. A much abbreviated summary is included in this volume [*9*]. We concluded that a variation of the instantaneous profile method [*10*] would be the most suitable, and only this method will be discussed.

In the instantaneous profile method a long cylindrical sample is confined in an impervious tube that is set in a horizontal position. Suctions are measured at a series of locations along the length, using tensiometers or psychrometers as required. The pore air pressure is maintained at some suitable value, usually atmospheric pressure. Water flow is induced by altering hydraulic conditions at one end, for example, by wetting the inflow end. The opposite end of the permeameter is typically made impervious to water. Readings of suction are obtained as a function of time and are generally continued until water has advanced through the sample to the far end. The values of volumetric water content, Θ, are evaluated from the suction measurements and from curves of Θ versus p, which are either measured on a duplicate sample or are obtained from measurements of Θ taken at the beginning and end of a test and matched to the measured suctions.

There are four steps needed to calculate conductivity from the measurements. First, plots are prepared of suction versus horizontal position, and values of dp/dx (which is analogous to the hydraulic gradient) are computed for each probe and each time. Second, the volumetric water content is plotted versus the horizontal position, and the total volume of water downstream of each point is determined for each time by integration

$$V_w = \int \Theta A \, dx \tag{4}$$

Third, the rate of flow past a point, dV_w/dt, is computed for each probe location and each time. Finally, conductivity is computed from the equation

$$k = \frac{\gamma_w}{A} \frac{dV_w/dt}{dp/dx} \tag{5}$$

for each probe and each increment of time. From the calculations of conductivity, plots may be developed of k versus suction, water content, or degree of saturation.

The accuracy of the method hinges on obtaining an accurate water content versus suction relationship and accurate measurements of suction in the conductivity test. Most tests of this type have been performed using a constant head at the inflow end. Such tests may lead to the formation of sharp wetting fronts and make it difficult to define conductivities in the range of water contents between the initial and final values. We used a controlled inflow rate and input water at a rate slow enough to spread out the wetting front and prevent saturation. A constant inflow rate is not required. Finally, the relationship between water content and suction has hysteresis, so care must be taken not to allow decreases in water content at any probe. Note that hysteresis was not considered in this study but could be studied using a similar dcvice.

Apparatus

Our conductivity tests were performed on samples compacted into a plastic tube (Fig. 1) with an inside diameter of 50 mm (2 in.) and a length of 114 mm (4.5 in.). The tube was provided with seven ports, with a spacing of 13 mm (0.5 in.), for insertion of psychrometers. The ends were closed with plastic end plates with 0-ring seals. The downstream end was provided with a hypodermic needle to act as an air vent and a sheet of filter paper to help maintain a uniform head. The upstream end was provided with another needle for inflowing water and several disks of filter paper to help distribute the flow across the entrance face. The inflow came from a 5-cm^3 capacity syringe. The plunger of the syringe was pushed at a controlled rate using a standard mechanical loading press (screw jack driven by an electric motor through a variable-speed gear box). Flow rates of the order of 0.7 cm^3/day were typically used. The entire cell was placed in a vessel of high humidity and controlled temperature. Further details of the apparatus design are provided by Hamilton [*11*]. The instrumentation for the psychrometers is discussed by Daniel et al [*7*].

A similar cell, with a tube length of 25 mm (1 in.) and two psychrometers, was used to measure the relationship between water content and suction. About 24 to 48 h were required for equilibrium to be established after each injection of water.

Soil Tested

The apparatus has been used successfully with several soils. Data reported here were obtained using compacted samples of a fire clay known commercially as Goose Lake clay. It had liquid and plastic limits of 26 and 19 percent, respectively, and was composed of 19 percent sand, 62 percent silt, and 19 percent clay. The standard maximum dry density and optimum water

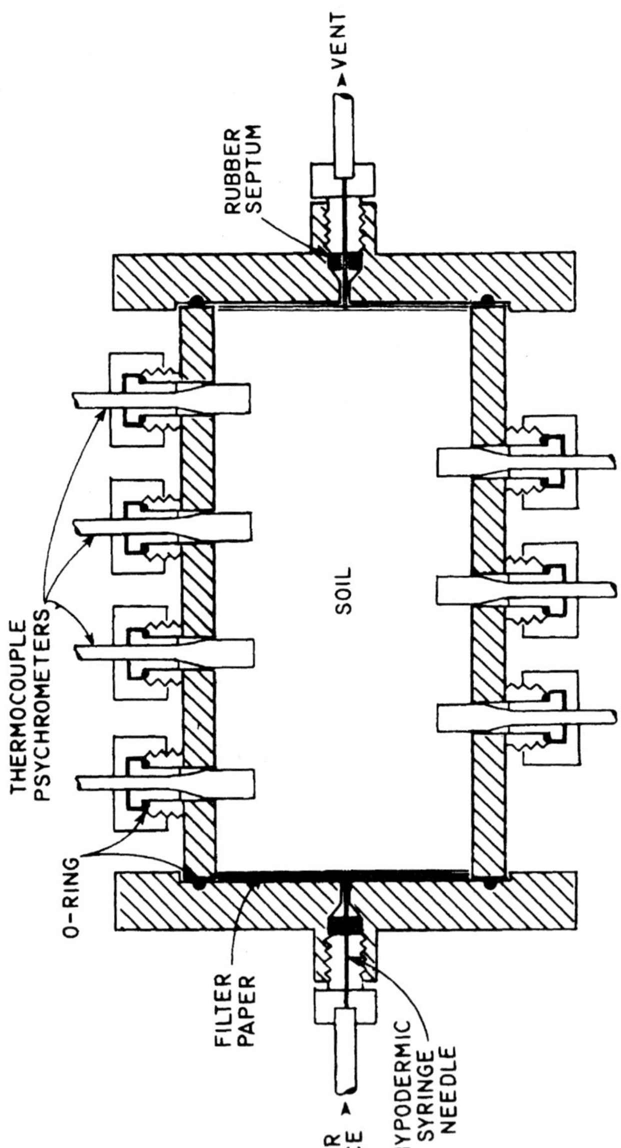

FIG. 1—*Schematic cross section through a permeameter.*

content were 1950 kg/m^3 (122 lb/ft^3) and 10.5 percent, and the modified values were 2130 kg/m^3 (133 lb/ft^3) and 9.5 percent, respectively.

Test Procedure

The soil was first sifted through a No. 40 sieve and then mixed with water to a gravimetric water content of 5.0 percent. The samples were compacted directly into the plastic tubes at a density of 1710 kg/m^3 (107 lb/ft^3). Holes were drilled in the sides of the soil and the psychrometers were inserted and sealed. Flow was then begun and readings were taken on the psychrometers, typically every 24 h. The flow was continued until the suction at the entrance end had decreased to about 4 atm, and the leading edge of the wetting front had at least reached the farthest psychrometer. Inflow was then stopped and the psychrometers were removed. A water content sample was taken from the soil around each psychrometer hole. The sample was then extruded and cut into disks to obtain another measurement of water content. A complete test for Goose Lake clay took about 20 days.

Results

Water Content-Suction Relationship

Data from six tests to measure the water content-suction relationship are shown in Fig. 2. The scatter in water contents is about ±0.5 percent from the mean curve. The curve becomes steep for suctions lower than about 6 atm and the psychrometers become less accurate in this range as well [*7*], thus leading to an increased scatter.

Conductivity Test Data

Measured curves of pore water pressure versus position in the sample are shown in Fig. 3 for one of the tests on Goose Lake clay. The water content profiles obtained by combining the pore pressures from Fig. 3 and the average curve in Fig. 2 are shown in Fig. 4. Note that a pumping rate was used which would prevent a sharp wetting front from developing.

The water contents were then integrated to obtain the quantity of water that has flowed past each probe as a function of time. Typical data for one probe are shown in Fig. 5.

The conductivities were then calculated for each probe at each time (excluding cases where an insignificant flow had occurred). The resulting conductivities are plotted versus suction in Fig. 6. The conductivity for $p = 0$ was obtained by compacting another sample at a dry density of 1660 kg/m^3 (104 lb/ft^3) and water content of 5 percent saturating it, and measuring conductivity by conventional means. No data were obtained in the range of suc-

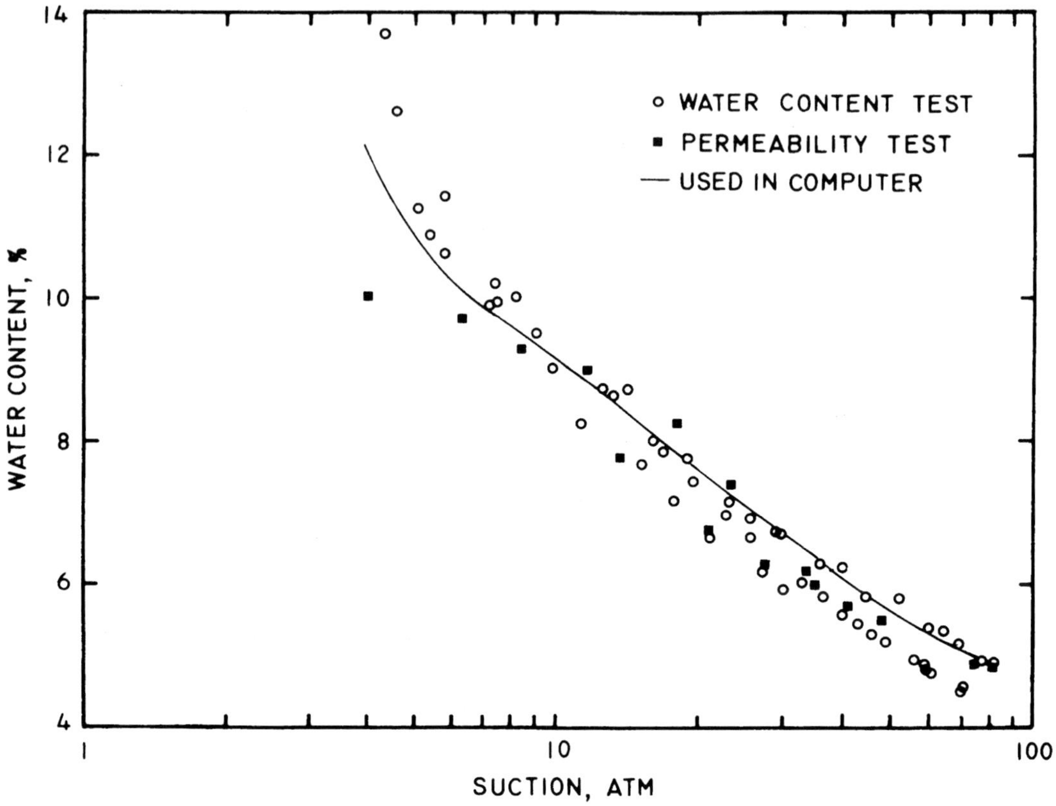

FIG. 2—*Relationship between water content and suction for Goose Lake clay.*

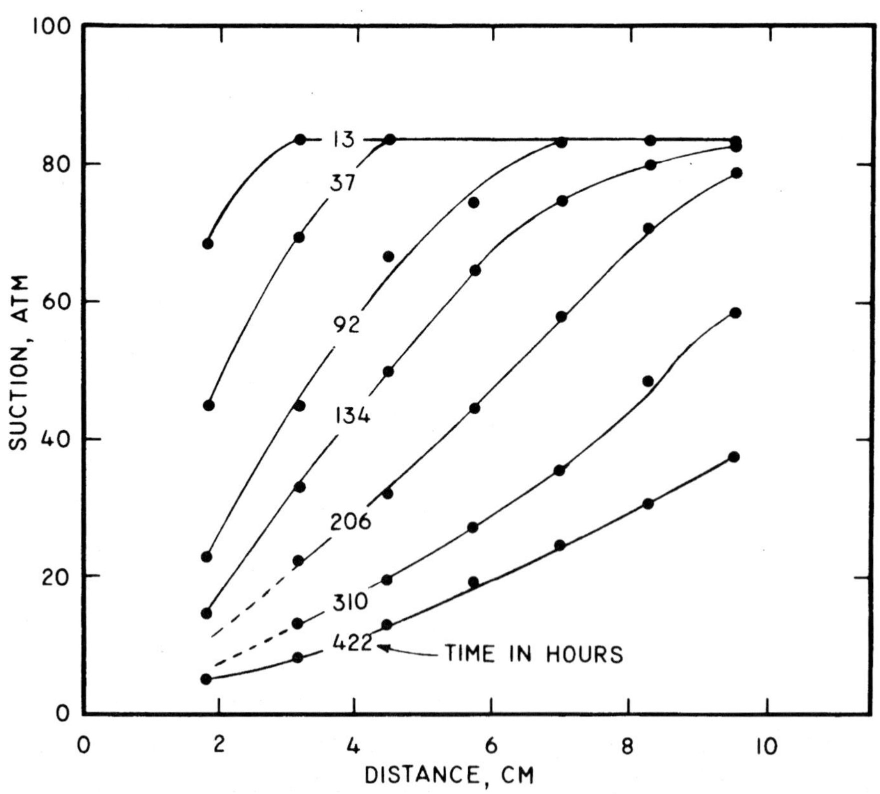

FIG. 3—*Measured curves of suction versus position in the sample at various times.*

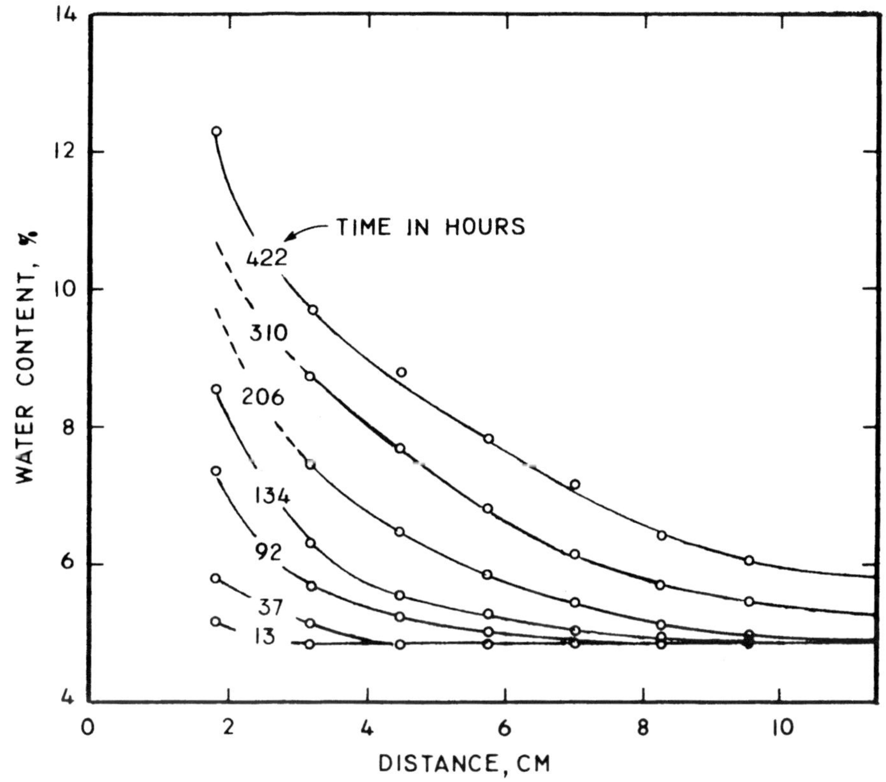

FIG. 4—*Water content profiles calculated from psychrometer readings.*

tions between about 0 and 1 atm because of the low accuracy of psychrometers in this range. It is striking that the conductivity changed by about 100 times for suctions between approximately 0 and 2 atm but only decreased by about 30 times for suctions from 2 to 80 atm. The exceedingly low conductivities, of the order 10^{-11} cm/s, for a soil that is a clayey silt are also striking.

A major part of the scatter in data in Fig. 6 is associated with Probes 1 and 2. Conductivities calculated at the Probe 1 position diverge from the bulk of the data between 40 and 10 atm of suction. The psychrometer in that position failed by corrosion due to attack from acids in the clay (pH = 2.9) and had to be replaced. The conductivities calculated based on the Probe 1 data are probably the result of the use of an inaccurate calibration for the failing psychrometer. Note that conductivities corresponding to readings by a replacement psychrometer (8 to 4 atm) agree with the rest of the data. Another possible cause of the scatter is a nonuniform distribution of water near the entrance end.

The large increase in conductivity between suctions 5 and 0 atm in Fig. 6 can be clarified by plotting log k versus degree of saturation (Fig. 7) from four tests on nearly identical samples of the same soil. The large reductions in conductivity correspond to substantial decreases in degree of saturation.

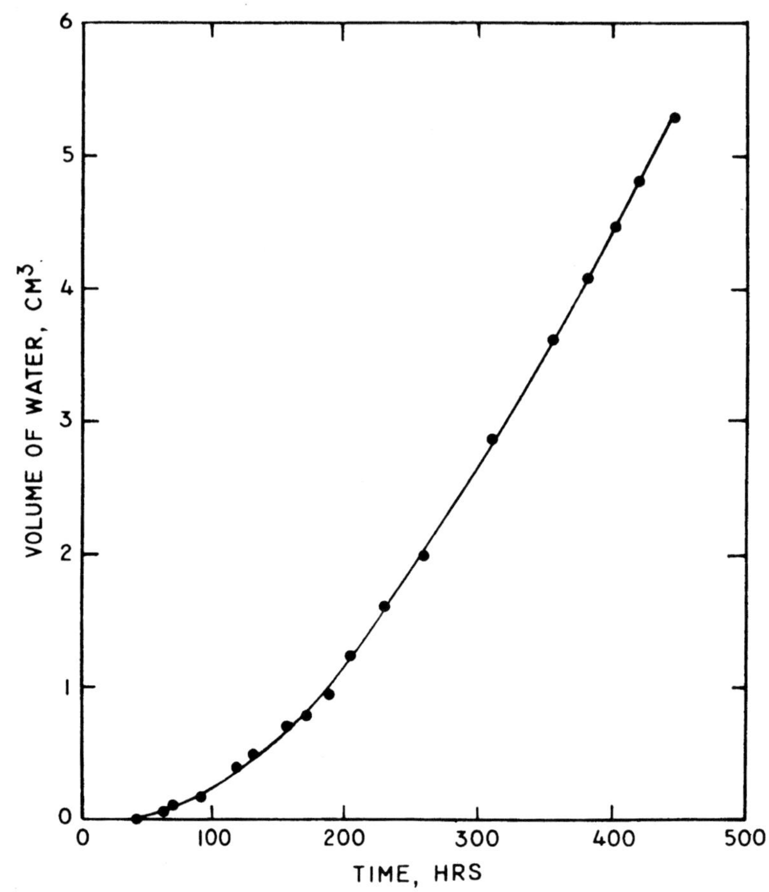

FIG. 5—*Calculated flow past a typical probe, as a function of time.*

At high degrees of saturation the bulk of the flow is probably in the middle of the voids, whereas at lower degrees of saturation it is in films on the particle surfaces where large drag forces should occur and where the total cross-sectional area through which flow occurs is reduced.

Numerical Analyses

The true test of the measurements reported here is to use the material properties in conjunction with water flow theory to predict water migration in a laboratory model. A Galerkin finite element computer program was written to solve the governing flow equation and to model one-dimensional, transient flow of water through an unsaturated soil. The problem is a nonlinear one because material properties are dependent upon suction. Details of the finite element analysis along with a Fortran listing of the computer program are given by Hamilton [*11*].

The computer program was used to model moisture movements in two conductivity tests. A comparison was drawn between the transient response of the soil as measured using thermocouple psychrometers and as predicted

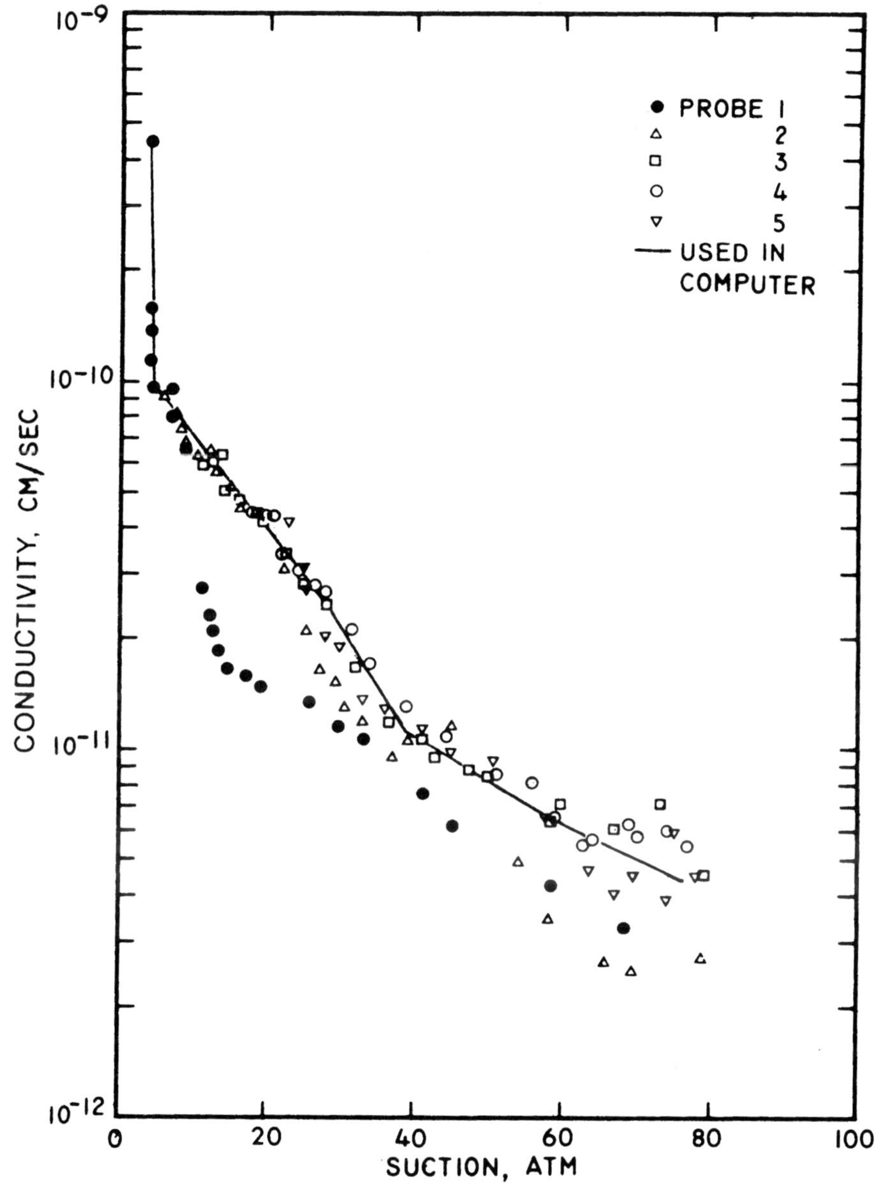

FIG. 6—*Calculated coefficients of hydraulic conductivity for Goose Lake clay.*

using the computer program. The comparison was made in an effort to answer two questions: (1) Are the material properties measured using the apparatus described earlier correct and sufficient to allow quantitative prediction of moisture migration in unsaturated soil under constant volume conditions? (2) Is the form of the flow equation most often used to predict flow in unsaturated soils capable of accurately modeling such a complex process?

It should be clearly asserted that the analysis did not provide a rigorous test of the applicability of the theory. The laboratory models are far removed from actual field conditions where anisotropy and inhomogeneity are ever-present. However, the results should not be taken casually, for if the theory is incapable of describing moisture migration under such ideal circumstances,

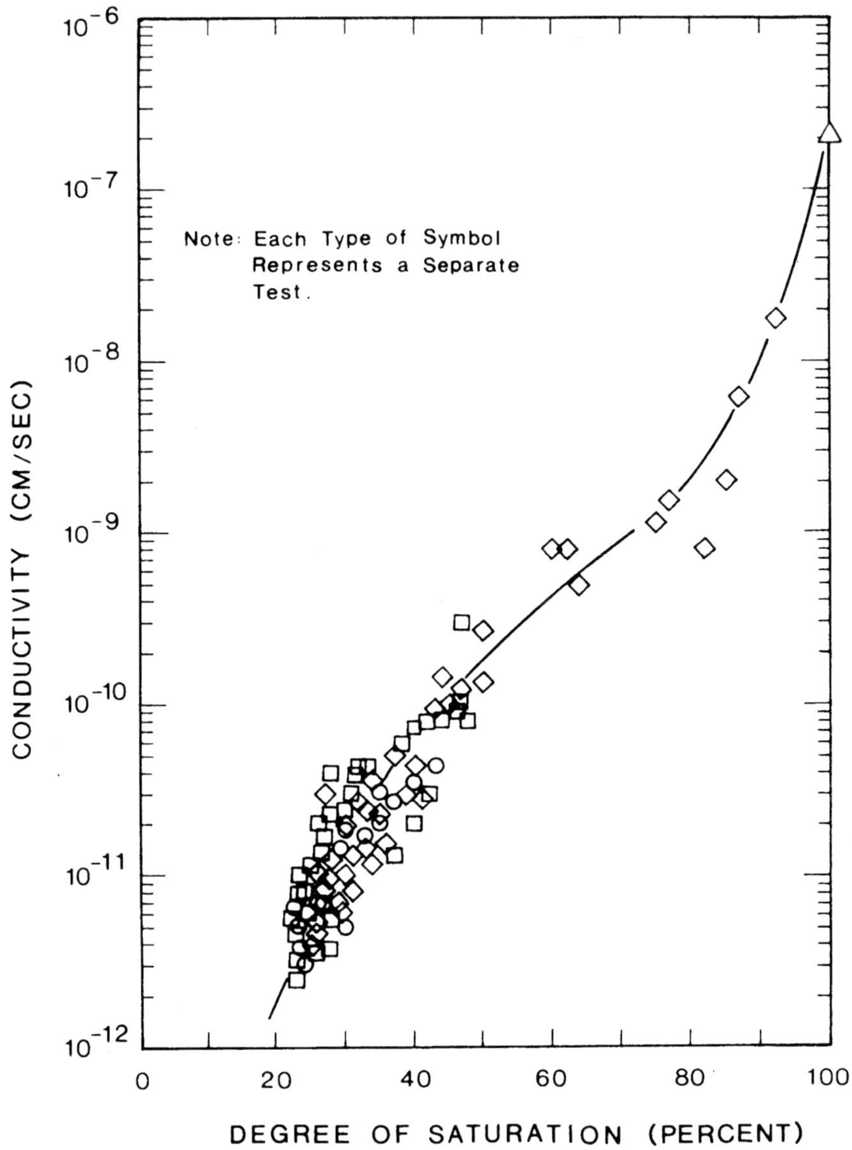

FIG. 7—*Effect of the degree of saturation on hydraulic conductivity for Goose Lake clay.*

the possibility of using that theory in conjunction with measured flow properties to model general field problems is remote.

The duration of the model test was 447 h. A constant inflow of 0.031 cm^3/h was delivered to the sample. The soil properties used in the numerical simulation of the test are shown in Figs. 5 and 6 as the solid lines.

In Fig. 8 suctions measured at seven probe locations at various times throughout the test are compared with those predicted using the computer program. The laboratory data are plotted as symbols, and the computer solution is plotted as solid lines. Agreement between measured and predicted results is excellent for all probe locations except the first one. Problems with conditions around the first probes were discussed earlier.

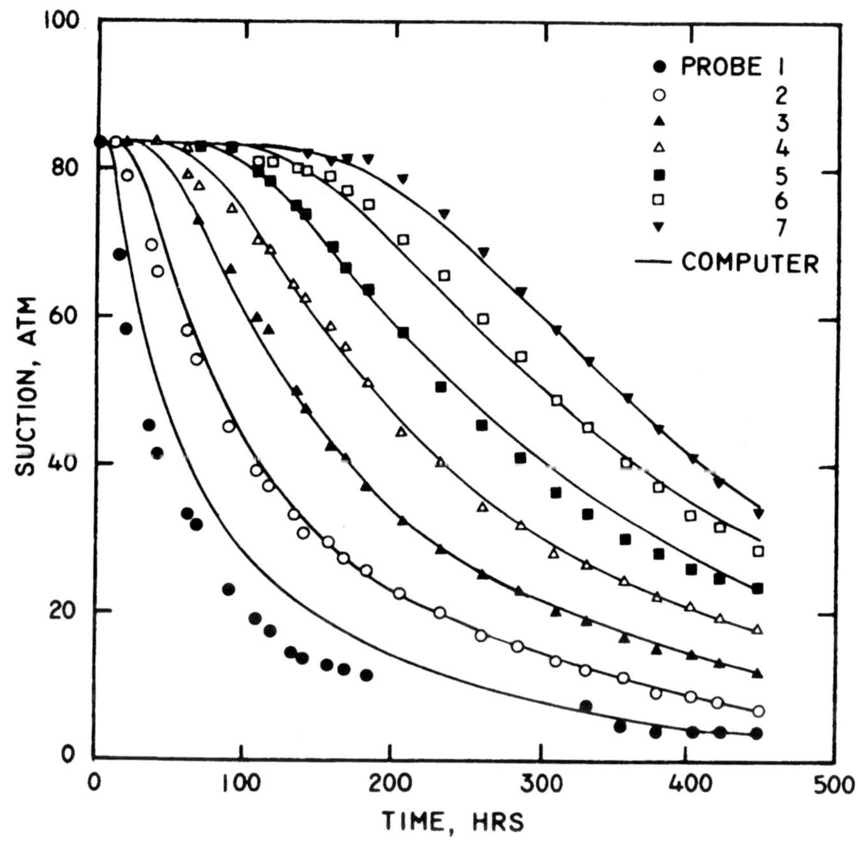

FIG. 8—*Comparison of measured suctions with values calculated using measured properties and a finite-element computer program.*

In Fig. 9, the predicted final moisture profiles, obtained from the theory and soil property measurements described previously, are compared with the profiles measured gravimetrically at the end of the test. There may have been an error in measurement of moisture content near the inlet boundary, or in the position of the suction-water content curve at low suctions. Generally, however, the comparison helps to substantiate the validity of the one-dimensional model and the correctness of the material properties that were measured.

Conclusions

A method has been presented whereby the relationship between permeability and suction for the wetting of an unsaturated soil can be measured for suctions between about 1 and 80 atm. A one-dimensional permeameter was instrumented with thermocouple psychrometers, from which both the hydraulic gradients and rates of flow could be estimated. The method is relatively convenient, at least compared with alternative methods [*8,9*], but requires about 3 weeks per test, not including calibration of psychrometers.

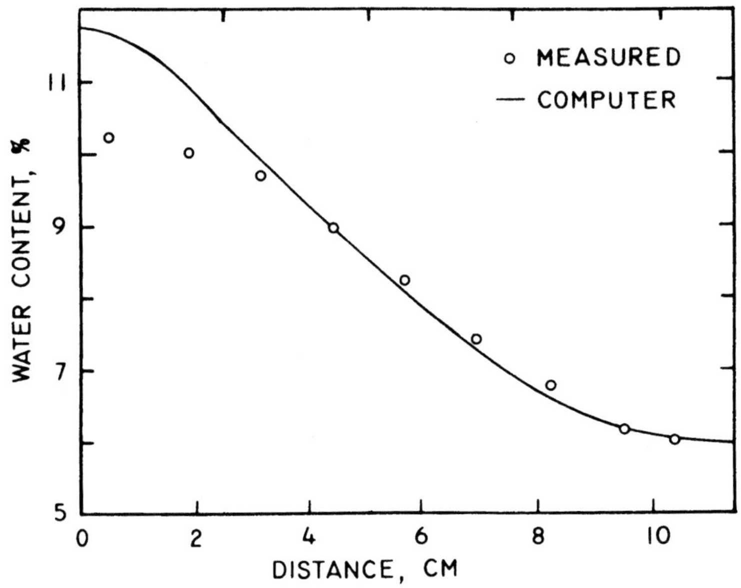

FIG. 9—*Comparison of measured and computed final water contents for the test with Goose Lake clay.*

Experiments have been performed on several soils, and the method has been found to work reasonably well for clays with degrees of saturation between about 30 and 90 percent, and for sands with degrees of saturation less than 50 percent. For higher degrees of saturation, the same method may be used, except that tensiometers would be used in place of psychrometers.

Experiments on a compacted silty clay indicate that the conductivity below a degree of saturation of 50 percent (suctions below about 5 atm) is low and ranges from 4×10^{-12} to 2×10^{-10} cm/s. Between 50 and 100 percent saturation, the conductivity increases by about 3 orders of magnitude to a value of about 10^{-7} cm/s.

A finite element solution to the equation described by Darcy's law was written for the case of one-dimensional wetting of a horizontal column of partially saturated soil, taking into account the dependence of soil properties on suction (or water content). Comparisons of predicted and measured profiles of suction at various times were good, which indicates that Darcy's law was generally valid for the soil examined. The measured material properties, including conductivity, were used to predict the final profile of water content within the sample of silty clay. This profile was compared with measured water contents obtained by slicing the sample into thin disks at the end of the test. Good agreement was found, which substantiates the correctness of the conductivity measurements for this soil. Although the methodology described for measuring conductivity and predicting water flow in partially saturated soils has only been investigated for a relatively small number of compacted soils, the results seem sufficiently promising to warrant further investigation.

Acknowledgments

This research was supported by grants from the Los Alamos Scientific Laboratory. Appreciation is expressed to Lamar Johnson, Merlin Wheeler, and Jim Steger for their support. The project at the University of Texas at Austin was administered by Prof. Joe O. Ledbetter.

References

[*1*] Olson, R. E. and Langfelder, L. J., *Journal of Soil Mechanics and Foundation Division, American Society of Civil Engineers*, Vol. 91, No. SM4, 1965, pp. 127-150.
[*2*] Letey, J., Kemper, W. D., and Norman, L., *Soil Science Society America Proceedings*, Vol. 33, 1969, pp. 16-18.
[*3*] Richards, L. A., *Soil Science*, Vol. 53, 1942, pp. 241-248.
[*4*] Richards, L. A. and Gardner, W., *Journal of the American Society of Agronomy*, Vol. 28, 1936, pp. 352-358.
[*5*] Richards, L. A., *Soil Science*, Vol. 68, 1949, pp. 95-112.
[*6*] Croney, D. and Coleman, J. D. in *Pore Pressure and Suction in Soils*, British Geotechnical Society, 1960, pp. 31-37.
[*7*] Daniel, D. E., Hamilton, J. M., and Olson, R. E., this publication, pp. 84-100.
[*8*] Klute, A., *Soil Science*, Vol. 113, No. 4, 1972, pp. 264-276.
[*9*] Olson, R. E. and Daniel, D. E., this publication, pp. 18-64.
[*10*] Richards, S. J. and Weeks, L. V., *Soil Science Society America Proceedings*, Vol. 17, 1953, pp. 206-209.
[*11*] Hamilton, J. M., "Measurement of Permeability of Partially Saturated Soils," M.S. thesis, the University of Texas at Austin, Austin, Tex., 1979.

FAILURE OF AN ANCHORED BULKHEAD

By David E. Daniel,[1] A. M. ASCE and Roy E. Olson,[2] F. ASCE

ABSTRACT: An 800-ft (240-m) long anchored bulkhead with a height of 65 ft (20 m) was designed more or less in accord with published recommendations by several acknowledged experts. Nevertheless, the wall failed. Post-failure analysis indicated that the source of error was a lack of understanding of soil behavior under conditions involving significant excavation near the bulkhead. The design was adequate for undrained conditions but was inadequate for fully drained conditions.

INTRODUCTION

In the early 1970's, a large anchored bulkhead was designed for installation along the Houston, Texas, ship channel. In performing the analyses, the designer consulted textbooks and papers written by generally acknowledged experts. The designer was a registered civil engineer with over 20 years of experience in designing waterfront structures. The site was explored using a reasonable number of borings and laboratory strength tests. Soils at the site were described using such terms as "stiff to hard clay" and "very dense fine sand." No soft clays or loose sands were found. Based on his best estimate of undrained shearing strengths, the designer calculated factors of safety in excess of two for the considered modes of failure. Nevertheless, soon after construction was completed, the wall began to move and, despite efforts to unload the wall, the horizontal movements reached 10 ft (3 m) within 3-1/2 months. When it became apparent that movements could not be stopped, the bulkhead was dismantled and a multimillion dollar litigation ensued.

Both sides retained consultants to conduct post-failure investigations. Taken collectively, the post-failure investigations included further soil borings, hundreds of laboratory tests, installation of seven slope indicator casings, in-situ pressuremeter tests, field measurement of the shape and location of the deformed bulkhead, and a general review of design methodology. Because detailed data are available, the failure affords an opportunity to evaluate the causes of failure, to determine if existing methods of analysis and design seem adequate, and to ex-

[1]Asst. Prof., Dept. of Civ. Engrg., Univ. of Texas, Austin, Tex. 78712.

[2]Prof., Dept. of Civ. Engrg., Univ. of Texas, Austin, Tex. 78712

Note.—Discussion open until March 1, 1983. To extend the closing date one month, a written request must be filed with the Manager of Technical and Professional Publications, ASCE. Manuscript was submitted for review for possible publication on October 21, 1981. This paper is part of the Journal of the Geotechnical Engineering Division, Proceedings of the American Society of Civil Engineers, ©ASCE, Vol. 108, No. GT10, October, 1982. ISSN 0093-6405/82/0010-1318/$01.00.

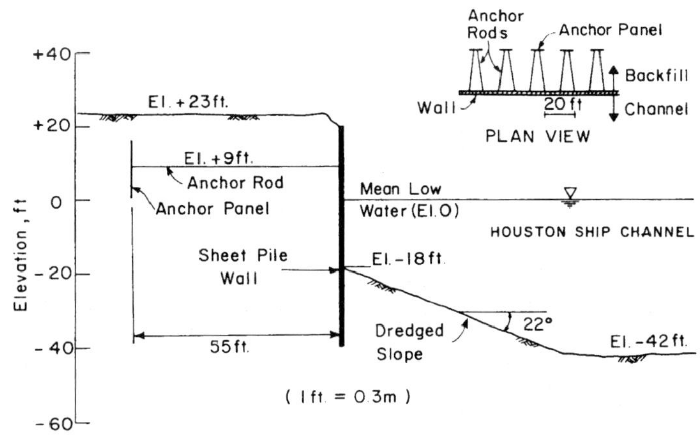

FIG. 1.—Cross Section of Bulkhead Before Failure

amine how practicing engineers might avoid future failures of this type.

Bulkhead.—The anchored bulkhead was constructed using 60-ft-long (18-m) Larssen 3-S interlocking steel sheet piles (Grade 60) that were driven to a tip elevation of −40 ft (−12 m), mean low water datum (Fig. 1). The wall was tied back using 2.5-in. (6.4-cm) diam Grade A36 steel anchor rods installed at elevation 9 ft (2.7 m), spaced 10 ft (3 m) on centers, and extended 55 ft (17 m) back to anchor panels (Fig. 1). The 15-ft-high by 9-ft-wide (4.6 by 2.7 m) anchor panels were constructed from PZ27 interlocking steel sheetpiles. Two anchor rods were connected to each anchor panel. Final ground elevation behind the wall was 23 ft (7.0 m). The dredgeline intersected the wall at elevation −18 ft (−5.5 m) but sloped down to elevation −42 ft (−13 m) at an angle of 22 degrees.

The original topography of the site is not known precisely. However, it is known that essentially no new fill was placed behind the wall, but up to about 40 ft (12 m) of excavation was required in front of the wall. The sequence of construction is summarized in Table 1.

The original design included plans for construction of a pile-supported dock on the dredged side of the bulkhead. Some lateral thrust from the bulkhead was to be taken by the dock at the elevation of the anchor rods. The designer considered the bulkhead to be a "temporary" structure. A schedule of construction

TABLE 1.—Sequence of Construction

Dates (1)	Activity (2)
September, 1973–March, 1974	sheet piles for main wall were driven
March, 1974–August, 1974	anchorage system was installed by excavating preexisting backfill, driving sheet-pile anchor panels, installing anchor rods, and backfilling to elevation 23 ft (7.0 m)
September, 1974–November, 1974	dredging in front of wall
November, 1974	construction complete

was not included in the design drawings. Presumably, the bulkhead would have to support itself for a period of at least several months while the dock was being constructed.

Failure.—In December, 1974, less than a month after construction of the bulkhead was finished, the backfill between the main wall and anchor panels began developing cracks. By February, 1975, even more cracks had developed and movement of the wall toward the ship channel had become noticeable. In an attempt to stabilize the bulkhead, approx 10 ft (3 m) of backfill was excavated between the wall and the anchor panels. The excavation did not stop the movements. On March 11, 1975, some 3-1/2 months after construction was complete, portions of the wall had displaced nearly 10 ft (3 m) toward the ship channel. Scarps were noticed just behind the anchor panels in some sections.

Between March 14, 1975 and April 4, 1975, seven slope indicator casings were installed in regions of current or potential movement. The slope indicator casing installed on April 4, 1975 was in the zone of maximum movement and was about 20 ft (6 m) behind the wall. Movements of this casing (Fig. 2) indicate that major shearing strains occurred between elevations −20 and −43 ft (−6 to −13 m). Movements were referenced to the bottom of the casing. Additional strains presumably occurred at greater depths, based on the observed gradient in displacements near the bottom of the casing.

The wall continued to deform through May and June of 1975. Field measurements, made in June, 1975 (Fig. 3), indicate that the failure involved both lateral movement and rotation of the base toward the ship channel.

By late summer 1975, it became clear that the wall would be unusable and, beginning in September, 1975, the wall was removed and a new structure was made.

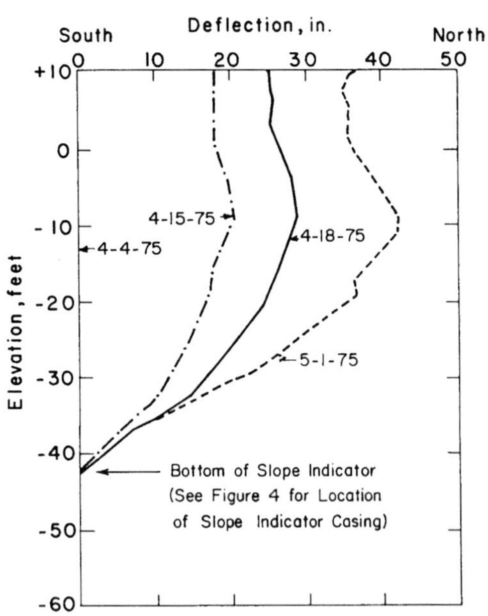

FIG. 2.—Horizontal Deflections Measured with Slope Indicator Installed 20 ft Behind Wall in Zone of Maximum Movement

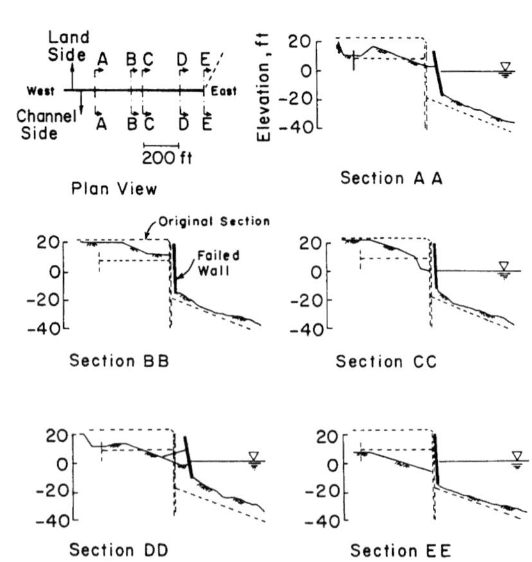

FIG. 3.—Topographic Cross Sections of Failed Bulkhead in June, 1975

Subsurface Conditions

Stratigraphy.—Twelve borings were drilled before the failure and fifteen more were drilled afterwards. Typical depth of borings was 80 to 100 ft (24 to 30 m). Subsurface conditions found in the borings are characterized by the profile shown in Fig. 4. The site is underlain predominately by stiff to hard clay with layers and lenses of sand and silt.

The stratum of main interest is a stiff-to-hard, fissured, slickensided clay found between average elevations of −15 and −43 ft (−5 and −13 m) and known locally as Beaumont clay. Essentially, all of the measured shearing strain appears to have occurred within the Beaumont clay (Fig. 2).

Index Properties.—Profiles of water content, Atterberg limits, dry unit weight, and standard penetration resistance (determined only for zones of cohesionless soil) are plotted in Fig. 5. The water contents were near the plastic limit, which is typical of stiff clays in the Houston area.

Shearing Strengths.—Soil samples were recovered from borings using 3-in.-diam (8-cm) thin-walled sampling tubes. Soil samples were extruded in the field, sealed, and transported to the laboratory for testing. Several pocket penetrometer tests were performed in the field immediately after the soil samples were ex-

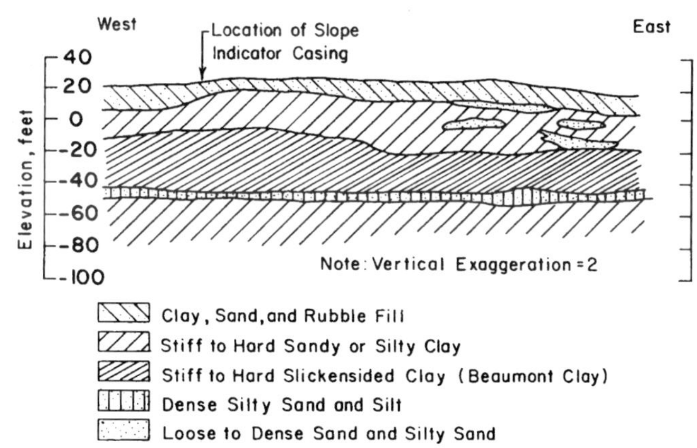

FIG. 4.—Section Along Length of Bulkhead

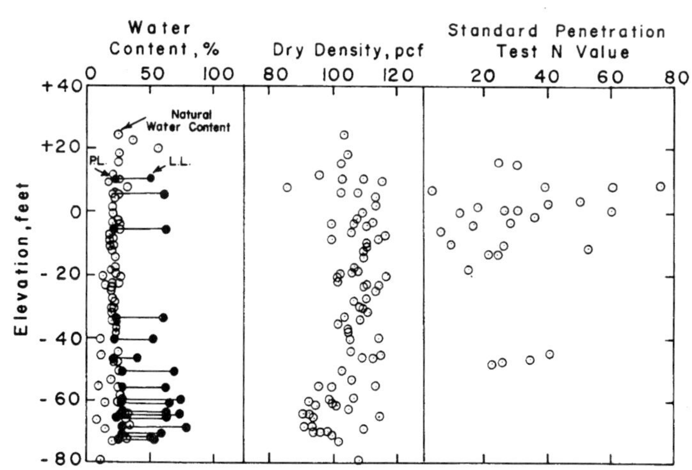

FIG. 5.—Index Properties

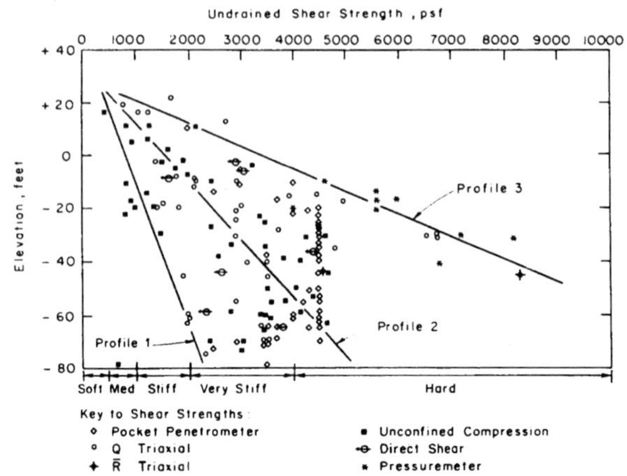

FIG. 6.—Undrained Shear Strength Plotted as a Function of Elevation

truded. Laboratory shear tests included unconfined compression tests and unconsolidated-undrained (Q-type) triaxial compression tests on the cohesive soils in general, and a few consolidated-undrained (R̄-type) triaxial compression tests on samples of silt from near elevation −46 ft (−14 m). The samples used for the R̄ tests were consolidated to the effective overburden pressure. A few direct shear tests were performed on 1-in. (25-mm) thick samples of clay, with short consolidation times (of the order of one minute) and shearing times of 8–42 minutes and with normal stresses near the effective field overburden pressure. Data from these tests, plus data obtained using a self-boring pressuremeter (7), are plotted in Fig. 6. The scatter is so great that definitive conclusions about the probable field strengths are difficult to draw. However, on the average, unconfined compressive strengths are the lowest, strengths measured with the pressuremeter are the highest, and strengths from triaxial tests are in between.

The three profiles of undrained shear strength shown in Fig. 6 were used for the analyses examined later. Profile 1, which is thought to provide an unrealistically low estimate of field strength, was used to examine the sensitivity of the analysis to the assumed soil properties.

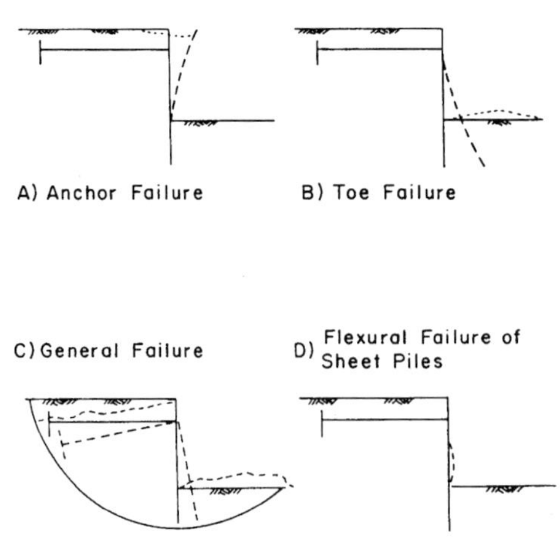

FIG. 7.—Possible Modes of Failure

Design of Bulkhead

Copies of the original design calculations were obtained as part of the litigation. Stresses applied to the wall were calculated using values of c in the range of 500–3,370 psf (24–161 kN/m^2) and ϕ in the range of 0–30°. The values of ϕ appear to have been based on Q-type triaxial compression tests. Total unit weights were used above the water table and submerged unit weights below. Two sets of analyses were performed, one attributed to Tschebotarioff and one to Terzaghi. In response to interrogatories, the designer indicated that the design was performed in accord with recommendations in Refs. 8 and 14–18.

The modes of failure usually considered in design (Fig. 7) include failure of the tie rods in tension, excessive movement of the anchors, toe failure, general failure, and flexural failure (14,17). Factors of safety against these modes of failure appear to have been calculated in the original design and were found to be adequate (in excess of two). The sheeting was designed using a hinge at the dredge line and using Rowe moment reduction factors.

We have rechecked the design in general accord with the assumptions believed used by the designer, but using somewhat more sophisticated methods. We will consider only general failure and toe failure because field observations eliminate other failure modes.

General Failure.—The possibility of general failure was analyzed assuming limit equilibrium and using a computer program developed for slope stability analyses on layered soils (19). The total unit weight of all strata was assumed to be 130 pcf (20.4 kN/m^3). Circular failure surfaces were assumed and the critical one was found by trial. In all cases, the critical circle passed just behind the anchor panels at one end and near the toe of the dredged slope at the other end. The factor of safety against general failure was defined as the ratio of the maximum possible resisting moment to the driving moment. Results of the analysis for the backfill at its full elevation of 23 ft (7 m), for strength profiles 1, 2, and 3 (Fig. 6), yielded factors of safety of 1.1, 2.2, and 5.1, respectively. The factor of safety for a reasonable estimate of undrained strength for the clay, e.g., profile 2, is in general accord with good engineering practice.

Toe Failure.—To analyze the possibility of a toe failure, the free-earth-support method of analysis was used (3,15–18). Soil behind the wall was assumed to be in a state of active failure and soil in front of the wall to be in a state of passive failure. Active and passive earth pressures were calculated using Rankine's theory (12,15) with $\phi = 0$. The effect of a sloping dredge line was taken into account (15). The factor safety against a toe failure was computed by summing moments about the anchor elevation and taking the ratio of maximum possible resisting moment to driving moment.

For the backfill at its final elevation, the factors of safety against a toe failure, for strength profiles 1, 2, and 3, were 1.2, 7.6, and infinity. Thus, again, use of reasonable estimates of undrained strength would lead to an acceptable margin of safety.

Post-Failure Analysis

Immediately after the failure, some of the consultants considered that there might exist a thin stratum of distinctly weaker soil that had not been discovered in the original soil exploration program. However, no such stratum was found in subsequent soil exploration and the existence of such a stratum would be incompatible with the measured deformation pattern (Fig. 2).

The possibility of serious errors in the calculations can probably be eliminated because our independent analyses yielded factors of safety apparently similar to those used in design. Factors of safety approaching one could be obtained using the minimum measured undrained strengths (Fig. 6). However, most of the minimum strengths were measured using unconfined compression tests. Such tests should underestimate field strengths of stiff fissured clays.

The most likely explanation of the failure would seem to lie in the use of undrained shearing strengths. Generally, loading a soil leads to generation of positive pore water pressures. Subsequent time-dependent dissipation of these excess pore water pressures leads to time-dependent increases in effective stress and strength. Anchored bulkheads have generally been constructed by backfilling behind the wall, which leads to time dependent increases in strength.

When soils are unloaded, negative pore water pressures are generated. The time-dependent increase in these pore water pressures leads to decreases in effective stress, and, thus, strength. Such an explanation is at least implied in Terzaghi's paper on stability of natural slopes (13) and a definitive statement is to be found in the U.S. Steel Sheet Piling Design Manual (Ref. 11, p. 25) and in Lambe and Whitman (Ref. 6, p. 472).

To evaluate the stability of the bulkhead under fully-drained conditions, the analyses were repeated using effective-stress strength parameters. The soils at the site can be divided into two groups, upper miscellaneous fill, sand, and clay, and the underlying Beaumont formation. The upper soils exerted little influence on stability; we have assumed average values of $\bar{c}$ and $\bar{\phi}$ for these soils of 0 and 25 degrees, respectively. The designer did not recognize the need for drained tests and none were performed. Most of the post-failure consultants also apparently considered the failure to have occurred under undrained conditions because the only effective-stress tests were for a "silt layer," at elevation about −45 ft (−14 m), which was considered a possible weak zone ($\bar{c}$ = 0, $\bar{\phi}$ = 33°). Our investigation began after other consulting efforts had been completed and no soil samples were available for testing. When preliminary drained analyses indicated that failure would be predicted if any rational parameters were used, we made use of Al-Layla's (1) values (Table 2), which were obtained using samples of similar Atterberg limits from the same formation. In the analyses, the water table was assumed to be at the average elevation measured in the boreholes (elevation 10 ft or 3 m). Although use of Al-Layla's values introduces a degree of uncer-

TABLE 2.—Effective Stress Strength Parameters for Beaumont Clay Reported by Al-Layla (1)

Type of test (1)	$\bar{\phi}$, in degrees (2)	$\bar{c}$, in pounds per square inch (3)
Drained direct shear (peak strengths)	13.0	3.6
Drained direct shear	8.0	1.0

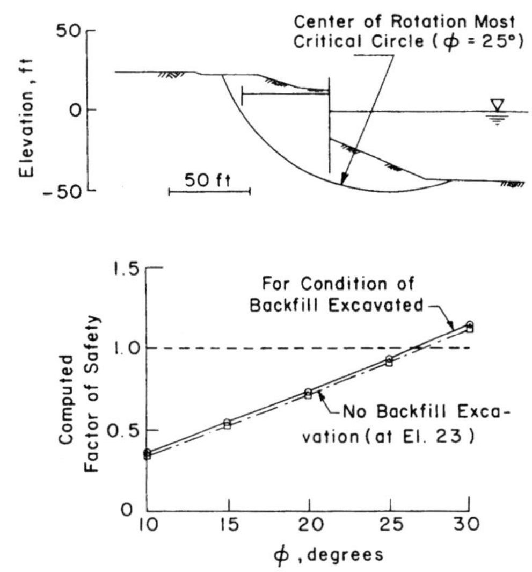

FIG. 8.—Sensitivity of Factor of Safety to Assumed Uniform $\bar{\phi}$

tainty into the analyses, sensitivity studies to be reported later show that the same conclusions would result from use of any reasonable set of effective-stress shearing parameters.

Analyses of overall stability were performed using the computer program referenced earlier. When applied to effective stress analyses, Spencer's (10) method was used to account for side forces between slices. For the fill at its full height (elevation 23 ft or 7 m) the calculated factors of safety for a general shear failure were 1.0 if peak strengths were used and only 0.5 for residual strengths. These values were increased by only about 10% for the case of backfill cut to elevation +13 ft (4 m). The critical failure surface intersected the ground surface in the general vicinity of the observed scarp at sections AA and DD (Fig. 3).

A toe failure was analyzed assuming logarithmic spiral failure surfaces and angles of wall friction of $0.5\bar{\phi}$ for both the active and passive conditions. The effects of upward seepage near the toe were taken into account by reducing the submerged unit weight as recommended by Terzaghi (15). Numerical values for active and passive earth pressure coefficients were obtained using tables in the U.S. Steel Manual (11). For the backfill at full elevation, the calculated factors of safety for a toe failure were 0.5 using peak strengths and 0.2 using residual strengths. Even with backfill at elevation +13 ft (4 m) these factors of safety were 0.8 and 0.4, respectively.

Analyses of overall stability were then performed using $\bar{c} = 0$ and a range in values of $\bar{\phi}$. They show (Fig. 8) that long term failure could be expected for any reasonable value of $\bar{\phi}$.

Analysis

Although the exact failure mode is unknown, post-failure analyses show that an appropriate design analysis would have warned of failure either by the toe pushing outward or by a general shear failure. The fact that failure occurred soon after construction was finished is in accord with the fact that the clay is fissured. Opening of the fissures would allow for rapid inflow of water.

The critical error in design appears to have been in lack of recognition that in soils subject to unloading the long-term or drained strengths are critical. This fact gradually became obvious for slope stability problems in the late 1940's and during the 1950's (4,9) but has apparently not yet thoroughly permeated design methods elsewhere in geotechnical engineering. This seems particularly true for cases involving design of anchored bulkheads because a number of experts on the subject have either implied that only undrained strengths need to be considered or have made definitive statements to that effect. For instance, Terzaghi has stated that "in the course of time, the shearing resistance of cohesive soils, such as silt or clay, increases because of progressive consolidation" (15, p. 1264). For cohesive soils, he recommended use of unconfined compression tests for design of anchored bulkheads (15, p. 1265). Tschebotarioff also recommended use of unconfined compression tests but he recommended increasing the unconfined strengths by 40% "because of inevitable weakening of such small specimens" (18, p. 483). Peck, Hanson, and Thornburn state that both the active and passive pressures acting on anchored bulkheads should be determined on the premise that undrained conditions prevail (18, p. 451).

Recommendations that undrained strengths be used for analysis of anchored bulkheads in clay are valid for the majority of bulkheads because the usual construction sequence involves implacement of large quantities of fill behind the wall and only moderate dredging in front of the wall. In such cases, consolidation will cause the clay to strengthen with time. It seems probable that all of the authors quoted in the previous paragraph had this typical condition in mind.

This case history was somewhat unusual in the sense that essentially no new fill was placed, but deep excavation was made. Under such circumstances, drained conditions should be critical.

Conclusions

A large anchored bulkhead, with a length of 800 ft (240 m) and a height from the top of the backfill to the bottom of the dredged channel of 65 ft (20 m), was designed more or less in accord with published recommendations by several acknowledged experts, and it failed, leading to a multimillion dollar law suit. Analysis of conditions at the site demonstrate that the failure was caused by a general shear failure that took out both the wall and the anchorage system. In addition, there may have been a toe failure that caused the wall to rotate into the adjacent ship channel.

The source of the design error was the lack of understanding of soil behavior, specifically a lack of realization that under the circumstances of this project the fully drained strength would be less than the undrained strength and, thus, would govern design. Similar circumstances may exist for other anchored bulkheads in which the quantity of soil excavated significantly exceeds the quantity of backfill.

Appendix.—References

1. Al-Layla, M. T. H., "Studies of Certain Geotechnical Properties of Beaumont Clay," thesis presented to Texas A&M University, at College Station, Tex., in 1970 in par-

tial fulfillment of the requirements for the degree of Doctor of Philosophy.

2. Alperstein, R., *Proceedings of the Sixth International Conference on Soil Mechanics and Foundation Engineering*, Vol. 3, 1965, pp. 379–380.
3. "Design Manual, Soil Mechanics, Foundations, and Earth Structures," *NAVFAC DM-7*, United States Naval Facilities Command, Washington, D.C., 1971.
4. Henkel, D. J., "Investigations of Two Long-Term Failures in London Clay Slopes at Wood Green and Northolt," *Proceedings of the Fourth International Conference on Soil Mechanics and Foundation Engineering*, Vol. 2, 1957, pp. 315–320.
5. Huntington, W. C., *Earth Pressures and Retaining Walls*, Article 13, John Wiley and Sons, Inc., New York, N.Y., 1957.
6. Lambe, T. W. and Whitman, R. V., *Soil Mechanics*, John Wiley and Sons, Inc., New York, N.Y., 1969.
7. Mast, K. H., "Field Testing with a Self-Boring Pressuremeter in Stiff Clay," thesis presented to the University of Texas at Austin, Tex., in 1977, in partial fulfillment of the requirements for the degree of Master of Science.
8. Peck, R. B., Hanson, W. E., and Thornburn, T. H., *Foundation Engineering*, 2nd Edition, John Wiley and Sons, Inc., New York, N.Y., 1974.
9. Skempton, A. W., "Time Rate of Softening in Stiff Fissured Clays, with Special Reference to London Clay," *Proceedings of the Second International Conference on Soil Mechanics and Foundational Engineering*, Vol. 2, 1948, pp. 50–53.
10. Spencer, E., "A Method of Analysis of the Stability of Embankments Assuming Parallel Inter-Slice Forces," *Geotechnique*, Vol. 15, No. 3, Sept., 1967, pp. 221–242.
11. *Steel Sheet Piling Design Manual*, U.S. Steel Company, Pittsburgh, Pa., 1972.
12. Taylor, D. W., *Fundamentals of Soil Mechanics*, John Wiley and Sons, Inc., New York, N.Y., 1948.
13. Terzaghi, K., "Stability of Slopes of Natural Clay," *Proceedings of the First International Conference on Soil Mechanics and Foundation Engineering*, Vol. 1, 1936, pp. 54–56.
14. Terzaghi, K., *Theoretical Soil Mechanics*, John Wiley and Sons, Inc., New York, N.Y., 1943.
15. Terzaghi, K., "Anchored Bulkheads," *Transactions*, ASCE, Vol. 119, 1954, pp. 1243–1324.
16. Terzaghi, K. and Peck, R. B., *Soil Mechanics in Engineering Practice*, John Wiley and Sons, Inc., New York, N.Y., 1967.
17. Tschebotarioff, G. P., *Soil Mechanics, Foundations, and Earth Structures*, McGraw-Hill Book Co., Inc., New York, N.Y., 1951.
18. Tschebotarioff, G. P., *Foundation Engineering*, Chapter 5, Leonards, G. A., (ed.), McGraw-Hill Book Co., Inc., New York, N.Y., 1962, pp. 438–524.
19. Wright, S. G., "SSTAB1—A General Computer Program for Slope Stability Analyses," *Research Report GE-74-1*, University of Texas, Department of Civil Engineering, Austin, Tex.

AXIAL CAPACITY OF STEEL PIPE PILES IN CLAY

By Norman D. Dennis, Jr., [1] M. ASCE, and Roy E. Olson, [2] F. ASCE

INTRODUCTION

Development of theories for the prediction of the axial compressive and tensile load capacities of driven piles is made difficult by a variety of factors, e.g., disturbance of the soil and changes in the state of stress caused by pile driving, dragdown of soil from one layer into another, and pile/soil interaction during loading. In the absence of a well developed theory supported by field evidence, engineers have developed a series of empirical design rules utilizing coefficients which are evaluated using full scale pile load tests. The size of data bases used in these correlations has generally been small and the methods have been applied in cases where the supporting data have been inadequate.

In the terrestrial environment it is generally possible to overcome uncertainties in design by driving piles to refusal or by using pile load tests to verify design assumptions. In the case of piles driven offshore, load tests are prohibitively expensive and the use of excessively conservative design assumptions may lead to installation problems. There is, therefore, a continuing need for improvement in design methods.

The improvements are needed in the complementry areas of improving the fundamental theories and improving the supporting data base and empirical correlations. We have increased the size of the readily available data base, examined the accuracy of several existing design methods, and developed improved correlations which have led to a revised design procedure.

GENERAL EQUATIONS

For vertical piles subject to axial compressive and tensile loads, the pile capacity (Q) may be written as:

$$Q = Q_s + Q_p \pm W_p \qquad (1)$$

1. Department of Engineering, U.S. Military Academy, West Point, NY

2. Professor, Department of Civil Engineering, University of Texas Austin, Texas

where Q_s and Q_p are the loads transferred from pile to soil in side shear and end bearing, respectively, and W_p is the weight of the pile beneath the ground surface (use plus for tension tests and minus sign for compression tests). If Q_p is the net tip capacity then W_p becomes the weight of pile minus the weight of soil displaced, and thus becomes sufficiently small compared with $Q_s + Q_p$ that it is generally ignored. For tension loadings Q_p is assumed to be negligible.

Load transferred in side shear is taken as:

$$Q_s = \int f_s dA_s = \overline{f}_s A_s \qquad (2)$$

where f_s and $\overline{f}_s$ represent local and average side shearing stresses and A_s is pile side area in contact with soil.

The tip capacity is taken as:

$$Q_p = q_p\ A_p = 9\ c_u A_p \qquad (3)$$

where q_p is the net tip bearing capacity, A_p is the tip area, and c_u is the undrained shearing strength of the clay. For open ended pipe piles the tip capacity was taken as the smaller of: (1) the tip capacity of the wall of the steel pipe plus the side shear capacity of a full-length plug, (2) the tip capacity calculated as if the pile had a cover plate.

CALCULATION OF SIDE SHEAR CAPACITY USING EXISTING METHODS

Analysis of load transfer between a pile and soil, in side shear, involves consideration of soil-structure interaction as suggested by Coyle and Reese (2), with local transfer dependent on the relative movement of the pile and soil. Load transfer is also influenced by such effects as pile whipping during driving, drag-down of soil between layers, preboring and jetting, use of oversized cover plates, pile taper, and time-dependent reconsolidation of the disturbed soil. In the empirical methods of analysis considered here these effects must be included in the empirical factors. The methods considered in this study are summarized below. The names assigned to each method are ones we used in a computer program, and are used here only to simplify discussion.

ALP1 Method

In the ALP1 method (10), the average side shear, $\overline{f}_s$, is calculated using:

$$\overline{f}_s = \overline{c}_a = \alpha \overline{c}_u \qquad (4)$$

where $\bar{c}_u$ is the average undrained shearing strength over the embedded length of the pile and α is an empirical coefficient, termed the "adhesion factor," which was made a function of $\bar{c}_u$ and pile material. The relationship between unit adhesion ($\bar{c}_a$) and undrained shearing strength used by Tomlinson (10) is illustrated in Fig. 1.

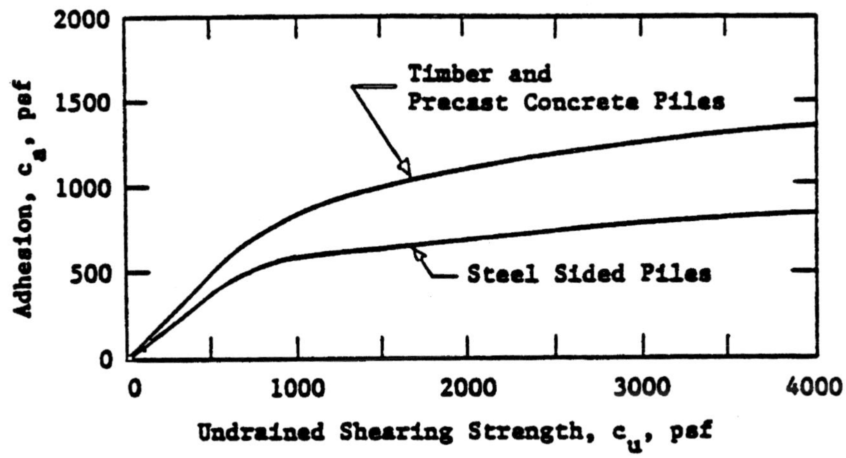

Fig. 1 Relationship between unit adhesion and undrained shearing strength after Tomlinson (10)

ALP2 Method

Based in part on field evidence obtained by excavating soil from around piles, Tomlinson (11) concluded that soil is dragged down from an overlying layer into the interface between the pile and soil in the next layer. He considered piles in three profiles: I, sand over "stiff clay:" II, soft clay over stiff clay, and III, stiff clay. He used Eq. 4 to calculate load transfer in the stiff clay, and made α a function of soil profile, average undrained strength in the stiff clay, and penetration of the pile into the stiff clay (Fig. 2). Tomlinson (11) calculated load transfer in an upper sand layer using Nordlund's (7) method and in an upper soft clay using α = 1 and Eq. 4. He concluded that pile material exerted a comparatively minor effect on side shear and used data from 93 load tests on timber, precast concrete, cast-in-place concrete, and steel pipe piles in his correlations. In our application of the ALP2 method, clays were considered "soft" if c_u was less than 500 psf, and we used the API (1) method for overlying sand layers.

APIC Method

The APIC method is essentially the American Petroleum Institute (1) method. The side shear capacity in clay, Q_s, is calculated using

$$Q_s = \sum_{\ell=1}^{N} \bar{f}_{s\ell} \, \Delta A_{s\ell} = \sum_{\ell=1}^{N} \alpha_\ell \, c_{u\ell} \, \Delta A_{s\ell} \qquad (5)$$

where $\bar{f}_{s\ell}$ is the average pile-soil shearing strength, $\Delta A_{s\ell}$ is the side area, α_ℓ is the adhesion factor, and $c_{u\ell}$ is the average undrained shearing strength of the clay, all in layer ℓ, and N is the number of layers penetrated by the pile.

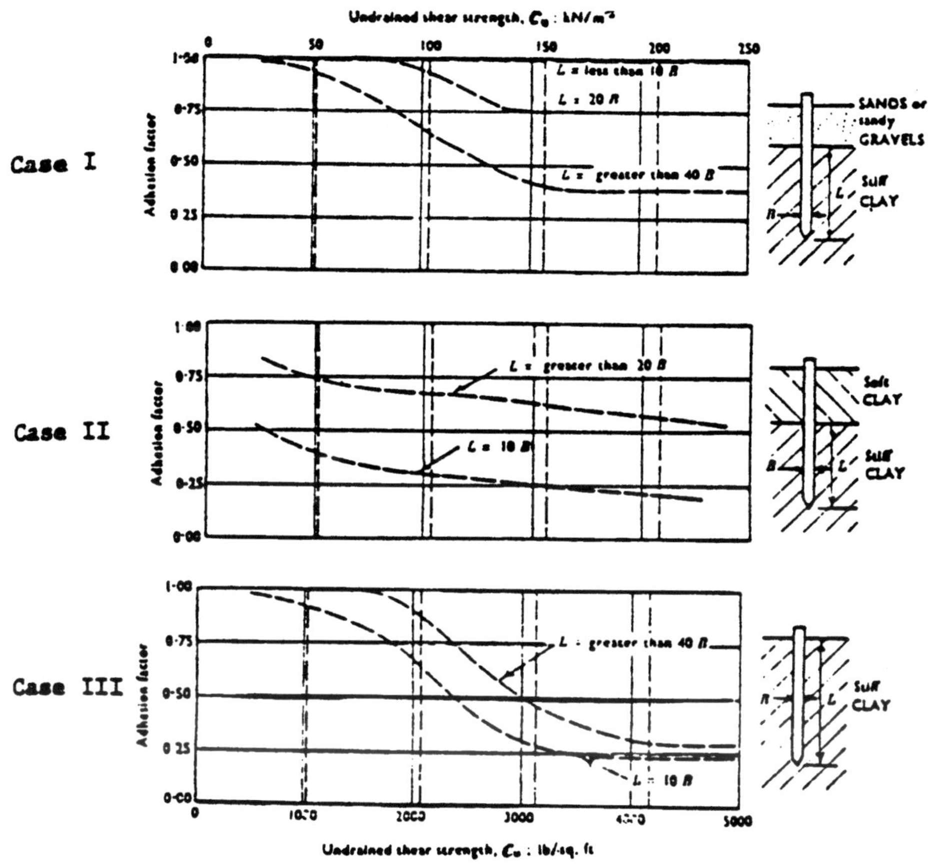

Fig. 2 Adhesion Factors Recommended by Tomlinson (11) and used in the ALP2 Method.

In the APIC method, clays are divided into two categories. Category 1 clays, here termed GULF clays (following the symbolism used in the computer program), are highly plastic clays, here including all clays with liquid limits and plasticity indices exceeding 50% and 35%, respectively. Category 2 clays are all other clays, here termed OTHR clays.

For normally consolidated and underconsolidated GULF clays, α is taken as one. For overconsolidated GULF clays α is taken as one but $\bar{f}_s$ is limited to the larger of 1000 psf or the c_u of a normally consolidated clay, whichever is larger.

For OTHR clays, α is 1.0 for $c_u \leq$ 500 psf, 0.5 for $c_u \geq$ 1500 psf, and varies linearly between those limits.

LAM1 Method

In the LAM1 method the side shear capacity is (12):

$$Q_s = \lambda(\sigma_m + 2c_m)A_s \tag{6}$$

where σ_m and c_m are the mean value of effective overburden pressure and undrained shearing strength, over the embedded length of the pile, and λ is an empirical factor which was found to depend on pile penetration (Fig. 3).

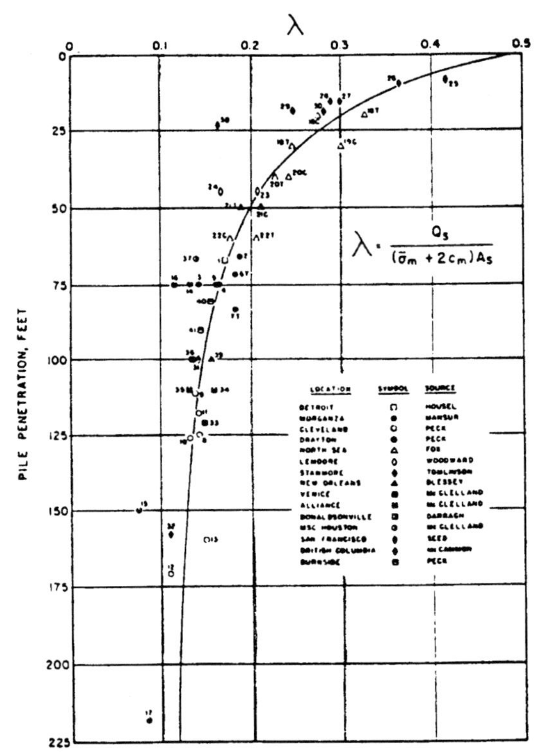

Fig. 3 Values of Lambda recommended by Vijayvergiya and Focht (12)

LAM2 Method

In the LAM2 method Eq. 6 is used but λ is given as (5):

$$\lambda = 0.178 - 0.016 \, \ell n \, \pi_3 \tag{7a}$$

for normally consolidated soils and:

$$\lambda = 0.232 - 0.032 \, \ell n \, \pi_3 \tag{7b}$$

for overconsolidated soils where:

$$\pi_3 = \pi B f_{max} L_e / AEu \tag{8}$$

and B = pile diameter, f_{max} is the peak pile-soil friction (taken as the mean undrained shearing strength), L_e is the embedded pile length, A is the cross-sectional area of pile material of modulus E, and u is the pile displacement needed to develop the side shear (taken as 0.1 inch). In the absense of consolidation data, Kraft et al. (5) considered a clay to be overconsolidated if $c/p \geq 0.4$. They set a lower limit of 0.14 on λ.

DATA BASE

A critical check on the accuracy of the predictive methods is a comparison of predicted (Q_c) and measured (Q_m) pile capacities using as large a data base as possible or, for a specific design application, a data base with conditions closely approximating those in the design problem. For this study, a data base of 1004 tests was developed using test results from the literature and unpublished data from government agencies, oil companies, and consulting firms.

Only data for tests carried essentially to plunging failure were used. Data used in this report were at a "defined" failure point, as opposed to plunging failure, because some piles appear not to reach plunging failure even at large displacements. Davisson's criterion (8, p. 215) was used to define the point of "failure."

The quality of collected data ranged widely. In the best cases, piles were fully instrumented, soil borings were near the test pile, soil properties were relatively uniform, soil properties were well defined (many measurements, reliable testing technique) and there seemed to be a reasonable consolidation period between pile driving and testing. For tests deemed good enough to be included in the data bank (about one test out of eight for which we had data), the quality of a test was denoted by an arbitrary "data quality factor" which ranged from 1 for the poorest acceptable test to 5 for the best tests. In the analyses reported here, only tests with data quality factors in the range of 2 to 5 were used.

The "strength" of the clay at a given site was measured, using a wide variety of methods, ranging from sophisticated laboratory and field tests to standard penetration tests. The most common measure was unconfined compression tests on 2-inch diameter thin-walled tube samples but Q-type and R-type triaxial compression tests and field vane tests were also often used. Lab vanes, Swedish fall cone, double ring shear, and other tests were also encountered. Reported strengths were used directly in analyses even though it is well known that they yield different values. In the case of the ALP1 and ALP2 methods the proper strength to use in the analysis would be the strength used in developing the original correlations (of the 93 tests used in the ALP2 correlations, 33 utilized unconfined compression, 11 field vane, and the technique was unknown for the other tests). Unconfined compression and miniature vane tests are specified in the API procedure and the engineer's interpreted strength is to be used in the LAM1 and LAM2 methods.

The data set used in this investigation was obtained by deleting from the master set of 1004 load tests all data for piles other than steel pipe piles, for piles with oversized cover plates or piles in prebored or jetted holes, tests with a data quality factor of one, tests where a load-settlement curve was unavailable, and tests where in situ vertical effective stresses could not be estimated with acceptable accuracy. The axial load capacities of the remaining piles were calculated using the methods discussed in this paper for clay layers and the API method (1) for sand layers, and tests were rejected if either the pile tip was in sand or a major part of the side capacity came from sand.

The final data set contained 84 load tests. Of these, 57 were entirely in clay, 11 derived 1-10% of their capacity in sand, 15 derived 10-20% of their capacity in sand, and one derived 29%, averaging 4% in compression tests and 3% in tension tests (based on analyses using the API methods in clay and sand). Of the 84 tests, 27 were loaded in tension and 57 were in compression. For the piles in compression, the calculated tip capacity was less than 10% of the calculated total capacity in 43 tests, 11-20% in 8 tests, 21-30% in 3 tests, and 31-40% in 3 tests, averaging 9%.

For the final data set, the average ratio of the peak reported load (often the plunging load) to the defined failure load was 1.06 with a range from 0.93 (3 tests had the load-settlement curve extrapolated to reach a defined failure) to 1.75.

It should be emphasized that calculated capacities are to be compared with "measured" capacities that are potentially in error. Measured capacities may be too high because of jack friction when loads are determined from hydraulic pressure in the jack and may be too low because of inadequate set-up time and because the long term static capacity may exceed the short term capacity measured in a load test, particulary for the pile tip. The measured capacity is influenced by the choice of failure criterion. Some of the measured capacities are potentially in error due to blunders, e.g., improper conversion of units, typographical errors, or use of incorrect calibration curves.

Q_c - Q_m COMPARISONS USING EXISTING METHODS

To accomodate load tests on piles of disparate sizes in a single diagram, Q_c and Q_m were plotted on log scales (Fig. 4). Lines were plotted with $Q_c/Q_m = \frac{1}{2}$ and 2 to help indicate scatter. The computer was programmed to prepare frequency diagrams for the ratio Q_c/Q_m on a log scale so that tests where say $Q_c = 2\ Q_m$ and $Q_c = \frac{1}{2}\ Q_m$ would plot the same distance from the $Q_c = Q_m$ point. For any test, a "factor of

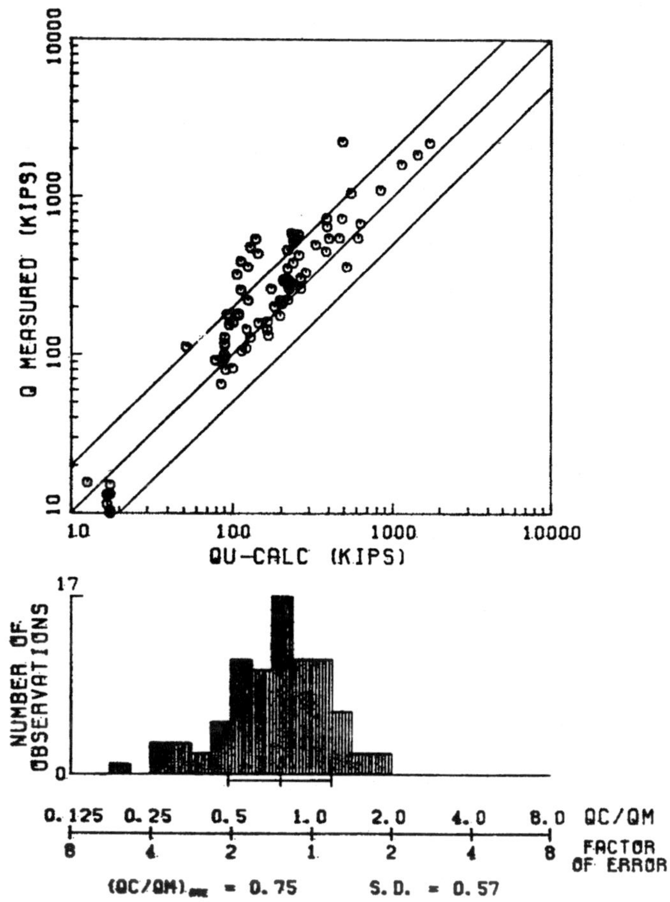

Fig. 4 Q_c - Q_m data, Method ALP1 (10)

error" was defined as Q_c/Q_m or Q_m/Q_c, whichever exceeded one. The mean value of Q_c/Q_m was calculated as:

$$(Q_c/Q_m)_{ave} = 10^{**}[\ \frac{1}{N} \sum_{j=1}^{N} \log\ (Q_c/Q_m)_j] \tag{9}$$

where ** denotes exponentiation, N is the number of tests, and $(Q_c/Q_m)_j$ is any one observation. The standard deviation, SD, was calculated as:

$$SD = 10^{**}\{\ \frac{1}{N-1} \sum_{j=1}^{N} [\ \log\ (\frac{Q_c}{Q_m})_{ave} - \log\ (\frac{Q_c}{Q_m})_j\ ]^2 \}^{\frac{1}{2}} - 1 \tag{10}$$

<u>Comparison of Methods</u>

Measured and computed capacities are compared in Figs. 4-8 for the ALP1, ALP2, LAM1, LAM2, and APIC methods, respectively. Statistical data are presented in Table 1.

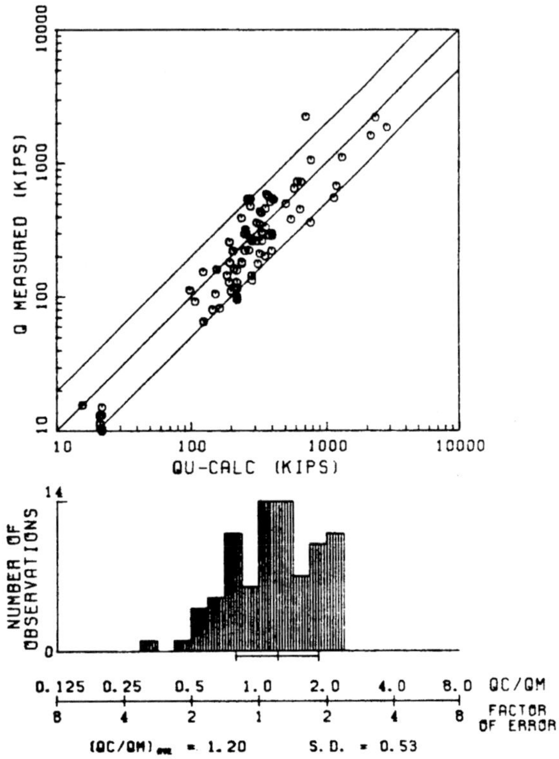

Fig. 5 Q_c - Q_m Data, ALP2 Method (10)

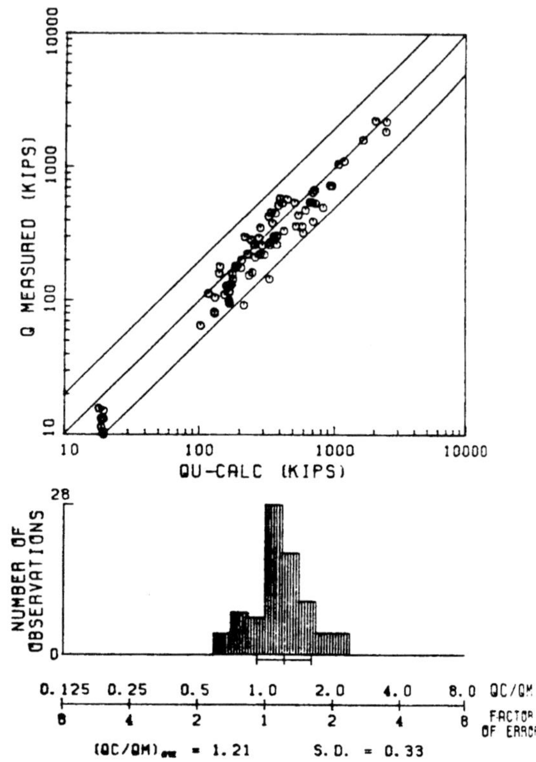

Fig. 6 Q_c - Q_m Data, LAM1 Method (12)

The ALP1 method leads to a significant underprediction of pile capacities for steel pipe piles and has the largest scatter. The ALP2 method leads to overprediction of capacities and a similar scatter. For both methods it was noticeable that piles with the largest values of Q_c/Q_m involved use of field vane shear tests or laboratory tests using samples of high quality (trimmed from 3-inch diameter, or larger, samples taken by pushing thin walled samplers) whereas tests with low values of Q_c/Q_m were often for cases where laboratory tests were performed on samples taken using smaller diameter, driven samplers. Data for tensile and compressive pile load tests were not significantly different and the existing differences were explainable in terms of sample and test quality.

Both lambda methods led to overprediction of pile capacities, with the LAM2 method being slightly better. The scatter in predicted capacities was much less for the lambda methods than for the two alpha methods. One reason for the reduced scatter in the lambda methods is that the term (σ_m + $2c_m$) in Eq. 6 is less effected by scatter in values of c_u than is the $\bar{c}_u$ term in the alpha methods (Eq. 4). Examination of the figures shows that the lambda methods are significantly more accurate for piles with capacities beyond 1000 kips, and none of those piles were in the original lambda

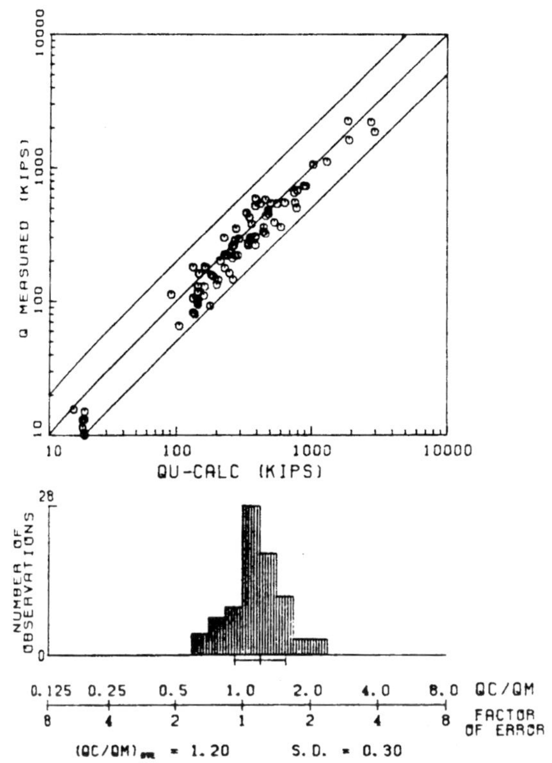

Fig. 7 Q_c - Q_m Data, LAM2 Method (5)

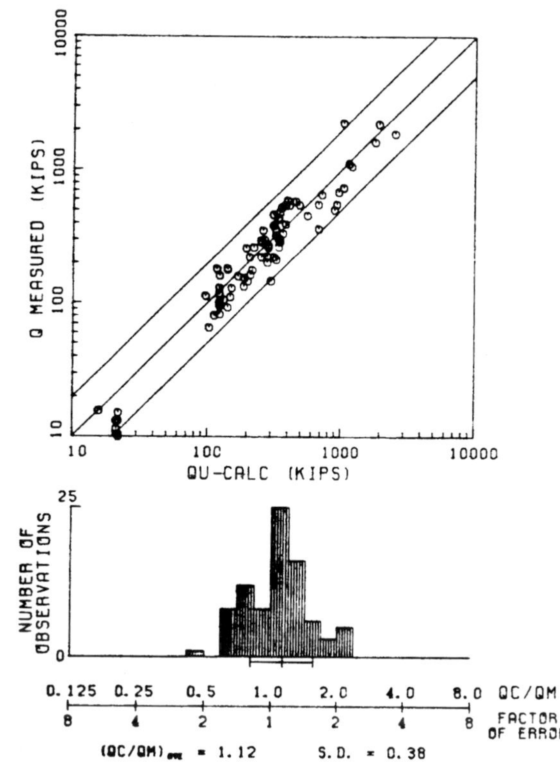

Fig. 8 Q_c - Q_m Data, APIC Method (1)

correlation so they constitute a valid test of the method.

The APIC method tended to overpredict capacities, but by less than did the lambda methods. However, there was more scatter in predictions, with the scatter again tending to correlate with sampling and testing quality.

DEVELOPMENT OF IMPROVED CORRELATIONS

Attempts could be made to improve correlations between measured and computed pile capacities using any of the existing formulations (alpha methods, beta methods, lambda methods) or new ones. Because of its simplicity, the alpha method was selected.

The data show an important effect of sampling and testing quality and thus indicate that a standard sampling and testing procedure should be used. We have chosen to define the standard as unconsolidated - undrained triaxial compression tests, performed using a confining pressure equal to the total overburden pressure, on samples taken using pushed thin-walled samplers, 3 inches in

Table 1 Statistical Summary of Analyses, using Existing Predictive Methods.

			Regression Line				Q_c/Q_m Data			
Method	Loading	N	a	b kips	r	SEE kips	Min	Max	Avg	SD
ALP1	T/C	84	1.30	400	.88	209	0.22	1.86	0.75	0.57
	COMP	57	1.30	450	.96	128	0.26	1.75	0.74	0.48
	TENS	27	3.00	-160	.80	260	0.22	1.86	0.79	0.75
ALP2	C/T	84	0.78	62	.85	231	0.32	2.32	1.20	0.53
	COMP	57	0.72	76	.93	161	0.49	2.17	1.13	0.45
	TENS	27	2.10	-240	.81	255	0.32	2.32	1.36	0.66
APIC	C/T	84	0.90	25	.90	186	0.46	2.32	1.12	0.38
	COMP	57	0.84	43	.95	131	0.65	2.17	1.07	0.37
	TENS	27	1.20	-54	.81	254	0.46	2.32	1.22	0.40
LAM1	C/T	84	0.91	-8	.97	103	0.66	2.30	1.21	0.33
	COMP	57	0.87	27	.98	92	0.66	2.30	1.11	0.32
	TENS	27	1.00	-75	.97	108	0.89	2.10	1.45	0.22
LAM2	C/T	84	0.81	30	.96	123	0.65	2.14	1.20	0.30
	COMP	57	0.74	68	.97	102	0.65	2.00	1.12	0.29
	TENS	27	1.10	-79	.97	100	0.81	2.14	1.38	0.25

diameter or larger. No suggestion is made here that the standard procedure be followed. However, if a different procedure is used, then the strengths should be adjusted to the standard strength.

To begin the process of improving the correlations, values of α were obtained from load tests on steel pipe piles by subtracting the calculated tip capacity (Eq. 3) from the measured pile capacity and dividing by the side area and average uncorrected undrained shearing strength to obtain a value for α. Because the most common method of estimating shearing strengths was unconfined compression tests, only load tests where unconfined compression tests were available were used. The resulting (first iteration) values of α are plotted versus c_u in Fig. 9, where solid symbols denote cases where samplers were generally 2¼-inch diameter or smaller, were driven, and often had large area ratios (often wire line samplers), and the hollow symbols denote cases where a 3-inch diameter, pushed, thin walled sampler, or better, was used.

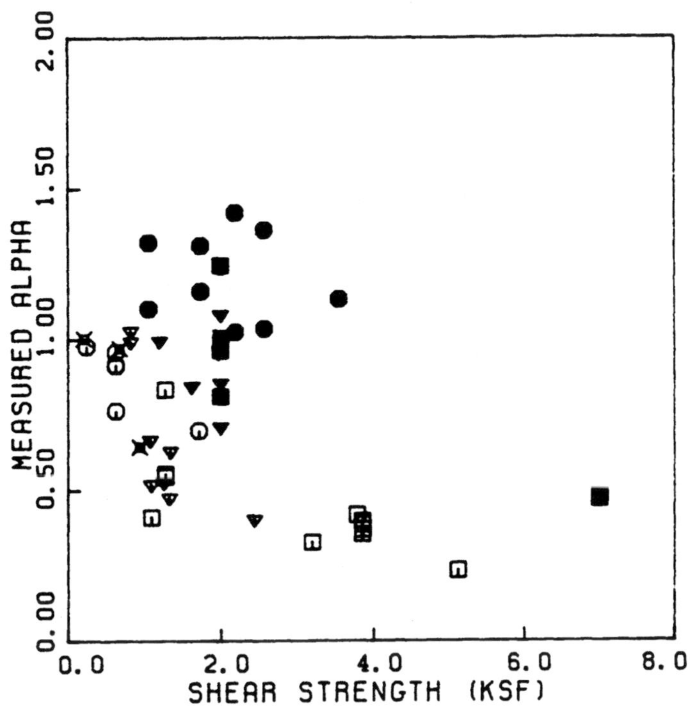

Fig. 9 Developed α Versus $\bar{c}_u$, with c_u Measured using Unconfined Compression Tests. The hollow symbols denote tests where the soil sampler was a 3-inch-diameter Shelby tube or better.

To increase the size of the data bank for tests with the better quality samplers, values of α were also determined for tests on untapered precast concrete piles. All values of α, for the better quality samples, were plotted versus c_u (Fig. 10) and a first iteration curve of α versus c_u was plotted. The present API curve for OTHR clays is shown for comparison.

The adhesion factors from Fig. 10 were then applied to the load tests on steel pipe piles for which strengths were measured using either laboratory tests (unconfined compression, triaxial compression) on high quality samples or in situ tests (vane, pressuremeter) and the ratio of Q_c/Q_m was plotted versus pile penetration (Fig. 11). Capacities seemed to be predicted tolerably well for piles of up to 100-foot penetration but the capacity of longer piles was underpredicted. The line in Fig. 11 could be used as a correction factor for pile length. The length effect is probably the result of interaction of a number of factors which cannot be separated using the data now available. Most of the short piles are in stiff clay where lateral movements of the piles during driving pushes the clay away and leads to reduced pile/clay contact stresses (10) and thus to reduced values of Q_m. All seven tests for piles with penetrations exceeding 250 feet were for piles in normally consolidated or lightly overconsolidated clays where good pile/clay contact is likely.

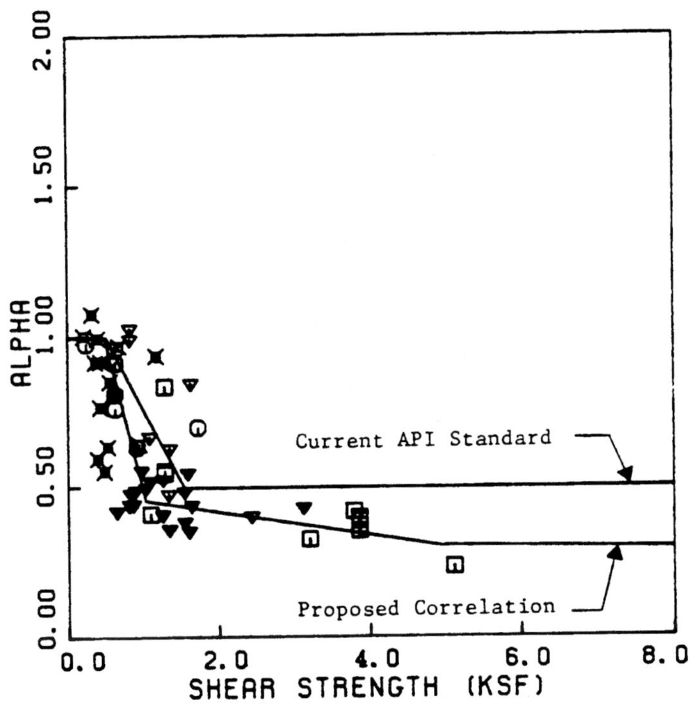

Fig. 10 Proposed Relationship Between α and c_u.

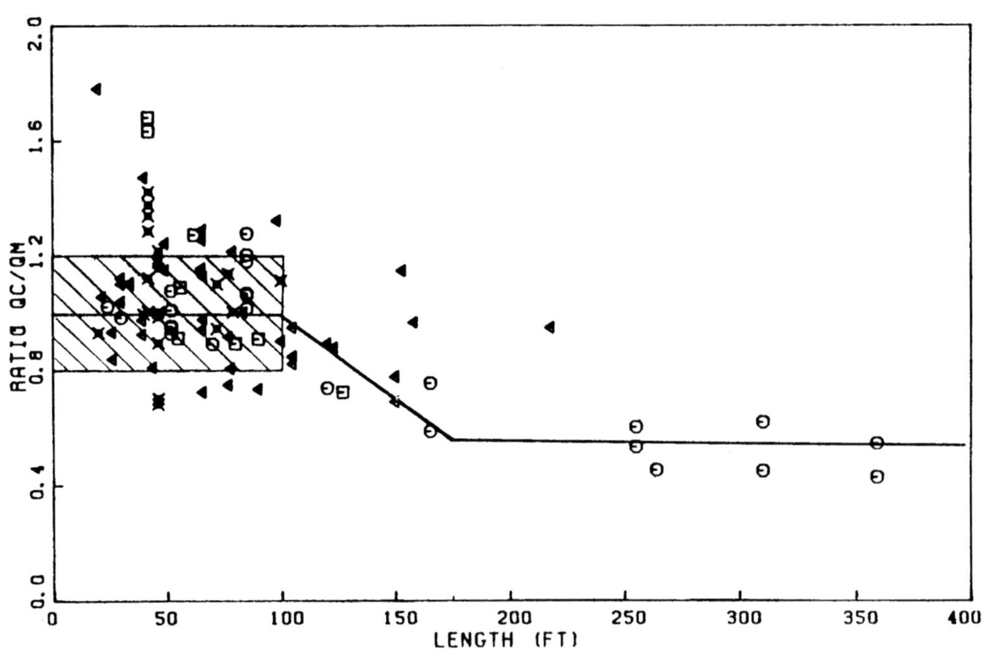

Fig. 11 Effect of Pile Penetration on the Ratio of Calculated to Measured Pile Capacity

Other effects that may have influenced the relationship include load shedding due to pile compressibility (2, 3, 6, 9), and dragdown of soil from one layer into the next (11). It is not clear, therefore, that the length correction shown in Fig. 11 has general applicability. It may also be noted that use of a dimensionless depth, e.g., the ratio of pile penetration to pile diameter, led to an appreciably poorer correlation.

Based on data shown in Fig. 9, and a point-by-point study of data in Fig. 11, it became apparent that the predictions would be improved by correcting the strengths to a standard value. It was found that on 13 sites where pile load tests were performed, the shearing strengths of the clays were measured using two or more techniques, thus allowing a limited amount of cross correlation. For these sites, the most often used measure of shearing strength was the unconfined compression test so other strengths were initially expressed in terms of the unconfined compressive strengths. On the average, the ratio of the unconfined compressive strength on high quality samples to another strength was 0.6 for field vane tests, 0.9 for Q-type triaxial compression tests on samples of high quality and 1.65 for laboratory unconfined compressive strengths of samples taken with driven samplers. Although the number of observations was small and the correction factors probably depend on various factors (plasticity, sensitivity, overconsolidation ratio, ...), the last factor is in agreement with a value reported by Emrich (4).

A revised method, called NCL1 (for new clay method 1), is now applied as follows. Calculate the tip capacity as:

$$Q_p = 9\, c_u F_c A_p \tag{11}$$

and the side capacity as:

$$Q_s = \alpha \bar{c}_u F_C F_L A_S \tag{12}$$

where c_u and $\bar{c}_u$ are the measured shearing strengths near the pile tip and averaged over the length of the pile, respectively, A_p and A_s are the areas of the pile tip and side, respectively, α is the adhesion factor, and F_c and F_L are correction factors for soil strength and pile penetration, respectively. The adhesion factor varies linearly as follows:

$\bar{c}_u F_c$ (psf)	0	600	1200	5000	∞
α	1.0	1.0	0.5	0.3	0.3

The correction factor for strength, F_c, is the ratio of the shearing strength determined using unconsolidated-undrained triaxial compression tests on samples of high quality to the shearing strength measured using some other technique. Values for F_c should be obtained from local experience. In the absence of such data, the engineer may use values found from the correlations reported earlier. These values would be 1.1 for unconfined compression tests on samples of high quality, 1.8 for unconfined compression tests on samples taken with typical driven samplers, and 0.7 for in situ vane shear tests. It should be noted that these values of F_c are based on a very small data set and do not show some well known effects, e.g., F_c should be larger for unconfined compression tests on stiff fissured and soft gaseous samples.

The correction factor for pile penetration, F_L, is:

L (feet)	0	100	175	∞
F_L	1.0	1.0	1.8	1.8

It should be re-emphasized that the data for long piles are for normally consolidated or lightly overconsolidated clays and may not be applicable for other soil conditions.

When the method was applied to open ended pipe piles, the side shear capacity was calculated as discussed above. The tip capacity was taken as the smaller of: (1) the tip capacity of a pile with a cover plate, and (2) the sum of the tip capacity of the base of the hollow pile and the side shear capacity of a full plug.

The NCL1 method was then applied to the data in the same data set used to study the existing methods. The method led to reasonably good correlations (Fig. 12) with a mean Q_c/Q_m of 1.0 and a scatter significantly smaller than found with the other methods. Statistical data are summarized in Table 2.

DISCUSSION

The recommended method is easy to apply and yields calculated capacities that compare well with measured capacities in the data bank used in this study. However, the method is empirical and is based on a relatively small number of tests. Developed pile capacities depend on a number of conflicting factors. The scatter in the values of Q_c/Q_m, found in this study, is unlikely to show the full scatter from all combinations of the conflicting errors. The corrections for method of measuring undrained shearing strength, and for pile length effects, are particularly suspect because of the small size of the relevant data bank. A long-term goal should be the development of rational methods of design.

Finally, it must be emphasized that the data set used in this study specifically excluded tapered piles, piles with oversized cover plates, and piles in prebored or jetted holes.

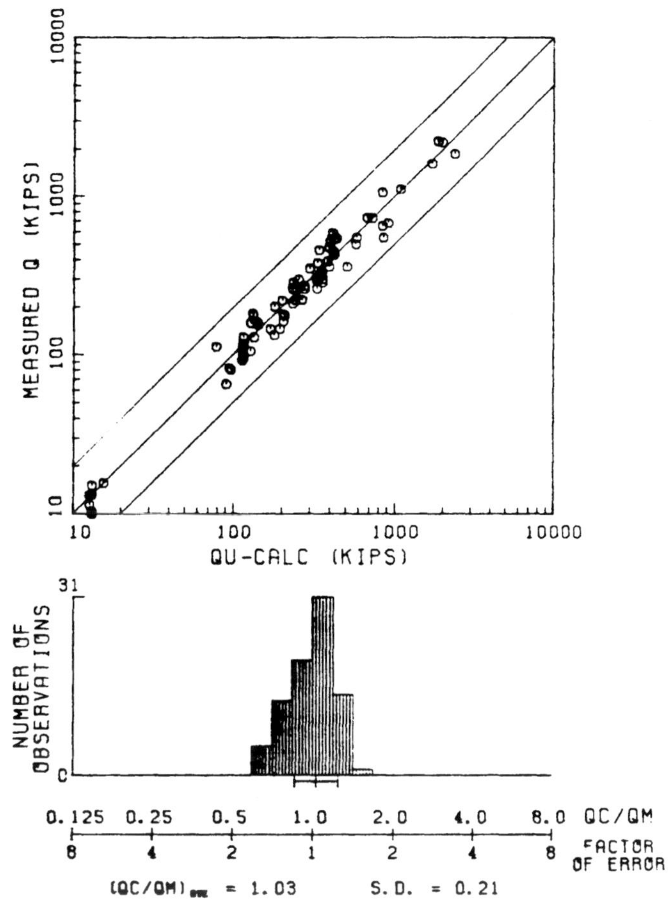

Fig. 12 Q_c - Q_m Data, Method NCL1

Table 2 Statistical Summary of Analyses using Method NCL1

Method	Loading	N	Regression Line a	Regression Line b kips	Regression Line r	SEE kips	Q_c/Q_m Data Min	Q_c/Q_m Data Max	Q_c/Q_m Data Avg	Q_c/Q_m Data SD
NCL1	Tens/Comp	84	0.98	6	0.97	104	0.70	1.55	1.03	0.21
NSA1	COMP	57	0.92	31	0.97	108	0.70	1.55	1.00	0.22
	TENS	27	1.20	-50	0.99	50	0.70	1.65	1.10	0.20

SUMMARY AND CONCLUSIONS

A data base of 84 pile load tests was developed for steel pipe piles, driven in clay (small amounts of sand near the surface for a few tests), where there was sufficient data to estimate effective overburden pressures with reasonable accuracy, where soil strengths were well defined, and where a load-settlement curve was available for the load test. Measured pile capacities were compared with values predicted using five methods that have been, are being, or potentially may be used in engineering practice. The two lambda methods had the least scatter but tended to overpredict capacities by about 20%. The current API method had slightly more scatter but overpredicted capacities by only about 13% on the average. The two alpha methods had the largest scatter and the greatest differences between measured and predicted capacities.

A major cause of scatter in prediction, for all five methods, was the different shearing strengths obtained using various methods of sampling and testing. To reduce the scatter, the analyses should utilize a standard strength which we have here taken as the unconsolidated-undrained triaxial compressive shearing strength measured on samples of high quality, preferably samples trimmed from ones taken using a pushed thin walled sampler of diameter 3 inches or more. An alpha-type method was developed using standardized strengths, a new α-c_u relationship, and a length correction factor. The resulting design method yielded an average ratio of calculated to measured capacities near one and the least scatter of any of the methods examined here.

ACKNOWLEDGMENTS

A major part of this research was performed with financial support from the American Petroleum Institute. The project advisory committee consisted of J.D. Allen (ARCO), D.D. Engle (Union Oil), L.M. Kraft, Jr. (McClelland Engineers), J.D. Murff (Exxon), B.S. Murphy (Amoco), J.R. Ruser (Chairman, Shell), G.E. Strickland (Chevron), and R.L. Thomas (Phillips). The guidance from the committee during the project and the permission from the American Petroleum Institute to publish the results are appreciated.

A major part of the literature survey and much of the early work on the computer program was carried out by D.G. Winter (Hart, Crowser, and Assoc.). Finally, we acknowledge the helpful critiques offered by the reviewers of the manuscript.

REFERENCES

1. American Petroleum Institute (1981), API Recommended Practice for Planning, Designing, and Constructing Fixed Offshore Platforms, API, Production Division, 211 North Ervay, Suite 1700, Dallas, Texas 75201.
2. Coyle, H. M., and L. C. Reese (1966), "Load Transfer for Axially Loaded Piles in Clay," Journ., Soil Mech. and Found. Div., ASCE, Vol. 92, SM2, pp. 1-26.
3. D'Appolonia, E., and I. P. Romualdi (1963), "Load Transfer in End-Bearing Steel H-Piles," Jour., Soil Mech. and Found. Div., ASCE, Vol. 89, No. SM2, pp. 1-25.

4. Emrich, W. J. (1970), "Performance Study of a Soil Sampler for Deep-Penetration Marine Borings, "Sampling of Soil and Rock, ASTM STP 483, pp. 30-50.
5. Kraft, L. M., J. A. Focht, Jr., and S. F. Amerasinghe (1981), "Friction Capacity of Piles Driven into Clay," Journ. Geot. Engr. Div., ASCE, Vol. 107, No. GT11, pp. 1521-1541.
6. Kraft, L. M., R. P. Ray, and T. Kagawa (1981), "Theoretical T-Z Curves," Jour., Geot. Engr. Div., ASCE, Vol. 107, No. GT11, pp. 1543-1561.
7. Nordlund, R. L. (1983), "Bearing Capacity of Piles in Cohesionless Soils," Journ. Soil Mech. and Found. Div., ASCE, Vol. 89, No. SM3, Paper 3506, pp. 1-35.
8. Peck, R. B., W. E. Hanson, and T. H. Thornburn (1974), Foundation Engineering, John Wiley and Sons, New York, NY.
9. Randolph, M. F., and C. P. Wroth (1978), "Analyses of Deformation of Vertically Loaded Piles," Jour., Geot. Engr. Div., ASCE, Vol. 104, No. GT12, pp. 1465-1488.
10. Tomlinson, M. J. (1957), "The Adhesion of Piles in Clay Soils," Proc. Fourth Intern. Conf. on Soil Mech. and Found. Engr., Vol. 2, pp. 66-71.
11. Tomlinson, M. J. (1971), "Some Effects of Pile Driving on Skin Friction," Behavior of Piles, Instn. of Civil Engrs., London, pp. 107-114. Response to Discussions on pp. 149-152.
12. Vijayvergiya, V. N., and J. A. Focht, Jr. (1972), "A New Way to Predict the Capacity of Piles in Clay," Offshore Technology Conference, Houston, Texas, Vol. 2, pp. 865-874.

LIST OF SYMBOLS

A	area (L^2)
a	slope of the linear regression line
A_p	total end area of the pile tip (L^2)
A_s	area of the pile side in contact with soil (L^2)
B	pile tip diameter (L)
b	intercept of the regression line (F)
$\bar{c}_a$	average pile soil adhesion along the pile side (F/L^2)
COMP	compressive load test
c_u	local undrained shearing strength (F/L^2)
$\bar{c}_u$	average undrained shearing strength along the pile side (F/L^2)
E	Youngs Modulus (F/L^2)
F_c	correction factor for shearing strength
f_{max}	peak pile soil friction (F/L^2)
F_L	correction factor for pile penetration
f_s	local unit shear at any point along the pile
L	total pile length (L)
Le	embedded pile length (L)
N	number of observations
q_p	unit tip resistance (F/L^2)
Q_c	calculated pile capacity (F)
Q_m	measured pile capacity (F)
Q_p	pile tip capacity (F)
Q_s	pile side capacity (F)

r	correlation coefficient
SD	standard deviation of Q_c/Q_m
SEE	standard error of the estimate (F)
T/C	tension and compression tests combined
TENS	tensile load test
u	displacement (L)
w_p	submerged pile weight (F)
α	adhesion factor, relating pile-soil adhesion to undrained shearing strength
λ	adhesion factor, relating pile-soil adhesion to undrained shearing strength and freefield effective overburden pressure.
π_3	pile-soil stiffness factor
σ_m	mean value of the free field effective overburden pressure (F/L^2)

Proceedings: Third International Conference on Case Histories in Geotechnical Engineering, St. Louis, Missouri, June 1-4, 1993, Paper No. 5.32

Failure of a Large Circular Excavation

R. E. Olson
L. P. Gilvin Professor of Civil Engineering, University of Texas, Austin, Texas

R. E. Heuer
Geotechnical Consultant, McHenry, Illinois

SYNOPSIS

A circular excavation, 117 feet (36 m) in diameter by 90 feet (27 m) deep, was designed by an experienced engineering firm and construction was performed by an experienced contractor. Nevertheless, the excavation support system suffered a complete collapse long before the maximum depth was reached. The failure is described and its causes are discussed.

INTRODUCTION

A buried tank, with an inside diameter of 105 feet (32 m), a wall thickness of 4 feet (1.2 m), and a depth to its bottom of 88 feet (27 m), was designed for a new sewage treatment plant at a site in Central Texas. The designer made a single soil boring at the center of the excavation and found that there was about 34 feet (10 m) of cohesive soil overlying a clay shale. A well known engineering firm, with wide international experience, designed a support system consisting of steel ring beams with vertical timber lagging. When the excavation reached a depth of about 42 feet (13 m) (Fig. 1), the support system collapsed and was a total loss.

In the resulting litigation, there was an opportunity to consider the design assumptions and the construction method, and to participate with experts in several fields in an effort to determine the cause(s) of the collapse. As usual in such cases, the various parties disagreed as to matters of fact as well as matters involving interpretation. Relevant parts of these considerations are reviewed in this manuscript and conclusions drawn regarding the causes of the failure.

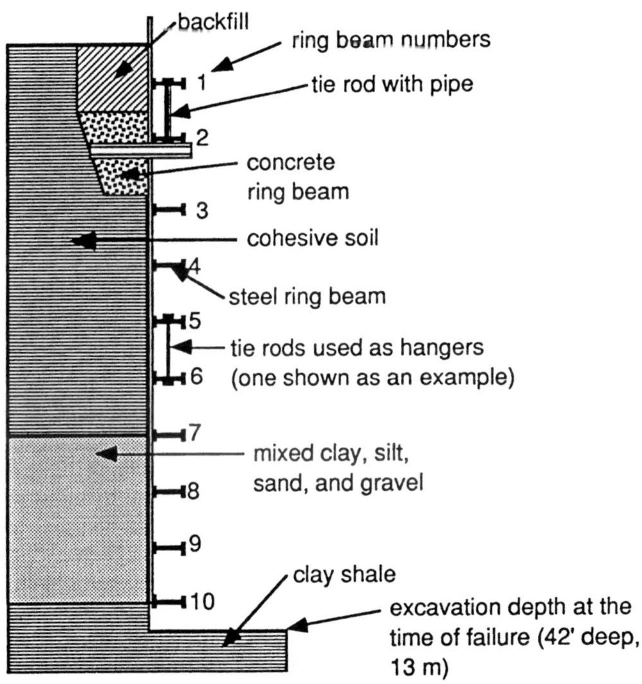

Fig. 1 Cross Section Through the West Side of the Excavation at the Time it Failed

SITE CONDITIONS

The site is flat and located in the flood plain of a river. The single soil boring specified by the design engineer, located at the center of the shaft, indicated the profile shown in Table 1. The soil boring was made during a dry period in the summer and no water table was encountered. The closest additional boring was 450 feet (137 m) away. That boring, and others at a greater distance, showed the presence of 19 to 35 feet (6-11 m) of cohesive soil over 7 to 12 feet (2-4 m) of sand and gravel. The designer estimated that the sand and gravel layer terminated about 200 feet (60 m) away from the edge of this shaft and thus estimated that there would be only minor seepage of water into the excavation. To be on the safe side, the designer assumed the possible presence of 3 feet (1 m) of relatively cohesionless soil just above the claystone.

DESIGN

The designers considered the site an obvious one for use of ring beams and timber lagging but indicated that liner plates

Table 1 Soil Profile from the Single Soil Boring

Depth feet (metres)	Description
0-12 (0-4)	SILTY CLAY, dark brown to brown, calcareous, organic material at top, sandy at bottom
12-34 (4-10)	SANDY SILTY CLAY, tan to rusty tan, brown, orange brown, calcareous, frequent shell fragments, some gravel and iron staining at the bottom
34-41 (10-12.5)	CHALKY CLAYSTONE, tan and light gray, weathered, occasional fossils and pyrite, upper few feet are fractured
> 41 (>12.5)	CHALKY CLAYSTONE, mostly gray, slickensided and fractured

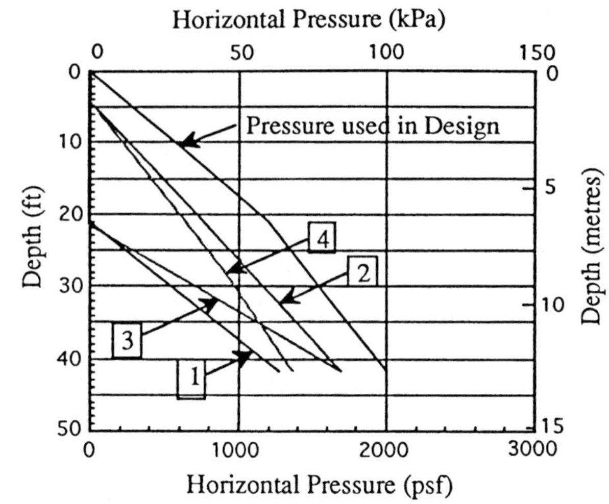

Fig. 2 Various Distributions of Earth Pressure

might be needed occasionally. The cohesive soils should stand unsupported for reasonable periods of time to allow placement of the supports. The lagging would provide needed local support for the fractured claystone.

A post-failure deposition by one of the design engineers led to the inference that the designer had not given serious consideration to any other support system in the overlying soils but had considered a variety of support systems in the underlying fissured claystone. In the deposition, the engineer claimed that steel sheeting could not be driven at this site, and that most other types of support would be uneconomical.

No original design computations could be obtained for the steel ring beams. The pre-bid geotechnical report mentioned need for only "nominal support" in the overlying soils. Testimony from one structural engineer was that the designer considered the beams to be in pure compression and did not consider the possibility of buckling failure. Design drawings contained the design pressure distribution labeled "Pressure used in Design" in Fig. 2. To resist those pressures, the designers recommended use of steel ring beams and vertical wood lagging (Table 2), assuming that the claystone would be encountered at a depth of about 34 feet (10 m). The designer's pressure diagram has been extended to the actual excavation depth of about 42 feet (13 m) in Fig. 2.

Ring beam spacing was 4 ft (1.2 m), except that rings 2 and 3 were 5 ft (1.5 m) apart.

The ring beams were tied to beams above and below using one-inch (25 mm) diameter steel tie rods spaced every six feet (1.8 m) around the circumference of the excavation. Each tie rod was connected to the two adjacent ring beams. The tie rods between ring 1 and 2 were encased in pipe so ring 1 could be supported on ring 2, whereas other tie rods were only in tension and had no pipe covers.

Square frame blockouts in the ring beam system were present at several locations around the shaft, extending from rings 4 through 6. One was located at the top of the initial failure area. Ring 5 was not continuous across this area.

Table 2.-Ring Beams and Lagging Selected by Designer

Depth Range, feet (metres)	Steel Section	Vertical Spacing of Beams ft. (m.)	Thick-ness of Lagging in. (mm)	Lagging Cover-age min. %
0-25 (0-7.6)	W24x55	4 (1.2)	2 (50)	25
25-57.5 (7.6-17.5)	W24x68	4 (1.2)	3 (75)	50
57.5-89.5 (17.5-27.3)	W24x104	4 (1.2)	3 (75)	100

The construction documents provided that the contractor was to be responsible for ensuring that the soil conditions were in accord with design assumptions and that the ring beams were adequately designed. The contract was let on a fast track basis because the city was under heavy pressure to improve the quality of the effluent that it was pumping back into the adjacent river.

The contractor hired a structural engineer to examine the support system and subsequently increased the size of the top group of ring beams to W24x62.

CONSTRUCTION

The contractor began construction in November with an excavation about 11 feet (3.3 m) deep, without support, and constructed a concrete ring beam (Fig. 1). This beam was 6 feet deep (2 m), 5 feet (1.5 m) wide at the top, and 3 feet (1 m) wide at the bottom, with a vertical inner edge (Fig. 1). Shaft lining began with lagging boards that were 12 feet (4 m) long. Steel ring beam number 2 was placed first. Then the upper ring (number 1) was placed and the upper lining backfilled. Normal downward shaft excavation then began.

The ring beams arrived on site, precurved, with each representing 1/15th of the circumference. The contractor welded groups of three sections together on the ground surface and then lowered these units into position. Once the first sections were in place, excavation was carried out over the entire bottom to about one foot (0.3 m) below the bottom edge of the next ring beam. The excavation extended about one foot (0.3 m) outside the projected position of the outer edge of the next ring beam. The next section was then lowered into position and hung from the ring above using tie rods. The wood lagging was placed on the outside of the ring beams and temporarily supported from the steel by driving a nail into the lagging and bending it over the outer flange of a steel ring. Wood wedges and blocking were placed between the lagging and the natural ground to provide support. Straw was stuffed into the voids where water seeps were encountered.

Actual soil conditions were different from those predicted from the single soil boring. A panoramic sketch of the soils exposed by the excavation is shown in Fig. 3.

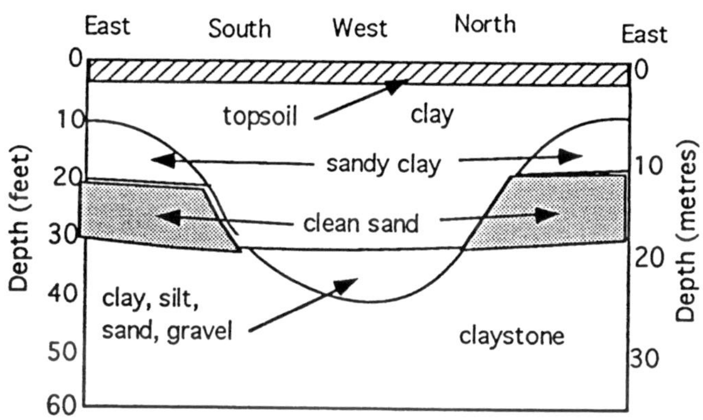

Fig. 3 Panoramic Sketch of the Excavation

On the east side there was about 13 feet (4 m) of clay, then 10 feet (3 m.) of sandy clay, and then 17 feet (5 m) of clean sand.

On the west side of the excavation, an apparent buried erosional surface cut down through the sandy clay and clean sand and into the underlying claystone. Overlying the erosional surface in the claystone was a layer of sand and gravel in turn overlain by mixed clay, silt, sand, and gravel. Above about 33 feet (10 m) depth, the "buried channel" was filled with the clay soils present at shallower depth around the entire excavation.

Raveling and small caveins occurred on the north and east sides early in December, apparently in the clean sand, which dried rapidly when exposed. The contractor used 3x6-inch (75 x 150 mm) wood cribbing and hay to backfill the voids initially. Several larger openings were filled with cement grout. Except for these problems, the project proceeded on schedule through the top forty feet (12 m).

On the west side of the excavation, when down to about ring beam 8, a fire, caused by welding activities, developed in the hay backfill, and forced the replacement of lagging from beams 3 and 6. The firemen used substantial amounts of water in fighting the fire. The new lagging was wedged into place as before but no special efforts were made to replace soil that may have been softened.

FAILURE

In early December, during the middle and end of the week of the failure, there were substantial rains in the area. On Monday, a block of soil, described as 3x4x4 feet (1x1.2x1.2 m), slid into the excavation on the west side, in the area of rib 9, apparently without causing concern on the part of anyone. It appears that routine hay and wood backfill operations were performed and work continued.

On Thursday evening, at about 5PM, workers were placing ring 11 at a depth just above the interface between the soils and the claystone. On the west side, water was flowing into the excavation through the lower gravel "channel filling" and two cavities were noted behind the rings and lagging. The contractor tried to use hay and timber cribbing to fill the cavities but the holes continued to enlarge. The construction superintendent for the day shift, which ended at 3PM, testified that the main void was about 4 feet (1.2 m) wide, 2 feet (0.7 m) deep, and 12 feet (3.7 m) high early in the day but had widened to about 12-15 feet (4 to 5 m) by the end of his shift. He was not concerned about the void although they were in the process of trying to stop its expansion. By about 9:30PM, the two voids had merged into one void along the gravel/claystone interface with a horizontal dimension along the circumference of the excavation estimated to be 50 feet (15 m), a maximum height of about 12 to 15 feet (4-5 m), and a radial dimension of about 5 feet (1.5 m). Workers changed tactics and welded some 2x2 (50x50 mm) steel angles to the outside of rings 8 through 10 and tried to construct angle iron bracing projecting into the void between the ring beams and the soil, upon which to support timber cribbing and hay filling of the void.

At about 12:30 AM, workers standing in the bucket of a front end loader, well up in the air, and working on the cribbing system, observed that they, and the loader, were slowly being pushed back toward the center of the excavation. They also heard loud popping noises and what sounded like an explosion in the general vicinity of ring beam 5.

All workers were removed from the excavation. The support system on the west side began to collapse (Fig. 1). The failure progressed downwards, and the entire west side of the support system had failed by 3:30 AM. Over the next few days, the entire support system unraveled and collapsed into the excavation as a twisted pile of steel (Fig. 4).

Eyewitness accounts and the locations of the steel rings immediately after the initial stages of the failure, indicate that the failure involved an inward buckling of rings starting with ring #5 at the top.

Fig. 4 Photograph of Collapsed Support System on the West Side

The time sequence of initial failure movements of rings is uncertain because the evidence was obliterated by movements in later stages of the failure.

POST MORTEM BY THE DESIGNER AND STRUCTURAL CONSULTANTS

Subsequent to the failure, the contractor filed a claim for changed conditions on the basis that the thick sand layer on the east side had not been shown in construction documents, nor had the thick gravel seam in the "buried channel" on the west side. The design engineer rejected the claim, pointing out that the contractor was responsible for ensuring the accuracy of the soil profile and performing the structural design. Further, the design engineer claimed that the contractor should have been able to stabilize the cavein more quickly.

A structural engineer, representing the designer, rejected eyewitness accounts of inwards movements of upper beams and concluded that the beams buckled outwards. He also argued that ring #5 did not bear properly on the blockout and was thus subjected to a twist for which it had not been designed. A structural engineer representing the contractor testified that the rings were unsafe for buckling failure.

SUBSEQUENT SITE INVESTIGATIONS

Site Visits

The first author visited the site on the following Monday afternoon (about 70 hours after the failure began) and on a number of other occasions during the next month. At the time of the first visit, water was flowing into the excavation through the debris on the west side. Soil samples were taken and field conditions were photographed. Much of the site, including the failure zone, was inaccessible.

Laboratory Tests

During the early phases of the investigation it was believed that there may have been an earth pressure problem in the cohesive soils at shallow depths and that this problem may have led to buckling of the beams. Accordingly, hand carved samples of some of the shallow soils were obtained and one each, drained and undrained, stage-type triaxial compression tests were performed. The failure envelopes had the following intercepts and slopes:

undrained: $c = 900$ psf = 43 kPa $\phi=19$ deg.

drained: $\bar{c} = 150$ psf = 11 kPa $\bar{\phi}= 27$ deg.

The finite value of ϕ for the undrained tests makes it clear that the samples were not saturated. No Atterberg limit tests were performed on the specific samples used for triaxial tests. However, numerous tests with adjacent samples showed that LL ranged from 31 to 38%, PI from 16 to 24%, water contents from 16 to 24%, with one sample classified as ML and the others as CL. The sand/silt/clay fractions were generally about 15/55/30 %.

Failure Area

Attempts by the first author to investigate conditions in the failure zone, several days after the failure, were only partially successful because the lower rings were buried in debris and the upper soils had slid down over the soils that were adjacent to the excavation in the failure zone.

However, during reconstruction, the steel and some of the soil debris was removed, and it was possible to access part of the west side where the failure apparently began. In that area, the top forty feet (12 m) (estimated depth) was cohesive but the top ten feet (3 m) (estimated) of supposed claystone was actually a mixture of soils ranging from clay to gravel. A sketch of the zone where failure was believed to have begun is shown in Fig. 5. The "red bed" was a red, highly plastic clay, with LL=72%, and PI=50%. Water was still flowing out of the lower gravel layer a month after the failure. The gravel layers may have been channel deposits with high pore water pressures resulting from exposure of the gravel at a higher elevation off site.

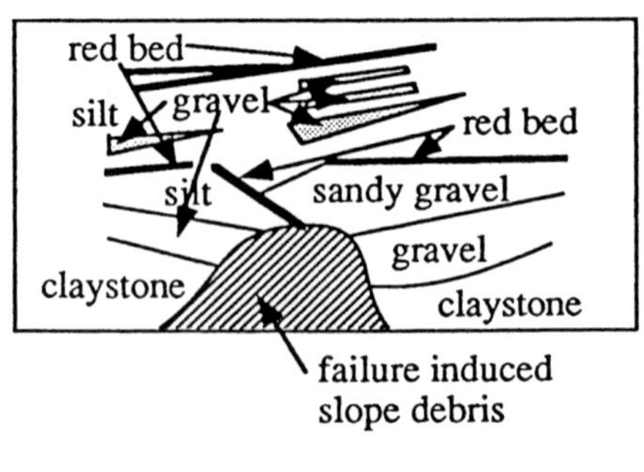

Fig. 5 Sketch of Exposed Soil in the Area of the Failure (Section about 35 feet=11 m wide and 15 feet=4.5 m high)

Earth Pressure Computations

For confining pressures approximating those in the field, the triaxial samples failed at low axial strains (1% undrained to 4% drained). In consideration of the construction method, which left significant thicknesses of soil unsupported before installation of the rings and lagging, and the large diameter-to-depth ratio at the time of failure, it is believed that earth pressures may be closer to classical Rankine values rather than those appropriate for prestressed, braced, or tied back, excavations.

A Rankine-type analysis was performed for active earth pressures using the soil profile assumed by the design engineer. Under undrained conditions, using the failure envelope reported earlier, and a total unit weight of 120 pcf (18.8 kN/m^3) the earth pressure would be as shown by curve 1 in Fig. 2 (no pressure to a depth of 21 feet (6.4 m) and the pressure at a depth of 42 feet (12.8 m) would be about 1280 psf (61.3 kPa). Vertical scarps of the order of this height were observed in the clays on the west side during the early stages of the clean up so the computations appear realistic.

Under drained conditions, the earth pressure depends on the assumed water conditions above the water table. If the water table is set at a depth of 42 feet (12.8 m) at the top of the claystone, and the water pressures above the water table are zero, then curve 2 (Fig. 2) is calculated, with a free standing height of only 4 feet (1.2 m) and an earth pressure of 1210 psf 57.9 kPa) at 42 feet (12.8 m).

If the pore water pressures above the water table are the no-flow negative values, then curve 3 (Fig. 2) is obtained. There is no earth pressure down to 22 feet (6.6 m) and the earth pressure is 1710 psf (81.9 kPa) at 42 feet (12.8 m).

Analyses under drained conditions for the actual soil profile on the west side are shown as curve 4 in Fig. 2.

Soil conditions at the time of the failure are unclear. Although it had rained periodically for the preceding six days, some eyewitnesses claimed to see dust blown into the air during parts of the collapse.

A comparison of the earth pressures diagrams suggests that the pressures used by the design engineer were reasonable and conservative and that problems with the earth pressure computations was not the cause of the failure.

ANALYSES OF RING BEAM DESIGN

According to conventional design procedures, the steel ring beams must have adequate capacity to support the external soil pressures while acting as pure compression members. The design condition is given by:

$$T = p\,R\,s = f_s\,A \qquad (1)$$

where T=compression thrust in the ring, p=design soil pressure, R=radius of excavation, s=ring spacing (4 feet = 1.2 m here), f_s=allowable steel stress, and A is the ring cross sectional area. If blocking space is not very close, a different analysis may be used which accounts for the bending moments induced in the curved ring between blocking points. Calculations suggest that the ring beams (Table 2) have adequate capacity to support the "Pressure used in Design" shown in Fig. 2.

Complete buckling failure and collapse of the underground support systems is uncommon, and generally not well understood by the profession. Two different structural consultants argued, after the failure, that the ring beams were inadequate because they did not have adequate capacity according to a buckling equation of the form:

$$p_{cr} = \frac{3EI}{R^3} \qquad (2)$$

where p_{cr}=critical external pressure on the support system, at which failure occurs, E and I are the lining material and sectional properties, and R=radius as before.

An equation of the form of Eq. 2 is not applicable in the present case, nor is it applicable to most conditions in underground construction. This equation is based on the assumption that the support is subjected to a uniform external fluid pressure. Failures of steel casings in slurry-filled holes, or surrounded by fluid grout, have occurred as predicted by this equation, or by similar equations which account for initial out-of-roundness of the cylinder due to manufacturing tolerances.

These boundary conditions do not apply to underground linings which are in contact with the surrounding soil (or rock) around their entire perimeter. Such linings include shaft and tunnel ring beams or continuous shells (steel pipe, steel liner plate, thin concrete or shotcrete shells) which are well blocked or grouted in place. Total collapse of such linings, by buckling failure, is rare because of soil-structure interaction effects. Whenever an inward deforming bulge begins to form in the lining, due to external soil pressure, arching tends to develop in the soil to relieve the driving pressure. Simultaneously, adjacent outward bulges of the lining induce passive soil pressures which tend to restrain further movement.

Accordingly, buried cylinders, properly installed, have much higher buckling strength than defined by the fluid buckling case of Eq. 2. Buckling equations of forms such as the following (Allgood, 1972) are applicable to most underground cases:

$$p_{cr} = \bar{c}\sqrt{E_s\left(\frac{EI}{D^3}\right)} \qquad (3)$$

where $\bar{c}$ is a parameter determined by boundary conditions such as depth of burial and Poisson's ratio of soil and lining materials, E_s=deformation modulus of the soil, E and I are as defined before, D=2R=diameter of the excavation. Boulson (1985) summarized the theoretical basis of buckling strength of underground structures, and various analytical and experimental studies which have been made. Note that the controlling parameter of the lining is the ring stiffness EI/R^3 or EI/D^3. Bucking failures of underground support systems can generally occur only in unusual conditions, such as extremely loose soils which permit large lining deformations (diametral deformations in the range of 15% to 20% can be tolerated before snap-through buckling failure in the first mode), highly concentrated loadings, or unusually low ring stiffness of the support or lining. Although the ring stiffness of the W24x68 ring beams for this case, about their strong (x-x) axis, is relatively low, about EI/D^3=0.42 psi (2.9 kPa), the second author has reported successful use of linings with ring stiffnesses as low as EI/D^3=0.05 psi (0.35 kPa) in shafts of approximately the same dimensions as this case (Heuer et al., 1983). Hence, the problem here was not simply use of a lining outside of previous experience.

The buckling strength of ring beam support systems, as defined by equations such as Eq. 3, is based on fundamental assumptions about construction methods and structural details. These are:

a. The lining is assumed to be in relatively continuous contact with the surrounding ground around the entire perimeter, and

b. The ring beams are restrained and supported such that they do not twist or buckle about their weak (Y-Y) axis.

Construction methods must ensure that these conditions are satisfied. For linings of low ring stiffness, assumption "a" is best satisfied by grouting the annulus between lining and excavated structure. It may possibly be satisfied by well developed wood blocking and wedging in relatively stable soils. Assumption "b" is normally satisfied by supporting the ring beams partially by the tie rods and partially by friction between soil and the ring system. Soil pressure is normally essentially horizontal against the lining, because soil weight is supported by the continuous column of confined soil outside the lining.

Both of these conditions were violated when the large void developed outside the west side of the shaft in the hours before the failure.

EXPLANATION OF THE FAILURE

Conditions which existed on the west side of the shaft just before failure were approximately as shown in Fig. 6.

There had been one or more voids on the west side for several days. Water was apparently flowing from the gravel layers into the excavation, eroding soil which was then being removed by the front end loaders as if it were soil from the bottom of the excavation. Routine efforts to support the cavities had not been successful. In the gravel, the flow of water rapidly enlarged the cavities to the point that use of hay and wood cribbing was inadequate. Soil contact outside rings 7 through 10 had been lost for several dozen feet around the perimeter.

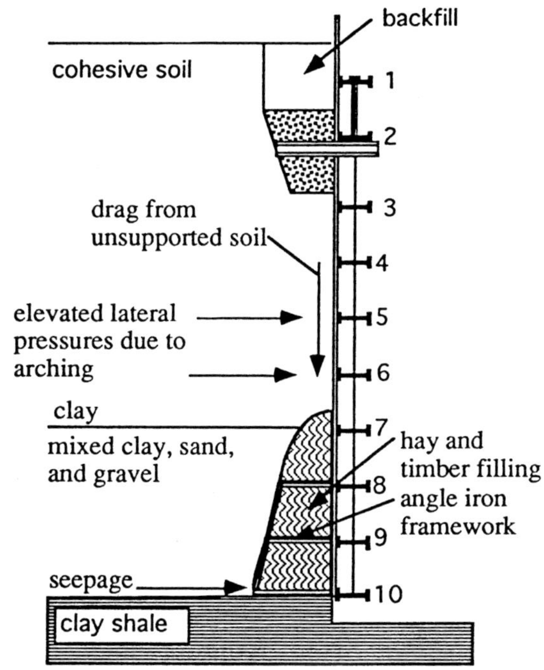

Fig. 6 Conditions In the West Wall at the Time of Failure

These rings may have deformed outwards sufficiently to relieve the load they were carrying. This movement was not observed, perhaps because the deformation was small, occurred over a substantial length, and no plastic hinge formed. The earth pressure that was originally carried by rings 7 through 10 must have been transferred through arching, partially into the floor of the excavation, and partially into the above ring beams.

Accurate calculation of the loads in the upper remaining ring beams is not possible due to lack of information. Various estimates of the loads can, however, be made. For example, if the soils are all assumed to be drained with a pore water pressure of zero above the water table, and a Rankine active condition is assumed, then the ring thrust loads would be about as shown in Fig. 7. It might be assumed that loads in rings 9 and 10 would be arched into the unexcavated "rock" beneath the excavation. The loads in rings 7 and 8 can be assumed to be arched into upper rings. If loads in rings 7 and 8 are distributed to rings 4 through 6, the possible ring loads might be as shown in Fig. 7. Peak ring loads would be in the range of 350 to 400 kips (1.5 to 1.8 MN).

However, use of Eq. 3 indicates that the W24x68 members in the failure area should have been adequate to support thrust loads of this magnitude.

The growing void in the lower soils undercut the overlying clay soils. These upper soils would tend to settle downwards, creating downward drag on the lagging behind rings 5 and 6, as illustrated in Fig. 6. This downward soil

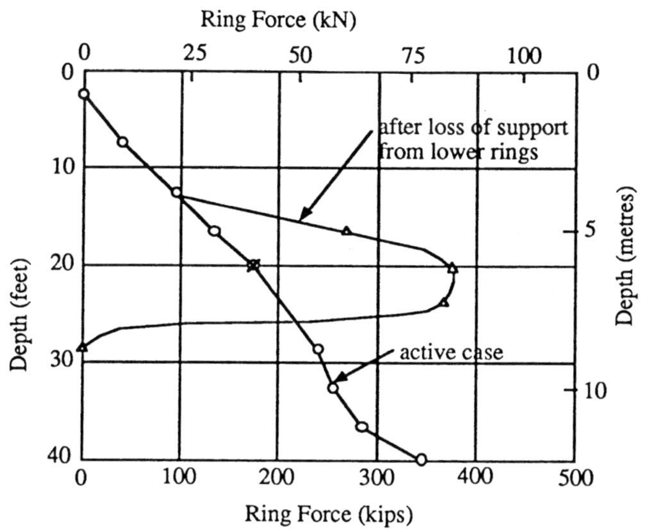

Fig. 7 Possible Ring Beam Thrust Loads Before and After Void Formation

movement may have been facilitated by overload and developing failure in ring 5 at the blockout (discussed below), which led to increased loading and yielding of ring 6 as well. Soil softening by water used to extinguish the earlier fire in the hay may have contributed to downwards soils movements into the void.

This vertical soil load, plus the weight of rings 5 through 10, would have subjected the tie rods between rings 4 and 5 to loads between 10 and 20 tons (88-176 kN). Resulting tension failure in the tie rods would allow the entire ring beam structure, from rings 5 to 10, to buckle downwards and into the excavation, buckling about the weak (Y-Y) axis. This scenario is in accord with observed positions of the members after the initial stages of the failure although the steel was so twisted after the large movements, and covered with soil, that the detailed conditions at the moment of failure could not be ascertained.

Some experts theorized that the failure was initiated because structural detailing at the blockout between rings 4 and 6 led to the line of thrust from the rings coming slightly to the inside of the center of the frame for the blockout, producing a twist in the side members of the blockout. The authors believe that the failure was initiated by load transfer from the lower rings because of the voids which formed outside the lower rings, and by downward drag of the soil. Structural detailing at the blockout may have led to failure initiating at that particular location, rather than elsewhere, but the basic cause of the failure was formation of voids outside the lower rings.

LESSONS LEARNED

The engineers on the two sides of the litigation may not agree on the lessons learned. The view of the authors (who represented opposing parties) is as follows:

1. The minimal soils investigation may have resulted from fast tracking and economic constraints imposed by the owner. The owner's attempt to transfer the responsibility for an inadequate soils investigation to the contractor did not succeed because there was insufficient time for an additional soils investigation and because such additional investigations are not in accord with usual practice.

2. Responses from the design supervising engineer during a deposition, led to the belief that the designer had performed the "design" in accord with their previous experience without giving much thought to alternatives, to the relevance of their previous experience, or to the consequences of having field conditions different from those assumed. Had the presence of clean sand and gravel been suspected, steel sheeting and ring beams, or some other system with full surface support, could have been used in the upper soils and should have prevented the failure.

3. The failure would not have occurred if the large voids outside the lining had not formed. Where granular soils were encountered, particularly with moderate seepage, the problem might have been avoided if a thick grout had been placed behind tightly placed wood lagging, or if liner plate had been installed and backgrouted, to fill any initial voids behind the support system. Short radial drains of slotted PVC pipe might have been placed at selected locations through the lining to permit controlled drainage of water without soil erosion, to prevent later formation of large voids. The site personnel seem not to have been aware of the potential serious consequences of such voids and did not take the necessary steps to control the problem at an appropriate time.

4. As occurs so often, this failure resulted because details of the design and construction process were not adequate for the ground conditions actually encountered.

REFERENCES

Allgood, J. R. (1972), *Summary of Soil-Structure Interaction*, Tech. Report R-771, U. S. Naval Civil Engineering Lab., Port Hueneme, CA. (Available as NTIS Report AD 748581).

Boulson, P. S. (1985), *Buried Structures, Static and Dynamic Strength*, Chapman and Hall, London and New York.

Heuer, R. E., W. C. Cox, and J. M. Loignon (1983), "Excavation and Support of Gatineau Shaft", *Proc., Rapid Excavation and Tunneling Conference*, ASCE and AIME, Vol. 1, Chap. 28, New York.

Proceedings: Third International Conference on Case Histories in Geotechnical Engineering, St. Louis, Missouri, June 1-4, 1993, Paper No. 5.31

Failure of a Twenty-Foot High Retaining Wall

R. E. Olson
L. P. Gilvin Professor of Civil Engineering, University of Texas, Austin, Texas

SYNOPSIS A cantilever retaining wall, designed in apparent accord with provisions in a civil engineering handbook, failed soon after construction. Analyses of the causes of the failure are presented.

INTRODUCTION

Cantilever retaining walls are typically designed using either the Rankine or Coulomb equations to estimate applied forces and considering failure modes involving sliding, overturning, toe failures, and overall failure. The standard design techniques are in civil engineering handbooks and have been in use for decades. Designers would seem to have the right to expect that a wall designed in accord with handbooks would perform adequately well.

A cantilever retaining wall with a maximum height exceeding twenty-six feet (eight metres) was designed by a registered professional engineer, in accord with a standard handbook. The designer was a generalist who performed design work in most areas of civil engineering and was thus not trained specially in geotechnical engineering. The designer used factors of safety in excess of two against the usual failure modes. The wall collapsed and subsequent litigation led to losses in excess of $1 million for the designer.

The purpose of this paper is to present the case history, examine the original design calculations, present a more appropriate, but still simple, set of analyses, and draw relevant conclusions.

CASE HISTORIES OF RETAINING WALL FAILURES

Anecdotal evidence of unacceptable performance of retaining walls, can be heard but well-defined case histories seem to be few. Ireland (1964) made a survey of walls used to support railway cuts. He categorized the backfill and subsoil simply as "sand" (including gravel) and "clay". The distribution of data for walls that had not performed properly are shown in Table 1.

These cases confirm the general view that the major problems are associated with cohesive soils.

Table 1 Summary of Cases of Retaining Walls Engaging in Unacceptable Behavior (Ireland, 1964)

Subsoil	Backfill	% of Walls
clay	clay	43
clay	sand	17
clay	unknown	8
sand	clay	8
unknown	unknown	24
sand	sand	0

DEVELOPMENT OF SITE

A developer selected a site in Central Texas for a shopping center. The site fronted on a major street and was generally flat until near the back edge and left side where the surface began to slope downwards into a large tract of land that was left wild and used as a nature preserve. The ground in the preserve sloped down at about 15 to 20 degrees.

Economic considerations dictated a certain floor area for the shopping center, and city ordinances then required a certain area for parking spaces and driveways. It was discovered that it would be necessary to provide parking and driveways to a line near the edges of the property and thus that it would be necessary to use a 1300-foot (400-m) long retaining wall along the left side and the rear, to support fill. The maximum height of the actual wall, measured from the top of the stem to the bottom of the keyway, was 26 feet 9 inches (8.2 m).

City environmentalists objected to the presence of a high concrete wall at the upper edge of the nature preserve so a solution was worked out such that there would be a relatively low single wall over much of the length and a double wall (Fig. 1) over a 700-foot (210-m) section where the wall was higher than about fifteen feet (4.6 m). Vegetation below the lower wall would obscure that wall from view from below, and vegetation in the flat area

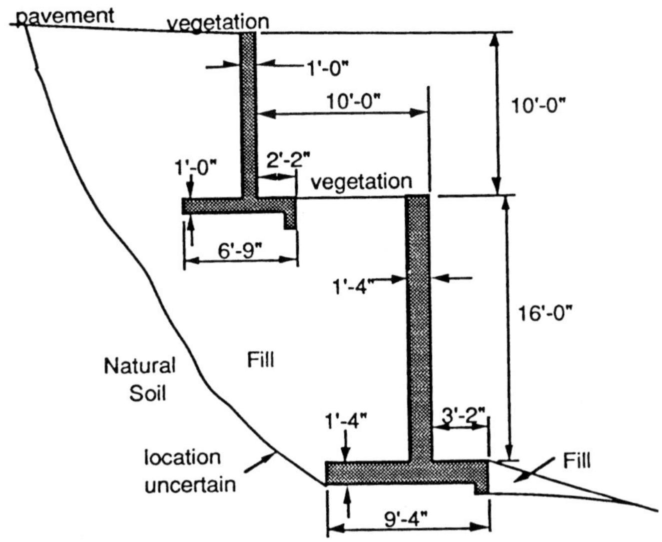

Fig. 1 Wall Configuration in the Area of the Failure

between the walls and above the top wall would obscure the upper wall and shopping center. An irrigation system was planned at the top and at the mid-height.

SITE INVESTIGATION AT DESIGN TIME

About thirty soil borings were made at the site as part of several engineering studies. However, discussion in the engineering reports was restricted to the buildings and to pavements. The wall may not have been contemplated during the main design time. The boring closest to the wall was one-hundred feet (30 m) away. Over much of the site there was one to four feet (1 to 1.2 m) of rocky clay on top of limestone. Several borings penetrated the limestone and encountered a layer of clay shale at a depth of fifteen to twenty feet (4.5 to 6 m). The limestone layer was missing over part of the site, with the clay shale exposed. The clay shale was also exposed in the slope behind the structure and that slope was probably at the angle of repose. No water table was encountered in any soil boring.

The boring closest to the wall showed a soil profile of four feet (1.2 m) of clay over limestone with the limestone described as weathered near its surface but very hard at a depth of six feet (1.8 m).

In accord with usual practice, most of the soil tests were of the classification type. Atterberg limit tests indicated that most the the cohesive soil had liquid limits in the range of 30 to 50% and plasticity points plotted above the A line. Unconfined compressive strengths of shallow clays were generally 1 to 3 tsf (100-300 kPa), with strengths up to 7 tsf (700 kPa) at greater depths in fissured clay shale.

DESIGN OF THE WALL

In view of the fact that subsurface materials were generally limestone or stiff to hard clay shale, that there was an interest in getting the project in operation quickly, and the wall may have been a late addition to the design, the designer apparently saw little need to perform expensive soil tests. The designer attempted to follow recommendations in the Standard Handbook for Civil Engineers (Merritt, 1983), literally tracking through each successive calculation using the same symbols as in the handbook. The designer selected an apparently conservative active earth pressure coefficient of 0.4 (ϕ~25 degrees) and a total unit weight of 100 pcf for the backfill.

The factor of safety against overturning was calculated as the ratio of the potential resisting moment to the apparent overturning moment, about the toe. The designer analyzed the two wall sections separately and ignored the effect of the upper wall on the lower one. The calculated factor of safety of the lower wall was 2.4.

For the sliding mode of failure, a wall-surface friction coefficient of 0.62 was used (δ~32 deg.), and the factors of safety defined using forces, were inexplicably around 1.0.

The designer estimated the toe stress by taking moments about the heel. In consideration of the presence of limestone and clay shale, the designer set a limit on the calculated toe stress of 5000 psf (240 kPa) based on an undefined handbook and a description of the soil as "clay shale". The actual calculated toe stress was 4.3 ksf (206 kPa).

No slope stability type analyses were performed to examine the possibility of an overall failure.

The designer specified that 2-inch (50-mm) diameter PVC drainage tubes be cast into both the upper and lower walls, near the base of their stems, on 20-foot (6-m) centers. A one-cubic foot (0.03 cu.m.) bag of clean 3/4-inch (18-mm) gravel was to be placed on the soil side of each drainage hole to provide filtration.

No specifications were provided for the backfill but it was believed that local practice, and city specifications were followed. The city specifications set no limit on the maximum particle size but specified that "The percentage of fines shall be sufficient to fill all voids and insure a uniform and thoroughly compacted mass of proper density". Construction records show that several density tests were performed. The backfill apparently ranged from tan, weathered, limestone gravel, to a brown gravelly clay. Field density tests indicated 95% to 100% of standard Proctor compaction.

BEHAVIOR OF THE WALL

Construction History

Construction records were not entirely available but it is known that backfilling of the wall occurred during April and it is believed that the wall was completely backfilled in May.

Wall Failure

In late July, after a period of heavy rain, a section of the lower rear wall, with a length of about 135 feet (41 m), suddenly began to displace horizontally. My first site inspection was about seven hours after the failure apparently began and, at that point, the lower wall had displaced horizontally about fifteen feet (5 m) (Fig. 2). A substantial vertical scarp had formed in the fill because the upper wall had dropped into the void formed when the lower wall displaced. The lower wall continued to displace horizontally at a slow rate. Some months later, when the wall approached the property line, the damaged portion of the wall was removed.

Fig. 2 View of the Failed Section of Wall

Site Observations

The exposed fill was mostly limestone gravel but there were several layers of clay. Water was leaking out of the most pervious layers in the fill.

Vegetation in front of the lower wall was dry and dead but vegetation that had been between or above the walls was green. The soil directly in front of the wall, and the slumped fill, was soft. Broken irrigation pipes could be seen in the fill between and above the walls.

Based on the lack of stains, many of the weep holes had apparently not functioned. Attempts to drive rods through them often failed. It appeared that the fill side of some of the holes was blocked with concrete. Water was spurting out of the top part of one of the weep holes, in a stream with a diameter about the size of a pencil. Apparently, the back side of the weep hole was blocked except for one small hole, and there must have been a substantial water head behind the wall.

Examination of intact portions of the wall showed the presence of rust stains along vertical joints, to a height of about six feet (two metres) above toe slab of the lower wall.

POST-FAILURE SOILS INVESTIGATION

Post failure site access was difficult because of the debris. We made two borings, using a light, portable rig, about thirty feet outside the original position of the lower wall and one in the fill just behind the scarp.

One boring in front of the wall encountered four feet (1.2 m) of medium to stiff black clay, grading rocky near the bottom, and met refusal at four feet (1.2 m), probably due to limestone. The other boring encountered four feet (1.2 m) of soft, wet, rocky clay (probably fill), over three feet (1.0 m) of tan clay (LL=54%, PI=33%, w=31%), and then refusal. The third boring was in the fill that had not yet failed. It encountered four feet (1.2 m) of pavement, base, and a rocky clay; two feet (0.6 m) of dry granular fill; 3.5 feet 1.1 m) of soft, wet, gravelly clay (LL=30%, PI=13%, w=22%); one foot (0.3 m) of very soft tan clay; five feet (1.5 m) of very soft, wet, gravelly clay (LL=28%, PI=11%, w=35%), two feet (0.6 m) of medium tan clay, and four feet (1.2 m) of dry, medium to stiff, black clay. This boring confirmed the heterogeneous nature of the fill.

The fact that an essentially vertical scarp, with a height of about six feet (2 m), stood for some days after the failure, seemed to indicate that the fill was reasonably well compacted.

Because of the heterogeneous nature of the fill, we did not perform any shearing tests on fill material. We performed four drained direct shear tests on samples of natural clay taken from a depth of four to six feet (1.2-2.0 m) in one of the borings in front of the wall (Fig. 3).

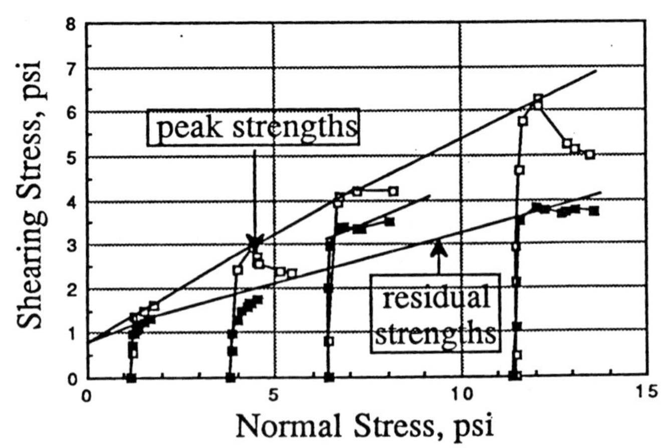

Fig. 3 Results of Direct Shear Tests on Natural Clay Shale

The "residual" friction angle was estimated simply by trimming a smooth failure surface in a direct shear sample and shearing the sample back and forth several times on that surface. Only the final stress path curves are shown in Fig. 3.

The estimated value of $\bar{c}$ was 100 psf (4.8 kPa) for peak and residual, but $\bar{\phi}$ was about 25 deg. for peak and 16 deg. for residual conditions (Fig. 3). The measured shearing properties agreed well with a number of other shearing tests performed in our laboratory on samples from this formation, or from other similar formations in the area, except that the peak friction angle for the other tests was usually closer to 21 deg. and the residual friction angle generally was in the range of 12 to 15 degrees.

ANALYSIS OF STABILITY

The estimated wall configuration in the area of the failure is shown in Fig. 1. The location of the contact between fill and natural soil is uncertain. The actual location of the bottom of the lower wall is unknown because design drawings indicated that the elevation of that base slab would be decided in the field. It is probable that the slab in the middle part of the failed zone was on natural soil because that was the apparent intent of the specifications, but it may also have been on fill.

The uncertainties in wall geometries and soil conditions are such that a sophisticated analysis is not warranted. Instead, the analysis is of the type that a practicing engineer might have made during the design phase. The purpose of the analysis is to determine if a relatively routine approach, properly applied, would have predicted failure.

The analysis was performed in two phases. In the first, the upper wall and soil below it (Fig. 4) were taken out as a free body. Equations for force equilibrium in the vertical and horizontal direction make it possible to calculate the force P that this upper wall and associated soil applies to the vertical plane through the heel of the lower wall. The orientation of the potential shear surface (θ in Fig. 4) was obtained by trial. Because of the tedious nature of the computations, a computer program was written that allowed the user to vary the parameters (β, c, ϕ, H_w, q, δ, γ, and W) as required.

In the second phase, the force P was applied to a free body composed of the lower wall and soil above its heel slab (Fig. 5) and factors of safety were calculated for various failure modes.

The analysis was of the Coulomb type and thus moment equilibrium was not satisfied. However, for active conditions, the Coulomb analysis seems to yield forces that are comparable to those calculated using more sophisticated approaches (Morgenstern and Eisenstein, 1970). The location of P and its obliquity were unknown. Trial analyses were performed to estimate the importance of these uncertainties.

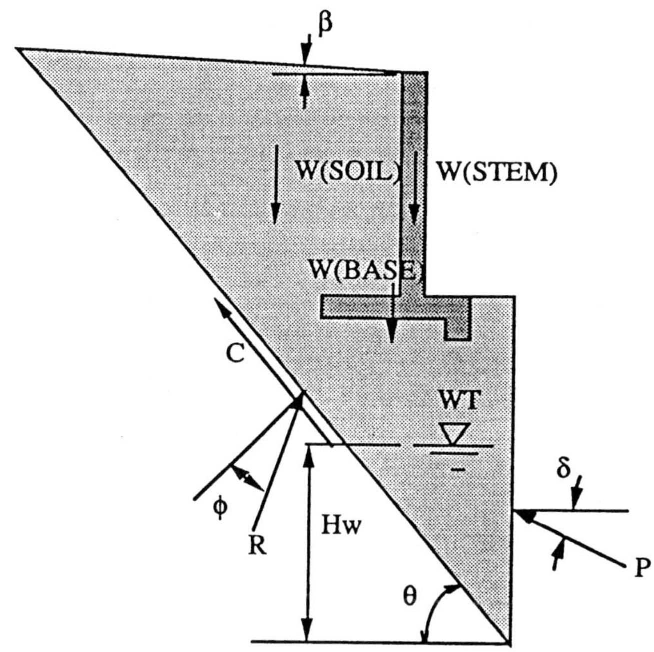

Fig. 4 Assumed Wedge of Soil Sliding Toward the Lower Wall

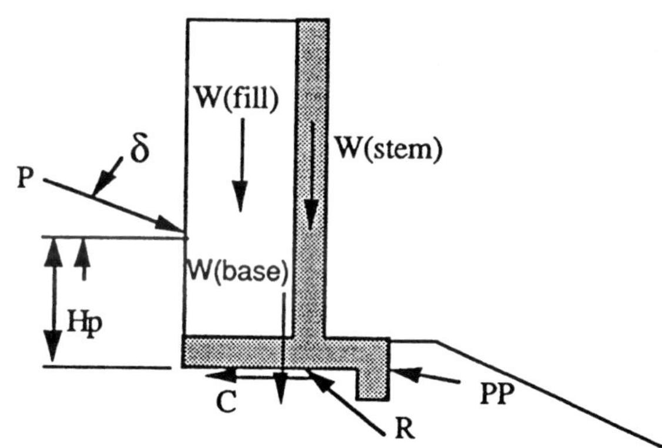

Fig. 5 Assumed Conditions for the Lower Wall Analyses to Obtain P

Assume the soil is saturated and undrained. Practicing engineers often assume that clays are saturated and undrained because then they can assume that ϕ=0 and the analysis is greatly simplified. The presence of gravel layers in the backfill, with water flowing through them, makes it clear that this condition could not have existed throughout the backfill. However, the designer performed some computations based on this assumption so it will be used here as one extreme case. In any case, the wall did stand for some period of time and that stability needs to be explained.

Analyses using δ=0 (Fig. 4) indicated that P dropped from 39 kips/foot when c=100 psf (4.8 kPa), to 0 when c=720 psf (34 kPa). Measured undrained strengths were in excess of 720 psf (34 kPa) for the natural soils and estimated undrained strengths for the exposed clay layers in the backfill also exceeded 720 psf (34 kPa). The analysis confirms the field observation that a substantial height of the clayey part of the fill would stand unsupported as long as it was undrained. The assumption of no drainage could not apply to the fill prior to failure, because of the gravel layers, but it may be relevant for the cohesive parts of the backfill immediately after failure.

Assume the soil is fully drained. Because of the variation in fill material both vertically and horizontally, it seemed most reasonable to assume a homogeneous fill but do a sensitivity analysis for a reasonable range in properties. The cohesionless layers probably had friction angles of at least 35 degrees, and no cohesion. The clay layers probably had friction angles close to that of the natural soil, 25 degrees, with no cohesion. The results of part of the sensitivity study are shown in Table 2.

Table 2-Sensitivity Study for the Upper Wall

c psf	ϕ psf	H_w ft	δ deg.	P kips/foot
100	25	0	0	12
		10		14
100	25	0	0	12
			15	11
0	25	0	0	15
100				12
200				8
300				4
0	15	0	0	23
	20			19
	25			15
	30			12

The effect of depth of water (H_w) and obliquity of the resulting earth force (δ) are minor compared with the effects of mean c and ϕ. It would appear that a reasonably conservative design would use about 15 kips/foot for P.

Local Stability of the Lower Wall

Stability of the lower wall (Fig. 5) was analyzed for overturning, sliding, and bearing capacity. Equations for overturning and sliding are found in most books on foundation engineering and need not be repeated.

Some engineers, including the designer of this wall, do not calculate bearing capacity but rather calculate the tip stress assuming a linear variation in contact stress across the base. Such a computation makes little sense for a surface footing on a cohesionless subsoil because the tip bearing capacity depends on footing width and has the irrational value of zero for an imaginary strip footing of differential width at the toe. The more rational approach is to calculate the bearing capacity of the base slab. The bearing capacity of a strip footing subjected to an inclined and eccentric load, and located on a sloping ground surface, is given by:

$$p_f = cN_c d_c i_c g_c + \gamma D_f N_q d_q i_q g_q + 0.5B'\gamma N_\gamma d_\gamma i_\gamma g_\gamma \quad (1)$$

where N is a dimensionless bearing capacity factor (depends on ϕ), d, i, and g are dimensionless factors to account for footing depth, inclined load, and effects of local ground slope, c is soil cohesion, γ is the soil unit weight above footing depth (q term) or below footing depth (γ term). The bearing capacity, p_f, is the vertical component of the applied force, divided by footing width. A variety of equations have been proposed for the various factors. Equations used here were as follows:

$$N_c = (N_q-1)\cot(\phi) \quad \text{(Prandtl, 1920, 1921)} \quad (2)$$

$$N_q = \exp(\pi \tan \phi)\tan^2(45+\frac{\phi}{2}) \quad \text{(Reissner, 1924)} \quad (3)$$

$$N_\gamma = (N_q-1)\tan(1.4\phi) \quad \text{(Meyerhof, 1961)} \quad (4)$$

$$d_q = 1 + 0.1\frac{D}{B} \quad \text{(Hansen, 1970)} \quad (5)$$

$$d_c = \frac{d_q s_q N_q - 1}{N_q - 1} \quad \text{(deBeer, 1970)} \quad (6)$$

$$d_\gamma = 1 \quad (7)$$

$$s_q = 1+\frac{B}{L}(\tan \phi_{triaxial}) \quad \text{(deBeer, 1970)} \quad (8)$$

$$i_c = i_q = (1 - \frac{\alpha}{90})^2 \quad \text{(Meyerhof, 1953)} \quad (9)$$

where α is the inclination of the resultant force on the base relative to a line normal to the plane of the base,

$$i_\gamma = (1 - \frac{\alpha}{\phi})^2 \quad \text{(Meyerhof, 1953)} \quad (10)$$

$$g_c = (1 - \frac{\beta}{147}) \quad \text{(Hansen, 1970)} \quad (11)$$

where β is the slope of the ground surface measured positively downwards from the horizontal (degrees)

$$g_q = g_\gamma = [1 - 0.5\tan(\beta)]^5 \quad \text{(Hansen, 1970)} \quad (12)$$

$$B' = B-2e \quad \text{(Meyerhof, 1953)} \quad (13)$$

and e is the eccentricity of the resultant force applied to the subsoil by the base.

Analyses were performed to determine the factors of safety against failure of the lower wall. A sensitivity study was again performed because of the uncertainties in the various input variables. The results of a part of those computations are shown in Table 3. For all analyses, it was assumed that there was 1.3 feet (0.4 m) of soil producing passive resistance at the toe.

Table 3 Sensitivity Study for the Lower Wall

P kpf	H p ft	δp deg.	δB deg.	aB psf	φ deg.	c psf	F_{sbc}	F_{so}	F_{ss}
15	5	0	25	0	25	10	0.08	1.13	0.47
13						0	0.19	1.31	0.55
11							0.31	1.55	0.65
15	3	0	25	0	25	10	0.49	1.89	0.47
	5					0	0.08	1.31	0.47
	6						0.00	-.--	-.--
15	5	0	25	0	25	10	0.08	1.13	0.47
		15				0	0.33	2.35	0.61
15	5	0	30	0	30	10	0.12	1.13	0.58
			25		25	0	0.08	1.13	0.47
15	5	0	25	0	25	10	0.08	1.13	0.47
						0	0.03	1.13	0.46
						0			
15	5	0	0	10	0	10	0.16	1.13	0.76
				00		00			

The first group of three analyses with variable force P shows that the lower wall was unstable in bearing capacity and sliding modes for any reasonable forces that would be predicted from the previous analyses.

The second set of analyses show that even a small increase in the assumed height of application of force would result in a bearing capacity failure because of the resulting eccentricity of resultant force on the base.

The third set of analyses shows that increasing the assumed obliquity of the applied force (positive for a downwards vertical component) results in an increase in the factors of safety, mostly because the horizontal component of the force is then reduced.

The fourth set of analyses shows that increasing the friction angles in the reasonable range had only a small numerical influence on the factors of safety.

Similarly, the sixth set of analyses showed that changing the cohesion, through a reasonable range, also had a small effect on the factors of safety.

The final analysis was for the undrained case and used the smallest measured strengths. In that case the wall was unstable for bearing capacity. The largest measured strengths were required to bring the bearing capacity factor of safety up to 1.0. These analyses utilized the value of P corresponding to a drained condition of the backfill. Previous analyses showed that under undrained condition there would be essentially no applied force and the wall would then clearly be stable.

Overall Stability

Overall stability analyses were performed using a slope stability program but the value of those analyses was limited by the uncertainties in soil properties once the clay shale became relatively hard. If the clay shale was assumed to possess a large effective cohesion at shallow depth, then the slope stability analysis gave factors of safety comparable to those obtained for bearing capacity and sliding.

CAUSES OF THE INADEQUATE DESIGN

The designer's computations were examined as part of the discovery phase prior to trial. In general, the main source of the problem was that the designer based the design on a standard engineering handbook without understanding the limitations of such an approach. On a more technical level:

1. the designer followed the common, but irrational, practice of limiting the apparent tip stress rather than estimating the bearing capacity of the base slab. The bearing capacity equations for cases involving inclined and eccentric load, and for footings on a slope, have a substantial degree of uncertainty in them but at least they provide a rational form of analysis.
2. the designer apparently did not understand the difference between the strengths of the soil under drained and undrained conditions.
3. for reasons that were never explained, the designer ignored the effect of the upper wall on the loads applied to the lower wall.

WHY DID THE WALL STAND UP AT ALL?

Based on the low calculated factors of safety, one wonders why the wall stood up at all. Some of the reasons include:

1. The maximum height of wall existed over a length of only about one-hundred feet (30 m). Stability increases rapidly as the wall height decreases so only a small portion of the wall was in danger. The unstable section of the wall probably derived limited amounts of support from adjacent stable portions.
2. Subsoil conditions were uncertain. Some of the wall may have been supported by a thin layer of limestone that was above the clay shale.
3. The clay shale became considerably harder with depth (could not be penetrated with thin-walled steel tubes) and may have been cemented, thus causing the factor of safety for bearing capacity to be higher than predicted.
4. Negative pore water pressures in the backfill during much of its life would have reduced applied forces greatly, perhaps essentially to zero. The rains that immediately preceded the failure apparently raised the pore water pressures in the fill and caused increased earth forces.
5. The drainage system, though widely used, was ineffective and large water pressures apparently built up on the wall after the rain. Because of the irregular nature of the fill, the water pressure was probably not hydrostatic but, instead, developed in the more pervious layers of fill.

CONCLUSIONS (RE)LEARNED

This case history demonstrates lessons that are generally taught in college but seem often forgotten in practice, including:

1. Engineers should not practice outside of their areas of training and experience.
2. Engineers must resist pressure from clients who want designs turned out immediately because of economic considerations. A careless design, made in haste, may cost the engineer dearly.
3. Design recommendations made in standard civil engineering handbooks, for geotechnical problems, often over simplify the problems and do not provide the kind of technical advice that is required for successful design.
4. Designers should consider both short term (undrained) and long term (drained) conditions unless a technical understanding of the problem makes it clear that one or the other is the critical case.
5. Positive means of controlling water pressures on walls is critical.

LIST OF VARIABLES

a_B adhesion of base slab on subsoil
c cohesion
F_{sbc} factor of safety against a bearing capacity failure
F_{so} factor of safety against an overturning failure
F_{ss} factor of safety against a sliding failure
H_p vertical distance from the heel to the point of application of the earth force
H_w height of ponded water
P force applied to the vertical plane through the heel of the lower wall (force/length)
δ obliquity of resultant force P (degrees)
δ_P obliquity of P on the vertical plane through the heel (degrees)
δ_B obliquity of resultant force applied to the subsoil by the base of the lower wall (degrees)

REFERENCES

de Beer, E. E. (1970), "Experimental Determination of the Shape Factors and Bearing Capacity Factors of Sand," *Geotechnique*", Vol. 20, No. 4, pp. 387-411.

Hansen, Brinch (1970), "A revised and extended formula for Bearing Capacity", Danish Geotechnical Institute, Bull. No. 28.

Ireland, H. O. (1964), "Design of Structures to Resist Earth Pressures", Soil Mechanics Lecture Series, Illinois Section, ASCE.

Merritt, F. S. (1983), *Standard Handbook for Civil Engineers*, McGraw-Hill Book Co., NYC.

Meyerhof, G. G. (1953), "The Bearing Capacity of Foundations Under Eccentric and Inclined Loads," *Proceedings*, Third International Conference on Soil Mechanics and Foundation Engineering, Zurich, Vol. 1, pp. 440-445.

Meyerhof, G. G. (1961), discussion, Proc., Fifth Intern. Conf. on Soil Mech. and Found. Engr., Vol. 3, pp. 193-194.

Morgenstern, N. R. and Z. Eisenstein (1970), "Methods of Estimating Lateral Loads and Deformations", *Proc., ASCE Specialty Conf. on Lateral Stresses and Earth-Retaining Structures*, Cornell Univ., pp. 51-102

Prandtl, L. (1920), "Uber die Harte platischer Korper", Nach. Gesell. Wiss. Gottingen. Math-Phys. Kl. p. 74-85.

Prandtl, L. (1921), "Uber die Eindringungsfestigkeit (Harte) Plastischer Baustoffe und deren Festigkeit von Schneiden", Zeitschrift fur Angewandte Mathematik und Mechanik.

Reissner, H. (1924), "Zum Erddruckproblem", *Proc., First Intern. Congress of Applied Mechanics*, Delft, pp. 295-311.

AXIAL LOAD CAPACITY OF PIPE PILES IN SANDS

by
Roy E. Olson and Magued G. Iskander‡

ABSTRACT

The axial load capacity of driven, open-ended, steel pipe piles in cohesionless soils, is often estimated using a method such as the American Petroleum Institute Recommended Practice 2A (API RP 2A). This method is empirical and is used mainly as a scheme for applying the results of pile load tests to new design problems. Unfortunately, the database upon which it is based is sparse and thus the uncertainties are large. In this paper, an expanded database is used to compare capacity predictions based on API RP 2A (1993) and Olson (1990) methods with measured values. A new method that utilizes continuous functions to represent design parameters is also presented and compared with measured values. It is found that the Olson (1990) and the "continuous parameter" methods are both superior to the API RP 2A (1993) method but that all of the methods overpredict capacities of long piles of the sizes used to support offshore structures.

INTRODUCTION

Driven pipe piles are widely used for the support of structures in both terrestrial and offshore environments. Typically, piles are designed using static methods with many designers of offshore structures following the API recommended practice (API RP 2A section 2.4.3) (Lacasse and Goulois, 1989; Focht and O'Neill, 1985). Studies of pile load test databases have shown that the margin of error in the prediction of the static axial load capacities of piles in cohesionless soils is larger than that of comparable piles in cohesive soils (Dennis and Olson, 1983a, 1983b). The margin of error in the prediction is particularly high for open-ended pipe piles, such as used to support of offshore oil production platforms.

The wide use of API RP 2A has brought along numerous criticisms (Iskander and Olson, 1992; Pelletier et. al, 1993). Many of the criticisms are valid but resolution of the criticisms is currently hindered by the profession's lack of good understanding of pile behavior. The API model is a simple one that is used to generalize pile load test data so those data can be used for design purposes. Many factors complicate such use. For example, for open-ended pipe piles in sand, the database consists of only a few tests and the design method is heavily influenced by measurements for closed ended pipe piles. The load test data are for piles with smaller diameters and lengths than typically used offshore. Most of the tests are in sands of medium densities whereas many offshore piles are in dense sands, sometimes in cemented sands.

---

‡ L. P. Gilvin Professor of Civil Engineering and Graduate Research Assistant, respectively, The University of Texas at Austin, ECJ 9.227, Austin, TX 78712, USA

The simple API model has a number of well known difficulties. For example, stresses around piles are greatly influenced by the presence of the pile. Soil is dragged down from one layer into the interface between the pile and lower layers (Tomlinson, 1971). Particle sizes of sands are reduced by the high normal stresses and large shearing deformations around a pile. Time-dependent changes in pile capacities in sand apparently occur over substantial periods of time even after excess pore pressures have been dissipated (Tavanas, 1971; Armishaw and Cox, 1979). High radial gradients in sand density probably result from lateral movements of the pile during driving (pile shaking, Poisson ratio effects, etc.) (Szechy, 1961; Robinsky and Morrison, 1964). Exterior stresses are influenced by the behavior of the plug inside the pile but that behavior involves a dynamic soil-structure interaction that has not been quantified (Ugaz, 1988; Raines et al. 1992).

Attempts can be made to analyze the behavior of piles using more generalized methods. However, these methods are currently all too simplified to model the problem realistically. Individual dynamic analyses have involved super-computer execution times as long as forty hours (El-Mabsout, 1991). The cost of measuring the numerous required soil properties is prohibitive. The analyses have not been supported by full scale field load tests and thus the accuracy of the final result is difficult to determine. The existence of these, and other, problems provides a challenge for researchers but does not provide a reasonable current design method.

Dynamic analyses based on the wave equation approach also hold promise. However, such analyses also suffer from a number of known problems. The analyses are oversimplified. They do not account for time-dependent changes in capacity although restrike analyses are helpful. The analyses involve low deformation measurements whereas pile capacities in sands involve large deformations. Finally, the dynamic analyses are used at driving time whereas the need for the analyses is during design time.

The result of these problems is that the best predictions of capacities in any region is made by people who are experienced in that region. Unfortunately, experience typically does not involve full scale load tests so experienced personnel must rely on less quantitative aspects of experience, e.g., driving records and the fact that structures designed using existing methods have not failed.

While supporting continued research in both dynamic and static methods, the authors believe that design will continue to be based mainly on a database of load test results and the experience of knowledgeable local engineers. This paper is concerned with the existing database for pile load tests in sand and with its use for prediction of pile capacities. Consideration is restricted to the API approach because of the paucity of data, available to us, involving cone, pressuremeter, dilatometer, and other methods, that are in use for terrestrial piles in some geographic regions.

API APPROACH TO PREDICTION OF STATIC CAPACITY OF PILES IN SAND

In the API approach, the static axial load capacity (Q_c) of a pile in cohesionless soil is estimated as:

$$Q_c = Q_s + Q_p \pm W_p \tag{1}$$

where Q_s and Q_p are the loads transferred to the soil in side shear and end bearing, respectively, and W_p is the weight of the pile submerged in soil (positive for tensile loading and negative for compressive loading).

Side shear is calculated using:

$$Q_s = \sum_{L=1}^{N} f_s \; C \; \Delta L = \sum_{L=1}^{N} K \; \bar{\sigma} \tan(\delta) \; C \; \Delta L \qquad (2)$$

where f_s is the local side shear between the pile and the surrounding cohesionless soil (limited to a value f_{lim}), C is pile circumference, ΔL is an increment of pile length in the Lth layer, K is the earth pressure coefficient, $\bar{\sigma}$ is the free-field vertical effective stress in the middle of the Lth layer, and δ is the friction angle between the pile and the surrounding soil.

Similarly, the tip capacity is calculated using:

$$Q_p = q_p \, A_p = \bar{\sigma} \; N_q \, A_p \qquad (3)$$

where q_p is the net pressure between the pile tip and the subsoil (limited to q_{lim}), A_p is the area of the pile tip, $\bar{\sigma}$ is the free field vertical effective stress at the pile tip, and N_q is a dimensionless bearing capacity factor.

The properties currently recommended by API are shown in Table 1. The earth pressure coefficient is assumed to be 1.0 for closed-ended pipes and 0.8 for open-ended pipes, with tension and compression treated as the same.

TABLE 1 - API RP 2A Design Values for Cohesionless Soils

Relative Density	*Soil Description*	*δ (deg.)*	*f_{lim} (ksf)*	*N_q*	*q_{lim} (ksf)*
very loose loose medium	sand sand-silt silt	15	1.0	8	40
loose medium dense	sand sand-silt silt	20	1.4	12	60
medium dense	sand sand-silt	25	1.7	20	100
dense very dense	sand sand-silt	30	2.0	40	200
dense very dense	Gravel sand	35	2.4	50	250

Problems involved in the use of such an approach include the following. The vertical stress near a pile is not the free field value so the error involved in use of the free field stress must be made up by a counterbalancing error elsewhere. The limitations on side (f_{lim}) and tip (q_{lim}) stresses represent a simplified effort to prevent overprediction of capacities of long piles but have little support from a rational point of view. The friction angle between the pile and sand (δ) is doubtless effected by grain size comminution during driving. Use of this approach thus requires that the parameters be ones selected to match a database. The parameters cannot be determined using current theories. For example, the tip bearing capacity is not likely to be given using classical bearing capacity theory. The equations thus rely on a database rather than on a theoretical foundation.

DATABASE

A database of about one thousand load tests on piles of a variety of types, and in a variety of soils, was developed during the early 1980's. and a method for analysis of pipe piles in sand was recommended (Olson and Dennis, 1982; Dennis and Olson, 1983b). API subsequently adopted a revised method for use in sand. Later efforts were made to improve on both the size of the database and on its interpretation (Olson and Al-Shafei, 1988; Olson, 1990). That effort has continued and is the basis for this paper.

Despite the size of the database, there are relatively few tests on steel pipe piles. The effort in 1990 used data for only four open-ended pipe piles in sand. It also used data for thirty-one closed-ended pipe piles in sand. The current study uses an updated database of forty eight load tests, of which thirteen were for open-ended pipes. The piles used in this study ranged in length from 3 - 43 m and in diameter from 200 - 800 mm.

In applying the API recommended practice to predict the capacity of the piles in the database, it is necessary to adopt a standard method to define relative density. In the absence of API guidelines for relative density, we followed the classification recommended by Peck et. al (1974), where relative density was defined using standard penetration resistance, N in blows per 30 cm (blows per foot). The N values were corrected for the effects of overburden pressure using the recommendation of Peck, et al. (1974).

Generally, piles in sand do not experience a sudden "plunging" failure. Several definitions of "failure" can be adopted under these circumstances. For the analyses reported herein, failure was defined as the peak applied load during the test, but tests were not accepted unless loads reached at least the value defined by Davisson (1973).

An examination of the existing database discloses limitations on the accuracy of analysis of the static type. First, the data are concentrated in piles in medium to dense clean sands. Based on the 1990 method, we estimate that for the database of forty eight tests, taken collectively, 63% of the side shear was in clean sand, and 28% in silty sands and sandy silts, with 4% in silt, 3% in sand/gravel mixtures, and 1% in clay. For the tip capacity, 89% was estimated to come from clean sand, 8% from sand/gravel mixtures, 1% from gravel, 2% from sand/silt mixtures, and none from silt or clay. Silty soils are not well represented because, geologically, the finer soils tend to occur at shallow depth, with sands at greater depth. The low effective stresses at shallow depth then lead to reduced frictional stresses. Piles are typically driven to firm material and thus logically tend to derive most of their side shear and tip capacities from deeper, thus coarser, soils. Second, significant vertical and lateral variations in the properties of cohesionless soils may result in soil borings not accurately reflecting the soil properties at the location of the load test. When standard penetration resistance are used for soil classification, variations in the efficiency of the hammers leads to scatter, especially when testing methods differ with respect to time or regionally. In a few cases, some of the soil was excavated from the site after soil borings were made and no borings were made after the excavation.

ANALYSIS USING THE 1993 API RP 2A

Analyses were performed for piles in the database using the 1993 API RP 2A recommendations. Measured and predicted capacities are compared in Fig. 1, with hollow symbols used tests in which the quality of the source data was judged to be deficient and the solid symbols for cases involving better data. The ratio of calculated-to-measured axial load capacity (Q_c/Q_m) ranged from 0.13 to 3.57. The ratios Q_c/Q_m tend to be normally distributed when $\log(Q_c/Q_m)$ is used. The mean log(Qc/Qm) was -0.129 (antilog is 0.74), and the standard deviation was 0.354. The median Q_c/Q_m of 0.74 indicates that the API method tended to underpredict capacities. The large scatter indicates that the method can be unsafe in some cases and uneconomical in others. It may be noted, however, that "unsafe" for individual piles in sands implies unexpectedly large, but stable, deformations, not collapse.

Use of the API method led to underprediction of capacities of short piles (Fig. 2) and overprediction of long piles. The overpredictions tend to be large for piles of lengths often used offshore. The number of observations for long piles is small and the quality of the data is generally suspect, but the consistency of the data is cause for concern.

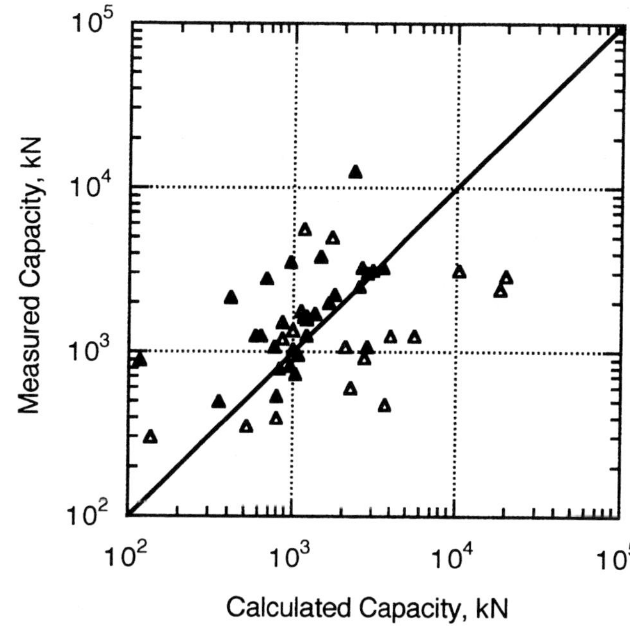

FIG. 1. Comparison of Measured Capacities to Capacities Predicted Using API RP 2A

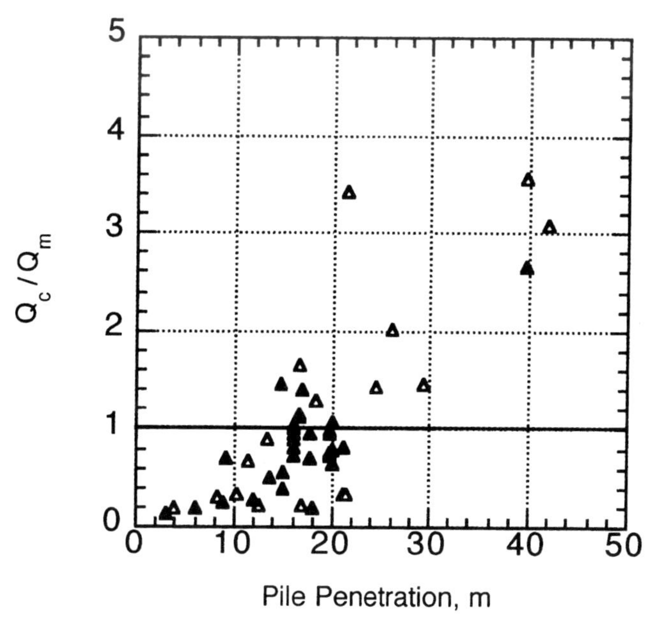

FIG. 2. Effect of Penetration on the Capacities Computed with API RP 2A

ANALYSIS USING OLSON 1990 METHOD

Olson's (1990) method uses the same framework as the API method. However, the design parameters (Table 2), are expressed as functions of the standard penetration resistance (corrected for the effect of overburden) instead of using more qualitative descriptive terms. Also, the lateral coefficient of earth pressure is made a function of the standard penetration resistance as follows:

$K = 0.16 + 0.015\,N$ for partial displacement piles

$K = 0.70 + 0.015\,N$ for full displacement piles

The properties shown in Table 2 were selected to fit the then-existing database. Such a procedure inevitably leads to a relatively good comparison of the method with the measured capacities because the method is compared with the data from which it came. If no effort is made to force properties to be rational functions of relative density, then excellent comparisons of measured and computed capacities result, especially for cases where there are only a few tests in a given relative density range, but the apparent accuracy is an illusion brought about by the fitting method and the small size of the database.

Analyses were performed for piles in the current database using the Olson (1990) method. The scatter between measured and predicted capacities is still significant (Fig. 3) but is less than for the API method. The logarithmic mean Q_c/Q_m was 0.064 (antilog was 1.16) and the standard deviation was 0.22. The range in Q_c/Q_m was from 0.36 to 4.43. This method also overpredicted the capacities of long piles (Fig. 4).

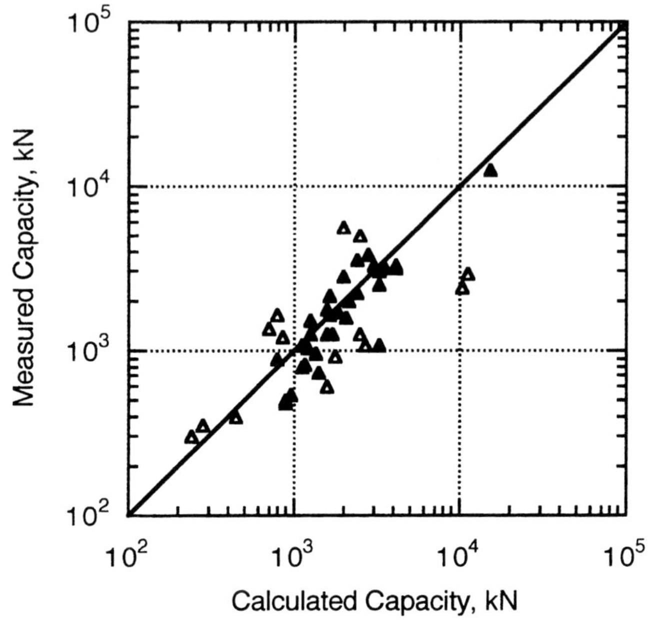

FIG. 3. Comparison of Measured Capacities to Capacities Predicted Using Olson, 90 Method

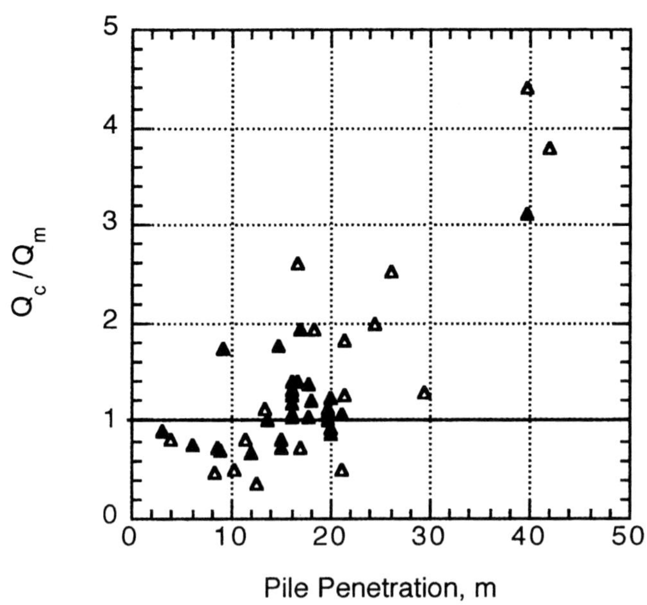

FIG. 4. Effect of Penetration on the Capacities Computed with Olson, 90 Method

TABLE 2 -- Soil Properties Used in Olson (1990) Method

Soil Type	*Range in N Values, bpf*	*δ (deg)*	*f_{lim} (ksf)*	*N_q*	*q_{lim} (ksf)*
Gravel	0-4	(20)	(1.4)	(12)	(60)
	5-10	(25)	(1.7)	(20)	(100)
	11-30	(30)	(2.0)	(40)	(200)
	over 30	(35)	(2.4)	(60)	(250)
Sand/ Gravel	0-4	(20)	(1.4)	(12)	(60)
	5-10	(25)	(1.7)	(20)	(100)
	11-30	(30)	(2.0)	(40)	(200)
	above 30	(35)	(2.4)	(60)	(250)
Sand	0-4	(20)	(1.0)	(50)	(40)
	5-10	30	1.1	120	120
	11-30	35	1.9	120	190
	31-50	40	2.6	120	190
	51-100	40	3.7	130	200
	over 100	40	3.8	220	530
Sand/ Silt	0-4	10	(1.0)	(10)	(10)
	5-10	10	(1.0)	(20)	(40)
	11-30	15	(1.4)	50	110
	31-50	20	2.0	100	160
	51-100	(30)	(2.0)	(100)	(200)
	over 100	(34)	(20.0)	(100)	(200)
Silt	0-4	(10)	(1.0)	(10)	(40)
	5-10	15	(1.0)	(10)	(40)
	11-30	20	(1.4)	(10)	(40)
	31-50	20	(1.4)	(12)	(60)
	over 50	(25)	(1.4)	(12)	(60)
	over 50	(25)	(1.4)	(12)	(60)

N.B: (numbers in parentheses were not used in the analyses)

TRIAL ANALYSIS USING CONTINUOUS FUNCTIONS

Instead of using cumbersome tables of parameters, it makes more sense to seek rational continuous functions to fit the data. The starting point is logically a log function of the form:

$$y = A + B \log(x) \tag{4}$$

where y is the dependent variable, in this case being δ, f_{lim}, N_Q, or q_{lim}, x is the independent variable, here the standard penetration resistance N, and A and B are parameters whose values depend on the parameter being evaluated. This function did not fit the design parameters used in either API (1993) or Olson (1990) but the reason may be the small size of the database and the fact that unreasonable parameters can occur in tables when there are only a few observations in some category.

The parameters used in the Olson (1990) method were used as a starting point and an effort was made to find the best fit values of A and B for each of the four parameters and for five soil classifications (gravel, sand/gravel, sand, silt/sand, and silt). However, the paucity of data for many cases led to the conclusion that, as a starting point, all cohesionless soils would be lumped into a single group. Several hundred trials were made and the best fit values were as follows:

$$\delta = 6 + 14 \log(N) \quad (5a)$$

with δ in degrees and N in blows/30 cm.;

$$f_{lim}\,(kPa) = 0 + 140 \log(N) \quad (5b)$$

$$N_Q = 60 + 50 \log(N) \quad (5c)$$

$$q_{lim}\,(kPa) = 1400 + 1900 \log(N) \quad (5d)$$

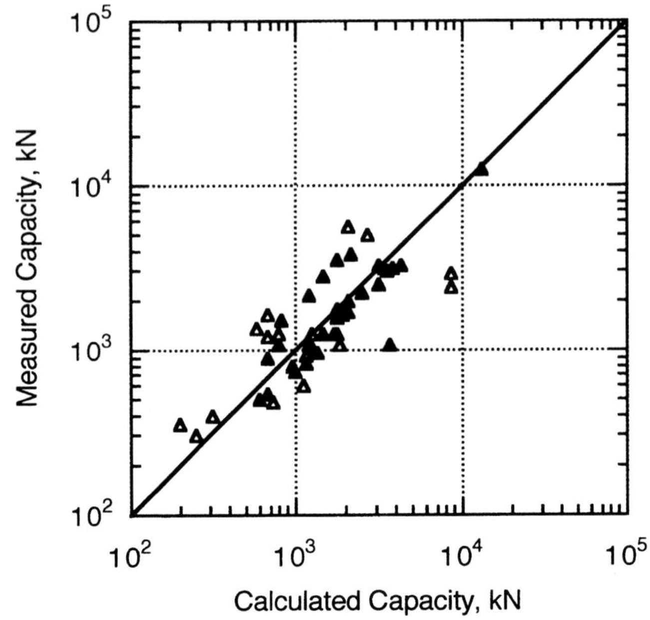

FIG. 5. Comparison of Measured Capacities to Capacities Predicted Using Continuous Functions

Almost all of the piles in the database encountered the limit on tip capacity so the equation for N_q has an unclear accuracy. Only one pile encountered the limit on side shear so the accuracy of values of f_{lim} is also questionable.

In all analyses, the values of K recommended by Olson (1990) were used and side shear in tension and compression were assumed to be the same (in accord with API RP 2A). Since the design parameters used in both RP 2A (1993) and Olson (1990) methods were always less for silty soils than for clean sands and gravels, an attempt was made to use reduction factors for silty soils but the effort failed because of data involving high measured capacities for cemented silty sands. These cemented silty sands were just classified as silty sands in this study but probably should be in some new classification.

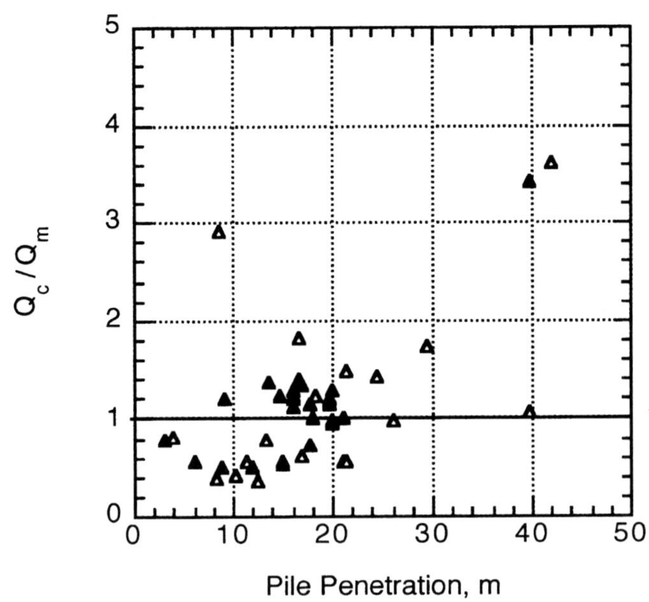

Fig. 6 - Effect of Penetration on the Capacities Computed with Continuous Functions

Analyses using Eqs. 5a-5d (Fig. 5) led to a mean log(Qc/Q_m) of 0 (antilog=1.00) and a standard deviation of 0.22. The ratio of calculated-to-measured axial load capacity (Q_c/Q_m) ranged from 0.37 to 3.63. The use of continuous functions again led to underprediction of capacities of short piles (Fig. 6) and overprediction of capacities of long piles.

Our current view is that the scatter between measured and computed pile capacities is partially due to defective data in the database, for example, due to conditions at the pile location differing from those where the borings were made, and to differences in the way that N values were determined. Part of the problem appears to be fundamentally in the way the API method is configured. For shallow piles, K seems to be large, whereas for deep piles K appears to be small near the surface and large near the base (Gregerson et al., 1973). Simple design equations can probably be configured to reflect such a condition but lack of definitive field data makes an empirical approach difficult.

It also appears that new soil classifications are needed to include cemented sands and silts such as found near the Arabian Gulf.

TENSION VERSUS COMPRESSION

Based on the general empirical observations of the database in the early 1980s (Olson and Dennis, 1982) and field data from Vesic (1967, 1970), side shear in tension and compression were assumed to be the same. It is possible that curve fitting could be performed using side shears in tension of only say 70% to 80% of those in compression, and equally good fits would be obtained, but with different parameters. Such fittings have not been attempted recently.

END EFFECTS

A different kind of end effect develops when the inside of an open-ended steel pipe pile is drilled out. Two case histories were presented by Olson and Al-Shafei (1988) to show that the unstressing of the pile tip results in the need for considerably larger tip settlements to develop tip capacity so a greatly reduced tip capacity is developed for deformations of the magnitudes usually encountered in load tests.

EFFECT OF CLAY LAYERS

Tests were excluded from the database used here if more than 15% of the calculated side shear came from clay layers. However, evidence exists in the database that suggests that clay layers cause a significant reduction in side shear in lower sand layers, in agreement with Tomlinson (1971), so the soil properties shown in this paper should be used with caution when clay layers are present.

CONCLUSIONS

The following conclusions are drawn from these analyses:

- The number of useful tests on steel pipe piles in predominantly cohesionless soils is small, especially for steel pipe piles. The number of load tests is less than the number of degrees of freedom in the system, for the API (1993) and Olson (1990) methods, so there are various combinations of properties that yield equivalent results.

- If the soil properties in the 1993 API RP 2A are used, the capacities of piles penetrating only a short distance into cohesionless soils is underpredicted. Use of the properties recommended by Olson (1990), and those in the continuous function method, resulted in reduced scatter and a smaller effect of penetration on the Q_c/Q_m ratio.

- Capacities of the long piles in the database are substantially overpredicted using all of the methods considered in this study.

- Sets of properties were found that lead to a reasonable prediction of pile capacities for terrestrial-sized piles, under the assumption that the side shearing stresses are the same in tension and compression. It is possible that similarly good predictions could be developed for the assumption that tensile side shear is some fraction of the value in compression.

ACKNOWLEDGMENTS

The work presented herein is part of an ongoing study supported by the Offshore Technology Research Center (NSF Engineering Research Centers Program, grant No. CDR8721512). The authors wish to thank the Offshore Technology Research Center for its financial support of this work.

REFERENCES

- American Petroleum Institute (1993), "API Recommended Practice for Planning, Designing, and Constructing Fixed Offshore Platforms - Load and Resistance Factor Design RP 2A," API, Production Div., 211 North Ervay, Suite 1700, Dallas, TX 75201

- Armishaw, J. W., and D. W. Cox (1979), "The Effects of Changes in Pore Water Pressures on the Carrying Capacities and Settlements of Driven Piles End Bearing in a Sand and Gravel Stratum", Recent Developments in the Design and Construction of Piles, Instn. of Civil Engrs., London, pp. 227-236.

- Davisson, M. T. (1973), "High Capacity Piles," Soil Mech. Div., Ill. Section, Am. Soc. of Civil Engrs., Chicago, pp. 81-112.

- Dennis, N. D., Jr., and R. E. Olson (1983a), "Axial Capacity of Steel Pipe Piles in Clay," *Proceedings,* Geotechnical Practice in Offshore Engineering, ASCE, Geotechnical Engr. Div. Specialty Conf., University of Texas at Austin, pp. 370-388.

- Dennis, N. D., Jr., and R. E. Olson (1983b), "Axial Capacity of Steel Pipe Piles in Sand," *Proceedings,* Geotechnical Practice in Offshore Engr., ASCE Geotechnical Engr. Div. Specialty Conf., Univ. of Texas at Austin, pp. 389-402.

- El-Mabsout, Mounir (1991), "A Finite Element Model for the Analysis of Pile Driving", Ph.D. dissertation, University of Texas at Austin.

- Focht, J.A. and M. O'Neill (1985), "International State of The Practice for Design and Installation of Axially Loaded Piles," *Proceedings*, 11th International Conference on Soil Mechanics and Foundation Engineering, vol. 1, pp. 187-209

- Gregorson, O. S., G. Aas, and E. DiBiagio (1973), "Load Tests on Friction Piles in Loose Sand", Proc., Eighth Intern. Conf. on Soil Mech. and Found. Engr., Moscow, Vol. 2, pp. 109-117.

- Iskander, M.G. and R.E. Olson (1992), "Review of API Guidelines for Pipe Piles in Sand," *Proceedings*, Civil Engineering in the Oceans V(CEOV), ASCE, pp. 798-812

- Lacasse, S. and A. M. Goulois (1989), "Uncertainty in API Parameters for Prediction of Axial Capacity of Driven Piles in Sand," *Proceedings*, 21st Offshore Technology Conference, Houston, OTC paper 6001 pp. 353-358

- Olson, R. E. (1990), "Axial Load Capacity of Steel Pipe Piles in Sand", Offshore Technology Conference, Houston, Paper 6419.

- Olson, R. E., and Dennis, N. D.(1982), "Review and Compilation of Pile Test Results, Axial Pile Capacity", final report to American Petroleum Institute on Project PRAC 81-29

- Olson, R. E., and K. S. Al-Shafei, (1988), "Axial Load Capacities of Steel Pipe Piles in Sand", Proc., Second Intern. Conf. on Case Histories in Geot. Engr., Univ. of Missouri, Rolla, Vol. 3 , pp. 1731-1738.

- Peck, R. B., W. E. Hanson, and T. H. Thornburn (1974), *Foundation Engineering*, John Wiley and Sons p. 114.

- Pelletier J. H., J. D. Murff and A. C. Young (1993), "Historical Development and Assessment of the Current API Design Methods for Axially Loaded Pipes" *Proceedings,* Offshore Technology Center, Houston, TX, pp. 253-282, OTC paper 7157

- Raines, R. D., O. G. Ugaz, and M. W. O'Neill (1992), "Driving Characteristics of Open-Toe Piles in Dense Sand", Jour of Geotechnical Engr., ASCE, Vol. 118, No. 1, pp. 72-88.

- Robinsky, E. I., and C. F. Morrison (1964), "Sand Displacement and Compaction Around Model Friction Piles", Canadian Geotechnical Journal, Vol. 1, No. 2, pp. 81-93.

- Szechy, C. (1961), "The Effects of Vibration and Driving upon the Voids in Granular Soil Surrounding a Pile", Proc., Fifth Intern. Conf. on Soil Mech. and Found. Engr., Vol. 2, pp. 161-164.
- Tavanes, F. (1971), "Load Test Results on Friction Piles in Sand", Canadian Geotechnical Journal, Vol. 8, pp. 7-22.
- Tomlinson, M. J., (1971), "Some Effects of Pile Driving on Skin Friction", *Proceedings*, Behaviour of Piles, Instn. of Civil Engrs., London pp. 149-152.
- Ugaz, O. G. (1988), "An Experimental and Numerical Study of Impact Driving of Open-Ended Pipe Piles in Dense Saturated Sand", Ph.D. Dissertation, Univ. of Houston.
- Vesic, A., (1970),"Load Transfer in Pile-Soil Systems", *Proc.,* Conf. on Design and Installation of Pile Found. and Cellular Structures, Lehigh Univ. pp. 47-73.
- Vesic, A. S. (1967), "A Study of Bearing Capacity of Deep Foundations", Final Report on Project B-189, Georgia Tech University, Atlanta.

Behaviour of open-ended pipe piles in sand during installation and loading

Le comportement des pieux tubulaires à extrémité ouverte dans le sable pendant l'installation et le chargement

M.G. Iskander – *Polytechnic University, Six Metro Tech Center, Brooklyn, N.Y., USA*
R. E. Olson – *The University of Texas, Austin, Tex., USA*

SYNOPSIS: An experimental study was performed to investigate the physical processes which influence the axial load capacities of open-ended steel pipe piles in sand. A double-walled pipe pile was developed to allow independent determination of the stresses on its inner and outer surfaces. An electronically operated single acting air hammer was developed for pile driving. Measurements were made of pore water pressures adjacent to the inner and outer walls, as well as side shear on the inside and outside.

INTRODUCTION

One of the major difficulties in predicting the axial load capacity of pipe piles in sand is the lack of understanding of the physical processes which control the behavior of piles during installation and loading. Variations between measured and predicted pile capacities arise due to a number of factors such as: severe changes in soil properties and state of stress caused by pile driving; difficulties in measuring field soil properties, particularly in the marine environment; variations in loading details and pile installation procedures; uncertainties in the behavior of the soil core (plug) during driving and subsequent loading; and soil structure interaction (Iskander and Olson, 1992, and Lehane and Jardine, 1994).

Designers account for these factors by calibrating simple theoretical models using empirical factors. The accuracy of this approach has been evaluated by comparing the predicted pile capacities to the measured pile capacities of known case histories. The scatter between measured and predicted pile capacities is significantly larger for piles in sand than piles in clay (Dennis and Olson, 1983a and 1983b). The primary inadequacy of this empirical approach is the data base used to calibrate the design method (Pelletier *et al*., 1993 and Olson, 1990). At the present time, the data base has fewer load tests than the number of degrees of freedom of the pile-soil system, which precludes determining a unique set of calibration factors (Olson and Iskander, 1994).

Analytical studies, alone, cannot resolve the uncertainties in axial capacity predictions due to the complex nature of pile behavior. An experimental study of the physical processes which control the behavior of piles during installation and loading was performed in order to resolve some of the issues involved (Iskander, 1995). A brief description of the developed apparatus and test results is presented in this paper.

EXPERIMENTAL SETUP

Piles were installed and load tested in a pressure chamber which was filled with fine, Oklahoma sand using dry pluviation. The sand had uniformity and curvature coefficients of 1.5 and 1.15, respectively; and was placed at a density of 17.6 kN/m3, which corresponds to a 94% relative density. The chamber is 0.90 m in diameter and 1.05 m high. Horizontal and vertical pressures, up to 275 kPa, can be applied to the sand. The sand can be saturated with de-aired water which is supplied at the base of the chamber.

A load-controlled MTS feedback control system was used to apply static and cyclic loads via a hydraulic actuator. The system has a compressive capacity of 260 kN, and a tensile capacity of 200 kN. The pressure chamber is located on a loading frame, centered below the hydraulic actuator. The frame has a dynamic design capacity of 270 kN.

This investigation utilized a number of experimental techniques which are different from previous studies, and are discussed next.

Double-Walled Pipe Pile

Attempts to measure the load transfer along the length of open-ended single-walled pipe piles are complicated by the inability to differentiate the shearing stresses acting inside and outside the pile. A double-walled steel pipe pile was developed to allow independent measurements of stresses acting on its inner and outer surfaces, as well as pore water pressures at the internal and external at the interfaces (Fig. 1). The pile is 0.87 m long, 89 mm in outside diameter, and has an 8 mm wall thickness. The pile is instrumented with 40 strain gages and eight pore-pressure transducers, which were designed to survive pile driving

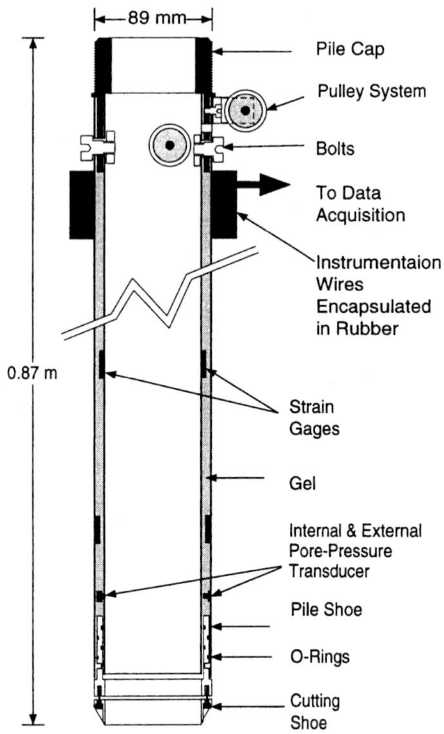

Figure 1. The instrumented double-walled pipe pile

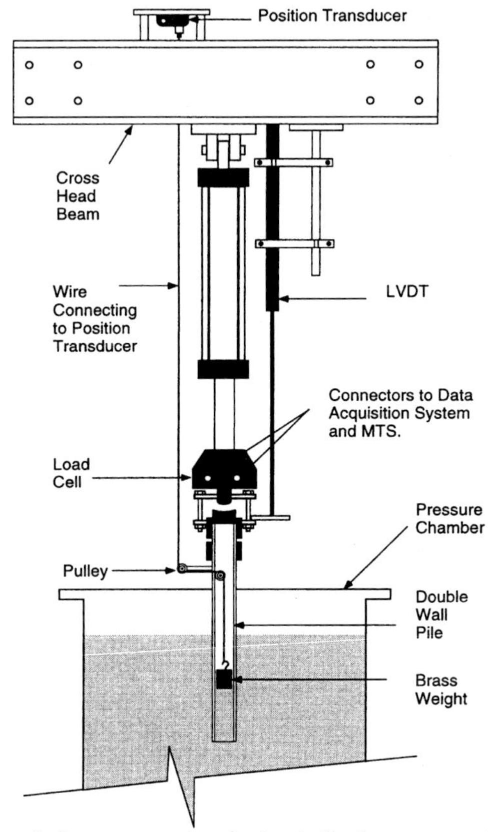

Figure 2. Instrumentation for load, displacement and plug movement

Electro-Pneumatic Pile Hammer

During installation, the dynamic interaction between the pile and surrounding soil may greatly influence the state of stress and soil conditions near the installed pile (Nunez *et al.,* 1988). Both dynamic and diffusive mechanisms occur when piles are driven. Dynamic effects deal with the inertial forces developed in the soil to resist pile penetration. Diffusive effects deal with the buildup and dissipation of pore water pressures during driving.

A fast electro-pneumatic hammer capable of representing the phenomena which occur during pile driving was built (Iskander, 1995). The hammer delivers a rated energy of 212 Joules with 57% efficiency. The hammer can operate with the full design energy at frequencies up to 1.2 Hz.

Elevation of The Soil Core Inside the Pile

The elevation of the soil core inside the pile was measured using a position transducer (Fig. 2). The transducer consists of a cable wound around a spring loaded shaft which is connected to a rotary optical encoder. When the cable is pulled out of the transducer, the shaft rotates and t he magnitude of the rotation is registered by the optical encoder. The transducer has a simulated wheatstone bridge which converts the cable displacement into a voltage change.

Data Acquisition System

A MASSCOMP 5400 UNIX work station capable of sampling at 1 MHz was used for data acquisition. Analog Devices signal conditioning modules were used to amplify, filter, and isolate the sensors. A total of 31 channels were recorded as follows: axial and circumferential strains at 5 elevations on the internal and external walls of the pile (20 channels); pore pressures at four elevations on the internal and external walls of the pile (8 channels); and axial load, pile displacement, and elevation of the soil core inside the pile (1 channel each).

TYPICAL EXPERIMENTAL RESULTS

Approximately 30 load tests were performed using both instrumented and uninstrumented pipe piles. Due to space limitation, only a typical load test on a driven pile in saturated sand is shown here.

Driving Stage

Approximately 90 blows were required for installation. The pile did not plug at any stage of the driving process.

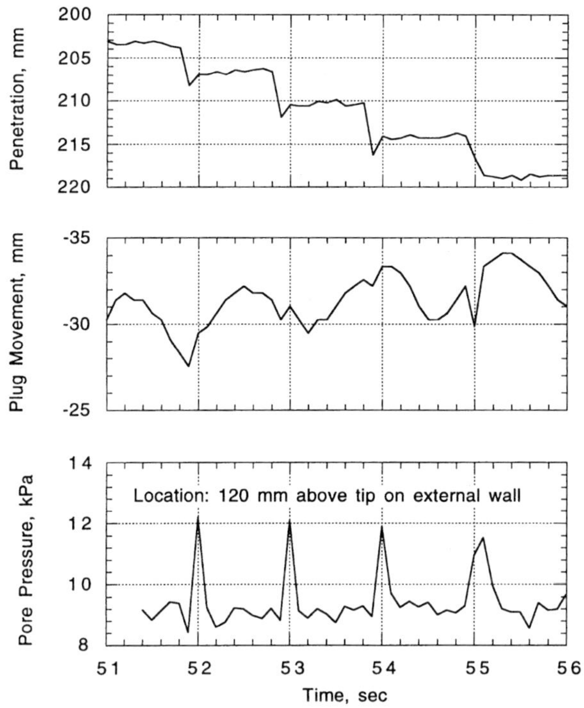

Figure 3. Penetration, plug movement, and pore pressure in four consecutive hammer blows

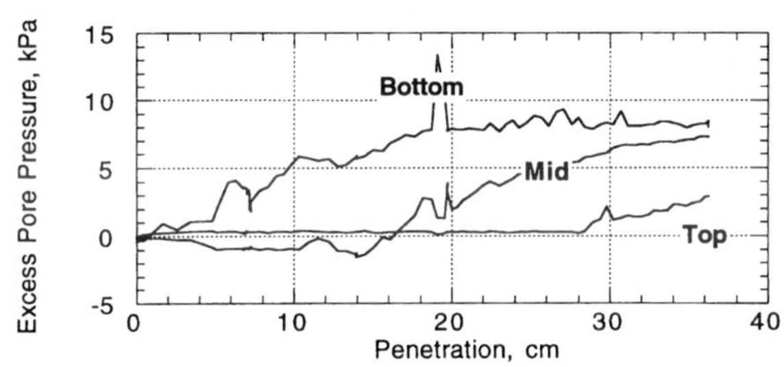

Figure 4. Build-up of pore pressures outside the pile during driving

The response of the pile-soil system during four consecutive blows in the middle of driving is shown in Fig. 3. The set per blow was approximately 4 mm. The frequency response of the position transducer used to monitor plug movements was slower than desired. However, it is clear that the plug continued to rise inside the pile after each blow. At the time of hammer impact, the measured pore water pressures rose by about 3 kPa in about 100 ms, and mostly dissipated in the next 100 ms. However, a gradual elevation in the pore water pressures existed between blows (Fig. 4). These residual pore water pressures rose to values of about 9 kPa (6% of the initial overburden pressure) at a height of 120 mm above the tip, on the outside, and up to about 10 kPa on the inner wall of the pile. Residual pore water pressures dissipated in about 20 minutes after driving ceased.

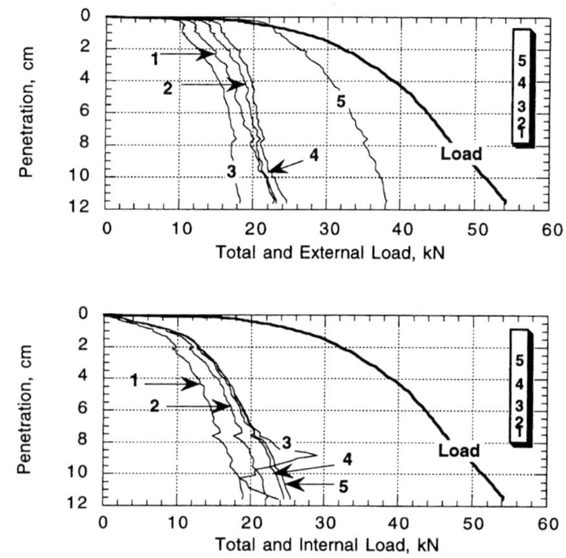

Figure 5. Load-settlement curve showing axial load transfer (a) outside and (b) inside the pile, during load test stage. Labels on contour lines represent elevations.

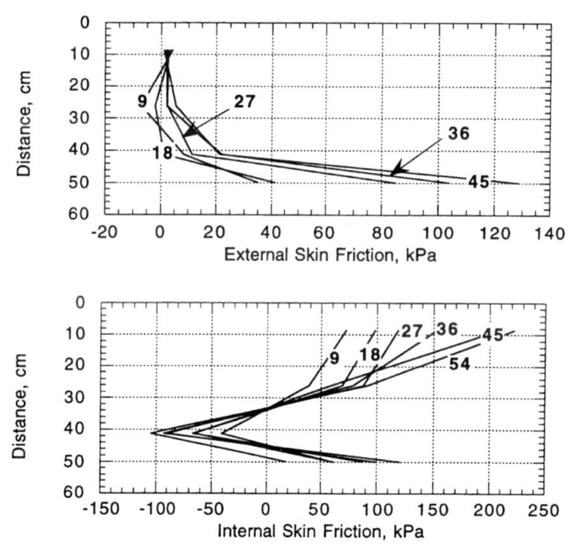

Figure 6. Skin friction distribution (a) outside, and (b) inside the pile during load test stage. Labels on contour lines represent applied load in kN

Load Test

The pile plugged during load testing and the surface of the soil core (plug) moved downward at approximately the same rate of penetration. Axial load transfer was continuously recorded during the load test (Fig. 5). The skin friction distributions shown in Fig. 6 were computed using the measured axial load transfer. Little skin friction developed in the top half outside the pile. Most of the load was transferred in end bearing which was transferred to the inner pile wall, by arching, in a

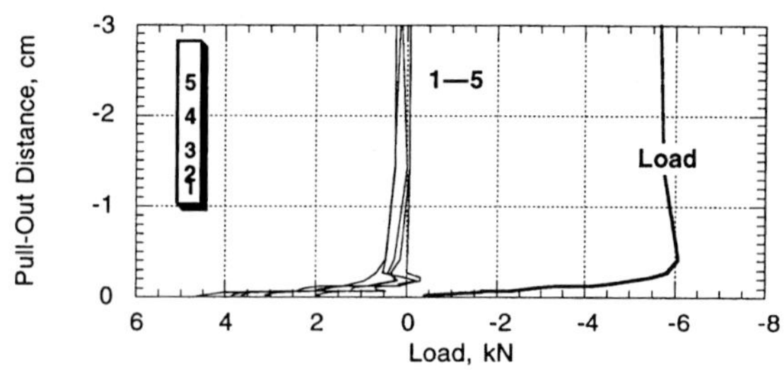

Figure 7. Pull-out curve showing axial load transfer inside the pile

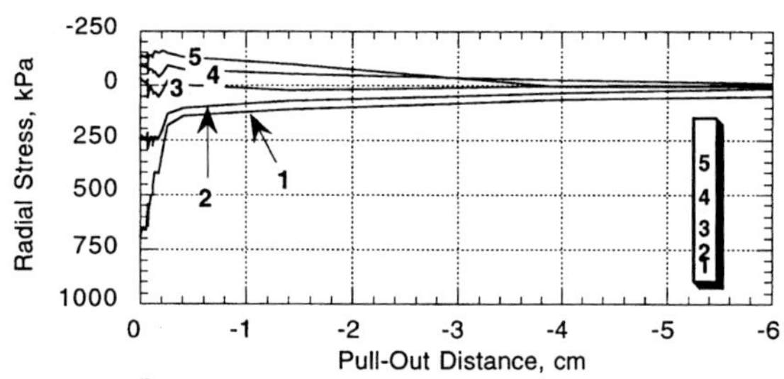

Figure 8. Radial stresses acting inside the pile during pull-out test

length equal to two pile diameters. Arching was accompanied by the development of large radial stresses (1000-1400 kPa) inside the pile (Iskander, 1995). A small build up of excess pore pressures occurred during the load test. Outside the pile, a maximum excess pore pressure of 5 kPa was recorded 120 mm above the tip. Excess pore pressures decreased with distance away from the tip.

Pull-Out Test

The pile's pull out capacity represented 20% of the capacity in compression and was developed entirely from friction outside the pile. The soil core (plug) was retained inside the pile during pull out, and essentially no plug slippage occurred. The plug contained large residual stresses which were released within 2 mm of the start of pull-out (Fig. 7). This observation was also supported by the measured radial pressure distribution inside the pile (Fig. 8). The pore pressures measured during pull out remained relatively unchanged.

CONCLUSIONS

The results of the testing program validate the performance of the developed apparatus, and provide unique insights into soil-structure interaction during pile driving and subsequent loading. Piles plugged during load tests, but not during driving. Plugging results in the formation of large radial stresses inside the pile. During pull out, only small pile movements were required to release the plug stresses. Pile driving resulted in a pore-pressure surge near the tip. The pressure surge resulted in a small build up of ambient pore pressure during driving. Excess pore pressures dissipated within 20 minutes from installation. A small build up of pore-pressure was also measured during load tests.

ACKNOWLEDGMENTS

Portions of this study were funded by NSF's Offshore Technology Research Center, The American Petroleum Institute, and Fugro McClelland Marine Geosciences. The authors are grateful to the three sponsors for their financial support.

REFERENCES

Dennis, N., Jr., and R. Olson (1983a) Axial Capacity of Steel Pipe Piles in Clay, Proc. Geotechnical Practice in Offshore Engineering, ASCE, Austin, TX, 370-388

Dennis, N., Jr., and R. Olson (1983b) Axial Capacity of Steel Pipe Piles in Sand, Proc., Geotechnical Practice in Offshore Engineering, ASCE, Austin, TX, 389-402

Iskander, M (1995) An Experimental Facility to Model the Behavior of Steel Pipe Piles in Sand, Dissertation, The University of Texas at Austin

Iskander, M. and R. Olson (1992) Review of API Guidelines for Pipe Piles in Sand, Proc. Civil Engineering in the Oceans V, ASCE, 798–812

Lehane, B. and R. Jardine (1994) Shaft Capacity of Driven Piles in Sand: A New Design Approach, Proc. Behavior of Offshore Structures (BOSS'94) 23-35

Nunez I., P. Hoadley, M. Randolph, and J. Hulett (1988) Driving and Tension Loading of Piles in Sand on a Centrifuge, Proc. Centrifuge 88, 353-362, Balkema

Olson, R. (1990) Axial Load Capacity of Steel Pipe Piles in Sand, Proc. OTC, Houston, TX, 17–24

Olson, R. and M. Iskander, (1994) Axial Load Capacity of Pipe Piles in Sands, Proc. First Regional Conference and Exhibition on Advanced Technology in Civil Engineering, Manama, Bahrain, ASCE–SAS, 383–394

Pelletier J., D. Murff, and A. Young (1993) Historical Development and Assessment of the Current API Design Methods for Axially Loaded Piles, Proc. OTC, Houston, TX, 253-282

Performance of suction caissons in sand and clay

Magued Iskander, Sherif El-Gharbawy, and Roy Olson

Abstract: The use of suction caissons (suction piles) in marine environments has been increasing in the last decade. A suction caisson is a steel pipe with an open bottom and a closed top that is inserted into the ground by pumping water out of it. Pumping creates a differential pressure across the caisson's top that pushes it into place, thus eliminating the need for pile driving. There are a number of uncertainties in the design of suction caissons. First, the state of stress and soil conditions adjacent to a suction caisson differs from those around typical driven piles or drilled shafts. Second, the axial load capacity of suction caissons depends on the rate of loading, hydraulic conductivity, drainage length, as well as the shearing strength properties of the foundation material. Finally, during pullout, volume change characteristics of the surrounding soils may change the theoretical suction pressures. A review of the existing knowledge relating to the design and construction of suction caissons is presented in this paper along with the results of a laboratory study on model caissons in sand and clay. Test results indicate that the use of suction pressure for installation of caissons is a viable alternative to conventional methods. Suction was also shown to resist some axial tensile loads.

Key words: suction, pile, bucket, foundation, anchor, capacity.

Résumé : L'utilisation de caissons à succion (pieux à succion) dans des environnements marins a augmenté au cours des dix dernières années. Un caisson à succion est un tuyau d'acier avec un fond ouvert et le dessus fermé qui est inséré dans le sol en extrayant l'eau de l'intérieur du pieu par pompage. Le pompage crée une pression différentielle sur la tête du caisson et le pousse en place, éliminant ainsi le besoin de battre le pieu. Il y a un certain nombre d'incertitudes dans la conception de caissons à succion. Premièrement, l'état des contraintes et les conditions du sol adjacent à un caisson à succion diffèrent de celles retrouvées autour de pieux battus ou forés typiques. Deuxièmement, la capacité portante axiale des caissons à succion dépend de la vitesse de chargement, de la conductivité hydraulique, de la longueur de drainage, de même que des propriétés de résistance au cisaillement du matériau de fondation. Finalement, durant l'arrachement, les caractéristiques de changement de volume du sol environnant peuvent changer les pressions théoriques de succion. On présente dans cet article une revue des connaissances existantes se rapportant à la conception et la construction de caissons à succion de même que les résultats d'une étude en laboratoire sur des modèles de caissons dans le sable et l'argile. Les résultats d'essai indiquent que l'utilisation de la pression de succion pour l'installation de caissons est une alternative viable aux méthodes conventionnelles. On a aussi démontré que la succion résiste à des chargements axiaux de traction.

Mots clés : succion, pieux, piston à clapet, fondation, ancrage, capacité.

[Traduit par la Rédaction]

Introduction

Foundations of offshore structures have undergone a steady evolution since the first offshore structure was constructed in 1947 (McClelland 1974). The first offshore structures were in shallow water, and their foundations were a natural extension of foundations used on land. Initially, piles driven with hammers operating above the water level were used. With time, structures migrated to deeper waters and became larger. Piles also became larger and special hammers, including ones operated under water were developed. For example, piles used for the Cognac platform in 313 m (1025 ft) of water were 2.15 m (7 ft) in diameter and 190 m (625 ft) long. The piles penetrated approximately 137 m (450 ft), with each pile weighing 450 t Lee and Light 1992).

In the last decade, the use of suction pressure to install foundations of offshore structures has progressed from a novelty concept to a viable alternative to pile driving. This effort has been driven by the discovery of large oil reserves in 2000 m (6500 ft) deep waters and the need to install offshore oil platforms at these depths. The cost of traditional steel pile jacket (SPJ) platforms, which resemble a pyramidal space frame, increases exponentially with depth due to the exponential increase in material and labor costs. Bullwinkle, which is the largest SPJ platform, was installed in 410 m (1350 ft) of water using 77 000 t of steel (Finn

Received 20 December 2000. Accepted 28 January 2002. Published on the NRC Research Press Web site at http://cgj.nrc.ca on 9 May 2002.

M. Iskander.[1] Polytechnic University, Six Metrotech Center, Brooklyn, NY 11201, U.S.A.
S. El-Gharbawy. The Petroleum Projects and Technical Consultations Company (Petrojet), Joseph Tito Street, Haikstep, Cairo, Egypt.
R. Olson. University of Texas at Austin, ECJ 9.227, Austin, TX 78712, U.S.A.

[1]Corresponding author (e-mail: iskander@poly.edu).

 DOI: 10.1139/T02-030

1992). Floating structures, such as tension leg platforms (TLP), are becoming, in some cases, the only economically viable design alternative. For example, the estimated cost of using an SPJ in 872 m (2860 ft) of water at the site of the Auger TLP would have exceeded the estimated value of the oil reserves. Auger, ASCE 1995 Civil Engineering Project of the Year, was installed at a cost of $1.1 billion (US) using 39 000 t of steel (Shell 2000).

TLP foundation loading consists of static uplift due to buoyancy and dynamic loading due to environmental conditions. The foundations are typically installed at great depths where suction caissons may realize economical advantages over traditional driven piles due to the speed of installation, elimination of the pile driving process, and reduction in material costs (Hogervorst 1980). In some cases, suction pressures may also contribute to the axial pullout capacity of the caissons.

The state of stress and soil conditions adjacent to a suction caisson are different from those around a typical driven pile or drilled shaft; thus typical experience-based design methods may not work well with suction caissons. It is for this reason that a laboratory study of model caissons in sand and clay was carried out. The results thereof are presented in this paper, along with a review of the knowledge gained from full-scale installation of suction caissons.

Suction caissons

Suction caissons have also been known as suction piles, bucket foundations, caisson anchors, and skirt foundations. They can be distinguished from traditional piling in a number of ways. Geometrically, suction caissons are larger in diameter and shorter in length than traditional piling. Length-to-diameter (L/D) ratios of typical suction caissons range between one and five, compared to a typical L/D ratio of 30–60 for traditional piling. Diameters of suction caissons range between 4 and 20 m (12–66 ft), compared with 0.3–3 m (1–10 ft) in traditional piling. For example, Snorre TLPs suction caissons are 20 m (66 ft) in diameter and penetrate 14.7 m (48 ft) below the seabed (Fines et al. 1991).

Suction caissons are sealed at the top. They are installed by lowering the caisson (or caissons in a group) to the sea floor, allowing them to settle under their own weight, with internal water draining through a valve in the top plate. Next, water is pumped out of the caisson using submersible pumps (Cuckson 1981). Pumping of water from within the caisson creates a differential pressure across the sealed top, thus resulting in a net downward hydrostatic force acting on the caisson's top. The developed hydrostatic force overcomes the soil resistance and pushes the caisson to the design depth. Once the design depth is reached, the pumps are disconnected and retrieved, the top of the caisson is sealed, and the space between the caisson's top and the soil plug is typically grouted. Pile hammers are not utilized in the installation.

When a tensile pullout load is applied to a suction caisson, suction pressure is developed beneath the sealed top, thus providing resistance to pullout. Given enough time, suction water pressures will dissipate, and prolonged pullout loading may result in caisson withdrawal. Suction caissons thus have greater capacity to resist short-term rather than long-term loading.

Suction anchors, should not be confused with suction caissons. Suction anchors are small inverted cups or hemispherical shaped anchors installed by pumping water from within, as with suction caissons (Brown and Nacci 1971; Sahota and Wilson 1982). The primary distinction between the anchors and caissons is that suction anchors are subject to a continuously applied suction and are typically much smaller than suction caissons. Since constant pumping is required to maintain their holding capacity, suction anchors are typically used for the temporary mooring of small rigs (Wang et al. 1978; Helfrich et al. 1976).

Historical background

The use of suction for mooring began almost two decades ago offshore of Norway. At that time, the cost of suction caissons was in excess of comparative foundation systems. It was only after significant experience had been gained with permanent structures, such as concrete gravity based structures (GBSs) and SPJs, that the use of suction re-emerged for the mooring of floating structures.

In 1989, Gullfaks C was the first structure to employ suction caisson foundations (Tjelta et al. 1990). The platform was the largest GBS ever built; it was installed in normally consolidated soft clay in 220 m (720 ft) of water. A soil drain system was utilized to accelerate consolidation thereby increasing strength and speeding primary settlement (Tjelta et al. 1992). During the design of the platform, two large-scale offshore installation trials were conducted using instrumented steel cylinders that penetrated 22 m (72 ft) into the seabed. The tip penetration resistance in the clay layers was approximately equal to the undrained bearing capacity. The applied suction decreased wall friction inside the skirt compartments in the sand layers and had no influence on wall friction or tip resistance in the clay layers. (Tjelta et al. 1986). The actual foundation consisted of 16 concrete skirt compartments that were 22 m (72 ft) high, 28 m (92 ft) in diameter and soil strength improvement by means of accelerated consolidation below the 22 m (72 ft) depth.

Snorre was the first TLP to implement the suction caisson concept (Fines et al. 1991). Suction caissons were installed in soft clay at a water depth of 320 m (1050 ft). The installation was preceded by extensive design verification and field load testing. Comparison between predicted and measured pullout loads from the model tests showed an agreement within 6% (Stove et al. 1992). The actual foundation consisted of four concrete foundation templates (CFT). Each CFT consisted of a concrete tricell that was 20 m (65.6 ft) high and 17 m (56 ft) in diameter.

Europipe (Draupner E), SPJ, was the first platform to utilize suction caissons in sand. Suction caissons were installed in very dense sand at a depth of 70 m (230 ft) (Tjelta 1994). Installation was also preceded by a series of field tests on an instrumented model caisson, due to the limited experience with suction caissons in sand. Relationships were developed between the weight of the structure, applied suction, and penetration resistance and also between applied suction and soil heave inside the caisson. (Tjelta 1995; Erbrich and Tjelta 1999). The actual foundation consisted of four suction caissons fixed to the structure, one in each corner. The caissons were 12 m (40 ft) in diameter with an overall height of 9.45 m (31 ft) and the penetration was set at 6 m (20 ft).

At the present time, suction caissons have been used to support other platforms in the North Sea, Africa, and South America. They have also been used for a variety of subsea systems such as underwater storage and offloading tanks, subsea manifolds, and anchor points for pipelines. Suction caissons also have proven to be a competitive alternative to driven piles and drag anchors for the temporary mooring of mobile offshore drilling units (MODUs) and floating, production, and storage offshore (FPSOs). A variety of soil types have been encountered at the installation sites. Layered soils, especially ones with stiff clays and dense sands, present more difficult conditions for penetration compared with homogeneous soils. The shape and geometry of the suction caissons is evolving to conform with specific localized soil conditions, water depths, installation methods, and types of loading. In each case, a solid understanding of the geotechnical aspects of caisson penetration is required for the successful installation of the foundation.

Pullout capacity of suction caissons

Drained pullout

The drained pullout capacity, Q, of a cylindrical caisson can be expressed as

$$[1] \quad Q = W_c + Q_s$$

where W_c is the sum of the submerged weights of the caisson, templates, and ballast; and Q_s is the side shear capacity (frictional resistance) on the inside and outside of the caisson wall. Generally Q_s can be calculated using an effective stress approach, as follows:

$$[2] \quad Q_s = \pi(D_o + D_i)\int_0^H \bar{\gamma}\, z\, K \tan(\delta)\,dz$$

where D_o and D_i are the outside and inside diameters of the caisson, respectively, H is the tip penetration of the caisson, $\bar{\gamma}$ is the appropriate effective unit weight of the soil, z is depth, K is the earth pressure coefficient, and δ is the friction angle of soil on caisson. Reduced values of $\bar{\gamma}$ and K should probably be used to account for perturbing the soil, particularly in sand.

Undrained pullout

The undrained pullout capacity of a cylindrical suction caisson can be expressed as

$$[3] \quad Q = Q_{so} + Q_b + W_c + W_s$$

where Q_{so} is the side shear on the outside wall of the caisson, Q_b is the tensile bearing capacity of the foundation soil, and W_c and W_s are the submerged weights of the caisson and soil plug, respectively.

In sand, an effective stress approach can be used to calculate Q_{so} as follows:

$$[4] \quad Q_{so} = \pi D_o \int_0^H (\bar{\gamma}\, z + i\gamma_w)\, K \tan(\delta)\,dz$$

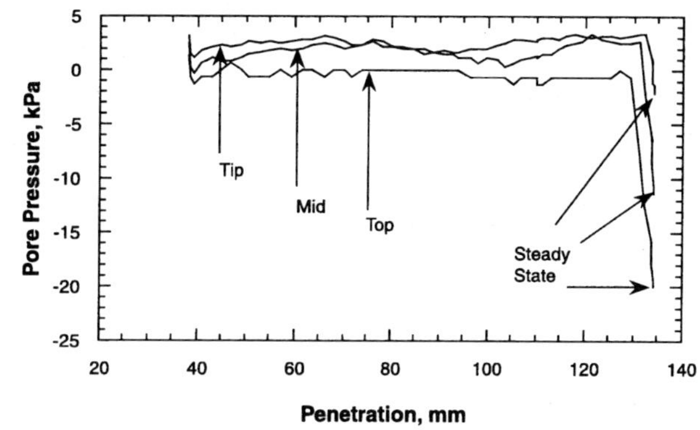

Fig. 1. Installation of the suction caisson using maximum vacuum in sand. Tip, Mid, and Top refer to the locations of porepresssure measurement inside the caisson.

where $i\gamma_w$ is the seepage force, and all other parameters are as defined above.

In clays, a total stress approach can be used to calculate Q_{so} and Q_b, as follows:

$$[5] \quad Q_{so} = \int_0^H \alpha\, S_u \pi\, D_o\,dz$$

$$[6] \quad Q_b = S_u N_c f\left(\frac{\pi}{4}\right) D_o^2$$

where S_u is the undrained shear strength of the soil, α is a friction factor, N_c is a bearing capacity factor usually taken as 9, f is a bearing capacity correction coefficient, and all other terms are as defined above.

Behavior in sands

A testing program was performed to study the generation and dissipation of suction pressures during the installation and pullout of suction caissons in sand (Pavlicek 1993; Jones 1994). Tests were performed using three different small-scale caissons with diameters ranging between 100 and 150 mm (4–6 in.), and L/D ratios of 2:1. The caissons were installed using a vacuum in fine dense sand having a saturated unit weight of 20.7 kN/m^3 (132 pcf). Loads, displacements, and pore-water pressures inside the model caissons were measured and recorded using a high-speed data acquisition system. The results of the testing were reported by Iskander et al. (1993) and Jones et al. (1994), and are summarized in this section along with relevant case histories.

Installation

The tests shown in Figs. 1–5 were performed using a 100 mm (4 in.) diameter, 200 mm (8 in.) long caisson (Table 1). Initially, caissons were seated by pushing them for approximately 30 mm (1.2 in.) to create a "seal." In practice, the self-weight of the caisson and ballast pushes the caisson and provides the seal. Next, the caisson was connected to a 550 mm (22 in.) of mercury vacuum source, to model a suction installation. Caisson insertion using suction draws water from the sea floor down through the soil outside of the caisson

Fig. 2. Installation of the suction caisson using minimum vacuum in sand.

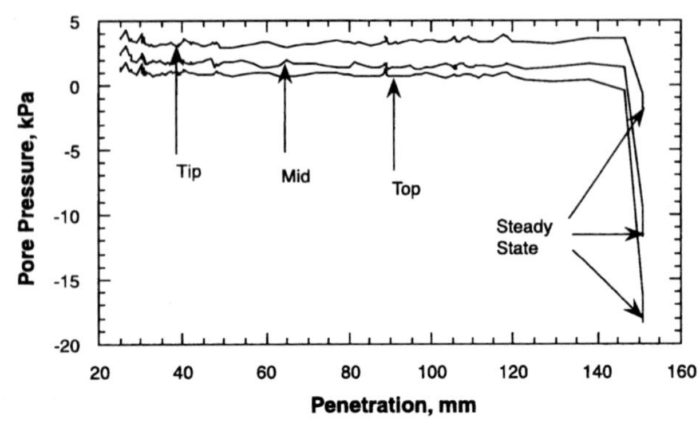

Fig. 3. Installation of the suction caisson by pushing in sand.

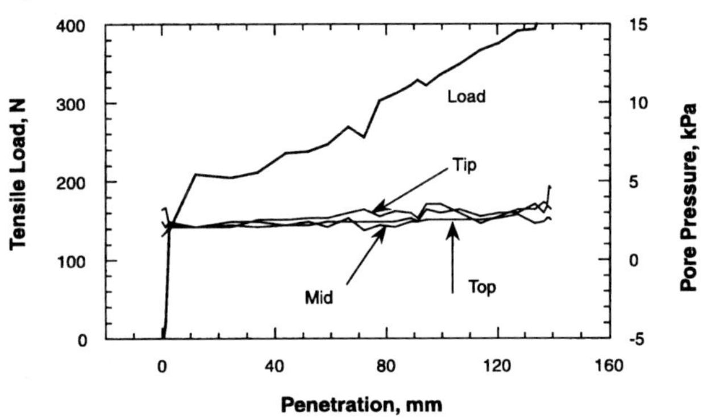

Fig. 4. Drained pullout of the suction caisson in sand.

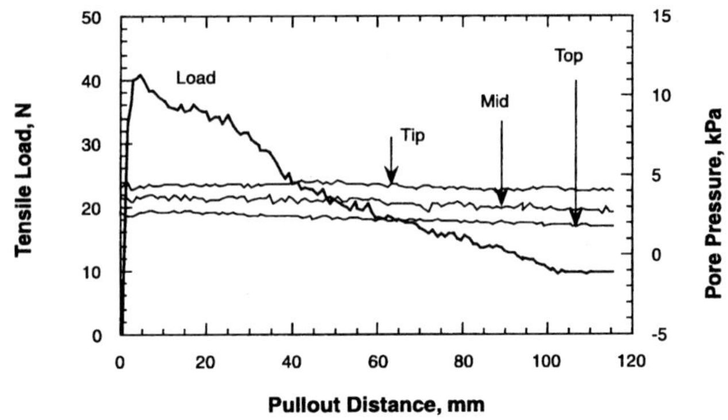

Fig. 5. Undrained pullout of the suction caisson in sand.

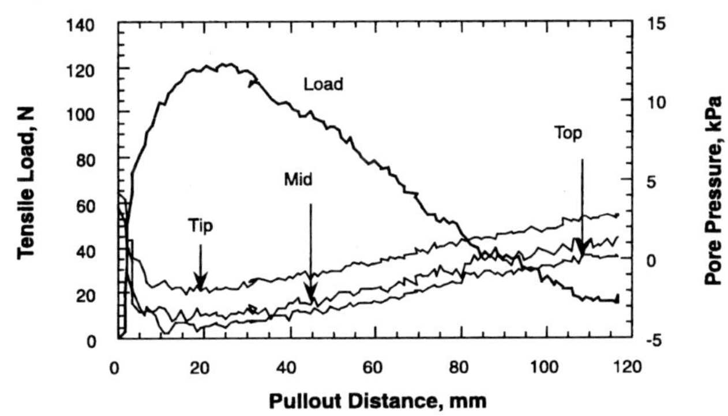

and up through the soil plug within the caisson. During this process, the upward hydraulic gradients (i = 20–40) exceed the critical gradient, resulting in a quick condition and liquefaction of the soil within the caisson and beneath its tip. The difference in the pressure head measured by the top and bottom pore-water pressure transducers corresponds to the head of the fluidized soil. In Fig. 1, the top and bottom transducers registered 3.5 kPa (0.5 psi) difference in head. For the given spacing of the transducers that head difference corresponds to a fluid weight of 20.9 kN/m^3 (133 pcf), which compares remarkably well with the measured total unit weight of the soil of 20.7 kN/m^3 (132 pcf). The occurrence of a quick condition in the model tests appears similar to that observed in the field by Senpere and Auvergene (1982) and Cuckson (1981). It also explains piping that was reported by Larsen (1989) and Brown and Nacci (1971) during the installation of suction caissons and anchors, respectively.

During installation, the level of soil inside the caisson always rose by approximately 50 mm (2 in.) above that in the tank, and a small conical depression formed surrounding the caisson. The difference in elevations between the soil inside the caisson and that in the test tank was termed the excess soil plug. Caisson penetration continued until the caisson top came in contact with the excess soil plug, at which point the pore-water pressures reached their steady state values (Fig. 1).

Several attempts to install the same suction caisson without developing a quick condition were performed by using the minimum suction pressure that caused caisson penetration. The pore pressures measured in reduced suction installations (Fig. 2) were similar to those measured in full suction installations (Fig. 1). It was not possible to install model caissons in sand without liquefying the soil, as evidenced by the measured pore pressure during the installation (Fig. 2) and the formation of an excess soil plug. Reduced suction installations were slower than full vacuum installations, but pore pressures were similar (Table 1).

The penetration force during installations (pressure differential times the area of caisson plus the submerged weight of caisson) was 45–50 N (10–11.3 lb) for full suction installations and approximately 34 N (7.7 lb) for reduced vacuum installations (Table 1). A number of caissons were pushed into position to provide a reference for comparison with caissons inserted using suction. Typical curves of force and measured pore-water pressure are shown in Fig. 3, as a function of penetration depth. The force required to push the caisson to the design depth was nearly one order of magnitude larger than that required for suction installation, which is consistent with field experience (Tjelta et al. 1986; Senpere and Auvergne 1982). The measured pore-water pressures were just slightly greater than the static values, indicating the development of a small positive excess pore-water pressure, due to installation by pushing.

Occurrence of a quick condition has two important practical implications. First, the force required for suction installation is substantially smaller than that needed for pushed or driven installations. The second is that the occurrence of an excess soil plug reduces the effective length of the suction caisson. In practice, formation of an excess soil plug may be

Table 1. Summary of installation and pullout data of caissons in sand.

	Installation						Pullout tests				
Test series	Method	Pressure differential across top (kPa) [psi]	Penetration force[a] (N) [lb]	Penetration distance (mm) [in.]	Penetration rate (mm/min) [in./min]	Excess soil plug, height (mm) [in.]	Method	Max load (N) [lb]	Displacement at max load (mm) [in.]	Suction force (N) [lb]	Side shear (kPa) [psf]
1	Max vacuum	4.75 [0.69]	50.3 [11.3]	138 [5.4]	152 [6.0]	59 [2.33]	Sealed top	131.7 [29.6]	22 [0.88]	74.7 [16.8]	0.78 [16.3]
2	Max vacuum	4.07 [0.59]	44.5 [10]	145 [5.7]	137 [5.4]	52 [2.05]	Open top	40.9 [9.2]	4.3 [0.17]	na	0.31 [6.4]
3	Push	na	445[b] [100]	178 [7.0]	533 [21.0]	17 [0.67]	Sealed top	149.7 [33.7]	18 [0.71]	68 [15.3]	1.02 [21.2]
4	Min vacuum	2.83 [0.41]	34.2 [7.7]	150 [5.9]	16.5 [0.65]	47 [1.85]	Sealed top	156.1 [35.1]	17 [0.68]	81 [18.2]	1.08 [22.5]

[a]Penetration force is equal to the pressure differential multiplied by the area plus the submerged caisson weight.
[b]From load cell measurement.

mitigated using dredging pumps. Senpere and Auvergne (1982) employed jetting inside the caisson to prevent the formation of an excess soil plug. However, jetting may reduce caisson capacity.

The observed experimental behavior was in agreement with later field tests conducted by NGI and Fugro-McClelland. In these tests, the penetration resistance was greatly reduced due to the use of suction (Tjelta 1995). Soil heave (excess soil plug) was also observed inside the caisson.

Static pullout

Rapid pullout tests were performed on the model caissons after the dissipation of the installation suction pressures. The caissons were pulled out at a constant displacement rate of 76 mm/s (0.3 in./s). Model caissons were designed such that they can be pulled out with a closed top, thus representing typical undrained behavior of suction caissons in pullout, or with their top opened to provide a comparison with the drained pullout capacity of conventional piles. Sealed-top tests resulted in a creation of a suction pressure beneath the caisson's top, which caused the soil plug to be pulled out with the caisson. Open-top tests tend to cause the caisson to pullout empty, leaving the soil plug behind. Plugs of soil were also observed to remain inside of suction anchors during pullout (Wang et al. 1978 and Helfrich et al. 1976).

The development of suction resulted in a marked increase in the pullout capacity. Sealed-caissons had approximately three times the pullout resistance of open-top caissons, as indicated in Figs. 4 and 5. The differential suction pressure is the main factor contributing to the increase in capacity (Table 1). The developed suction in sealed-top caissons depends on the rate of loading, drainage length, hydraulic conductivity, and shearing strength properties of the foundation material. A comprehensive relationship between the various parameters affecting the pullout capacity and the rate of loading of suction caissons is not available.

The displacement at the maximum load of open-top caissons was 4 mm (0.157 in.) (Fig. 4), consistent with typical displacements required to mobilize side shear of driven pipe piles in sand. The maximum suction developed for closed-top caissons peaked at displacements of about 20 mm (0.8 in.) (Fig. 5). Negative pore-water pressure tended to peak sharply and dissipate slowly. The development and dissipation of these pressures was a function of the rate of pullout loading, hydraulic conductivity, and the drainage length. The measured negative pressures support the contention of Albert et al. (1989) that negative pressures are sufficient to hold the soil plug inside the caisson during pullout, thus adding the submerged weight of the plug to the dead weight of the caisson but eliminating internal side shear.

In open top tests, the side shear (f_s) developed during pullout tests was calculated in Table 1 by rearranging eqs. [1] and [2], as follows:

$$[7] \quad f_s = \frac{Q - W_c}{\pi (D_o + D_i) z} \quad \text{(open top)}$$

where Q is the pullout tensile load, W_c is the submerged weight of the caisson, z is the caisson penetration, and D_o and D_i are the outer and inner diameters of the caisson, respectively. The side shear measured in open top tests where high vacuums were used during installation was 0.31 kPa (6.4 psf). This value is an average of internal and external side friction, which could be different.

In sealed top tests, the value of the side shear developed during pull (f_s) depends on the magnitude of end bearing, Q_b, which cannot be directly measured. Therefore, f_s was calculated in Table 1, using eq. [3] and assuming that the tensile end bearing, Q_b, was approximately equal to the suction force acting on the caisson's top, as follows:

$$[8] \quad f_s = \frac{Q - pa - W_c - W_s}{z \pi D_o} \quad \text{(sealed top)}$$

where p is the pressure differential across the caisson top, a is the internal area of the caisson, W_s is the submerged weights of the soil plug, and all other variables are as defined above. The side shear measured in sealed top tests where high vacuums were used during installation was 0.78 kPa (16.3 psf), which is 25% lower than that of caissons installed by pushing or using a low vacuum (1.05 kPa, 22 psf).

The use of $Q_b = pa$ can be justified by the fact that the maximum suction is directly related to the developed tensile end bearing (reverse bearing capacity). It is understood that the magnitude of the developed suction and tensile end bearing depends entirely on the rate of loading, the caisson geometry, and the hydraulic conductivity of the soil. The maximum developed suction is limited by the cavitation pressure and the soil's tensile capacity. In sand, the tensile capacity depends on the development of negative pore pressures during pullout loading.

The use of $Q_b = pa$ was verified by comparing the measured skin frictions to their theoretical values. The external side friction (f_s) can be calculated as follows:

$$[9] \quad f_s = \frac{z}{2}(\bar{\gamma} + i\gamma_w)K \tan\delta$$

f_s (series 1) = 0.075 × [(20.7 − 9.81) + (2.1 × 9.81)] × 0.8 × tan(25) = 0.85 kPa (sealed top)

f_s (series 2) = 0.075 × [(20.7 − 9.81) × 0.8 × tan(25) = 0.30 kPa (open top)

where z is the caisson penetration (caisson length minus height of excess soil plug), $\bar{\gamma}$ is the effective unit weight of the soil, $i\gamma_w$ is the increase in vertical stress due to the downward hydraulic gradient, i is the hydraulic gradient, γ_w is the unit weight of water, K is the coefficient of lateral earth pressure taken per API (1992) as 0.8 for nondisplacement piles, and δ is the friction angle between the caisson and soil. The computed skin frictions for drained and undrained conditions (0.30 and 0.85 kPa, respectively) are in agreement with the measured values (0.31 and 0.78 kPa, respectively). The gradient associated with pullout suction was responsible for the increase in frictional resistance.

Cyclic pullout

Tests were performed to determine the cyclic pullout capacity of suction caissons. In these tests, the reference pullout capacity was that corresponding to fully drained pullout, i.e., slow pullout with an open caisson top. Sinusoidal cyclic loads ranging between 4 and 15 Hz were applied with a variety of amplitudes. Cyclic capacity tended to decrease with the increase in the frequency of the applied load. Caisson pullout occurred in cases where the peak sinusoidal load exceeded the drained reference load. Pullout occurred in less than 5 s when the peak cyclic load exceeded 125% of the reference load. Pullout occurred over several hours when peak cyclic loads ranged between 100 and 110% of the reference load. When peak cyclic loads were maintained, even slightly, below the reference load, pullout did not occur in agreement with Allersma et al. (2000). For example, one test conducted with a peak cyclic load corresponding to 90% of the reference load was terminated after 160 000 cycles without any pullout.

The cyclic pullout load capacity of suction caissons depends on the same parameters as for static load. The small size and the fast rate of loading in the model tests obviously contributed to the lack of cyclic capacity of model suction caissons. However, it is probably imprudent, at this time, to depend on suction to resist any cyclic loading in excess of the drained pullout capacity.

Behavior in clays

The focus of this part of our investigation has been the behavior of suction caisson foundations for tension leg platforms in water depths of 2000–3000 m (6 500 – 10 000 ft) in the Gulf of Mexico. At these depths, the foundation material is predominantly normally consolidated clay. Traditionally, suction caissons have been built in the North Sea with 0.5–2 L/D ratios, e.g., Gullfaks C, Snorre, and Europipe. However, in Gulf of Mexico clays, capacity is derived mainly from increased soil shear strength with depth. It was therefore necessary to examine the possibility of extending the range of L/D ratios of suction caissons beyond that used in the North Sea.

A multidirectional displacement-controlled loading device capable of applying static and cyclic, axial and inclined tensile loading as well as combined vertical and horizontal cyclic tension loading representative of actual field loading conditions was developed. An experimental program was designed to test a set of model foundations under simulated TLP static and dynamic loads (Goldberg 1994; El-Gharbawy 1998). The foundation material was normally consolidated kaolinite clay consolidated using a vacuum. Model caissons with L/D ratios of 2–12 were used to study the feasibility of using suction to install caissons with L/D ratios in that range and the effect of increasing length on penetration resistance and pullout capacity of the caissons. Each of the instrumented caissons consisted of a 50–100 mm (2–4 in.) outside diameter tube fitted with a cap. Loads, displacements, and pore-water pressures inside and in the vicinity of the model caissons were measured and recorded using a high-speed data acquisition system.

Installation

The tests shown in Figs. 6–8 were conducted using a 100 mm (4 in.) diameter, 600 mm (24 in.) long caisson. Tests began with seating the caisson on the surface of the soil and allowing it to settle under its own weight, thereby creating the "seal" required for the application of suction. A pump was then fitted to the top of the caisson and used to

Fig. 6. Installation of the suction caisson in clay. (*a*) Penetration. (*b*) Pore-water pressure. (Solid lines are measurements inside the caisson. Dashed lines are measurements outside of the caisson).

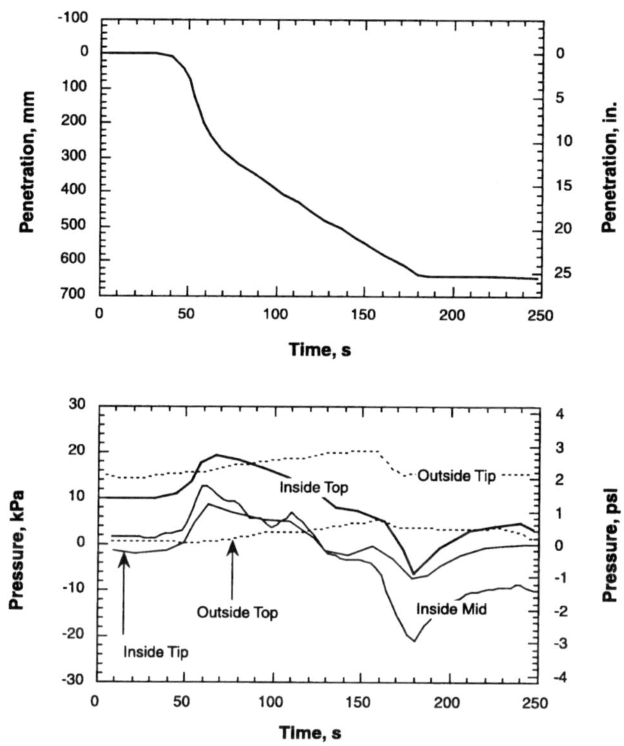

Fig. 7. Undrained pullout of the suction caisson in clay.

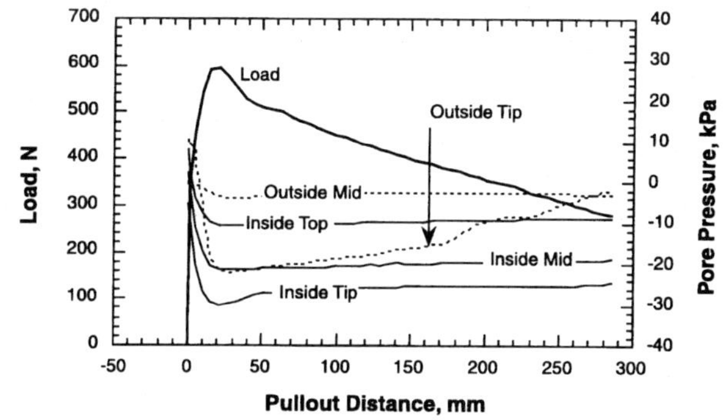

Fig. 8. Incremental, drained pullout of the suction caisson in clay.

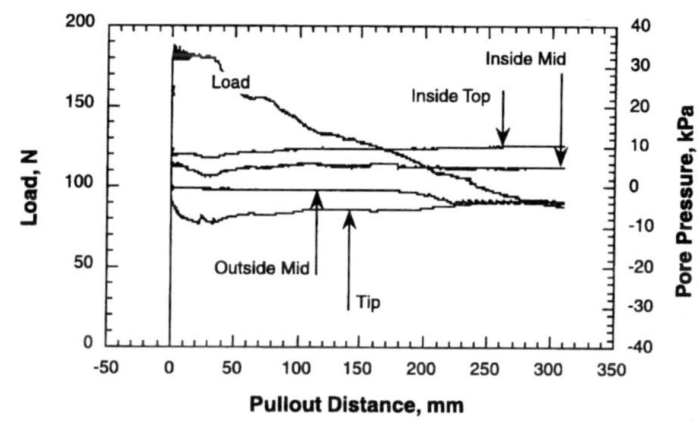

create a reduced pressure zone inside the caisson. The pressure differential across the top of the caisson worked to push the caisson into the soil to full penetration. As shown in Fig. 6, caisson installation was composed of two distinct phases: self-weight penetration followed by suction installation. Pore-water pressures inside and around the caisson increased during the first phase due to the weight and volume displacement of the caisson. In the second phase, pressures within the caisson decreased sharply in response to the application of suction, while those outside the caisson were little affected or had proceeded to increase with the increased penetration. Model suction caissons with L/D ratios ranging between 2 and 12 were installed successfully, in agreement with the theoretical predictions of House et al. (1999). Self-weight penetration accounted for approximately a third to half the penetrable length of the caisson, in agreement with later full-scale installations in 2100 m (7000 ft) water depths in the Gulf of Mexico (El-Gharbawy et al. 1999).

Static pullout

Pore-water pressures induced during the installation process were allowed to dissipate before caissons were extracted. The capacities of these caissons were similar to those of piles having similar dimensions and pushed into clay (El-Gharbawy and Olson 1999). Undrained pullout resulted in larger capacities than drained pullout. For example, tests run on a 100 mm (4 in.) diameter caisson with a L/D ratio of 6 indicated that undrained pullout had approximately three times the capacity of drained pullout (Figs. 7 and 8). Undrained loading involved pulling the model caissons out of the soil fast enough, in a displacement controlled mode, so as to prevent pore-water pressure dissipation. Typical pullout speeds were on the order of 25–50 mm/s (1–2 in./s). When the caisson was withdrawn under undrained conditions, the soil plug came out with the caisson and the outer surface of the caisson was covered with about 3 mm of clay. In the drained tests, loading was applied in increments, in a load-controlled mode, to allow the pullout suction pressures to dissipate. Under drained conditions, the caisson came out with no soil adhering to either the inner or the outer surfaces.

Negative pore-water pressures developed within the caisson during undrained pullout (Fig 7) and remained largely unchanged during drained pullout (Fig. 8). The increase in pullout capacity in the undrained condition was easily attributed to these pressures (El-Gharbawy and Olson 1998). Negative pressures also resulted in the soil plug being retained within the caisson during undrained pullout. These results are in agreement with Finn and Byrne (1972), Byrne and Finn (1978), and Clukey and Morrison (1993). Accordingly, the weight of the plug could be counted upon to resist the fast pullout load, typical of environmental effects.

Cyclic pullout

Stress-controlled cyclic tests were conducted by applying and maintaining a bias load on the caisson overnight, then superimposing vertical cyclic pullout loads with different peak values. The bias load was chosen as a function (60–75%) of the long-term capacity (LTC) of the caisson. Increasing the peak cyclic load resulted in an increase in displacement after a given number of cycles (Fig. 9). A peak cyclic load

Fig. 9. Axial cyclic loading of suction caissons in clay. LTC, long term capacity.

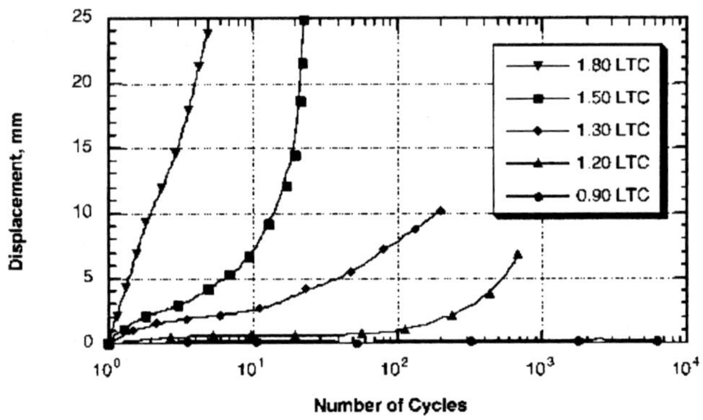

equivalent to 85% of the LTC of the 600 mm (24 in.) long caisson could be sustained for at least 10 000 cycles with no significant cap displacement. However, peak cyclic loads greater than the LTC resulted in the caisson pulling out at rates that were directly proportional to the magnitude of the load. For example, a peak cyclic load reaching twice the magnitude of the bias load, the 400 mm (16 in.) long caisson (L/D = 4) pulled out after 6 cycles, which is in general agreement with the Gullfaks C field test results. When the same loading was applied to the 60 mm (24 in.) long caisson (L/D = 6), the caisson withstood 80 cycles prior to failure. The number of cycles before failure increased with decreasing peak cyclic load or with increasing caisson length.

Conclusions and recommendations

Utilizing suction pressures for installation and development of pullout capacity of caissons is a viable concept. Successful model suction caisson installations were performed with ease. The following observations were made during the installation and pullout of model suction caissons:

(1) A quick condition occurs during the installation of suction caissons in sand, which influences the behavior in several ways. First, the penetration resistance is reduced. Second, the penetration distance is reduced due to the formation of an excess soil plug. Third, suction installation results in a reduction in the frictional capacity of the caissons.

(2) Model suction caissons with L/D ratios ranging between 2 and 12 were installed successfully in clay.

(3) Suction, and the associated tensile bearing capacity, are the dominant parameters contributing to the short-term (undrained) pullout capacity of sealed top suction caissons in both sand and clay. Additionally, the suctions developed during pullout were sufficient to retain the soil plug inside the caisson, thereby increasing the dead weight of the caisson and eliminating the internal side shear.

(4) During pullout, the gradient associated with the suction pressures resulted in a marked increase in the frictional capacity of caissons in sand.

(5) Suction caissons derive their vertical pullout capacities from one or more of the following components: (*a*) weight of the suction caisson, foundation templates, and ballast loads; (*b*) submerged weight of the soil plug captured within the caisson; (*c*) shearing resistance between the soil and the caisson wall; and (*d*) a reverse bearing capacity mechanism developed due to suction.

(6) The long-term drained capacity of the caisson was found to represent an upper limit to the peak cyclic load that could be withstood by the caisson before significant displacement occurred. The number of cycles before failure decreased with an increase in the peak cyclic load. However, suction pressure can be used to resist transient events.

Acknowledgments

We are grateful to the Offshore Technology Research Center (National Science Foundation (NSF) Engineering Research Centers program grant no. CDR8721512) for funding portions of this work. Our research partners Robert Pavlicek, Casey Jones, Aaron Goldberg, Wade Williams, Jody Kerwin, and Pawel Zieminski were instrumental in conducting the experimental work and provided us with valuable insights and suggestions.

References

Albert, L., Holtz, R., and Magris, E. 1989. Super pile system: a feasible alternate foundation for tension leg platform in deep water. Marine Geotechnology, **8**: 133–158.

Allersma, H.G.B, Kierstein, A., and Maes, D. 2000. Centrifuge modelling on suction piles under cyclic and long term vertical loading. *In* Proceedings of the 10th International Offshore and Polar Engineering Conference, (ISOPE-2000), Seattle, WA, 28 May – 2 June 2000. Vol. 2, pp. 334–341.

API. 1992. American Petroleum Institute (API) recommended practice for planning, designing and constructing fixed offshore platforms, RP2A. API Production Div., 211 North Ervay, Suite 1700, Dallas, TX 75201.

Brown, G., and Nacci, V. 1971. Performance of hydrostatic anchors in granular soils. *In* Proceedings, Offshore Technology Conference, OTC 1472, pp. 533–542.

Byrne, P., and Finn, W. 1978. Breakout of submerged structures buried to shallow depth. Canadian Geotechnical Journal, **15**: 146–154.

Clukey, E., and Morrison, M. 1993. A centrifuge and analytical study to evaluate suction caissons for TLP applications in the Gulf of Mexico. *In* Design and performance of deep foundations, piles and piers in soil and soft rock, American Society of Civil Engineers, Geotechnical Special Publication GSP No. 38, pp. 141–156.

Cuckson, J. 1981. The suction pile finds its place. Offshore Engineer, April, pp. 80–81.

El-Gharbawy, S. 1998. The pullout capacity of suction caisson foundations for tension leg platforms. Ph.D. dissertation, The University of Texas at Austin, Austin, TX.

El-Gharbawy, S., and Olson, R. 1998. The pullout capacity of suction caisson foundations for tension leg platforms. *In* Proceedings of the 8th International Offshore and Polar Engineering Conference, Montreal, PQ, 24–28 May 1998.

El-Gharbawy, S., and Olson, R. 1999. Suction caisson foundations in the Gulf of Mexico. *In* Analysis, Design, Construction, and Testing of Deep Foundations, Proceedings of the OTRC '99 Conference, Reston, VA. *Edited by* J. Roesset. American Society of Civil Engineers, Geotechnical Special Publication No. 88, pp. 281–295.

El-Gharbawy, S., Olson, R., and Scott, S. 1999. Suction anchor installation for deep Gulf of Mexico applications. *In* Proceedings, Offshore Technology Conference, OTC 10992, pp. 747–754.

Erbrich, C., and Tjelta, T. 1999. Installation of bucket foundations and suction caissons in sand – geotechnical performance. *In* Proceedings, Offshore Technology Conference, OTC 10990, pp. 725–735.

Fines, S., Stove, O., and Guldberg, F. 1991. Snorre TLP tethers and foundation. *In* Proceedings, Offshore Technology Conference, OTC 6623, pp. 587–597.

Finn, L. 1992. Offshore structures — past, present and future. *In* Proceedings Civil Engineering in the Oceans V. College Station, Texas, 2–5 November. *Edited by* R. Hudspeth. American Society of Civil Engineers (ASCE), pp. 218–242.

Finn, W., Byrne, P. 1972. The evaluation of the breakout force for a submerged ocean platform. *In* Proceedings Offshore Technology Conference, OTC 1604.

Goldberg, A. 1994. Design and construction of a model suction pile. M.Sc. thesis, The University of Texas at Austin, Austin, TX.

Helfrich, S., Brazil, R., and Richards, A. 1976. Pullout characteristics of a suction anchor in sand. *In* Proceedings, Offshore Technology Conference, OTC 2469, pp. 501–506.

Hogervorst, J. 1980. Field trials with large diameter suction piles. *In* Proceedings, Offshore Technology Conference, OTC 3817, pp. 217–224.

House, A., Randolph, M., and Borbas, M. 1999. Limiting aspect ratio for suction caisson installation in clay. *In* Proceedings of the 9th International Offshore and Polar Engineering Conference (ISOPE 99), Brest, 30 May – 4 June 1999, Vol. 1, pp. 676–683.

Iskander, M., Olson, R., and Pavlicek, R. 1993. Behavior of suction piles in sand. *In* Design and performance of deep foundations, piles and piers in soil and soft rock, American Society of Civil Engineers, Geotechnical Special Publication GSP No. 38, pp. 157–171.

Jones, W. 1994. Axial capacity of suction piles in sand. M.Sc. thesis. The University of Texas at Austin, Austin, TX. 90 pp.

Jones, W., Iskander, M., Olson, R., and Goldberg, A. 1994. Axial capacity of suction piles in sand. *In* Proceedings of the 7th International Conference on the Behavior of Offshore Structures (BOSS'94). *Edited by* C. Chryssostomidis. Pergamon, pp. 63–75.

Lee, G., and Light, L. 1992. Fixed offshore platforms – design and construction considerations. *Notes for short course* Design of Fixed Offshore Platforms. University of Texas, Austin, TX.

Larsen, P. 1989. Suction. Anchors as an anchoring system for floating offshore constructions. *In* Proceedings, Offshore Technology Conference, OTC 6029, pp. 535–540.

McClelland, B. 1974. Design of deep penetrating piles for ocean structures. Journal of Geotechnical Engineering Division, ASCE, **100**(GR7): 705–748.

Pavlicek, R. 1993. Axial tensile load capacity of suction piles. M.Sc. thesis, The University of Texas at Austin, Austin, TX. 165 pp.

Sahota, B., and Wilson, Q. 1982. The breakout behavior of a suction anchor embedded in submerged sand. *In* Proceedings, Offshore Technology Conference, OTC 4175, pp. 117–132.

Senpere, D., and Auvergne, G. 1982. Suction anchor piles – a proven alternative to driving or drilling. *In* Proceedings, Offshore Technology Conference, OTC 4206, pp. 483–493.

Shell Oil. 2000. Auger tension leg platform fact sheet [online]. Available from Shell Exploration and Production Company Web Site, http://www.shell.com [cited 22 March 2002].

Stove, O.J., Bysveen, S., and Christophersen, H. 1992. New foundation systems for the Snorre Development. *In* Proceedings, Offshore Technology Conference, OTC 6882, pp. 75–83.

Tjelta, T. 1994. Geotechnical aspects of bucket foundations replacing piles for the Europipe 16/11-E jacket. *In* Proceedings, Offshore Technology Conference, OTC 7397, pp. 73–82.

Tjelta, T. 1995. Geotechnical experience from the installation of the Europipe jacket with bucket foundations. *In* Proceedings, Offshore Technology Conference, OTC 7795, pp. 897–908

Tjelta, T., Guttormsen, T., and Hermstad, J. 1986. Large scale penetration test at a deep water site. *In* Proceedings, Offshore Technology Conference, OTC 5103, pp. 201–212.

Tjelta, T., Aas, P., Hermstad, J., and Andenaes, E. 1990. The skirt piled Gullfaks C platform installation. *In* Proceedings, Offshore Technology Conference, OTC 6473, pp. 453–462.

Tjelta, T.I., Janbu, N., and Grande, L. 1992. Observations on drainage control effects on Gullfaks C structure. *In* Proceedings of International Conference on Behavior of Offshore Structures (BOSS 92), Vol. 1, pp. 703–720.

Wang, M., Demars, K., and Nacci, V. 1978. Applications of suction anchors in offshore technology. *In* Proceedings, Offshore Technology Conference, OTC 3203, pp. 1311–1320.

Geotechnical Testing Journal, March 2003, Vol. 26, No. 1
Paper ID GTJ200310648_261
Available online at: www.astm.org

Robert C. Pedersen,[1] *Roy E. Olson,*[2] *Alan F. Rauch*[3]

Shear and Interface Strength of Clay at Very Low Effective Stress

ABSTRACT: Thin-specimen direct shear (TSDS) tests were conducted to measure the shearing strength of kaolinite, and interface strengths between kaolinite and acrylic plastic and anodized aluminum, at normal effective stresses from 1 to 2400 Pa (0.02 to 50 lb/ft^2). At the lowest effective normal stresses, curved strength envelopes fitted through the data exhibited no cohesion and high secant friction angles. Accurate information on the behavior of soil in this low-pressure range is needed to properly interpret the behavior of prototype foundations in laboratory-scale model tests.

KEYWORDS: shear strength, low stress, interface friction, direct shear, tilt-table

Knowledge of the shearing resistance between soil and concrete, steel, geosynthetics, and wood is needed to evaluate the stability of many types of geotechnical structures. Numerous studies have produced useful data on the shearing resistance of interfaces involving sand on steel or concrete, and clay on polymer geosynthetics, but much less information is available on the strength of interfaces between clays and rigid foundation materials. Further, most interface shear tests have been conducted at stress levels that are convenient for laboratory testing and appropriate for typical field conditions. In contrast, there is a lack of experimental data on soil interface strengths at low stress levels.

At the University of Texas at Austin, scale models of offshore foundations are being tested in tanks of normally and overconsolidated kaolinite (El-Gharbawy and Olson 1999; Olson et al. 2001; Rauch et al. 2001). The behavior of suction caissons is being studied using acrylic and aluminum models measuring up to 100 mm in diameter by 1 m long, in a 1-*g* environment. Suction caissons are large-diameter steel "piles" that are used in the deep marine environment to anchor floating structures. They are installed using a combination of dead weight and internal suction. Early suction caissons had relatively large diameters and shallow depths, but the soft clays found in the Gulf of Mexico require deeper penetration to develop the desired capacity. Knowledge of the shearing resistance at the caisson/clay interface is particularly important for design, especially for vertical loading.

Laboratory-scale model tests are being used to study the behavior of suction caisson foundations, with the findings used to support the development of improved analytical and design methods. To interpret these laboratory tests, the interface shearing behavior of clay at low effective normal stresses must be characterized. Thin-specimen direct shear (TSDS) tests were undertaken to characterize the drained strength of kaolinite interfaces at low effective normal stresses.

Received July 17, 2001; accepted for publication April 5, 2002; published Jan. 17, 2003.

[1] Graduate research assistant, Department of Civil Engineering, The University of Texas at Austin, Austin, TX 78712.
[2] L. P. Gilvin Professor, Department of Civil Engineering, The University of Texas at Austin, Austin, TX 78712.
[3] Assistant professor, Department of Civil Engineering, The University of Texas at Austin, Austin, Texas, 78712.

Previous Studies of Soil and Interface Shear Strength

Many engineers accept the presence of an effective stress cohesion intercept for a "cohesive" soil, but not for a "cohesionless" soil. However, clean sands can exhibit cohesion intercepts if straight lines are fitted to strength envelopes that are actually curved. Test results dating back to at least Casagrande (1940) indicate that effective stress failure envelopes for sands are curved. Curved failure envelopes for sands have been widely reported (Vesić and Clough 1968; Lee and Seed 1967; Ponce and Bell 1971). De Beer (1970) demonstrated that large errors can result from using failure envelopes measured in the usual range of effective stress to interpret small-scale laboratory tests in sand.

For clays, early shear tests of normally consolidated Boston Blue clay indicated no cohesion intercept (Jurgenson 1934). Casagrande and Hirschfeld (1960) assumed that the cohesion intercept for clay was zero, but the friction angle decreased with increasing normal stress. Jacobson (1953) found an "origin cohesion" for normally consolidated natural clays. Bjerrum (1954) also found a cohesion intercept, which he attributed to attractive forces between the clay particles, while Kenney at al. (1967) and Tavenas et al. (1971) demonstrated cohesion due to cementation. Bjerrum (1973) indicated that strengths could increase due to the effects of aging. However, the curvature of failure envelopes for normally consolidated clays is apparent in the usual stress range (Olson 1974) as well as at elevated stresses (Bishop et al. 1965; Brouillette et al. 1993).

Duncan et al. (1980) found that curved failure envelopes for many soils could be conveniently modeled in the typical pressure range using a friction angle that decreases linearly with the logarithm of effective normal stress. Following this suggestion with the secant friction angle (ϕ_s'), a non-linear envelope can be represented (Fig. 1) with:

$$\phi_s' = \phi_{s1}' - \Delta\phi_s' \log(\sigma_n') \tag{1}$$

where σ_n' is the effective normal stress, and ϕ_{s1}' and $\Delta\phi_s'$ are parameters used to fit the measured strength data. Like the internal strength of soil, failure envelopes for soil sheared along various interfaces are likely curved and can be approximated in a similar manner:

$$\delta_s' = \delta_{s1}' - \Delta\delta_s' \log(\sigma_n') \tag{2}$$

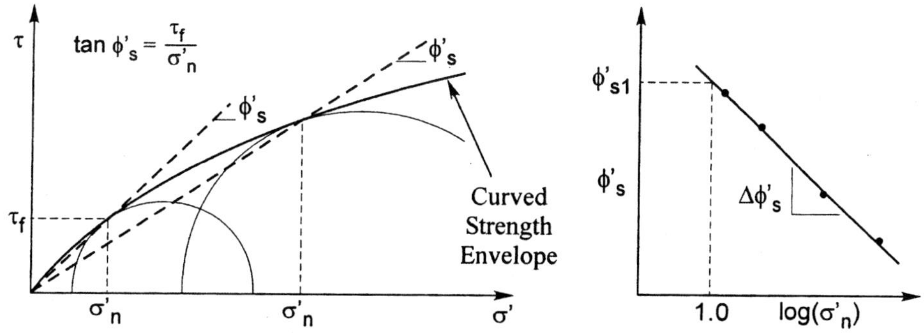

FIG. 1—*Model of curved strength envelope using the secant friction angle.*

where δ_s' is the secant interface friction angle and δ_{s1}' and $\Delta\delta_s'$ are fitting parameters.

Potyondy (1961) measured interface friction angles in the normal-stress range of 50–150 kPa using direct shear tests. It is likely that Potyondy's tests with clayey soils involved partial drainage, making it impossible to interpret the interface shearing resistance in terms of effective stress. Direct shear tests on interfaces between sand and wood, concrete, or metal have also been reported by Acar et al. (1982), Bosscher and Ortiz (1987), Reddy et al. (2000), and others.

Relying mostly on direct simple shear tests of sand on steel or concrete, Uesugi and Kishida (1986a) investigated the relationships between interface shear resistance and particle size, particle shape, surface roughness, and normal stress. They found that the interface friction varied with surface roughness up to a critical value; on even rougher surfaces, failure occurred in the soil and the interface strength was determined by ϕ'. The interface friction angle was found to decrease slightly between normal stresses of 100 to 4000 kPa. They also related interface friction to particle roundness and a surface roughness index (Uesugi and Kishida 1986b; Kishida and Uesugi 1987). For effective normal stresses in the range of 100 to 500 kPa, Uesugi et al. (1988, 1990) found that little sliding occurs on the interface at shear stresses below failure. At the peak shear stress, sand slipped along a smooth interface, while shearing was concentrated in the sand on a rough interface.

Fewer interface shear tests have been conducted with cohesive soils on concrete or steel. In drained direct shear tests, Clark and Meyerhof (1972) and Littleton (1976) found contacts between steel and clays to be entirely frictional (no cohesion). Tsubakihara and Kishida (1993) used direct simple shear tests and found that, for clay-steel interfaces, shearing deformations occur in the soil up to the peak shear stress, followed by sliding along a smooth interface. On a surface rougher than the critical value, deformation and failure occur completely in the soil.

Lehane and Jardine (1996) summarized the results of interface shear tests performed with clays, silts, and sands on steel. They showed that the interface friction angle decreased as the soil became coarser, consistent with the findings of Uesugi and Kishida (1986b). Tsubakihara et al. (1993) observed three failure modes: shearing within the soil placed against a rough surface, sliding of the soil along a smooth interface, and, for clay-sand mixtures, partial sliding in localized areas of the interface contact area.

Direct shear tests of clays on geomembranes have been reported by Koerner et al. (1986), Seed and Boulanger (1991), Fishman and Pal (1994), and others. Direct shear tests have also been performed on geosynthetic clay liners, some involving normal pressures less than 10 kPa (Gilbert et al. 1996; Merrill and O'Brien 1997; Fox et al. 1998). Tests of sand on polymer membranes show that the interface friction angle increases with higher normal stresses due to particles becoming embedded in the membrane (Girard et al. 1990), an effect that changes with the hardness of the polymer (O'Rourke et al. 1990).

Most of the interface shear tests described in the literature were conducted using either direct shear or simple shear devices. Torsional shear devices are sometimes used to achieve larger shearing deformations (Yoshimi and Kishida 1981; Stark and Poeppel 1994). Almost all of the interface tests cited here were conducted at normal stresses exceeding about 15 kPa.

Direct Shear Tests with Tilt Tables and Thin Specimens

A simple means of conducting shear tests at very low stress levels involves the use of a tilt-table, where shear stresses are generated by inclining the specimen. Tilt-table devices have been used to measure low-pressure friction along rock joints (Cawsey and Farrar 1976; Hencher 1976; Bruce et al. 1989) and geosynthetic interfaces (Girard et al. 1990; Shan 1993; Lalarakotoson et al. 1999). Tilt-table shear tests have the following disadvantages (Hencher 1976; Shan 1993; and Gilbert et al. 1995): displacements are not controlled and the post-peak response cannot be measured, test pressures are limited by toppling of the surcharge weights, and nonuniform normal stresses develop on the interface as the device is inclined. Advantages include the elimination of internal machine friction, not forcing failure to occur along the interface, and the ability to perform tests under low normal stresses.

Thin soil specimens can be used in tests to measure the strength of soils. Terzaghi (1925) performed direct shear tests on specimens as thin as 3 mm and found that ϕ_s' for his "yellow clay" decreased from 38 to 15° as the normal stress rose from 3 to 90 kPa. respectively. Over the same range of pressures, he measured δ_s' for clay on glass of 33 to 12°. Although Terzaghi's samples were initially consolidated, full drainage may not have occurred during shearing. Jurgenson (1934) performed direct shear tests on remolded clay specimens as thin as 2 mm and measured friction angles that were independent of specimen thickness. Casagrande and Hirschfeld (1960) reported obtaining the same failure envelope using standard thick specimens and specimens as thin as 3 mm, in the stress range of 10 to 30 kPa. Kenney et al. (1967) performed simple shear tests on specimens as thin as 1 mm and found similar friction angles from thin or thick specimens. Chandler and Hardie (1989) studied residual friction angles using specimens as thin as 2 mm.

Although not suitable for testing undisturbed specimens, using thin specimens in direct shear tests on remolded soil is an established technique with many advantages: thinner specimens reduce the time needed to dissipate excess pore pressures and eliminate the need for lateral confinement.

Thin-Specimen Direct Shear Tests on Kaolinite

In this study, submerged kaolinite specimens, 0.3 to 0.5 mm thick, were sheared to failure in what will be called "thin-specimen direct shear" (TSDS) tests. The tests were conducted using a tilting table apparatus in a water bath (Fig. 2). Specimens were prepared by pouring kaolinite slurry on top of a 13-mm-thick by 150-mm-diameter porous stone that had a temporary slurry retention band around the sides. A top platen fabricated from acrylic, aluminum, or porous stone was then placed on top. The slurry was allowed to consolidate, with drainage through the lower stone, while the band was allowed to slide down over the lower stone. Consolidation pressures ranged from 1 to 2400 Pa (0.02 to 50 lb/ft^2). The tests were set up, then submerged in the water bath, and the retaining band was removed to leave the specimen unconfined.

Shearing stresses were applied by elevating one end of the table (Fig. 2) in increments of one degree, with the apparatus locked in position for 1 min after each increment to allow for full dissipation of excess pore water pressures (Pedersen 2001). The angle of inclination (θ) was determined optically using a cathetometer. Shearing displacements were not measured.

The average normal (σ) and shear (τ) stresses were:

$$\sigma = (W / A) \cos \theta \qquad (3)$$

$$\tau = (W / A) \sin \theta \qquad (4)$$

where A is the area of the shear plane, W is the submerged weight of the top platen and surcharge weights, and θ is the angle of rotation. The estimated accuracies of the test measurements are summarized in Table 1. Rotation of the assembly of soil, platens, and weights converts the applied normal force to an inclined and eccentric load (Hencher 1976; Shan 1993; Pedersen 2001). Flat weights were used to minimize this problem. The maximum inclination of the test was limited by the tendency for the weights to tip.

A 150-mm-diameter porous stone was used as the top platen in obtaining measurements of the kaolinite's internal shear strength (Jurgenson 1934; Chandler and Hardie 1989). Interface tests were conducted with a top platen of smooth, unaltered acrylic or anodized aluminum, the two materials used to make the model suction caissons. The interface platens were larger in diameter (160 mm) than the test soil (150 mm) to provide for a constant contact area as shearing displacements developed.

Failure involved the top platen and weights sliding down the inclined table. Whether failure occurred within the kaolinite or by sliding along the interface was judged by observing the exposed surface of the lower porous stone. For sliding along the interface, the top platen should slide off cleanly and the exposed part of the lower stone should remain covered with kaolinite. When failure occurs in the soil, some of the soil should adhere to the top platen and some to the bottom platen, leaving some areas on each that have no adhered soil.

All test specimens were overconsolidated to some degree at failure, because both apparatus submergence and rotation during testing reduced the normal stress on the sample (Pedersen 2001). The normal stress decreases with increasing specimen inclination (θ) and the overconsolidation ratio (OCR) becomes $1/\cos\theta$. In these tests, the maximum inclination achieved was almost 60°, which resulted in an OCR of 2 from this effect. In addition, because the tests were conducted by consolidating the soil under the applied weights before submerging the apparatus, buoyancy reduced the effective weight of the surcharge and the applied effective stress. For most of the tests, the combined effects of submergence and rotation lead to an OCR at failure between about 1.2 and 10.

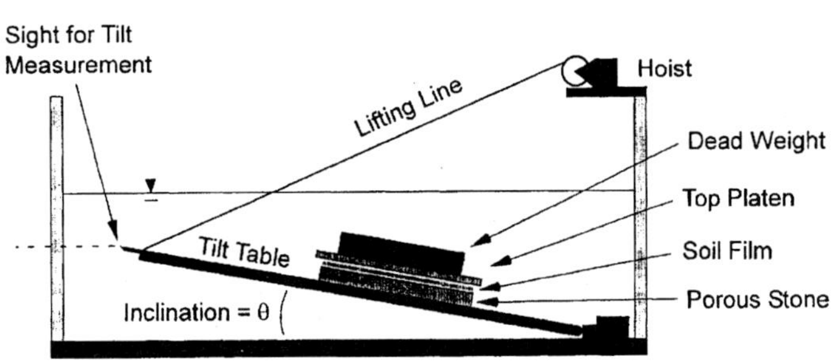

FIG. 2—*Tilt-table apparatus used in thin-specimen direct shear (TSDS) tests.*

TABLE 1—*Estimated accuracy of TSDS test measurements.*

Measurement	Accuracy	Contributing Factors
Inclination (θ)	± 0.1°	Accuracy/precision of cathetometer
Submerged surcharge weight	± 2 %	± 0.03 % in dry weight, ± 2 % in specific gravity
Normal effective stress (σ'_n)	± 4 %	± 2 % in submerged weight of surcharge, ± 2 % in specimen area
Secant friction angle (ϕ'_s or δ'_s)	± 0.6° at 30°	Average over tilt increment of 0.8–1.0°

Results from TSDS Tests on Kaolinite

The kaolinite clay tested was "Hydrite R," purchased as a dry powder from the Dry Branch Kaolin Co. It has a liquid limit in the range of 54–58 %, a plastic limit in the range of 31–34 %, and a specific gravity of the solids of 2.59–2.63 (El-Gharbawy 1998). El-Gharbawy (1998) used conventional direct shear tests on 100-mm-diameter by 13-mm-thick specimens of this soil and found that $\phi' = 28°$ and $c' = 140$ Pa in the pressure range of 200–1600 Pa (Fig. 3).

The first series of TSDS tests with Hydrite R kaolinite used a porous stone on top of the kaolinite (Table 2), except for two tests

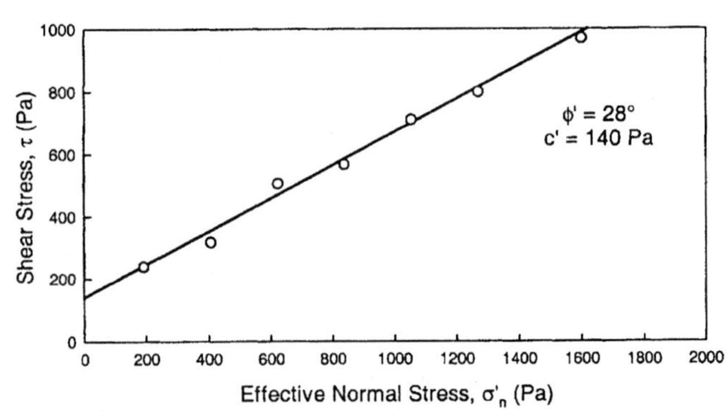

FIG. 3—*Failure envelope from drained direct shear tests on kaolinite (El-Gharbawy 1998).*

TABLE 2—*Summary of TSDS tests conducted on kaolinite.*

		Normal Stress, Pa		OCR at Failure	
Top Platen	Number of Tests	Minimum	Maximum	Minimum	Maximum
Stone	15	1	2320	1.4	115.1
Acrylic	15	10	1980	1.2	17
Aluminum	11	40	1840	1.2	2.1

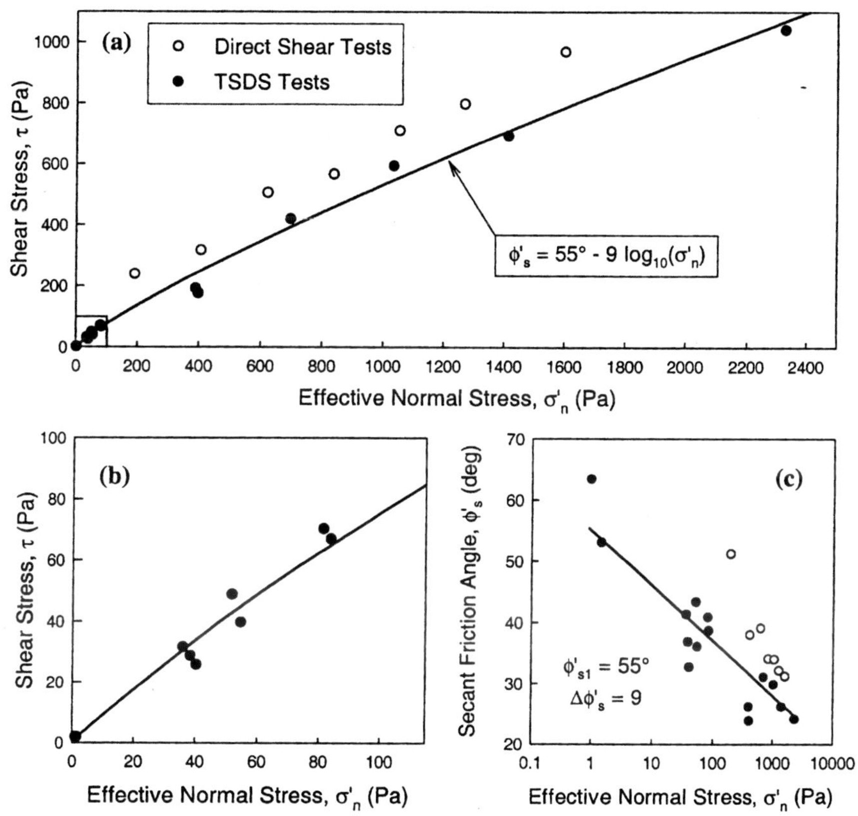

FIG. 4—*Drained strength of kaolinite measured in TSDS and direct shear tests.*

that involved a layer of kaolinite with no stone on top. It was difficult to determine by observation whether failure occurred at the stone/soil interface or in the soil in these tests, but it appears that the difference was insignificant. The failure envelope (Figs. 4*a* and 4*b*) was curved and passed through the origin. The previous direct shear tests (El Gharbawy 1998) gave a slightly steeper failure envelope than the TSDS tests, possibly because of friction in the direct shear apparatus.

The measured secant friction angles decreased from about 60° at low normal stresses to about 23° at the highest normal stresses (Fig. 4*c*). The semi-logarithmic envelope (Eq 1) was fit to the data and is shown as a solid line in all three plots of Fig. 4. A linear strength envelope passing through the TSDS data would have given a cohesion intercept of about 50 Pa.

When an acrylic top platen was used, failure typically involved a gradual displacement of the acrylic platen with shearing displacements up to about 6 mm. The leading portion of the acrylic platen, which moved beyond the edge of the lower stone, usually appeared to be clean and transparent, whereas the clay exposed by the sliding top plate remained along the upper edge of the lower porous stone, suggesting that failure occurred at the interface between the kaolinite and the acrylic plastic. After failure, rapid displacement occurred, and parts of the acrylic plate that had sheared beyond the underlying kaolinite had small amounts of adhered kaolinite, and patches of bare porous stone could be seen along the trailing edge where the clay had apparently adhered to the acrylic.

The measured shearing resistances from the acrylic interface tests (Fig. 5) are lower than the internal drained strength of the soil. Large friction angles and no intercept were observed at low normal stresses, and the failure envelope could be represented using a semi-logarithmic relationship (Eq 2).

For the interface tests with kaolinite and anodized aluminum, the self-weight of the aluminum platen limited the lowest effective normal stress to about 40 Pa. The measured interface strength for aluminum (Fig. 6) was lower than that of kaolinite. Hence, even though about half of the tests visually seemed to involve failure through the soil, the measured strengths suggest that failure occurred by sliding at the interface.

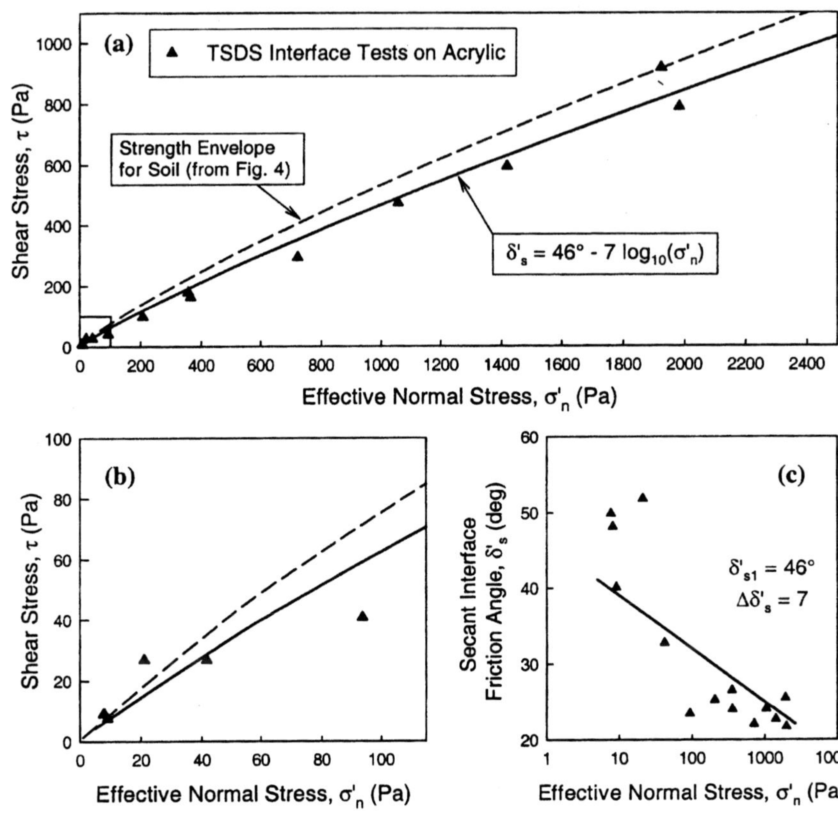

FIG. 5—*Drained strength of kaolinite-acrylic interface measured in TSDS tests.*

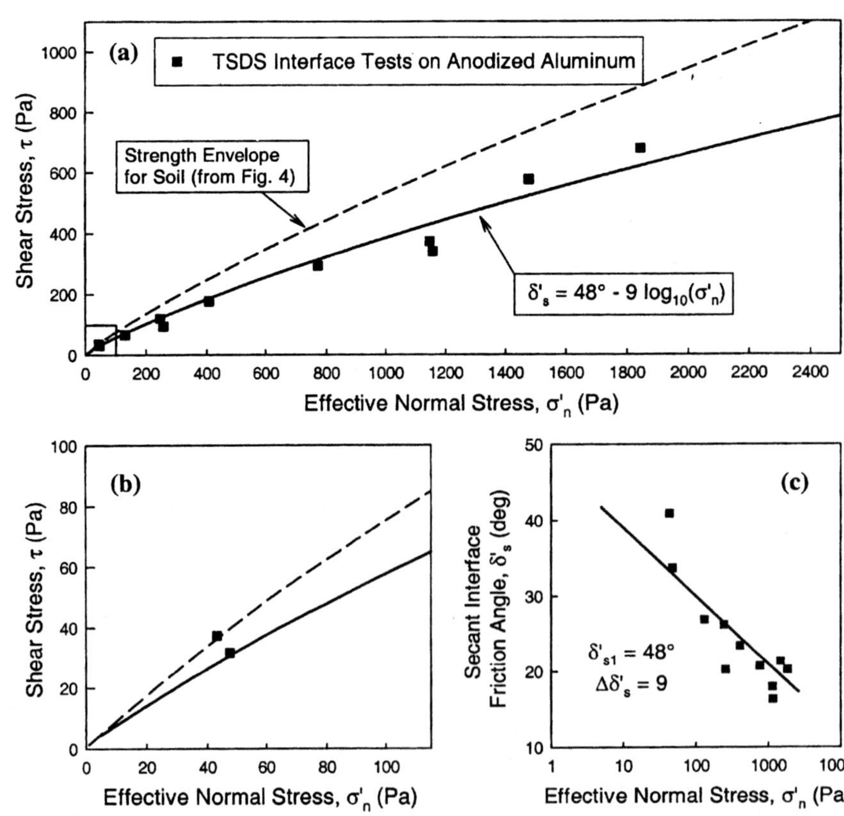

FIG. 6—*Drained strength of kaolinite-anodized aluminum interface measured in TSDS tests.*

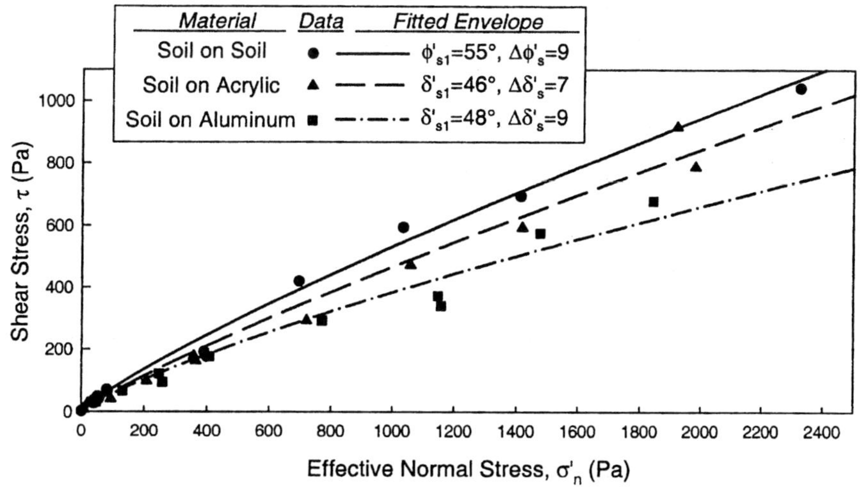

FIG. 7—*Comparison of kaolinite strengths measured in TSDS tests.*

Data from all of the TSDS tests are plotted together in Fig. 7, along with the fitted curved strength envelopes. Even with the scatter in this very low pressure test data, it is clear that the interface shear strengths for both acrylic and anodized aluminum are lower than the drained shear strength of kaolinite. Ratios of interface strength to kaolinite strength ranged from 0.7 to 0.9 for acrylic and 0.6 to 0.8 for aluminum. The TSDS tests confirm observations previously made during pullout tests of suction caisson models in the laboratory (El-Gharbawy 1998; El-Gharbawy and Olson 1999) that failure occurred at the caisson interface under drained loading conditions.

Summary and Conclusions

Low-pressure direct shear tests were performed using thin specimens of saturated kaolinite on a tilt-table apparatus. These "thin-specimen direct shear" (TSDS) tests were conducted to characterize the drained internal strength of kaolinite, as well as the shearing resistance of kaolinite on interfaces of smooth acrylic and anodized aluminum, at normal effective stresses less than 2.4 kPa (50 lb/ft^2). Observations from these test results include:

- The TSDS apparatus functioned well, although it was difficult to ascertain from visual observations whether failure occurred at the interface or in the clay.
- The effective-stress failure envelope of kaolinite is curved for pressures below 2.4 kPa. At the lowest test pressures, the measured secant friction angles for the soil and the interfaces, with values approaching 60°, were well in excess of the values that would be expected based on tests at higher pressures.
- The interface shearing resistance was generally less than the drained shear strength of the kaolinite.
- Whereas best-fit linear strength envelopes through the TSDS data give a cohesion or adhesion intercept, the curved envelopes pass through the origin. There was little opportunity for aging effects in these tests.

Analyses of laboratory-scale model tests should use soil properties determined in the appropriate stress range rather than properties extrapolated from tests at higher pressures. The difference between observed and extrapolated properties would be considerable for the kaolinite tested. To allow constitutive modeling of the soil response in model tests, research is needed to develop a test device that would allow measurement of stresses and strains in very soft soil specimens confined at low effective stresses.

Acknowledgments

This work was funded by the Offshore Technology Research Center, which is supported by the U.S. Minerals Management Service and the OTRC Industry Consortium.

References

Acar, Y. B., Durgunoglu, H. T., and Tumay, M. T., 1982, "Interface Properties of Sand," *Journal of the Soil Mechanics and Foundations Division*, American Society of Civil Engineers, Vol. 108, No. GT4, pp. 648–654.

Bishop, A. W., Webb, D. L., and Skinner, A. E., 1965, "Triaxial Tests on Soil at Elevated Cell Pressures," *Proceedings, 6th International Conference on Soil Mechanics and Foundation Engineering*, Vol. 1, pp. 170–174.

Bjerrum, L., 1954, "Theoretical and Experimental Investigations on the Shear Strength of Soils," Ph.D. dissertation, Federal Institute of Technology at Zurich, reprinted as Pub. 5 of the Norwegian Geotechnical Institute.

Bjerrum, L., 1973, "Problems of Soil Mechanics and Construction on Soft Clays and Structurally Unstable Soils," *Proceedings, 8th International Conference on Soil Mechanics and Foundation Engineering*, Moscow, Vol. 3, pp. 111–159.

Bosscher, P. J. and Ortiz, C., 1987, "Frictional Properties Between Sand and Various Construction Materials," *Journal of Geotechnical Engineering*, American Society of Civil Engineers, Vol. 113, No. 9, pp. 1035–1039.

Brouillette, R. P., Olson, R. E., and Lai, J. R., 1993, "Stress-Strain Characteristics of Eagle Ford Shale," *Proceedings, International Symposium on Hard Soils-Soft Rocks*, Athens, Greece, Vol. 1, pp. 397–404.

Bruce, I. G., Cruden, D. M., and Eaton, T. M., 1989, "Use of a Tilting Table to Determine the Basic Friction Angle of Hard Rock Samples," *Canadian Geotechnical Journal*, Vol. 26, No. 3, pp. 474–479.

Casagrande, A., 1940, "First Progress Report to the Waterways Experiment Station," Harvard University.

Casagrande, A., and Hirschfeld, R. C., 1960, "Stress-Deformation and Strength Characteristics of a Clay Compacted to a Constant Dry Unit Weight," *Proceedings, Research Conference on the Shear Strength of Cohesive Soils*, American Society of Civil Engineers, Boulder, Colorado, pp. 359–417.

Cawsey, D. C., and Farrar, N. S., 1976, "A Simple Sliding Apparatus for the Measurement of Rock Joint Friction," *Géotechnique*, Vol. 26, No. 2, pp. 382–386.

Chandler, R. J., and Hardie, T. N., 1989, "Thin-Sample Technique of Residual Strength Measurement," *Géotechnique*, Vol. 39, No. 3, pp. 527–531.

Clark, J. I., and Meyerhof, G. G., 1972, "The Behavior of Piles Driven in Clay: Part I, An Investigation of Soil Stress and Pore Water Pressure as Related to Soil Properties," *Canadian Geotechnical Journal*, Vol. 9, No. 4, pp. 351–373.

De Beer, E. E., 1970, "Experimental Determination of the Shape Factors and Bearing Capacity Factors of Sand," *Géotechnique*, Vol. 20, No. 4, pp. 387–411.

Duncan, J. M., Byrne, P., Wong, K. S., and Mabry, P., 1980, "Strength, Stress-Strain and Bulk Modulus Parameters for Finite Element Analyses of Stresses and Movements in Soil Masses," Report No. UCB/GT/80-01, University of California, Berkeley.

El-Gharbawy, S. L., 1998, "The Pullout Capacity of Suction Caisson Foundations for Tension Leg Platforms," Ph.D. dissertation, The University of Texas at Austin.

El-Gharbawy, S. and Olson, R. E., 1999, "Suction Caisson Foundations in the Gulf of Mexico," *Analysis, Design, Construction, and Testing of Deep Foundations*, American Society of Civil Engineers, Geotechnical Special Pub. 88, pp. 281–295.

Fishman, K. L., and Pal, S., 1994, "Further Study of Geomembrane/Cohesive Soil Interface Shear Behavior," *Journal of Geotextiles and Geomembranes*, Vol. 13, No. 9, pp. 571–590.

Fox, P. J., Rowland, M. G., and Scheithe, J. R., 1998, "Internal Shear Strength of Three Geosynthetic Clay Liners," *Journal of Geotechnical and Geoenvironmental Engineering*, American Society of Civil Engineers, Vol. 124, No. 10, pp. 933–944.

Gilbert, R. B., Liu, C. N., Wright, S. G., and Trautwein, S. J., 1995, "A Double Shear Test Method for Measuring Interface Strength," *Proceedings, Geosynthetics '95*, Vol. 3, Nashville, pp. 1017–1029.

Gilbert, R. B., Fernandez, F., and Horsfield, D. W., 1996, "Shear Strength of Reinforced Geosynthetic Clay Liner," *Journal of Geotechnical Engineering*, American Society of Civil Engineers, Vol. 122, No. 4, pp. 259–266.

Girard, H., Fischer, S., and Alonso, E., 1990, "Problems of Friction Posed by the Use of Geomembranes on Dam Slopes—Examples and Measurements," *Journal of Geotextiles and Geomembranes*, No. 9, pp. 129–143.

Hencher, S. R., 1976, "Discussion on: A Simple Sliding Apparatus for the Measurement of Rock Joint Friction," *Géotechnique*, Vol. 26, No. 4, pp. 641–644.

Jacobson, B., 1953, "Origin Cohesion in Clays," *Proceedings, 3rd International Conference on Soil Mechanics and Foundation Engineering*, Vol. 1, pp. 35–37.

Jurgenson, L., 1934, "Shearing Resistance of Soils," *Journal of the Boston Society of Civil Engrs.*, July, reprinted in *Contributions to Soil Mechanics, 1925–1940*, Boston Society of Civil Engineers, 1940.

Kenney, T. C., Moum, J., and Berre, T., 1967, "An Experimental Study of Bonds in a Natural Clay," *Proceedings Geotechnical Conference*, Oslo, Norway.

Kishida, H., and Uesugi, M., 1987, "Tests of the Interface Between Sand and Steel in the Simple Shear Apparatus," *Géotechnique*, Vol. 37, No. 1, pp. 45–52.

Koerner, R. M., Martin, J. P., and Koerner, G. R., 1986, "Shear Strength Parameters Between Geomembranes and Cohesive Soils," *Journal of Geotextiles and Geomembranes*, Vol. 4, No. 1, pp. 21–30.

Lalarakotoson, S., Villard, P. and Gourc, J. P., 1999, "Shear Strength Characterization of Geosynthetic Interfaces on Inclined Planes," *Geotechnical Testing Journal*, Vol. 22, No. 4, pp. 284–291.

Lee, K., and Seed, H. B., 1967, "Drained Strength Characteristics of Sands," *Journal of Soil Mechanics and Foundation Division*, American Society of Civil Engineers, Vol. 93, No. SM6, pp. 117–141.

Lehane, B. M., and Jardine, R. J., 1996, "Shaft Capacity of Driven Piles in Sand: A New Design Approach," *Proceedings, Conference on Behavior of Offshore Structures*, pp. 23–36.

Littleton, I., 1976, "An Experimental Study of the Adhesion Between Clay and Steel," *Journal of Terramechanics*, Vol. 13, No. 3, pp. 141–152.

Merrill, K. S., and O'Brien, A. J., 1997, "Strength and Conformance Testing of a GCL in a Solid Waste Landfill Lining System," *Testing and Acceptance Criteria for Geosynthetic Clay Liners, ASTM STP 1308*, L. W. Weill, Ed., pp. 71–88.

Olson, R. E., 1974, "Shearing Strengths of Kaolinite, Illite, and Montmorillonite," *Journal of Geotechnical Engineering Division*, American Society of Civil Engineers, Vol. 100, No. GT 11, pp. 1215–1229.

Olson, R. E., Rauch, A. F., Gilbert, R. B., Tassoulas, J. L., Aubeny, C. P., and Murff, J. D., 2001, "Toward the Design of New Technologies for Deep-Water Anchorages," *Proceedings, 11th International Offshore and Polar Engineering Conference, ISOPE-2001*, Stavanger, Norway, June.

O'Rourke, T. D., Druschel, S. J., and Netravali, A. N., 1990, "Shear Strength Characteristics of Sand-Polymer Interfaces," *Journal of Geotechnical Engineering*, American Society of Civil Engineers, Vol. 116, No. 3, pp. 451–469.

Pedersen, R. C., 2001, "Model Offshore Soil Deposit: Design, Preparation, and Characterization," M.S. thesis, The University of Texas at Austin, May.

Ponce, V. M., and Bell, J. M., 1971, "Shear Strength of Sand at Extremely Low Pressures," *Journal Soil Mechanics and Foundation Division*, American Society of Civil Engineers, Vol. 97, No. SM4, pp. 625–638.

Potyondy, J. G., 1961, "Skin Friction Between Various Soils and Construction Materials," *Géotechnique*, Vol. 11, No. 4, pp. 339–353.

Rauch, A. F., Olson, R. E., Mecham, E. C., and Pedersen, R. C., 2001, "A Laboratory Facility for Testing Model Suction Caissons," *Proceedings, OTRC 2001 International Conference—Geotechnical, Geological and Geophysical Properties of Deepwater Sediments*, Houston, April, pp. 198–216.

Reddy, E. S., Chapman, D. N., and Sastry, V. V. R. N., 2000, "Direct Shear Interface Test for Shaft Capacity of Piles in Sand," *Geotechnical Testing Journal*, ASTM, Vol. 23, No. 2, pp. 199–205.

Seed, R. B., and Boulanger, R. W., 1991, "Smooth HDPE-Clay Liner Interface Shear Strengths: Compaction Effects," *Journal of Geotechnical Engineering*, American Society of Civil Engineers, Vol. 117, No. 4, pp. 686–693.

Shan, H.-Y., 1993, "Stability of Final Covers Placed on Slopes

Containing Geosynthetic Clay Liners," Ph.D. dissertation, University of Texas at Austin, August.

Stark, T. D. and Poeppel, A. R., 1994, "Landfill Liner Interface Strengths from Torsion-Ring-Shear Tests," *Journal of Geotechnical Engineering,* American Society of Civil Engineers, Vol. 120, No. 3, pp. 597–615.

Tavenas, F., Chagnon, J.-Y., and La Rochelle, P., 1971, "The Saint-Jean-Vianney Landslide, Observations and Eyewitnesses Accounts," *Canadian Geotechnical Journal*, Vol. 8, No. 3, pp. 463–478.

Terzaghi, K., 1925, "Principles of Soil Mechanics: VII—Friction in Sand and Clay," *Engineering News-Record,* Vol. 95, No. 26, pp. 1026–1029.

Tsubakihara, Y., and Kishida, H., 1993, "Frictional Behavior Between Normally Consolidated Clay and Steel by Two Direct Shear Type Apparatuses," *Soils and Foundations,* Vol. 33, No. 2, pp. 1–13.

Tsubakihara, Y., Kishida, H., and Nishiyama, T., 1993, "Friction Between Cohesive Soils and Steel," *Soils and Foundations,* Vol. 33, No. 2, pp. 145–156.

Uesugi, M., and Kishida, H., 1986a, "Influential Factors of Friction Between Steel and Dry Sands," *Soils and Foundations,* Vol. 26, No. 2, pp. 33–46.

Uesugi, M., and Kishida, H., 1986b, "Frictional Resistance at Yield Between Dry Sand and Mild Steel," *Soils and Foundations,* Vol. 26, No. 4, pp. 139–149.

Uesugi, M., Kishida, H., and Tsubakihara, Y., 1988, "Behavior of Sand Particles in Sand-Steel Friction," *Soils and Foundations,* Vol. 28, No. 1, pp. 107–118.

Uesugi, M., Kishida, H., and Uchikawa, Y., 1990, "Friction Between Dry Sand and Concrete Under Monotonic and Repeated Loading," *Soils and Foundations,* Vol. 30, No. 1, pp. 115–128.

Vesić, A. S., and Clough, G. W., 1968, "Behavior of Granular Materials Under High Stresses," *Journal of Soil Mechanics and Foundation Division*, American Society of Civil Engineers, Vol. 94, No. SM3, pp. 661–688.

Yoshimi, Y. and Kishida, T., 1981, "A Ring Torsion Apparatus for Evaluating Friction Between Soil and Metal Surfaces," *Geotechnical Testing Journal,* ASTM, Vol. 4, No. 4, pp. 145–152.

AXIAL LOAD CAPACITY OF PILES IN CALIFORNIA IN COHESIONLESS SOILS

Roy E. Olson, Life Member, ASCE, P.E.[1] and Thomas J. Shantz, Member, ASCE, P.E.[2]

ABSTRACT: The literature abounds with methods to estimate the axial tensile or compressive load capacity of driven piles. Most of these methods have disadvantages associated with undue complexity, non-linear relationships, lack of sufficient supporting load test data, use of out-of-date technology, or simply lack of desired accuracy. The California Department of Transportation performed a series of load tests on driven piles for bridges, over a period of years, but lacked the kinds of soil information needed to extend their experience to new sites, and did not have the data in easily accessible forms. A project was initiated to make new soil borings and cone soundings at a number of sites and to incorporate data into an electronic database for ready access. There was a further interest in examining the accuracy of existing methods for predicting capacity and of developing estimation procedures that were fitted to experience in California. This paper is concerned with pile tests in cohesionless soils, ranging from silts through clean sands, but excluding sites with soils classified as gravelly sands or coarser. Experience indicates that the margins of error for predictions in cohesionless soils is substantially in excess of that for cohesive soils. The considered pile types are open-end and closed-end steel pipe piles and precast concrete piles. H piles were included in the investigation but they all encountered gravels and are thus excluded from this paper.

INTRODUCTION

Over the last 40 years, the California Department of Transportation (Caltrans) has performed several hundred load tests on driven piles. These load tests were largely "proof" tests that were performed to verify whether or not the installed pile met

[1] Emeritus Professor, Department of Civil Engineering, University of Texas, Austin, TX 78712, olson@mail.utexas.edu

[2] Senior Research Engineer, Caltrans, 5900 Folsom Blvd., Sacramento, CA 95819, tom.shantz@dot.ca.gov

design load requirements. A smaller subset of these tests was performed as part of preconstruction load-test programs that usually consisted of several load tests at one or more project locations. These load-test programs typically considered multiple tip elevations and also evaluated construction issues such as predrilling or use of vibratory hammers. On occasion, different pile types or end condition (open or closed) were evaluated. In all cases, Caltrans' load-testing measurements were limited to the top of pile. No instrumentation was included that allowed inference of tip loads or distributions of shaft load with depth.

Engineers at Caltrans recognized the potential value of these tests in evaluating existing procedures to estimate pile capacity as well as possible usefulness in developing improved procedures. A project was devised to develop an electronic database that would facilitate analysis of the load-test data. It was recognized that a key shortcoming of the load-test data was generally inadequate site characterization information. Caltrans practice in the past has generally down-played the importance of obtaining accurate soil properties as part of the site exploration program. It was rare, for example, for load-test reports to have strength tests on clay samples beyond Standard Penetration Test (SPT) N-values and estimates of undrained shearing strengths from pocket penetrometers or Torvanes. Furthermore, project sites were often large, resulting in borehole densities of the order of one per hundred meters. It was not uncommon for the nearest boring to the load-tested pile to be in excess of 30-meters. Thus, it was decided that an important component of the project would be site investigations. These investigations would include "undisturbed" sampling and laboratory testing of clay layers along with use of an energy-calibrated automatic SPT hammer for characterization of cohesionless layers. Frequent use of the cone penetrometer would also be made.

Caltrans contracted with University of Texas at Austin to perform three major components of the project: database development and population, site investigation and laboratory testing, and data analysis. Additional support contracts with a California drilling firm and a Texas-based soil testing lab were also utilized.

The load-test dataset is here discussed, along with some of the many issues one faces in evaluating and analyzing load-test data. Initial findings from the authors' on-going analytical work is also included.

CALTRANS DATASET

The Caltrans dataset consists of 319 static load-tests on 227 test piles at 75 bridge locations. Some bridge locations had tests at multiple footing (pile cap) locations. In all, 160 different footing locations were tested. The oldest tests were performed in the early 1960's and were compression tests. More recent testing frequently included both uplift and compression loading (considered here to be separate tests) or just uplift loading.

After a critical review of the data, only 130 tests were accepted for further analysis. Of these, 89 were steel pipe piles (23 closed-end and 66 open-end), 24 were precast concrete, and 6 were H piles. Ten tests were in profiles that were entirely clay, 26 in entirely cohesionless soils, and 81 were in mixed profiles. There were 55 compression tests and 64 tension tests. The piles ranged in diameter/width from 250

FIG. 1. Test Sites

mm (10 inches) to 1.8 m (72 inches), and penetrations ranged from 7.3 m (24 feet) to 35 m (114 feet). For piles in predominantly cohesive soils, reported set-up times ranged from 4 days to 76 days but some times were unreported.

Most of the test sites were in heavily populated areas along the coast (Fig. 1). Test sites with mostly cohesionless soils were concentrated around Los Angeles and San Diego but with one site in Oakland. Clay sites were mostly from the San Francisco Bay Area. The majority of soil profiles included layers of both sand and clay. Overall, the dataset tends to be dominated by test-sites in Los Angeles, San Diego, and the San Francisco Bay area.

CHALLENGES IN WORKING WITH PILE LOAD-TEST DATA

Failure Criteria

It is common for load-tests to be terminated before the load reaches a well-defined maximum (failure) value. Typical reasons are lack of jack or reaction frame capacity and the fact that many tests are terminated once design capacity is verified. Furthermore, piles in mostly cohesionless soils may not achieve plunging failure at any rational butt settlement.

Many definitions of pile capacity have been suggested. The four definitions of "failure" used in developing the Caltrans database, were load at a butt movement of 12.5 mm (0.5 inch) (Caltrans criterion), Davisson's (1972) load, the extrapolated failure load (Brown, 2001), and the peak applied load. A restricted form of Brown's (2001) method was used here.

Brown used a variation on Fleming's (1992) hyperbolic method in which:

$$Q = \frac{S_p}{mS_p + b} \tag{1}$$

where Q is applied load, S_p is plastic pile head movement, and m and b are fitted parameters. The plastic movement is the total head movement minus the elastic movement of the pile acting as a free-standing column. Then:

$$Q_m = \frac{1}{m} \tag{2}$$

and Q_m is the "measured" pile capacity. Once m and b were evaluated for each test, the fitted curve was plotted on top of the actual measured curve. In most cases, the extrapolated load and the peak applied load were within 1% of each other because data were excluded at an early stage if the load-settlement curve did not show clear evidence of failure. In a few cases in which there was a significant thickness of clay, the applied load dropped off substantially after the peak, and then the extrapolated load exceeded the clearly defined peak load. In those cases, the peak load was used.

For the tests used in the current set of analyses, the Caltrans load averaged 93% of the failure load (range 81% to 100%) and Davisson's load averaged 96% of the failure load (range was 86% to 100%).

Evaluation of Data Quality

Older load-tests often utilized jack pressures to estimate the applied load, a practice recognized to overestimate actual loads (Davisson, 1958; Dunnicliff, 1988). Fortunately, it has been Caltrans' practice, for at least 25 years, to use calibrated electronic load cells and redundant LVDTs for the development of the load-deflection data. All but the oldest available load-test reports are thought to have accurate load-deformation curves.

Site characterization is a more serious challenge. Available boring or sounding information may be at such a distance from the test pile that they do not accurately reflect site conditions. Even when nearby borings are available, issues of inadequate material characterization remain. Laboratory strength testing of clays was unusual in the original load-test data. For cohesionless soils, Caltrans performed standard penetration tests using mostly safety hammers but hammer energies still probably varied among different operators.

Construction practices can also lead to uncertainty. A significant percentage of the tested piles were installed in predrilled holes 1- to 2-meters deep. Others may have been installed in shallow predrilled holes but the fact was not documented. Fortunately, the effect of shallow predrilling on capacity is small However, some piles were installed in deeper predrilled holes. These holes were usually drilled to diameters less than the effective diameter of the pile and in some cases the holes were reported to be partially collapsed. In these cases, predrilling may have had a significant impact on capacity.

It was also common for piles to be installed in the bottom of footing excavations. We used Boussinesq stress distribution to account for the associated reduction in vertical stress for pile analysis. The N values used in our analyses were the measured ones corrected for hammer energy, e.g., N_{60}. If a particular analytical method utilized N values corrected to a standard pressure, e.g., $N_{1,60}$, then that correction was applied as well.

SUPPLEMENTAL SITE EXPLORATION PROGRAM

Fifty borings and 58 cone soundings were made at various test locations as part of the current investigation. These locations were selected based on the number and quality of load tests at that site, the availability of existing site information, and practical issues such as site access.

Field and laboratory testing methods were meant to reflect typical practices employed by consulting firms or DOT's for the design of bridge foundations. Samples of cohesive soils were obtained using 75 mm (3-inch) diameter thin-walled soil samplers (Shelby tubes). Standard penetration tests using an energy calibrated hammer were performed in cohesionless soils. Sampling or testing normally occurred at 1.5 m (5-foot) intervals. Original plans called for a large number of piezo-cone soundings. Unfortunately, at some sites the cones could not be pushed to the depth of the pile tip due to the presence of dense gravel layers or cobbles (particularly in Southern California), or obstruction-laden urban fill (particularly in San Francisco).

The 75 mm (3-inch) Shelby tube samples were sealed in the tubes in the field, placed in custom-built, padded, wood boxes that maintained the vertical orientation of the tubes. Boxes were stored in Caltrans facilities for up to about a week and were then driven back to UT for testing. A portion of the testing was subcontracted to a commercial soils laboratory.

The laboratory testing of cohesive soils involved unconsolidated-undrained triaxial compression tests on samples with trimmed ends but no trimming on the sides. This procedure is in accord with usual engineering practice and it was mandated by both budget and time considerations. Water contents and Atterberg limits were also measured, and a few one-dimensional consolidation tests were performed. For cohesionless soils, a number of grain size analyses were performed.

ANALYSIS - INITIAL FINDINGS

The emphasis in this paper is on piles in cohesionless soils because experience shows that uncertainties are substantially larger for piles in cohesionless soils as opposed to piles in cohesive soils (Pelletier et al., 1993). To be of immediate practical application, the plan was to focus on methods based on standard *in situ* tests, i.e., standard penetration and piezo-cone tests. Only the standard penetration data are used in this paper.

Meyerhof (1976) used standard penetration resistances in predicting pile capacities and we have used them in unpublished work since about 1980, and they have been used in other published work (Decourt, 1982). Meyerhof assumed linear relationships between both side shear and end bearing with the N value but Decourt included an intercept. The formulation considered here is:

$$f_s=A_s+B_sN_{60} \tag{3}$$

where f_s is the pile/soil shearing stress at failure, N_{60} is the standard penetration resistance at a hammer efficiency of 60%, and A_s and B_s are fitted parameters for side shear, the numerical values of which depend on the selected system of units. For piles in compression:

$$q_p=A_p+B_pN_{60} \tag{4}$$

where q_p is the pile tip stress, and A_p and B_p are fitted parameters, with numerical values that depend on the selected system of units. We have used other forms of the equations, e.g., using log(N) (Olson and Iskander,1984) but only the simple linear relationship is considered here.

In the analyses, we made corrections for pile weight. Further, Eq. 4 was applied to the net tip capacity, i.e., q_p is the stress in addition to the total free-field vertical stress in the soil at that depth. For piles in tension, if the tip was below the water table, we assumed that there would be an upwards water pressure on the tip but that the cohesionless soil would just fall into any water-filled void left when the tip displaced upwards.

Scatter

In evaluating methods for estimating pile capacity it is usual practice to compare the calculated capacity, Q_c, with the measured capacity, Q_m, usually in the form of the ratio Q_c/Q_m. The mean and scatter of Q_c/Q_m can be expressed using a "normal distribution" but the distribution is actually log normal. Hereafter, we will use the terms Gaussian mean and Gaussian standard deviation (SD) when that distribution is used and log-normal mean and log normal standard deviation (SDL) when a log normal distribution is used. When the scatter is small, the two means and standard deviations become essentially the same. With a Gaussian distribution, it is convenient to use the coefficient of variation (COV) to provide a dimensionless distribution. To make the means appear in more familiar form, the log-normal mean Q_c/Q_m will be exponentiated so the sought value is 1.0 instead of 0.0, but it will still be referred to as the log-normal mean. For use with values of Q_c/Q_m, the log-normal mean is always less than the Gaussian mean but the difference is small when the scatter is small, e.g., in the best cases the log-normal mean was 1.00 and the Gaussian mean was 1.01.

Restrictions on Soil Layers

Analyses were performed with the requirement that at least 60% of the side capacity had to come from cohesionless soils. Tests were not included in current analyses if even one soil layer was classified as gravelly sand or coarser.

Restrictions on Pile Installation Methods

Tests were excluded if the pile installation involved use of vibration, jetting, or relief drilling.

Restrictions on Pile Types

All piles used in the study were steel pipe piles, solid precast concrete piles, or H piles, but all of the H piles penetrated gravel layers and are thus not included in this report. No piles with diameters in excess of 1 m (3 feet) were included. None of the piles were tapered. There were no timber or cast-in-place concrete piles. There were several Raymond cylinder piles but they were installed using jetting and were thus excluded as well.

Analyses in Clay Layers

The majority of tests in the Caltrans database are for piles in mixed soil profiles so any attempt to estimate capacities in cohesionless soils requires estimates for layers of clays as well. A single, relatively unbiased method of analysis was required for use in clay layers. Several methods were considered. The details of each method were presented by Brown (2001).

Two independent studies were performed, using load tests in which at least 80% of the capacity came from clays. In the first, Brown (2001) compared measured capacities with predictions from eight methods in clay and five in sand, in all combinations. In that study, the coefficient of variation of Q_c/Q_m was least for the method of Dennis and Olson (1983a) but most of the other methods were close. The mean of Q_c/Q_m was closest to one for the Dennis/Olson method. A subsequent study, performed after a more extensive study of the data, and development of a somewhat different database, led to the same conclusion (Table 1). In these studies, the method used in sand layers was of limited importance because so little of the calculated capacity came from sand. The SDL was lowest for the Dennis/Olson method, which was also among the easiest to use, and it was adopted for use with clay layers.

Table 1. Comparisons of Mean Values of Q_c/Q_m for Different Combinations of Methods used to Estimate Axial Capacities in Cohesive and Cohesionless Soils, using the Caltrans Database. The method for clay layers is across the top row and the methods in sand are shown in column two.

Loading (1)	Method in Sand (Below) (2)	API (1990) (3)	Dennis and Olson (1983a) (4)	Tomlin-son (1971) (5)	Kraft et al. (1981) (6)	Karlsrud (1999) (7)
Tension	Nordlund (1979)	1.31	1.04	1.20	1.35	1.18
	API (1990)	1.30	1.03	1.19	1.34	1.17
	Decourt(1982)	1.31	1.04	1.21	1.35	1.19
	Olson (1990)	1.31	1.04	1.20	1.35	1.18
Comp-ression	Nordlund (1979)	1.17	0.95	1.12	1.23	1.11
	API (1990)	1.16	0.94	1.11	1.22	1.10
	Decourt (1982)	1.20	0.97	1.14	1.26	1.13
	Olson (1990)	1.19	0.97	1.14	1.25	1.13

Analyses at "Sand Sites"

Values were sought for A_s, B_s, A_t, and B_t in Eqs. 3 and 4. It was necessary to subdivide load test data, e.g., to separate open-end and closed-end steel pipes, and to separate tension tests from compression tests. However, excessive subdivision tended to lead to unreasonably small data sets and sometimes data from only one or two sites, in which case irrational coefficients could be found that fit the small data set well but would not apply to more typical datasets.

TYPICAL DATA

The first need was to establish some sort of basis for comparative purposes. The authors have performed studies of this type in the past and numerous dataset and analytical methods can be used with varying results. For current purposes, three methods were used. The API (1990) method was selected because it is a well known standard that has been used throughout the world, but it is applicable mainly to open-end steel pipe piles. Caltrans uses steel pipe piles for many bridge structures and thus the API method applied to studies reported here. The method published by Nordlund (1963) and apparently modified in an unpublished set of class notes in 1979, was selected because it has been recommended for use by the FHWA. Finally, the method of Olson and Dennis (1983b) was used because we used it extensively with a larger database in the 1980s.

None of the tests in the Caltrans database were used in development of any of the three methods and thus the analyses are unbiased. The Gaussian distribution was used because it was used when the method was developed. The database used for this comparison contained only 35 tests (Table No. 2) because of the restrictions previously discussed.

Table 2. Comparison of Three Analytical Methods for Piles in Compression and Tension, using the Caltrans Database

Load (1)	Num. of Tests (2)	API (1990) Q_c/Q_m (3)	API (1990) COV (4)	Nordlund (1963) Q_c/Q_m (5)	Nordlund (1963) COV (6)	Olson (1990) Q_c/Q_m (7)	Olson (1990) COV (8)
Comp	18	0.60	0.24	0.80	0.46	0.67	0.42
Tens	17	0.68	0.34	0.99	0.37	0.66	0.38

These data indicate a general underprediction of pile capacity. A possible explanation is that these methods were all based on use of standard penetration resistances but for almost all of the source tests, the hammer energies were unknown and probably less than 60%. Hammers were generally manually operated and in many cases the rod was 19-mm (¾-inch) heavy duty water pipe, which flexed considerably during driving and for deep measurements probably struck the side walls and dissipated energy. Because the methods chose fitting parameters to make Q_c/Q_m equal to one, the lowered hammer energy led to increased N values and thus lowered fitting coefficients. When these lowered coefficients are used with current data, where the hammer efficiency was 60%, they result in underprediction of pile capacity.

In addition to ensuring a mean Q_c/Q_m close to one and a small COV, it is also important to demonstrate that the value of Q_c/Q_m does not vary systematically with other variables. Our experience is that most analytical methods underpredict the capacities of short piles and overpredict the capacities of long piles. The API method is typical in that respect (Fig. 2) indicating a fundamental bias in the model used.

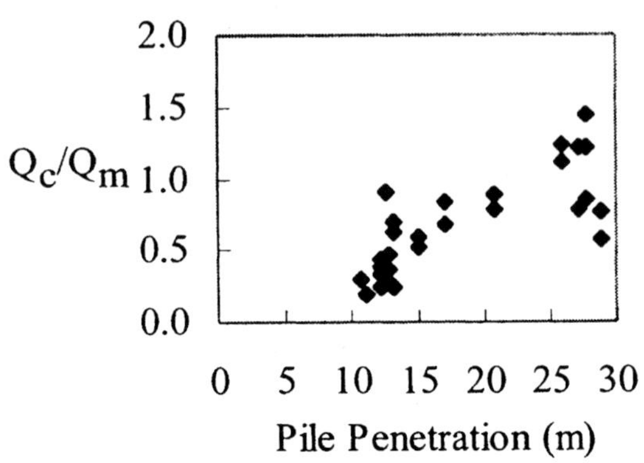

FIG. 2. Effect of Pile Penetration on Q_c/Q_m for Open-End Piles, Mostly in Sand, Based on API(1990)

Plots similar to Fig. 2 were generated for the other methods and for almost all subsets, i.e., steel pipes and precast concrete piles separately, for compression and tension separately, and closed-end and open-end piles separately.

For the fitting used in this investigation, piles were separated into groups based on pile type (steel pipe, concrete), end condition (open or closed/solid), and direction of loading (tension or compression) and attempts were made to find the best fit values of A_s, B_s, A_t, and B_t for each group (Table No. 3). The fitting was performed by

Table 3. Best-Fit Parameters (Trial is a number to tie data in this table to data in the next table, SP denotes a steel pipe pile, PC denotes a precast concrete pile)

Trial	Pile Type (1)	Load Dir. (2)	End Cond. (3)	Num of Tests (4)	Equation *(kPa)* (5)	SDL (6)	Num. of Sites (7)
22	SP	comp	closed	1	f_s=31+3.11N	0.15	3
22	PC	comp	solid	6	q_p=36+312N		
23	PC	tens	solid	2	f_s=11+2.30N	0.24	2
24	SP	tens	closed	3	f_s=14+3.50N	0.17	1
25	SP	tens	open	10	f_s=9+2.39N	0.23	5
26	SP	comp	open	14	f_s=24+3.59N q_p=77+57.5N	0.22	6

selecting a range in each parameter, plus the number of trial values in that range, and performing a full set of analyses for each combination of the trial values. Data were output to a spreadsheet and sorted, first using Q_c/Q_m and then SDL. In most cases there were numerous combinations of parameters in which Q_c/Q_m was 1.000 but there would be one set of parameters with the minimum SDL. A high degree of accuracy cannot be ascribed to the results because the number of sites in each category was small. A partial compensation for the small number of sites was the fact that special borings were made near each load test and considerable effort was made in interpretation of the soil profiles using visual descriptions, driving records, and cone soundings.

There are some striking features of the data in Table 3, including:

1. The number of tests in each group was small considering that 108 tests satisfied our quality criteria. The small number is the result of the application of the

restrictions discussed previously. For tests that were carried reasonably to failure, the largest cause of rejection here was the presence of too much clay.

2. A second feature was the small logarithmic standard deviation (SDL). The small value is often associated with having a group of tests at the same site, or in similar deposits in the same city but it also suggests that the assumed functional relationship between side shear or end bearing and the standard penetration resistance was reasonable.
3. A third feature of the data is the presence of an intercept in all cases. Meyerhof (1986) interpreted his data to indicate a linear relationship between f_s or q_p and N, with no intercept. However, it seems that there must be an intercept. Even if N approaches zero, there is still a vertical effective stress, there is still a value of the lateral earth pressure coefficient K, and the effective friction angle of the sand is unlikely to drop below some minimum value, e.g., 29 deg. for loose fine siliceous sand. Thus, there must be both a side and tip capacity even for a loose sand with N near zero.

Considering the relatively small number of tests involved in the correlations, and the even smaller number of sites, a question arises as to whether the differences are real or just represent conditions at different sites that cannot be quantified using just the standard penetration resistance. The answer to this question will always be vague but it will be clarified as additional data, of comparable quality, are added to the database. However, ranges in pile penetration, pile diameter, N values, and fraction of the calculated side capacity in cohesionless soil (silt through clean sand but no coarser soil) are included in Table No. 4 to aid potential users in deciding whether or not a new site is tolerably represented in this data set.

Table 4. Details of Pile Penetrations (L), Diameter (D), N Value (N), and Fraction of Calculated Side Capacity from Cohesionless Soil (Sand)

Trial	L (m)			D (mm)			N (b/30 cm)			Sand (%)		
	Min	Max	Ave	Min	Max	Ave	Min	Max	Ave	Min	Max	Ave
22	7.3	14.0	10.7	280	380	356	4	51	17	77	100	97
23	7.9	12.5	10.1	356	356	356	6	52	23	100	100	100
24	10.7	13.1	11.9	610	610	610	86	86	86	80	87	84
25	12.2	29.3	20.7	356	610	483	8	90	28	75	100	92
26	11.0	29.3	18.9	356	610	508	8	167	36	73	100	93

It is possible to calculate side shears and end bearing using the different coefficients in Table No. 3 and make direct comparisons to get estimates of the effect of having open or closed ends, or tension versus compressive side capacities. Caution is required, however, because the number of sites represented by individual formulas is typically small, and there are likely to be factors built into empirical correlations that are not known. Furthermore, the closed-end pipe piles all had oversized cover plates which probably cause a reduction in side shear, at least near the pile tip. If the Q_c/Q_m ratio varies with depth, as for many methods, then comparisons between two sites where the pile penetrations vary will lead to distorted conclusions about some compensating effect. The number of tests is insufficient to sort out such diverse effects so we tend to emphasize effects where we have previous experience in support of a certain view.

Compare Tensile Side Shears for Open-End and Closed-End Piles

Consider open-end steel pipe piles in tension and compression. Values of calculated side shear can be tabulated as a function of N, using equations in Table No. 3, and the ratio of side shears in open-end piles to closed-end piles can be compared. O'Neill (2001) summarized data from recent years and found that some researchers or designers used a ratio of 1.0 whereas others argued for ratios around 2/3's. From the equations in Table No. 3, the ratio of the calculated side shear in tension to that in compression ranges from 0.67 for N=10 bpf to 0.68 for N=100 bpf, which supports the argument that the proper ratio is around 2/3's.

Compare Side Shear in Compression for Open-End and Closed-End Piles

The ratio of the compressive side shear for open-end steel pipe piles (14 tests) to that of closed-end steel pipe piles (1 test) plus precast concrete piles (6 tests), ranges from 0.96 for N=10 bpf to 1.12 for N=100 bpf. It seems very unlikely that side shear in compression will be more for open-end steel pipes than it is for closed-end steel pipes. Data summarized in Table No. 4 indicate that there are significant differences between piles in the two groups but the influence of these differences on results is unclear.

Compare End Bearing for Open-End and Closed-End Piles

The data just discussed can also be used to compare end bearing pressures for open-end and closed-end piles but with the same problems. The ratio of the calculated end bearing stress for open-end piles to that of closed-end piles ranged from 0.21 for N=10 bpf to 0.19 for N=100 bpf. That ratio seems excessively low. It would be possible to increase the end bearing of the open-end piles and reduce the side shear and still maintain a mean Q_c/Q_m close to one but we chose not to make such subjective adjustments to the data. There was one test on a 1.8-m (72-inch) diameter open-end steel pipe pile (not used here) where the data suggest that there was a negligible end bearing and that same situation may apply to open-end pipe piles of smaller diameter but further analysis, and hopefully additional load tests, may be needed to clarify this issue.

Compare Side Shear for Open-End Piles - (Tension/Compression)

In this case, there were 10 tests on open-end pipes in tension and 14 tests on open-end pipes in compression so the comparison may be more meaningful. The ratio of calculated side capacity in tension to that in compression ranges from 0.55 for N=10 bpf to 0.65 for N=100 bpf. The lower side shear in tension is what many investigators have suggested.

Compare Side Shear for Closed-End Piles - (Tension/Compression)

There were only three tests on open-end pipe piles in tension (Trial 25) and six tests on open-end pipe piles in compression (Trial 26) and the N values were considerably higher for the tests in tension (Table No. 4) so the comparison has a reduced probability of being valid. The ratio of calculated side shear in tension to that in compression ranged from 0.79 at N=10 bpf to 1.06 for N=100 bpf.

EFFECT OF PILE PENETRATION ON Q_c/Q_m

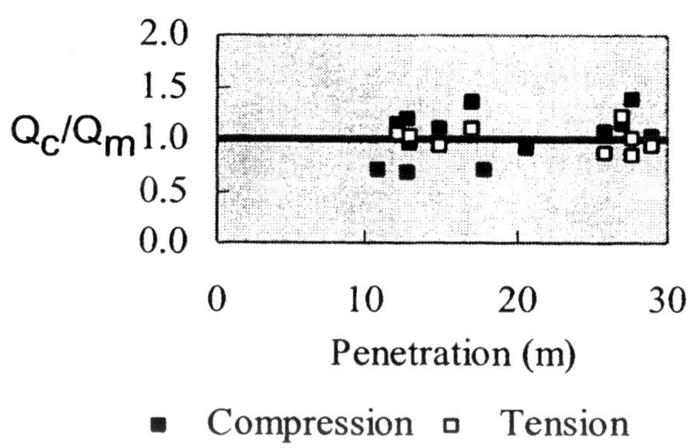

FIG. 3. Effect of Pile Penetration on Q_c/Q_m when the NVal Method is Used

When the fitted equation, using N values, is used for open-end steel pipes in tension and compression, the ratio Q_c/Q_m does not vary with pile penetration (Fig. 3) in contrast to our experience with most other methods (Fig. 2). This observation suggests that the formulation of the method is likely to prove useful.

SUMMARY AND INFERENCES

Databases for pile load tests are expensive and time consuming to develop but they are essential if experience from tests at one site are to be useful in preliminary designs at a different site. When existing load test data are used, the number of close soil borings may be insufficient for more careful consideration so new soil boring programs or cone soundings may be required for development of a rational database. It is also sometimes found that tests that seemed acceptable during construction are less useful for current purposes and it would have been better if load tests were carried reasonably to plunging failure. Plunging failures do not usually occur for compression tests on piles in cohesionless soils but tests with butt movements of up to say 50 mm (2 inches) usually suffice.

As a result of the required effort and expense, the number of tests of sufficient quality to be included in the data analysis was limited to 108. Furthermore, when efforts are made to separate out the effects of the multitude of variables that are believed to influence pile capacity, the number of tests in any one logical grouping tends to be small and comparisons between different groups may be strongly influenced by variations in pile penetration, pile diameter, mean N values, clay content, or other factors not necessarily understood currently. Grouping tests leads to larger samples but it obscures the effects being sought. With the sample sizes available in this study, it is believed that the ability of Caltrans to estimate pile capacities is enhanced significantly. The analytical approach used here may also prove useful elsewhere. However, the applicability of the predictive methods to diverse sites remains to be examined.

Most researchers consider the standard penetration test to be crude, perhaps too crude to be useful in estimating pile capacity. Indeed, we found wide ranges in N values within single borings and between adjacent borings. However, the N values for most of these tests were carefully measured and the hammers were calibrated for energy so most of the scatter is believed to result from inherent variations in sand density. Further, there is an averaging effect because the piles penetrate through a number of zones with different N values. In spite of the criticisms of researchers, there are advantages to the standard penetration test in that it yields samples for visual classification and, we could use it at depths and sand densities that did not allow penetration by a cone. We would thus argue that the standard penetration test in cohesionless soils is a useful and widely available test.

Engineers must recognize that correlations presented here were for siliceous sands. These correlations will not apply to calcareous sands, nor to volcanic sands, and probably not to residual sands formed by *in situ* weathering of igneous rocks.

In this presentation, we have assumed a linear relationship between side shear or end bearing and the N value. We have previously considered using log(N) and other functional relationships may also warrant trials.

In spite of the size of the database, and the restricted geologic environments, some aspects of the correlations look promising. The ratio of side shears for open-end pipe piles in tension to the values in compression were similar to the 2/3's value often assumed. The tip capacities that appear to be smaller for open-end piles compared to closed-end piles also fit with previous experience but the differences found here seem excessive and may result, in part, from use of a small database

The fact that the ratio Q_c/Q_m typically varies substantially with pile penetration for various other methods, but is acceptably constant when N values are used, is a further indication that the proposed approach has promise.

Finally, potential users are warned against a slavish use of the coefficients. In many cases, the analyses yielded numerous combinations of factors, all of which yielded a mean Q_c/Q_m of 1.000 (exponentiated natural logarithmic mean). The values selected here were the ones yielding the smallest logarithmic standard deviations (SDL) but the small size of the data sets makes it probable that more consistent values could be obtained had they been carefully selected so as to maintain preconceived relationships between sets. We chose not to enforce our preconceived ideas on the data. Nevertheless, inclusion of other tests with similarly well-defined values of N should lead to somewhat different coefficients and larger amounts of scatter that may be more representative of the actual state of the art.

ACKNOWLEDGEMENTS

Financial support and direction were provided by Caltrans and Caltrans personnel. Two students worked on the project. Cem Akguner was heavily involved in all aspects of the field and laboratory work. Rollins Brown performed initial interpretations of the data and developed an electronic database. The project had an "Oversight Panel" composed of Dr. Michael W. O'Neill, Dr. Bengt H. Fellenius, Dr. Paul W. Mayne, and Dr. Donald G. Anderson. Their constructive comments helped guide the project and kept it on track. Finally, several reviewers of the first draft of this manuscript brought up important points that were addressed in the final manuscript. The contributions of all were essential for this investigation to be successful, and are appreciated.

REFERENCES

American Petroleum Institute (1990). "Recommended Practice for Planning, Designing and Constructing Fixed Offshore Platforms – Working Stress Design: API Recommended Practice 2A-WSD (RP 2A-WSD)." Washington, DC.

Brown, R. P. (2001). "Predicting the Ultimate Axial Resistance of Single Driven Piles." *Ph.D. Dissertation*, University of Texas, Austin, TX.

Davisson, M. T. (1958). personal communication.

Davisson, M.T. (1972). "High Capacity Piles." Proc., *Lecture Series on Innovations in Foundation Construction*, Illinois Section, ASCE, Chicago, 81-112.

Decourt, L. (1982). "Prediction of the Bearing Capacity of Piles Based Exclusively on N Values of the SPT." *Proc. 2nd European Symposium on Penetration Testing, ESOPT-II*, Amsterdam, 1, 29-34.

Dennis, N. D. and Olson, R. E. (1983a). "Axial Capacity of Steel Pipe Piles in Clay." *Proc. Geotechnical Practice in Offshore Engineering*, ASCE, Austin, Texas, 370-388.

Dennis, N. D. and Olson, R. E. (1983b). "Axial Capacity of Steel Pipe Piles in Sand," *Proc. Geotechnical Practice in Offshore Engineering*, ASCE, Austin, Texas, 389-402.

Dunnicliff, J. (1988). "Geotechnical Instrumentation for Monitoring Field Performance." *John Wiley & Sons, Inc.*, pp. 301-303.

Fleming, W. G. K. (1992). "A New Method for Single Pile Settlement Prediction and Analysis." *Geotechnique*, 42, No. 3, 411-425.

Karlsrud, K. (1999). "Lesson Learned from Instrumented Axial Pile Load Tests in Clay," distributed as separate, OTRC Conference, *Analysis, Design, Construction and Testing of Deep Foundations*, Austin, Texas.

Kraft, L. M., Focht, J. A., Amerasinghe, S. F. (1981). "Friction Capacity of Piles Driven into Clay." *J. of the Geotech. Engrg Division*, ASCE, 107, No. 11, 1521-1541.

Meyerhof, G. G. (1976). "Bearing Capacity and Settlement of Pile Foundations." *J. Geotech. Engrg Div.*, ASCE, 102, No. 3, 195-228.

Nordlund, R. L. (1963). "Bearing Capacity of Piles in Cohesionless Soils." J. Soil Mechanics and Foundations Div., ASCE, 89, No. 3, 1-35.

Olson, R. E. (1990). "Axial Load Capacity of Steel Pipe Pile in Sand." *Proc., Offshore Technology Conference*, Houston, OTC Paper No. 6419, 4, 17-24.

Olson, R. E., and M. G. Iskander (1994). "Axial Load Capacity of Piles in Sand." *Advanced Technology in Civil Engineering*, ASCE-Saudi Arabian Section, Manama, Bahrain, 383-394.

O'Neill, M. (2001). "Side Resistance in Piles and Drilled Shafts." *J. of Geot. and Geoenvironmental Engr.*, ASCE, 127, No. 1, pp. 1-16.

Pelletier, J. H., Murff, J. D. and Young, A. C. (1993). "Historical Development and Assessment of the Current API Design Methods for Axially Loaded Pipes." *Proc., Offshore Technology Conference*, Houston, OTC Paper No. 7517, 2, 253-282.

Tomlinson, M. J. (1971). "Some Effects of Pile Driving on Skin Friction." *Proc., Conference on Behavior of Piles*, Instn. of Civil Engineers, London, 107-114, 149-152.

ELSEVIER

Available online at www.sciencedirect.com
SCIENCE DIRECT

Ocean Engineering 32 (2005) 878–891

OCEAN ENGINEERING

www.elsevier.com/locate/oceaneng

Components of suction caisson capacity measured in axial pullout tests

Adam M. Luke[a], Alan F. Rauch[b], Roy E. Olson[c,*], Elliott C. Mecham[d]

[a]*Groundwater Services, Inc., 2211 Norfolk, Suite 1000, Houston, TX 77098, USA*
[b]*Fuller Mossbarger Scott & May Engineers, Inc., 1409 North Forbes Road, Lexington, KY 40511, USA*
[c]*Department of Civil Engineering, The University of Texas at Austin, 1 University Station C1792, Austin, TX 78712, USA*
[d]*Foundation Engineering, 7420 SW Hunziker Road, Suite A, Portland, OR 97223, USA*

Available online 15 December 2004

Abstract

Suction caissons are the foundation of choice for offshore structures in deep water. Systematic study of caisson behavior is relegated to the laboratory due to the high cost of full-scale testing. Our laboratory caisson was installed in normally consolidated clay using dead weight and suction. Tensile axial capacity was measured with the top cap vented or sealed, and with the soil undrained or drained. For the common case of rapid pullout with a sealed top, the test results indicate an external side resistance factor (α) of 0.5–0.8 and a reverse end bearing factor (N_c) of 13–21.

Keywords: Suction caissons; Offshore foundations; Pullout capacity; Tensile loading; Prototype tests; Normally consolidated soil

1. Introduction

A suction caisson is a closed-top steel tube that is lowered to the seafloor, allowed to penetrate the bottom sediments under its own weight, and then pushed to full depth with differential pressure produced by pumping water out of the interior. Suction caissons are an attractive alternative for providing anchorage for offshore facilities because they can be

* Corresponding author.
E-mail addresses: amluke@gsi-net.com (A.M. Luke), arauch@fmsmengineers.com (A.F. Rauch), olson@mail.utexas.edu (R.E. Olson), mechamec@yahoo.com (E.C. Mecham).

0029-8018/$ - see front matter © 2004 Elsevier Ltd. All rights reserved.
doi:10.1016/j.oceaneng.2004.10.007

installed reliably in deep water with good precision and at relatively low cost, have a large axial and lateral holding capacity, and can be recovered for re-use.

Early experience with this technology often involved relatively stiff soils and axial compressive loads applied at the top center of the caisson. As a result, relatively short caissons with large diameters were used. Designs for tension leg platforms in the normally consolidated or underconsolidated, highly plastic clays in the Gulf of Mexico led to the need for greater penetrations. Later designs for floating structures in deeper water, where horizontal or inclined mooring lines are attached to caissons, led to the need for increased lateral capacity. Although the offshore industry is deploying suction caissons in these configurations, a number of design issues remain unresolved (Clukey, 2001; Gilbert and Murff, 2001). In particular, there is a significant body of field data on the installation of suction caissons but still a limited amount of information on the behavior under long-term loading, especially for long, slender caissons.

This paper is concerned with the axial pullout capacity of relatively long suction caissons in normally consolidated clay. Although taut mooring systems for floating platforms in deep water can impart large lateral forces, axial capacity typically controls caisson design when the load is applied at inclinations exceeding about 30° above horizontal (Clukey, 2001).

Only a few field tests of suction caissons have been reported in the open literature (e.g. Tjelta et al., 1986). A number of investigators have tested scale models of suction caissons in geotechnical centrifuges (e.g. Clukey et al., 1995). Centrifuge tests are 'model' tests in that the results can be scaled up to the size of full-scale caissons. Testing in a centrifuge produces results quickly, but some measurements are difficult on small models and scaling laws for different aspects of behavior are in conflict. Field and centrifuge tests are expensive, making it difficult to perform systematic studies.

A wider range of conditions can be investigated at less expense using laboratory tests on small-scale prototype caissons in a 1-*g* environment, although the time needed for soil preparation is much longer than for centrifuge testing. While the results of 1-*g* tests do not scale up to field dimensions, the test data can be used to verify numerical analyses, which can then be used to predict the behavior of full-size caissons. Tests on axially loaded, small scale, prototype suction caissons have been reported by Steensen-Bach (1992), Rao et al. (1997), Whittle et al. (1998), El-Gharbawy and Olson (1999), and others.

In this investigation, laboratory tests using a 100 mm (4 in.) diameter suction caisson were conducted in tanks of normally consolidated clay (Olson et al., 2003a; Rauch et al., 2001, 2003, Luke et al., 2003). Here, components of the vertical pullout resistance are studied using data from nine tests, where the prototype caisson was inserted using dead weight and/or suction pressures, extracted rapidly versus slowly (soil was undrained or drained), and pulled with a vented or sealed top cap.

2. Test methods

2.1. Instrumentation

The prototype caisson (Fig. 1) was constructed from a 100-mm (4-in.) diameter anodized aluminum tube with a wall thickness of 0.81 mm (0.032 in.), giving a wall

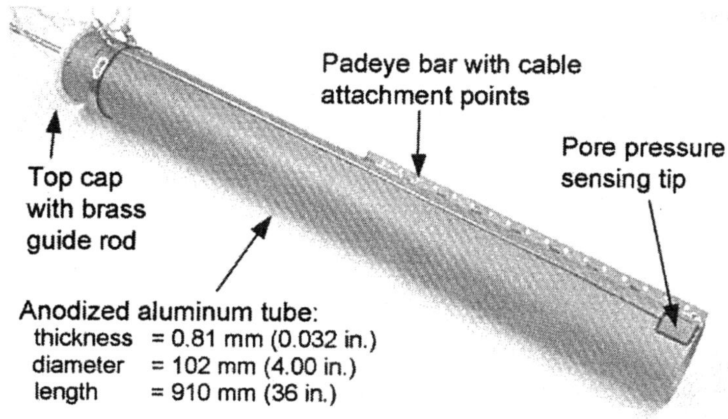

Fig. 1. Prototype caisson used in the model tests.

thickness-to-diameter ratio similar to that of typical full-scale caissons. The prototype caisson can be inserted to depths up to about 860 mm (2.8 ft.). An aluminum bar, 3.2× 2.7 mm (1/3×1/2 in.) in cross-section, was welded to the lower half of the caisson to provide a series of cable attachment points for lateral load tests. No internal stiffeners were present.

Tests were conducted by inserting the caisson into soil deposits prepared in steel tanks. The rate of insertion and removal was controlled using a winch and wire cable. Caisson displacements were recorded using a linear displacement transducer, while axial forces were measured using a 110-N (25-lb) load cell. Water pressures were measured inside the top cap and at five locations on the caisson wall in selected tests. The pore pressure sensing tips (Fig. 1) consisted of porous bronze patches epoxied to the caisson wall and covering the ends of the sensing lines, which were connected to pressure transducers mounted separately from the caisson.

2.2. Test bed soil

Tests were conducted in normally consolidated clay to represent conditions at many offshore sites. After considering various soils, it was apparent that only kaolinite would yield an acceptable consolidation time and final thickness of the test bed deposits. The selected kaolinite was 'Hydrite R' (Dry Branch Kaolin Co., Dry Branch, Georgia). Physical properties included a median particle size of 0.8 mm, a specific gravity of solids of 2.58, liquid limits between 54 and 58%, plastic limits between 31 and 34%, and effective angles of internal friction that decreased from about 55 to 28° as the effective normal stress increased from 1 Pa to 1 kPa (Pedersen et al., 2003).

The test bed was prepared by allowing a slurry of 'Hydrite R' to consolidate under self weight (Olson et al., 2003b). The slurry was mixed to an initial water content of about 164% (liquidity index of 5.6) inside a steel tank measuring 1.2×2.4 m (4×8 ft.) in plan and 1.8 m (6 ft.) high. With double drainage, about 8 months were required to consolidate

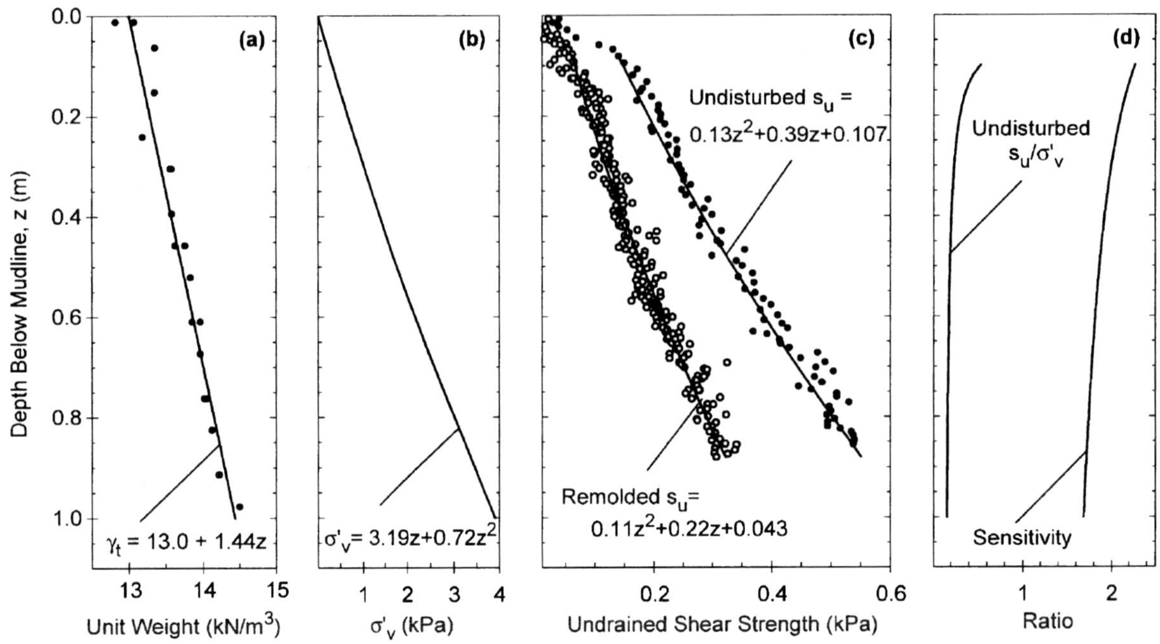

Fig. 2. Profiles of: (a) total unit weight, (b) vertical effective stress, (c) undrained shearing strength and (d) sensitivity and s_u/σ'_v, in the consolidated test bed.

the soil from an initial slurry thickness of 1.6 m (61 in.) to a final thickness of about 1.1 m (43 in.). Soil water contents, determined from samples acquired with a small diameter piston sampler, were used to establish profiles of total unit weight (Fig. 2a) and vertical effective stress (Fig. 2b).

Undrained shearing strengths in the test bed soil were estimated using T-bar penetration tests (Stewart and Randolph, 1994) conducted with a 25-mm (1-in.) diameter by 100-mm (4-in.) long, smooth acrylic T-bar. Undrained shearing strengths (Fig. 2c) were calculated using a bearing factor of 10.5 (Stewart and Randolph, 1994). Undisturbed strengths were obtained from an initial penetration, while the remolded strength was measured after repeated penetrations at the same location. The s_u/σ'_v ratio of the undisturbed soil was about 0.2 over most of the depth, while the sensitivity varied between 2.3 and 1.7 (Fig. 2d).

2.3. Caisson tests

Nine axial load tests were conducted (Table 1). The center-to-center spacing between adjacent test locations, as well as the distance between the side walls of the test tank and the center of each test location, was three caisson diameters (300 mm or 12 in.).

The caisson was pushed to full penetration with dead weight in three tests. In six tests, dead weight was used to install the caisson to a tip depth of about 250 mm (10 in.) and then suction was used to cause penetration to the final depth (Rauch et al., 2003).

Table 1
Summary of test conditions for axial pull out tests

Maximum depth of insertion by		Conditions during pullout test	
Dead weight (mm)	Suction (mm)	Top cap	Rate of pullout
854	none	Vented	Rapid
833	none	Vented	Rapid
279	821	Vented	Rapid
279	808	Vented	Rapid
858	none	Sealed	Rapid
279	819	Sealed	Rapid
279	808	Sealed	Rapid
229	794	Sealed	Rapid
229	812	Vented	Slow

Pore water pressures measured on caisson wall.

Based on measurements of the pore water pressures (Olson et al., 2003a), all tests included a minimum set-up time of 48 h following insertion, to ensure dissipation of excess pore water pressures on the exterior caisson wall.

For extraction tests with a vented top, the unused pressure ports in the cap were left open. Water (or air) could then flow into the caisson during pullout and the plug of soil inside the caisson was left behind in the soil deposit. For the sealed-top tests, the ports were closed; suction then developed inside the top cap during withdrawal, resulting in the interior soil plug coming out with the caisson.

Rapid extraction tests (soil assumed to be undrained) involved pulling the caisson out of the soil at a rate between 5 and 20 mm/s (0.2–0.8 in./s). In tests with slow extraction, the caisson was displaced upwards in steps of about 1 mm, with successive steps applied only when the measured excess pore water pressures had decreased to the measurement accuracy of the transducers (about 0.3 kPa). On average, the caisson was extracted in steps lasting about 30 min each.

3. Analysis of pullout resistance

The forces acting on the caisson during extraction are indicated in Fig. 3. For both the vented and sealed conditions, the resistance to axial pullout (Q) can be written

$$Q = W_{\mathrm{caisson}} + F_{\mathrm{press}} + W_{\mathrm{soil}} + Q_{\mathrm{side}} + Q_{\mathrm{tip}} \quad (1)$$

where W_{caisson} is the self weight of the caisson (in air) plus any added surcharge weight, F_{press} is the force on the caisson due to water pressures (including hydrostatic pressure on the top of the caisson cap, pressure inside the top cap during vented pullout, and buoyancy effects), W_{soil} is the weight of soil being pulled out with the caisson, Q_{side} is the total shear resistance along the interior (vented case only) and exterior sides of the caisson, and Q_{tip} is the total end bearing force, with all terms on the right side positive when acting downwards.

When the top cap is vented, Q_{side} develops on both the interior and exterior surfaces and W_{soil} is the weight of soil adhering to both surfaces. When the top is sealed, Q_{side} acts only

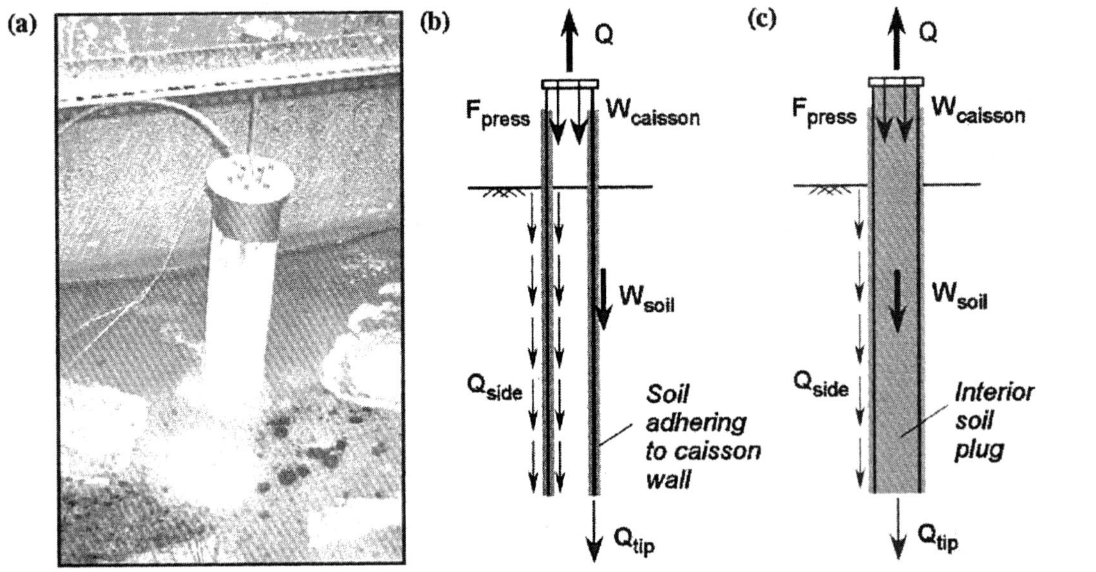

Fig. 3. Caisson pullout: (a) photograph of test, (b) free body diagram for extraction with vented top and (c) free body diagram for extraction with sealed top.

on the caisson exterior and W_{soil} includes the weight of the soil plug. If the caisson is vented, Q_{tip} acts on the thin cross-sectional area of the caisson and is negligible.

The total end bearing force (Q_{tip}) on the caisson can be estimated from bearing capacity theory. For the case of a saturated soil that is loaded so quickly that there is essentially no drainage, the appropriate expression for the total end bearing force is

$$(Q_{tip})_{extraction} = (N_c s_u - \sigma_v)A_{tip} \quad (2)$$

where N_c is the bearing capacity factor including appropriate corrections for depth and shape, s_u is the undrained shear strength of the soil at the tip, σ_v is the total vertical stress at the tip elevation, and A_{tip} is the cross-sectional area of the tip. For normally consolidated soil, σ_v generally exceeds $N_c s_u$ so the actual tip stress is negative (acts upwards on the caisson). The net end bearing resistance (Q'_{tip}) is defined as

$$Q'_{tip} = N_c s_u A_{tip} \quad (3)$$

In analyzing the pullout data, σ_v was calculated by integrating the equation for the unit weight of the test bed soil (Fig. 2a) and s_u was obtained from the undisturbed strength profile (Fig. 2c) at the elevation of the caisson tip. Consistent with design practice, Q_{tip} was computed using s_u of the undisturbed soil. Once the caisson capacity is exceeded, the tip is pulled through a zone of partially remolded soil, resulting in a reduced post-failure tip resistance.

The values of N_c and A_{tip} depend on whether the caisson is sealed or vented during extraction. The internal soil plug was not pulled out when the caisson was vented, so the reverse end bearing resistance can be modeled by treating the caisson tip as a strip footing (shape factor = 1) having a width equal to the caisson wall thickness. Since, the wall thickness is much less than the depth for even shallow penetrations, the depth factor can be assumed to have a maximum value of 1.5 (Skempton, 1951) throughout the pullout test.

Table 2
Capacity measured in rapid pullout tests when caisson top was (a) vented, and (b) sealed

Test ID	Installed by[a]	W_{soil} (*N*)	Q'_{tip} (*N*)	Q_{side} (*N*)	α
(a) Pullout with vented top cap (no soil plug)					
1-022702	DW	16	–	–	–
1-040802	DW	9	1	101	0.67
1-030502	DW+S	16	1	82	0.55
1-041002	DW+S	7	1	82	0.57
(b) Pullout with sealed cap (soil plug remains in place)					
1-030102	DW	107	40	78	0.97
1-030802	DW+S	103	38	77	1.04
1-031002	DW+S	102	38	73	1.01
1-041502	DW+S	98	37	82	1.17

[a] DW, dead weight; S, suction.

The bearing capacity factor is thus $N_c=(5.14)(1.0)(1.5)=7.7$. Because of the small value of A_{tip} (cross-sectional area of the caisson tube), Q_{tip} contributes little to the pullout resistance of the vented caisson.

When the internal soil plug is pulled with the sealed caisson, end bearing pressure develops over the end area of the caisson tube plus the end area of the soil plug. The effective area of the caisson tip is now large enough to contribute significantly to the pullout resistance (Table 2b). Calculations were made using a shape factor of 1.2 for a circular tip. With the depth exceeding 2.5 times the diameter, a depth factor of 1.5 was used (Skempton, 1951). The bearing factor is then $N_c=(5.14)(1.2)(1.5)=9.3$.

Given the measured or computed values of $W_{caisson}$, F_{press}, W_{soil}, and Q_{tip}, the side shear resistance (Q_{side}) was estimated using Eq. (1). To compare the maximum capacities from different rapid pullout tests, the unit side resistance was normalized with respect to the shear strength of the soil to obtain the ratio α

$$\alpha = \frac{Q_{side}}{A_{side} s_{u,avg}} \tag{4}$$

where A_{side} is the surface area of the caisson wall, and $s_{u,avg}$ is the average undrained shear strength, both defined in the zone of caisson–soil contact.

4. Measured pullout capacity

4.1. Rapid pullout with a vented top cap

The caisson was pulled rapidly from the soil with a vented top cap in four tests. Values of W_{soil}, Q'_{tip}, Q_{side}, and α for the ultimate capacity are given in Table 2a. In all cases, the caisson came out of the test bed covered with a thin layer of soil on both the inside and outside surfaces (Fig. 3a), indicating that failure was in the soil, not at the interface.

The side resistances measured during vented pullout reached a peak at displacements of 5–13 mm (5–13% of the caisson diameter) (Fig. 4a). The weight of adhered soil (W_{soil}), which was assumed to be constant during pullout, was determined by weighing the caisson

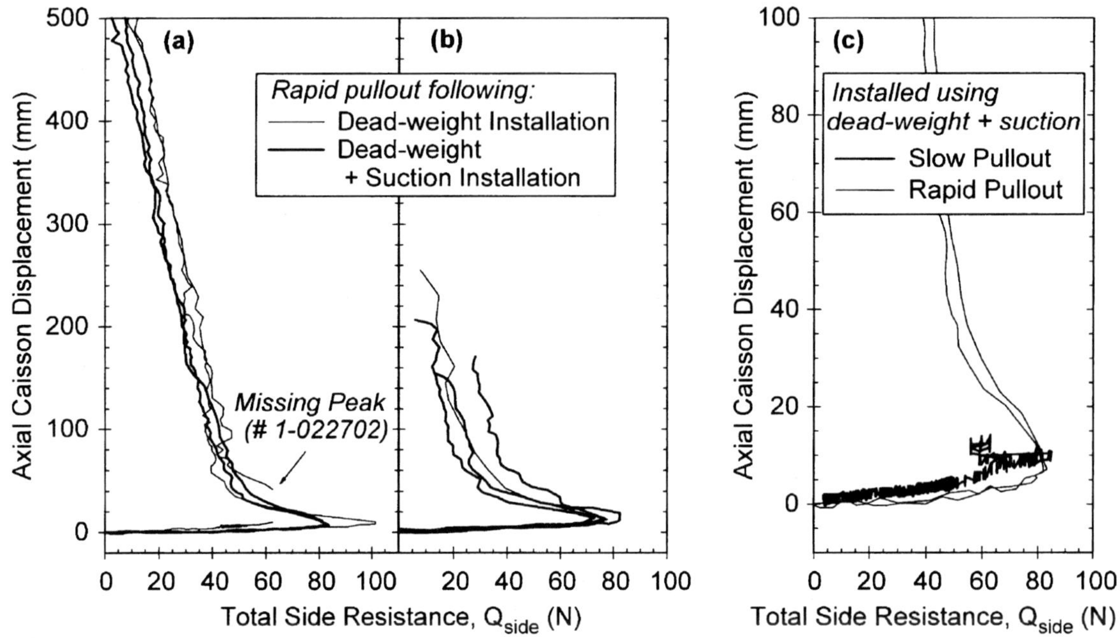

Fig. 4. Measured side resistance: (a) rapid pullout with a vented top, (b) rapid pullout with a sealed top and (c) comparison of forces for slow and rapid pullout with a vented top.

after it was fully extracted. After failure, the pullout resistance decreased by about 50% (Fig. 4a), which is in keeping with partial remolding of the soil (sensitivity of about two) coupled with a slight reduction in contact area.

The load cell was over-ranged in the first test (1-022702) and the peak load was not recorded; data from this test are reported because it shows the weight of soil extracted with the caisson, as well as the consistency in the measurements after failure (Fig. 4a). For vented pullout, W_{soil} and Q'_{tip} were relatively small (Table 2a) and most of the capacity developed from side resistance. An average value of α is reported because the side shear forces on the interior and exterior could not be separated.

The peak side resistance was about 20% lower for the caissons installed partially using suction compared with the single caisson completely installed using dead weight (Table 2a and Fig. 4a). The post-peak side resistance of the suction-installed caisson was also less during pullout (Fig. 4a). The possibility of having lower side shear resistance following suction installation has been the subject of recent discussion (Clukey, 2001; Gilbert and Murff, 2001).

4.2. Rapid pullout with a sealed top cap

In four tests, the caisson was pulled rapidly from the soil with a sealed top. Values of W_{soil}, Q'_{tip}, Q_{side}, and α, for the measured ultimate capacity in each test, are given in Table 2b. The internal soil plug was pulled out of the test bed with the caisson (Fig. 3c), greatly increasing W_{soil} and Q_{tip}. Side resistance developed only on the caisson exterior when the interior soil plug was extracted with the caisson, so α was computed with A_{side} equal to the outside surface area.

In a preliminary test with a sealed top, the caisson was fully extracted and the soil plug was weighed, but the hole left in the test bed collapsed and disturbed a large zone of soil. To preserve adjacent test locations, all remaining tests with a sealed caisson were halted after about 150–250 mm of displacement, well after the peak pullout capacity was recorded. At this point, the caisson was vented and the interior soil was allowed to slide back into the deposit. This procedure minimized the area disturbed in the test bed, but also prevented the direct measurement of W_{soil} in subsequent tests.

In all four tests with a sealed top (Fig. 4b), the maximum pullout capacity developed in about 13 mm of displacement, or 13% of the caisson diameter. The pullout resistance then dropped off (Fig. 4b), but by more than observed in the vented caisson tests, and probably more than can be attributed to soil disturbance and reduced contact area.

4.3. Slow pullout with a vented top cap

One pullout test was performed with a vented top cap at a rate such that measured excess pore water pressures were zero. The peak load was reached in 5.5 h after about 11 mm (0.43 in.) of axial displacement (Fig. 4c). The calculated value of α was 0.58. Essentially no soil adhered to the caisson during withdrawal, in agreement with previous tests with overconsolidated kaolinite (El-Gharbawy and Olson, 1999).

The observed side resistances under drained and undrained soil conditions are similar (Fig. 4c). Excess pore water pressures were measured only on the sides of the caisson, not under the tip, and it is not clear that full dissipation occurred under the tip. Dissipation of positive excess pore water pressures along the sides should have increased the side capacity. However, unloading the tip would generate reduced water pressures and partial dissipation would lead to a weakening of the soil and increased uplift on the tip, thus counterbalancing the increased side capacity and possibly leading to essentially the same capacity for rapid and slow tests.

The side shear capacity of a deep foundation is sometimes predicted using effective stress and the ratio β

$$\beta = \frac{Q_{side}}{A_{side}\sigma'_{v,avg}} \tag{5}$$

where $\sigma'_{v,avg}$ is the average vertical effective stress along the length of the caisson. For a maximum side resistance of 85 N, the value of β was 0.12.

5. Components of axial capacity for sealed, rapid pullout

The values of α for rapid pullout (Table 2a and b) differ substantially for the vented and sealed caisson. Interpretation of the sealed caisson tests are complicated by the uncertainty associated with the values of W_{soil}, N_c, and α. The total side resistances determined from the vented pullout tests are believed to be more accurate because the contributions of W_{soil} and Q_{tip} are relatively small.

Note that the α values reported for the vented tests (Table 2a) are average values for the interior and exterior surfaces. In reality, the internal and external side shear forces may differ

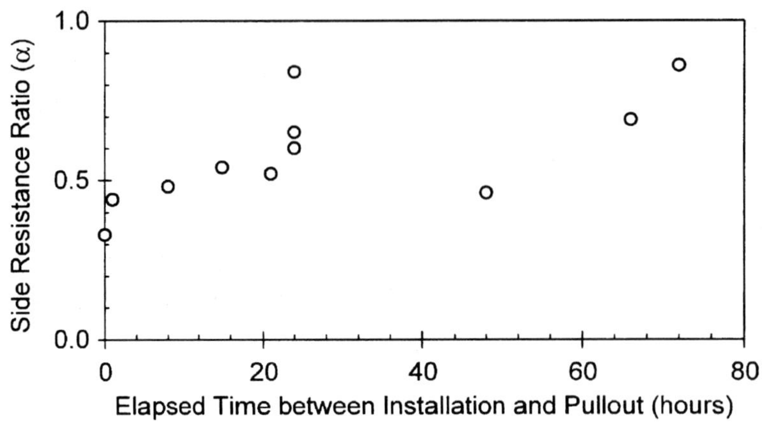

Fig. 5. Side resistance measured during rapid pullout of a thin plate.

due to differences in the radial effective stresses on the inside and outside surfaces of the caisson. However, values of the average α for the vented caisson are consistent with those measured in tests with a thin aluminum plate inserted vertically into the soil (Fig. 5). Eleven such tests were conducted in the same test bed using a plate that was 100 mm wide by 800 mm long (4×32 in.) and 0.8 mm (0.03 in.) thick. While considerable scatter was observed (Fig. 5), the results indicate that α scattered between about 0.5 and 0.9 after 4 h of set-up. The values of α determined in the vented pullout tests (Table 2a) fall in the middle of this range.

Given these results, it is reasonable to assume that α is between 0.5 and 0.8 for the external side resistance on the sealed caisson. This is smaller than the values reported in Table 2b, indicating that the value of N_c or the weight of the soil plug may be larger than

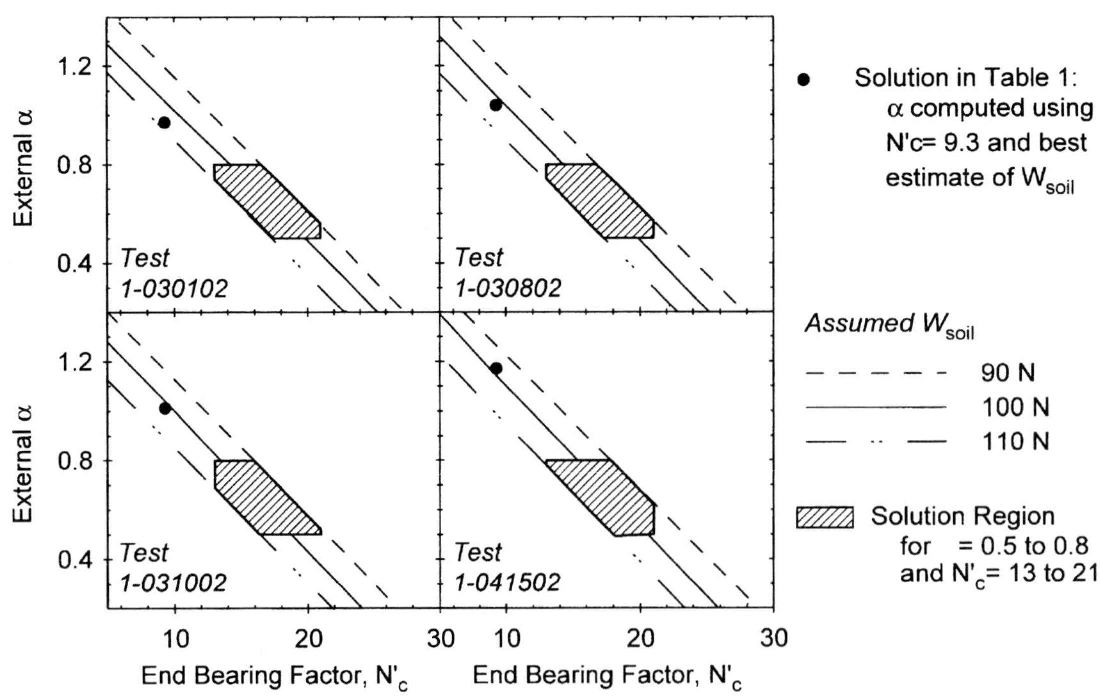

Fig. 6. Combination of soil weight, side resistance, and reverse end bearing that give the peak capacity measured in four rapid pullout tests with a sealed top.

assumed. Data from centrifuge model tests (House and Randolph, 2001) also suggest N_c is greater than 9.3 for axial pullout of a sealed suction caisson.

Different combinations of soil weight, end bearing, and side resistance that would produce the capacity measured in the four tests with the sealed caisson were investigated (Fig. 6). The solutions given in Table 2b, for $N_c=9.3$, are also indicated. It seems reasonable to assume that the soil weight (interior plug plus soil adhering to the caisson exterior) was between 90 and 110 N (20–25 lb) in each test. For values of α in the range of 0.5–0.8, reasonable solutions can be found for all four tests using values of N_c between 13 and 21. This region of likely solutions in indicated by the hatched boxes in Fig. 6.

6. Summary and conclusions

The behavior of suction caissons was investigated using a 100 mm (4 in.) diameter caisson prototype installed in a 1.1 m (43 in.) thick laboratory deposit of normally consolidated kaolinite. Nine axial pullout tests were performed in this phase of the study.

Tests involved caisson installation using suction, as well as insertion to full depth using dead weight. The soil was allowed to re-consolidate after caisson insertion, to simulate field conditions involving full set-up. Pullout tests were then conducted with a sealed or vented top cap, with rapid extraction (undrained soil conditions) in eight tests and slow extraction (drained soil conditions) in one test. In all cases, the peak extraction force was reached within an axial movement of about 13 mm (13% of the diameter) and there was a substantial loss in capacity after failure.

In pullout tests with a vented caisson, less side resistance was observed following suction installation compared with dead weight installation. The test results suggest that side resistance is about the same for withdrawal under drained or undrained conditions.

Axial capacities were considerably larger when the caisson was sealed during extraction, because of the large end bearing capacity and the weight of the interior soil plug, which was pulled out with the caisson. Reasonable predictions of the measured capacities could be obtained using a side resistance factor (α) between 0.5 and 0.8 and an end bearing capacity factor (N_c) between 13 and 21.

Results from these laboratory tests help to elucidate the behavior of suction caissons, and the data can be used to support the development of simplified design models and sophisticated computer simulations. However, these 1-g model test results should not be applied directly to the design of full-scale suction caissons.

Acknowledgements

This study was supported by the Offshore Technology Research Center at the University of Texas and Texas A&M University, with funding from the US Minerals Management Service and the OTRC Industry Consortium.

References

Clukey, E.C., 2001. Suction caisson design issues, Proceedings, OTRC 2001 International Conference, Geotechnical, Geological and Geophysical Properties of Deepwater Sediments, Houston, April. pp. 163–181.

Clukey, E.C., Morrison, M.J., Gariner, J., Corté, J.F., 1995. The response of suction caissons in normally consolidated clays to cyclic TLP loading conditions, Proceedings, Offshore Technology Conference, OTC 7796. pp. 909–918.

El-Gharbawy, S., Olson, R., 1999. Suction caisson foundations in the Gulf of Mexico, analysis, design, construction, and testing of deep foundations. ASCE Geotechnical Special Publication 88, 281–295.

Gilbert, R.B., Murff, J.D., 2001. Design methodologies and criteria for suction caissons for deepwater mooring applications, Workshop Report, Offshore Technology Research Center, June.

House, A.R., Randolph, M.F., 2001. Installation and pull-out capacity of stiffened suction caissons in cohesive sediments, Proceedings of 11th International Offshore and Polar Engineering Conference, Stavanger, Norway, June.

Luke, A.M., Rauch, A.F., Olson, R.E., Mecham, E.C., 2003. Behavior of suction caissons measured in laboratory pullout tests. Paper no. OMAE2003-37023, Proceedings of 22nd International Conference on Offshore Mechanics and Arctic Engineering, ASME, Cancun, Mexico, June.

Olson, R.E., Rauch, A.F., Luke, A.M., Maniar, D.R., Tassoulas, J.L., Mecham, E.C., 2003a. Soil reconsolidation following the installation of suction caissons, Proceedings, Offshore Technology Conference, OTC 15263, May.

Olson, R.E., Rauch, A.F., Mecham, E.C., Luke, A.M., 2003b. Self-weight consolidation of large laboratory deposits of clay, Proceedings, 12th Pan-American Conference on Soil Mechanics and Geotechnical Engineering, Cambridge, MA, June.

Pedersen, R.C., Olson, R.E., Rauch, A.F., 2003. Shear and interface strength of clay at very low effective stress. ASTM Geotechnical Testing Journal 26 (1).

Rao, S.N., Ravi, R., Ganapathy, C., 1997. Pullout behavior of model suction anchors in soft marine clays, Proceedings, Seventh International Offshore and Polar Engineering Conference, May, Honolulu.

Rauch, A.F., Olson, R.E., Mecham, E.C., Pedersen, R.C., 2001. A laboratory facility for testing model suction caissons, Proceedings, OTRC 2001 International Conference, Geotechnical, Geological and Geophysical Properties of Deepwater Sediments, Houston, April. pp. 198–216.

Rauch, A.F., Olson, R.E., Luke, A.M., Mecham, E.C., 2003. Measured response during laboratory installation of suction caissons, Proceedings, 13th International Offshore and Polar Engineering Conf., Honolulu, Hawaii, May. pp. 780–787.

Skempton, A.W., 1951. The bearing capacity of clays, Proceedings, Building Research Congress, London. pp. 180–189.

Steensen-Bach, J.O., 1992. Recent model tests with suction piles in clay and sand, Proceedings, Offshore Technology Conference, OTC 6844. pp. 323–330.

Stewart, D.P., Randolph, M.F., 1994. T-bar penetration testing in soft clay. Journal of Geotechnical Engineering, ASCE 118 (12), 2230–2235.

Tjelta, T.I., Guttormsen, T.R., Hermstad, J., 1986. Large-scale penetration test at a deepwater site, Proceedings, Offshore Technology Conference, OTC 5103. pp. 201–212.

Whittle, A.J., Germaine, J.T., Cauble, D.F., 1998. Behavior of miniature suction caissons in clay, Offshore Site Investigation and Foundation Behaviour '98, SUT. pp. 279–300.

Adam M. Luke received a BS degree in Civil Engineering from Texas A & M University in 2000, and an MS degree in Civil Engineering from the University of Texas at Austin in 2002. For about a year afterward, he worked at the World Trade Center site in New York for RJLee Group, Inc. He is currently working for Groundwater Services, Inc. in Houston, TX, USA.

Alan F. Rauch received his PhD in Civil Engineering in 1997 from Virginia Polytechnic Institute and State University. He was an assistant professor of civil engineering at the University of Texas at Austin, from 1997 to 2004. He is currently a Senior Geotechnical Engineer with Fuller, Mossbarger, Scott, and May Engineers in Lexington, KY, USA.

Roy E. Olson received BS and MS degrees in Civil Engineering from the University of Minnesota, and a PhD degree with a major in geotechnical engineering and minors in geology, agronomy, and chemistry, from the University of Illinois. He was on the faculty at the University of Illinois for a decade and the University of Texas for three decades, both in Civil Engineering, and is now semi-retired. His interests were originally in physico-chemical aspects of soils, and in consolidation of soft clays. His major fields of interest in recent years have been in design of foundations to support bridges, buildings, and offshore structures. He is a member of the National Academy of Engineering.

Elliott C. Mecham received his BS degree in Civil Engineering from Utah State University in 1999, and his MS degree in Civil Engineering from the University of Texas at Austin in 2001. He was an engineer with GRL Inc. in Cleveland for several years, engaged in the analysis and testing of deep foundations. He is currently employed by Foundation Engineering, Inc., in Portland, OR, USA, where he is a foundation engineer.

A Perspective on the Levee Failures in New Orleans from Hurricane Katrina

David E. Daniel[1], P.E., NAE, Dist.M.ASCE

[1]President, Univ. of Texas at Dallas, 800 W. Campbell Rd, Richardson, TX 75080; dedaniel@utdallas.edu

ABSTRACT: The fundamental cause of failure of the levee system in New Orleans following Hurricane Katrina was poor decision making. The design hurricane was not updated over time as new information became available. The inevitable settlement of levees founded on compressible soil was not taken into account. Benchmarks used to establish elevations settled over time, but adjustments were not made. Many sections of levee were built of fine sand and silt materials that were not capable of withstanding erosion from overtopping. Unconservative design assumptions contributed to failure of levees that were not overtopped. Warning signs of imminent failure from a full-scale I-Wall test did not trigger re-consideration of design parameters. These observations lead to the conclusion that the failure of the New Orleans flood protection system was the result of a failure of judgment and decision-making on the part of engineers and the authorizing/funding agencies. The New Orleans levees provide an example of the importance of the first canon of ASCE's code of ethics: "Engineers shall hold paramount the safety, health and welfare of the public."

INTRODUCTION

The author had the privilege of chairing ASCE's External Review Panel, which investigated causes of the failure of levees in New Orleans from Hurricane Katrina. The Panel's report, "The New Orleans Hurricane Protection System: What Went Wrong and Why" (ASCE, 2007) documents findings. This paper draws heavily on the Panel's report and findings.

On the morning of August 29, 2005, Hurricane Katrina stuck and devastated New Orleans and nearby coastal regions. One thousand one hundred eighteen (1,118) people lost their lives, and 135 more are missing and presumed dead (ASCE, 2007). An estimated 49% of those who died were age 75 or greater (Brunkhead et al., 2008).

The peak water levels from storm surge varied from approximately 3 to 6 m above sea level in the New Orleans area (USACE, 2006, Vol. 4). Storm surge elevations were higher than the levee crest elevations in many sections of the levee system, resulting in overtopping. Some failures occurred at sections of levee that were not overtopped and should not have failed.

In this paper, findings from the Panel's study are knitted together to paint a picture of poor decision-making that led, collectively, to the levee breaches and to the resulting catastrophe. Several excellent analyses of geotechnical conditions associated with specific failures and causes of failure have been published. The reader is directed to a special issue of the ASCE *Journal of Geotechnical and Geoenvironmental Engineering* (May, 2008), which contains articles about various aspects of the failures. Of particular relevance to topics covered in this paper are articles on overall conditions (Seed et al., 2008a), levee erosion (Briaud et al., 2008), failure of the 17th St. Canal levee (Sasanakul et al., 2008; and Seed et al., 2008c); and stability of I-Walls (Duncan et al., 2008; and Seed et al., 2008b).

As an engineering educator, the author has spent most of his career teaching students the tools and methodologies of sound engineering analysis and design. But as this case history illustrates, there is more to good engineering than just good analysis – sound judgment and decision making play a key role, as well. It is this aspect of engineering – decision making under constraints of budget and other factors – that is the focus of this paper.

DESIGN HURRICANE

Perhaps the most fundamental challenge in designing the Hurricane Protection System for New Orleans is selecting the design hurricane. The larger the design hurricane, the greater the storm surge, the higher the required crest elevations of levees, and the greater the cost.

As wind from a hurricane whips across the Gulf of Mexico toward the U.S. coast, wind shear pushes water toward the shore and creates storm surge. The height of storm surge is a function of wind speed and direction as well as other characteristics of the hurricane and of the coast line. Storm surges from extreme hurricanes can be as high as 10 m above normal sea level.

Criteria for the design hurricane in New Orleans are established in Federal legislation. The U.S. Army Corps of Engineers (USACE) was directed by this legislation to design the New Orleans Hurricane Protection System for "the most severe combination of meteorological conditions that are considered 'reasonably characteristic' of the region." The terminology "reasonably characteristic" is viewed as synonymous with "maximum probable." Note that "maximum probable" is not as extreme as "maximum possible." Congress did not specify probabilistic parameters – just the words "reasonably characteristic."

Starting in the mid 1960s, the USACE relied upon a 1959 study of hurricanes published by the U.S. Weather Bureau (Graham and Nunn, 1959) for selecting the design hurricane, which is termed the "standard project hurricane (SPH)." The 1959 study documented the characteristics of hurricanes that had impacted the U.S East Coast and Gulf Coast over a 56-year period between 1900 and 1956. This period was not necessarily a period in which the worst that could happen did in fact happen. To the contrary, there would be no reason to expect that the most extreme event possible would have occurred during this particular 56-year period. But at the time (1950s), this was the best information available. Graham and Nunn (1959) indicate that the maximum wind speed for a SPH in New Orleans is 101 to 111 mph (162 to 197 kmh). The SPH used by the Corps to design New Orleans' levees had a maximum wind speed of 100 mph (160 kmh) and a recurrence interval of approximately 100 years (USACE, 2006, V. 3). It is not clear why the maximum wind speed used by the USACE is slightly less than the minimum of the range reported by Graham and Nunn (1959).

Not surprisingly, in the ensuing years, even larger hurricanes struck the U.S. Gulf Coast, e.g., Carla (1961) and Camille (1969). In 1979 the National Oceanic and Atmospheric Administration issued a report that updated hurricane information (NOAA, 1979). The maximum wind speeds vary with factors such as forward speed of the hurricane but are about 100 mph (160 kmh) and 133 mph (214 kmh) for SPH and probable maximum hurricane, respectively.

Over time, the USACE did not upgrade its SPH to incorporate new information about extreme hurricanes. Keeping the design hurricane the same over a period of many years simplified the process for the UASCE. Had the Corps increased the magnitude of its SPH, the Corps would not only have been required to construct higher, more expensive levees, but it would also have been pressured to raise crest elevations for older levees, too, creating a complex situation.

Some have said that the essential problem with Hurricane Katrina is that it exceeded Congressional authorization. Katrina was a massive storm. When the eye of the Hurricane was 300 km offshore, the maximum wind speed was 260 kmh (160 mph). Just a few hours before landfill, as the surge was building along the coast, the maximum wind speed was approximately 225 kmh (140 mph). By landfall, maximum sustained wind speed had dropped to 185 kmh (115 mph) (USACE, 2006, Vo. 4) to 204 kmh (127 mph) (NOAA, 2005), well above the SPH maximum wind speed of 160 kmh (100 mph). As a result, storm surge was significantly higher than design criteria at many locations in New Orleans.

It is disappointing that so little attention was given to the inevitability of New Orleans being hit by a hurricane larger than the standard project hurricane. No matter what the choice for SPH, the only question was "when" not "if" a more extreme hurricane would strike.

DESIGN FOR OVERTOPPING

In general, the levees were of two types: (1) earthen levees, and (2) I-Walls placed within earthen levees (Fig. 1). Levees breached at approximately 50 locations in the flood protection system. Most but not all of the failures were caused by overtopping and erosion.

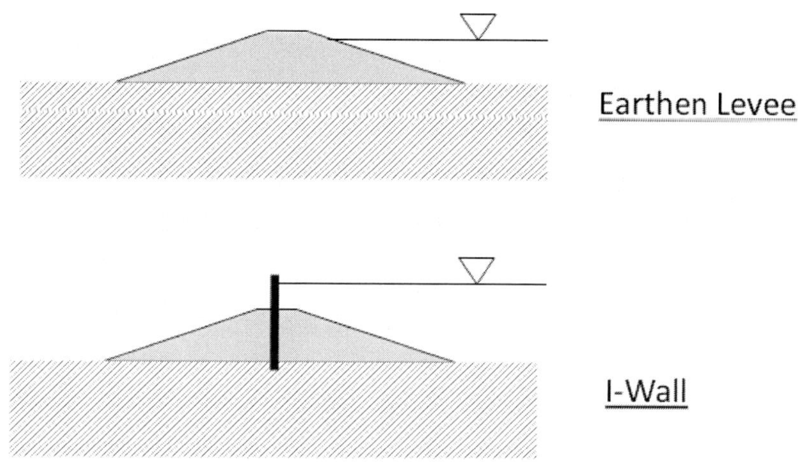

Figure 1. Two Types of Levee: Earthen Levee and I-Wall.

Many overtopped earthen levees were washed away because they were constructed of highly erodible materials such as fine sand and silt. The author has not located any explanation for why levees were constructed of highly erodible materials. Such materials were presumably used because they were readily available and were the least expensive and most convenient materials available. Some members of the Corps' post-failure investigation team described the levees as "disposable levees" because it was obvious that the fine sands and silts would be washed away if/when the levees were overtopped.

It can be argued that construction of levees from more expensive, erosion-resistant materials would have exceeded Congressional authorization because it would have required the engineer, in effect, to design the levees for a hurricane larger than the standard project hurricane, which was not authorized by Federal law. The author found no record of any dialogue about whether use of erodible material might be a poor design choice, perhaps because the engineers believed they were not authorized to consider hurricanes larger than the SPH. Whatever the reasons, the choice to construct levees from highly erodible materials was a tragic one for which failure was entirely foreseeable. If the levees had been constructed of erosion-resistant materials, levees would have been overtopped for a few hours during peak storm surge and significant flooding would have occurred, but catastrophic consequences would have been lessened or avoided.

LEVEES WERE TOO LOW

Subsidence. It is well known that New Orleans is undergoing regional subsidence. The subsidence is estimated to be occurring at an average rate of about 5 mm (0.2 in.) per year (ASCE, 2007). Subsidence was not taken into account in the design. Therefore, to an increasing extent over time, the levees sunk below design crest elevation.

Subsidence could easily have been taken into account. For example, for a design life of 100 years and 5 mm of subsidence per year, levees would need to be built 500 mm (1.6 ft) higher to account for subsidence. Accounting for subsidence would have been tractable, but costly.

The author is at a loss to understand how one might rationalize ignoring subsidence. Storm surges are referenced to sea level. Levees are intended to last for decades. In the author's opinion, the engineer should and must consider subsidence.

Benchmarks. The USACE referenced its survey datums for construction projects to land-based benchmarks that were themselves undergoing subsidence. The Corps recommends that a survey datum be evaluated and corrected prior to use on any new construction project. It appears, however, that over a period of many years, adjustments in the survey datums were not made. The result was that critical sections of levees were constructed approximately 600 mm (2 ft) too low (USACE, 2006, V. 2).

Consolidation Settlement from Weight of Levee. Newly constructed or expanded earthen levees are expected settle as a result of consolidation of underlying, compressible soils. Even if a levee is constructed at the correct elevation, the crest will soon settle below the design elevation independent of regional subsidence. The author found no evidence of any attempt to correct for this predictable settlement, i.e., construct the levees at in initial crest elevation that ensures the crest will remain above the minimum design elevation after consolidation settlement of subsoils. Again, no explanation of justification was found in the record. In the author's opinion, there simply is no justification for ignoring this type of predictable settlement.

If adjustments had been made for settlement, various sections of levee constructed at different times likely would have been constructed at different crest elevations. This might have caused confusion among members of the public wondering why the levee next to their home was at a different elevation than the levee next to a home down the street. Also, building levees to account for settlement would have cost more. In addition, if at any point in time the engineer had decided to account for settlement, that decision might have led to a massively open-ended problem of going back to all existing levees and correcting for settlement. The simplest thing to do was not account for settlement. However, it was not the safest thing to do.

17th STREET CANAL

The 17th Street Canal levee failure has received much attention. The levee failed when the water level was 0.6 to 1.5 m below the top of the I-Wall (USACE, 2006; Duncan et al., 2008; and Seed et al., 2008c). The 17th Street Canal levee failure was particularly catastrophic in terms of great loss of life in the adjacent, heavily populated area (Seed et al., 2008c).

Because the water level was below the top of the levee at the time of failure, the levee should not have failed, but did. The question is "why?" The answer to this question provides insights into the flawed decision making process.

A cross-sectional view of the 17th Street Canal is shown in Fig. 2, along with a potential slip surface that includes a "crack" behind the I-Wall. The author brings to question five aspects of decision-making.

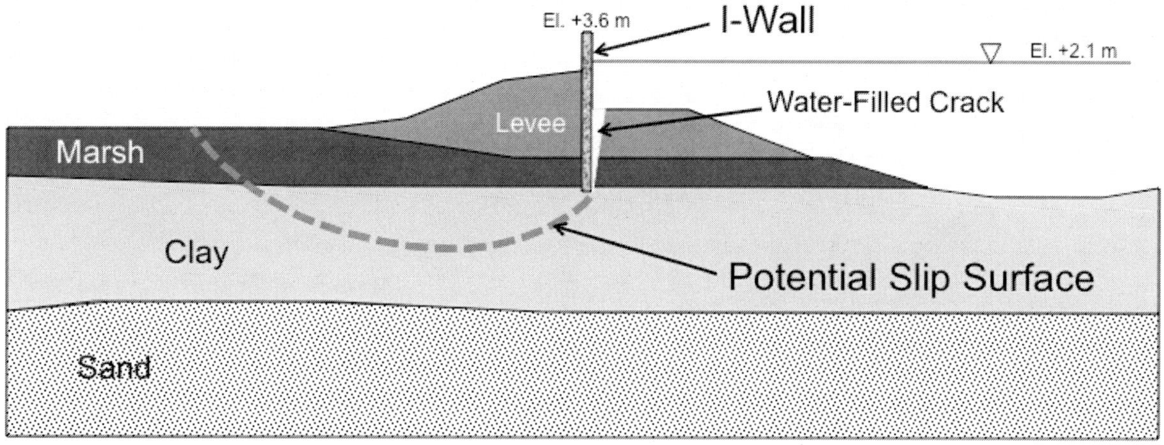

Figure 2. Cross Section of 17th Street Canal Levee Showing Stratigraphy (from USACE, 2006) and Potential Slip Surface Incorporating a Water-Filled Crack Behind the Wall.

1. *Selection of the Factor of Safety*. Stability analyses were performed to support design with a minimum factor of safety of 1.3 against slope stability failure using the method-of-planes technique (Eustis Engineering, 1988). Typical minimum factors of safety recommended by the Corps for slope stability design in the late 1980s appear to be as follows: (1) for earth- and earth-rock dams: 1.3 for end of construction, 1.5 for long-term conditions, and 1.4 for maximum storage pool with a footnote stating, "If consequences of … failure are great … resulting in potential catastrophic failure, higher factors of safety should be considered" (UASCE, 2003); and (2) for levees: 1.3 for end of construction and 1.4 for long-term conditions (USACE, 2000).

The Corps' Engineer Manual for Slope Stability (UASCE, 2003) notes that the two most important considerations in selecting a factor of safety for design are uncertainties in the conditions being analyzed (especially shear strength) and the consequences of failure. In the author's opinion, for the 17th Street Canal, the uncertainties in shear strength were significant considering the variable geology in the region. The consequences of failure were extreme: numerous people living immediately adjacent to the levee at elevations below sea level.

The author believes that safety of levees was thought of in too generic a way. The Corps' levee manual, in recommending factors of safety, does not discuss consequences of failure. Most levees in the U.S. protect flood plains next to rivers. Rivers typically crest with ample warning after heavy rains or snowmelt, and land adjacent to levees is mostly farmland. The consequences of failure are dramatically different for such situations compared to New Orleans. And yet it appears that factors of safety used in design were essentially the same for levees protecting farmland adjacent to rivers and levees protecting New Orleans. The selection of a factor of safety of 1.3, with the benefit of hindsight, did not adequately account for either uncertainties or consequences of failure.

2. *Shear Strength Selected Unconservatively*. Subsoil conditions were investigated in the early 1980s when vertical expansion of the 17th Street Canal levees using I-Walls was considered. Soil borings were drilled, samples obtained, and shear strength tests (typically unconfined compression tests) performed (USACE, 1990).

The shear strength test results were plotted versus elevation to select the design strength profile. The critical elevation is near and slightly below the base of the I-Wall (El. -18 ft, or -5.5

m), where a slip surface would pass. The plot of shear strength vs. elevation from the original design memorandum for that critical elevation range is shown in Fig. 3 (see Seed et al., 2008c, Fig. 19, for the complete graph).

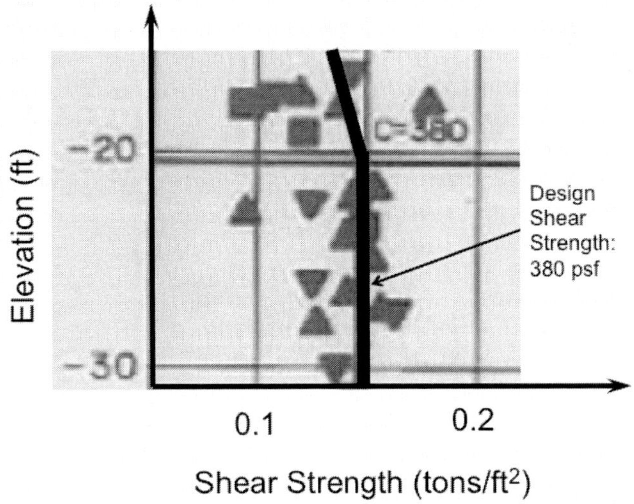

Figure 3. Shear Strength vs. Elevation from Design Memorandum [1 tsf = 2000 psf = 96 kPa] (from U.S. Army Corps of Engineers, 1990, Plate 56).

The most surprising observation from Fig. 3 is that more data points lie to the left of the design shear strength curve than to the right. This means that the design curve is an unconservative interpretation of the available shear strength data, with no explanation or justification offered in the design report. It is the author's experience that, unless there is a clear reason to interpret the data differently, the usual practice is to draw a conservative design curve from the available data, i.e., if anything, bias the design curve so that a majority of the shear strength test results are greater than the design values. The U.S. Army Corps of Engineers Engineer Manual for Retaining and Flood Walls (1989) recommends "...selection of design strengths at or below the thirty-third percentile of test results..." The actual selection was inconsistent with the Corps-recommended practice. The Chief of the Engineering Division, New Orleans District, reviewed the design and wrote a review letter stating that the shear strength trend that was drawn should not be used but, rather, a strength trend should be selected such that two-thirds of the test values exceed the design curve (Chatry, 1986). It remains a mystery why the engineer unconservatively interpreted the test data or why the reviewer's recommendation for a more conservative interpretation was not heeded.

3. *Shear Strength Averaged from Widely Spaced Borings*. The data points in Fig. 3 represent laboratory strength test results from borings spaced about 50 m apart, covering a length of approximately 2.4 km along the levee alignment. Lumping together data in this manner is not uncommon, but the prudent engineer looks for local anomalies, i.e., zones of lower strength. Two exploratory soil borings, No. 62 and 64, were drilled within the failure zone. Table 1 summarizes strength test results in the critical elevation zone (slightly below the base of the I-Wall). The average shear strength in the critical zone was 12.5 kPa (262 psf) for the two borings. This average strength was 31% less than the average used for design (18.2 kPa or 380 psf). Seed et al. (2008c) report that an even weaker seam of material was found in the post-failure investigation.

Table 1. Shear Strength Data from Boreholes Closest to 17th Street Canal Failure (USACE, 1990).

Borehole	Elevation (m below mean sea level)	Shear Strength (kPa)
62	-5.2	14.6
62	-8.2	12.4
64	-4.6	4.9
64	-8.1	18.3
AVERAGE		**12.5**

The design memorandum provides no explanation for why so many borings were lumped together in considering design shear strength or, more importantly, why local areas of lower shear strength were not considered. The unconservative interpretation of data (Fig. 3) combined with a low factor of safety (1.3) are especially perplexing considering the obviously variable soil conditions in New Orleans. The decision-making process that resulted in unconservative assumptions clearly did not make safety the top priority.

4. *Shear strengths measured from borings taken beneath crest of levees*. Drilling and soil sampling were conducted at the crest of existing levees, before the I-Walls were constructed, as indicated in Fig. 4. However, a failure surface, such as the one sketched in Fig. 1, intersects soil beneath and beyond the limits of the levee.

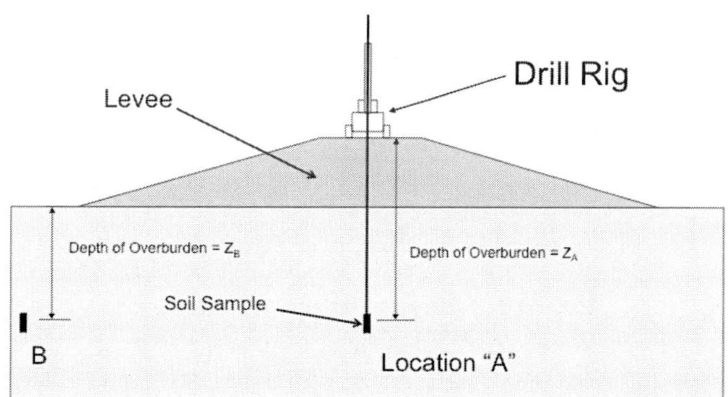

Figure 4. Exploratory Drill Holes Located at Crest of Levee Prior to Construction of I-Wall.

Soil sampled at Location "A" in Fig. 4 is consolidated to a larger vertical effective stress than soil located at "B" and, thus, should be stronger. No account was made for the fact that soil at the same elevation but at locations off the centerline along a potential slip surface should not be as strong as beneath the centerline.

Ladd (2009) provides a detailed discussion of some of the critical elements of shear strength testing and effects of vertical effective stress. This discussion and the closure by Duncan et al. (2009) make interesting and thought-provoking reading. But, nevertheless, the designers of levees along the 17th Street Canal made decisions that were unconservative, including the choice to measure shear strength at the single location (centerline of levee) where the soil was strongest.

5. *Warning signs from tests not heeded.* The U.S. Army Corps of Engineers performed a full-scale test of the sheet piling system for I-Walls termed the "E-99" test (USACE, 1988) at about the same time that design of I-Walls was being finalized for the 17th Street Canal. The test was performed at a site in the Atchafalaya Basin south of Morgan City, LA, where soil conditions were similar to those at the 17th Street Canal site. The purpose of the test was to confirm structural design criteria for sheet piling. The wall was loaded incrementally to a water depth of 2.4 m (8 ft). A PCV membrane separated the water from the sheet piling and subsoil. The calculated minimum factor of safety against slope-stability failure was 1.47 using the same methods-of-plane technique that was used for the 17th Street Canal. With a calculated factor of safety of 1.47, one would expect the wall to be stable.

Deformation at the top of the wall was measured and is plotted as a function of depth of impounded water in Fig. 5. The graph turns nearly horizontal as the water depth approaches 8 ft. The UASCE (1988) report states, "Although the test wall was not loaded to "failure," i.e., structural failure of the steel sheet piling or overturning of the wall, the plot … indicates failure may have been imminent as the head on the wall approached and exceeded 8 ft."

What is remarkable is what happened next: nothing. The reality of this test was that the theoretical factor of safety was 1.47 but the real factor of safety was essentially 1.0. Analysis was not converging with reality. Since reality is inalterable, one would have thought that these findings would have triggered a large-scale reassessment of the analysis and design criteria, but they evidently did not.

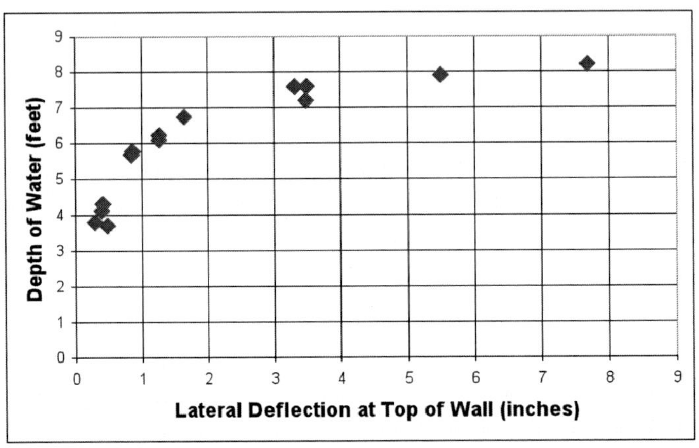

Figure 5. Deflection of E-99 Test Wall as a Function of Water Depth behind Wall (after UASCE, 1988).

Some years later, Oner et al. (1999) performed further analysis and concluded that "As the water level rises, the increased loading may produce separation of the soil from the pile on the flooded side (i.e., a "tension crack" develops behind the wall)." Indeed, a crack, such as sketched in Fig. 2, seems all but certain to have occurred at the 17th Street Canal (USACE, 2006; and Duncan et al., 2008). The author has not been able to locate any literature prior to Oner et al. (1999) providing specific warnings to consider cracks for design of I-Walls or similar structures, although the effect of vertical cracks on slope stability analysis in general has been known for decades. What the author has trouble understanding is why the E-99 test results, which showed conditions of imminent failure despite a calculated factor of safety of 1.47, did not trigger a pause and further analysis to consider the implications. In addition, once the problem of tension cracking behind I-Walls was clearly known and disseminated (Oner et al., 1999), there was no systematic effort to reanalyze the existing I-Walls to determine if they would be stable with this new, more limiting (but verified) design criterion.

LONDON AVENUE CANAL

The southern failure of the London Avenue Canal provides further insights into what went wrong with the levees in New Orleans. A cross section of the levee is shown in Fig. 6. The predominant underlying materials were sand. Failure occurred as a result of massive erosion of sand from beneath the levee and deposition on the landside – an obvious piping failure as sand was eroded from beneath the levee (UASCE, 2006). The question is: why?

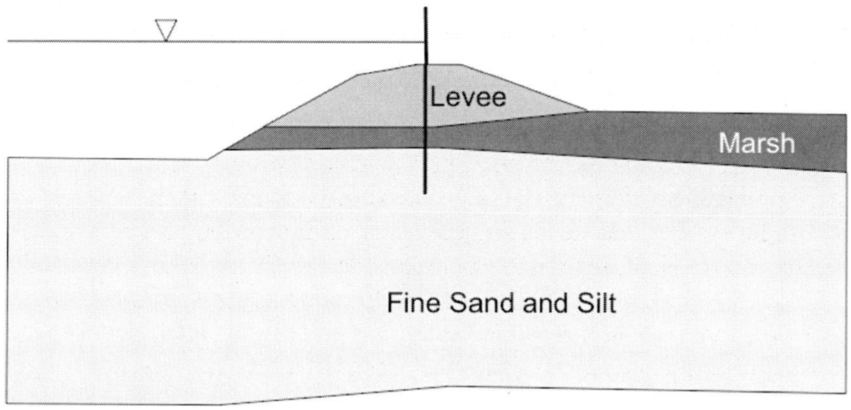

Figure 6. Cross Section beneath Southern Breach of London Avenue Canal (after USACE, 2006).

The author searched design documents for the flow net used in design. The labeling was not specific, but it appears that the flow net shown in Fig. 7 was the basis for analysis and design. The critical flow-net element at the toe of the levee is labeled by the author.

This is a poorly drawn flow net because: (1) the critical element is not square (i.e., it cannot be subdivided into square sub-elements); and (2) flow and equipotential lines are not perpendicular. But there is an even worse mistake: the low-permeability marsh layer at the surface is omitted! If the flow net had been correctly drawn with an impermeable layer covering the fine sand and silt to the right of the toe of the levee, the discharge point would be pushed further to the right in Fig. 7. The result would be uplift pressures beneath the marsh layer that were in excess of the weight of the marsh layer, pointing to instability involving uplift.

The flow net was a poorly drawn flow net for the wrong problem.

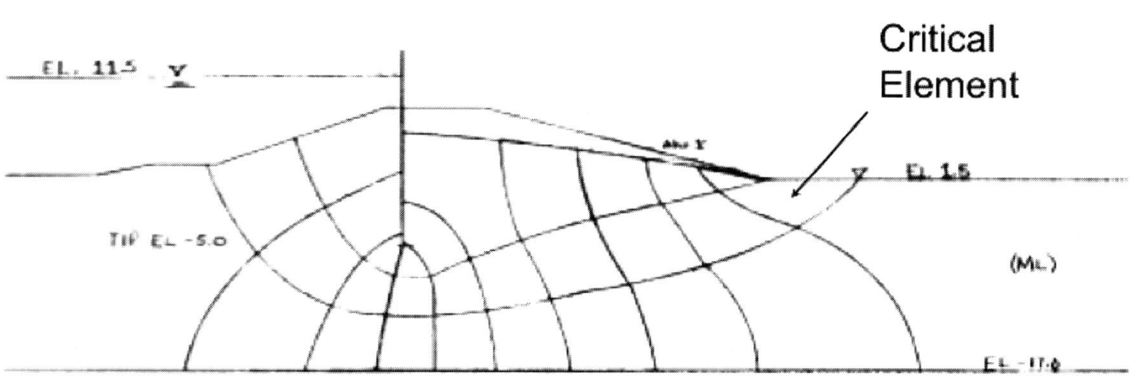

Figure 7. Flow Net Used for Design of London Avenue Canal Levee (USACE, 1989).

OTHER FACTORS

Other factors contributed to the inadequacies of the hurricane protection system in New Orleans (ASCE, 2007). The hurricane protection system was not a system at all but, rather, a collection of individual projects. A variety of local, city, state, and federal officials had certain responsibilities for components of the levee system, but no one was responsible for the whole system. In that sense, no one was in charge. Inspection and maintenance were inadequate. There was far too little external peer review that might have identified and corrected some of the flawed decisions.

DISCUSSION AND CONCLUSIONS

Engineers responsible for design of levees in New Orleans made deeply flawed decisions from the smallest level of detail to the largest scale of design parameters. Examples of flawed decisions made at the level of detailed design include selection of unconservative design shear strengths for the 17th Street Canal and use of an inadequate and inappropriate flow net to analyze under-seepage beneath the London Avenue Canal. Several higher-level decisions are also difficult to understand, such as why the alarming results from the E-99 test did not trigger wholesale re-analysis of I-Walls, why the levees were not designed to maintain their desired crest elevations after foreseeable settlement had occurred, or why survey datums that were not corrected for settlement. On an even larger scale of decision making, one wonders why new information about stronger hurricanes did not trigger an upgrading of the design hurricane. Perhaps the decision that is most difficult for the author to comprehend was the decision to construct levees out of highly erodible material that could not possibly survive the inevitable overtopping, dooming levees to destruction and adjacent areas to devastation when levees were overtopped.

How would one describe organizations that consistently, over a period of years, from lowest to highest levels, make critical decisions that trend toward the unsafe rather than safe? The author would describe such organizations as having a problem with safety culture. The author believes that the safety culture degraded because the funding agency (U.S. Congress) wanted to minimize the amount of money spent on levees, pressuring the engineer to develop less costly designs. Securing any funding for levees in New Orleans was a challenge. Hurricanes strike infrequently and memories are short. Over time, there can be a temptation to compromise on safety to meet the cost expectations of the funding agency. The challenge for the engineer is to maintain a healthy, functional safety culture despite these pressures.

At some point, the engineer may be asked to (or feel that he or she must) cross a line separating safe from un-safe design because of lack of funding. It is at that point that an organization's engineering leaders have a particular responsibility. One wonders what would have happened if the Chief Engineer of the Corps had gone before the U.S. Congress and said, "I will not design this project at this level of funding – it cannot be done safely."

The first canon of ASCE's code of ethics states: "Engineers shall hold paramount the safety, health and welfare of the public." Had the safety, health and welfare of the people of New Orleans always been the top priority, the decisions might have been different. Regardless, the failure of levees in New Orleans provides a valuable and sobering reminder of the responsibility that all engineers share to protect public welfare above all else.

ACKNOWLEDGMENTS

First and foremost, the author wishes to acknowledge the enormous human tragedy resulting from the levee failures in New Orleans and to express his sympathies to those whose lives were lost and whose friends and families suffered.

This paper is based upon the findings from an ASCE (2007) report that was written by a dedicated group of talented individuals: Christine F. Anderson, Jurjen A. Battjes, Billy Edge, William Espey Jr., Robert B. Gilbert, Thomas L. Jackson, David Kennedy, Dennis S. Mileti, James K. Mitchell, Peter Nicholson, Clifford A. Pugh, George Tamaro, and Robert Traver. Lawrence H. Roth provided invaluable assistance.

Numerous engineers not associated with the ASCE Panel performed exhaustive investigations individually and collectively. The ASCE Panel relied especially heavily on the Interagency Performance Evaluation Task Force (IPET), whose report was published by UASCE (2006) and that was chaired by Lewis B. Link. The dedicated work of individuals serving on the IPET, National Science Foundation, and various other teams contributes significantly to the body of knowledge. On behalf of the entire profession, this work is gratefully acknowledged.

Finally, as the author reflects on how he got to a point in his career where synthesis of such a complex project was even possible, he owes a special debt of gratitude to his graduate advisor, Prof. Roy E. Olson. If the author has any skills at all in geotechnical engineering, the root of such skill is traceable to the countless hours that Prof. Olson spent with the author teaching, challenging, and showing. He set the highest possible standard for dedication to his profession and for ethical conduct. Most importantly, he taught the author a way of thinking. It is for this gift, along with his guidance and friendship, that the author humbly extends to Prof. Olson his deepest possible expression of appreciation, gratitude, and respect.

REFERENCES

American Society of Civil Engineers (2007). "The New Orleans hurricane protection system: what went wrong and why," ASCE, Reston, VA.

Briaud, J.-L., Chen, H.-C., Govindasamy, A.V., and Storesund, R. (2008). "Levee erosion by overtopping in New Orleans during the Katrina Hurricane," *J. Geotech. & Geoenv. Eng.*, 134 (5), 618-632.

Brunkhead, J., Namulanda, G., and Ratard, R. (2008). "Hurricane Katrina Deaths, Louisiana, 2005," *Disaster Medicine and Public Health Preparedness,* http://new.dhh.louisiana.gov /assets/docs/katrins.deceasedreports/KatrinaDeaths 082008.pdf

Chatry, F. M. (1986). Letter Report to Mr. John Holtgreve, Design Engineering Inc., February 10, Department of the Army, New Orleans District Corps of Engineers.

Eustis Engineering (1988). Letter Report to Mojeski and Masters, 31 August, Matairie, LA.

Sasanakul, I., Vanadit-Ellis, W., Sharp, M., Abdoun, T., Ubilia. J., Steedman, S., and Stone, K. (2008). "New Orleans levee system performance during Hurricane Katrina: 17th street canal and Orleans canal north," *J. Geotech. & Geoenv. Eng.*, 134 (5), 657-667.

Duncan, J.M., Brandon, T.L., Wright, S.G., and Vroman, N. (2008). "Stability of I-walls with gaps between the I-wall and the levee fill," *J. Geotech. & Geoenv. Eng.*, 134 (5), 681-691.

Duncan, J.M., Brandon, T.L., Wright, S.G., and Vroman, N.D. (2009). "Closure to 'Stability of I-Walls in New Orleans during Hurricane Katrina'," *J. Geotech. & Geoenv. Eng.*, 135 (12), 2002-2006.

Graham, H.E., and Nunn, D.E. (1959). "National Hurricane Research Project Report No. 33, Meteorological conditions pertinent to standard project hurricane, Atlantic and Gulf Coasts of the United States," U.S. Dept. of Commerce, Weather Bureau, Washington, DC.

Ladd, C.C. (2009). Discussion to 'Stability of I-Walls in New Orleans during Hurricane Katrina'," *J. Geotech. & Geoenv. Eng.*, 135 (12), 1999-2002.

National Oceanic and Atmospheric Administration (1979). "Meteorological criteria for standard project hurricane and probable maximum hurricane windfields, Gulf and East Coasts of the United States," Washington, DC.

National Oceanic and Atmospheric Administration (2005). "Hurricane Katrina, a climatological perspective," Washington, DC.

Seed, R.B., Bea, R.G., Abdelmalek, R.I., Althanasopoulos-Zekkos, A., Boutwell, G.P., Briaud, J.-L., Cheung, C., Cobos-Roa, D., Ehrensing, L. Govindasamy, A.V., Harder, J.F. Jr., Inkabi, K.S., Nicks, J., Pestana, J.M., Porter, J., Rhee, K., Riemer, M.F., Rogers, J.D., Storesund, R., Vera-Grunauer, X., and Wartman J. (2008a). "New Orleans and Hurricane Katrina. I: Introduction, overview, and the east flank," *J. Geotech. & Geoenv. Eng.*, 134 (5), 701-717.

Seed, R.B., Bea, R.G., Althanasopoulos-Zekkos, A., Boutwell, G.P., Bray, J.D., Cheung, C., Cobos-Roa, D., Ehrensing, L. Harder, J.F. Jr., Pestana, J.M., Riemer, M.F., Rogers, J.D., Storesund, R., Vera-Grunauer, X., and Wartman J. (2008b). "New Orleans and Hurricane Katrina. II: The central region and the lower ninth ward," *J. Geotech. & Geoenv. Eng.*, 134 (5), 718-739.

Seed, R.B., Bea, R.G., Althanasopoulos-Zekkos, A., Boutwell, G.P., Bray, J.D., Cheung, C., Cobos-Roa, D., Harder, J.F. Jr., Ross, R.E.S., Pestana, J.M., Riemer, M.F., Rogers, J.D., Storesund, R., Vera-Grunauer, X., and Wartman J. (2008c). "New Orleans and Hurricane Katrina. III: The 17th Street drainage canal," *J. Geotech. & Geoenv. Eng.*, 134 (5), 740-761.

U.S. Army Corps of Engineers (1988). "E-99 sheet pile wall field load test report," Vicksburg, MS.

U.S. Army Corps of Engineers (1989). "London Avenue outfall canal," Design Memorandum No. 19A, Vol. II, New Orleans District, LA.

U.S. Army Corps of Engineers (1990). "Design memorandum No. 20, general design, 17th Street Outfall Canal (Matairie Relief)," Vol. 1, New Orleans, LA.

U.S. Army Corps of Engineers (2000). "Design and construction of levees," Engineer Manual EM110-2-1913, Washington, DC.

U.S. Army Corps of Engineers (2003). "Slope stability," Engineer Manual EM110-2-1902, Washington, DC.

U.S. Army Corps of Engineers (2006). "Performance evaluation of the New Orleans and southeast Louisiana hurricane protection system," Washington, DC.

Insights into the Performance Reliability of Offshore Piles Based on Experience in Hurricanes

Jiun-Yih Chen[1], A.M. ASCE and Robert B. Gilbert[2], M. ASCE, P.E.

[1]Principal Technical Professional – Marine, Energo Engineering (A KBR Company), 1300 W. Sam Houston Pkwy. S., Suite 100, Houston, TX 77042
[2]Brunswick-Abernathy Regents Professor, The University of Texas at Austin, 301 E. Dean Keeton St. Stop C1792, Austin, TX 78712

ABSTRACT: Between 2004 and 2009, five major hurricanes in the Gulf of Mexico loaded numerous offshore platforms and their pile foundations to near or beyond their capacities. The performances of these foundation systems provide a tremendous opportunity to learn about their reliability and to improve design practice. A set of thirteen platforms are analyzed in detail based on hindcast data for the hurricane loads. The foundation systems for these platforms represent a range of characteristics with non-redundant and highly redundant systems, systems governed by the lateral capacities and systems governed by the axial capacities of the piles, and systems with piles driven primarily in normally consolidated marine clays and those with piles driven in interbedded sands and clays. An upper-bound plasticity model, calibrated to match a three-dimensional finite element method model, is used to predict the performance of these foundation systems. A probabilistic approach is then used to update information about the systematic bias in predicting the capacities of the foundation systems and hurricane loads. Findings from this work include (a) actual and predicted performances of the thirteen foundation systems agree reasonably well, (b) the use of nominal versus expected yield stress for steel affects the predicted performance and reliability, and (c) the robustness and redundancy of a foundation system have a significant effect on its reliability.

INTRODUCTION

Between 2004 and 2009, five major hurricanes in the Gulf of Mexico loaded numerous offshore platforms and their pile foundations to near or beyond their capacities. The performances of these foundation systems provide a tremendous opportunity to learn about their reliability and to improve design practice. A set of thirteen offshore platforms were analyzed in detail based on hindcast data for the hurricane loads. The foundation systems for these platforms represent a range of characteristics with non-redundant and highly redundant systems, systems governed by the lateral capacities and systems governed by the axial capacities of the piles, and systems with piles driven primarily in normally consolidated marine clays and those with piles driven in interbedded sands and clays (Table 1).

Offshore pile foundation design involves both geotechnical and structural engineers. The geotechnical engineer establishes ultimate axial pile capacities as a function of pile diameter and length following the normal industry practice for offshore platforms in the Gulf of Mexico (API 2000 and 2011). The geotechnical engineer also provides non-linear relationships between unit side shear and relative shear displacement between the pile wall and soil (t-z curves), end bearing resistance and displacement of the pile tip (Q-z curves), and lateral resistance and pile deflection (p-y curves) for structural analyses.

The structural engineer then performs structural analyses incorporating these soil springs with different pile diameters and wall thicknesses using very long piles. Several combinations of dead loads, live loads and environmental loads (e.g., wind, wave, current, etc.) are used to check the structural capacity of all steel tubular members and joints including the piles according to API (2000). After pile diameters and wall thicknesses are selected, a structural analysis is performed on the final configuration of the structure to select pile lengths. The design axial loads of the piles are typically determined by applying the 100-year metocean combination of wind, wave and current loads to the structure. These design axial loads are increased by a factor of safety of 1.5 to determine the required ultimate axial capacities of piles according to API (2000). The pile lengths are then selected to satisfy the required ultimate axial capacities.

Table 1. Foundation characteristics of 13 case study platforms

Platform No.	Water Depth (m)	No. of Piles	No. of Wells	Hurricane Exposure	Foundation Failure Mechanism	Soil Profile
1	42	8	18	Katrina	Axial/Lateral	Sand and Clay
2	42	6	12	Katrina	Axial	Sand and Clay
8	66	4	12	Katrina	Axial	Sand and Clay
9	19	4	1	Katrina	Axial	Sand and Clay
10	109	3	1	Ike	Axial	Clay
11	36	4	4	Katrina	Lateral	Sand and Clay
12	57	4	12	Rita	Lateral	Sand and Clay
22	34	4	None	Rita	Axial/Lateral	Sand and Clay
25	28	4	4	Katrina	Axial/Lateral	Sand and Clay
27	92	4	2	Rita	Lateral	Clay
29	47	8	12	Katrina	Axial/Lateral	Sand and Clay
30	47	6	12	Katrina	Lateral	Sand and Clay
31	30	8	None	Ike	Axial/Lateral	Sand and Clay

FOUNDATION SYSTEM CAPACITY MODEL

An upper-bound plasticity model (Murff and Wesselink 1986, Tang and Gilbert 1993, Chen et al. 2009 and Murff 2012), calibrated to match a three-dimensional finite element method model, was used to back analyze the performances of these foundation systems. This model, schematically shown in Figure 1(a), assumes a plastic collapse mechanism, where all elements of resistance are characterized as rigid and perfectly plastic. This model incorporates structural capacities of piles and well conductors as well as axial and lateral soil resistance acting on the piles and conductors. The structure supported by the piles is assumed to be perfectly rigid and infinitely strong so that it can distribute loads as necessary to develop a full foundation collapse mechanism. Well conductors are modeled as piles that are connected to the structure with rollers to constrain them horizontally. Each pile or conductor in the system collapses when one hinge forms at the mudline and a second hinge forms at some depth below the mudline. Individual piles can fail either structurally due to the combination of axial load and bending moment or by plunging or pulling out from the soil axially (note that axial capacities in compression versus tension are distinguished). Collapse of the entire system occurs when double hinges form in each of the piles and conductors, while mobilizing the axial and lateral soil resistance along the piles and conductors. Therefore, the capacity of the foundation system depends on the structural capacity for individual piles, the lateral and axial soil resistance along the piles, and the geometry of the pile system including the direction and magnitude of batter for individual piles.

Figure 1(b) presents a schematic of the foundation system capacity interaction diagram. The interaction curve is an envelope; for hurricane loads located within the envelope, the foundation system is expected to be stable. The potential failure mechanism depends on the resultant of the environmental loads. In the shear failure mechanism, the environmental loads are applied relatively close to the mudline (i.e., with a small moment arm), whereas in the overturning failure mechanism, they are applied at a higher location. These distinct regions of foundation system behavior are shown in Figure 1(b). In the shear failure mechanism, structural capacities of the piles and conductors contribute primarily to the system capacity. In the overturning failure mechanism, axial soil resistances on the piles contribute primarily to the system capacity. The base shear capacity in the shear failure mechanism increases with increasing overturning moment (or moment arm) because of pile batters.

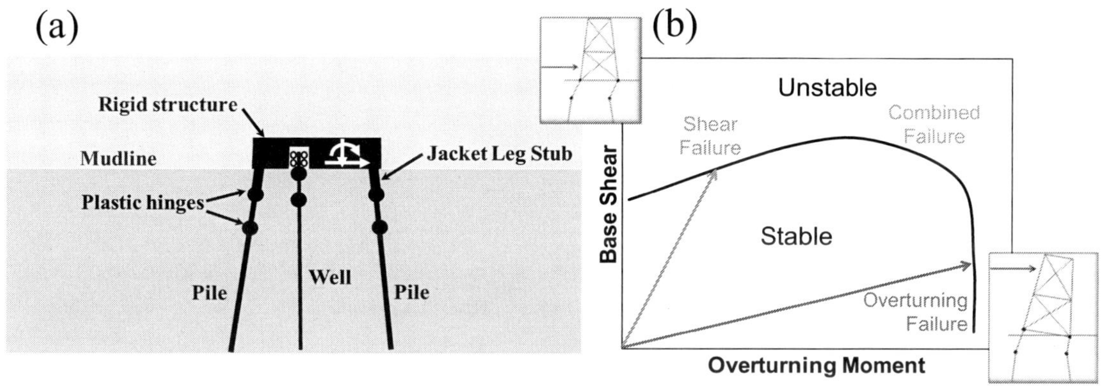

Figure 1. Schematics of (a) foundation system capacity model and (b) foundation system capacity interaction diagram

COMPARISON OF PREDICTED AND OBSERVED PERFORMANCES

Gilbert et al. (2010) compared the predicted and observed performances of the thirteen platform foundations shown in Table 1. The upper-bound plasticity model was used to estimate the foundation system capacity based on available information (e.g., geotechnical reports, structural drawings, etc.). The hurricane load, consisting of base shear and overturning moment, on the foundation system was estimated based on the hindcast wind, wave and current conditions at the platform location using a 3-dimensional (3-D) structural model. For example, Figure 2 shows the foundation system capacity interaction diagram of Platform 9 when loaded by Hurricane Katrina. The potential failure mechanism of this foundation system is overturning. The reserve strength ratio (RSR) at the peak of hurricane load was estimated by dividing the base shear capacity (R) by the base shear load (S) at the moment arm of the environmental load (indicated by the dashed line) to be 1.24. As such, this foundation system is not expected to fail and it did survive the hurricane. The predicted and observed performances are consistent.

Results from back analyses of all thirteen platform foundations are summarized in Table 2. As shown, for the first nine platforms, the predicted and observed performances are consistent. When the foundation system is not expected to fail, it survived the hurricane loading (e.g., Platforms 8, 22, 25, 27, 29). In some cases (e.g., Platforms 9, 12 and 30), the shear and moment capacities contributed by the well conductors need to be included to explain the survival of the foundation system, which underscores the importance of considering the capacity of the well conductors in addition to the environmental loads they attract in an ultimate-strength assessment. In one case (e.g., Platform 10), the foundation system is expected to fail in overturning and it did fail in Hurricane Ike due to the pull-out failure of the most critically loaded pile and bending failures of the other two piles (Chen et al. 2013).

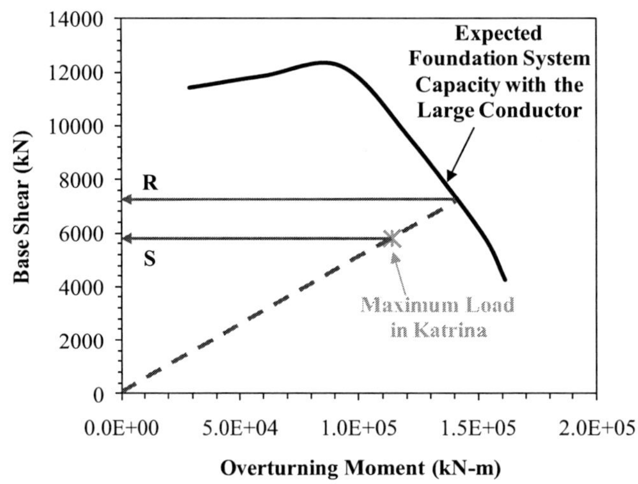

Figure 2. Foundation system capacity interaction diagram of Platform 9 showing base shear capacity (R) and load (S)

Table 2. Hurricane performance of 13 case study platforms

Platform No.	RSR= R/S*	Predicted Performance	Observed Performance	Comments
8	1.22	Foundation Capacity Not Exceeded	Survival	Predicted and Observed Performances Consistent
22	1.07			
25	1.77			
27	1.22			
29	1.42			
9	1.24	Foundation Capacity with Conductors Not Exceeded	Survival	Predicted and Observed Performances Consistent
12	1.45			
30	1.29			
10	1.00	Potential Overturning Failure	Failure	Foundation Failure Predicted and Observed
11	***1.00***	Potential Shear Failure	Survival	More Realistic Lateral Capacity Assumptions Explain Survival
31	***0.84***			
1	***0.63***	Potential Overturning Failure	Survival	Lack of Site-Specific, Modern Soil Borings and CPTs
2	***0.64***			

*Base shear capacity (R) includes the contribution of well conductors, if present.

Platforms 11 and 31 survived potential foundation shear failures. Structural factors and more realistic lateral capacity assumptions can explain the survival of these foundation systems (e.g., Chen et al. 2010). For Platform 11, the higher expected (vs. nominal) yield stress of the piles alone is sufficient to show that the actual foundation system capacity is higher than the hurricane load; hence, the survival of this foundation system is not a surprise. For Platform 31, more realistic lateral capacity assumptions, such as static (vs. cyclic) p-y curves, expected (vs. nominal) yield stress of the piles and modeling the additional lateral soil resistance due to the presence of leg stubs penetrating below the mudline, which push the plastic hinges to form at greater depths, help explain the survival of this foundation system.

Platforms 1 and 2 survived potential foundation overturning failures, where the capacity is dominated by the axial capacities of piles. The piles supporting these bridge-connected platforms are embedded in interbedded layers of sands and clays corresponding to alluvial/fluvial geologic settings. Their axial capacities can be highly variable due to the spatial variability in not only the depths and thicknesses of the sand layers but also the density of sands. A site-specific soil boring was not available for the assessment of these platforms. Hence, a boring located approximately 300 m away from the platform locations was used. Further, a Driven Penetration Test versus a modern Cone Penetration Test (CPT) was used to characterize the density of sands in this available boring. Plenty of evidence suggests that the density of the sands at the pile tips is much higher than indicated in the available boring (Gilbert et al. 2010). Sensitivity analyses show that with higher density of sands at the pile tips and more realistic lateral capacity assumptions, the survival of these foundation systems can be explained.

UNCERTAINTIES IN HURRICANE LOAD AND FOUNDATION CAPACITY

Chen (2011) assessed the uncertainties in the estimated hurricane load and foundation system capacity of Platform 10. This 3-pile foundation system failed in Hurricane Ike due to the pull-out failure of the most critically loaded pile and bending failure of the other two piles. Back analyses confirmed that the failure is expected (Chen et al. 2013). The RSR of this foundation system at the peak of hurricane load was approximately 1.0 (Figure 3).

The uncertainty in the estimated load on the foundation system during the peak of Hurricane Ike from the hindcast data was assessed based on the methodology presented in ABS Consulting (2004). This methodology is based on the variability of wave heights during a hurricane (Forristall 1978) and the relationship between wave height and platform loading (e.g., Manuel et al. 1998 and Bea et al. 1999). Chen (2011) shows with this methodology that the total uncertainty in the estimated base shear load is represented approximately by a lognormal distribution with a coefficient of variation (c.o.v.) of 0.35. Similarly, the estimated overturning moment has a c.o.v. of 0.37 and the correlation coefficient between base shear and overturning moment is nearly one.

This information about uncertainty in the hindcast load on the foundation system is depicted in Figure 3a. The reliability contours represent combinations of applied base shear and overturning moment for comparison with the system capacity interaction curve (Winterstein and Kumar 1995). Each reliability contour corresponds to a constant probability of exceedance; two contours are shown for probabilities of exceedance equal to 0.1 and 0.25 (Figure 3a). The probability that load on the system exceeds a given capacity is obtained from where the interaction curve for the system capacity is tangent to the reliability contour. For example in Figure 3a, if the foundation system is dominated by its base shear capacity (i.e., a horizontal interaction curve), then there is a 25-percent probability that the load in Hurricane Ike would have exceeded a base shear capacity of 6,800 kN and a 10-percent probability that the hurricane load would have exceeded a base shear capacity of 8,400 kN (Figure 3a).

Likewise, if the foundation system is dominated by its overturning capacity (i.e., a vertical interaction curve), then there is a 25-percent probability that the hurricane load would have exceeded an overturning capacity of 6.2×10^5 kN-m and a 10-percent probability that it would have exceeded an overturning capacity of 7.7×10^5 kN-m (Figure 3a).

The uncertainty in the estimated foundation system capacity was also assessed. Variables affecting foundation system capacity include axial soil resistance (T), lateral soil resistance (P) and yield stress of pile steel (F). These three variables were assumed to be independent of one another. While both axial and lateral soil resistances are related to the shear strength of soils, the justification for assuming independence between them is that the soils near the pile tips generally affect axial capacity more, while those at shallower depths affect lateral capacity. The c.o.v. in T, P and F were assumed to be 0.3, 0.3 and 0.15, respectively. Using the approach presented in Chen (2011), the mean plus and minus one standard deviation capacity envelopes were developed (Figure 3a). As shown, the foundation system capacity in the overturning failure mechanism is more variable than the capacity in the shear failure mechanism; the c.o.v.'s of base shear capacity in overturning and shear failure mechanisms are 0.25 and 0.15, respectively.

Figure 3(b) shows the probability distribution of RSR for the foundation system of Platform 10 at the peak of hurricane load. This probability distribution is obtained by combining the models accounting for uncertainty in the hurricane load and uncertainty in the foundation system capacity. The shaded area represents the likelihood that this foundation system would fail in Hurricane Ike (i.e., RSR< 1.0), if there is no systematic bias in the estimated capacity or load; the likelihood is approximately 47 percent.

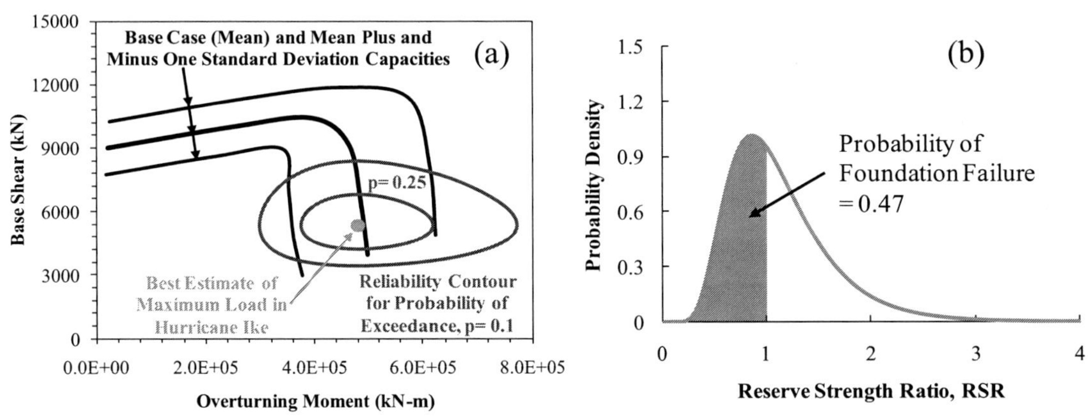

Figure 3. (a) Foundation system capacity interaction diagram of Platform 10 with reliability contours of hurricane load and mean plus and minus one standard deviation capacities, and (b) probability distribution of RSR

BAYESIAN UPDATING BASED ON HURRICANE PERFORMANCE

The Bayesian approach (Ang and Tang 2007) was used to update information about the systematic bias in predicting the capacities of the foundation systems and hurricane loads based on the performances of selected platforms. The reserve strength ratio (RSR) is then formulated as:

$$RSR = B \times (R/S) \tag{1}$$

where B is the systematic bias, R is the estimated foundation system capacity and S is the estimated hurricane load on the foundation.

A B-value greater than 1.0 indicates a conservative bias in the approaches used to estimate foundation system capacity and load. A diffuse prior uniform distribution was used for the bias. The likelihood function was developed as the probability of observing the performance given a range of B-values (e.g., 0.01 to 4.0). For example, the likelihood that the foundation system of Platform 10 would fail due to the loading in Hurricane Ike if there is no systematic bias (i.e., B= 1.0) is 0.47 (Figure 3b). Therefore, the likelihood function at a B-value of 1.0 is 0.47. Note that this result is consistent with earlier work comparing measured and predicted axial capacities for driven piles in normally to slightly overconsolidated clays (e.g., Olson and Dennis 1982). This earlier work, based on pile load tests, indicated that the bias on average was about one and there was roughly a 50-50 chance for any given measured capacity to be less than or greater than the predicted capacity. This new data point complements the existing databases by providing information of full-scale piles under in-service loading conditions.

The updated distribution of B in light of the performance of the foundation system supporting Platform 10 in Hurricane Ike was developed using the Bayesian approach (Figure 4). This foundation system failed in Hurricane Ike due to the pull-out failure of the most critically loaded pile embedded in normally consolidated to slightly overconsolidated marine clays. Because the observed performance is a foundation failure, the bias is more likely to be a value less than 1.0, as shown in Figure 4.

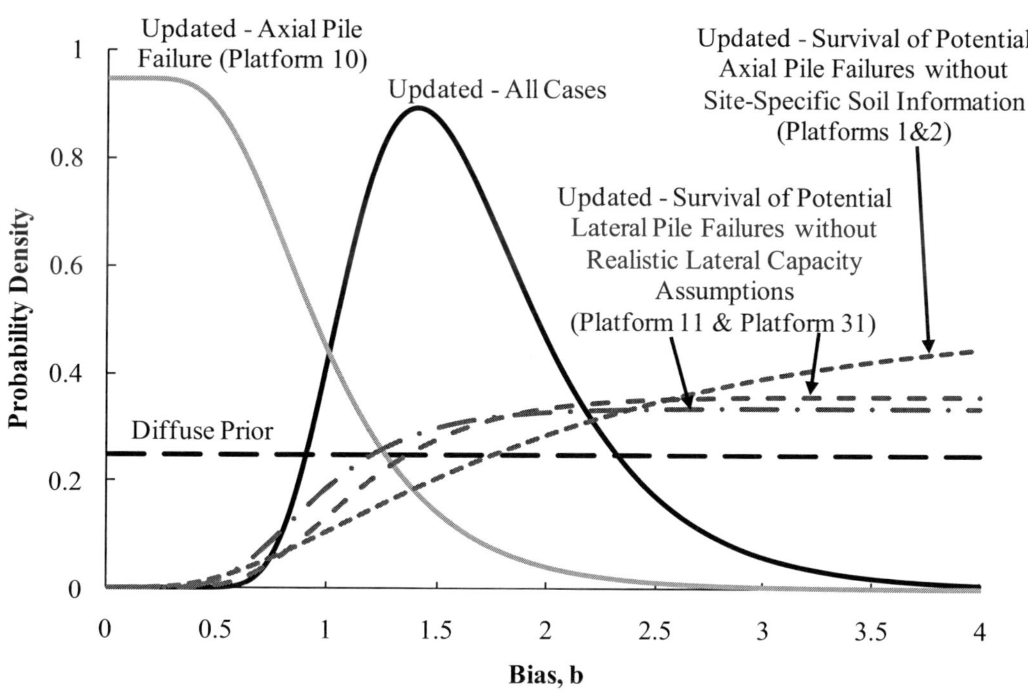

Figure 4. Prior and updated distributions of bias according to the performances of Platform 10, Platforms 1 & 2, Platform 11, and Platform 31

Likewise, the updated distributions of B according to the performances of Platforms 1 and 2, Platform 11, and Platform 31 were developed (Figure 4). The foundation systems supporting the bridge-connected Platforms 1 and 2 survived potential overturning failures in Hurricane Katrina. Their piles are embedded in interbedded layers of sands and clays corresponding to alluvial/fluvial geologic settings. Additionally, because there is no site-specific soil boring or modern CPT data for the assessment of these platforms, the uncertainty in foundation system capacity is relatively high with a c.o.v. on the order of 0.7. As a result, the updated distribution is relatively flat. Also, because the observed performances are unexpected survivals, the bias is more likely to be a value greater than 1.0, as shown in Figure 4. Note that because these platforms are located very close to each other and loaded by the same storm, the performances of these platforms essentially represent one proof load test.

The foundation systems supporting Platforms 11 and 31 survived potential shear failures. Because the foundation system capacity in shear is less variable than in overturning, the updated distributions according to the performances of these platforms are steeper than the updated distribution according to the performances of Platforms 1 and 2. Similarly, because the observed performances are unexpected survivals, the bias is more likely to be a value greater than 1.0.

The updated distribution according to all of the above observations can also be developed assuming these observations are independent of one another. The justification of assuming independence is that these cases are located in different regions in the Gulf of Mexico, their geologic settings are different (e.g., marine clay vs. alluvial/fluvial), they were loaded by different storms and, lastly, their potential failure mechanisms are different (e.g., shear vs. overturning). The updated distribution (Figure 4) indicates a conservative bias in estimating the capacities of the foundation systems and hurricane loads. This result is driven primarily by the observed performances of Platforms 1 and 2, where site-specific soil data were not available for a case of potential overturning failure in alluvial/fluvial geologic settings, and Platforms 11 and 31, where the shear capacities of the foundation systems were underestimated because more realistic lateral capacity assumptions were not incorporated in the assessment practice.

One final observation on the Bayesian updating of bias using the diffuse uniform prior distribution is that both cases of foundation failures and survivals are necessary in order to achieve an updated probability distribution of bias that is approximately bell-shaped. In the case of observed foundation failure, the probability decreases as the bias increases, whereas in the case of observed foundation survival, the probability increases with increasing bias.

SYSTEM ROBUSTNESS AND REDUNDANCY

The foundations of Platforms 10 and 30 represent two very different systems. Platform 10, supported by three piles and one well, is a non-redundant system, while Platform 30, supported by 6 piles and 12 wells, is a more redundant system. The foundation of Platform 10 failed in overturning in Hurricane Ike, whereas the foundation of Platform 30 survived a potential shear failure in Hurricane Katrina. Sensitivity analyses were performed by increasing and decreasing the axial and lateral soil resistance of each of the piles by 30 percent to establish the possible bounds for the variations of foundation system capacity due to uncertainty in soil resistance. Figure 5 shows that the foundation system capacity of a non-redundant system (Platform 10) is more affected by the possible variations in soil resistance than a more redundant system (Platform 30). The shear capacities of both systems are less affected by the variations in soil resistance than their overturning capacities. These observations show that system redundancy contributes to built-in robustness of the foundation system. A component-based design approach (e.g., API 2000), where individual piles are designed to have a minimum factor of safety of 1.5 under the design environmental events (e.g., 100-year storm), will result in different levels of performance and reliability.

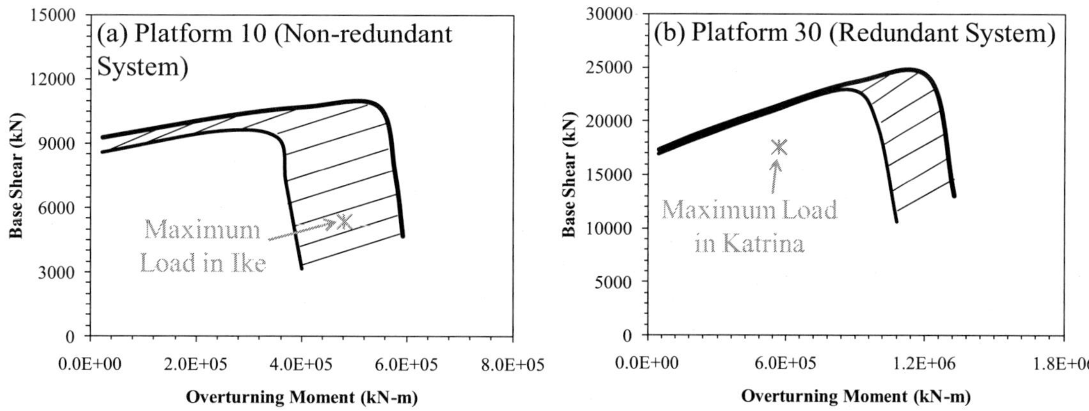

Figure 5. Comparison of foundation system robustness for (a) three-pile system and (b) six-pile system (after Chen et al. 2010)

CONCLUSIONS

The primary findings and conclusions based on the performances of all thirteen case study platforms in recent hurricanes are as follows:

1. The actual performances for nine of the thirteen foundation systems match the predicted performances. In the one case where a foundation failure is predicted to be likely, it occurred. While this data set is small, these results are promising in supporting the current practice for foundation design.
2. The nominal biases included in modeling structural properties in design, such as the use of nominal (vs. expected) yield stress for steel, affect the predicted versus actual performances and, consequently, the performance reliabilities of the foundation systems when uncertainties in the loads and capacities are considered.
3. The robustness and redundancy of a foundation system can have a significant effect on its reliability. Current design practice, which is based on evaluating individual piles, can be improved with system-level checks, such as an ultimate-strength pushover analysis.

ACKNOWLEDGMENTS

The authors appreciate the financial support of the former U.S. Minerals Management Service (MMS) and American Petroleum Institute (API) through the Offshore Technology Research Center (OTRC) for research projects from which this paper is derived. Frank Puskar, Sean Verret, Aditya Hariharan and Matthew Garcia of Energo Engineering were helpful in providing data and structural analyses of the case study platforms. Britain Materek of HNTB and Justin Carpenter of Shell, former graduate students of UT-Austin, were helpful in the analyses of several case study foundations. Don Murff, Alan Young of GEMS and Philippe Jeanjean of BP provided valuable review and guidance throughout the research projects. The first author acknowledges funding from the KBR TIIP program, which supports the preparation of this paper. The views and opinions contained in this paper are those of the authors alone and do not necessarily reflect those of our employers, sponsors or collaborators.

REFERENCES

ABS Consulting, Inc. (2004), *Hurricane Lili's Impact on Fixed Platforms*, Final Report

Prepared for Minerals Management Service, June.

Ang, A.H-S. and Tang, W.H. (2007), *Probability Concepts in Engineering, Emphasis on Applications to Civil and Environmental Engineering*, 2nd Edition, John Wiley & Sons, Inc., 406 pp.

API (2000), *Recommended Practice for Planning, Designing and Constructing Fixed Offshore Platforms – Working Stress Design*, American Petroleum Institute, API Recommended Practice 2A-WSD, 21st Edition.

API (2011), *Geotechnical and Foundation Design Considerations*, American Petroleum Institute, ANSI/API Recommended Practice 2GEO, 1st Edition.

Bea, R.G., Xu, T., Ramos, R. and Valle, O. (1999), "Reliability-based design criteria for floating drilling & production structures in the Bay of Campeche," *Proceedings of Offshore Technology Conference*, Paper Number OTC 11065, May 3-6, Houston, Texas.

Chen, J.Y. (2011), *Analysis of Performance and Reliability of Offshore Pile Foundation Systems based on Hurricane Loading*, Ph.D. Dissertation, The University of Texas at Austin, 287 pp.

Chen, J.Y., Gilbert, R.B., Murff, J.D., Young, A.G. and Puskar, F.J. (2010), "Structural factors affecting the system capacity of jacket pile foundations," *Proceedings of the Second International Symposium on Frontiers in Offshore Geotechnics*, November 8-10, Perth, Australia.

Chen, J.Y., Gilbert, R.B., Puskar, F.J., Verret, S. (2013), "Case study of offshore pile system failure in Hurricane Ike," *Journal of Geotechnical and Geoenvironmental Engineering*, ASCE, Vol. 139, No. 10.

Chen, J.Y., Materek, B.A., Carpenter, J.F., Gilbert, R.B., Verret, S. and Puskar, F.J. (2009), *Analysis of Potential Conservatism in Foundation Design for Offshore Platform Assessment*, Final Project Report Prepared for the Minerals Management Service, MMS Project Number 612.

Forristall, G.Z. (1978), "On the statistical distribution of wave heights in a storm," *Journal of Geophysical Research*, Vol. 83, pp. 2353-2358.

Gilbert, R.B., Chen, J.Y., Materek, B., Puskar, F., Verret, S., Carpenter, J., Young, A. and Murff, J.D. (2010), "Comparison of observed and predicted performance for jacket pile foundations in hurricanes," *Proceedings of Offshore Technology Conference*, Paper No. OTC 20861, May 3-6, Houston, Texas.

Manuel, L., Schmucker, D.G., Cornell, C.A. and Carballo, J.E. (1998), "A reliability-based design format for jacket platforms under wave loads," *Marine Structures*, Vol. 11, pp. 413-428.

Murff, J.D. (2012), "Estimating capacities of offshore foundations," The McClelland Lecture.

Murff, J.D. and Wesselink, B.D. (1986), "Collapse analysis of pile foundations," *Proceedings of the 3rd International Conference on Numerical Methods in Offshore Piling*, Nantes, France, 445-459.

Olson, R.E. and Dennis, N.D. (1982), *Review and Compilation of Pile Test Results, Axial Pile Capacity*, Final Report to American Petroleum Institute, Project PRAC 81-29, The University of Texas at Austin.

Tang, W.H. and Gilbert, R.B. (1993), "Case study of offshore pile system reliability," *Proceedings of Offshore Technology Conference*, Paper No. OTC 7196, May 3-6, Houston, Texas.

Winterstein, S.R. and Kumar, S. (1995), "Reliability of floating structures: extreme response and load factor design," *Proceedings of Offshore Technology Conference*, Paper No. OTC 7758, May 1-4, Houston, Texas.

Long Term Field Measurement of Micro Meter Crack Response to Climatological and Blast Vibration Induced Effects

Charles H. Dowding[1], Ph.D., P.E., M.ASCE, and
Pierre-Alexandre Abeel[2], M.ASCE

[1]Professor of Civil and Environmental Engineering, Northwestern University, 2145 Sheridan Road, Evanston, IL 60208; c-dowding@northwestern.edu

[2]Graduate Student of Civil and Environmental Engineering, Northwestern University, 2145 Sheridan Road, Evanston, IL, 60208; abeel.pa@u.northwestern.edu

ABSTRACT: This paper compares more than two years of micro meter crack displacement response and structural velocity response of a wood frame house adjacent to an operating aggregate quarry. The same transducers that monitor long term, climatologically induced micro inch crack response also measure dynamic responses induced by blast-induced ground motions, occupant activities and wind gusts. The two story house has been expanded several times, is founded on an irregular basement, framed in wood and clad with wood siding and clapboard. This project is part of a larger research effort of structural health monitoring via autonomous crack monitoring [ACM]. The unusually long monitoring period allowed observation of two maximum, climatologically induced peak crack responses, and as such represents one of the longest periods of continuous observation in the literature. These once a year peak climatological responses are compared with unusually intensive ground motions in excess of that allowed by regulation. Intense ground motions were legally possible because the test house is owned by the quarry and located on the quarry property. These comparisons show that climatic and environmental variations cause greater crack response than ground motions that exceed regulatory limits.

INTRODUCTION

This paper presents structural response and the resulting crack response of a two-story residential structure situated on the property of an aggregate quarry in Sycamore, Illinois.

Focus of Study

This paper compares the structural response and resulting crack response produced by:

- Ground motions from blasting
- Environmental conditions (long-term changes in temperature and humidity)
- Occupant activity
- Wind gusts

The instrumented house is a two-story wood-framed structure with a basement foundation. It is located approximately 300 feet (91m) away from the edge of the quarry as shown in Figure 1.

The autonomous crack monitoring (ACM) system was installed on June 16th and June 17th, 2010. Data collection began on July 2nd, 2010, and has continued since. The air overpressure transducer and the indoor temperature and humidity gauge were installed on July 21st, 2010. The air overpressure transducer did not begin recording properly until November of 2010.

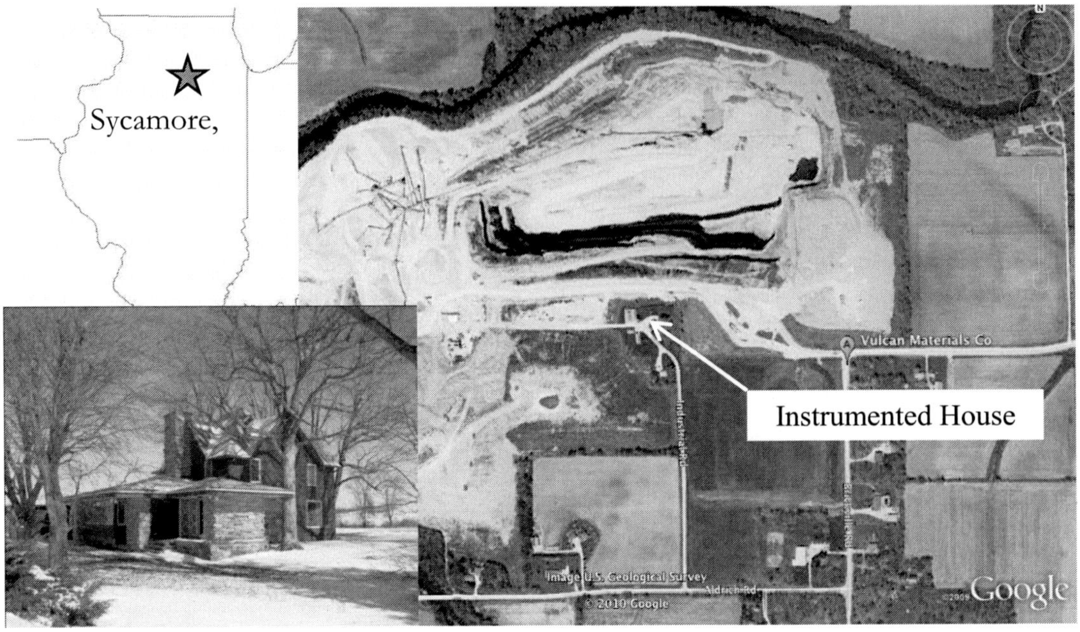

Figure 1. Overall view of Sycamore quarry and photograph and location of the instrumented house: some 300 ft (100m) from the southern edge of the quarry.

During the study period (from July 1st, 2010 to October 27th, 2012), the quarry generated blasts 36 times. Table 1 describes the blast vibration environment. Blasts were initiated at varying distances from the house: between 300ft and 1400ft (90-425m) away. These distances were evaluated by triangulation using distances between the blasts and surrounding houses, provided by Vulcan Materials. Example ground motion time histories shown in Figure 7. On some occurrences (indicated in Table 1 by a star), data were not recorded by the ACM system. However, Vulcan Materials compliance monitoring would have recorded ground motions where necessary.

Instrumentation

Structural and crack response is autonomously measured by the combination of sensors listed in Table 2. The crack sensors are described in more detail below. Other sensors are described in the Installation Report (Meissner, 2010). Although the house was equipped with both wired and wireless monitoring systems, the present paper is based on data from the wired system. Wireless response is documented in two reports by Koegel (2011) and Dowding et al. (2012). Reports are downloadable thrugh; http://iti.northwestern.edu/acm/publications.html.

The sensor installation plan is shown in Figure 2. Photographs of the crack sensors and their orientation on the walls are shown in Figure 3. Details of the installation of the velocity transducers are documented in the Installation Report (Meissner, 2010).

Two types of data are recorded: long-term and dynamic:

- Long-term response is obtained by measuring crack response as well as temperature and humidity for comparison once every hour. These single points averages of 1000 samples obtained at 1000 samples per second (sps)
- Dynamic response is obtained by measuring crack response, ground and structural velocity, and air overpressure at 1000 sps for 3 seconds when triggered during dynamic events, with a 0.5 second pre-trigger.

LVDT displacement sensors have been installed to monitor the in-plane responses of three cracks. Table 2 describes the locations, purposes and sensors used for each crack.

Table 1 Characteristics of blasts producing the vibrations throughout the study period.

Date	Recorded PPV (in/s)			Distance to blast (ft)	Air overpressure		Crack response (μin)		
	[L]	[T]	[V]		[10^-4 psi]	[dB SPL]	Shear	Seam	Ceiling
09/01/10	0.108	0.134	0.239	1200	-	-	53	59	215
09/13/10	0.172	0.16	0.262	1300	-	-	82	82	427
09/20/10	0.157	0.12	0.219	1300	-	-	255	84	459
09/22/10 (*)	-	-	-	-	-	-	47	54	205
09/30/10	0.389	0.307	0.371	1300	-	-	401	242	509
10/08/10	0.452	0.34	0.636	900	-	-	274	211	444
10/19/10	0.16	0.128	0.224	1400	-	-	218	102	76
04/20/11	1.341	0.491	0.976	400	48.2	127.6	1120	532	486
04/22/11	0.817	0.792	0.646	300	50.2	128.0	2050	448	1260
05/11/11	0.197	0.189	0.201	600	27.2	122.7	154	86	201
05/19/11	0.516	0.523	0.417	400	34.4	124.7	807	439	1094
05/26/11	1.642	0.905	0.797	700	33.1	124.4	3277	1666	2162
06/03/11	0.454	0.275	0.301	1200	30.3	123.6	283	213	567
06/15/11 (*)	-	-	-	-	-	-	-	-	-
06/30/11	0.61	0.529	0.389	900	31.7	124.0	387	243	648
07/12/11 (*)	-	-	-	-	-	-	-	-	-
07/22/11	-	-	-	-	-	-	-	-	-
08/02/11	0.701	0.758	0.741	-	23.4	121.4	253	329	3032
08/10/11 (*)	-	-	-	-	-	-	-	-	-
08/18/11 (*)	-	-	-	-	-	-	-	-	-
09/07/11 (*)	-	-	-	-	-	-	-	-	-
09/14/11	1.137	0.749	0.665	400	36.6	125.3	1107	550	2363
09/22/11 (*)	-	-	-	-	-	-	-	-	-
10/03/11 (*)	-	-	-	-	-	-	-	-	-
10/10/11	0.443	0.291	0.215	-	64.3	130.1	279	170	536
06/28/12	0.847	0.452	0.662	-	65.7	130.3	918	414	2125
07/17/12	1.011	0.750	0.554	-	40.8	126.2	591	382	1630
07/23/12	1.023	0.683	0.559	-	41.5	126.3	433	314	1812
07/27/12	0.760	0.465	0.730	-	52.6	128.4	754	284	946
08/13/12	0.514	0.507	0.752	-	63.4	130.0	335	324	1203
08/17/12	0.384	0.349	0.495	-	45.1	127.1	327	200	1135
08/18/12	0.570	0.367	0.373	-	30.4	123.6	495	256	797
08/27/12	0.719	0.621	0.539	-	42.5	126.5	580	515	1059
09/05/12	0.746	0.598	0.671	-	55.5	128.9	426	483	2108
09/19/12	0.614	0.539	0.638	-	100.0	134.0	413	387	1058
09/24/12	0.468	0.319	0.233	-	37.3	125.4	512	579	1141

Data were not recorded by the ACM system on instances shown by an asterisk;
PPV = Peak Particle Velocity (1in/s = 25.4mm/s, 1μin = 0.0254μm):
L= longitudinal, T= Transverse, V= vertical

Table 2: Sensor Description, identification, and location as shown in Figure 3.

Sensor	Channel Name	Description/Location
Triaxal Geophone (buried)	Geo_L_1 Geo_T_2 Geo_V_3	Buried outside the house - south-east corner
Air Overpressure	Air_Blast_4	Located on the outside wall
Temperature and Humidity	InTemp_5 InHumid_6	Indoor climatic data
Linear Variable Differential Transformer (LVDT)	LVDT_9_Shear	Shear crack, South wall, 1st floor
	LVDT_10_Null	Null sensor, uncracked area adjacent to Shear crack
	LVDT_11_Seam	Addition Seam crack, South wall, 1st floor
	LVDT_12_Ceil	Ceiling crack, Bedroom ceiling, 2nd floor
Horizontal Wall Geophone	HG_13_Bot1f HG_14_Top1f HG_15_Top2f HG_16_MidW	Wall-mounted

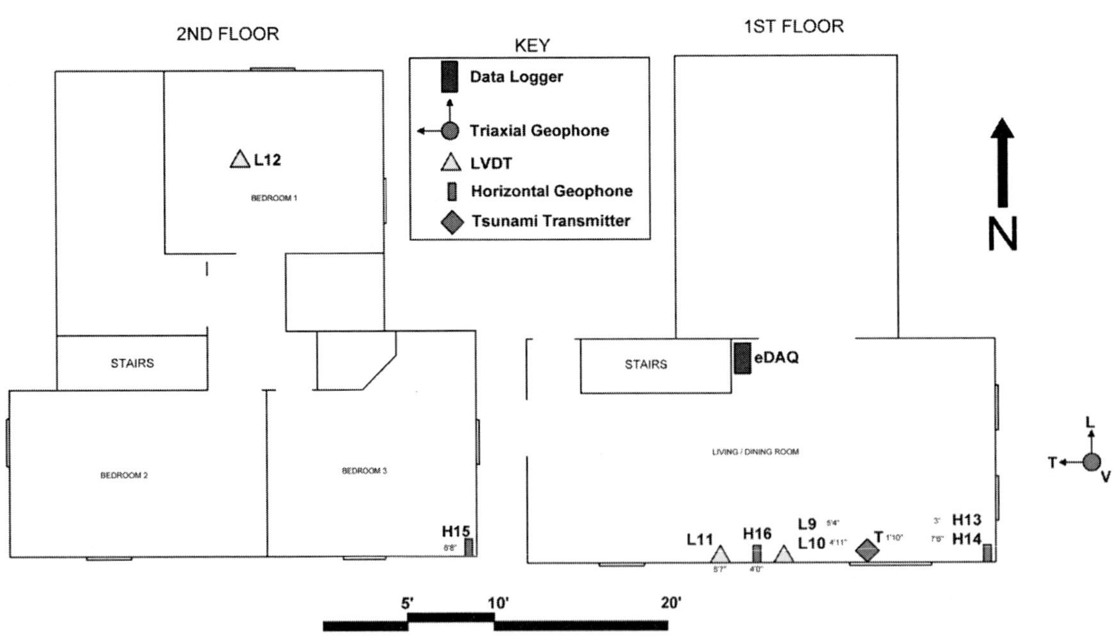

Figure 2.Plan view of sensor and equipment locations in the house

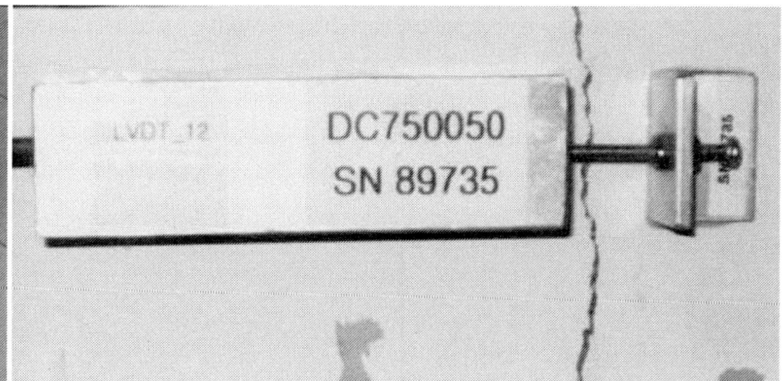

↑ a - Close-up of ceiling crack monitored by LVDT_12

← b – Upstairs bedroom ceiling showing the location of the crack and the transducer LVDT-12

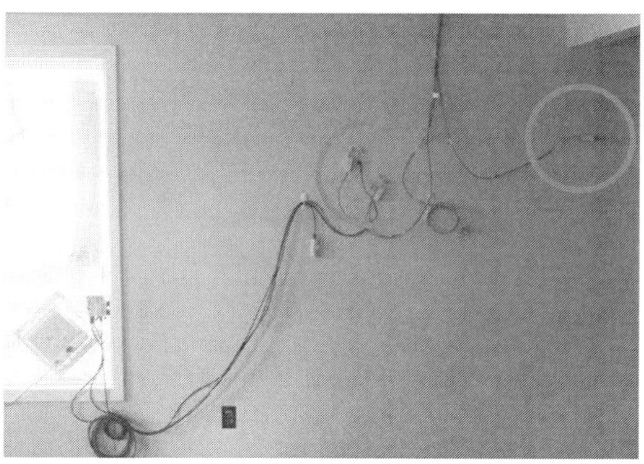

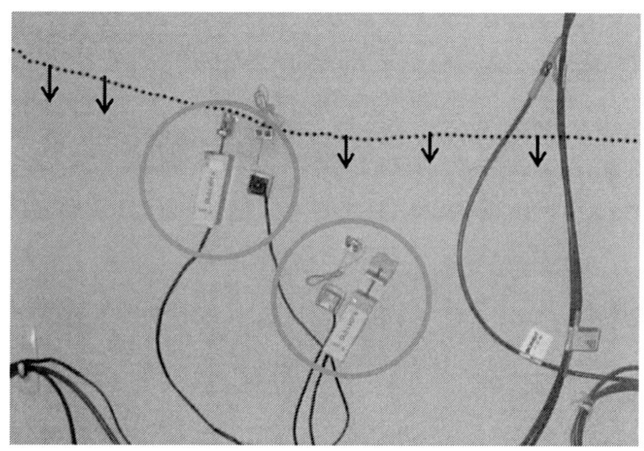

↑ c - Overall view of sensor suite LVDT-9, 10 & 11 on south wall (first floor)

↗ d - Close-up of Shear crack monitored by LVDT_9 and Null gauge LVDT_10.

→ e - Close-up of Addition Seam Crack monitored by LVDT_11

(Cracks identified by arrows from the offset dotted crack paths)

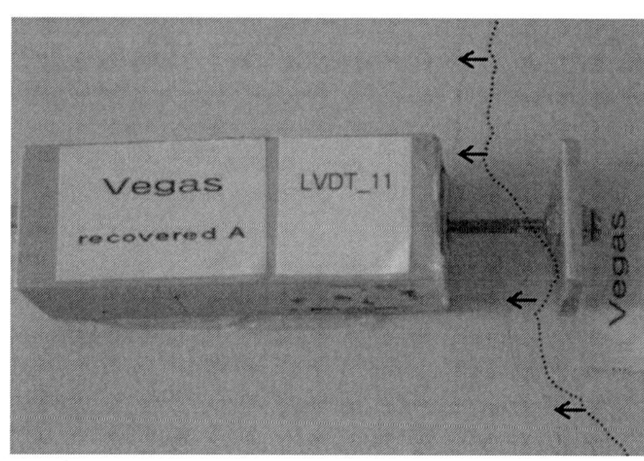

Figure 3. Overall views of the wall-mounted crack sensors

Crack response is the *change* in crack width, not total crack width, and is called *response* in this report. Figure 4 (Siebert, 2000) illustrates this definition. All transducers have been installed so that positive response indicates crack opening and negative indicates crack closing. All measurements are made in micro-inches. A null sensor, which is placed on an adjacent uncracked area, provides a record of any drift or thermal effects on sensor metal or electronics. It has been shown that the null sensors' response is small relative to the cracks' responses (Kosnik, 2008).

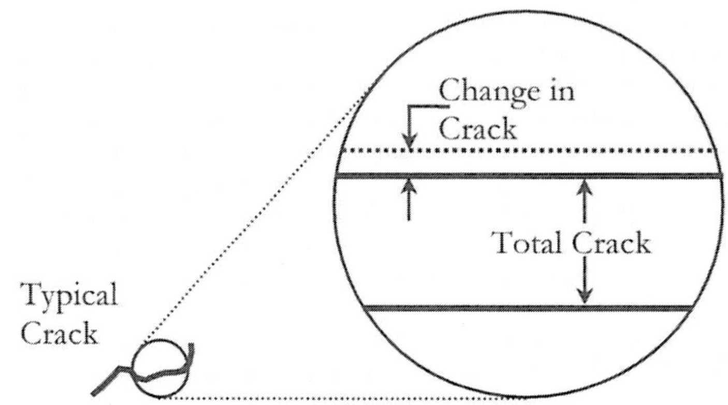

Figure 4. Crack response is the change in crack width, not total crack width.

RESULTS – CRACK RESPONSE

Long Term Climatological Effects

Long-term crack response is measured hourly by averaging 1000 samples obtained at 1000 samples per second to eliminate noise spikes. Figure 5 presents a two-month fragment of the whole project that will be seen again later in this paper. Long-term response to date (28 months) is compared for all three cracks in Figure 6 with indoor temperature and humidity. This hourly data is represented by the red, most highly variable line in Figures 5 and 6. The less variable blue line is a 24-hour central moving average (CMA) of the hourly data, which shows the response to weather fronts. The even less variable black line is a 30-day CMA of the hourly data, which shows the response to seasonal trends as it varies about the 2 year average green horizontal line.

Climatic responses are defined in Figure 5 to clarify the time scales of these influences. Daily response is defined as the difference between the hourly data and the 24-hour CMA (red arrow). Frontal response is defined as the difference between the 24-hour CMA and the 30-day CMA (blue arrow). Seasonal response is the difference between the black, 30-day CMA curve and the green overall average curve (green arrow). Finally, the maximal response is the sum of the daily, frontal and seasonal effects, which is the difference between the red hourly data and the overall average curve (black arrow). The maximum values from Figure 6 are listed in Table 3.

Table 3. Maximum crack response to weather effects (1μin=0.0254μm). All measurements are zero-to-peak

Response (μin, zero-to-peak)	Shear crack	Null	Seam	Ceiling crack
Max Daily	3276	258	2791	1849
Max Frontal	3840	122	3459	5182
Max seasonal	10485	285	2757	9157
Maximal	12064	745	4861	11726

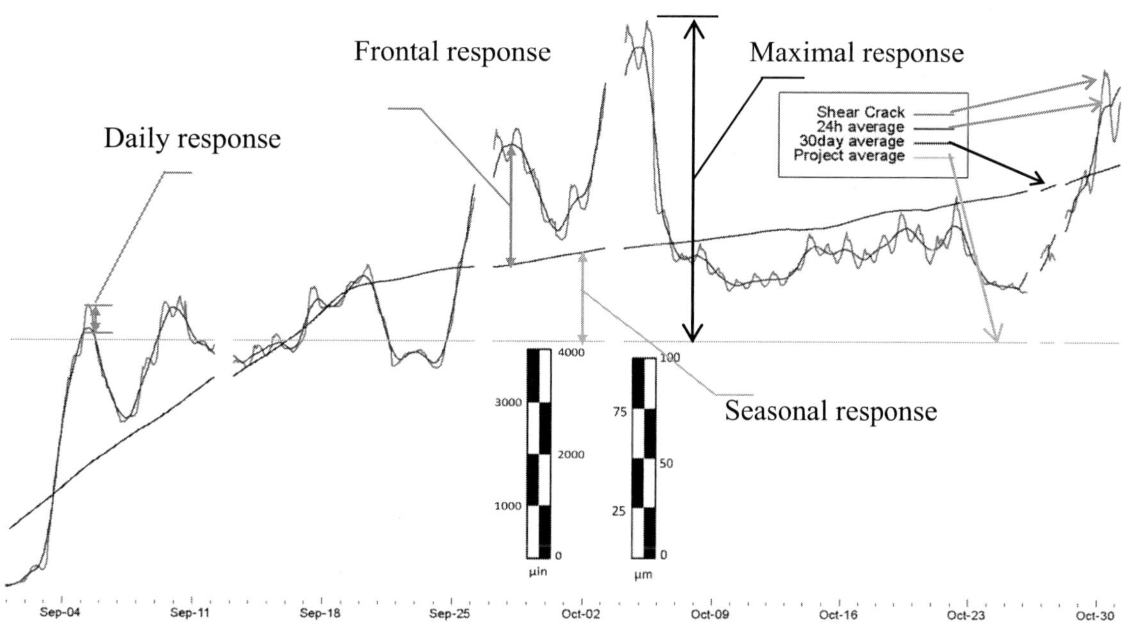

Figure 5. Crack response caused over the course of two months as shown in Figure 6

The unusual longevity of this study (28 months) allows measurement and observation of yearly responses. As described elsewhere (Dowding, 2008), such long term obersvations show the dominance of seasonal response, which produce the largest of the responses. The yearly maximum shown by the black arrows on Figure 6 occurs once per year for a brief period when all three factors (daily, frontal and seasonal) combine.

Blast-induced (Dynamic) Ground Motion Crack Response

Table 1 compares the zero-to-peak crack responses with the maximum peak particle velocities and the air overpressures over the course of the 28 month study. The ceiling crack response is larger than the south wall responses (shear and seam). This difference is a function of surface orientation with respect to vertical, location of the surface, construction details such as spacing of studs or joists, and construction materials (drywall, plaster, and lath). The seam has the least response with the thinnest crack width but located across a transition between an addition and the original structure.

Time histories for both crack and structural responses are presented in Figure 7 for two blasts. The first (top graph) was recorded on September 20, 2010 with a maximum Peak Particle Velocity (PPV) of 0.219ips (5.56 mm/s) in the vertical direction. The second (bottom graph) was recorded on May 19, 2011 with a maximum PPV of 0.523ips (13.3 mm/s) in the transverse direction. The air overpressure was recorded only for the May event and is shown in the figure.

As shown in Figure 8, the air overpressure can cause distinct crack response, especially when the ground vibrations are low. The later arriving air overpressure produces a larger ceiling crack response than the ground vibrations for the May 11, 2011 blast (Max PPV = 0.201ips = 5.16mm/s). Even so, it is still much smaller than the climatologically-induced response.

High PPVs in the ground are recorded because the house is located on mine property and will eventually be dismantled as the rock is mined out underneath it. As will be discussed later, even peak particle velocities (PPVs) above 0.5ips (12.7mm/s) produce crack responses that are only fractions of the seasonal, frontal, or even the daily "maximum" responses.

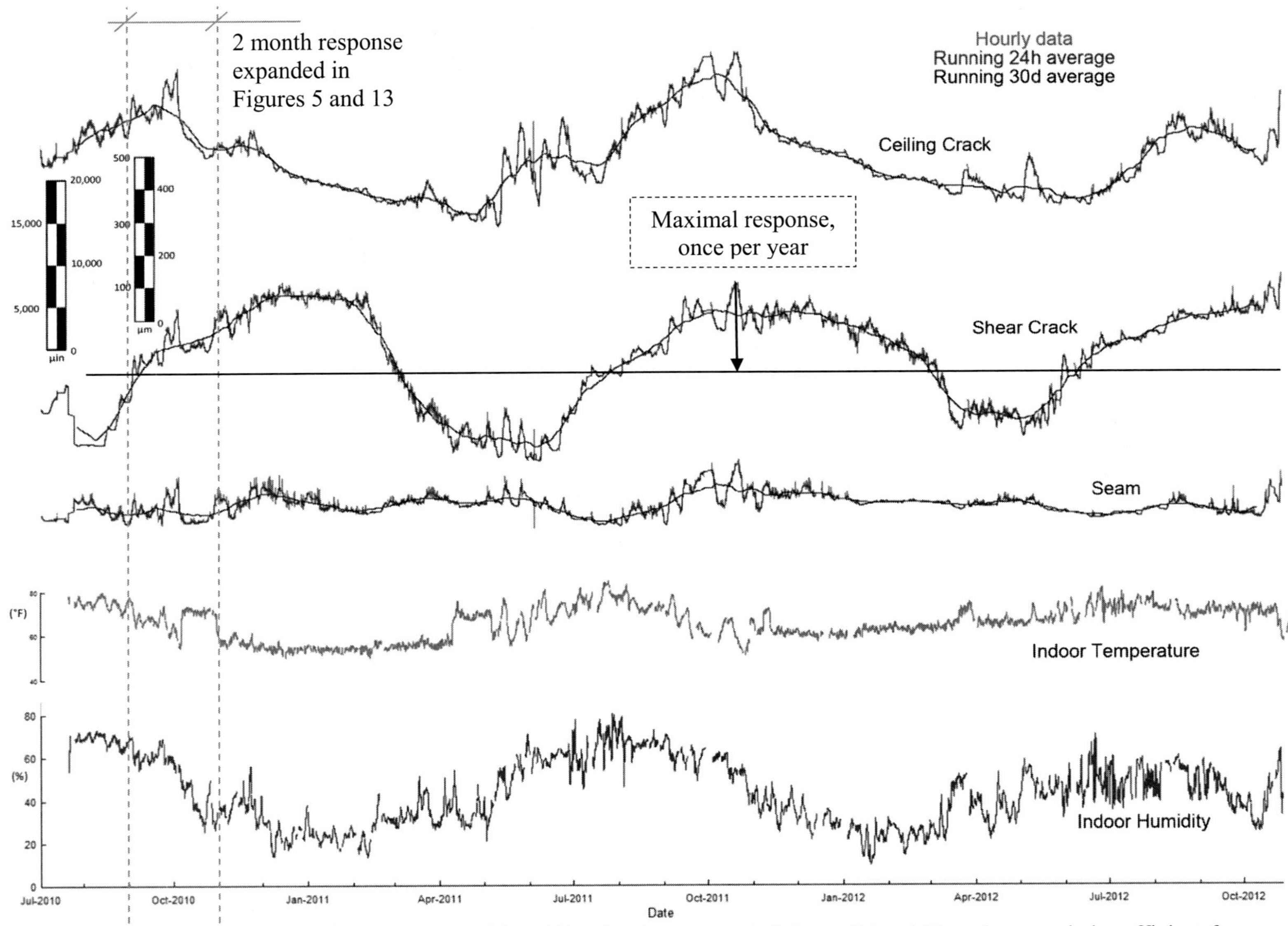

Figure 6. Crack response compared w/ temperature and humidity Crack response is “change” in width and one scale is sufficient for comparison

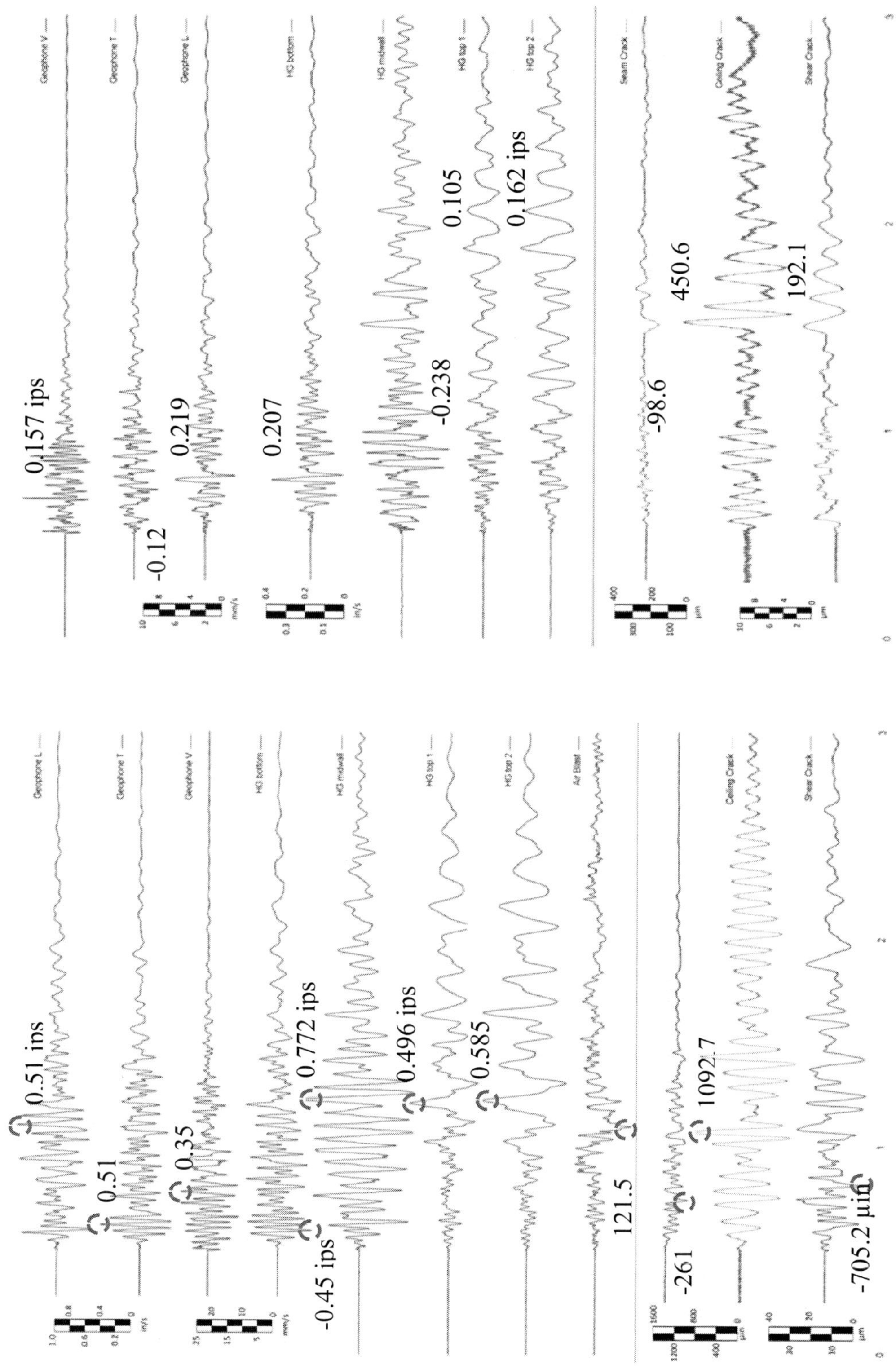

Figure 7. Time related structural (velocity) and crack response (displacement) time histories to blast events on Sept. 20, 2010 (right) and May 19, 2011 (left). All peak values in black are zero-to-peak. (1μin = 0.0254μm, 1ips = 25.4mm/s). One amplitude scale is applicable for each measurement type.

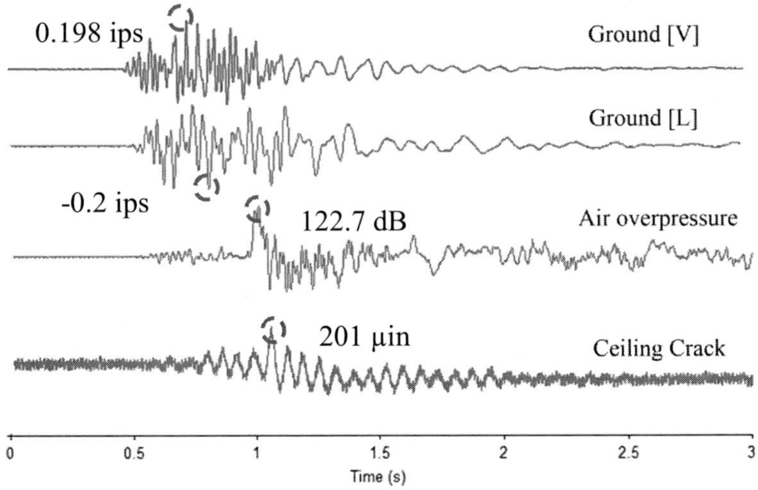

Figure 8. Influence of late arrival of air overpressure – Blast event on May 11, 2011. Measurements are zero-to-peak.

Occupant Activity

Crack responses to the unplanned entrance of someone into the house are shown in Figure 9. Comparison with planned events (Meissner, 2010) shows that the entrance was probably by the front door. Maximum values for the crack responses to this door opening are tabulated in Table 4, along with structural and crack responses to 6 different planned occupant induced activities. Their time histories can be found in the Installation Report (Meissner, 2010). The signal to noise ratio is high for the ceiling crack because of line losses resulting from the longer distance between the data logger and the transducer. One single event such as closing or slamming the bedroom door can induce an important response from the ceiling crack, located on the bedroom ceiling. Slamming the bedroom door produces a response of 2036μin (51.7μm).

Blast events must produce PPVs greater than 0.75ips (19.1mm/s) to produce larger ceiling crack response than the one produced when slamming the bedroom door. Moreover, the unplanned front door opening produced crack responses that are on the same order of magnitude of the responses produced by blast events such as those that occurred on Sep 01, 2010 or on May 11, 2011. These blast events produced ground motions slightly above 0.2ips (5.08mm/s).

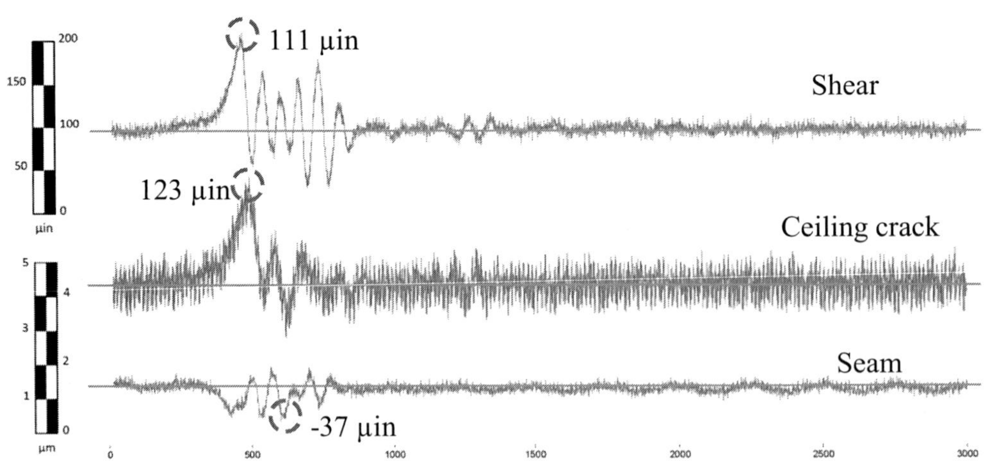

Figure 9. Crack response from opening the front door on the first floor

Table 4. Crack and structural responses to 7 occupant activities (1μin = 0.0254μm)

	Crack response (μin)				Structural response (in/s)			
Event	Shear	Null	Seam	Ceiling	HG 13 (Bottom)	HG 14 (Top 1)	HG 15 (Top 2)	HG 16 (Midwall)
Unplanned front door opening	111	18	37	123	0.008	0.02	0.03	0.12
Slam front door	18	14	39	190	0.013	0.02	0.031	0.128
Run down stairs	17	14	13	30	0.005	0.01	0.013	0.055
Close bedroom door	18	15	13	518	0.005	0.007	0.012	0.015
Slam bedroom door	13	14	13	2036	0.015	0.028	0.055	0.043
Slam garage door	15	14	23	150	0.009	0.025	0.038	0.123
Close window	17	13	12	38	0.028	0.014	0.01	0.07
Heel drop	18	20	15	39	0.003	0.003	0.004	0.025

Influence of Inside Temperature Regulation

On October 25th, 2012, natural gas supply was terminated for future demolition of the structure, which ended the temperature regulation inside the house (heating). As shown in Figure 10, the inside temperature dropped 6°C (13°F) in one day on October 26th. After that, the temperature kept varying following a daily cycle, but at values much lower than when the heat was on.

Peak values of the crack response during large fall temperature swings are compared to average values during the 5 week period in Figure 10. They show that the absence of inside temperature regulation caused a response comparable to the largest seasonal response observed during the 30 month study, as shown in Table 5 and Figure 12.

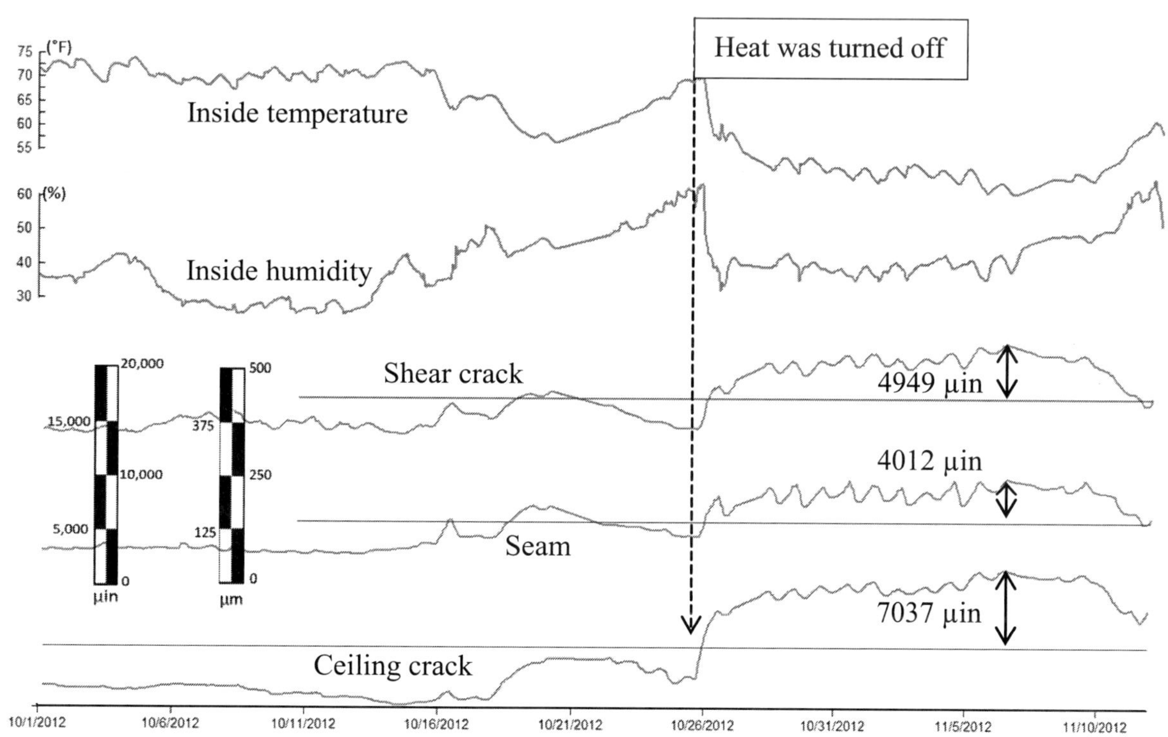

Figure 10. Influence of the inside temperature regulation during the fall of 2012

Wind Response

Weather patterns, particularly involving high-velocity wind gusts, induced crack responses that could be as strong as the responses induced by blasting. Wind gusts triggered disturbances in the air overpressure sensor. Figure 11 shows an example of a response in the air overpressure sensor and the simultaneous crack displacements for all 3 cracks.

This wind gust event produced crack responses that are significant. As was observed with the occupant activity, the crack responses are on the same order of magnitude of those produced by blast events such as those which occurred on Sep 01, 2010 or on May 11, 2011. These blast events produced ground motions slightly above 0.2ips (5.1mm/s).

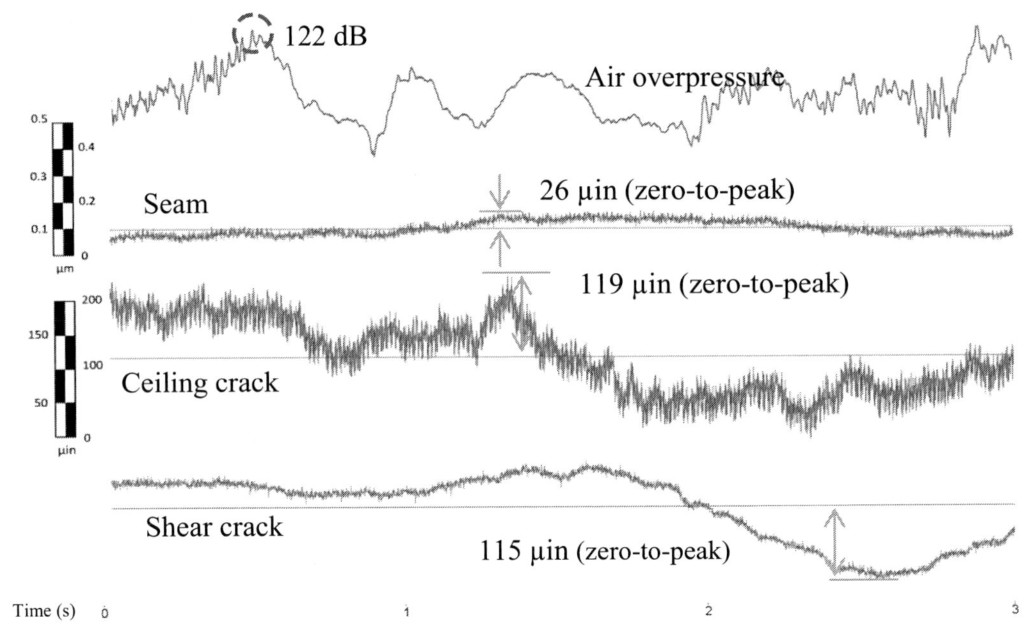

Figure 11. Example wind event on May 15, 2011 showing Air Overpressure and Crack Responses (1µin = 0.0254µm). Measurements are zero-to-peak

ANALYSIS

Comparison of Crack Response to Climatological and Vibration Effects

Blast-induced crack responses are compared to long-term environmental effects and occupant induced activities in Table 5. Long-term response is at least an order of magnitude larger than any of the dynamic responses, even those produced by ground motions as high as 0.5ips (12.7mm/s). Figure 12 compares the tabulated responses in graphical form.

In general, the greater the climatologically induced long term response, the greater the dynamic response. Overall, the shear crack and the ceiling crack respond more than the seam. This difference is shown by the overall maximum response of the seam, which is less than half that of the shear or ceiling cracks. These ratios are consistent regardless of the source: occupant or blast, or the magnitude of ground motion: low (0.219ips, 5.56mm/s) or high (0.523ips, 13.28mm/s).

Time Histories of Daily Climatological and Blast-induced Response

Time histories of crack responses to the 0.219ips (5.56mm/s) blast in Figure 13 can be compared with the long-term climatic response in Figure 6. This comparison is made with the two-month timespan bounded by the vertical dashed lines in Figure 6. Two months of data illustrate the large effect of the passage of large weather systems/fronts. Blast responses are compared to the week of climatic response delimited by the dashed blue rectangle on Figure 13 (top). This week-long comparison illustrates the daily environmental fluctuations that are superimposed over the longer term effects.

Long-term and dynamic responses are plotted on the same vertical scale for both comparisons. The small black vertical bar on September 20th represents the maximum magnitude of the dynamic response, which is expanded in the rectangular box below the graph. Figure 13 shows that the environmentally induced response during the week surrounding the blast event is approximately an order of magnitude larger than the blast-induced response. A similar conclusion was reached by Kosnik (2008) and Meissner and Dowding (2009).

Table 5. Maximum crack response to all observed sources of vibrations. All measurements are zero-to-peak (1μin = 0.0254μm)

Response (μin)	Shear crack	Null	Seam	Ceiling crack
Max Daily	3276	258	2791	1849
Max Frontal	3840	122	3459	5182
Max seasonal	10485	285	2757	9157
Maximal	12064	745	4861	11726
Ground motion (09/20/11) Max PPV = 0.219ips (5.56mm/s)	192	22	99	451
Ground motion (05/19/11) Max PPV = 0.523ips (13.3mm/s)	705	30	261	1093
Occupant activity (front door open)	111	18	37	123
Wind event (May 15, 2011)	115	20	27	119
Turning off the heat inside the house (on Oct. 26, 2012)	4949	117	4012	7037

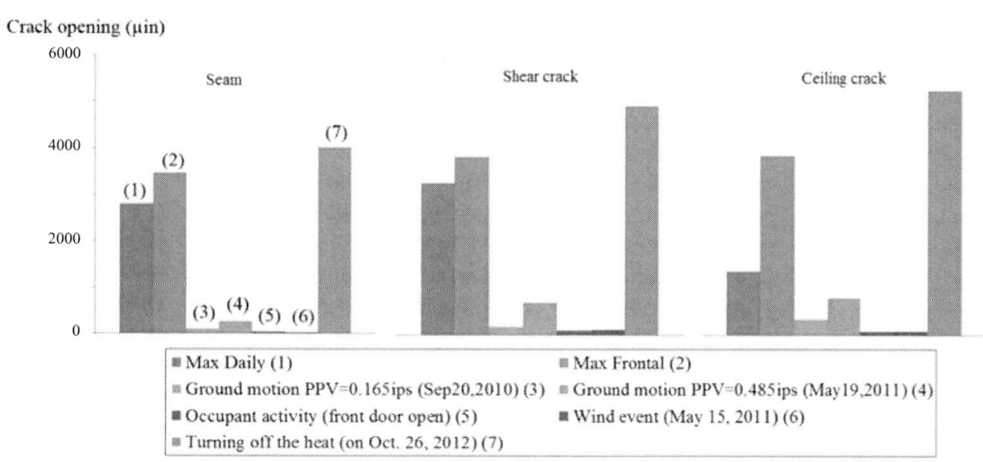

Figure 12. Comparison of crack response magnitudes as presented numerically in Table 5

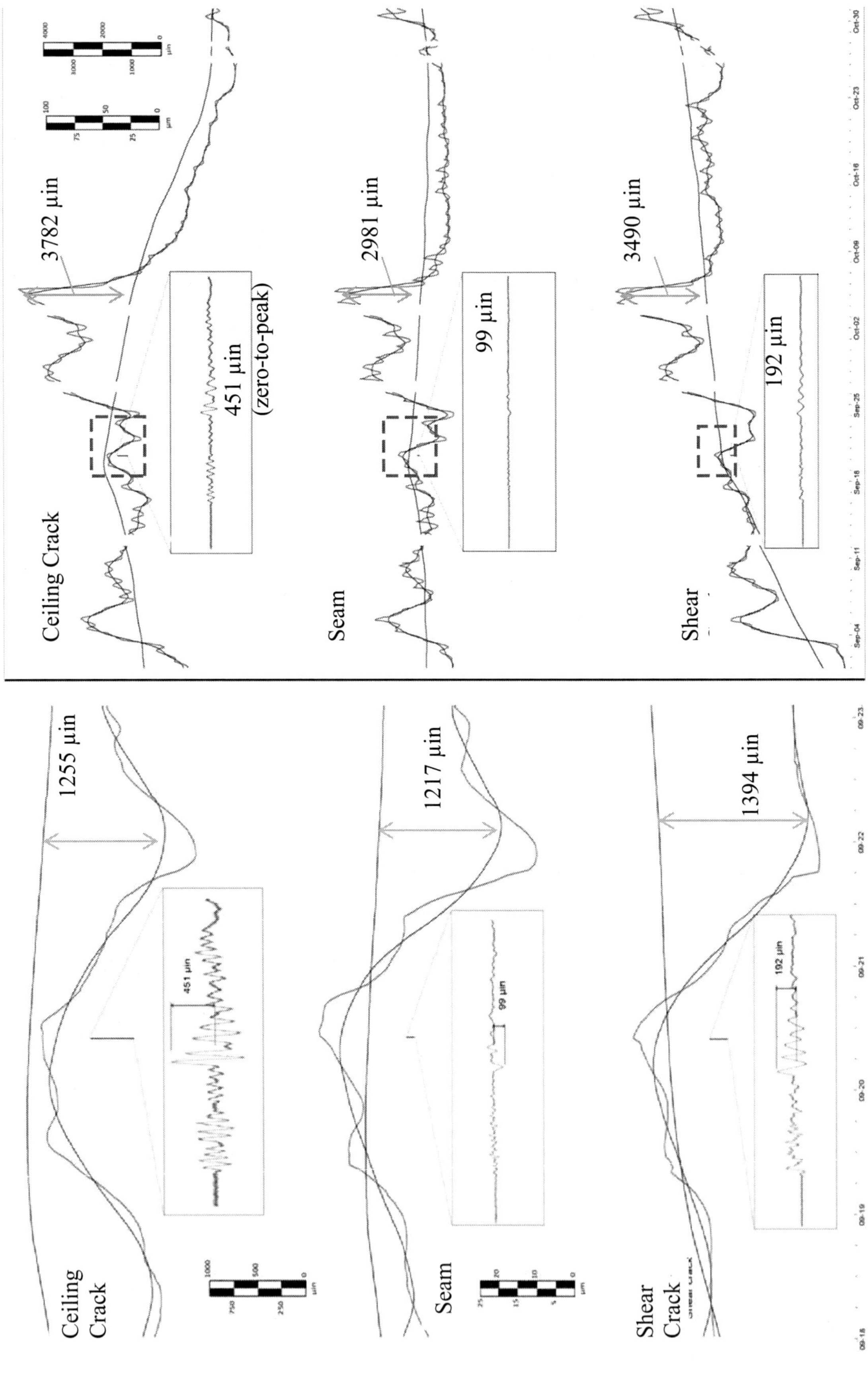

Figure 13. Comparison of zero-to-peak crack response to environmental effects over the course of two months (right) and one week (left, delimited in dashed boxes on right) with that produced by the PPV=0.219in/s event. (1µin=0.0254µm, 1in/s=25.4mm/s)

CONCLUSIONS

Data provided herein are a compilation of one of the longest continuously recorded crack responses to date. More than 28 months of continuous crack response have been recorded, including several periods of many months without blasting, thus showing that large crack response occurs without blasting. The unusually long period of observation provided the opportunity to observe response to two seasonal variations.

As has been observed before, crack response to environmental variations is overwhelmingly larger than that produced by blast-induced ground motion and associated air overpressure pulses. Seasonal variations and even the passing of weather fronts can produce crack response that is larger by at least an order of magnitude. Turning off the heat inside the house in the fall can cause crack response of that order of magnitude as well, but over periods of time as short as a week.

Observation of occupant activity and wind gust events shows that both can produce crack response as large as that produced by blast-induced ground motions.

ACKNOWLEDGEMENTS

This paper is the result of research sponsored and supported by the Infrastructure Technology Institute (ITI) at Northwestern University, which is funded by a block grant from the US Department of Transportation. The authors wish to thank Vulcan Materials Corp. for supplying the test house and blast environment, as well as ITI research engineers for their guidance and contribution.

REFERENCES

Abeel, P.-A. (2012). *Building and Crack Response to Blasting, Construction Vibrations and Weather Effects.* MS Thesis, Northwestern University, Evanston, IL.

Dowding, C. H. (2008). *Micrometer Crack Response to Vibration and Weather.* International Society of Explosives Engineers.

Dowding, C. H., Kotowsky, M., & Koegel, T. (2012). *Field Qualification of Inexpensive Wireless System to Monitor Micro-Inch Crack Response for Structural Health Monitoring.* Technical report, ITI - Northwestern University, Evanston, IL.

Koegel, T. (2011). *Comparative Field Qualification of ACM and ACSM Systems at Sycamore, IL.* Honors Thesis, Dept. of Civil and Environmental Engineering, and Internal Report, ITI - Northwestern University, Evanston, IL.

Kosnik, D. E. (2008). *Autonomous Crack Displacement Monitoring of a Residence Near a Limestone Quarry, Naples, Florida.* Technical Report, ITI - Northwestern University, Evanston, Illinois.

Meissner, J. E. (2010). *Autonomous Crack Monitoring of Residential Structure – Sycamore, IL.* Installation Report, ITI - Northwestern University, Evanston, Illinois.

Meissner, J. E., & Dowding, C. H. (2009). *Response of a Concrete Block Structure to Quarry Induced Ground Motion and Weather.* Technical report, ITI - Northwestern University, Evanston, Illinois.

Siebert, D. (2000). *Autonomous Crack Comparometer.* MS Thesis, Northwestern University, Evanston, IL.

The $\phi = 0$ Concept: Review of its Theoretical Basis and Pragmatic Issues with Implementation

Russell A. Green[1], P.E., M.ASCE, and William F. Marcuson, III[2], P.E., Hon.M.ASCE

[1]Professor, Department of Civil and Environmental Engineering, 120B Patton Hall, Virginia Tech, Blacksburg, VA 24061; rugreen@vt.edu
[2]Director Emeritus, Geotechnical Laboratory, U.S. Army Engineer Research and Development Center (ERDC), 3909 Halls Ferry Road, Vicksburg, MS 39180-6199; wfm3@att.net

ABSTRACT: The "$\phi = 0$ Concept" is central to the specification of the undrained shear strength of saturated clays in engineering practice. While the theoretical basis for the Concept lies in Mohr's rupture hypothesis, its pragmatic value to geotechnical engineers lies in Terzaghi's effective stress principle. Specifically, the $\phi = 0$ Concept circumvents the onerous need for estimating (or measuring) excess pore water pressures in situ during construction, etc., which is required to perform an effective stress analysis. Despite the significance of the $\phi = 0$ Concept, a survey of geotechnical engineering textbooks published over the last 60+ years showed that most opt to just state the $\phi = 0$ Concept without putting it into proper theoretical or contextual perspective. The authors view this as a missed opportunity to put the various rules of thumb and sets of governing principles for the shear strength of soil into a consistent framework. Accordingly, the objectives of this paper are to review the theoretical basis of the $\phi = 0$ Concept in the context of Mohr's rupture hypothesis and Terzaghi's effective stress principle and to discuss some of the pertinent pragmatic issues with implementing the Concept. In relation to the latter objective, the procedure for multi-staged rapid drawdown analysis for earthen dams proposed by Duncan et al. and the procedure for determining the strength of liquefied soil proposed by Poulos et al. are reviewed. It is shown that while Duncan et al. fully address all the pragmatic issues in implementing their procedure, Poulos et al. do not.

INTRODUCTION

A colleague who was preparing to write a new introductory textbook on geotechnical engineering conducted an informal survey among geotechnical engineering faculty in the US asking about their likes and dislikes of existing texts. The responses to the survey showed an almost ubiquitous dislike for the coverage of shear strength (Fiori, 2008). The reason for this is likely because, in many ways, the texts mimic research papers and reports. And, shear strength research in the US largely evolved following a semi-empirical approach wherein each soil type-stress state-drainage condition combination was treated as being unique and having its own set of rules of thumb and governing principles. Few texts put these rules of thumb and sets of principles into the context of a consistent mechanics-based framework. One example of this is the "$\phi = 0$ Concept" (e.g., Skempton, 1948a), which is commonly invoked when specifying the undrained strength of saturated clays, wherein the soil is assumed to be non-frictional and purely cohesive (i.e., Tresca material). Accordingly, the objectives of this paper are to review the theoretical basis of the $\phi = 0$ Concept in the context of Mohr's rupture hypothesis and Terzaghi's effective stress principle and to discuss some of the pertinent pragmatic issues with implementing the Concept.

Arguably the most significant theory in all of soil mechanics is Terzaghi's effective stress principle. This principle has two tenets: (1) deformation (i.e., volumetric strain) in soil is a function of effective stress, and (2) shear strength of soil is a function of effective stress (Terzaghi, 1936). Although various researchers had identified aspects of the effective stress principle prior to Terzaghi (e.g., Lyell, 1871; Reynolds, 1886; Fillunger, 1915), Terzaghi is credited with fully developing the effective stress principle and with being the first to use it in geotechnical engineering practice (e.g., Skempton, 1960; de Boer, 2005). The first tenet was postulated in the 1920s and underlies consolidation theory (e.g., Terzaghi, 1924 and 1925). The second tenet evolved with time, and to quote Skempton (1960): "Although this principle is of paramount importance, its development and implications did not occur at once but over a period of several years and, wholly or in part, in many different publications." Even at the time Skempton wrote this statement, the implications of the second tenet were not fully realized in the context of plasticity theory that was just being developed and applied to soil mechanics (e.g., Drucker and Prager, 1952; Mroz, 1963). And, even today, few geotechnical engineering textbooks mention plasticity theory in relation to Terzaghi's effective stress principle, despite the fact that relevant aspects of this theory have been experimentally verified, as discussed later in this paper.

The first investigations of the $\phi = 0$ Concept for geotechnical engineering applications actually predate Terzaghi's effective stress principle and were completely based on empirical laboratory observations (Bell, 1915; Westerberg, 1926; Fellenius, 1927). However, by the mid-1930s and throughout the 1940s, the pragmatic issues of implementing the Concept for specifying the undrained strength of saturated clays were seemingly a hot topic of research (e.g., Terzaghi, 1936; Skempton, 1948a, b; Golder and Skempton, 1948; Skempton and Golder, 1948). Focus was primarily on the relationship, and seeming inconsistencies, among the $\phi = 0$ Concept, Terzaghi's effective stress principle, and the orientation of the slip surface observed both in the laboratory and in the field. From reviewing numerous papers written at this time, it became obvious that the effective stress principle was not universally appreciated, as exemplified by the frequent lack of distinction between ϕ and ϕ' (i.e., total and effective stress angles of internal friction, respectively). Additionally, during the 1940's and 50's, the use of a failure envelope that was independent of the applied [total] normal stress was referred to as a "$\phi = 0$ Analysis", with the moniker "$\phi = 0$ Concept" not coming until later (e.g., Lambe and Whitman, 1969). Although the change in the name from "$\phi = 0$ Analysis" to "$\phi = 0$ Concept" may seem trivial, it clearly signifies the evolving understanding of the relationship between the $\phi = 0$ Concept and Terzaghi's effective stress principle.

To better gage the evolution of the $\phi = 0$ Concept, the first author performed a survey of geotechnical engineering textbooks published over the last 60+ years (i.e., Table 1: Note that no attempt was made to perform an exhaustive review of all geotechnical engineering textbooks published since the 1930s; rather, the first author only reviewed the texts in his personal library). The review showed that most of the texts just state the $\phi = 0$ Concept without putting it into proper theoretical or contextual perspective. (It is worth noting that in preparation of the final draft of this paper Professor Mick Pender, University of Auckland, performed a similar review of several geotechnical engineering textbooks in his personal library that were not reviewed by the authors, with his conclusion of the coverage of the $\phi = 0$ Concept being the same as that of the authors (Pender, 2013)). While this approach to covering the $\phi = 0$ Concept is not necessarily incorrect and, by all means, the authors view many of these texts as excellent, in the authors' opinion it is a missed opportunity to help put the various rules of thumb and sets of governing principles for the shear strength of soil into a consistent framework. This is just one example of a potential plethora of reasons for the results of our colleague's survey which showed dissatisfaction among academicians with the coverage of shear strength by existing geotechnical engineering textbooks.

In the subsequent sections of this paper, the theoretical basis for the $\phi = 0$ Concept is first presented. This is followed by a discussion of the pragmatic issues on implementing the Concept. Finally, the procedure for multi-staged rapid drawdown analysis for earthen dams proposed by Duncan et al. (1990) and the procedure for determining the strength of liquefied soil proposed by Poulos et al. (1985) are reviewed in the context of the $\phi = 0$ Concept.

Table 1. Geotechnical engineering textbooks reviewed as part of this study

Year Published	Author(s)	Title
1948	D.W. Taylor	Fundamentals of Soil Mechanics
	K. Terzaghi and R.B. Peck	Soil Mechanics in Engineering Practice, 1st ed.
1951	G.P. Tschebotarioff	Soil Mechanics, Foundations, and Earth Structures
1962	G.A. Leonards, ed.	Foundation Engineering
1963	R.F. Scott	Principles of Soil Mechanics
1967	K. Terzaghi and R.B. Peck	Soil Mechanics in Engineering Practice, 2nd ed.
1969	T.W. Lambe and R.V. Whitman	Soil Mechanics
1991	M.D. Bolton	A Guide to Soil Mechanics
2002	V.N.S Murthy	Geotechnical Engineering: Principles and Practices of Soil Mechanics and Foundation Engineering
2005	J.K. Mitchell and K. Soga	Fundamentals of Soil Behavior, 3rd ed.
2006	B.M. Das	Principles of Geotechnical Engineering, 6th ed.
2007	M. Budhu	Soil Mechanics and Foundations, 2nd ed.
	R.L. Handy and M.G. Spangler	Geotechnical Engineering: Soil and Foundation Principles and Practice, 5th ed.
	D.F. McCarthy	Essentials of Soil Mechanics and Foundations: Basic Geotechnics, 7th ed.
2008	R. Salgado	The Engineering of Foundations
2011	R.D. Holtz, W.D. Kovacs, and T.C. Sheehan	An Introduction to Geotechnical Engineering, 2nd ed.
	I. Ishibashi and H. Hazarika	Soil Mechanics Fundamentals

THEORETICAL BASIS FOR THE $\phi = 0$ CONCEPT

The theoretical basis for the $\phi = 0$ Concept actually lies in Mohr's rupture hypothesis, which is not unique to soil mechanics but rather applies to a range of materials (Mohr, 1882). Mohr hypothesized that an envelope of plastic states (i.e., failure envelope) exists in which the shear stress mobilized on the slip surface is a function of the normal stress on the slip surface. The Mohr-Coulomb failure envelope that is commonly used in geotechnical engineering is a specific form of the more general Mohr hypothesis, where the Mohr-Coulomb envelope assumes the shear stress mobilized on the slip surfaces is a linear function of the normal stress acting on the slip surface. If the Mohr hypothesis is correct, then a stress circle that is tangent to the failure envelope represents a limit state (i.e., failure). Although the stress vectors corresponding to the point of tangency between the failure circle and failure envelope represent the stress components acting on the Coulomb slip surface (Coulomb, 1776), it is important to note that the entire failure circle represents the stresses in the element at failure. As such, the stress vectors acting on any plane (other than planes parallel to the principal stress planes) cutting through the element are unique to that plane at failure under the given stress conditions. This notion is central to the $\phi = 0$ Concept (and to the shear strength of all soils in general, as discussed subsequently), but of the textbooks listed in Table 1, only Lambe and Whitman (1969) emphasize this point. Their stated rule in this regard is that consistency is required between the assumed failure envelope and orientation of the slip surface, where the slope of the assumed failure envelope can be arbitrary.

This rule is illustrated in the retaining wall example shown in Figure 1. In this figure, a 6 m high retaining wall has a dry sand backfill with $\phi' = 30°$. A wedge analysis is performed to determine the lateral force imposed on the wall by the backfill for active conditions. In Figure 1b, the assumed failure envelope is as shown (i.e., $\phi' = 30°$ and $c' = 0$ kPa, where c' is the effective cohesion of the soil) and the corresponding slip surface forms a 60° angle with the horizontal (i.e., $45° + \phi'/2 = 60°$). The stresses in a soil element at mid-depth in the backfill are taken as being representative of the stress state of the entire backfill and the stress circle for this soil element is shown in Figure 1b. From a force balance of the wedge, the lateral force per unit width into the page imposed on the wall by the backfill is computed to be 113.4 kN/m.

The wedge analysis is repeated for the same retaining wall system in Figure 1c, this time defining the failure envelope by an arbitrary angle of $\phi = 15°$. Because the assumed failure envelope has to be tangent to stress circle representing limit state of our soil element in the backfill (i.e., the same stress circle used in Figure 1b), for $\phi = 15°$, c has to be equal to 9.44 kPa, where c is the "apparent" cohesion of the soil. In this case the slip surface corresponding to the assumed failure envelope has an inclination of 52.5° (i.e., $45° + \phi/2 = 52.5°$). Again, from a force balance of the wedge, the lateral force per unit width into the page imposed on the wall by the backfill is computed to be 113.4 kN/m; the same as computed in Figure 1b! (Note the notation change from ϕ' and c' to ϕ and c used to define the Mohr-Coulomb failure envelopes in Figures 1b and 1c, respectively. In this paper, ϕ' and c' are used exclusively to represent the Mohr-Coulomb envelope used in conventional effective stress analyses. In contrast, ϕ and c are used to define a failure envelope having an arbitrary slope.).

Finally, the wedge analysis is repeated one more time for the same retaining wall system in Figure 1d, this time defining the failure envelope by an arbitrary angle of $\phi = 0°$. Again, because the assumed failure envelope has to be tangent to the stress circle representing limit state of our soil element in the backfill (i.e., the same stress circle used in Figures 1b and 1c), for $\phi = 0°$, c has to be equal to 18.9 kPa. In this case the slip surface corresponding to the assumed failure envelope has an inclination of 45° (i.e., $45° + \phi/2 = 45°$). And, once again from a force balance of the wedge, the computed lateral force per unit width into the page is 113.4 kN/m; the same as computed in Figures 1b and 1c! Moreover, using Eqn (1), it can be shown that $P_a = 113.4$ kN/m for any arbitrarily assumed failure envelope that is tangent to the failure circle shown in Figure 1 and that has $-90° < \phi < 90°$, with the corresponding inclination of the slip surface between 0° and 90° (i.e., $0° < \alpha_f < 90°$). (Note that in using Eqn (1), either ϕ or α_f is assumed and the remaining unknown parameters are then computed.)

$$P_a = R \cdot \cos(90° - \alpha_f + \phi) - C \cdot \cos(\alpha_f) \quad \text{(1a)}$$

$$R = \frac{W - C \cdot \sin(\alpha_f)}{\sin(90° - \alpha_f + \phi)} \quad \text{(1b)}$$

$$W = \frac{1}{2} \cdot \frac{H^2}{\tan(\alpha_f)} \cdot \gamma \quad \text{(1c)}$$

$$C = c \cdot L \quad \text{(1d)}$$

$$c = \frac{H \cdot \gamma}{4} \left\{ tan\left(45° - \frac{\phi}{2}\right) - \frac{tan^2\left(45° - \frac{\phi'}{2}\right)}{tan\left(45° - \frac{\phi}{2}\right)} \right\} \quad \text{(1e)}$$

$$L = \frac{H}{\sin(\alpha_f)} \tag{1f}$$

$$\alpha_f = 45° + \frac{\phi}{2} \tag{1e}$$

where: P_a = lateral force imposed on the wall by the backfill for active conditions.
R = resultant force acting between the failure wedge and the stable backfill.
W = weight of the failure wedge.
C = force acting on the assumed slip surface that is associated with the apparent cohesion of the backfill.
γ = unit weight of the backfill.
L = length of the assumed slip surface.
H = height of the wall.
c = intercept of the assumed failure envelope.
ϕ = angle of the assumed failure envelope.
α_f = angle of inclination of the assumed slip surface.

PRAGMATIC ISSUES WITH THE ϕ = 0 CONCEPT

As outlined in the preceding section, the assumed ϕ-c combination used to define the failure envelope and corresponding inclination of the slip surface can range widely, yet still yield the same result (e.g., Figure 1). However, given the semi-empirical path that the evolution of shear strength concepts has followed, the ϕ-c combinations used to define the failure envelopes in geotechnical engineering practice are often determined from various laboratory tests performed on soil. While this is surely reasonable and preferred in many cases, the thorough understanding of the relationship between Terzaghi's effective stress principle and Mohr's rupture hypothesis that was required to properly interpret the laboratory data and use it to make design decisions was still evolving in the formative days of geotechnical engineering. Of particular issue was the lack of distinction between effective stress and total stress strength parameters (i.e., ϕ' and c' versus ϕ and c). For highly permeable soils (i.e., sandy or coarse grained soils), this was not much of an issue, but it was a significant issue for clays (e.g., Bell, 1915; Westerberg, 1926; Fellenius, 1927).

By the late 1940s, the significance of the change in water content (or lack thereof) of a saturated clay sample during the application of both confining stress and shearing stress was generally understood (e.g., Skempton, 1948a,b). Specific to the ϕ = 0 Concept, it was understood that for a saturated clay specimen under undrained conditions increases in confining stress were carried by the pore water in the sample, with the effective stress in the sample remaining unchanged. And, consistent with the second tenet of Terzaghi's effective stress principle, if the effective stress in a sample does not change, the deviatoric stress required to cause failure in the sample does not change. This results in total stress failure circles having equal diameters, but having different total major and minor principal stresses at failure. The failure envelope drawn tangent to the total stress failure circles is characterized by ϕ = 0° and c = S_u, where S_u is the undrained shear strength of the soil. This is conceptually illustrated in Figure 2, where the total stress failure circles from three unconsolidated-undrained (UU) triaxial tests performed on identical, normally consolidated saturated clay samples are shown. For truly identical samples, the pore pressures developed during the shearing phase of the UU tests would be the same for all tests, resulting in all the tests having the same effective stress failure circle, also conceptually illustrated in Figure 2.

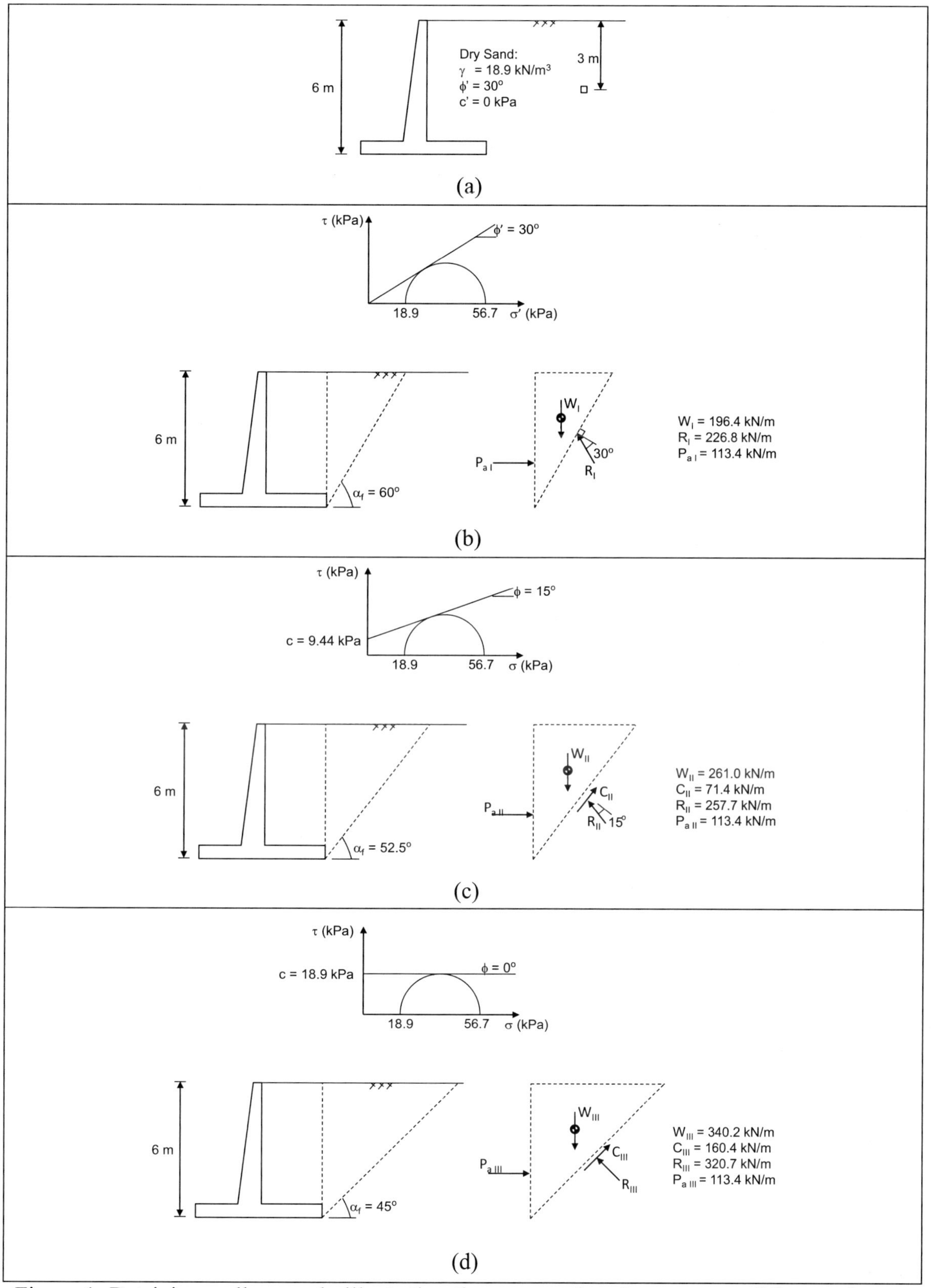

Figure 1. Retaining wall example illustrating the concept that the stresses at failure acting on any plane cutting through the element (other than planes parallel to the principal stress planes) are unique to that plane.

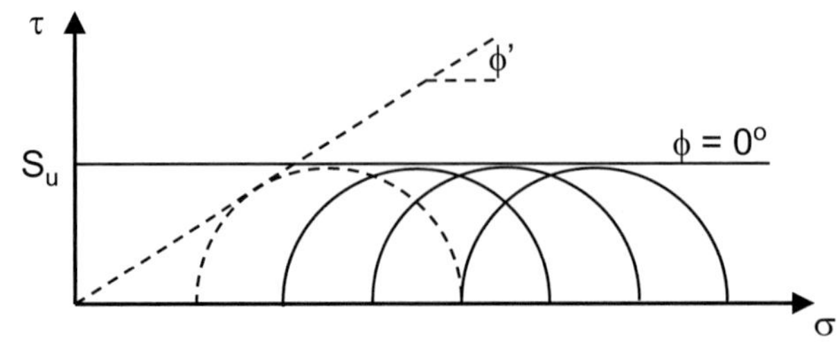

Figure 2. Conceptual illustration of the results from three UU triaxial tests performed on identical, normally consolidated saturated clay samples. The circles drawn with the solid lines are the total stress failure circles and the associated failure envelope identified by $\phi = 0°$. The circle drawn with the dashed line is the effective stress failure circle and the associated failure envelope identified by ϕ'. (e.g., Bishop and Bjerrum, 1960).

While an understanding of the $\phi = 0°$ failure envelope in the context of Terzaghi effective stress principle was fairly well established by the late 1940's, a full understanding of the relationship among the $\phi = 0$ Concept, Terzaghi's effective stress principle, and Mohr's rupture hypothesis was not. A lot of attention was focused on the observed orientation of the slip surface in both laboratory specimens and field slope failures. At the time, it was generally assumed that the slip surface corresponded to the plane having maximum obliquity in stresses (i.e., $45° + \phi'/2$ from the plane of major principal stress), as postulated by Coulomb (1776). And, the fact that the observed slip surfaces in saturated clay samples that were sheared undrained were closer to $45° + \phi'/2$ than to 45° was used to give credence to Terzaghi's effective stress principle (e.g., Skempton, 1948a). However, the main issue of contention seemed to be that specifying the undrained strength of saturated clay by a failure envelope having $\phi = 0°$ inherently implied a 45° slip surface orientation, which differed from the orientations of observed slip surfaces. This point of contention may have stemmed from a lack of appreciation for Mohr's rupture hypothesis (i.e., the importance is to have consistency in the failure envelope and the assumed slip surface used in the analysis, with the orientation of the actual slip surface being irrelevant, barring any significant issues due to geometrical and/or material anisotropy in the soil).

To some extent, the general lack of appreciation of Mohr's rupture hypothesis continues today (e.g., Ladd and DeGroot, 2003). This is exemplified by only one of the texts listed in Table 1 (viz., Lambe and Whitman, 1969) emphasizing the significance of this hypothesis to geotechnical engineering analyses. Additionally, two of the texts in Table 1 discuss the errors associated with quantifying the shear strength of the soil using S_u versus $S_u \times \cos(\phi')$, where the latter is the shear stress mobilized on the Coulomb slip surface (i.e., τ_{ff}). Their contention is that the shear strength should be quantified by τ_{ff}, not by S_u. However, they go on to state that because of sampling disturbance, etc., S_u for a disturbed sample is approximately equal to τ_{ff} for an undisturbed sample, which offsets the error from using S_u as opposed to τ_{ff} (i.e., "compensating errors"). The compensating error postulate is conceptually illustrated in Figure 3 for unconfined compression test results. The premise that using τ_{ff} is more correct than using S_u is contrary to Mohr's rupture hypothesis, inherent to which using either τ_{ff} or S_u will result in the same answer for a uniform soil profile as long as there is consistency in the assumed failure envelopes and slip surfaces (e.g., Figure 1). Furthermore, any issues related to sampling disturbance, etc. are completely separate from whether τ_{ff} or S_u are used to quantify shear strength, and the authors believe that such issues should be accounted for by the minimum factors of safety required by the various design specifications.

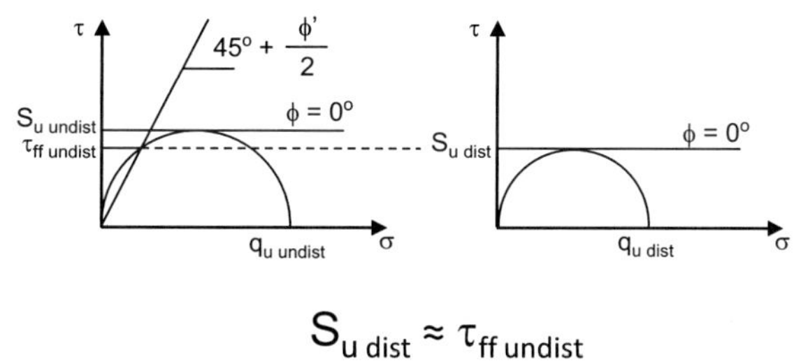

Figure 3. Conceptual illustration of the compensating error postulate proposed by two of the texts listed in Table 1.

It needs to be emphasized that the inherent assumption in the $\phi = 0$ Concept that the slip surface is oriented 45° from the plane of major principal stress by no means violates Terzaghi's effective stress principle. This is because the shear strength of the soil (i.e., S_u) that is assumed to be mobilized along this slip surface is indeed a function of effective stress, not total stress. This can be seen in the following equation, which was derived from the geometry of an effective stress failure circle.

$$S_u = \frac{\sigma'_{1f}}{2} \cdot \left\{1 - tan^2\left(45° - \frac{\phi'}{2}\right)\right\} \tag{2}$$

where σ'_{1f} = the major principal effective stress in the element at failure. However, from examination of data from isotropically consolidated-undrained compression (CIUC) triaxial tests performed on normally consolidated saturated clay samples, it was observed that $\sigma'_{1f} \approx \sigma'_c$, where σ'_c = consolidation stress (Rutledge, 1944). This allows Eqn (2) to be written in a much more useful form:

$$\frac{S_u}{\sigma'_c} = \frac{1}{2} \cdot \left\{1 - tan^2\left(45° - \frac{\phi'}{2}\right)\right\} \tag{3}$$

Skempton and Bishop (1954) further generalized these equations for saturated clay samples that are initially consolidated to K_o conditions (i.e., at rest conditions) and sheared undrained in triaxial compression and extension (i.e., CK_oUC and CK_oUE, respectively):

$$\frac{S_u}{\sigma'_c} = \frac{[K_o + (1-K_o)\cdot\bar{A}_f]\cdot\sin(\phi')}{1 + (2\cdot\bar{A}_f - 1)\cdot\sin(\phi')} \qquad (CK_oUC) \tag{4a}$$

$$\frac{S_u}{\sigma'_c} = \frac{[1 - (1-K_o)\cdot\bar{A}_f]\cdot\sin(\phi')}{1 + (2\cdot\bar{A}_f - 1)\cdot\sin(\phi')} \qquad (CK_oUE) \tag{4b}$$

where: K_o = coefficient of lateral earth pressure at rest, and $\bar{A}_f$ = Skempton's pore pressure coefficient at failure (Skempton, 1954), which is an index of the soil's dilative tendencies among other things.

Note that S_u/σ'_c is rarely computed using Eqns (3) and (4), but rather, S_u/σ'_c is measured directly in the laboratory by subjecting the soil to a stress path representative of that expected in the field (e.g., Ohta et al., 1985). However, the importance of Eqns (3) and (4) is that they show that the undrained shear strength of normally consolidated saturated clay can be specified without the explicit quantification of the effective stress state in the sample at failure, which is required if shear strength is specified in terms of the Mohr-Coulomb failure criterion. *This alleviates the* *onerous task of having to estimate (or measure) excess pore pressures in situ during construction, etc., and lies at the heart of the pragmatic value of the $\phi = 0$ Concept for engineering practice (italics used for emphasis).* In this vein, the majority of the textbooks listed in Table 1 have statements to the effect that "… the $\phi = 0$ Concept is applicable to only saturated clays and silts." This statement is technically incorrect. A more correct statement would be that the pragmatic value of the $\phi = 0$ Concept applies only to saturated clays and silts. As discussed above and as illustrated in Figure 1d, the $\phi = 0$ Concept can be used to specify the shear strength of any soil. However, excess pore pressures generated during construction processes rapidly dissipate in highly permeable, saturated soils. As a result, the pore pressures at failure in these soils are hydrostatic or in steady state seepage conditions, and thus the effective stresses at failure can be readily computed, allowing shear strength to be determined using the Mohr-Coulomb failure criterion.

Finally, as stated above, the inclination of the observed slip surfaces in saturated clay samples that were sheared undrained were closer to 45° + $\phi'/2$ from the plane of major principal stress than to 45°, which was used to give credence to Terzaghi's effective stress principle (Terzaghi, 1936). However, the observed slip surfaces sometimes varied rather significantly from 45° + $\phi'/2$. Based on these observations, it was then hypothesized that the slip surfaces were oriented 45° + $\phi'_e/2$ from the plane of major principal stress rather than 45° + $\phi'/2$ (e.g., Terzaghi, 1936; Gibson, 1953), where ϕ'_e is the Hvorslev effective friction parameter (Hvorslev, 1960). Consequently (and among other reasons), ϕ'_e was often referred to as the "true angle of internal friction" (e.g., Leonards, 1962). However, the results of a detailed laboratory study performed by Rowe (1962, 1963) showed that this was not the case. Although this may have seemed at the time to be a chink in the armor of Terzaghi's effective stress principle, it is not. Starting in the 1950's – 1960's, plasticity theory was being developed and applied to soil mechanics (e.g., Drucker and Prager, 1952; Mroz, 1963). Per this theory, the predicted orientation of the slip surface is 45° + $\psi/2$ from the plane on which the major principal stress acts, where ψ is the dilatancy angle given by Eqn (5) and shown schematically in Figure 4 (e.g., Davis, 1968; Davis and Selvadurai, 2002; Salgado, 2008).

$$\tan(\psi) = -\frac{d\varepsilon_v}{|d\gamma^z|} \tag{5}$$

where: ε_v = volumetric strain; γ^z = shear strain in the x_1^z-x_3^{zp} plane (Figure 4); x_1^z is the direction of zero normal strain, and x_3^{zp} is the direction normal to x_1^z (e.g., Salgado, 2008). Accordingly, the Coulomb slip surface equals that predicted by plasticity theory only if $\psi = \phi'$. However, for most clays ψ is considerably less than ϕ'. Arthur et al. (1977) extended the plasticity theory to take into account strength anisotropy in soil and showed that the inclination of the slip surface typically varied between 45° + $\phi'/2$ and 45° + ¼($\psi+\phi'$). Of the textbooks listed in Table 1, only Mitchell and Soga (2005) and Salgado (2008) discuss that the actual slip surfaces differ from the Coulomb slip surface, which is commonly assumed for effective stress analyses in engineering practice. Fortunately, however, as discussed above, from an engineering perspective, the correct solution to an analysis can be achieved as long as there is consistency in the assumed failure envelope and analyzed slip surface, regardless of whether the analyzed slip surface is the actual slip surface or not.

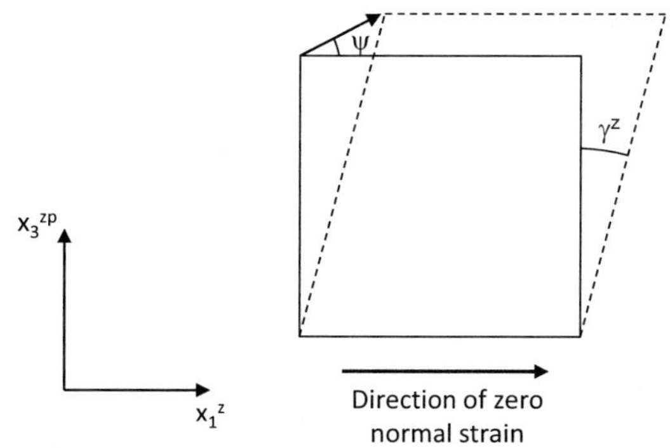

Figure 4. State of strain visualized from an element with one side aligned with the direction of zero normal strain (after Salgado, 2008).

ANALYSIS PROCEDURES OF DUNCAN ET AL. AND POULOS ET AL.

In this section, the procedure for multi-staged rapid drawdown analysis for earthen dams proposed by Duncan et al. (1990) and the procedure for determining the strength of liquefied soil proposed by Poulos et al. (1985) are reviewed in the context of the $\phi = 0$ Concept (note it is not the intent of the authors to comprehensively present and/or review these procedures in their entirety, but rather, the intent is to review the portions of these procedures that are relevant to the objectives of this paper). Both of these procedures entail using undrained strengths of soil and are generally used only in high-end geotechnical engineering practice.

Duncan et al. (1990)

The rapid drawdown analysis procedure proposed by Duncan et al. (1990) consists of three stages and is an extension of the two stage procedure proposed by Lowe and Karafiath (1960). The first stage of the procedure proposed by Duncan et al. entails performing a pre-drawdown, effective stress slope stability analysis, wherein the pore water pressures are determined assuming steady state seepage conditions. The purpose of this stage is to identify the critical slip surface and to estimate the effective stresses acting along it. The second stage entails performing a post-drawdown, total stress slope stability analysis. In this analysis, undrained shear strengths are used for soils having a low permeability and are related to the consolidation stresses acting along the critical slip surface, which were determined in Stage 1. Finally, the third stage entails a comparison of the drained and undrained strengths of the soil along the slip surface, and the stability of the slope is re-analyzed using the lower of the strengths.

Of specific interest herein is the way in which Duncan et al. quantify the undrained shear strength of soils having low permeability in their Stage 2 slope stability analysis, which is given by Eqn (6) and shown in Figure 5a.

$$\tau_{ff} = \frac{\sigma_{1f}-\sigma_{3f}}{2} \cdot \cos(\phi') \qquad (6)$$

where σ_{1f} and σ_{3f} are the total major and minor principal stresses at failure, respectively. As can be seen from Eqn (6)/Figure 5a, Duncan et al. do not define the undrained shear strength of the soil by the stress at the top of the failure circle. Rather, they define it as the shear stress mobilized on the Coulomb slip surface, consistent with an effective stress analysis. The reason for this is that the critical slip surface used in the total stress analysis in Stage 2 is the surface determined from the effective stress slope stability analysis performed in Stage 1. As a result, the procedure proposed by Duncan et al. is completely consistent in its determination of the critical slip surface and its definition of undrained shear strength.

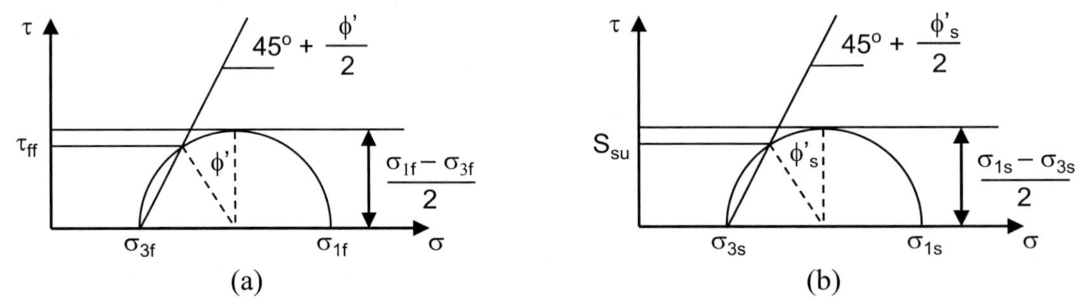

Figure 5. Undrained strengths as used in the (a) Duncan et al. procedure and (b) Poulos et al. procedure.

Poulos et al. (1985)

The procedure proposed by Poulos et al. (1985) for determining the undrained strength of liquefied soil for evaluating flow liquefaction potential is based on the steady state concept initially proposed by Casagrande (1936, 1938). Per Poulos et al. (1985), "The steady state of deformation for any mass of particles is that state in which the mass is continuously deforming at a constant volume, constant normal effective stress, constant shear stress, and constant rate of shear strain." Hence, steady state deformation is an undrained phenomenon. Flow liquefaction is predicted to occur when the driving shear stresses exceed the steady state strength (S_{su}) of the soil, where S_{su} is given by Eqn (7) and shown in Figure 5b.

$$S_{su} = \frac{\sigma_{1s}-\sigma_{3s}}{2} \cdot \cos(\phi'_s) \qquad (7)$$

where σ_{1s} and σ_{3s} are the total major and minor principal stresses at steady state, respectively, and ϕ'_s is the steady state friction angle (in terms of effective stress). As can be seen from Eqn (7)/Figure 5b, Poulos et al. define the undrained shear strength analogously to Duncan et al. (1990). However, unlike Duncan et al., they do not give explicit directions on how to determine the critical slip surface but, rather, just state that it should be determined by "... conventional methods of stability analysis." Accordingly, it is left up to the user of Poulos et al.'s steady state strength to know to use an effective stress analysis to determine the critical slip surface, such that consistency is maintained between the definition of S_{su} and how the critical slip surface is determined. Fortunately, however, if the critical slip surface is determined using a total stress analysis (as is likely to be the case), unforeseen failure is not imminent; rather, the factor of safety against flow failure will be underestimated by $\{1/\cos(\phi'_s) - 1\}\times 100\%$ (i.e., ~10% - 15% for loose sands).

CONCLUSIONS

As discussed above, the "$\phi = 0$ Concept" is widely used in practice in conjunction with total stress analyses of saturated soils that have low permeability. The Concept circumvents the onerous need for estimating (or measuring) excess pore pressures in situ during construction, etc., which is required for performing effective stress analyses. However, while its pragmatic value is widely understood and appreciated, its relationship to Mohr's rupture hypothesis and Terzaghi's effective stress principle is not. This is illustrated by the results of a review of geotechnical engineering textbooks published over the last 60+ years, which showed that most texts opt only to state the $\phi = 0$ Concept and do not put it into proper theoretical or contextual perspective. The authors view this as a missed opportunity to tie the various rules of thumb and sets of governing principles for the shear strength of soil together by presenting them within a consistent framework. Additionally, two procedures that are used in high-end geotechnical engineering practice that entail using undrained strengths of soil were reviewed in the context of the $\phi = 0$ Concept. It was shown that while the Duncan et al. (1990) procedure for rapid drawdown analyses fully addresses all the pragmatic issues for implementing their procedure, the Poulos et al. (1985) procedure for determining the strength of liquefied soil does not.

ACKNOWLEDGMENTS

The authors gratefully acknowledge the insightful review comments by Professors Chris Baxter, University of Rhode Island, Radoslaw Michalowski, University of Michigan, Tom Brandon, Virginia Tech, Mick Pender, University of Auckland, and Brady Cox, University of Texas at Austin, as well as those by Dr. Kord Wissmann, Geopier Foundation Company, and the two anonymous reviewers. Moreover, the authors acknowledge the many significant contributions made by Professor Emeritus Roy E. Olson, University of Texas at Austin, to the field of geotechnical engineering and his impact on our professional practice.

REFERENCES

Arthur, J.R.F., Dunstan, T., Al_Ani, Q.A., and, Assadi, A. (1977). "Plastic Deformation and Failure in Granular Media", *Geotechnique*, 27(1), 53-74.

Bell, A.L. (1915). Lateral Pressure and Resistance of Clay", *Minutes of the Proceedings of the Institute of Civil Engineers*, 199, 233-272.

Bishop, A.W. and Bjerrum, L. (1960). "The Relevance of the Triaxial Test to the Solution of Stability Problems", *Research Conference on Shear Strength of Cohesive Soils*, American Society of Civil Engineers, 437-501.

Bolton, M.D. (1991). *A Guide to Soil Mechanics*, Universities Press (India) Private Limited, Hyderguda, Hyderabad, India.

Budhu, M. (2007). *Soil Mechanics and Foundations*, 2nd edition, John Wiley & Sons, New York.

Casagrande, A. (1936). "Characteristics of Cohesionless Soils Affecting the Stability of Slopes and Earth Fills", *Journal of the Boston Society of Civil Engineers*, 23(1), 13-32.

Casagrande, A. (1938). "The Shearing Resistance of Soils and Its Relation to the Stability of Earth Dams", *Proc. Soils and Foundation Conf. of the U.S. Engineer Department*, 17-21 June, Boston, MA, A-1 – A-20.

Coulomb, C.A. (1776). "Essai sur une application des regles des maximis et minimis a quelques problemes de statique relatifs a l'architecture", *Memoires de Mathematique & de Physique, presents a l'Academie Royale des Sciences par divers Savans, & lus dans ses Assemblees*, 7, 1773, 343-382, Paris.

Das, B.M. (2006). Principles of Geotechnical Engineering, 6th edition, Thomson.

Davis, E.H. (1968). "Theories of Plasticity and the Failure of Soil Masses", *Soil Mechanics: Selected Topics* (I.K. Lee, ed.), America Elsevier Publishing Co., Inc., New York, 341-380

Davis, R.O. and Selvadurai, A.P.S. (2002). Plasticity and Geomechanics, Cambridge University Press, New York.

de Boer, R. (2005). *The Engineer and the Scandal*, Springer, New York

Drucker, D.C. and Prager, W. (1952). "Soil Mechanics and Plastic Analysis of Limit Design", *Quarterly of Applied Mathematics*, 10(2),157-175.

Duncan, J.M., Wright, S.G., and Wong, K.S. (1990). "Slope Stability during Rapid Drawdown", *Proc. H. Bolton Seed Memorial Symposium* (J.M. Duncan, ed.), BiTech Publishers, LTD, 2, 253-272.

Fellenius, W. (1927). *Erdstatische Berechnungen*, Berlin

Fillunger, P. (1915). "Versuche uber die Zugfestigkeit bei allseitigen Wasserdruck", *Osterreichische Wochenschrift offentlichen Baudienst (OWOB)*, 29, 443-448.

Fiori, C. (2008). Personal communication with R. Green

Gibson, R.E. (1953). "Experimental Determination of the True Cohesion and True Angle of Internal Friction in Clays", *Proc. 3rd Intern. Conf. Soil Mechanics and Foundation Engineering*, 1, 126-130.

Golder, H.Q. and Skempton, A.W. (1948). "The Angle of Shearing Resistance in Cohesive Soils for Tests at Constant Water Content", *Proc. 2nd Intern. Conf. on Soil Mechanics and Foundation Engineering*, 1, 185-192.

Handy, R.L. and Spangler, M.G. (2007). *Geotechnical Engineering: Soil and Foundation Principles and Practice*, 5th edition, McGraw Hill, New York.

Holtz, R.D., Kovacs, W.D., and Sheehan, T.C. (2011). *An Introduction to Geotechnical Engineering*, 2nd edition, Prentice Hall, New York.

Hvorslev, M.J. (1960). "Physical Components of the Shear Strength of Saturated Clays", *Proc. ASCE Research Conf. on the Shear Strength of Cohesive Soils*, Boulder, CO, 169-173.

Ishibashi, I. and Hazarika, H. (2011). *Soil Mechanics Fundamentals*, CRC Press, Taylor & Francis Group, Boca Raton, FL.

Ladd, C.C. and DeGroot, D.J. (2003). "Recommended Practice for Soft Ground Site Characterization: Arthur Casagrande Lecture", *Proc. 12th Panamerican Conference on Soil Mechanics and Geotechnical Engineering*, 1, 3-57 (*revised May 9, 2004*)

Lambe, T.W. and Whitman, R.V. (1969). *Soil Mechanics*, John Wiley & sons, Inc., New York.

Leonards, G.A. (1962). *Foundation Engineering* (G.A. Leonards, ed.), McGraw-Hill Book Company, New York.

Lowe, III, J. and Karafiath, L. (1960). "Stability of Earth Dams Upon Drawdown", *Proc. 1st Panamerican Conf. on Foundation Engineering*, 2, 537-552.

Lyell, C. (1871). Elements of Geology, The Student's Series, London.

McCarthy, D.F. (2007). *Essentials of Soil Mechanics and Foundations: Basic Geotechnics*, Prentice Hall, New York.

Mitchell, J.K. and Soga, K. (2005). *Fundamentals of Soil Behavior*, 3rd edition, John Wiley & Sons, Inc., New York.

Mohr, O. (1882). "Ueber die Darstellung des Spannungszustandes und des Deformationszustandes eines Korperelementes und uber die Anwendung derselben in der Festigkeitslehre", *Civilingenieur*, 28, 113-156.

Mroz, Z. (1963). "Non-associated Flow Laws in Plasticity", *Journal de Mecanique*, 2(1), 21-42.

Murthy, V.N.S (2002). *Geotechnical Engineering: Principles and Practices of Soil Mechanics and Foundation Engineering*, Marcel Dekker, Inc., New York.

Ohta, H., Nishihara, A., and Morita, Y. (1985). "Undrained Stability of K_o-Consolidated Clays", *Proc. 11th Intern. Conf. on Soil Mechanics and Foundation Engineering*, San Francisco, 2, 613-616.

Pender, M. (2013). Personal communication with R. Green

Poulos, S.J., Castro, G., and France, J.W. (1985). "Liquefaction Evaluation", ASCE, *Journal of Geotechnical Engineering*, 111(6), 772-792.

Reynolds, O. (1886). "Experiments Showing Dilatancy, a Property of Granular Material, Possibly Connected with Gravitation", *Proceedings of the Royal Institute of Great Britain*, 11, 354-363.

Rowe, P.W. (1962). "The Stress-Dilatancy Relation for Static Equilibrium of an Assembly of Particles in Contact", *Proceedings of the Royal Society of London, Series A*, 269, 500-527.

Rowe, P.W. (1963). "Stress-Dilatancy, Earth Pressures, and Slopes", ASCE, *Journal of Soil Mechanics and Foundation Engineering*, 89(SM3), 37-62.

Rutledge, P.C. (1944). "Relation of Undisturbed sampling to Laboratory Testing", *Transactions of the American Society of Civil Engineering*, 109, 1155-1183.

Salgado, R. (2008). *The Engineering of Foundations*, McGraw Hill, New York.

Scott, R.F. (1963). *Principles of Soil Mechanics*, Addison-Wesley Publishing Company, Inc., Reading, MA.

Skempton, A.W. (1948a). "The $\phi = 0$ Analysis of Stability and Its Theoretical Basis", *Proc. 2nd Intern. Conf. on Soil Mechanics and Foundation Engineering*, 1, 72-78.

Skempton, A.W. (1948b). "A Study of the Immediate Triaxial Test on Cohesive Soils", *Proc. 2nd Intern. Conf. on Soil Mechanics and Foundation Engineering*, 1, 192-196.

Skempton, A.W. (1954). "The Pore-Pressure Coefficients A and B", *Geotechnique*, IV, 143-147.

Skempton, A.W. (1960). "Significance of Terzaghi's Concept of Effective Stress", *From Theory to Practice in Soil Mechanics, Selections from the Writings of Karl Terzaghi* (L. Bjerrum, A. Casagrande, R.B. Peck, and A.W. Skempton, eds.), John Wiley & Sons, New York, 42-53.

Skempton, A.W. and Bishop, A.W. (1954). "Soils", *Building Materials, Their Elasticity and Inelasticity* (M. Reiner, ed.), North-Holland, Amsterdam, Chapter X, 417-482.

Skempton, A.W. and Golder, H.Q. (1948). "Practical Examples of the $\phi = 0$ Analysis of Stability of Clays", *Proc. 2nd Intern. Conf. on Soil Mechanics and Foundation Engineering*, 2, 63-70.

Taylor, D.W. (1948). *Fundamentals of Soil Mechanics*, Wiley, New York.

Terzaghi, K. (1924). "Die Theorie der hydrodynamischen Spannungserscheinungen und ihr erdbautechnisches Anwendungsgebiet", *Proc. Intern. Congress of Applied Mechanics*, 288-294.

Terzaghi, K. (1925). *Erdbaumechanik auf bodenphysikalischer Grundlage*, Franz Deuticke, Leipzig/Wien.

Terzaghi, K. (1936). "The Shearing Resistance of Saturated Soils", *Proc. 1st Intern. Conf. Soil Mechanics and Foundation Engineering*, 1, 54-56.

Terzaghi, K. and Peck, R.B. (1948). *Soil Mechanics in Engineering Practice*, 1st edition, John Wiley & sons, Inc., New York.

Terzaghi, K. and Peck, R.B. (1967). *Soil Mechanics in Engineering Practice*, 2nd edition, John Wiley & Sons, Inc., New York.

Tschebotarioff, G.P. (1951). *Soil Mechanics, Foundations, and Earth Structures*, McGraw-Hill Book Company, Inc., New York.

Westerberg, N. (1926). "Jordtryck i Kohesionara jordarter", *Tek. Tidsk. Vag. o. Vatten.*, 51, 25-29

Consolidation of Soils

G. Mesri[1], M. ASCE and T.W. Feng[2], M. ASCE

[1]Ralph B. Peck Professor of Civil Engineering, University of Illinois at Urbana-Champaign, IL 61801; gmesri@illinois.edu
[2]Professor of Civil Engineering, Chung Yuan Christian University, Chung-li, Republic of China; twfeng@cycu.edu.tw

ABSTRACT: Primary compression and secondary compression of saturated soils are consistent with the Terzaghi effective stress principle. Both primary compression and secondary compression, and associated settlement and increase in undrained shear strength, result from the rearrangement of soil particles to more compact packing, assisted by soil particle deformation, soil particle compression, and soil particle damage. The secondary compression behavior of soils is well explained and predicted by the C_α/C_c law of compressibility. The debate on the independence of end-of-primary (EOP) void ratio versus effective vertical stress relation from the duration of primary consolidation continues in spite of the following series of convincing evidence (a) EOP e versus σ'_v relation from laboratory consolidation tests is independent of the duration of primary consolidation, (b) preconsolidation pressure mobilized in the field is equal to EOP preconsolidation pressure determined from laboratory oedometer tests on 20 mm thick undisturbed soil specimens, (c) surface and subsurface settlements measured in the field during primary consolidation of soil subjected to embankment loading, are equal to the settlement predicted using the EOP e versus σ'_v relation from laboratory oedometer tests on 20 mm thick undisturbed soil specimens, and (d) data on compressibility with time, $(\partial e / \partial t)_{\sigma'_v}$, provide an explanation for the independence of EOP e versus σ'_v relation from the duration of primary consolidation. There is no direct observed evidence supporting the "extreme" hypothesis B.

INTRODUCTION

Consolidation is the time-dependent compression of saturated soils. Consistent with the Terzaghi effective stress principle (Terzaghi 1923), compression begins with an increase in effective stress; however, compression continues indefinitely at decreasing rate after the effective stress increase comes to an end at elapsed time t_p. The compression that takes place during the increase in effective stress is termed primary compression, and the compression that follows at constant effective stress is termed secondary compression. The practical consequences of consolidation are an increase in stiffness and strength of soil and settlement of the soil surface as water is squeezed out of the soil, and total volume decreases.

Compression of soils mainly results from the rearrangement of soil particles to a more compact packing; however, it may also result from deformation of soil particles as in micaceous silts, compression of soil particles as in fibrous peats, and from soil particle damage as in fracturing of quartz sand particles. Secondary compression phenomenon is not inconsistent with the effective stress principle as the definition of the Terzaghi effective stress is in terms of total stress and porewater pressure, both of which are external to the soil structure. Secondary compression is the manifestation of continued readjustments of the internal components of effective stress toward a more stable fabric and interparticle forces. The subsequent interpretation in this paper is for horizontally constrained vertical compression, and consolidation is in response to an increase in effective vertical stress:

$$\sigma'_v = \sigma_v - u \qquad (1)$$

where σ_v is the total vertical stress and u is porewater pressure.

Secondary Compression and Effective Stress

The C_α/C_c law of compressibility (Mesri and Godlewski 1977, Mesri 1987, Mesri and Castro 1987, Mesri 2001, Mesri and Ajlouni 2007, Mesri and Vardhanabhuti 2009), where $C_\alpha = \Delta e/\Delta \log t$ is the slope of e versus log t relation at any σ'_v and time greater than t_p, and $C_c = \Delta e/\Delta \log \sigma'_v$ is the slope of e versus log σ'_v relation at any time equal or greater than t_p, explains and predicts for a wide variety of soils:

1. Secondary settlement after loading to the recompression, to the transition from recompression to compression (near σ'_p), or to the compression range (Mesri et al. 1994, Mesri and Ajlouni 2007).
2. Secondary settlement after the removal of surcharge (Mesri 1986, Mesri and Feng 1991, Mesri et al. 1997, Mesri et al. 2001, Mesri and Ajlouni 2007).
3. Shape of the settlement versus log t curve in the transition from primary compression to secondary compression (Mesri and Godlewski 1977, Mesri and Choi 1980, Mesri 1987).
4. Strain rate effect on e versus log σ'_v relation, preconsolidation pressure, and undrained shear strength (Mesri and Choi 1979, 1984, Mesri 1987).
5. Imposed strain rate for CRS oedometer test to obtain the EOP e versus log σ'_v relation (Mesri and Feng 1992, Mesri et al. 1994).
6. Preconsolidation pressure, shear modulus, and undrained shear strength resulting from secondary compression (Mesri and Choi 1979, 1983, Mesri 1987).
7. Decrease in effective stress resulting from the arresting of secondary compression (Mesri 1987).
8. Increase in coefficient of earth pressure at rest resulting from secondary compression (Mesri and Hayat 1993).
9. Post-densification penetration resistance of granular soils (Mesri et al. 1990).

One remarkable aspect of the C_α/C_c law of compressibility, shown in Table 1 (Terzaghi et al. 1996) is that the magnitude of C_α/C_c for each soil type, such as inorganic soft clays and silts, amorphous and fibrous peats, and granular soils, has a narrow range of values; and even the total range for all soils, 0.01 to 0.07, is remarkably small in comparison to the observed ranges for other soil properties. This result, plus the general applicability of the compressibility law to all particulate materials, suggests a more fundamental meaning for C_α/C_c than merely a compressibility ratio. An examination suggests that the magnitude of C_α/C_c is actually a measure of the increase in internal effective stress with time, as follows:

$$\Delta \log \sigma'_v = (C_\alpha/C_c)\, \Delta \log t \qquad (2)$$

Thus, it is not surprising that C_α/C_c has a narrow range of values because a limited set of interparticle primary valence bonding, attraction, and repulsion mechanisms control the internal interparticle stress in all soil compositions. The total range in the magnitude of C_α/C_c reflects the difference in importance of specific particle interaction mechanisms among different soil compositions ranging from granular space lattice silicates to organic fibrous peats (Mesri and Ajlouni 2007, Mesri and Vardhanabhuti 2009).

Table 1. Values of C_α/C_c for Geotechnical Materials.

Material	C_α/C_c
Granular soils including rockfill	0.02 ± 0.01
Shale and mudstone	0.03 ± 0.01
Inorganic clays and silts	0.04 ± 0.01
Organic clays and silts	0.05 ± 0.01
Amorphous and fibrous peats	0.06 ± 0.01

After an effective vertical stress increase, $\Delta\sigma'_v = \Delta\sigma_v - \Delta u_s$, is realized in primary consolidation time period t_p, at any time t equal or greater than t_p, the internal effective vertical stress from Eq. 2 is:

$$\sigma^*_{vf} = \sigma'_{vf}\left(\frac{t}{t_p}\right)^{C_\alpha/C_c} \qquad (3)$$

where $\sigma'_{vf} = \sigma'_{vo} + \Delta\sigma'_v$. In general, in normally consolidated soil deposits of age t, the internal effective vertical stress is

$$\sigma^*_v = \sigma'_v\left(\frac{t}{t_p}\right)^{C_\alpha/C_c} \qquad (4)$$

where $\sigma'_v = \sigma_v - u_s$, and t_p is the end-of-primary consolidation time required to realize the last increment of σ'_v, and u_s = static or steady state porewater pressure. According to this general principle of effective stress, compression of normally consolidated saturated soils depends on the consolidation pressure which is an internal effective stress defined by Eq. 4.

For a normally consolidated young deposit that experiences a $\Delta\sigma'_v$ in a time period t_p, the settlement at elapsed time t equal or greater than t_p is:

$$S = \frac{C_c}{1+e_o} L_o \log\frac{\sigma^*_{vf}}{\sigma'_{vo}} \qquad (5)$$

or

$$S = \frac{C_c}{1+e_o} L_o \log\left[\frac{\sigma'_{vf}}{\sigma'_{vo}}\left(\frac{t}{t_p}\right)^{C_\alpha/C_c}\right] \qquad (6)$$

which may be rewritten as:

$$S = \frac{C_c}{1+e_o} L_o \left[\log\frac{\sigma'_{vf}}{\sigma'_{vo}} + \frac{C_\alpha}{C_c}\log\frac{t}{t_p}\right] \qquad (7)$$

Equation 7 is the well-known expression for primary settlement plus secondary settlement of a normally consolidated young soil with a linear EOP e versus log σ'_v relation in the range of σ'_{vo} to σ'_{vf}, and a linear e versus log t relation in the elapsed time range t_p to t (Terzaghi et al. 1996, Art. 16.6).

For a normally consolidated young deposit that experiences a $\Delta\sigma'_v$ in a primary consolidation time period t_p, undrained shear strength at elapsed time t equal or greater than t_p is:

$$s_u = \left[\frac{s_u}{\sigma'_v}\right]\sigma^*_{vf} \tag{8}$$

or

$$s_u = \left[\frac{s_u}{\sigma'_v}\right]\sigma'_{vf}\left[\frac{t}{t_p}\right]^{C_\alpha/C_c} \tag{9}$$

where s_u = undrained shear strength, and σ'_{vf} = consolidation pressure reached at time t_p.

In summary, secondary compression is consistent with the Terzaghi principle of effective stress, and the C_α/C_c law of compressibility explains and predicts observed implications of secondary compression in terms of settlement and undrained shear strength behavior.

Primary Compression and Duration of Primary Consolidation

The most commonly used procedure for computing settlement resulting from primary compression is implicitly based on the assumption of end-of-primary (EOP) void ratio versus effective vertical stress relation independent of the duration of primary consolidation (e.g., Leonards 1972, Ladd et al. 1977, Mesri 1977, Mesri and Choi 1985a, Jamiolkowski et al. 1985, Mesri et al. 1994, Leonards and Deschamps 1995, Mesri et al. 1995). The significant implication of this assumption is that the EOP e versus σ'_v relation determined from oedometer testing of 20 mm thick undisturbed soil specimens with duration of primary consolidation in minutes or days is used for settlement analysis of field soil layers with duration of primary consolidation in months or years.

Mesri and co-workers have proposed an EOP compression independent of t_p based on the following observations (Mesri 2001):

1. Primary compression of thin and thick specimens:
   Based on consolidation tests on thin and thick undisturbed specimens of five soft clay deposits, Mesri and Choi (1985a) and Mesri et al. (1995) have demonstrated an EOP void ratio versus effective vertical stress relation independent of duration of primary consolidation (Fig. 1). The typical duration of primary consolidation, t_p, in Fig. 1, for the pressure increments in the compression range, for specimens with maximum drainage distance of 125 mm is in the range of 4 to 25 days, whereas the same range for maximum drainage distance of 500 mm is 54 to 354 days. Similarly reliable data on thin versus thick specimens have been published by Aboshi (1973).
2. Preconsolidation pressure mobilized in the field compared to σ'_p from oedometer tests:
   Using the interpretations proposed by Hoeg et al. (1969) and Sällfors (1975), porewater pressure measurements under embankments have been used to determine preconsolidation pressure mobilized in the field, which is then compared to the preconsolidation pressure obtained from the EOP e versus log σ'_v relation of 20 mm thick undisturbed laboratory specimens. Mesri et al. (1995) and Mesri (2001) have summarized data from the literature for over 70 separate measurements on some 25 soft clay and silt deposits that are shown in Fig. 2. That σ'_p/σ'_{vo} determined from oedometer tests on 20 mm thick specimens correctly defines the σ'_p in the field, is most fortunate because the preconsolidation pressure from the

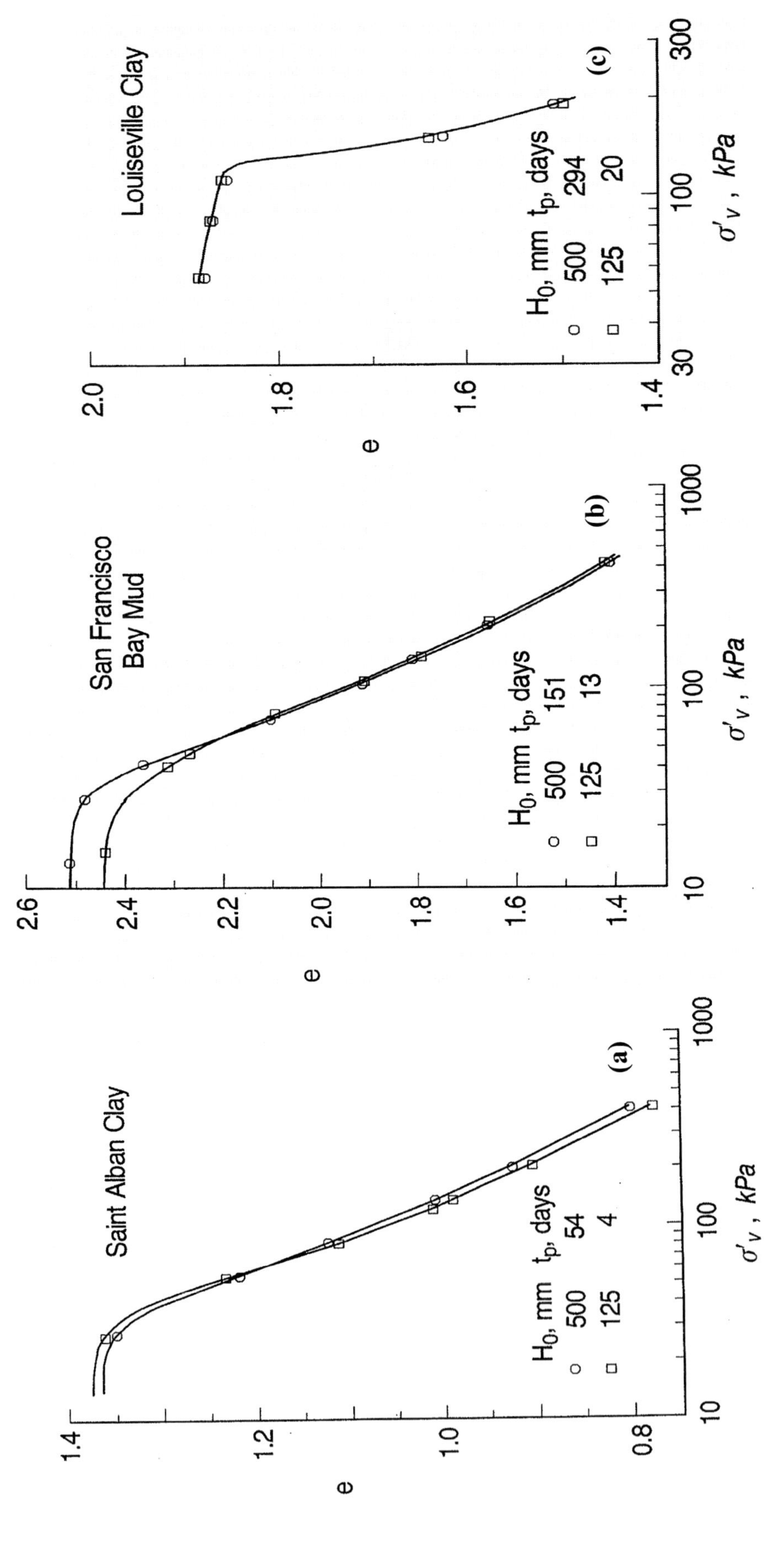

Figure 1. EOP e versus log σ'_v relations of a thin and thick layer: **(a)** Saint Alban Clay, **(b)** San Francisco Bay Mud, and **(c)** Louiseville Clay (Mesri and Choi 1985a)

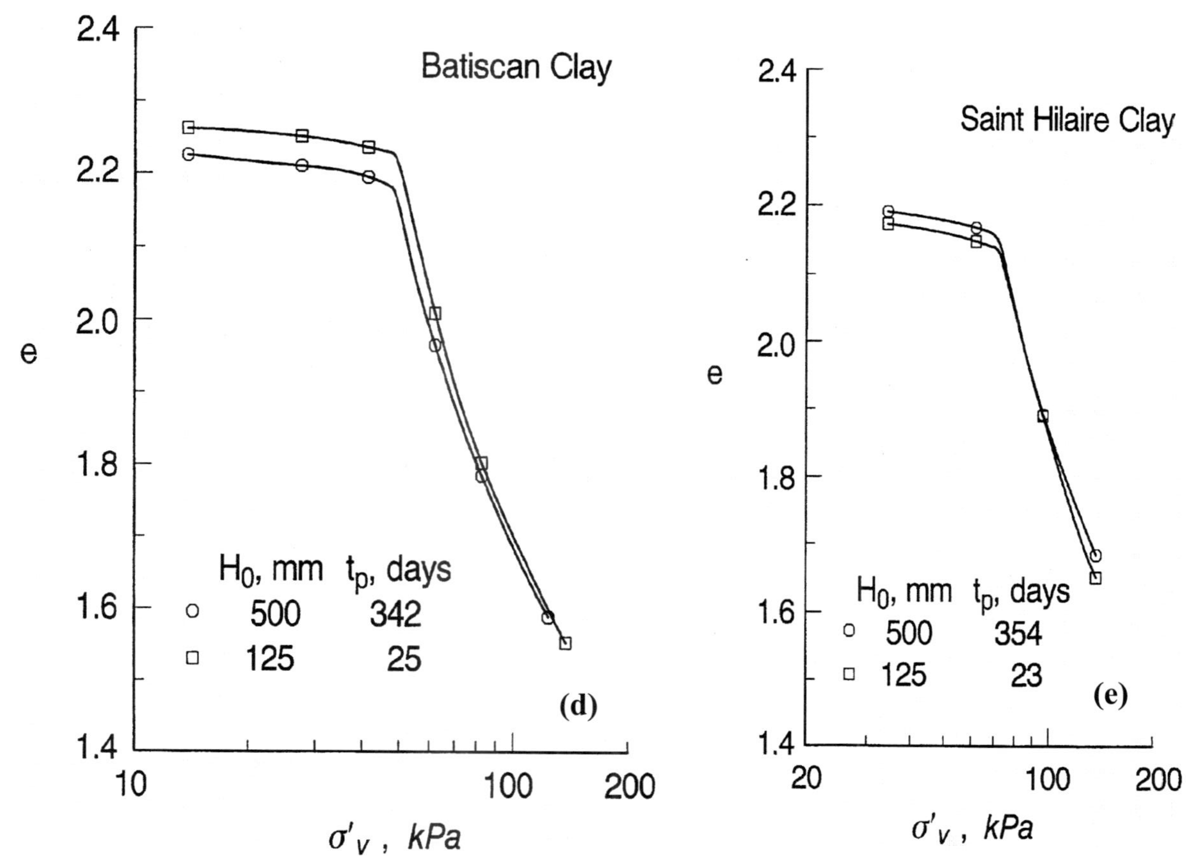

Figure 1 (cont.). EOP e versus log σ'_v relations of a thin and thick layer: **(d)** Batiscan Clay, and **(e)** Saint Hilaire Clay (Mesri et al. 1995).

laboratory tests has been utilized not only for settlement analysis, but also for interpretation of other important soil properties, especially the undrained shear strength of soft clay and silt deposits (Terzaghi et al. 1996).

3. Settlement analysis based on EOP e versus σ'_v relation from laboratory oedometer tests:
Primary settlement of soft clay and silt deposits is computed for embankment loading, using the EOP e versus log σ'_v relation from oedometer tests on 20 mm thick specimens, and is compared with the field observations of settlement (Table 2). The details of the ILLICON procedure for settlement analysis have been described by Mesri and Choi (1985b) and Mesri et al. (1994), with the results shown in Fig. 3. The computed surface and subsurface settlements during and at the end of primary consolidation are in excellent agreement with observations.

Table 2. Case Histories of Embankment Construction on Soft Clay Deposits.

Case Record	w_o %	w_ℓ %	σ'_p/σ'_{vo}	C_α/C_c	SS	SUS
Olga B & C, Canada	30-92	30-68	2.0-3.0	0.033	2	-
Rupert 7, Canada	27-62	30-36	1.6-2.6	0.030	3	-
Vasby, Sweden	43-122	40-125	1.1-1.3	0.060	1	5
Gloucester, Canada	25-90	37-64	1.4-1.9	0.031	1	4
Ellingsrud, Norway	30-45	24-27	1.1-2.4	0.044	1	1
Skå-Edeby, Sweden	50-130	50-150	1.2-1.3	0.050	7	21
Changi, Singapore	30-75	35-95	1.5-3.1	0.035	11	26
ChekLapKok, Hong Kong	25-125	40-100	1.1-6.6	0.03-0.065	7	28
LaGrande, Canada	34-62	25-43	1.3-2.5	0.054	2	6
Bangkok, Thailand	50-120	72-100	1.1-2.0	0.050	2	-

SS = number of surface settlement observations
SUS = number of subsurface settlement observations

4. Compressibility parameters during primary consolidation:
Reasoning either implicitly, or explicitly in terms of the following general equation for primary compression:

$$\int_o^{t_p} de = \int_o^{t_p} \left[\left(\frac{\partial e}{\partial \sigma'_v} \right)_t \frac{d\sigma'_v}{dt} + \left(\frac{\partial e}{\partial t} \right)_{\sigma'_v} \right] dt \tag{10}$$

one might suppose that with increasing t_p, the compressibility of soil structure with time, $(\partial e / \partial t)_{\sigma'_v}$, may contribute more to primary compression. However, such a reasoning is based on the implicit assumption that $(\partial e / \partial t)_{\sigma'_v}$ is independent of t_p. In fact $(\partial e / \partial t)_{\sigma'_v}$ is related to the rate of effective stress increase and thus t_p. Mesri et al. (1995) have shown that a combination of a small value of t_p together with large values of $(\partial e / \partial t)_{\sigma'_v}$ leads to the same EOP e versus σ'_v relation as the combination of a large t_p together with small values of $(\partial e / \partial t)_{\sigma'_v}$.

In spite of these observed data favoring the EOP void ratio versus effective stress relation independent of the duration of primary consolidation, the uniqueness concept (Mesri and Choi 1985a) has been challenged. The debate has continued in terms of the

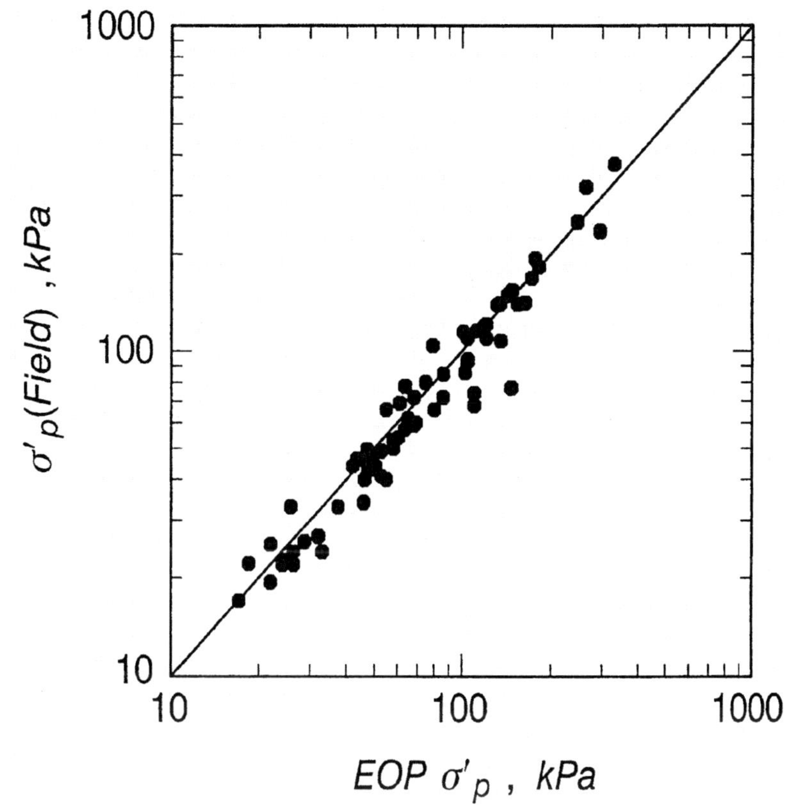

Figure 2. Preconsolidation pressure mobilized in the field compared with EOP σ'_p from 20 mm thick oedometer specimens (Mesri et al. 1995).

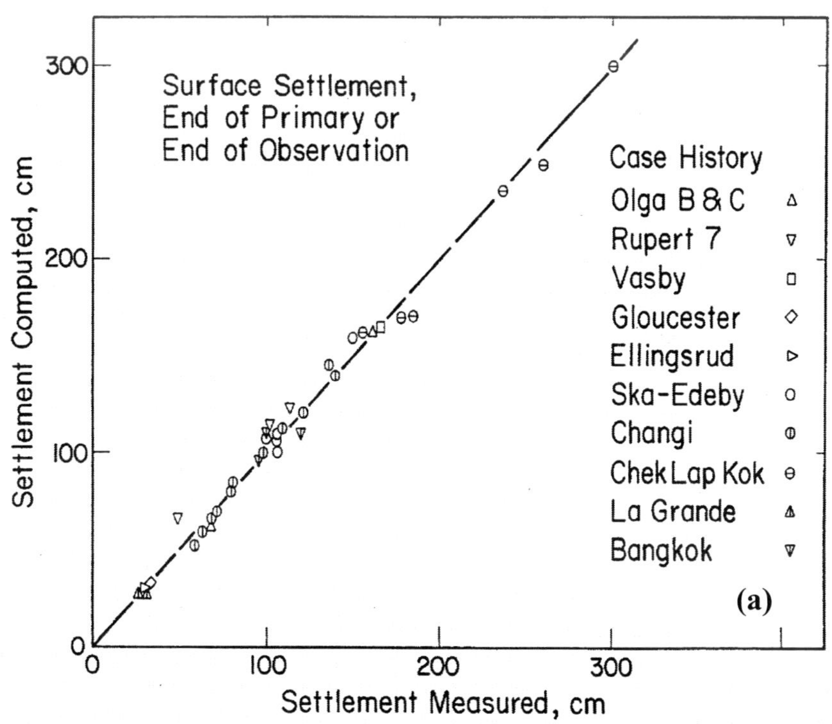

Figure 3. **(a)** Computed and measured EOP or EOO surface settlement of embankments on soft clay and silt deposits (Mesri et al. 1994).

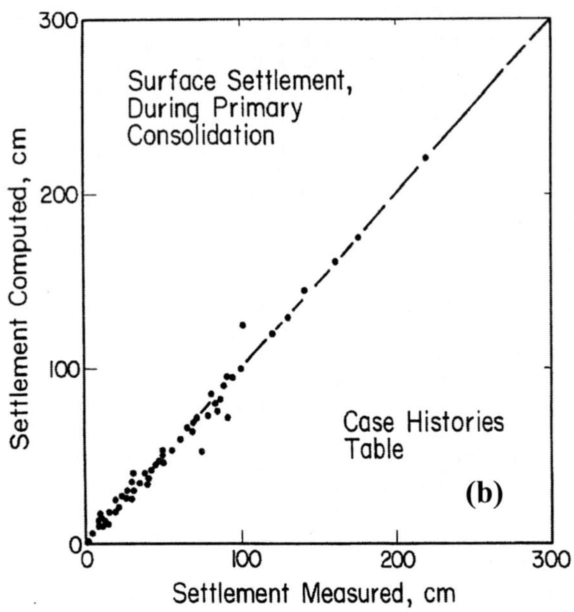

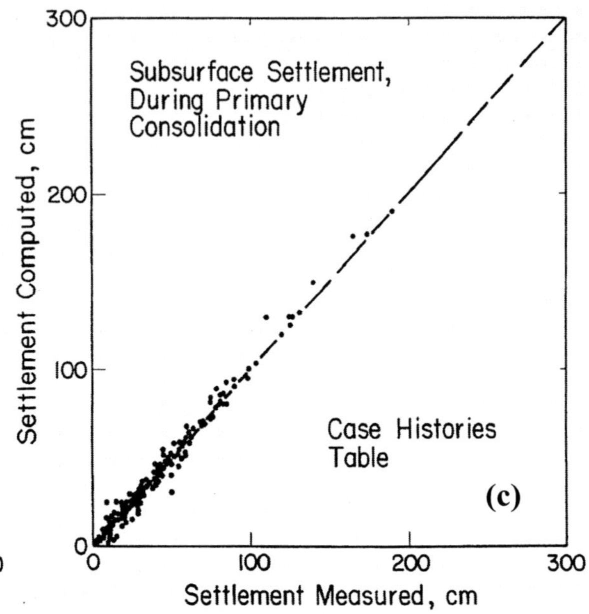

Figure 3 (cont.). Computed and measured surface settlement **(b)** and subsurface settlement **(c)** during primary consolidation for embankments on soft clay and silt deposits (Mesri et al. 1994).

two alternative hypotheses A (EOP e-σ'_v independent of t_p) and B (EOP e-σ'_v dependent on t_p) defined by Ladd et al. (1977) for the possible relation of primary settlement to the duration of primary consolidation, illustrated in Fig. 4. If in fact hypothesis B were true, then primary settlement observed in the field would have been significantly larger than computed using EOP e versus σ'_v relation from 20 mm thick undisturbed specimen in oedometer tests. This has not been the general experience (e.g., Leonards 1972, Mesri 1977, Ladd et al. 1977, and Jamiolkowski et al. 1985), and the data in Fig. 3.

The most recent objection to the uniqueness concept has come from Degago et al. (2009, 2011). Degago et al. (2009) reinterpreted laboratory data in Feng (1991), and using a computer program based on Soft Soil Creep (Stolle et al. 1999) predicted consolidation behavior of laboratory specimens. Degago et al. (2009) used the C_α/C_c law of compressibility, intended for secondary compression, to define their so-called creep parameter during primary consolidation. Having decided that the pressure increment durations in the Feng (1991) consolidation tests for the 500 mm thick specimens were not long enough, Degago et al. (2009) subtracted an increment of void ratio from the EOP void ratio of the 125 mm specimens at consolidation pressures of 62.1 kPa and 96.6 kPa, respectively, for Batiscan clay and St. Hilaire clay (Figs. 1d and 1e). In other words, for the 125 mm thick specimens, the EOP void ratio was incorrectly defined by Degago et al. (2009) at an elapsed time significantly less than t_p. More importantly, the additional compression that took place at 62.1 kPa and 96.6 kPa, respectively, for Batiscan clay and St. Hilaire clay, were not included in the next EOP compression at 96.6 kPa and 138.0 kPa, respectively, of Batiscan clay and St. Hilaire clay.

The fundamental question for Degago et al. (2009) was "... whether or not creep acts as a separate phenomenon during primary compression, while excess pore pressure dissipates." Creep as a separate phenomenon originates from the concept of springs and

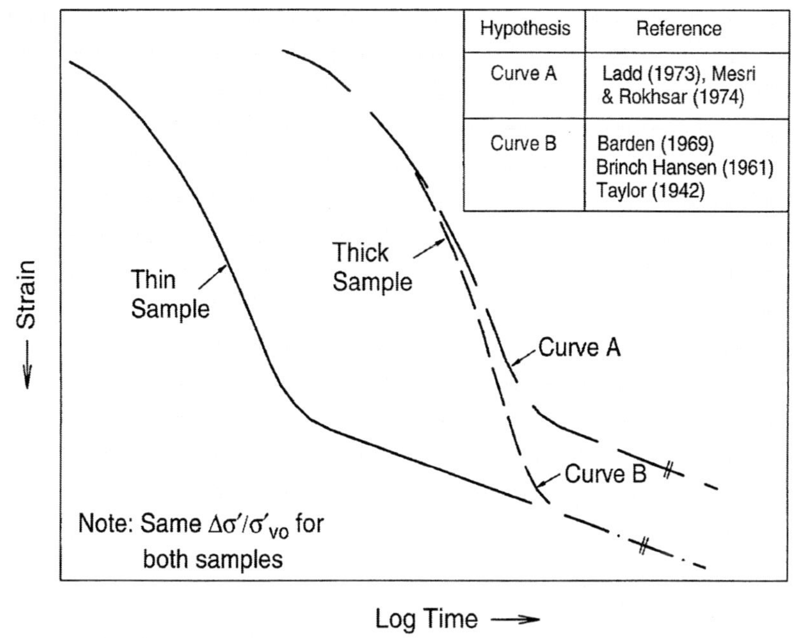

Figure 4. Hypotheses A and B proposed by Ladd et al. (1977).

dashpots – i.e., one spring associated with effective stress increase and a separate spring, completely unaware of the first spring, associated with progress of time. In fact, compressibility with time contributes during primary compression; however, soil has only one structure which controls both compressibility with effective stress and compressibility with time.

Degago et al. (2009) propose that "strictly speaking, it is the degree of dissipation of the excess pore pressure that should establish a true EOP criterion." Degago et al. (2009) do not seem to be aware of Mesri et al. (2005): "Excess porewater pressure during secondary compression." Two of the conclusions of that article are: a) The value of excess porewater pressure at the beginning of secondary compression stage (i.e., end of primary consolidation) corresponds to u'_m/σ'_v values in the range of 1 to 3%, where u'_m is the maximum excess porewater pressure within the layer and σ'_v is the consolidation pressure, and b) For soft clay deposits, with u'_m/σ'_v near 2%, EOP u'_m is often near 1 kPa and is not expected to exceed 10 kPa. The important implication of these conclusions is that it is not always straight forward to use the measurements of excess porewater pressure to "… establish a true EOP criterion", especially for pressure increments spanning the preconsolidation pressure of the highly structured soft clays of Eastern Canada such as the Batiscan clay and St. Hilaire clay.

Degago et al. (2009) then proceeded to examine the excess porewater pressure at EOP compression of the 125 mm and 500 mm specimens. The excess porewater pressures were in the range of 0.1 to 0.8 kPa for Batiscan clay and 1.0 to 2.2 kPa for St. Hilaire clay. Degago et al. (2009) concluded that primary consolidation "… time for the 500 mm thick specimen has been too short …". For the pressure increment from 62.1 kPa to 82.8 kPa after about 382 days the excess porewater pressure at the impermeable boundary (i.e., maximum excess porewater pressure) is less that 1 kPa.

Thus, having satisfied themselves that "… creep acts as a separate phenomenon …" during primary consolidation, Degago et al. (2009) proceeded to make predictions of compression rate "... using concepts derived from the Isotache concept ...". Degago et al. (2009) report that "the material model used is the Soft Soil Creep (SSC) ... the Model incorporates creep during the consolidation process and yields a non-unique EOP strain for different specimen thickness." However, Degago et al. (2009) do not illustrate as to how non-unique is the prediction by the SSC model. In other words, for a field situation, does it predict EOP vertical strain 1 percent greater than the EOP strain of a 20 mm oedometer specimen or does it predict 100% greater vertical strain? Therefore, two main conclusions of Degago et al. (2009) are quite unjustified and troubling. These are: (a) Re-evaluation of the measurements gave "... results that support Hypothesis B" and (b) "According to the isotache concept ... any distinction between thick and thin specimen is revealed through the respective preconsolidation stress ...". The "extreme" hypothesis B as originally defined (Ladd et al. 1977) could predict settlements for field conditions that are as much as 100 percent larger than those predicted using the uniqueness of EOP compression. The observed behaviors in Figs. 1, 2 and 3 dismiss the main conclusions of Degago et al. (2009).

Degago et al. (2009) does not include an application of the SSC model to field situations. However, Neher et al. (2001) use SSC to predict settlement of two test embankments. In connection with one embankment they conclude that "... SSC-model is not so useful." It is the settlement predictions by Neher et al. (2001) for the second embankment – Skå-Edeby Area IV - that is quite revealing about the SSC model.

Mesri and co-workers have carried out settlement analyses using the ILLICON computer program for all of the test fills at the Skå-Edeby test site in Sweden (Mesri et al. 1994). For example, for the 12 m thick soft clay deposit at Skå-Edeby test field, subjected to the same embankment load, however with duration of primary consolidation, for different drainage boundaries by vertical drains, of about 2, 4, 8, and 38 years, the measured EOP settlements are, respectively, 114, 100, 108, and 107 cm (Mesri et al. 1994; Terzaghi et al. 1996, Fig. 25.16; Mesri and Huvaj-Sarihan 2009). Neher et al. (2001) concluded that "... the SSC-model and measured data agree fairly well." However, the SS-model (i.e., SSC-model without "creep") predicted, for an elapsed time of 20 years, a settlement of 15 cm as compared to the observed settlement of 90 cm. The incredible implication of this result is that using hypothesis A one may underpredict settlement by a factor of 6. No such experience assuming hypothesis A has ever been reported in the geotechnical literature (see, e.g., Leonards 1972). In order to understand the unusual prediction by Neher et al. (2001), one needs only to examine their input data on $C_c/(1+e_o)$ of the Skå-Edeby clay. It turns out that $C_c/(1+e_o)$ magnitudes in Table 3 of Neher et al. (2001) are about 1/6 of the true values for the Skå-Edeby clay. In other words, in order for the SSC-model and measured data to agree, Neher et al. (2001) used $C_c/(1+e_o)$ values about 1/6 of the true magnitudes, and this resulted in the unusual prediction of 15 cm settlement for the SS-model. If Neher et al. (2001) had used the true values of $C_c/(1+e_o)$, their SSC-model would have overpredicted settlements by a huge factor (Mesri and Feng 2009).

Degago et al. (2011) conclude that "... the isotache approach can capture the main characteristics of the time-dependent compressibility of clays during both the primary and secondary consolidation phases". They admit, however, that "... when the isotache concept was first presented (Suklje 1957), it was meant to describe compressibility of soils subjected to long-term creep, such that the primary consolidation phase plays a minor role in the whole compression process". The linear secondary compression-log time relations in the Degago et al. (2011) "simulation" illustrate the limitations of the isotache concept, as only in certain cases $C_\alpha/(1+e_o)$ may remain constant with time when $C_c/(1+e_o)$ remains constant with effective vertical stress (Mesri 1987,Mesri and Castro 1987.

Degago et al. (2011) adapted the explanations by Mesri et al. (1995) of the uniqueness of EOP void ratio-effective vertical stress relation, or hypothesis A, by stating that for stress increments in the compression range beyond σ'_p, "... the thin specimen deforms with a higher creep rate for a shorter consolidation period, whereas the thick specimen deforms with a lower creep rate for a longer consolidation period. Consequently, the incremental EOP strains, in the normally consolidated regime, will be almost the same for both specimens". Among the objections to this rewording of the Mesri et al. (1995) explanation is that Mesri and Choi (1985a) and Mesri et al. (1995) did not use the so-called "incremental EOP strains". Mesri and co-workers have obtained EOP e versus log σ'_v relations independent of duration of primary consolidation not by adding primary compression of pressure increments, but rather by computing EOP void ratio at each consolidation pressure, thus taking into account previous compression history. In other words, we have always used the so-called "nominal strain". For a further clarification of the confusion associated with Degago et al.'s (2011) "nominal strain" as opposed to "incremental EOP strain", the reader is referred to Mesri (1986).

The 125 mm and 500 mm thick undisturbed specimens of five soft clay deposits, including the Batiscan clay and St. Hilaire clay, were subjected to equal all around pressure (to avoid side friction and to maintain constant applied pressure over periods in excess of one year) and volumetric compression of the specimen was measured by the volume of water that flowed out. The volumetric compression data were used to compute the EOP void ratio-consolidation pressure relations. The results on undisturbed specimens of five soft clay deposits support hypothesis A, and they indicate EOP void ratio versus consolidation pressure relation independent of duration of primary consolidation (Fig. 1).

The porewater pressure at the bottom and axial compression at the top of each 125 mm sublayer of the 500 mm thick specimens of Batiscan clay and St. Hilaire clay were also measured (see e.g., Figs. 8 and 9 in Mesri et al. 1995 or Fig. 3 in Mesri 2001). The observed data in Feng (1991) that Degago et al. (2011) selected to justify their preference for hypothesis B are the 125 mm sublayer compression measurements for the pressure increment spanning the preconsolidation pressure for Batiscan and St. Hilaire clays. The authors' reinterpretation of the alternative methods examined by Feng (1991) only for the computing of fundamental compressibility parameters $(\partial e/\partial\sigma'_v)_t$ and $(\partial e/\partial t)_{\sigma'_v}$, exaggerates the importance they attach to these data. The alternative assumptions examined by Feng (1991) to compute local change in void ratio had no effect whatsoever on the EOP e versus σ'_v relations which were computed using the volumetric compression of the entire specimen. Disregarding existing evidence, Degago et al. (2011) generalize that the behavior of clays "... agrees w ell with hypothesis B". There is no justification to conclude even the Batiscan and St. Hilaire clay behavior observed for the pressure increment spanning the preconsolidation pressure, supports the "extreme" hypothesis B as it is defined in Fig. 4. Further interpretation of the sublayer compression measurements for a highly structured soft clay from Eastern Canada as it is destructured by a small equal all around pressure increment spanning the preconsolidation pressure, is beyond the scope of this paper (see e.g., Mesri and Choi 1979, Mesri 1981, Mesri et al. 2005, etc.)

Degago et al. (2011) also conclude that "… the in situ EOP preconsolidation stress is lower than that determined from EOP laboratory tests." This serious claim is obviously ignoring the data in Fig 2 that includes over 70 separate measurements on some 25 soft clay and silt deposits. It is suggested that the potential of the isotache concept and "extreme" hypothesis B, for describing consolidation of soils should be substantiated by conducting settlement analyses and comparing with observations, for an adequate number of full-scale cases of embankments on soft clay and silt deposits. Leonards (1972) and Ladd (see Ladd et al. 1977) were "... biased toward hypothesis A ..." based on significant experience with consolidation of soils.

CONCLUSIONS

The following conclusions are based on the analyses, data, and interpretation presented in this paper:

1. The Terzaghi effective stress principle, when interpreted in terms of the internal interparticle stress in soils, is consistent with both primary compression and secondary compression.
2. Both primary compression and secondary compression, and associated settlement and the increase in undrained shear strength, result from the rearrangement of soil particles to more compact packing, assisted by soil particle deformation, soil particle compression, and soil particle damage.
3. The secondary compression behavior which results from the continued internal redistribution of interparticle stress, is completely explained and predicted by the C_α/C_c law of compressibility.
4. The debate on the independence of EOP void ratio versus effective vertical stress from the duration of primary consolidation continues in spite of the following series of convincing evidence (a) EOP e versus σ'_v from laboratory consolidation tests is independent of the duration of primary consolidation, (b) preconsolidation pressure mobilized in the field is equal to the preconsolidation pressure determined from laboratory oedometer tests on 20 mm thick undisturbed soil specimens, (c) surface and subsurface settlements measured in the field during primary consolidation of soil subjected to embankment loading, are equal to the settlement predicted using the EOP e versus σ'_v relation from laboratory oedometer tests on 20 mm thick undisturbed soil specimens, and (d) data on compressibility with time, $(\partial e/\partial t)_{\sigma'_v}$, provide an explanation for the independence of EOP e versus σ'_v relation from the duration of primary consolidation.
5. The existing direct reliable laboratory and field evidence support primary consolidation in accordance with hypothesis A; there is no direct convincing evidence supporting the "extreme" hypothesis B.
6. The proponents of hypothesis B need to verify the claim in terms of settlement analyses together with comparison with measurements, for a sufficient number of embankment loading case histories of soft clay and silt deposits.

ACKNOWLEDGMENT

The first author's introduction to consolidation testing was in Professor Roy Olson's laboratory, and together they explored solutions of the Terzaghi theory of consolidation for a wide range of initial and boundary conditions.

NOTATION

The following symbols are used in this paper:

C_c = compression index = $\Delta e/\Delta \log \sigma'_v$
C_α = secondary compression index = $\Delta e/\Delta \log t$
e = void ratio
e_o = insitu void ratio under effective overburden pressure σ'_{vo}
L_o = preconstruction thickness of compressible layer with void ratio e_o
S = settlement
s_u = undrained shear strength
t = elapsed time
t_p = duration of primary consolidation
u = porewater pressure
u_m = maximum porewater pressure within a layer
u_s = reference porewater pressure
w_ℓ = liquid limit
w_o = in situ water content under effective overburden pressure σ'_{vo}
σ'_p = preconsolidation pressure determined from EOP e versus log σ'_v relation
σ_v = total vertical stress
σ'_v = effective vertical stress
σ^*_v = internal effective vertical stress
σ'_{vf} = postconstruction effective overburden pressure
σ'_{vo} = preconstruction effective overburden pressure
EOO = end-of-observation
EOP = end-of-primary consolidation

REFERENCES

Aboshi, H. (1973). "An experimental investigation on the similitude in the consolidation of a soft clay, including the secondary creep settlement." *Proc. 8th Int. Conf. soil Mech. Found. Engrg.*, Moscow 4(3), 88.
Degago, S.A., Grimstad, G., Jostad, H.P., and Nordal, S. (2009). "The non-uniqueness of the end-of-primary (EOP) void ratio-effective stress relationship." *Proc. 17th Int. Conf. Soil Mech. Geotech. Engrg.*, Alexandria 1, 324-327.
Degago, S.A., Grimstad, G., Jostad, H.P., Nordal, S., and Olsson, M. (2011). "Use and misuse of isotache concept with respect to creep hypotheses A and B." *Géotechnique*, 61(10), 898-908.

Feng, T.W. (1991). "Compressibility and permeability of natural soft clays and surcharging to reduce settlements." PhD thesis, University of Illinois at Urbana-Champaign, Urbana, Illinois.

Hoeg, K., Andersland, O.B., and Rolfsen, E.N. (1969). "Undrained behavior of quick clay under load tests at Asrum." *Géotechnique,* 19(1), 101-115.

Jamiolkowski, M., Ladd, C.C., Germaine, J.T.,, and Lancellotta, R. (1985). "New developments in field and laboratory testing of soils." *Proc. 11th Int. Conf. Soil Mech. Found. Engrg.*, San Francisco, 1, 57-153.

Ladd, D.C., Foott, R., Ishihara, K., Schlosser, F., and Poulos, H.G. (1977). "Stress-deformation and strength characteristics." General Report, *Proc. 9th Int. Conf. Soil Mech. Found. Engrg*, 2, 421-494.

Leonards, G.A.. (1972). "Shallow foundations." Discussion, *Proc. Spec. Conf. on Performance of Earth-Supported Structures, ASCE*, Lafayette, Indiana, 3, 169-173.

Leonards, G.A., and Deschamps, R.J. (1995). "Origin and significance of the quasi-preconsolidation pressure in clay soils." *Proc. Compression and Consolidation of Clay Soils*, Hiroshima, Japan, 1, 341-347.

Mesri, G. (1977). "Stress-deformation and strength characteristics." Discussion, *Proc. 9th Int. Conf. Soil Mech. Found. Engrg.*, Tokyo, 3, 354-355.

Mesri, G. (1981). "Groundwater and seepage problems." Discussion, *10th Int. Conf. Soil Mech. Found. Engrg.*, Stockholm, 4, 659.

Mesri, G. (1986). "Post construction settlement of an expressway built on peat by precompression." Discussion, *Canadian Geot. J.*, 23(3), 403-407.

Mesri, G. (1987). "Fourth law of soil mechanics: A law of compressibility." *Proc. Int. Symp. Geot. Engrg. Soft Soils*, Mexico City, 179-187.

Mesri, G. (2001). "Primary compression and secondary compression." In *Soil behavior and soft ground construction*, (eds. J.T. Germain, T.C. Sheahan and R.V. Whitman), *Geotechnical Special Publication, ASCE*, 119, 122-166.

Mesri, G., and Ajlouni, M. (2007). "Engineering properties of fibrous peats." *J. Geotech. Engrg., ASCE,* 133(7), 850-866.

Mesri, G., Ajlouni, M.A., Feng, T.W., and Lo, D.O.K. (2001). "Surcharging of soft ground to reduce secondary settlement." *Proc. 3rd Int. Conf. Soft Soil Engrg.*, Hong Kong, 55-65.

Mesri, G., and Castro, A. (1987). "The C_α/C_c concept and K_o during secondary compression." *J. Geotech. Engrg., ASCE*, 113(3), 230-247.

Mesri, G., and Choi, Y.K. (1979). "Strain rate behavior of Saint-Jean-Vianney clay." Discussion, *Canadian Geotech. J.,* 16(4), 831-834.

Mesri, G., and Choi, Y.K. (1980). "Excess pore water pressure and preconsolidation effect developed in normally consolidated clays of some age." Discussion, *Soils and Foundations,* 20(4), 131-136.

Mesri, G., and Choi, Y.K. (1983). "Dynamic properties of soft clays for wide strain range." Discussion, *Soils and Foundations*, 23(1), 125-127.

Mesri, G., and Choi, Y.K. (1984). "Time effects on the stress-strain behavior of natural soft clays." Discussion, *Géotechnique*, 34(3), 439-442.

Mesri, G., and Choi, Y.K. (1985a). "The uniqueness of the end-of-primary (EOP) void ratio-effective stress relationship." *Proc. 11th Int. Conf. Soil Mech. Found. Engrg.*, San Francisco, 2, 587-590.

Mesri, G., and Choi, Y.K. (1985b). "The behavior of embankments on clay foundation." Discussion, *Canadian Geotech. J.*, 18(3), 460-462.
Mesri, G., and Feng, T.W. (1991). "Surcharging to reduce secondary settlement." *Proc. Int. Conf. Geotech. Engrg. For Coastal Development*, Yokohama, Japan, 1, 359-364.
Mesri, G., and Feng, T.W. (1992). "Constant rate of strain consolidation testing of soft clays." *Proc. Marsal Symposium*, Mexico City, 49-59.
Mesri, G., and Feng, T.W. (2009). "The non-uniqueness of the end-of-primary (EOP) void ratio-effective stress relationship." Discussion, *Proc. 17th Int. Conf. Soil Mech. Geotech. Engrg.*, Alexandria, 3559-3561.
Mesri, G., Feng, T.W., and Benak, J.M. (1990). "Postdensification penetration resistance of clean sands." *J. Geotech. Engrg. Div., ASCE*, 116(7), 1095-1115.
Mesri, G., Feng, T.W., and Shahien, M. (1995). "Compressibility parameters during primary consolidation." *Proc. Int. Symp. Compression and Consolidation of Clayey Soils*, Hiroshima, 2, 1021-1037.
Mesri, G., and Godlewski, P.M. (1977). "Time and stress compressibility interrelationship." *J. Soil Mech. Found. Div., ASCE,* 103, GT5, 417-430.
Mesri, G., and Hayat, T.M. (1993). "The coefficient of earth pressure at rest." *Canadian Geotech. J.*, 30(4), 647-666.
Mesri, G., and Huvaj-Sarihan, N. (2009). "The Asaoka Method revisited." *Proc. 17th Int. Conf. Soil Mech. and Geotech. Engrg.*, Alexandria, Egypt, 1,131-134.
Mesri, G., Huvaj. N., Vardhanabhuti, B., and Ho, Y-H. (2005). "Excess porewater pressure during secondary compression." *Proc. 16th Int. Conf. Soil Mech. Geotech. Engrg.*, Osaka 2, 1087-1090.
Mesri, G., Lo, D.O.K., and Feng, T.W. (1994). "Settlement of embankments on soft clays." *ASCE Conf. on Vertical and Horizontal Deformations of Foundations and Embankments*, College Station, Texas, 8-56.
Mesri, G., Stark, T.D., Ajlouni, M.A., and Chen, C ..S. (1997). "Secondary compression of peat without or with surcharging." *J. of Geotech. Engrg. Div., ASCE*, 411-421.
Mesri, G., and Vardhanabhuti, B. (2009). "Compression of granular materials." *Canadian Geot. J.*, 46, 369-392.
Neher, H.P., Wehnert, M., and Bonnier, P.G. (2001). "An evaluation of soft soil models based on trial embankments, computer methods and advances in geomechanics." *Balkema*, 373-378.
Sållfors, F. (1975). "Preconsolidation pressure of soft high plastic clays." PhD thesis, Chalmers University of Technology, Göteborg, Sweden.
Stolle, D.F.E., Vermeer, P.A., and Bonnier, P.G. (1999). "Consolidation model for creeping clay." *Canadian Geotech. J.*, 36(4), 754-759.
Suklje, L. (1957). "The analysis of the consolidation process by the isotache method." *Proc. 4th Int. Conf. Soil Mech. Found. Engrg.*, London, 1, 200-206.
Terzaghi, K. (1923). "Die Berechnung der Durchlässigkeitsziffer des Tones aus dem Verlauf der Hydrodynamischen Spannungserscheinungen." *Sitz. Akad. Wissen. Wien Mathnaturw Kl Abt. Ila*, 132, 105-124.
Terzaghi, K., Peck, R.B. Mesri, G. (1996). *Soil Mechanics in Engineering Practice, Third Edition*, John Wiley & Sons, New York, 549 p.

Variability and Uncertainty in Consolidation and Settlement Parameters from Different Sampling and Testing Methods

Dan Ding[1], S.M.ASCE, William J. Likos[2], M.ASCE, J. Erik Loehr[3], M.ASCE

[1]Ph.D. Candidate, Dept. of Civil & Environ. Engineering, University of Missouri, Columbia, MO, 65211; ddn56@mail.missouri.edu
[2]Assoc. Professor, Dept. of Geological Engineering, University of Wisconsin, Madison, WI, 53706; likos@wisc.edu
[3]Assoc. Professor, Dept. of Civil & Environ. Engineering, University of Missouri, Columbia, MO, 65211; eloehr@missouri.edu

ABSTRACT: A comprehensive research program was recently completed to evaluate variability and uncertainty in geotechnical design parameters, including those used for consolidation and settlement analyses. Numerous one-dimensional consolidation tests were conducted following incremental load and constant rate of strain loading procedures for specimens obtained from four different sites using "state of the practice" and "state of the art" sampling techniques. The four sites included two "soft clay" sites, one "soft silt" site, and a "stiff silt/clay" site. Laboratory consolidation tests were conducted on specimens from each site to determine compression and recompression indices, preconsolidation stress, and coefficients of consolidation. Results from different sampling and testing methods are compared to draw conclusions regarding the variability and uncertainty associated with consolidation parameters obtained following different sampling and testing protocols. "State of the art" boring, sampling, and testing methods were found to produce lower values of preconsolidation stress, greater values of compression index, and similar values of recompression index and coefficient of consolidation compared to results from "state of the practice" methods for the three soft soil sites. Variability and uncertainty in consolidation parameters determined from state of the art methods were also found to be notably less than the variability and uncertainty established using state of the practice methods for the soft soil sites while the variability and uncertainty in consolidation parameters for the stiff soil site were similar for both state of the art and state of the practice methods.

INTRODUCTION

Site characterization programs are commonly used in practice to gather information for estimating soil engineering parameters. Compressibility parameters measured from one-dimensional consolidation tests are generally used to estimate the amount and rate of settlement for geotechnical design at cohesive soil sites. It is widely known that these measurements are affected by disturbance induced during the process of sampling and testing that may lead to bias and variability in laboratory measured consolidation parameters (Ladd & Lambe 1963). It is also generally believed that the potential bias and variability introduced into design by sample disturbance will lead to conservative estimates of settlement, and thus, conservative designs that may be more costly than necessary. The primary motivation for improving site characterization practices is, therefore, to reduce bias and uncertainty in estimated soil parameters, which may lead to more cost effective designs.

Obtaining truly undisturbed samples is unfortunately impossible (Olson 1986). Sample disturbance of varying degree inevitably occurs during the process of boring, sampling, transportation, storage, specimen preparation and testing (Santagata & Germaine 2002). The degree of disturbance generally depends on the magnitude of stress relief due to removal of overlying soils, the amount of vibration during handling, sample distortion during extrusion and specimen preparation, the magnitude of temperature changes, and even the amount of time that samples are stored (Olson 1986, 1998). Significant research conducted on sample disturbance (Santagata & Germaine 2002, Hight 2003, Ladd & DeGroot 2003, DeGroot et al. 2005) suggests that "state of the art" sampling to minimize disturbance should involve:

- Using heavy drilling mud to maintain borehole stability and reduce the magnitude of stress relief during boring and sampling;
- Using large diameter, thin-walled, fixed-piston samplers with sharp cutting edges;
- Careful handling, transportation, and storage;
- Transporting and storing sampling tubes vertically;
- Storing samples in a high humidity environment with controlled temperature;
- Careful extraction of samples from sampling tubes that generally includes cutting of the tubes, breaking bonds at the soil-tube interface, and extruding the sample in the same direction as sampling in the field;
- Trimming of specimens to remove soil that has been distorted; and
- Testing specimens as soon after sampling as possible.

Despite this knowledge, however, such practices are not commonly followed in routine practice (DeGroot 2003). Samples acquired using "state of the practice" techniques are generally expected to have more disturbance and show more bias and variability in measured soil parameters. For one-dimensional consolidation tests, this disturbance generally leads to a decrease in the compression index, C_c, preconsolidation stress, σ'_p, and coefficient of consolidation, c_v, and an increase in the recompression index, C_r (Simons 1974, Janbu et al. 1981, Leroueil et al. 1983, Leroueil & Samson 1983, Crawford 1986, Kabbaj et al. 1986, Olson 1986, Leroueil 1996, Santagata & Germaine 2002).

Two forms of one-dimensional consolidation tests are commonly used for measuring compressibility parameters. These tests include conventional incremental load (IL) tests, where the applied load is sequentially increased and specimens are allowed to fully consolidate under each applied load, and constant rate of strain (CRS) tests that utilize continuously increasing loading. The IL consolidation tests are the more conventional type of consolidation test in practice and are most familiar to practicing engineers. However, IL consolidation tests are substantially more labor intensive and time consuming than CRS tests, which are automatically controlled during testing (Ladd & DeGroot 2003). In contrast, CRS tests require more elaborate analysis and interpretation and require careful selection of strain rate values to allow excess pore water pressures to dissipate while simultaneously avoiding secondary consolidation (Tanaka et al. 2000). CRS type tests with an appropriate strain rate are generally considered to be preferable to IL tests (Olson 1986, Ladd & DeGroot 2003) and have become increasingly common in practice.

To evaluate bias and variability in compressibility parameters, a comprehensive site characterization program was conducted in the State of Missouri at four soil sites. Both "state of the art" and "state of the practice" boring and sampling techniques were used at each site. Constant rate of strain consolidation tests (ASTM-D4186) were conducted on specimens obtained using state of the art sampling techniques to produce results that are subsequently referred to as "state of the art measurements", or SOA measurements. Incremental load consolidation tests (ASTM-D2435) were conducted for specimens obtained using state of the practice sampling techniques to produce results that are referred to as "state of the practice measurements", or SOP measurements. In this paper, SOA and SOP measurements from each site are reported along with calculated values representing the variability and uncertainty in consolidation parameters. The results are then compared and conclusions are derived regarding the bias, variability and uncertainty associated with consolidation parameters obtained from the two different boring, sampling and testing protocols.

FIELD TEST SITES

Four soil sites were selected for boring and sampling as shown in Figure 1. General soil profiles and measured natural water contents (w_n) and Atterberg limits (w_P and w_L) are shown in Figure 2 for each of the four sites. Table 1 provides the mean and ranges of bulk mineralogical characteristics for the soils at each site measured from X-Ray Diffraction (XRD) tests.

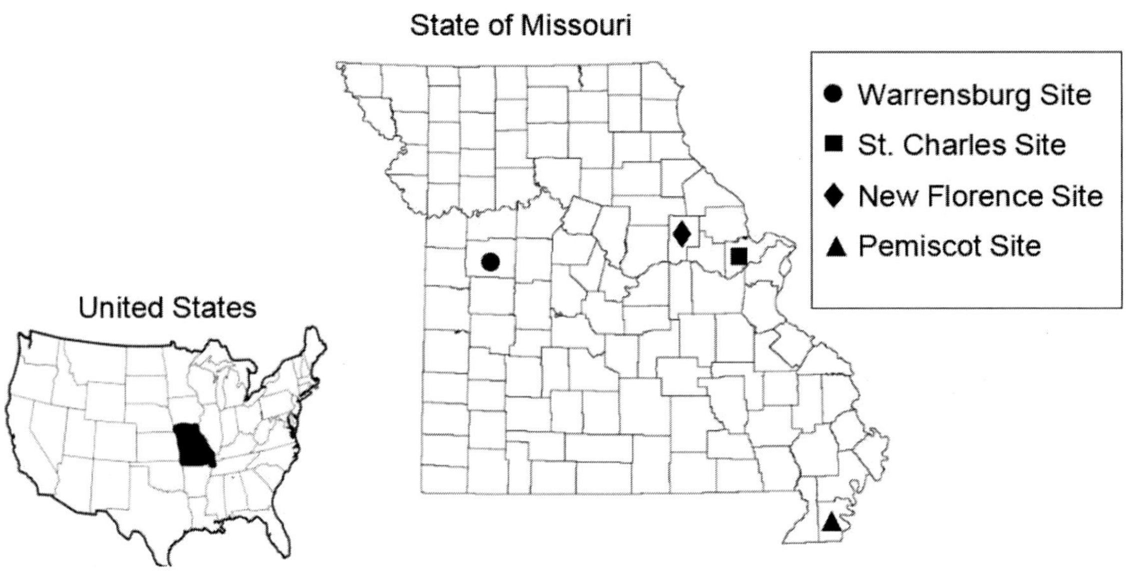

Figure 1. Locations of four test sites.

The Warrensburg site is composed of generally soft, silty soils overlying shale, sandstone, or limestone bedrock. In the immediate vicinity of the borings, the depth of the overburden soils is approximately 14 m. In situ moisture contents and Atterberg limits are generally consistent throughout the upper 14 m of the profile, except that an increase in the liquid limit is observed in some samples from 4 to 7 m below ground. The ground water table was located at a depth of approximately 3 m. The mean specific gravity measured throughout the 14 m of overburden by water pycnometry (ASTM-D854-10) is 2.71. The mean total unit weight is 18.4 kN/m^3.

Table 1. Summary of bulk powder fractions from XRD tests for soils from four test sites.

Test Site	Relative Abundance (%)			
	Total Clay	Quartz	Plagioclase/Feldspar	Calcite
Warrensburg	46.8±10.1	38.0±8.8	10.5±2.4	4.8±0.2
St. Charles	70.6±8.4	20.1±9.0	6.3±2.9	3.0±1.9
New Florence	50.3±12.0	33.4±4.2	8.5±8.0	7.9±7.8
Pemiscot	66.9±7.9	19.2±5.3	10.1±3.7	3.8±1.4

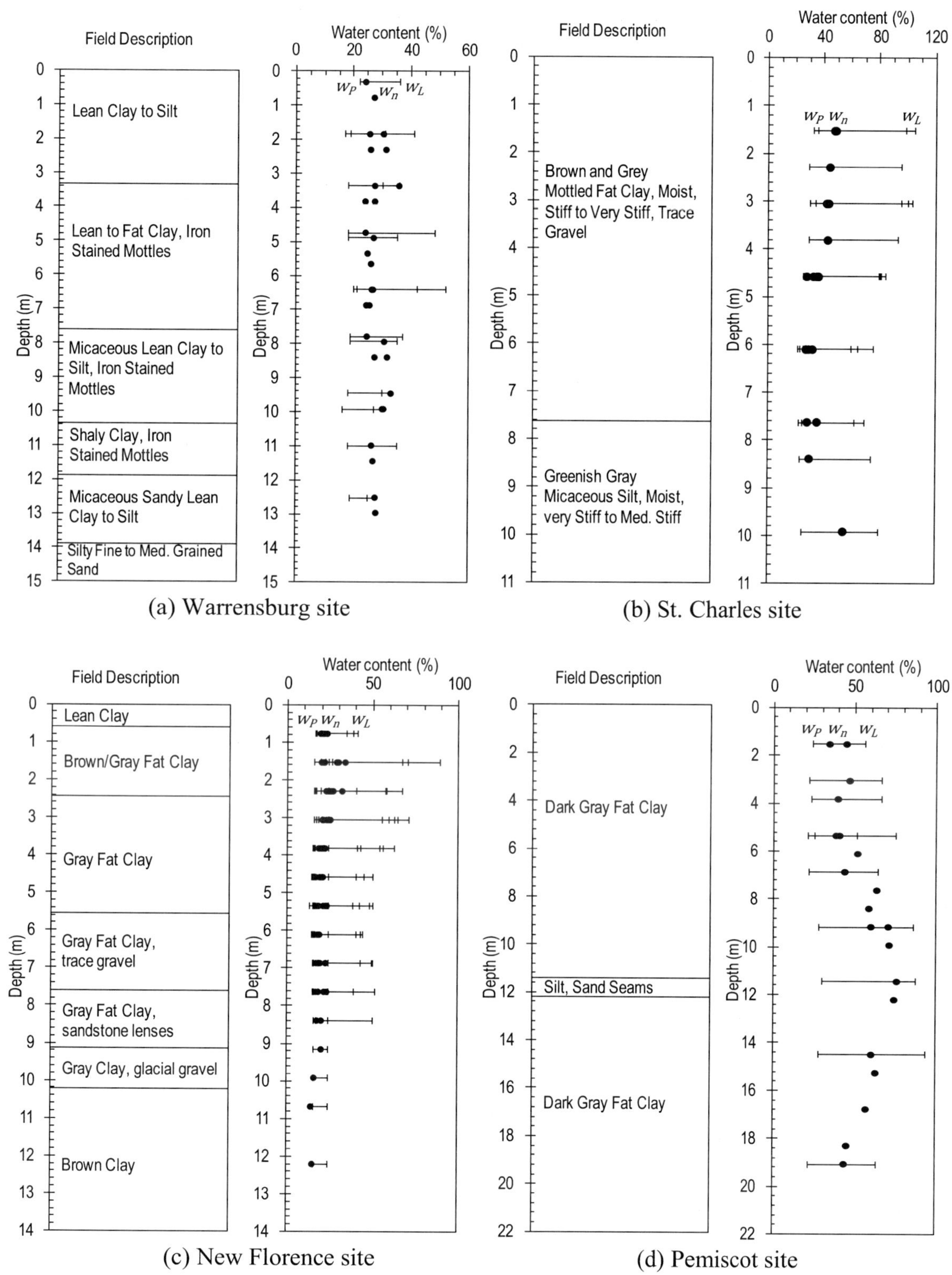

Figure 2. General soil profiles for selected research sites: (a) Warrensburg site, (b) St. Charles site, (c) New Florence site, and (d) Pemiscot site.

The St. Charles site is an alluvial site located near the confluence of the Missouri and Mississippi rivers. The subsurface is composed of approximately 11 m of soft, fat clay overlying alluvial sands. The ground water table was located at a depth of approximately 7 m. The mean specific gravity from samples taken throughout the 11 m sampling depth is 2.74 and the mean total unit weight over the same depths is 17.2 kN/m^3.

The New Florence site is an "upland" site composed of relatively stiff clay overlying sandstone, limestone and/or shale bedrock. Liquid limits (w_L) for the stiff clay generally decrease with depth from a high of 89 at 1.5 m depth to approximately 25 at 9 m depth and below. Plastic limits (w_P) are practically uniform throughout the profile. The ground water table was located below 14 m, beyond the depths of sampling and testing conducted for the present study. The mean specific gravity from all samples is 2.63 and the mean total unit weight is 19.3 kN/m^3.

The Pemiscot site is an alluvial site composed of approximately 20 m of soft, fat clay overlying alluvial sands. Both Atterberg limits and natural water contents are observed to increase between depths of 8 to 16 m, with the moisture content approaching the liquid limit at about 12 m. The ground water table was located near the ground surface at a depth of approximately 1 m. The mean specific gravity is 2.72 and the mean total unit weight is 16.2 kN/m^3.

BORING, SAMPLING AND TESTING METHODS

Comparisons are made among site interpretations derived from measurements on samples obtained from "research" borings, "sister" borings, and "historical" borings. Research borings are those that were drilled and sampled using "state of the art" drilling, sampling, and testing procedures. Sister borings were drilled and sampled in the immediate vicinity and within several days time of the research borings, but were drilled and sampled using current (2010) Missouri Department of Transportation (MoDOT) state of the practice techniques. Results from historical borings are those obtained from MoDOT's historical records. These borings and test results also generally represent current MoDOT practice and provide for additional comparison to the research borings. Only boreholes considered representative of present subsurface conditions at each site were included in this category.

Research borings were generally advanced using rotary wash boring methods with samples obtained using an Osterberg hydraulic piston sampler. Rotary wash boring methods were used to reduce disturbance resulting from stress relief and to facilitate removal of cuttings. Drilling slurries consisting of mixtures of water and powdered bentonite were used to produce drilling slurry with suitable viscosity and density. Periodic measurements of Marsh Funnel viscosity and slurry density were taken for quality control of the drilling fluid consistency. A 7.6-cm (3 in) diameter Osterberg hydraulic piston sampler was used to obtain samples with a 0.75 m (2.5 ft) push followed by cleanout with a tri-cone bit before attempting a subsequent sample. At the New Florence site, where the soil profile includes stiff clays and silts, use of rotary wash boring methods proved to be problematic, and thus hollow stem auger boring methods were applied at this particular site. When soil at the New Florence site became too stiff for sampling with the Osterberg sampler, samples were acquired by pushing conventional 7.6-cm (3 in) diameter Shelby tubes. In the stiffest soils, a 7.6-cm (3 in) diameter Pitcher sampler was utilized. Drilling and sampling at each of the four sites was completed within a two-week time interval such that samples could generally be considered time invariant.

Sister borings were advanced following state of the practice drilling and sampling techniques consistent with current (2010) MoDOT practices that employ a hollow stem auger boring method. Samples were obtained by first pushing a conventional 7.6-cm (3 in) diameter Shelby tube sampler, followed by a Standard Penetration Test (SPT), followed by cleanout of the borehole over the 1.5-m (5 ft) interval encompassing the Shelby tube and SPT sampling intervals. Sister borings at the Warrensburg, St. Charles, New Florence, and Pemiscot sites were drilled and sampled in the immediate vicinity and within several days time of the research borings.

Comparisons among results obtained using samples from the research and sister borings are thus considered time invariant.

Historical borings refer to site exploration information obtained from MoDOT's historical records at locations at or nearby each of the four sites. Practice for drilling and sampling the historical borings is generally consistent with techniques used for sister borings. Laboratory test results reported from the historical borings were thus used to supplement results from sister borings, and the combined sister/historical results representing the state of the practice were used for comparison with the results obtained from the research borings representing the state of the art. While sampling and testing for the historical borings was completed in the past, the influence of time is not explicitly considered in this paper.

Sample tubes acquired using the state of the art boring and sampling techniques from the research borings were carefully handled and double-sealed on both ends with O-ring compression seals and plastic caps to prevent moisture loss during transportation and storage. All sample tubes were stored vertically in padded containers for transport to the laboratory for testing. Upon arrival at the laboratories, sample tubes were stored in a refrigerated and humidified environment until specimens were cut from the tubes and trimmed for testing.

Samples acquired using state of the practice boring and sampling techniques from the sister borings were extruded in the field, wrapped with aluminum foil, sealed in paraffin wax, and stored in cardboard cases for transportation to a different testing laboratory. Once at the laboratory, samples were stored in a humidified environment until trimming and testing.

Specimens for CRS tests were carefully trimmed into a 6.4-cm (2.5 in) diameter sharp-edge steel consolidation ring. Strain rate was adjusted to approximately 1.5 x 10^{-6} m/s to keep a base excess pore water pressure ratio ($\Delta u/\sigma_v'$) between 3% and 15% throughout loading. For IL consolidation tests, specimens were carefully trimmed into a 6.4-cm (2.5 in) diameter, thick-wall brass consolidation ring. Loads were incrementally applied by doubling the previous load to levels needed to identify the preconsolidation stress followed by an unloading cycle. Individual load increments were generally held for 24 hours before applying the next load increment.

ANALYSIS AND INTERPRETATION

Values of preconsolidation stress, σ_p', were estimated from results of the CRS and IL tests using Casagrande's standard graphical method, which is the most widely used technique for estimating preconsolidation stress. Values of compression index, C_c, and recompression index, C_r, were interpreted from the slopes of constructed compression and recompression curves. Values for coefficient of consolidation, c_v, were calculated at the estimated σ_p' for each test.

Profiles for each consolidation parameter (σ_p', C_c, C_r, and c_v) with depth were developed for each site (Figures 3-6). Linear least squares regression of the σ_p' datasets was used to develop linear subsurface profiles of mean preconsolidation stress, written in general form as:

$$E(y) = \beta_0 + \beta_1 z \tag{1}$$

where $E(y)$ is the estimated mean value of a parameter at depth z and β_0 and β_1 are intercept and slope values obtained from linear regression of the experimental data over various depth intervals. The standard deviation at a given depth is calculated as

$$\sigma_{total} = \sqrt{Var(y)} = \sqrt{\left[MSE \times \left(1 + \frac{1}{n} + \frac{(z-\bar{z})^2}{\sum(z_i-\bar{z})^2}\right)\right]} \tag{2}$$

where σ_{total} is the total standard deviation of the parameter, which incorporates both variability from the measurements and uncertainty in the linear regression model. $Var(y)$ is the total variance of the parameter calculated at depth z; $\bar{z}$ is the average depth for all measurements in the dataset; z_i is the depth of the ith measurement and n is the total number of measurements. The mean square error (MSE) in Eq. 2 is computed as:

$$MSE = \sum(y_i - \hat{y}_i)^2/(n-2) \tag{3}$$

where y_i is the ith measurement and $\hat{y}_i$ is the estimated value of y at x_i (from Eq. 1).

Subsurface profiles of compression index, C_c, recompression index, C_r, and coefficient of consolidation, c_v, were evaluated by fitting a least squares regression model with $\beta_1 = 0$, or a constant model. The mean in this case is calculated by averaging all measurements:

$$E(y) = \frac{\sum y_i}{n} \tag{4}$$

The standard deviation for the constant model is computed as

$$\sigma_{total} = \sqrt{Var(y)} = \sqrt{\left(\frac{\sum(y_i-\hat{y}_i)^2}{n-1}\right)\left(1+\frac{1}{n}\right)} \tag{5}$$

Variability and uncertainty for each of the consolidation parameters may also be quantified by the coefficient of variation, COV, which is the ratio of σ_{total} to the estimated mean, $E(y)$. Note that σ_{total} and COV are independent of depth in cases where a constant model is used. However, both σ_{total} and COV vary with depth when a linear model is used, and tend to be least at intermediate depths and greatest at the upper and lower extreme depths. Values of σ_{total} and COV reported subsequently are thus given for a specified intermediate depth in the respective profiles.

RESULTS AND DISSUSSION

Warrensburg Site

At the Warrensburg site, there are 17 SOA measurements and 27 SOP measurements. Results of these measurements are plotted in Figure 3 with interpreted models reflecting the mean value of the respective consolidation parameters based on the SOA and SOP measurements. Calculated mean, standard deviation, and COV values for consolidation parameters at a depth of 7.6 m are summarized in Table 2.

The σ'_p profiles respectively interpreted from SOA and SOP measurements are in close agreement (Figure 3a) except at shallow depth. Both profiles suggest that the clay is lightly overconsolidated over much of the profile, becoming normally consolidated at a depth of approximately 13 m. However, the SOA measurements indicate lower preconsolidation stress than SOP measurements at depths less than 3 m. The scatter in measured values for σ'_p is noticeably smaller for the SOA measurements than for the SOP measurements. Accordingly, the calculated COV at a depth of 7.6 m from SOP measurements (approx. 50%) is substantially greater than the COV determined from SOA measurements (approx. 30%).

Results from both SOA and SOP measurements suggest that the compression index, C_c, is essentially constant over the depth range of all measurements. Compression indices from SOA measurements have a mean value of 0.238, which is approximately 20 percent greater than the mean value determined from SOP measurements (0.185). At least part of this difference can be attributed to results from four SOP measurements that produce C_c-values that are noticeably lower than indicated by the remaining tests. The variability and uncertainty in C_c from both sets of measurements are similar, with a COV of approximately 20 percent.

Measured values for the recompression index, C_r, and the coefficient of consolidation, c_v, show large variability with COV's being greater than 40 percent for both the SOA and SOP measurements. However, the mean values for C_r and c_v from the two sets of measurements are generally consistent, with the SOA measurements indicating slightly lower mean values than the SOP measurements for both parameters.

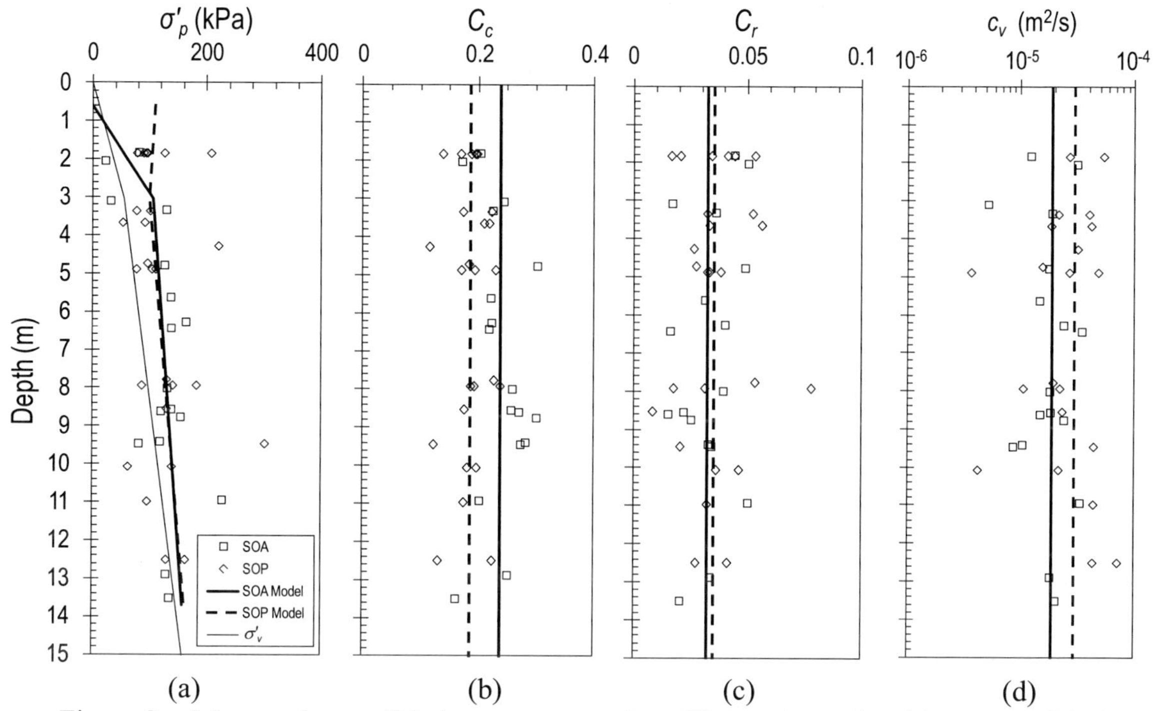

Figure 3. Measured consolidation parameters from Warrensburg site: (a) preconsolidation stress; (b) compression index; (c) recompression index; and (d) coefficient of consolidation.

Table 2. Summary of interpreted consolidation parameters at a depth of 7.6 m from Warrensburg site.

Parameter	Method	μ	σ_{total}	COV (%)
σ'_p (kPa)	SOA	129	42	32
	SOP	126	60	48
C_c	SOA	0.238	0.042	18
	SOP	0.185	0.032	17
C_r	SOA	0.032	0.012	39
	SOP	0.036	0.015	42
c_v ($10^{-5}\,m^2/s$)	SOA	1.9	0.86	45
	SOP	3.0	1.75	58

St. Charles Site

Test results from 19 SOA measurements and 30 SOP measurements from the St. Charles site are compared in Figure 4 and Table 3. The preconsolidation stress profile (Figure 4a) interpreted from SOA measurements suggests that the soil is overconsolidated to depths of approximately 6 m where the soil becomes normally consolidated. At greater depth, the SOA measurements indicate that σ'_p increases more than the in situ vertical effective stress with depth, suggesting that the soil is again lightly overconsolidated at depths greater than 6 m. In contrast, the SOP measurements suggest increasing values of preconsolidation stress with depth that are generally greater than σ'_p values suggested by SOA measurements at depths less than 8 m. The variability in σ'_p from SOP measurements is substantially greater than determined from the SOA measurements. The variability in σ'_p from both SOA (COV=19%) and SOP (COV=27%) measurements is less that observed for the Warrensburg site.

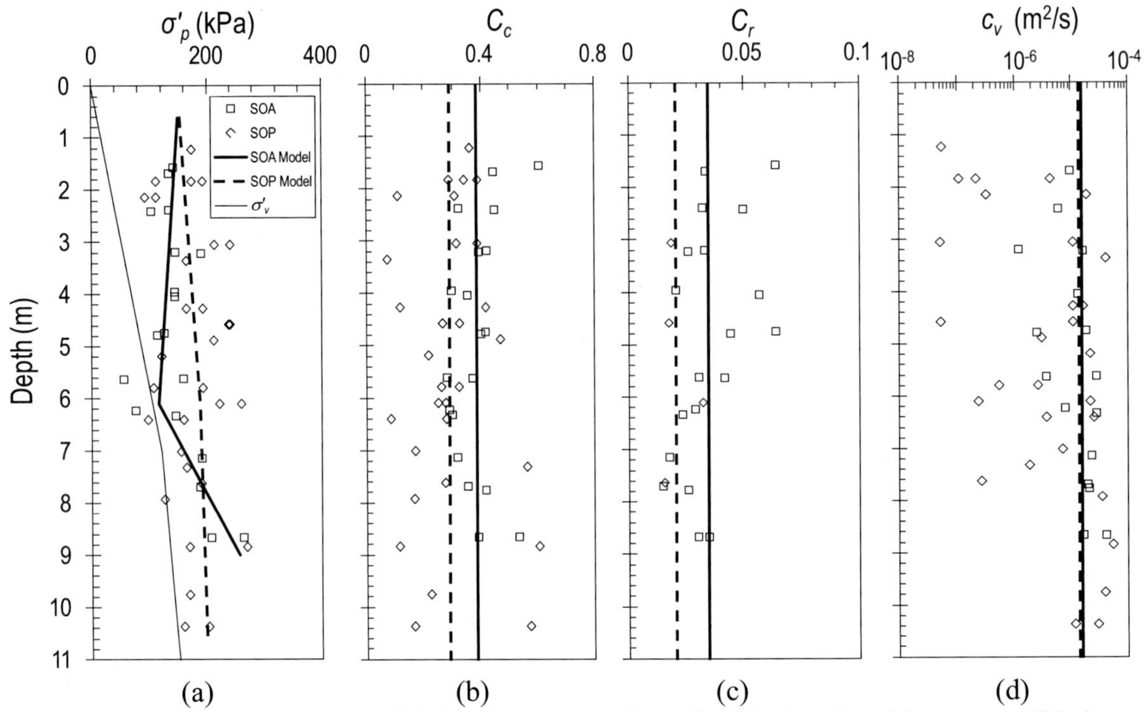

Figure 4. Measured consolidation parameters from St. Charles site: (a) preconsolidation stress; (b) compression index; (c) recompression index; and (d) coefficient of consolidation.

Results shown in Figure 4 indicate that the compression index (C_c), recompression index (C_r), and coefficient of consolidation (c_v) can all be practically considered to be constant over the depth range of the measurements. The established mean values for C_c from SOA and SOP measurements vary substantially, from 0.289 for SOP measurements to 0.384 for SOA measurements. However, the mean value established from the SOP measurements again is influenced by a series of measurements with exceptionally low values for C_c (measurements of $C_c \approx 0.1$). If these measurements are neglected, the remaining measurements are more consistent with SOA measurements, but still produce a mean value for C_c that is less than obtained for the SOA measurements as shown by the values in parentheses in Table 3. The mean value for C_r from SOA measurements is also significantly greater than the mean value from SOP measurements while the mean values for c_v are similar for both sets of measurements.

The variability present in C_c and c_v for the St. Charles site is generally greater than observed for the Warrensburg site while the variability in C_r is similar to that observed for the Warrensburg site for both SOA and SOP measurements. The variability in C_c and c_v from SOP measurements is also substantially greater than observed for the SOA measurements while the variability of C_r is similar for the SOA and SOP measurements.

Table 3. Summary of interpreted consolidation parameters at a depth of 5 m from St. Charles site.

Parameters	Methods	μ	σ_{total}	COV (%)
σ'_p (kPa)	SOA	141	26	19
	SOP	185	51	27
C_c	SOA	0.384	0.087	23
	SOP	0.289 (0.327)[†]	0.142 (0.123)[†]	49 (38)[†]
C_r	SOA	0.035	0.015	43
	SOP	0.021	0.009	43
c_v ($10^{-5}\,m^2/s$)	SOA	1.6	1.1	70
	SOP	1.4	1.7	119

[†] values in parentheses reflect computations that neglect unusually low measured values

New Florence Site

Because of sampling difficulties at the New Florence test site (stiff clays and silts), only five SOA measurements are available for comparison with 25 SOP measurements. Comparisons of these measurements are shown in Figure 5 and Table 4. Figure 5 shows substantial differences in preconsolidation stress from SOA and SOP measurements, with SOA measurements indicating σ'_p is substantially greater than indicated by SOP measurements. These differences are perplexing. On one hand, the SOP measurements follow a reasonably consistent trend that suggests decreasing overconsolidation with depth. However, such trends are not common or expected for "stiff clay" sites, which raises questions about the magnitude of the preconsolidation stress from the SOP measurements. In contrast, results of the SOA measurements are less consistent, but indicate greater values of preconsolidation stress that are expected for stiff clays. The magnitude of σ_{total} from SOA measurements is substantially greater than from SOP measurements, but both types of measurements have similar COV's of approximately 40 percent.

As observed at the Warrensburg and St. Charles sites, SOA measurements again produced greater values of C_c than SOP measurements. The mean value of C_r from SOA measurements is slightly less than from SOP measurements. The variability in C_c and C_r from both SOA and SOP measurements is greater than observed for the Warrensburg and St. Charles sites and the variability in c_v from SOA measurements is seen to be exceptionally large. No c_v values from SOP measurements were available.

Pemiscot Site

Consolidation parameters from 26 SOA measurements and 41 SOP measurements for the Pemiscot site are compared in Figure 6 and Table 5. Preconsolidation stress models for the SOA and SOP measurements are similar in form, with essentially constant σ'_p to a depth of 9 m, followed by slightly decreasing values of σ'_p to depths of 18 to 20 m where the soil is normally consolidated. However, the SOA measurements consistently indicate lower σ'_p and lower variability and uncertainty in σ'_p than the SOP measurements.

Compression indices (C_c) from both SOA and SOP measurements suggest different mean values are appropriate for depths less than 9 m and for depths between 9 and 22 m with C_c for the shallower range being typical for clay soils while C_c for the deeper range is greater than observed for most clays. In contrast, values of C_r and c_v are generally consistent throughout the profile.

Table 4. Summary of interpreted consolidation parameters at a depth of 7.6 m from New Florence site.

Parameters	Methods	μ	σ_{total}	COV (%)
σ'_p (kPa)	SOA	1316	503	38
	SOP	209	86	41
C_c	SOA	0.300	0.123	41
	SOP	0.159	0.067	42
C_r	SOA	0.018	0.010	53
	SOP	0.030	0.018	60
c_v $(10^{-4}\,m^2/s)$	SOA	1.7	3.9	224
	SOP	--	--	--

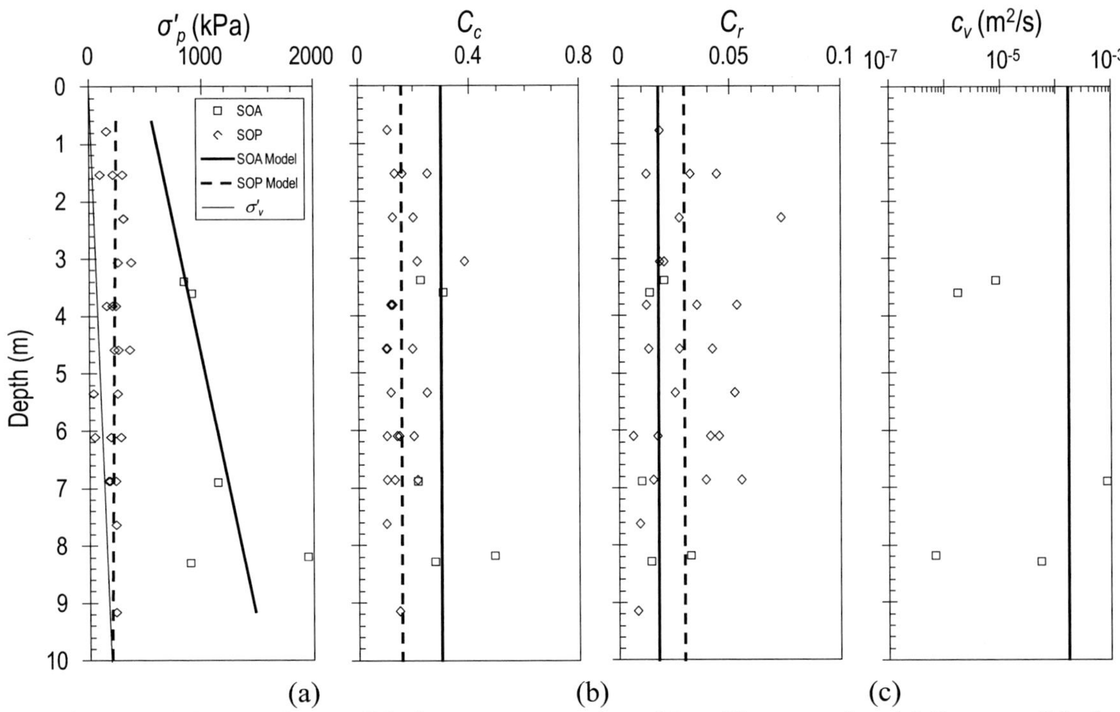

Figure 5. Measured consolidation parameters from New Florence site: (a) Preconsolidation stress; (b) Compression index; (c) Recompression index.

The mean values of C_c, C_r, and c_v from SOA measurements are all slightly greater than determined for the SOP measurements. The variability of C_c and c_v are both observed to be significantly lower from SOA measurements while the variability of C_r from the SOP measurements is lower. The variability and uncertainty observed for all parameters generally falls within the bounds of results from the Warrensburg, St. Charles, and New Florence sites except for the variability of c_v from the SOP measurements, which is exceptionally large.

Discussion of Collective Results

The results presented for the four different test sites lead to several general observations about differences between consolidation parameters established using SOA and SOP measurements. For the Warrensburg, St. Charles, and Pemiscot sites (the "soft soil sites"), the mean preconsolidation stress established from SOA measurements is generally less than or similar to the mean preconsolidation stress from SOP measurements whereas the preconsolidation stress established from SOA measurements for the New Florence site (the "stiff soil site") is substantially greater than that from SOP measurements. The coefficient of variation for σ_p' from SOA measurements is also consistently lower than that from SOP measurements, especially for the soft clay/silt sites.

Table 5. Summary of interpreted consolidation parameters at a depth of 8 m from Pemiscot site.

Parameters		Methods	μ	σ_{total}	COV (%)
σ_p' (kPa)		SOA	109	36	33
		SOP	179	80	45
C_c	Above 9 m	SOA	0.455	0.105	23
		SOP	0.405	0.167	41
	Below 9 m	SOA	0.808	0.232	29
		SOP	0.692	0.246	36
C_r		SOA	0.063	0.033	52
		SOP	0.036	0.009	26
c_v $(10^{-6}\,m^2/s)$		SOA	5.26	4.44	84
		SOP	3.23	7.89	244

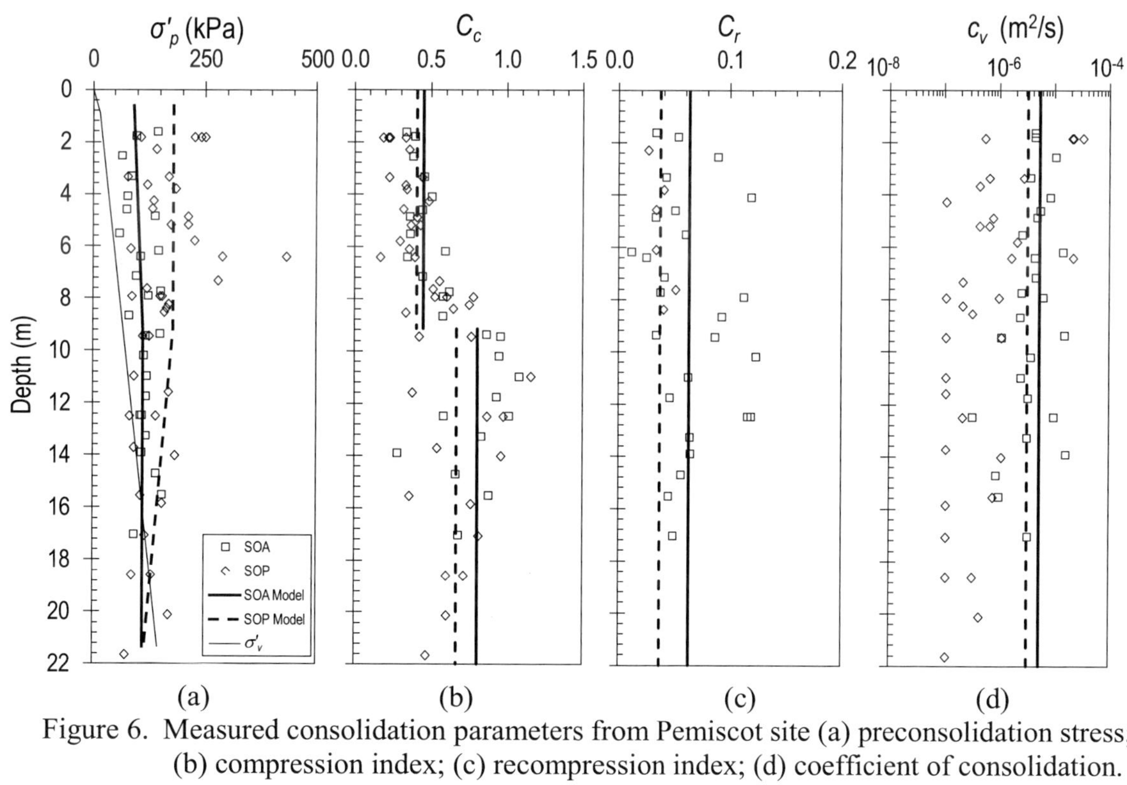

Figure 6. Measured consolidation parameters from Pemiscot site (a) preconsolidation stress; (b) compression index; (c) recompression index; (d) coefficient of consolidation.

The mean values of C_c from SOA measurements are greater than the mean values determined from SOP measurements for all sites. The differences in the mean value of C_c can be at least partially attributed to select SOP measurements that produce values of C_c that are exceptionally low, especially for the soft clay/silt sites. Coefficients of variation for C_c were similar for SOP and SOA measurements at the Warrensburg and New Florence sites, where the soils are silty and have lower mean values for C_c. For the St. Charles and Pemiscot sites, the COV for C_c from SOA measurements is substantially lower than from SOP measurements.

No consistent trend is observed in comparing the mean values for C_r from SOP and SOA measurements at the different sites with SOA measurements producing mean C_r-values that are greater than the mean C_r from SOP measurements at two sites and less than the mean C_r from SOP measurements at two sites. The variability of C_r from SOA and SOP measurements, as represented by the reported COV-values, is generally seen to be similar, but substantially greater than the variability observed in both σ_p' and C_c.

Mean values for c_v from SOP and SOA measurements are generally similar for each of the four sites. Coefficients of variation for c_v are generally quite large, and greater than the COV-values determined for the other consolidation parameters, with the COV-values established from SOA measurements generally being smaller than those established from SOP measurements.

CONCLUSIONS

Comparison of consolidation parameters established using "state of the art" and "state of the practice" techniques for boring, sampling, and testing at four different sites leads to the following conclusions:

- Use of state of the art boring, sampling, and testing methods at the three "soft soil" sites tended to produce lower values of preconsolidation stress, greater values of compression index, and similar values of recompression index and coefficient of consolidation when compared to values established from state of the practice methods.
- For the three soft soil sites, the variability and uncertainty in consolidation parameters established using state of the art methods was generally less than the variability and uncertainty in consolidation parameters established using state of the practice methods.
- The variability and uncertainty in consolidation parameters established for a "stiff soil" site using state of the art and state of the practice techniques were generally similar.
- Coefficients of variation for the recompression index and coefficient of consolidation were generally observed to be substantially greater than coefficients of variation for the preconsolidation stress and compression index.

These conclusions suggest that state of the art techniques can lead to substantial improvements to the efficiency of foundation designs, particularly for soft soil sites.

ACKNOWLEDGMENTS

The work presented was funded by the Missouri Department of Transportation (MoDOT), the Center for Transportation Infrastructure and Safety at Missouri University of Science and Technology (MS&T), and the University of Missouri. This support is gratefully acknowledged. The opinions, findings, and recommendations in this publication are not necessarily those of MoDOT or the U.S. Federal Highway Administration. This document does not constitute a standard, specification or regulation.

REFERENCES

ASTM-D854-10. 2010. "Standard Test Methods for Specific Gravity of Soil Solids by Water Pycnometer." *American Society for Testing and Materials (ASTM)*:DOI: 10.1520/D0854-10.

ASTM-D2435. 2011. "Standard Test Methods for One-Dimensional Consolidation Properties of Soils Using Incremental Loading." *American Society for Testing and Materials (ASTM)*:DOI: 10.1520/D2435_D2435M-11.

ASTM-D4186. 2012. "Standard Test Method for One-Dimensional Consolidation Properties of Saturated Cohesive Soils Using Controlled-Strain Loading." *American Society for Testing and Materials (ASTM)*:DOI: 10.1520/D4186_D4186M-12.

Crawford, C.B. 1986. "State of the Art: Evaluation and Interpretation of Soil Consolidation Tests." *Consolidation of Soils: Testing and Evaluation. ASTM STP 892, R.N. Yong and F.C. Townsend, Eds.*:71-103.

DeGroot, D.J. 2003. "Laboratory Measurement and Interpretation of Soft Clay Mechanical Behavior." *Soil Behavior and Soft Ground Construction, American Society of Civil Engineers, Geo-Institutue Geotechnical Special Publication* no. 119:167-200.

DeGroot, D.J., S.E. Poirier, and M.M. Landon. 2005. "Sample Disturbance — Soft Clays." *Studia Geotechnica et Mechanica* no. XXVII (3-4):107-120.

Hight, D.W. 2003. "Sampling Effects in Soft Clay: An Update on Ladd and Lambe (1963)." *Soil Behavior and Soft Ground Construction, ASCE GSP 119, MIT,*:86-122.

Janbu, N., O. Tokheim, and K. Senneset. 1981. "Consolidation Tests with Continuous Loading." *Proc. 10th Int. Conf. Soil Mech. and Foundation Engineering, Stockholm* no. 1:645-654.

Kabbaj, Mouslim, Fusao Oka, Serge Leroueil, and Francois Tavenas. 1986. "Consolidation of Natural Clays and Laboratory Testing." *Consolidation of Soils: Testing and Evaluation. ASTM STP 892, R.N. Yong and F.C. Townsend, Eds.*:378-404.

Ladd, C.C., and D.J. DeGroot. 2003. "Recommended practice for soft ground site characterization: The Arthur Casagrande Lecture." *Proceedings of the 12th Panamerican Conference on Soil Mechanics and Geotechnical Engineering* no. 1:3-57.

Ladd, C.C., and T.W. Lambe. 1963. "The Strength of Undisturbed Clay Determined from Undrained Tests." *Laboratory Shear Testing of Soils, ASTM STP 361*:342-371.

Leroueil, S, and L. Samson. 1983. "Laboratory and Field Determination of Preconsolidation Pressures at Gloucester." *Canadian Geotechnical Journal* no. 20 (4):477-490.

Leroueil, S. 1996. "Compressibility of Clays: Fundamental and Practical Aspects." *Journal of geotechnical engineering* no. 122 (7):534-543.

Leroueil, S., F. Tavenas, L. Samson, and P. Morin. 1983. "Preconsolidation Pressure of Champlain Clays. Part II. Laboratory Determination." *Canadian Geotechnical Journal* no. 20 (4):803-816.

Olson, R.E. 1986. "State of the Art: Consolidation Testing." *Consolidation of Soils: Testing and Evaluation. ASTM STP 892, R.N. Yong and F.C. Townsend, Eds.*:7-70.

Olson, R.E. 1998. "Settlement of Embankments on Soft Clays (The Thirty-First Terzaghi Lecture)." *Journal of Geotechnical and Geoenvironmental Engineering*, 124 (4):278-288.

Santagata, M.C., and J.T. Germaine. 2002. "Sampling Disturbance Effects in Normally Consolidated Clays." *Journal of Geotechical and Geoenvironmental Engineering* no. 128 (12):997-1006.

Simons, N.E. 1974. "Normally Consolidated and Lightly Overconsolidated Cohesive Materials." *Proc. Conf. Settlement of Structures, Cambridge*:500-530.

Tanaka, H., O. Mishima, and M. Tanaka. 2000. "The Rate Effect on P_c Value for Nine Undisturbed Marine Clays." *Tenth International Offshore and Polar Engineering Conference*:649-652.

Determination of the Time Rate of Radial Consolidation Using Laboratory Inflow Consolidation Testing

Cem Akgüner[1], A.M.ASCE

[1]Lecturer, Civil Engineering Department, Davutpasa Campus, Yildiz Technical University, Esenler, Istanbul, Turkey 34220; akguner@yildiz.edu.tr

ABSTRACT: A simple, easily applicable correlation is proposed for the relationship between the degree of radial consolidation (U_r) and the time factor (T_r) based on the free-strain assumption and drainage to a central drain (inflow radial consolidation). The suggested correlation is successfully applied to the results of a series of tests on soft clayey soils from Louisiana using the incremental radial inflow test equipment developed and utilized by a series of R. E. Olson's former graduate students in the late 70's and early 80's. The proposed correlation is practical, yet powerful, and can easily/quickly be applied using spreadsheets. One can also conveniently and consistently determine radial consolidation coefficient (c_r) without the conventional graphical construction.

INTRODUCTION

Vertical drains (sand drains, nowadays prefabricated vertical drains (PVDs) or "wicks") are commonly used with preloading to accelerate the rate of consolidation in improving the properties of soft, compressible soils, which may be susceptible to excessive settlements under working loads. The horizontal (radial) hydraulic conductivity is usually significantly higher than the vertical. Thus, the excess pore-water pressures generated due to surcharge loads can dissipate faster through radial flow and the time to complete consolidation is cut-down by shorter drainage distances, further leading to reduced time and potential lower costs in improvement efforts.

The determination of the time factor (T_r) – degree of consolidation (U_r) relationship is critical for the design of cases involving soil improvement by preloading and vertical drains. A radial consolidation coefficient for each loading increment is required, which can be obtained through laboratory and/or in-situ testing. There are two theoretical approaches in calculating the time factor for radial consolidation of a cylindrical sample draining to an inner or outer drain (well), each satisfying a different set of vertical boundary conditions within the soil sample being tested: free-strain and equal-strain theories (Barron, 1948; Da Silviera, 1953; Escario & Uriel, 1961; McKinlay, 1961; Barry & Wilkinson, 1969; Hansbo, 1981; and Onoue, 1988).

In earlier studies a laboratory setup to conduct incremental, radial consolidation tests which considers inflow (water collecting to a central drain within the sample) and/or outflow (water accumulating to an outer drain) was developed, utilized and critically evaluated (Thomas, 1978; Collins, 1979; Trautwein, 1980; Trautwein et al., 1981; and Olson, 1986). It was determined that inflow testing should be preferred over outflow tests because of the formation of a smear zone in the soil next to the drain in the outflow tests resulting in lower values of c_r. Furthermore, measured and back-calculated (from consolidation theory) hydraulic conductivities are in closer agreement when free-strain conditions are assumed. However, along with most practitioners, some researchers, such as Kjellman (1948) and Leo (2004), prefer equal-strain theory due to its simplicity in applying. Barron (1948) and Hansbo (1981) conclude that both approaches result in similar outcomes. On the other hand, Fox et al. (2003) indicated that neither method works well when large strains are involved. Nevertheless, in general for laboratory and field applications the free-strain approach is preferable as it involves fewer approximations in-spite being more complex.

In this paper, the free-strain approach is assumed in deriving a correlation for the time factor (T_r) – radial consolidation (U_r) relationship during incremental radial consolidation. Also the application of the suggested correlation is shown using the laboratory measurements obtained for one of the loading increments of an inflow radial consolidation test conducted on a soft clayey soil sample from Southern Louisiana.

SUGGESTED CORRELATION FOR TIME RATE VERSUS DEGREE OF CONSOLIDATION

The differential equation governing the rate of change of excess pore water pressures along the depth of the sample which can then be used to calculate vertical settlements (deformations) involving radial drainage is as follows:

$$\frac{\partial u}{\partial t}=c_r\left[\frac{1}{r}\frac{\partial u}{\partial r}+\frac{\partial^2 u}{\partial^2 r}\right] \tag{1}$$

where, c_r is the radial coefficient of consolidation defined as:

$$c_r=\left[\frac{k_r(1+e)}{a_v\gamma_w}\right]=\frac{T_r r_{sample}^{\;2}}{t} \tag{2}$$

(k_r: radial hydraulic conductivity; e: void ratio; a_v: coefficient of compressibility; γ_w: unit weight of water; T_r: radial time factor; r_{sample}: radius of consolidating sample in laboratory; t: time).

The solution for inflow using the free-strain theory involves Bessel functions of first and second kind of order zero and one, and it is given elsewhere (Barron, 1948; Moran et al., 1958). The free-strain approach can easily be programmed using readily available spreadsheet software.

In this paper, a practical and easily adaptable correlation is proposed for the relationship between the degree of radial consolidation and the time factor for a cylindrically shaped laboratory sample based analytically on the free-strain assumption and drainage to a central drain (inflow radial consolidation):

$$U_{r,i}=\left(\frac{T_{r,i}^{\;2.58}}{T_{r,i}^{\;2.58}+0.7}\right)^{0.26} \tag{3}$$

for $N=d_{sample}/d_{drain}=5$, and i (%) varying from 0 to100. The equation is similar to the suggestion by Brinch-Hansen (1961) for vertical flow, which is also used by Chan (2003) for a least squares approach.

The U_r - T_r values by the suggested correlation and Barron's equation are compared in Figure 1 where a very good match is obtained with a maximum error of 1.1% in the average degree of radial consolidation.

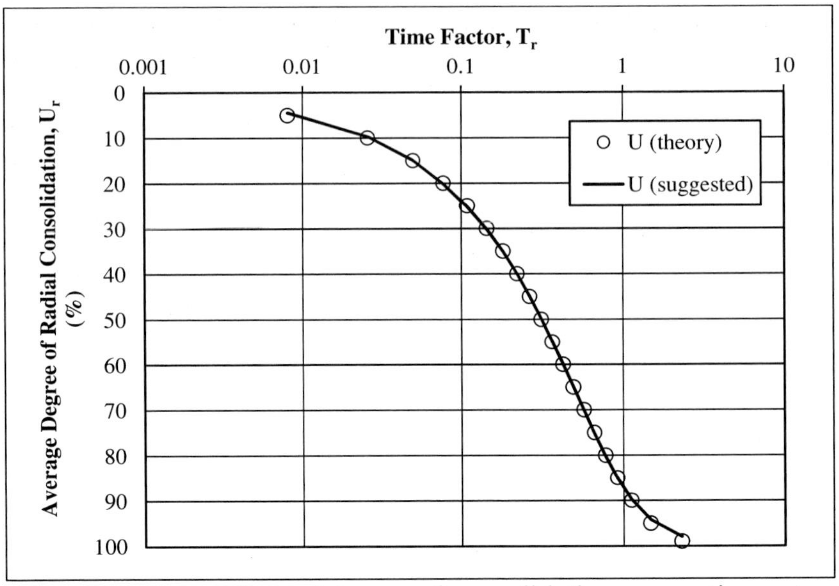

Figure 1. Comparison of degree of consolidation versus time rate.

LABORATORY TESTING

Incremental consolidation tests with inflow towards a central drain were conducted by the author in the laboratory using an electronic data acquisition system. The loading was started at 125 psf (6 kPa) until a pressure of 32000 psf (1530 kPa) was reached with a load increment ratio (LIR) of two and then it was reduced by a LIR of four in three stages to 500 psf (24 kPa). The tested samples had a height and diameter of 51 mm (2 inches) and 64 mm (2.5 inches), respectively. A central drain with a diameter of 12.7 mm (0.5 inches) was constructed, which consisted of medium to fine, poorly graded sands (SP). The ratio of the drain area to the total area is 4% so that any differences in the compressibility between sand drain and clayey soil can be neglected. Top and bottom boundaries are impervious. Therefore, the water flow is horizontal and only towards the central drain, i.e., no drainage occurs towards the outside boundaries (top, bottom, side) of the sample. The sides of the consolidation cells are lined with polytetrafluoethylene (Teflon®) to reduce side friction which is greater than in standard one-dimensional vertical flow tests because of the larger ratio of height to diameter for radial flow tests.

A slight modification of the conventional data reduction was applied according to the suggestion by Trautwein (1980), which is similar to Taylor's (1942) approach for vertical consolidation data. In vertical consolidation testing the square root time – settlement measurements are linear for approximately 60% of the settlements. In radial consolidation tests, instead of square root of time, 0.67 root of time ($T^{0.67}$) versus settlement measurements are graphed for the linearity approximately within the same percentage of settlements. Then a straight line is drawn having an abscissae 1.31 times the corresponding abscissae on the linear part of the plotted curve rather than 1.15 suggested by Taylor's method for vertical consolidation to obtain the end-of-primary settlements during each loading increment. T_{50} is 0.312 as compared to 0.197 for calculating the coefficient of radial consolidation from the laboratory data. Swelling may occur in the initial phases of the testing where the applied compressive loads are less than the field effective stresses for samples. However, no swelling or compression beyond the determined end-of-primary consolidation was allowed. The suggested correlation (3) allows for the whole time-settlement curve to be individually represented for each loading stage.

The "Solver" function in Excel may be referred to as a what-if analysis tool for optimization where an optimal value for a formula in a target cell by adjusting the values of other specified cells that are related to the formula in the target cell (Microsoft, 2013). The "Solver" can conveniently be utilized to find the values for each stage such that the beginning of the primary settlements S_0 and end-of-primary settlement S_{100} can be adjusted along with the time to reach 50% of radial consolidation (t_{50}) (given in Figure 2 with $A$3:$A$5 respectively) by minimizing the difference between the measured and calculated settlements (shown as $P$25 in Figure 2), which are found by the time factor, $T_{r,i}$ (= $c_{r,i}t/r_{sample}^2$) for the stage and consolidation percentage based on the suggested correlation in Eq. (3). Typically the solution can be obtained after a single run within seconds although the results should be checked for reasonableness. An initial fitting with assumed values for a loading stage using the above described methodology is presented in Figure 3 where the significant difference between the measurements and fitting can be seen.

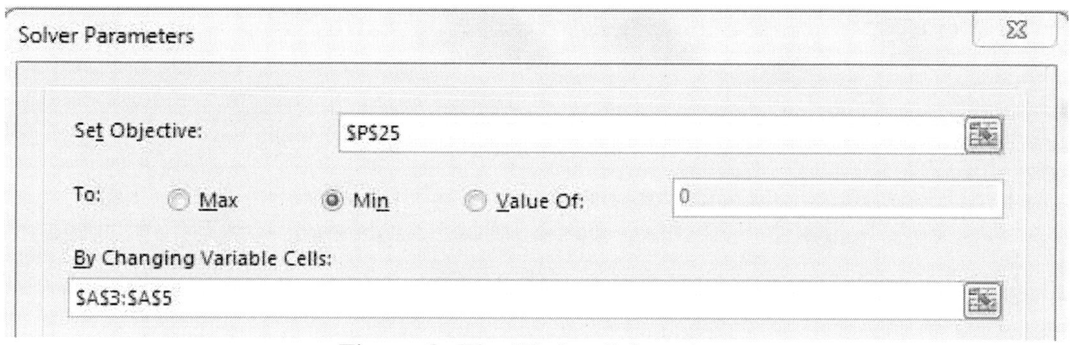

Figure 2. The "Solver" function.

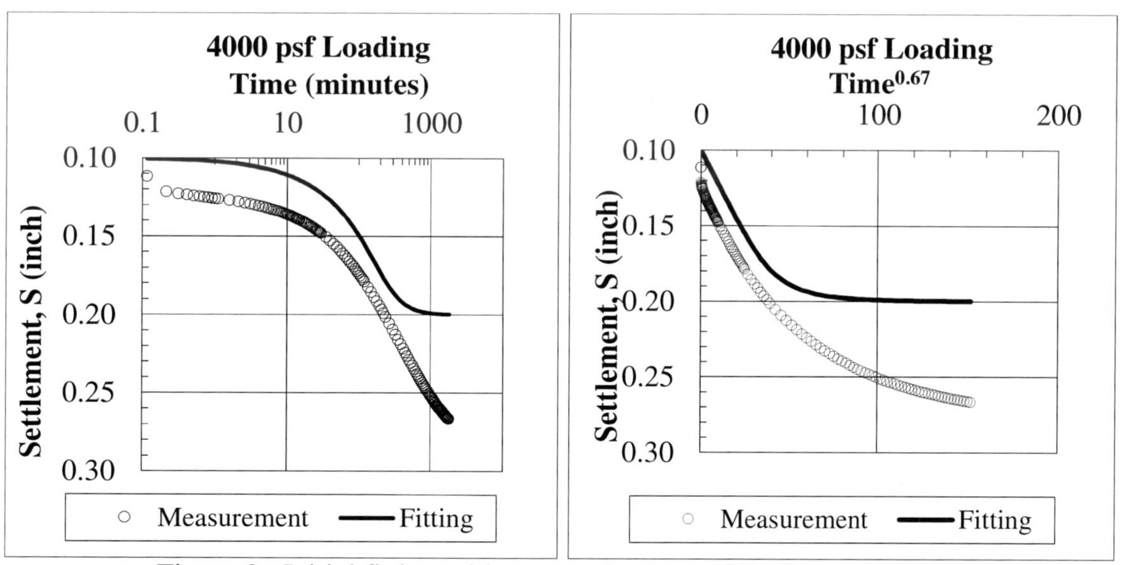

Figure 3. Initial fitting with assumed values of S_0, S_{100} and T_{50}.

The suggested correlation is successfully applied to a series of nine incremental radial inflow consolidation tests on high plasticity soft and at times organic clayey samples from Southern Louisiana. An example of the good agreement between measurements and fitting is presented for one of the incremental loading stages of a sample obtained from an average depth of 21 feet (6.4 m) (LL=120%; PI=88%; water content=76%; s_u=440-600 psf (21-29 kPa)) in Figure 4. Similar outcomes can be seen for all of the stages of this sample as well as the eight other samples. As expected from laboratory measurements, the match is not exact, but very good considering many possible sources of error in laboratory testing in geotechnical engineering, such sample disturbance, thin stratification, limited sampling procedures, and measurement/personnel errors.

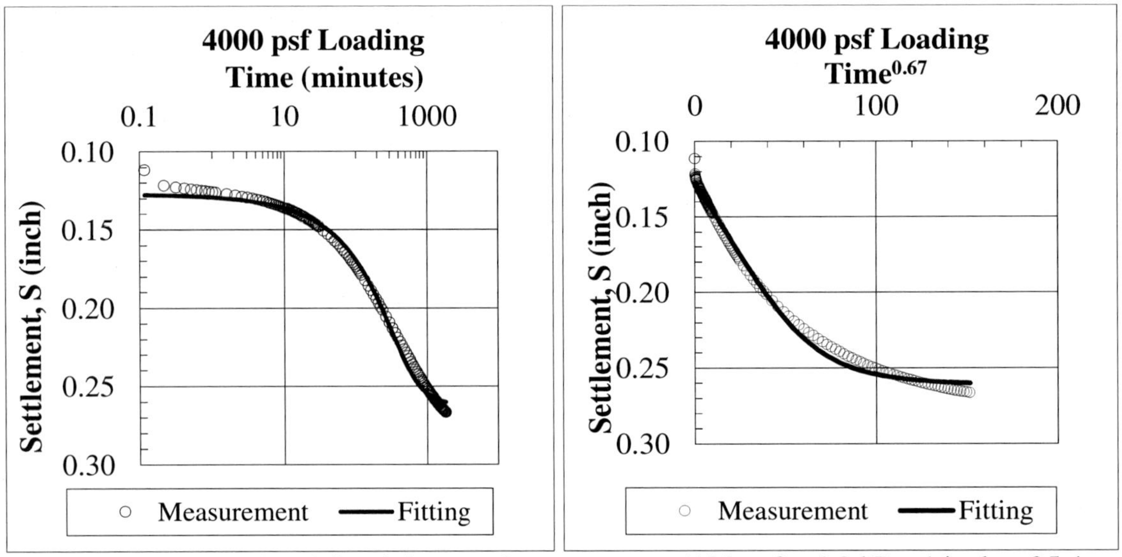

Figure 4. The measured and fitted time-settlement curves (100 psf = 4.8 kPa; 1 inch = 25.4 mm; calculated radial coefficient of consolidation for this stage, $c_r = 2.3 \times 10^{-3}$ m^2/day)

CONCLUSIONS

Determining the coefficient of radial consolidation is the critical parameter when estimating the displacements and the necessary time to reach the expected soil improvement of soft, highly compressible clays with vertical drains. A correlation and methodology to calculate the radial consolidation coefficient based on the free-strain approach for incremental radial consolidation testing with a central drain is suggested and successfully applied to a series of samples. The correlation involves the relationship between the radial time factor (T_r) and degree of consolidation (U_r), and matches very well with the theoretical values. Furthermore, it is quick and easy to apply minimizing any errors related to reduction of collected consolidation data.

ACKNOWLEDGMENTS

I was supervised by Professor Roy Olson as Ph.D. student who gave me my first and last jobs during my extended time at UT-Austin. We had long, entertaining, hearty, and at times heated conversations (or should I call them experiences) on a multitude of subjects, freely ranging from geotechnical engineering, Longhorn sports, food, cooking, restaurants, trips, photography, local and world politics, family, relationships, history, WW1 and WW2, commercial and military planes, race relations, past and present professors/students. The author is thankful for the experience. Most of the work for this paper was conducted by the author as part of a consulting work for Rickey P. Brouillette, also an Olson protégé.

REFERENCES

Barron, R.A. (1948). "Consolidation of fine-grained soils by drain wells". *Transactions of the American Society of Civil Engineers*, Vol. 113, 718-754.

Berry, P. L. and Wilkinson, W.B. (1969). "The radial consolidation of clay soils". *Géotechnique*, Vol. 19, No. 2, 253-284.

Brinch Hansen, J. (1961). "A model law for simultaneous primary and secondary consolidation". *Proceedings of 5th International Conference on Soil Mechanics and Foundation Engineering*, Paris, Vol. 1, 133-136.

Chan, A.H.C. (2003). "Determination of the coefficient of consolidation using a least squares method". *Géotechnique*, Vol. 53, No. 7, 673–678.

Collins, J.T. (1979). *Laboratory measurements of soil response during consolidation with sand drains*. MS thesis, The University of Texas at Austin.

Da Silviera, I. (1953). "Consolidation of a cylindrical clay sample with external radial flow of water". *Proceedings of 3rd International Conference on Soil Mechanics and Foundation Engineering*, Vol. 1, 55-56.

Escario, V. & Uriel, S. (1961). "Determining the coefficient of consolidation and horizontal permeability by radial drainage". *Proceedings of 5th International Conference on Soil Mechanics and Foundation Engineering*, Paris, Vol. 1, 83-87.

Fox, P., Di Nicola, M., & Quigley, D. (2003). "Piecewise-linear model for large strain radial consolidation". *ASCE Journal of Geotechnical and Geoenvironmental Engineering*, Vol. 129, No. 10, 940–950.

Hansbo, S. (1981). "Consolidation of fine-grained soils by prefabricated drains". *Proceedings of 10th ICSMFE*, Stockholm, Vol. 3, 677-682.

Kjellman, S. (1948). Accelerating consolidation of fine grain soils by means of cardboard wicks. *Proceedings of 2nd International Conference on Soil Mechanics and Foundation Engineering*, Vol. 2, 302-305.

Leo, C.J. (2004). "Equal strain consolidation by vertical drains". *ASCE Journal of Geotechnical and Geoenvironmental Engineering*, Vol. 130, No. 3, 316-327.

McKinlay, D.G. (1961). "A laboratory study of rates of consolidation in clays with particular reference to conditions of radial porewater drainage". *Proceedings of 5th International Conference on Soil Mechanics and Foundation Engineering*, Paris, Vol. 1, 225-228.

Microsoft Co. (2013). *Office 2010 – Excel help pages: about Solver*. Retrieved on August 30, 2013 from http://office.microsoft.com/en-us/excel-help/about-solver-HP005198368.aspx.

Moran, Proctor, Meuser & Rutledge (1958). *Study of deep soil stabilization by vertical sand drains*. Publication No. PB151692, U.S. Department of Commerce.

Olson, R. E. (1986). "State of the art: consolidation testing". *Consolidation of Soils: Testing and Evaluation*, ASTM STP 892, R. N. Yong and F. C. Townsend, Eds., American Society for Testing and Materials, Philadelphia, 7-70.

Onoue, A., (1988). "Consolidation by vertical drains taking well resistance and smear into consideration". *Journal of Soils and Foundations*, Vol. 28, No. 4, 165-174.

Taylor, D.W. (1942). *Research on consolidation of clays*. Massachusetts Institute of Technology. Department of Civil and Sanitary Engineering, Publication 82.

Thomas, R.L. (1978). *A new radial drainage consolidation cell*. MS thesis, The University of Texas at Austin.

Trautwein, S.J. (1980). *Practical measurements of the radial flow consolidation properties of clays*. MS thesis, The University of Texas at Austin.

Trautwein, S.J., Olson, R.E. & Thomas, R.L. (1981). "Radial flow consolidation testing". *Proceedings of 10th International Conference on Soil Mechanics and Foundation Engineering*, Stockholm, Vol. 1, 811-814.

Consolidation of estuarine marine clays for coastal reclamation using vacuum and surcharge loading

Buddhima Indraratna[1], FASCE, Cholachat Rujikiatkamjorn[2], A.S. Balasubramaniam[3]

[1]Professor and Research Director, Centre for Geomechanics and Railway Engineering, Faculty of Engineering, University of Wollongong, Wollongong, NSW 2522, Australia (corresponding author). E-mail: indra@uow.edu.au

[2] Senior Lecturer, Centre for Geomechanics and Railway Engineering, Faculty of Engineering, University of Wollongong, Wollongong, NSW 2522, Australia.

[3] Adjunct Professor, School of Civil Engineering Griffith University, Gold Coast, Queensland, Australia

ABSTRACT: Soft clays in coastal areas have low shear strength and high compressibility. Consequently, certain construction activities for infrastructure developments in these deposits often pose geotechnical problems due to large time dependent settlements and lateral movements. Ground improvement techniques are adopted in such terrains to reduce the water content of soft clays by preloading with surcharge fill over vertical drains. Depending on the magnitude of the surcharge load used substantial immediate settlement with lateral movements can take place during preloading, leading to undrained stability problems in various parts of the clay foundation. Therefore, the use of vacuum assisted preloading has now become a popular method in ground improvement works where substantial loads need to be carried out to meet a desired rate of settlement and mitigate undrained failure by controlling lateral displacements. To assist the vacuum propagation to significant depths, vertical drains are used in tandem At the Port of Brisbane, Australia, vacuum assisted surcharge preloading and conventional surcharge preloading schemes were adopted to reduce the consolidation time and long term settlement in soft Holocene clays in 2009. It is shown that a combined vacuum surcharge loading system with a standard surcharge fill highlights the obvious benefits of vacuum consolidation in reducing long term settlement and enhanced stability.

INTRODUCTION

The Port of Brisbane is recognized as Australia's third largest container port located between the mouth of the Brisbane River and Fisherman Islands (Indraratna et al. 2011). With rapid demand growth in trading activities, a new outer area (235ha) adjacent to the current port facilities is being reclaimed to provide the additional berths suitable for cargos and containers. In this area, soil profile comprises a highly compressible dredged mud and clay layer over 20m in thickness with an undrained shear strength of less than 15 kPa near the surface. The shear strength of the dredged mud for reclamation has much lower shear strength depending on the time of placement and the duration the capping material (surcharge) placement. Without any appropriate ground improvement scheme (Indraratna et al. 2001), it is expected that the consolidation period will be more than 50 years with vertical deformations of 2.5-4.0m under the actual service loadings (60 kPa). Therefore, vacuum consolidation with prefabricated vertical drains (PVDs) was recommended to accelerate the consolidation process and to limit horizontal deformation for the site located immediately adjacent to the Moreton Bay Marine Park.

Chu et al. (2000), Indraratna and Chu (2000) and Chai et al. (2005) also showed that using a system vacuum preloading combined with PVDs, suction pressure can be transmitted to a significant depth in the soft subsoil. Also, prolonged consolidation time due to staged embankment construction can be significantly reduced (Indraratna et al. 2005). The surcharge fill height may be lowered by several meters, if a vacuum pressure of at least 70% the atmospheric pressure is maintained (Rujikiatkamjorn et al. 2008). In addition, the embankment construction rate can be increased with the reduction in the number of construction stages. Once the soil shear strength increases due to consolidation, the post-construction settlement can be significantly less, thereby reducing risk of long term differential settlement (Shang et al. 1998; Yan and Chu 2003). The ground improvement provided by PVDs combined with vacuum pressure may be an economically attractive alternative in deep soft clay sites. Olson (1977), Olson et al. (1974) and Geng et al. (2012) introduced a solution for a single ramped load, for both vertical and radial drainage. To date, there is no comprehensively reported case history in which both the conventional surcharge preloading and vacuum technique are applied in the same area. In this paper, the performance comparison between the vacuum and non-vacuum methods has been made on the basis of measured vertical deformations, excess pore pressures and horizontal displacements. The effects of improvement techniques on the long term settlement and excess pore pressure are also elucidated.

VACUUM PRELOADING SYSTEMS

Currently, there are two main types of vacuum preloading systems adopted in the field (Geng et al. 2011):

A. Membrane system: After PVDs are installed and the sand blanket is placed with horizontal perforated pipes, the membrane is laid on the top and its borders are flooded under a bentonite slurry channel (Fig. 1a). The vacuum pumps are then connected to the discharge system. A major advantage here is that the vacuum can propagate within the sand platform, along the soil surface and down the PVDs. An obvious disadvantage is that the efficiency relies mainly on the ability of the airtight system to prevent any air leaks over a significant period of time.

B. Membraneless system: When an area has to be subdivided and progressed individually, the vacuum preloading can only be conducted one section after another and therefore the membrane system may not be a reasonable solution. In order to avoid this problem, the vacuum pipes are joined directly to each individual PVD using a tubing system (Fig. 1b). In contrast to the membrane system where any air leak can affect the entire system, each drain acts independently. However, the requirement of significant tubing for hundreds of drains can affect the installation time and cost.

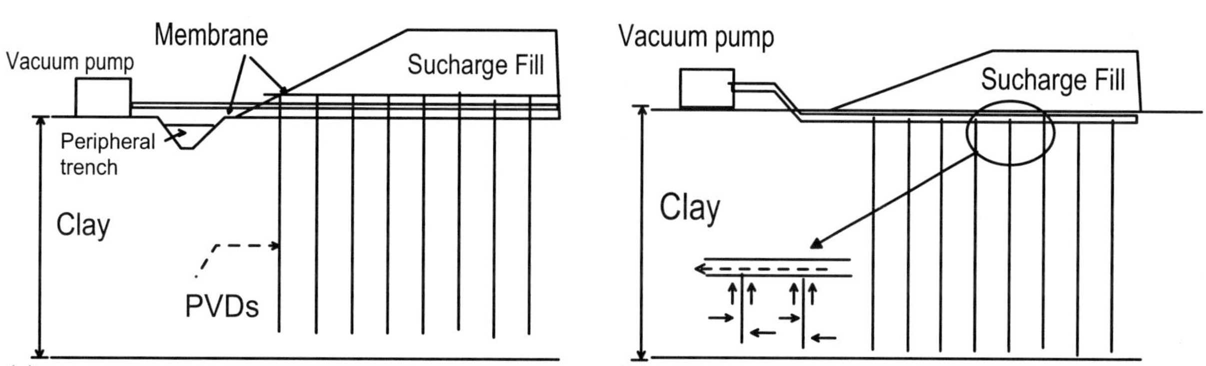

(a) (b)

Figure 1. Types of vacuum preloading systems (a) Membrane system and (b) Membraneless system

GENERAL DESCRIPTION OF EMBANKMENT CHARACTERISTICS AND SITE CONDITIONS

In 2003, the Port of Brisbane Corporation started to recover a sub-tidal land area of 235ha at Fisherman Islands near the mouth of the Brisbane River (Fig. 2). The aim is to provide additional berths and associated infrastructure to accommodate the future demand of the Port. To compare the performance of the vacuum system with the non-vacuum system (PVD and surcharge load), there were 3 contractors selected to carry out this trial. Each contractor was allocated a trial area of about 3000m^2. The aim was to compare performance based on their design and construction work.

Contractor A: Involved in 8 trial zones in Area S3a, designated as WD1, WD2, WD3, WD4, WD5a, WD5b, VC 1 and VC 2. Areas WD1 to WD5a and WD5b had surcharge only, while VC1 and VC2 had surcharge and vacuum consolidation with membrane system.

Contractor B: The trial area T11 consisted of seven zones, of which 5 had surcharge and different drain types. Two sub-areas had surcharge in conjunction with the membraneless system.

Contractor C: This site T11 was subdivided into trial Areas 4, 5 and 6. Areas 4 and 5 had a surcharge period of a year, whereas Area 6 had a surcharge period of 0.5 years. Vertical drains with 1.4m spacing were adopted in Areas 4 and 5 while Area 6 used vertical drains with a spacing of 1m.

The upper Holocene sand underneath the dredged mud was about few meters thick, and overlaid the Holocene clay layer having a thickness from 5m to 24m. The highly soft and compressible Holocene clay layer had relatively low shear strength (Ameratunga et al. 2010). The Holocene layer overlies a Pleistocene deposit comprising of highly over-consolidated clay. Site investigations e.g. cone penetration/piezocone tests, dissipation tests, boreholes, field vane shear tests and oedometer tests were carried out to quantify the consolidation and stability design parameters. The soil parameter variations are shown in Fig. 3. Groundwater level was at +3.5m RL. The water contents of the soil layers were at or beyond their liquid limits. The field vane tests indicate that the undrained shear strength of the dredged mud and the Holocene clays varied from 5 to 60 kPa. The compression index (C_c) varied from 0.1-1.0. The coefficient of consolidation in vertical direction (c_v) was about the same as that in horizontal direction (c_h) for the remoulded dredged mud layer, while the ratio c_v/c_h was about 2 for the Holocene clay layer. This is based on laboratory testing on samples obtained in vertical and horizontal directions to the vertical drains. The coefficient of consolidation of remolded samples was very similar to c_v of the undisturbed sample obtained in Holocene layer.

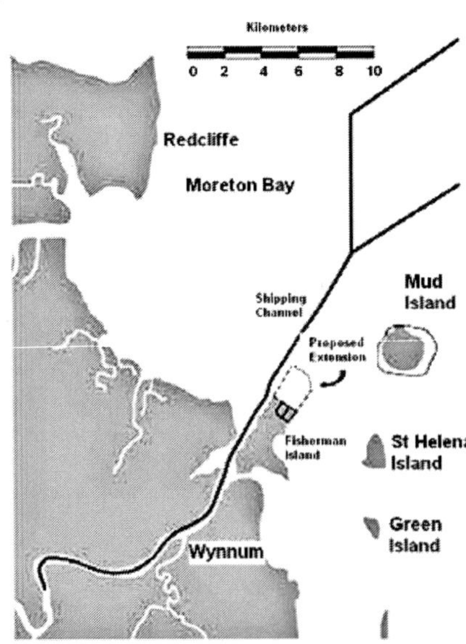

Figure 2. Map of the proposed extension area at the Port of Brisbane (adopted from Indraratna et al. 2011)

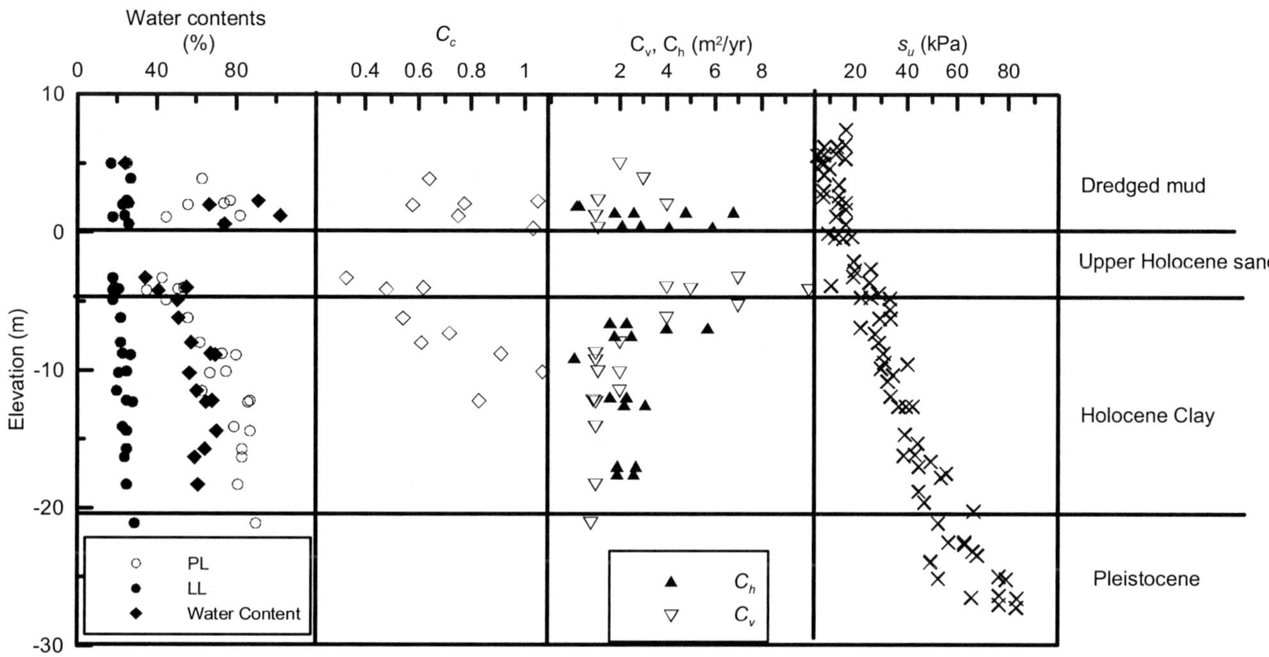

Figure 3. Soil properties and profile at S3A, Port of Brisbane (Indraratna et al. 2011)

As the Holocene clay layer is rather thick (15m or more), two preloading approaches were used to reduce the long term settlement including conventional surcharge preloading system and the vacuum consolidation system both applied in conjunction with PVDs. Rigorous design specifications were considered for the design and construction of fill embankments and vacuum application over the soft Holocene deposits: (a) Service load of 15-25 kPa, (b) maximum residual settlement of not more than 250 mm over 20 years after the application of service load.

ASSESSMENT OF RELATIVE EFFECTIVENESS OF THE TRIAL SCHEMES

Degree of Consolidation (DOC) with Time

The numerically determined DOC using measured settlements via Asaoka's method with time plots via spreadsheet software are shown in Fig. 4 for an array of locations, and they all show a comparable behaviour, irrespective of the treatment technique (S3A and T11) and the type of applied loading (i.e. vacuum vs. surcharge). A relatively high DOC has been achieved after 12 months, and all plots provide DOC > 80%. In order to separate the 'clustering' especially after one year, the DOC is normalized by the dimensionless factor, β (Indraratna et al. 2012) .

The β factor is independent of the soil properties and is designed to capture the drain and site loading conditions, and comprises the favorable effects of: (i) the drain length (l_d), (ii) the drain spacing (s_d) and its pattern (α = 1.05 for triangular and 1.13 for square spacing), and (iii) the surcharge load height (H) to consolidate the given clay thickness (h_c), represented by the ratio (H/h_c).

In this respect, β can be defined as:

$$\beta = (l_d/\alpha s_d) \times (H/h_c) \tag{1}$$

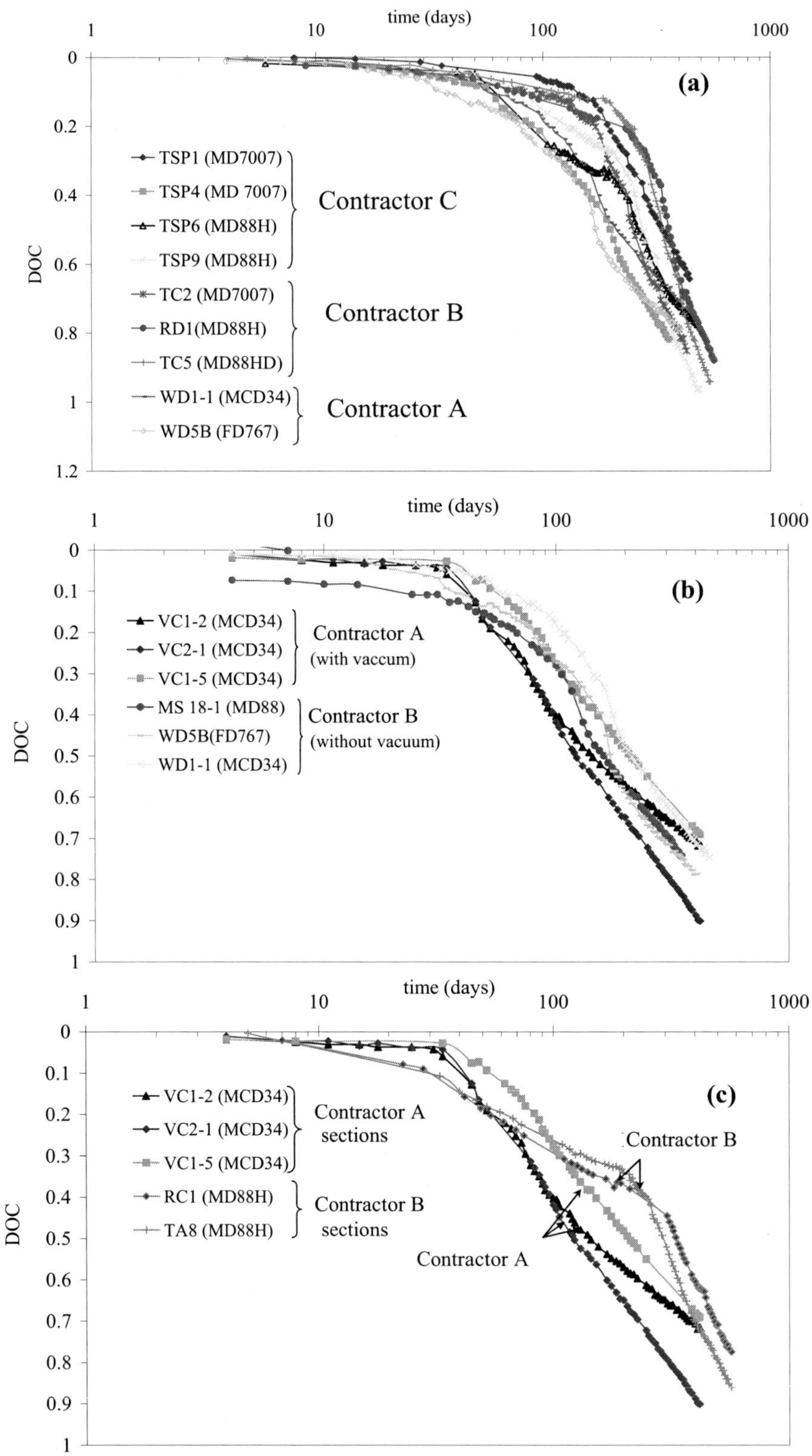

Figure 4. Analytically computed DOC with time for (a) non-vacuum in S3A and T11, (b) treatment in S3A only and (c) vacuum areas in S3A and T11

Based on the magnitude of β determined at each settlement plate location for S3A and T11, the drain and site conditions at the 3 trial areas can be differentiated as:

(i) Low β impact: 2-6 (for S3A area under Contractor A),

(ii) Moderate β impact: 8-12 (for T11 area under Contractor B), and
(iii) High β impact: 12 -18 (for T11 area under Contractor C).

Although the value of β has no specific relationship to the converging target of DOC intended to be attained at the date of fill removal by all contractors, it can act as a 'filter' in distinguishing the relative performance in S3A and T11, by dividing the DOC by β. Figure 5 shows the variation of DOC divided by β plotted against time. This results in a separation between vacuum and non-vacuum areas, and also separates the vacuum consolidation effects of Contractors A and B. When considering all 3 sets of plots (Figs. 4a-c), the relative consolidation performance seems more superior in the case of Contractor A treatment areas when using vacuum consolidation, in comparison with all other locations in S3A and T11.

Excess Pore Water Pressure Dissipation

Figure 6 shows the reduction in measured pore water pressure with time, and it is shown that VC2 in S3A shows the largest reduction closely followed by Section IV in T11 (Contactor C). However, due to the varying fill heights and clay thickness in S3A and T11 paddocks these plots cannot be directly compared and most of them are clustered together during the first 3 months showing little differences. Figure 7a indicates the rate of change of excess pore water pressure for the same locations, and it is observed that VC2, VC1 and WD1 indicate the highest rate of change of excess pore pressure at the start, with VC1 maintaining a steady state over a long period of time. The membraneless systems by Contractor B do not seem to indicate a high rate of excess pore pressure dissipation in comparison with VC1 and VC2 areas. When these plots are normalized by β (Figure 7b), it is shown that VC1 and VC2 provide greater benefit in view of excess pore pressure dissipation, compared to all other areas. While the fill height is reduced in VC areas of S3A thereby involving less mucking operations, the applied suction (-70 kPa) more than compensates for accelerated excess pore pressure dissipation rates, confirming the effective performance of membrane-type vacuum consolidation technique.

Long Term Settlements

All contractors have proposed measures that are anticipated to control the long term settlements within 25 years, either to be less than 150mm or 250mm depending on the clay thickness and anticipated service loads in the respective areas. In Figure 8, the values of long term settlements for both S3A and T11 areas are calculated based on methods provided by Terzaghi et al. (1996); and then plotted with the β-factor; the observations suggest that the critical long term settlement occurs in the range $4 < \beta < 16$. In this critical zone, that includes locations from all 3 contractors from both S3A and T11 paddocks, the long term settlements are close to the permissible limits. At low values of $\beta < 4$, the residual settlements are much smaller mainly because of vacuum consolidation. In relation to the clay thickness, at very high values of $\beta > 16$ (T11), the long term settlements tend to decrease, mainly because of the high fill surcharge levels (i.e. relatively high H/h_c ratio).

Figure 9 provides approximately linear relationships between the long term settlement and clay thickness for a range of OCR from 1.1 to 1.4 for DOC > 80%. As expected, it is observed that when the OCR increases the residual settlement (RS) decreases substantially. In general, as the total Holocene clay thickness increases, the RS also increases, and the corresponding regression lines and best–fit equations are also provided on Figure 9. In particular, the vacuum consolidation locations of S3A (VC1-2, VC2-2 and VC2-3) show considerably reduced long term settlements at OCR approaching 1.4, well below the permissible limit. At an OCR of approximately 1.3, the long term settlements associated with membraneless consolidation (TA8,) and VC1-5 (S3A) are also small. Based on Figure 9, a lower bound and upper bound for long term settlement in terms of clay thickness (h_c) can be obtained as follows for the entire range of over-consolidation upon fill removal:

Lower Bound: Long term settlement = 3.8 h_c – 27 (vacuum consolidation in S3A at OCR = 1.4)
Upper Bound: Long term settlement = 14.3 h_c + 34 (surcharge only sites at OCR = 1.1)

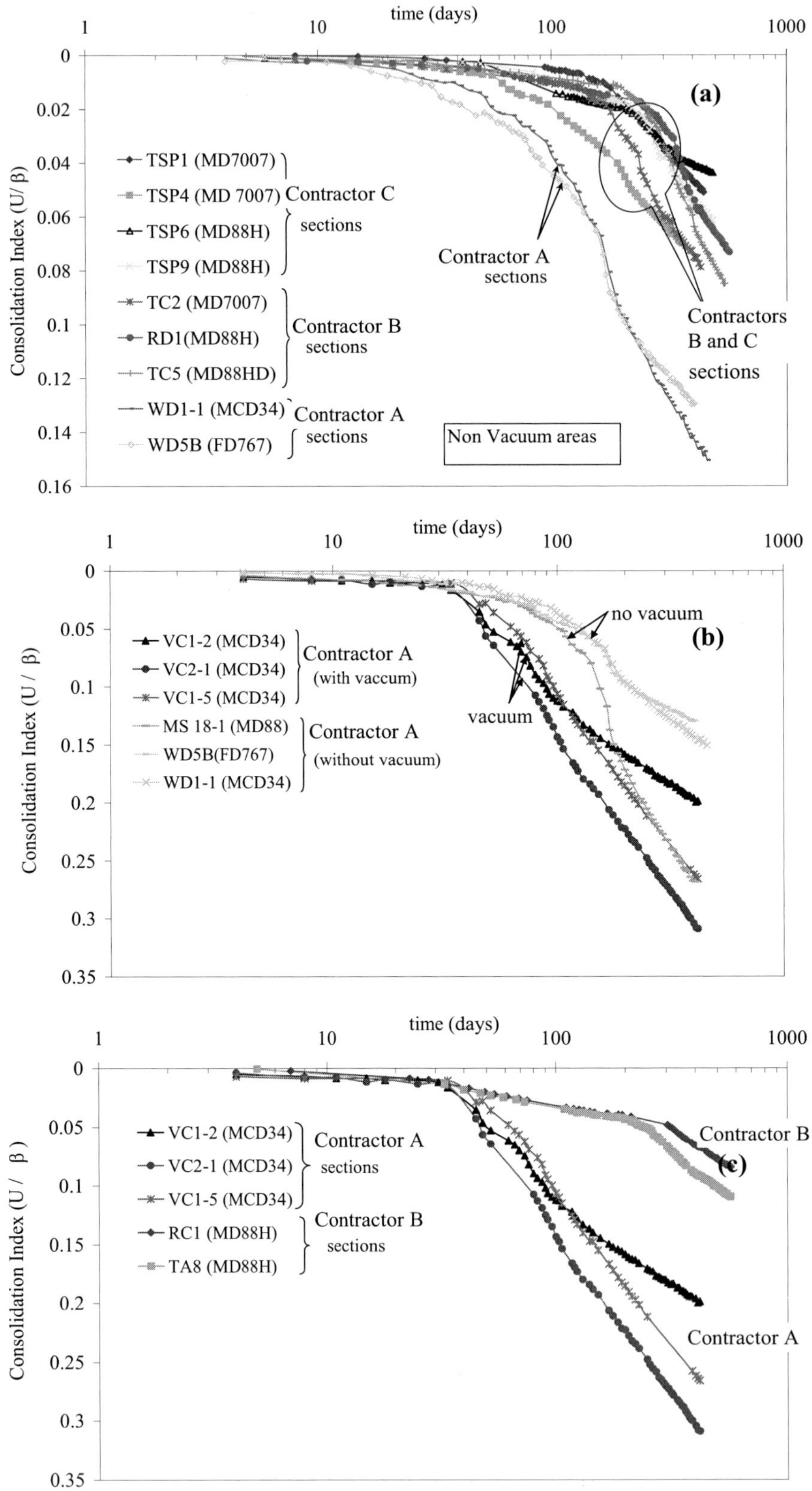

Figure 5. Computed DOC/β with time for (a) non-vacuum in S3A and T11, (b) treatment in S3A only and (c) vacuum areas in S3A and T11

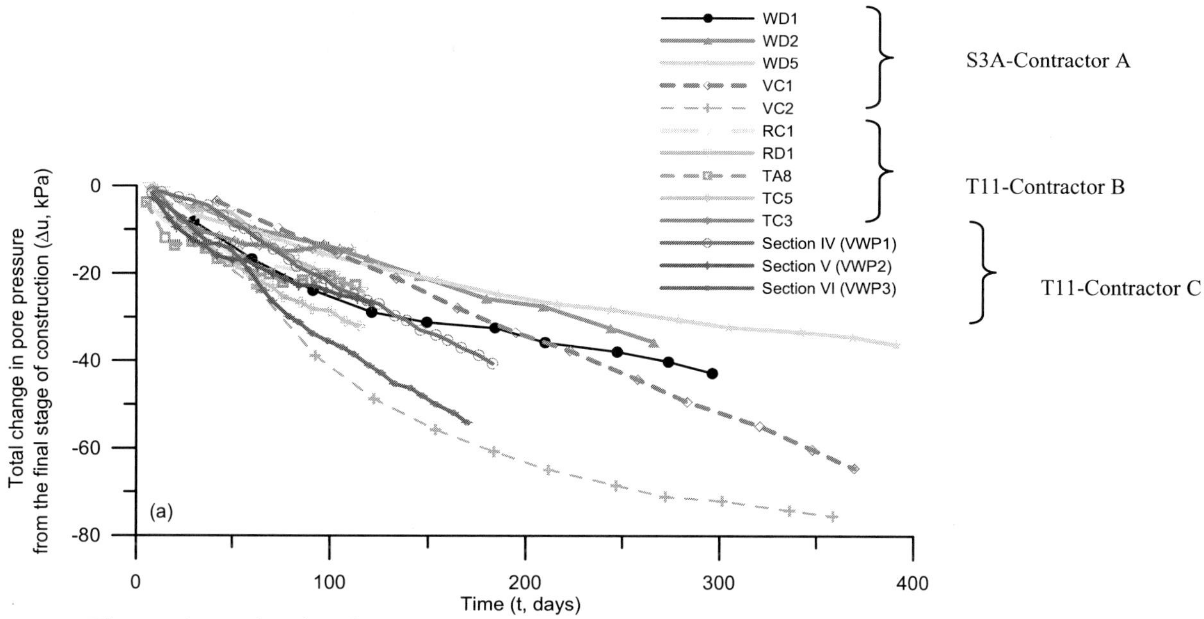

Figure 6. Reduction in Excess Pore Water Pressure with Time in S3A and T11 areas

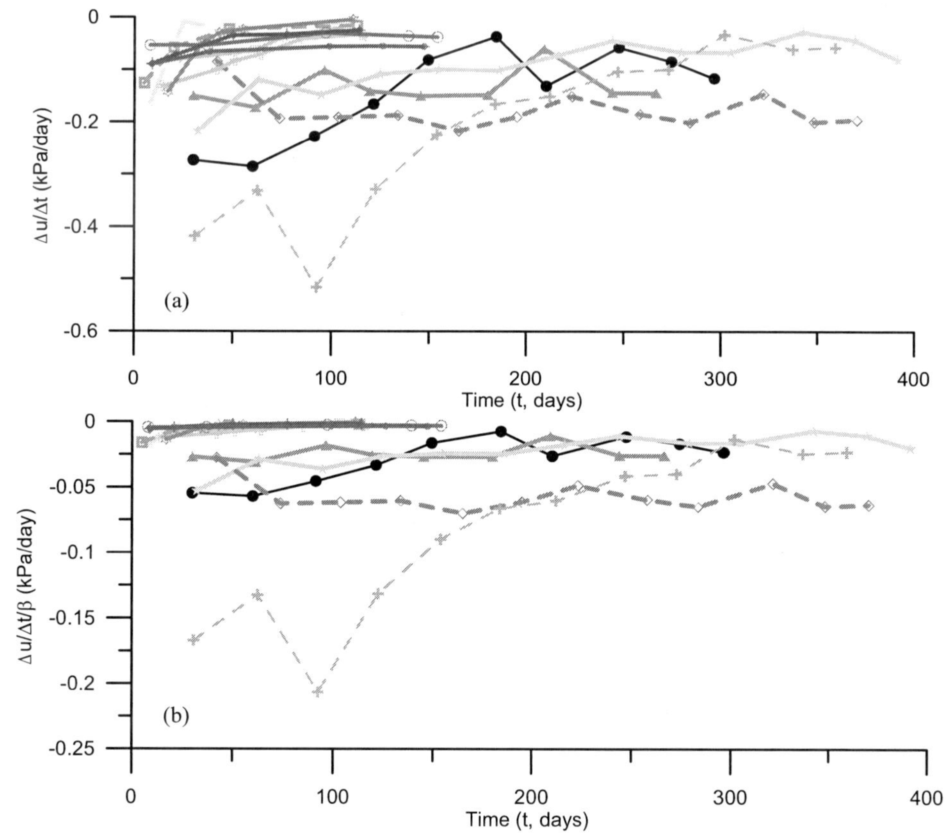

Figure 7. Comparison of excess pore pressure dissipation between S3A and T11 (a) Rate of dissipation of Excess pore pressure, (b) Excess pore pressure dissipation rate normalised by β

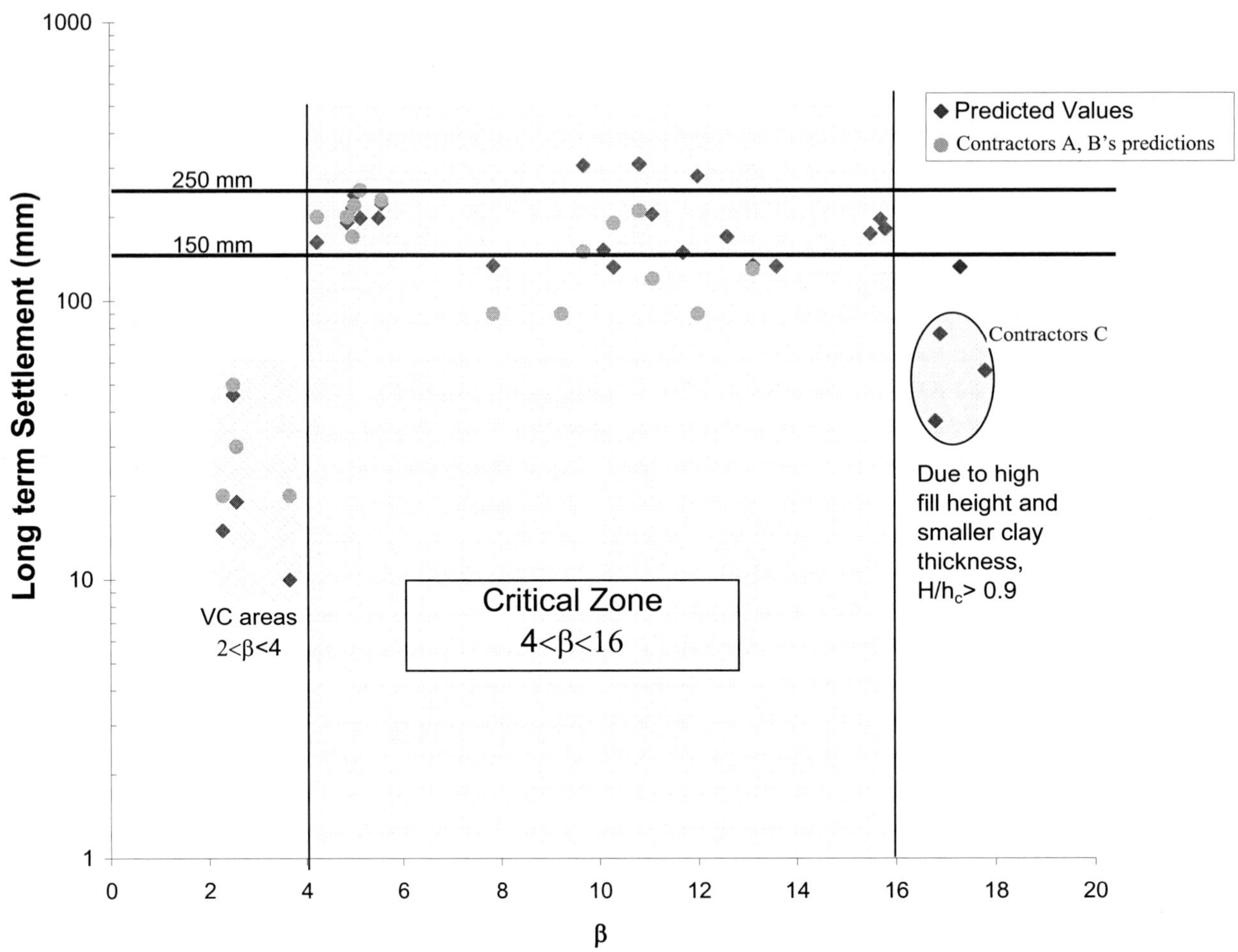

Figure 8. Critical β values for permissible Residual Settlement in S3A and T11

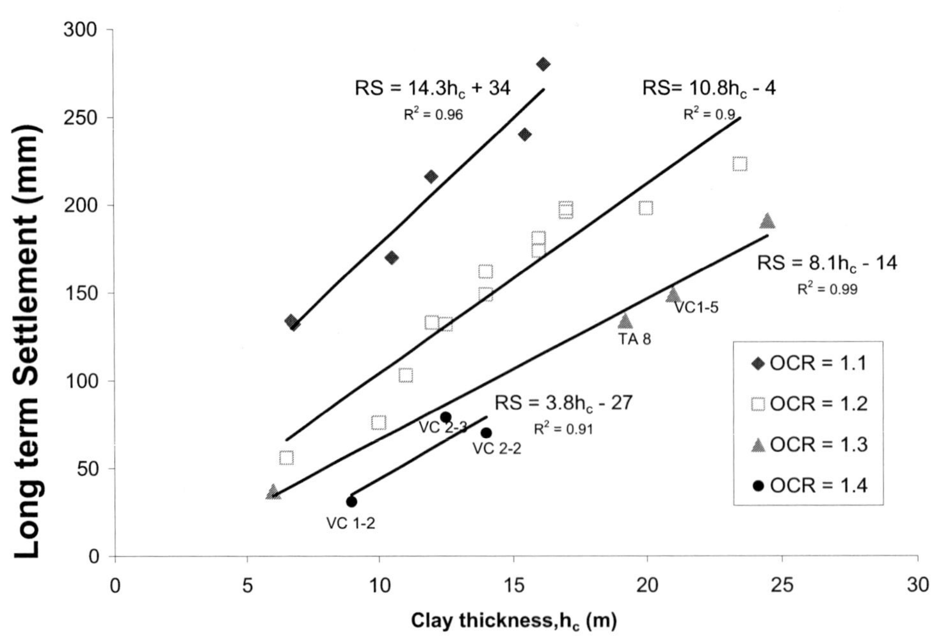

Figure 9. Effect of OCR and Clay Thickness on Residual Settlement

Controlling the Lateral Displacements

It is well-known that the vertical drains provide the advantage of reducing the lateral yield in soil and that the application of vacuum pressure further curtails the lateral movement, and in some cases may even make the lateral movements go inwards rather than outwards from the centerline of the embankment (Indraratna et al. 2005). The use of vacuum pressure to control of lateral displacements is very important in sensitive areas such as in the vicinity of marine parks. At this site, only very limited field data has been available from a few inclinometers. Nevertheless, in order to compare the lateral movements of selected vacuum and non-vacuum areas that have very different soil profile and surcharge load conditions, the lateral displacement can be divided by the applied effective stress at the same depth.

The normalized lateral displacement profiles with depth for the limited data sections are shown in Figure 10. These plots clearly indicate that while vacuum consolidation is definitely beneficial for controlling the lateral movement, the membrane system with 70 kPa suction further demonstrates the most significant reduction in the normalized lateral displacement (i.e. compare VC1-MS28 with WD3-MS27). In the Membraneless vacuum system with 50 kPa suction, while a reduction in the lateral movement is definitely achieved (i.e. compare MS24 with MS34), the amount of this reduction is not as significant as that of Membrane system. The shape of the lateral displacement curves suggest that in all vacuum areas the suction head propagates significantly with depth such that both the LHC and UHC layers are favorably influenced.

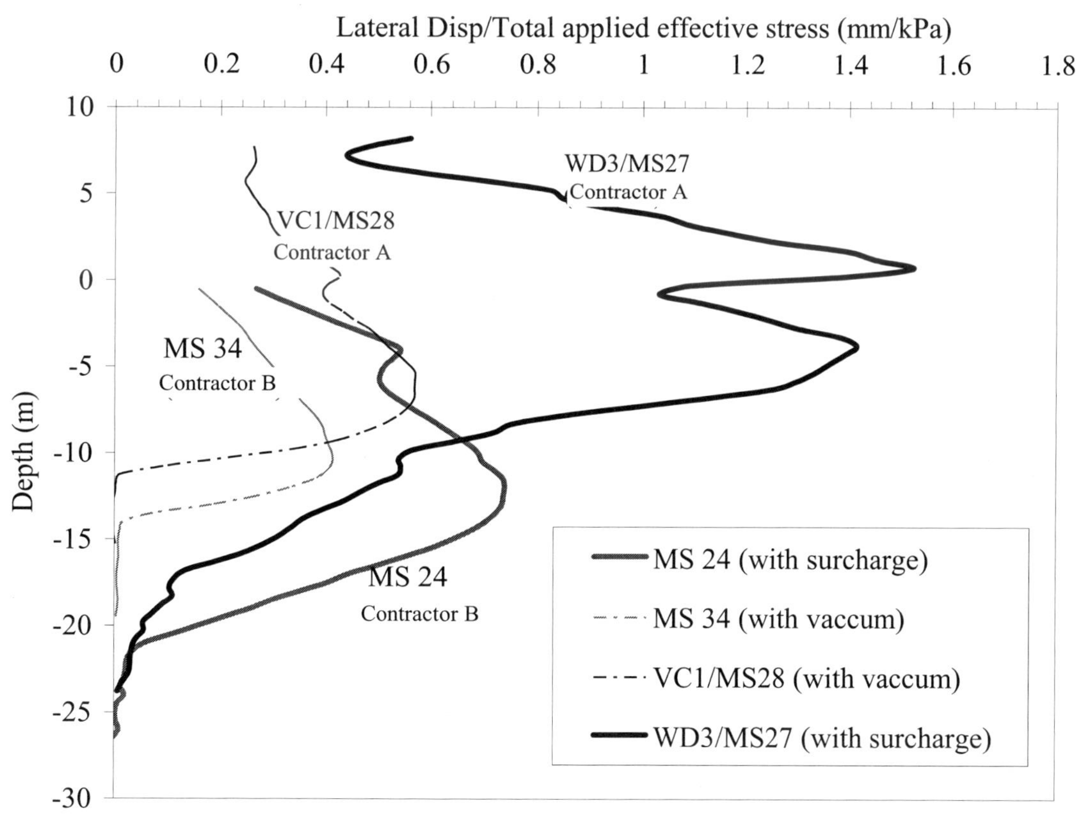

Figure 10. Role of Vacuum Consolidation on Lateral Displacement

CONCLUSIONS

The performance of soil consolidation at the Port of Brisbane was analyzed and discussed. The technical contents reported in this study are not intended to compare the capabilities of any contractors, but solely focused on the relative efficiency of different methods applied at the Port of Brisbane reclamation sites. The land was reclaimed using mud dredged from the seabed of shipping channels and berths. A total of 3 trial areas were selected to investigate the behavior of surcharge and vacuum consolidation. In general, a system of vertical drains with vacuum preloading is an effective method for accelerating soil consolidation, however, purely on the basis of Degree of Consolidation (DOC), it was not possible to compare the benefits of relative treatments applied in the trial areas. This is because, in all cases, albeit different time scales, the desired or target DOC could be attained irrespective of the type of drains and their installation pattern, nature of surcharge loading (with or without vacuum) and clay thickness. In order to distinguish the differences between the trial locations, a drain and site representation factor totally independent of soil consolidation properties was defined as the β-factor, designed to capture the drain and site loading conditions. It comprises the effects of: (i) i drain length (l_d), (ii) drain spacing (s_d) and its pattern (α = 1.05 for triangular and 1.13 for square spacing), and (iii) the surcharge load height (H) in relation to a given clay thickness (h_c).

Normalizing the DOC, settlement and lateral displacement/settlement ratio by proposed β-factor, provides a performance indicator that represents the returns per unit value of β. In such a comparison, the vacuum consolidation applied in S3A seems to be the most beneficial. The membraneless vacuum system application is also effective in terms of controlling lateral displacements, however, the field inclinometer data is very limited to formulate overall conclusions. Control of lateral displacement effectively in sensitive areas such as marine parks would benefit immensely by the application of vacuum pressure and thereby decreasing the required fill heights on the surface.

While a distinct relationship between the DOC and long term settlement is difficult to determine for the given conditions, there is no doubt that the long term settlement decreases almost linearly with the increase in the over-consolidation ratio, and also the long term settlement tends to become closer to the prescribed 150mm limit for the critical range $4 < \beta < 16$. The minimum long term settlement is attained in the vacuum consolidation sites in S3A when the OCR exceeds 1.3. The long term settlement tends to become critical when the OCR is close to or less than 1.1, and this situation mainly occurs for surcharge only sites with large clay thickness, where the treatment is not as effective as when vacuum pressure is applied. It verifies that a large surcharge fill height becomes necessary in the absence of vacuum pressure in order to keep the long term settlement less than the prescribed limit, and the need to remove a large amount of fill in order to achieve a significant OCR can be a cumbersome process in the field. Higher the service load, the greater will be the advantage of vacuum application as a means of reducing the need for excessive fill heights as well as lateral displacement control. In view of stringent residual settlement and lateral displacement control plan, the application of sufficiently high vacuum pressure in tandem with some surcharge fill to achieve a relatively high DOC (i.e. > 85%) and subsequent unloading for attaining an OCR > 1.3 would be the optimum choice for the site characteristics and loading conditions encountered here.

ACKNOWLEDGEMENTS

Writers acknowledge the support of the Port of Brisbane Corporation and Coffey Geotechnics. The research funding from the Australia Research Council is acknowledged. The assistance of Prof Harry Poulos, Dr Jay Ameratunga, Cynthia De Bok, Tine Birkemose and Chamari Bamunawita of Coffey Geotechnics is appreciated. More elaborate details of the contents discussed in the paper can also be found in previous publications of the first Author and his research students in ASCE and Canadian Geotechnical Journals, since the mid 1990's.

REFERENCES

Ameratunga, J., Boyle, P., De Bok, C. and Bamunawita, C. (2010). "Port of Brisbane (PoB) Clay Characteristics and Use of Wick Drains to Improve Deep Soft Clay Deposits." *Proceedings 17th Asian Geotechnical Conference*, Taipei, Vol. I, pp 116-119.

Chai, J.C., Carter, J.P., and Hayashi, S. (2005). "Ground deformation induced by vacuum consolidation". *Journal of Geotechnical and Geoenvironmental Engineering*, 131(12):1552-1561.

Chu, J. Yan, S.W., and Yang, H. (2000). "Soil improvement by the vacuum preloading method for an oil storage station". *Geotechnique*, 50(6): 625-632.

Geng, X. Y., Indraratna, B. and Rujikiatkamjorn, C. (2012). "Analytical solutions for a single vertical drain with vacuum and time-dependent surcharge preloading in membrane and membraneless systems." *International Journal of Geomechanics*, ASCE, 12(1), 27-42.

Geng, X. Y., Indraratna, B. and Rujikiatkamjorn, C. (2011). "The effectiveness of partially penetrating vertical drains under a combined surcharge and vacuum preloading", *Canadian Geotechnical Journal*, 48(6), 970-983.

Indraratna, B., Sathananthan, I., Rujikiatkamjorn C. and Balasubramaniam, A. S. (2005). "Analytical and numerical modelling of soft soil stabilized by PVD incorporating vacuum preloading." *International Journal of Geomechanics*, 5(2). 114-124.

Indraratna, B., Rujikiatkamjorn, C., Kelly, R. and Buys, H. (2012). "Soft soil foundation improved by vacuum and surcharge loading". *Ground Improvement,* 165 (2): 87-96.

Indraratna, B., Rujikiatkamjorn, C., Ameratunga, J., and Boyle, P. (2011) "Performance and Prediction of Vacuum Combined Surcharge Consolidation at Port of Brisbane". *J. of Geotechnical & Geoenvironmental Engineering,* ASCE, 137 (11), 1009-1018.

Olson, R. E. (1977). "Consolidation under time-dependent loading." *J. Geotech. Engrg.* Div. , 103 (1), 55–60.

Olson, R. E., Daniel, D. E., and Liu, T. K. (1974). "Finite Difference Analyses for Sand Drain Problems," *Proceedings, Conference on Analysis and Design in Geotechnical Engineering*, ASCE, 85-110.

Rujikiatkamjorn C., Indraratna, B. and Chu, J. (2008). "2D and 3D Numerical Modeling of Combined Surcharge and Vacuum Preloading with Vertical Drains". *International Journal of Geomechanics*, ASCE, 8(2), 144-156.

Shang, J.Q., Tang, M., and Miao, Z. (1998) "Vacuum preloading consolidation of reclaimed land: a case study", *Canadian Geotechnical Journal*, 35, 740-749.

Terzaghi, K., Peck, R. B. and Mesri, G., (1996). *Soil Mechanics in Engineering Practice*, 3rd Ed. Wiley-Interscience.

Yan, S.W. and Chu, J. (2003). "Soil improvement for a road using a vacuum preloading method." *Ground Improvement*, 7(4): 165-172.

Hydraulic Conductivity and Compatibility of Bentonite for Hydraulic Containment Barriers

Charles D. Shackelford[1], P.E., M.ASCE, and Kristin M. Sample-Lord[2], P.E., S.M.ASCE

[1]Professor, Department of Civil and Environmental Engineering, 1372 Campus Delivery, Colorado State University, Fort Collins, CO 80523-1372; shackel@engr.colostate.edu
[2]PhD Candidate, Department of Civil and Environmental Engineering, 1372 Campus Delivery, Colorado State University, Fort Collins, CO 80523-1372; ksample@engr.colostate.edu

ABSTRACT: High-swelling sodium bentonite (Na-bentonite) commonly is used in engineered containment barriers for hydraulic control and containment of liquid-phase contaminants. Such barriers include compacted liners comprised of bentonite amended natural soils (i.e., sands, silts, and clays), manufactured geosynthetic clay liners (GCLs) comprised of thin (~ 10 mm) sheets of Na-bentonite sandwiched between two porous geotextiles, highly compacted bentonite buffers for high-level nuclear waste disposal, and soil-bentonite (SB) vertical cutoff walls. The primary role of the bentonite in all of these barriers is to serve as a high swelling, low hydraulic conductivity, k, material ($k < 10^{-10}$ m/s) to minimize seepage of liquids (groundwater, leachates in landfills, pure-phase organic liquids, etc.) and migration of contaminants (e.g., heavy metals, radionuclides, organic compounds, etc.) through the barriers. A primary concern with bentonite-based barriers used for such containment applications is the potential for incompatibility between the bentonite and the liquid being contained, resulting in increase in k and concomitant decrease in the performance of the barrier. With this concern in mind, the purpose of this paper is to provide a review of the factors and potential impacts resulting from such incompatibility ($\Delta k > 0$) in several different types of bentonite based-barriers, including traditional GCLs that rely solely upon bentonite as the primary hydraulic resistant material, compacted soil-bentonites (CSBs), and soil-bentonite backfills (SBBs) used in vertical cutoff walls for control of contaminated groundwater. The results show that the potential for significant incompatibility in bentonite-based barriers (i.e., Δk ranging from one to three or more orders of magnitude) generally increases with decrease in the amount of bentonite in the barrier, increase in the initial void ratio of the barrier, and/or increase in the composition and concentration of dissolved chemical species (solutes) contained in the permeant liquid. Values of k back-calculated from the results of consolidation tests based on Terzaghi's theory typically are unconservatively low and, therefore, should not be relied upon for assessing the performance of bentonite-based barriers.

INTRODUCTION

Compatibility refers to the impact of a liquid on the engineering properties of a material, such as the stress-strain behavior, shear strength, compressibility, and hydraulic conductivity of a soil (e.g., clay) or a geosynthetic (polymer-based) material (e.g., geomembrane, geotextile). If exposure of the material to a liquid results in little or no change in an engineering property of interest, the liquid and material are deemed to be compatible. In contrast, if exposure to the liquid results in a significant change in an engineering property of interest, the material is considered to be incompatible with the liquid.

Incompatibility in engineering practice typically represents an issue only when the change in the material property results in a negative impact in performance. For example, exposures of clays to high concentrations of salt solutions have been shown to result in an increase in strength, decrease in compressibility, and increase in permeability (e.g., Mitchell et al. 1965, Mesri & Olson 1970, 1971a,b). If none of these changes in properties are deemed undesirable from the perspective of the use of the clay for engineering applications, then such incompatibility obviously would not be an issue.

However, within the context of clays used as engineered barriers for hydraulic containment applications, such as landfill liners and covers, surface impoundments (e.g., ponds and lakes, aeration lagoons, fly ash lagoons, and other surface impoundments), canals, storage tanks, and secondary containment of above-grade fuel storage tanks, significant increases in the permeability of the clays due to exposure to chemicals would be considered undesirable, whereas an increase in strength and decrease in compressibility generally would be considered desirable. Thus, the primary compatibility issue with respect to clays used as hydraulic containment barriers is with respect to the permeability or, more specifically, the hydraulic conductivity, k, of the clay.

In this regard, bentonite is a naturally occurring, high swelling clay that is commonly used wholly or partially in a variety of engineered barriers to control liquid flow (seepage) and contaminant transport (e.g., Gates et al. 2009). Such barriers include compacted liners comprised of bentonite amended natural soils (i.e., sands, silts, and clays), manufactured geosynthetic clay liners (GCLs) comprised of thin (~ 10 mm) sheets of sodium bentonite sandwiched between two porous geotextiles, highly compacted bentonite buffers for high-level nuclear waste disposal, and soil-bentonite vertical cutoff walls. Bentonite is composed primarily of the mineral montmorillonite, a member of the smectite group of clay minerals, which is characterized by large surface area (100-800 m^2/g), a net negative charge typically on the order of 80 to 150 $cmol_c/kg$ (or 80 to 150 meq/100 g), and exchangeable surface cations (Grim 1968, Mitchell & Soga 2005). These characteristics impart a strong affinity by bentonite for water, resulting in high swelling and low k in the presence of water (Eisenhour & Brown 2009, Gates et al. 2009).

In all of these applications, sodium bentonite (Na-bentonite) is preferred, meaning that sodium (Na^+) is the cation that dominates the exchange sites of the individual bentonite particles. The preference for Na-bentonite stems from desirable engineering properties, such as low hydraulic conductivity to water, k_w (typically < 10^{-10} m/s), and the existence of semipermeable membrane behavior, the latter giving rise to hyperfiltration, chemico-osmotic flow, and reduced diffusion (Shackelford 2013). Unfortunately, Na-bentonite is thermodynamically unstable in environments where multivalent cations (e.g., Ca^{2+}, Mg^{2+}) are present, including most naturally occurring pore waters as well as waste streams containing heavy metals (e.g., landfill leachates, acid mine drainage) or radionuclides. Under such conditions, multivalent cations exchange for the Na^+ initially present on the exchange sites of the bentonite particles, resulting in increase in permeability and potentially a negative impact (e.g., Gates et al. 2009). With these considerations in mind, the purpose of this paper is to provide a review of the factors and potential impacts resulting from chemical incompatibility in several different types of bentonite based-barriers, including traditional GCLs that rely solely upon bentonite as the primary hydraulic resistant material, compacted soil-bentonites (CSBs), and soil-bentonite backfills (SBBs) used in vertical cutoff walls for control of contaminated groundwater.

BACKGROUND

Hydraulic Conductivity

The ability of a porous medium to conduct seepage is controlled in accordance with Darcy's law, as follows (e.g., Olson & Daniel 1981, Daniel 1994, Shackelford 1994):

$$q = kiA \tag{1}$$

where q is the volumetric flow rate (L^3T^{-1}, L = units of length, T = units of time), i is the hydraulic gradient (LL^{-1} or dimensionless) or driving force for seepage, and A is the total (solids plus voids) cross-sectional area perpendicular to the direction of seepage (L^2). The proportionality constant, k, in Eq. 1 is a function of the properties of both the porous medium and the liquid in accordance with the following expression (e.g., Shackelford 1994):

$$k = K \cdot \left(\frac{\rho g}{\mu}\right) = K \cdot \left(\frac{g}{\nu}\right) \tag{2}$$

where K is the intrinsic, absolute, or specific permeability of the porous medium (L^2), ρ is the mass density of the liquid (ML^{-3}, where M = mass units), g is the acceleration due to gravity (LT^{-2}), μ is the absolute or dynamic viscosity of the liquid ($ML^{-1}T^{-1}$), and ν is the kinematic viscosity of the liquid (L^2T^{-1}).

Changes in Hydraulic Conductivity

In accordance with Eq. 2, changes in the hydraulic conductivity, Δk, can occur via three processes. For example, changes in k can result if the structure of the porous medium changes independent of the permeant liquid. In this case, changes in k will result only from changes in the intrinsic permeability of the porous medium, or

$$\Delta k = \Delta K \cdot \left(\frac{\rho g}{\mu}\right) \tag{3}$$

An example of this process is the change in k of a highly compressible soil, such as a silt or clay, to water resulting from consolidation of the soil (Lambe & Whitman 1969). Since the permeant liquid is the same regardless of the void ratio, e, the liquid properties remain constant such that the increase in k with increasing e is due entirely to an increase in K (i.e., Δe reflects ΔK).

A second process causing Δk involves a change in the permeant liquid in the absence of any change in intrinsic permeability. This process can be represented with respect to Eq. 2 as follows:

$$\Delta k = K \cdot \Delta\left(\frac{\rho g}{\mu}\right) \tag{4}$$

An example of this process involves measuring the k of a relatively incompressible, clean (no fines) coarse-grained soil (e.g., sand or gravel) with water, and then repeating the measurement on the same coarse-grained soil with a liquid other than water, such as oil. In this case, assuming no change in the structure of the network of pores due to the low compressibility of the soil, and lack of interaction between the chemical liquid and the solid particles of the soil, the magnitude of the ratio of the k of the porous medium with respect to the chemical, k_c, relative to that with respect to water, k_w, or k_c/k_w, will be reflected directly by the magnitude of the ratio of the properties of the chemical relative to those of the water as follows (e.g., Shackelford et al. 2000):

$$\frac{k_c}{k_w} = \frac{\left(\frac{\rho g}{\mu}\right)_c}{\left(\frac{\rho g}{\mu}\right)_w} = \left(\frac{\rho_c}{\rho_w}\right)\left(\frac{\mu_w}{\mu_c}\right) = R_l \tag{5}$$

Values of R_l are shown in Fig. 1 for several different liquids. For inorganic salt solutions (Fig. 1a), values of R_l decrease to below unity as concentration increases. For these solutions, both ρ_c and μ_c increase as the concentration increases. The change in μ_c (relative to μ_w) in the denominator is more significant than the change in ρ_c (relative to ρ_w), such that the overall value of R_l decreases. For example, μ_c for 5,000 mM NaCl solution is 1.68 times greater than μ_w, while ρ_c is only 1.16 times greater than ρ_w, resulting in an R_l of 0.69. However, for KCl, as the concentration increases, μ_c increases less significantly than ρ_c, such that R_l increases to values greater than 1.0.

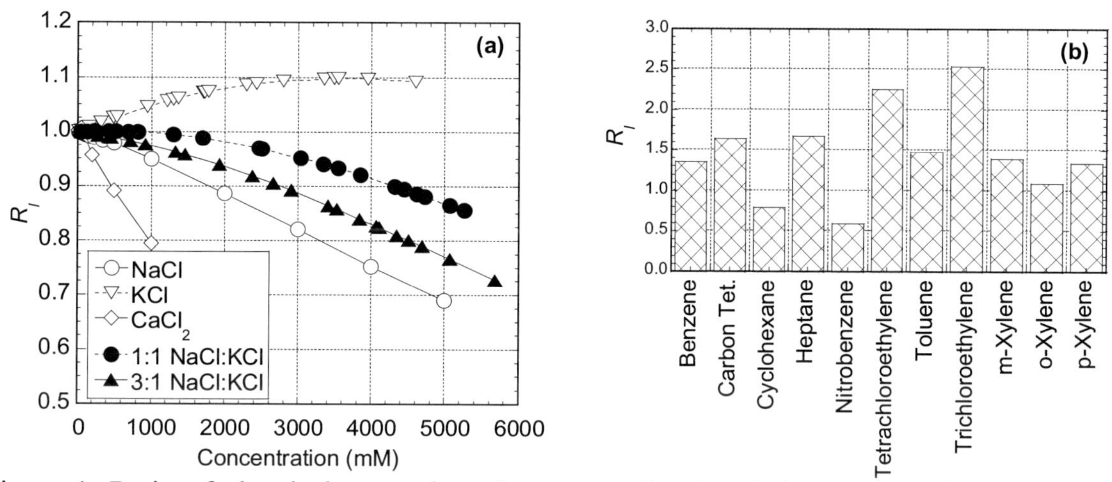

Figure 1. Ratio of chemical properties of permeant liquids relative to those for water, R_l, in accordance with Eq. 5: (a) R_l for simple inorganic salt solutions and mixtures of salt solutions (molar basis) as a function of concentration at 20 or 25°C; (b) R_l for non-aqueous phase liquids (NAPLs) at 20°C [data from Shackelford 1994, Hai-Lang & Shi-Jun 1996, Ozbek 2010, Malusis & McKeehan 2013].

For nonaqueous phase liquids, or NAPLs (Fig. 1b), the values of R_l are essentially independent of concentration since these organic compounds are essentially immiscible in water (although the aqueous solubility for most of these compounds still is sufficiently high to make the compounds problematic with respect to the potential for an adverse environmental impact). With the exceptions of cyclohexane ($R_l = 0.8$) and nitrobenzene ($R_l = 0.6$), all of the R_l values for the other NAPLs are greater than unity, with $1.1 \leq R_l \leq 2.5$. Thus, for these other NAPLs, k_c can be expected to be greater than k_w simply on the basis of differences in liquid properties, i.e., all other conditions being equal, whereas the opposite situation (i.e., $k_c < k_w$) can be expected in the case of cyclohexane and nitrobenzene.

The third process in which changes in k can occur is when the properties of the permeant liquid are such that the liquid interacts with the solid particles of the porous medium, causing a change in the intrinsic permeability of the porous medium. With respect to Eq. 2, this effect of interaction between the liquid and the porous medium can be represented as follows:

$$\Delta k = \Delta\left(K \cdot \frac{\rho g}{\mu} \right) \tag{6}$$

The effect on k represented by Eq. 6 is in fact the compatibility effect. However, because the changes in K are the result of using permeant liquids different than water, the effect of the change in the chemical properties of the permeant liquid on k must be distinguished from the effect of the change in K on k.

For example, consider the data shown in Fig. 2 based on the effect of ethanol concentration of the permeant liquid on the k of a needle-punched GCL as reported by Petrov et al. (1997b). As shown in Fig. 2a, the k of the GCL actually decreases for ethanol concentrations $\leq$ 50 %, whereas the k progressively increases as the ethanol concentration increases above 50 %. However, as shown in Fig. 2b, there is virtually no change in intrinsic permeability for ethanol concentrations less than 50 %, but significant increases in intrinsic permeability occurred with increasing ethanol concentrations higher than 50 %. Thus, for ethanol concentrations less than 50 %, the changes in k are attributed to changes in the properties of the permeant liquid, principally the viscosity, in accordance with Eq. 4, whereas the significant increases in k at the higher ethanol concentrations are attributable to an incompatibility effect (i.e., ΔK) resulting from interactions between the ethanol used as the permeant liquid and the bentonite comprising the GCL.

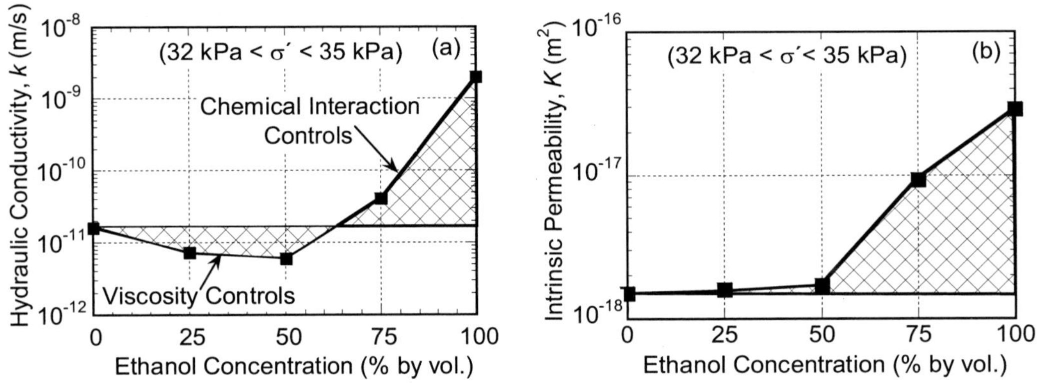

Figure 2. Hydraulic conductivity (a) and intrinsic permeability (b) of a GCL permeated with ethanol solutions (Note: σ′ = effective stress) [modified after Shackelford et al. 2000].

Although determination of the intrinsic permeability is required to discern mechanistically if changes in k are due to chemical attack of the soil or simply to differences in liquid properties, determining the density and viscosity of permeant liquids generally is only feasible in the case where such permeant liquids have relatively simple chemistries, such as the simple salt solutions and pure solvents illustrated in Fig. 1. For many practical applications, the chemistries of the liquids to be contained can be complex, such as landfill leachates. In these cases, the measured value of k_c based on permeation with the actual liquid should be sufficient to estimate the liquid flux through the barrier, i.e., assuming the conditions upon which k_c is measured are the same as those in the field. In this regard, several additional considerations should be taken into account in interpreting measured values of k of bentonites used in hydraulic containment applications.

Additional Factors Affecting Hydraulic Conductivity and Compatibility

Several factors in addition to pore-water chemistry can affect measurement and/or interpretation of hydraulic conductivity of bentonite permeated with chemical solutions (e.g., see Shackelford 1994, Shackelford et al. 2000, Mitchell et al. 2007). Among these factors, probably the five most significant factors are the quality of the bentonite, the effect of prehydration, the type of permeameter (rigid-wall vs. flexible-wall), the use of high hydraulic gradients to reduce testing duration and the importance of achieving chemical equilibrium between the effluent and influent prior to terminating a hydraulic conductivity test.

Bentonite Quality. The variability in the quality of bentonites can affect the hydraulic performance of bentonites used for hydraulic containment applications (e.g., Shackelford et al. 2000, Ashmawy et al. 2002, Lee and Shackelford 2005a, Shackelford 2008, Guyonnet et al. 2009). In general, the quality of a bentonite is affected by several factors, including the mineralogical composition of the bentonite (e.g., montmorillonite or smectite content), the surface area of the bentonite (e.g., particle-size distribution of the bentonite), the surface charge deficiency of the bentonite (e.g., amount and type of isomorphic substitution), and the composition of the exchange complex (e.g., amount of exchangeable Na^+ relative to exchangeable Ca^{2+}) (e.g., Shackelford et al. 2000, Ashmawy et al. 2002, Lee and Shackelford 2005a, Guyonnet et al. 2009, Shackelford 2008). As illustrated in Fig. 3, the quality of the bentonite generally increases with increases in montmorillonite content, surface area (i.e., decrease in particle size), surface charge deficiency, and/or percentage of sodium (Na^+) on the exchange complex. The effects of these factors generally is indicated macroscopically in terms of an increase in cation exchange capacity (*CEC*), increases in plasticity and swell capacity in the presence of water, and a decrease in hydraulic conductivity based on permeation with water (k_w).

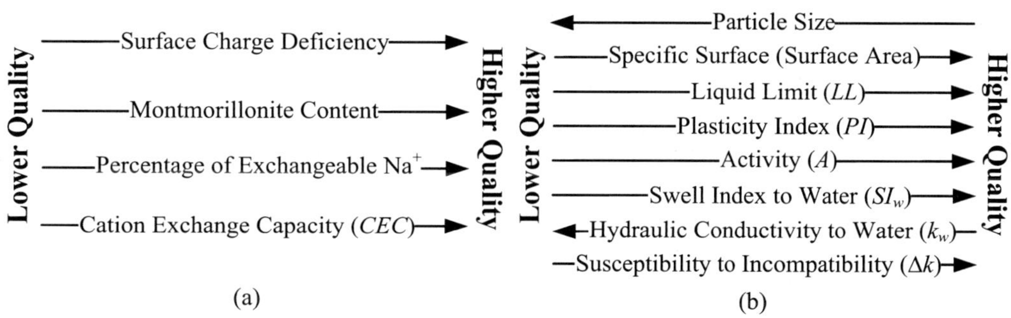

Figure 3. Relative quality of bentonite as reflected by relative magnitudes in (a) mineralogical and chemical properties and (b) physical properties of bentonites. Arrows indicate directions of increasing magnitudes of properties [redrawn after Shackelford 2008].

For example, Haug & Bolt-Leppin (1994) evaluated the effect of polymer addition on the hydraulic conductivity of what was referred to as a "marginal quality bentonite" used in sand-bentonite mixtures. They measured the physico-chemical properties of two bentonites referred to as "high quality bentonite" and "marginal quality bentonite", and distinguished the "high quality bentonite" as the bentonite with a higher specific surface (668 vs. 490 m^2/g), a higher *CEC* (90 vs. 69 meq/100 g), and a greater liquid limit (590 vs. 425), although the percentages of exchangeable Na^+ and the plastic limits for the two qualities of bentonite were similar.

However, as noted by Lee and Shackelford (2005a), even though higher quality bentonites typically result in lower k_w, such higher quality bentonites also are typically more susceptible to chemical attack upon permeation with non-standard liquids (i.e., liquids other than water), resulting in greater increases in k (i.e., $\Delta k > 0$) for higher quality bentonites relative to those that occur with lower quality bentonites. For example, Lee & Shackelford (2005a) permeated two GCLs containing different quality bentonites with calcium chloride ($CaCl_2$) solutions. Even though k_w for the GCL with the higher quality bentonite (HQB) was about 3.5 times lower than that for the GCL with the lower quality bentonite (LQB), i.e., 2.4×10^{-11} m/s vs. 7.0×10^{-12} m/s, the GCL with the HQB was 40 times more permeable to a 500 mM $CaCl_2$ solution than the GCL with the LQB (i.e., 6.0×10^{-7} m/s vs. 1.5×10^{-8} m/s). Relative to permeation with water, the k of the GCL with the HQB increased by almost 5 orders of magnitude (i.e., log 85,700 = 4.93) upon permeation with the 500 mM $CaCl_2$ solution, whereas the k of the GCL with the LQB increased by only about 3 orders of magnitude (i.e., log 625 = 2.80) upon permeation with the 500 mM $CaCl_2$ solution. Thus, higher quality bentonites generally are more reactive, i.e., more susceptible to chemical attack and incompatibility, than lower quality bentonites.

Prehydration. Several studies have reported that prehydration of bentonite barriers with water prior to permeation with actual permeant liquids can have a profound effect on k (Shackelford 1994, Lee & Shackelford 2005b). For example, Shackelford (1994) reported the results of two tests involving permeation of identical specimens of compacted sand-bentonite mixtures containing 16 % of the same sodium bentonite by dry weight (w/w) with a calcium saturated tailings solution. The first test involved permeation with water resulting in a value k_w of 3.4 x 10^{-10} m/s, followed by permeation with the saturated calcium solution on the same specimen resulting in a final steady-state value for k_c of 1.7 x 10^{-9} m/s. Thus, the hydraulic conductivity of the specimen increased by a factor of five (i.e., $k_c/k_w \cong 5$) upon permeation with the calcium solution relative to initial permeation with water. The second test involved permeation of an identical, but separate specimen directly with the calcium solution (i.e., without prehydration), resulting in a final steady-state value for k_c of 2.3 x 10^{-7} m/s. Thus, although both tests indicated Δk, the value of k_c for the non-prehydrated specimen was approximately 135 times greater than that for the prehydrated specimen (i.e., 2.3 x 10^{-7} m/s $\div$ 1.7 x 10^{-9} m/s $\cong$ 135).

In general, the degree of prehydration may be complete or partial, depending on the degree of saturation upon prehydration. Prehydration typically has been imposed in laboratory testing by permeating the specimens with water, although prehydration by spraying or soaking of specimens with water also has been undertaken (e.g., Lee & Shackelford 2005b). In the field, prehydration may occur naturally, such as by precipitation or absorption of water from adjacent soils, or may be imposed either by spraying water or, in the case of GCLs, by installing commercially available prehydrated GCLs (e.g., see Lee & Shackelford 2005b).

Type of Permeameter. The effect of type of permeameter on the measurement of k of clay barriers has been well established (e.g., Daniel et al. 1985, Daniel 1994, Shackelford 1994, Petrov et al. 1997b, Shackelford et al. 2000). In general, there are two categories of permeameters, viz. flexible-wall or triaxial-type permeameters and rigid- or fixed-wall permeameters, and both types of permeameters have relative advantages and disadvantages (e.g., Daniel et al 1985). In terms of compatibility testing, the primary advantage of flexible-wall permeameters is the ability of the flexible membrane to maintain contact with the testing specimen despite any volume change in the specimen that may occur upon permeation with non-standard liquids (i.e., liquids other than water). In the case of rigid-wall specimens, permeation with non-standard liquids may result in significant shrinkage of the specimen, such that the specimen contracts away from the walls of the cell, resulting in side-wall leakage or liquid flow around, instead of through, the specimen, and measurement of unrealistically high k_c values (e.g., see Daniel et al. 1985). As a result, rigid-wall cells have been criticized on the basis that the measured k values may represent system, rather than specimen, values in that the true k of the specimen will not be reflected by the measured k value when side-wall leakage is present. However, in such cases, one also may argue that such measured values may be indicative of the potential for shrinkage in the field and, therefore, may provide useful information of a potential compatibility issue. Also, rigid-wall cells have been adapted with double outflow rings to account for side-wall leakage and/or vertical loading pistons to minimize or prevent side-wall leakage by maintaining specimen contact with the cell walls during permeation (Daniel 1994, Shackelford 1994).

Hydraulic Gradient. The impact resulting from the use of high hydraulic gradient to shorten the test durations of hydraulic conductivity tests also has been well established (e.g., Daniel et al. 1985, Daniel 1994, Shackelford 1994, Shackelford et al. 2000). In general, the use of hydraulic gradients much higher than those expected in the field can result in a measured hydraulic conductivity that is too low due to consolidation by high seepage forces. However, Shackelford et al. (2000) showed that the primary consideration is the effective stresses imposed on the specimen due to the application of a hydraulic gradient, i.e., as opposed to the actual magnitude of the hydraulic gradient, and that the use of elevated hydraulic gradients on the order of 200 to 500 may not appreciably affect the hydraulic conductivity of thin, bentonite-based testing specimens such as GCLs. Regardless, consideration should be given to the potential impact of high hydraulic gradients on the magnitude of any measured value of k.

The situation involving two-phase flow, or the migration of mixtures containing two separate liquid phases in soil, requires special consideration when the interaction between the two liquid phases involves permeating a barrier soil at an initially high degree of water saturation with nonpolar, hydrophobic organic liquids such as the aforementioned NAPLs indicated in Fig. 1b (e.g., Shackelford 1994, Mitchell et al. 2007). In this case, the existence of surface and interfacial tension effects dictates that the non-polar liquid (a non-wetting fluid) can displace the water (a wetting fluid) in the pore space of the soil only after a minimum pressure (known as the entry pressure) for the non-wetting fluid is reached. Thus, although some studies have shown that permeation of clay soils by hydrophobic organic liquids has resulted in significant increases in *k*, the hydraulic gradients required to force these liquids through the soils typically have been quite large (> 100), such that the relevance of these results to field situations where much smaller hydraulic gradients are applicable is questionable (Shackelford 1994, Mitchell et al. 2007).

Chemical Equilibrium. In the case of permeation with non-standard liquids, the importance of performing hydraulic conductivity tests with sufficient durations to achieve chemical equilibrium between the effluent and influent has been demonstrated for both prehydrated and non-prehydrated bentonite based specimens (e.g., Shackelford et al. 2000, Mitchell et al. 2007). Unconservative (low) values of hydraulic conductivity may be measured if the test is terminated before complete chemical equilibrium has been established. Such unconservative *k* values may not reflect the true long-term performance of the barrier in the field (e.g., Mitchell et al. 2007). For example, Shackelford (2005) performed an analysis showing that substantially lower hydraulic gradients in the field can result in substantially longer times required to achieve chemical equilibrium in the field relative to those in the laboratory, such that the overall long-term hydraulic performance of the GCL in the field may not be realized for tens of years.

Conceptual Interpretations of Incompatibility

Two theories have been promulgated to provide explanations for the behavior of clays in terms of incompatibility in hydraulic conductivity upon permeation with chemical solutions, viz., the diffuse-double layer (DDL) theory and the theory related to swell. The origin of both of these theories is in the field of clay colloid chemistry (e.g., van Olphen 1977), as the theories were developed primarily to explain the behavior of clay particles suspended in liquids with different chemical properties, e.g., compositions and concentrations of solutes.

The DDL theory pertains to the influence of the thickness of the DDL or adsorbed layer of ions, predominantly cations in the case of clays with negative surface charges such as bentonites, and associated water molecules on the interaction between adjacent clay particles, and the associated effect of these interactions on the micro-fabric and *k* of the clay. Based on the classical DDL theory, an increase in the concentration and/or valence (charge) of the ions in the pore water and/or a decrease in the dielectric constant of the pore water results in individual clay-particle orientations within the overall clay mass, whereby the positive edges (+) of individual clay particles are associated with the negative faces (-) of adjacent clay particles (e.g., Mitchell & Soga 2005). This electrostatic attraction between adjacent particles results in a relatively open or flocculated micro-fabric for the clay and a corresponding increase in *k*, whereas the opposite trends in concentration, valence, and dielectric constant result in the formation of a relatively dispersed micro-fabric and a corresponding decrease in *k*. The DDL theory also includes an effect of temperature, but this effect generally is minor due to compensation by an opposite effect of temperature on the dielectric constant (Mitchell & Soga 2005).

Although the DDL theory has been used extensively to explain the results of compatibility testing (e.g., Shackelford 1994), the theory presumes the ability of individual particles of clay to rearrange themselves within a complex arrangement of particles in response to changes in the chemistry of the pore water. In the case of load bearing barriers, such as GCLs and CSBs, this particle rearrangement ability is suspect, but may be more accurate in the case of SBBs, where the high slump of the backfill suggests a consistency closer to that of a clay suspension. Nonetheless, the DDL theory generally has provided reasonable, albeit qualitative, interpretations of the results of compatibility studies (Shackelford 1994).

In terms of swell, the swelling of bentonites generally is known to be inversely related to *k*, i.e., the greater the swell, the lower the *k* (Shackelford et al. 2000, Ashmawy et al. 2002, Jo et al. 2004, Kolstad et al. 2004). Thus, factors that promote swelling of bentonites also promote low *k*, and vice versa. In general, appreciable swelling and lower *k* occur when the interlayer cations in montmorillonite are monovalent, whereas very little swelling and higher *k* occur when the cations are multivalent (Kolstad et al. 2004).

RESULTS AND DISCUSSION

Geosynthetic Clay Liners

Results from the literature regarding the relationship between *k* and *e* for bentonites and GCLs comprised of bentonite are presented in Fig. 4. Regardless of differences in the types of bentonites, test specimen preparations, or permeant liquids, all of the results in Fig. 4 indicate increasing *k* with increasing *e*, as expected.

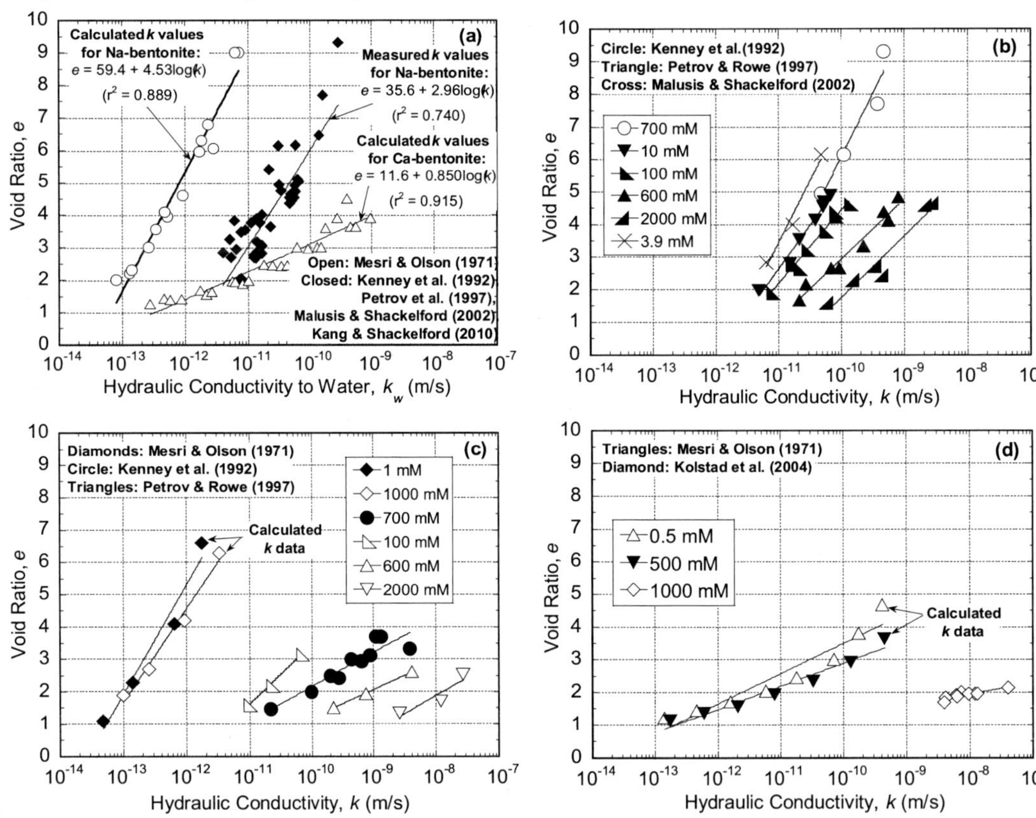

Figure 4. Effects of different testing conditions on *k* versus *e* for GCLs and other bentonites: (a) calculated and measured *k* values to water; (b) prehydrated and permeated with NaCl solutions; (c) & (d) calculated and measured *k* to NaCl and $CaCl_2$ solutions, respectively, for non-prehydrated specimens.

The data in Fig. 4a represent test conditions where water was used both as the liquid in preparing the test specimens and as the permeant liquid. The closed symbols represent measured values reported in the literature for *k* of Na-bentonites permeated with water. The open symbols represent values of *k* that were calculated by Mesri & Olson (1971) from consolidation test results for Na- and Ca-bentonites using Terzaghi's one-dimensional theory of consolidation. As expected on the basis of DDL theory, for a specific void ratio (*e*) the calculated *k* values for the Ca-bentonites were higher than the *k* values for the Na-bentonites by a factor ranging from 120 to 1,860, with the difference increasing as *e* increases from 2.0 to 4.0. Also, for the Na-bentonite, the calculated k values were lower than the measured k values reported in the literature by a factor ranging from 22 to 98. Mesri & Olson (1971) suggested that the calculated k values for Na- and Ca-bentonites were likely lower than measured values by 5 – 20%. This difference was attributed to Terzaghi's theory assuming that permeability was the sole cause of delay in compression, whereas the actual time dependent resistance to compression has been attributed by others to both permeability and a structural resistance component of the clay (Mesri & Olson 1971).

The data in Fig. 4b are from test specimens of bentonites that were prehydrated prior to permeating with an NaCl solution. With the exception of the data from Kenney et al. (1992), the relative results from the literature indicate an increase in k with an increase in NaCl concentration. The lower values of k observed by Kenney et al. (1992) for 700 mM NaCl are likely attributable, in part, to higher bentonite quality relative to the results from other studies shown in Fig. 4b. For example, the *CEC* of the bentonite used by Kenney et al. (1992) was 100 $cmol_c/kg$ (meq/100 g), whereas the *CEC* of the bentonite used by Petrov & Rowe (1997), who reported higher k values, was only 86 $cmol_c/kg$.

The relationships between k and e for non-prehydrated bentonites are shown in Figs. 4c and 4d for NaCl and $CaCl_2$, respectively. Note that the specimens in the Mesri & Olson (1971) study were not permeated with either salt solution, but rather were prepared with the salt solutions as the pore water in the specimens and then consolidated, and that the k values were calculated on the basis of Terzaghi's consolidation theory as opposed to being measured. Similar to the results shown in Fig. 4a, the resulting calculated values of k based on NaCl and $CaCl_2$ are significantly lower than measured k values reported in Figs. 4c and 4d, respectively. Also, similar to the prehydrated specimens, k of the nonprehydrated specimens generally increases with increasing salt concentration.

As noted by Shackelford et al. (2000), the effect of chemical interaction on k can be isolated from the effect of e on k by plotting the values of k as a function of the salt concentration for a given value of e. Examples of the results of such an analysis are shown in Fig. 5, where values of the ratio of k_c relative to k_w, or k_c/k_w, based on the data from Petrov & Rowe (1997) for both prehydrated GCLs (Fig. 5a) and non-prehydrated GCLs (Fig. 5b) are plotted as a function of the NaCl concentration for different void ratios. Note that the data from Petrov & Rowe (1997) were reported with respect to a bulk void ratio, e_b, which included the geotextile portion of the GCL.

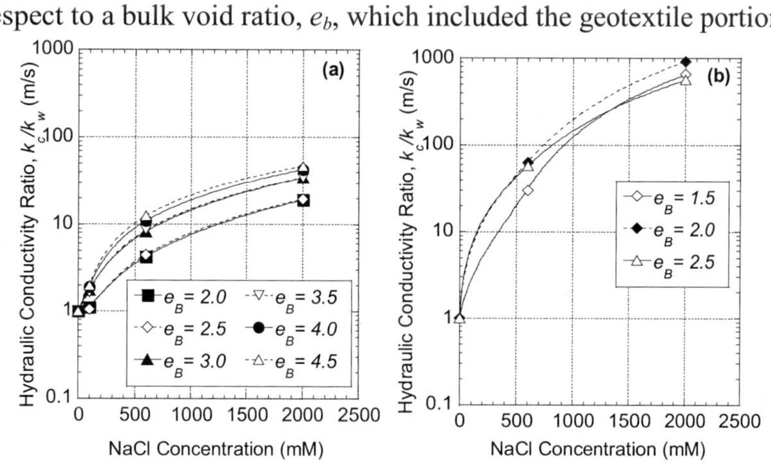

Figure 5. Effect of salt (NaCl) concentration on the ratio of the hydraulic conductivity of GCL specimens to the chemical, k_c, relative to that based on permeation with water, k_w: (a) prehydrated specimens; non-prehydrated specimens (data from Petrov & Rowe 1997).

At least three observations are apparent from the results shown in Fig. 5. First, the effect of incompatibility (i.e., $k_c/k_w > 1$) increases with increasing salt concentration, and can be substantial even for solutions of simple salts, such as NaCl (5a,b), i.e., when the concentration of the salts is sufficiently high. Both of these observations are consistent with expected behavior based on the

DDL and swelling behavior. In addition, these increases in k_c relative to k_w cannot be attributed to the effects of liquid properties, since the R_l values for both NaCl and $CaCl_2$ decrease with increasing salt concentrations (see Fig. 1a). Second, all other conditions being the same, incompatibility generally is greater the higher the void ratio, which is consistent with studies indicating that an increase in effective stress and, therefore, decrease in void ratio, can be used to offset incompatibility (e.g., Fernandez & Quigley 1991). Third, incompatibility effect tends to be more significant in the case of non-prehydrated specimens (Fig. 5b) relative to prehydrated specimens (Fig. 5a), which is consistent with the aforementioned first exposure or prehydration effect.

The potential influence of prehydration or first exposure is illustrated further in Fig. 6, where the value of the first exposure ratio, or *FER*, defined by Stern & Shackelford (1998) as the measured value of k_c for a non-prehydrated specimen relative to that for an identical specimen that is prehydrated, is plotted as a function of $CaCl_2$ concentration of the permeant liquid based on the results of several studies. Note that all of the data in Fig. 6 pertain to solutions of $CaCl_2$ except the aforementioned data from Petrov & Rowe (1997), which pertain to NaCl solutions. For relatively dilute salt concentrations (< 20 mM), there is little if any effect, whereas the prehydration or first exposure effect can become significant (i.e., $1 < FER < 10$) at higher salt concentrations ranging from about 100 to 300 mM, and substantial (i.e., $10 < FER < 100$) for salt concentrations greater than about 600 mM. For intermediate concentrations between about 20 mM and 100 mM, the results apparently can be mixed.

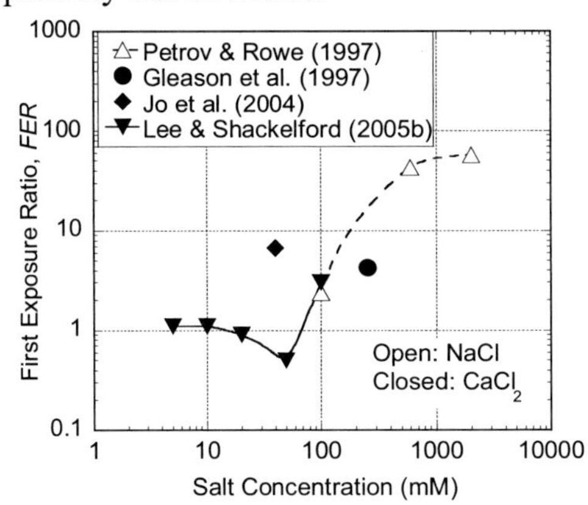

Figure 6. Effect of first exposure or prehydration on the hydraulic conductivity of GCLs permeated with different salt solutions (modified after Lee & Shackelford 2005b).

Compacted Soil-Bentonite Barriers

Values of k_w for compacted soil-bentonite (CSB) mixtures are shown in Fig. 7. All of the specimens for the data shown in Fig. 7a were compacted mixtures of a basalt base soil (LL = 52, 12 % clay (< 2 μm) content) amended with 17 % (dry wt. basis) of four different bentonites containing different percentages of montmorillonite (MMT) and, therefore, different qualities of bentonite (e.g., LL = 311, 391, 511, and 616 for bentonites A, B, C, D, respectively). Also, the values of k_w in Fig. 7a were calculated from the results of consolidation tests based on Terzaghi's theory and, therefore, are likely somewhat lower than what would be expected for the true measured values of k_w (Mesri & Olson 1971). The trends in e versus log k_w are similar to those for the GCLs previously shown (Fig. 4a). Also, for a given value of e, the value of k_w tends to decrease the higher the quality of the bentonite (i.e., the higher the MMT content), which is consistent with the description in Fig. 3.

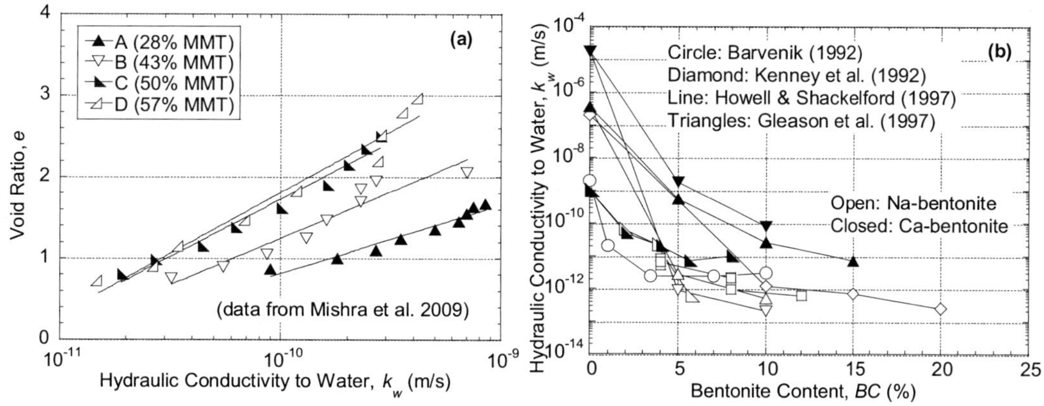

Figure 7. Hydraulic conductivity to water for compacted soil-bentonite specimens as a function of (a) void ratio and (b) bentonite content (data from Mishra et al. (2009) calculated from consolidation test results).

The effect of bentonite content, *BC*, for both Na-bentonite and Ca-bentonite on k_w for a variety of CSB specimens is shown in Fig. 7b. In all cases, a significant decrease in k_w occurs with increasing *BC* such that extremely low k_w values ($\leq 10^{-10}$ m/s) were achieved for $BC \geq 10$ %. Also, in general, the amount of bentonite required to achieve a given k_w value tends to be lower when Na-bentonite was used in the CSB relative to Ca-bentonite, as expected. Both of these trends are attributable to the high swelling behavior of bentonites, in general, and Na-bentonite, in particular. Note that, for the data from Gleason et al. (1997), measureable differences in *k* values occurred between specimens with the same bentonite content and type. These differences were attributed, in part, to the variations in the sand used as the base soil in the compacted CSBs. For example, the highest values of k_w occurred for mixtures with uniform medium Ottawa sand, whereas the lowest values of k_w values occurred for a silty sand, with the values of k_w for a widely-graded sand being intermediate between these two extremes.

The effects of salt concentration on the hydraulic conductivity of CSB specimens are shown in Fig. 8. The data in Fig. 8a from Gleason et al. (1997) represent specimens of compacted sand-bentonite that were prepared with 5 % and 15 % bentonite, whereas the data from Stern & Shackelford (1998) pertain to compacted sand-bentonite mixtures containing 10, 15, or 20 % Na-bentonite. The specimens from Gleason et al. (1997) either were permeated directly with $CaCl_2$ solutions (nonprehydrated), or were prehydrated by permeation with water and then permeated with $CaCl_2$ solutions, whereas those from Stern & Shackelford (1998) were permeated directly with a solution of 500 mM $CaCl_2$. As expected, the Gleason et al. (1997) data demonstrate an increase in *k* of the CSB by two orders of magnitude as the $CaCl_2$ concentration increased from 0 to 250 mM. In addition, three times more Ca-bentonite than Na-bentonite was required (15 % vs. 5 %) to obtain approximately the same *k*. Interestingly, Gleason et al. (1997) did not measure a significant difference in the *k* values of the prehydrated and nonprehydrated specimens for $CaCl_2$ solutions up to concentrations of 250 mM.

In terms of the data from Stern & Shackelford (1998), doubling the bentonite content from 10 % to 20 % resulted in an almost two order-of-magnitude decrease in *k* to the same 500 mM $CaCl_2$ solution. These results illustrate that, for a given chemical loading, the negative impact on *k* can be overcome to some extent by increasing bentonite content, although bentonite contents for CSB barriers in excess of 10 % likely are cost prohibitive in most practical applications.

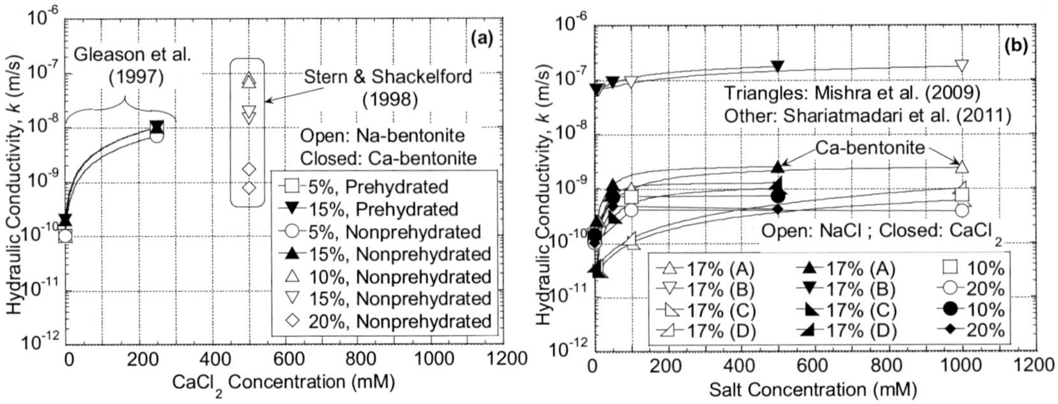

Figure 8. Hydraulic conductivity to salt solutions for compacted soil-bentonite specimens as a function of bentonite content: (a) measured values; (b) calculated values.

In contrast to the results shown in Fig. 8a, which were based on measured values of k on specimens compacted with water, all of the values of k in Fig. 8b were calculated on the basis of the results of consolidation testing in accordance with Terzaghi's theory, and all of the specimens were prepared with the salt solutions as opposed to water and, therefore, represent non-prehydrated specimens. The sand-bentonite specimens for the Mishra et al. (2009) data set were prepared with four different commercial bentonites, denoted as A through D in the legend (the same bentonites shown in Figure 7a). Except for the data from Mishra et al. (2009) for the B bentonite, the resulting calculated values of k (Fig. 8b) generally are lower than measured values of k for similar bentonite contents (Fig. 8a), which is consistent with the previously noted relative differences for GCLs in Figure 4. However, this relative difference persists despite the preparation of specimens with salt solutions that would be expected to result in a relatively flocculated micro-fabric of the bentonite and higher k. In addition, the trends with respect to bentonite quality are not entirely consistent with those suggested in Fig. 3, since bentonites C and D contained higher montmorillonite contents and exhibited higher LL values relative to those for bentonites A and B, but bentonites C and D exhibited lower k values to salt solutions than bentonites C and D. This inconsistency in behavior suggests that caution should be exercised in using values of k back-calculated from consolidation test results on specimens prepared with salt solutions, as such values could be unconservatively low.

The data in Fig. 8b from Shariatmadari et al. (2011) were based on specimens comprised of Varamin clay soil (93 % fines, 7 % sand) mixed with 10-20 % bentonite. The significant portion of fines in the base soil likely accounts for the low k values even for specimens with as low as 10 % bentonite content. Consistent with the results from Stern & Shackelford (1998) shown in Fig. 8a, the mixtures in the Shariatmadari et al. (2011) study with higher bentonite contents (20 %) resulted in lower values of k for a given concentration of NaCl or $CaCl_2$ relative to the mixtures with 10 % bentonite.

Soil-Bentonite Backfills

Values of k_w for various soil-bentonite backfills (SBBs) are shown as a function of void ratio in Fig. 9. The values of k_w shown in Fig. 9 were measured either in oedometer cells at the end of each loading stage of a consolidation test (Yeo et al. 2005) or in flexible-wall permeameters (Malusis & McKeehan 2013).The results shown in Fig. 9a correspond to values of k_w for sand-bentonite backfills as a function of the bentonite content in the backfills prepared at slumps ranging from 100 to 125 mm. For these results, the bentonite contents include both the dry bentonite added to the sand and the bentonite imparted to the backfill from achieving the desired slump using bentonite-water slurry. The results in Fig. 9b, extracted from the study by Yeo et al. (2005), pertain to backfills comprised from mixing sand in different proportions with a natural clay (Nelson Farm Clay) containing 89 % fines to achieve backfills without any dry bentonite amendment. However, the fines contents shown in Fig. 9b do not include any bentonite imparted to the backfills from the use of the bentonite slurry required to achieve the desired slump of 100 mm, which ranged from 1.2 to 2.1 % (Yeo et al. 2005).

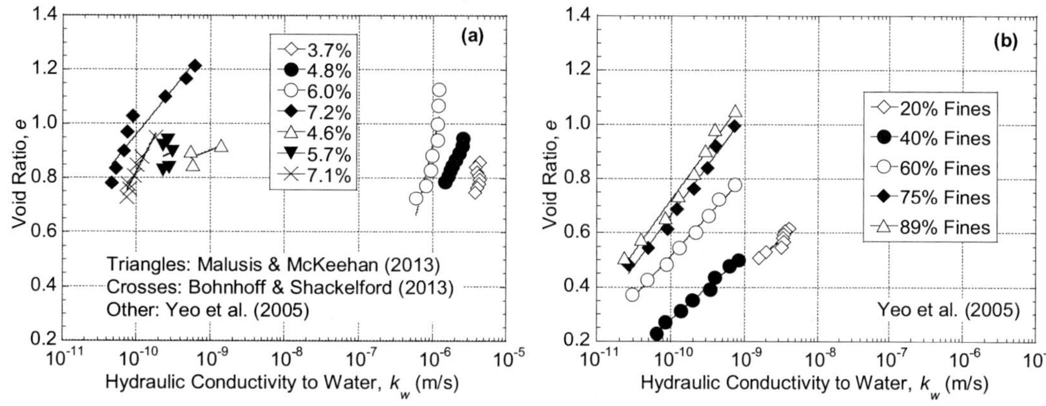

Figure 9. Hydraulic conductivity to water for soil-bentonite backfills as a function of void ratio: (a) effect of bentonite content for sand-bentonite backfills prepared at slumps ranging from 100 mm to 125 mm; (b) effect of backfill fines contents for various sand-natural clay backfills prepared at a slump of 100 mm.

Two observations are apparent from the results shown in Fig. 9. First, most of the data in Fig. 9 illustrate the expected correlation between e and k_w, although the slopes in e-log k_w data are much steeper and the ranges in e are much smaller than those shown for either GCLs (Fig. 4) or CSBs (Fig. 7). The differences in behavior are attributable to relatively incompressible behavior of SBBs relative to either GCLs or CSB resulting primarily from the dominating influence of the sand in the SBBs. Second, the data fall into two distinct groups, those data with relatively low k_w values less than about 10^{-9} m/s versus those data with relatively high k_w values greater than about 5 x 10^{-7} m/s. This approximately 2.5 orders-of-magnitude difference in k_w values points to the critical importance of the roles of the type and amount of bentonite in the SBBs, the type (e.g., (fine, medium, coarse) and gradation of the sand (e.g., well graded vs. poorly graded), and the amount of fines in the sand, if any, prior to bentonite amendment in achieving a suitably low k_w. As an example of the importance of these factors, the data from Malusis & McKeehan (2013) in Fig. 9a exhibit lower k_w values than would be expected on the basis of bentonite content alone compared with the other data. These lower k_w values from Malusis & McKeehan (2013) were attributed to the use of a finer sand relative to the other studies, and the likely use of a higher quality (swelling) bentonite based on Atterberg limits. Also, as clearly indicated in Fig. 9b, base backfill soils with higher fines contents generally results in lower k_w values, and are able to achieve similarly low k_w values relative to those for the sand-bentonite backfills (Fig. 9a) despite much lower bentonite contents (i.e., 1.2 - 2.1 % for Fig. 9b data vs. 3.7 - 7.2 % for Fig. 9a data).

The effect of compatibility in terms of the values of k based on permeation of both prehydrated and non-prehydrated sand-bentonite backfills to $CaCl_2$ solutions in flexible-wall permeameters is shown in Fig. 10. The data in Figure 10 indicate an initial increase in k as $CaCl_2$ concentration increases from 0 to 50 mM, followed by values of k that are relatively insensitive to $CaCl_2$ concentrations greater than 50 mM. However, the increase in k based on permeation with 50 mM $CaCl_2$ in the case of the backfill tested by Bohnhoff & Shackelford (2014) is substantially greater than that for the backfill tested by Malusis & McKeehan (2013), which is only marginal. This difference in results occurred even though the total bentonite content of the sand-bentonite mixtures used by Bohnhoff & Shackelford (2014) was greater, and the quality of the bentonite apparently was higher in the case of Malusis & McKeehan (2013). For example, the *LL* and *PI* of the bentonite from Malusis & McKeehan (2013) were 583 and 530, respectively, whereas the *LL* and *PI* of the bentonite in Bohnhoff & Shackelford (2014) were 420 and 381, respectively. Thus, differences in bentonite quality relative to the expected trends illustrated in Fig. 3 cannot explain the differences in the measured *k* values. However, two other factors favor the lower measured *k* values reported by Malusis & McKeehan (2013), viz., the use of a finer sand in the case of Malusis & McKeehan (2013), and the application of somewhat higher average effective stress during the testing (i.e., 34.5 kPa vs. 20 kPa).

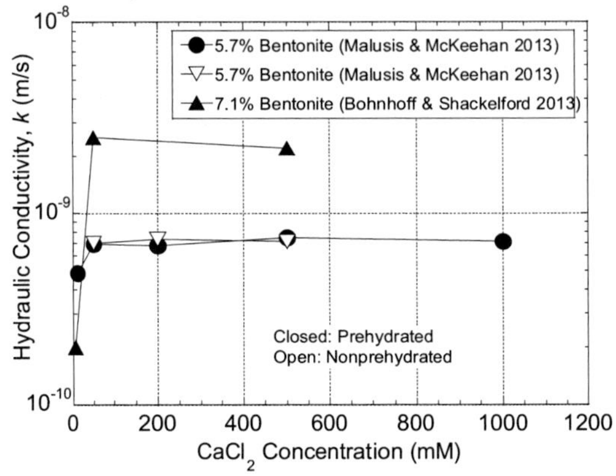

Figure 10. Measured hydraulic conductivity for prehydrated and non-prehydrated sand-bentonite backfill specimens permeated with $CaCl_2$ solutions.

Two other observations can be made from the results shown in Fig. 10. First, based on the data from Malusis & McKeehan (2013), there was virtually no effect resulting from prehydration, even for the results based on permeation with relatively high $CaCl_2$ concentrations (i.e., $\geq$ 200 mM). This result contradicts that previously shown for GCLs in Fig. 6. Malusis & McKeehan (2013) attributed this lack of a prehydration effect to the likelihood that specimens of SBB, unlike those of GCLs, likely are inherently prehydrated as a result of mixing the backfill with bentonite slurry to achieve a desired slump, such that even the non-prehydrated specimens are essentially prehydrated upon permeation with chemical solutions. Second, the sharp increase in *k* for the backfill from Bohnhoff & Shackelford (2014) at the relatively low $CaCl_2$ concentration of 50 mM also seemingly contradicts results for GCLs where such low concentrations of $CaCl_2$ typically have resulted in little, if any, effect on *k* (e.g., Lee & Shackelford 2005a,b). However, this difference between the compatibility behaviors of GCLs for SBBs again can be attributed, in part, to the much lower bentonite contents (typically $\leq$ 7 %) in SBBs relative to GCLs, containing essentially 100 % bentonite, and to the relatively loose structure (high *e*) associated with SBBs.

CONCLUSIONS

Five potentially significant factors affecting the measurement and interpretation of *k* and Δk of bentonite-based barriers are quality of the bentonite, effect of prehydration, type of permeameter (rigid-wall vs. flexible-wall), hydraulic gradients used during testing, and achievement of chemical equilibrium between the effluent and influent prior to terminating a hydraulic conductivity test. Higher quality bentonites typically result in lower *k* to water, k_w, but also are more susceptible to chemical attack and incompatibility (i.e., $\Delta k > 0$).

For GCLs or specimens containing 100 % bentonite at a specified void ratio, *e*, values of *k* for Ca-bentonites were higher than those for Na-bentonites, as expected from theoretical considerations. Values of *k* calculated from the results of consolidation tests in accordance with Terzaghi's theory generally were lower than measured *k* values. Also, whether *k* is calculated or measured, *k* increased as NaCl or $CaCl_2$ concentrations increased, for both prehydrated and nonprehydrated specimens. Therefore, the effect of incompatibility generally increases with increasing salt concentration, and can be substantial (1 to 3 orders of magnitude) even for simple salt solutions (e.g., NaCl or KCl), if the salt concentration is sufficiently high. All other factors being equal, incompatibility generally is greater the higher *e*, which is consistent with studies indicating that an increase in effective stress, σ', can be used to offset incompatibility. Finally, incompatibility effects tend to be more significant for non-prehydrated specimens relative to prehydrated specimens for salt concentrations greater than about 100 to 300 mM.

For compacted soil-bentonite (CSB) mixtures, the value of k_w tends to decrease the higher the quality of the bentonite, all other factors being equal. A significant decrease in k_w occurs with increasing bentonite content, *BC*, such that extremely low k_w values ($\leq 10^{-10}$ m/s) can be achieved for $BC \geq 10$ %. A lower amount of Na-bentonite relative to Ca-bentonite is required to achieve a given k_w. Generally, *k* increased with increase in salt concentration, and substantial increases in *k* (2 to 3 orders of magnitude) can occur at $CaCl_2$ concentrations as low as 200 to 500 mM. Such magnitude of increase in *k* generally is greater than that for GCLs at similar concentrations due, in part, to the lower *BC* of CSBs relative to GCLs. This observation is consistent with data showing that the adverse impact of a given chemical solution on the *k* of a CSB can be overcome somewhat simply by increasing the *BC* of the CSB. Finally, caution should be exercised in using values of *k* back-calculated from consolidation test results on specimens prepared with salt solutions, as such values likely will be unconservatively low by one to two orders of magnitude.

Finally, compatibility data for soil-bentonite backfills (SBBs) are more limited than those for GCLs or CSBs. However, important factors for achieving suitably low *k* for SBBs include the type and amount of bentonite in the SBBs, the type and gradation of the sand, and the amount of fines in the sand prior to bentonite amendment. The results of compatibility tests on specimens of sand-bentonite backfills permeated with $CaCl_2$ solutions are mixed, indicating little or no incompatibility to significant incompatibility at $CaCl_2$ concentrations as low as 50 mM, with the significant incompatibility at low $CaCl_2$ concentrations being attributed to the relatively loose structure of SBBs and the relatively small amounts of bentonite ($\leq$ 7 %) commonly associated with SBBs. Finally, all SBB specimens are essentially prehydrated upon permeation with chemical solutions as a result of mixing the backfill with bentonite slurry to achieve a desired slump.

ACKNOWLDEGEMENTS

Financial support for portions of the work presented herein was provided by the U.S. National Science Foundation (NSF), Arlington, VA, under grants MSS-9257305, CMS-9634649, CMS-9820863, CMS-0099430, and CMMI-0757815. The opinions expressed in this paper are solely those of the authors and are not necessarily consistent with the policies or opinions of the NSF.

REFERENCES

Ashmawy, A. K., El-Hajji, D., Sotelo, N. & Muhammad, N. (2002). "Hydraulic performance of untreated and polymer-treated bentonite in inorganic landfill leachates." *Clays and Clay Minerals*, 50(5), 546-552.

Bohnhoff, G.L., & Shackelford, C.D. (2014). "Hydraulic conductivity of polymerized bentonite amended backfills." *J. Geotech. & Geoenv. Eng.*, in press.

Daniel, D.E. (1994). "State-of-the-art: Laboratory hydraulic conductivity test for saturated soils." *Hydraulic Conductivity and Waste Contaminant Transport in Soil (STP 1142)*, ASTM, West Conshohoken, PA, 30-78.

Daniel, D.E., Anderson, D.C., & Boynton, S.S. (1985). "Fixed-wall versus flexible-wall permeameters." *Hydraulic Barriers in Soil & Rock (STP 874)*, ASTM, West Conshohoken, PA, 107-126.

Eisenhour, D.D. & Brown, R.K. (2009). "Bentonite and its impact on modern life." *Elements* 5

(2), 83-88.

Fernandez, F. & Quigley, R.M. (1991). "Controlling the destructive effects of clay – organic liquid interactions by application of effective stresses." *Can. Geotech. J.*, 28 (3), 388-398.

Gates, W.P., Bouazza, A., & Churchman, G.J. (2009). "Bentonite clay keeps pollutants at bay." *Elements,* 5 (2), 105-110.

Gleason, M.H., Daniel, D.E., & Eykholt, G.R. (1997). "Calcium and sodium bentonite for hydraulic containment applications." *J. Geotech. & Geoenv. Eng.*, 123 (5), 438-445.

Grim, R. (1968). *Clay Mineralogy*, McGraw-Hill, New York, NY.

Guyonnet, D., Touze-Foltz, N., Norotte, V., Pothier, C., Didier, G., Gailhanou, H., Blanc, P., & Warmont, F.(2009). "Performance-based indicators for controlling geosynthetic clay liners in landfill applications." *Geotextiles & Geomembranes*, 27 (5), 321-331.

Hai-Lang, Z., & Shi-Jun, H. (1996). "Viscosity and density of water + sodium chloride + potassium chloride solutions at 298.15° K." *J. Chem. & Eng. Data*, 41 (3), 516-520.

Haug, M.D. & Bolt-Leppin, B. (1994). "Influence of polymers on the hydraulic conductivity of marginal quality bentonite-sand mixtures." *Hydraulic Conductivity and Waste Contaminant Transport in Soil (STP 1142)*, ASTM, West Conshohoken, PA, 407-421.

Howell, J.L., & Shackelford, C.D. (1997). "Hydraulic conductivity of sand admixed with processed clay mixtures." *Proc. 14th Int.Conf.on Soil Mech. & Found. Eng.,* Hamburg, Germany, Sep. 6-12, 1977.

Jo, H.Y., Benson, C.H., & Edit, T.B. (2004). "Hydraulic conductivity and cation exchange in non-prehydrated and prehydrated bentonite permeated with weak inorganic salt solutions." *Clays & Clay Miner.*, 52 (6), 661-679.

Jo, H.Y., Katsumi, T., Benson, C.H., & Edil, T.B. (2001). "Hydraulic conductivity and swelling of non-prehydrated GCLs permeated with single-species salt solutions." *J. Geotech. & Geoenv. Eng.*, 127 (7), 555-567.

Kang, J-B., & Shackelford, C.D. (2010). "Consolidation of a geosynthetic clay liner under isotropic states of stress." *J. Geotech. & Geoenv. Eng.*, 136 (1), 253-259.

Kenney, T.C., van Veen, W.A., Swallow, M.A., & Sungaila, M.A. (1992). "Hydraulic conductivity of compacted bentonite-sand mixtures." *Can. Geotech. J.*, 29 (3), 364-374.

Kolstad, D.C., Benson, C.H., & Edil, T.B. (2004). "Hydraulic conductivity and swell of nonprehydrated geosynthetic clay liners permeated with multispecies inorganic solutions." *J. Geotech. & Geoenv. Eng.*, 130 (12), 1236-1249.

Lambe, T.W. & Whitman, R.V. (1969). *Soil Mechanics*, Wiley, New York, NY.

Lee, J.-M. & Shackelford, C.D. (2005a). "Impact of bentonite quality on hydraulic conductivity of geosynthetic clay liners." *J. Geotech. & Geoenv. Eng.*, 131 (1), 64-77.

Lee, J.-M. & Shackelford, C.D. (2005b). "Concentration dependency of the prehydration effect for a geosynthetic clay liner." *Soils & Found.*, 45 (4), 27-41.

Malusis, M. A. & McKeehan, M.D. (2013). "Chemical compatibility of model soil-bentonite backfill containing multiswellable bentonite." *J. Geotech. & Geoenv. Eng.*, 139 (2), 189-198.

Malusis, M. A. & Shackelford, C.D. (2002). "Chemico-osmotic efficiency of a geosynthetic clay liner." *J. Geotech. & Geoenv. Eng.*, 128 (2), 97-106.

Mishra, M., Ohtsubo, M., Li, L., Higashi, T., & Park, J. (2009). "Effect of salt of various concentrations on liquid limit and hydraulic conductivity of different soil-bentonite mixtures." *Environ.Geol.*, 57, 1145-1153.

Mesri, G. & Olson, R.E. (1970). "Shear strength of montmorillonite." *Géotech.*, 20 (3), 261-270.

Mesri, G. & Olson, R.E. (1971a). "Consolidation characteristics of montmorillonite." *Géotech.*, 21 (4), 341-352.

Mesri, G., & Olson, R.E. (1971b). "Mechanisms controlling the permeability of clays." *Clays & Clay Miner.*, 19, 151-158.

Mitchell, J.K., Alvarez-Cohen, L., Atekwana, E.S., Burns, S.E., Gilbert, R.B., Kavazanjian, Jr., E., O'Riordan, W.H., Rowe, R.K., Shackelford, C.D., Sharma, H.D., and Yesiller, N. (2007). *Assessment of the Performance of Engineered Waste Containment Barriers*. The National Academies Press, 500 Fifth Street, N.W., Washington, DC 20001.

Mitchell, J.K., Hooper, D.R., & Campanella, R.G. (1965). "Permeability of compacted clay." *J. Soil Mech. & Found. Div.*, 91 (SM4), 41-65.

Mitchell, J.K. & Soga, K. (2005). *Fundamentals of Soil Behavior*, 3rd ed., Wiley, New York, NY.

Olson, R.E. & Daniel, D.E. (1981). "Measurement of the hydraulic conductivity of fine-grained soils." *Permeability and Groundwater Contaminant Transport (STP 746)*, ASTM, West Conhohoken, PA, 18-64.

Ozbek, H. (2010). "Viscosity of aqueous sodium chloride solution from 0 to 150°C." *Proc. . Amer. Chem.l Soc. 29th Regional Meeting,* Tampa, FL, Nov. 9-11, 1977.

Petrov, R.J. & Rowe, R.K. (1997). "Geosynthetic clay liner (GCL) – chemical compatibility by hydraulic conductivity testing and factors impacting its performance." *Can. Geotech. J.*, 34, 863-885.

Petrov, R.J., Rowe, R.K., & Quigley, R.M. (1997). "Selected factors influencing GCL hydraulic conductivity." *J. Geotech. & Geoenv. Eng.*, 123(8), 683-695.

Petrov, R.J., Rowe, R.K., & Quigley, R.M. (1997b). "Comparison of laboratory-measured GCL hydraulic conductivity based on three permeameter types." *Geotech. Testing J.*, 20 (1), 49-62.

Shackelford, C.D. (1994). "Waste-soil interactions that alter hydraulic conductivity." *Hydraulic Conductivity and Waste Contaminant Transport in Soil (STP 1142)*, ASTM, West Conshohoken, PA, 111-168.

Shackelford, C.D. (2005). "Environmental issues in geotechnical engineering." *16th Int. Conf. on Soil Mech. & Geotech. Eng.*, Osaka, Japan, Sept. 12-16, 2005, Millpress, Rotterdam, The Netherlands, Vol. 1, 95-122.

Shackelford, C.D. (2008). "Selected issues affecting the use and performance of GCLs in waste containment applications." *Geosynthetics and Environment, Proc.s . Geotech. Eng.Conf.of Torino (XXI Edition)*, M. Manassero, and A. Dominijanni, eds., Politecnico di Torino, Torino, Italy, Nov. 27-28, 2007, Pàtron Editore, Bologna (CD-ROM; ISBN: 978-88-555-2953-2).

Shackelford, C.D. (2013). "Membrane behavior in engineered bentonite-based containment barriers: State of the art." *Coupled Phenomena in Environmental Geotechnics (CPEG 2013)*, July 1-3, 2013, Torino, Italy, CRC Press/Balkema, Taylor & Francis Group, London, 45-60.

Shackelford, C.D, Benson, C.H., Katsumi, T., Edil, T.B., & Lin, L. (2000). "Evaluating the hydraulic conductivity of GCLs permeated with non-standard liquids." *Geotext. & Geomem.*, 18 (2-4), 133-161.

Shariatmadari, N., Salami, M., & Fard, M. (2011). "Effect of inorganic salt solutions on some geotechnical properties of soil-bentonite mixtures as barriers." *Int. J. of Civil Eng.*, 9 (2), 103-110.

Stern, R.T. & Shackelford, C.D. (1998). "Permeation of sand-processed clay mixtures with calcium chloride solutions." *J. Geotech. & Geoenv. Eng.*, 124 (3), 231-241.

van Olphen, H. (1977). *An Introduction to Clay Colloid Chemistry, 2nd ed.*, Wiley, New York, NY.

Yeo, S., Shackelford, C.D., & Evans, J. (2005). "Consolidation and hydraulic conductivity of nine model soil-bentonite backfills." *J. Geotech. & Geoenv. Eng.*, 131 (10), 1189-1198.

Deformation and Shear Strength Behaviors of Overconsolidated Clay from Stress-Path Testing for a Deep Open Cut Excavation

Jeong-Yun Won[1], Ph.D., P.E., M.ASCE,
Blake Cotton[2], P.E., M.ASCE, and Byron W. Porter[3], P.E., M.ASCE

[1]Senior Engineer, Technip, 11700 Katy Freeway Suite 1200, Houston, TX 77079; jy9007@gmail.com
[2]Vice President, Fugro Consultants, Inc., 15 Veterans Blvd, Kenner, LA 70062; bcotton@fugro.com
[3]Senior Consultant, Fugro Consultants, Inc., 6100 Hillcroft Ave., Houston, TX 77081; bporter@fugro.com

ABSTRACT: A stress-path testing program was proposed and performed to investigate shear strength and deformation behaviors due to an open cut excavation in overconsolidated clay. Stress changes due to the staged excavation for different locations of the slope were estimated and applied on undisturbed specimens using a triaxial testing system. Axial load, chamber pressure, and back pressure were actively controlled to implement the various stress paths. Vertical and horizontal deformations were monitored during the stress paths, and then an undrained compression or extension shearing stage followed. Test results for three cases from the testing program: the crest and the toe of the slope, and the base of the excavation are presented. The clay deposit was found to be non-expansive, based on the deformation behaviors. Effective stresses appeared to govern the vertical and horizontal strains. Failure envelopes for compression and extension were established from the stress path testing. The mobilized undrained shear strength from the stress path approach is substantially lower than one from the conventional method.

INTRODUCTION

An open cut excavation of 24-29 m deep in stiff overconsolidated clay was under consideration for a river diversion channel. The overconsolidated clay is characterized as Pleistocene clay ranging from lean to fat clay with fine sand and silt sublayers. Swelling (deformation) and strength decrease due to the stress relief and softening were one of the main design concerns for slope stability of the open cut excavation.

Duncan and Dunlop (1969) pointed out some key aspects of slope stability for heavily overconsolidated clay, such as stress-strain and strength characteristics, effects of fissures, and effects of high lateral stress, based on the historical slope failures. The delayed slope failure of cut slopes in overconsolidated clay was explained by pore water pressure dissipation, swelling, release of strain energy, strength decrease with increasing loading time, and fissures (Vaughan and Walbancke 1973). Mesri et al. (1978) proposed several processes related to softening due to unloading. Softening of overconsolidated or swelling potential of clay has been an intense study area for slope stability problems. Recently, fully softened shear strength was recommended for repeated wet and dry processes in overconsolidated clay (Wright, 2005). On the other hand, Hird and Pierpoint (1997) and Ng (1999) performed stress path testing to investigate strength and deformation characteristic of clay under excavation conditions. In fact, degradation, progressive failure, softening, swelling, and stress relief are closely related to each other.

Soil elements along the open cut slope would experience complicated stress changes as the excavation progresses. Lateral or horizontal stress in the cut slope will decrease dramatically while the vertical stress remains practically the same. Lateral movement is expected due to the horizontal stress relief that is not directly related to a slope failure. Lateral slope movements due to an excavation was reported by Duncan and Dunlop (1969) and Hird and Pierpoint (1997). Because of the lateral movement, the ratio between the effective horizontal and vertical stresses, K, is not the same as the initial state (e.g. K_o) and will not be a K_o state any more. At the base of the cut slope, horizontal stress may increase because of the potential sliding force. Vertical stress, however, is reduced dramatically because of the cut, and vertical deformation (swelling) is expected. While these complicated stress changes occur in open cut excavations, comprehensive investigations to characterize the soil behaviors through stress-path testing are seldom performed.

A special testing program was proposed and carried out to investigate the effects of stress changes on deformation and undrained shear strength of the overconsolidated clay along the side and at the base of the excavation. The testing program comprised 34 stress path tests for various locations of the slope and the base of the proposed excavation. In the stress path tests, stresses for present and future conditions were applied on undisturbed specimens to model in-situ situations more realistically. Each test had two to nine stress path stages and different undrained shear modes, depending on the slope locations and possible stress changes. This paper presents three stress path test results from the testing program for a compressional failure zone (crest), an extensional failure zone (toe), and the base of the excavation.

TEST MATERIAL

The site investigation program consisted of 6 borings and continuous 127 mm (5-inch) diameter undisturbed sampling by a fixed piston sampler. The Pleistocene clayey layers at the site extend to a depth of at least 30 m. Numerous thin (1-3 m thick) sublayers of silt, sand and their mixtures, and abundant slickensides were the characteristic of the clay layers. Light brown and olive gray colored clay layers were classified as CL and CH following ASTM D2487. The ground water table was at the ground surface during the investigation. Figure 1 summarizes physical and mechanical properties of the clay layers studied.

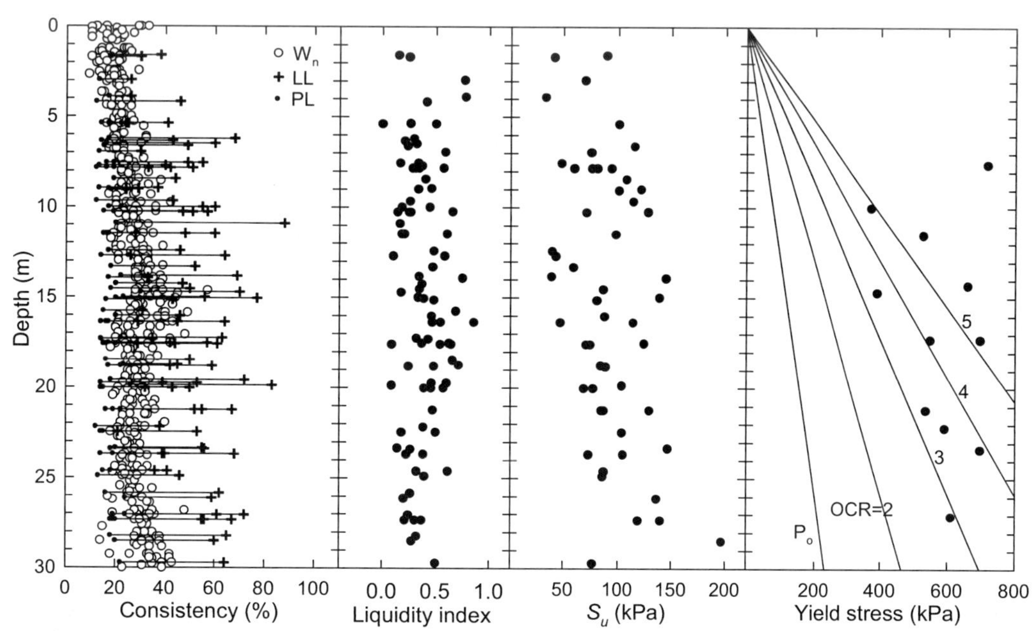

Figure 1. Consistency and mechanical properties of the clay layer from 6 borings.

Natural moisture contents of the clay ranged from 18 to 44 % and were generally close to the plastic limits (12 to 30 %). Liquid limits ranged from 24 to 88 % and the liquidity index ranged from 0.0 and 0.9. Undrained shear strengths measured from unconsolidated-undrained (UU) triaxial tests generally ranged from 40 to 150 kPa with an average of about 90 kPa. The overconsolidation ratios evaluated from 100 mm (4-inch) diameter one-dimensional consolidation tests varies from 3 to over 5, due to the complicated geological history associated with the adjacent Mississippi river.

STRESS CHANGES DUE TO AN EXCAVATION

Lambe (1967) provided a stress path example for a tank foundation with a variation in ground water table. He explained that K_o conditions exist because the dewatering occurs over a large lateral area therefore horizontal strain must be equal to zero. Furthermore, he has shown that vertical total stress remains constant while the lateral total stress drops during the dewatering. Lambe and Marr (1979) provided a stress path example for the stability of a dam, in which a steady-state seepage occurs within the dam when it was flooded, causing a possible instability of the downstream face. They demonstrated that the effective stress path extends horizontally, with a constant deviator stress (q=0.5(σ'_v-σ'_h)) and decreasing effective mean stress (p'=0.5(σ'_v+σ'_h)), when the pore pressures become positive, while total stresses remain constant (where σ'_v and σ'_h represent effective vertical stress and effective horizontal stress, respectively). Hird and Pierpoint (1997) implemented a stress path for swelling of an excavation as a constant deviator stress (q), while the effective mean stress (p') decreases. They presented an interesting vector plot for an excavation movement in Oxford Clay. It was clearly shown that there was a horizontal (lateral) movement at the excavation faces after the excavation was performed. A similar lateral behavior of a cut slope was noted by Duncan and Dunlop (1969). When, and if, a lateral movement occurs, a K_o condition adjacent to the excavation cannot be justified. Lambe (1970) pointed out important aspects that can occur in soil elements near an excavation: first, the reduction of total vertical stress and total horizontal stress, and second, the change in equilibrium pore water pressure. In fact, the stress paths for a braced cut in normally consolidated soil presented by Lambe (1970) have been referenced by many research papers (for example, Ng 1999). Mesri et al. (1978) suggested that a passive failure might have happened to an overconsolidated clay because the effective horizontal stress became so large relative to the effective vertical stress during the geological unloading. The delayed failures of slopes cut in overconsolidated clays have been explained in terms of delayed pore pressure equilibration and softening (Vaughan and Walbancke 1973 and Mesri et al. 1978). According to Mesri et al. (1978), it was assumed that failure was delayed by the slow rate of increase in pore water pressure (from negative excess pore pressure) following stress reduction and undrained deformation during excavation. Malandraki and Toll (2001) showed an example of vertical and horizontal stress changes in a soil element adjacent to an excavation. Bolton (1993) illustrated stress changes at the base of an excavation and suggested that the clay would behave like an undrained triaxial extension test (reducing vertical stress with a constant horizontal stress).

Soil elements along the excavation slope and around the excavation base would experience various stress changes and different shearing modes. During a staged excavation, as an example, the following vertical and horizontal stresses, and pore pressure changes are expected in the soil element, Point D in Figure 2. The stress changes and deformations that are expected in the soil element D in Figure 2 are summarized in Table 1. Ground water table is at the ground level. The total vertical stress (σ_v) is reduced by an amount of $\gamma_{total} \cdot h_{excavation}$, right after the 1st stage excavation, where γ_{total} is the saturated unit weight of the soil and $h_{excavation}$ is the excavation depth. The total horizontal stress (σ_h), however, remains constant initially. Since the excavation can be treated as rapid, in other words undrained condition, a negative excess pore pressure is expected at the Point D. The magnitude of the negative excess pore pressure may be estimated from K_o-consolidated triaxial extension tests. The effective vertical stress (σ'_v) would be approximately the same as the one before the excavation, because the decrease in the total vertical stress is compensated by the negative excess pore pressure. The effective horizontal stress (σ'_h) would increase substantially because of the negative excess pore pressure, while the total horizontal stress (σ_h) remains constant. During a waiting period, the negative excess pore pressure would dissipate bringing the pore pressure back to a hydrostatic equilibrium state. Since the excess pore pressure was negative, during the dissipation period (waiting period), effective vertical and horizontal stresses would decrease while total vertical and horizontal stresses remain constant. For the next excavation Stage 2, the soil element Point D in Figure 2 would experience a decrease in total horizontal stress (σ_h), while total vertical stress (σ_v) remains constant. Excess pore pressure is not likely to develop. Thus, the effective vertical stress (σ'_v) will not change. Now, because there is a stress increase due to the potential sliding action, the effective horizontal stress (σ'_h) may increase. The failure of the Point D in Figure 2 would be in compression or direct simple shear mode.

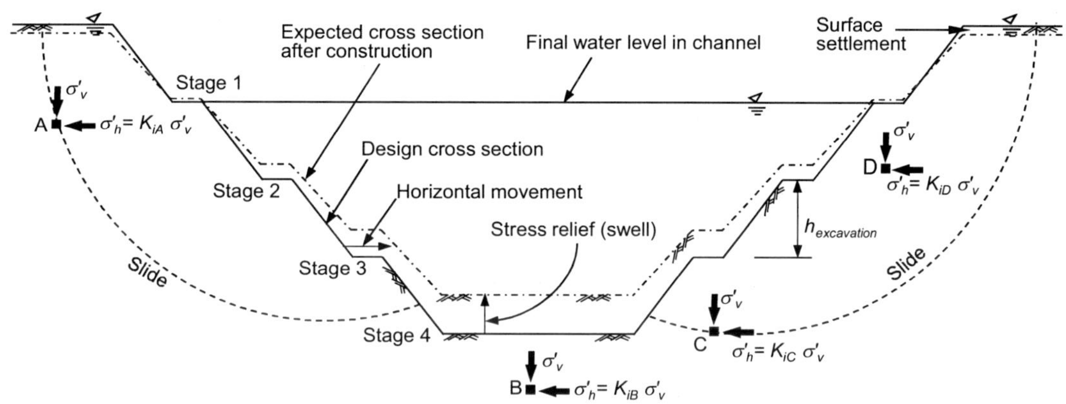

Figure 2. Schematic and conceptual drawing for a deep excavation.

Table 1. Changes in stress and deformation during excavation in the soil element D in Figure 2.

Excavation sequence		1st Stage	Waiting period	2nd Stage	Waiting period
Total stress	Vertical, σ_v	Decrease	Constant	Constant	Constant
	Horizontal, σ_h	Constant	Constant	Decrease	Constant
Excess pore pressure, u'		Decrease	Increase	-	-
Pore pressure, u		Decrease	Increase	Constant	Constant
Effective stress	Vertical, σ'_v	Constant	Decrease	Constant	Constant
	Horizontal, σ'_h	Increase	Decrease	Decrease	Increase
Deformation	Vertical, δ_v	Constant	Swell	Constant	Constant
	Horizontal, δ_h	Constant	Constant	Increase	Increase

Similar to the Point D in Figure 2, the Points B and C in Figure 2 would experience similar stress paths repeated successively, whilst the failure mode is extension. The Point B and C in Figure 2 would experience multiple stages (decrease in total vertical stress) before the total horizontal stress decrease. In fact, the swelling process under the base of excavation (Point B in Figure 2) may not be fully represented by the one-dimensional swelling condition, because the sliding force acts horizontally while the seepage force acts vertically. Examples of the complete set of stress changes in vertical and horizontal stresses, excess pore pressure, and stress paths for the Point C in Figure 2 are shown in Figure 3. Mean and deviator stress changes due to a staged excavation are shown in Figure 3 (a) and the changes in axial and radial stresses in terms of total and effective stresses during the stress paths are shown in Figure 3(b). Axis p and q in Figure 3 (a) are oriented at 45 degrees to σ_h and σ_v in Figure 3 (b) with a $\sqrt{2}$ scaling factor. The stress paths 1, 3, 5 and 7 replicate rapid excavation stages, whereas stress paths 2, 4, and 6 are for the swelling (waiting) stages. In the last stress path, 8, the pore pressure was raised without any change in effective stresses as shown in Figure 3 (c), to model a rapid water rise to the final water level, followed by the undrained extension shear. Until the fourth excavation, the stress paths in Figure 3 are similar to that from a typical triaxial extension test. In addition to the complicated stress changes due to the staged excavation, failure modes of each soil element would be either compressional, direct simple shear, or extensional. The soil elements near the top and possible failure surface would experience compression and/or direct simple shear modes, while elements around the toe and the base would go through an extension shear mode. The extension shear mode can either be a decrease in vertical stress with constant horizontal stress or an increase in horizontal stress with constant vertical stress. As for the Point A in Figure 2, a compression failure mode is expected because of the horizontal stress relief and increase in sliding force. In this study, the failure modes were idealized as either compression or extension, because the testing was performed in triaxial cells. In fact, vertical and horizontal stress controls as well as excess pore pressure control in a direct simple shear test apparatus are difficult. Unlike a conventional triaxial test, all the stresses and excess pore pressure were actively controlled to follow pre-determined stress paths during the stress path testing. The stress paths were idealized as a set of modules that have pre-determined stress-paths based on the locations of the soil element. Some of the stress paths from the testing program, following the excavation process, resulted in a failure as a result of approaching the compression/extension failure envelopes.

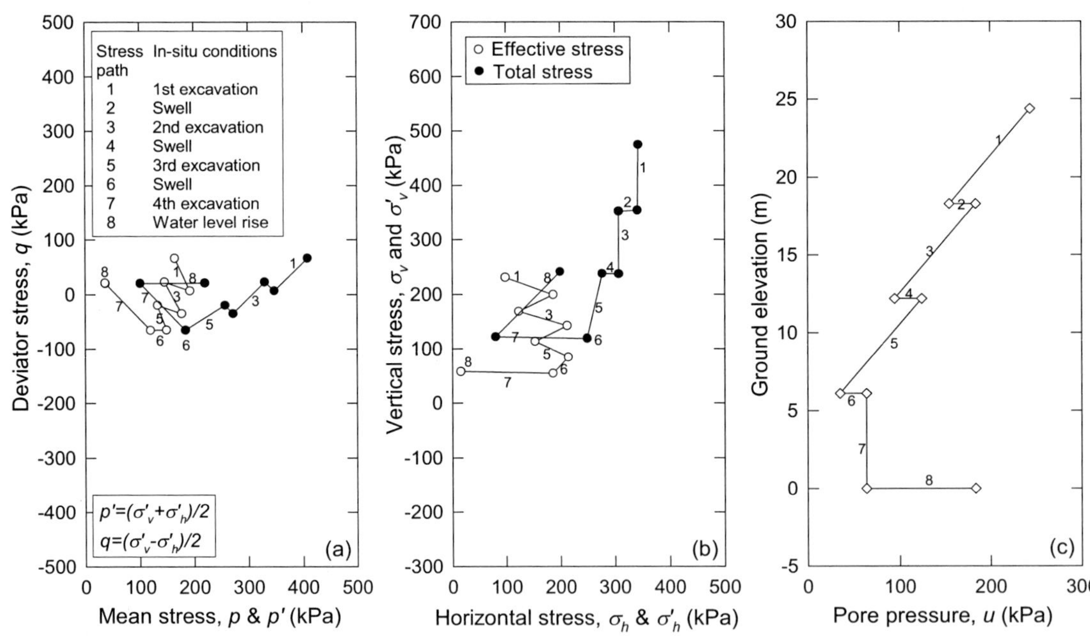

Figure 3. Examples of stress changes during excavation and swelling, Point C in Figure 2.

STRESS-PATH TESTING

Multi-stage stress path tests were carried out in a triaxial cell on 50 mm diameter specimens trimmed by hand from undisturbed samples of a nominal height of about 120 mm. The same procedures described by Won et al. (2010) were followed for sampling, handling, transportation, extruding, and storage. More than 10 triaxial testing units were used simultaneously for the stress path testing program. Each stress path triaxial testing unit was comprised of three independent computer controlled pressure systems: axial load control, chamber pressure control, pore water pressure control as shown in Figure 4. The axial load was controlled mainly by an internal load cell and an external load cell was monitored as a backup system. The internal load cell removes the needs of corrections for the friction between the piston rod and the top cap of the triaxial cell. The bottom plate was raised and lowered for axial loading and unloading while the frame was fixed. Pore pressure was measured from the top and the bottom of the specimens.

Specimen height and diameter right after the trimming were used as the reference values to calculate any vertical and radial deformations during saturation, K_o-consolidation, stress paths and undrained shearing stages. Saturated porous stones were placed at both ends after placing filter paper discs. Side filter strips were not used. Nominal back pressure (u_o) and chamber pressure (σ_{cell}) applied for the saturation stage were 482 kPa and 496 kPa, respectively. During a B-value check, an isotropic pressure of 34 kPa was applied to measure excess pore pressure at the bottom of the specimen. Typically within about 5 minutes a B-value of 0.95 or more was achieved. In fact, the calculated initial degrees of saturation of the tested specimens were over 95 %. At least an overnight waiting time was given after the completion of the saturation process. Cho et al. (2007) reported a reduction in initial stiffness of triaxial specimens due to alteration of the soil structure when the swelling during saturation was not controlled. However, in this testing program, the swelling during saturation was not controlled. After the saturation process, the specimens were consolidated in a K_o-state to an estimated in-situ effective vertical stress level, following the recompression concept. After the end of K_o-consolidation, the axial stress, radial stress, and pore pressure were recorded, before initiating a series of stress paths.

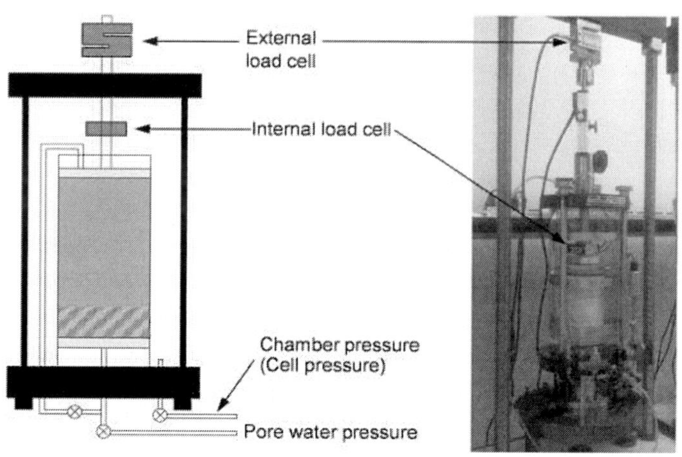

Figure 4. Triaxial stress path cell.

The stress change rate was one of the critical testing considerations because it governs testing time and pore pressure equalization within the testing specimens. From the literature review, a variety of stress change rates were observed. For example, Hird and Pierpoint (1997) used a stress change rate of 1.0-1.5 kPa/hr during stress paths on heavily overconsolidated clay. Hird and Pierpoint (1997) pointed out, even with the aide of side filter paper for drainage, there must have been some excess pore pressure in the central region of the specimen. Ng (1999) applied stress change rates of 1.0-1.5 kPa/hr for drained tests and 5-10 kPa/hr for undrained tests. Malandraki and Toll (2001) applied a 50 kPa/hr stress change rate for stress path testing. A rate of 1.4 kPa/hr was selected as the maximum stress change rate in this testing program. With the stress change rate selected, a full multi-stage (9 stages) stress path test took about 31 days.

For the stress path testing, three stresses were controlled simultaneously. The vertical (axial) and horizontal (radial) stresses were controlled by the axial load cell and chamber pressure. Unlike typical triaxial tests, pore pressure was actively controlled by the back pressure. The three stresses were simultaneously ramped to reach each target value. Some of the stresses, especially pore pressure were changed at a very slow rate when the pressure difference was small. Since the stress path testing was a stress controlled test, axial and radial strains were allowed to develop during the stress paths. Volume changes from the axial pressure, chamber pressure, and pore pressure control systems were logged and converted to axial and radial specimen deformations. In the undrained extension shearing, the axial stress was decreased while the radial stress was maintained constant.

Because of the active control of each stress component including pore pressure, more refined definitions for each pressure were necessary. The term "vertical stress" was used in lieu of "axial stress" and "horizontal stress" was interchanged with "radial stress".

Effective horizontal stress: $\sigma'_h = \sigma'_{radial} = \sigma_{cell} - u_{applied}$ (1)

Effective vertical stress: $\sigma'_v = [(F_{ILC} - F_{mc} + F_{tc}) / A_s] + \sigma'_h$ (2)

Applied pore pressure: $u_{applied} = u_{measured} = u' + u_o = u' + u_{hs} + u_{base}$ (3)

Hydrostatic pore pressure, variable: $u_{hs} = z_{depth} \cdot \gamma_w$ (4)

Excess pore pressure: $u' = u_{measured} - u_o$ (5)

Where,

F_{ILC} is the load from the inner load cell
F_{mc} is the membrane correction
F_{tc} is the top cap weight
A_s is the specimen area = specimen volume / (h_o - δ_v)
h_o is the initial specimen height
$\delta_v = \delta_a$ is the axial deformation
z_{depth} is the depth of soil element at a given excavation stage
γ_w is the unit weight of water
u_o is the back pressure at the start of the first stress path
u_{base} is arbitrary base (allowance) pore pressure for testing

For example, if pressures measured at a given time are $u_{measured} = u_{applied} = 418.2$ kPa, hydrostatic pore pressure, $u_{hydrostatic} = 244.2$ kPa, and initial back pressure, $u_o = 482.1$ kPa, then excess pore pressure, $u' = -63.9$ kPa, which means that a soil element is experiencing a negative excess pore pressure (-63.9 kPa) due to an unloading. The vertical and horizontal stresses were then, calculated accordingly based on the pore pressure calculated.

RESULTS

Stress path test results for three cases are shown in Figure 5 in *p-q* planes. Axial strains due to deviator stress changes are also plotted. Effective stress paths (ESP) and total stress paths (TSP) are presented in the same *p-q* plots. Figures 5 (a) and (b) are for the Point A in Figure 2, e.g. near the crest of the cut slope, where a compression failure is expected. Figure 5 (c) and (d) are for the Point C in Figure 2, e.g. near the toe of the slope, where an extension failure is expected. In fact, the failure mode of the Point C could be a combination of reduction in vertical stress and increase in horizontal stress. Figures 5 (e) and (f) are for the Point B in Figure 2, e.g. at the bottom of the excavation, where an extension failure is expected. For the compression failure case (Point A in Figure 2), a stress path stage for a drought condition was included before the undrained compression shearing. For the extension failure case, a stress path for a water level rise condition was implemented before the undrained extension shearing. Sign convention for axial deformation followed the conventional geotechnical practice, i.e. negative for swelling (vertical upward deformation) and positive for compression (vertical downward deformation). Similarly, negative horizontal deformation is for horizontal movement to the excavated side and positive is for horizontal movement to the back of the cut slope.

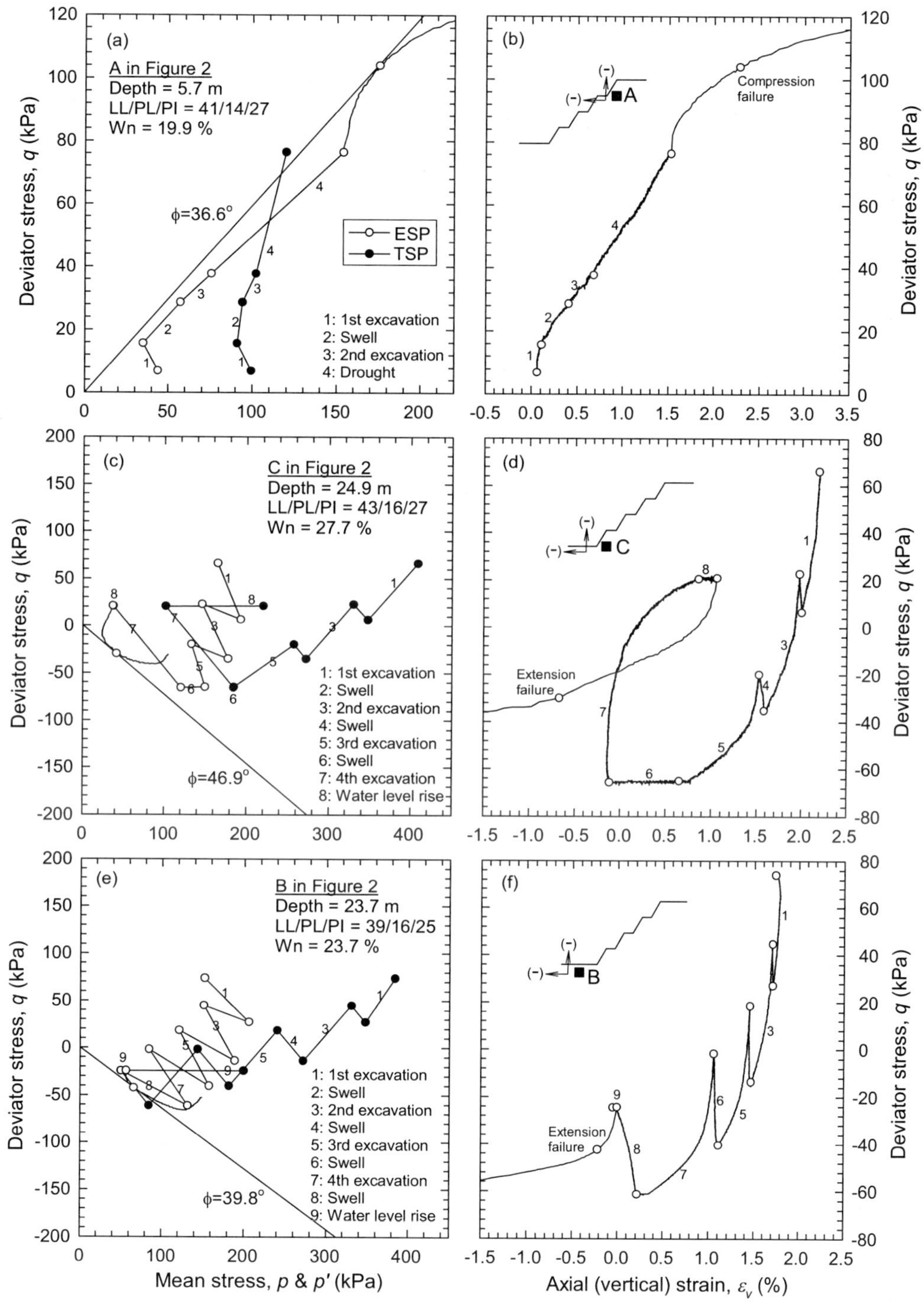

Figure 5. Mean and deviator stress changes, undrained shearing, and axial strain due to deviator stress changes: (a) and (b) for near the crest of slope (Point A in Figure 2), (c) and (d) for near the toe of slope (Point C in Figure 2), and (e) and (f) for base the of excavation (Point B in Figure 2).

Undrained failures both in compression and extension were defined as the maximum obliquity (σ'_v / σ'_h) points for a consistent comparison. Thus, friction angles and undrained shear strengths at failure would be different when a different failure definition is used. The starting points of the axial strains are positive because of strains occurred during seating, saturation and K_o consolidation. The axial strains that occurred between the trimming and the end of saturation phase (before K_o consolidation) were -0.15, 0.33, and 0.41 percent for the Points A, C, and B in Figure 2, respectively. Sampling depths, natural moisture contents, and Atterberg limits are presented in Figure 5.

As for the crest case (Point A in Figure 2), stability after the 1st excavation stage seems to be critical, whereas the compressional axial deformation continued to occur due to the increase in deviator stress q. The friction angle of 36.6° was estimated assuming zero cohesion. Four stages of excavation and water level rise were followed by an undrained extension shear for the Point C in Figure 2, as shown in Figure 5 (c). The swelling stage (stage 6) after the 3rd excavation stage, when negative excess pore pressure (u') dissipates, in turn, causing σ'_v to decrease appeared to be critical. In that stage, the changes in σ'_h would be minor, because the increase in the sliding force would compensate the decrease in σ'_h due to the negative u' dissipation. During the water level rise, no changes in σ'_v and σ'_h occurred thus no change in q was implemented. The friction angle from the Point C in Figure 2 was exceptionally high, indicating that certain cohesion could be anticipated. The base of the excavation, the Point B in Figure 2 approaches the failure envelope as the excavation process proceeds as shown in Figure 5 (e). The stress changes in σ'_h were different from Figure 5 (c) because the bottom point was assumed to be far from the toe of the slope so that the increase in σ'_h due to the sliding force can be ignored. Axial strains in extension sides (Points B and C in Figure 2), shown in Figures 5 (d) and (f) showed that the axial strain (ε_v) responses due to deviator stress changes are not linear, partly because of the different rates of changes in and σ'_v and σ'_h.

DISCUSSION

In Figure 6, vertical (axial) and horizontal (radial) strains are plotted with pore pressure, u. Because the stress path testing was stress-controlled, vertical and horizontal strains were allowed through the in/out of water from the pressure control system. The starting points of the pore pressure (u) were the hydrostatic pore pressures of the Points A, C, and B before the excavation. Pore pressure changes during the undrained shearing were plotted continuously after the series of stress paths. The horizontal strains (ε_h) during the undrained compression and extension shearing (constant volume) were back calculated from the volume at the end of the series of stress paths.

The vertical strain (ε_v) at the crest (Point A in Figure 2), as shown in Figure 6 (a), is expected to be positive, e.g. the slope surface will trend down. However, horizontal strain (ε_h) is expected to be negative, which means the slope face will move toward the excavated side (Figure 6 (b)). This horizontal strain is not necessarily related to a slope failure. The horizontal strain, in fact, can be interpreted as volumetric adjustment to accommodate the stress changes due to the excavation. Due to the increase in sliding force (in compression) and the decrease in pore pressure, the deviator stress q was increased, causing a vertical strain increase.

In the meantime, the vertical strain (ε_v) at the toe (Figure 6 (c)) is expected to be negative, e.g. vertical upwards as the excavation proceeds. Until the swelling stage after the 3rd excavation, stages 1 through 6 in Figure 6 (d), horizontal strain (ε_h) moved in a positive direction (decrease in specimen diameter) and vertical strain (ε_v) moved negative direction (increase in specimen height). During the 4th excavation (stage 7), due to the substantial decrease in effective horizontal stress (σ'_h), horizontal strain (ε_h) moved in a negative direction, e.g. increase in diameter. Accordingly, the vertical strain (ε_v) moved in a positive direction, e.g. decrease in specimen height because of the substantial decrease in horizontal strain. During the water level rise (stage 8), the pore pressure was raised by increasing back pressure while the effective axial and radial stresses remained constant.

Because of the different stress paths for the bottom of the excavation (Point B in Figure 2), the vertical and horizontal strains during the stress path stages of 6, 7, 8, and 9 (Figures 6 (e) and (f)) are different from those of the Point C in Figure 2. Horizontal strains showed loops due to the pore pressure changes by the excavation in Figure 6 (f). The axial strains moved in a negative direction (increase in specimen height) when the pore pressure and deviator stress decreased. However, the axial strains did not fully recover when the pore pressure and deviator stresses increased (stress path 2, 4, 6, and 8 in Figure 6 (e)).

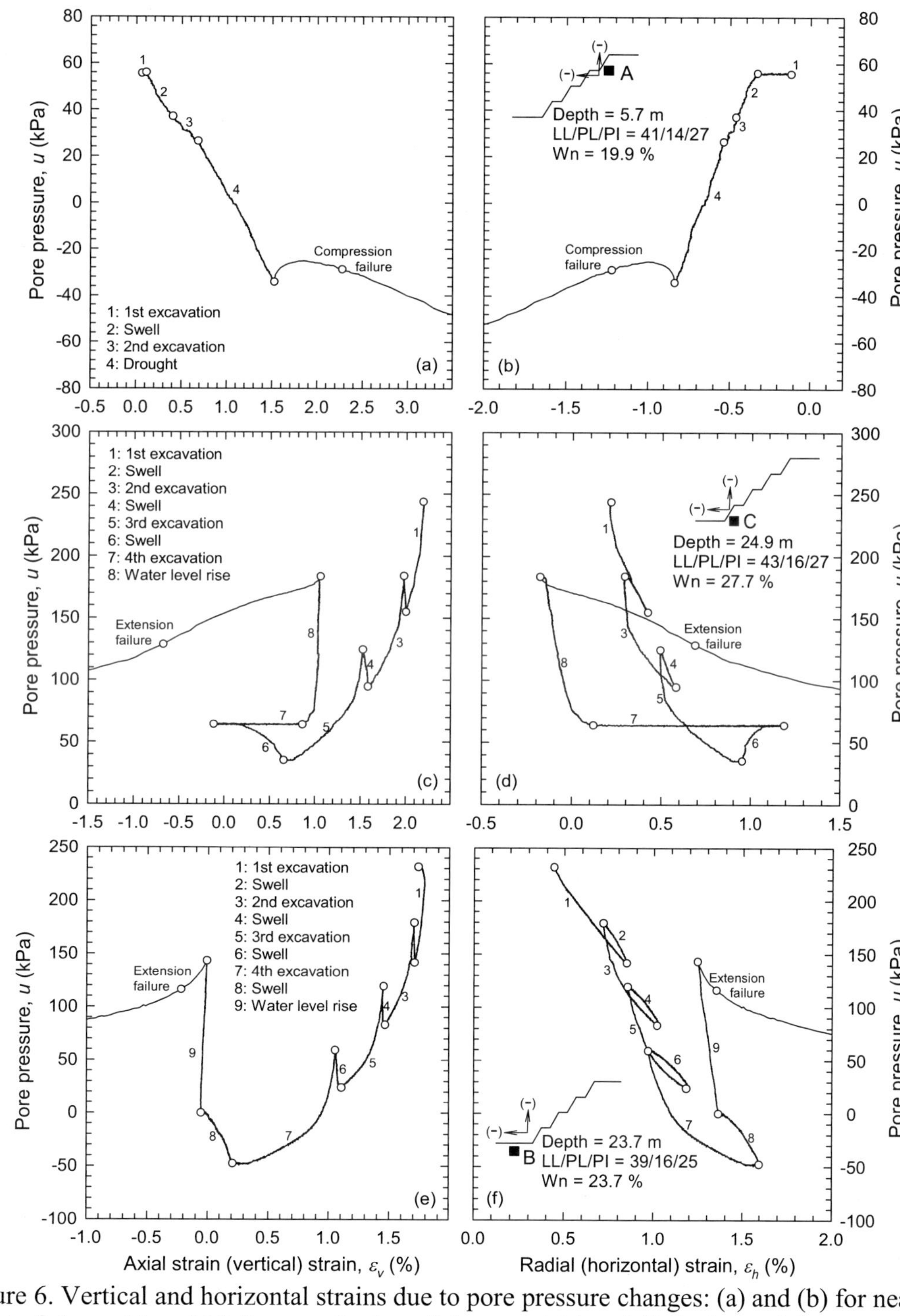

Figure 6. Vertical and horizontal strains due to pore pressure changes: (a) and (b) for near the crest of slope, (c) and (d) for near the toe of slope, and (e) and (f) for base of the excavation.

Unlike the drought condition as in the Point A, no significant changes in vertical and horizontal strains were observed during the water level rise stages for the Points C and B. Based on the minimal changes in the axial strains during the saturation stage and the strain behaviors during the water level rise stages, the clay layer could be treated as non-expansive soil. From Figure 6, the vertical and horizontal strains seemed to be controlled mainly by the effective stresses rather than the pore pressure changes. When the horizontal strain is constrained, for example the bottom of the excavation, the horizontal strain (or deformation) potential could be converted to vertical strain, causing swelling or upheaval.

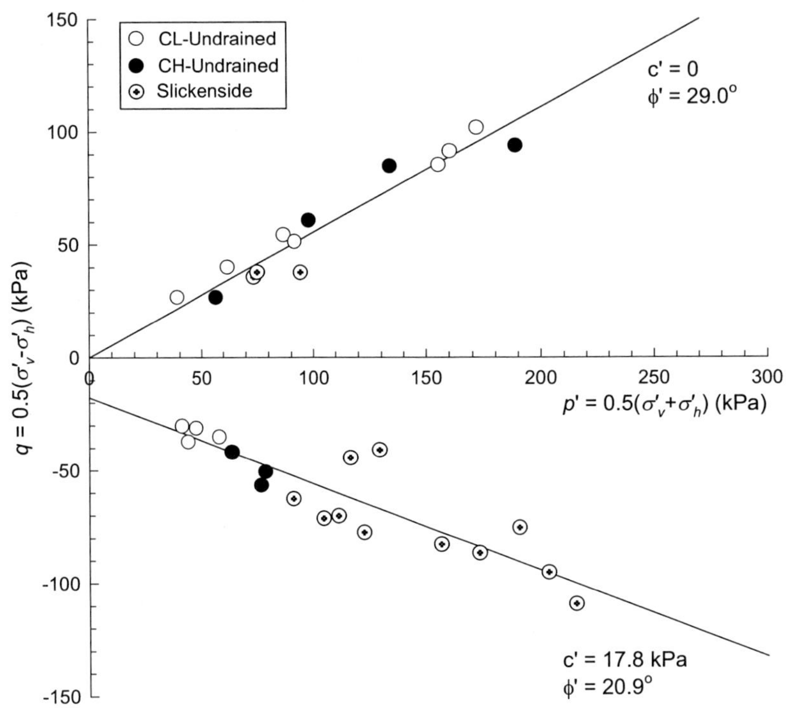

Figure 7. Failure envelopes for compression and extension shear

In Figure 7, the stress path test results from the entire testing program were summarized in a *p-q* plot for compression and extension cases. About 40 % of the stress path testing specimens failed during the stress paths before the undrained shearing stage. The pre-maturely failed specimens were typically slickensided, showing a brittle and distinctive failure surface. The slickensided failure results were also included in Figure 7 to establish failure envelopes. As for the compression shear, the clay layer studied could be represented by effective cohesion, c'=0 and effective friction angle (ϕ') of 29.0 degrees. In the extensional shear, effective cohesion, c'=17.8 kPa and effective friction angle (ϕ') of 20.9 degrees were estimated. Undrained shear strengths that mobilize under a given stress condition should be predicted by locating the soil element's stress path(s) in *p-q* plot. For instance, as for the Point C in Figure 2, undrained shear strength measured by an UU triaxial test was 86.6 kPa. The actual failure mode of the soil element, however, is extension shear. Under the expected stress condition after a series of stress paths shown in Figure 5 (c), the undrained extension shear strength of the soil element was 30.2 kPa. The undrained shear strength that can mobilize is only 35 % of the undrained shear strength from the UU test. This significant difference in undrained shear strengths between UU (Figure 1) and stress path results should be taken into consideration during excavation design. For slope stability, when a total stress analysis is performed, a proper undrained shear strength value that is expected to mobilize under a given stress condition should be selected.

CONCLUSIONS

A stress-path test program was proposed and carried out to investigate shear strength changes due to stress relief and swelling due to a deep open cut excavation in overconsolidated clay. Complicated stress changes and undrained failure modes for different locations along the excavation slope were estimated. Stress path test results for three locations were presented in this paper: near the crest and toe of the cut slope, and the base of the excavation. Each stress path testing consisted of a series of staged excavations and swelling, followed by an undrained compression or extension shearing. In the triaxial testing system, axial load, chamber pressure, and back pressure were actively controlled to implement the stress paths that modeled in-situ stress changes during the excavation. Based on the deformation behaviors from the stress path testing, the clay layer was characterized as non-expansive clay. The vertical and horizontal deformations that were monitored during each stress path were mainly controlled by the effective stress changes, rather than the pore pressure changes. Many of the specimens failed pre-maturely during the stress paths because of slickensides in the specimens. Failure envelopes for the compression and extension were developed based on the entire stress path test results including slickensided failures. It was shown that only 35 % of the undrained shear strength from UU test can be mobilized at a toe of the slope. Based on the stress path testing program, an effective friction angle of 29° with zero cohesion and an effective friction angle of 21° with an effective cohesion of 18 kPa were proposed for compression and extension shearing, respectively.

REFERENCES

Bolton, M.D. (1993). "Mechanisms of ground deformation due to excavation in clay". *Excavation in Urban Areas*, Adachi, T. ed., *Int. Geotech. Forum on Comparative Geotech. Eng.*, Kobe University, Japan, 1-33.

Cho, W., Holman, T.P., Jung, Y.H. & Finno, R.J. (2007). "Effects of swelling during saturation in triaxial tests in clays". *Geotech. Testing J.*, 30 (5), 378-386.

Duncan, J.M. & Dunlop, P. (1969). "Slopes in stiff-fissured clays and shales". *J. Soil Mech. & Fdns Div.*, 95 (SM2), 467-492.

Hird, C.C. & Pierpoint, N.D. (1997). "Stiffness determination and deformation analysis for a trial excavation in Oxford Clay". *Géotechnique*, 47 (3), 665-691.

Lambe, T.W. (1967). "Stress path method". *J. Soil Mech. & Fdns Div.*, 93 (SM6), 309-331.

Lambe, T.W. (1970). "Braced excavation". *Proc. Specialty Conf. Lateral Stresses in the Ground and Design of Earth-Retaining Structures*, ASCE, 149-218.

Lambe, T.W. & Marr, W.A. (1979). "Stress path method: second edition". *J. Geotech. Eng. Div.*, 105 (GT6), 727-738.

Mesri, G., Ullrich, C.R., & Choi, Y.K. (1978). "The rate of swelling of overconsolidated clays subjected to unloading". *Géotechnique*, 28 (3), 281-307.

Malandraki, V. & Toll, D.G. (2001). "Triaxial tests on weakly bonded soil with changes in stress path". *J. Geotech. & Geoenv. Eng.*, 127 (3), 282-291.

Ng, C.W.W. (1999). "Stress paths in relation to deep excavations". *J. Geotech. & Geoenv. Eng.*, 125 (5), 357-363.

Vaughan, P.R. & Walbancke, H.J. (1973). "Pore pressure changes and the delayed failure of cutting slopes in overconsolidated clay". *Géotechnique*, 23 (4), 531-539.

Won, J.Y., Porter, B. & Cotton, B. (2010). "End of primary consolidation for Gulf coast soils". *GeoFlorida*, GSP No.199, ASCE, West Palm Beach, FL, 1108-1115.

Wright, S.G. (2005). *Evaluation of Soil Shear Strengths for Slope and Retaining Wall Stability Analysis with Emphasis on High Plasticity Clays*, Report of Center for Transportation Research at The University of Texas at Austin.

Comparison of DMT- and CPT-Correlated Constrained Moduli in Clayey and Silty Sands

E. Gregory McNulty[1], Ph.D., P.E., P.G., M.ASCE and Michael D. Harney[2], P.E., M.ASCE

[1]Senior Engineering Manager, Parsons Government Services, 10010 Junction Drive, Suite 100N, Annapolis Junction MD 20701; egreg.mcnulty@parsons.com
[2]Senior Principal Engineer, Shannon & Wilson, 400 N 34th Street, Seattle, WA 98103, mdh@shanwil.com

ABSTRACT: This paper compares constrained modulus and related data from a geotechnical engineering investigation of a 40,000 m^2 site in the Southeastern United States that lies within the Atlantic Coastal Plain. This study includes 47 cone penetration tests (CPTs and SCPTs) and 13 flat-plate dilatometer tests (Seismic and non-seismic DMTs) and over 80 incremental and constant rate of strain consolidation tests. This paper compares the CPT- and DMT-correlated constrained moduli and laboratory test results from co-located borings. The comparative analyses show major differences and key similarities in the correlated values; sources for these differences and similarities are discussed. This paper concludes that in contrast to the DMT, CPT/SCPT soundings and often laboratory consolidation testing underestimate key settlement parameters because of disturbance. Correlations for the investigated soil types are also presented.

INTRODUCTION

An extensive geotechnical site characterization study has been undertaken on a 40,000 m^2 site located in the Southeastern United States within the Atlantic Coastal Plain. The study includes field explorations, laboratory testing, development of soil properties and stratigraphy, comparative and statistical analyses of data, engineering analysis to estimate static settlement and bearing capacity, and preparation of design and construction recommendations. The field explorations include standard penetration test (SPT) borings, Cone Penetration Tests (CPT) and Seismic Cone Penetration Test (SCPT) soundings, flat plate dilatometer (DMT) measurements, and cross-hole seismic tests. This paper presents some results of the CPT, SCPT and DMT measurements, and compares constrained moduli correlated from these tests to laboratory incremental (ICON) and constant rate of strain (CRS) consolidation tests.

FIELD EXPLORATIONS

The field explorations included drilling 15 standard penetration test (SPT) mud-rotary borings, advancing 47 CPT and SCPT soundings, making flat dilatometer (DMT) measurements in 13 soundings, measuring shear and compression wave velocities at two locations using cross-hole seismic methods, and collecting bulk samples from eight, 4.5-meter deep auger borings. The typical penetration depth for this study was approximately 45 m below ground surface. Including previous subsurface explorations at the site and those completed for this study, soundings and/or borings were generally spaced on an about 15-meter grid pattern across the project site. Borings, CPTs and SCPTs, and DMTs were co-located in selected areas to facilitate data comparisons.

Based on the results of the subsurface explorations, and laboratory program, the engineering stratigraphy of the site was interpreted. CPT and SCPT explorations and SPT boring data were used to determine the thickness and presence of the engineering layers. Patterns in tip stress, sleeve stress, ratio of tip stress to sleeve stress, and pore pressure responses were used in differentiating soil layers. The site subsurface interpretation is comprised of 13 identified engineering soil layers. These layers are generally horizontally-stratified at the site, and consist of very loose to dense silty and clayey sands. Interbeds of sandy silt and silty clay are scattered in some of the engineering layers. Table 1 summarizes some of the characteristics of the engineering layers.

Table 1. Representative characteristics of identified engineering soil layers. (McNulty and Harney, 2010)

Layer	Average Depth to Layer (m)	USCS	Average Fines Content (%)	Average Index Properties of Fines		Average CPT Results		
				LL	PI	Corrected Tip Stress (MPa)	Pore Pressure (MPa)	Friction Ratio
1D	0	SC	23	37	18	16	0.11	2.6
1C	4.7	SP, CL	24	53	30	11	0.13	2.9
1B	7.5	SC, CL	15	41	20	7.9	0.07	2.9
1A	9.5	SC	14	39	19	3.4	0.64	3.5
2C	11.4	SM, SC	13	30	9	10	0.02	1.7
2B	15.1	SM	11	24	9	13	-0.03	0.8
2A	18.7	SP-SM	5	29	16	22	0.07	0.5
3	24.5	SC, SM	11	45	26	6.9	0.45	1.1
4	28.9	SP-SM	7	36	17	18	0.23	0.6
5C	35.3	SC	13	42	21	3.0	0.78	2.2
5B	38.1	SP-SM	9	31	9	17	0.22	0.9
5A	40.2	SC, CL	13	40	16	3.8	1.1	1.9
6	42.9	SC	16	48	20	9.7	2.9	1.7

Interpretation of Test Results

Based on published relationships, several calculated and correlated soil parameters were determined as part of the data interpretation. The published correlations are generally based on a combination of theoretical and semi-empirical concepts, many of which are based on site- or region-specific information. Several published correlations for use with CPT/SCPT and DMT data were determined to be applicable to the site. For the project, correlations were selected that represent commonly-referenced procedures believed to be applicable across a wide range of subsurface conditions. Site-specific correlations and/or methods were used whenever available. The interpretation methods of constrained modulus and related parameters are presented in Table 2.

Table 2. Interpretation methods used.

Test Method	Parameter		Reference
CPT	Stress-normalized tip resistance	q_{t1N}	Schmertmann and Crapps (1988)
	Normalized friction ratio	FR	Wroth (1984)
	Pore pressure parameter	B_q	Wroth (1984)
	Soil behavior type	SBT	Robertson and Fear (1995)
	Soil behavior type index	I_c	Site-specific
	Vertical Constrained modulus	M	Mayne (2006)
DMT	Material Index	I_D	Marchetti (1980)
	Constrained modulus	M	Marchetti (1980)
	Overconsolidation ratio	OCR	Mayne (1995), Marchetti (1980)
	Effective preconsolidation Stress	σ'_p	Marchetti (1980)
	Initial vertical effective stress	σ'_{vo}	Marchetti (1980)
ICON	Incremental loading consolidation test		ASTM 2435
CRS	Constant rate of strain consolidation test		ASTM 4136

COMPARATIVE ANALYSES

Comparative analyses of field and laboratory-derived soil properties were performed to visualize and evaluate the soil properties measured, calculated, and correlated from the subsurface data collected at the site. The comparative analyses assisted in the assessment of accuracy and reliability of individual data sets. These comparisons consisted of graphs of soil parameters versus elevation and depth at paired exploration locations across the site. Nearly 250 of these comparative analyses were performed in total, for 14 different soil parameters, each at up to 35 paired exploration locations. An example of comparative analyses for M is presented in Figure 1. Some laboratory results are also presented.

STATISTICAL ANALYSES

Geotechnical engineering and index parameters for the soil were measured, calculated or correlated from field and laboratory test data. A statistical analysis of each data set and blended data set was performed to evaluate the central tendency and variability of the soil properties for each soil layer. Statistical analyses were performed for 19 different parameters for each of the 13 interpreted engineering soil layers.

The results of the statistical analyses were visualized using a series of "box-and-whisker" plots. Box-and-whisker plots were selected to compare relative distributions between engineering layers because measures of central tendency, variability, and overall range are apparent. The

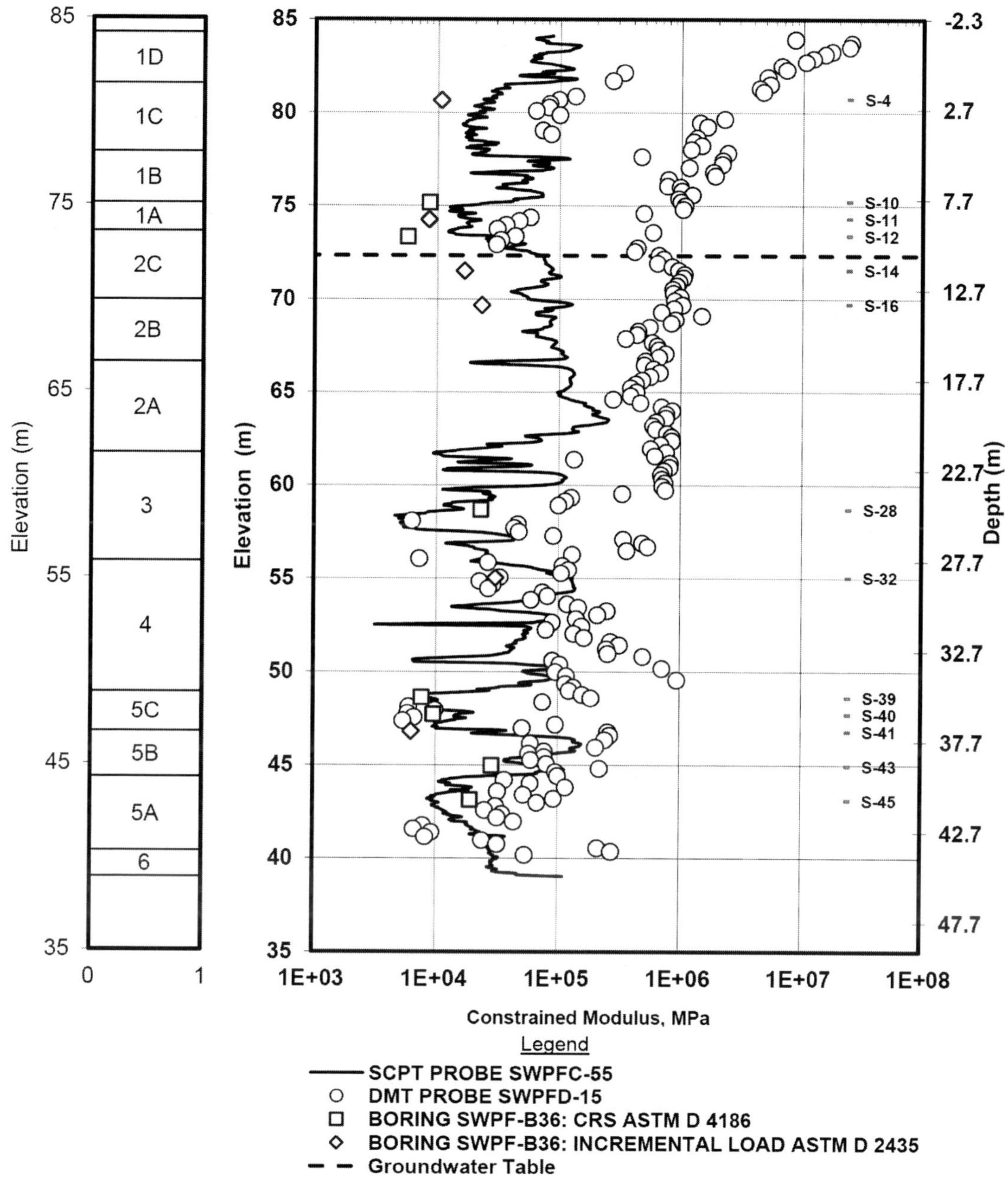

Figure 1. Comparative analysis of constrained modulus for a paired set of CPT, DMT, and laboratory consolidation results (Boring B36, SCPT C55, and DMT D15).

box-and-whisker plots show median (vertical bar), inter-quartile range (shaded boxes), and minimum and maximum data range (whiskers). The median is defined as the data point such that there are an equal number of values larger and smaller than that data point. For an odd number of data points, the medium is the middle value. For an even number of data points, the median is the mean of the two middle values. The inter-quartile range divides the data set into quarters. For example, the first quartile is a value that has 25 percent of the observations below its value (Johnson, 2000).

Confidence intervals that are based on the standard deviation do not provide a representative measure of spread about the central tendency if the underlying distribution is far from normal. As can be observed from the box plots, some of the distributions are skewed; therefore, other measures of variability or dispersion about the central tendency are more appropriate. For some data sets, the sample size was too small to calculate meaningful confidence limits.

Box-and-whisker plots were developed for (a) all data sources for each parameter for each soil layer; and (b) all soil layers for each soil parameter and data source. An example of a box-and-whisker plot for all data sources for the constrained modulus in a specific layer is shown in Figure 2. Similar plots were developed for every soil layer. Figure 2, CV denotes the coefficient of variation, N denotes the number of samples in the population, and XHL denotes crosshole sonic logging. Figures 3 and 4 present the constrained modulus for all layers using the CPT-based correlation and DMT-based correlation, respectively.

DISCUSSION

Potential sources of bias

In the selection of the design M's, the uncertainty in the data source from direct measurements, calculations or empirical correlations, was evaluated. In the comparative and statistical analyses, biases were observed in the data source or data sets for a given soil property. The observed differences or bias among data sets for M could be the scatter imbedded in the measurement technique, subsurface conditions, or the empirical correlations used to estimate some of the soil properties. These potential sources of bias tend to limit the ability to make conclusions regarding the relative accuracy and precision between two data sources or sets and the source of the bias.

The various data sources were assumed to be reliable as they all have been extensively used throughout the field of geotechnical engineering. Strict quality assurance procedures were followed during the field explorations and laboratory testing to increase the confidence in the reliability of these tests.

However, in the laboratory, two significant sources of biases were observed late in the CRS laboratory testing programs that were not part of or not controlled by the quality assurance program. First, upon extrusion, the soil specimens quite noticeably expanded and separated vertically into individual lens of material along thin layers of sand. Second, the CRS lab technician used a thick aluminum blade to trim the soil specimens and roughly pushed the consolidation ring into the soil. By contrast, the ICON lab technician used a wire soil cutter and gently handled the specimens. The failure to use a wire soil cutter and delicately treat the CRS soil specimens caused significant mechanical disturbance of the consolidation test specimens. Hence, all the CRS tests appeared to have significant laboratory induced disturbance as their stress-strain curves were more rounded than the ICON stress-strain curves. ICON test specimens likely incurred at least the disturbance produced by extrusion from the Shelby tubes.

The in situ soundings were advanced at distances of 1.5 to 8 meters or more apart. Changes in subsurface conditions over this distance could have contributed to the bias between two data sets of the same soil property.

Empirical correlation equations used for the project were developed by various researchers in the cited references. Empirical relationships are based on a database of soil properties that generally exhibit scatter. The final predictive correlation equation is only one best-fit estimate of that soil property, and there is a distribution (or scatter) of data about that estimate. The estimate could represent the average, lower bound, upper bound, or some other value depending on the interest of the author. For example, one empirical correlation equation using inputs from CPT estimates M lower than that from the DMT in part because CPT shears at the ultimate or critical shear strain while the DMT shears at a strain about three orders of magnitude lower as shown in Figure 5 (Mayne, 2001).

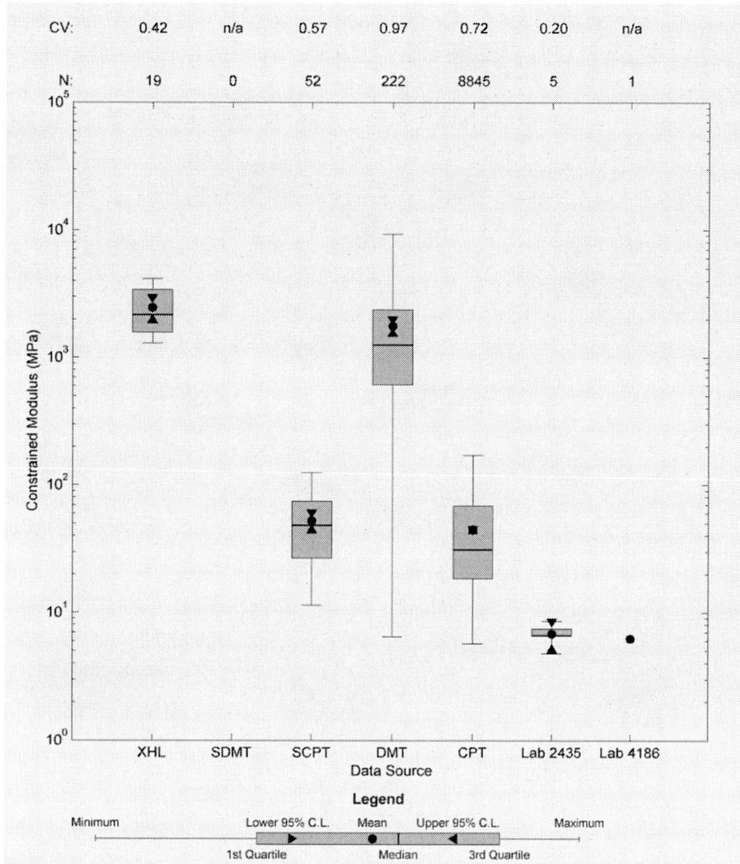

Figure 2. Box-and-whisker statistical visualization for constrained modulus correlated from all data sources for soil layer 1C, including ICON and CRS consolidation tests.

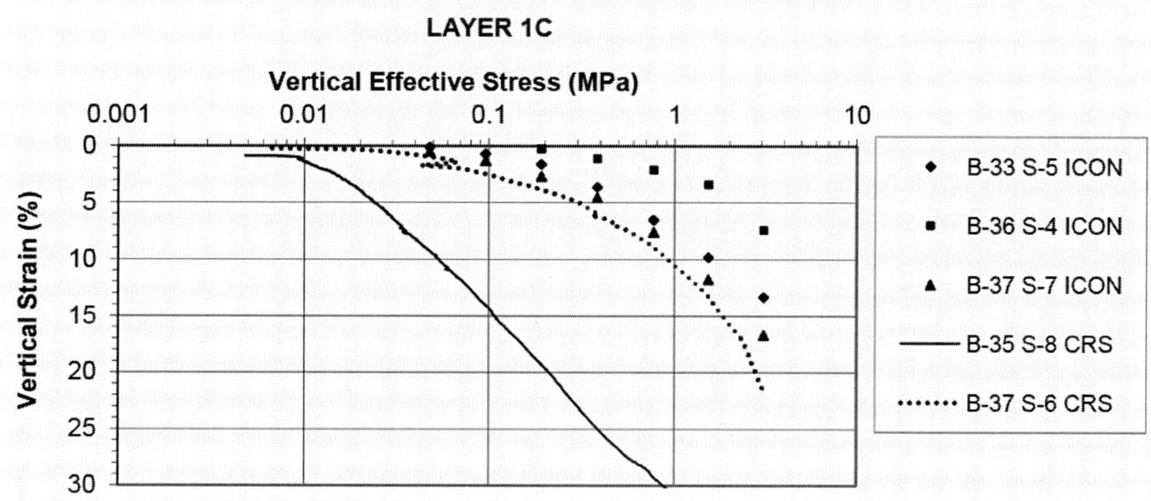

Figure 3. Laboratory ICON and CRS test stress-strain curves for soil layer 1C; CRS curves suggest significant disturbance, believed to be caused by extrusion and trimming of soil specimens.

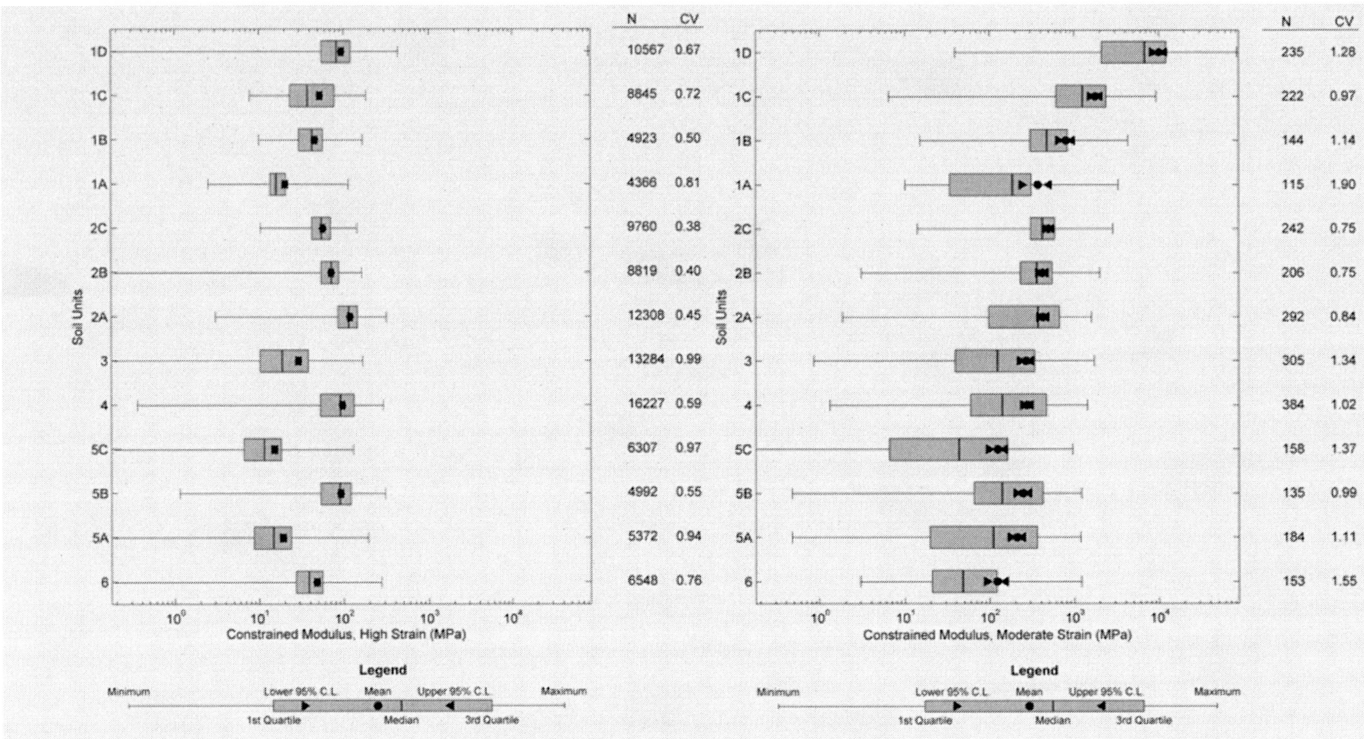

Figure 4. Box-and-whisker statistical visualization for constrained modulus from CPT (high shear strain) and DMT (moderate shear strain) for all soil layers.

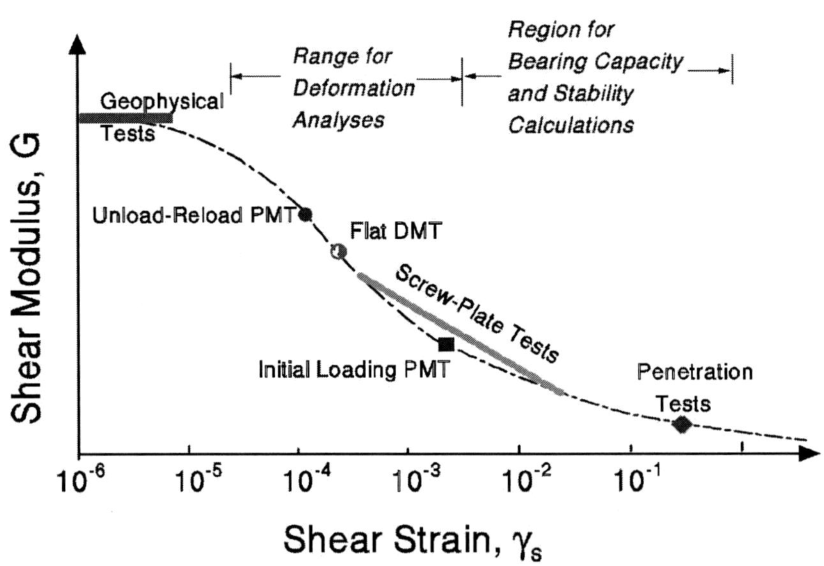

Figure 5. Conceptual illustration of shear modulus reduction, hence disturbance, incurred by various method of in – situ testing and their corresponding appropriate use for geotechnical analyses (Mayne, 2001).

Constrained modulus

At 14 paired test locations, M estimated from CPT has been compared with those determined from ICON and CRS consolidation tests performed on undisturbed samples from the nearest SPT borings. Direct-push Shelby tubes, pitcher barrel samplers, or hydraulically-operated fixed-piston samplers were used, depending on the soil type sampled. At six paired test locations, M estimated from DMT were compared with those determined from ICON and CRS tests performed on undisturbed samples from the nearest SPT borings.

The M estimated from the ICON and CRS laboratory tests compares favorably with the CPT-correlated values below the groundwater table, but generally is higher above the groundwater table. In particular, the high shear strains at which the CPT operates seems to confirm that both the ICON and CRS soil specimens experienced significant disturbance. The larger constrained moduli in the engineering layers above the groundwater table are likely due to the significant overconsolidation caused by desiccation in Layers 1D and 1C.

The DMT generally measures constrained moduli an order of magnitude greater than those from the other data sets; M reduces only slightly for the relatively loose engineering layers 3, 5A, and 5C, and is more sensitive, as indicated by the larger coefficients of variation, to changes in the stiffness of the soil (Schmertmann et al,, 1986). Unit weight tests on samples retrieved from these layers indicate relatively loose soils, and other in-situ tests (e.g., CPT) and laboratory consolidation tests indicate lower M in these layers. Because of the apparent amount of in situ destruction of the soil matrix by the CPT (particularly at the hard bottom of engineering layer 4 before entering the "soft zones" starting at 5C) and disturbance in the laboratory due to sample extrusion, the DMT M were used in the engineering analyses to compute settlement.

Preconsolidation stress and overconsolidation ratio

Figure 6 compares the preconsolidation stress and overconsolidation ratios determined by the DMT versus those derived from ICON and CRS tests using the Becker and Casagrande methods to determine ranges of minimum and probable estimates. These data suggest that the DMT estimates a preconsolidation stress that equals or often greatly exceeds those estimated by the Casagrande or Becker methods; note that the Casagrande method depends on the selection of point of maximum curvature that often does not exist or is difficult to determine. Typically, the data suggest that the Becker probable stress agrees better with the DMT estimates of P_c. By contrast, the CPT significantly underestimates the overconsolidation ratio because the high shear strain at which the CPT operates erases all stress history and obliterates the structure of the soil.

CONCLUSIONS

Constrained moduli correlated from multiple data sources at a site composed of silty and clayey sands have been compared. Comparisons were performed among results correlated from different in-situ techniques (i.e., CPT and DMT) and among correlated M and laboratory ICON and CRS consolidation tests. Because of the observed disturbance incurred during extrusion and trimming of the consolidation test specimens, the CPT's high strain/high disturbance M correlated M well with laboratory measurements. In particular, the flattened stress strain curves from the CRS consolidation tests suggest that their soil specimens experienced the most disturbance. In addition, engineering management leadership in instituting consistent training of lab technicians in proper equipment and specimen handling could help reduce these disturbance issues.

The DMT-correlated M for all soil layers, particularly above the water table, exceeded the CPT and consolidation test values by an order of magnitude or more. DMT-correlated M's generally measured the highest OCR, even in the looser layers. In the looser layers, the DMT-correlated M and OCR did not exhibit a reduction, as would be expected for a technique that protects soil structure and stress history in the surrounding in-situ. It appears that the DMT, not the CPT, should be used to estimate M in-situ and be used for settlement design analyses.

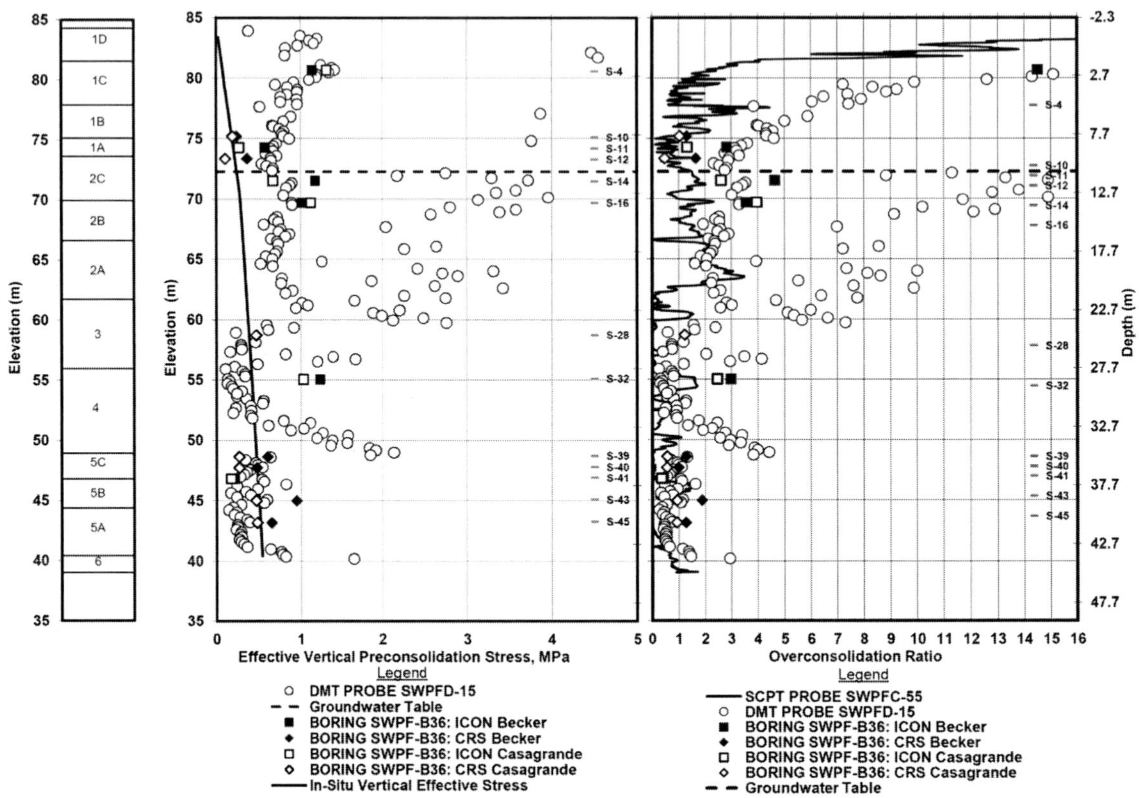

Figure 6. Comparison of preconsolidation stress and overconsolidation ratios from DMT, ICON and CRS tests, and CPTs.

REFERENCES

Becker, D.E., Crooks, J.H., Been, K., and Jefferies, M.G. (1987). "Work as a criterion for determining in situ and yield stresses in clays. *Can. Geotech. J., 24(4), 549-564.*

Casagrande, A. (1936). "The Determination of the Pre-Consolidation Load and Its Practical Significance," Discussion D-34, *Proceedings of the First International Conference on Soil Mechanics and Foundation Engineering, Cambridge,* Vol. III, pp. 60– 64.

Johnson, R.A., 2000, Miller & Freund's probability and statistics for engineers, 6th ed.: New Jersey, Prentice hall.

Marchetti, Silvano, 1980, "In situ tests by flat dilatometer." *J. Geotech. & Geoenv. Eng* Journal of the Geotechnical Engineering Division, v. 106, no. GT3, p. 299-321

Marchetti, Silvano, 1997, The flat dilatometer: Design applications, keynote lecture, in Third International Geotechnical Engineering Conference, Cairo, Egypt, 1997, *Proceedings: Cairo, Egypt, Cairo University, p. 421-448,* available: http://www.marchetti-dmt.it/.

Mayne, P., 2001, Stress-strain-strength-flow parameters from enhanced in-situ tests. Proceedings, *International Conference on In-Situ Measurement of Soil Properties & Case Histories In-Situ 2001, Bali, Indonesia, May 21-24, 2001, p. 27-48.*

Mayne, P., 2006. Interrelationships of DMT and CPT readings in soft clays, *Proceedings of the 2nd International Flat Dilatometer Conference.*

McNulty, E.G. and Harney, M.D., 2010. "Comparison of CPT- and DMT-correlated effective friction angle in clayey and silty sands," *Proceedings of the 2nd International Conference on Cone Penetration Testing,* Huntington Beach California, May 10-12.

Robertson, P.K. and Campanella, R.G., 1983, Interpretation of cone penetration tests, part I: sand: *Canadian Geotechnical Journal, v. 20, no. 4, p. 718-733.*

Robertson, P.K. and Fear, C.E., 1995, Liquefaction of sands and its evaluation, in Ishihara, Kenji, ed., Earthquake geotechnical engineering, *The First International Conference on Earthquake Geotechnical Engineering, IS-Tokyo '95, Tokyo, 1995, Proceedings: Rotterdam, A. A. Balkema.*

Schmertmann, J.H., 1975, Measurement of in-situ shear strength, *in ASCE Specialty Conference on In-Situ Measurement of Soil Properties, v. 2, pp. 57-138.*

Schmertmann, J.H., , W. Baker, R. Gupta, and K. Kessler,1986, CPT/DMT QC of Ground Modification at a Power Plant, *Proceedings of In Situ '86 GT Div., ASCE, June 23-25, 1986 Blacksburg, VA*

Schmertmann, J. H., and D. Crapps, 1988, Guidelines for Using the CPT, CPTU and Marchetti DMT for Geotechnical Design: DMT test methods and date reduction, Rep. No. FHWA-PA-87-014-84-24, *Pennsylvania. Dept. of Transportation. Office of Research and Special Studies, United States. Federal Highway Administration.*

Senneset, K.; Sandven, R.; and Janbu, N., 1989, Evaluation of soil parameters from piezocone tests: *Transportation Research Record no. 1235, p. 24-37.*

Wroth, C.P., 1984, The interpretation of in situ soil tests: *Geotechnique, v. 34, no. 4, p. 449-489.*

Estimating van Genuchten Parameters α and n for Clean Sands from Particle Size Distribution Data

By Craig H. Benson[1], I. Chiang[2], Tanit Chalermyanont[3], and Auckpath Sawangsuriya[4]

[1]Wisconsin Distinguished Professor and Chair, Dept. of Geological Engineering, Dept. of Civil and Environmental Engineering, University of Wisconsin, Madison, WI 53706, USA, chbenson@wisc.edu
[2]Former Graduate Student, Dept. of Civil and Environmental Engineering, University of Wisconsin, Madison, WI 53706, USA
[3]Associate Professor, Dept. of Civil Engineering, Faculty of Engineering, Prince of Songkla University, Songkla 90112, Thailand, tanit.c@psu.ac.th
[4]Researcher, Bureau of Road Research and Development, Dept. of Highways, Bangkok 10400, Thailand, sawangsuriya@gmail.com

ABSTRACT: The effect of median particle size and breadth of particle sizes in clean sands on the SWCC and the van Genuchten parameters α and n is described. For uniform sands, the air entry suction increases (α decreases) and the steepness diminishes slightly (n decreases) for the drying SWCC as the particles become finer. Increasing the breadth of particle sizes also increases the air entry suction (lower α) and reduces the steepness (lower n) of the SWCC. Breadth of the particle size distribution has a larger effect on steepness and n of the SWCC, and median particle size has a larger effect on the air entry suction and α. Similar effects occur for the wetting SWCC, although the impact on the water entry suction is less than the impact on the air entry suction. The systematic effects of median particle size and breadth of particle sizes on the SWCC are captured in a pedotransfer function (PTF) to estimate α and n for clean sand based on d_{60} and C_u. Comparison with other SWCC data shows that α estimated with the PTF typically is within a factor of 2 of the actual α and n estimated with the PTF is within ± 2 of the actual n. These findings are applicable to α for sands with non-plastic fines provided the fines content is less than 20%, but apply to n only for sands with negligible fines content. A procedure to convert between van Genuchten and Brooks-Corey parameters for sands is also presented.

Key words: unsaturated soil, soil water characteristic curve, sand, van Genuchten parameters, Brooks-Corey parameters, pedotransfer function

INTRODUCTION

The soil water characteristic curve (SWCC) defines the relationship between volumetric water content (θ) and matric suction (ψ) and is used in essentially all analyses related to unsaturated soil. An example of a SWCC for sand is shown in Fig. 1. The SWCC generally is defined using a functional relationship, the most common being the van Genuchten equation (van Genuchten 1980):

$$\Theta = \frac{\theta - \theta_r}{\theta_s - \theta_r} = \left[\frac{1}{1 + (\alpha\psi)^n}\right]^{1-n^{-1}} \qquad (1)$$

where θ_s is the saturated water content, θ_r is the residual water content, Θ is the effective saturation, and α and n are the van Genuchten parameters. The parameters α, n, and θ_r generally are obtained by non-linear regression of Eq. 1 on SWCC data, whereas θ_s is obtained by weight-volume relationships (θ_s = porosity). The α parameter is related inversely to the air entry suction for drying (ψ_a, suction at which soil begins to de-saturate, Fig. 1) or the water entry suction for wetting (ψ_w, Fig. 1), the n parameter describes the slope of the SWCC for $\psi > \psi_a$ or ψ_w, and θ_r describes the lowest practically attainable water content. For a properly measured SWCC, θ_r is close to or equals zero.

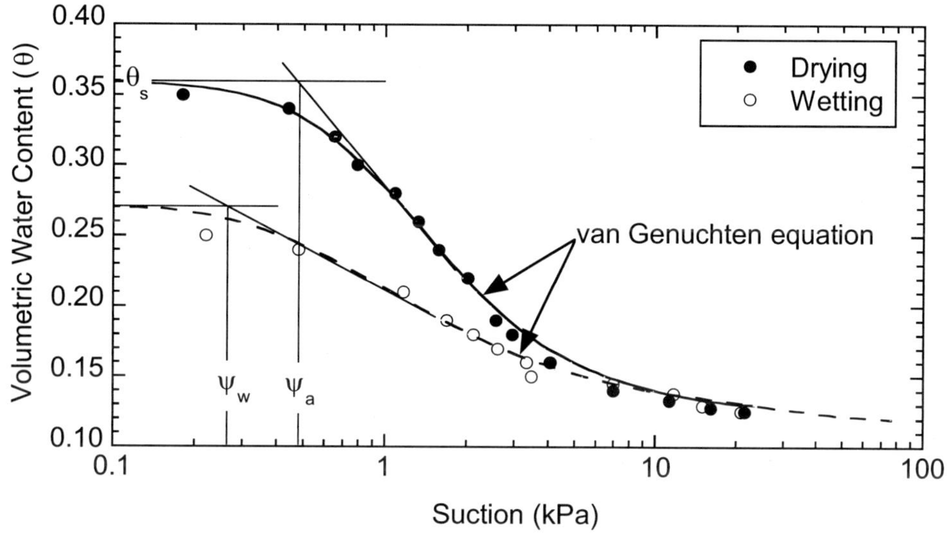

Figure 1. Example of SWCC showing fit of van Genuchten equation and air entry suction (ψ_a), water entry suction (ψ_w), and saturated water content (θ_s).

Most SWCCs are measured in the laboratory using procedures such as those described in ASTM D6836 (*Standard Test Methods for Determination of the Soil Water Characteristic Curve for Desorption Using a Hanging Column, Pressure Extractor, Chilled Mirror Hygrometer, and/or Centrifuge*). Measurement of SWCCs is time consuming (weeks to months) and expensive (US $850-1500 per test in 2013). Thus, alternative methods to estimate the SWCC are sought (Aubertin et al. 2003, Perera et al. 2005, Chin et al. 2010), particularly during preliminary design. Empirical functional relationships based on the particle size distribution and other compositional factors, know as "pedotransfer functions" (PTFs), are the most common method to estimate the SWCC (van Genuchten et al. 1992). Examples for coarse-grained soils based on particle size data include the modified Kovács model by Aubertin et al. (2003) and the NCHRP model by Perera et al. (2005).

This paper describes a PTF method to estimate the van Genuchten parameters α and n for clean sands that is intuitive, simple, and quick and relies only on information typically reported for a particle size analysis (d_{60} and the coefficient of uniformity, C_u). The method was developed based on a series of experiments where SWCCs were measured for a collection of sands prepared systematically with a range of median particle size and breadth of particle sizes. The method is validated using a database of SWCCs for sands from other sources and studies.

BACKGROUND

The distribution of pore sizes in soil governs the SWCC (Lu and Likos 2004). In accordance with the Young-Laplace equation, soils having larger pores de-saturate at lower suctions (lower

ψ_a) and retain less water at a given matric suction compared to soils with smaller pores (Fig. 2a). The water content of soils with a broader distribution of pore sizes varies more gradually as the matric suction varies (Fig. 2b) (Hillel 1998). In terms of α and n, soils with larger pores have larger α and soils with a broader distribution of pore sizes have lower n. Soils comprised of minerals that have lower surface charge and less affinity for hydration also have lower water content at a given matric suction and larger α (Tinjum et al. 1997).

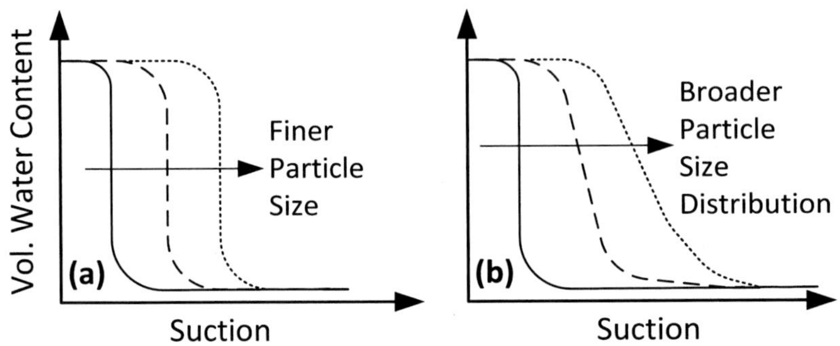

Figure 2. Conceptual diagram of effect of median particle size of uniform sand (a) and breadth of particle size distribution (b) on shape of the SWCC of sand.

Clean sands and other coarse materials with limited fines exhibit a strong relationship between particle size distribution, pore size distribution, and the SWCC (Yang et al. 1994, Aubertin et al. 2003, Jaafar and Likos 2011). As the particle sizes become finer, the pore sizes become smaller (Juang and Holtz 1986), and in accordance with the Young-Laplace equation, higher suctions are required to de-saturate pores. Similarly, sands with a broader distribution of particle sizes typically have finer pores because of the packing efficiency afforded by a broad particle size distribution (Juang and Holtz 1986). Thus, more broadly graded sands retain more water for a given matric suction. The diversity of pore sizes also increases as the particle size distribution becomes broader, resulting in a SWCC with more gradually varying slope. Consequently, α and n should vary systematically with the median particle size and the breadth of particle sizes in sand: α should decrease as the median particle size decreases and breadth of the particle sizes increases (Fig. 3a) and n should decrease as the breadth of particle sizes increases (Fig. 3b).

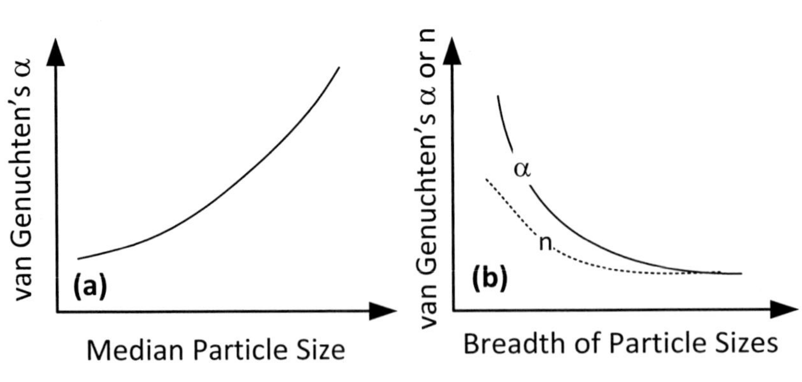

Figure 3. Conceptual diagram of effect of median particle size of uniform sand (a) and breadth of particle size distribution (b) on α and n.

The objective of this study was to develop a simple, intuitive, and quick quantitative method to estimate α and n based on particle size distribution data that captures the conceptual relationship between α and n and the size and distribution of particles in sand. Sands with a range of median particle size (e.g., fine to coarse sand) and distribution of particles were custom blended using sand and gravel-size particles from three coarse-grained soils. The SWCC of each custom-blended sand was measured using the hanging column procedure in ASTM D6836 (Method A) and subsequently fit with Eq. 1 using a non-linear optimization procedure. During optimization, θ_s was set at the porosity when fitting the wetting curve and allowed to vary when fitting the drying curve. The parameters α and n for these sands were related to parameters describing the center of the particle size distribution and the breadth of particle sizes.

MATERIALS AND METHODS

Soils

Three soils were used in this study: Torpedo sand, Wisconsin River stone, and Verona gravel. Custom-blended sands were prepared in groups by blending select particle size fractions from the four soils. The d_{50} for each group was fixed, and C_u was varied from 1 to 43 to simulate conditions varying from uniformly graded to broadly graded. Particle sizes were proportioned so that the particle size curve varied smoothly (none of the sands were gap graded).

Torpedo sand, a clean and broadly graded sand (Fig. 4), was the primary source of sand-size particles. A 100-L drum of torpedo sand was washed with tap water, air-dried, and segregated into groups of different size particles using sieves. Wisconsin River stone and Verona gravel were sources of larger gravel particles (Wisconsin River stone: 9.5-12.7 mm; Verona gravel: 12.7-19.7 mm) not available in the Torpedo sand. An example of particle size distributions that were created is shown in Fig. 5. Particle size distributions for the Wisconsin River stone and Verona gravel are not shown in Fig. 4 because only a fraction of the large particles from these soils were used in the experiments.

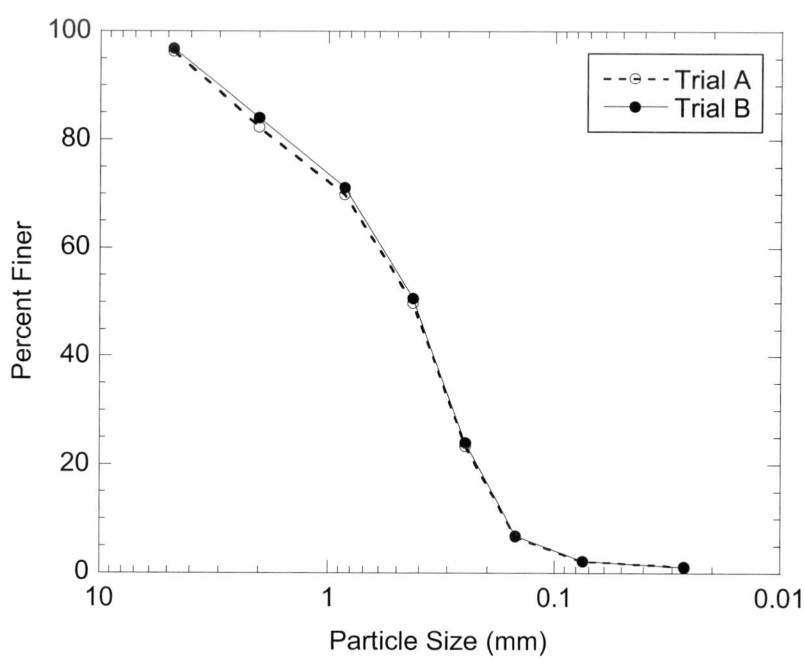

Figure 4. Particle size distribution of Torpedo sand.

All of the tests described in this paper were conducted on sands with no fines. However, sands blended with non-plastic fines were evaluated as a separate part of this study (Chiang 1998). Chiang (1998) shows that the findings for α reported in this paper are applicable to sands with non-plastic fines provided the fines content is less than 20%. In contrast, the findings for n apply only to sands with negligible fines.

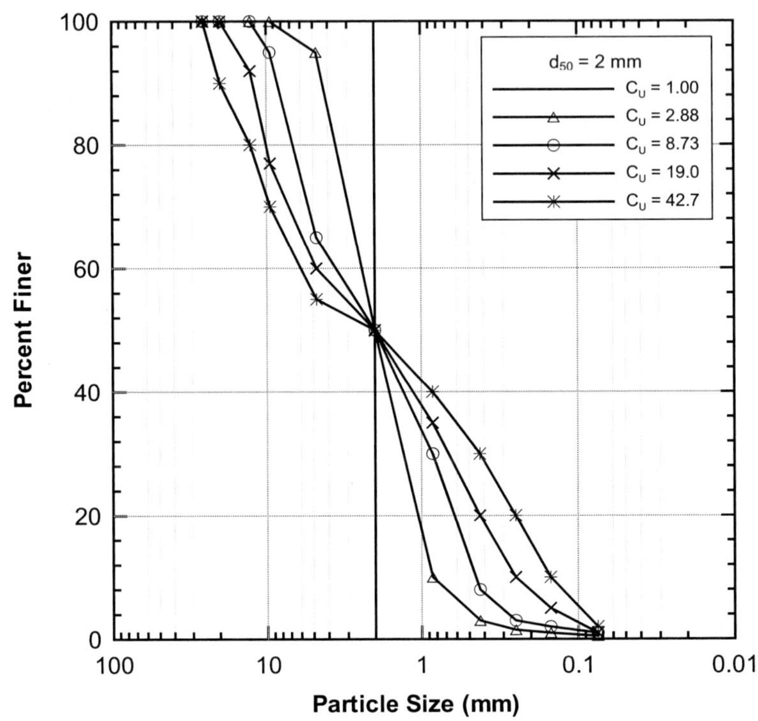

Figure 5. Example of particle size distributions of custom-blended sands for d_{50} = 2 mm.

SWCC

A hanging column apparatus identical to the apparatus described in ASTM D6836 was used to measure the SWCC of the sands. This apparatus is the same as the apparatus described in Wang and Benson (2004). Specimens were prepared in a volume-calibrated Büchner funnel in a single lift compacted using an aluminum tamper consisting of a 12-mm-diameter disk attached to an aluminum rod. Tamps were applied across the surface until the density ceased changing. All specimens had a diameter of 87 mm and a thickness of 15-20 mm.

Specimens were inundated with de-aired tap water after compaction and allowed to soak for 24 hr. After soaking, free water was removed with a syringe, the outflow apparatus was attached and filled with water, and the specimen and apparatus were allowed to equilibrate at zero suction for 48 hr. Matric suction was then applied incrementally in accordance with the procedure in ASTM D6836-Method A. Drying and wetting curves were measured for each specimen, with the wetting curve measured by incrementally decreasing the suction after the drying curve was completed.

INFLUENCE OF PARTICLE SIZE DISTRIBUTION ON THE SWCC

Effect of Median Particle Size

SWCCs for the uniformly graded blends are shown in Fig. 6 for drying and wetting in terms of effective saturation (Θ) vs. suction. Effective saturation is used in lieu of θ to eliminate the influence of porosity on θ_s as the particle size distribution varies. Each of the SWCCs in Fig. 6 corresponds to a sand with $C_u = 1$. Each of the SWCCs can be found in Chiang (1998)

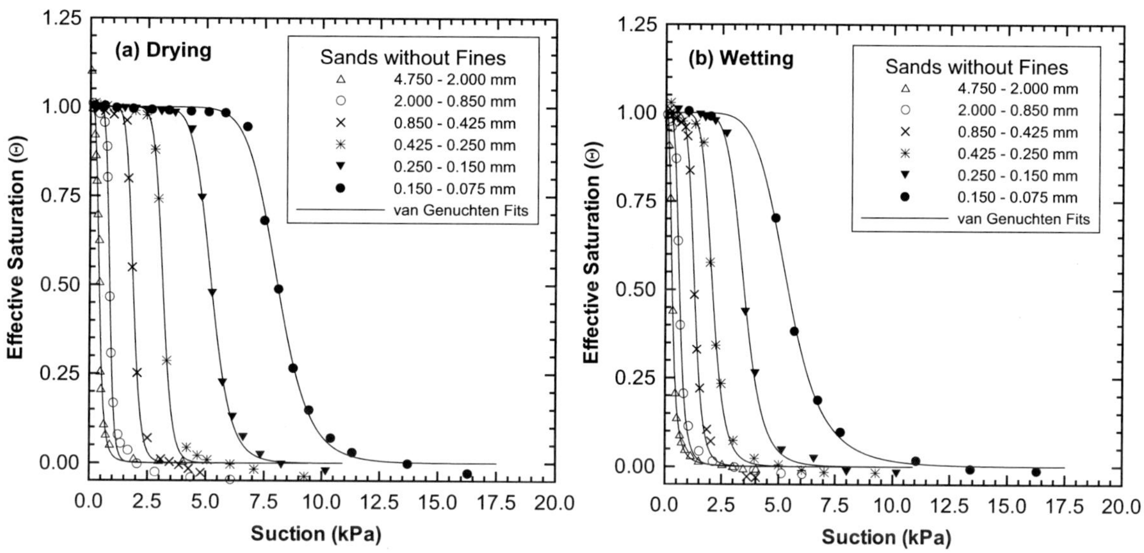

Figure 6. SWCCs of clean uniform sands ($C_u = 1$) for (a) drying and (b) wetting.

Desaturation of the uniformly graded sands occurs at higher suctions (i.e., larger ψ_a) as the sand becomes finer, and the finer sands retain more water at a given suction. The effective saturation also drops dramatically after desaturation is initiated during drying (i.e., the SWCC has a steep slope, Fig. 6a) and increases rapidly once re-saturation begins during wetting (Fig. 6b). The suction at which saturation is reached on wetting (i.e., the water entry suction, ψ_w) is less sensitive to particle size than ψ_a, and ψ_w is less than ψ_a for each sand. Steepness of the SWCC diminishes modestly as the sand becomes finer for both the drying and wetting curves.

The increase in ψ_a as the particle size becomes finer is consistent with the sensitivity of the SWCC described conceptually with Fig. 2. Finer sands have smaller pores that remain saturated at higher suctions. The very steep SWCCs reflect the relatively uniform pore size distribution in uniformly graded sands. Increasing the suction slightly following desaturation of the largest pore ($\psi > \psi_a$) results in drainage of nearly all of the pore space. The modest reduction in steepness as the sand becomes finer suggests that the finer sands had a slightly broader pore size distribution relative to the coarser sands, even though all of the sands in Fig. 6 were uniformly graded with $C_u = 1$.

Effect of Breadth of Particle Sizes

The effect of breadth of particle sizes on the SWCC is shown in Fig. 7 for sands with a median particle size of 2 mm (coarse sands) and C_u ranging from 1 to 43 (uniformly graded to very broadly graded). SWCCs with similar shape were obtained for each d_{50} (Chiang 1998). As the breadth of particle sizes increases, steepness of the SWCC diminishes because the distribution of pore sizes increases as the sand becomes more broadly graded. More broadly graded sands also have larger ψ_a and ψ_w because they have smaller pores overall due to more efficient packing.

However, breadth of particle sizes has more impact on steepness of the SWCC than on ψ_a, and less influence on ψ_a than median particle size (Fig. 6).

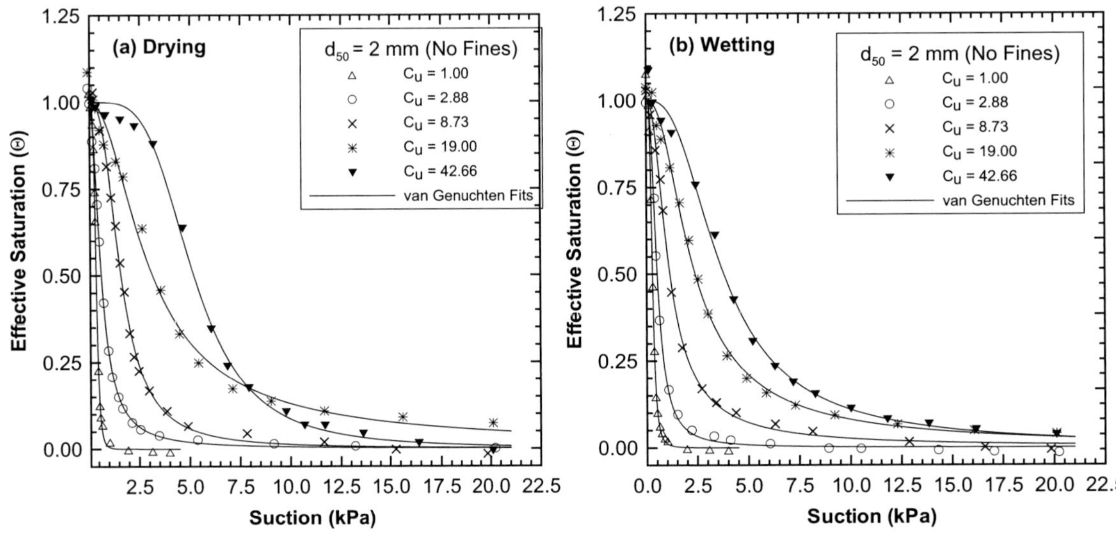

Figure 7. SWCCs of clean sands with a median particle size of 2 mm and C_u ranging from 1 to 43 for (a) drying and (b) wetting.

van Genuchten's α and n Parameters

The effect of median particle size and breadth of particle sizes on α for uniformly graded sands is shown in Fig. 8 for drying and wetting. Breadth of the particle size distribution was described by the coefficient of uniformity (C_u=d_{60}/d_{10}). The center of the particle size distribution was defined by d_{60} (rather than the median, d_{50}) for consistency with C_u and to promote the simplicity of the method, as d_{60} is frequently stated in laboratory reports when soils are classified (and d_{50} is not). In addition, for most sands, d_{50} and d_{60} are similar and the relationship between them is monotonic. Thus, d_{60} can be used in lieu of d_{50} in a PTF without loss of generality.

For the uniform sands (C_u = 1), α varies linearly with d_{60} for wetting and drying (Fig. 8a), which reflects the systematic variation in pore size with particle size for uniformly graded sands. The influence of breadth of particle sizes on α is shown in Fig. 8b in terms of normalized α (N_α), which is defined as the ratio of α for a given C_u to α for C_u = 1 (i.e., $N_\alpha = \alpha_{Cu} \div \alpha_{Cu=1}$). N_α decreases systematically as C_u increases, which reflects the systematic decrease in pore size as the breadth of particle sizes increases and the sand packs more efficiently. The variation in N_α with C_u is unique regardless of drying or wetting and can be described with the power function shown in Fig. 8b.

The effect of d_{60} and C_u on the n parameter for uniformly graded sands is shown in Fig. 9a for drying and wetting conditions. The systematic variation in n with d_{60} (Fig. 9a) is not as strong as observed for α (Fig. 8a), but the decrease in n for drying and wetting with increasing d_{60} is consistent with the diminished steepness of the SWCC evident in Fig. 6 as the sand becomes coarser. The n parameter for wetting is also smaller than the n parameter for drying for any d_{60}, which is consistent with the SWCCs for wetting of uniform sand (Fig. 6b) being less steep than the SWCCs for drying (Fig. 6a) for uniform sands.

The effect of breadth of particle sizes on n is shown in Fig 9b in terms of N_n, which is a normalized n defined analogous to N_α. As the sand becomes more broadly graded (C_u increases), N_n diminishes. The drop in n is substantial with a small increase in breadth of particles (i.e., C_u increasing from 1 to 2), which reflects the filling of larger pores by slightly smaller particles as the breadth of particle sizes increases. A more subtle decrease in n occurs as C_u increases beyond 2.

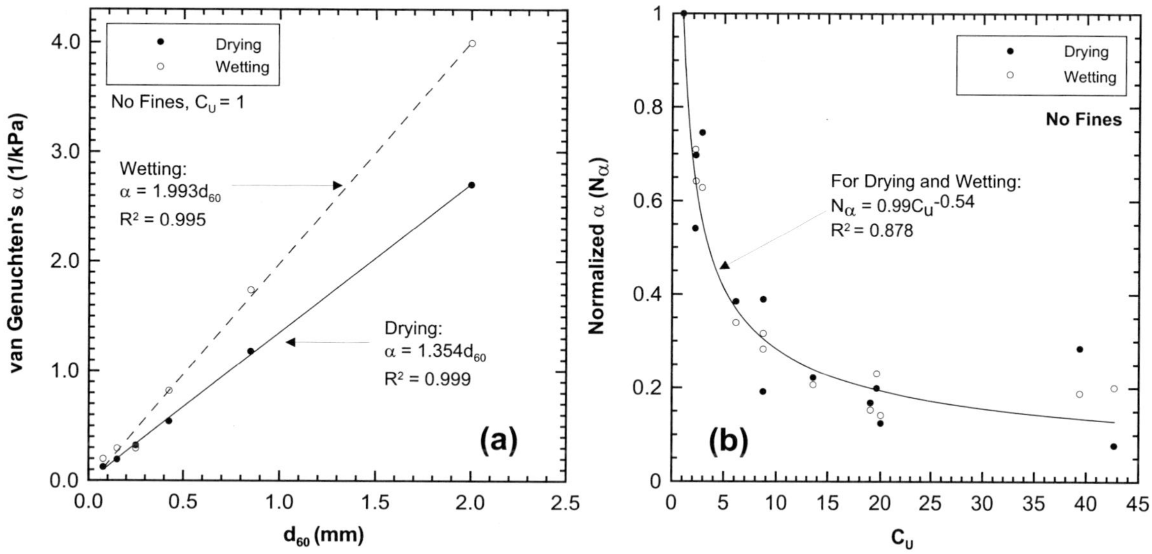

Figure 8. Effect of d_{60} on α of clean and uniform sand (C_u=1) (a) and effect of breadth of particle sizes (in terms of Cu) on α of clean sand (b).

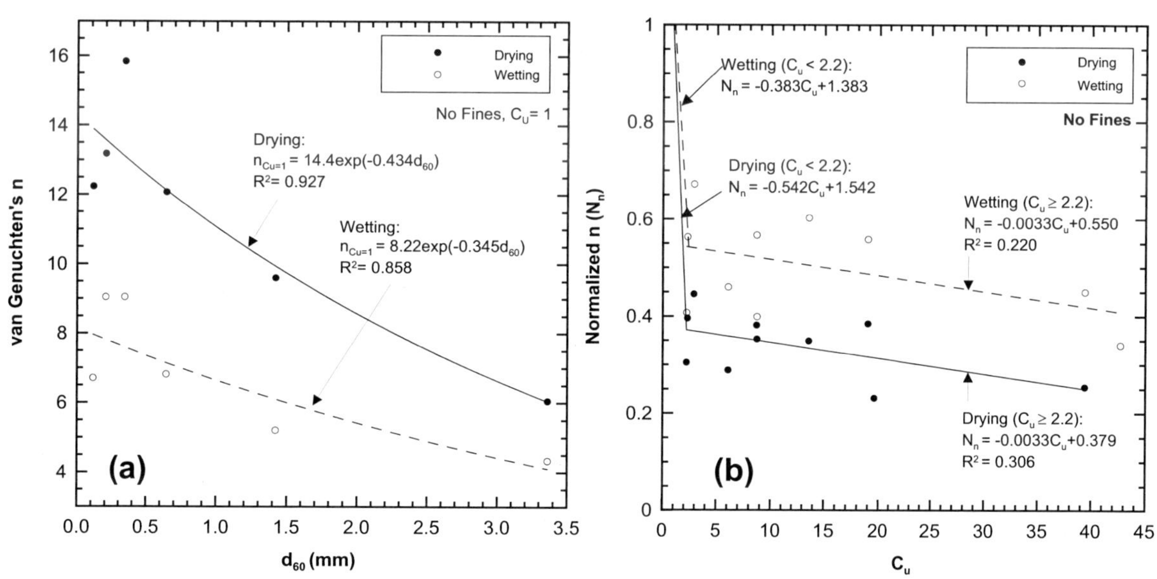

Figure 9. Effect of d_{60} on n of clean and uniform sand (C_u=1) (a) and effect of breadth of particle sizes (in terms of C_u) on n of clean sand (b).

PEDOTRANSFER FUNCTION (PTF) FOR CLEAN SANDS

The regression lines in Figs. 8 and 9 were used to create a PTF method for α and n for clean sands. The following procedure can be used:

1. Estimate α for wetting or drying assuming that the sand is uniformly graded ($C_u = 1$) using the regression equations relating α and d_{60} in Fig. 8a. These estimates are referred to as α_{1w} and α_{1d} (i.e., α for $C_u = 1$ for wetting and drying).

2. Correct the estimates of α from Step 1 to account for the breadth of particle sizes in the sand. Calculate the parameter N_α using the equation on Fig. 8b and C_u of the sand. Compute the corrected α for wetting (α_w) or drying (α_d) as:

$$\alpha_w = \alpha_{1w} N_\alpha \quad (2a)$$

$$\alpha_d = \alpha_{1d} N_\alpha \quad (2b)$$

3. Estimate n for wetting or drying assuming that the sand is uniformly graded ($C_u = 1$) using the regression equations relating n and d_{60} in Fig. 9a. These estimates are referred to as n_{1w} and n_{1d} (i.e., n for $C_u = 1$ for wetting and drying).

4. Correct the estimates of n made in Step 3 for gradation. Calculate N_n for wetting and drying using the known C_u of the sand and the equations on the graph on Fig. 9b. Compute the corrected n for wetting (n_w) and drying (n_d) as:

$$n_w = n_{1w} N_{nw} \quad (3a)$$

$$n_d = n_{1d} N_{nd} \quad (3b)$$

Experience using this method indicates that n obtained from Eqs. 3 can be too low for very large C_u, and the data collected in this study suggest that n for sands levels out at approximately 2.0 for large C_u. (Fig. 10). Thus, if n computed with Eq. 3 is unreasonable, then n can be set at 2.0.

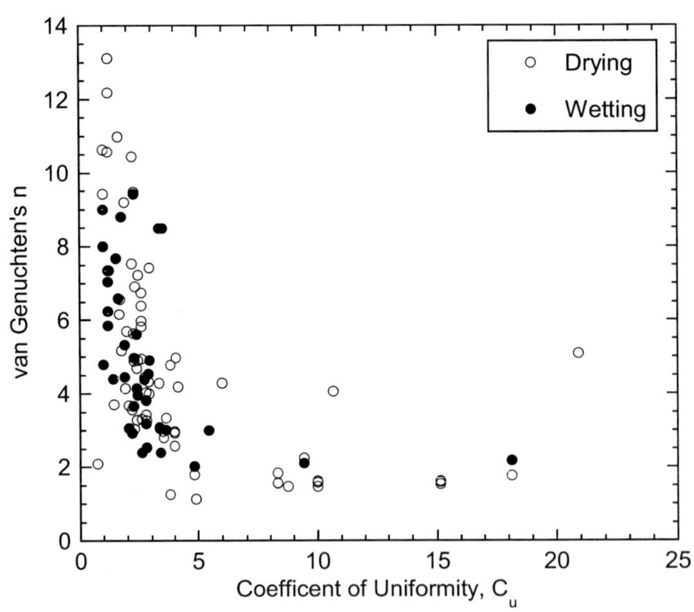

Figure 10. van Genuchten's n for clean sands vs. C_u for drying and wetting.

The α and n estimated using Eqs. 2 and 3 can be used with θ_s and θ_r to predict the SWCC From weight-volume calculations, θ_s is calculated as:

$$\theta_s = 1 - \frac{\gamma_d}{\gamma_s} \tag{4}$$

where γ_d is the dry unit weight and γ_s is the unit weight of solids. For the SWCCs in this study, θ_r obtained by fitting Eq. 1 to the SWCC data ranged from 0.030 to 0.045, and averaged 0.040. These θ_r represent the best fit of Eq. 1 to data, but should not be interpreted as being phenomenological. From a practical perspective, θ_r can be assumed to be 0.040 (or 0 without significant error).

The efficacy of this method was evaluated by applying it to sands from other sources and studies for which the SWCC and particle size distribution are known. Particle size characteristics and SWCC parameters for these sands are summarized in Table 1. Comparisons of α and n estimated using the aforementioned PTF procedure to the measured α and n are shown in Fig. 11. The estimated and predicted α exhibit essentially no bias and generally fall within a factor of 2 (dashed lines parallel to the 1:1 line). The estimated and predicted n exhibit more scatter than α, with n being predicted within ± 2 (dashed lines parallel to the 1:1 line) for most sands. This good comparison suggests that the PTF procedure provides a reliable estimate of α and n for clean sands. As an alternative to Eqs. 3, n can also be estimated directly from Fig. 10.

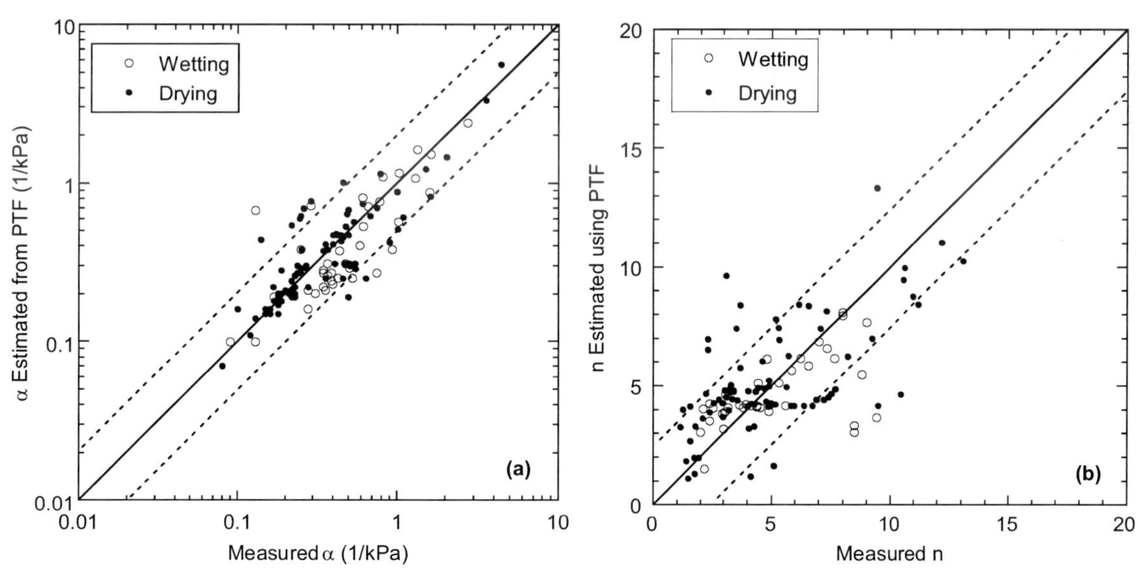

Figure 11. Comparison of α estimated from PTF to measured α (a) and n estimated from PTF to measured n (b).

Table 1. Particle size characteristics and measured and estimated van Genuchten parameters for sands evaluated in the Wisconsin Geotechnics Laboratory (WGL) or from the literature.

Source	Sample Name	d_{60} (mm)	d_{10} (mm)	C_u	% Fines	θ_s	θ_r	Measured				Estimated by PTF			
								Wetting		Drying		Wetting		Drying	
								α_w (kPa^{-1})	n_w	α_d (kPa^{-1})	n_d	α_w (kPa^{-1})	n_w	α_d (kPa^{-1})	n_d
Stormont & Anderson (1999)	Concrete Sand	0.68	0.20	3.40	1.0	0.350	0.050	0.13	2.40	-	-	0.67	3.50	-	-
Tanahashi et al. (2000)	Field Sand	0.93	0.17	5.47	0.0	0.333	0.069	0.66	3.00	-	-	0.71	3.17	-	-
	Glass Beads (0.05 mm)	0.05	0.05	1.00	0.0	0.379	0.033	0.09	8.00	-	-	0.10	8.08	-	-
	Glass Beads (0.1 mm)	0.10	0.10	1.00	0.0	0.402	0.025	0.17	8.00	-	-	0.19	7.94	-	-
	Glass Beads (0.2 mm)	0.20	0.20	1.00	0.0	0.366	0.022	0.25	9.00	-	-	0.38	7.67	-	-
She & Sleep (1998)	F75 Silica Sand	0.20	0.11	1.87	3.0	0.367	0.045	0.75	5.33	0.50	9.20	0.27	5.12	0.19	6.98
Kamon et al (1998)	Toyoura Sand	0.20	0.12	1.68	0.0	0.383	0.054	-	-	0.21	6.57	-	-	0.20	8.36
	Silica Sand #7	0.18	0.09	1.98	3.0	0.419	0.084	-	-	0.10	5.70	-	-	0.16	6.27
Simon (2001)	Yahara Sand	2.90	0.16	18.1	6.0	NA	0.120	1.03	2.17	1.61	1.76	1.15	1.48	0.82	1.31
Baker and Hillel (1990)	Sublayer (1.0-2.0)	1.52	1.07	1.41	0.0	0.374	0.000	2.74	4.40	-	-	2.39	4.10	-	-
	Sublayer (0.71-1.0)	0.87	0.73	1.19	0.0	0.433	0.000	1.63	5.85	-	-	1.51	5.65	-	-
	Sublayer (0.5-0.71)	0.62	0.52	1.19	0.0	0.415	0.000	1.30	6.23	-	-	1.07	6.16	-	-
	Sublayer (0.355-0.5)	0.44	0.37	1.19	0.0	0.446	0.000	0.78	7.35	-	-	0.76	6.57	-	-
	Sublayer (0.25-0.355)	0.31	0.26	1.19	0.0	0.441	0.000	0.62	7.04	-	-	0.53	6.85	-	-
	Sublayer (0.106-0.25)	0.18	0.12	1.54	0.0	0.444	0.000	0.35	7.68	-	-	0.27	6.15	-	-
	Sublayer (0.5-0.71)	0.62	0.52	1.19	0.0	0.400	0.000	1.30	6.23	-	-	1.07	6.16	-	-
Baker (2002)	NW1 - Coarse Sand	1.10	0.33	3.33	1.0	0.404	0.098	0.82	8.50	0.29	4.28	1.09	3.03	0.77	3.29
	NW2 - Coarse Sand	0.84	0.24	3.50	2.0	0.392	0.067	0.61	8.50	0.54	2.97	0.81	3.31	0.57	3.68
	NW4 - Coarse Sand	0.46	0.21	2.19	1.5	0.374	0.031	1.02	2.93	0.40	3.57	0.57	3.81	0.41	4.39
	NW5 - Gravel	1.07	0.22	4.86	4.0	0.340	0.018	1.59	2.02	1.09	1.80	0.87	3.03	0.61	3.29
	NW7 - Fine Sand	0.25	0.09	2.78	7.0	0.442	0.068	0.39	3.20	0.23	4.03	0.27	4.08	0.19	4.78
	NW8 - Fine Sand	0.26	0.09	2.89	7.0	0.370	0.099	0.35	4.54	0.18	4.33	0.28	4.06	0.20	4.75
	NW9 - Fine Sand	0.37	0.11	3.36	6.0	0.347	0.049	0.44	3.06	0.23	3.09	0.37	3.90	0.26	4.51
	NW10- Fine Sand	0.28	0.10	2.80	6.0	0.423	0.051	0.48	2.54	0.28	3.27	0.31	4.04	0.22	4.72
	NW11 - Silty Sand	0.19	0.07	2.71	11.0	0.434	0.098	0.28	4.39	0.16	4.44	0.21	4.17	0.15	4.91

Table 1. Particle size characteristics and measured and estimated van Genuchten parameters for sands evaluated in the Wisconsin Geotechnics Laboratory (WGL) or from the literature (continued).

Source	Sample Name	d_{60} (mm)	d_{10} (mm)	C_u	% Fines	θ_s	θ_r	Measured				Estimated by PTF			
								Wetting		Drying		Wetting		Drying	
								α_w (kPa^{-1})	n_w	α_d (kPa^{-1})	n_d	α_w (kPa^{-1})	n_w	α_d (kPa^{-1})	n_d
Baker (2001)	NW12 - Silty Sand	0.20	0.08	2.38	8.0	0.445	0.077	0.39	4.14	0.18	4.69	0.24	4.16	0.17	4.90
	NW16 - Fine Sand	0.17	0.07	2.43	13.0	0.434	0.067	0.31	3.96	0.13	4.90	0.20	4.20	0.14	4.96
	NW17 - Fine Sand	0.18	0.08	2.25	8.0	0.423	0.050	0.35	3.67	0.15	5.63	0.22	4.19	0.16	4.95
	NW18 - Fine Sand	0.18	0.09	2.05	7.0	NA	NA	0.40	3.06	0.16	3.68	0.23	4.63	0.16	5.77
	NW19 - Fine Sand	0.14	0.05	2.59	22.0	0.415	0.000	0.28	2.41	0.12	3.31	0.16	4.24	0.11	5.02
	NW20 - Fine Sand	0.18	0.10	1.76	5.0	0.404	0.030	0.43	8.80	0.17	5.18	0.25	5.46	0.18	7.80
	NW21 - Fine Sand	0.19	0.10	1.88	4.0	0.415	0.077	0.37	4.45	0.19	5.33	0.26	5.10	0.18	6.93
	NW22 - Silty Sand	0.17	0.08	2.27	10.0	0.430	0.095	0.36	4.97	0.15	4.90	0.21	4.21	0.15	4.97
	NW23 - Fine-Coarse	0.21	0.09	2.39	7.0	0.381	0.110	0.53	5.61	0.18	3.29	0.25	4.15	0.18	4.88
	NW24 - Fine-Coarse	0.26	0.09	2.77	6.0	0.377	0.097	0.51	3.82	0.23	3.42	0.29	4.07	0.20	4.76
	NW25 - Fine-Coarse	0.40	0.11	3.64	4.0	0.385	0.070	0.94	3.02	0.25	3.35	0.38	3.85	0.27	4.44
	NW27 - Fine-Coarse	0.38	0.13	2.92	1.0	0.366	0.041	0.59	4.90	0.25	7.42	0.40	3.90	0.29	4.51
RMA (1997)	Howe Concrete Sand	1.09	0.22	4.92	1.1	0.349	0.004	-	-	0.25	1.13	-	-	0.62	3.26
	Squeegee	3.50	1.83	1.91	0.3	0.461	0.003	-	-	3.56	4.14	-	-	3.32	1.18
	Pea Gravel	3.50	4.83	0.72	0.2	0.442	0.003	-	-	4.40	2.10	-	-	5.61	3.62
	Mk96TP07GH 19-20	0.65	0.17	3.82	6.5	0.340	0.015	-	-	0.90	1.26	-	-	0.42	3.98
	Mk96TP15GH 15-16	3.50	0.40	8.75	4.0	0.341	0.009	-	-	2.03	1.48	-	-	1.46	1.10
Benson and Gribb (1997)	Sand #1	0.58	0.23	2.57	0.0	NA	NA	-	-	0.42	5.81	-	-	0.47	4.15
	Sand #2	0.58	0.23	2.57	0.0	NA	NA	-	-	0.40	6.40	-	-	0.47	4.15
	Sand #3	0.58	0.23	2.57	0.0	NA	NA	-	-	0.50	6.74	-	-	0.47	4.15
	Sand #4	0.58	0.23	2.57	0.0	NA	NA	-	-	0.45	4.94	-	-	0.47	4.15
	Sand #5	0.58	0.23	2.57	0.0	NA	NA	-	-	0.44	5.98	-	-	0.47	4.15
Khire et al. (1999)	Wenatchee Sand	0.58	0.20	2.90	1.0	0.400	0.010	-	-	0.14	4.00	-	-	0.44	4.14
Schroth et al. (1995)	Accusand grade 12/20	1.01	0.82	1.23	0.0	NA	0.012	-	-	1.51	7.35	-	-	1.22	8.13
	Accusand grade 20/30	0.72	0.61	1.19	0.0	NA	0.016	-	-	1.00	10.57	-	-	0.88	9.45
	Accusand grade 30/40	0.51	0.42	1.21	0.0	NA	0.018	-	-	0.68	13.10	-	-	0.62	10.25
	Accusand grade 40/50	0.35	0.29	1.20	0.0	NA	0.020	-	-	0.045	12.18	-	-	0.043	11.03

Table 1. Particle size characteristics and measured and estimated van Genuchten parameters for sands evaluated in the Wisconsin Geotechnics Laboratory (WGL) or from the literature (continued).

Source	Sample Name	d_{60} (mm)	d_{10} (mm)	C_u	% Fines	θ_s	θ_r	Measured				Estimated by PTF			
								Wetting		Drying		Wetting		Drying	
								α_w (kPa^{-1})	n_w	α_w (kPa^{-1})	n_w	α_w (kPa^{-1})	n_w	α_w (kPa^{-1})	n_w
Rassam and Williams (1999)	Kidston Gold Mine	0.17	0.02	9.44	28.0	0.395	0.050	0.13	2.11	0.08	2.24	0.10	4.022	0.07	4.65
Topp and Miller (1966)	Glass-Bead	0.18	0.18	1.00	0.0	0.326	NA	-	-	0022	9.42	-	-	0.24	13.32
Gurdal et al. (2003)	Omaha (DC1 L2 D2 33)	0.45	0.19	2.32	0.4	NA	NA	-	-	0.37	6.92	-	-	0.38	4.40
	Omaha (DC1 L2 D1 23)	0.44	0.18	2.44	2.1	NA	NA	-	-	0.35	7.22	-	-	0.37	4.41
	MPL1-L1-D1-01	0.50	0.12	4.17	7.7	NA	NA	-	-	0.41	4.18	-	-	0.31	4.23
	MPL1-L1-D2-02	0.46	0.12	3.83	8.7	NA	NA	-	-	0.51	4.78	-	-	0.30	4.32
	MPL1-L1-D3-03	0.48	0.08	6.00	9.0	NA	NA	-	-	0.46	4.29	-	-	0.25	4.20
	MPL1-L1-D4-04	0.49	0.12	4.08	8.6	NA	NA	-	-	0.47	4.98	-	-	0.31	4.26
	MPL2-L1-D1-05	0.48	0.12	4.00	8.9	NA	NA	-	-	0.49	2.58	-	-	0.31	4.28
	MPL2-L1-D2-06	0.48	0.12	4.00	8.4	NA	NA	-	-	0.51	2.93	-	-	0.31	4.28
	MPL1-L1-D3-07	0.43	0.12	3.54	8.9	NA	NA	-	-	0.55	2.81	-	-	0.29	4.40
	MPL2-L1-D4-08	0.48	0.12	4.00	9.3	NA	NA	-	-	0.55	2.98	-	-	0.31	4.28
Lu et al (2006)	Well-graded Sand	0.51	0.08	6.40	9.0	0.299	0.093	-	-	0.64	1.56	-	-	0.25	4.13
	Esperance Soil	0.30	0.16	1.90	0.6	0.390	0.111	-	-	0.19	2.32	-	-	0.28	6.51
	Poorly-graded Sand	0.68	0.25	2.70	0.0	0.369	0.030	-	-	0.22	3.23	-	-	0.54	3.96
Likos et al (2010)	Sand	0.25	0.14	1.80	1.0	0.400	0.000	-	-	0.36	3.51	-	-	0.25	7.42
WGL	Portage Sand	0.85	0.85	1.00	0.0	NA	0.059	1.34	4.80	0.79	10.63	1.62	6.13	1.14	9.96
	Foundry Base Sand	0.21	0.13	1.62	2.0	0.403	0.068	0.37	6.59	0.23	10.98	0.31	5.84	0.22	8.76
	Fine Sand	0.59	0.26	2.27	NA	NA	0.385	0.29	9.42	1.01	9.48	0.72	3.64	0.51	4.14
	Brazil Sand	0.33	0.15	2.20	0.1	NA	0.049	-	-	0.27	10.45	-	-	0.29	4.64
	730-Group A	1.00	0.09	10.6	0.0	NA	0.059	-	-	0.25	4.05	-	-	0.38	3.21
	730-Group B	2.30	0.11	20.9	0.0	NA	0.064	-	-	0.25	5.10	-	-	0.60	1.65

Table 1. Particle size characteristics and measured and estimated van Genuchten parameters for sands evaluated in the Wisconsin Geotechnics Laboratory (WGL) or from the literature (continued).

Source	Sample Name	d_{60} (mm)	d_{10} (mm)	C_u	% Fines	θ_s	θ_r	Measured				Estimated by PTF			
								Wetting		Drying		Wetting		Drying	
								α_w (kPa^{-1})	n_w	α_w (kPa^{-1})	n_w	α_w (kPa^{-1})	n_w	α_w (kPa^{-1})	n_w
WGL	730-Group C	0.63	0.44	1.43	0.0	NA	0.147	-	-	0.75	3.70	-	-	0.70	8.39
	Monona Sand	0.25	0.11	2.27	1.6	NA	0.046	-	-	0.17	3.05	-	-	0.22	4.80
	Sand 1	0.20	0.12	1.67	0.0	0.421	0.029	-	-	0.21	6.16	-	-	0.20	8.43
	Wraith Soil 2122	0.24	0.11	2.18	1.0	0.450	0.130	-	-	0.22	7.53	-	-	0.21	4.66
	Boise Sand	1.50	0.20	7.50	2.3	0.373	0.021	-	-	0.50	1.56	-	-	0.68	2.66
	Phoenix Sand	0.20	0.10	2.00	2.5	0.373	0.027	-	-	0.18	4.62	-	-	0.19	6.05
	Vail Sand	0.55	0.22	2.50	0.9	0.412	0.030	-	-	0.46	5.14	-	-	0.45	4.21
	Dominican Sand	0.22	0.10	2.20	0.1	0.559	0.053	-	-	0.22	7.69	-	-	0.19	4.87
	Tampa Bay Sand	0.20	0.12	1.67	0.0	0.426	0.036	-	-	0.19	11.21	-	-	0.20	8.43
	Green Roof Sand	2.20	0.15	14.67	5.0	0.544	0.000	-	-	0.26	1.39	-	-	0.69	1.83
	Newcastle Sand	0.31	0.17	1.82	0.1	0.421	0.015	-	-	0.24	2.33	-	-	0.30	6.97
	Pattaya Sand	0.50	0.27	1.85	0.0	0.443	0.050	-	-	0.42	8.20	-	-	0.48	6.24
	San Diego Sand	0.30	0.17	1.76	0.1	0.429	0.033	-	-	0.27	7.09	-	-	0.30	7.40
	Suvarnphumi Sand	0.70	0.15	4.67	2.4	0.347	0.046	-	-	0.36	2.42	-	-	0.41	3.86
	Turtle Bay Sand	0.70	0.45	1.56	0.0	0.456	0.034	-	-	0.61	5.29	-	-	0.74	7.43
	WisDOT Sand	0.60	0.27	2.22	0.3	0.378	0.033	-	-	0.48	4.86	-	-	0.53	4.13
	Weimar Sand	2.20	0.30	7.33	3.1	0.367	0.000	-	-	0.46	1.74	-	-	1.01	1.97
	Kegonsa Sand	0.25	0.17	1.47	0.4	0.410	0.020	-	-	0.24	3.11	-	-	0.27	9.62
	Tampa Sand	0.25	0.10	2.50	1.7	0.405	0.035	-	-	0.20	3.40	-	-	0.21	4.79
	Copper Mtn Sand	3.00	0.10	30.00	5.4	0.385	0.074	-	-	0.49	1.91	-	-	0.64	1.97
	Bluebench Sand	0.17	0.08	2.13	6.6	0.420	0.030	-	-	0.18	4.90	-	-	0.15	5.22

DETERMINING BROOKS-COREY PARAMETERS

In some cases the Brooks-Corey equation is used to describe the SWCC (Brooks and Corey 1966). The Brooks-Corey equation is

$$\Theta = \frac{\theta - \theta_r}{\theta_s - \theta_r} = \left(\frac{\psi_{a,w}}{\psi}\right)^{\lambda} \tag{5}$$

where $\psi_{a,w}$ is the air entry suction (ψ_a) or the water entry suction (ψ_w) for the drying and wetting curves and λ is the pore size distribution index. Eq. 5 applies for $\psi > \psi_{a,w}$; otherwise $\Theta = 1.0$.

Because Eq. 1 and Eq. 5 are parametric representations of SWCC data, their parameters should be related systematically. To determine this relationship, the Brooks-Corey parameters $\psi_{a,}$, ψ_{w}, and λ were obtained for the SWCC data presented in this paper and the other SWCC data for sands with fines reported in Chiang (1998) using the same non-linear least-squares optimization method employed to obtain the van Genuchten parameters. The Brooks-Corey parameters are graphed vs. the corresponding van Genuchten parameters in Fig. 12. There is a strong relationship between $\psi_{a,w}$ and α and between λ and n. Equations to convert between the parameters are shown in Fig. 12. Of particular importance is the relationship between $\psi_{a,w}$ and α; these parameters are inversely related, but are not reciprocals as is commonly (and mistakenly) assumed.

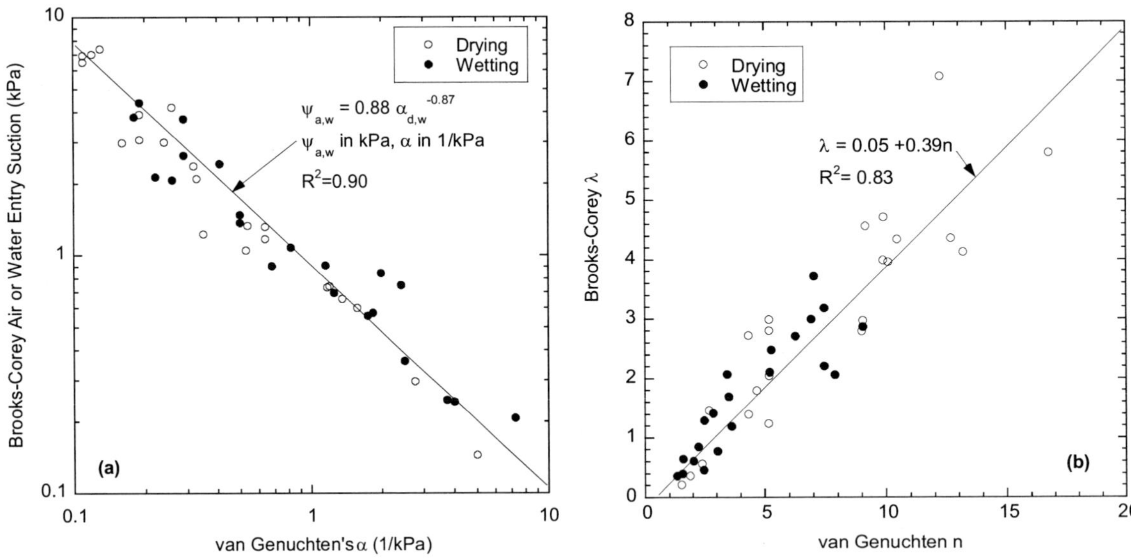

Figure 12. Relationship between Brooks-Corey air (ψ_a) or water (ψ_w) entry suction (a) or Brooks-Corey pore size distribution index (λ) (b) and van Genuchten parameters α and n.

SUMMARY AND CONCLUSIONS

This paper has described how the SWCC of clean sand, and the van Genuchten parameters used to describe the SWCC, vary with median particle size and breadth of particle sizes in clean sands. For uniform sands (C_u=1), the air entry suction increases (α decreases) and the steepness diminishes slightly (n decreases) for the drying SWCC as the particles become finer. Increasing breadth of particle sizes also increases the air entry suction (lower α) and reduces the steepness (lower n) of the SWCC. Breadth of the particle size distribution has a larger effect on steepness and n, and median particle size has a larger effect on air entry suction and α. The wetting SWCC exhibits similar sensitivity to particle size characteristics, although the impact on the water entry suction is less than the impact on the air entry suction.

These effects of median particle size and breadth of particle sizes on the SWCC reflect how particle size characteristics affect the pore size distribution. Sands with finer particles, or a broader distribution of particle sizes, have smaller pores and therefore higher air entry suction and lower α. Sands with a broader distribution of particle sizes also have a broader distribution of pore sizes, and thus a more gradually varying SWCC with lower n.

The systematic effects of median particle size and breadth of particle size on the SWCC were captured in a pedotransfer function (PTF) that can be used to estimate α and n for clean sand if the dry unit weight, d_{60}, and C_u of the sand are known. Comparison with SWCC data from the literature and determined in the Wisconsin Geotechnics Laboratory shows that α estimated with the PTF typically is within a factor of 2 of the actual α and n estimated with the PTF is within $\pm$ 2 of the actual n. These findings are applicable to α for sands with non-plastic fines provided the fines content is less than 20%, but apply to n only for sands with negligible fines content. A method to estimate parameters for the Brooks-Corey SWCC function from the van Genuchten parameters has also been provided.

ACKNOWLEDGEMENT

This paper was prepared in honor of Professor Emeritus Roy E. Olson, a mentor and friend of the senior author for nearly three decades (as of 2014). Professor Olson has dedicated his career to nurturing the knowledge and success of young people. He has had a profound impact on the senior author, and many other geotechnical engineers throughout his career. His support and guidance of the senior author over many years is gratefully acknowledged.

Financial support for this study was provided by the US Environmental Protection Agency (EPA) through the Alternative Cover Assessment Program, by the US Department of Energy (DOE) under Cooperative Agreement Number DE-FC01-06EW07053 entitled "The Consortium for Risk Evaluation with Stakeholder Participation III," and by Benson's Wisconsin Distinguished Professorship. The opinions, findings, conclusions, and recommendations expressed herein are those of the authors and do not necessarily represent the views of the US EPA or US DOE.

REFERENCES

Aubertin, M., Mbonimpa, M., Bussière, B., and Chapuis, R. (2003), A Model to Predict the Water Retention Curve from Basic Geotechnical Properties, *Canadian Geotechnical J.*, 40, 1104-1122.

Baker, K. (2001), Investigation of Direct and Indirect Hydraulic Property Laboratory Characterization Methods for Heterogeneous Alluvial Deposits: Application to the Sandia-Tech Vadose Zone Infiltration Test Site, *MS Thesis*, New Mexico Institute of Mining and Technology, Socorro, NM

Baker, R. and Hillel, D. (1990). Laboratory Test of a Theory of Fingering During Infiltration into Layered Soils, *Soil Science Society of America J.*, 54, 20-30.

Benson, C. and Gribb, M. (1997), Measuring Unsaturated Hydraulic Conductivity in the Laboratory and Field, *Unsaturated Soil Engineering Practice*, GSP No. 68, ASCE, S. Houston and D. Fredlund, eds., 113-168.

Brooks, R. and Corey, A. (1966), Properties of Porous Media Affecting Fluid Flow, *J. Irrigation and Drainage Division*, ASCE, 92 (2), 61-88.

Chiang, I. (1998), Effect of Fines and Gradation on Soil Water Characteristic Curves of Sands, MS Thesis, University of Wisconsin, Madison, WI.

Chin, K., Leong, E., and Rahardjo, H. (2010), A Simplified Method to Estimate the Soil-Water Characteristic Curve, *Canadian Geotechnical J.*, 47, 1382–1400.

Gurdal, T., Benson, C., and Albright, W. (2003), Hydrologic Properties of Final Cover Soils from the Alternative Cover Assessment Program, Geo Engineering Report 03-02, Geo Engineering Program, University of Wisconsin-Madison.

Hillel, D. (1998), *Environmental Soil Physics*, Academic Press, New York, 771 p.

Jaafar, R. and Likos, W. (2011), Estimating Water Retention Characteristics of Sands from Grain Size Distribution Using Idealized Packing Conditions, *Geotechnical Testing J.*, 34(5), 1-14.

Juang, C., and Holtz, R. (1986). Fabric, Pore Size Distribution and Permeability of Sandy Soils, *J. Geotechnical Engineering*, ASCE, 112(9), 855–868.

Khire, M., Benson, C., and Bosscher, P. (1999), Field Data from a Capillary Barrier in Semi-Arid and Model Predictions with UNSAT-H, *J. Geotech. Geoenvironmental Eng.,* 125(6), 518-528.

Kamon, M., Endo, K., and Inui, T., Kawabata, J., and Katsumi, T. (1998), Tank Modeling Test for the Investigation of DNAPL Migration Processes, Environmental Geotechnics, *Proc. 4th International Conference on Environmental Geotechnics*, Swets and Zeitlinger, Lisse, 89-94.

Likos, W., Wayllace, A., Godt, J., and Lu, N. (2010), Modified Direct Shear Apparatus for Unsaturated Sands at Low Suction and Stress, *Geotechnical Testing J.*, 33(4), 286-298.

Lu, N. and Likos, W. (2004), *Unsaturated Soil Mechanics*, John Wiley, Hoboken, NJ, 556 p.

Perera, Y., Zapata, C., Houston, W., and Houston, S. (2005), Prediction of the Soil-Water Characteristic Curve Base on Grain-Size-Distribution and Index Properties, *Advances in Pavement Engineering*, GSP No. 130, C. Schwartz, E. Tutumuluer, and L. Tashman, eds., ASCE, Reston, VA, 1-12.

Rassam, D. and Williams, D. (1999), A Numerical Study of Steady State Evaporative Conditions Applied to Mine Tailings, *Canadian Geotechnical J.*, 36, 640-650.

RMA (1997), Rocky Mountain Arsenal, Final RCRA-Equivalent Cover Demonstration Project Comparative Analysis and Filed Demonstration Scope of Work, Dec. 1997, Rocky Mountain Arsenal Remediation Venture Office, Commerce City, CO.

Schroth, M., Ahearn, S., Selker, J., and Istok, J. (1995). Characterization of Miller-Similar Sands for Laboratory Hydrologic Studies, *Soil Science Society America J.*, 60, 1331-1339.

She, H. and Sleep, B. (1998), The Effect of Capillary-Pressure-Saturation Relationships for Air-Water and Perchloroethylene-Water Systems, *Water Resources Research*, 34(10), 2587-2597.

Simon, D. (2001), Comparison of Three Geophysical Imaging Techniques for Characterization of an IAS Plume, MS Thesis, University of Wisconsin, Madison, WI.

Stormont, J. And Anderson, C. (1999), Capillary Barrier Effect From Underlying Coarser Soil Layer, *J. Geotechnical Geoenvironmental Engineering*, ASCE, 125(8), 641-648.

Tinjum, J., Benson, C., and Blotz, L. (1997), Soil-Water Characteristic Curves for Compacted Clays, *J. Geotechnical Geoenvironmental Engineering*, ASCE, 123(11), 1060–1069.

Tanahashi, H., Sato, T., and Konishi, J. (2000), Degree of Imbibition and Residence Oil Saturation in Porous Media, *Proc Creation of a New Geoenvironment*, Fourth Kansai International Geotechnical Forum, Kansai Branch of the Japanese Geotechnical Society, Kyoto, Japan, 167-172.

Topp, G. and Miller, E. (1966), Hysteretic Moisture Characteristic and Hydraulic Conductivity for Glass Media, *Soil Science Society America Proc.*, 30, 156-162.

van Genuchten, M.. (1980). A Closed-Form Equation for Predicting the Hydraulic Conductivity of Unsaturated Soils, *Soil Science Society of America J.*, 44(5), 892-898.

van Genuchten, M., Leij, F., and Lund, L. (1992), *Indirect Methods for Estimating the Hydraulic Properties of Unsaturated Soils*, University of California, Riverside and US Department of Agriculture Salinity Laboratory, Riverside, CA, 718 p.

Wang, X. and Benson, C. (2004), Leak-Free Pressure Plate Extractor for Measuring the Soil Water Characteristic Curve, *Geotechnical Testing J.*, 27(2), 1-10.

Yang, H., Rahardjo, H., Leong, E., and Fredlund, D. (1994), Factors Affecting Drying and Wetting Soil-Water Characteristic Curves of Sandy Soils, *Canadian Geotechnical J.*, 41, 908–920.

Pile Damage Prevention and Assessment Using Dynamic Monitoring and the Beta Method

Garland Likins, P.E., M.ASCE[1] and Frank Rausche, P.E., PhD., D.GE., M.ASCE[2]

[1] President, Pile Dynamics, Inc., 30725 Aurora Road, Cleveland, OH 44139, glikins@pile.com
[2] Partner, Pile Dynamics, Inc., 30725 Aurora Road, Cleveland, OH 44139, frausche@pile.com

ABSTRACT

Driven piles are subjected to high stresses during installation. It is, therefore, important not to exceed acceptable stresses along the pile shaft and at the toe to prevent damage. Dynamic monitoring has been used for decades to evaluate not only the installation stresses, but also check test piles for signs of structural damage. The Beta Method (ß-Method) for evaluation of the *location and extent of a potential damage* was developed over thirty years ago and has proven effective as a QC and QA tool. As an aid in the process of pile rejection or acceptance, the ß-Method also offers a rating scale which translates the automatically determined ß-number into a helpful pile integrity assessment tool. The reliability of this algorithm has been proven by numerous extracted piles. However, one limitation of the ß-method concerned detection of damage near the pile toe where high toe resistance effects and/or stress wave reflections reduce the effectiveness of the traditional ß-Method. In the past, therefore, near-toe damage was determined by the testing engineer, not only by visual inspection of the dynamic monitoring data, but also by reviewing the pile toe compressive stresses throughout the monitored driving history and the strength and stiffness of the soil response from the pile toe. This approach has now been automated and subjected to tests on existing data. After a review of the existing methods of pile stress and damage calculations, the paper presents the new method, illustrating its effectiveness by examples from measurements on both concrete and steel piles.

INTRODUCTION

Driven pile foundations are very popular because they are both cost effective deep foundation solutions and provide by their installation history assurance that the pile has been installed to sufficient bearing with good structural integrity. Usually the maximum loads that can be applied to a driven pile foundation are limited by the resistance that can be applied to the supporting soil in shaft resistance and end bearing, although any weakness in the structural strength of the shaft may then be the limiting factor. Under very favorable conditions, such as when the pile toe is driven to a hard rock or the pile is long and attains significant resistance through soil setup, the maximum load can be limited by the structural strength of the pile. The pile shaft may be subjected to various loading conditions, including lateral and uplift which impose additional structural strength requirements.

Dynamic monitoring of driven piles on a routine basis has its origin as a research project at Case Western Reserve University (Goble et al., 1975) with the goal to replace static load testing by determining the pile capacity using the hammer impact. The basic method employed by the commonly used Pile Driving Analyzer® (PDA) system measures strain and acceleration with reusable sensors making the cost to instrument a pile very economical.

These measurements were analyzed in closed form solutions called the Case Method and by signal matching in rigorous numerical models called CAPWAP® (Rausche et al., 1972). Numerous correlations were made with static load tests over many decades of testing with generally good agreement when the times after installation for both tests were similar (Likins and Rausche, 2004).

Early tests by the researchers at Case Western Reserve University were generally done on closed-end steel pipe piles considered as friction piles and no pile damage was observed. A later research project was performed on H piles driven to end bearing on rock which did identify damage at the pile toe and noted "If the velocity increases sharply relative to the force at any point earlier than the 2L/c time it indicates damage has weakened the pile" (Goble et al. 1977). The test piles were extracted and visual inspection confirmed the pile damage. An example of the pile toe damage for these research project piles is shown in Figure 1. Although pile damage was detected, no specific attempt was made to estimate the magnitude of the damage, other than to note the capacity was greatly reduced in both dynamic and static load tests after damage occurred when compared with undamaged piles in the study. However, Teferra (1977) analyzed the force

Figure 1. Damaged H pile.

and velocity records of the test piles in closed form, yielding pile toe resistance versus pile toe displacement and, therefore, pile toe stiffness. Today this analysis method is incorporated in CAPWAP as the Pile End Bearing Wave Analysis Program (PEBWAP). For end bearing piles it shows very clearly the reduction of stiffness for piles with toe damage.

The ß-Method for evaluation of the *extent of damage* was developed and published (Rausche and Goble 1979) after dynamic measurements at a large commercial construction site clearly indicated pile distress on prestressed concrete piles of up to 240 ft (73 m) length. The piles were assembled using multiple mechanical splices, and detected damages were located at or near the splices. Rejecting piles with indications of only minor damage was considered unreasonable, particularly since the damage could have been merely a sheared mechanical connector, allowing the pile sections to slightly separate at the joint. But some rating scale was needed. An equation was derived from wave propagation theory which related the stresswave reflection magnitude from the potential damage to the magnitude of the impact force, taking into consideration the effects of soil resistance above the damage. This approach resulted in an integrity factor β indicating the theoretically remaining cross section (100% representing no damage and 0% a completely missing lower section). It was recommended that if, in theory, the thus estimated remaining cross section was greater than 80%, that then the piles were "slightly damaged" and would be accepted. If ß was between 80 and 60%, then the pile was rated as "damaged" and all information available would be reviewed before rejecting the pile. For a β less than 60%, reflections from the pile toe were usually not apparent from the dynamic records and the pile was considered "broken" and rejected. Rausche and Goble (1979) also gave an example of calculating the gap at the splice between sections. These methods were applied to many test piles during a restrike test program where as many as 44 piles were tested in one day, and a correlation was established between the ß-Method and the penetration per blow. Later installed production piles used this β-value versus pile penetration resistance relationship during pile restrikes for a quick decision on pile acceptance or rejection. All parties involved in the project accepted this approach and the project was completed successfully. While no piles were extracted for verification of the β-Method at this particular site, the subsequent 35 years have produced a wealth of independent assessments of detected damaged piles which were positively confirmed by extraction of those piles. For example, Rausche et al. (1988), Hussein and Morgano (1993) and Goble and Hussein (1994) described several case studies. Also, Verbeek and Middendorp (2011) reviewed the β-Method approach and generally agreed with it. They also proposed replacing the rating scale terminology of "damaged" or "broken" with "issues" of various significance.

Implementation of the Case Method

Conventionally, only selected test piles are monitored, and this information provides the basis for an economical and safe installation criterion for production piles.

While the basic methods of data interpretation are based on solid theory, the variety of hammer, pile and soil conditions require that the user has a thorough understanding of foundation construction methods, geotechnology, and the basic theory behind the various dynamic monitoring results. Therefore, training and continued education are important to correctly implement and use the dynamic monitoring results; this is certainly true when evaluating pile damage.

Allowable Driving Stresses

The pile installation process generally results in the highest stresses the pile will experience. Since pile stresses during installation are key to limiting damage to driven piles, determining allowable stresses and then confirming they are not exceeded is critical to preventing pile damage. Measuring the strain and acceleration during pile driving is now a routine practice, and by using these measurements the critical stresses in the pile are either directly obtained or can be calculated from the measurements and wave propagation theory. Recommended limits on tension and compression driving stresses have been established based on decades of experience and observations of the conditions which lead to structural failures of driven piles. These limits relate to the structural strength of the pile material, e.g., the yield strength of steel, and the prestress and strength of the concrete. Typical limits, such as proposed by AASHTO (2010) are as follows.

- 90% of the steel yield strength for steel piles in either tension or compression
- 85% of the concrete compressive strength minus the effective prestress for prestressed concrete piles in compression, and
- 100% of the effective prestress plus an allowance for the concrete tensile strength which is typically 0.2 to 0.3 ksi (1.4 to 2.1 MPa) for prestressed concrete piles in tension.

It should be noted however that once the cross section has suffered a crack the tension strength component of the concrete will be lost. . Limiting the tension stress to the prestress level would therefore be a more conservative approach.

Dynamic monitoring of the stresses in the piles, and developing a driving protocol which maintains production pile stresses below these recommended limits, are the most effective means of reducing the likelihood of pile damage. Such a protocol would specify maximum hammer fuel setting or energy output to blow count. For example, to limit tension stresses for concrete piles the hammer output is often reduced at low blow counts. Limiting toe compressive stresses would potentially require reduced energies at high blow counts.

Case Method Closed-Form Solutions

As mentioned, measuring the strain and acceleration during pile driving using reusable bolt-on sensors is now routine practice. Any pile can be thus tested if installation records raise doubts about pile integrity without the need for preplanning specific pile instrumentation. Strain measurements are converted to force, F(t), by multiplication with the material modulus, E, times the pile area, A. Velocity, v(t), is obtained by integrating the measured acceleration. Until reflections occur from soil resistance or a change in pile cross section, the force and velocity are proportional by the pile impedance, Z, which is the product of modulus E and cross sectional area A divided by the material wave speed, c. Stress waves due to hammer impact can be separated from reflections by separating the measurements into downward and upward travelling wave components. The force in the downward wave can be obtained from

$$Fd(t) = \frac{1}{2} [F(t) + Z\, v(t)] \tag{1}$$

And the force in the upward wave is

$$Fu(t) = \frac{1}{2} [F(t) - Z\, v(t)] \tag{2}$$

Considering that superposition of the forces in the stress waves yields the total force at any point, the above two expressions can be conveniently used to derive a variety of soil resistance, stress and other results of interest. The following provides a brief summary of what can be readily calculated for uniform piles by a PDA system during pile installation or restrike of any pile type using only the standard bolt-on sensors.

Bearing capacity. Based on the above basic equations, the closed-form capacity solution is computed from the downward wave at time t and the upward wave at time t + 2L/c, where L is the pile length below sensors and c the material wave speed. The basic Case Method equation for the maximum total resistance RTL which includes static and dynamic resistance components of shaft resistance and end bearing is

$$RTL = Fd(tx) + Fu(tx + 2L/c) \tag{3a}$$

Where tx is a time chosen to yield the maximum total resistance value. Static capacity RMX can then be calculated under the assumption of a damping factor J (Rausche et al. 1985).

$$RMX = (1 - J)\, Fd(tx) + (1 + J)\, Fu(tx + 2L/c) \tag{3b}$$

The component and distribution of resistance acting on shaft can be evaluated from the upward wave as a function of time during the initial 2L/c time period. The toe resistance can be deduced from subtracting the shaft resistance from RMX.

Stress extrema. Although dynamic testing was originally developed to evaluate pile capacity, it became apparent that other uses were also beneficial. Using only the standard force and velocity measurements, usually taken near the pile head, methods were developed from wave propagation theory to evaluate the driving stresses in compression at the generally critical pile head and toe locations, and maximum tension somewhere along the pile, which is generally of great importance for concrete pile installations where tension stresses can cause cracks that may develop into more serious pile damage. At the pile head, the maximum stress can be directly obtained from the measurements. The maximum pile toe force is

$$Rt,toe = RTL - Rt,shaft \tag{4}$$

The maximum tension force requires that the record is searched for the minimum of Fu(t) (which is the peak of the upward travelling tension wave) and superimposing it with an earlier occurring minimum value of Fd(t).

Hammer performance. The performance of the hammer can be assessed by measuring the energy transferred to the pile from

$$E(t) = \int F(t)\, v(t)\, dt \tag{5}$$

Considerable variability was found by comparing the maximum energy transferred to the manufacturer's energy rating (Rausche et al. 1986). Identifying poorly performing hammers is important as the energy transferred influences the blow count, which generally is used as a driving criterion.

The β-Method

The β-Method looks for an early tension reflection caused by a reduced cross section along the pile shaft. The maximum local reduction in the upward wave from this early reflection is related to the extent of the damage. The local reflection caused by damage along the pile shaft can be computed as

$$\Delta = Z\, v(t3) - F(t3) + R \tag{6}$$

where t3 is a time (between the initial impact and reflection from the toe) of the local minimum in upward wave that corresponds to the damage and R is the total shaft resistance, including damping, above the damage location. From wave theory, Rausche and Goble (1979) showed that the term

$$\alpha = 0.5\, \Delta / [\, F(t1) - R\,] \tag{7}$$

where t1 is the time of the initial force input peak and α can be used to quantify the section reduction by inclusion in the expression

$$\beta = (\, 1 - \alpha\,) / (\, 1 + \alpha\,) = Z_2 / Z_1 \tag{8}$$

where Z_1 is the original cross section and Z_2 is the reduced section at the damage. Thus, β expresses the percentage of remaining section compared to the original pile cross section. If F(t1) and Fd(t1) are nearly identical, as is usually the case, this can be put in simplest terms as

$$\beta = [\, Fd(t1) - 1.5\, R + Fu(t3)\,] / [\, Fd(t1) + 0.5\, R - Fu(t3)\,] \tag{9}$$

It is generally more accurate to look at the time between the beginning of the input wave and the beginning of the local minimum in the upward wave and calculate the damage location x from the wave speed c. The wave speed is well known for steel piles (16,800 ft/s or 5,123 m/s), but for concrete piles the wave speed depends on the concrete mix design and the age of the concrete. It typically ranges between 10,500 and 15,500 ft/s (3,250 and 4,800 m/s). While this is a large range of potential wave speeds, it should be understood that, within certain geographic areas, concrete wave speeds typically vary only 5 to 10% because of the predominate and consistent use of certain types of aggregates and mix designs by the local concrete pile suppliers. In any event, for concrete piles, the best estimate of wave speed comes from evaluating the time of the toe reflection from the earliest easy driving when the pile is known to be undamaged.

The closer the apparent impedance reduction is to the pile head, the earlier the reflection from the damage will arrive. Although the engineer should visually inspect the data for a premature tensile reflection, the PDA system automatically does such a search and calculates β and the distance below sensors, if the reflection occurs at a time more than one rise time (defined as the time between initial onset of the impact and the first input peak) before the pile toe reflection. For tensile reflections occurring later, but before the toe reflection, a toe damage is identified and a ß-evaluation is not performed. Further discussion on toe damage assessment is given below.

Example of a Damage Evaluation Along the Shaft for a Spliced Pile

Figure 2 shows measurements of force (F) and velocity times impedance (Z v), and the calculated upward wave force (Fu) and the top displacement (D) as a function of time for a 16.5 inch (420 mm) octagonal prestressed concrete pile. The pile has total length below the sensors of 134 ft (40.9 m) with a splice located 64 ft (19.5 m) below the sensors. The two arrows indicate the beginning of the rise time at impact and the beginning of the expected toe reflection. The time period between the arrows is equal to the wave travel time, 2L/c. After 202 blows the PDA system first automatically detected a very small tensile reflection (β of 96%) occurring just above the splice.

It should be noted that, because of the needed development length, the prestress is not developed at the pile segment ends, so the tension at the splice has to be carried by the dowel bars that connect the sections. In such case a tensile failure therefore happens most likely at the end of either the upper or lower dowel bars, or due to a failure of the splice connection if it is a mechanical connection. Figure 2 shows that the damage had gradually progressed to a β of 81% by blow number 692; the damage is marked by the vertical dashed line about midway between the two solid vertical lines marking the beginning of initial impact and 2L/c later. Even though the prestress was designed to be 1.0 ksi (6.9 MPa), the computed tension from the measurements of only a maximum of 0.65 ksi (4.5 MPa) was sufficient to start the damage since, most likely, the full prestress was not available at the end of the dowel bars. After blow number 692, the ß-value then rapidly regressed to 64% by blow number 700 (Figure 3) and to 50% by blow number 703 (Figure 4). Driving was halted at blow number 713. The damage location was indicated 2 ft (0.6 m) above the splice location. It can be observed in Figures 2 and 3 that the reflection from the pile toe is still apparent (strong increase in velocity relative to force, and a reduction in Fu, after the second vertical line). In Figure 4 the toe reflection (after the third vertical line) is vastly different from that in Figure 3 and it is not certain that the toe reflection can be reliably ascertained.

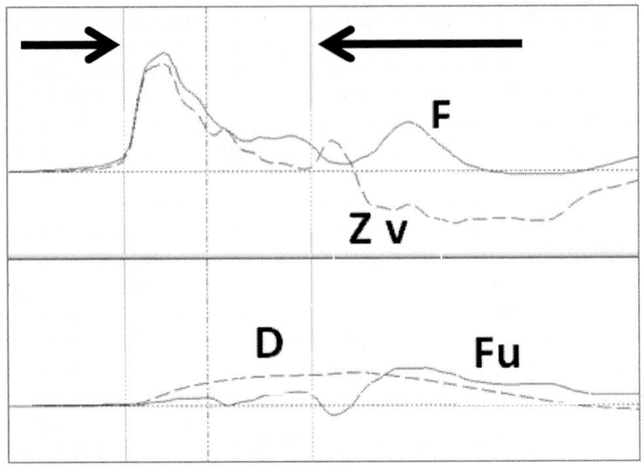

Figure 2. Spliced concrete pile, blow number 692.

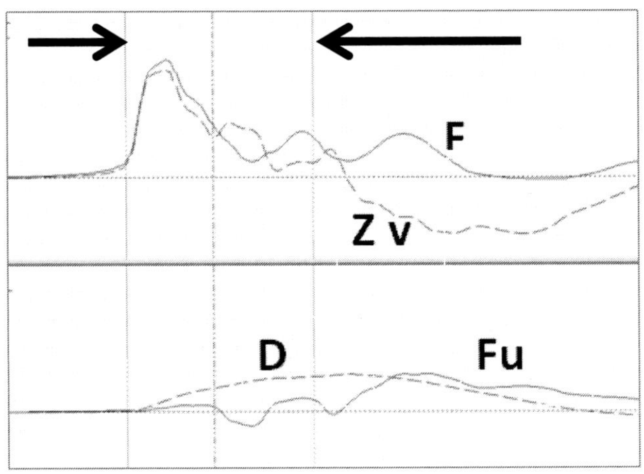

Figure 3. Spliced concrete pile at blow number 700.

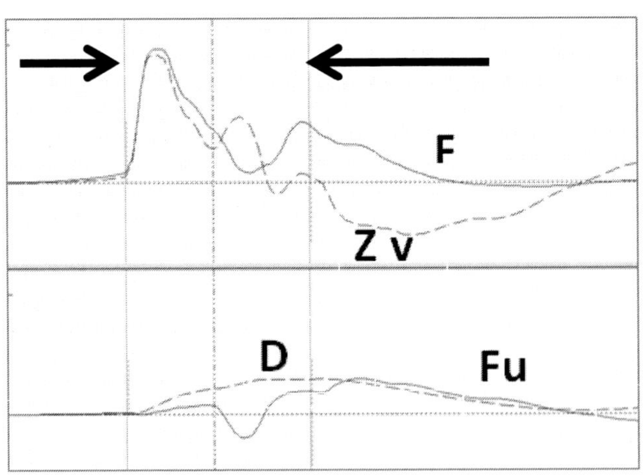

Figure 4. Spliced concrete pile at blow number 703.

It should be noted that there had been ample opportunity to cease pile driving when the damage was still relatively moderate (β above 81%), and even during the last few blows before β dropped below 60% when the pile broke completely. Such a broken pile is unreliable in the service condition. Many extracted piles show a shear failure with the upper section likely sliding off the lower section. The resistance against the broken end might include forces associated with the prestressing strands which are unreliable in the long-term service condition because of probably high stress concentrations and corrosion potential.

Damage Near the Pile Toe

A major defect near the pile toe is classified as toe damage, although there is no strict definition as to the distance of the damage from the toe. However, in such a case, the damage reflection will occur only slightly earlier than the toe reflection and careful observation is required to differentiate between them. Monitoring the pile installation continuously helps determine the time of pile toe reflection of the undamaged pile from early driving and, therefore, allows for toe damage detection should it actually happen. However, for restrike testing, if the concrete wave speed is not accurately known, toe damage assessment may not be possible.

If the damage is located within a distance above the toe that corresponds to one rise time, the reflection will include tension waves from both damage and pile toe and compression waves from the end bearing plus shaft resistance near the toe. Unfortunately these effects all superimpose, and it is not possible to sort out the three individual components. So while it is possible, given an accurate wave speed, to detect the early wave arrival, and thus diagnose existence of pile damage near the toe, it is not possible to clearly assess the magnitude of such a defect. Instead, the pile tester has to evaluate the toe condition from the pile driving behavior and the strength and stiffness of the toe resistance, as may be apparent from soil information and static and dynamic analysis. In addition, a signal matching analysis, such as CAPWAP, can help shed further light on the extent of the distress.

Example of Near-Toe Damage for a Concrete Pile

A 24 inch square solid section pile with length 91 ft (27.7 m) was driven by an APE D50-42 (5 metric ton ram weight) single acting diesel hammer. For hammer blow No. 1160 at a penetration depth of 68.5 ft (20.9 m), Figure 5 shows measured time histories of force and velocity times impedance, Z, plus the force in the upward traveling wave, Fu, together with the pile head displacement D. The toe reflection is obvious from either the relative increase of

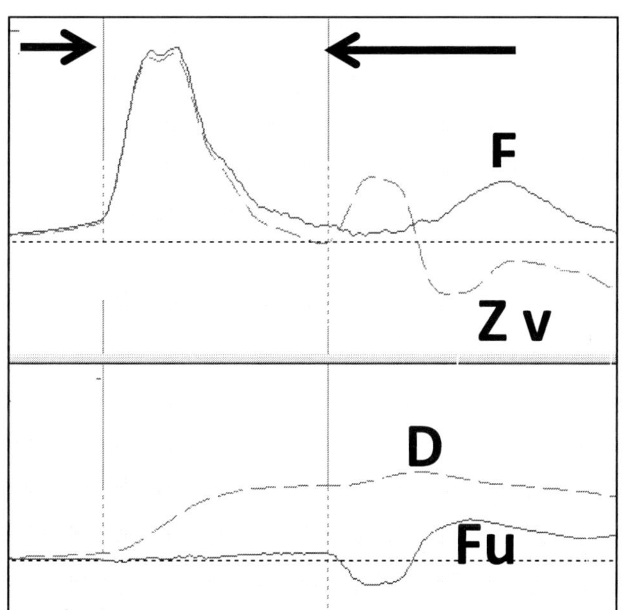

Figure 5. Early record for concrete pile, blow number 1160.

velocity compared to force, or the local reduction in the upward wave, Fu. The material wave speed can be determined from the measured time and known pile length as 13,500 ft/s (4,110 m/s). Figure 6 shows a record taken on the same pile a few blows later. The extra continuous dashed line, pointed to by the dashed arrow, just prior to the 2L/c reflection shows the automatically detected near-toe defect and its location is 2 ft (0.6 m) above the undamaged toe position. Significantly, the critical compressive stress condition in prestressed concrete piles is not at the toe but rather where the full prestress is first achieved, which is some small distance above the toe since right at the pile toe the prestress is zero. Due to the development length for the prestress, this critical distance is typically at two to three feet (0.6 to 1.0 m) above the toe, making it possible to detect near-toe damage for concrete piles using the standard pile top measurements. That distance will be longer for piles where the "stress transfer length" has become longer due to improper strand release in the casting yard.

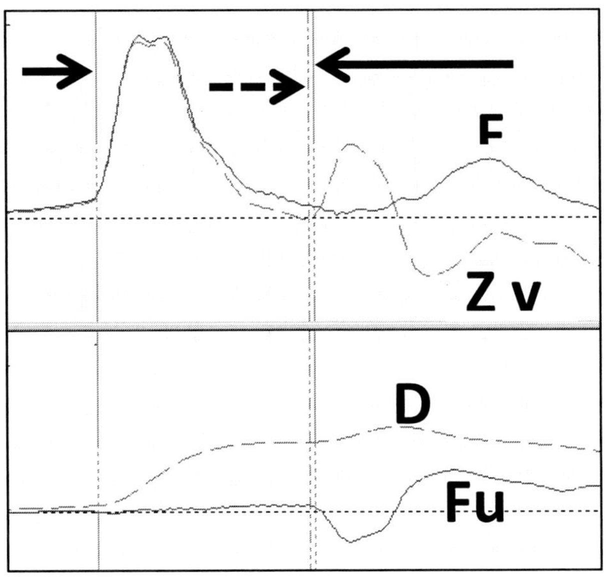

Figure 6. Prestressed concrete pile, a few blows after first damage detection.

The damage is detectable, yet minimal and, if it met other driving criteria, the pile would almost assuredly perform satisfactorily in the service condition if the driving had been terminated at this first detection.

Figure 7 shows a record for the same pile taken about 300 blows later. The damage is now more progressed and clearly indicated by the early reflection, most easily observed in the reduction in upward wave (Fu) just prior to the time of wave return (2L/c). The damage location is now 3 ft (0.9 m) above the original toe. This prestressed concrete pile was extracted and Figure 8 shows the toe condition of this extracted pile. The bottom of the pile is basically still intact, and connected to the main pile shaft by the prestressing strands. It is therefore obvious that the damage did not start at the bottom and progress up the pile shaft, but rather the damage began at the critical section about two feet above the bottom, as shown by the standard measurements at the pile head. It is also likely that the extraction process has caused additional harm to the pile.

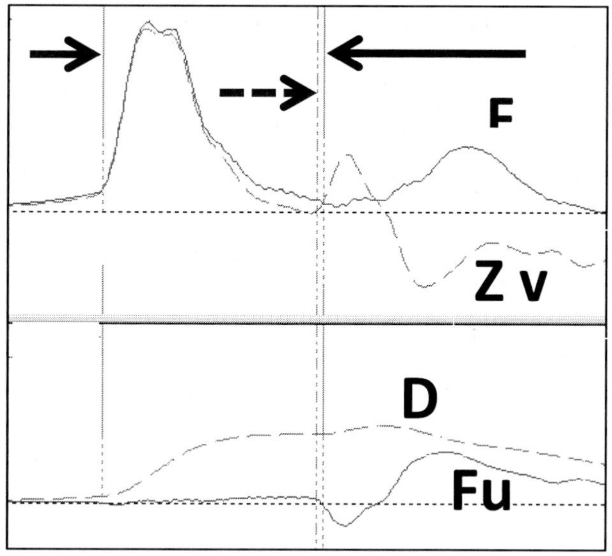

Figure 7. Concrete pile at blow number 1420,
300 blows after first damage detection.

Figure 8. Extracted concrete pile
showing the pile damage.

Example of Steel Pile Damage

The original ß-Method is useful for any pile type which can be dynamically tested, including concrete piles, prestressed or not, timber piles, and all steel pile types. Steel piles may suffer defective splices, bends, tears or twists when encountering obstructions during driving. Similarly near-toe damage can be detected from pile head measurements for any pile type. Steel piles, for example, can suffer collapsed pile toes when improperly trying to seat them with too many hammer blows into a hard material or on bedrock. Exceptions are near-toe deformations that leave the pile impedance virtually unchanged. An example for this type of damage is an ovalization of an open-end pipe.

Large 14x117 (360x174) H-piles were driven by a Delmag D46-32 (4.6 metric ton ram) to a karst limestone. Pile lengths for two test piles were 90 ft (27.4 m) and were equipped with high strength cast-iron shoes. Although the piles were driven in relatively close proximity to each other, the final elevations and the installation records were quite different. Because the shaft of an H pile cannot be visually inspected after installation, dynamic testing was requested to explain the differences. Figure 9 shows measured pile head force, velocity and the associated forces in the downward and upward traveling waves for a blow when one of the piles reached refusal at 59 ft (18 m) penetration. The toe reflection comes at exactly the time predicted based on pile length and steel wave speed. The force F after 2L/c is quite large, and the upward wave Fu after 2L/c is also in compression and a significant percentage of the input downward wave. The dynamic measurements clearly show that this pile has high end bearing and is not damaged.

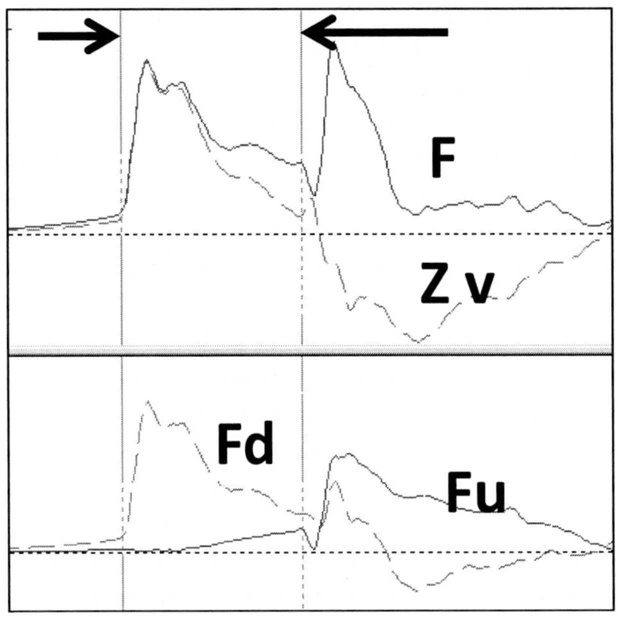

Figure 9. H pile driven to refusal at penetration of 59 ft (18 m).

Figure 10 shows the record of a second pile which was driven on the same project site to 78 ft (23.8 m) penetration. The early portion of the upward wave Fu which is a reflection from the shaft resistance is similar to the upward wave from the previous pile, a similarity that might be expected for neighboring piles in the same soil profile. By contrast, prior to the time of wave return from the toe, both the force F and the upward wave force Fu show an obvious and early reduction. The beginning of the reflection, defined by the relative increase of velocity to force or the beginning of the reduction in upward wave Fu, and depicted by the early full dash-dot line prior to 2L/c, starts too early when compared with the expected time of wave return. This is a clear indication of damage; a photograph of the extracted pile is shown in Figure 11. The pile also originally had a cast-iron shoe that is now missing. The sharp bend about 8 ft (2.4 m) above the pile toe causes a reduced axial stiffness from this near-toe damage. The compressive force in the upward wave after 2L/c is significantly smaller indicating considerably reduced capacity compared to the undamaged pile.

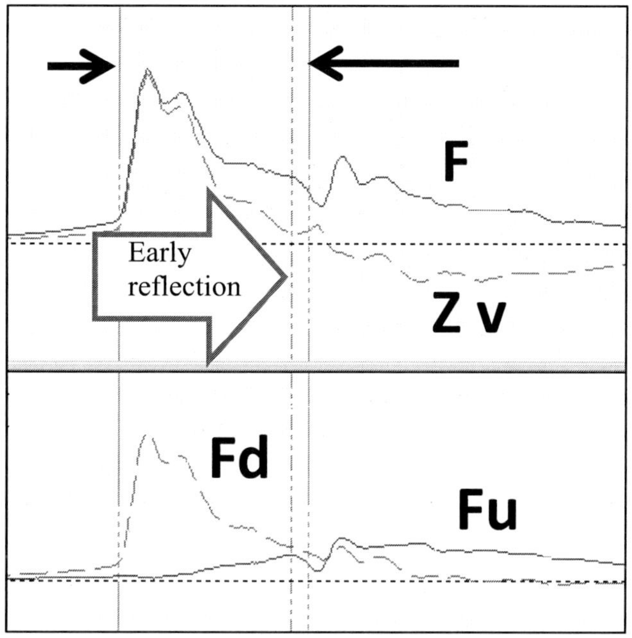

Figure 10. Pile driven to 78 ft (23.8 m) demonstrates near-toe damage.

Figure 11. Extracted pile with near-toe damage.

Discussion of Pile Toe Damage Assessment

Considering the many different pile types and failure modes, an automatic assessment of the severity of damage is always problematic. For this reason, advice has always been given to pile testers to visually inspect the data for an early stress wave arrival which could indicate near-toe pile damage. As mentioned earlier, this has been the advice for decades as noted in the early research project report that included H-pile damage: "If the velocity increases sharply relative to the force at any point earlier than the 2L/c time it indicates damage has weakened the pile" (Goble, et al. 1977).

Over more than three decades, numerous independent engineers all over the world have used the ß-method to detect damaged piles and the major damage has been independently verified by extraction, confirming the validity of the original recommendations derived from theory. In general, the reflection from the pile toe and thus a positive indication of a continuous pile over its full length cannot be observed when the β-value is less than 60%. This is a further justification of the aforementioned damage scale. Since frequently the bearing capacity and pile integrity are closely related, the Case Method bearing capacity determination combined with the β-Method has proven to be a very powerful tool.

For dynamic measurements, the typical data sampling frequency of 10,000 Hz gives a resolution in distance to damage location evaluation of 0.2 m. Given the wave speed from pre-damage pile monitoring, near-toe damage can be investigated automatically by a PDA system by checking for relatively small local reductions in the upward traveling compression wave just prior to the expected time of the wave return from the toe. While this method can detect toe damage and its distance above the toe, it cannot assess the extent of the damage. However, both the traditional β-Method and the near-toe damage detection are applicable not only to prestressed concrete piles but also to all other pile types including steel H-piles and pipes. In general, the only piles that need to be instrumented and tested are test piles. Production piles that do not meet the installation criteria should also be instrumented at their pile head and tested during a restrike. Conveniently, this can be done remotely, i.e., by having field personnel attaching the sensors to the pile and sending the data from the PDA via internet to the experienced test engineer.

To evaluate toe damage in concrete piles, Verbeek and Middendorp (2011) propose to measure changes in strains in all piles near their toe with expendable sensors. Although this promises a direct measurement of pile integrity, diagnosing significant pile damage from an apparent strain variation requires establishing an arbitrary criterion which could lead to rejecting piles that meet the driving criteria and would perform acceptably in long-term application. Experience shows that slipping prestressing strands frequently produce some prestress loss near the pile ends, even though the concrete of the pile has not been damaged.

It is important to consider the day-to-day operations on a pile driving site and the consequences of pile acceptance and rejection. A pile which has suffered significant damage will not likely achieve the same blow count as a neighboring undamaged pile and therefore its installation criterion will not be met. A damaged pile which does not meet the driving criteria, and has been dynamically tested and confirmed damaged, is most likely rejected and so are untested production piles which do not achieve the driving criterion. Test piles with damage near the toe, as determined by dynamic testing, exhibiting high resistance from the bearing layer may be analyzed with a slightly shorter length and may still be deemed acceptable as are all untested piles which meet the driving criterion. Similarly, a closed-end pipe pile with a collapse in the vicinity of the toe may still be filled with concrete and become serviceable.

Best practice is to perform initial dynamic pile tests with the actual driving equipment intended for the project and for all contemplated pile types to establish a driving sequence and installation criteria with acceptable stresses that avoid damage. This is a time honored procedure which provides a known quality foundation while being both economical and construction time sensitive. It should be noted that the initial test piles on a site are often driven harder, deeper, and to higher loads than required, resulting in a higher chance for pile damage. This is particularly true for concrete piles when the reason for the test program is to establish casting lengths for the piles. Further, the driving system and in particular the plywood cushion for the pile head are usually selected by a wave equation analysis. However, after measuring the stresses by dynamic monitoring, the cushion thickness might be increased or the energy output of the hammer reduced to decrease the stresses for subsequent production piles.

The greatest problems on a site occur when excessive time is spent on meetings trying to decide if an apparent pile damage is serious enough to require pile extraction or replacement. If the damage cannot be detected from the dynamic pile head measurements alone and the pile meets the installation criteria, then it is unlikely that the pile's ability to sustain the required load is compromised. With the availability of automatic methods of damage detection the pile testing engineer is well equipped to make the proper judgment within a short time after installation.

CONCLUSIONS

Dynamic pile monitoring according to the Case Method, as a quality control and assurance method, has gained worldwide acceptance because it provides construction professionals with a complete evaluation of a foundation's geotechnical and structural adequacy. Together, calculated bearing capacity, pile stresses, hammer performance, and the ß-integrity factor provide the basis for rational decisions for pile acceptance or rejection. The ß-Method was developed for detection and estimation of damage along the pile shaft and, over the subsequent more than three decades, numerous independent engineers have used the ß-method to detect damaged piles. Major damage has then been independently verified by extraction, confirming the validity of the original recommendations derived from theory.

The ß-Method only requires instrumentation near the pile head to detect potential damage at any location below this point of measurement. This single point of measurement makes instrumentation reusable and therefore economical since it can be employed for any driven pile, even after-the-fact, and avoids extra instrumentation permanently embedded in the pile. Additional economy can be achieved with remote testing technology. A further advantage of the ß-Method is its applicability to any uniform cross section pile, including steel and concrete piles, covering the vast majority of all driven piles. For non-uniform piles, unless continuous monitoring is performed, which would show from record characteristics developing damage from blow to blow, a more rigorous analysis such as CAPWAP is needed to evaluate pile integrity.

Where near-toe pile damage is indicated, the ß-Method cannot evaluate the extent of the damage due to superposition of the closely following reflection from the pile toe. However, damage near the pile toe can be detected automatically by the PDA system by looking for a reflection occurring too early for the correct 2L/c. In any event, near-toe pile damage should always be evaluated along with an assessment of pile bearing and its stiffness. Dynamic testing, including a rigorous signal matching analysis, provides information on the magnitude and stiffness of the pile toe response which are invaluable when deciding on the acceptance of a pile with near-toe damage.

Pile damage is best prevented by establishing an installation protocol based on a wave equation analysis and then confirmed by dynamic measurements. Such a driving prescription may, for example, require a minimum cushion thickness and replacement schedule and hammer energy setting as a function of driving resistance.

Based on measurements near the pile head alone, piles that lack any sign of damage either along the shaft or near the pile toe will perform satisfactorily in service conditions provided they also meet the established driving criteria.

REFERENCES (many are available on www.pile.com/Reference)

AASHTO. (2010). "LRFD Bridge Design Specifications". *5th Edition. American Association of State Highway and Transportation Officials*, Washington D.C.

Goble, G. G., Likins, G. E., Rausche, F. (1975). "Bearing Capacity of Piles from Dynamic Measurements". *Final Report, Ohio Department of Transportation.* Cleveland, OH.

Goble, G. G., Likins, G. E., Teferra, W. (1977). "Piles and Pile Driving Hammer Performance for H-Piles Driven to Bedrock". *Ohio Department of Transportation and Federal Highway Administration.* Cleveland, OH.

Goble. G., and Hussein, M. H. (1994). "Dynamic Pile Testing in Practice". *13th Int. Conf. on Soil Mechanics and Foundation Engineering*. New Delhi, India, 713-716.

Hussein, M.H., and Morgano, C. M. (1993). "Structural Integrity Evaluation Of Concrete Piles From Stress Wave Measurements". *Society for Experimental Mechanics, Conf. on Nondestructive Testing Of Concrete In The Infrastructure.* Dearborn, MI.

Likins, G.E. and Rausche, F. (2004). "Correlation of CAPWAP with Static Load Tests". *Proc. of the 7th Int. Conf. on the Application of Stresswave Theory to Piles.* Petaling Jaya, Malaysia, 153-165.

Rausche, F., Goble, G. G. (1979). "Determination of Pile Damage by Top Measurements". *American Society for Testing and Material, STP 670, January 1979.* Philadelphia, PA; 500-506.

Rausche, F., Goble, G., Likins, G. and Miner, R. (1985). "The Performance of Pile Driving Systems". *Federal Highway Administration Contract DTFH61-82-1-00059*, (four volumes).

Rausche, F., Likins, G.E., and Hussein, M.H. (1988). "Pile Integrity by Low and High Strain Impacts". *Proc. of the 3rd Int. Conf. on the Application of Stress Wave Theory to Pile.* Ottawa, Canada, 44-55.

Rausche, F., Moses, F., Goble, G. G. (1972). "Soil Resistance Predictions From Pile Dynamics". *Journal of the Soil Mechanics and Foundations Division, American Society of Civil Engineers*. Vol 98, SM9, Proc. Paper 9220, September, 1972 (Reprinted in Current Practices and Future Trends in Deep Foundations, Geotechnical Special Publication No. 125, August, 2004. American Society of Civil Engineers: Reston, VA; 418-440.)

Teferra, W. (1977), "Dynamic Studies of Steel Piles with High Driving Stresses," Unpublished Master's Thesis, Case Western Reserve University, Cleveland, Ohio.

Verbeek, G., and Middendorp, P. (2011). "Determination of Pile Damage in Concrete Piles". *DFI Journal Vol. 5, No 2.* December 2011.

Capacities and Resistance Factors for Driven Piling in Illinois

James H. Long[1], A.M.ASCE, Andrew Anderson[2], M.ASCE, and William Kramer[3], M.ASCE

[1]Emeritus Faculty, University of Illinois at Urbana, Champaign, 205 North Mathews, Urbana, IL 61801, jhlong@illinois.edu

[2]Graduate Research Assistant, University of Illinois at Urbana, Champaign, 205 North Mathews, Urbana, IL 61801, acandrsn@illinois.edu

[3]Head, Foundations and Geotechnical Unit, Illinois Department of Transportation, Springfield, IL, William.Kramer@illinois.gov

ABSTRACT: The Illinois Department of Transportation (IDOT) and the University of Illinois collected information on driven piling to improve predictions for pile capacity and to develop resistance factors appropriate for LRFD-based design. Several methods were investigated including static methods, dynamic formula, WEAP, PDA, and CAPWAP.

Pile dynamic testing was conducted on 37 piles from 19 different construction sites throughout Illinois. Measurements of penetration resistance were recorded for end of driving (EOD), and beginning of restrike (BOR) conditions. Estimates for static capacity were based on CAPWAP analyses of pile penetration at beginning of restrike (BOR) conditions. All piles were steel H-piles or closed-ended pile piles and driven with open-ended diesel hammers.

Comparisons of measured and predicted capacity were made for several different predictive methods. The agreement between measured and predicted capacity for dynamic formulas, WEAP, and PDA are significantly better than observed in previous studies. The reason agreement improved is due primarily to the limited types of piles and hammers used for driving piles for common IDOT bridges. Improved reliability for predicting capacity can justify higher resistance factors and more efficient pile design.

INTRODUCTION

The Illinois Department of Transportation (IDOT) spends more than thirteen million dollars per year on driven piling for repair, expansion, and new construction for support of bridge structures. The majority of these structures are modest-sized bridges with two abutments, and two piers or less. Each substructure is typically supported on six to twelve driven piles. Tight construction schedules typically do not accommodate the additional time necessary to restrike the piles, therefore, IDOT is particularly interested in improving and quantifying the accuracy of pile capacity methods based on EOD conditions. An increase in the precision of estimated capacity allows use of higher resistance factors, which can result in significant savings due to the large volume of piles driven by IDOT per year.

Pile driving was monitored for 37 piles at 19 different sites in Illinois in which driven piles were installed for bridge foundations. Typical records of pile driving (pile penetration rate and hammer stroke) were accompanied with dynamic measurements using a Pile Driving Analyzer (PDA) for both end of driving and beginning of restrike conditions. All piles were H-piles or closed-ended pipe piles driven with a single-acting diesel hammer. Soil profiles varied, however, there was no site in which soft fine grained soil was predominant, and most piles were driven into stiff fine grained soil, till, or granular soil.

Estimates of driven pile capacity were made using static methods, dynamic formulas, wave equation analysis, and methods that require dynamic monitoring during driving. Static methods investigated include IDOT, Olson, and DRIVEN. Methods based on a pile driving formula include FHWA-modified Gates, Minnesota modified Gates, and the Washington State DOT methods. Capacities are also estimated using PDA and CAPWAP for both end of driving and beginning of restrike conditions. The estimate for static capacity was based on CAPWAP results for beginning or restrike (BOR) conditions. CAPWAP (BOR) capacities are compared with prediction of pile capacity using the other methods. Only one static load test was performed during this study, and the agreement between the capacity based on the static load test (using Davisson's method) and CAPWAP (BOR) was within 3 percent. Statistical parameters for the ratio of measured capacity to predicted capacity are used to assess the accuracy and precision of the methods. The statistical parameters are used to develop resistance factors for these methods.

COLLECTION OF DRIVEN PILE DATA

Pile testing was performed on 37 piles at 19 sites throughout Illinois; however, the majority of tests were clustered in the middle half and western portion of Illinois. Typically three pile tests were performed at a particular site with a range of 1 to 5 piles at each site.

Of the 37 piles, 26 were closed-ended pipe (CEP) piles and 11 were H-piles. The predominant H-pile section was the HP 12×53 (305x79) section, whereas the metal shell section MS 14×0.25 inch (356x6.35mm) was the most common CEP pile. Driven pile lengths ranged from 5.5 to 31m with an average length of about 15m.

Piles were driven with 12 different open-ended diesel hammers from 5 manufacturers. The most commonly used hammers were the Delmag D19-32 and D19-42. Piles were driven into a variety of soil types. Fine-grained soil profiles were predominantly stiff clay. CEP piles driven into mixed soils accounted for 30 percent of the cases, while the fewest cases were for H-piles driven into clay (4 percent). Thirty-six percent of the piles were driven into sand, 49 percent of piles were driven into mixed soil profiles, while only 15 percent were driven into primarily clay soil. Axial capacity for the driven piles ranged from 685kN to 3200 kN with an average of about 1500kN.

METHODS FOR PREDICTING CAPACITY

Pile capacities were estimated using static methods, dynamic formulas, wave equation analyses (WEAP), and from measurements of pile dynamic monitoring (PDA, and CAPWAP).

Static Methods

Static methods are typically used during initial design to calculate pile capacity and estimate pile size and length. Static methods use results from a subsurface investigation and laboratory testing such as soil description, visual classification, Atterberg limits, standard penetration tests, and undrained strength to calculate unit side resistance and unit end bearing. The capacity of the pile is determined from the dimensions of the pile and the mode of failure (plugged, unplugged).

Three static methods were investigated: the IDOT method, DRIVEN, and Olson's method. Details for each method are too lengthy to describe within the limits of this paper, therefore a brief description of each is given along with appropriate references.

The IDOT method was developed by the Illinois Department of Transportation and has been modified several times since its original development. Unit side resistance and end bearing for piles driven into coarse-grained soil is determined on the basis of standard penetration test results (N_{60}). Standard penetration tests conducted by Illinois DOT use automatic trip hammers. Unit side resistance for fine-grained soil is based on the unconfined compressive strength of the soil.

Further adjustments are applied depending on pile type (H or CEP) and on predominant soil type (fine-grained or coarse grained). A more detailed description for the method can be found in Long, et al, (2012).

Olson's method for coarse grained soils (Olson, 1990) is based on SPT N-values for coarse-grained soils with recommendations for limiting values of unit side resistance and end bearing. Olson's method for clay (Dennis and Olson, 1983) relates the unit side resistance to the soil's undrained strength.

DRIVEN is a program written for FHWA in 1998 to determine pile capacity. The program is described by Hannigan et al. (2006). The program has several options for selecting unit side resistance and unit end bearing. In this study we selected options to use Nordlund (1963, 1979) and Thurman (1964) for coarse-grained soil and Tomlinson (1980) for fine grained soils.

Dynamic Formulas

Three dynamic formulas were examined for predicting pile capacity. These methods use information recorded during the driving of a pile to determine its capacity. The most important parameters for these methods are the energy delivered to the pile due to the weight and drop of the pile hammer and the number of blows to drive the pile a given distance at the end of driving. Investigated were the Washington State DOT formula (WSDOT), the FHWA-Gates formula, and the Minnesota State DOT formula (MnDOT).

The WSDOT (Allen, 2005) uses the following formula to determine pile capacity:

$$R_n = 21.6 * F_{eff} * WH * \ln(10N_b) \tag{1}$$

where R_n is the ultimate pile capacity (kN), F_{eff} is the hammer efficiency factor based on hammer and pile type, W is the weight of the ram (kN), H is the observed hammer stroke (m), and N_b is the average penetration resistance [blows per 25mm]. The hammer efficiency factor, F_{eff} is 0.55 for air/steam hammers with all pile types, F_{eff} is 0.37 for open-ended diesel hammers with concrete or timber piles, F_{eff} is 0.47 for open-ended diesel hammers with steel piles, and F_{eff} is 0.35 for closed-ended diesel hammers with all pile types.

The Federal Highway Administration has modified Gates' original equation and recommends the following (Hannigan et al. 2006):

$$R_n = 212 * \sqrt{WH} * \log(10N_b) - 445 \tag{2}$$

where R_n is the pile capacity in kN, W is the weight of the ram in kN, H is the observed hammer stroke in meters, and N_b is the driving resistance in blows per 25mm. This method is referred to as the FHWA-modified Gates equation.

The third dynamic formula is referred to as the Minnesota Gates equation (Paikowsky, 2009) where pile capacity is determined according to the formula below:

$$R_n = 134 * \sqrt{WH} * \log(10N_b) \tag{3}$$

where R_n is the pile capacity in kN, W is the weight of the ram in kN, H is the observed hammer stroke in meters, and N_b is the driving resistance in blows per 25mm. This method is referred to as the MNDOT-modified Gates equation.

Wave Equation, PDA, and CAPWAP

Wave equation analyses use the one-dimensional wave equation to estimate pile stresses and pile capacity during driving (Goble and Rausche 1986). The analyses model details of the hammer, hammer cushion, pile, and soil. Analyses used in this study were conducted with the software GRLWEAP (PDI, 2010). Modeling of the side resistance distribution with depth and the portion of total capacity carried in side resistance was determined from the IDOT static method. No setup factors were used in the WEAP analyses.

The PDA method is a procedure for determining pile capacity based on values of pile head force and velocity with time. Accordingly, estimates of capacity using the PDA require the pile to be instrumented during driving. The PDA monitors instrumentation attached to the pile head and measurements of strain and acceleration are recorded versus time. Strain measurements are converted to pile force, and acceleration measurements are converted to velocities. A simple dynamic model (Case model) is applied to estimate the pile capacity. The calculations for the Case model are simple enough for static pile capacity to be estimated during pile driving operations. Several versions of the Case method exist, and each method will yield a different static capacity. All capacities in this study used the RMX method with a damping factor of 0.6. A more detailed presentation of the PDA and Case methods is presented in Hannigan, et al., 2006.

CAPWAP employs dynamic measurements obtained with the PDA during driving with more realistic modeling of the pile and the static and dynamic soil resistance along the side and end of the pile. The method uses the acceleration history measured at the top of the pile as a boundary condition. The result of the analyses is a predicted force versus time response at the top of the pile. Comparison of predicted and measured force response allows the user to determine the accuracy of the solution, and model parameters are modified until the measured and predicted force versus time plots are in close agreement (Hannigan, et al., 2006). While CAPWAP does not provide a unique solution, solutions with good match quality generally result in similar capacities.

PERFORMANCE OF PREDICTIVE METHODS

The ratio of measured capacity (Q_M) to predicted capacity (Q_P) is the metric used to quantify how well a predictive method estimates capacity. Values of Q_M/Q_P close to one represent good agreement between measured and predicted capacity. Values of Q_M/Q_P less than unity indicate the predictive method overestimates capacity whereas values of Q_M/Q_P greater than unity indicate the predictive method underestimates capacity. The average of Q_M/Q_P values for all piles is referred to herein as the bias (λ), and represents, on the average, how well the method predicts capacity. The standard deviation (σ) of Q_M/Q_P for all piles is a convenient metric to quantify the scatter associated with the predictive method, and finally, the coefficient of variation (cov) is a normalized metric for scatter (cov = σ/λ).

The bias is useful to assess whether a predictive method tends to, on the average, over- or under-predict capacity. Correcting for bias can be easily accomplished for a predictive method by applying a constant to compensate for the bias. However, the normalized scatter (cov) associated with a predictive method is independent of a correction factor. Accordingly, the cov is a useful parameter for quantifying which predictive methods exhibit the least scatter. Smaller cov's reflect less scatter and better ability to predict capacity consistently and efficiently. Statistical parameters for all methods are provided in Table 1.

The agreement between predicted capacity and measured capacity is shown in Figures 1a, 1b, and 1c for the three static methods IDOT, DRIVEN, and Olson. Measured capacity was estimated using CAPWAP for beginning of restrike conditions (CAPWAP BOR). Significant scatter is exhibited for all the static methods. Values of cov for the three methods exceed 0.49 which represents considerable scatter. Several factors contribute to the high degree of scatter observed for these methods such as: the inherent variability associated with use of SPT to define soil parameters, and the difference between the soil profile estimated from the nearby boring and the soil profile at the location of the pile contributes significantly to the scatter observed.

Dynamic formulas predicted capacities with greater consistency than static methods as shown in Figure 2. WSDOT and FHWA-modified Gates methods exhibited a small degree of scatter for all piles (cov approximately 0.25) while the MnDOT Gates methods was only slightly larger (cov equal 0.275). Accordingly, the dynamic formula provide significantly more reliable results than static methods, and reinforce the value of using driving resistance as a measure for predicting pile capacity.

Results comparing WEAP predictions are given in Figures 3a and 3b for both BOR and EOD conditions, respectively. Predicted capacities for BOR conditions are in very good agreement. Scatter for BOR conditions is less than for the static and dynamic formulas. Predictions for capacity based on EOD driving behavior however, exhibits slightly more scatter. There is a consistent trend for WEAP EOD to either predict capacities well, or to underpredict capacity. Under-predictions occurred most commonly for CEP piles in mixed soil. Comparisons of predicted and measured capacities using the PDA are shown in Figures 4a and 4b for BOR and EOD conditions, respectively. The cov for PDA BOR was the smallest of all methods, and the cov for PDA EOD was the second smallest for all predictive methods.

Values of cov for H piles are consistently less than the cov for CEP piles for the dynamic formulas, WEAP, and PDA. This observation suggests that the capacity of H piles may be predicted with greater reliability than CEP piles. Furthermore, the difference in cov's is significant enough to warrant separate treatment of H and CEP piles. However, the number of H piles tested (11) is too small to reliably assess cov. Efforts to collect more H pile cases in Illinois are currently in progress. MnDOT (Pakowski, 2009) used a different collection of pile data and found cov's for H piles and CEP piles varied enough to treat each separately.

Table 1. Statistical parameters for Q_M/Q_P for all methods.

Method	Pile	n	λ	cov
Static IDOT	H	11	1.04	0.326
	CEP	26	0.78	0.514
	All Piles	37	0.86	0.492
Static - DRIVEN	H	11	1.34	0.698
	CEP	26	1.65	0.553
	All Piles	37	1.55	0.604
Static - Olson	H	11	1.24	0.627
	CEP	26	2.24	0.808
	All Piles	37	1.92	0.802
Dynamic Formula WSDOT	H	11	0.87	0.171
	CEP	26	0.93	0.283
	All Piles	37	0.91	0.252
Dynamic Formula MnDOT	H	11	1.01	0.192
	CEP	26	1.01	0.308
	All Piles	37	1.01	0.275
Dynamic Formula FHWA	H	11	0.78	0.149
	CEP	26	0.82	0.287
	All Piles	37	0.81	0.251
WEAP EOD	H	11	1.04	0.171
	CEP	26	1.42	0.328
	All Piles	37	1.31	0.316
WEAP BOR	H	11	1.04	0.171
	CEP	26	1.01	0.233
	All Piles	37	1.02	0.238
PDA EOD	H	11	1.01	0.139
	CEP	24	1.12	0.270
	All Piles	35	1.09	0.237
PDA BOR	H	11	0.86	0.191
	CEP	26	0.82	0.183
	All Piles	37	0.83	0.184

DETERMINATION OF RESISTANCE FACTORS

Load Resistance Factor Design (LRFD) is being increasingly used in geotechnical design and is part of standard practice in structural engineering. The primary advantage of using LRFD is the ability to separate the uncertainty in the loading from the uncertainty in the resistance.

Resistance factors are determined using FOSM (First Order Second Moment) and the correct representation for the covariance of the total load. Calculations were performed to compare the results obtained from FOSM (with corrected covariance of total load) and FORM (First Order

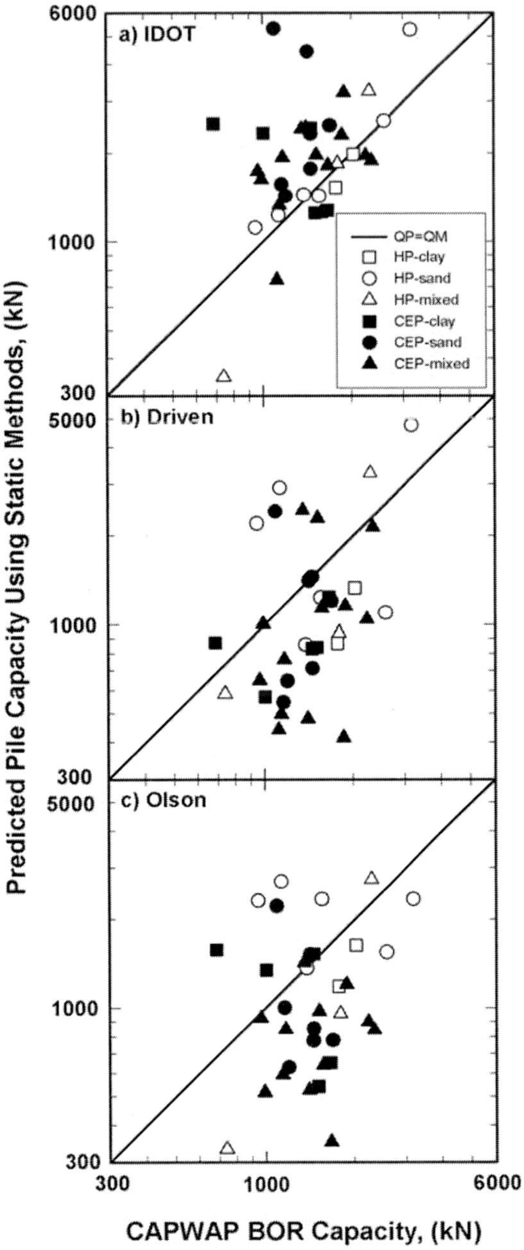

Figure 1. Predicted versus measured capacity for static methods.

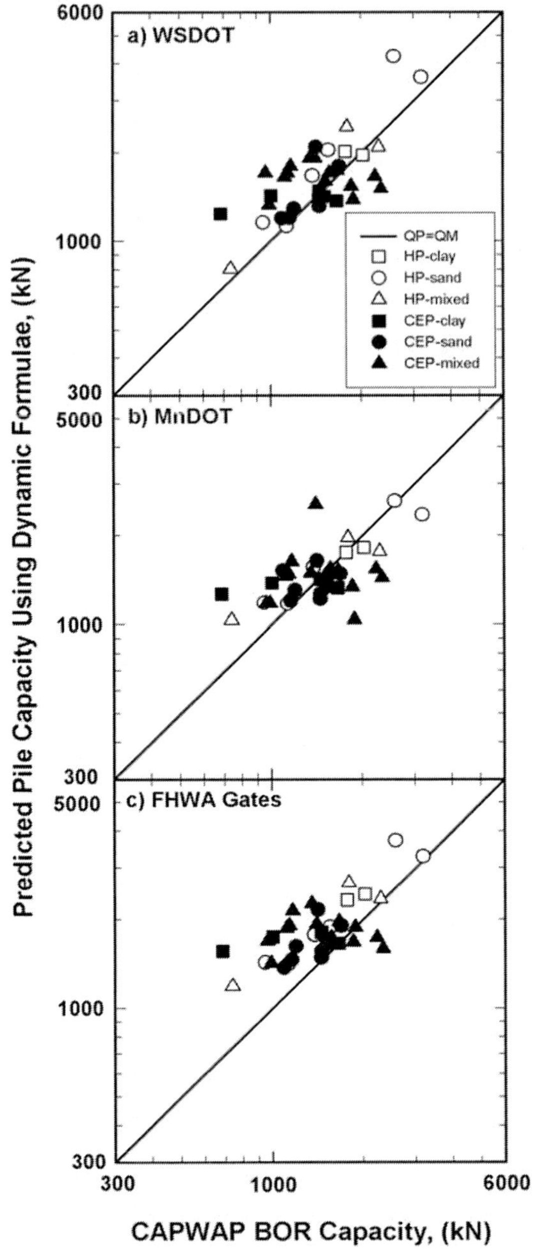

Figure 2. Predicted versus measured capacity for dynamic formulas.

Reliability Method) and found the difference negligible. The FOSM approach was applied in this study for ease of calculation due to FORM requiring an iterative solution process. Additional detail in the LRFD calculation procedure and other methods to determine the resistance factor are presented in Long et al. 2009.

The FOSM method (Paikowsky 2004) can be used to determine the resistance factor, φ. The FOSM method was modified to include the Bloomquist et al. (2007) modification for a more accurate covariance of the total load (dead load and live load) resulting in the following equation

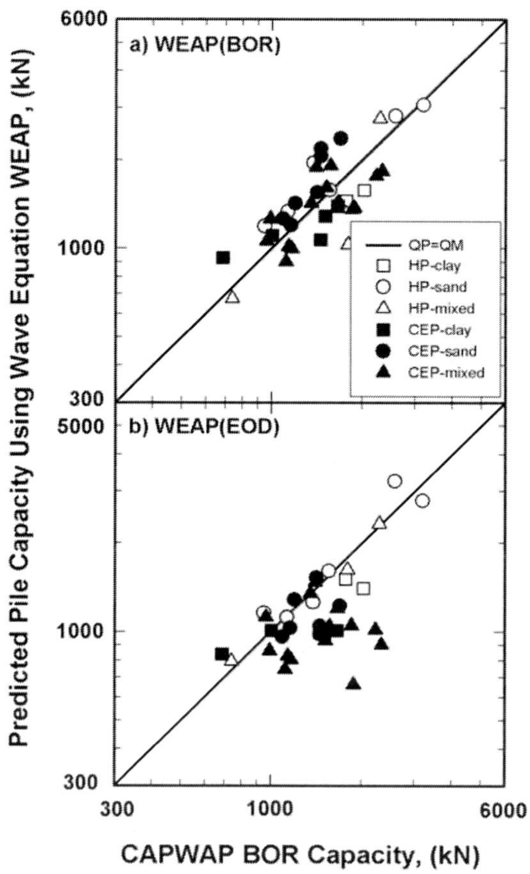

Figure 3.Predicted versus measured capacity for WEAP.

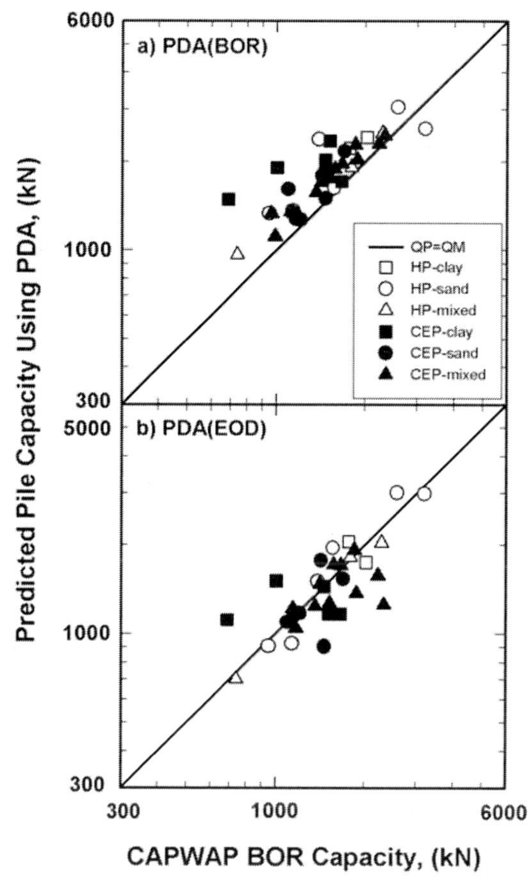

Figure 4. Predicted versus measured capacity for PDA

for determination of the resistance factor, ϕ.

$$\phi = \frac{\lambda_R \cdot \left(\gamma_D \cdot \frac{Q_D}{Q_L} + \gamma_L\right)\sqrt{\frac{\frac{\frac{Q_D^2}{Q_L^2} \cdot \lambda_{QD}^2 \cdot COV(QD)^2 + \lambda_{QL}^2 COV(QL)^2}{\left(\frac{Q_D^2}{Q_L^2} \cdot \lambda_{QD}^2 + 2\frac{Q_D}{Q_L} \cdot \lambda_{QD}\lambda_{QL} + \lambda_{QL}^2\right)}}{(1 + COV(R)^2)}}}{\left(\lambda_{QD} \cdot \frac{Q_D}{Q_L} + \lambda_{QL}\right) \cdot e^{\beta\sqrt{ln\left\{(1+COV(R)^2)\cdot\left(1+\frac{\frac{Q_D^2}{Q_L^2}\cdot\lambda_{QD}^2\cdot COV(QD)^2+\lambda_{QL}^2 COV(QL)^2}{\left(\frac{Q_D^2}{Q_L^2}\cdot\lambda_{QD}^2+2\frac{Q_D}{Q_L}\cdot\lambda_{QD}\lambda_{QL}+\lambda_{QL}^2\right)}\right)\right\}}}} \quad (4)$$

where λ_R is the bias factor for resistance, COV(QD) is the coefficient of variation for the dead load (0.1), COV(QL) is the coefficient of variation for the live load (0.2), COV(R) is the coefficient of variation for the resistance, β_T is the target reliability index (2.33), γ_D is the load factor for dead loads (1.25), γ_L is the load factor for live loads (1.75), Q_D/Q_L is the ratio of dead load to live load (2.0), λ_{QD} is the bias factor for dead load (1.05), and λ_{QL} is the bias factor for live load (1.15). Comparisons of Eqn. 4 with FORM methods show the results to be in good agreement.

Resistance factors were determined for several predictive methods using Eqn. 4 and the statistics reported in Table 1. Resistance factor for the IDOT method, which exhibited the least scatter of the static methods, was determined to be 0.33. The resistance factor for the WSDOT method, which is currently used by IDOT, was 0.62. WEAP resistance factors were calculated as 0.77 and 0.72 for EOD and BOR conditions. One would expect the resistance factor to be greater for BOR than for EOD because the cov is smaller for BOR conditions and therefore more precise. However, the resistance factor (Eqn. 4) also includes the effect of bias of the predictive method. The higher resistance factor observed for WEAP EOD is because WEAP under-predicts capacity for EOD conditions, and the higher resistance factor results from that bias. Accordingly, it is important to note that comparing cov's, not resistance factors, is an effective means to identify which methods exhibit greater precision. Resistance factors for the PDA resulted in 0.77 and 0.66 for EOD and BOR, respectively. Again, the high resistance factor for PDA BOR is compensating for the difference in bias between the two conditions.

This study uses CAPWAP (BOR) results as a measure of static pile capacity instead of static load test results. All estimates of bias and cov in Table 1, as well as estimates for resistance factors, are based on the agreement between the predicted capacity and the capacity predicted with CAPWAP(BOR). The bias, cov, and estimates for resistance factor presented above can be adjusted to represent estimates based on static load test results by using statistics that relate CAPWAP(BOR) results to static load test results. Rausche et al. (1996) used a collection of 99 load tests on driven piling and reported values of 0.92 and 0.22 for the mean and cov for the ratio of capacity determined from CAPWAP(BOR) to that determined with a static load test. These statistics correspond to values of 1.14 and 0.22 for bias and cov, respectively. Values of bias and cov for all predictive methods were adjusted by combining the two distributions, CAPWAP(BOR)/Predicted Capacity and Static Load Test Capacity/CAPWAP(BOR). Resistance factors based on static load tests were slight less than estimates that used CAPWAP(BOR) as the measure of capacity (Table 2).

Previous studies with a different pile database resulted showed cov's for dynamic formulas, WEAP, and PDA to be higher (Long, et al, 2006) than this study. Coefficient of variation for WSDOT methods were 0.45 in the previous study compared to 0.25 in this study. Other predictive methods exhibited similar reduction in cov values. A reason that contributes significantly to the reduction in scatter is that this study only considered a small subset of the pile types, hammer types, and soil conditions. This study included only steel H and CEP piles driven

Table 2. Comparison of resistance factors based on CAPWAP(BOR) with factors based on static load test.

Predictive Method	ϕ_1	ϕ_2
Static - IDOT	0.33	0.33
Dynamic Formula – WSDOT	0.62	0.58
WEAP EOD	0.77	0.74
WEAP BOR	0.72	0.67
PDA EOD	0.77	0.71
PDA BOR	0.66	0.59

Note:

ϕ_1 - uses CAPWAP(BOR) as measure of static capacity

ϕ_2 - adjusted for static load test as measure of static capacity with single acting diesel hammers. Soil conditions were mostly stiff clays and granular soils. Soft soils were not a significant part of any soil profile.

SUMMARY

The accuracy of several methods for predicting driven pile capacity was investigated. Predictive methods included static methods, dynamic formulas, wave equation, PDA and CAPWAP. CAPWAP BOR results were used as a measure of the pile capacity and adjustments were made to determine resistance factors based on static load tests. Thirty-seven cases were collected from 19 sites in Illinois. These cases were representative of soil, pile, and installation conditions commonly encountered by the Illinois Department of Transportation when installing driven piling for typical bridge foundations. Static methods predicted capacity with the least consistency. The coefficients of variation associated with the methods were high, and the resistance factor for the IDOT static method was determined to be 0.33. Dynamic formulas performed considerably better than the static methods. Coefficient of variation for WSDOT was about 0.25 and a resistance factor of between 0.58 and 0.62 was determined using a FOSM method. WEAP (BOR) exhibited greater precision than static methods and the dynamic formulas. WEAP EOD exhibited slightly more scatter than the dynamic formula, however, incorporating setup into the estimates would improve EOD estimates. PDA BOR and EOD exhibited the least amount of scatter for all the methods.

Resistance values determined from this study are greater than resistance values recommended by AASHTO(2010). The resistance factors were developed for a limited collection of driven piles. All piles were either steel H or closed end pipe piles driven with single acting diesel hammers and located within the State of Illinois. Furthermore, resistance factors determined using WEAP and PDA are also a result of the methodology used to select and use parameters for each predictive method. It is likely that other driving equipment, soil conditions, piles, and different methodologies for selecting parameters will result in different resistance factors than presented here. Furthermore, methods that use EOD behavior to estimate static pile capacity implicitly assume a value for setup. The magnitude of setup is affected by both pile and soil conditions and as a result, these methods may overpredict capacity for cases with little setup and underpredict capacity for cases with significant setup.

ACKNOWLEDGMENTS

This work is based on the results of ICT-R27-69, Improved Design for Driven Piles on a Pile Load Test Program in Illinois. The work was conducted in cooperation with the Illinois Center for Transportation; the Illinois Department of Transportation, Division of Highways; and the U.S. Department of Transportation, Federal Highway Administration.

The contents of this report reflect the view of the authors, who are responsible for the facts and the accuracy of the data presented herein. The contents do not necessarily reflect the official views or policies of the Illinois Center for Transportation, the Illinois Department of Transportation, or the Federal Highway Administration. This report does not constitute a standard, specification, or regulation.

REFERENCES

AASHTO (2010), *Standard Specifications for Highway Bridges*, 15th Addition, American Association of State Highway and Transportation Officials, Washington, D.C.

Allen, T.M. 2005. *Development of the WSDOT Pile Driving Formula and Its Calibration for Load and Resistance Factor Design (LRFD)*. Prepared for Washington State Department of

Transportation and in cooperation with the U.S. Department of Transportation, Federal Highway Administration.

Bloomquist,D., McVay,M. and Hu, Z. (2007) *Updating Florida Department of Transportation's (FDOT) Pile/Shaft Design Procedures Based on CPT & DTP Data.* BD-545, RPWO #43, UF Project #00005780. Florida Department of Transportation.

Dennis,N.D.Jr. and Olson,R.E. (1983). "Axial Capacity of Steel Pipe Piles in Clay." *Proceedings of the Conference on Geotechnical Practice in Offshore Engineering*, S.G. Wright, ed., ASCE, New York, 370-388.

Goble,G.G. and Rausche,F. (1986). "*Wave Equation Analysis of Pile Driving – WEAP86 Program, 1 IV.*" US. Dept. of Trans. FHWA, Implementation Division, McLean, VA.

Hannigan,P.J., Goble,G.G., Likins,G.E., and Rausche,F., 2006. *Design and Construction of Driven Pile Foundations, Volume 1*, Report No. FHWA-NHI-05-042, Washington, D.C.

Long,J.H., Hendrix, J., and Baratta, A. (2009). *Evaluation/Modification of IDOT Foundation Piling Design and Construction Policy*, Research Report FHWA-ICT-09-037. Rantoul: Illinois Center for Transportation

Long,J.H. and Anderson, A.C. (2012). *Improved Design for Driven Piles on a Pile Load Test Program in Illinois*, Research Report FHWA-ICT-12-011. Rantoul: Illinois Center for Transportation

Nordlund,R.L. (1963). "Bearing Capacity of Piles in Cohesionless Soils," *ASCE, SM&F Journal SM-3*.

Nordlund,R.L. (1979). "Point Bearing and Shaft Friction of Piles in Sand," *5th Annual Fundamentals of Deep Foundation Design*, University of Missouri-Rolla.

Olson,R. E. (1990), "Axial Capacity of Steel Pipe Piles in Sand," *Proceedings of the Offshore Technology Conference*, Houston, Tx, pp. 17-24.

Paikowsky,S.G.,(2004). *Load and Resistance Factor Design (LRFD) for Deep Foundations. NCHRP Report 507*, Washington, DC: Transportation Research Board.

Paikowsky,S.G., (2009). *Developing a Resistance Factor for Mn/DOT's Pile Driving Formula.* St. Paul: Minnesota Department of Transportation.

Pile Dynamics, Inc. (2010). *GRLWEAP: Wave Equation Analysis of Pile Driving, Procedures and Models, Version 2010*, Cleveland, OH

Rausche,F., Hendean,G., Abou-matar,H., Likins,G.E. and Goble,G.G. (1996), *Determination of the Driveability and Capacity from Penetration Tests*, Final Report, U.S. Department of Transportation, Federal Highway Administration Research Contract DTFH61-91-C-00047.

Thurman, A.G. (1964). "*Computed Load Capacity and Movement of Friction and End-Bearing Piles Embedded in Uniform and Stratified Soil*," Ph.D. Thesis, Carnegie Institute of Technology.

Tomlinson, M.J. (1980), *Foundation Design and Construction*, Pitman Advanced Publishing, Boston, MA, 4th Edition.

Pile Capacity in a Bouldery Stratum

P.V. Chandramohan, Dr

President (Technical), Navayuga Engineering Company Ltd, Plot 379, Rd #10, Jubilee Hills, Hyderabad – 500 033, India, pvcmohan@gmail.com

ABSTRACT: A bridge was to be constructed across the Lohit River in India. The subsoil consisted of cobbles (80 to 300mm size) and boulders (>300mm size) with interspaces loosely packed with gravel (4.75 to 80mm) and sand (0.075 to 4.75mm). A soil boring could not be carried out due to the difficulty in making a 150mm borehole and also as it would not yield any reasonable data. A shaft bore of 1700mm diameter was made. A test shaft was provided with a steel liner for the full length of the shaft. Four reaction shafts (tension) were cast with liners only for the upper part of the shaft. During the first load test, there was excessive settlement. At the same time, the reaction shafts experienced very little movement. It was inferred that, the steel liner prevented friction from being developed in this medium. The liner was required to prevent collapse of the soil during boring, but not necessarily to the full depth. A second load test was conducted with a reduced length of liner and there was a significant increase in the capacity of the shaft. The load test also showed plastic behavior of the medium. However the plastic was minimal in the vicinity of the design loads.

KEY WORDS: cobbles, boulders, liner, drilled shafts, Load test, shaft shear, end bearing

INTRODUCTION

Soil strata with a substantial quantity of 'large particles' are found at certain locations in rivers. These large particles are in the form of rounded rock masses. Rivers originate at the mountains and flow down the slopes. Near the origin, slope of the river is much steeper than that in the lower reaches. Steeper slopes will result in higher current velocities with lower flow depths. These high velocities have high carrying capacities. This means that higher current velocities will be able to transport heavier bodies, like rock masses. As the river comes down to the lower reaches, the slope becomes milder and flow velocities are reduced. The carrying capacity of the current reduces and the larger particles settle out. As the river goes into the plains beyond the mountain slopes, the size of particles deposited will further reduce. The material deposited in the plains will be predominantly sand. Upstream of the sand deposits, sand will be mixed with small cobbles and further upstream, the particle size increases eventually reaching boulder size material with rounded corners and a smooth surface.

BOULDERY STRATUM

A bridge was to be constructed across River Lohit at Alubarighat in India. The bed of the river was typical of a low depth-high velocity flow regime. The river is susceptible to flash floods and most of the bed is visible during low water conditions. The bed is made up of cobbles of more than 100mm size. Please see figure 1 and 2.

An open excavation was made to investigate the distribution of grain sizes with depth. The depth of excavation was limited to about 3m due to collapse of the side walls of the excavation. The material consisted mainly of rounded particles of varying sizes ranging up to 600mm diameter boulders. Please see figure 3 and 4.

FIG 1 Wide river bed – low water

FIG 2 Closer look at the bed

FIG 3 Inside the excavation

FIG 4 The odd big one

The bridge would be simply supported with a span of 50m. A drilled shaft foundation was selected for support of the bridge piers during the conceptual stage. There were two primary questions to be answered before proceeding with the shaft foundations. 1) Determine the nature of the stratum at least down to the founding level of the shafts and 2) determine whether a shaft bore could be completed successfully in this type of strata.

Normally, soil exploration would be carried out, by drilling a 100 to 150mm borehole. Undisturbed and disturbed samples would be collected for laboratory tests. In-situ tests, likely standard penetration tests, would also be performed. However, the trench excavation revealed that it would be very difficult to provide borings due to large boulders. Even if borings could be successfully performed, the information would be of limited value no information would be gained outside the 150mm diameter boring limits. The second question was whether it would be possible to provide a satisfactory excavation for the bored shafts. A test bored shaft was selected to determine their feasibility.

TEST BORED SHAFT

The test trench showed that particle sizes up to 600mm could be encountered during foundation excavation. Note that the boulder size could be larger than 600mm below the 3m test excavation depth. Generally, larger diameter particles are easier to excavate in larger diameter boreholes. Well foundations (sunken caissons) are commonly used in India for bridge foundations and were considered in the early stages of planning. However, they were eliminated from consideration due to their high cost, longer time for construction and difficulties offered by the embedded boulders in maintaining the verticality of the caisson during construction and due to the difficulty of correcting shift or tilt in such materials. Bored shafts appeared to be a viable alternate. However, there was no local experience in the area with bored shafts. Therefore, it was important to work out the design and construction procedures for this bridge and possibly others with similar conditions. A 1700mm diameter was selected for the test bore mainly to accommodate boulders larger than the 600mm encountered in the test excavation. This diameter was much larger than required for structural considerations.

Difficulties Encountered In Test Boring

As previously mentioned, a 1700 diameter test boring was selected for the first alternate boring size to determine the feasibility of bored shafts and to obtain details of the medium through which the shafts must pass. (The next alternate size test boring would have been 2000mm before again considering well foundations.) Initially, a 20m deep borehole was made. A Bauer BG36 drill rig was used. The excavation was very difficult as expected. There were many interruptions due to mechanical problems. Many bullet teeth and/or buckets had to be replaced (23 teeth in the test hole; but, an average of 70 teeth per shaft during production shaft excavation). Two buckets were replaced during the test hole excavation. (With the benefit of experience, no buckets required replacement during production shaft construction.) The test hole took 11 days to complete including three days lost to flooding of the river. The actual time required to complete the boring was only 100 hours of this total time.

From 0 to 13m, the strata consisted of 100 to 400mm boulders mixed with coarse sand. The drilling in the upper 13m required high torque due to the large size of the boulders. Thereafter, the size of the boulders reduced to 50 to 200mm. Another major problem encountered was the collapse of the sides of the boring. Steel liner (casing) of 1700mm diameter and 12mm thickness was used to retain the sides of the borehole. As all these contingencies had to be addressed, the actual time spent on the boring came down to only 22 hours. Up to a depth of 8m, the liner was pushed down using the drill rig. But thereafter, a vibro-hammer (ICE make, model 815c) was required to drive the casing. Liner was required for a depth of 10 to 12m. Thereafter, drilling mud was used for stabilizing the sides. The density of the bentonite solution (drilling mud) was maintained at a specific gravity of 1.2 to retain the sides below the liner . Consumption of bentonite was high in the test boring – 200 bags. Photos of the soil collapse during boring and the in-situ material are given in Figure 5 to 8.

FIG 5 Collapse around bucket

FIG 6 Collapsed soil

FIG 7 Bored out material-1

FIG 8 Bored out material-2

To supplement the data from visual observation, samples were collected and a sieve analysis was carried out. Samples were collected at various depths. Results of the particle size analysis are given in figure 9 and 10.

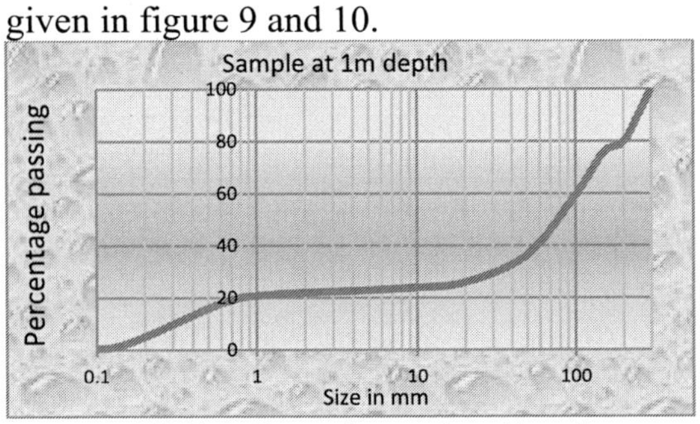

FIG 9 Sample at 1m depth

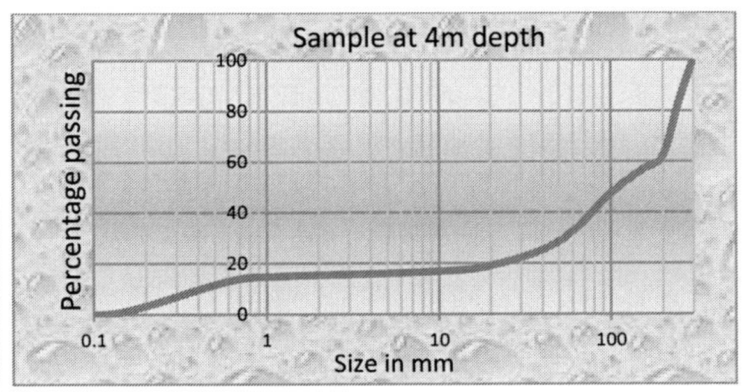

FIG 10 Sample at 4m depth

The above curves are different from those of a normal soil like sand or clay. The sample at 1m depth shows a sharp rising tendency from 60mm size at 40% upwards. The sample at 4m depth also showed the climb. The weight of fines is less in the mixture. The interspaces of the boulders were loosely filled up with cobbles and coarse sand. The material of the boulders was identified to have sandstone origin. This material was found to have a specific gravity of 2.54 to 2.68 gm/cc. A compressive strength test yielded a value of 17.36 MPa.

DESIGN PARAMETERS TAKEN

For the purpose of designing a shaft foundation, certain shear parameters and deformation characteristics of the soil are required. Here, the bulk medium consisted of various sizes of cobbles and boulders at random with the interspaces filled with gravel at various levels of compaction. It was inferred that this medium will have a high degree of permeability. A conceptual diagram of the composition of the medium is given in figure 11.

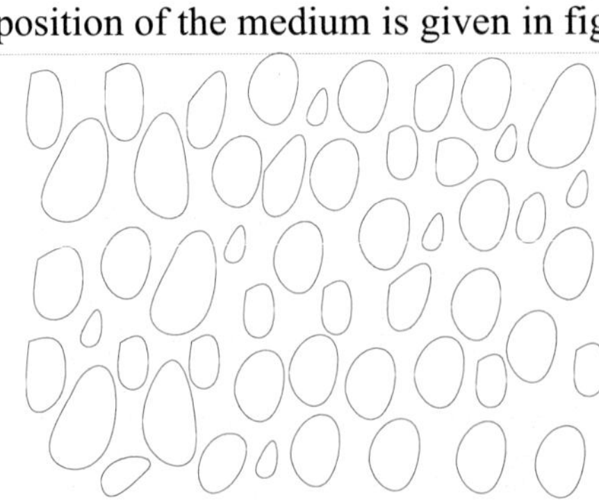

FIG 11 Conceptual diagram of the medium

Particle size distribution is the only test that was feasible. Since soil boring could not be performed in this stratum, standard penetration tests could not be done. The large boulders would provide very high blow counts which could be misleading with regard to the strength and compressibility of the medium. However, calculations were necessary to provide an estimated shaft depth. (See Bureau of Indian Standards 1979 and 1981 for design guidelines for bored shafts and shallow foundations.) N-values could be correlated with the angle of internal friction by using graphs given by the Indian Standards. Considering the N-value to be 100, from graphs an angle of internal friction (φ) of 43° was arrived at. Computations were made for φ=42° and angle of wall friction (δ)=42°. Submerged unit weight of the medium was assumed as 12kN/m^3. For φ=42°, the code, gave bearing capacity factors as N_γ=174.35 and N_q=92.47. Coefficient of earth pressure at rest for bored cast-in-situ shafts was taken as 1.5 from Indian code. With the above parameters, depth of shaft below scour level was arrived at as 9m only. Ultimate side friction came to only 3507kN but ultimate end bearing worked out as 23008kN. All this resulted in a safe carrying capacity of 10606kN. It is evident that end bearing constituted the major component mainly because the shaft is short and bearing capacity factors were are high. The design load was 8816kN. At the location of the test, total shaft depth was only 17m. With a diameter of 1.7m, length to diameter ratio was only 10. There should not be any difficulty in obtaining good quality of concrete. But core drilling and grouting was carried out on the test shaft to ensure quality of concrete.

LOAD TESTING

The shaft had to be load tested (see Bureau of Indian Standards (1985) for guidelines). The maximum test load would be 22041kN. Four reaction shafts (tension shafts) were provided because the test load was comparatively high. Each reaction shaft would provide one-fourth of the test load. Bearing would not contribute to the capacity of the reaction shafts. Many engineers provide a higher factor of safety for uplift. The length of tension shaft was computed to be 22m. A smaller diameter is often used for reaction shafts. However, a diameter of 1700mm was selected for the reaction shafts at this site due to the boulders. Minimum distance between shafts is 3 times the diameter for non-interference during loading. With this distance, there was no chance of boring in one affecting the other during excavation as well. Arrangement for test loading is shown in figure 12 and 13.

High tensile strands, embedded in the shafts, were anchored on top of steel plate girders (secondary girders). They spanned between the reaction shafts. The main girder spanned between the secondary girders and was located below them, supported from top by the bottom flanges of the secondary girders. Loading jacks were located below the main girder; directly over the test shaft. For abundant precaution against borehole sides collapse during excavation, a full length liner was provided for the test shaft. The liner was limited to the upper 12m for the reaction shafts. An outer liner, down to the scour level, was provided around the test shaft to simulate scour. In this particular case, this was provided for a depth of 8m from the cut off level. Figure 14 and 15 shows the photographs of the test set up. The outer casing may be seen in Figure 14.

FIG 12 3-D view of test set up-1

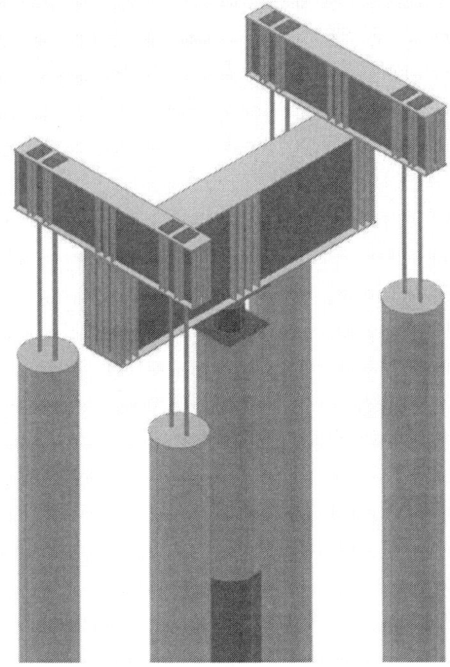

FIG 13 3-D view of test set up-2

FIG 14 Photo of test set up-1

FIG 15 Photo of test set up-2

Initial Load Test - 1

Indian Code (see Bureau of Indian Standards, 1985) requires the test loads to be applied in increments of one-fifth the design load. The first load was to be 1763kN. Loads were rounded off to the nearest figure in the pressure gauge. The first load then came to 2220kN. For this load, the settlement was only 0.72mm. The next load was 3330kN. The settlement went up to 4.06mm.

Please see the load settlement graph in figure 16. The settlement then followed almost a linear behavior with almost equal increase in settlement for every increase in load. At a load of 9434kN, the shaft settled by 38.1mm. Please note that this was only marginally more than the design load of 8816kN. It is needless to say, that this performance was not acceptable and so, loading was stopped. The rebound was small during unloading. This proved to be another surprise. Kindly see the load-settlement graph. The rebound was a negligible 1.42mm. The rebound curve hardly rose from the bottom. The initial settlement was almost equal to the permanent settlement for all practical purposes which provides evidence of plastic behavior.

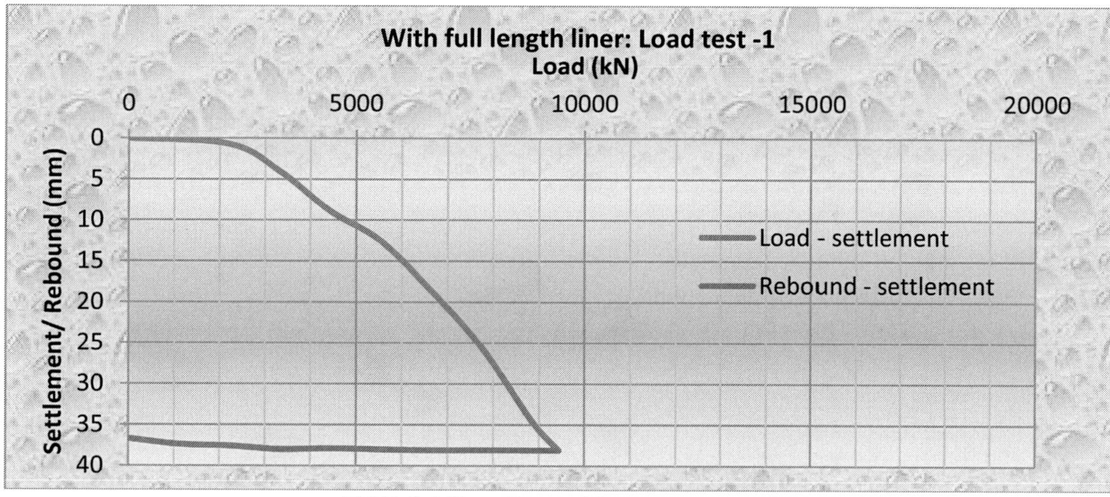

FIG 16 Load – settlement graph

Test Shaft Behavior

The estimated ultimate end bearing was 23008kN. It was clear that mobilization of end bearing was far from this value. The estimated ultimate value of side friction was 3507kN. Here also, the shaft was slipping through the medium slowly. One salient feature of the load-settlement curve was that there was no clear indication of a failure mode. The straight line portion of the curve had a constant slope which continued till the termination of the test. The settlement was not elastic. The toe of the shaft was supported on boulders and cobbles with lose gravel packing in the interspaces. When the load was introduced, the boulders adjusted during loading of the test shaft. In doing so, they moved downward slightly and mobilized a higher resistance. This was a continuing process. The test shaft had steel liner over its shaft throughout its length. The fact that side friction also did not mobilize were attributed to the fact that the rounded boulders and cobbles were acting like point contacts facilitating the shaft to move down without much resistance. Such behavior from the shaft was not expected. The feasibility of shafts for the foundation of the bridge was in question.

Tension Shaft Behavior

It was also observed that while the test shaft was settling slowly, no distress was noticed in the reaction shafts. The reaction (tension) shaft did not have an end bearing component. This shaft was resisting the uplift purely by friction. At the final stage of the test, each reaction shaft was carrying 2359kN. The estimated ultimate value of skin friction was 14668kN. Even though, no measurement of the movement of reaction shaft was made, no visible movement was noticed. For the shaft to move, it had to separate out from the surrounding soil. This was not the case. The length of the shaft shaft was 22m. Steel liner was provided for a depth of 12m, the minimum depth to prevent collapse of the sides. Steel is expensive and there is always an effort made to minimize the depth of liner in any piling operation. While trying to sort out the likely cause of the failure of the test shaft, parameters of the tension shaft, which performed satisfactorily, were compared with those of the test shaft.

Inference

The only difference was that the test shaft had full length liner. If boulders were acting as pointed contacts and reducing the side friction to low values in the test shaft, then the sizeable value of friction in the reaction shafts was from the side friction created by the shaft that was not covered with steel liner. This triggered a discussion about the interface between the shaft bore and the shaft. Please see figure 17 for illustration of the interface with full length liner. As discussed earlier, the boring operations were tough, as the tool was making its way down through boulders and cobbles. Sides of the drilling tool hit against the projecting boulders during its rotation. If a liner was present in the bore, concrete was confined to its precise cylindrical shape. Volume of concrete in this case was the theoretical one. In the case of tension shafts, which had partial liner cover only, consumption of concrete exceeded by one cubic metre. But in reaches where there was no liner, poured concrete would have intrusions of boulders into it. Part of the boulders would be inside the shaft and the other part outside would project out into the medium. The boulders that projected outside the shaft concrete into the supporting soil medium are also part of the shaft. These projections will participate in load transfer.

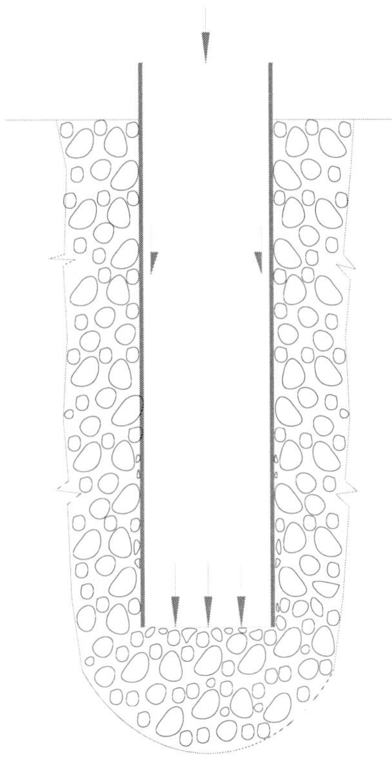

FIG 17 Liner-bore interface and load transfer

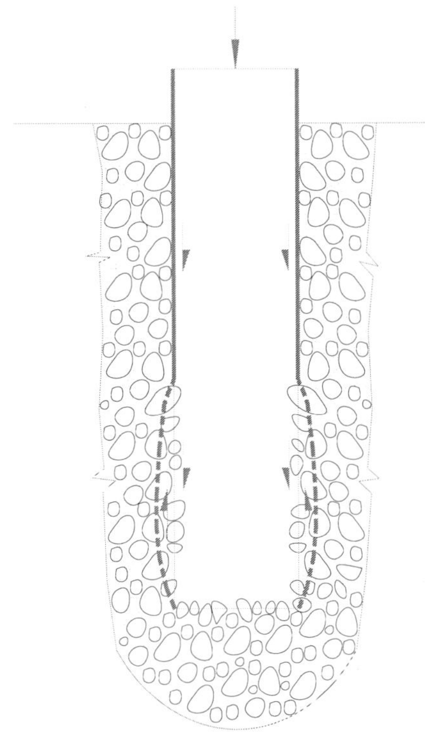

FIG 18 Shaft-bore interface without liner

Kindly see FIG 18. The figure illustrates the portion of the shaft shaft that is not covered by liner. Effect of the projecting boulders is to provide a larger diameter of resistance with effectively a high angle of wall friction (a rougher shaft sidewall than the cased portion). The conclusion was that though liner was required for retaining soil, it was the culprit in preventing the side friction from being developed, in a bouldery stratum. So, if shaft capacities were to be enhanced, liner lengths were to be kept to a minimum.

INITIAL LOAD TEST – 2

If one were confident about this hypothesis regarding the effect of liner upon shaft shear, the next load test would be conducted with a reduced liner length to provide the required resistance. This would have required installation of one more test shaft and two more reaction shafts. The load test would require another one and half months. Since there were no references about such an experience, this would have been a gamble. But the stakes were very high. The Project Team decided to go ahead with this expensive and time consuming experiment. Proportioning of the shaft geometry was a matter of arriving at the length of the shaft without liner, as the diameter was fixed. End bearing was neglected. The side shear value that would be mobilized was not known. A safe length of shaft without liner was chosen as 16m. This 16m uncased length later became the reference length for future structures in the region. The stipulation was that site engineers could provide as much length of liner as required by them for stability of the shaft bore. Requirement from capacity angle was that below the liner, the shaft should have a minimum length of 16m. This was the length that would provide resistance against the load. Two reaction shafts of the old test were retained and a new test shaft was installed as a mirror image of the old one. Two additional reaction shafts were installed directly opposite the old ones which were retained for the new test. Girders were placed in position and the second test was conducted. There was initial apprehension about the result. But the load test was a great success. The result is given in figure 19.

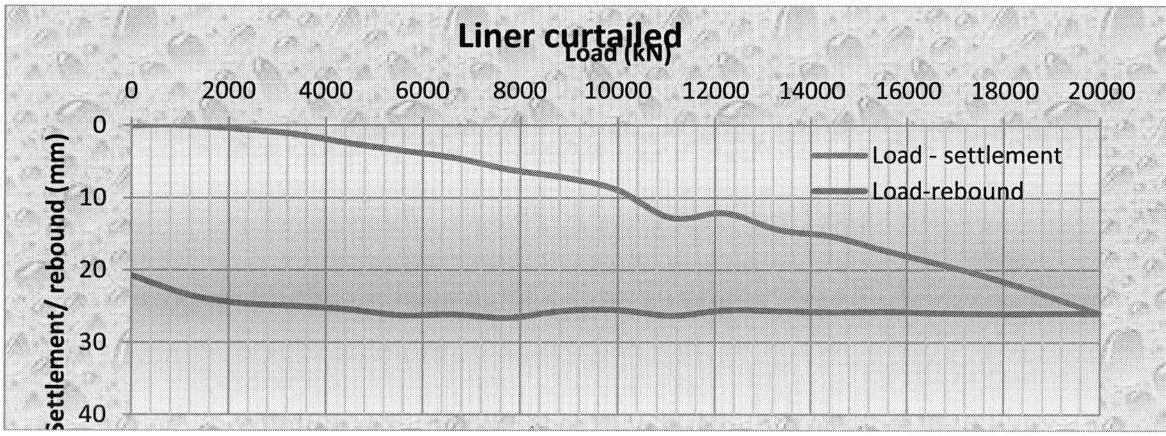

FIG 19 Load-settlement graph with curtailed liner

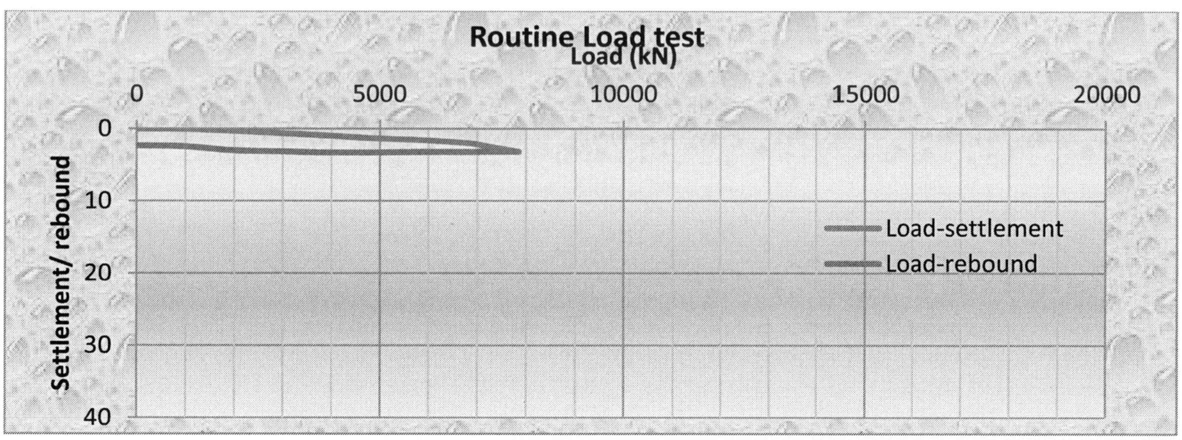

FIG 20 Load-settlement graph – routine test

Though the test load was 22041kN, the test was terminated at 19979kN when the piston of the jack reached its limit of extension. This was due to the elongation of the high tensile strands and the deflection of the girders which occurred as the load caused the test shaft to settle. At the maximum load applied, the settlement was 26mm. The settlement did not increase even after retaining the load for 24 hours. The rebound was of a concern because the shaft top rebounded very little, only about 5mm. Analyses of the test data justified a design load of 8816kN for a 2-shaft arrangement at a pier location and a design load of 4896kN for a 4-shaft alternate. The estimated settlement for a 4-shaft alternate was only 2.5mm. Considering the host of uncertainties, the 4-shaft arrangement was selected for this bridge.

ROUTINE LOAD TEST

One could not be sure of the load carrying capacity in a stratum like this with the results of only one load test. A couple of routine load tests were conducted during the execution of the project. The result of one of the routine tests is presented in figure 20. The shaft was to be loaded to one and a half times the design load. This came to 7344kN for the 4-shaft group. The settlement came to less than 3.5mm but for the design load it was only 1.5mm. This result was even better than the initial results. However, the plastic nature of the movement was observed in this load test also.

Lateral load test

The site was classified as highly seismic in India. The shaft would be subjected to a horizontal load of 850kN. A lateral load test was conducted on a working shaft. The test was done for 1.5 times the design load. Resistance to lateral movement of the shaft was remarkable. The results are given in figure 21. Ultimate movement of the shaft head was 9.19mm. The lateral load test had a free-head condition. The service condition would be a fixed head. Hence movement in fixed head condition would be only 2.30mm for the test load. For the design load, this value would be 1.0mm. Allowable movement would be 17mm. Unlike the vertical load tests, the plastic property was not pronounced in the lateral case. This is partly due to the structural rebound of the shaft as a cantilever instead of elastic compression in the vertical case.

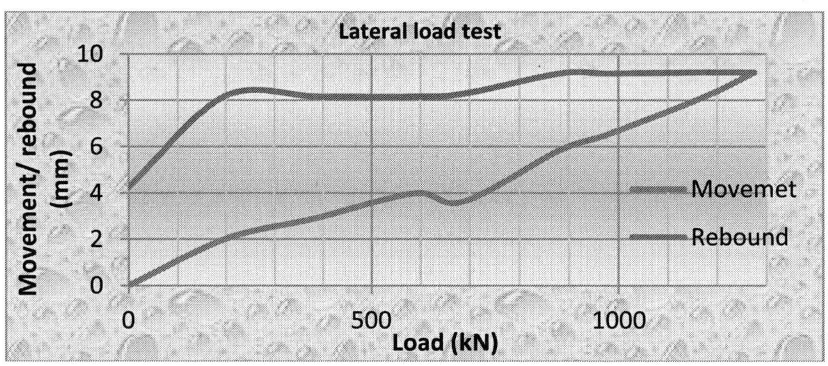

FIG 21 Lateral load test result

CONCLUSIONS

The medium in which the test and reaction shafts were installed could hardly be described as soil. It consisted of sand, cobbles, gravel and boulders. The shear strength of the medium could not be described by the normal parameters like angle of internal/ wall friction. Terzaghi's bearing capacity factors also were not applicable for capacity computations. The medium has highly plastic properties. A few load tests were conducted on shafts in the medium. It was found that rounded cobbles and boulders act as pointed lateral supports on the liner and offer negligible resistance against vertical loads. On the other hand, if liner were not present, portion of boulders projecting into the shaft bore, become embedded in pile shaft (concrete). The other portion jetting out into and interacting with the medium would provide high resistance against vertical loads. This fact was used to provide a high shaft shear component. The end bearing component underwent large settlements in the process of resisting loads. Therefore, bearing was neglected in the design. The tests revealed a large plastic component to the settlement with the permanent settlement approximately equal to the settlement of the test shaft during loading. However, a short length of uncased shaft provided a high shear which helped to compensate for the lower than expected bearing.

REFERENCES

Bureau of Indian Standards (1979). *IS:2911(Part I/Sec 2)-1979 Code of Practice for Design and Construction of Pile foundations Part I Concrete Piles Section 2 Bored Cast-in-situ Piles, (First revision).* Bureau of Indian Standards, New Delhi, India

Bureau of Indian Standards (1985). *IS:2911(Part4)-1985 Code of Practice for Design and Construction of Pile foundations Part 4 Load test on Piles, (First revision).* Bureau of Indian Standards, New Delhi, India

Bureau of Indian Standards (1981). *IS:6403-1981 Code of Practice for Determination of Bearing Capacity of Shallow Foundations, (First revision).* Bureau of Indian Standards, New Delhi, India

Time-dependent Capacity Increase of Piles Driven in the Puget Sound Lowlands

Seth C. Reddy[1], E.I., SM.ASCE and Armin W. Stuedlein[2], Ph.D., P.E., M.ASCE

[1] Graduate Research Asst., School of Civil and Construction Engineering, Oregon State University, 101 Kearney Hall, Corvallis, OR, 97331, reddyse@onid.orst.edu

[2] Assistant Professor and Loosley Faculty Fellow, School of Civil and Construction Engineering, Oregon State University, 101 Kearney Hall, Corvallis, OR, 97331, armin.stuedlein@oregonstate.edu

ABSTRACT: Piles driven in the Puget Sound Lowlands frequently exhibit a high magnitude of soil setup or gain in capacity over time. In fact, the energy associated with production pile driving hammers is often insufficient to fully mobilize pile toe bearing resistance during dynamic restrike testing. Due to the potential for significant increases in capacity, practitioners often must consider the uncertain trade-off between the potential for setup and time-to-restrike. This study presents the results of 76 pile case histories from the Puget Sound Lowlands that were dynamically monitored during end-of-drive (EOD) and begin-of-restrike (BOR). The setup time (i.e. time between EOD and BOR) ranged from 5.5 to 312.5 hours. Signal matching using CAse Pile Wave Analysis Program (CAPWAP) indicated that the majority of the piles experienced significant increases in shaft resistance. In order to assess setup, piles were grouped into three categories: closed-end steel pipe piles, open-end steel pipe piles, and solid concrete piles. The accuracy and uncertainty of the well-known Skov and Denver (1988) setup prediction model was evaluated using the mean bias, defined as the ratio of measured and predicted shaft resistance, and the coefficient of variation (COV), defined as the ratio of the standard deviation of the point biases to the mean bias. The Skov and Denver model exhibited considerable uncertainty (COV = 45 to 59 percent) when evaluated using the local database. Recalibration efforts made using a hyperbolic function and least squares regression resulted in a reduction of the COV to 29 to 32 percent. Based on considerable scatter in the setup ratio (i.e. ratio of BOR to EOD shaft resistance), it was determined that the behavior of setup over time could not be adequately captured by considering total shaft resistance. Therefore, setup due to the overall increase in shaft resistance between EOD and BOR was characterized in an initial effort to provide guidance for incorporating setup into pile design. Closed-end concrete piles exhibited the largest average setup ratio (3.92), followed by closed-end steel piles (2.37) and open-ended steel piles (2.26).

INTRODUCTION

It is widely recognized that driven piles typically undergo a time-dependent increase in capacity following initial installation due to "setup" or "freeze" in the remolded soil surrounding the pile shaft. Although rare, "relaxation," or a decrease in capacity has been observed for piles installed in dilative soils or when lateral stress is decreased after pile installation (York et al. 1994). In some cases, pile capacity may increase by up to twelve times the end-of-driving (EOD) capacity (Komurka 2003). Setup is initially associated with an increase in effective stress due to the dissipation of positive porewater pressure during primary consolidation. However, numerous case histories (e.g. Tavenas and Audy 1972, Samson and Authier 1986, among others.) indicate that pile capacity will often continue to increase long after effective stresses have equilibrated. Although not well understood, setup from soil "aging" has been observed in all soil types and can comprise a significant portion of the total capacity gained from setup (Tan et al. 2004).

Accounting for setup in pile design can result in a more cost-effective solution by reducing the number and length of piles and the required installation time. Despite the availability of relatively quick procedures for estimating the static capacity of piles post-installation (e.g. dynamic load tests or DLTs), pile capacity from setup is often under-estimated due to demanding construction schedules which often inhibit long-term load test programs (Bullock 2008). Additionally, it appears that there is no clear consensus regarding the appropriate delay between EOD and restrike. Hannigan et al. (2006) suggested that five to seven days was adequate to assess setup in granular soils and at least two weeks in fine grained soil; however, Chen et al. (1999) observed steady setup for 80 days following end-of-driving, and Fellenius (2006) reported continued porewater pressure dissipation eight months following installation of piles in soft marine clays.

Long delays between end-of-driving and restrike testing are often impractical. As a result, numerous empirical, semi-empirical, and analytical models to estimate pile capacity from setup have been proposed (Skov and Denver 1988, Tan et al. 2004, etc.). However, most models exhibit significant uncertainty without site-specific calibration. More advanced models (e.g. Steward and Wang 2011) have been proposed, but are often developed using geologic-specific databases and require specific soil properties (e.g. friction angle, overconsolidation ratio). A simple, reliable methodology for estimating pile capacity from soil setup can reduce foundation construction costs and aid in decisions regarding planning of restrikes. This paper evaluated the accuracy and variability of one well-known setup prediction model using a regional geologic-specific DLT database. In order to reduce uncertainty, model recalibration was performed using a hyperbolic function and least squares regression. The magnitude and variability of setup in the Puget Sound Lowland was assessed for different pile types (closed-end concrete, closed and open-end steel piles) in an effort to provide a guide for incorporating setup into pile design.

BACKGROUND

Pile Setup and Aging

Setup takes place following pile installation as the disturbed soil regains strength. During installation, the soil adjacent to the pile shaft experiences larger shear strains and restructuring over a larger surface area compared to soil beneath the pile toe (Bullock 2008); therefore, setup is primarily associated with an increase in shaft resistance (Bullock et al. 2005). Dynamic load tests reported by Samson and Authier (1986), Skov and Denver (1988), Kehoe (1989), and others have indicated minimal changes in toe bearing resistance following pile installation.

Setup is associated with multiple interrelated mechanisms acting within the region of disturbed soil. In soft to medium-stiff fine-grained and loose granular soils, setup is initially associated with the recovery of effective stress during primary consolidation (Seed and Reese 1955, Vesic 1977, Karlsrud and Haugen 1985). Consolidation is usually assumed to occur within 24 hours in highly permeable sands (Tan et al. 2004), whereas it may take several months in cohesive soils (Fellenius 2006). Large diameter and closed-end piles (i.e. displacement piles) may induce significant porewater pressure changes within the large volume of displaced soil; therefore, the consolidation period may be longer for displacement piles compared to non-displacement piles (e.g. open-ended pipe piles, H-piles) (Randolph 2003). In addition, loose, saturated granular soils may liquefy due to the dynamic motion of the pile (Rausche et al. 2004); therefore, the difference in pile capacity between EOD and BOR can be significant.

Driven piles often continue to gain capacity long after effective stresses have stabilized (Huang 1988). Schmertmann (1991) presented several examples of long-term setup in different soil types which he associated with mechanical aging due to soil restructuring. Bullock (2008) suggested that consolidation and aging effects coincide in cohesive soils, resulting in continuous setup; whereas aging in sands may not begin for several weeks following installation. Karlsrud and Haugen (1985) observed a 30 percent increase in shaft resistance 24 days after consolidation in sensitive overconsolidated Norwegian clay. Tavenas and Audy (1972) observed a 70 percent capacity increase within the first 15 to 20 days for piles installed in clean sands, and Chow et al. (1998) associated an 85 percent capacity increase with the relaxation of a temporary radial soil arch formed adjacent to the pile wall during driving. Aging may also be associated with soil fatigue in stiff cohesive soils (Rausche et al. 2004), corrosion of steel piles (Chow et al. 1998), clay dispersion, thixotropy, and secondary compression (Steward and Wang 2011), and re-cementation in calcareous materials following installation (Mitchell and Solymar 1984).

Estimation of Setup

It is usually impractical to conduct a long-term load test program, thus several methodologies for predicting pile capacity due to setup have been proposed. Many researchers have observed a linear correlation between setup and the logarithm of time (Bullock et al. 2005). Based on concrete piles driven in predominately cohesive soils, Skov and Denver (1988) proposed a dimensionless setup factor, A, to predict pile capacity with the contribution of setup at a specific time:

$$\frac{Q_t}{Q_o} = 1 + A \cdot \log\left(\frac{t}{t_o}\right) \tag{1}$$

where A represents the capacity increase per log cycle of time, Q_t = pile capacity at time t after driving, and Q_o = pile capacity at the reference time t_o. The authors originally recommended A = 0.6 for clay with t_o = 24 hours, and A = 0.2 for sand with t_0 = 12 hours, where A includes all sources of capacity increase (e.g., consolidation, aging). Skov and Denver (1988) selected t_o as the beginning of semi log-linear capacity increase; however, determining t_o and the associated Q_o is subjective, and its selection impacts the log cycle capacity factor A (Bullock et al. 2005). Due to variable site and ground conditions, different pile types and materials, and the inherent variability of setup, pile case histories reported by Svinkin et al. (1994), Axelsson (1998), Bullock et al. (2005), and others have indicated considerable scatter in A and t_o. Therefore, pile capacity predictions made using Eq. 1 without site-specific data will likely include significant uncertainty.

Steward and Wang (2011) recognized that Eq. 1 describes the relationship between capacity gain and setup time, but does not provide an estimate of ultimate pile resistance (i.e., capacity). They studied load tests on heavily instrumented concrete piles installed in predominately clayey soils conducted by Bullock (1999), and developed a model to predict long-term capacity using the remolded friction angle and over-consolidation ratio. Based on the limited number of examples presented, the proposed model provided a more accurate estimation of setup compared to Skov and Denver (1988). Other setup prediction methods have been proposed by Svinkin (1996) (power law), Bogard and Matlock (1990) (hyperbolic function), and Wang et al. (2010) which used a variable growth rate approach.

GEOLOGY OF THE PUGET SOUND LOWLAND

This study presents an analysis of a database of piles driven in the Puget Sound Lowlands, a region that has been subjected to repeated glaciation over the last 2.4 million years (Troost and

Booth 2008). Based on previous research, Easterbrook (1994) suggested that glaciation occurred at least six times during the Pleistocene. The most recent period was during the Vashon Stade of the Fraser Glaciation in which the Puget Lobe of the Cordilleran Ice Sheet advanced into the Puget Sound lowlands (Borden and Troost 2001). As the ice moved southward, free-flow of seawater into the Puget Sound ceased, causing a series of proglacial lakes to form. Glacial lacustrine deposited in these lakes is locally known as the Lawton formation which consists of laminated silts and clays. As the ice continued southward, advance outwash was deposited over the Lawton formation and subsequently consolidated. Borden and Troost (2001) describe a typical glacial soil formation from bottom to top as: (1) fine grained, glaciolacustrine sediments, (2) coursing-upward advance outwash, (3) till and poorly sorted ice-contact deposits, and (4) fining upward recessional outwash locally interbedded with dead-ice deposits. During nonglacial periods, lacustrine and marine deposits filled in the center of the lowland (Borden and Troost 2001). At present, erosion and deposition is largely confined to local alluvial, deltaic, and estuarine environments.

PILE DATABASE

Piles in the database were installed in soil stratigraphy typical of the Puget Sound Lowlands. The database assessed herein was developed from 76 driven pile case histories gathered from the project records of several consulting firms with experience in the Puget Sound Lowlands. Data gathered included geographic and geologic data, borings and in-situ tests, pile types and materials, driving system information, pile installation logs, and CAse Pile Wave Analysis Program (CAPWAP) records. Pile diameters and embedded lengths ranged from 356 to 914 millimeters (14 to 36 inches), and 8.7 to 48.8 meters (29 to 160 feet), respectively. Measured total static capacities determined using signal matching (i.e. CAPWAP) at BOR ranged from 1,023 to 10,675 kN (230 to 2,400 kips), with shaft capacities of 623 to 4,937 kN (140 to 1,110 kips). Figure 1a through c presents the histograms and fitted normal and lognormal distributions of total static pile capacity (sum of shaft and toe resistance) at EOD and BOR, and setup time, respectively. Pile types included open and closed-end steel pipe and concrete piles (square and octagonal), and observed setup times ranged from 5.5 hours to 312.5 hours. The majority of piles in the database were installed with open-ended diesel hammers, although other hammer types included hydraulic, external combustion, and steam.

Each pile in the database was associated with a single DLT and subsequent signal matching analysis at EOD and beginning of restrike (BOR) performed by a single, local engineer specializing in dynamic testing. Because a significant portion of the piles were driven to bear on very dense, deep glacially overridden soils, toe bearing resistances were only partially mobilized at BOR for some pile cases. Signal matching analysis revealed that toe bearing remained constant, or in some cases decreased due to insufficient hammer energy, between EOD and BOR. Therefore, this study considered setup resulting from shaft resistance only. Although not explicitly addressed, it is also possible that the shaft resistance along the lower portion of the pile shaft was not fully mobilized in some cases resulting in a more conservative estimate of soil setup.

SETUP FOR PILES DRIVEN IN THE PUGET SOUND LOWLAND

The Skov and Denver (1988) Setup Prediction Model

Compared to other setup models, the Skov and Denver (1988) method is popular because it is simple and requires minimal data; however, many researchers have reported significant error between measured and predicted capacity using conventional values of A and t_o. As a result, designers often neglect additional pile capacity gained after the final load test (Steward and Wang

2011). The wide range of A and t_o values reported in literature further reinforces the lack of confidence in Eq. 1.

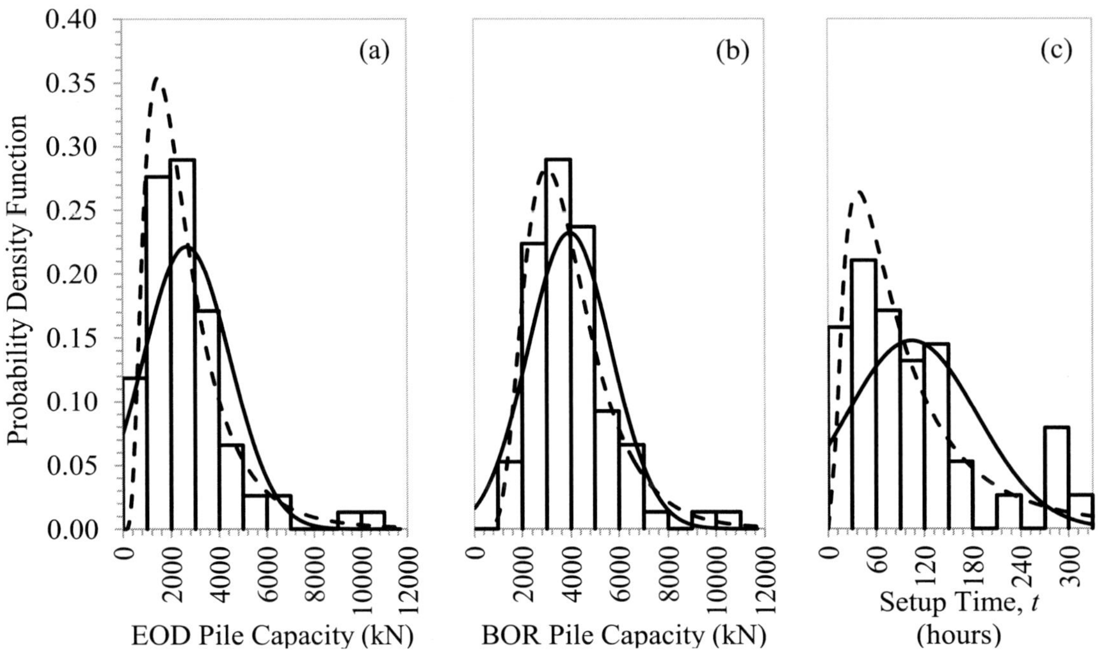

FIG 1. Histograms and fitted normal and lognormal probability density functions of (a) total static (sum of shaft and toe) pile capacity at end-of-driving and (b) beginning-of-restrike, and (c) setup time.

The database in this study was used to assess the accuracy and uncertainty associated with the Skov and Denver (1988) model and to determine the range of A; subsequently, recalibration efforts were made in order to increase the accuracy and reduce variability of this model. Since setup occurs primarily along the pile shaft, Q_o and Q_t in Eq. 1 are set equal to shaft resistance at EOD and BOR, respectively. For simplicity, t_o was assumed to be equal to one hour. Provided that the depth of embedment does not change between EOD and BOR, piles of different size can be assessed together in Eq. 1. Piles were grouped into three categories: closed-end steel pipe piles, open-end steel pipe piles, and solid concrete piles. Figure 2 illustrates a relatively poor correlation between setup ratio, $\eta = Q_t / Q_o$, and time, t, along with the average setup factor, $\overline{A}$, for each pile grouping and the total range of A (0.05 to 6.33). With the exception of two piles, η is bounded between one and seven, with an average of three. The high variability associated with A is reflected in setup capacity predictions made using Eq. 1. The accuracy and uncertainty of the setup prediction model can be evaluated using the mean bias, defined as the ratio of measured and predicted shaft resistance, and the coefficient of variation (COV), defined as the ratio of the standard deviation of the point biases to the mean bias. Using $\overline{A}$ for each pile grouping, Q_t was calculated using Eq. 1. Figure 3a shows that the Skov and Denver model exhibits considerable uncertainty (COV = 45 to 59 percent) for the database considered. Although the mean bias for each pile grouping is near unity, some dependence may be present between bias (i.e. the tendency to over or under-predict capacity) and capacity predicted using the Skov and Denver model (Figure 3a). At smaller capacities, the model appears to under-predict capacity, whereas the opposite may be true for larger capacities.

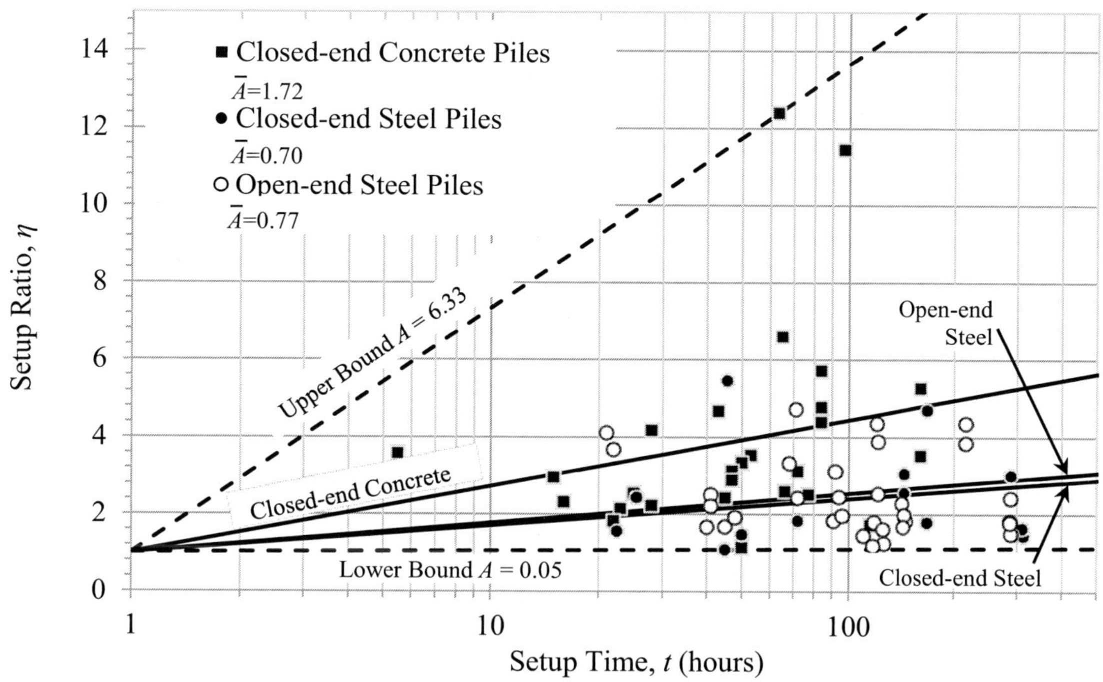

FIG 2. Average, and upper and lower bounds of the setup parameter A for each pile group for the Skov and Denver (1988) prediction model using the Puget Sound Lowland database. Note the setup ratio represents the increase in shaft resistance only.

Recalibration of the Skov and Denver (1988) Setup Prediction Model

Recalibration efforts were made to reduce the uncertainty in the Skov and Denver setup prediction model. Based on a series of load tests following EOD, Bogard and Matlock (1990) and Tan et al. (2004) showed that the increase in pile capacity due to setup could be approximated using a hyperbolic curve; applying this approach to Eq 1. yielded:

$$Q_t = \frac{Q_o \cdot A \cdot \log\left(\frac{t}{t_o}\right)}{k_1 + k_2 \cdot Q_o \cdot A \cdot \log\left(\frac{t}{t_o}\right)} + Q_o \qquad (2)$$

where k_1 and k_2 are fitting parameters determined using least squares regression, and govern the rate of gain in capacity. The shape of the hyperbolic curve is also affected by Q_o, where a larger Q_o indicates a higher initial setup rate and a better defined horizontal asymptote.

Figures 3b through 3d present the accuracy (mean bias) and uncertainty (COV) of shaft resistance from setup predicted using Eq. 2 compared to the CAPWAP capacity at BOR and k_1 and k_2 for closed-end concrete piles, and closed and open-ended steel piles, respectively. The solid line represents a one-to-one line between the measured and predicted shaft resistance, whereas the dashed lines represent one standard deviation of the model residuals added and subtracted from the one-to-one line. These bounds indirectly represent the level of uncertainty associated with the setup prediction model; narrow bands indicate a more accurate model. Most of the points fall within one standard deviation of the residuals, which indicate a reasonably accurate setup prediction model as compared to the use of a single average fitting parameter for each pile group (e.g., Figure 3a).

Because each pile is associated with only two load tests (i.e., the EOD and a single BOR capacity), the actual shape of the setup curve between EOD and BOR for piles in this database is unknown; therefore, Eq. 2 should be considered as an approximation of setup behavior over time. This modified version of the Skov and Denver setup prediction method is a purely empirical model and developed from a distinct set of pile case histories, and therefore it should only be used with reasonable confidence if the design case under consideration falls within the bounds of the dataset (i.e. geologies, pile sizes and range in capacities).

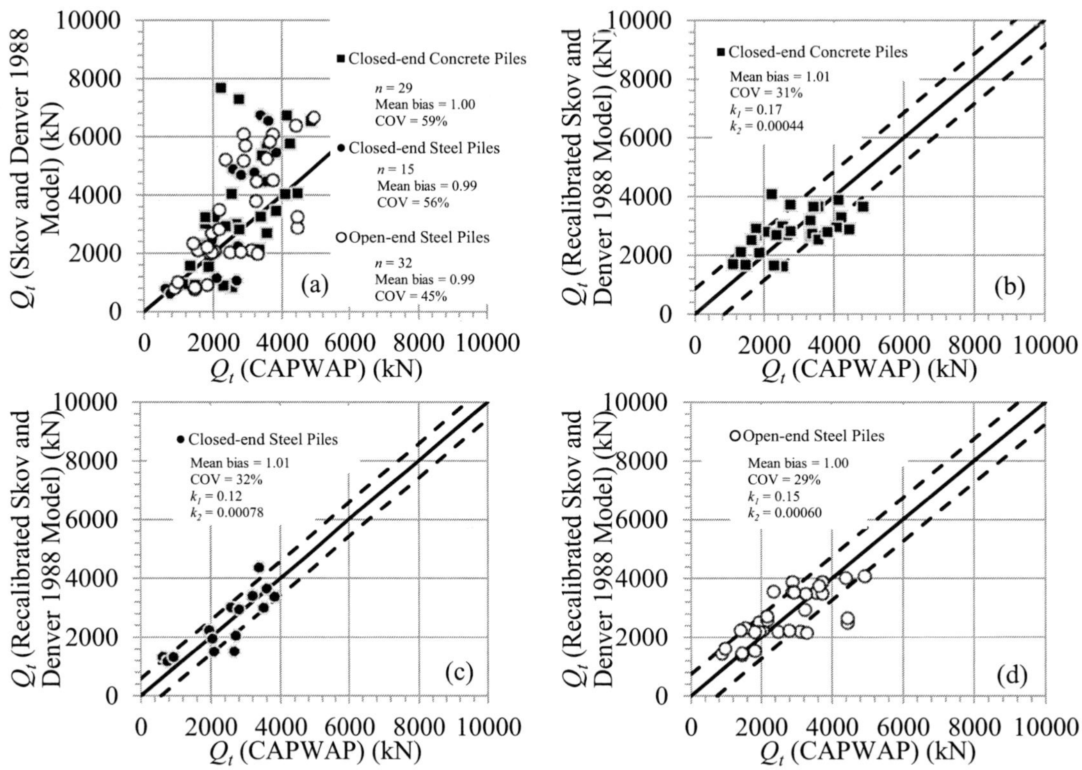

FIG 3. Accuracy and uncertainty of: (a) the Skov and Denver model using $\overline{A}$ for each pile group, and fitting parameters of the recalibrated Skov and Denver model for (b) closed-end concrete piles, (c) closed-end steel piles, and (d) open-end steel piles. Note, the dashed lines represent one standard deviation of the model residuals added and subtracted from the one-to-one line.

Characteristics of Setup in the Puget Sound Lowland

This study investigated setup in the Puget Sound Lowlands using a DLT database, where pile capacity was determined using CAPWAP. It is well known that setup is time dependent, where the rate of setup is largest immediately following EOD before decreasing to zero over time. However, the relative shaft resistance gained from setup (i.e. Q_t / Q_o) was found to be independent of setup time for the database considered. For example, a pile restuck after 10 hours exhibited a similar relative capacity gain compared to a pile tested after 200 hours. Based on the considerable scatter presented in Figure 2, it is clear that setup behavior in the Puget Sound Lowlands cannot be adequately captured by considering total shaft resistance. Instead, setup should be assessed for individual soil layers that are characteristic of exhibiting significant setup, which falls beyond the scope of this paper. This study intends to provide a summary of the degree of setup observed in the Puget Sound Lowlands, and offer intermediate guidance for incorporating setup into pile design until such time that a more detailed evaluation of each pile case history can be made.

Nonetheless, it is helpful to provide typical values of setup observed in the Puget Sound Lowlands and its variability. Figures 4a through 4c present the empirical, fitted normal, and fitted lognormal cumulative distribution functions (CDF) of the setup ratio, η, for closed-end concrete, and closed and open-end steel piles, respectively. The Anderson-Darling goodness-of-fit test (Anderson and Darling 1952) provided a means for determining the distribution of η for each pile grouping. For closed-end concrete and steel piles, the null hypothesis of normality is rejected in favor of a lognormal distribution at a 5 percent level of significance (i.e. p-value > 0.05). However, there was convincing evidence against normality and lognormality for open-end steel piles (i.e. p-value < 0.05) due to the poor fit of the upper tail. The database, representative of conditions encountered in the Puget Sound Lowlands and as presented in Figure 4, can provide an indirect level of confidence for incorporating setup into pile design. Based on the empirical distributions in Figures 4a through 4c, approximately 90 percent of the closed-end concrete piles in the database exhibited a setup ratio of two or greater; compared to 52 and 44 percent for open and closed-end steel piles, respectively. The range of η was 1.1 to 12.4, 1.1 to 5.5, and 1.2 to 4.7 for closed-end concrete and closed and open-end steel piles, respectively. Overall, closed-end concrete piles exhibited the largest average setup ratio, although this pile subset also exhibited the largest variability.

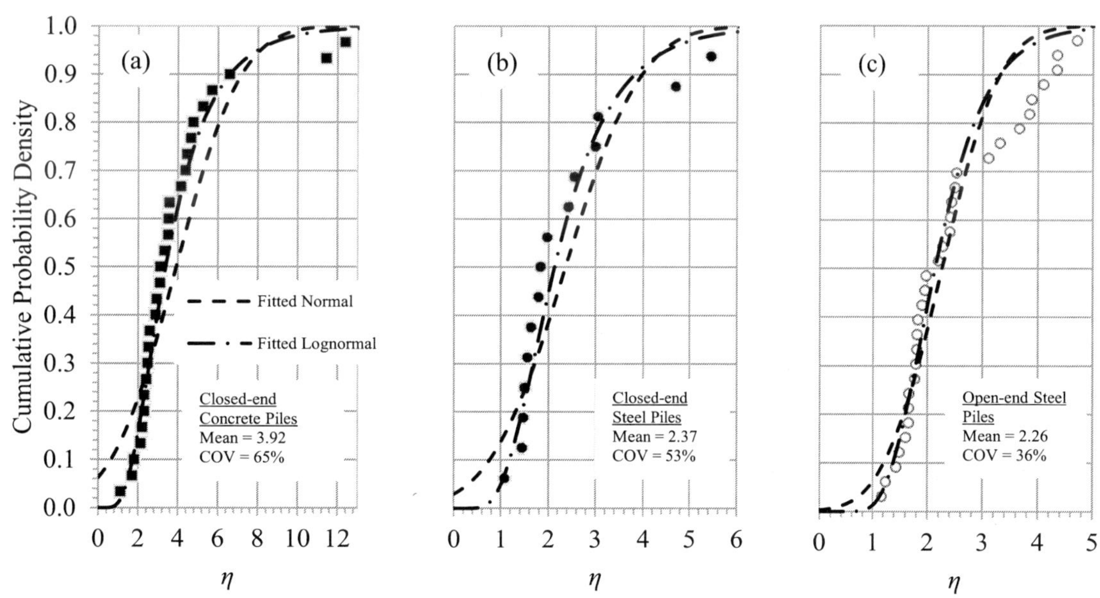

FIG 4. The empirical, fitted normal and lognormal cumulative distribution functions and relevant statistics of the setup ratio for (a) closed-end concrete piles, (b) closed-end steel piles, and (c) open-end steel piles. Note, the setup ratio represents the increase in shaft resistance only.

SUMMARY AND CONCLUSIONS

A large, regional geologic-specific dynamic load test (DLT) database was analyzed to investigate the time-dependent gain in capacity for piles installed in the Puget Sound Lowlands. Pile types represented in the database included open and closed-end steel pipe and closed-end concrete piles. Each pile in the database was associated with a single DLT at end-of-driving (EOD) and beginning-of-restrike (BOR) and a corresponding capacity determined using signal-matching (i.e. CAPWAP). Based on CAPWAP records, toe bearing resistance remained constant, or in some cases decreased due to insufficient hammer energy, between EOD and BOR. Therefore, this study considered setup resulting from shaft resistance only. Because each pile was associated with only two DLTs, there was insufficient information to assess the behavior of the rate of setup over time for individual piles; therefore, piles were grouped together based on pile type (closed-end concrete, closed-end steel, open-end steel).

Owing to its wide usage, the accuracy and uncertainty in the Skov and Denver (1988) setup prediction model was investigated using the local pile database. This prediction model produced unacceptably high uncertainties (coefficient of variation or COV = 45 to 59 percent) upon calibration when compared to shaft resistance at BOR due to significant scatter in setup ratio (i.e., the ratio of BOR and EOD shaft resistance) and a poorly defined relationship with time. Based on past research, a hyperbolic function was used to recalibrate the Skov and Denver model; recalibration provided considerable improvement and yielded an unbiased model, where COVs ranged from 29 to 32 percent. According to the Anderson-Darling goodness-of-fit test, the setup ratio for closed-end concrete and steel piles could be adequately described with a lognormal distribution, whereas open-end steel piles did not follow a normal or lognormal distribution. Overall, closed-end concrete piles exhibited the largest average setup ratio, followed by closed and open-end steel piles. Approximately 90 percent of closed-end concrete piles had a setup ratio of two or greater, compared to 52 and 44 percent for open and closed-end steel piles, respectively.

REFERENCES

Anderson, T.W. and Darling, D.A. (1952) "Asymptotic Theory of Certain Goodness-of-Fit Criteria Based on Stochastic Processes," *The Annals of Mathematical Statistics,* Vol. 23, No. 2, June, 1952, pp. 193-212.

Axelsson, G. (1998). "Long-term setup of driven piles in noncohesive soils evaluated from dynamic tests on penetration rods." *Proc., 1st Int. Conf. on Site Characterization,* P.K. Robertson and P.W. Mayne, eds., Balkema, Brookfield, VT., 2, 895-900.

Bogard, J.D. and Matlock, H. (1990). "Application of model pile tests to axial pile design." *Proc. of the 22nd Annual Offshore Technology Conference,* Houston, Texas, 3, 271-278.

Borden, R.K. and Troost, K.G. (2001). "Late Pleistocene stratigraphy in the south-central Puget Lowland, Pierce County, Washington." Washington State Department of Natural Resources: Division of Geology and Earth Resources, Report No. 33, 40 pp.

Bullock, P. J. (1999). "Pile friction freeze: A field and laboratory study." PhD dissertation, Dept. of Civil Engineering, Univ. of Florida, Gainesville, Fla.

Bullock, P.J. (2008). "The easy button for driven pile setup: dynamic testing." *Proc. of the Symposium Honoring Dr. John H. Schmertmann: From Research to Practice in Geotech. Engrg. Congress,* New Orleans, Louisiana, GSP 180, 471-488.

Bullock, P.J., Schmertmann, J.H., McVay, M.C., and Townsend, F.C. (2005). "Side shear setup. I: test piles driven in Florida." *J. Geotech. Geoenviron. Eng..* ASCE, Reston, VA , 131(3), 292-300.

Chen, C.S., Liew, S.S., and Tan, Y.C. (1999). "Time effects of bearing capacity of driven piles." *Proc. of 11th Asian Regional Conferences,* Seoul, South Korea, 50-64.

Chow, F. C., Jardine, R. J., Brucy, F., and Nauroy, J. F. (1998). "Effects of time on capacity of pipe piles in dense marine sand." *J. Geotech. Geoenviron. Eng.*, 124(3), 254–264.

Easterbrook, D. J. (1994). "Chronology of pre-late Wisconsin Pleistocene sediments in the Puget Lowland, Washington." *In* Lasmanis, Raymond; Cheney, E. S., convenors, Regional geology of Washington State: Washington Division of Geology and Earth Resources Bulletin 80, p. 191-206.

Fellenius, B.H., (2006). "Results from long-term measurement in piles of drag load and downdrag." *Canadian Geotechnical Journal*, 43(4), 409-430.

Hannigan, P.J., Goble, G.G., and Likins, G.E. (2006). "Design and construction of driven pile foundations - Reference manual Vol. II." Federal Highway Administration, Pub. No. FHWA NHI-05-043, April, 2006.

Huang, S (1988). "Application of dynamic measurements on long H-pile driven into soft ground in Shanghai." *Proc. of the 3rd International Conference on the Application of stress wave theory to piles*, B H Fellenius, Ed., Ottawa, Canada, 635-643.

Karlsrud, K., and Haugen, T. (1985). "Axial static capacity of steel model piles in overconsolidated clays." *Proc., 11th Int. Conf. on Soil Mechanics and Foundation Engineering*, Balkema, Brookfield, Vt., 3,1401–1406.

Kehoe, S. P. (1989). "An analysis of time effects on the bearing capacity of driven piles." *Master's Rep.* Dept. of Civil Engineering, Univ. of Florida, Gainesville, Fla.

Komurka, V.E. (2003). "Incorporating setup into driven pile design and installation." *Piledriver,* 4(1), 13-20.

Laprade, W.T. (1982). "Geologic implications of pre-consolidated pressure values, Lawton clay, Seattle, Washington." *Proc. of the 19th Annual Engineering Geology and Soils Engineering Symposium*,Pocatello, Idaho, 303-321.

Mitchell, J.K. and Solymar, Z.V. (1984). "Time-dependent strength gain in freshly deposited or densified sand." *J. Geotech. Engrg.,* ASCE, 110(11), 1559-1576.

Randolph, M.F. (2003). "Science and empiricism in pile foundation design." *Geotechnique,* 53(10), 847-875.

Rausche, F., Robinson, B., and Likins, G.E. (2004). "On the prediction of long term pile capacity from end-of-driving information." *Proc. of GeoTrans 2004: Current Practices and Future Trends in Deep Foundations,* GSP 125, Los Angeles, California, 77-95.

Samson, L., and Authier, J. (1986). "Change in pile capacity with time: Case histories." *Can. Geotech. J.*, 23(2), 174–180.

Schmertmann, J. H. (1991). "The mechanical aging of soils." *J. Geotech. Eng.*, 117(9), 1285–1330.

Seed, H. B., and Reese, L.C. (1955). "The action of soft clay along friction piles." *ASCE No. 81, Paper 842.* American Society of Civil Engineers.

Skov, R., and Denver, H. (1988). "Time-dependence of bearing capacity of piles." *Proc., 3rd Int. Conf. on the Application of Stress-Wave Theory to Piles*, B. G. Fellenius, ed., BiTech Publishers, Vancouver, BC, 879–888.

Steward, E.J. and Wang, X. (2011). "Predicting pile setup (freeze): A new approach considering soil aging and pore pressure dissipation." *Proc. of GeoFrontiers 2011: Advances in Geotechnical Engineering,* GSP 211, Dallas, Texas, 11-19.

Svinkin, M. (1996). "Discussion – Setup and relaxation in glacial sand." *J. of Geotech. Engrg.,* 122(4), 319-321.

Svinkin, M. R., Morgano, C. M., and Morvant, M. (1994). "Pile capacity as a function of time on clayey and sandy soils." *Proc., 5th Int. Conf. on Piling and Deep Foundations*, Deep Foundations Institute, Englewood Cliffs, N.J., 1.11.1–1.11.8.

Tan, S.L., Cuthbertson, J., and Kimmerling, R.E. (2004). "Prediction of pile setup in non-cohesive soils." *Proc. of GeoTrans 2004: Current Practices and Future Trends in Deep Foundations,* GSP 125, Los Angeles, California, 50-65.

Tavenas, F. and Audy, R. (1972). "Limitations of the driving formulas for predicting the bearing capacities of piles in sand." *Canadian Geotechnical Journal.* National Research Council of Canada, Ottawa, Ontario, 9(1), 47-62.

Troost, K. G., and Booth, D. B., (2008). "Geology of Seattle and the Seattle area, Washington, in landslides and engineering geology of Seattle, Washington, Area." R. L. Baum, J. W. Godt, and L. M. Highland (eds.), Geological Society of America Review in Engineering Geology XX, 35 pp.

Vesic, A.S. (1977). *Synthesis of highway practice 42: Design of Pile Foundations*, TRB, National Research Council, Washington, D.C.

Wang, X., Verma, N., Tsai, C., and Zhang, Z. (2010). "Setup prediction of piles driven into Louisiana soft clays." *Proc.of GeoFlorida 2010: Advances in Analysis, Modeling and Design.* GSP 199, West Palm Beach, Florida, 1573-1582.

York, D., Brusey, W., Clémente, F., and Law, S. (1994). "Setup and relaxation in glacial sand." *J. Geotech. Engrg.*, 120(9), 1498–1513.

Bearing Capacity Reduction of Vibratory Installed Large Diameter Pipe Piles

Ethan C. Lamiman[1], EIT, A.M.ASCE, and Brent Robinson[2], PE, M.ASCE

[1]Staff Engineer, Froehling & Robertson, Inc., 833 Professional Place W., Chesapeake, Virginia 23320; ELamiman@fandr.com
[2]Senior Engineer, GRL Engineers, Inc., 30725 Aurora Road, Cleveland, Ohio, 44129; BRobinson@grlengineers.com

ABSTRACT: This paper presents a case history where 91.4-centimeter (36-inch) diameter open-end pipe piles were installed using both impact and vibratory installation techniques. Thirteen dynamically tested piles were installed along a new sheet pile containment wall located in the Southern Branch of the Elizabeth River, bounded by Chesapeake and Portsmouth, Virginia. The soil conditions encountered generally consisted of interbedded layers of silt, sand, and clay forming the Alluvium and the Norfolk Formations. The piles were advanced into the underlying Yorktown Formation bearing stratum consisting of clayey to silty sand with varying amounts of marine shell fragments. At two test pile locations, impact driven test piles were extracted and relocated by vibratory hammer and subjected to restrike driving with dynamic analysis to assess bearing capacity. Seven to fourteen day restrikes were performed on the thirteen hammer driven test piles. Restrikes at one or more months were performed on one impact driven pile, and both vibratory installed test piles. Signal matching analyses of restrike driving events indicate an approximate 50% reduction in overall bearing capacity of the vibrated piles compared with the driven piles. Additionally, long-term restrike driving of vibrated piles did not continue to gain capacity akin to driven piles.

INTRODUCTION

The installation of deep foundations has been loosely documented since the 12th century, and observed through archeology into prehistoric times (Boyer 1985). Beginning with human powered drop hammers to early steam machines and into today's diesel and hydraulic impact hammers, the understanding, design and specification of deep foundations is based upon the breadth of knowledge derived from the long history of impact driving. With the development of modern vibratory hammers come the trepidations of designers and engineers to specify their use for deep foundations requiring axial bearing capacities. The reservations for use in design and construction are mainly consequent to the lack of understanding of the soil-pile interaction during and subsequent to installation (Viking 2005).

Previous studies comparing measured bearing capacities of impact driven and vibratory installed methods have indicated differences between impact and vibratory driven piles. O'Neill et al. (1990) installed 102 mm diameter instrumented displacement piles in sands with relative densities of 65 and 90% using a vibrator and an impact hammer. For these displacement piles installed in a pressure chamber, pile bearing capacity was higher for impact driven piles in the medium dense sand and higher for vibratory installed piles in dense sand. Mosher (1987) reported on a number of case studies using a variety of pile types and hammer types. In a majority of cases summarized, piles installed in sand by impact hammer exhibited higher axial capacities than those installed solely with a vibratory hammer. This study summarizes a case history in which two open end pipe piles were installed with both an impact and vibratory hammer.

SITE DESCRIPTION

The construction site, administrated by the Environmental Protection Agency (EPA) Region 3, is under the construction jurisdiction the United States Army Corps of Engineers (USACE) Norfolk District. The project is identified as the Atlantic Wood Industries Super Fund Site, Off Shore Sheet Pile Containment Wall. Construction of the new containment wall phase of the project was accomplished by McLean Contracting Company. The project site is located on the western shore of the Southern Branch of the Elizabeth River, in Portsmouth, Virginia. The site is generally bounded to the north and west by private industrial properties and the Norfolk Naval Shipyard, to the south by the United States of America South Gate Annex property, and aerially bisected by the new South Norfolk Jordan Bridge.

Figure 1. Project site (starred)
(Source: google.com, 2013)

The project generally consisted of a 550-meter (1,180-foot) steel combined pile bulkhead, with 35 meters (120 feet) of tied-back steel combination pile bulkhead along the river shoreline, and 112 meters (370 feet) of steel sheet pile cutoff wall constructed onshore. This construction phase of the containment wall was one of the preliminary stages of the remediation of the Atlantic Wood Industries Superfund Site.

Deep Foundation Elements

The foundation elements of the new sheet pile containment wall consist of a land based dead man wall at the southwest portion of the wall utilizing 610-mm (24-inch) square, precast, prestressed concrete (PPC) piles and 610 mm diameter spiral welded, open-end pipe piles driven alternately on a 4 Horiz: 12 Vert batter, 26.5 meters (87 feet) and 35.1 meters (115 feet) in length, respectively. The containment wall piles consist of plumb 1219-mm (48-inch) diameter open-end steel pipe piles at the southwest portion of the containment wall. The offshore portion of the containment wall consists of 914-mm (36-inch) diameter steel pipe piles with a 15.875-mm (0.625-inch) wall thickness battered 4:12 along the exterior and 1219-mm diameter open-end steel pipe piles along the interior driven plumb. The battered exterior 914-mm pipe piles were initially designed with an open-end condition, and later changed to closed-end pipe piles during the test pile program to achieve design axial compressive capacity requirements.

SUBSURFACE CONDITIONS

The project site lies within the Coastal Plain physiographic province of Virginia, which extends from the Fall Zone eastward to the Atlantic Ocean. Numerous transgressions and regressions of the Atlantic Ocean have deposited marine, lagoonal, and fluvial (stream lain) sediments. The regional geology is very complex, and generally consists of interbedded layers of varying mixtures of sands, silts, and clays.

As stated by Barker and Bjorken (1978), the site subsurface soil strata beneath the water generally consists of Alluvium of sand and marsh sediment, estuarine-beach, tidal marsh, and fluvial silt, sand and clay with organic material (peat); underlain by the Norfolk Formation of brackish marine silty sand and fluvial estuarine, silty sand. The Norfolk Formation is subdivided into the Upper and Lower member. The upper member is typically composed of brackish marine silty sand and fluvial estuarine clayey silty sand and may also include marine silt and clay deposits present in areas where the underlying Yorktown Formation has been eroded. The lower member of the Norfolk Formation generally consists of fine to coarse quartz sands with varying amounts of silt. Underlying the Norfolk Formation is the Yorktown Formation consisting of near shore marine fossiliferous, silty, coarse sand and coquina.

The Yorktown Formation is the typical bearing stratum for deep foundations in the Tidewater, Virginia, area and is considered to be over-consolidated. Very limited consolidation data indicates an average over-consolidation range of 2.1 to 3.2 (Martin, James, Powell & Bertoulin 1987).

The following generalized subsurface information shown in Table 1 contains a composite profile of collected site data and was derived from multiple subsurface exploration programs made public in the project's Contract Documents (USACE 2011). The depth of water to the Alluvium varied from five to twelve feet across the length of the wall.

Table 1. Generalized subsurface profile.

Depth (m)	Depth (ft)	Material Description	Range of N_{60} (blows per foot)	Stratum/ Formation
0–1.5	0–5	Water, Southern Branch of the Elizabeth River	–	–
1.5–4.6	5–15	Very Soft, ORGANIC SOIL, trace fine sand (OL, OH)	0	ALLUVIUM
4.6–13.7	15–45	Very Soft, FAT CLAY, trace fine sand (CH)	0 to 1	UPPER NORFOLK
13.7–21.3	45–70	Very Loose to Medium Dense, Fine to Coarse SAND, varying amounts of silt (SM, SP-SM)	0 to 13	LOWER NORFOLK
21.3–33.5^{+}	70–110^{+}	Medium Dense to Dense, Silty to Clayey, Fine to Coarse SAND, contains varying amounts of marine shell fragments	12 to 40	YORKTOWN

TEST PILE PROGRAM

Test pile installation was performed by McLean Contracting Company, based out of Glen Burine, Maryland, with regional operations in Chesapeake, Virginia. The impact driven test pile program included high-strain dynamic pile analysis during initial drive and pile restrike for two 610-mm PPC piles, seven 914-mm diameter open-end, spiral welded pipe piles, and six 914-mm diameter closed-end, spiral welded pipe piles. One 1219-mm diameter open-end, spiral welded pipe pile was dynamically monitored during initial installation.

A static load test of a 914-mm diameter closed-end, spiral welded pipe pile was performed at the southeast corner of the wall, where a member of the reaction frame, one 610-mm diameter vibratory installed, spiral welded pipe pile, was dynamically tested during restrike to confirm uplift capacity of the reaction frame.

During the test pile program, two previously impact driven 914-mm diameter test piles (66A and 111A) were extracted with a vibratory hammer following restrike, relocated in the vicinity of the initial installation, and vibratory installed to the same approximate toe elevation. These two test piles were later dynamically tested through impact restrike driving.

For the purposes of this publication, the discussion herein will consider the comparison of the 914-mm open-end pipe piles installed by impact and vibratory driving.

Hammer Assemblies

The 914-mm impact driven test piles were installed with an American Piledriving Equipment (APE) model D80-23 open ended diesel (OED) hammer assembly. The APE D80-23 has a ram weight of 78.47 kN (17.64 kips) and a rated energy of 287.7 kN-m (212.2 kip-ft) at the maximum 3.51-meter (11.5-foot) stroke. The hammer assembly was installed in 1320-mm (52-inch) Gage Swinging Leads for pipe pile installations.

Vibratory installation of 914-mm diameter piles was performed utilizing an ICE model 4450 vibratory hammer with HPSI model 500 power unit. The ICE 4450 has a dynamic force of 1619.2 kN (364 kips), an eccentric moment of 51 kg-m (4400 in-lbs), a frequency of 1,600 vpm, and an amplitude of 30 mm (1.17 inches).

Impact Driven Test Pile Installations

The 914-mm diameter impact driven test piles were installed plumb at approximate test pile locations identified alpha numerically by line and bent. The 914-mm piles were located along line A, with a bent spacing of 269.6 centimeters (106.125 inches) on center. From the southeast corner of the combination wall at pile location 45A, 914-mm test piles were linearly aligned to the wall terminus at pile location 146A.

The 914-mm test piles were initially installed and restruck with the APE D80-23 hammer assembly. Following the impact driven test pile program, two piles at location 66A and 111A were extracted and relocated at a distance of approximately 3 to 4.5 meters with an ICE 44-50 vibratory hammer.

A preconstruction wave equation analysis and driveability study was performed for the APE D80-23 hammer assembly installing the 914-mm open-end pipe piles. The preconstruction drivability study indicated a driving resistance of 40 to 50 blows per 0.3 meters considering a 90 percent partially plugged condition at the toe. Using a coring model of toe resistance, the driveability study predicted driving resistances of 13 to 14 blows per 0.3 meters. Of the test piles measured following installation, the depth of soil within the pile was observed to approximate the mud line.

Impact driven test piles were installed with the APE D80-23 hammer assembly using the lowest fuel setting (fuel setting 1). As shown by Figure 2 for impact driven open-end pipe piles, there was consistency across the site with regards to driving resistances. Open-end test piles were driven to an approximate embedment depth of 31.7 m (104 feet) below the water line.

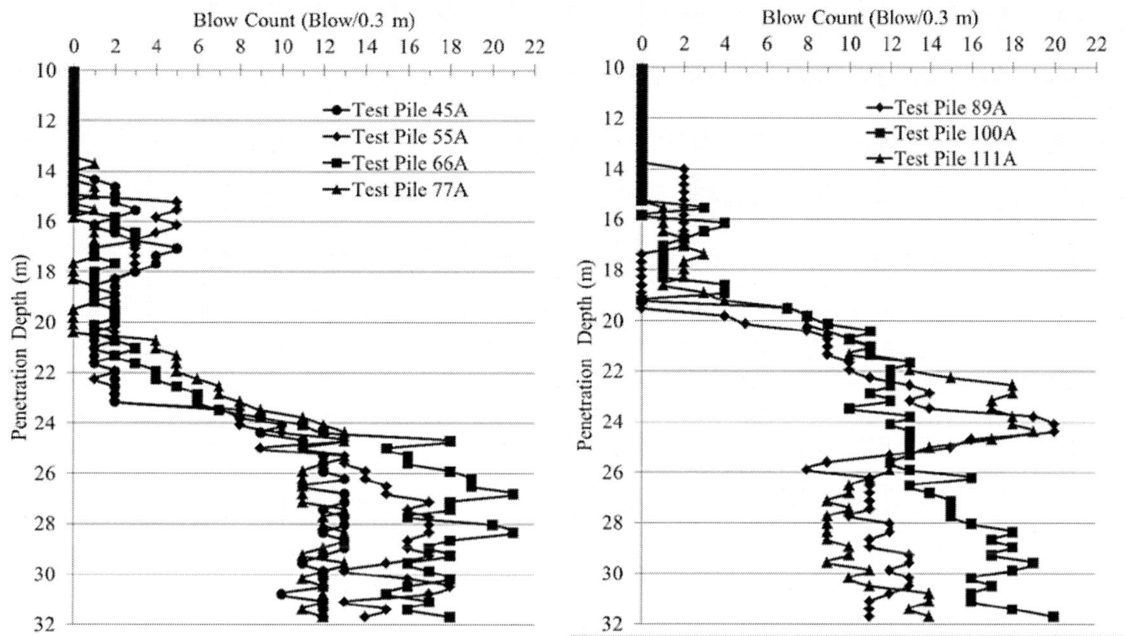

Figure 2. Impact driven pile installation summary.

Restrike Driving of Test Piles

Impact driven and vibratory installed test piles were each subjected to restrike driving using the APE D80-23 hammer assembly in conjunction with dynamic analysis. Restrike driving was performed following a range of setup periods from 5 to 47 days for impact driven piles and 26 to 68 days for vibratory installed piles.

The first test pile was restruck on the maximum fuel setting 4 with the APE D80-23 following a 13 day setup period. It was observed that the static capacity was much lower than anticipated where the signal matching with CAPWAP® software (Pile Dynamics, 2006) indicated an ultimate compressive capacity of 3200 kN (720 kips) at a driving resistance of approximately three blows per inch. Restrike driving was then attempted on fuel setting 1 on the same pile following an additional 6 day setup period after the first restrike where the signal matching analysis indicated an ultimate static capacity of 4220 kN (950 kips) at a driving rate of approximately four blows per inch. The total shaft resistance was observed to increase slightly from the reduction of fuel settings from 2800 to 3035 kN (630 to 680 kips), but the static end bearing resistance increased substantially from 405 to 1185 kN (90 kips to 265 kips). A summary of these results are presented in the following Table 2 for capacity evaluation at the end of initial installation (EOID) and successive high and low energy restrike (BOR) driving events.

Table 2. Pile number 55A energy variation during restrike driving results.

Driving Event	Date of Event	Fuel Setting	Static Pile Capacities			Hammer Performance	
			Shaft (kN)	End (kN)	Total (kN)	Max. Force FMX (kN)	Energy Transfer EMX (kN-m)
EOID	Dec-21-11	1	2180	378	2558	6272	75.5
BOR1	Jan-3-12	4	2802	400	3203	7922	104.5
BOR2	Jan-9-12	1	3025	1179	4204	5551	55.0

Following these observations, subsequent test piles were restruck utilizing fuel setting 1. Based on these results, it is likely that the reduced fuel setting and associated energy transfer into the pile allowed for any soil mass plugged onto the pile to move with the pile.

Vibratory Installed Test Piles

Test pile installation by vibratory hammer was performed with the ICE 44-50 hammer hydraulic vibratory driver for test piles 66A and 111A. This hammer has an eccentric moment of 51 kg-m (4400 in-lbs). Vibratory installation was timed at approximately nine minutes each to an embedment depth of 31.7 m (104 feet) below the water line. Dynamic measurements were not obtained during vibratory installations. Wave equation analyses were performed with dynamic parameters recommended by Rausche (2002), and indicated predicted driving times of 3 or 6 minutes, depending on whether the N-value method (SA) or the general soil type (ST) static analysis option, respectively, was used to predict static capacity.

RESULTS AND CONCLUSIONS

In general, the total static bearing capacity of the vibratory installed 914-mm diameter, 15.875-mm walled, open-end, spiral welded pipe piles was observed to be approximately 50 percent of a sister pile installed by impact driving. Additionally, the loss of bearing capacity attributed to shaft resistance was observed to vary between the individual soil strata. The comparison of signal matching results from restrike testing of sister piles initially installed with the vibratory hammer and impact driven is shown in Table 3.

Table 3. Static resistances of impact and vibratory installed sister piles.

	Pile No.		66A	66A	111A	111A
	Installation Method[1]		IMP	VIB	IMP	VIB
	Installation Date		Jan-4-12	Feb-8-12	Jan-9-12	Mar-23-12
	Setup Period (days)		5	26	10	69
Stratum	Depth (ft)	Depth (m)	Unit Resist. (kPa)	Unit Resist. (kPa)	Unit Resist. (kPa)	Unit Resist. (kPa)
Upper Norfolk	17	5.2	6.70	1.92	5.27	0.00
	24	7.3	21.55	3.83	12.45	4.31
	30	9.1	25.86	5.75	21.07	10.05
	37	11.3	30.16	7.18	29.69	12.93
Lower Norfolk	44	13.4	54.58	14.36	33.52	17.24
	50	15.2	54.58	16.76	40.22	18.67
	57	17.4	54.58	28.73	46.44	25.86
	64	19.5	58.89	32.56	58.89	25.86
Yorktown	70	21.3	62.24	30.16	61.29	25.86
	77	23.5	67.03	33.04	60.81	28.73
	84	25.6	67.03	33.04	58.89	32.56
	90	27.4	67.03	36.39	56.98	32.56
	97	29.6	67.03	36.39	56.98	35.91
	104	31.7	67.03	19.15	50.27	32.56
Static End Bearing Resistance (kN)			1468	685	1201	672
Total Static Capacity (kN)			5605	2713	4715	2469

1 – IMP is Impact hammer, VIB is vibratory hammer

Averaged between the two vibratory installed piles, the highly plastic clays of the Upper Norfolk Formation underwent an approximate reduction of 73 percent of the shaft resistance as compared with the two impact driven sister piles. The shaft resistance within the loose sands of the Lower Norfolk Formation were observed to be reduced by approximately 55 percent under vibratory installation; and the medium dense to dense silty, and clayey sands of the Norfolk Formation lost approximately 49 percent of the shaft resistance during vibratory installation.

The average end bearing resistance of the vibratory piles was observed to be 50 percent of the impact driven sister piles, and 50 percent of the total static capacity. Given the SPT N-values and general description of the sands in the Norfolk Formation as loose to medium dense, the lower capacity in the vibratory hammer driven piles is consistent with the lower capacities observed by O'Neill et al. (1990) in their model piles driven to medium dense sands with relative density of 65%.

ACKNOWLEDGMENTS

The authors would like to express their gratitude to McLean Contracting Company who made this data possible and available for use through their dedication to the advancement and understanding of pile foundation installation methods.

REFERENCES

Boyer, N.B. (1985). "Resistance to technological innovation: the history of the pile driver through the 18th century". Technology and Culture Vol. 26, No. 1.

Viking, K. (2005). "The vibratory pile installation technique, part I". PDCA Pile Driver Magazine, Winter Issue.

Google.com,(2013). https://maps.google.com/maps?q=richmond,+va&hl=en&ll =36.807086,-76.294212&spn=0.216059,0.482712&sll=36.849543,-76.298161&sspn=0.107695,0.085316&hnear=Richmond,+Virginia&t=m&z=12

Barker, W.J. & Bjorken, E.D. (1978). "Geology of the Norfolk South quadrangle, Virginia: Virginia Department of Mineral Resources Publication 9". Text and 1:24,000 scale map.

Martin, R.E., James, J.S., Powell, G.W. & Bertoulin, M. (1987). "Concrete pile design in Tidewater Virginia". ASCE Journal of Geotechnical Engineering, Vol. 113.

United States Army Corps of Engineers (2011). "Remedial design phase 1C – offshore sheet pile containment wall Atlantic Wood Industries Superfund Site". Unrestricted Construction Solicitation and Specifications W91236-11-R-0020 Volume 2 of 2.

Rausche, F. (2002). "Modeling of vibratory pile driving". Proceedings of the International Conference on Vibratory Pile Driving and Deep Soil Compaction: Louvain-La Neuve, Belgium.

O'Neill, M.W., Vipulanandan, C., Wong, D. (1990). "Laboratory modeling of vibro-driven piles", *Journal of geotechnical engineering*, v 116, n 8, p 1190-1209, Aug 1990.

Mosher, R. (1987), "Comparison of axial capacity of vibratory-driven piles to impact-driven piles." Final Report, Department of the Army.

Pile Dynamics, Inc. (2006). "CAPWAP® Background Report" Pile Dynamics, Inc. Cleveland, Ohio.

O-Cell Tests on Two 70-m Long Bored Piles in Vietnam

Nguyen Minh Hai[1)] and Bengt H. Fellenius[2)] M.ASCE, P.Eng.

[1)] Faculty of Civil Engng, Thu Dau Mot Univ., Binh Duong Province, Vietnam; haitdmu@gmail.com
[2)] Consulting Engineer, 2475 Rothesay Avenue, Sidney, BC, Canada, V8L 2B9; Bengt@Fellenius.net

ABSTRACT Static loading tests using single-level O-cells were carried out in mid-2008 on two bored piles constructed at the Binh Loi Bridge, crossing the Saigon River at Ho Chi Minh City, Vietnam. The soils consist of surficial layers of soft clay and silt to 11 and 15 m depth on compact silty sand to about 55 m depth followed by dense to very dense silty sand. Two bored piles, 2.0 m and 1.5 m in diameter were installed to 73.5 m and 73.0 m depth, respectively, and constructed using reverse circulation technique and bucket drill with casing advanced ahead of the hole. Drilling slurry was composed of polymer slurry for the 2.0-m diameter pile and bentonite slurry for the 1.5-m diameter pile. For both piles, the O-cell assembly was attached to a reinforcing cage lowered into the stabilized hole to 11.4 and 11.6 m, respectively, above the pile toe, 5 and 2 days after completing drilling and cleaning, respectively. Each reinforcing cage was instrumented with two pairs of diametrically opposed vibrating wire strain-gages at three levels below and five levels above the O-cell level. The static loading tests were performed 49 and 28 days, respectively, after the piles had been concreted.

The strain-gage records, when based on the nominal pile diameters, indicated Young's modulus values of about 25 GPa for the nominal cross sections of the piles. Analysis of the records showed that residual load had developed during the wait time. The unit shaft resistance above the O-cell for the 2.0-m diameter, polymer-slurry constructed pile was moderately larger than that for the 1.5-m diameter, bentonite-slurry constructed pile. Detailed analysis showed the shaft resistance to be post-peak softening. The pile toe stress-movement responses were essentially linear and almost identical for the two piles.

INTRODUCTION

Static loading tests using single-level O-cells were carried out for two bored piles constructed using different drilling techniques at the Binh Loi Bridge, crossing the Saigon River at Ho Chi Minh City, Vietnam. The Binh Loi Bridge is a part of Tan San Nhat – Binh Loi Outer Ring Road Project connecting Tan Son Nhat International airport at Ho Chi Minh City to the city outer ring road. The Binh Loi Bridge will link up the Binh Thanh to Thu Duc district of Ho Chi Minh City across the Saigon River. It has a total length of 957 m with a 150 m long main span. The height above the river water level is 15 m.

The bridge is supported on long bored piles constructed through deltaic alluvial soils, consisting of an about 40 m thick deposit of alternating layers of soft clay, compact silty sand, and soft to firm sandy clay, underlain by compact silty sand, becoming very dense at 70 m depth. The main span is supported on 2,000 mm diameter piles and the approach spans on 1,500 mm diameter piles. The working loads were 10 MN and 6 MN, for the 2,000 mm and 1,500 mm diameter piles, respectively. As the assigned factor of safety was 2.5, the required pile capacities were 25 MN and 15 MN, respectively.

To verify that the design of the pile support is satisfactory, static loading tests were performed on one test pile of each diameter, constructed at opposite banks of the river, and 338 m apart, at a pier location nearest the river. The 2,000 mm diameter test pile, Pile D2000, was constructed on the east bank and the 1,500 mm diameter test pile, Pile D1500, on the west bank (Figure 1). The piles had almost the same embedment depth, 73.5 m and 73.0 m, respectively. Both piles were strain-gage instrumented. The static loading test was performed by means of the bi-directional O-cell test (Osterberg 1989; 1998). The results of the test are presented and correlated to the site conditions. The test data and back-analysis are considered to be of interest beyond the design of the foundations for the Binh Loi Bridge.

SUBSURFACE CONDITIONS

The Mekong River basin is composed of a Quaternary soil overlaying a Tertiary soil (Workman 1977). The upper part of the Mekong basin is filled in with recent deposits from the Mekong River and regional settlements occur in the general area. Pre-Holocene Cenozoic deposits frequently crop out, forming terraces along the Dong Nai and Saigon Rivers. The Saigon River has a total length of 256 km and cuts through an estuary delta stretching from Dau Tieng Lake to Dong Nai River confluence. The seasonal variation of the river water surface elevation at the site is about 3 m.

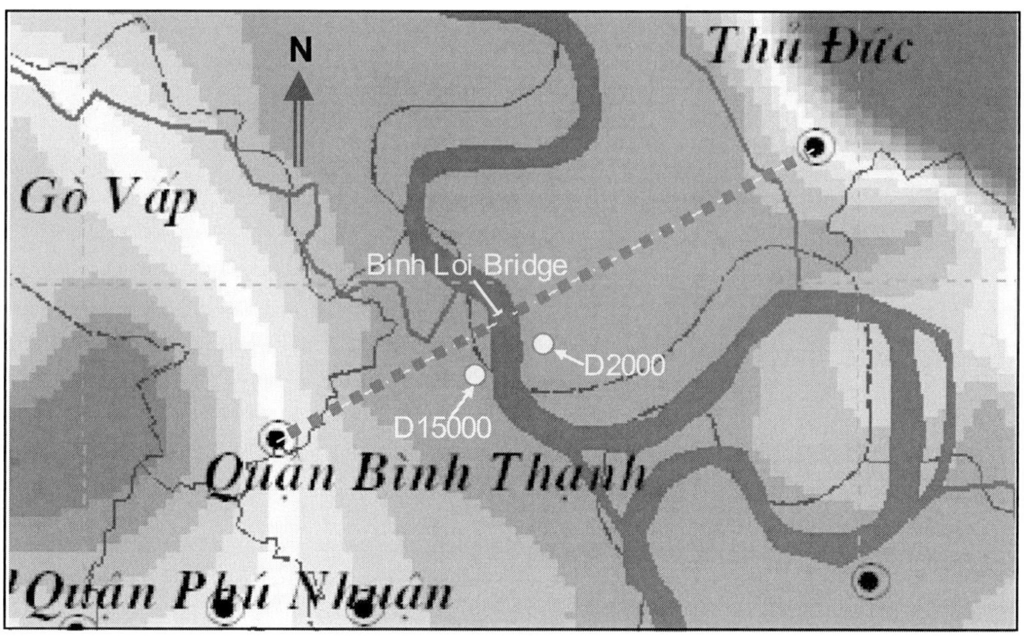

Fig. 1 Location of test piles and Boreholes

The soil profile at the locations of the test piles (Piles D2000 and D1500) consists of surficial layers of soft clay and silt to 11 m and 15 m depths, respectively, underlain by compact silty sand to about 55 m depth followed by dense to very dense silty sand (Figure 2). At Pile D2000, a zone of clayey silt was encountered between 30 to 40 m depth in the borehole. Table 1 summarizes the soil parameters for the soil layers. The pore pressure distribution is hydrostatic and corresponds to a groundwater table at 3 m depth. Figure 3 shows the measured distribution of cone stress, sleeve friction, and U2 pore pressure from two CPTU soundings pushed close to the test pile locations. The soundings were terminated once the cone encountered the compact silty sand. Figure 4 shows the distribution of SPT N-indices close to each test pile location

TEST PROGRAMME

Details of the Test Piles and Instrumentation

The two pre-production test piles, Pile D2000 and D1500, were drilled on May 20, 2008, and July 22, 2008, to 73.5 m and 73.0 m depth below grade, respectively. The diameter of Pile D2000 was 2,000 mm and the pile was supplied with a reinforcing cage of forty-four 32 mm bars, resulting in a steel reinforcement area of 350 cm^2 and a reinforcement ratio of 0.8 % of the 3.14 m^2 total nominal pile cross section. The diameter of Pile D1500 was 1,500 mm and the pile was supplied with a reinforcing cage of thirty-two 32 mm bars, resulting in a steel reinforcement area of 260 cm^2 and a reinforcement ratio of 1.5 %, of the 1.77 m^2 total nominal pile cross section. The concrete cylinder strength was determined one week after casting to 35 MPa and 37 MPa for the concrete used in Piles D2000 and D1500, respectively.

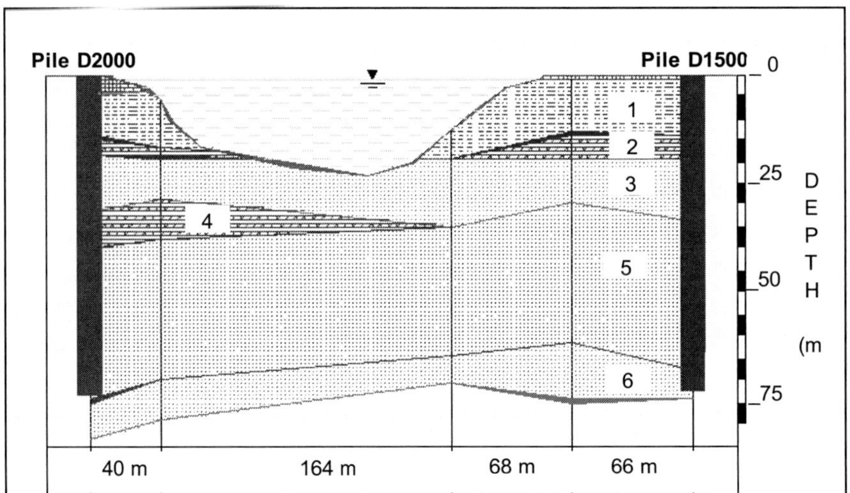

1. Soft clay and silt 2. Firm clay and silt 3. Compact silty sand some clay
4. Soft sandy clayey silt 5. Compact to very dense clayey sand, some silt
6. Dense to very dense silty sand, some silt

Fig. 2 Soil profile between the test piles

TABLE 1 Soil Parameters

Items		Layer 1	Layer 2	Layer 3	Layer 4	Layer 5	Layer 6
Water content (w_n)	%	92	29	22	43		
Saturated density (ρ_{sat})	kg/m^3	1,470	1,500	2,050	1,800	2,100	2,100
Void Ratio (e_0)	- - -	2.48	0.80	0.60	1.16	0.54	0.54
Liquid Limit (LL)	%	96	54			-	
Plastic Limit (PL)	%	34	23			-	
Clay Fraction (%)	(- -)	42	47	13	28	21	6
Silt Fraction (%)	(- -)	55	46	20	45	12	12
Sand Fraction (%)	(- -)	3	7	67	27	67	82
Friction Angle (ϕ)	^{0}	3°	13°	20°	13°	31°	23°
Vane Shear Strength (τ_{vb})	kPa	14	28		11		
SPT Index range	bl/0.3 m			11-30	7-15	18->50	>50

The test piles were constructed by first inserting 2,032 mm and 1,520 mm outer diameter temporary casings, respectively, to 10 m depth. Thereafter, the D2000 and Pile D1500 shafts were drilled to 73.5 m depth using reverse circulation drilling and to 73.0 m depth using a bucket drill, respectively. The test piles were constructed by two separate contractors, each with a preference for the type of slurry to use polymer slurry (Pile D2000) and bentonite slurry (Pile D1500). No records were kept on the procedures.

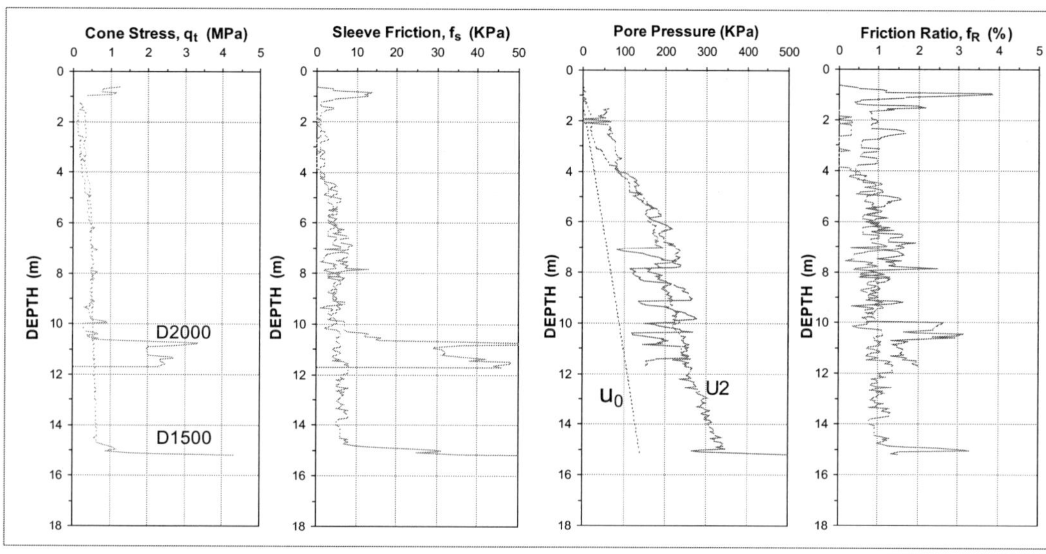

Fig. 3 CPTU diagrams from two soundings near Piles D2000 and D1500

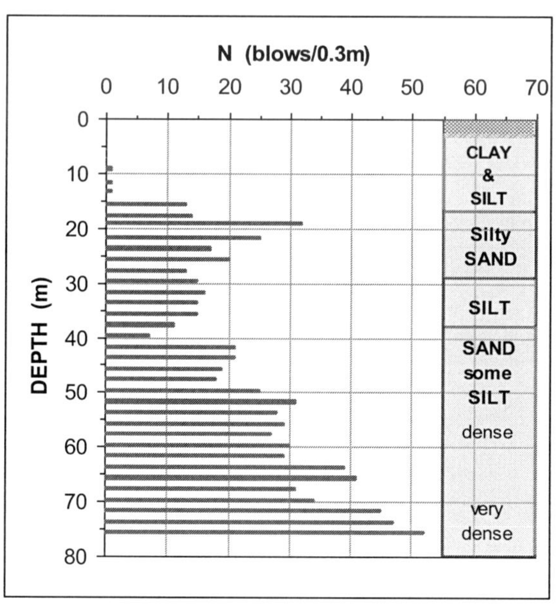

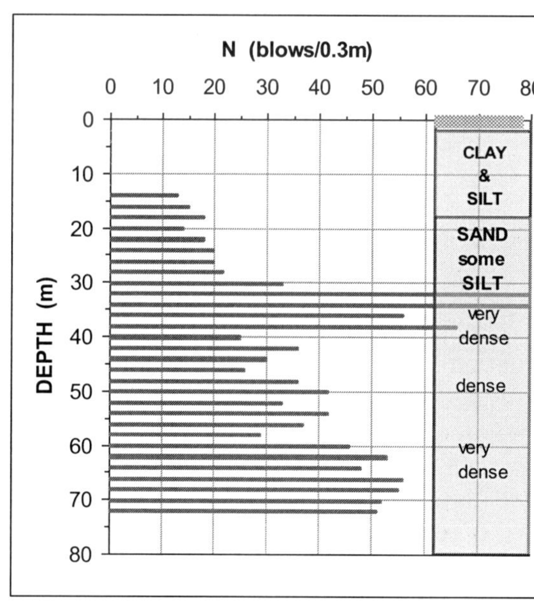

Fig. 4a SPT N-indices at Pile D2000 **Fig. 4b SPT N-indices at Pile D1500**

Before concreting each shaft, the shafts were cleaned and a reinforcing cage with the O-Cell assembly attached 11.4 and 11.6 m above the cage end, respectively, was lowered into the shaft. Pile D2000 O-cell level had a single 670 mm diameter O-cell and Pile D1500 O-cell level had three 330 mm cells.

On May 25, 2008, 5 days after drilling, the D2000 drilled shaft a 300 mm O.D. tremie pipe was inserted to the bottom of the shaft to start tremie placing concrete, displacing the polymer slurry. However, when the tremie pipe had reached the pile shaft bottom, it became blocked by debris that had accumulated on the shaft bottom. The pipe was extracted and the debris was removed from the shaft bottom. Thereafter, tremie placing of the concrete was resumed.

The bottom 1 m length of Pile D2000 had been equipped with a 114 mm O.D. coring tube attached to the reinforcing cage. When the core was core drilled after the concrete had cured, debris and slurry was found to exist below the pile toe. As a remedy, on July 5, 2008, the pile toe was pressure-grouted through the cored hole with a 1:1 cement-sand slurry. Grouting pressure was not recorded and no strain-gage-measurements were taken.

On July 24, two days after the drilling, tremie placing concrete displacing the bentonite slurry in Pile D1500 was carried out, without mishaps. Also Pile D1500 had been constructed with a coring tube. However, the tube turned out to be damaged at the O-cell level and while the tremie placing of the concrete appeared to have been successful, coring of the pile and the soil below the pile toe was not possible. Therefore, pressure-grouting of the soil below the pile toe could not be carried out.

The pile and instrumentation details are shown in Table 2. Each test pile was instrumented with two pairs of diametrically opposed vibrating wire strain-gages at three levels below and five levels above the O-cell, as indicated in Figure 5. The static loading test on Pile D2000 was performed on July 15, 49 days after concreting Pile D1500 was tested on August 21, 28 days after concreting. The complete test reports are available in Loadtest (2008).

Table 2. Pile and Instrumentation details

Item	Pile D2000	Pile D1500
Length (m)	73.5	73.0
Diameter (mm) to 10 m	2,032	1,520
Diameter (mm) 10 m to toe	2,000	1,500
Length (m) above O-cell	62.1	61.4
Length (m) below O-cell	11.4	11.6
Buoyant weight above O-cell (KN)	2,900	1,660
Rebar number and size	44xΦ32	32Φ32
Spiral size (75 m spacing)	Φ10	Φ10

The O-Cell Test

The test consists of applying load increments to the pile by means of incrementally increasing the pressure in the jack, which causes the O-Cell to expand, pushing up the pile shaft above the O-cell level and pushing down on the pile shaft below the O-cell. The measurements recorded are the O-Cell pressure (producing the O-cell load), the expansion of the O-cell, the shortening of the pile shaft above and below the O-cell level, and the movement of the pile head, which combine to provide the upward and downward movements of the top and bottom of the O-cell (Osterberg 1989; 1998).

Loading Procedure

The method of testing was in accordance with the "quick test" of ASTM Guidelines D1143-07. Pile D2000 was loaded in 23 equal increments of 790 KN to a maximum bi-directional O-Cell load of 18.14 MN and then unloaded in six decrements. Pile D1500 was loaded in 20 equal increments of 465 KN to a bi-directional O-Cell load of 9.31 MN and then unloaded in five decrements. Each successive load level was held constant for ten minutes by manually adjusting the O-Cell pressure as needed. Unloading waswith 5-minute load-holding.

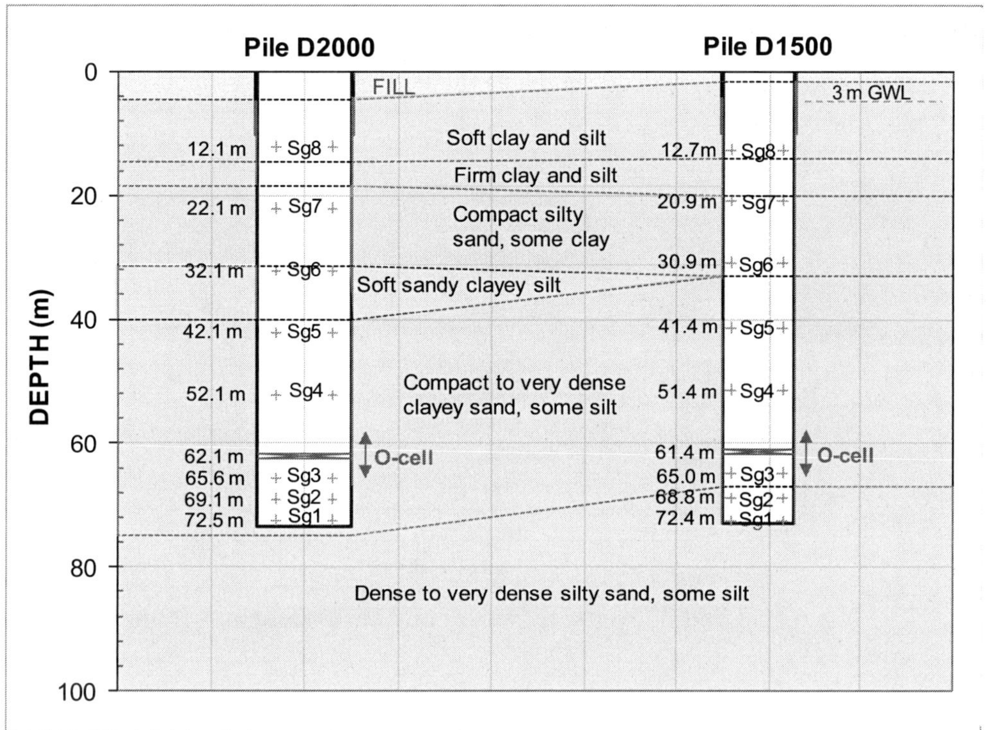

Fig. 5 Details of Instrumented Test Piles

TEST RESULTS

Load-Movement

The loads applied by the O-Cell act in two opposing directions, mobilizing resistance above and below the O-cell. Figure 6 shows the load-movement records of the two O-cell tests. At the maximum O-cell loads (18.14 and 9.31 MN), the O-cell downward movements were 77.8 and 123.0 mm, respectively. Theoretically, the O-Cell does not impose an upward movement until its expansion force exceeds the buoyant weight of the pile above the O-Cell plus the residual load, if any, acting at the O-cell level. The pile buoyant weights above the O-Cell for Piles D2000 and D1500 were 2.91 MN and 1.66 MN, respectively. Thus, the maximum upward resistances were 15.23 and 7.64 MN, respectively, generating O-cell upward movements of 5.3 mm and 8.5 mm, respectively. The maximum pile head movements (upward) were 0.5 and 2.3 mm, respectively.

The measurements of Pile D2000 do not indicate presence of any significant residual load at the O cell level. However, for Pile D1500, a small amount of residual load, about 1 MN, could be present at the Pile D1500 O-cell level, as indicated by the fact that significant upward movement did not start until the applied load was about 1 MN larger than the buoyant weight of the pile above the O-cell level.

Movement and Strain Measurements and Determining Axial Modulus

The upward shaft movements measured in the tests were small. At Gage Levels GL4 and GL5, the maximum relative movements between the pile shaft and the soil were no more than about 2 to 3 mm in both piles. This movement is not likely sufficient to have mobilized the shaft resistance fully. In contrast, the downward movements of 77 mm and 120 mm are larger than the minimum movement necessary to fully mobilize the shaft resistance along the pile portion below the O-cell levels. The full shaft resistance mobilization is reached where the slope (stiffness) of the load-strain curves (Figures 7a and 7b) become approximately linear.

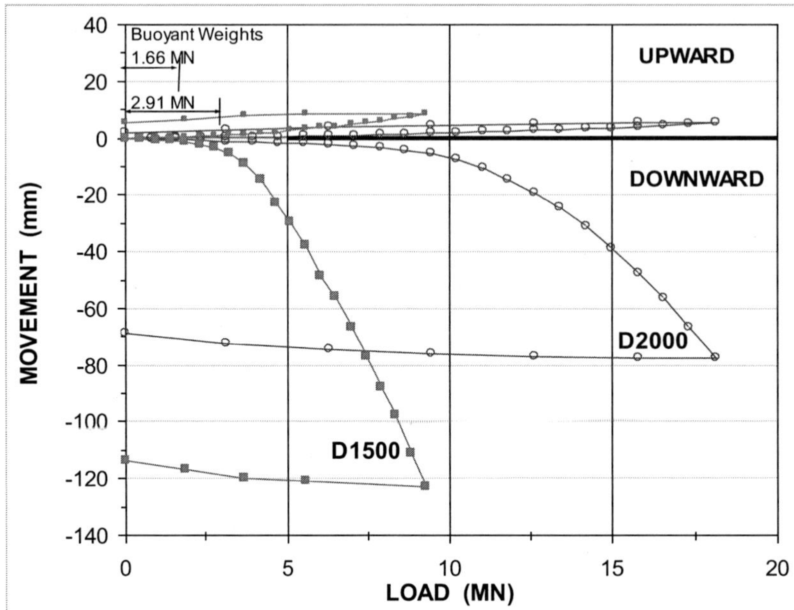

Fig. 6 Load-Movement Curves of Piles D1500 and D2000

Figures 7a and 7b report the measured load-strain curves. The values from Pile D2000 show that the slopes for the lowest, GL1, and highest, GL3, positioned gage levels below the O-cell level were approximately similar, whereas the GL2 curve deviated significantly from the other two. The deviation is considered due to the difficulties with the initial tremie and cleaning of the pile below the O-cell level. It is likely that the pile is bulging at GL2 resulting in a larger cross section and, therefore, a seemingly larger stiffness. Gage GL2 data have not been used in the further analysis.

Furthermore, both load-strain relations from Pile D2000 GL1 and GL3 plot along two approximately straight sloping lines. For each, one slope, called "early", was about 80 GN, appearing from the applied loads of about 10 and 8 MN, respectively, when the downward movements were 7 and 4 mm, respectively. The slopes then changed to a "late" slope of about 60 GN, appearing from the applied loads of about 13 and 10 MN, respectively, when the downward movements were about 30 and 7 mm, respectively.

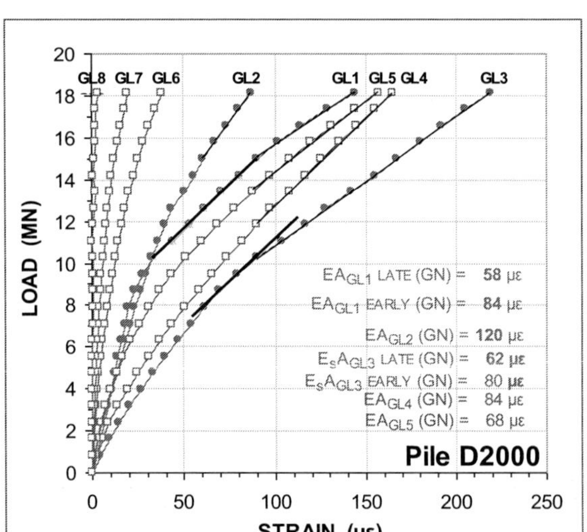

Fig. 7a Load versus strain, Pile D2000

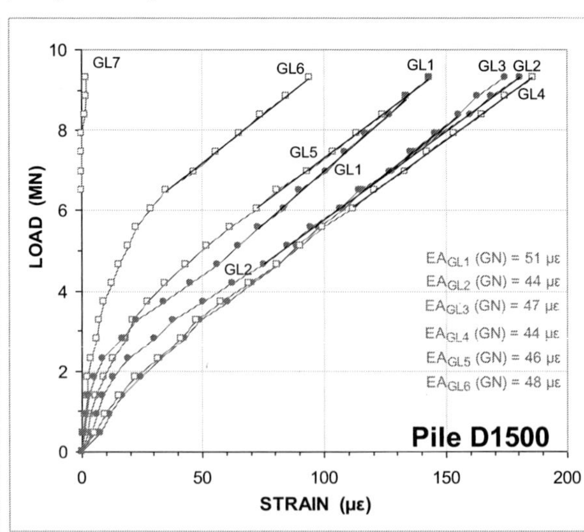

Fig. 7b Load versus strain, Pile D1500

The change of slope is considered due to the fact that the shaft shear response went into a strain-softening mode above the particular gage level, the load increment reaching the gage level became larger than the applied load increment, resulting in a larger strain change due to the softened shaft resistance.

The load-strain relations from Pile D1500 (Figure 7b) do not show a similar two-slope load-strain curves.

In order to convert the strain values to load, it is necessary to know the modulus of the concrete. Determining the modulus to use is simple in theory, but complex in practice. The modulus of concrete differs between different batches and it, therefore, needs to be determined for the actual batches used in the pile. Although there is a relation between concrete strength—cylinder or cube—and modulus, the relation is not sufficiently rigorous to provide accurate strain to load relations for the test data. The relation is best determined from studying the actual load-strain relations of the test. As indicated by Fellenius (1989; 2011), the load at a gage level is a function of the secant modulus or stiffness, E_s or AE_s, for the combination of the concrete and the steel reinforcement in the pile. The secant modulus of concrete is not a constant parameter, but reduces with increasing strain over the large strain range usually imposed in a static loading test. However, for the rather low strain range imposed in the subject tests—the maximum strains measured in the Piles D2000 and D1500 were 180 and 240 $\mu\varepsilon$—the modulus can be assumed to be essentially constant. Thus, a satisfactory accuracy can be obtained from the slope of the curves in a diagram showing the applied load and measured strains, as indicated in Figures 7a and 7b for Piles D2000 and D1500, respectively. Generally, once the shaft resistance acting along the pile between the load applications (the O-cell, in the subject tests) and the strain gage locations has been mobilized, the load-strain curve will become a practically straight line. That is, provided that the shaft resistance is plastic and not strain-hardening or strain-softening after reaching its peak strength. The slope of the line is the pile axial stiffness as a function of the induced strain.

Usually, the best way of determining the pile modulus is by means of a so-called "tangent modulus" or "incremental stiffness" plot (Fellenius 1989; 2011), that is, the applied increment of load over the induced increment of strain plotted versus the measured strain. This is shown in Figures 8a and 8b for Pile D2000 and D1500, respectively. The tangent modulus determined from the incremental stiffness procedure establishes the secant modulus for a modulus relation that reduces with strain. It is a differentiation process and, therefore, it is very sensitive to small variations of accuracy of both load and strain measurements. Moreover, it requires that the shaft response shows neither strain-hardening nor strain-softening trends.

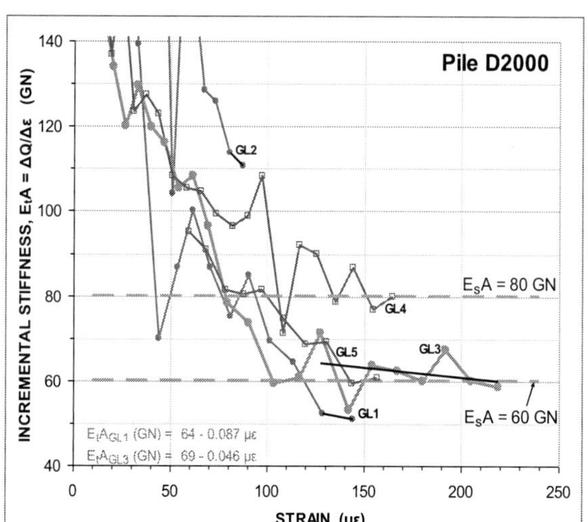

Fig. 8a Pile D2000 Stiffness Plot

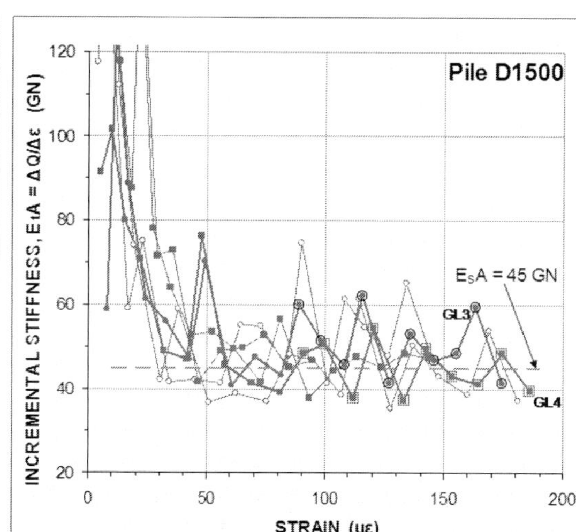

Fig. 8b Pile D1500 Stiffness Plot

The incremental stiffness method assumes that for load increments applied after the shaft resistance at the studied gage level has been fully mobilized, the continued incremental stiffness values will plot along a slightly sloping line. The line defines the tangent modulus relation for the pile cross section. The tangent modulus can then be converted to secant modulus as indicated by Fellenius (1989; 2011). Because of the combined effect of the strain-softening and the scatter of values, the incremental stiffness method did not provide reliable values for the tangent and secant stiffnesses. Therefore, for this case, the authors have preferred to rely on the linear portions of load-strain relations and convert the measured strains using constant stiffnesses, AE, of 80 GN and 45 GN for Piles D2000 and D1500, respectively. Correlated to the nominal cross sectional areas, both values indicate an E-modulus of 25 GPa.

Load Distributions

The evaluated stiffness (EΛ) valucs were used tu convert the ten-minute load-holding strain values to load as shown in Figures 9a and 9b. The minimal shaft resistance shown along the upper about 20 m for both piles could be due to either or both facts that the gages are in the soft clay and silty clay layers or that the shaft resistances were not fully mobilized. However, with regard to the relative change of shaft resistance of the load distribution further down the piles, it is neither logical that the resistance between GL6 and GL5 in Pile D2000 would be much larger than the resistance between GL5 and the O-cell, nor that the resistance in D1500 between GL4 and the O-cell is so small. Actually, as shown in Figure 2, the soil at the depth of GL6 to GL5 at Pile D2000 was a soft silt, not sand. It is a reasonable expectation that the load distributions would have shown a gentle shape with a decreasing slope toward depth, reflecting increasing shaft resistance as the effective overburden stress and the sand density both increase with depth. That this is not the case is considered due to varying effect of the slurry filter-cake and, possibly varying pile cross section.

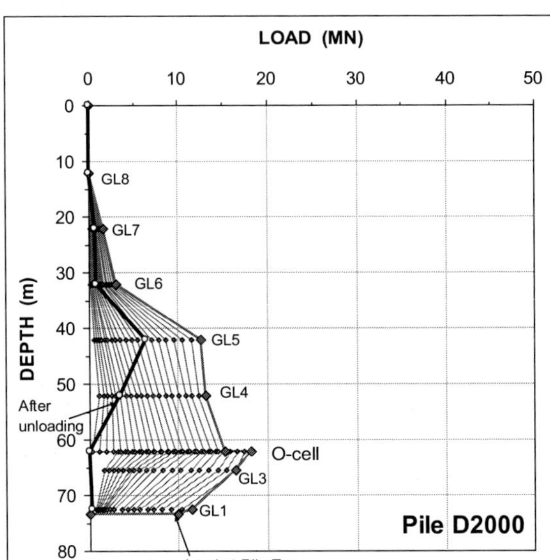

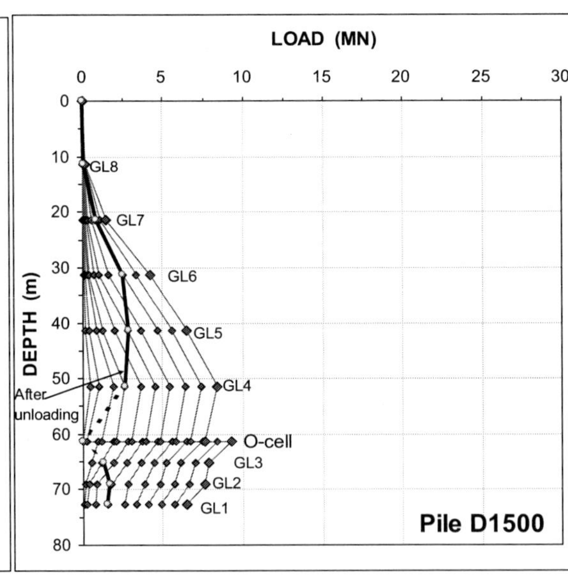

Fig. 9a Pile D2000 Load Distributions **Fig. 9b Pile D1500 Load Distributions**

Load and unit stress versus movement for shaft and toe resistances

Disregarding the strain-gage loads above the O-cell and assuming that the shaft resistance would show a uniform proportionality to effective overburden stress, Figures 10a and 10b show the actual and approximate equivalent head-down load-distribution curves. The latter are obtained by an effective stress calculation for ß-coefficients of 0.14 and 0.12 for Piles D2000 and D1500, respectively, throughout the pile lengths.

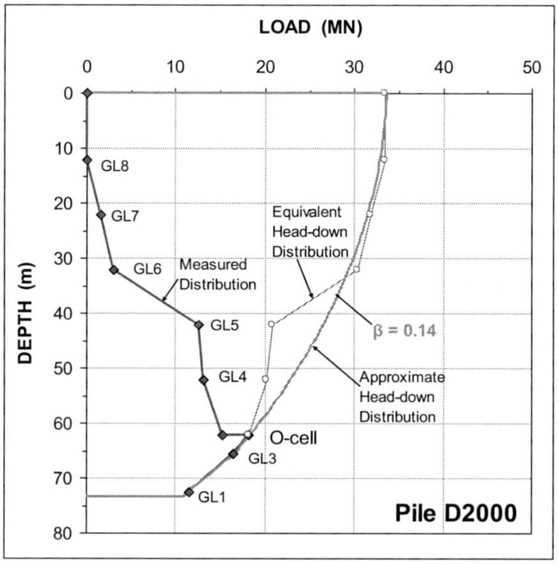

Fig. 10a Pile D2000 Load Distributions with Approximate Head-down Curve

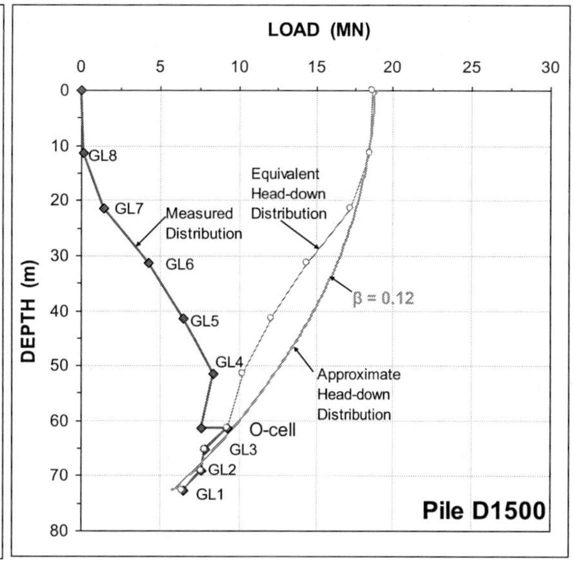

Fig. 10b Pile D1500 Load Distributions with Approximate Head-down Curve

The average unit shaft shear resistance between the gage levels can be determined as the difference in evaluated strain-gage load divided by the surface area between the gage levels. Figures 11a and 11b show those values plotted against movement for the particular gage levels. Thus, values between GL1 and GL2 are plotted against downward pile toe movement, values between GL2 and GL3 are plotted against the mean of movements of the pile toe and the downward O-cell movement, values between the O-cell level and GL3 are plotted against the downward O-cell level. Similarly, values between the O-cell level and GL 4 are plotted against the upward O-cell movements, etc.

The records of unit shear resistance between GL5 and GL6 in Pile D2000 were considered unreliable and are therefore omitted. (The shear force vs. movement showed a straight-line trend; the maximum unit resistance was more than four times the undrained shear strength and would correspond to a beta-coefficient of about 0.6, which is not consistent with the soft silt found in-between GL5 and GL6).

The uppermost gage level movements and loads were too small to produce meaningful curves in either pile. The peak resistance was reached at very small movements, a few millimetres only. The tendency to post-peak softening is noticeable in several of the curves.

The lowest diagrams in Figures 11a and 11b show the unit toe resistances, as determined by dividing the measured load in GL1 with the respective nominal pile cross section and plotting the stress values against the measured toe movements. For comparison, the figure includes also the shaft resistance curves. The toe curves show no tendency toward an ultimate resistance, confirming that ultimate toe resistance is a concept not existing in reality (Fellenius 1999; 2011).

The unit shaft resistances estimated from the load distributions are small, as demonstrated by the small beta-coefficients (Figures 10a and 10b). However, the distributions shown are those for the maximum load, which was applied when the peak resistance had been surpassed. Judging by the unit resistance versus movement curves, the peak shear resistance would be more than twice as large as that shown for the maximum applied load. What the peak pile shaft capacity would be is not important for assessing the pile in a design, however. Design by applying a factor of safety, or resistance factor, to a capacity is a crude approach, and, therefore, what working load to accept from the test results must be determined using conservative values of shaft resistance. In the subject case, the post-peak values are used to estimate the shaft resistances and beta-coefficients.

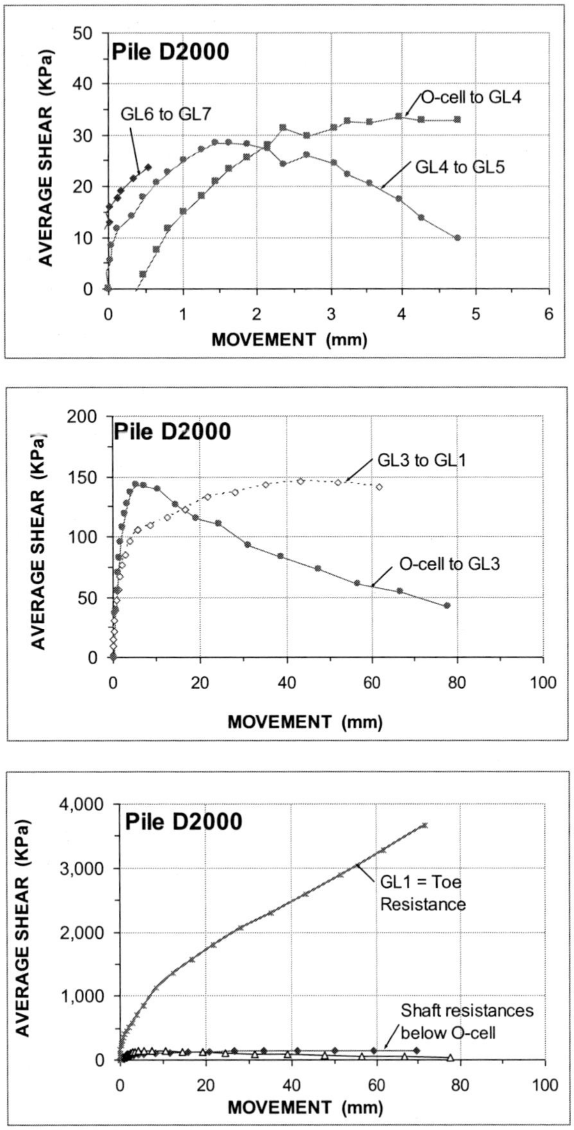

Fig. 11a Pile D2000 Unit Shaft and Toe Resistances

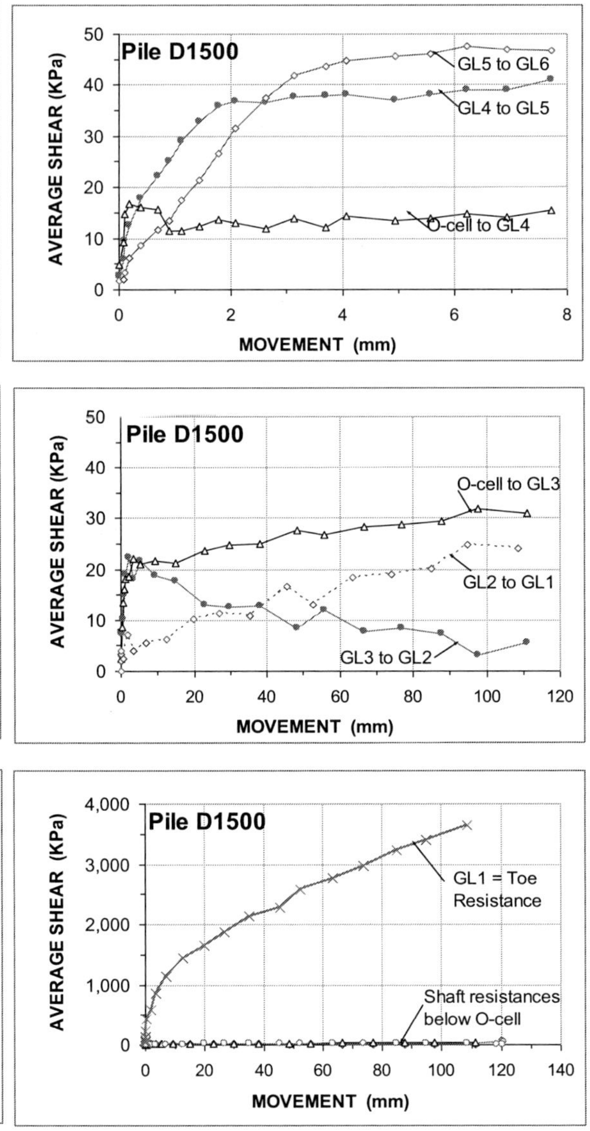

Fig. 11b Pile D1500 Unit Shaft and Toe Resistances

It is surprising that such distinct post-peak softening shaft resistance response would occur in a dense to very dense sand. As mentioned, it is therefore likely that the shaft resistance for both piles is affected by the presence of left-behind slurry filter cake between the pile and the soil. Furthermore, also the toe responses are very soft and not anywhere near what one would expect in a very dense sand. As mentioned, the toe responses are similar for the two test piles, which indicates that the pressure grouting at Pile D2000 failed to remedy the contamination of the pile toe.

Moreover, as the measured pile toe load-movement responses clearly show, the toe contribution to the pile capacity for use in a factor of safety or a LRFD approach must be defined as the toe resistance mobilized at some acceptable toe movement, often chosen to a values of about 10 to 15 mm—"a full half inch" and not some value of "pile toe capacity".

The total shaft resistances at maximum O-cell load were about 25,000 and 13,000 KN, for the two piles, respectively. The pile toe resistances at about 10 mm toe movement were about 4,000 and 2,000 KN, respectively. It was decided that the tests had established pile capacities that satisfied the specified values.

About ten months earlier, at a bridge site (Phu My) about 10 km downstream and south of from the Binh Loi bridge, a 1,800 mm diameter pile had been tested using two O-cell levels. The pile was drilled through an almost identical soil profile to a depth of 74.4 m using bentonite slurry. The only essential difference between the piles at the two sites was that the construction was executed by another contractor than the two who did the pile construction at the Binh Loi site. With the exception of one gage level, the strain-gage instrumentation in the Phu My test pile was not successful. However, the two-level O-cell test showed a unit toe resistance of about 8 MPa at a movement of about 70 mm, i.e., an about 2.5 to 3 times stiffer response than that shown for Pile D1500. Furthermore, the unit shaft resistance was at least about twice that of Pile D1500. The working load for the Phu My piles was 12 MN, twice that for the Binh Loi D1500 pile.

Moreover, about two years later, November 2010, an 1,500 mm diameter O-cell test pile was constructed at the Sunrise City complex, a series of office towers. The contractor was different to the contractors for the Phu My and Binh Loi sites. The Sunrise City site was located about 8 km west of the Binh Loi site in a very similar soil profile. However, the site investigation characterized the soil at the pile toe level as a very dense sandy silt. The construction method employed bentonite slurry and bucket drilling and the pile length was 76 m. The results showed the same low toe stiffness as found for the Binh Loi D1500 test pile and the unit shaft resistance correlated the same narrow range of beta-coefficients as for the subject case. The pile working load was 8 MN.

The back-calculated pile stiffness, EA, for the test piles at the Phu My and Sunrise sites was about the same as for the Binh Loi piles suggesting that neither site had piles constructed with a diameter that deviated much from the nominal diameter. That is, applying the nominal diameter when calculating the unit shaft resistance was reasonable for the analysis of the strain-gage data.

Modeling the long-term pile response

The results of an O-cell test, or head-down test for which the pile toe response has been determined, are eminently suitable for a more exacting analysis of the pile design: analysis based on settlement of the piled foundation, rather than pile capacity. The process for the subject piles and project is illustrated in Figures 12a and 12b. The process combines—unifies—design for capacity, drag load, settlement, and downdrag. It is, therefore, called "The Unified Pile Design" method (Fellenius 1984; 1988; 2011, Fellenius and Ochoa 2009). The main feature of the process is that the sustained working load, the shaft resistance distribution, and the soil settlement distribution are known entities for each case analyzed and they interact with the neutral plane location, the downdrag, and the pile toe load-movement response, which are interdependent variables. The calculations shown in Figures 12a and 12b are performed with the software UniPile5 (Goudreault and Fellenius 2013).

The Unified Method assumes that the long-term shaft resistance is fully mobilized; indeed, in the subject case, acting at the post-peak condition. As to pile toe resistance, it applies the load-movement curve for the pile toe (measured or according to an assumed q-z relation). The toe load is then determined by the correlation between the toe movement relative to the soil (toe penetration), that is, the difference for the soil settlement at the pile toe level and for the pile toe movement that generated the pile toe load.

The analysis results show that the piles will be having a drag load that is about equal to the sustained working load on the piles. However, the maximum axial loads will be well within the structural strength of the piles and is therefore of no concern.

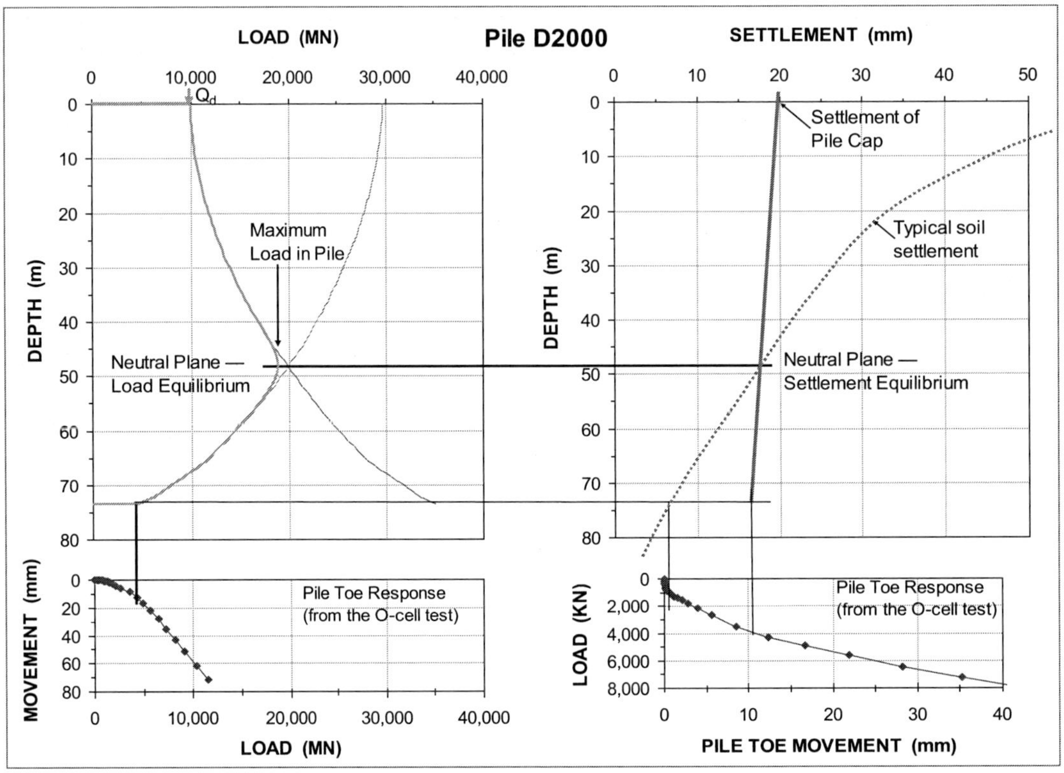

Fig. 12a Pile D2000 Load and Settlement Compilation for the Design

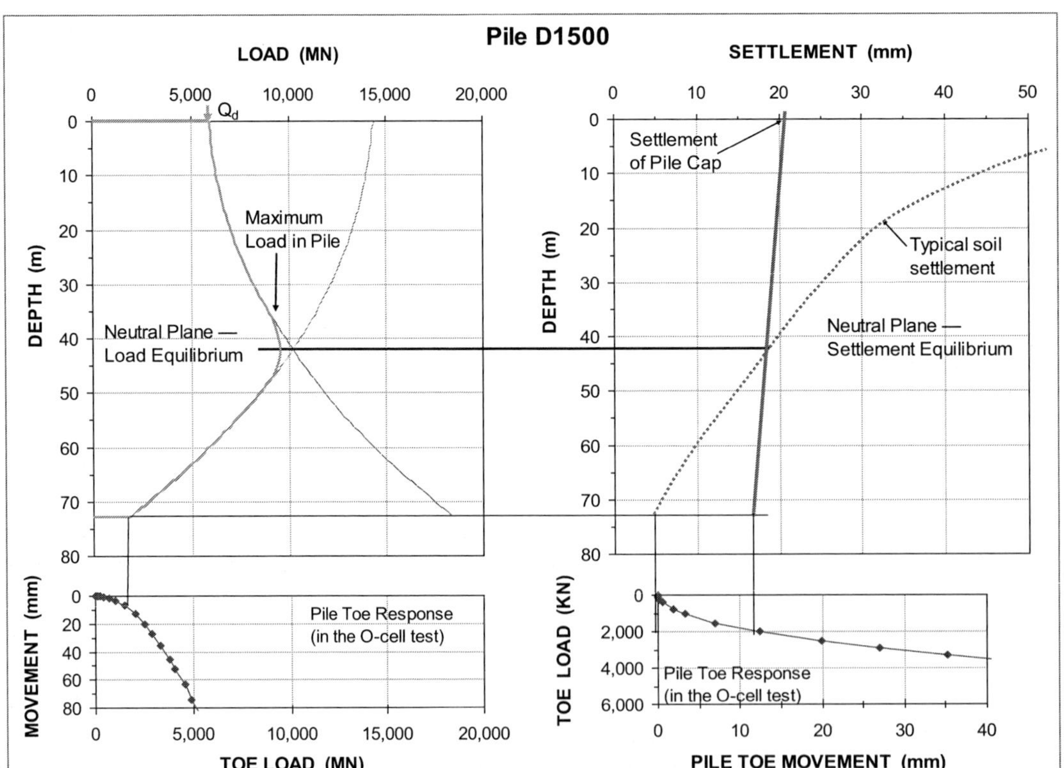

Fig. 12b Pile D1500 Load and Settlement Compilation for the Design

The analysis process is iterative. In a first attempt, an assumed pile toe load is chosen that will establish a force equilibrium—neutral plane—which is not likely to satisfy the settlement equilibrium, that is, show a location of the neutral plane which has a pile toe movement that agrees with that of the chosen pile toe load. However, after a couple of iterations, a load distribution will be found with a pile toe load that places the depth to the neutral plane that is the same magnitude as the pile toe load found for the settlement and toe movement diagram. The end result is the settlement at the pile head, which is the settlement of the structure supported by the pile(s). If that settlement is acceptable to the structure, the design requirements are reached.

Note, the load distribution shown rising from the pile toe load to a maximum value at the pile head is governed by the pile toe load, which in turn is determined by the soil settlement and pile toe movement. It is not the same load as that may have been chosen as the pile toe "capacity" used in a factor of safety or LRFD approach.

For the current piles, it is assumed that no residual load acts at the pile toe when the structure is built. In most cases, however, there would be some initial load present at the pile toe and the placement of the pile toe load-movement diagram in Figures 12a and 12b would have to be adjusted (moved to the left) accordingly. Such preloading condition would result in a reduced amount of toe movement, an increase of toe load, a lowering of the neutral plane, and a reduced pile cap settlement.

Unless extreme factors are applied, a design found to result in an acceptable settlement satisfactory by the Unified Method will also be shown to be safe in a factor of safety or LRFD. The reverse is not true, however.

SUMMARY AND CONCLUSIONS

The two O-cell tests on Piles D2000 and D1500 established that the pile capacities satisfied the specified values.

Analysis by the Unified Design method shows that if the structure can sustain long-term settlements of about 20 mm, the pile lengths and construction methods would be accepted.

The tests achieved maximum upward movements at the O-cell location of 5.3 mm and 8.5 mm, and maximum downward movements of 77.8 and 123.0 mm. The maximum upward movement of the pile heads were 0.5 and 2.3 mm, respectively, which means that the shaft resistance along the upper pile lengths were not quite fully mobilized.

The peak pile shaft resistance required only a few millimetres to develop and showed a significant post-peak softening. The shaft resistance at large movement was small and correlated to an effective stress analysis beta-coefficient of only 0.14 and 0.12 for Piles D2000 with polymer slurry and D1500 with bentonite slurry, respectively. The difference is too small to be significant.

The measured pile toe stress-movements for the piles were almost identical, showing about the same stress and essentially linear trend. The pile toe response was very soft and not representative for a pile in very dense sand.

The use of different slurry composition—bentonite and polymer—for the two test piles showed no difference in distribution of shaft resistance or collection of debris at the bottom of the shaft. Brown (2012) has indicated a series of other differences and factors of importance for such comparisons. However, no pertinent observations were recorded that could have provided such factors.

It appears that the shaft and toe resistances for both test piles were affected by presence of slurry filter cake between the concrete and the soil, and this to a similar degree for the piles.

The toe stiffness and shaft resistance found for the Binh Loi and Sunrise City test piles were very similar to each other, but remarkably different from the test pile at the nearby Phu My site, constructed under soil conditions very similar to the Binh Loi soil profile, but by a different contractor. This indicates that cause of the low values of shaft resistance and toe stiffness at the Binh Loi and Sunrise City sites were due to less than desirable construction procedures.

The evaluated pile Young's modulus, E, for the two piles correlated to 25 GPa for the nominal pile cross section of both piles, as well as for the test piles at the Phu My and Sunrise City sites.

No measurements were made of shaft diameter and results of the cleaning effort. While the comparison between the three identical test piles (similar, but not identical diameters) at the subject and the two nearby projects, showed about the same pile axial stiffness, this does not mean that the actual diameter was the same as the nominal. Therefore, when applying the evaluated unit shaft resistance values to other piles in the area, it should be recognized that a mere two inches of diameter widening would result in a 10 % larger shaft circumference of a 1,500 mm diameter, and a proportional difference in the evaluated beta-coefficients.

The pressure grouting of the toe of Pile D2000 was not successful.'

ACKNOWLEDGEMENTS

The bidirectional O-cell tests were performed by Loadtest Asia Pte. Ltd. under the direction of Mr. Khoo Han Sen. Thanks are owed to Dr. Mauricio Ochoa, PE, TWEI, Houston, TX, for his valuable review comments on the draft manuscript.

REFERENCES

Brown, D., 2012. Recent advances to the selection and use of drilled foundations. ASCE GeoInstitute Geo-Congress Oakland, March 25-29, 2012, Keynote Lectures, State of the Art and Practice in Geotechnical Engineering, ASCE, Reston, VA, Rollins, K. and Zekkkos, D., eds., Geotechnical Special Publication, GSP 226, pp. 519-548.

Fellenius, B.H., 1984. Negative skin friction and settlement of piles. Proceedings of the Second International Seminar, Pile Foundations, Nanyang Technological Institute, Singapore, 18 p.

Fellenius, B.H., 1988. Unified design of piles and pile groups. Transportation Research Board, Washington, TRB Record 1169, pp. 75-82.

Fellenius, B.H., 1989. Tangent modulus of piles determined from strain data. ASCE, Geotechnical Engineering Division, the 1989 Foundation Congress, F.H. Kulhawy, Editor, Vol. 1, pp. 500-510.

Fellenius, B.H., 1999. Bearing capacity—A delusion? Deep Foundation Institute, Hawthorne, NJ, Proc. of Annual Meeting, Dearborn, MI, Oct. 14-16, 1999, 17 p.

Fellenius, B.H., 2011. Basics of foundation design, a text book. Revised Electronic Edition, [www.Fellenius.net], 374 p.

Fellenius, B.H. and Ochoa, M., 2009. Testing and design of a piled foundation project. A case history. J. of the S-E Asian Geot. Soc., Bangkok, 40(3) 129-137.

Goudreault, P.A. and Fellenius, B.H., 2013. UniPile Version 5, Users and Examples Manual. UniSoft Geotechnical Solutions Ltd. [www.UniSoftLtd.com]. 99 p.

Loadtest International Pte. Ltd., 2008. Reports on Bored Pile Testing, Binh Loi Than Son Nhat, Vietnam, LTI-2620-1 and LTI-2620-2, 68 p and 104 p.

Osterberg, J.O., 1989. A new device for load testing driven piles and drilled shafts separates friction and end bearing. Proc. of Deep Foundations Institute, Int. Conf. on Piling and Deep Foundations, London, May 15-18, pp. 421-427.

Osterberg, J.O., 1998. The Osterberg load test method for drilled shaft and driven piles. The first ten years. 7th Int. Conf. on Piling and Deep Foundations, Deep Foundation Institute, Vienna, Austria, June 15-17, 1998, 17 p.

Workman, D.R., 1977. Geology of Laos, Cambodia, South Vietnam, and the eastern part of Thailand. Overseas geology and mineral resources, 50, Her Majesty's Stationery Office, London, 33 p.

Analysis of Settlement-Induced Bending Moments in Battered Piles

William M. Isenhower[1], P.E., M. ASCE, Luis G. Vasquez[2], P.E., M. ASCE, and Shin-Tower Wang[3], P.E., M. ASCE

[1] Project Manager, Ensoft, Inc./Lymon C. Reese & Associates, 3003 Howard Lane, Austin, TX, 78728, wmi@ensoftinc.com.
[2] Project Manager, Ensoft, Inc./Lymon C. Reese & Associates, 3003 Howard Lane, Austin, TX, 78728, gonzalo@ensoftinc.com.
[3] President, Ensoft, Inc./Lymon C. Reese & Associates, 3003 Howard Lane, Austin, TX, 78728, stw@ensoftinc.com.

ABSTRACT: The effects of settlement on the development of bending moments in battered piles were studied for a flood protection structure using nonlinear three-dimensional finite element model. The model was of one configuration for a flood protection structure consisting of an earth embankment and a pile-supported T-wall. The constitutive models used for the soil were the Mohr-Coulomb and the Bounding Surface Plasticity model. Both models were used for drained and undrained analyses. The drained analyses represented the long-term consolidation analysis of the embankments and the undrained analyses represented the short-term loading conditions present during a storm surge loading event. The time rate of consolidation and dissipation of pore water pressures in the embankment and supporting subsoil were not modeled in these analyses. The study found that different levels of bending moments were developed in the three rows of batter piles due to the settlement of the embankments and the maximum moments were developed in the flood side row of piles. The undrained analyses of storm surge loading event revealed that bending moments in the flood side row of piles increased and the moments in the protected side rows of piles decreased and the sign of the moment in the piles was reversed.

INTRODUCTION

The US Army Corps of Engineers and the State of Louisiana are expanding and improving the hurricane protection system for New Orleans, Louisiana. As part of this effort, various types of flood barriers have been designed and constructed. This paper presents analysis of one flood barrier design that incorporates a pile-supported flood wall constructed on an existing earthen embankment that is being raised and widened. The analysis presented in this paper was conducted to determine the magnitude of loading on the battered piles due to long-term settlement effects of the embankment prior to a raised water loading event and the change in loading due to two levels of elevated water loading.

A three-dimensional finite element model was used to analyze the section. The analysis consisted of drained analyses to model the long-term consolidation of the embankment and subsoils, followed by an undrained analysis to model the loading by the storm surge event. The finite element model was of sufficient mesh refinement to permit accurate evaluation of bending stresses and moments in the piles. In addition, special modeling techniques were employed to simulate pinned-head connection conditions between the pile head and the T-wall footing. In addition, two constitutive models were employed in the analyses. The constitutive models used in finite element model were the Mohr-Coulomb model (Simulia, Corp., 2011) and the Bounding Surface Plasticity (BSP) model developed by Dafalias and Herrmann (1986).

VALIDATION STUDIES

Four validation studies were made prior to performing the analyses presented in this paper. The purpose of the validation studies was to verify that the finite element program and user-defined constitutive model could capture the behavior of the soil and foundations.

The validation problems were:

1. Modeling of one-dimensional drained settlement. This validation problem was checked using a hand calculation of one-dimensional consolidation settlement.
2. Modeling of two-dimensional undrained loading and comparison to slope stability analysis. This validation problem compared the solution of bearing capacity failure load to that computed using the UTexas4 slope stability program (Wright, 1999).
3. Modeling of plastic flow around the corners of a square pile with chamfered corners. This analysis was performed to verify that soil could flow plastically around the corners of a pile with chamfered corners.
4. Modeling of a laterally loaded pile compared to an analysis using LPile 2012 (Ensoft, Inc., 2012). This analysis was used to compare the combined behavior of the finite element model with a nonlinear constitutive model, contact elements, and gapping behavior to an equivalent LPile analysis of a previously performed lateral load test in soils similar to that modeled in this study.

PROBLEM GEOMETRY, SOIL PROPERTIES, AND FINITE ELEMENT MODELING

The problem shown in Figure 1 represents a case where an existing embankment is being both raised and widened, piles are driven, and a flood wall is constructed on the crest of the embankment. The flood wall and piles were represented as solid elements and connected to the soil mesh through contact interfaces that allow both slip and separation. The finite element model was developed to capture the settlement from soil surcharge loading and flood-side loading from the hurricane storm surge. The finite element mesh is of sufficient size and refinement to permit accurate calculation of stresses, bending moments, and deformations at the locations of interest.

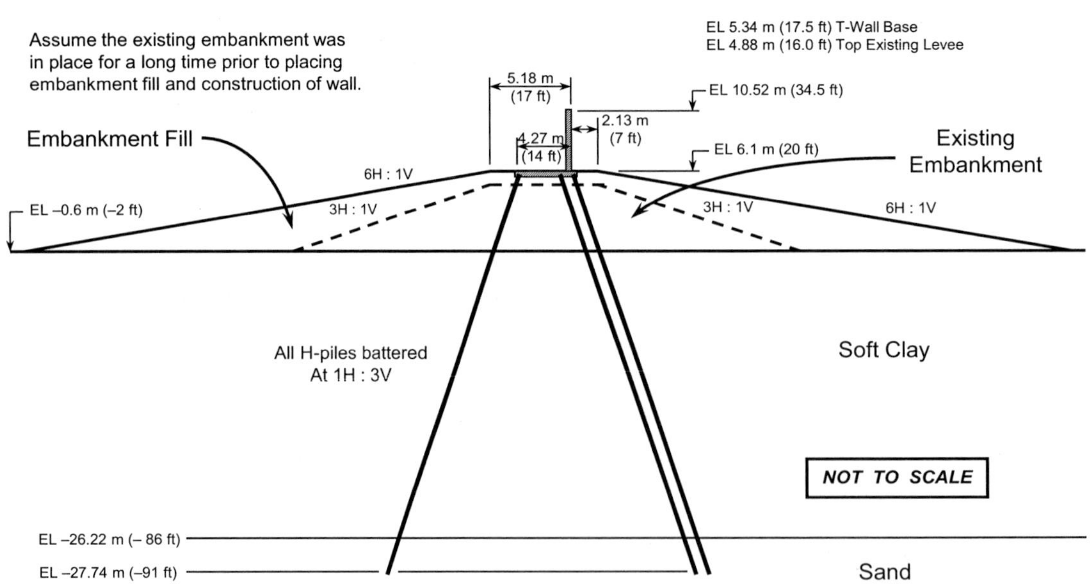

Figure 1. Cross-section of embankment and T-wall.

Pile Properties

The cross section and properties of the HP360×108 (14×73) piles are shown in Figure 2. Creating the finite element mesh for the steel flanges and web of the H-Pile posed a modeling problem due to the small dimensions of flange and web thicknesses. Instead, the H-Piles were modeled with a rectangular cross section with chamfered corners.

A finite element model with the actual H-Pile geometry requires a very fine and complex mesh to represent the flanges, web, and fillets between the flanges and web. The final mesh will require a very large number of finite elements that will be very difficult to be analyzed. An alternative approach is to model the H-Pile with a rectangular cross section using equivalent bending properties. Therefore, to keep the bending stiffness constant, the elastic modulus was adjusted to give the rectangular beam an equivalent bending stiffness to the H-pile. The procedures for deriving the equivalent unit weight are detailed in following paragraphs. For this study, it was assumed that the axial stiffness has less effect compared to the bending stiffness.

The other geometric change to the H-Pile is the use of chamfered corners. The geometric change tries to address the numerical issue of nodes "locking" at the corners of the pile in contact with the soil. In the initial case studies, the finite element analysis did not encounter nodes locking at the corners. The main computational issues encountered involved nodal penetration and high stress concentrations. Chamfering of the corners mitigated these effects.

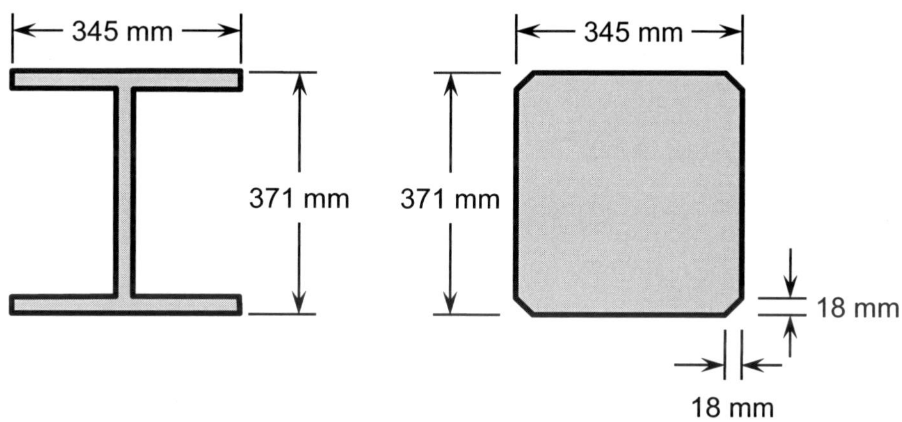

I = 729 in^4 = 3.021×10^8 mm^4
A = 21.4 in^2 = 13,810 mm^2
E = 29,000 ksi = 200,000 MPa

I = 3,477 in^4 = 1.447×10^9 mm^4
A = 197.6 in^2 = 127,480 mm^2
E_e = 6,053 ksi = 41,735 MPa

Figure 2. Cross sections and properties of H-pile and equivalent section pile.

For the equivalent section, the equivalent Young modulus is computed so the bending stiffness of the equivalent pile is equal to the bending stiffness of the actual pile, so:

$$E_e I_e = E_p I_p$$

where E_e is the equivalent Young's Modulus, I_e is the equivalent moment of inertia, E_p is the Young's modulus of the H-pile, and I_p is the moment of inertia of the H-pile. Solving for E_e

$$E_e = \frac{E_p I_p}{I_e} = 41{,}740 \text{ MPa} = 6.053 \times 10^6 \text{ psi}$$

For the equivalent section, the unit weight of the equivalent pile is computed so that the total unit weight of the equivalent pile is equal to the total unit weight of the actual pile plus the soil between the flanges. The equivalent unit weights for the piles used in the model are listed in Table 1.

Table 1. Equivalent unit weights used for pile.

Pile zone	γ (kN/m^3)	γ (pcf)
Above surface of soil	8.33	53
In embankment	23.74	151
In clay (submerged)	14.31	91
In sand (submerged)	15.72	100

Finite Element Modeling

The computer program used for the finite element analysis was Abaqus, version 6.11, developed by Simulia, Corp. (2011). This program was selected because of the robust performance of the program's contact element algorithms permitted modeling of large soil displacements along the pile surfaces and gapping around the piles. In addition, Abaqus permits the use of a user-defined constitutive models.

The three-dimensional finite element model is created in several stages. First a three-dimensional solid model is created. Next, material properties and contact surfaces are defined. Lastly, the finite element nodal points and elements are mapped to the solid model and refined in zones where needed.

A side view of the three-dimensional solid model is shown in Figure 3.

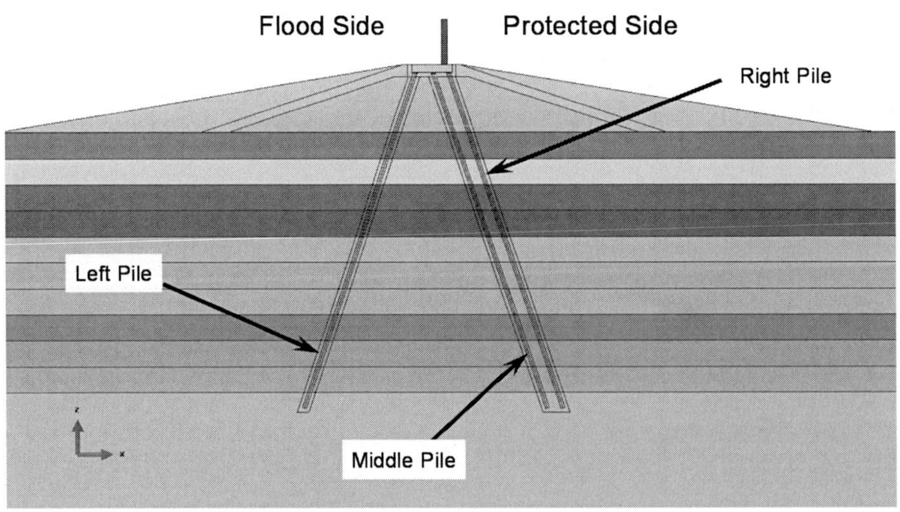

Figure 3. Side view of three-dimensional solid model.

The side view of the finite element mesh is shown in Figure 4 and a close up view of the T-wall and piles is shown in Figure 5. One feature of the model is the inclusion of wedges at the connection between the pile cap and the piles. The purpose of the wedges is to simulate a pinned pile-head connection between the pile cap and the piles. In reality, the embedment of the piles into the base slab is insufficient to develop pile head fixity, so a simulated pinned-head condition was desired. A low value of Young's modulus was assigned to the wedges within the piles to allow for relative rotation at the pile head and to minimize the bending moment developed in the connection. The finite element model consists of a total of 1,043,934 tetrahedral elements and 235,238 nodes. The piles are modeled using a maximum element size of 2.5 inches (63 mm). In addition, the soil close to piles is refined with a similar mesh size. Then the mesh size is increased toward the boundary of the model.

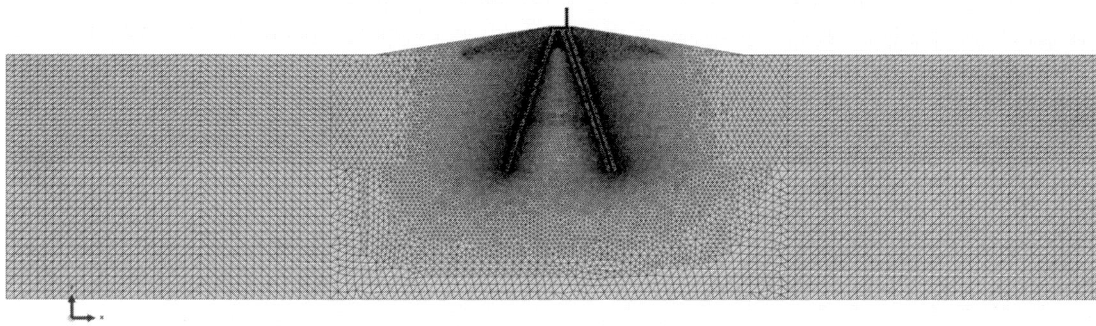

Figure 4. Side view of three-dimensional finite element model.

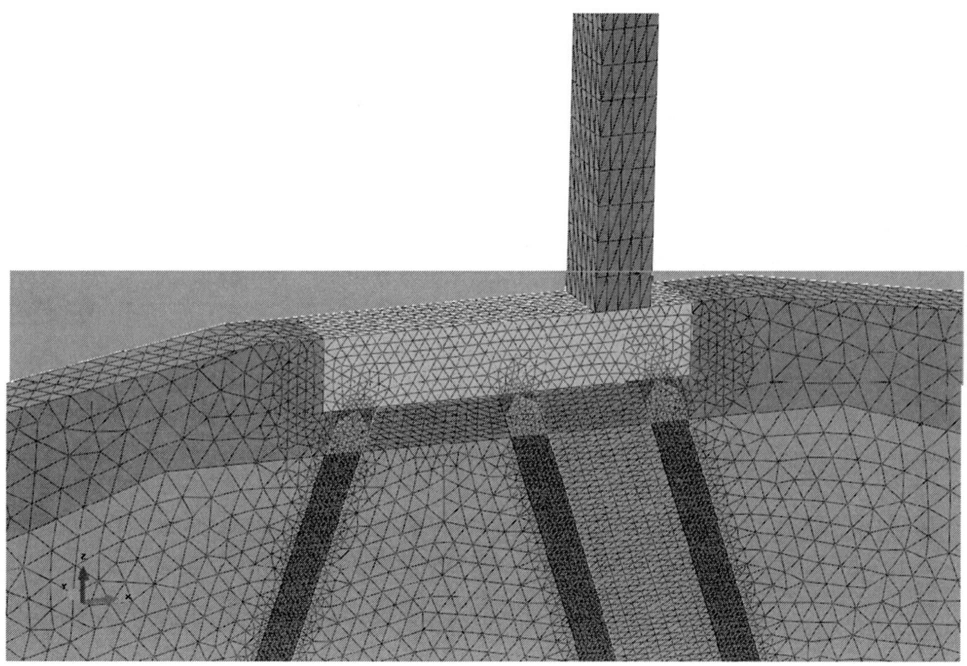

Figure 5. View through mid-plane of finite element model of embankment and pile heads.

The small size of the elements for piles was required in order to evaluate the bending moments developed in the piles. The bending moments were evaluated by integration of stress resultants across the section of the pile. Prior to performing the main analyses, a parameter study was made to determine the maximum size of elements that could be used. It was determined that an element size of 63 mm (2.5 inches) was needed in the cross-section of the piles to achieve the desired level of accuracy for evaluation of bending moments, shear forces, and axial thrust forces.

Constitutive Models

Three constitutive models were used in the analysis of the model. The T-wall and piles were modeled as elastic materials; the cohesive soils were modeled as a user-defined material based on the Bounding Surface Plasticity (BSP) model; and the cohesionless soils were modeled using an elastic Mohr-Coulomb material.

Loading Steps

The finite element model includes all the components, but it is being loaded in steps. These steps are illustrated in Figures 6 and 7. First, the initial stresses are applied. The initial stresses are taken into account for all the elements below the base level of –0.6 m (–2 ft) and the old embankment. The new embankment and all the structural elements above the reference level have zero initial stresses. Because of the presence of the old embankment, the initial stresses are estimated using a five-step process.

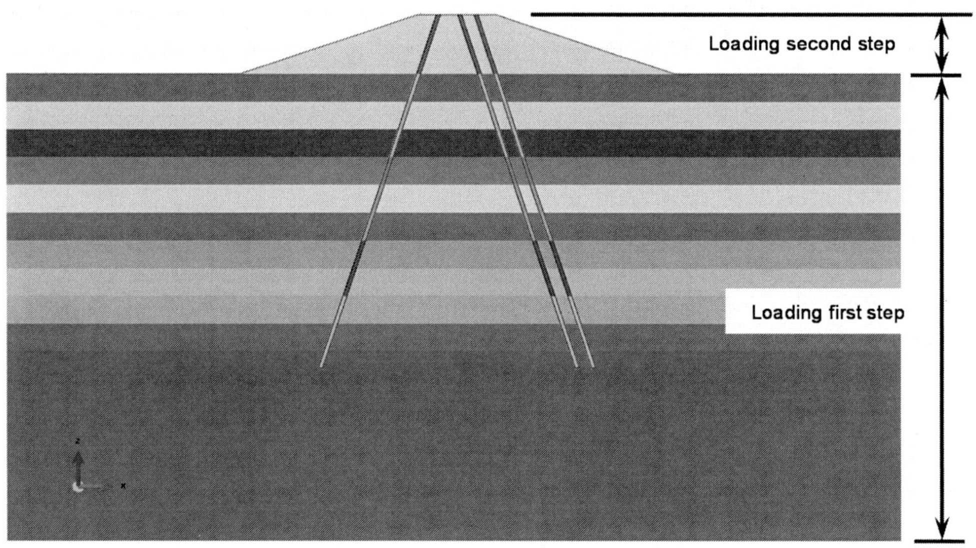

Figure 6. Parts of model loaded during loading steps 1 and 2 for computation of initial stresses.

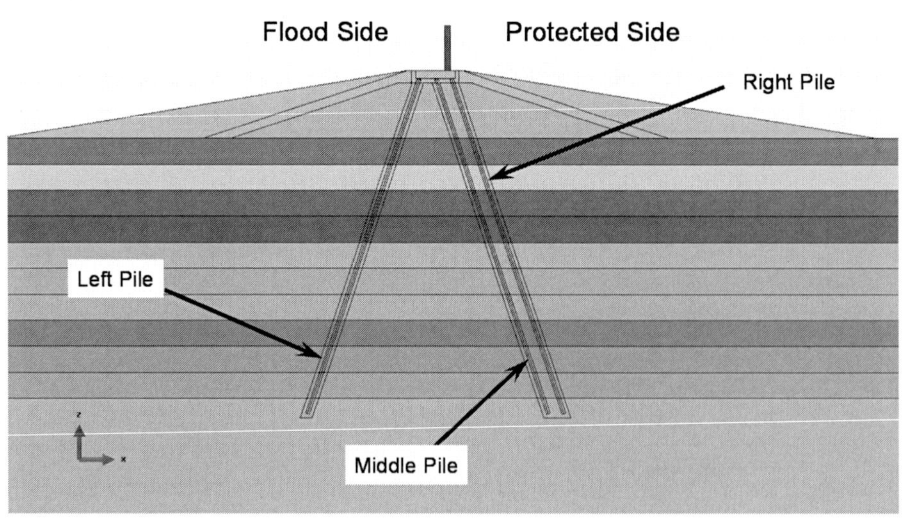

Figure 7. Parts of model loaded during loading step 3 for construction of raised and widened embankment.

In loading step 1, the soil below the base level of –0.6 m is considered. For this part of the soil, the initial stresses are estimated based on the vertical stresses and the coefficient of earth pressure.

In loading step 2, the old embankment is added. An analysis is performed considering the self-weight of the old embankment. The self-weight is applied in increments. This distribution of stresses at the end of the analysis is used as initial stresses for subsequent analysis. Also, as a part of the analysis the system will undergo some deformation, and the deformations from the initialization steps are not included in subsequent analyses.

In loading step 3, the load due to the self-weight of the embankment plus all the structural elements is added above the surface profile of the old embankment. The load is applied in increments by increasing the weight of the embankment from zero at the beginning of the step to the total weight at the end of the step. In addition, the difference of weight in the structural elements not considered during initial stresses is added at the beginning of this step. For this loading, drained properties for clay and sand are employed.

In loading steps 4 and 5, the storm surge loading is applied to the structure. The load is applied to the left hand (flood) side of the model. The load is applied in two steps, in loading step 4 from the reference elevation, −0.6 meters (−24 inches) to the top of the embankment and in loading step 5 up to the top of the T-wall. In each loading step, the load is applied in increments. For this loading, undrained properties for clay and drained properties for sand are employed.

LONG-TERM ANALYSES

Contours of initial vertical stresses are presented in Figure 8 (the negative sign for compressive stress is due to the sign convention used in Abaqus). In the analysis of initial vertical stresses, the initial vertical stresses are checked to be in equilibrium with gravitational loading. The effect of self-weight of the old embankment in the middle portion of the model can be seen.

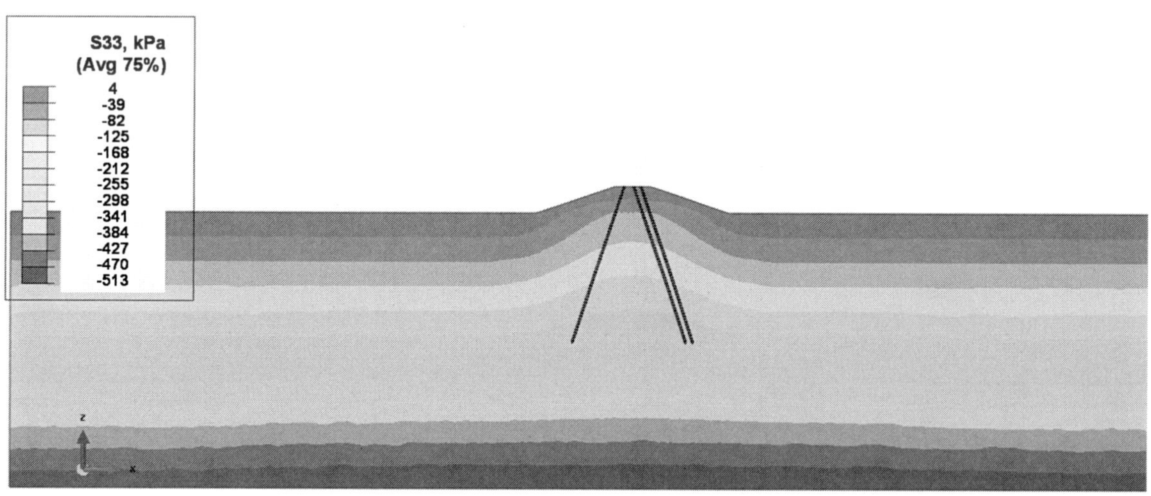

Figure 8. Contours of vertical stresses after step 2.

Loading due to Self-Weight of Embankment

Drained properties for the materials are used in the analysis during the application of the self-weight loading. These material parameters are listed in Table 2. It is important to note that in this analyses the embankment is modeled using a Mohr-Coulomb model, while all the clay layers are modeled using the BSP model and are numbered from top to bottom.

Table 2. Material properties for bounding surface plasticity model.

Parameter	Layer 1	Layer 2	Layers 3 to 10	Embankment
λ	0.179	0.179	0.179	
κ	0.0179	0.0179	0.0179	
v	0.380	0.380	0.380	0.380
M_c	0.890	0.890	0.890	
M_e	0.691	0.691	0.691	
R_c	1.870	2.070	2.170	
R_e	1.380	1.560	1.620	
A_c	0.010	0.010	0.010	
A_e	0.010	0.010	0.010	
T	0.010	0.010	0.010	
C	0	0	0	
S_p	1.200	1.200	1.200	
a	1.200	1.200	1.200	
w	5.00	5.00	5.00	
E				10,270. kPa
S_u				28.7 kPa

The deformed shape of the embankment (shown with a magnification factor of 5) is shown with contours of horizontal and vertical deformations in Figures 9 and 10. The maximum horizontal displacement of the soil due to the self-weight of the embankment is about 259 mm (10.2 inches) and is located at the top left of the embankment. The horizontal displacement at the right hand side is 229 mm (9 inches). In general, the soil is moving away from the base slab of the T-wall. The maximum vertical displacement of the soil is about 655 mm (25.8 inches) and is located near the bottom of the embankment faces on both sides of the T-wall. The soil at crest of embankment is moving downwards, so a gap under the pile cap is formed, thereby suspending the pile cap.

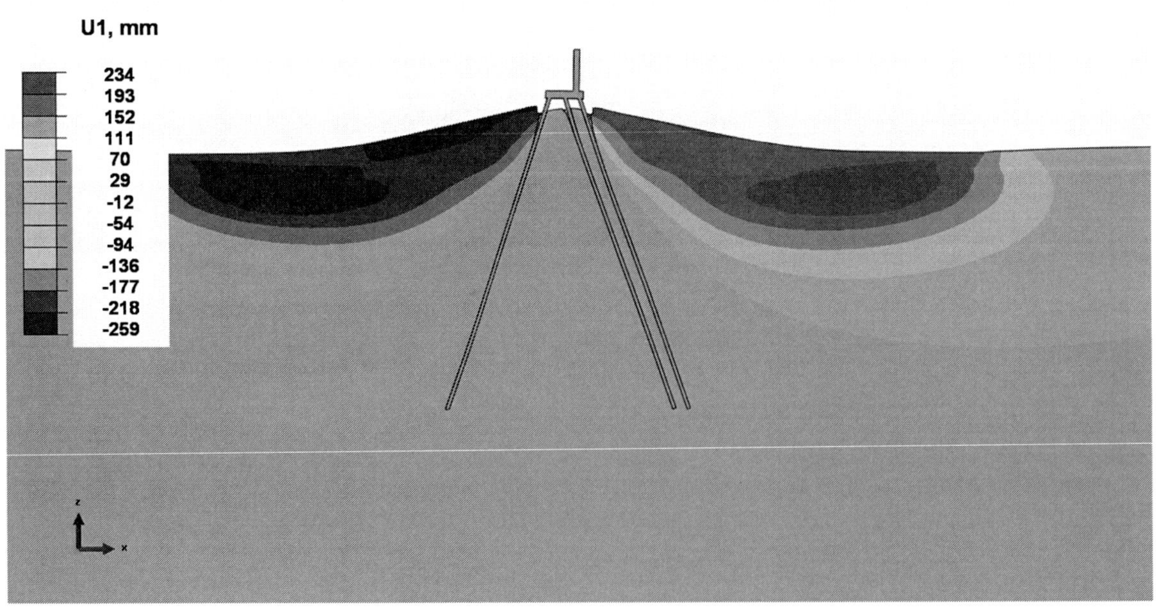

Figure 9. Contours of horizontal displacements (U1) due to self-weight of embankment.

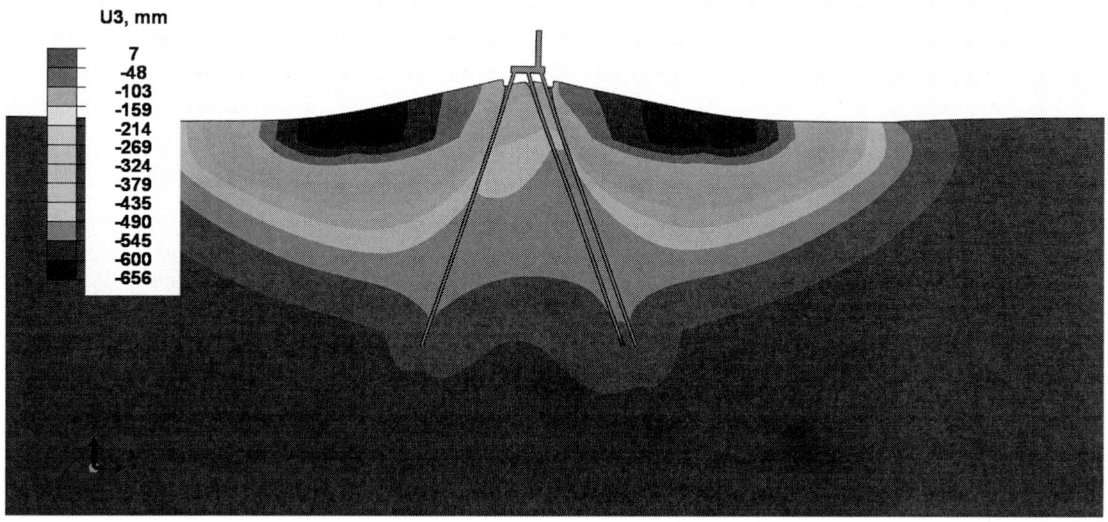

Figure 10. Contours of vertical displacements (U3) due to self-weight after raising and widening of embankment.

The deformed shape of the piles and T-wall (drawn with a magnification factor of 200) is shown with contours of horizontal and vertical deformations in Figure 11. A higher magnification factor was used in Figure 11 to aid in visualization of curvature in the piles induced by the soil settlement. The internal forces and transverse displacements developed in the piles are shown in Figure 12. A summary of the maximum displacements, bending moments, and axial thrust forces developed in the three piles is presented in Table 3.

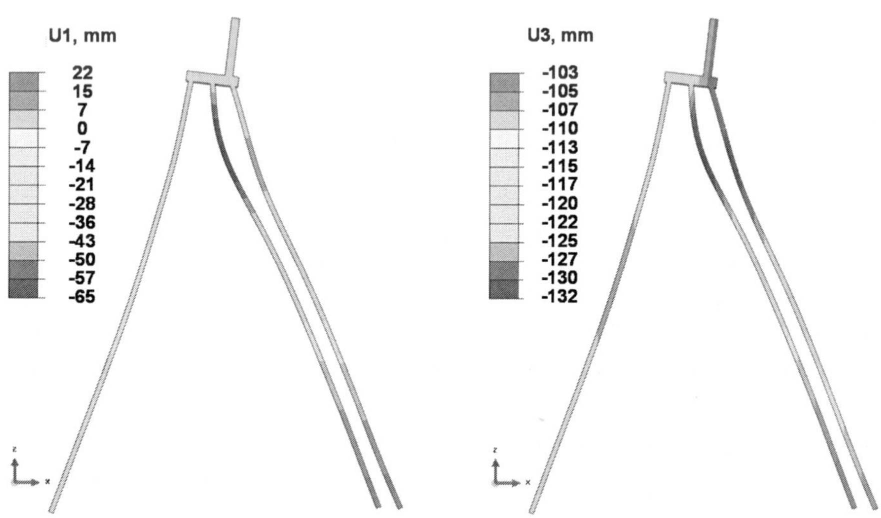

Figure 11. Contours of horizontal (U1) and vertical (U3) displacements of piles and T-wall due to self-weight of the embankment.

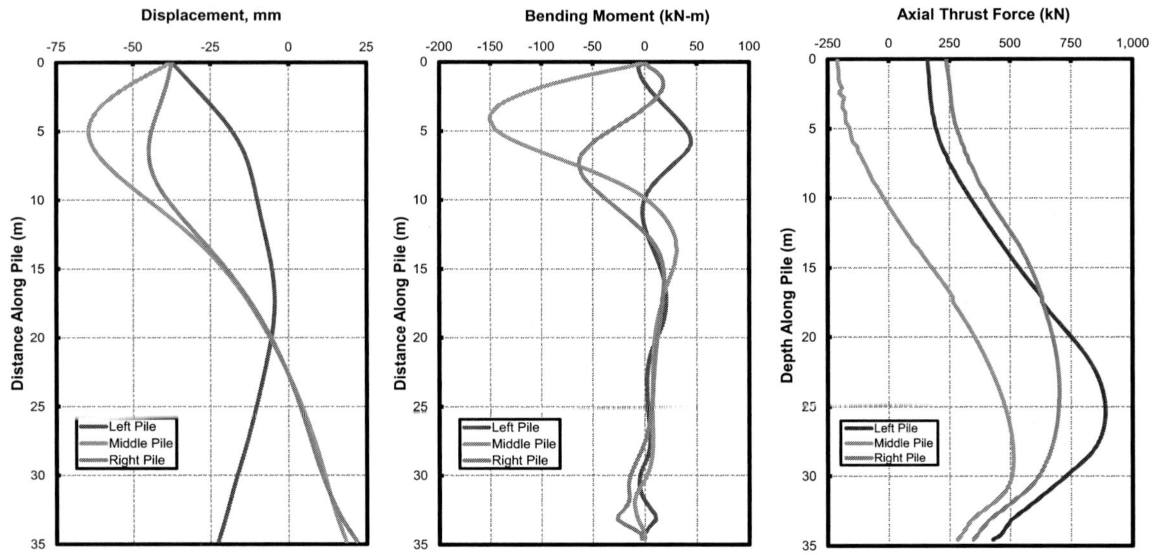

Figure 12. Transverse displacements and internal forces developed in piles due to self-weight of embankment.

Table 3. Maximum transverse displacements and internal forces developed in piles due to self-weight of embankment.

Pile response	Left pile	Middle pile	Right pile
Lateral displacement	−37.6 mm	−64.5 mm	−44.9 mm
Positive bending moment	44.5 kN-m	30.8 kN-m	18.6 kN-m
Negative bending moment	−6.10 kN-m	−149.1 kN-m	−63.3 kN-m
Shear force	16.5 kN	70.3 kN	26.5 kN
Axial thrust force	890. kN	516. kN	703. kN

The net effect of the settlement is to cause the batter piles to curve inward towards the centerline of the embankment. The greatest curvature and bending moment is developed in the middle pile, though smaller bending moments are also developed in the left and right piles. The initial movement of the T-wall is leftward is due to the unbalanced lateral stiffness of the foundation and the downdrag of the settling soils in and under the embankment. The largest bending moment of approximately −150 kN-m is developed in the middle pile about 4 meters below the pile head. This bending moment value is about 57 percent of the allowable bending moment in the piles.

ANALYSES OF STORM SURGE LOADING

The storm surge loading analyses were performed to assess the performance of the embankment and T-wall when loaded by two levels of storm surge. The area of concern was to assess if the additional loadings on the piles due to storm surge, when added to the pre-existing loads due to long-term settlement would overload the piles.

In the storm surge loading analyses, two levels of storm surge loading were evaluated. These storm surge elevations are illustrated in Figure 13. The first water level loads only the embankment and the second water level loads the embankment and T-wall to their maximum capacity.

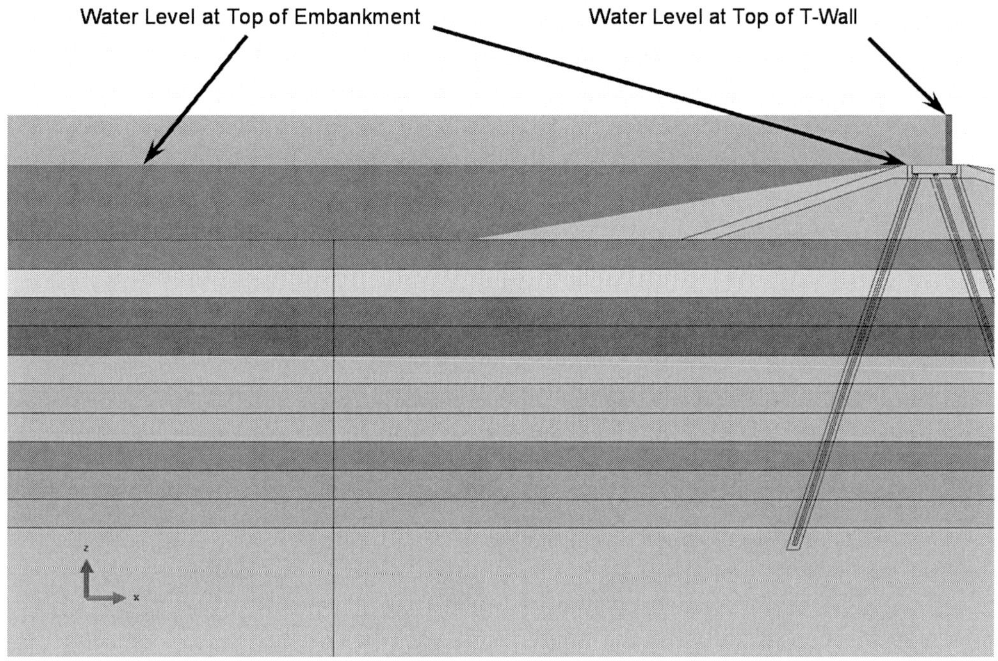

Figure 13. Water levels applied for storm-surge loading in steps 4 and 5.

The deformed shape of the piles and T-wall (shown with a magnification factor of 200) along with contour shading of horizontal and vertical deformations are shown in Figures 14 and 15 for the two levels of storm surge loading. Summary plots of pile deformation, bending moment, and axial thrust force developed in all levels of loading for the analyses are presented in Figures 16, 17, and 18, respectively. Note that the pile displacements shown in these figure are the cumulative displacements from all stages of the analysis. Tables 4 and 5 present summaries of the maximum effects of loading in the three piles for water levels at the top of the embankment and at the top of the T-wall.

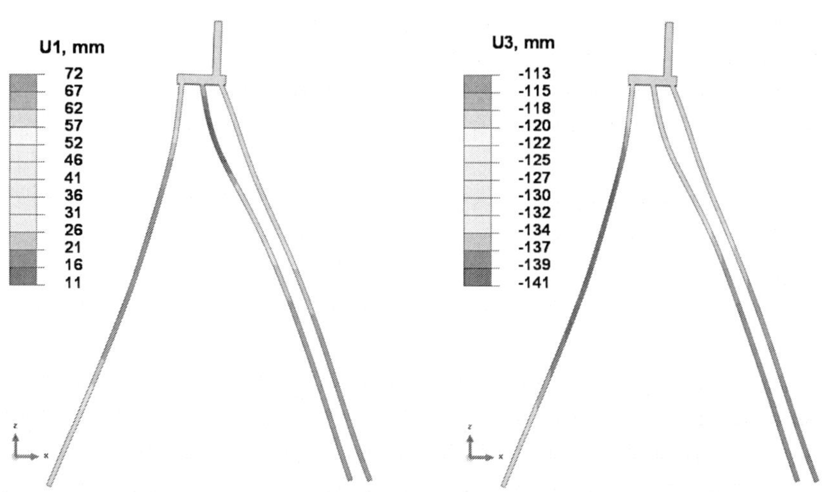

Figure 14. Contours of horizontal (U1) and vertical (U3) displacements of piles and T-wall due to storm surge loading with water level at top of embankment.

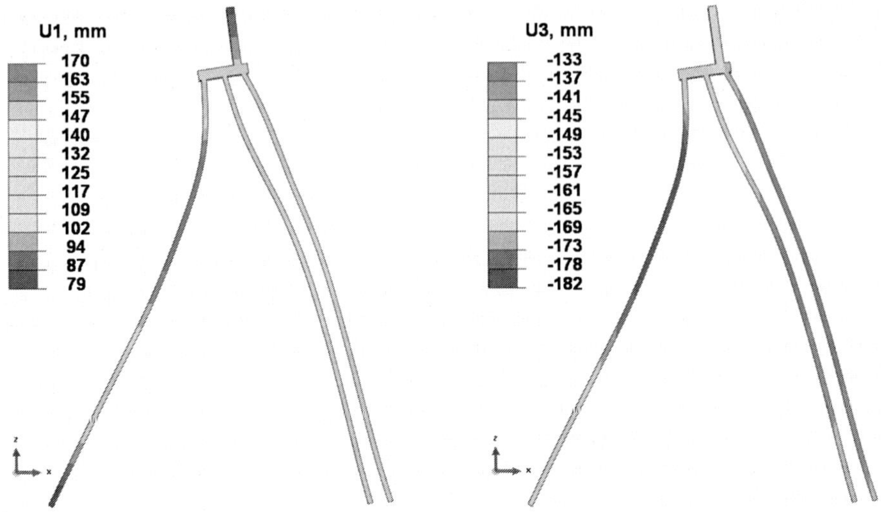

Figure 15. Contours of horizontal (U1) and vertical (U3) displacements of piles and T-wall due to storm surge loading with water level at top of T-wall.

An examination of transverse pile displacements shown in Figure 16 finds that all piles are initially displaced to the left by the application of self-weight of the embankment, then the piles are displaced to the right by movement of the embankment during the storm surge loadings. The initial movement to the left is due to the unbalanced axial and lateral stiffness on the left side of the foundation due to the unequal number of battered piles in the foundation. Attention should be drawn to the lateral movement of the pile tips. The non-zero displacement of the pile tips indicates that the whole model is being displaced towards the protected side of the embankment.

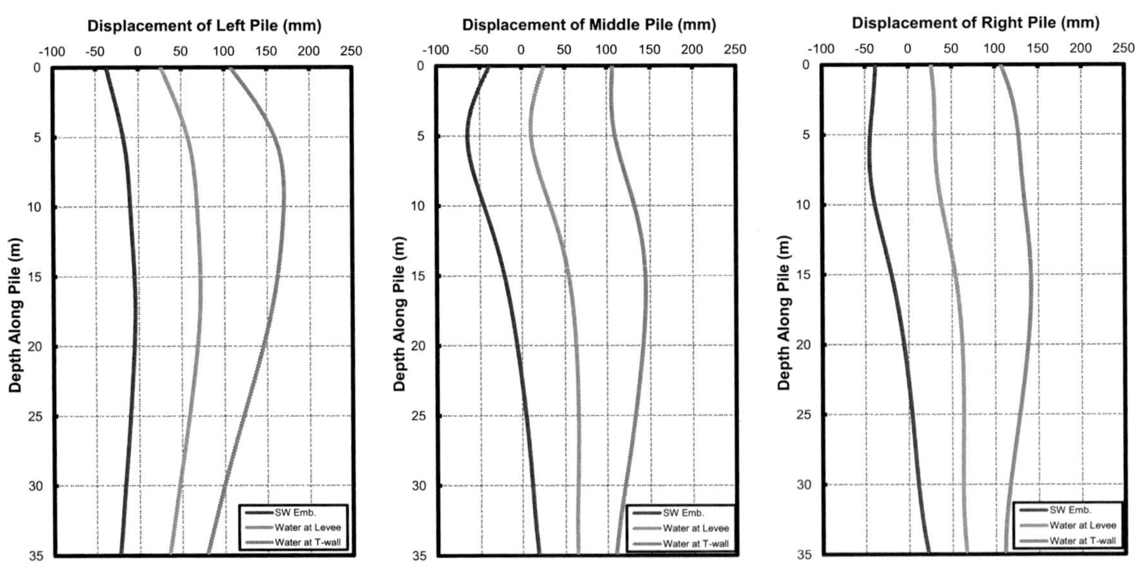

(a) Left Pile (b) Middle Pile (c) Right Pile

Figure 16. Lateral pile displacements developed in piles after embankment construction, water at top of embankment, and water level at top of T-wall.

The bending moments developed in the piles for all stages of loading are shown in Figure 17. Differences in moment were seen primarily in the upper 20 m of the piles. In the self-weight stage of loading, the maximum moment is developed in the middle pile. In the first storm surge loading, when the water level reaches the top of the embankment, the bending moment in middle pile is reduced, but is still the maximum moment developed in the three piles. In the second storm surge loading, the maximum moment is developed in the left pile, but is reduced by 11 percent from the moment developed in the middle pile in the self-weight loading.

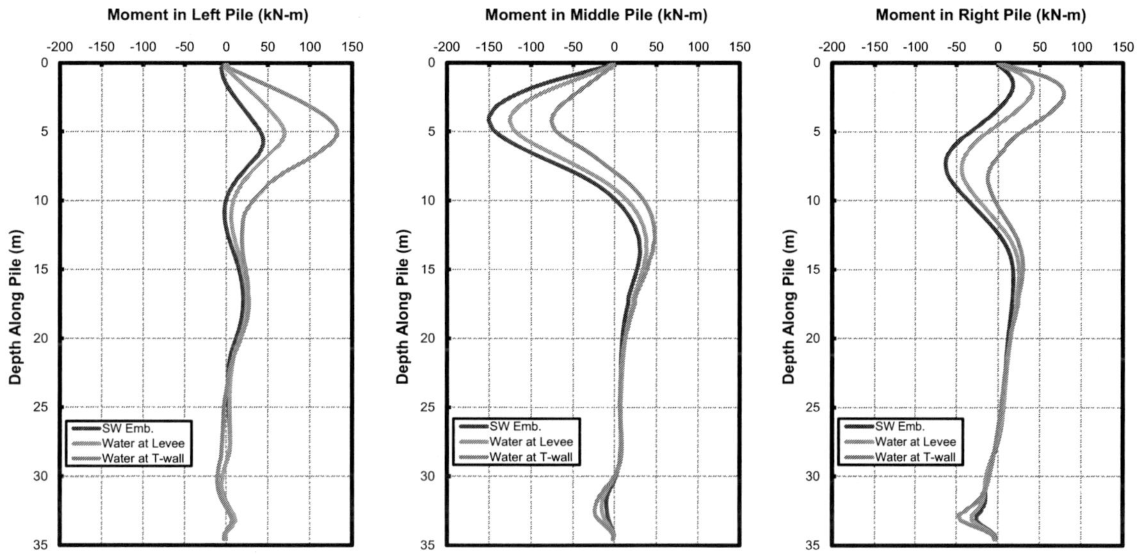

(a) Left Pile (b) Middle Pile (c) Right Pile

Figure 17. Bending moments developed in piles after embankment construction, water at top of embankment, and water level at top of T-wall.

Table 4. Maximum displacements and internal forces developed in piles due to storm surge loading with water level at top of embankment.

Pile response	Left pile	Middle pile	Right pile
Lateral displacement	72.1 mm	65.3 mm	66.3 mm
Positive bending moment	69.7 kN-m	38.4 kN-m	41.8 kN-m
Negative bending moment	−5.29 kN-m	−125.8 kN-m	−43.8 kN-m
Shear force	20.2 kN	46.3 kN	46.7 kN
Axial thrust force	394. kN	587. kN	618. kN

The axial thrust forces developed in the pile are presented in Figure 18. The maximum axial thrust developed during the self-weight stage of loading is developed in the left pile and is due to downdrag. Maximum axial thrust levels are reduced during the first storm surge loading, with the maximum thrust developed in the right pile. In the second storm surge loading, the maximum axial thrust is developed in the middle pile and is 4.5 percent higher than the axial thrust developed in the left pile in the self-weight stage of loading.

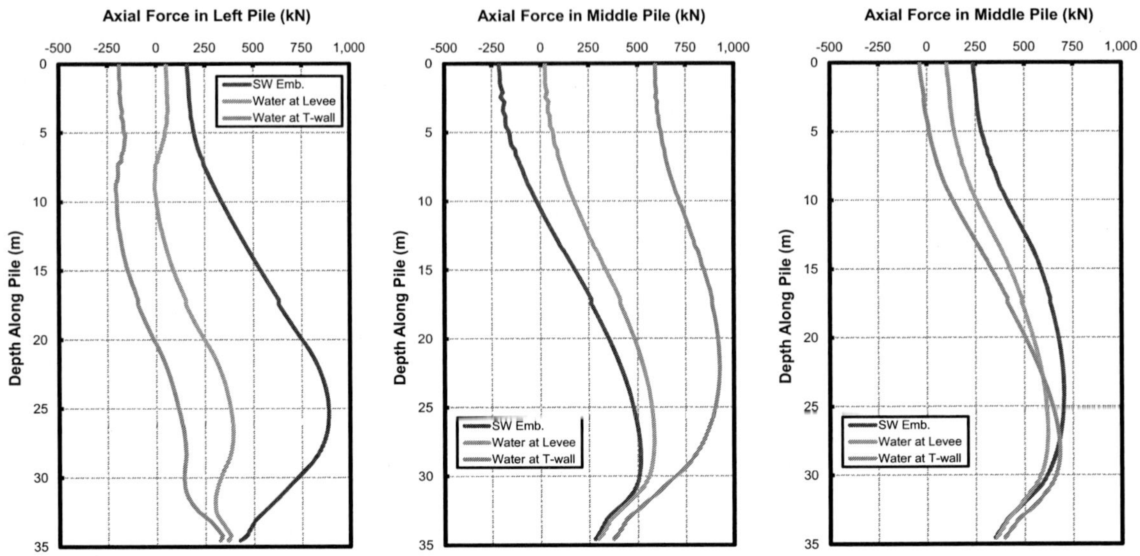

(a) Left Pile (b) Middle Pile (c) Right Pile

Figure 18. Axial thrust forces developed in piles after embankment construction, water at top of embankment, and water level at top of T-wall.

Table 5. Maximum displacements and internal forces developed in piles due to storm surge loading with water level at top of T-wall.

Pile response	Left pile	Middle pile	Right pile
Lateral displacement	170.2 mm	144.8 mm	144.2 mm
Positive bending moment	132.6 kN-m	48.3 kN-m	78.9 kN-m
Negative bending moment	−10.58 kN-m	−75.9 kN-m	−47.0 kN-m
Shear force	38.7 kN	29.4 kN	74.7 kN
Axial thrust force	336. kN	930. kN	681. kN

DISCUSSION

One goal for performing this study was to serve as an example of a type of analysis that might be needed for cases where the presence of additional features may need to be included. Examples of additional features are near changes in alignment of a flood barrier (curves or corners) or inclusion of underground utilities such as cables, pipelines, and sewers. The soil profile used for the analyses performed in this study is not representative of all soil profiles for this type of flood barrier. Consequently, no general conclusions can be drawn for flood barriers constructed on different soil profiles or constructed in areas with markedly different soil or foundation properties.

This particular study found that the magnitude of the maximum bending moments developed in the battered piles during the long-term settlement stage of the analysis was less than the maximum bending moments developed in the storm surge analyses. It is not possible to say whether this type of performance is typical of all pile-supported T-wall structures or was only the result for this analysis.

CONCLUSIONS

A study to estimate settlement induced bending moments in batter piles using nonlinear three-dimensional finite elements models was made. Four validation models and two T-wall foundation configurations were considered for this study. Two soil models were used, the Mohr-Coulomb model and the Bounding Surface plasticity model. In addition, contact surfaces and interface friction between the soil and piles and between the soil and T-wall were included in the model. All parts of the model were modeled using three-dimensional solid elements. Smaller elements were used in the piles in order to evaluate the bending moments developed in the piles.

Maximum moments developed at the middle of the piles or close to the top of piles (about 3 to 5 m below the pile head). There are more significant changes (increases or decreases) for the bending moment close to the top of the pile for storm surge loading. The moment at the middle of the pile does not change significantly under storm surge loading.

In general, bending moments at the middle of the pile are mostly affected for loading under self-weight. While moment close to the top of the pile is affected under both self-weight and storm loading.

In general, axial force in the middle pile is smaller for self-weight loading, while it increases significantly under storm surge loading. Axial forces in left and right piles are larger due to for self-weight loading, reducing under storm surge loading.

ACKNOWLEDGMENTS

The authors wish to thank the following individuals who participated in the performance or review of the analyses reported in this paper: Mr. Felix Chen and Dr. Daqing Xu of Ensoft, Inc./Lymon C. Reese & Associates; Mr. Neil T. Schwanz of the US Army Corps of Engineers; Professor John Tassoulas of the University of Texas at Austin, and Professor George M. Filz of Virginia Tech University.

The authors also wish to acknowledge the major influence of Professor Roy E. Olson on their education and practice as civil engineers. The authors of this paper were privileged to have taken 12 courses taught by Professor Olson during their student years at the University of Texas at Austin.

REFERENCES

Dafalias, Y.F. and Herrmann, L.R., (1986). "Bounding Surface Plasticity. II: Application to Isotropic Cohesive Soils," *J. Engineering Mechanics*, ASCE, Vol. 112 (12), 1263-1291.

Ensoft, Inc., (2012). *LPile 2012, A computer program for the analysis of laterally loaded piles*, Ensoft, Inc., Austin, Texas.

Lymon C. Reese & Associates, (2011). "Evaluation of Bending in Batter Piles Due to Settlement," Report to FFEB Joint Venture, October 13, 2011, 186 p.

Simulia, Corp., (2011). *Abaqus, Version 6.11 Documentation*, 2011.

Wright, S. G. (1999). *UTexas4: A computer program for slope stability calculations*, Shinoak Software, Austin, Texas.

Measurement of Ground Motion near Piles during Driving

Richard Woods[1], P.E., D.GE, Dist.M. ASCE, Adda Athanasopoulos-Zekkos[2], A.M. ASCE, Athena Gkrizi[3], S.M.ASCE, Anthony Pietrangelo[4], P.E., and Andrew Zimmerman[5], M. ASCE

[1]Professor Emeritus, Department of Civil and Environmental Engineering, University of Michigan, Ann Arbor, MI 48109-2125; rdw@umich.edu
[2]Assistant Professor, Department of Civil and Environmental Engineering, University of Michigan, Ann Arbor, MI 48109-2125; addazekk@umich.edu
[3]Graduate Student, Department of Civil and Environmental Engineering, University of Michigan, Ann Arbor, MI 48109-2125; agkrizi@umich.edu
[4]Geotechnical Construction Support Engineer, Michigan Department of Transportation, Lansing, MI 48906; pietrangeloa@michigan.gov
[5]CEO Civionics, LLC, Colorado Springs, CO, 80919; andy@civionics.com

ABSTRACT: Two types of vibration damage caused by driving piles have been reported in the literature: direct structural damage and damage due to settlement. Direct damage results from vibratory excitation of structures at amplitude exceeding the structural tolerance. Damage from settlement is a consequence from vibratory densification of loose soils resulting in total or differential settlement of structures. Problems of settlement due to pile driving have been experienced recently by the Michigan Department of Transportation (MDOT) during operations associated with replacement of deteriorating bridges. The work described here represents an attempt to understand the mechanisms of energy transfer from steel H-piles driven with diesel hammers to the surrounding soil and the energy attenuation through the soil by measuring ground motion in the near vicinity of the pile. The main feature of this study consisted of installing motion transducers very close, within 0.5 foot, to piles and measuring the resulting ground motion during pile driving. Selection, fabrication, and installation of the transducers and preliminary measured pile driving vibrations are presented.

BACKGROUND

Damage to structures caused by impact pile driving has been the topic of many published papers as in O'Neill (1971); Clough and Chameau (1980); Lacy and Gould (1985); Picornell and del Monte (1985); and Dowding (1991). Two types of vibration damage caused by driving piles have been identified in those papers: direct structural damage and damage due to settlement. Direct damage results from vibratory excitation of structures at amplitudes exceeding the structural tolerance to cracking or other damaging phenomena. Settlement damage results from vibratory densification of loose sands resulting in total or differential settlement of structures or other civil engineering features. Vibrations that cause direct damage are often considered to be in the range of 0.5 to 2 in/sec particle velocity at the structure (Woods, 1997). Problems of settlement due to pile driving required two conditions: presence of loose soil (usually loose sand) and ground vibration exceeding threshold shear strain amplitude of about 0.01 %. Depending on soil conditions, particle motions much less than those necessary for direct damage can cause shear strain amplitudes in excess of the 0.01% threshold and cause settlement of loose sand. Recent experience by the Michigan Department of Transportation (MDOT) during replacement operations associated with deteriorating bridges has identified shakedown settlement of loose sand from pile driving as an important problem. The writers have been working with MDOT to understand and predict the potential for settlement from driving piles.

The mechanisms of energy transfer from driven piles were postulated in the FHWA Synthesis # 253 by Woods (1997) as shown in Fig. 1. This idealized schematic assumes a half-space consisting of a homogeneous, isotropic, elastic material with a Poisson's ratio, μ. Primary waves (P-waves) radiate from the pile tip and cylindrical waves (S-waves) radiate from the pile shaft and from the developing Rayleigh waves on the surface. However, this hypothesis has not been proven with physical ground motion measurements. Likewise, vibration attenuation in three general soil behavior zones, plastic, non-linear, and nearly elastic, shown in Fig. 2, are also unsubstantiated. Shearing strain, γ, is suggested for defining boundaries of these zones. Shear strain in the ground can be calculated from particle velocity (which is the subject of this work) divided by shear wave velocity which can be measured or inferred from standard blow count. To verify these assumed soil behaviors, measurements of ground motion in the vicinity of a driven pile are required and the literature does not report on measurements of this kind. The work described here represents a first attempt to make ground motion measurements in close proximity to driven H-piles.

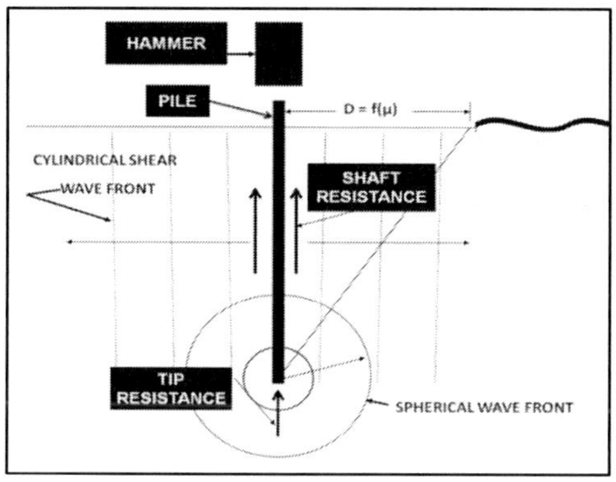

Figure 1 - Basic mechanisms of energy transfer from pile to soil (after Woods, 1997)

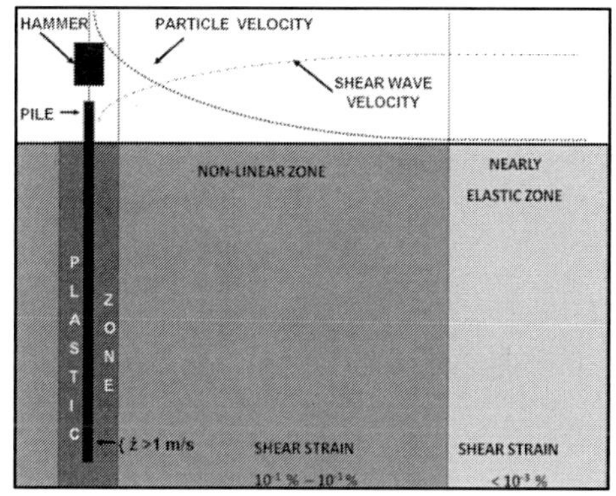

Figure 2 - Assumed soil behavior zones near driven pile (after Massarsch, 2002)

By measuring ground motion at three distances from the pile, it was expected that a vibration decay curve could be developed from which an attenuation coefficient or coefficients could be developed to define the soil behaviors in Fig. 2. It was not considered feasible to determine the ground vibration precisely at the pile-soil interface, but Massarsch and Fellenius (2008) present an equation for predicting the maximum amplitude of soil motion at the pile-soil interface that depends on shearing strength of the soil:

$$\dot{z} = \tau / V_s \, \rho \qquad (1)$$

where $\dot{z}$ is the peak particle velocity in the soil at the pile-soil interface, τ is the shearing strength of the soil, V_s is the shear wave velocity in the soil at the contact with the pile and ρ is the mass density of the soil. One preliminary goal of this research was to compare calculated values of the interface particle velocity from EQ 1 with projections back to the interface soil-pile interface from measured particle velocity from sensors embedded in the ground near the driven pile. In this way it was expected that the near pile attenuation of ground motion could be verified.

SACRIFICIAL GROUND MOTION TRANSDUCERS

Construction operations are not compatible with installation and recovery of buried transducers. Conductor cables are also vulnerable in the pile driving construction environment so it was determined to develop sacrificial transducer packages that could be pushed into the ground and not be recovered.

Two types of motion sensor cones were designed and fabricated to be pushed into the ground and not recovered. These sacrificial transducers consisted of single component (vertical) geophones and triaxial accelerometers. The least expensive approach was to use 4.5 Hz geophones (Fig. 3). The geophone cans, model RGI-20DX, were supplied by Racotech Geophysical Instruments. Since the range of ground motion was not known in advance, more expensive triaxial, MEMS based accelerometers were also prepared. Based on preliminary tests, it was estimated that ground acceleration could be as high as 5 g's so accelerometers with this capability were procured. Triaxial accelerometer units to meet this criterion were custom designed and fabricated by Civionics, LLC. Model MMA7361LCT triaxial, MEMS type accelerometers supplied by Freescale, Inc. with an acceleration range of 6 g's were the basis of Civionics' design. A 3.3 V voltage regulator that allowed any voltage source from 4V to 14 V was used to provide a zero level, 1.65V (0 – 3.3 volt range) output with a sensitivity of 206 mV/g. These components were mounted on a 1" x 1" PCB, as shown in Fig. 4.

SENSOR CONES

Sensor cones were machined from steel to be very robust so they could be pushed into the ground to their planned elevations. These cones had 60 degree tapered tips, a hollowed-out cylindrical center to house the sensors and a special adaptor that allowed downward pushing of the cone with rods from the drill rig, (Fig.5), but the adaptor could not apply a withdrawing force to the sensor cone so it was unrecoverable. Figure 6 shows a geophone potted into the push cone and Fig.7 shows the accelerometer chip being fitted into a cone cavity. The conducting cable from the sensors was fed upward through the adapter and hollow of the drill rods to the ground surface and to the data acquisition system, as shown in Fig. 8.

SACRIFICIAL SENSOR INSTALLATION

Pile driving contractors who had been awarded the contracts to build the replacement bridges at each of the selected sites recommended by MDOT were approached for their willingness to cooperate with the University of Michigan (U of M) research project. In each case the contractor agreed to drive an H-pile of the size typical for their site in a location where ground motion measurements could be made with minimal interruption of their operations but favorable for the U of M research.

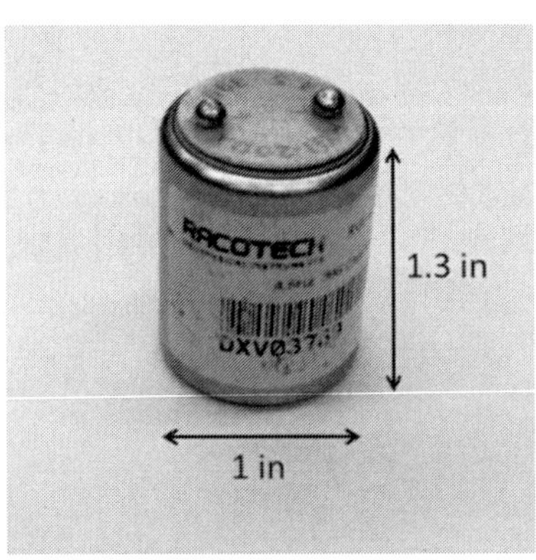

Figure 3 – 4.5 Hz geophone can

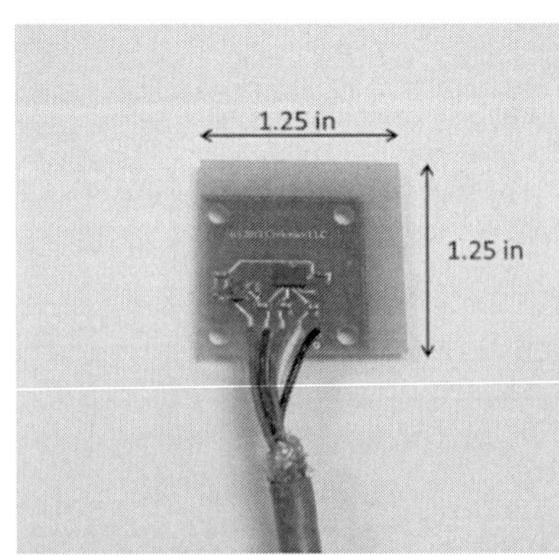

Figure 4 – Civionics Triaxial MEMS accelerometer on chip

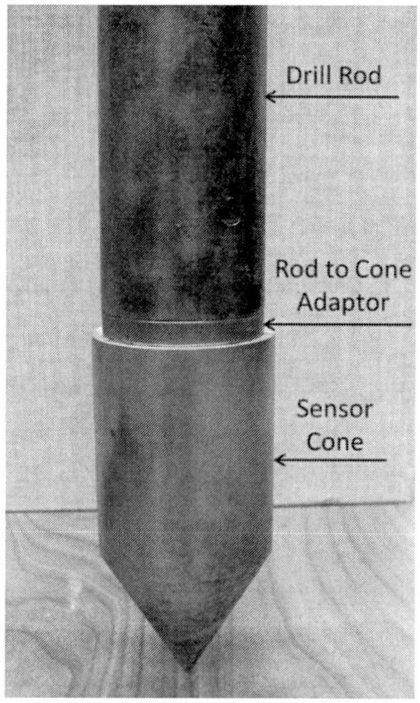

Figure 5 – Sacrificial sensor push cone, and drill rod

Figure 6 – Geophone potted in push cone cavity

Figure 7 - Accelerometer chip being fitted into push cone cavity

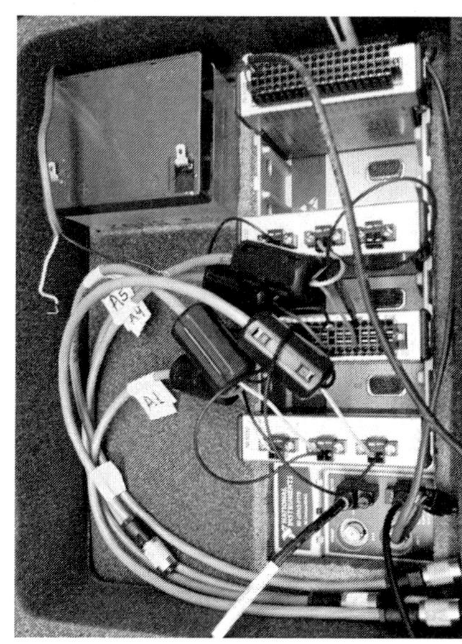

Figure 8 – Data acquisition system

The location of where the pile was to be driven was selected by the contractor and the plan positioning of the intended sensors was laid out on the ground by the U of M group. At each depth sensors were installed starting with the one closest to the pile and working outward from the pile for additional sensors. An MDOT drill rig was positioned over the closest to the pile sensor location and the sensor package pushed into the ground. Sensors were pushed using the hydraulic thrust capabilities of the drill rig (Fig. 9). When all sensors were installed, the pile was spotted over the planned location as shown in Fig. 10 and driven. Plastic bags containing coils of conductor cable on the ground surface in Fig. 10 identify the ground surface projection of embedded sensors.

Sites chosen for this work consisted generally of loose to medium dense sands so little difficulty was expected during installation of the sensors. However, initial attempts at installation of sensor cones resulted in breaking off of the conductor cables due to encountering unknown obstacles. Hydraulic crowd capabilities of the drill rigs and free, unsupported column action of the drill rod also limited how far the sensors could be pushed without bending the rods beyond the elastic range. As an alternate approach, installation of the sensor packages through the center of hollow stem flight augers was attempted. Difficulties were encountered with this approach as well because the sensor packages needed to be pushed below the lower end of the hollow stem augers to get the sensor cone below the zone of disturbance caused by the hollow stem auger's installation. Withdrawal of the center rod of the hollow stem auger and re-installation of the drill rod with the sensor package resulted in loss of either the sensor package or severing of the conductor cable. While each sensor's position in space was planned, in practice, the local ground conditions and installation process controlled the final depths of the sensors. A steep learning curve was experienced that resulted in successful installation of seven triaxial accelerometer sensor cones as shown in Fig. 11. A view of one of the sites prepared for collecting data from buried and surface sensors is shown in Fig. 12.

Figure 9 – All-terrain drill rig installing sensors using hydraulic thrust

Figure 10 – Spotting pile and showing plastic bags containing cables from buried sensors

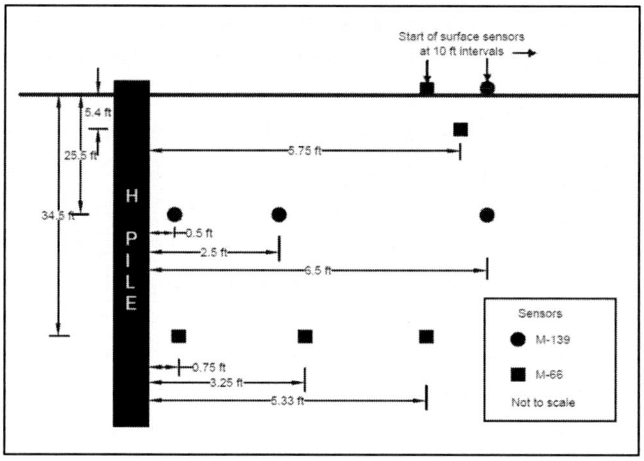

Figure 11- Schematic of embedded sensor cones at depth and with distance at two sites

Figure 12 – Ground surface geophones on line with pile being driven

SAMPLE OF RESULTS

Three H-pile driving sites were instrumented and pile driving vibrations recorded. Measured vibrations from one of the three sites will be presented in preliminary form. A plan view of the M-139 site near Niles, Michigan showing the location of buried sensors and surface geophones is presented in Fig. 13. At this site five triaxial accelerometer packages were installed in the ground near the pile and four surface geophone packages were set along the ground surface to record surface waves. Figure 14a shows a simplified representation of the soil profile with soil descriptions and blow counts of a boring performed after the pile had been driven and within 5 feet of the driven pile. Accelerations of sensors at a depth of 25.5 ft. (intended depth of sensors was 20 ft. but loose sand would not "grab" the sensor cone at that depth) and at three distances from the pile as the pile tip reached each depth are shown in Fig. 14b. At the depths where the pile tip is above the accelerometers, the three sensors show moderately increasing amplitude with pile tip depth to within about 5 feet above the sensors, but after the pile tip reaches the depth of the sensors, the accelerations increase dramatically particularly for the sensor nearest to the pile. It can be seen that there are different behaviors of the three sensors at their common depth as the pile approaches from the top and as the pile departs to greater depths. There is little or no shaft contribution to the ground motion as the pile tip is still above the sensor, but contributions from both shaft and tip combine constructively after the pile tip reached and passed the depth of the sensor.

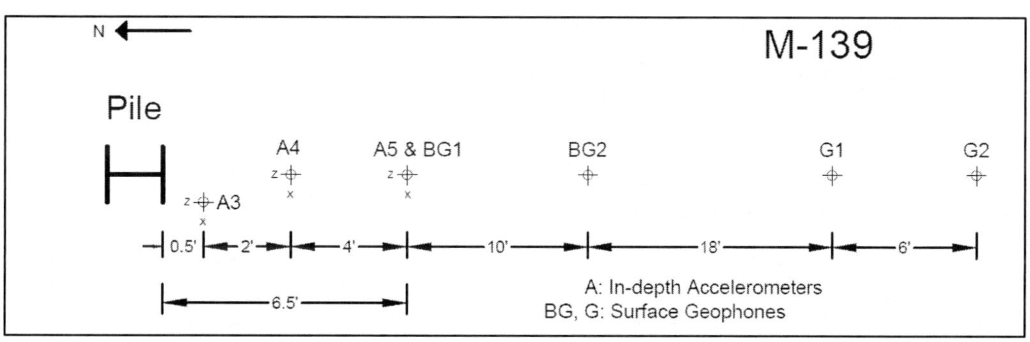

Figure 13 – Plan view of sensor locations at M-139 site, both buried and surface

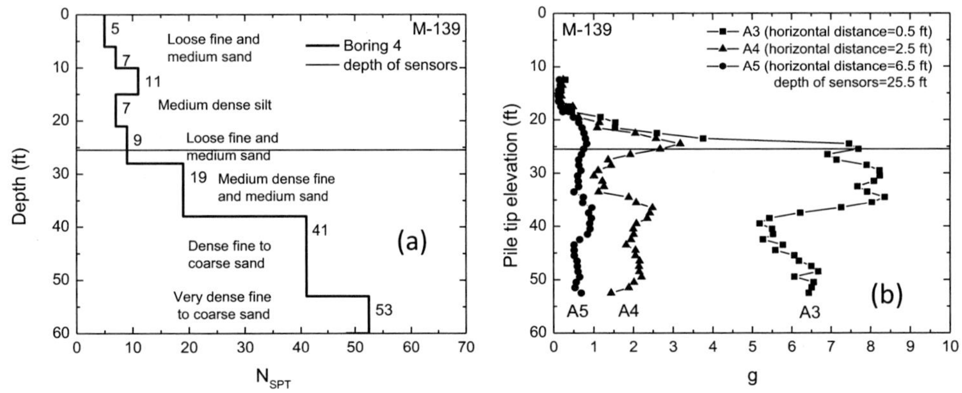

Figure 14 – (a) Boring log near pile after driving and (b) observed vertical accelerations at sensors at three depths as pile tip penetrates the ground

Another way of looking at the ground vibration amplitude as the pile tip penetrates the ground is shown in Figs. 15 a, b & c. In these figures the peak particle velocity is plotted vs. diagonal distance (resultant of vertical and horizontal distances) from the pile tip to the buried sensor. Furthermore, data is plotted in Figs. 15 for two regions of the drive: where the pile tip is above the elevation of the sensors and where the pile tip is below the sensors. To interpret these figures, start at the lower right portion of Fig. 15a, for example, where the pile tip is furthest above the sensors and proceed upward to the left as the pile tip approaches the depth of the sensor. The amplitude increases smoothly to the level where the pile tip is at the same elevation as the sensors. Continue to follow the amplitude of vibration curve as the pile tip descends below the sensors and the vibration amplitude first makes a dip then increases and remains at the higher level throughout the remainder of the driving.

Figures 16 a & b show attenuation curves for peak particle velocity in both arithmetic and logarithmic scales respectively developed from records of the three sensors at a common depth (25.5 ft.) and vibration amplitude at the pile interface (called 0.1ft for convenience on log scale) estimated from EQ 1. These curves show the high rate of attenuation very near the pile and lesser rates of attenuation with distance from the pile. Figures 16 c & d show similar curves for the surface geophones. As more data of this type gathered, coefficients of attenuation and power function equations will be developed for various soils encountered.

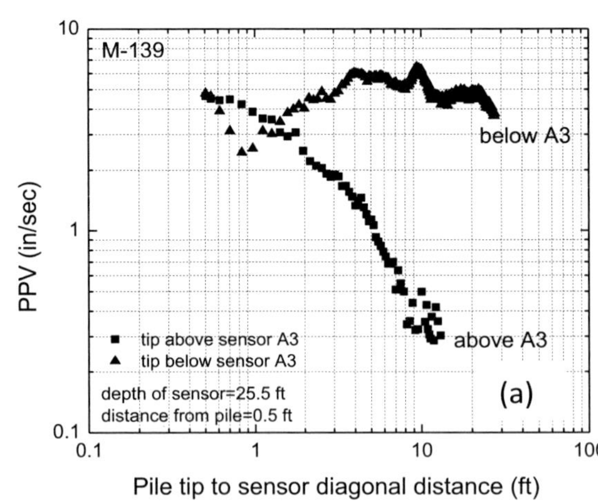

Figure 15a – Pile tip approaching from above and departing below depth of sensor A3 for vertical motion

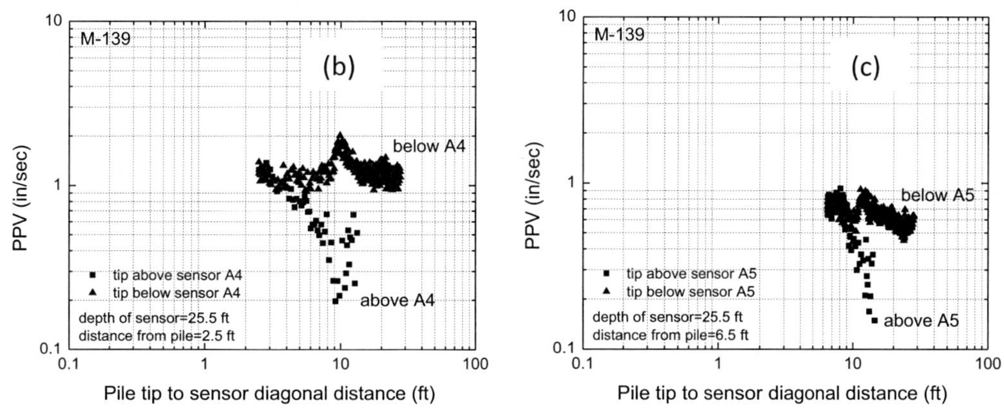

Figure 15 b & c – Pile tip approaching from above and departing below depth of sensor (b) A4 and (c) A5 all for vertical motion

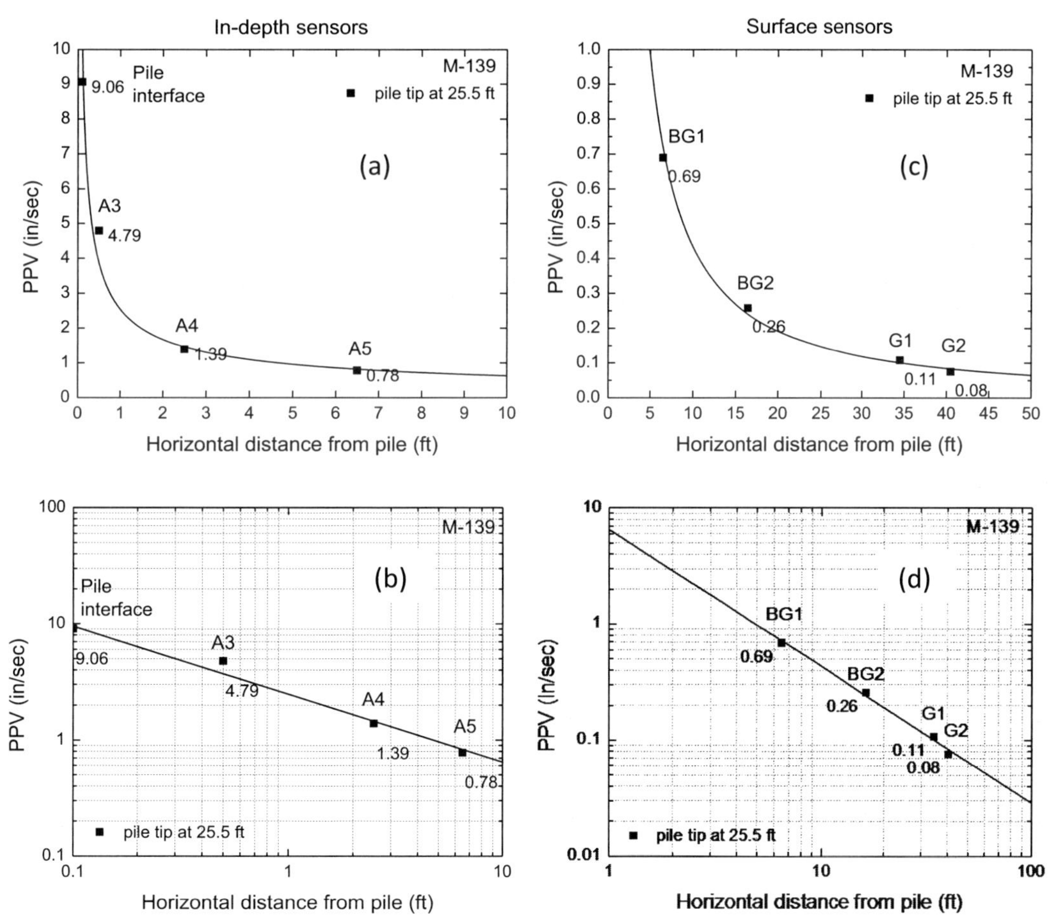

Figure 16 – Attenuation curves for vertical vibrations in the ground and on the surface

CONCLUSIONS

To the writers knowledge this work represents the first example of installation of motion sensors at varying depths and very near to driven piles for the expressed purpose of measuring vibrations as the piles were being driven. Ground motion sensor packages that could be pushed into the ground and sacrificed after their use were designed, fabricated and installed in the ground very near to driven piles. Piles were driven at three, loose sand sites after installation of sensor cones at three depths and three radial distances from the pile. Measured vibrations show that sensors at different distances from the pile at the same elevation, record different amplitudes of vibration as the pile tip approaches and departs from the elevation of the sensors. The data suggests to the writers that when the pile tip is far above the sensor, pile tip generated vibrations mainly affect the sensors but as the tip gets closer to the sensor level, contribution from the pile shaft also reaches the sensor. After the pile tip passes the depth of the sensor, both pile tip and shaft generated vibrations combine constructively for larger vibrations. The vibration levels after the tip passes the elevation of the sensors stays relative constant suggesting that the pile tip which is getting further from the sensor contributes less and less to the vibration than the shaft. This response is not contradictory to the hypothesis at the beginning of this paper as suggested in Figs.1 and 2. It is further anticipated that attenuation data will help define boundaries of vibration decay with radial distance from the driven pile. Additional sites will be studied and more exhaustive analyses made and reported in future publications.

ACKNOWLEDGEMENTS

This work was performed under contract 2010-0296 with the MDOT, project #114128, dates 10/1/2011-03/31/13

REFERENCES

Clough, G.W. and Chameau, J-L. (1980). "Measured Effects of Vibratory Sheetpile Driving." *Journal of the Geotechnical Engineering Division, ASCE*, 106(GT 10), 1081-1099.

Dowding, C.H. (1991) "Permanent Displacement and Pile Driving Vibrations." *Proceedings, 16th Annual Member Conference of the Deep Foundations Institute,* Chicago, Oct. 7-9, 67-84.

Lacy, H.S. and Gould, J.P. (1985). "Settlement From Pile Driving in Sands." *Vibration Problems in Geotechnical Engineering,"* Proceedings of a Symposium edited by Gazetas and Selig, sponsored by the Geotechnical Engineering Division in conjunction with the ASCE Convention, Detroit Michigan, October 22, 152-173.

Massarsch, K.R. (2002). "Effects of Vibratory Compaction." *TransVib 2002 – International Conference on Vibratory Driving and Deep Soil Compaction*, Louvain-la-Neuve, 33-42.

Massarsch, K.R. and Fellinius, B.H. (2008)." Ground Vibrations Induced by Impact Driving." *Proccedings of 6th International Conference on Case Histories in Geotechnical Engineering,* Edited by S. Prakash, University of Missouri-Rolla, Rolla, Mo, August 12-16, Arlington, VA, 38.

O'Neill, D.B. (1971). "Vibration and Dynamic Settlement from Pile Driving." *Proceedings of the Conference on Behavior of Piles*, Institute of Civil Engineers, London, 135-140.

Picornell, M. and del Monte, E. (1985). "Pile Driving Induced Settlements of a Pier Foundation. " *Vibration Problems in Geotechnical Engineering,* Proceedings of a Symposium edited by Gazetas and Selig sponsored by the Geotechnical Engineering Division, in conjunction with the ASCE Detroit, Michigan, Convention, October 22, 174-186.

Schwab, J.P. and Bhatia, S.K. (1985). "Pile Driving Influence in Surrounding Soil and Structures." *Civil Engineering for Practicing and Design Engineers*, Pergamon Press Ltd., 4(8), 641-684.

Woods, R.D. (1997). *Dynamic Effects of Pile Installations on Adjacent Structures, Synthesis of Highway Practice NCHRP#253,* Transportation Research Board, National Research Council, National Academy Press, 86 p.

Effective Mitigation of Underwater Piling Noise

Karl-Heinz Elmer

Dr.-Ing., OffNoise-Solutions GmbH, D-31535 Neustadt, Leinstrasse 36, Germany;
kh.elmer@offnoise-solutions.com

ABSTRACT: During the installation of offshore wind farm foundations and deep water foundations, the underwater piling noise of hydraulic impact hammers induces considerable hydro sound that is radiated from the piles into the surrounding water. These very high underwater noise emissions can reach dangerous levels and are potentially harmful to marine life in a large area.

Both knowledge about the complex mechanisms and dynamic interactions of the hammer and pile system and knowledge about the acoustic behaviour of the piles, the shallow water and the ground are necessary to understand the generation and radiation of underwater piling noise and to develop effective noise mitigation systems.

The new mitigation method of hydro sound dampers (HSD) uses curtains of robust air filled elastic bladders showing high resonant effects, similar to air bubbles, and special dissipative PE-foam elements to reduce impact noise. These HSD- elements are fixed to pile surrounding fishing nets.

HSD-systems are independent of compressed air and not influenced by currents. The theoretical background, numerical simulations, laboratory tests and offshore tests with noise mitigations between 17 dB to 35 dB (SEL) are described.

INTRODUCTION

Underwater noise from impact piling results in considerable sound emissions and is distinguishable above ambient noise over distances of several tens of kilometers from the source as given in Elmer et al. (2007). The construction noise of offshore wind turbines is potentially harmful to marine life, in particular to marine mammals as reported by Southall et al. (2007) and Lucke et al. (2009). Different zones of underwater noise immissions can be defined in the region surrounding the source of acoustic noise. The ranges of zones strongly depend on the kind of hydraulic hammers and possible noise mitigation methods.

The zone of audibility also depends on the kind of marine animals. Within several kilometers with moderate exposure levels, marine animals like harbour porpoises, harbour seals, grey seals and also fish will show some kind of reaction or change in their behaviour. At higher exposure levels, within the closer zone of several hundreds of meters from the source, underwater noise can induce temporary (TTS) or permanent (PTS) threshold shift or hearing loss. Important acoustic information might be masked by reductions in hearing sensitivity of an animal.

Close to the source of a very loud noise source like pile driving, extreme intensity levels of underwater noise can cause physical trauma or death. Even the lowest level of damage, which is a temporary threshold shift (TTS), is dangerous and must be avoided. Different injury ranges for different mammals are described by Thomsen et al. (2006), Southall et al. (2007) and Lucke et al. (2009).

Due to large piles that will require higher driving energies, high underwater noise levels are expected in future offshore projects. This is also accompanied by an increasing number of projects such as erected offshore wind turbines.

Effective noise reducing methods that reduce sound levels below recommended acoustic emission thresholds so they are no longer harmful and disturbing to marine mammals and other protected animals are in great demand.

UNDERWATER PILING NOISE

Numerical simulations and measurements of underwater noise during the construction of offshore foundations such as monopiles, tripods, tripiles and jackets like the first offshore research platform in the North Sea FINO1 after Figure 1 are used to study the generation, radiation and attenuation of underwater noise as published in Elmer et al. (2007) and Elmer (2010).

Hammer:	100 % impact energy
Pile:	60-90% impact energy
Water:	~ 1% impact energy
Sea ground:	rest of impact energy

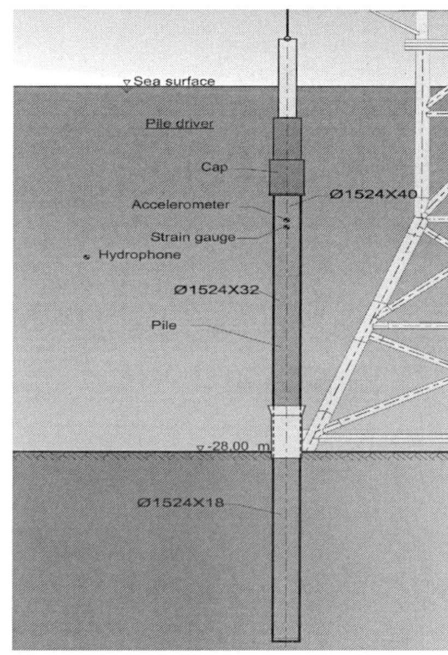

Figure 1. FINO1 foundation system in the North Sea and impact energy partitions after Elmer et. al. (2007).

Most of the piling energy is driven into the sea ground. As noted in Figure 1, only about 1% of the whole ram energy is radiated directly from the wet surface of the pile into the surrounding water inducing very high underwater sound levels after Elmer (2010). Depending on the properties of the ground material and depending on the propagation of vibrational waves in the ground a part of the impact energy is radiated indirectly from the sea ground into the water, resulting in additional underwater sound. These are especially the lower frequency vibrations that are travelling through the sea ground.

The short impact pulse of the hammer induces an impact wave in the pile with small radial displacements of the pile surface. Numerical simulations and measurements of the impact's wave in the pile after Figure 1 and Figure 2 show that the resulting travelling wave is reflected up to several times at stepped cross sections of the pile and at both ends of the pile until all the kinetic energy is damped out and radiated into the ground. These travelling waves induce sound waves in the surrounding water with an impact pressure of about 10-30 bar. This radiated underwater noise propagates at the speed of sound in water of about 1500m/s into all directions. The noise is also reflected at the free water surface and at the sea ground after Elmer et. al. (2007).

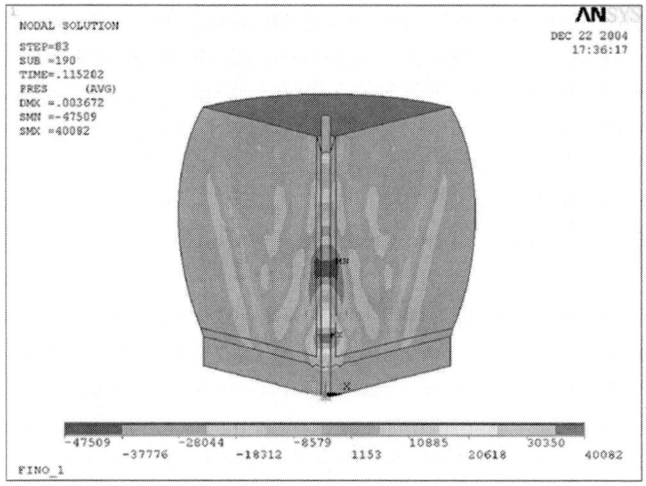

Figure 2. Travelling wave within the pile inducing under water sound waves and reflections of the sound waves at the free surface of the water and at the sea ground after Elmer et. al. (2007).

The underwater piling noise is usually described by two sound levels. The first level is the peak Sound Pressure Level (peak SPL) in decibels (dB) of the maximum instantaneous positive or negative sound pressure $|p_{peak}|$ of the measured impact noise relative to the reference underwater sound pressure of $p_0 = 1\mu Pa$.

$$peak\ SPL = 20\log\left(\frac{|p_{peak}|}{p_0}\right) \quad \text{in dB re:} 1\mu\text{Pa}. \tag{1}$$

The second quantity for describing pile driving underwater noise is the Sound Exposure Level SEL in decibels (e.g. dB re:$1\mu Pa^2 s$), which is an equivalent energy level of the noise of a single sound event between T_1 and T_2 like a single pile driving impulse. The impulse energy between T_1 and T_2 is spread over the period of $T_0 = 1s$ duration.

$$SEL = 10\log\left(\frac{1}{T_0}\int_{T1}^{T2}\frac{p(t)^2}{p_0^2}dt\right) \quad \text{in dB re:} 1\mu\text{Pa}^2\text{s}. \tag{2}$$

The SEL is the level of a theoretical continuous sound of 1s duration which has the same sound energy as the pile driving impulse, independent of the real impulse duration of $T_2 - T_1$.

Measurements of the underwater piling noise show peak levels of more than 210 dB (SPL) re $1\mu Pa$ and sound exposure levels of more than 180 dB (SEL) re $1\mu Pa^2 s$ at a distance of 750 m from pile driving sites, depending on ram energy and pile size.

UNDERWATER NOISE MITIGATION

Several underwater noise mitigation methods have been developed in the recent years. Nearly all of them use the underwater noise mitigation effects of air bubble curtains. Small air bubbles in the water are released from perforated tubes with compressed air at the bottom of the seabed. Surrounding the piles these air bubble curtains can reduce the radiated noise using the effects of acoustic impedance mismatching of the bulk acoustic impedance change and of frequency dependent resonance scattering effects of the bubbles.

Tests were first done by Würsig et. al. (2000) and extended tests were done by Reyff (2003) during the San Francisco Oakland Bay Bridge Demonstration Projects and by Reyff (2007) with very different noise mitigation results. Best results are achieved with pile sleeves or steel tubes as noise barrier methods and additional constrained bubble curtains or with coffer dams. But these experiments have been implemented at great expense on a limited number of test piles.

In Germany, the first offshore applications of air bubble curtains achieved noise reductions of up to 10-13dB during pile driving operations of the offshore platforms FINO3 and "alpha-ventus" in the North Sea as described by Betke (2008) and Betke & Matuschek (2010). The problems are sound leakage through the bubble curtain resulting from bubble drift with tidal currents. Therefore, the sizes of diameters of unconstrained bubble curtains around offshore piling sites are between 150m and 250m.

Spectral information of sound measurements from individual impact events are given by third-octave spectra of the sound exposure levels SEL of three different hammers in Figure 3. The highest spectral levels of the measured underwater ram noise of the hammers are shown in the low frequency range from 100 to 300 Hz, responsible for the high broadband level of piling noise.

Figure 4 demonstrates that the air bubble curtain attenuation of high frequency noise above 1 kHz is very high. But the broadband sound level of the piling noise mainly depends on the lower frequency noise, far below 1 kHz, where the attenuation from bubble curtains is poor.

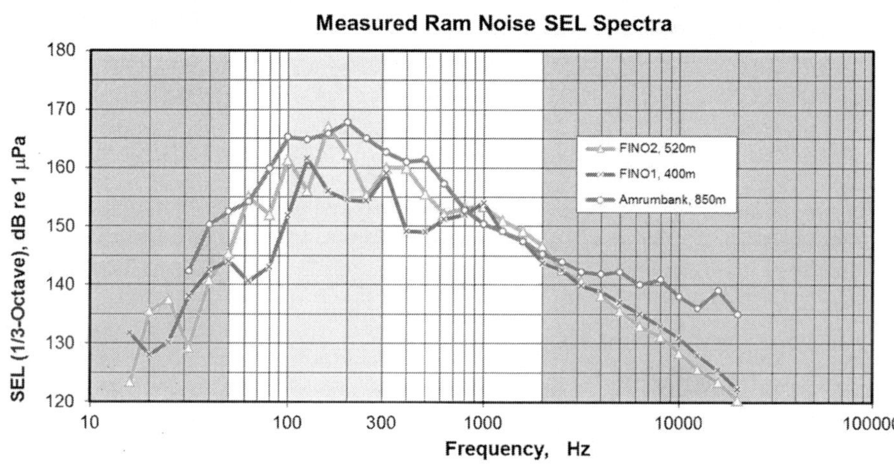

Figure 3. UW-noise spectra of FINO1, FINO2 and Amrumbank installations, Betke (2008).

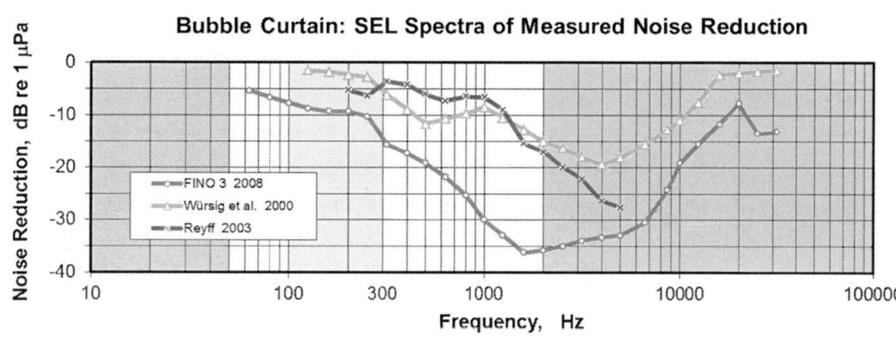

Figure 4. Noise reduction of bubble curtains; Wursig(2000), Reyff (2003) & Betke (2008).

The reasons for this are that large air bubbles (several cm) with low resonant frequencies are uncontrolled. They exhibit chaotic movements and divide as they slowly arise to the surface of the water. This implies that only constant shapes of air bubbles and nearly steady state resonant excitations are able to achieve high sound attenuations of resonant air bubbles. As a consequence there is no benefit from high theoretical underwater sound attenuation potential of large air bubbles with low resonant frequencies, when using conventional air bubble curtains. That is why the attenuation of high frequency noise is very high above 1 kHz, but only poor in the most important frequency range between 100Hz - 300Hz after Fig. 4.

Offshore applications of air bubble curtains are very expensive at great water depth and currents. The main problems are the supply with compressed air, the control of the bubble size or the control of bubble distribution, and the influence of the tide currents of the sea, together with slow ascent rates of the bubbles. The noise reducing potential of air bubble curtains is not sufficient to reduce the noise of modern large hydraulic hammers of the last years within the even lower most important frequency range between about 50Hz - 200Hz.

The German Federal Maritime and Hydrographic Agency (BSH) has set the standard sound exposure level of 160 dB (SEL) and the peak level of 190 dB (SPL) at 750m distance from offshore pile driving sites as part of the building permission of offshore wind farms.

Effective noise reducing methods are necessary to achieve these standard levels.

HYDRO SOUND DAMPERS (HSD)

To overcome these problems, a new underwater noise reducing method has been developed, using gas filled envelope bodies and PE-foam elements as hydro sound dampers, instead of free natural air bubbles as described by Elmer (2010).

The size of the bodies, the effective frequency range, the damping rate, the number and distribution of the hydro sound dampers (HSD) and the influence from hydrostatic pressure can be fully controlled, if the envelope bodies are fixed to a pile surrounding fishing net or to stiff frames.

Figure 5 shows HSD offshore applications as staggered HSD-grids or large fishing nets with HSD-elements around a pile. Systems of hydro sound dampers can also be fixed to the hammer, a piling frame or a gripper after Fig. 5. Covering the whole sea in the near-field of a pile to reduce the indirect noise, transmitted from the ground into the water is another HSD offshore application.

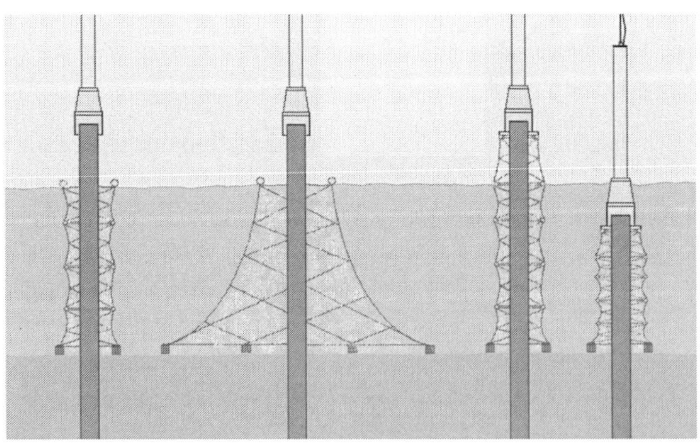

Figure 5. Staggered HSD-nets and large fishing nets with HSD-elements, Elmer (2010).

The HSD-system of Figure 6 is a donut-like container enclosing the HSD-net. It can be fixed below a piling frame, or below the hydraulic ram, or it may float around the pile as shown in Figure 6. The efficacy of HSD in reducing underwater noise depends on the frequency and the special density of the hydro sound dampers. Volume rates of about 1-2% of the HSD are sufficient to obtain good results. At these special densities vertical forces from buoyancy and horizontal forces from tide currents are still small.

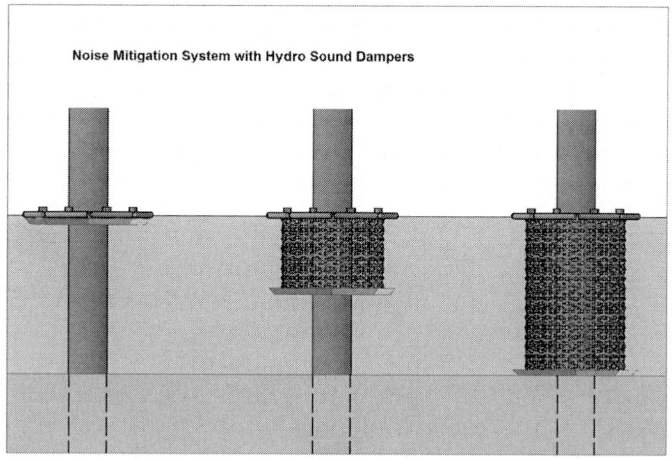

Figure 6. HSD-net enclosed in a donut-like container after Elmer et.el. (2011).

In contrast to the free air bubbles of conventional air bubble curtains, hydro sound dampers benefit from three different physical reasons for effective underwater noise attenuation:

- Resonant effects of small air filled balloons and robust PE-foam elements in water can reduce underwater sound up to 35 dB or more like small air bubbles. The resonance frequency of these HSD-elements is adjustable, even to very low frequency ranges below 100Hz, in contrast to free air bubbles. The resonance frequency is inversely proportional to the diameter of the elements. Other factors affecting the resonance frequency include: the gas pressure inside, the water depth and the stiffness of the envelope material.
- Dissipation and material damping effects according to the material damping potential of the envelope material and the filling material inside the HSD-elements. Maximum damping is obtained near the resonance frequency of a damped element, achieving noise reductions between 10 and 30 dB (SEL).
- Reflections of sound waves at impedance steps, as HSD-elements, like air bubbles in water, increase the compressibility of the mixed water-body, decrease the bulk modulus of the mixture and decrease the sound speed and the specific impedance of the mixture. These effects result in noise reductions between 5dB and 15dB (SEL).

The important resonant effect with high scattering, multiple reflections and effective absorption of sound waves in the water can be seen in Figure 7. The very strong interaction of a vibrating HSD-element and the surrounding water can be seen at the water surface in Figure 7. This interaction also takes place under water as shown in Figure 8, but it is not visible there.

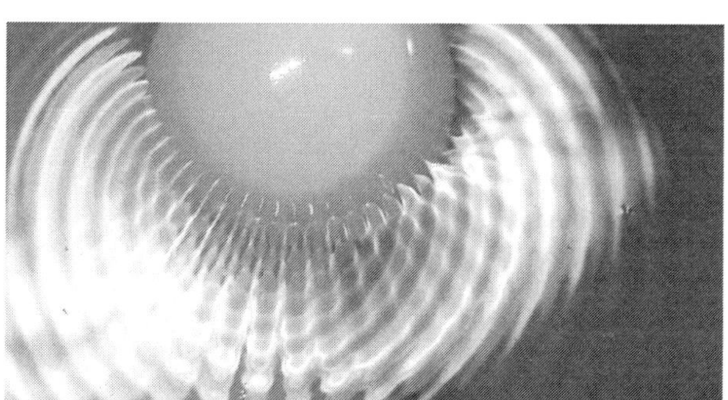

Figure 7. Scattering, radiation and strong interaction of a vibrating HSD, Elmer et.al. (2011).

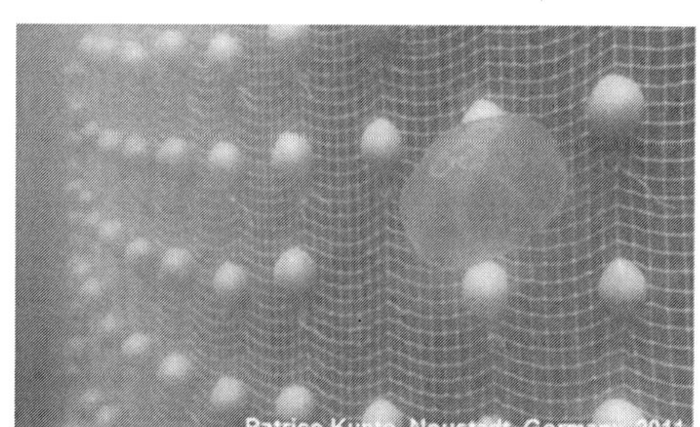

Figure 8. HSD-net and elements under water.

Hydro sound dampers are used in the whole frequency range of pile driving noise from 50 - 5000 Hz. It is possible to control the damping rate, the size, the number and the distribution of the HSD around the pile. Finally, HSD-systems don't need compressed air supply.

MEASURED OFFSHORE NOISE REDUCTIONS OF HSD-SYSTEMS

Experimental tests and measured results of offshore tests in the Baltic Sea confirm the high underwater sound attenuation potential of hydro sound dampers. For a first offshore test the designed prototype of the HSD-System was a floating installation with winches to lower 3 different types of HSD-nets to seabed after Bruns (2012).

Figure 9. HSD-system at ESRa test (2011) after Bruns et.al. (2012).

The total weight of the system is about 10t. Gas filled thin balloons together with robust PE-foam elements of high material damping rate are used as HSD-elements and attached to the nets. All elements are tuned to the resonant frequency of 120 Hz to effectively achieve reduced noise transmissions in a very effective way within the most important lower frequency range. The net layout with blue colored HSD-elements is shown in Figure 10. The radiated underwater sound pressure was measured at 4 m above the ground at a distance of about 6 m from the pile to get most of the directly radiated sound and to avoid influences of reflections from water surface and sea ground. Figure 11 shows the 1/3 octave spectrum of the SEL of the original piling noise and the reduced sound exposure level after applying HSD noise mitigation of both gas filled balloons and PE-foam elements.

There is a very broad noise reduction up to 23 dB (SEL) within the most important range between 100 Hz and about 600 Hz after Bruns et.al. (2012). Higher frequencies were not tested.

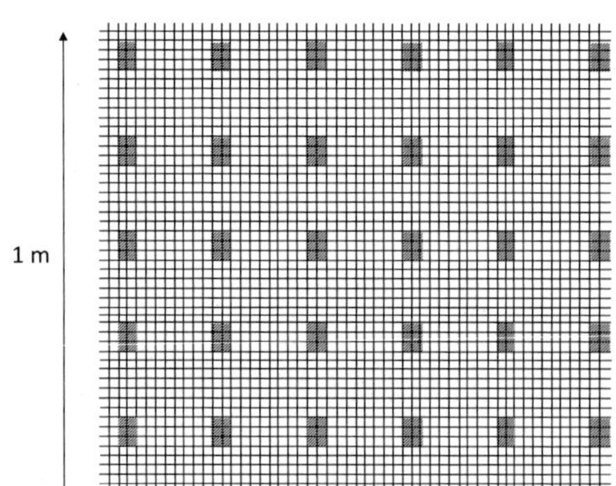

Figure 10. HSD net-layout at ESRa-test in the Baltic Sea in 2011.

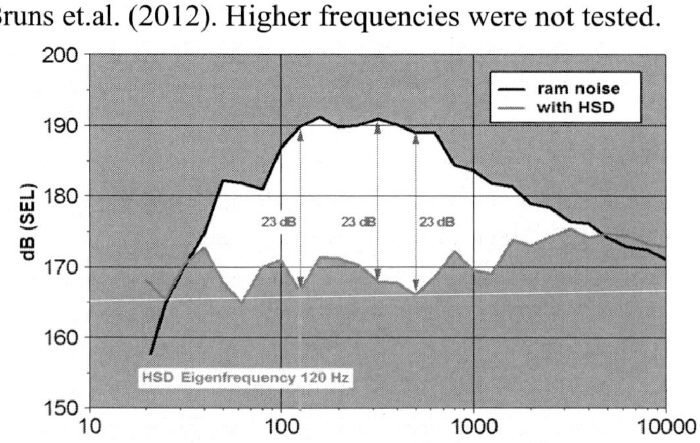

Figure 11. Third octave SEL spectra of underwater piling noise with and without HSD noise mitigation.

Another offshore test was done at the London Array (LA) wind farm in August of 2012 in the North Sea near the coast of the United Kingdom. The designed HSD was a self-expanding system with a total weight of 17t, and a size of 9 m in outer diameter after Figure 12. It consists of three parts: the buoyancy ring (floating on the water surface), the HSD net and the ballast box. The compressed HSD height was 1,8m and was applicable in variable water depth of up to 28m as described by Bruns et. al. (2012).

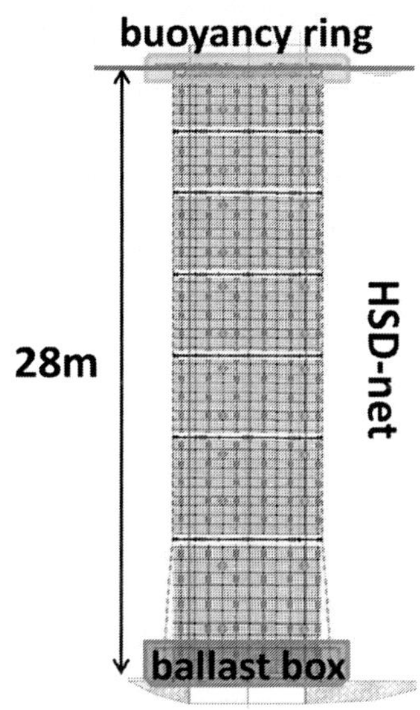

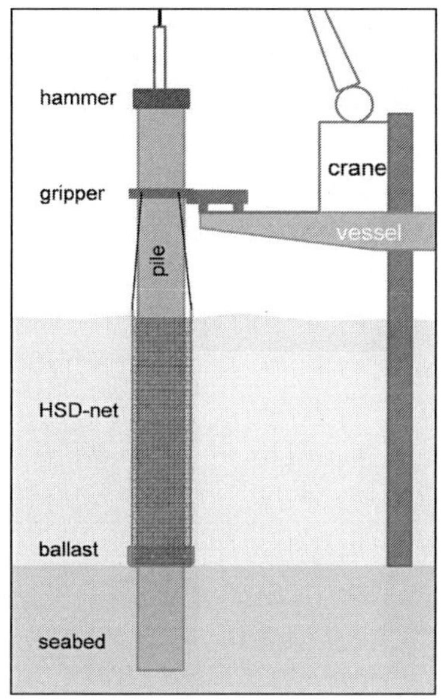

Figure12. HSD-net. Figure 13. Net below the gripper. Figure 14. London Array test.

The additional time required to install the HSD-system can be easily reduced to zero by installing the HSD-system below the gripper after Figure 13 and 14 as reported in Bruns (2012).

The net layout of LA in Figure 15 shows the same compilation of HSD elements as used at ESRa. In addition, smaller and larger elements were used to achieve improved noise reduction in the frequency range below 100 Hz and higher than 1000 Hz.

The underwater sound mitigation was measured at 1 m above the seabed at a distance of 15 m. Figure 16 shows the 1/3 octave spectrum of the SEL from the original piling noise and reduced sound level with HSD in use. The two spectra are from different piles.

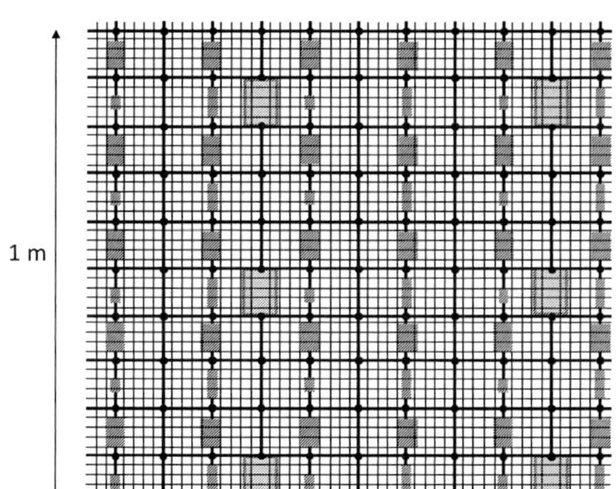

Figure 15. HSD net-layout at London Array test in the North Sea in 2012.

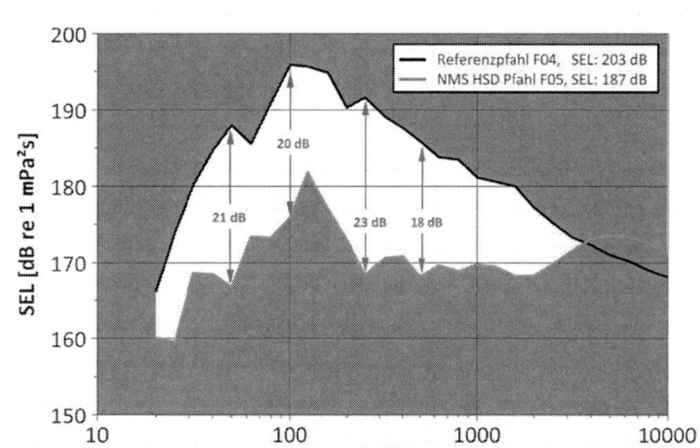

Figure 16. Third octave SEL spectra of underwater piling noise with and without HSD noise mitigation.

The impact of the additional applied HSD-elements causes, in comparison to the ESRa test results, increased reductions up to 21dB between 20Hz and 100 Hz and above 1000 Hz. Again a very broad noise reduction of up to 23 dB (SEL) was achieved using the HSD-noise mitigation.

CONCLUSIONS

During the construction of offshore wind turbines the use of hydraulic impact hammers induces considerable underwater sound emissions that are potentially harmful to marine life, in particular to marine mammals, in a large area. Due to larger piles requiring higher driving energies, even higher underwater noise levels are expected in future offshore projects such as an increasing number of erected offshore wind turbines. Effective noise mitigation methods are in great demand for the purpoise of reducing sound levels to below recommended acoustic emission thresholds that are no longer harmful and disturbing to marine mammals and other protected animals.

Air bubble curtains as first used in noise mitigation systems are very expensive. They are very effective above 1 kHz, but they only achieve poor sound attenuation in the decisive lower frequency range of impact hammers. Bubble curtains are also strongly influenced by tidal currents. Improved air bubble curtains with increased noise mitigation require heavily weighted additional pile sleeves or surrounding steel tubes as noise barriers.

The new underwater noise mitigation method of hydro sound dampers (HSD) overcome these problems. HSD-systems are very effective barriers to underwater piling noise between about 50Hz to 5000Hz. The HSD-elements can be tuned and optimized even to the very low frequency ranges of modern heavy weighted hydraulic hammers. Offshore noise reductions of up to 23 dB in 1/3 octave bands have been measured. This implies that 99.5% of the sound energy was damped out even though the HSD-net is only covered by HSD elements on less than 10% of its surface.

HSD-systems are light-weighted and small (the size is about 2 x 9 x 9 m). HSD-systems are independent of a compressed air supply, they are not expensive, and they are not influenced by the tide currents since water can flow through the open HSD-nets. Only the underwater noise is stopped. HSD-systems can be applied in different offshore configurations: floating around the pile, fixed to a piling frames or a gripper, or hanging below the hammer or fixed to pre-piling templates of jacket foundations.

HSD-systems are already patented in Germany, international patents are pending.

ACKNOWLEDGMENTS

All offshore test results and concepts of "London Array" wind farm were carried out within the research project "Hydro Sound Dampers" FKZ 0325365 of the Institute for Soil Mechanics and Foundation Engineering of the Technische Universität Braunschweig, Germany (IGB-TUBS) together with Arsleff Bilfinger Berger Joint Venture (ABJV) and OffNoise-Solutions GmbH. This research project is supported by the German Federal Ministry for the Environment, Nature Conservation and Nuclear Safety (BMU).

REFERENCES

Betke, K. (2008) "Mitigation of uw-noise during construction of OWF", (in German), *BSH-workshop FINO3*, Hamburg, Germany, October 8, 2008.

Betke, K. & Matuschek, R. (2010) „Messungen von Unterwasserschall beim Bau derWind-energieanlagen im Offshore-Testfeld "alpha ventus" - Abschlussbericht zum Monitoring nach StUK 3 in der Bauphase", *ITAP Report*, 2010.

Bruns, B., Kuhn, C., Stein, P., Gattermann, J., Elmer, K.-H. & Imhorst, F. (2012) "Underwater noise reduction from measurements and offshore tests using systems of hydro sound dampers (HSD)", *Proceed. DEWEK2012 Conference*, Bremen, Germany, Nov. 7/8, 2012.

Elmer, K.-H., Betke, K., & Neumann, T. (2007): "Standard Procedures for the Determination and Assessm. of Noise Impact on Sea Life by OWF", (in German), *BMU final report 0329947, 2007.*http://www.naturschutzstandards-erneuerbarer-energien.de/images/literatur/2007_Elmer_Schall 2_Offshore-Wind.pdf

Elmer, K.-H. (2010) "Pile driving noise reduction using new hydro sound dampers", *BSH-workshop on pile driving, ESC2010,* Stralsund, March 21, 2010. http://www.igb.tu-bs.de/veroeff/ELMER_Hydro_Sound_Dampers_ecs2010.pdf

Elmer, K.-H., Gattermann,J., Bruns, B., Kuhn, C. & Stahlmann, J. (2011) "Mitigation of underwater piling noise using new hydro sound dampers (HSD)", *Proceed. of the 8th FMGM (Field Measurement in GeoMechanics) International Symposium,* Berlin, Sept. 12-16, 2011.

Lucke, K., Siebert, U., Lepper, P. A. & Blanchet, M. A. (2009) "Temporary shift in masked hearing thresholds in a harbor porpoise (Phocoena phocoena) after exposure to seismic airgun stimuli*", J. of the Acoust. Society of America*, 125 (6), 4060-4070.

Reyff, J. A. (2003) "Underwater Sound Pressure Associated with the Restrike of the Pile Installation Demonstration Project Piles", *Final Report, State of California, Department of Transportation*, District 4, 2003.

Reyff, J. A. (2007) "Compendium of pile driving sound data", Report prepared by Illinworth & Rodkin for The California Department of Transportation, Sept. 2007.

Southall, B. L., Bowles, A. E., Ellison, W. T., Finneran, J. J., Gentry, R. L., Greene Jr., C. R., Kastak, D., Ketten, D. R., Miller, J. H., Nachtigall, P. E., Richardson, W. J., Thomas, J. A. & Tyack, P. L. (2007) "Marine Mammal Noise Exposure Criteria: Initial Scientific Recommendations"*, Aquatic Mammals,* 33 (4), 411-509.

Thomsen, F., Lüdemann, K., Kafemann, R. & Piper, W. (2006): "Effects of offshore wind farm noise on marine mammals and fish". BIOLA, Hamburg, Germ., on behalf of COWRIE Ltd.

Würsig, B., Greene, C. R. & Jefferson, T. A. (2000) "Development of an air bubble curtain to reduce underwater noise of percussive piling", *Marine Environm. Research*, 49 (1), 79-93.

Full Scale Field Study of Jack-in Piles

Siew-Ann Tan [1], Jie Sun[2], and K.S. Ng[3]

[1]Associate Professor, Dept. of Civil Engineering, National Univ. of Singapore, 1 Engineering Dr. 2, Singapore 117576; cvetansa@ nus.edu.sg
[2]Ph.D candidate, Dept. of Civil Engineering, National Univ. of Singapore, 1 Engineering Dr. 2, Singapore 117576; sunjie@ nus.edu.sg
[3]Lecturer, Faculty of Civil Engineering, Universiti Teknologi MARA, 40450 Shah Alam, Selangor, Malaysia. E-mail: kokshien@yahoo.com

ABSTRACT: The prediction of pile bearing capacity to include installation effects is a very complex problem, especially for jack-in piles. This paper describes a series of full scale pile load tests that shed light on the effects of jack-in force on the ultimate bearing capacity of jack-in piles. Three spun pile were installed in similar soil condition under different jack-in forces. The extensive in-situ investigations, including standard penetration test (SPT), pressuremeter test (PMT) and piezocone penetration test (CPTU), were conducted in the experimental site to get better understanding of the soil condition around the test piles. The results of the tests show that the different Jack-in forces did not affect the shaft friction significantly and that the difference in behaviors between test piles is caused by the difference in the toe stiffness response. The data show that the larger the jack-in force, the larger the stiffening effect, which is due mainly to the increase in volumetric compression of the bulb of soil below the toe of the piles.

INTRODUCTION

Pile foundations are one of the oldest topics in geotechnical engineering and rational design methods based on soil mechanics principles were established over 50 years ago. A great volume of field experience and empirical data on the performance of pile foundations have been published and an enormous amount of research has been carried out. However, the prediction of pile bearing capacity inclusive of installation effects is a very complex problem which is partially based on theoretical concepts derived from the sciences of soil mechanics, but is mainly based on empirical methods obtained from field experience until the present time and is arguably the area of the greatest uncertainly in pile foundation design (Randolph *et al.,* 1994). In addition, the effects of various methods of pile installation on the bearing capacity and deformation characteristics cannot be calculated by strict application of soil or rock mechanics theory (Tomlinson and Woodward, 2008), especially for displacement type piles.

In order to develop reliable and broadly applicable design method, clear existing experimental evidence should be evaluated first. It is suggested that carefully designed field tests with highly sensitive instrumented pile may provide the key to understanding the mechanisms that govern pile behavior and established valid design criteria. Several field tests have been conducted in the past that do provide some better insights into pile load transfer mechanisms. (Bond & Jardine 1991; Lehane *et al*,1993; Bond *et al* 1991; Lehane *et al*, 1994a and Lehane *et al* 1994b).

Static loading tests were performed on three 600mm diameter Jacked-in spun piles, constructed to depths of 28.7m (TP1), 29.9m (TP2) and 31.7m (TP3) at the same site in Singapore with different Jack-in forces. The primary objective was to investigate the effects of Jack-in force magnitude on the ultimate bearing capacity of Jack-in pile. The soil conditions at the testing site are first discussed based on in situ piezocone penetration tests (CPTU). Next the experimental set-up and testing procedures are described. Then, the results and analysis of the pile installation and static load test results are presented.

SOIL CONDITION

Site

The experimental site is located in Tuas South Ave 2, at the western end of Singapore. The stratigraphy comprises a 10m-12m of loose to medium dense reclaimed sand fill overlying approximately 4m soft marine clay; this deposit is underlain by stiff to very stiff sandy clay and completely weathered sandstone/mudstone (Jurong Formation). The land was reclaimed more than 20 years ago, and it is expected that the site had reached normal consolidation to slight over consolidation state. The water level is at an average depth of about 3m below the ground surface.

In-Situ Tests

The extensive in-situ investigations, including standard penetration test (SPT), pressuremeter test (PMT) and piezocone penetration test (CPTU), were conducted in the experimental site. This gave valuable data for better understanding of the soil condition around the test piles. The site layout map Figure 1, shows the location of borings, SPT, PMT and CPTU soundings relative to the test pile.

Total of three borings BH-1, BH-2 and BH-3, about 15m apart, were drilled for the purpose of collecting soil samples and conducting SPTs and PMTs at specific depths. Figure 1 presents the SPT results of three boreholes. The subject test piles, Piles TP1, TP2 and TP3, were constructed at the positions of BH-1, BH-2 and BH-3, respectively. As can be seen, the soil conditions for all test piles are very similar and do not vary significantly. Three PMTs were performed at similar depths below the test piles. Another three PMTs were performed at the depth where the soft Marine clay is. The results of total six PMTs are summarized in Table 1.

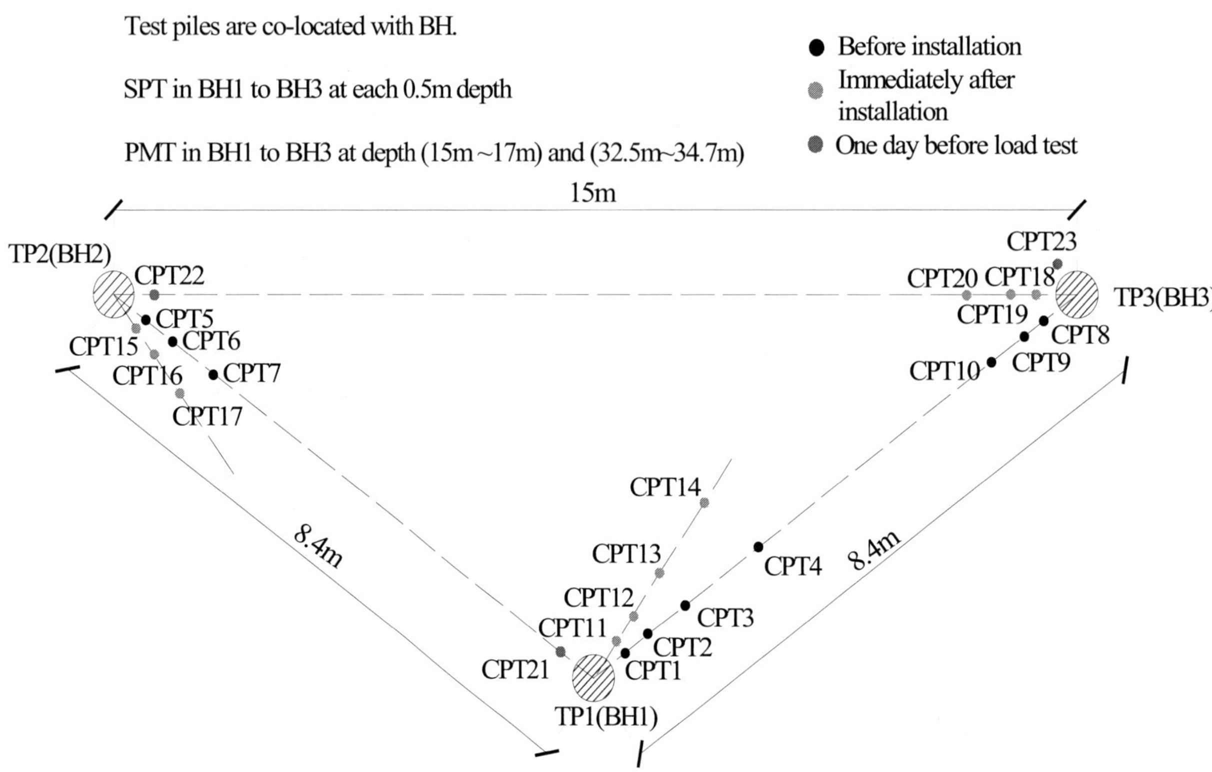

NOTE: CPT1 (CPT5, CPT8, CPT11, CPT15, CPT18):0.6m from the center of pile.
CPT2 (CPT6, CPT9, CPT12, CPT16, CPT19):0.9m from the center of pile.
CPT3 (CPT5, CPT8, CPT11, CPT15, CPT18):1.5m from the center of pile.
CPT 4 (CPT14):3m from the center of pile.

FIG.1 Soil stratigraphy of the experimental site.

TABLE 1. Summary of the pressuremeter test results

Borehole	*Depth(m)*	*Soil Type*	*SPT-N*	E_{PMT} *(kPa)	E_{ur}^{PMT} *(kPa)
BH1(TP1)	16.5	M*	10	13034	68911
BH1(TP1)	33.0	S V*	100	78778	303570
BH2(TP2)	15.0	M*	4	3518	24563
BH2(TP2)	32.5	S V*	100	44357	768739
BH3(TP3)	17.0	M*	10	20770	103936
BH3(TP3)	34.7	S V*	100	81119	525934

*M: Kallang Marine clay
S V: Completely Weathered sandstone/mudstone in Jurong Formation, consists of sandy silt
E_{PMT} : Loading stiffness
E_{ur}^{PMT} : Unloading stiffness

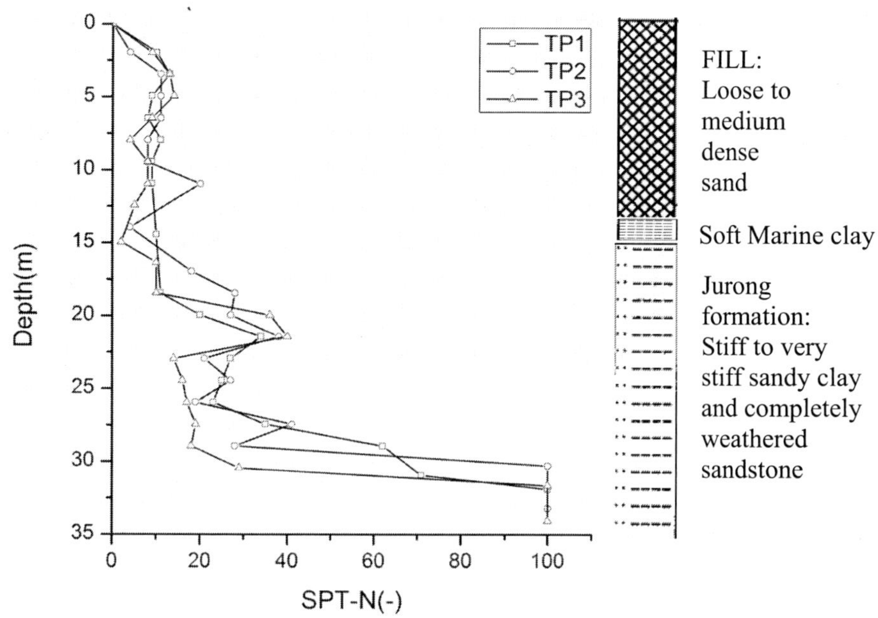

FIG.2 The profile of SPT-N value for BH1 to BH3.

In total, twenty CPTUs were conducted close to the pile tests locations, at distances from 0.6m to 3m relative to the center of each test pile. CPTUs were conducted in two stages: before and after the pile installation. They are designated as CPT1 to CPT10 for "before installation" period and CPT11 to CPT20 for "after installation" period. Figure 3 and Figure 4 show the penetration results of CPT1 to CPT10. It can be observed that despite the variation in q_t value between each CPTU, there is a good agreement in the trend of the penetration results which indicate the soils have similar geological profile. The results of the cone pore pressures indicated that sand type soil extends from ground to approximate 12m depth. Then significant excess pore pressures were observed from 12m depth to 30m which means clay type soil underlying the sandy fill layer. The negative pore pressures, found from 17m depth to 22m depth, indicate that the clay type soil in this layer may be over-consolidated.

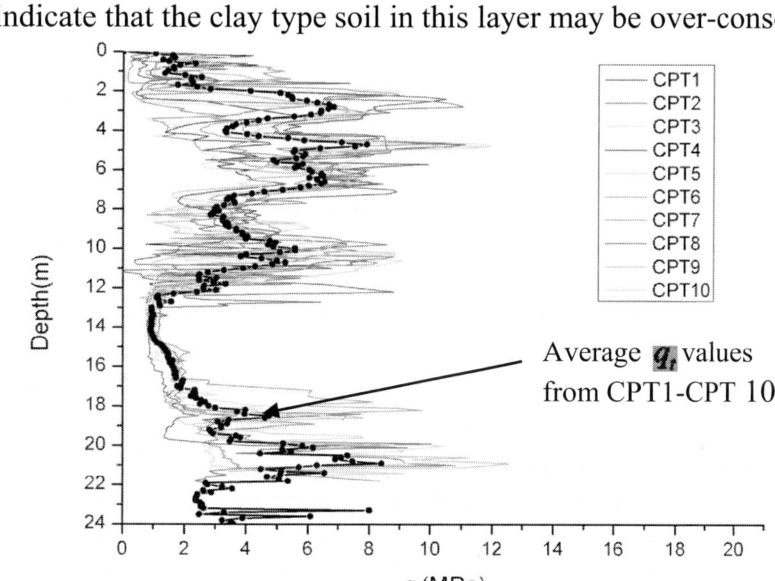

FIG.3 CPTU q_t profiles before the pile installation.

The comparison of the penetration results of the CPTU tests before and after the pile installation are shown in Figure 5. A noticeable increase in cone resistance was observed in the sand fill layer after pile installations. The magnitude of the increase of the cone resistance reduces as the normalized radii (r/r_0) increases, as shown in Figure 6. However, the cone resistance does not change significantly in the soft marine clay layer and the sandy silt layer.

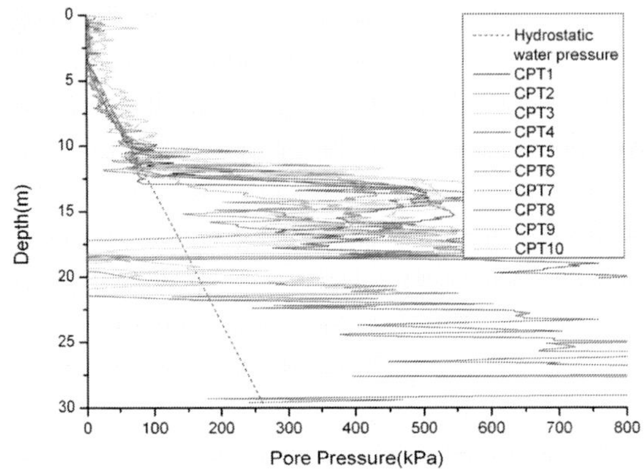

FIG.4 CPTU pore pressure profiles before the pile installation.

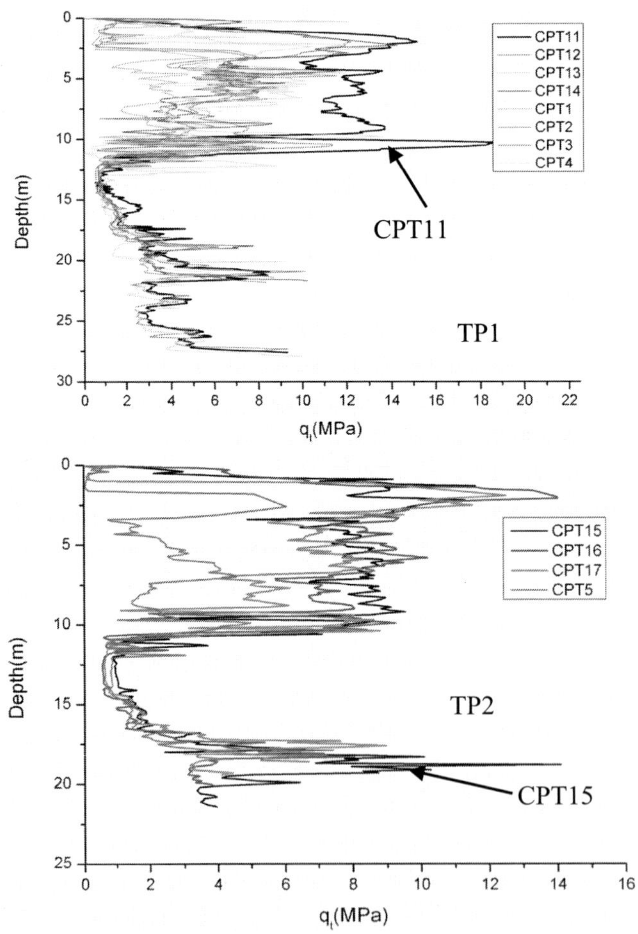

FIG.5 Compare CPTu qt profiles before and after pile installations.

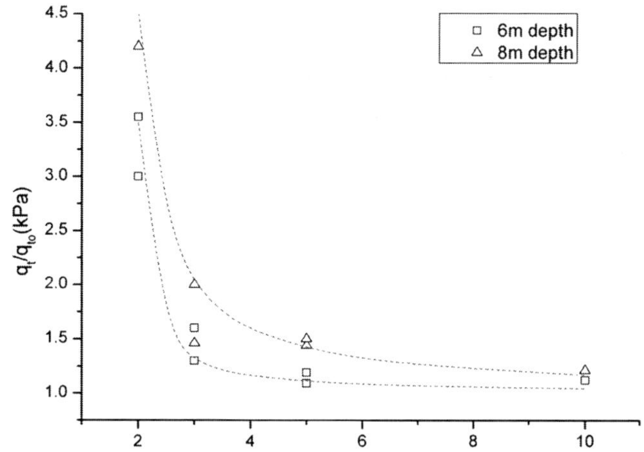

FIG.6 Ratio of q_t/q_{to} plotted against the normalized radii.

Laboratory Tests

Undisturbed samples were taken from the experimental site, in the three boreholes at specific depths. The laboratory tests comprised of index tests, six sets of CIU triaxial tests and three oedometer tests. The results of the laboratory tests are summarized in Table 2. The CIU triaxial tests are used to obtain the strength parameters for the design and the numerical study. The average c'and ϕ' for S VI and S V are 2.5 kPa, 28.5°, and 25kPa, 32.5°. High cohesion is usually found in the completely weathered sandstone/mudstone (S V). The over-consolidation ratio (OCR) can be obtained from the oedometer tests. The average values of OCR for Kallang Marine clay and Residual soil are 1 and 1.15 respective. Due to highly disturbed sample, no results are obtained from the oedometer tests for the completely weathered sandstone/mudstone (S V). In summary, all the borings as well as the CPTU tests showed very similar soil conditions around the three test piles.

TABLE 2 Summary of the laboratory test results

Bore hole	*Depth (m)*	*Soil Type*	*Water content (%)*	*c' (kPa)*	*ϕ' (Deg)*	*LL* (%)*	*PL* (%)*	*PI* (%)*	*Pre-consolidation pressure (kPa)*	*Compression Index*
BH1 (TP1)	14.5-15.4	M	27	5	27	40	21	19	211	0.147
BH1 (TP1)	30.5-31.5	S V	21	27	33	35	23	12	-	-
BH2 (TP2)	12.0-12.9	M	53	0	22	69	30	39	124	0.585
BH2 (TP2)	30.5-31.5	S V	22	-	-	43	28	15	-	-
BH3 (TP3)	15.0-15.5	M	28	0	30	43	20	23	184	0.153
BH3 (TP3)	30.5-31.5	S V	22	23	32	43	27	16	-	-

*LL: Liquid limit
*PL: Plastic limit
*PI: Plasticity index

TEST ARRANGEMENT AND TESTING PROGRAMME

Test Program

The experiment was conducted between October 22 and November 27, 2009 at Tuas View Ave 2. Three closed-ended spun piles (TP1, TP2 and TP3) were jacked into the ground under different Jack-in forces. A month later, three static load tests were conducted on the test piles. The relative positions of the test piles and CPTUs are shown in Figure 1. The three piles installed at the experimental site are 600 mm diameter prestressed high-strength concrete (PHC) spun piles. A steel plate was welded to the toe of all test piles to form the close-end.

Pile Installation and Instrumentations

The three test piles, Piles TP1, TP2 and TP3, were jacked-in to depths of 28.7m, 29.9m and 31.7m respectively by jacked-in rig, on October 23, October 22 and October 24, 2009, respectively. TP1 was jacked-into the completely weathered Jurong sandstone (SPT-N=70) at 4475 kN jack load (1.5×W.L.). TP2 and TP3 were jacked-into the completely weathered Jurong mudstone (SPT-N=100) by 5903 kN jack load (2×W.L.) and 6620 kN jack load (2.25×W.L.) respectively.

The conventional vibrating wire strain gage (VWSG) instrumentation method for driven and jack-in pile is done by installing either VSWG instrumented reinforcement cage or an instrumented pipe, into the hollow core of spun piles followed by cement grout in filling. However, the infilling of cement grout substantially alters the structural properties of the spun piles, thus rendering them to be different from the actual working spun piles, which are usually not grouted internally. The change in strain in the post-grouted core under the applied loading may also not be the same as the change in strain in the prestressed concrete wall of the pile because of the different stiffness of the two materials of different mix, strength and age. In order to overcome these obvious shortcomings, a new strain deformation monitoring system named Global Strain Extensometers (GLOSTREXT) was developed by Ali and Lee (2008) and was used in this study. The GLOSTREXT Method for static load testing on piles is a deformation monitoring system using advanced pneumatically anchored extensometers and a novel analytical technique for determining axial loads and movements at various levels down the pile shaft including the pile base level. The instrumentation is installed after the installation of the pile. With the instrumentation set-up, the state-of-the-art GLOSTREXT system is able to measure shortening and strains over an entire section of the test pile during each loading steps of a typical static pile load test.

Static Load Test

Three static load tests were conducted from 14th November to 27th November, 2009. A kentledge reaction system of approximately 10.3MN weight was set up to serve as reaction for TP1 and TP3 load tests. A jack-in rig counter-weights reaction system was used for TP2 load test. In the test set-up used, the test load was applied using a single hydraulic jack acting against the main beam. The jack was operated by an electric pump. The applied load was indicated by calibrated Vibrating Wire Load Cell as primary load measurement, while the pressure gauge reading was used as the secondary cross-checking purpose.

In each test pile, four numbers of Linear Vertical Displacement Transducers (LVDTs) were mounted to the reference beams; with plunger pressing vertically against glass plates fixed to pile top. Vertical scales were also provided on the reference beams to monitor frame movement during load testing for correction purposes; and vertical scale rules fixed to pile top sighted by a survey precise level instrument. All instrumentation was connected to a data logger (Micro-10x Data logger) and the test data were recorded and stored automatically.

The testing program consisted of testing TP1 and TP3 piles in four cycles applied using the Quick Load Test Method for individual pile (ASTM 1143-81). The applied load of 4 cycles were 2930 kN (1×W.L.), 5860 kN (2×W.L.), 7325 kN (2.5×W.L.) and 10255 kN (3.5×W.L.) or failure load. The testing program consists of testing TP2 in three cycles due to maximum capacity of the jack-in rig counter-weights reaction system is 8000 kN. As a result, TP2 was not tested to failure..

ANALYSIS OF TEST RESULTS

Load-Movement Behavior of Test Piles

The ultimate bearing capacity of TP1 and TP3 are 7690 kN and 8762 kN, respectively when jack-in force increases from 4475 kN (1.5×W.L.) to 6617 kN (2.25×W.L.). The ultimate bearing capacity of TP2 cannot be obtained from load-movement curve since TP2 was not loaded to failure due to insufficient capacity of the jack-in rig counter-weights reaction system. The relationship between ultimate bearing capacity of the test pile and Jack-in force is shown in Figure 7. It can be observed that the increment of ultimate bearing capacity is not in linear proportion to the increment of jack-in force. The ultimate bearing capacity of TP2 would be expected between 7690 kN and 8762 kN and closer to 8762 kN.

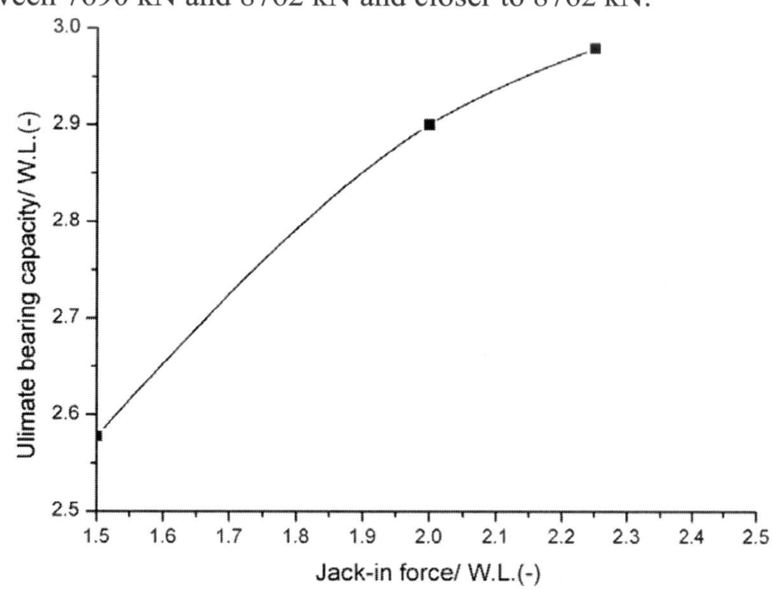

FIG. 7 The relationship between normalized ultimate bearing capacity of test pile and normalized Jack-in force. (Point for TP2 is estimated)

The comparison of Jack-in pile behavior due to different Jack-in force is shown in Figure 8. It is observed TP1, TP2 and TP3, showed almost the same load-movement curves up to 2 time working load (5860 kN) as in Figure 8 (a). Movement of TP1, TP2 and TP3 is 19.59mm, 18.73 mm and 18.22 mm at that load level respectively. Figure 8 (b) shows that TP1 which was jacked-in by 1.5×W.L. is softer compared to TP2 and TP3 which was jacked-in by 2×W.L. and 2.25×W.L. respectively. Since the soil conditions around the three piles are approximately the same, the different behaviors between test piles must be caused by differences in the toe stiffness responses of TP1, TP2 and TP3 which is an integral component of the pile head load-movement response and governs whether the test pile meets or fails the acceptance pile head settlement criteria. The differences in toe stiffness between the piles, TP1, TP2, and TP3 will be discussed later.

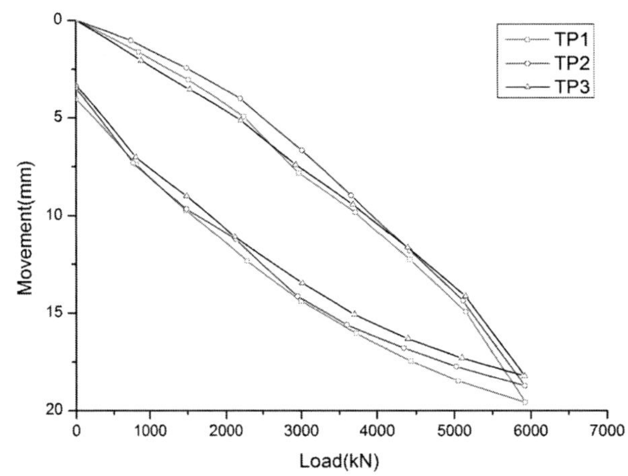

(a)

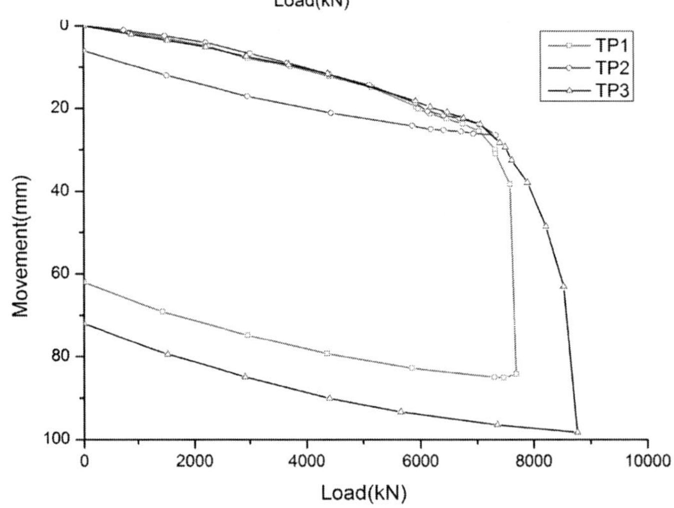

FIG.8 The comparison between three test piles (a) under 2 time working load and (b) at the ultimate bearing capacity.

Residual load and true load distribution in the pile

The load distribution along the pile shaft and the base can be derived from computations based on the measured changes of the global strain gauge readings and the estimated pile properties. Load transferred at each global strain gage can be calculated as:

$$P = (\varepsilon \times E_s)A_c \quad (1)$$

where ε is average change in global strain gauge readings; E_s is secant modulus of concrete; A_c is cross-section area of the pile.

For the manufactured spun pile, A_c does not vary much along the pile shaft. E_s becomes very important in computing the transferred load. Generally, the concrete modulus is not constant over the large strain range imposed in the static load test. It reduces with the increasing strain value. It is indicated that secant modulus can be best determined from a so-called "tangent stiffness" plot, which presents the applied increment of load over the induced increment of strain (Fellenius 1989). When the shaft friction fully mobilizes, the "tangent stiffness" will plot along a slightly sloping line and then the "tangent stiffness" is directly converted to secant modulus, E_s.

The "tangent stiffness" plots for the last cycle of each test pile are shown in Figure 9. For TP1 and TP2, the "tangent stiffness" lines in Figure 10 indicated that E_s= (50-0.005με GPa). The similar analysis for TP3 shows E_s=(55-0.006 με GPa). The high values of E_s were obtained is due to the high-strength concrete used for the manufactured spun piles.

Figure 10 shows the load distribution curves for the last cycle of each test pile. However, these load distributions are not true resistances. This is because of the existence of the residual load that it is well demonstrated to exist in several driven and jack-in pile tests. In the driven piles or Jack-in piles, residual load results from shear stress developed between the pile and the soil during the installation. The measured load distributions shown in Figure 10 were evaluated based on "zeroing" all gages immediately before the start of the load test. As a result, strain gauge readings did not provide any information about the residual loads, and the shaft resistance was overestimated and the base resistance underestimated based on the measured data.

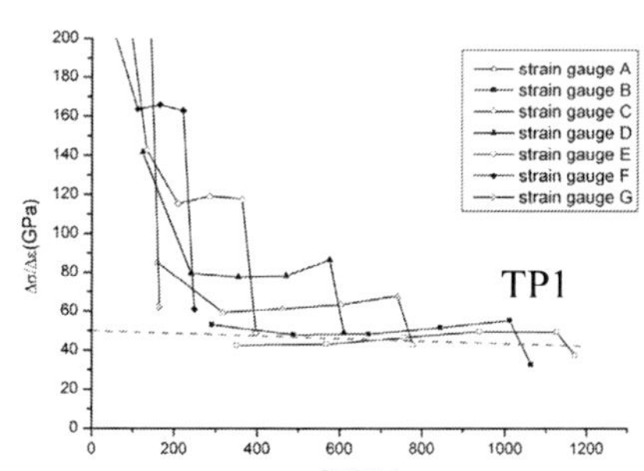

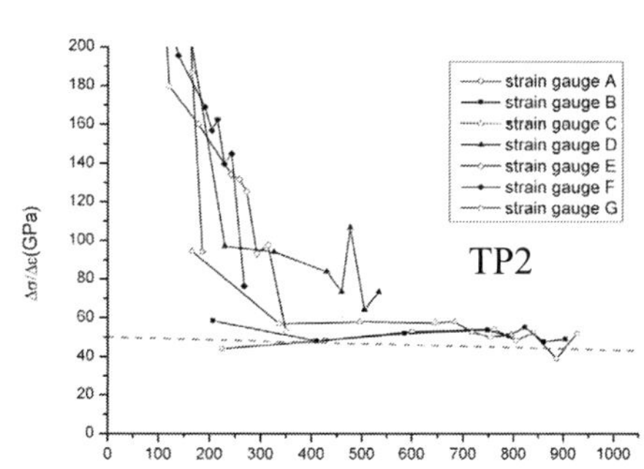

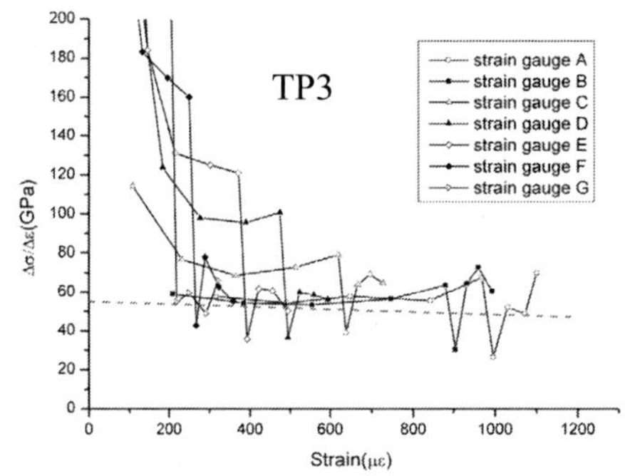

FIG.9 Secant modulus plotted against strain at each gage level for the last loading cycle of three test piles.

Fellenius proposed a method to determine the residual loads from the static load test , based on the effective stress method and the assumption that the shear resistance is independent of the direction of shear (Fellenius 2002). It can be assumed that the residual load is developed fully in the upper part of the pile. The residual load must be unloaded before the positive skin friction is mobilized during the static load test. Therefore, if the negative skin friction is fully mobilized, the reduction of load along the upper part of the pile is twice the true shaft resistance. It is also recognized that the soil becomes stiffer with depth (as indicated by the SPT-N value). As a result, the full shaft friction may not be mobilized in the lower proportion of the pile. Because of the loading/unloading actions, the evaluation of the residual load needed some judgments. The distributions of the residual load for the test piles shown in Figure 10 are considered the most probable distribution. The unit negative skin friction along the upper shaft of 15 m length of the piles corresponds to a beta-coefficient of 0.55 in sand fill layer and 0.35 in soft clay layer in the effective stress analysis. As can be seen in the figure, the residual load develops fully in the upper part of the pile, and then the residual load tends to reduce with depth. The evaluated residual loads at the pile toe are 618kN, 500kN and 335kN for TP1, TP2 and TP3, respectively. The residual load at the pile toe reduces as the pile is seated deeper in the completely weathered Jurong mudstone. This may have caused an increase in the positive shaft resistance to be mobilized in the lower part of the pile.

It would be expected that the total ultimate shaft friction should be approximately proportional to the pile depth since the soil condition does not vary very much between the three piles. But the total ultimate shaft friction for TP1 (28.7m), TP2 (29.9m) and TP3 (31.7m) are 6200 kN, 6020 kN and 6190 kN respectively from the measured data. These are not consistent with the expected behavior. When the residual load is considered, the total ultimate shaft friction for TP1 (28.7m), TP2 (29.9m) and TP3 (31.7m) are 5400 kN, 5662 kN and 5900 kN respectively. These results showed better agreement with the expected pile shaft resistance behavior.

As discussed in the previous section, the different behaviors between test piles must be caused by differences in the toe stiffness response of piles TP1, TP2 and TP3. The differences in toe stiffness among the piles, TP1, TP2, and TP3 are due to the different jack-in forces applied during pile installation. The difference in toe stiffness, considering the residual load, between TP1 and TP3 is shown in Figure 11. From Figure 11 we can see, when jack-in force increases, the behavior of the pile toe becomes stiffer and if the ultimate end bearing resistance q_b as defined by settlement criterion of 10% of the pile diameter, q_b is increased from 4900 kPa to 8900 kPa when jack-in force increases from 4475 kN (1.5×W.L.) to 6617 kN (2.25×W.L.). This toe resistance behavior is consistent with the Jack-in construction method, which prestresses the soil bulb below the pile toe and ensures a larger stiffness of the toe response with larger Jack-in force. At start of the static load test, the toe was already subjected to the Jack-in load and the load-movement curve displays an initial stiffer reloading portion. When the toe load-movement curve for TP1 was shifted a bit to the left, the two responses for TP1 and TP3 seem to have the same virgin compression line, as in Figure 12. The larger the Jack-in force, the larger volumetric compression of the bulb of soil below the toe of the piles. As a result, the larger reloading portion causes the larger stiffening effect when the pile is loaded after the installation.

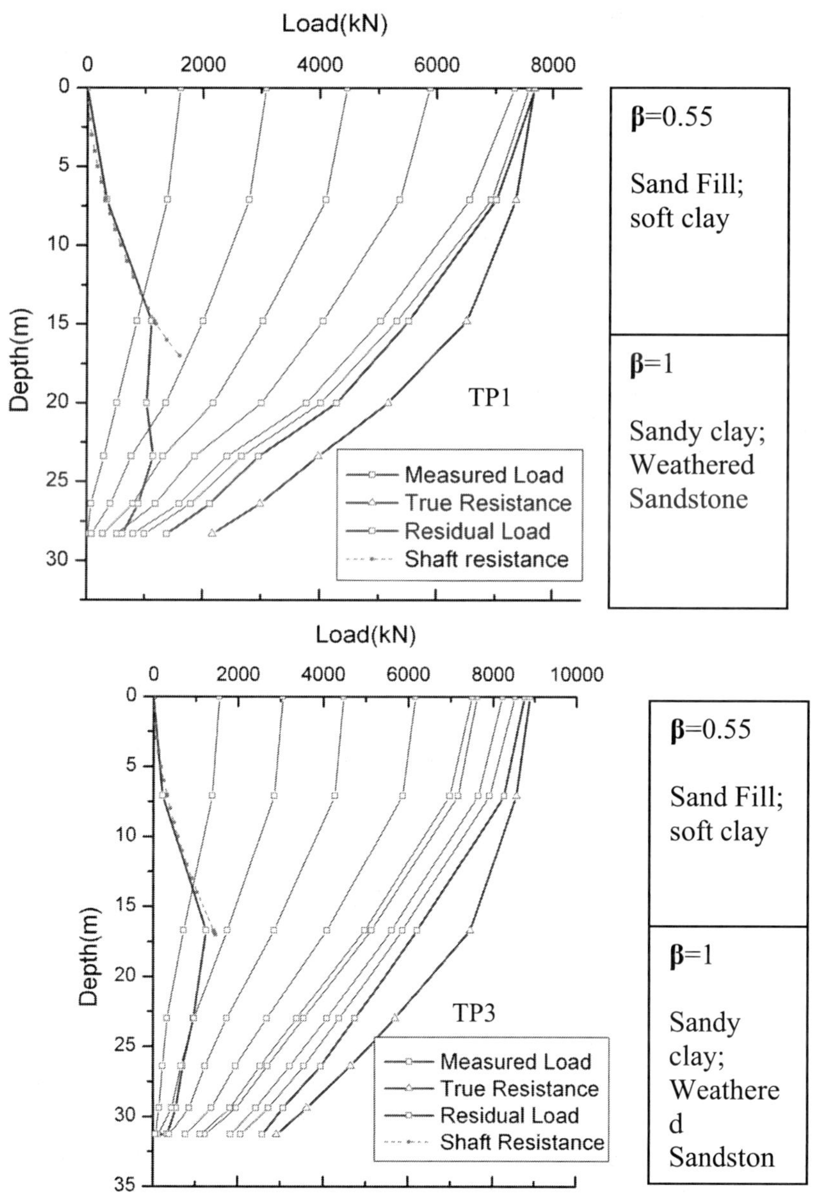

FIG.10 Evaluated distributions of measured load, residual load, load corrected for residual loads, and shaft resistance based on beta method for TP1 and TP3.

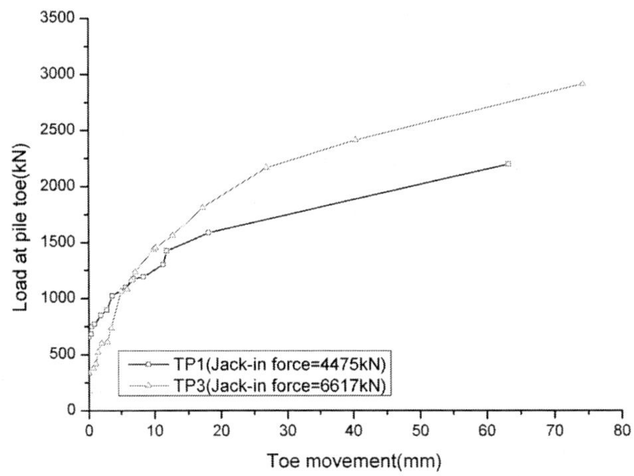

FIG.11 Toe load plotted against toe movement.

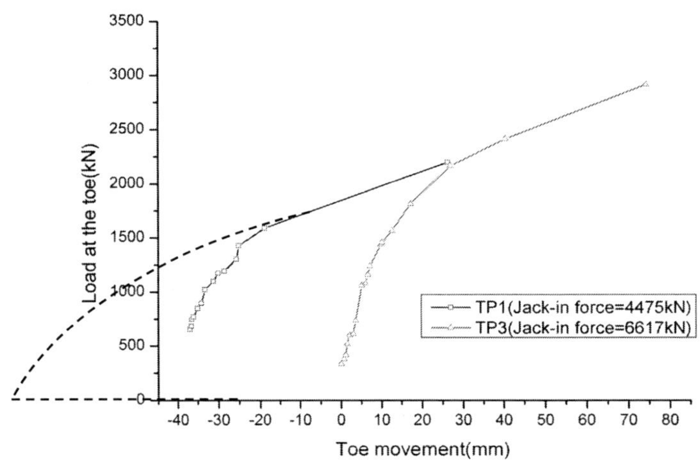

FIG.12 Virgin compress curve for pile toe.

CONCLUSIONS

In order to investigate the effects of Jack-in force magnitude on the ultimate bearing capacity of Jack-in pile, a comprehensive in-situ investigation and laboratory testing of soils surrounding three static pile load test on Jacked-in spun piles were conducted.

In general, very similar soil conditions were observed around the three test piles from both borehole and CPTU results. Similar soil parameters were evaluated from well documented empirical methods on CPTU data. A significant increase in the cone resistance was observed in the reclaimed sand fill layer after pile installation. The magnitude of the increase of the cone resistance reduces with distance away from the center of the pile. However, the cone resistance did not change significantly in the soft marine clay layer and the sandy silt layer after installation due to its undrained response.

The three test pile load tests indicated that the ultimate bearing capacity of Jack-in pile increases as the Jack-in force increases. Generally, the pile developed greater stiffness response in the load-movement curve when it is installed with larger jack-in force. However, the three test piles showed approximately the same load-movement curves under 2 times working load. The difference in behaviors between test piles is caused by the difference in the toe stiffness response of TP1, TP2 and TP3, as their shaft responses are nearly the same for similar soil conditions around the test piles. Due to the Jack-in construction method, at the start of the static load test, the toe was already subjected to initial lock-in residual loads from the jack-in installation process, and the load-movement curve displays an initial steep (stiffer) reloading portion when the pile is loaded after installation. The larger the Jack-in force, the larger the volumetric compression of the bulb of soil below the toe of the piles. As a result, the larger reloading portion was observed that caused the larger stiffening effect in the early reload portion of the pile load tests.

ACKNOWLEDGMENTS

The authors appreciate CS Construction & Geotechnic Pte Ltd and Soil Investigation Pte Ltd for the opportunity to conduct field testing.

REFERENCES

Ali, F.H. and Lee, S.K. (2008). “A new instrumentation method for driven prestressed spun concrete piles.” *Electronic Journal of Geotechnical Engineering* 13.

Bond, A.J. and Jardine, R.J. (1991). “Effects of installing displacement piles in a high OCR clay.” *Geotechnique* 41(3): 341-363.

Bond, A.J. and Jardine, R.J. (1995). “Shaft capacity of displacement piles in a high OCR clay.” *Geotechnique* 45(1): 3-23.

Lehane, B.M., Jardine, R.J., Bond, A.J., and Frank, R. (1993). "Mechanisms of shaft friction in sand from instrumented pile tests." *Journal of geotechnical engineering* 119(1): 19-35.

Lehane, B.M. and Jardine, R.J. (1994 a). "Displacement-pile behaviour in a soft marine clay." Canadian Geotechnical Journal 31(2): 181-191.

Lehane, B.M. and Jardine, R.J. (1994 b). "Displacement pile behaviour in glacial clay." *Canadian Geotechnical Journal* 31(1): 79-90.

Eslami and Fellenius, B.H. (1997). "Pile capacity by direct CPT and CPTU methods applied to 102 case histories." *Canadian Geotechnical Journal* 34(6): 888-904.

Fellenius, B.H. (1989). “Tangent modulus of piles determined from strain data.” 1989 Foundation Congress.

Fellenius, B.H. (2002). “Determining the Resistance Distribution in Piles.” *Geotechnical News Magazine* 20(3): 25-29.

Randolph, M.F., Dolwin, J., and Beck, R. (1994). “Design of driven piles in sand.” *Geotechnique* 44(3): 427-448.

Tomlionson, M. and Woodward, J. (2008). “Pile Design and Construction Practice.” London and New York, *Taylor&Francis*.

The Mechanics of Friction Fatigue in Jacked Piles Installed in Sand

Prasenjit Basu[1], A.M.ASCE, Dimitrios Loukidis[2], Monica Prezzi[3], A.M.ASCE, and Rodrigo Salgado[4], F.ASCE

[1]Assistant Professor, Department of Civil and Environmental Engineering, The Pennsylvania State University, University Park, PA 20191-4400; pbasu@engr.psu.edu
[2]Lecturer, Department of Civil and Environmental Engineering, University of Cyprus, Nicosia, Cyprus; loukidis@ucy.ac.cy
[3]Professor, School of Civil Engineering, 550 Stadium Mall Drive, Purdue University, West Lafayette, IN 47907-2051; mprezzi@ecn.purdue.edu
[4]Professor, School of Civil Engineering, 550 Stadium Mall Drive, Purdue University, West Lafayette, IN 47907-2051; rodrigo@ecn.purdue.edu

ABSTRACT: The limit shaft resistance of displacement piles decreases with the number of axial loading cycles applied during pile installation (e.g., jacking, driving) or due to the nature of superstructure loading. This study examines the mechanisms governing the phenomenon of friction fatigue along piles jacked in sandy soils. Finite element analysis (FEA) of a thin disk of soil surrounding the pile shaft is performed. The pile installation process is modeled as sequential combination of cylindrical cavity expansion and vertical shearing along the pile-soil interface. The soil disk is subjected to several vertical shearing cycles in order to simulate the successive application and removal of jacking loads. An advanced soil constitutive model based on two-surface plasticity and critical state theory is used in the FE simulations. FEAs examine the development and evolution of the normal stress acting on the pile shaft, which multiplied by an appropriate friction coefficient, yields the value of mobilized shaft resistance. Results show that, in the absence of loading cycles, the normal stress attains values considerably larger than the in situ vertical stress. However, with subsequent application of loading cycles, the normal stress decreases at rates that increase with decreasing relative density and in situ vertical stress.

INTRODUCTION

The installation of displacement piles often causes complex changes to the soil surrounding the piles. As a result of these changes, the response of piles to structural loads varies greatly depending on the installation method used. Thus, for reliable estimation of pile capacity, design methods should account for the effect of the installation technique. Numerous recent experimental investigations point to the fact that the limit shaft resistance in displacement piles decreases with the number of applied axial loading cycles. These loading cycles may be imposed on piles as a consequence of the installation process (e.g., jacking, driving) or due to transient superstructure loading. The decrease in limit shaft capacity as a result of successive loading-unloading cycles is commonly referred to as friction fatigue.

Jacking of precast concrete or steel piles into ground often provides an effective alternative to the conventional dynamic pile driving methods (using hammers or vibrators) that produces noise, vibration and air pollution. Pile jacking was primarily used in the past for generating additional foundation capacity or to reduce settlement of existing foundations (Babushkin & Ginzburg 1984; Yang et al. 2006). However, piles installed by jacking (or jacked piles) are now being increasingly used as foundations of new structures in different parts of the world. By jacking a pile into the ground, it is possible to get an estimate of the ultimate (and, indeed, of the limit) load capacity of the pile based on the measured jacking load during installation (Yang et al. 2006). Empirical or semi-empirical equations, based on results of load tests performed on driven piles, are often used to calculate the capacity of jacked piles. However, the mechanics involved in pile driving (a dynamic process) differ from those prevalent in pile jacking (a quasi-static process). Therefore, it is important to investigate whether the existing design methods for driven displacement piles are applicable to jacked piles as well.

This study investigates the development of shaft resistance for piles jacked in sandy soils focusing on the mechanisms governing the phenomenon of friction fatigue. This is achieved through axisymmetric finite element analysis (FEA) of a thin disk of soil surrounding the pile shaft at a given elevation. An advanced two-surface plasticity-based constitutive model is used to simulate the mechanical behavior of sand under multi-axial loading. Based on the FEA results, equations are established for the calculation of the coefficient of lateral earth pressure K acting on the pile shaft. Friction fatigue is observed through degradation of K as a function of the number of jacking strokes.

IDEALIZATION OF PILE JACKING

In this study, pile jacking is modeled using simplified modes of deformation, which allow adequate treatment of the problem in a one-dimensional (1D) analysis framework. The jacking of a pile into the ground is assumed to load the soil mass through a sequential combination of cylindrical cavity expansion and vertical shearing. As this study deals with shaft resistance, the evolution of stress states within a thin disk of soil around the pile shaft is examined (Figure 1). Similar approaches were followed by Randolph et al. (1979) and Potts and Martins (1982) for the analysis of shaft resistance of displacement piles in clays. In reality, as the pile tip approaches a point in the soil mass, the vertical stress increases substantially. The high vertical stress on the tip is also evident in cone penetration testing. According to Salgado et al. (1997) and Salgado & Prezzi (2007), the cone resistance q_c is significantly larger than the horizontal radial stress (the cylindrical cavity expansion limit pressure) in the vicinity of the cone shoulder. As the pile base passes the depth under consideration, there are significant reversals in all strain components (Randolph & Wroth 1978), causing major unloading of the vertical stresses. The 1D approach used in this paper cannot capture the exact stress path and the large vertical stress in a soil element just below the pile tip; however, the following analysis results show that the development of limit shaft resistance and friction fatigue in jacked piles can be adequately captured using the simple approach of combining cylindrical cavity expansion with vertical shearing.

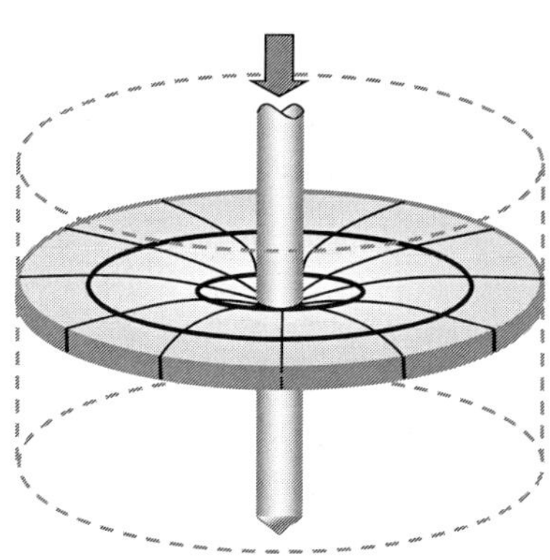

Figure 1. Soil disk around the pile shaft

Different analysis stages used to model the jacking of a pile into the ground are shown in Figure 2. Stage 1 corresponds to the penetration of the pile in a virgin soil mass and is modeled as a combination of two distinct phases: the cavity expansion (CE) phase and the primary vertical shearing phase. The first phase (i.e., CE) represents the creation of a cylindrical space (occupied by the pile) within the ground as the pile tip pushes the soil away from the path of the pile. The cylindrical cavity expansion stops when the cavity radius becomes equal to the pile radius, i.e., as the pile tip clears element "A" and the left face of element "A" becomes aligned with the pile shaft. CE is followed by a vertical shearing (primary shearing) phase, which continues until a limit condition (critical state) is reached along the pile shaft. The primary shearing phase is essential for the estimation of limit shaft capacity. Azzouz et al. (1990) showed that modeling pile installation using only cavity expansion leads to an over prediction of the effective lateral stress on the pile shaft by a factor of 2 when compared with predictions using the strain path method (SPM). The addition of the primary shearing phase eliminates this over prediction by introducing reversal in the radial stresses and strains, leading to substantial reduction in radial stress by the end of Stage 1.

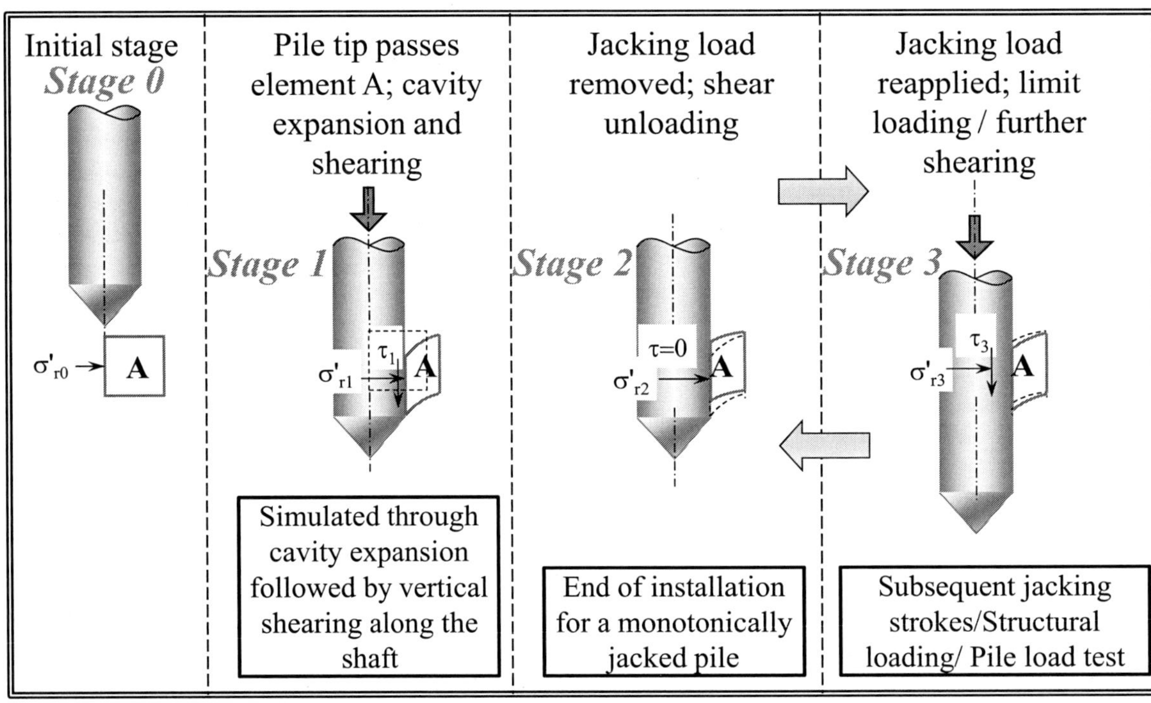

Figure 2. Idealization of pile jacking into the ground

Stage 2 represents the removal of the jacking load. This occurs either after the maximum jacking stroke length s of the hydraulic jack is reached or at the end of the pile installation. During the unloading stage, the shear stress τ reduces until it reaches a residual value τ_{res} and the normal stress σ'_r acting on the pile shaft reduces due to contraction of the soil elements in its vicinity. It is well known that soils exhibit strongly contractive response upon shearing reversals, even if they are dilative immediately before reversal. This exactly happens in the vicinity of the pile shaft during transition from Stage 1 to Stage 2. Soil in the vicinity of the pile shaft (which is dilative at the end of Stage 1) exhibits purely contractive behavior when the direction of shear stress increment is reversed on removal of the jacking load (Stage 2). The magnitude and direction of the residual shear stress τ_{res} acting on the shaft of a displacement pile installed in sand vary with depth and depend on many factors, such as relative pile-soil stiffness, pile geometry and base-to-shaft load ratio (Briaud & Tucker 1984; Darrag & Lovell 1989). Field and laboratory test results on model jacked piles (Lehane et al. 1993; Colombi 2005) show that τ_{res} is negative (acting downward) in the upper part of the pile and positive (acting upward) near the bottom of the pile, while in some portion of the pile τ_{res} is equal to zero. In the present study, the value of τ_{res} at the end of unloading is assumed to be equal to zero. During Stage 3, the load is reapplied to the pile until a limit condition is once again reached. For a pile installed monotonically, Stage 3 represents the application of the structural load (or the performance of a static pile load test). For typical jacked piles, which are installed with multiple strokes, Stage 3 represents the second jacking stroke. Subsequent jacking strokes are represented by successive repetition of Stages 2 and 3 (multiple removals and re-applications of the jacking load at the pile head). At each load increment reversal, there is a renewed phase of soil contraction in the vicinity of the pile, which results in net loss of confining stress, and, consequently, reduction in unit shaft resistance (commonly known as friction fatigue). The last reloading stage always corresponds to the application of the load from the superstructure or the load applied during a field load test.

FRICTION FATIGUE FOR JACKED PILES IN SAND

White & Bolton (2002) and White (2005) have shown that the horizontal effective stress σ'_h (= $K\sigma'_v$; σ'_v is the vertical effective stress) acting on the shaft of a pile jacked into sand at a given depth decreases gradually as the pile tip penetrates deeper past that depth. In addition, results from field and laboratory experiments indicate that the lateral stress acting on the pile shaft decreases significantly when a pile is installed in sand using multiple jacking strokes (Lehane 1992, Chow 1997, Leung et al. 1998, White & Lehane 2004). The successive removal and reapplication of jacking load during pile installation using multiple jacking strokes, induces shearing cycles to the soil surrounding the pile shaft. Therefore, the number of jacking strokes applied to the pile needs to be a key factor in determining the design capacity of jacked piles.

Recent studies on this topic relate friction fatigue to the contraction of the soil elements in the vicinity of the pile shaft resulting from successive cycles of vertical shearing (White & Bolton 2002, Randolph 2003, White 2005). Figure 3 schematically shows the evolution of dilation and contraction zones surrounding the pile shaft during Stage 1 (i.e. a cylindrical cavity expansion phase followed by a primary shearing phase) of pile jacking. Soil near the pile shaft dilates (loosens) due to the intense shearing in the plane of the soil disk caused by the expansion of a cylindrical cavity within the ground. This dilated zone is surrounded by a zone of contracted (densified) soil. During the primary vertical shearing phase that follows cavity expansion, the dilated zone in the vicinity of the pile shaft has the tendency to contract, and similarly, the contracted zone away from the pile has the tendency to expand elastically. The tendency for contraction of soil in the vicinity of the pile shaft results in a loss of normal stress acting on the pile shaft. Cycles of vertical shearing (i.e., unloading-reloading cycles) contribute positively to the loss of normal stress on the pile shaft, and consequently, aggravate friction fatigue.

White & Lehane (2004), through a series of centrifuge tests, showed that the reduction in lateral stress can be correlated with the number of loading cycles N_L applied during installation and subsequent loading of the pile. Observing the reduction in shaft resistance for (i) a multi-sleeve CPT probe (Dejong 2001), (ii) a jacked instrumented model pile (Chow 1997), and (iii) a long offshore pile (Fugro 1996), White & Lehane (2004) concluded that friction fatigue does not occur in the absence of loading cycles. White (2005) summarized the results of field tests for both silica and carbonate sands (Fugro 1996, Chow 1997) and showed that N_L plays a role in reducing the normalized shaft resistance (defined as the ratio of the average shaft resistance to the average CPT sleeve friction resistance along the embedded pile length). However, currently available design equations for displacement piles, which were developed mostly based on field or experimental data, either do not consider the effect of loading cycles (e.g., Nordlund 1979,

Fleming et al. 1992) or consider the effect indirectly by expressing K as a decreasing function of the distance from the pile tip (e.g., Randolph et al. 1994, Jardine et al. 1998, Randolph 2003, Lehane et al. 2005). The latter consideration is based on the fact that the soil at the pile tip elevation experiences only one cycle, while the final number of loading cycles experienced by soil elements along the shaft of jacked and driven piles increases the farther they lie from the pile tip.

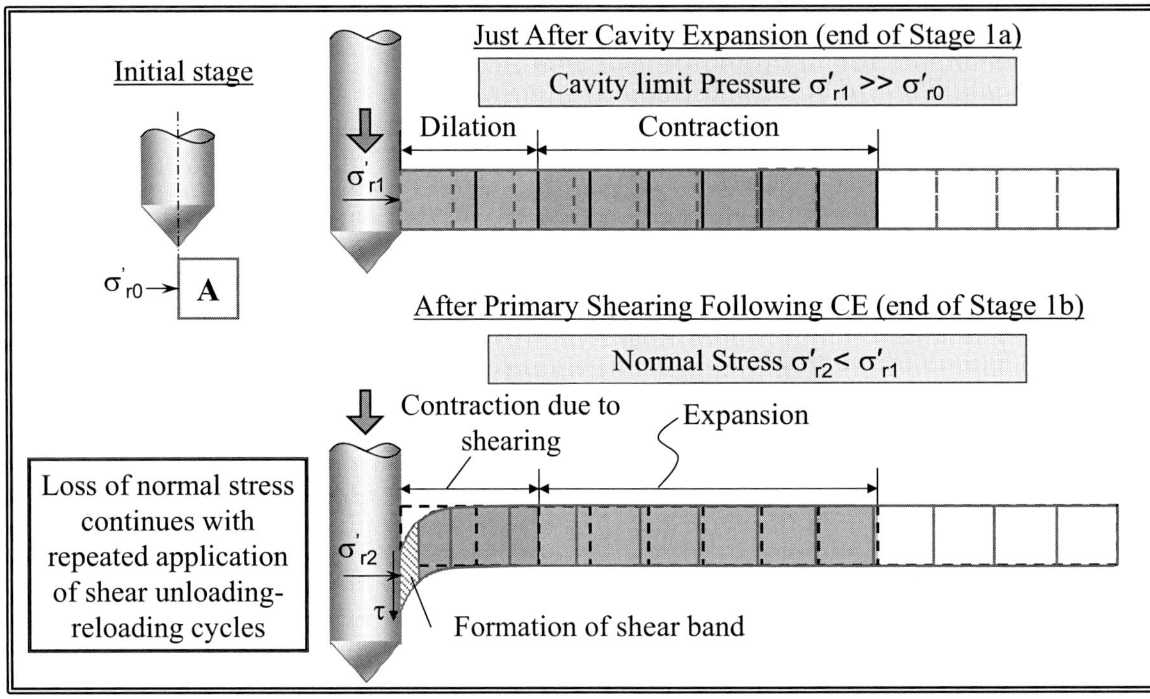

Figure 3. Evolution of dilative and contractive zones surrounding the pile shaft

FINITE ELEMENT MODEL

The pile shaft-soil interaction is modeled using a disk of soil around the pile shaft. It is assumed that, at a depth far from pile head and pile tip, the vertical normal strain is negligible and that there is no bending deformation. Such assumptions make the analysis independent of the thickness of the soil disk and thus, one-dimensional. The FE code SNAC (Solid Nonlinear Analysis Code; Abbo & Sloan 2000) is used for the analyses. For proper simulation of a large deformation problem like the one addressed in this paper, a simple updated Lagrangian approach is used by updating the nodal coordinates after each solution increment.

Mesh, Boundary Conditions and Applied Displacement Increments

The FE mesh used in this study consists of a row of eight-noded rectangular axisymmetric elements (total 125 elements) with four Gauss-quadrature points. Figure 4 shows a schematic representation of deformed FE mesh, boundary conditions and prescribed displacement increments at different stages of analyses. The thickness t_s of the first (leftmost) element (which is adjacent to the pile shaft) is selected in a way that at the end of the CE phase t_s falls in the range of 5 to 20 times the mean particle diameter D_{50} , as typically observed in sands (Uesugi et al. 1988; Oda et al. 2004). The formation of a shear band along the pile shaft can be correctly represented by keeping the thickness of the leftmost element consistent with the expected shear band thickness in sand. The degrees of freedom (in the vertical and radial directions) of all nodes lying along any vertical line are tied together to enforce the condition of zero normal vertical strain and deformation with no bending (i.e., no rotation of the vertical sides of the elements). Imposition of these constraints guarantee that shearing takes place only in the vertical direction and makes the analysis results independent of the thickness of the soil disk (i.e., the height of the rectangular elements). The nodes at the far-field (rightmost) boundary of the soil disk are fixed in both vertical and radial directions.

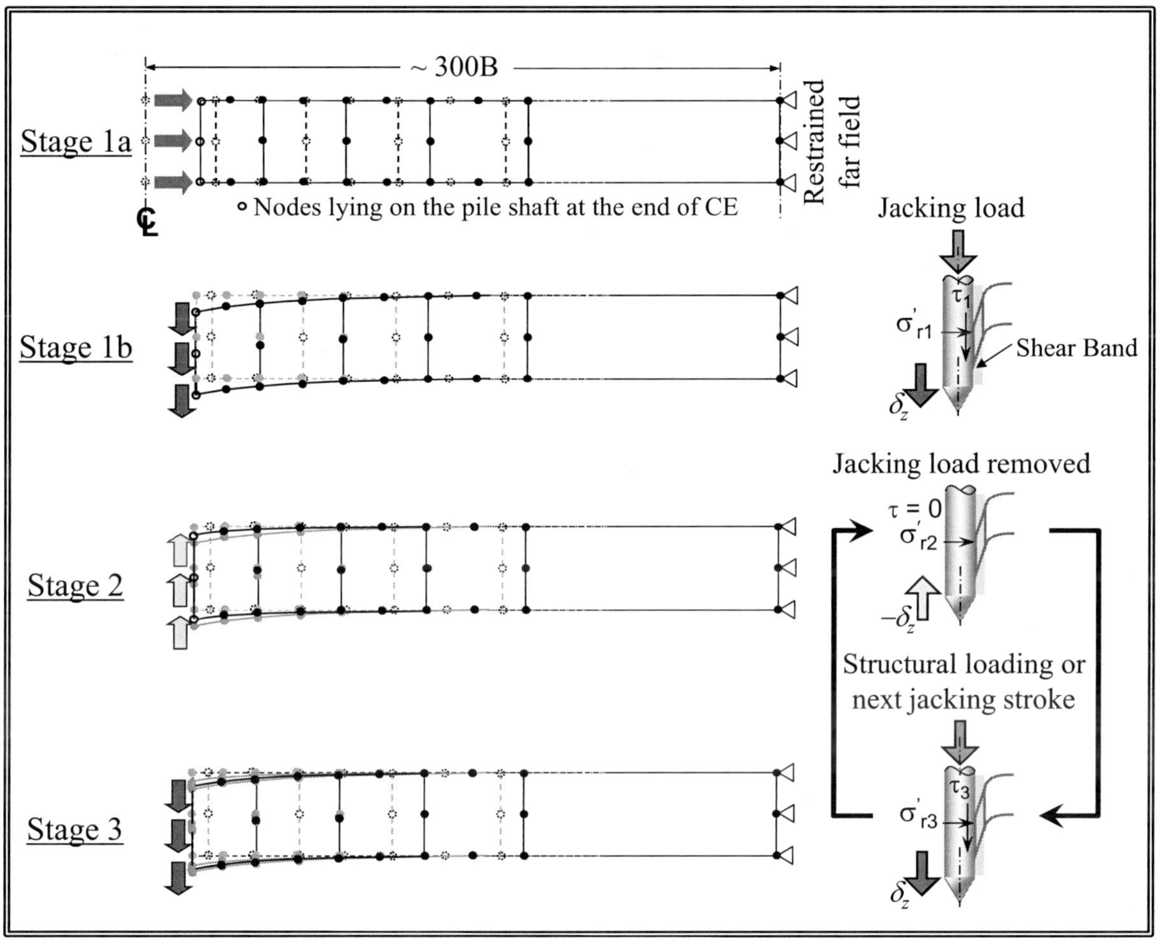

Figure 4. Finite element mesh, boundary conditions and applied displacement increments during different analysis stages

Displacement-controlled analyses are performed; the displacement increments are applied at the nodes on the left boundary of the domain (the contact surface between soil and pile shaft) and the corresponding reactions are monitored at these nodes. In particular, horizontal displacements are applied during the CE phase (Stage 1a; Fig. 4). The CE phase starts from a very small initial cavity radius r_0 (= 0.015 m) and ends when the cavity radius r becomes equal to the pile radius R, which in the present analyses is equal to 0.165m (pile diameter B equal to 0.33m). In reality, a cavity is created (expanded from zero radius) in the soil during pile installation. However, with a sufficiently small initial cavity radius (compared to the final radius), the analyses yield cavity pressure at the end of the cavity expansion process (i.e., when r/r_0 = 11) that matches the limit cavity pressure. This was confirmed by closely predicting the analytical and semi-analytical results of cavity creation problems (Yu & Houlsby 1991; Collins et al. 1992; Salgado & Randolph 2001) using the present FE formulation.

Vertical displacement increments are applied directly at the nodes of the leftmost boundary to simulate the vertical shearing (Stages 1a, 2 and 3; Fig. 4). No interface elements are placed at the left boundary (shaft wall). Therefore, any slippage between pile and soil is simulated by the formation of a vertical shear band inside the soil adjacent to the pile shaft. This condition corresponds to a perfectly rough interface expected in the case of a precast concrete pile in fine and medium sands and steel piles in fine sands (Lings & Dietz 2005).

Each vertical loading stage (primary shearing and subsequent reloading stages) is stopped when both normal and tangential reactions at the left boundary of the domain stabilize and, thus, the limit shaft resistance is reached. Negative (upward) vertical displacement increments are applied to simulate the removal of the jacking load (unloading stage) between successive jacking strokes (Stage 2; Fig. 4). Each unloading stage is stopped when the shear stress along the left boundary of the domain becomes equal to zero.

Constitutive Model for Sand

A two-surface, plasticity-based constitutive model is used to simulate the mechanical response of sand. This constitutive model, which is based on the critical-state soil mechanics framework, was originally proposed by Manzari & Dafalias (1997) and subsequently modified by Li & Dafalias (2000), Papadimitriou & Bouckovalas (2002), Dafalias et al. (2004), Loukidis (2006) and Loukidis & Salgado (2009). The model parameters used in the present analyses are calibrated for dry-deposited/air-pluviated Toyoura sand. This constitutive model is particularly suitable for the analysis of pile jacking because it can simulate sand response accurately from the early stages of each shearing cycle, the transition from contractive to dilative behavior, all the way to critical state. The model realistically captures sand response during shear stress reversals by using the small-strain shear modulus and predicting a renewed contractive phase at the beginning of each shearing cycle. The constitutive model also takes into account the inherent anisotropy due to the initial fabric, but not the potential evolution of this anisotropy upon cyclic shearing along the pile shaft. The constitutive model equations are integrated using a semi-implicit algorithm adapted with sub-incrementation and error control (Loukidis 2006). The relative stress error tolerance used is equal to 10^{-4}.

ANALYSIS RESULTS

Finite element analyses (FEAs) were performed for Toyoura sand with different values of relative density (D_R = 30, 45, 60, 75 and 90%), different values of initial vertical effective stress (σ'_{v0} = 25, 50, 100, 200 and 400 kPa) and for in situ lateral earth pressure coefficient K_0 = 0.45 (a value appropriate for normally consolidated sand). To assess the effect of inherent (fabric) anisotropy, which is due to the preferred orientation of the sand particles, on the pile-soil load transfer behavior, two sets of analyses were performed: i) considering fabric anisotropy as observed in laboratory experiments on Toyoura sand, and ii) ignoring fabric anisotropy (this is achieved by switching off the constitutive model components pertaining to fabric anisotropy). These two sets of analyses are referred to as *anisotropic* analysis and *isotropic* analysis, respectively.

Evolution of Normal and Shear Stress

The evolution of the normal and shear stresses acting on the shaft of a pile subjected to 30 jacking strokes (i.e., shear unloading-reloading cycles) is shown in Figure 5. In this figure, the initial values of the normal (radial) stresses are equal to those at the end of the CE phase (i.e., the limit cavity pressure). At the end of the CE phase, the elements very close to the shaft reach critical state, and elements not far from it are on their way to critical state.

As the primary vertical shearing phase starts, the sand behavior in the soil disk becomes predominantly contractive, leading to loss of radial stress. This happens because the direction of loading changes from horizontal during the CE phase to vertical during the primary shearing phase. This is consistent with model pile experiments by White & Bolton (2004), who observed with the aid of photogrammetry that sand particles move away from the path of the penetrating pile, but after the pile tip passes the given elevation, the direction of particle movement reverses, signifying the onset of contractive behavior in the vicinity of the shaft wall and the loss of the high stress values developed during creation of the cavity. A small gain in radial stress is subsequently observed just after the local peak in the shear stress response (Figure 5). This marginal gain in radial stress is due to dilation towards critical state inside the shear band; nevertheless, the net volumetric response during the primary shearing phase remains contractive, resulting in a net loss of radial stress. Results show that once limit shaft resistance is reached at the end of the primary shearing phase, almost half of the initial limit cavity pressure is lost due to net contractive behavior. This observation is consistent with the observations made by Azzouz et al. (1990) for displacement piles in clays. Subsequent unloading/reloading cycles lead to further net reduction in normal stress. The degradation of normal stress acting on the pile shaft is larger in case of initial anisotropic sand fabric than for an initial isotropic sand fabric.

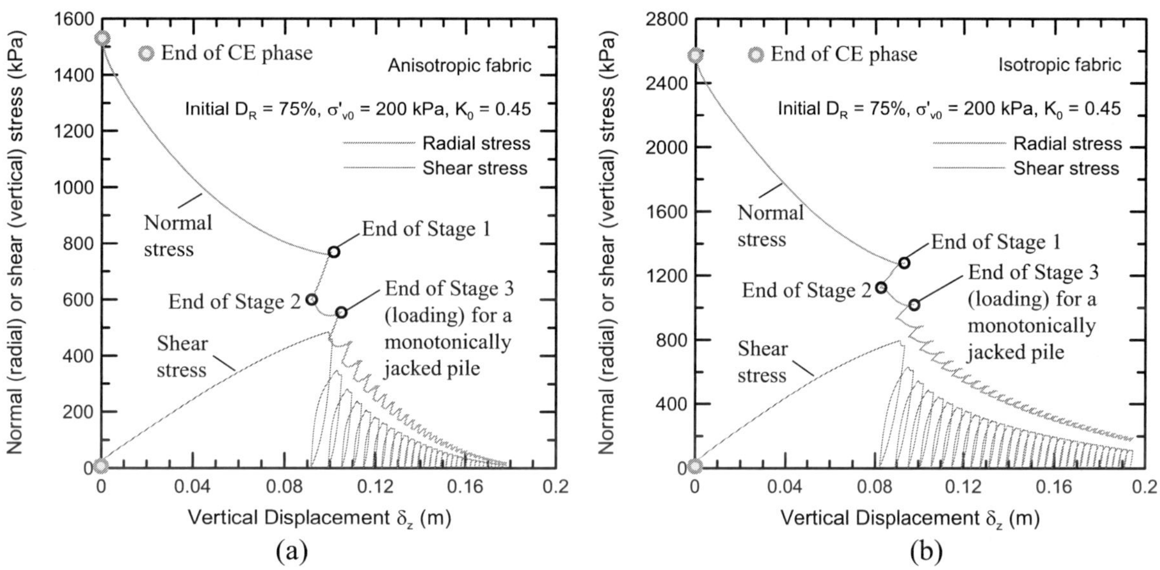

Figure 5. Evolution of normal and shear stresses on the shaft of a pile jacked in sand with (a) initially anisotropic fabric and (b) initially isotropic sand fabric

Influence of Pile Jacking in the Surrounding Soil

Figure 6 shows the distributions of radial effective stress σ'_r and relative density D_R in the surrounding soil at different stages of installation of a monotonically jacked pile. At the end of primary shearing phase, the radial stress σ'_r decreases significantly throughout the soil disk from its large value at the end of the CE phase (Fig. 6a). Within an annular zone of soil surrounding the pile with thickness approximately equal to $2B$, D_R is approximately 33% lower than its in-situ (initial) value as a result of the cavity expansion (Fig. 6b). On the contrary, the relative density outside this annular inner zone becomes slightly larger than the in-situ D_R. Subsequent removal of shearing load along the pile shaft (end of installation of a monotonically jacked pile) leads to a slight increase in D_R, particularly inside the inner annular region.

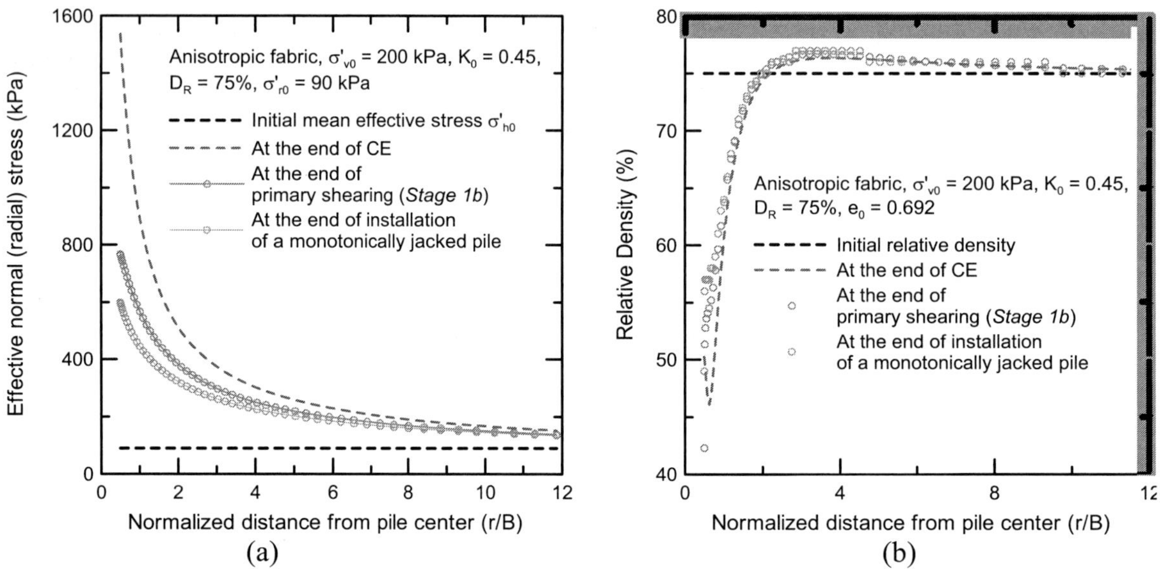

Figure 6. Influence of pile jacking in the soil surrounding the pile (a) distribution of normal stress and (b) distribution of relative density

Lateral Earth Pressure Coefficient

The ratio K/K_0 is calculated from the recorded normal reaction at the shaft wall at the end of each jacking stroke. Based on the FEA results, the following mathematical expressions are proposed to estimate K/K_0 for limit shaft resistance calculations for a monotonically installed jacked pile (i.e., number of jacking strokes $N = 1$):

$$\left(\frac{K}{K_0}\right)_{N=1} = 2.27\exp\left[\frac{D_R}{100}\left\{2.5 - 0.55\ln\left(\frac{\sigma'_{v0}}{p_A}\right)\right\}\right] \quad (1)$$

for an initially isotropic sand fabric, and

$$\left(\frac{K}{K_0}\right)_{N=1} = 0.93\exp\left[\frac{D_R}{100}\left\{2.8 - 0.45\ln\left(\frac{\sigma'_{v0}}{p_A}\right)\right\}\right] \quad (2)$$

for an initially anisotropic sand fabric, where D_R is expressed as a percentage (%) between 0 and 100 and p_A is a reference stress (= 100 kPa or equivalent in other units). It is observed that K/K_0 increases with increasing relative density and decreasing initial confinement (a direct consequence of the increased dilatancy of the soil). The K/K_0 for the anisotropic case is always smaller than the one for the isotropic case because the installation of a jacked pile involves stress paths and loading modes in which the direction of the principal stress increment is closer to the horizontal plane (which the preferred orientation of the particle long axis for sands deposited under the action of gravity) than to the vertical plane.

The coefficient of lateral earth pressure calculated using Eqs. 1 and 2 can be used to estimate the limit shaft resistance q_{sL} of axially loaded jacked piles as:

$$q_{sL} = (K\tan\delta)\sigma'_{v0} \quad (3)$$

where δ = friction angle mobilized at the pile-soil interface. The value of δ in Eq. 3 depends on the critical-state friction angle ϕ_c of the soil and the relative roughness of the pile-soil interface and falls within a well-defined range; in the order of $0.95\phi_c$ for precast concrete piles and 0.85 to $0.9\phi_c$ for steel piles (Salgado 2008). Thus, unlike K, δ is independent of the soil state prior to pile installation.

Effect of Number of Jacking Stroke N

After establishing (K/K_0) for a monotonically jacked pile (N = 1), the ratio K/K_0 for a pile installed using multiple jacking strokes (N > 1) can be estimated by multiplying $(K/K_0)_{N=1}$ by a degradation ratio (or friction fatigue coefficient) μ_N. Based on the FEA results, the following equation is established:

$$\mu_N = 0.02 + 0.98\exp\left\{-C_1(N-1)^{C_2}\right\} \tag{4}$$

where C_1 and C_2 are coefficients that depend on the initial soil state.

$$C_1 = 0.2\left(\frac{p_a}{\sigma'_{v0}}\right)^{0.3} \text{ and } C_2 = 1 - 0.4\left(\frac{D_R}{100}\right)^{1.1} \text{ (for initially isotropic fabric)} \tag{5}$$

$$C_1 = 0.25\left(\frac{p_a}{\sigma'_{v0}}\right)^{0.3} \text{ and } C_2 = 1 - 0.3\left(\frac{D_R}{100}\right)^{2} \text{ (for initially anisotropic sand fabric)} \tag{6}$$

Figure 7 shows the variation of the friction fatigue coefficient μ_N with the number of jacking strokes N for anisotropic and isotropic sand fabric and D_R = 75%. For a particular number of jacking strokes and initial soil state, μ_N is always lower for the anisotropic analysis than for the isotropic analysis. The reduction of μ_N with N is faster for low than for high confining stresses.

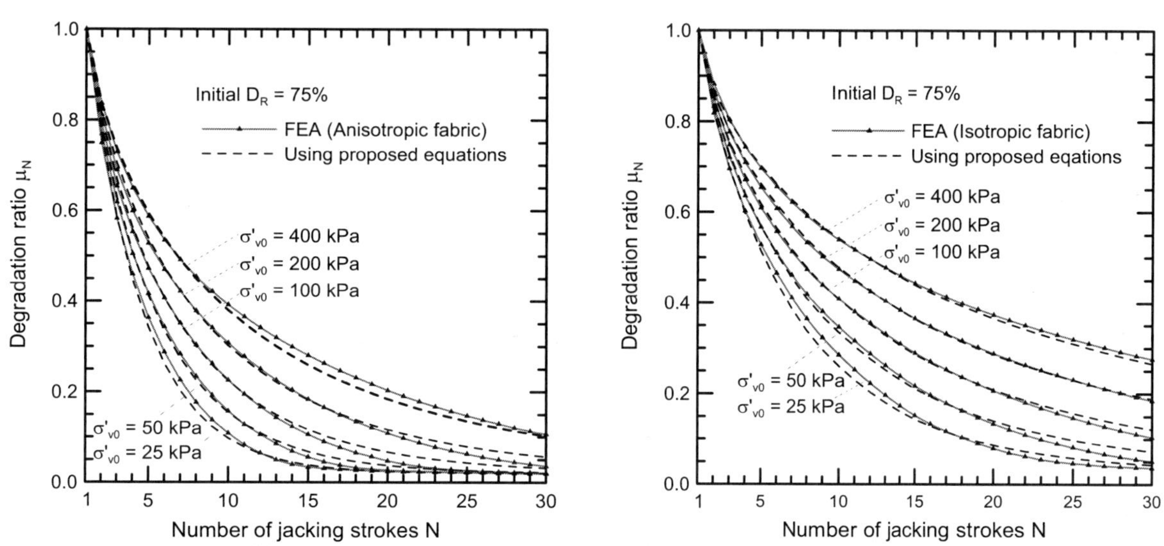

Figure 7. Friction fatigue coefficient μ_N for (a) anisotropic analysis and (b) isotropic analysis

CONCLUSIONS

Axisymmetric finite element analyses are performed to investigate the mechanism of friction fatigue in piles jacked in sand. The number of jacking strokes used for jacking a pile in sand plays an important role in determining the overall shaft resistance, as cycles of shearing along the pile shaft during pile installation cause significant degradation of the normal stress acting on the pile shaft-soil interface. For monotonic jacking, the ratio K/K_0 increases with increasing relative density D_R and decreasing initial vertical effective stress σ'_{v0}. The friction fatigue, i.e. the degradation of K/K_0 with increasing number of jacking strokes N becomes more intense for large D_R and small σ'_{v0} values.

REFERENCES

Abbo, A.J. & Sloan, S.W. (2000). "*SNAC user manual version 2.0*". Dept. of Civil, Surveying and Environmental Engineering, University of Newcastle, Australia.

Azzouz, A.S., Baligh M.M. & Whittle, A.J. (1990). "Shaft resistance of piles in clay". *J. Geotech. Eng.*, 116 (2), 205–221.

Babushkin, G.U. & Ginzburg, L.K. (1984). "Strengthening of foundations on collapsible soils by means of jacked piles". *Soil Mechanics and Foundation Engineering*, 21(1), 14–20.

Briaud, J-L & Tucker, L. (1984). "Piles in sand: a method including residual stresses". *J. Geotech. Eng.*, 110(11), 1666–1680.

Chow, F.C. (1997). "*Investigations into the behavior of displacement piles for offshore foundations*". Ph.D. thesis, University of London (Imperial College).

Collins, I.F., Pender, M.J. & Yan, W. (1992). "Cavity expansion in sands under drained loading conditions". *Int. J. Num. & Ana. Meth. in Geomech.*, 16(1), 3-23.

Colombi, A. (2005). "*Physical modeling of an isolated pile in coarse grained soil*". Ph.D. Thesis. University of Ferrara.

Dafalias, Y.F., Papadimitriou, A.G. & Li, X.S. (2004). "Sand plasticity model accounting for inherent fabric anisotropy". *J. Eng. Mech.*; 130(11), 1319–1333.

Darrag, A.A. & Lovell, C.W. (1989). "A simplified procedure for predicting residual stresses for piles". *Proc., 12th Intl. Conf. on Soil Mech. & Foundation Eng.*, Rio de Janeiro, Brazil, 1127–1130.

DeJong, J.T. (2001). "*Investigation of particulate-continuum interface mechanisms and their assessment through a multi-friction sleeve penetrometer attachment*". Ph.D. thesis, Georgia Institute of technology, USA.

Fleming, W.G.K., Weltman, A.J., Randolph, M.F. & Elson, W.K. (1992). *Piling Engineering*, Surrey University Press, Surrey.

Fugro. (1996). Euripides database report, Fugro Engineers BV, The Netherlands and Geodia SA, 1–5.

Jardine, R. & Chow, F.C. (1998). *Research into the behavior of displacement piles for offshore foundations*. Health and Safety Executive, OTO 98 833.

Lehane, B.M. (1992). *Experimental investigations of pile behavior using instrumented field piles*, Ph.D. Thesis, University of London (Imperial College).

Lehane B.M., Jardine, R.J., Bond, A.J. & Frank, R. (1993). "Mechanisms of shaft friction in Sand from instrumented pile tests". *J. Geotech. Eng.*, 119(1), 19–35.

Lehane B.M., Schneider J.A. & Xu, X. (2005). "The UWA-05 method for prediction of axial capacity of driven piles in sand". *Proc., Int. Sym. on Frontiers in Offshore Geotechnics*, Perth, Australia, 683-689.

Leung, C.F, Lee, F.H, & Yet, N.S. (1998). "Behavior of pile subjected to lapses during installation". *Centrifuge 98*, Balkema, Rotterdam, 491-494.

Li, X.S & Dafalias, Y.F. (2000). "Dilatancy of cohesionless soils". *Géotechnique*, 50(4), 449-460.

Lings M.L. & Dietz, M.S. (2005). "The peak strength of sand–steel interfaces and the role of dilation". *Soils and Foundations*, 45(6),1-14.

Loukidis, D. (2006). "*Advanced constitutive modeling of sands and applications to foundation engineering*". Ph.D. thesis, Purdue University, USA.

Loukidis, D. & Salgado, R. (2009). "Analysis of the shaft resistance of non-displacement piles in sand". *Géotechnique*, 58(4), 283-296.

Loukidis, D. & Salgado, R. (2009). "Modeling sand response using two-surface plasticity". *Comp. & Geotechnics*, 36(1–2), 166–186.

Manzari, M.T. & Dafalias, Y.F. (1997). "A critical state two-surface plasticity model for sands". *Géotechnique*, 47(2), 255-272.

Nordlund, R.L. (1979). "Point Bearing and Shaft Friction of Piles in Sand". *Proc., 5th Annual Fundamentals of Deep Foundation Design*, University of Missouri-Rolla.

Oda, M., Takemura, T. & Takahashi, M. (2004). "Microstructure in shear band observed by microfocus X-ray computed tomography". *Géotechnique*, 54(8), 539-542.

Papadimitriou, A.G. & Bouckovalas, G.D. (2002). "Plasticity model for sand under small and large cyclic strains: a multiaxial formulation". *Soil Dyn. & Earthquake Eng.*, 22(3), 191-204.

Potts, D.M. & Martins, J.P. (1982). "The shaft resistance of axially loaded piles in clay". *Géotechnique*, 32(4), 369-386.

Randolph, M.F. (2003). "Science and empiricism in pile foundation design". *Géotechnique*, 53(10), 847-875.

Randolph, M.F., Carter, J.P. & Wroth, C.P. (1979). "Driven piles in clay—the effects of installation and subsequent consolidation". *Géotechnique*, 29(4), 361-393.

Randolph, M.F, Dolwin, J., & Beck, R. (1994). "Design of driven piles in sand". *Géotechnique*, 44(3), 427-448.

Randolph, M.F. & Wroth, C.P. (1978). "Analysis of deformation of vertically loaded piles". *J. Geotech. Eng.* 104(12), 1465-1488.

Salgado, R., Mitchell, J.K. & Jamiolkowski, M. (1997). "Cavity expansion and penetration resistance in sands". *J. Geotech. & Geoenv. Eng.*, 123(4), 344-354.

Salgado, R. & Randolph, M.F. (2001). "Analysis of cavity expansion in sand". *Int. J. Geomech.*, 1(2), 175-192.

Salgado, R. & Prezzi, M. (2007). "Computation of cavity expansion pressure and penetration resistance in sands". *Int. J. Geomech.* 7(4), 251-265.

Salgado, R. (2008). *The Engineering of Foundations*. Mc Graw-Hill: New York.

Uesugi, M., Kishida, H., and Tsubakihira, Y. (1988). "Behavior of sand particles in sand-steel friction". *Soils and Foundations*, 28(1), 107-118.

White, D. J. & Bolton, M. (2002). "Observing friction fatigue on a jacked pile". *Constitutive and Centrifuge Modeling: Two Extremes*, Balkema, Rotterdam, 347–354.

White, D.J. & Bolton, M. (2004). "Displacement and strain paths during plane-strain model pile installation in sand". *Géotechnique*, 54(6), 375-397.

White, D.J. & Lehane, B.M. (2004). "Friction fatigue on displacement piles in sand". *Géotechnique*, 54(10), 645-658.

White, D.J. (2005). "A general framework for shaft resistance on displacement piles in sand. Frontiers in offshore geotechnics", *Proc. of ISFOG 2005*, London, 697–703.

Yang, J., Tham, L. G., Lee, P.K.K., & Yu, F. (2006). "Observed performance of long steel H-piles jacked into sandy soils". *J. Geotech. & Geoenv. Eng.* 132(1), 24-35

Yu H. S. & G. T.Houlsby, 1991. Finite cavity expansion in dilatant soils: loading analysis. Géotechnique; Vol. 41, No. 2:173–183.

A Framework for Analysis of Piles with Rectangular Cross Section

Dipanjan Basu[1], C Eng., M.ASCE, Rodrigo Salgado[2], P.E., F.ASCE, and Monica Prezzi[3], A.M.ASCE

[1]Assistant Professor, Department of Civil and Environmental Engineering, University of Waterloo, 200 University Avenue West, Waterloo, N2L 3G1, Canada; dipanjan.basu@uwaterloo.ca

[2]Professor, School of Civil Engineering, Purdue University, 550 Stadium Mall Drive, West Lafayette, IN 47907; rodrigo@purdue.edu

[3]Professor, School of Civil Engineering, Purdue University, 550 Stadium Mall Drive, West Lafayette, IN 47907; mprezzi@purdue.edu

ABSTRACT: An analysis framework is presented for piles with rectangular cross section. Existing analysis and design methods are mostly applicable for piles with circular cross section. For piles with cross section other than a circle, an equivalent circle is often assumed and the methods for circular piles are applied. The newly developed framework explicitly takes into account the rectangular cross section and produces the response of piles under axial and lateral loads. A rational soil displacement field surrounding the rectangular cross section of the pile is assumed, and the total potential energy of the loaded pile-soil system is considered. The potential energy is minimized using calculus of variations to obtain the differential equations governing the pile and soil displacements. Closed-form solutions are obtained for pile settlement and axial force in the case of axially loaded piles. For laterally loaded piles, closed-form solutions are obtained for lateral deflection, slope of the deflected curve, pile bending moment and shear force. The input parameters needed for the analysis are the pile geometry, applied load and the elastic constants of the soil and pile. The new analysis framework produces results in seconds and has the accuracy of equivalent three-dimensional finite element analysis.

INTRODUCTION

Existing analysis and design methods for piles are mostly developed for circular cross sections (Dennis and Olson 1983a, b). Although piles with circular cross sections (e.g., drilled shafts, pipe piles, and CFA piles) are common, there are a large number of piles used in practice that have rectangular cross sections (e.g., precast concrete piles and barrettes). It is a common practice to convert the rectangular cross section into equivalent circle for the purpose of analysis and design. However, such conversions are approximate and more accurate results can be obtained if the actual rectangular cross section of the pile can be retained in the analysis. In this paper, a framework for analysis of axially loaded and laterally loaded piles with rectangular cross sections is presented. The analysis is developed following the variational principles of mechanics for an assumed soil displacement field surrounding the rectangular cross section. Closed-form solutions are obtained for pile settlement and axial force in the case of axially loaded piles. For laterally loaded piles, closed-form solutions are obtained for lateral deflection, slope of the deflected curve, pile bending moment and shear force.

Settlement analysis of axially loaded piles can be broadly classified into load-transfer methods and continuum-based methods. The load-transfer methods (Seed and Reese 1957, Coyle and Reese 1966, Randolph and Wroth 1978, Armaleh and Desai 1987, Guo and Randolph 1997,

Mylonakis 2001) essentially model the soil as a series of Winkler springs with spring constants defined in a variety of ways (e.g., t-z curves). Approximate analytical or numerical solutions of pile settlement have been obtained using this approach. In the continuum approach, the soil is represented as a continuum, and numerical techniques such as the boundary integral method, the finite layer method or the finite element method are used to obtain solutions (Poulos and Davis 1968, Butterfield and Banerjee 1971, Randolph and Wroth 1978, Rajapakse 1990).

Analysis of laterally loaded piles can also be classified into load-transfer methods and continuum-based methods. The load-transfer methods typically include the p-y method and other similar approaches in which the lateral soil resistance is represented by Winkler springs, and approximate analytical or numerical solutions of pile response are obtained (Reese and Matlock 1956, Broms 1964, Matlock 1970, Reese et al. 1974, 1975, O'Neill et al. 1990). The continuum-based methods typically include the boundary element method, the finite difference method or the finite element method (Poulos 1971a, b; Banerjee and Davies 1978, Randolph 1981, Verruijt and Kooijman 1989, Trochanis et al. 1991, Ng and Zhang 2001).

The load-transfer (spring-based) methods mentioned above are typically used in routine projects because of their ease of use and because of the speed with which results are obtained. However, these methods may not always produce accurate results because they are developed by calibrating numerical results with the results of a limited number of field tests performed mostly on circular cross section piles, and do not explicitly take into account the pile cross section shape. The continuum-based numerical methods, on the other hand, are theoretically more rigorous and can explicitly take into account the shape of the pile cross section. In spite of that, the use of continuum-based methods is rather limited, both in research and in practice, because they require specially trained personnel and are often time consuming. Recently, Basu et al. (2008) and Basu and Salgado (2008) developed continuum-based analysis of rectangular cross section piles subjected to axial and lateral loads, respectively, which are integrated in a unified framework and presented in this paper. The advantages of this analysis framework are that it is rigorous, explicitly captures the rectangular pile cross section, and produces results in seconds.

ANALYSIS FRAMEWORK

Problem Definition

A pile with a rectangular cross section with dimensions $2a \times 2b$ and length L_p is considered embedded in a layered soil medium containing n layers (Figure 1). Each soil layer extends to infinity in all horizontal directions, and the bottom (n^{th}) layer extends to infinity in the downward direction. The depth to the base of any intermediate layer i is H_i, which implies that the thickness of the ith layer is $H_i - H_{i-1}$ with $H_0 = 0$. The top of the pile (head) is at the level of the ground surface, and the bottom (base) is embedded in the nth layer. The pile is subjected to a vertical axial force F_v at the head for the case of an axially loaded pile (Case 1) and to a horizontal transverse force F_h and moment M_o at the head for the case of a laterally loaded pile (Case 2). F_h and M_o are orthogonal as vectors (Figure 1); the vector F_h acts perpendicular to the face with a width of $2b$, and the vector M_o acts perpendicular to the face with a width of $2a$. A rectangular Cartesian coordinate system is chosen with its origin located at the center of the pile head such that the positive z-axis points downward and coincides with the pile axis and the vertical force F_v. The positive x-axis is aligned in the direction of the horizontal force vector F_h, and the moment vector M_o acts in the direction of the negative y-axis (Figure 1).

The soil medium is assumed to be elastic, isotropic and homogeneous within each layer, with Lame's constants λ_{si} and G_{si} for the ith layer. There is no slippage or separation between the pile and the surrounding soil or between the soil layers. For the case of axially loaded pile (Case 1), the pile behaves as a column (rod) with an axial stiffness E_pA_p (E_p is the Young's modulus of pile and $A_p = 2a \times 2b$ is the area of pile cross section). For the case of laterally loaded pile (Case

2), the pile behaves as an Euler-Bernoulli beam with a flexural rigidity E_pI_p ($I_p = 4ba^3/3$ is the second moment of inertia of the pile cross section for bending in the x-z plane).

Soil Displacement, Strain and Stress Fields

A soil displacement field, compatible with the rectangular geometry of the pile cross section, is assumed around the pile (Figure 2). A separable variable technique is used to describe the soil displacement so that the variations of the displacement in each of the coordinate directions are uncoupled. For the axially loaded pile (Case 1), it is assumed that the horizontal soil displacements u_x and u_y are negligible (i.e., $u_x = 0$ and $u_y = 0$) and that the vertical soil displacement u_z is given by

$$u_z = u(z)\,\phi_x^{(v)}(x)\,\phi_y^{(v)}(y) \tag{1a}$$

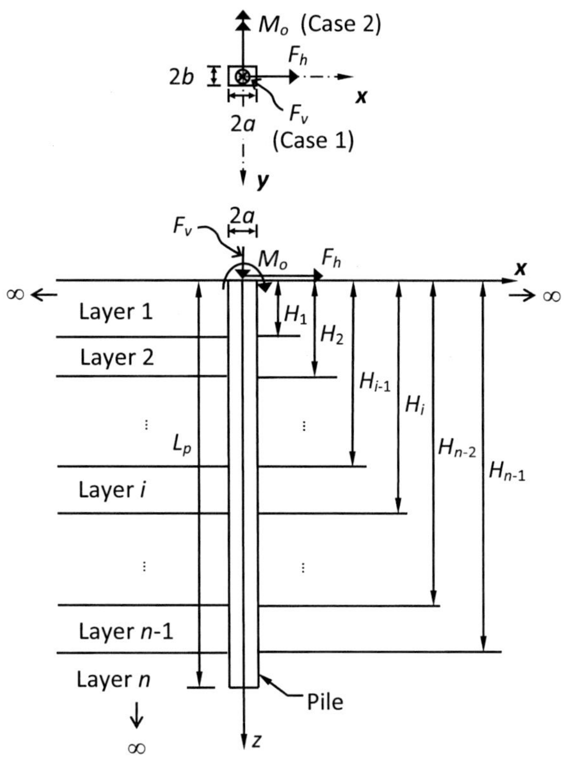

Figure 1. Axially and laterally loaded pile.

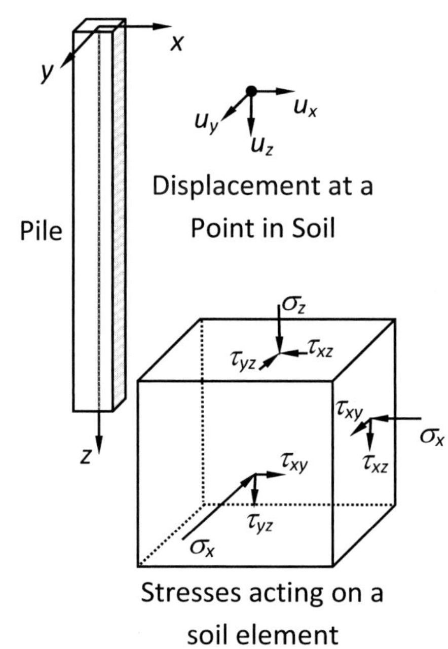

Figure 2. Displacement and stresses in soil.

where $u(z)$ is a displacement function (with a dimension of length) varying with depth z, which describes the axial pile settlement; $\phi_x^{(v)}(x)$ is a dimensionless displacement function varying along x; and $\phi_y^{(v)}(y)$ is a dimensionless displacement function varying along y. These functions are set equal to one at the pile-soil interface (so that the soil displacement u_z becomes equal to the pile displacement u at the interface), which ensures displacement continuity (strain compatibility) between the pile and soil. The dimensionless functions are set equal to zero at infinite horizontal distance from the pile, which ensures that the soil displacements due to pile movement decrease with increase in horizontal distance from the pile and become negligible at sufficiently large horizontal distance.

For the laterally loaded pile (Case 2), it is assumed that the displacements in the z (vertical) and y directions are negligible (i.e., $u_z = 0$ and $u_y = 0$), and the horizontal displacement in the x direction is given by

$$u_x = w(z)\,\phi_x^{(h)}(x)\,\phi_y^{(h)}(y) \tag{1b}$$

where $w(z)$ is a displacement function (with a dimension of length) varying with depth z, which describes the lateral pile deflection; and $\phi_x^{(h)}(x)$ and $\phi_y^{(h)}(y)$ are dimensionless displacement functions varying along x and y that ensures perfect contact between the pile and soil at the interface (displacement continuity) and describe how the displacements in the soil mass (due to pile deflection) decrease with an increase in horizontal distance from the pile ($\phi_x^{(h)}(x)$ and $\phi_y^{(h)}(y)$ are set equal to one at the pile-soil interface and equal to zero at an infinite horizontal distance from the pile).

The strain fields in the soil can be derived from the displacement fields (with the compression positive sign convention) as

$$\begin{bmatrix}\varepsilon_{xx} & \varepsilon_{yy} & \varepsilon_{zz} & \varepsilon_{xy} & \varepsilon_{xz} & \varepsilon_{yz}\end{bmatrix} = \begin{bmatrix}0 & 0 & -\dfrac{du}{dz}\phi_x^{(v)}\phi_y^{(v)} & 0 & -\dfrac{1}{2}u\dfrac{d\phi_x^{(v)}}{dx}\phi_y^{(v)} & -\dfrac{1}{2}u\phi_x^{(v)}\dfrac{d\phi_y^{(v)}}{dy}\end{bmatrix} \tag{2a}$$

for axially loaded piles and as

$$\begin{aligned}&\begin{bmatrix}\varepsilon_{xx} & \varepsilon_{yy} & \varepsilon_{zz} & \varepsilon_{xy} & \varepsilon_{xz} & \varepsilon_{yz}\end{bmatrix}\\ &= \begin{bmatrix}-w(z)\dfrac{d\phi_x^{(h)}(x)}{dx}\phi_y^{(h)}(y) & 0 & 0 & -\dfrac{1}{2}w(z)\phi_x^{(h)}(x)\dfrac{d\phi_y^{(h)}(y)}{dx} & -\dfrac{1}{2}\dfrac{dw(z)}{dz}\phi_x^{(h)}(x)\phi_y^{(h)}(y) & 0\end{bmatrix}\end{aligned} \tag{2b}$$

for laterally loaded piles, where ε_{pq} (= $[u_{p,q} + u_{q,p}]/2$) is the strain tensor (the indices p and q can be x, y or z and, q in the subscript means differentiation with respect to q). The corresponding stress fields can be obtained by using the elastic constitutive relationship, expressed in terms of indicial notation as

$$\sigma_{pq} = \lambda_s \delta_{pq}\varepsilon_{kk} + 2G_s\varepsilon_{pq} \tag{3}$$

where σ_{pq} is the stress tensor, δ_{pq} is the Kronecker's delta (summation is implied by repetition of index).

Potential Energy of Pile-Soil System and its Minimization

Using the above stress-strain fields, the potential energy Π of the pile-soil system can be quantified as

$$\begin{aligned}\Pi &= \frac{1}{2}E_pA_p\int_0^{L_p}\left(\frac{du}{dz}\right)^2 dz + \frac{1}{2}\int_{\Omega_0}\sigma_{pq}\varepsilon_{pq}d\Omega - F_v u\big|_{z=0}\\ &= \frac{1}{2}E_pA_p\int_0^{L_p}\left(\frac{du}{dz}\right)^2 dz + \frac{1}{2}\int_{\Omega_0}\left[(\lambda_s + 2G_s)\left(\frac{du}{dz}\right)^2\left(\phi_x^{(v)}\right)^2\left(\phi_y^{(v)}\right)^2 + G_su^2\left(\frac{d\phi_x^{(v)}}{dx}\right)^2\left(\phi_y^{(v)}\right)^2\right.\\ &\quad\left. + G_su^2\left(\phi_y^{(v)}\right)^2\left(\frac{d\phi_y^{(v)}}{dy}\right)^2\right]d\Omega - F_a u\big|_{z=0}\end{aligned} \tag{4a}$$

for axially loaded piles, and as

$$\begin{aligned}\Pi &= \frac{1}{2}E_pI_p\int_0^{L_p}\left(\frac{d^2w}{dz^2}\right)^2 dz + \frac{1}{2}\int_{\Omega_0}\sigma_{pq}\varepsilon_{pq}d\Omega - F_h w\big|_{z=0} + M_o\frac{dw}{dz}\bigg|_{z=0}\\ &= \frac{1}{2}E_pI_p\int_0^{L_p}\left(\frac{d^2w}{dz^2}\right)^2 dz + \frac{1}{2}\int_{\Omega_0}\left[(\lambda_s + 2G_s)w^2\left(\frac{d\phi_x^{(h)}}{dx}\right)^2\left(\phi_y^{(h)}\right)^2 + G_sw^2\left(\phi_x^{(h)}\right)^2\left(\frac{d\phi_y^{(h)}}{dy}\right)^2\right.\end{aligned}$$

$$+G_s\left(\frac{dw}{dz}\right)^2\left(\phi_x^{(h)}\right)^2\left(\phi_y^{(h)}\right)^2\Bigg]d\Omega - F_h\, w\big|_{z=0} + M_o\left.\frac{dw}{dz}\right|_{z=0} \tag{4b}$$

for laterally loaded piles. In the right hand side of equations (4a) and (4b), the first integral represents the internal potential energy of the pile, the second integral represents the internal potential energy of the soil, and the remaining term(s) represent(s) the external potential energy. In the second integral, the soil domain Ω_0 extends from $-\infty$ to $+\infty$ in the x and y directions and from 0 to $+\infty$ in the z direction, but excludes the volume $L_p \times 2a \times 2b$ occupied by the pile.

The total potential energy is minimized to obtain the equilibrium configuration of the pile-soil system under the externally acting forces. This is achieved by setting the first variation of the total potential energy equal to zero (i.e., $\delta\Pi = 0$) and manipulating the equation using calculus of variations, which produces an equation of the form:

$$\int_{\Upsilon_0}\left[A(v)\delta v + B(\phi_x)\delta\phi_x + C(\phi_y)\delta\phi_y\right]d\Upsilon + D(v)\delta v\big|_{\Gamma_i} + E(\phi_x)\delta\phi_x\big|_{\Gamma_j} + F(\phi_y)\delta\phi_y\big|_{\Gamma_k} = 0 \tag{5}$$

where the function v is to be replaced by u for axially loaded pile and by w for laterally loaded piles, ϕ_x is $\phi_x^{(v)}$ or $\phi_x^{(h)}$ and ϕ_y is $\phi_y^{(v)}$ or $\phi_y^{(h)}$, Υ_0 represents the domain and Γ_i represents a boundary. In the above equations, $A(\cdot)$, $B(\cdot)$, $C(\cdot)$, $D(\cdot)$, $E(\cdot)$ and $F(\cdot)$ are differential operators. In order for the above equation to be valid, the individual terms must be equal to zero, i.e., $A(v)\delta v = 0$, $B(\phi_x)\delta\phi_x = 0$, $C(\phi_y)\delta\phi_y = 0$, $D(v)\delta v = 0$, $E(\phi_x)\delta\phi_x = 0$ and $F(\phi_y)\delta\phi_y = 0$. These individual terms can each be zero if the functions v, ϕ_x and ϕ_y are known a priori (in which case, their respective variations are zero, i.e., $\delta v = 0$, $\delta\phi_x = 0$ and $\delta\phi_y = 0$) or if the differential operators are each equal to zero, leading to equations of the form: $A(v) = 0$, $B(\phi_x) = 0$, $C(\phi_y) = 0$, $D(v) = 0$, $E(\phi_x) = 0$ and $F(\phi_y) = 0$. Within the domain Υ_0 and over part of the boundaries Γ_i (called the natural boundaries), the functions v, ϕ_x, and ϕ_y are not known a priori and, therefore, the equations represented by $A(v) = 0$, $B(\phi_x) = 0$ and $C(\phi_y) = 0$ give the differential equations for the domain Υ_0 and the equations $D(v) = 0$, $E(\phi_x) = 0$ and $F(\phi_y) = 0$ give the natural boundary conditions for the natural boundaries. At the remaining portions of the boundaries Γ_i (called the essential boundaries), v, ϕ_x, and ϕ_y are known or prescribed and automatically satisfy equation (5). Since these differential equations and boundary conditions of the displacement functions $v(z)$, $\phi_x(x)$ and $\phi_y(y)$ are obtained by satisfying the minimum potential energy, solutions of these equations give the equilibrium configuration of the pile-soil system.

Differential Equations and Solutions of Soil Displacement Functions

The differential equations governing the functions $\phi_x^{(v)}$ and $\phi_x^{(h)}$ have the same mathematical form and can be generically written as

$$\frac{d^2\phi_x}{dx^2} - \left(\frac{\gamma_x}{r_p}\right)^2\phi_x = 0 \tag{6}$$

where $\phi_x = \phi_x^{(v)}$ and $\gamma_x = \gamma_x^{(v)}$ for axially loaded piles, $\phi_x = \phi_x^{(h)}$ and $\gamma_x = \gamma_x^{(h)}$ for laterally loaded piles, and $r_p = \sqrt{ab}$. The differential equations governing the functions $\phi_y^{(v)}$ and $\phi_y^{(h)}$ also have the same mathematical form as that of equation (6) and can be generically represented by replacing the variables ϕ_x and γ_x by ϕ_y and γ_y, respectively. The boundary conditions of ϕ_x are: $\phi_x = 0$ at $x = \pm\infty$ and $\phi_x = 1$ for $-a \le x \le a$. The boundary conditions of ϕ_y are also mathematically similar: $\phi_y = 0$ at $y = \pm\infty$ and $\phi_y = 1$ for $-b \le x \le b$. Therefore, the solutions of ϕ_x and ϕ_y have the same mathematical form, and are given by

$$\phi_x = \begin{cases} = e^{\gamma_x \frac{x+a}{r_p}}; & -\infty < x \le -a \\ = 1; & -a \le x \le a \\ = e^{-\gamma_x \frac{x-a}{r_p}}; & a \le x < \infty \end{cases} \tag{7a}$$

$$\phi_y = \begin{cases} = e^{\gamma_y \frac{y+b}{r_p}}; & -\infty < y \le -b \\ = 1; & -b \le y \le b \\ = e^{-\gamma_y \frac{y-b}{r_p}}; & b \le y < \infty \end{cases} \tag{7b}$$

The dimensionless parameters $\gamma_x^{(v)}$ and $\gamma_y^{(v)}$ for axially loaded piles are given by

$$\left(\gamma_x^{(v)}\right)^2 = \frac{1}{\psi^2 \left[\left(\frac{r_p}{\gamma_y} + 2b \right) \sum_{i=1}^{n} G_{si} \int_{\tilde{H}_{i-1}}^{\tilde{H}_i} \tilde{u}_i^2 d\tilde{z} + \left(\frac{r_p}{\gamma_y} + 2b \right) G_{sn} \left[\tilde{u}_n^2 \big|_{\tilde{z}=1} \right] \sqrt{\frac{\tilde{t}_{n+1}^{(v)}}{2\tilde{k}_n^{(v)}}} \right]} \times \\ \left[\left(\frac{r_p}{\gamma_y} + 2b \right) \left\{ \sum_{i=1}^{n} (\lambda_{si} + 2G_{si}) \int_{\tilde{H}_{i-1}}^{\tilde{H}_i} \left(\frac{d\tilde{u}_i}{d\tilde{z}} \right)^2 d\tilde{z} + (\lambda_{sn} + 2G_{sn}) \left[\tilde{u}_n^2 \big|_{\tilde{z}=1} \right] \sqrt{\frac{\tilde{k}_n^{(v)}}{8\tilde{t}_{n+1}^{(v)}}} \right\} \right. \\ \left. + \frac{L_p^2 \gamma_y^{(v)}}{r_p} \left\{ \sum_{i=1}^{n} G_{si} \int_{\tilde{H}_{i-1}}^{\tilde{H}_i} \tilde{u}_i^2 d\tilde{z} + G_{sn} \left[\tilde{u}_n^2 \big|_{\tilde{z}=1} \right] \sqrt{\frac{\tilde{t}_{n+1}^{(v)}}{2\tilde{k}_n^{(v)}}} \right\} \right] \tag{8a}$$

$$\left(\gamma_y^{(v)}\right)^2 = \frac{1}{\psi^2 \left[\left(\frac{r_p}{\gamma_x^{(v)}} + 2a \right) \sum_{i=1}^{n} G_{si} \int_{\tilde{H}_{i-1}}^{\tilde{H}_i} \tilde{u}_i^2 d\tilde{z} + \left(\frac{r_p}{\gamma_x^{(v)}} + 2a \right) G_{sn} \left[\tilde{u}_n^2 \big|_{\tilde{z}=1} \right] \sqrt{\frac{\tilde{t}_{n+1}^{(v)}}{2\tilde{k}_n^{(v)}}} \right]} \times \\ \left[\left(\frac{r_p}{\gamma_x^{(v)}} + 2a \right) \left\{ \sum_{i=1}^{n} (\lambda_{si} + 2G_{si}) \int_{\tilde{H}_{i-1}}^{\tilde{H}_i} \left(\frac{d\tilde{u}_i}{d\tilde{z}} \right)^2 d\tilde{z} + (\lambda_{sn} + 2G_{sn}) \left[\tilde{u}_n^2 \big|_{\tilde{z}=1} \right] \sqrt{\frac{\tilde{k}_n^{(v)}}{8\tilde{t}_{n+1}^{(v)}}} \right\} \right. \\ \left. + \frac{L_p^2 \gamma_x^{(v)}}{r_p} \left\{ \sum_{i=1}^{n} G_{si} \int_{\tilde{H}_{i-1}}^{\tilde{H}_i} \tilde{u}_i^2 d\tilde{z} + G_{sn} \left[\tilde{u}_n^2 \big|_{\tilde{z}=1} \right] \sqrt{\frac{\tilde{t}_{n+1}^{(v)}}{2\tilde{k}_n^{(v)}}} \right\} \right] \tag{8b}$$

where $\psi = L_p / r_p$. The dimensionless parameters $\gamma_x^{(h)}$ and $\gamma_y^{(h)}$ for laterally loaded piles are given by

$$\left(\gamma_x^{(h)}\right)^2 = \frac{1}{\psi^2 \left[\left(\frac{r_p}{\gamma_y^{(h)}} + 2b \right) \sum_{i=1}^{n} (\lambda_{si} + 2G_{si}) \int_{\tilde{H}_{i-1}}^{\tilde{H}_i} \tilde{w}_i^2 d\tilde{z} + \left(\frac{r_p}{\gamma_y^{(h)}} + 2b \right) (\lambda_{sn} + 2G_{sn}) \left[\tilde{w}_n^2 \big|_{\tilde{z}=1} \right] \sqrt{\frac{\tilde{t}_{n+1}^{(h)}}{2\tilde{k}_n^{(h)}}} \right]} \times \\ \left[\left(\frac{r_p}{\gamma_y^{(h)}} + 2b \right) \sum_{i=1}^{n} G_{si} \int_{\tilde{H}_{i-1}}^{\tilde{H}_i} \left(\frac{d\tilde{w}_i}{d\tilde{z}} \right)^2 d\tilde{z} + \frac{L_p^2 \gamma_y^{(h)}}{r_p} \sum_{i=1}^{n} G_{si} \int_{\tilde{H}_{i-1}}^{\tilde{H}_i} \tilde{w}_i^2 d\tilde{z} \right. \\ \left. + G_{sn} \left[\tilde{w}_n^2 \big|_{\tilde{z}=1} \right] \left\{ \left(\frac{r_p}{\gamma_y^{(h)}} + 2b \right) \sqrt{\frac{\tilde{k}_n^{(h)}}{8\tilde{t}_{n+1}^{(h)}}} + \frac{L_p^2 \gamma_y^{(h)}}{r_p} \sqrt{\frac{\tilde{t}_{n+1}^{(h)}}{2\tilde{k}_n^{(h)}}} \right\} \right] \tag{9a}$$

$$\left(\gamma_y^{(h)}\right)^2 = \frac{1}{\psi^2\left[\left(\frac{r_p}{\gamma_x^{(h)}}+2a\right)\sum_{i=1}^{n} G_{si}\int_{\tilde{H}_{i-1}}^{\tilde{H}_i}\tilde{w}_i^2 d\tilde{z}+\left(\frac{r_p}{\gamma_x^{(h)}}+2a\right)G_{sn}\left[\tilde{w}_n^2\big|_{\tilde{z}=1}\right]\sqrt{\frac{\tilde{t}_{n+1}^{(h)}}{2\tilde{k}_n^{(h)}}}\right]}\times$$
$$\left[\left(\frac{r_p}{\gamma_x^{(h)}}+2a\right)\sum_{i=1}^{n} G_{si}\int_{\tilde{H}_{i-1}}^{\tilde{H}_i}\left(\frac{d\tilde{w}_i}{d\tilde{z}}\right)^2 d\tilde{z}+\frac{L_p^2\gamma_x^{(h)}}{r_p}\sum_{i=1}^{n}(\lambda_{si}+2G_{si})\int_{\tilde{H}_{i-1}}^{\tilde{H}_i}\tilde{w}_i^2 d\tilde{z}\right.$$
$$\left.+\left(\frac{r_p}{\gamma_x^{(h)}}+2a\right)G_{sn}\left[\tilde{w}_n^2\big|_{\tilde{z}=1}\right]\sqrt{\frac{\tilde{k}_n^{(h)}}{8\tilde{t}_{n+1}^{(h)}}}+\frac{L_p^2\gamma_x^{(h)}}{r_p}(\lambda_{sn}+2G_{sn})\left[\tilde{w}_n^2\big|_{\tilde{z}=1}\right]\sqrt{\frac{\tilde{t}_{n+1}^{(h)}}{2\tilde{k}_n^{(h)}}}\right] \quad (9b)$$

In the above equations, the pile displacements $u(z)$ and $w(z)$ are expressed in terms of the normalized variables $\tilde{u}=u/L_p$, $\tilde{w}=w/L_p$ and $\tilde{z}=z/L_p$ with the subscript i associated with $\tilde{u}$ and $\tilde{w}$ describing the respective normalized functions within the ith layer; $\tilde{H}_i = H_i / L_p$; and $\tilde{k}_n^{(v)}$, $\tilde{t}_{n+1}^{(v)}$, $\tilde{k}_n^{(h)}$ and $\tilde{t}_{n+1}^{(h)}$ are normalized soil parameters defined later. Note that, for the purpose of the analysis, the nth (bottom) layer is split into two parts, with the part below the pile denoted by the subscript $n+1$; therefore, $H_n = L_p$ (i.e., $\tilde{H}_n = 1$) and H_{n+1} (or $\tilde{H}_{n+1}$) $\to \infty$.

Differential Equations and Solutions of Pile Displacement

The differential equation governing the normalized axial pile displacement is given by:

$$\left(1+2\tilde{t}_i^{(v)}\right)\frac{d^2\tilde{u}_i}{d\tilde{z}^2}-\tilde{k}_i^{(v)}\tilde{u}_i=0 \quad (10)$$

with the following boundary conditions:

$$-\left(1+2\tilde{t}_1^{(v)}\right)\frac{d\tilde{u}_1}{d\tilde{z}}=\tilde{F}_v \quad \text{(at the pile head, i.e., at } z=\tilde{z}=0\text{)} \quad (11)$$

$$\tilde{u}_i=\tilde{u}_{i+1} \quad \text{(at the interface between adjacent layers, i.e., at } z=H_i \text{ or } \tilde{z}=\tilde{H}_i\text{)} \quad (12a)$$

$$\left(1+2\tilde{t}_i^{(v)}\right)\frac{d\tilde{u}_i}{d\tilde{z}}=\left(1+2\tilde{t}_{i+1}^{(v)}\right)\frac{d\tilde{u}_{i+1}}{d\tilde{z}} \quad \text{(at the layer interfaces, i.e., at } z=H_i \text{ or } \tilde{z}=\tilde{H}_i\text{)} \quad (12b)$$

$$\left(1+2\tilde{t}_n^{(v)}\right)\frac{d\tilde{u}_n}{d\tilde{z}}+\sqrt{2\tilde{k}_n^{(v)}\tilde{t}_{n+1}^{(v)}}\,\tilde{u}_n=0 \quad \text{(at the pile base, i.e., at } z=L_p \text{ or } \tilde{z}=1\text{)} \quad (13)$$

The differential equation governing the normalized lateral pile displacement is given by:

$$\frac{d^4\tilde{w}_i}{d\tilde{z}^4}-2\tilde{t}_i^{(h)}\frac{d^2\tilde{w}_i}{d\tilde{z}^2}+\tilde{k}_i^{(h)}\tilde{w}_i=0 \quad (14)$$

The corresponding boundary conditions are:

$$\frac{d^3\tilde{w}_1}{d\tilde{z}^3}-2\tilde{t}_1^{(h)}\frac{d\tilde{w}_1}{d\tilde{z}}=\tilde{F}_h \quad (15a)$$

and

$$\frac{d\tilde{w}_1}{d\tilde{z}}=0 \quad \text{(applicable if a cap is present at the pile head)} \quad (15b)$$

or

$$\frac{d^2\tilde{w}_1}{d\tilde{z}^2}-\tilde{M}_o=0 \quad \text{(applicable for a free-head pile)} \quad (15c)$$

at the pile head (i.e., at $z=\tilde{z}=0$);

$$\tilde{w}_i=\tilde{w}_{i+1} \quad (16a)$$

$$\frac{d\tilde{w}_i}{d\tilde{z}} = \frac{d\tilde{w}_{i+1}}{d\tilde{z}} \tag{16b}$$

$$\frac{d^3\tilde{w}_i}{d\tilde{z}^3} - 2\tilde{t}_i^{(h)} \frac{d\tilde{w}_i}{d\tilde{z}} = \frac{d^3\tilde{w}_{i+1}}{d\tilde{z}^3} - 2\tilde{t}_{i+1}^{(h)} \frac{d\tilde{w}_{i+1}}{d\tilde{z}} \tag{16c}$$

$$\frac{d^2\tilde{w}_i}{d\tilde{z}^2} = \frac{d^2\tilde{w}_{i+1}}{d\tilde{z}^2} \tag{16d}$$

at the interface between two adjacent layers (i.e., at $z = H_i$ or $\tilde{z} = \tilde{H}_i$); and

$$\frac{d^3\tilde{w}_n}{d\tilde{z}^3} - 2\tilde{t}_n^{(h)} \frac{d\tilde{w}_n}{d\tilde{z}} - \sqrt{2\tilde{k}_n^{(h)}\tilde{t}_{n+1}^{(h)}}\,\tilde{w}_n = 0 \tag{17a}$$

$$\frac{d^2\tilde{w}_n}{d\tilde{z}^2} = 0 \tag{17b}$$

at the pile base (i.e., at $z = L_p$ or $\tilde{z} = 1$).

The normalized applied forces and moment in the above equations are defined as $\tilde{F}_v = F_v/(E_pA_p)$, $\tilde{F}_h = F_hL_p^2/(E_pI_p)$ and $\tilde{M}_o = M_oL_p/(E_pI_p)$. The normalized soil parameters are defined as

$$\tilde{t}_i^{(v)} = \begin{cases} \dfrac{1}{2}\dfrac{(\lambda_{si} + 2G_{si})r_p^2}{E_pA_p}\left(\dfrac{1}{\gamma_x^{(v)}\gamma_y^{(v)}} + \dfrac{2b}{\gamma_x^{(v)}r_p} + \dfrac{2a}{\gamma_y^{(v)}r_p}\right); & i = 1,\ 2,\ ...,\ n \\ \dfrac{1}{2}\dfrac{(\lambda_{si} + 2G_{si})}{E_pA_p}\left(\dfrac{r_p}{\gamma_x^{(v)}} + 2a\right)\left(\dfrac{r_p}{\gamma_y^{(v)}} + 2b\right); & i = n+1 \end{cases} \tag{18a}$$

$$\tilde{k}_i^{(v)} = \frac{G_{si}L_p^2}{E_pA_p}\left[\frac{\gamma_x^{(v)}}{r_p}\left(\frac{r_p}{\gamma_y^{(v)}} + 2b\right) + \left(\frac{r_p}{\gamma_x^{(v)}} + 2a\right)\frac{\gamma_y^{(v)}}{r_p}\right] \tag{18b}$$

$$\tilde{t}_i^{(h)} = \begin{cases} \dfrac{r_p^2G_{si}L_p^2}{2E_pI_p}\left(\dfrac{1}{\gamma_x^{(h)}\gamma_y^{(h)}} + \dfrac{2b}{\gamma_x^{(h)}r_p} + \dfrac{2a}{\gamma_y^{(h)}r_p}\right); & i = 1,\ 2,\ .....n \\ \dfrac{G_{si}L_p^2}{2E_pI_p}\left(\dfrac{r_p}{\gamma_x^{(h)}} + 2a\right)\left(\dfrac{r_p}{\gamma_y^{(h)}} + 2b\right); & i = n+1 \end{cases} \tag{19a}$$

$$\tilde{k}_i^{(h)} = \frac{L_p^4}{E_pI_p}\left[(\lambda_{si} + 2G_{si})\frac{\gamma_x^{(h)}}{r_p}\left(\frac{r_p}{\gamma_y^{(h)}} + 2b\right) + G_{si}\frac{\gamma_y^{(h)}}{r_p}\left(\frac{r_p}{\gamma_x^{(h)}} + 2a\right)\right] \tag{19b}$$

The governing differential equation for axially loaded pile resembles that of a column (or rod) supported by an elastic foundation (i.e., elastic soil mass). The soil parameter $t^{(v)}$ accounts for the resistance of the soil due to vertical compression, while the parameter $k^{(v)}$ represents the resistance of soil against shearing in the vertical direction. The boundary condition at the head ensures that the external vertical force is balanced by the resistive axial force in the pile. The boundary conditions at the interfaces of the adjacent layers ensure the continuity of displacements and axial forces. The boundary condition at the pile base ensures that the axial force in the pile infinitesimally above the base is equal to that infinitesimally below it.

The governing differential equation for laterally loaded pile resembles that of an Euler-Bernoulli beam supported by an elastic foundation (i.e., elastic soil mass). The parameter $k^{(h)}$ is analogous to the Winkler spring constant and determines the compressive resistance of soil against pile (beam) movement. The parameter $t^{(h)}$ determines the soil resistance due to shearing in the horizontal direction. The boundary conditions at the pile head ensures that the applied horizontal force is equal to the shear force of the pile (which comprises of the shear resistance of the pile section and the horizontal shear resistance of soil) and that either the slope is zero (which is a reasonable assumption if a pile cap is present) or the applied moment is equal to the pile bending moment (if the head is free to rotate). The boundary conditions at the interface of any two layers imply that the pile deflection, slope, bending moment and shear force in the adjacent layers have the same magnitude at the interface. The boundary conditions at the pile base states that the pile bending moment is zero and that the shear force infinitesimally above the pile base is equal to that infinitesimally below it.

The general solution of equation (10) for the axially loaded pile is given by:

$$\tilde{u}_i(\tilde{z}) = C_1^{(i)} \sinh\sqrt{\frac{\tilde{k}_i^{(v)}}{1+2\tilde{t}_i^{(v)}}}\tilde{z} + C_2^{(i)} \cosh\sqrt{\frac{\tilde{k}_i^{(v)}}{1+2\tilde{t}_i^{(v)}}}\tilde{z} \tag{20}$$

The integration constants $C_1^{(i)}$ and $C_2^{(i)}$ for any layer i can be determined using the relevant boundary conditions.

The general solution of equation (14) for the laterally loaded pile is given by:

$$\tilde{w}_i(\tilde{z}) = C_1^{(i)}\Phi_1 + C_2^{(i)}\Phi_2 + C_3^{(i)}\Phi_3 + C_4^{(i)}\Phi_4 \tag{21}$$

where, $C_1^{(i)}$, $C_2^{(i)}$, $C_3^{(i)}$ and $C_4^{(i)}$ are integration constants of the ith layer; $\Phi_1 = \sinh(\alpha_i\tilde{z})\cos(\beta_i\tilde{z})$, $\Phi_2 = \cosh(\alpha_i\tilde{z})\cos(\beta_i\tilde{z})$, $\Phi_3 = \cosh(\alpha_i\tilde{z})\sin(\beta_i\tilde{z})$ and $\Phi_4 = \sinh(\alpha_i\tilde{z})\sin(\beta_i\tilde{z})$ ($\alpha_i = \sqrt{[\tfrac{1}{2}\{\sqrt{\tilde{k}_i^{(h)}} + \tilde{t}_i^{(h)}\}]}$ and $\beta_i = \sqrt{[\tfrac{1}{2}\{\sqrt{\tilde{k}_i^{(h)}} - \tilde{t}_i^{(h)}\}]}$) if $\sqrt{\tilde{k}_i^{(h)}} > \tilde{t}_i^{(h)}$, and $\Phi_1 = \sinh(\alpha_i\tilde{z})$, $\Phi_2 = \cosh(\alpha_i\tilde{z})$, $\Phi_3 = \sinh(\beta_i\tilde{z})$ and $\Phi_4 = \sinh(\beta_i\tilde{z})$ ($\alpha_i = \sqrt{[\tilde{t}_i^{(h)} + \sqrt{\{(\tilde{t}_i^{(h)})^2 - \tilde{k}_i^{(h)}\}}]}$ and $\beta_i = \sqrt{[\tilde{t}_i^{(h)} - \sqrt{\{(\tilde{t}_i^{(h)})^2 - \tilde{k}_i^{(h)}\}}]}$) if $\sqrt{\tilde{k}_i^{(h)}} < \tilde{t}_i^{(h)}$. The integration constants can be determined using the boundary conditions.

Adjustment of Soil Modulus

Modifications of the Lame's constants of the elastic soil are necessary to counterbalance the excessively stiff pile response obtained by solving the above equations because the assumption of zero horizontal displacements in the case of axially loaded piles and that of zero displacements in the y and z directions in the case of laterally loaded piles artificially restrain the pile movement. It was found by comparing the results of this analysis with those of equivalent finite element analysis that setting $\lambda_{si} = 0$ (which is equivalent to setting the soil Poisson's ratio $\upsilon_{si} = 0$) removed the artificial stiffness in the system and that the effect of soil Poisson's ratio can be indirectly taken into account by replacing the shear modulus G_{si} by $G_{si}^* = 0.6G_{si}\left(1+1.25\upsilon_{si}^2\right)$ for the case of axially loaded piles and by $G_{si}^* = 0.7G_{si}\left(1+0.75\upsilon_{si}\right)$ for the case of laterally loaded piles. These adjustments were made before the final results were obtained.

Solution Algorithm

The dimensionless parameters γ_x and γ_y are required to be known in order to estimate the parameters k_i and t_i without which pile displacements cannot be determined. The parameters γ_x and γ_y, in turn, depend on the pile displacement. Hence, an iterative solution scheme is required. In the first iteration, initial guesses on γ_x and γ_y are made and, using these guessed values, k_i and t_i are determined and pile displacements are calculated. Using the calculated pile displacements, new values of γ_x and γ_y are calculated and compared with the assumed initial values of γ_x and γ_y, respectively. If the differences are greater than a tolerance limit, iterations are continued with the calculated values of γ_x and γ_y taken as the new guesses. Successive iterations are continued until the values of both γ_x and γ_y obtained from two consecutive iterations fall below the tolerance limit. A value of one is chosen as the initial guess for both γ_x and γ_y, and the tolerance limit prescribed on both γ_x and γ_y between the ith and $(i+1)^{th}$ iteration is $|\gamma_{x/y}^{(i+1)} - \gamma_{x/y}^{(i)}| < 0.001$ (this tolerance limit was determined by a systematic convergence study). In most problems, solutions are obtained in three iterations.

RESULTS

The use of the analysis framework is demonstrated with the use of examples. In the first example, a 15-m-long axially loaded precast pile with a square cross section of 0.5 m × 0.5 m is considered. The pile is embedded in a four-layer soil deposit with H_1 = 2 m, H_2 = 5 m and H_3 = 10 m; E_{s1} = 10 MPa, E_{s2} = 15 MPa, E_{s3} = 30 MPa and E_{s4} = 100 MPa; υ_{s1} = 0.4, υ_{s2} = 0.35, υ_{s3} = 0.3 and υ_{s4} = 0.15 (E_{si} and υ_{si} are the Young's modulus and Poisson's ratio of the ith soil layer, respectively; $G_{si} = E_{si}/2(1+\upsilon_{si})$, $\lambda_{si} = E_{si}\upsilon_{si}/[(1+\upsilon_{si})(1-2\upsilon_{si})]$). The pile is subjected to an axial force F_v = 3000 kN at the head. Figure 3(a) shows the pile deflection as a function of depth, as obtained from our analysis and equivalent three-dimensional (3D) linear finite element (FE) analysis performed using Abaqus. It is evident that the match between the results of the present analysis and that of the FE analysis is good — the difference in the pile settlement at the head is about 0.5%.

For the second example, a laterally loaded rectangular barrette, 40 m long, 2.8 m × 0.8 m in cross section with $E_p = 25 \times 10^6$ kPa is considered embedded in a four-layer soil profile with H_1 = 1.5 m, H_2 = 3.5 m, and H_3 = 8.5 m; E_{s1} = 20 MPa, E_{s2} = 25 MPa, E_{s3} = 40 MPa and E_{s4} = 80 MPa; and υ_{s1} = 0.35, υ_{s2} = 0.3, υ_{s3} = 0.25 and υ_{s4} = 0.2. A horizontal force F_h = 3000 kN acts at the pile head along the longer dimension of the pile cross section (i.e., $2a$ = 2.8 m and $2b$ = 0.8 m). Figure 3(b) shows the pile deflection profiles, as obtained from the present analysis and 3D linear FE analysis performed using Abaqus. The deflection profiles match well; the difference in the head deflection is equal to 5%.

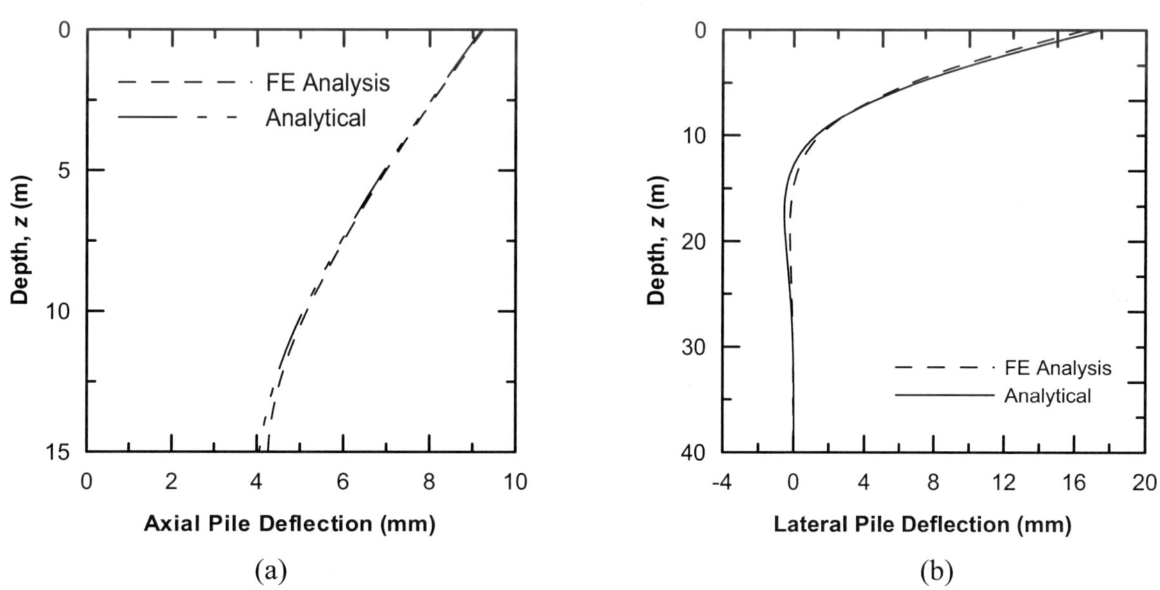

Figure 3. (a) Settlement of a 15-m long precast pile and (b) lateral deflection of a 40-m long barrette.

Several other examples have been solved and compared with equivalent 3D FE analysis, and the ability of the analysis framework to predict accurate pile response have been established — the details are available in Basu et al. (2008) and Basu and Salgado (2008). Moreover, comparisons with the response of piles with circular cross section obtained using a similar analytical framework developed by the authors (Basu et al. 2009, Seo et al. 2009, Basu and Salgado 2007) have shown that conversions of rectangular cross section into equivalent circles may produce reasonably accurate results if the equivalent circle has the same area of cross section as that of the actual rectangular cross section in the case of axially loaded pile and if the equivalent circle has the same second moment of inertia as that of the actual rectangular cross section in the case of laterally loaded pile. However, such conversions into equivalent circular cross sections may not produce accurate results when soil nonlinearity is considered because, as demonstrated in the analysis, pile geometry has a role in determining the soil strain and stress fields, which are different for circular and rectangular cross sections.

It is also important to note that the inputs required for the analysis — pile geometry and Young's modulus, soil layering and elastic constants, and magnitude of applied forces — can be given in the form of a simple text file and no elaborate numerical mesh generation is required. Moreover, because closed-form solutions are used, the results are obtained within a few seconds.

CONCLUSIONS

An analysis framework for axially and laterally loaded piles with rectangular cross section embedded in multi-layered elastic media is developed. The analysis produces the pile settlement and axial resistance force for axially loaded piles, and lateral deflection, slope of the deflected curve, bending moment and shear force for laterally loaded piles. The inputs required are the pile cross sectional dimensions and length, thicknesses of the soil layers, Young's modulus of the pile material, Young's modulus and Poisson's ratio (or any pair of elastic constants) of the soils in the various layers, and the magnitudes of the applied forces and moment.

The governing differential equations for the pile deflections are obtained using the principle of minimum potential energy. Closed-form solutions are obtained. The solution depends on two parameters, γ_x and γ_y, that describe the rate at which the deflections in the soil medium decrease with an increase in the distance from the pile. These parameters are not known a priori and must be determined iteratively. The method produces pile responses in seconds.

The shape of the pile cross section has a bearing on the pile response. The analysis presented in the paper can be used to make realistic predictions of the response of axially and laterally loaded rectangular piles.

ACKNOWLEDGEMENTS

This material presented in this paper is based upon work supported by the National Science Foundation under Grant No. 0556347. The authors are thankful for this support.

REFERENCES

Armaleh, S. and Desai, C. S. (1987). "Load-deformation response of axially loaded piles." *J. Geotech. Engng.*, 113(12), 1483-1500.

Banerjee P. K., and Davies T. G., (1978). "The behaviour of axially and laterally loaded single piles embedded in nonhomogeneous soils.", *Geotechnique*, 28(3), 309-326.

Basu, D., and Salgado, R. (2007). "Elastic analysis of laterally-loaded pile in multilayered soil." *Geomech. Geoeng.: An Int. J.*, 2(3), 183-196.

Basu, D., and Salgado, R. (2008). "Analysis of laterally loaded piles with rectangular cross sections embedded in layered soil media." *Int. J. Numer. Anal. Meth. Geomech.*, 32(7), 721-744.

Basu, D., Prezzi, M., Salgado, R. and Chakraborty, T. (2008). "Settlement analysis of piles with rectangular cross sections in multi-layered soil." *Computers and Geotechnics*, 35, 563-575.

Basu, D., Salgado, R. and Prezzi, M. (2009). "A continuum-based model for analysis of laterally loaded piles in layered soils." *Géotechnique*, 59(2), 127-140.

Broms, B. B. (1964). "Lateral resistance of piles in cohesionless soils." *J. Soil Mech. Fdn. Div.*, 90(SM3), 123-156.

Butterfield, R. and Banerjee, P. K. (1971). "The elastic analysis of compressible piles and pile groups." *Géotechnique*, 21(1), 43-60.

Coyle, H. M. and Reese, L. C. (1966). "Load transfer for axially loaded piles in clay." *J. Soil Mech. Fdn Div.*, 92(SM2), 1-26.

Dennis, N. D. and Olson, R. E. (1983a) "Axial capacity of steel pipe piles in clay." *Proc. Conf. Geotech. Practice Offshore Eng.*, New York, 370-388.

Dennis, N. D. and Olson, R. E. (1983b) "Axial capacity of steel pipe piles in sand." *Proc. Conf. Geotech. Practice Offshore Eng.*, New York, 389-402.

Guo, W. D. and Randolph. M. F. (1997). "Vertically loaded piles in non-homogeneous media." *Int. J. Num. Analyt. Meth. Geomech.*, 21(8), 507-532.

Matlock, H. (1970). "Correlations for design of laterally loaded piles in soft clay." *Proc. 2nd Offshore Tech. Conf.,* Houston, Texas, 1, 577-594.

Mylonakis, G. (2001). "Winkler modulus for axially loaded piles." *Géotechnique*, 51(5), 455-461.

Ng, C. W. W. and Zhang, L. M. (2001). "Three-dimensional analysis of performance of laterally loaded sleeved piles in sloping ground." *J. Geotech. Geoenv. Engng.*, 127(6), 499-509.

O'Neill, M. W., Reese, L. C., and Cox, W. R. (1990). "Soil behavior for piles under lateral loading." *Proc. 22nd Offshore Tech. Conf.,* Houston, Texas, 3, 279-287.

Poulos, H. G. (1971a). "Behavior of laterally loaded piles: I – single piles." *J. Soil Mech. Fdn Div.*, 97(SM5), 711-731.

Poulos, H. G. (1971b). "Behavior of laterally loaded piles: III – socketed piles." *J. Soil Mech. Fdn Div.*, 98(SM4), 341-360.

Poulos, H. G. and Davis, E. H. (1968). "The settlement behavior of single axially loaded incompressible piles and piers." *Géotechnique*, 18(3), 351-371.

Rajapakse, R. K. N. D. (1990). "Response of an axially loaded elastic pile in a Gibson soil." *Géotechnique*, 40(2), 237-249.

Randolph, M. F. and Wroth, C. P. (1978). "Analysis of deformation of vertically loaded piles." *J. Geotech. Div.*, 104(GT12), 1465–1488.

Randolph, M. F. (1981). "The response of flexible piles to lateral loading." *Geotechnique*, 31(2), 247-259.

Reese, L. C. and Matlock, H. (1956). "Non-dimensional solutions for laterally loaded piles with soil modulus assumed proportional to depth." *Proc. 8th Texas Conf. on Soil Mech. Fdn. Engng.*, Austin, Texas, 1-41.

Reese, L.C., Cox, W. R. and Koop, F. D. (1974). "Analysis of laterally loaded piles in sand." *Proc. 6th Offshore Tech. Conf.*, Houston, Texas, 2, 473-483.

Reese, L.C., Cox, W. R. and Koop, F. D. (1975). "Field testing and analysis of laterally loaded piles in stiff clay." *Proc. 7th Offshore Tech. Conf.,* Houston, Texas, 2, 671-690.

Seed, H. B. and Reese, L. C. (1957). "The action of soft clay along friction piles." *Transactions* ASCE, 122.

Seo, H., Basu, D., Prezzi, M. and Salgado, R. (2009). "Load-settlement response of rectangular and circular piles in multilayered soil." *J. Geotech. Geoenviron. Eng.*, 135(3), 420-430.

Trochanis, A. M., Bielak, J. and Christiano, P. (1991). "Three-dimensional nonlinear study of piles." *J. Geotech. Engng.*, 117(3), 429-447.

Verruijt, A. and Kooijman, A. P. (1989). "Laterally loaded piles in a layered elastic medium." *Geotechnique*, 39(1), 39-46.

Three-Dimensional Finite Element Modeling of Pile and Pile Group System Response

Ning Wang[1], Ahmed Elgamal[2], M.ASCE, and Jinchi Lu[3], P.E., M.ASCE

[1]Graduate Student Researcher, Dept. of Structural Engineering, Univ. of California, San Diego, La Jolla, CA 92093-0085; E-mail: n9wang@ucsd.edu
[2]Professor, Dept. of Structural Engineering, Univ. of California, San Diego, La Jolla, CA 92093-0085 (corresponding author); E-mail: elgamal@ucsd.edu
[3]Assistant Project Scientist, Dept. of Structural Engineering, Univ. of California, San Diego, La Jolla, CA 92093-0085; E-mail: jinlu@ucsd.edu

ABSTRACT: Advancements in high-speed computing and developments in material modeling techniques are increasingly allowing for large-scale three-dimensional (3d) linear and nonlinear geotechnical finite-element (FE) simulations. The open-source computational platform OpenSees provides such 3d simulation capabilities for static/seismic ground-foundation analyses. With the aid of the graphical user interface OpenSeesPL (http://cyclic.ucsd.edu/openseespl), 3d computations for pile and pile group studies are greatly facilitated. Using this FE framework, an effort is made to capture key features associated with axial and lateral loading on piles and pile groups. For that purpose, the behavior of a laterally loaded 3×3 pile group at different pile-spacing scenarios is presented first, and contrasted with that of the corresponding single pile scenario. Response under linear and nonlinear soil behavior representations is discussed. Lateral pile-group performance and efficiency considerations are investigated. Thereafter, the behavior of a large (8×4) pile group foundation supporting a high-clearance (for marine navigation) bridge deck is addressed. For seismic response evaluation, push-over combined axial/lateral loads along with a substantial moment are imposed. All studies are presented within a scope that highlights current modeling capabilities and directions for further research.

INTRODUCTION

Soil-structure interaction (SSI) plays a major role in the response of structures subjected to earthquakes (e.g., Wolf 1985; Mylonakis and Gazetas 2000). For piles and pile groups, numerical FE investigations have been conducted to further understand the involved SSI mechanisms within an environment that attempts to represent the actual involved geometric configuration (Brown and Shie 1990; Chae et al. 2004; Dodds 2005; Elgamal et al. 2008, 2009a; Elgamal and Lu 2009; Lu et al. 2010a; Maheshwari et al. 2004a,b; Maki et al. 2006; Muqtadir and Desai 1986; Peng et al. 2004; Uzuoka et al. 2007; Wakai et al. 1999; Yang and Jeremic 2003, 2005).

For seismic pile-ground analyses, advanced 3d nonlinear constitutive material models that are applicable to arbitrary cyclic loading scenarios are needed (e.g., Yang 2000, Yang and Elgamal 2003). In this regard, the open-source platform OpenSees (Open System for Earthquake Engineering Simulation) of the Pacific Earthquake Engineering Research (PEER) Center (Mazzoni et al. 2006) allows for such capabilities and is employed herein. Developed for 3d pile-ground analyses, the user friendly interface OpenSeesPL is utilized to facilitate the pre- and post-processing efforts (Lu et al. 2010b, http://cyclic.ucsd.edu/openseespl/).

Using this numerical analysis framework, this paper aims to present a number of pilot investigations related to the response of piles and pile groups. Soil is represented by three alternative idealizations, in the form of linear as well as nonlinear (undrained clay-type) models, with and without a tension cut-off logic.

Additional details of the employed analysis framework are presented briefly below. Influence of pile group spacing on basic system behavior mechanisms is then illustrated using a relatively simple 3×3 configuration. Thereafter, an analysis of a large 8×4 pile group system under push-over combined axial/lateral loads and overturning moment is performed. Distribution of load between the piles is analyzed and the group interaction effects are discussed. As such, it is important to note that FE mesh refinement is among the important considerations behind overall accuracy. In this regard, effort has been made throughout to develop and employ adequately refined FE meshes for the purpose of the above-mentioned pilot investigations. Along with the insights gained from this study, the reported results highlight the current analysis framework capabilities and range of potential applications.

COMPUTATIONAL FRAMEWORK

A graphical user interface (GUI) OpenSeesPL (http://cyclic.ucsd.edu/openseespl, Elgamal et al. 2009b) is employed throughout this study. This interface allows for: i) convenient generation of the soil mesh with an embedded single pile or pile group and associated boundary conditions, ii) execution of numerical simulation under seismic excitation as well as under static/dynamic push-over loading in prescribed displacement or prescribed force modes, iii) study of various ground modification scenarios by appropriate specification of the material within the pile zone (Elgamal et al. 2009a, Nguyen et al. 2013, Rayamajhi et al. 2012), and iv) graphical display of the results for the pile and the ground system (Elgamal et al. 2009b). As mentioned earlier, the FE analysis engine for this interface is OpenSees (http://opensees.berkeley.edu, Mazzoni et al. 2006).

LATERAL PILE RESPONSE CALIBRATION

Elgamal and Lu (2009) employed the analytical elastic solution of Abedzadeh and Pak (2004) to develop a satisfactory 3d FE model for the purpose of lateral pile response studies. Details of this calibration effort are described briefly as follows:

Configuration of the Pile-Soil System

A circular steel-pipe free-head pile with steel Young's Modulus $E_p = 199.95$ GPa, outer radius (a) of 203.20 mm and wall thickness (h) of 20.32 mm was studied (cross-sectional moment of inertia $I = \pi a^3 h = 5.36\times10^{-4}$ m^4). The pile length (l) was selected to be 10.15 m in order to numerically model the analytical elastic solution (case of l/a=50). The pile was fully embedded in a homogeneous isotropic linearly elastic soil domain with submerged unit weight $\gamma' = 1005.96$ kg/m^3, shear modulus of soil $G_s = 5.50\times10^4$ kPa, and Poisson's ratio $\nu_s = 0.25$ (i.e., Bulk modulus $B_s = 9.17\times10^4$ kPa). In this linear-response calibration scenario, a horizontal load H of 140.12 kN (31.5 kips according to Elgamal and Lu 2009) was applied to the free pile head (at ground surface).

Finite Element Model and Results

In view of symmetry, a half-mesh was employed as shown in Figure 1 (lateral load was applied along the longitudinal direction of this mesh). Both 8-node and 20-node brick elements were used to model the soil domain. The pile was modeled by beam-column elements, in order to easily view axial loads, moments, and shear forces. Within this configuration, essentially rigid beam-column elements (links) were used to represent the pile cross-sectional spatial domain (diameter) and the interface with the surrounding 3d soil elements. As such, these links were defined to be 10^4 times stiffer (both axially and flexurally) than the pile elements. Away from the pile centerline beam-column elements, the edge nodes of these rigid links are tied to the coincident nodes of the adjacent 3d soil brick element in the three translational degrees of freedom (Elgamal et al. 2008; Law and Lam 2001; Rollins and Brown 2011). In total there were 2900 brick elements, 20 beam-column elements, and 189 rigid links in this FE model (Figure 1).

The boundary conditions of the model were defined as: i) Base of the ground domain was fixed in the longitudinal (x), transverse (y), and vertical (z) directions, ii) Left, right and back sides of the mesh were fixed in the x and y directions, and iii) Plane of symmetry was fixed in the transverse (y) direction. A satisfactory numerical solution was obtained for a mesh of 158.19 m in lateral extent and 20.12 m in depth (i.e., the bottom of the soil domain was 9.97 m below the pile tip). The numerical pile deflection profile obtained using OpenSeesPL is shown in Figure 2 along with the analytical elastic solution (Abedzadeh and Pak 2004).

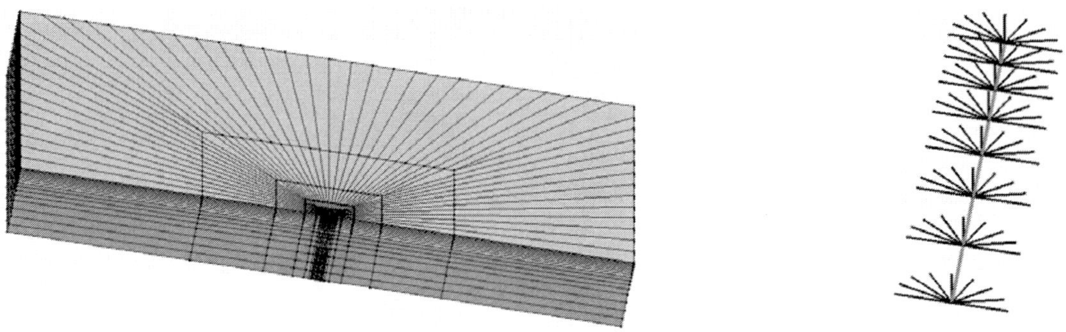

(a) (b)

Figure 1. Employed half mesh configuration (due to symmetry) for the single pile simulation (a) 3d isometric view (b) Illustration of pile model with rigid links.

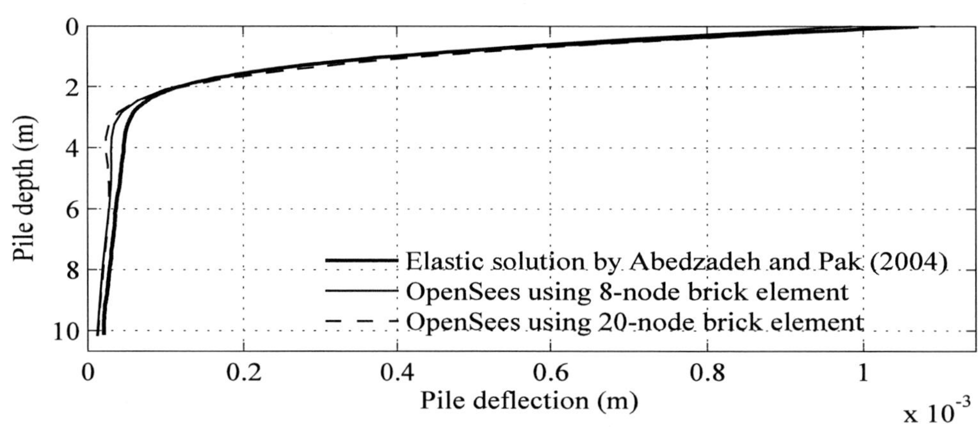

Figure 2. Comparison of pile deflection profiles for the linear-response calibration scenario (Elgamal and Lu 2009).

NONLINEAR SOIL MODELING

In addition to the elastic soil simulations, a multi-surface pressure-independent J_2 plasticity model (Von-Mises) was employed (undrained clay-type soil material) to demonstrate the influence of nonlinear ground response (Figure 3). Response using this nonlinear model was also explored in the presence of a tension cut-off mechanism where shear strength vanishes upon occurrence of tensile effective confinement (Figure 3c). Along with the elastic properties of the calibrated model, cohesion c= 40.68 kPa in the range of a Medium Clay, was specified. Using a hyperbolic backbone curve model, the maximum shear strength was specified to occur at an octahedral shear strain (γ_{max}) of 3%.

MODELING OF PILE GROUP EFFECTS

Benchmark Lateral Response of Single Pile

Numerical single pile simulations for linear, nonlinear, and nonlinear with tension cut-off soil cases were conducted first to provide a reference for the pile group behavior comparison. A lateral load (H) was applied to the pile head (free head condition), located essentially at the ground surface, in 40 increments up to a final load of 420.36 kN (94.5 kips according to Elgamal and Lu 2009) to clearly demonstrate the effects of nonlinear soil response.

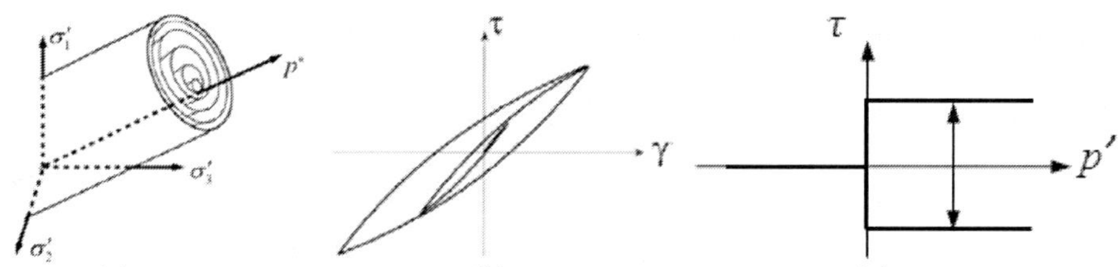

Figure 3. Multi-surface plasticity J_2 model: (a) Von Mises multi-surfaces, (b) Hysteretic shear response (Elgamal et al. 2008), and (c) Illustration of tension cut-off logic.

Figure 4 illustrates the influence of nonlinear soil response on the pile head computed deflections. As the load increases, the difference in pile head deflection becomes more pronounced. At maximum applied load, the nonlinear soil tension cut-off scenario appears to be nearing ultimate capacity. In this case, the back-soil zones (adjacent soil that the pile is moving away from at any depth) provide minimal support, leading to a marked reduction in overall lateral resistance.

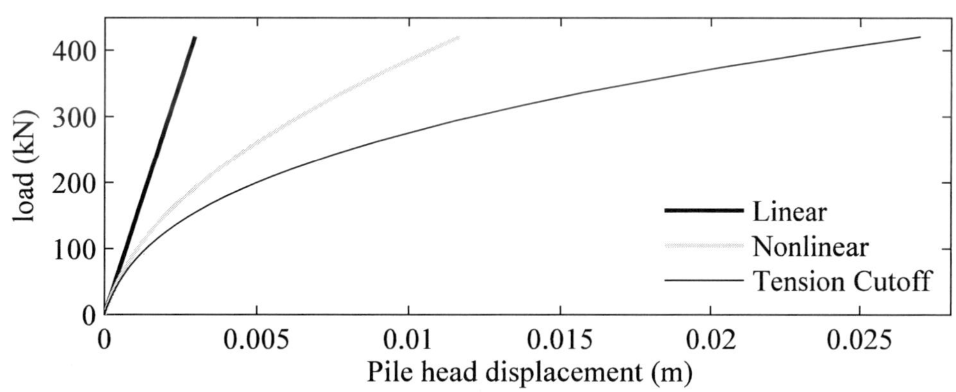

Figure 4. Computed pile head load-deflection curves for the single pile simulation.

The corresponding pile deflection, bending moment, and shear force profiles are shown in Figure 5. Compared to the linear case, the nonlinear scenarios displayed: i) a significant increase in lateral pile head displacement and maximum moment, and ii) a clear increase in depth to the location where maximum moment occurs. The soil nonlinear yielding response and the tension cut-off logic (when activated) decrease lateral soil resistance, and thus account for these mechanisms.

3×3 Pile Group Configuration

Lateral resistance may be significantly reduced for a closely spaced pile group (Rollins et al. 2006). In order to model this mechanism and explore the resulting pile group efficiency, the behavior of a 3×3 pile group (Figure 6) at different pile-spacing configurations was studied (based on above single pile scenario). The 3×3 pile group (free head condition rigidly connected by a pile cap) was subjected to a horizontal load (*H*) of 3,783.21 kN, nine times that of the single pile case.

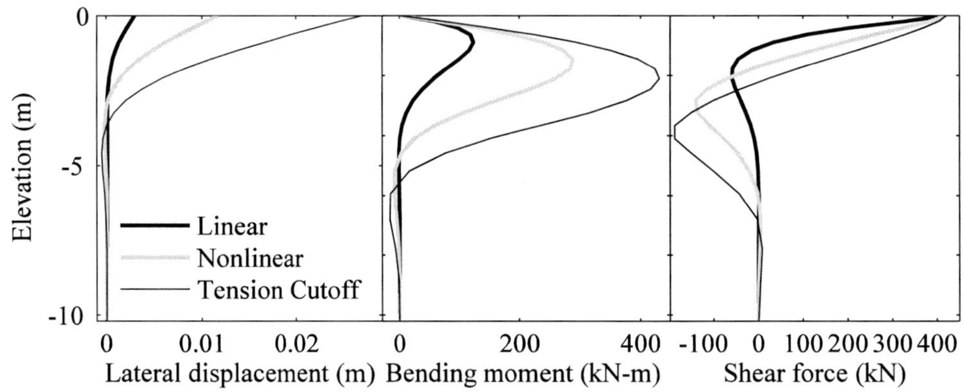

Figure 5. Pile deflection, bending moment and shear force profiles for the linear, nonlinear, and nonlinear with tension cut-off soil cases at the maximum applied lateral load of 420.36 kN.

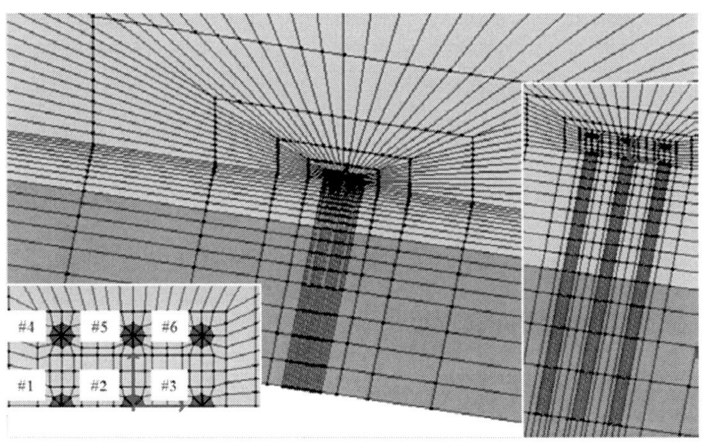

Figure 6. Finite element mesh (half-mesh configuration) employed for 3×3 pile group (3D spacing case) with pile numbering scheme and close-up of the pile group.

The pile group functionality implemented in the user-interface OpenSeesPL allows for convenient study of pile spacing effects. Up to about 6 hours of computational time was needed using a 3GHz CPU PC Desktop for the more challenging nonlinear soil cases with the tension cut-off logic activated (in terms of the needed number of iterations for convergence).

3×3 Pile Group Response

For representative pile spacing configurations (the tension cut-off case at maximum applied load), Figure 7 depicts the attained stress ratio (J_2 computed shear stress divided by failure stress counterpart). As such, the pile-group interaction mechanism is illustrated in terms of influence on the surrounding soil. At close pile spacing (3D, where D is the pile diameter), it is evident that the entire soil between the piles is highly stressed with more noticeable overall lateral translation (Lu et al. 2011). As the spacing increases, each pile is surrounded by an adequate soil domain (independent from the influence of surrounding piles).

Based on this logic, the peak-load displacement efficiency (deflection of single pile / deflection of pile group) as a function of pile spacing (Figure 8) displays higher values with the increase in pile spacing. When the spacing reaches 10 D, the pile group efficiency approaches its maximum (1.0) for the tension cut-off soil condition. At close pile spacing, it is of interest to note that linear case shows the highest displacement efficiency (Figure 8).

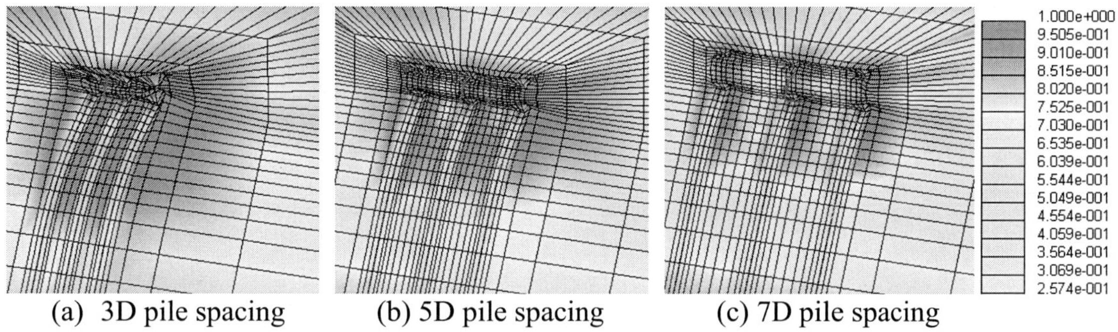

Figure 7. Tension cut-off case: stress ratio for 3×3 pile group at the maximum lateral load (displacements as shown are magnified by a factor of 8).

As the pile spacing increases, this efficiency increases for all cases. However, the rate of this increase appears to be highest for the nonlinear case with tension cut-off logic activated. On this basis, it may be inferred that (at the presented maximum load situation of Figure 8): a) closely spaced piles intensify nonlinear soil response in the immediate surrounding soil, further reducing the group efficiency compared to their single pile counterpart, and b) as spacing increases, nonlinear soil behavior tends to localize the influence of each pile on the surrounding soil domain, gradually leading to increased pile-group efficiency compared to the linear soil case (i.e., liner behavior tends to engage a larger domain of the ground around each pile).

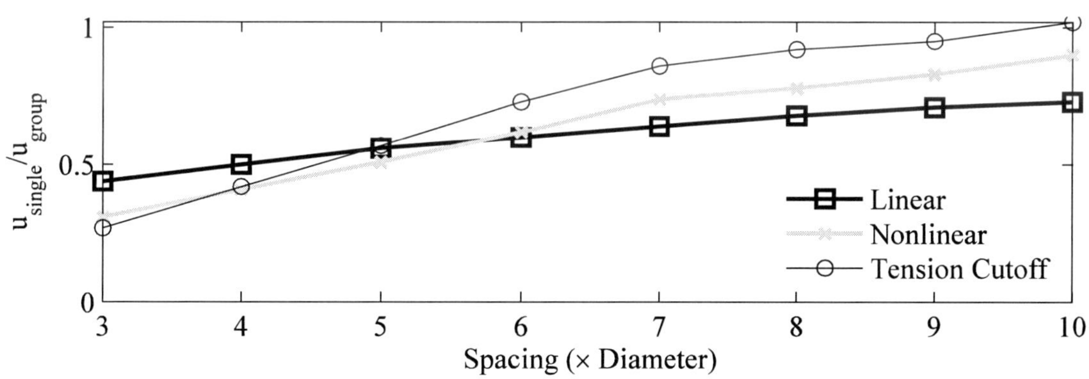

Figure 8. Pile head displacement efficiency of the 3×3 pile group for different spacing under the maximum applied lateral load of 420.36 kN per pile (u_{single}, the single pile head deflection is 0.003 m, 0.012 m, and 0.027 m for the linear, nonlinear and nonlinear with cut-off cases respectively as shown in Figure 4).

Figure 9 displays salient response characteristics as a function of the level of applied load and resulting pile head displacement. Load-deflection curves (Figure 9a) clearly show that the pile group (PG) undergoes considerably more displacement than the single pile (SP) under the same average load/pile (Brown et al. 1987; McVay et al. 1995). This is particularly evident (Figure 9a) in the highly nonlinear tension cut-off case. As spacing increases, the pile group response tends toward the single pile situation (Figure 9a).

In Figure 9b, pile group displacement efficiency is seen to change with the level of applied lateral load (the nonlinear soil situations). At close pile spacing, efficiency decreases as the load increases, as the influence zone of each pile overlaps with the other thus causing higher levels soil shear stress and yielding (Figure 7). As the pile spacing increases, the opposite occurs, and group efficiency is seen to increase with the level of applied load. Apparently, for the larger pile spacing scenarios, further localization of the pile influence on the surrounding soil occurs with the increase in nonlinear soil response, resulting in a corresponding increase in efficiency. As such, it may be concluded that group efficiency is dependent on the level of applied load, soil nonlinearity, and pile group spacing (McVay et al. 1998). It may increase or decrease as the level of applied load increases depending on spacing.

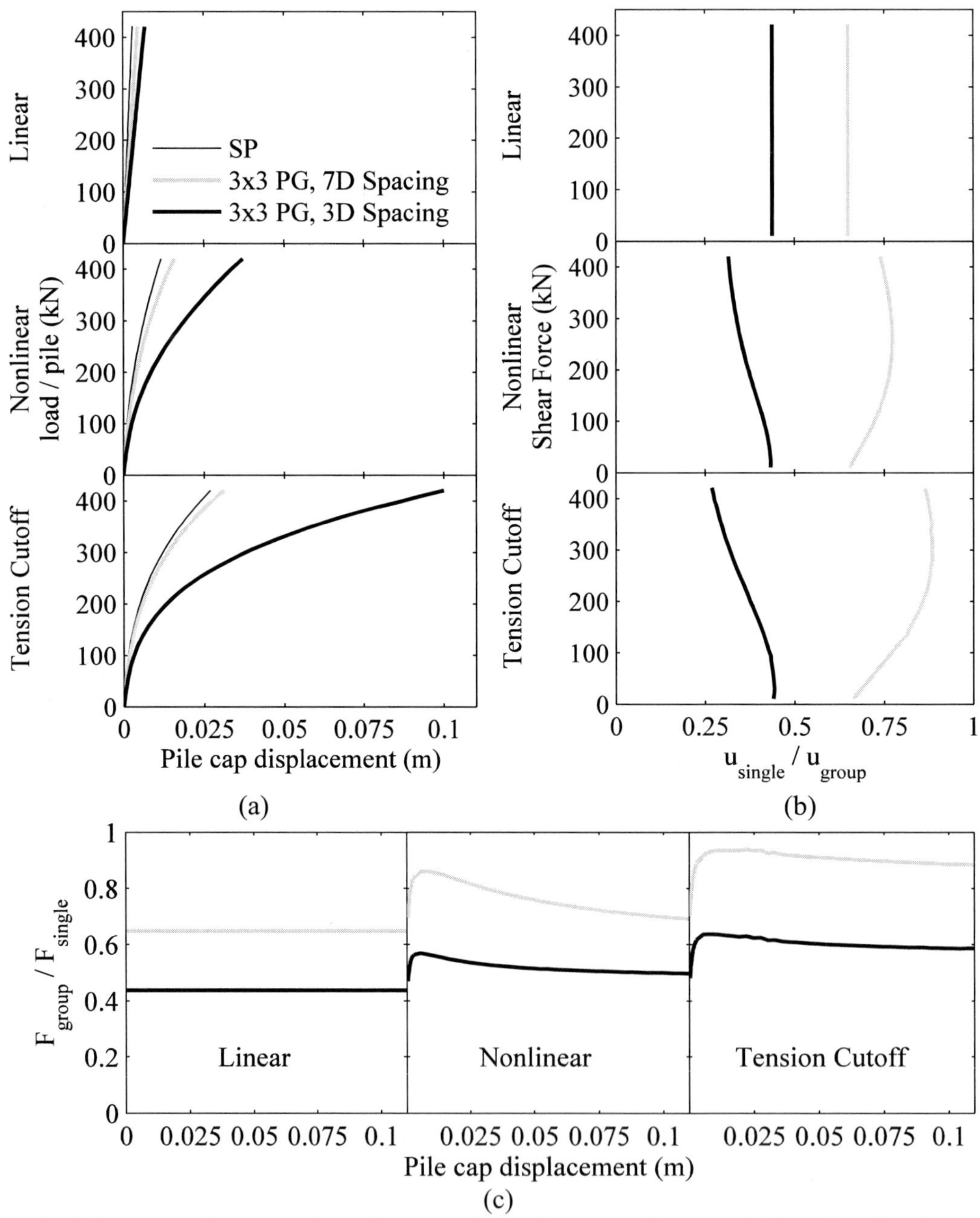

Figure 9. Linear, nonlinear, and nonlinear with tension cut-off cases: (a) Load-deflection curves, (b) corresponding displacement efficiency, and (c) load or force-based efficiency.

Efficiency may be assessed alternatively based on sustained force by the single pile and the pile group at equal levels of lateral displacement. For the case under discussion, this force-based efficiency (Average load per pile in group / Single pile load) as shown in Figure 9c reveals that initially, the group efficiency increases markedly as deflection increases (the nonlinear soil scenarios). At such relatively low levels of deflection, compared to the pile group with overall large lateral dimensions, the single pile appears to cause higher levels of stress concentration and soil yielding in the immediate surrounding ground. For large spacing (the shown 7D scenario), this effect remains evident as deflection continues to increase with the efficiency sustained at high values above about 0.9. For close spacing (the shown 3D scenario), efficiency shows a more noticeable gradual decrease with the increase in deflection due to the increased soil nonlinearity and yielding around the entire pile group.

Lateral loads carried by each pile are shown in Figure 10 for the 3D and 7D spacing scenarios, and bending moment profiles are presented in Figure 11. For the linear and nonlinear soil cases, the front (3 and 6) and back (1 and 4) piles show similar load patterns. Being on the outer edges, away from the influence of other piles, the corner piles (4 and 6) are seen to shoulder a higher portion of the load (Figure 10). Upon activation of the tension cut-off mechanism, the front corner pile (6) shoulders a higher portion of the overall load compared to the back corner pile (4). This is especially obvious in the closely-spaced 3D pile group scenario. As the pile spacing increases, a more uniform loading sharing pattern takes place.

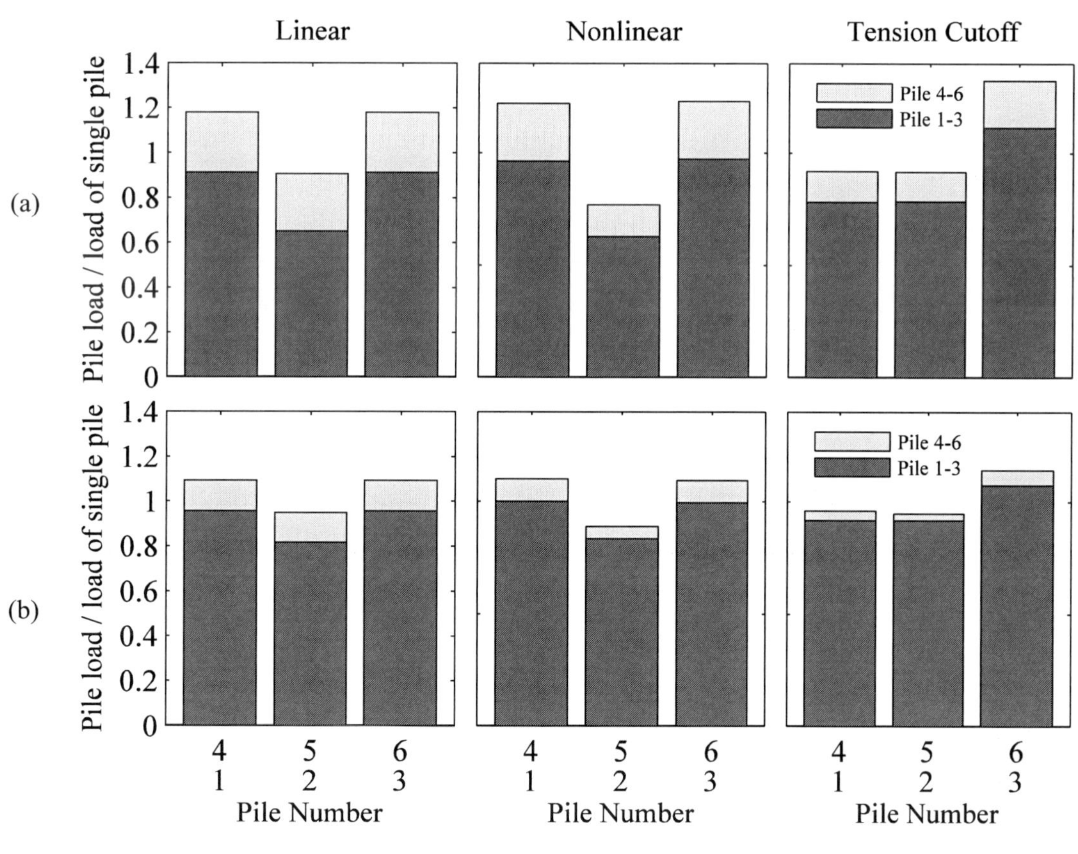

Figure 10. Distribution of pile head lateral force at the maximum applied load for (a) 3D pile spacing and (b) 7D pile spacing (Pile numbers are shown in Figure 6).

From the results of this 3x3 pile-group simulation, the following response patterns were observed:

1. Displacement efficiency of the pile group dropped to as little as about 27% (Figure 9b) for

close pile-spacing scenarios (3D spacing tension cut-off case). Such decrease in efficiency would be expected as the immediate soil regions providing resistance around each pile heavily overlap.

2. For such close spacing, displacement efficiency of the nonlinear and tension cut-off soil cases was lower than that of the linear counterpart. For these nonlinear cases, the overlapping influence of each closely-spaced pile results in increased loads and yielding of the immediate surrounding soil.

3. As spacing increased, group displacement efficiency becomes higher and more rapidly so in the nonlinear soil cases. This observation implies that the loaded zone of influence around each pile decreases with the level of nonlinearity in soil response.

4. The center pile (pile 2 in Figure 10 and Figure 11) shoulders a lower share of the overall load. Soil within this zone translates laterally due to the action of all surrounding piles, a mechanism that reduces available resistance to the center pile translation (Scott 1981).

5. The mechanism of tension cut-off plays a major role in dictating the analysis outcome. Along with reduced overall lateral stiffness, the front piles shoulder a higher portion of the overall load, and experience higher bending moments and shear forces (Figure 10 and Figure 11).

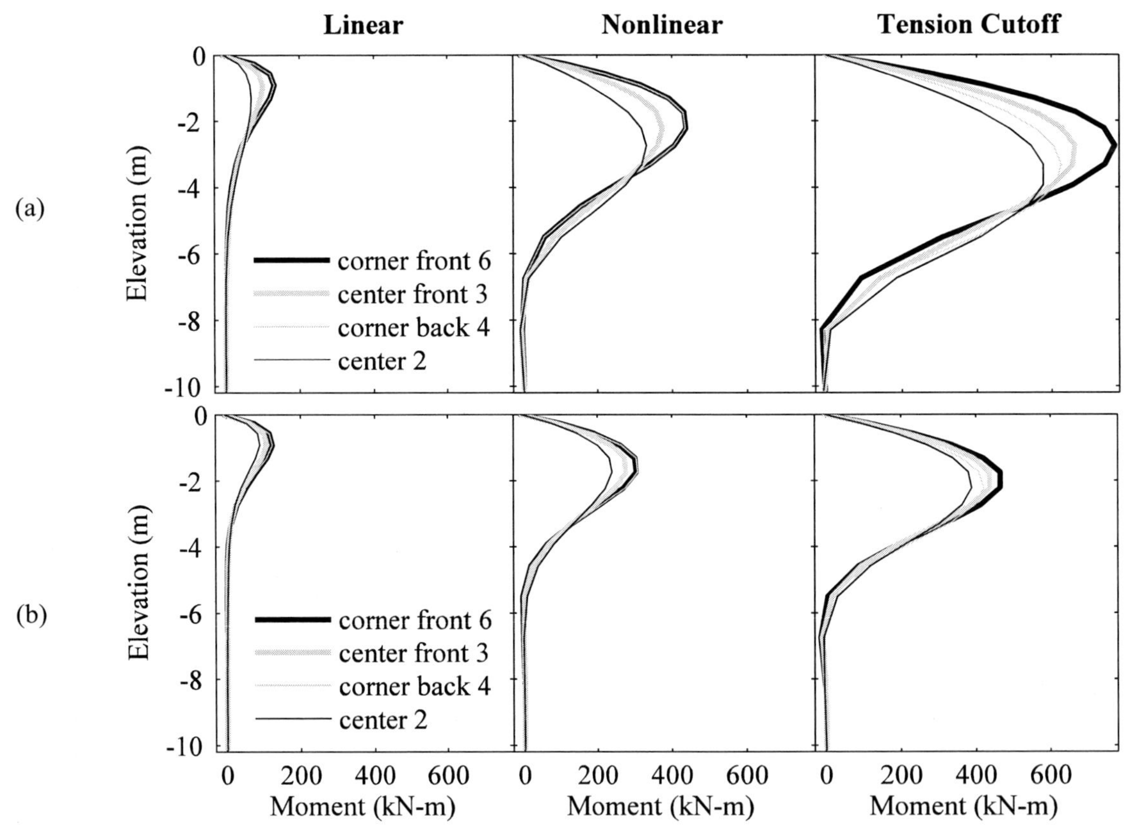

Figure 11. Bending moment profiles at maximum load in the 3×3 pile group of Figure 6 for (a) 3D pile spacing and (b) 7D pile spacing.

PILE GROUP CONFIGURATION AT DUMBARTON BRIDGE PIER 23

In the following section, a model that represents salient characteristics of the Dumbarton Bridge (California) Pier 23 pile-foundation geometric configuration was studied. A seismic evaluation of this bridge was conducted recently by Ke et al. (2010) in which conventional *p-y* springs were used to represent the involved kinematic soil-pile interaction mechanisms at selected piers.

The Pier 23 pile-group foundation is configured in an 8 × 4 arrangement (Figure 12) with a longitudinal (the 8 piles direction) and transversal (the 4 piles direction) spacing of 2 and 2.15 pile diameters on center, respectively. Each pile is 1.37 m in diameter and 30.8 m long. The group is essentially rigidly connected at the pile cap, 14.3 m above the mudline (Figure 12).

Each concrete pile is encased by a prestressed concrete shell (wall thickness $h = 0.1778$ m) and filled with cast-in-place concrete up to 19.8 m below the ground surface (Ke et al. 2010). To simplify the FE model, pile response is assumed to remain linear with a bending stiffness of $EI = 2 \times 10^6$ kN-m^2 for each pile (average value of actual stiffness along the pile height).

For the purpose of this study, three soil strata were considered (Table 1) in which the upper 2 layers were 6.7 m each and the bottom layer was 30.5 m in thickness (Figure 12a). The pressure-independent (J_2) multi-yield surface plasticity model (Elgamal et al. 2008) was employed in which a hyperbolic relationship described the soil shear stress-strain backbone response. A Poisson's ratio of 0.4 was specified for all layers. In this section of the study, the tension cut-off logic was activated (Figure 5c).

Pushover computational analysis of this large pile group system was performed under combined: a) bridge tributary self weight vertical load, and b) lateral load along with overturning moment in the 8-pile longitudinal direction of the FE mesh. In the sections below, distribution of load within the pile group is analyzed and the group interaction effects are examined.

Table 1 Soil Model Properties

Material Property	Top layer	Middle layer	Bottom layer
Thickness (m)	6.7	6.7	30.5
Mass density (kg/m^3)	1300	1500	1800
Shear wave velocity (m/s)	120	250	300
Shear strength (kPa) at a specified octahedral shear strain $\gamma_{max} = 3\%$	25	60	75

Finite Element Model

In the employed ½ mesh configuration of Figure 12b and Figure 12c, length of the mesh in the longitudinal 8-pile direction was 393 m, with 191 m transversely (resulting in a 393 m x 382 m soil domain in Plan view). Total thickness of the soil layers was 43.9 m (i.e., the base of the soil domain located 27.4 m below the pile tip). The soil domain was modeled by 8-node brick elements (23,040 in total), the piles were modeled by beam-column elements (512 in total), and the pile cross-sectional spatial domain (diameter) was modeled by lateral rigid beam-column elements as discussed earlier (1,664 in total). Boundary conditions for this mesh (Figure 12b) were defined as described earlier in this document.

Imposed initially, a vertical dead load of 28,900 kN was estimated to represent the tributary self weight of the bridge deck at Pier 23 (after applying the soil domain self weight). In order to clearly demonstrate the effects of nonlinear soil response and the associated permanent displacement effects of the foundation, a lateral load of 11,677.6 kN (approximately 40% of the tributary bridge-deck self weight) was applied at the bridge deck level (27.6 m above the pile cap) in the 8-pile direction (Figure 12). As such, both lateral loading along with a substantial bending moment of 321,800 kN-m was imposed on the pile group at the pile cap elevation (Figure 12). In this lateral direction, load was applied gradually in 40 equal increments, and thereafter removed in the same fashion (i.e., loading followed by un-loading was explored).

Summary of Main Numerical Results

Under the maximum applied lateral load, the corresponding shear force and bending moment distribution between piles in the pile group is shown in Figure 13a and Figure 13b. The front corner pile 16 carries the highest portion of shear force (-416.23kN) and bending moment (4028.04 kN-m at the pile cap). The edge front pile 8, and back piles 1, 9 also sustain relatively high levels of load. Due to the soil tension cut-off mechanism, the back piles 1, 9 were found to sustain a slightly lower portion of shear force and bending moment compared to their front pile 8, 16 counterparts. The inner piles 3-6 carry the least burden (each, about 80% of the share of pile 16).

Upon removal of the applied lateral load, distribution of residual shear forces and bending moments are shown in Figure 13c and Figure 13d. While relatively small (about 4% of the maximum), residual shear forces are seen to vary in sign between the front and back sectors of the pile group (Figure 13c). In the back sector, the piles return towards their original position a bit more than the surrounding soil, resulting in this positive/negative distribution (Figure 13c). Residual moments in the range of 5-10% of the maximum also remained (Figure 13d).

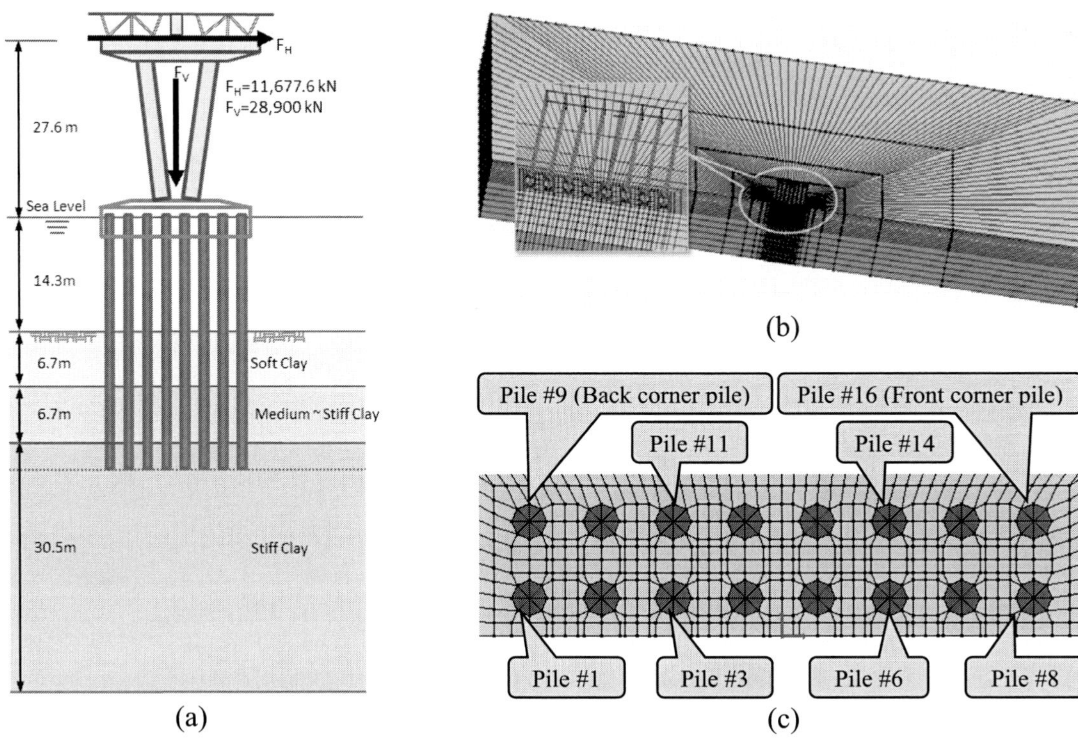

Figure 12. Large 8x4 pile group: (a) Soil profile used for the FE simulation with 8 piles in the direction of applied lateral load, (b) The finite element mesh employed for the simulation (1/2 mesh configuration due to symmetry of the geometry and the applied load), and (c) Plan view of pile group layout (1/2 mesh configuration due to symmetry of the geometry and the applied load).

Axial force distribution between piles in the pile group is shown in Figure 14. Even in the initial axial load static state (bridge-deck self weight), the share of each pile varied in a wide range. Piles along the circumference carry most of the load and the corner piles shoulder the biggest burden (Scott 1981). At the prescribed 11,677.6 kN peak lateral load, the compressive axial forces increase dramatically in the front piles 6-8 and 13-16. Conversely, the back piles experience tensile forces in the range of 0.2-2.4 of the initial static compressive force. Such lower compressive or even tensile axial forces in the back piles may greatly weaken the confinement-dependent structural reinforced concrete properties (Priestley et al. 1994). Upon removal of the applied lateral load, a noticeable redistribution of axial forces is observed with the center piles carrying more load compared to corner and edge piles (Figure 14a and 14c).

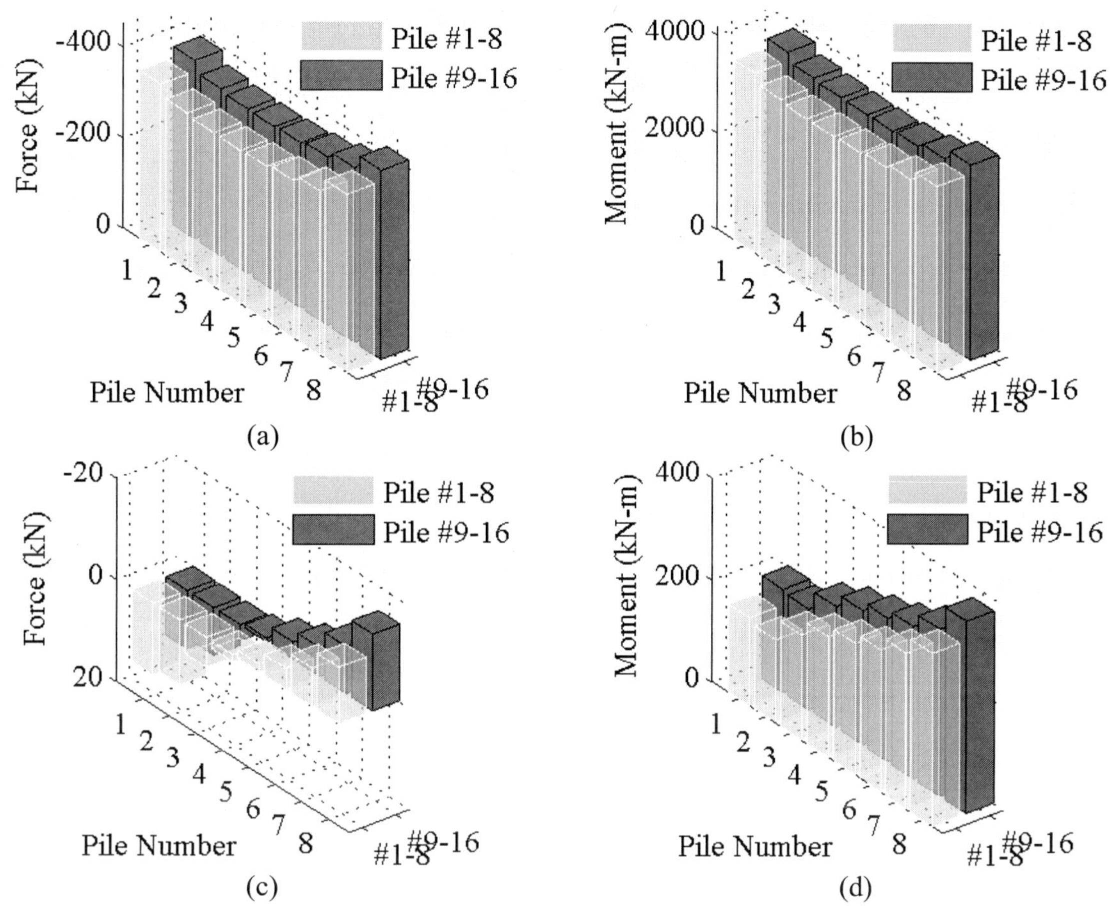

Figure 13. Shear forces and moments in each pile: (a) Peak shear force, (b) Peak bending moment, (c) Residual shear force (after removal of lateral load), and (d) Residual bending moment (after removal of lateral load).

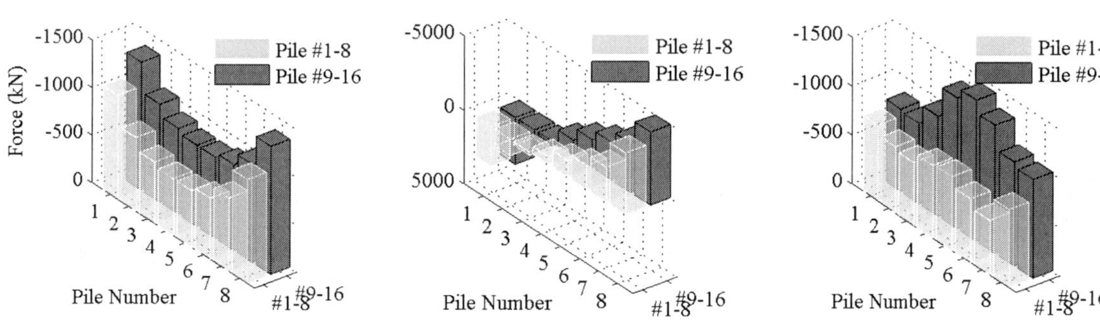

Figure 14. Axial force distribution: (a) under gravity load (b) under the applied peak lateral load and (c) after removal of the applied lateral.

As for peak lateral displacements at the mudline, inner piles 2-6 experienced the most lateral movement, whereas the corner front pile 16 translated the least due to the resistance provided by the surrounding soil (Figure 15). Compared to their front pile counterparts, the back piles 1, 9 moved a bit easier with the movement of the preceding piles 2-6 and 10-14. Overall the inner piles moved as much as 25 % more than the front corner pile. Upon removal of the applied lateral load, about 40 % of the peak displacement remains (Figure 15b).

Peak and residual vertical displacements are shown in Figure 16 (at the maximum lateral load and upon removal). Variation of the pile and surrounding soil displacements highlight the mechanism of load transfer from the pile group to the surrounding soil.

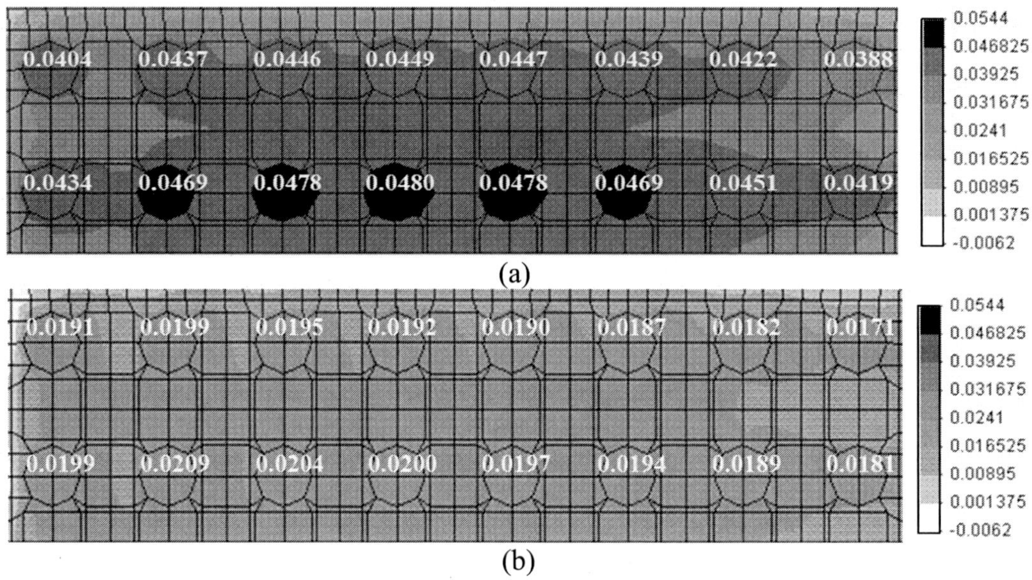

Figure 15. Lateral pile displacements (m) at the mudline: (a) at the maximum lateral load, and (b) upon removal of lateral load.

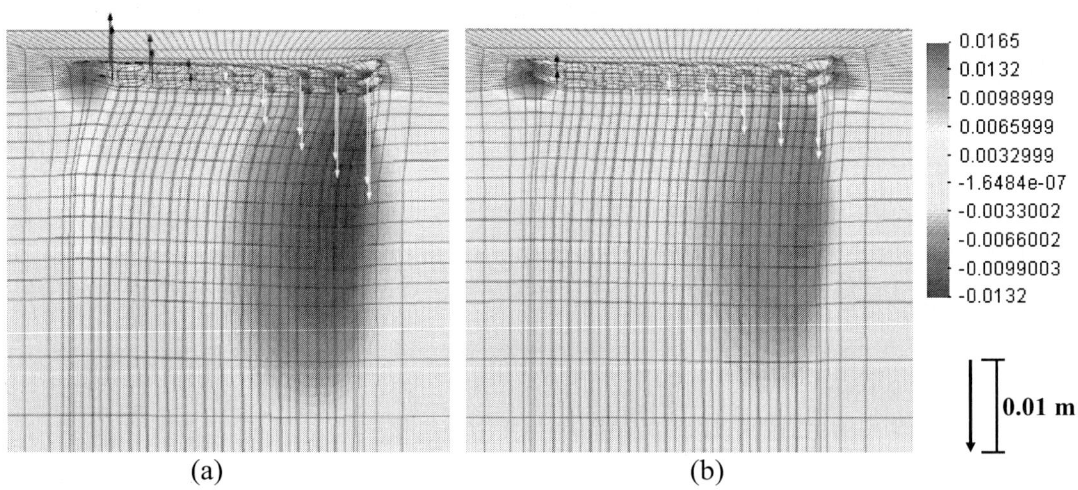

Figure 16. Vertical displacement (m) contour (magnified by a factor of 50): (a) at the maximum lateral load, and (b) upon load removal.

CONCLUSIONS

OpenSeesPL, a robust and versatile framework for 3d computational analysis of pile-ground systems was presented in this paper. A series of numerical analyses on a 3×3 pile group subjected to lateral loading in linear, nonlinear and nonlinear with tension cut-off ground system idealizations were conducted with the aid of this framework. Lateral pile group behavior was compared with that of its single pile counterpart. In addition, a pilot computational study of a large pile group system (Dumbarton Bridge Pier 23 geometric configuration) under combined axial/lateral loads and an associated moment was discussed.

The results of this research bring insights into the response of pile groups at different spacing scenarios, where theoretical solutions and/or experimental/field data may be scarce or nonexistent today. Lateral and axial load distribution among piles in the pile group was examined. Further studies, to be conducted efficiently with the aid of OpenSeesPL, may be directed towards effects of soil/pile stiffness properties, mesh refinement, pile geometric parameters (diameter, length, pile spacing and pile cap contribution), as well as soil-pile boundary conditions. Validation and refinement of this numerical analysis framework can result in higher fidelity and more robust foundation analysis/design principles.

ACKNOWLEDGMENTS

The authors are grateful for the funding provided by the National Science Foundation Grant No. CMMI-1201195.

REFERENCES

Abedzadeh, F., and Pak, R. Y. S. (2004). "Continuum Mechanics of Lateral Soil-Pile Interaction". *Journal of Engineering Mechanics*, 130(11), 1309-1318.

Brown, D. A., Reese, L. C., and Oneill, M. W. (1987). "Cyclic Lateral Loading of a Large-Scale Pile Group". *Journal of Geotechnical Engineering*, 113(11), 1326-1343.

Brown, D. A., and Shie, C.-F. (1990). "Numerical experiments into group effects on the response of piles to lateral loading". *Computers and Geotechnics*, 10(3), 211-230.

Chae, K. S., Ugai, K., and Wakai, A. (2004). "Lateral Resistance of Short Single Piles and Pile Groups Located Near Slopes". *International Journal of Geomechanics*, 4(2), 93-103.

Dodds, A.M. (2005). "A numerical study of pile behavior in large pile groups under lateral loading". *PhD Dissertation,* University of Southern California, Los Angeles.

Elgamal, A., Yan, L., Yang, Z., and Conte, J. P. (2008). "Three-Dimensional Seismic Response of Humboldt Bay Bridge-Foundation-Ground System". *J. Struct. Eng.*, 134(7), 1165.

Elgamal, A., and Lu, J. (2009). "A Framework for 3D Finite Element Analysis of Lateral Pile System Response". *GeoFlorida 2009,* ASCE GSP 186, 616–623.

Elgamal, A., Lu, J., and Forcellini, D. (2009a). "Mitigation of liquefaction-induced lateral deformation in a sloping stratum: 3D numerical simulation". *J. Geotech. Geoenviron. Eng.*, 135(11), 1672-1682.

Elgamal, A., Lu, J., Yang, Z., and Shantz, T. (2009b). "Scenario-focused three-dimensional computational modeling in Geomechanics". *Alexandria, Egypt, 4 iYGEC'09 - 4th Intl Young Geotech. Engineers' Conference*, 2-6 October, ISSMGE.

Ke, T.-C., Law, H., Lam, P., Maroney, B., Mohan, S., and Weinstein, J. (2010) "Ground motion study on Dumbarton toll bridge". *Proc., 5th International Conf. on Recent Advances in Geotech. Earthq. Eng. and Soil Dyn.*, San Diego, CA.

Law, H., and Lam, I. (2001). "Application of Periodic Boundary for Large Pile Group." Journal of Geotechnical and Geoenvironmental Engineering, 127(10), 889-892.

Lu, J., Elgamal, A., Sikorsky, C., and Shantz, T. (2010a). "Computational Modeling of a Large Pile Group under Lateral Load". *Proc., 5th International Conf. on Recent Advances in Geotech. Earthq. Eng. and Soil Dyn.*, San Diego, CA.

Lu, J., Yang, Z., and Elgamal, A. (2010b). "OpenseesPL: three-dimensional lateral pile-ground interaction". *User's Manual*, Dept. of Structural Engineering, Univ. of California, San Diego.

Lu, J., Elgamal, A., Yan, L., Law, K., and Conte, J. (2011). "Large-Scale Numerical Modeling in Geotechnical Earthquake Engineering". *International J. of Geomechanics*, 11(6), 490-503.

Maheshwari, B. K., Truman, K. Z., El Naggar, M. H., and Gould, P. L. (2004a). "Three-dimensional finite element nonlinear dynamic analysis of pile groups for lateral transient and seismic excitations". *Canadian Geotechnical Journal*, 41(1), 118-133.

Maheshwari, B. K., Truman, K. Z., El Naggar, M. H., and Gould, P. L. (2004b). "Three-dimensional nonlinear analysis for seismic soil-pile-structure interaction". *Soil Dynamics and Earthquake Engineering*, 24(4), 343-356.

Maki, T., Maekawa, K., and Mutsuyoshi, H. (2006). "RC pile-soil interaction analysis using a 3D-finite element method with fibre theory-based beam elements". *Earthquake Engineering & Structural Dynamics*, 35(13), 1587-1607.

Mazzoni, S., McKenna, F., Scott, M. H., and Fenves, G. L. (2006). "Open System for Earthquake Engineers Simulation User Manual". *Pacific Earthquake Engineering Research Center,* University of California, Berkeley, CA (http://opensees.berkeley.edu/).

McVay, M., Casper, R., and Shang, T. I. (1995). "Lateral Response of Three-Row Groups in Loose to Dense Sands at 3D and 5D Pile Spacing". *J. of Geotechnical Engineering*, 121(5), 436-441.

McVay, M., Zhang, L. M., Molnit, T., and Lai, P. (1998). "Centrifuge testing of large laterally loaded pile groups in sands." *J. of Geotech. and Geoenviron. Eng.*, 124(10), 1016-1026.

Muqtadir, A., and Desai, C. S. (1986). "Three-dimensional analysis of a pile-group foundation". *International Journal for Numerical and Analytical Methods in Geomechanics*, 10(1), 41-58.

Mylonakis, G., and Gazetas, G. (2000). "Seismic soil structure interaction: Beneficial or Detrimental?". *Journal of Earthquake Engineering*, 04(03), 277-301.

Nguyen, T., Rayamajhi, D., Boulanger, R., Ashford, S., Lu, J., Elgamal, A., and Shao, L. (2013), "Design of DSM Grids for Liquefaction Remediation". *J. of Geotech. and Geoenviron. Eng.*.

Peng, J., Lu, J., Law, K. H., and Elgamal, A. (2004). "ParCYCLIC: Finite element modelling of earthquake liquefaction response on parallel computers". *International Journal for Numerical and Analytical Methods in Geomechanics*, 28(12), 1207-1232.

Priestley, M., Verma, R., and Xiao, Y. (1994). "Seismic Shear Strength of Reinforced Concrete Columns". *Journal of Structural Engineering*, 120(8), 2310-2329.

Rayamajhi, D., Nguyen, T.V., Ashford, S.A., Boulanger, R.W., Lu, J., Elgamal, A., and Shao, L. (2012). "Effect of discrete columns on shear stress distribution in liquefiable soil". *Geo-Congress 2012*, ASCE GSP 225, 1908-1917.

Rollins, K. M., Olsen, R. J., Egbert, J. J., Jensen, D. H., Olsen, K. G., and Garrett, B. H. (2006). "Pile Spacing Effects on Lateral Pile Group Behavior: Load Tests." *J. of Geotech. and Geoenviron. Eng.*, 132(10), 1262-1271.

Rollins, K., and Brown, D. (2011). "Design Guidelines for Increasing the Lateral Resistance of Highway-Bridge Pile Foundations by Improving Weak Soils."

Scott, R.F. (1981). *Foundation analysis*. Prentice Hall, Englewood Cliffs, NJ.

Uzuoka, R., Sento, N., Kazama, M., Zhang, F., Yashima, A., and Oka, F. (2007). "Three-dimensional numerical simulation of earthquake damage to group-piles in a liquefied ground". *Soil Dynamics and Earthquake Engineering*, 27(5), 395-413.

Wakai, A., Gose, S., and Ugai, K. (1999). "3-D elasto-plastic finite element analyses of pile foundations subjected to lateral loading". *Soils and Foundations*, 39(1), 97-111.

Wolf, J. P. (1985). *Dynamic soil-structure interaction*. Prentice-Hall, Englewood Cliffs, NJ.

Yang, Z. (2000). "Numerical Modeling of Earthquake Site Response Including Dilation and Liquefaction". *PhD*, Columbia University, New York.

Yang, Z., and Elgamal, A. (2003). "Application of unconstrained optimization and sensitivity analysis to calibration of a soil constitutive model". *International Journal for Numerical and Analytical Methods in Geomechanics*, 27(15), 1277-1297.

Yang, Z., and Jeremic, B. (2003). "Numerical study of group effects for pile groups in sands". *International J. for Numerical and Analytical Methods in Geomech.*, 27(15), 1255-1276.

Yang, Z., and Jeremic, B. (2005). "Study of Soil Layering Effects on Lateral Loading Behavior of Piles". *Journal of Geotechnical and Geoenvironmental Engineering*, 131(6), 762-770.

Finite Element Analyses of Offshore Monopile Deflection Accumulation Under Harmonic Loading

Deepak Pappusetty[1], S.M.ASCE and Miguel A. Pando[2], Ph.D., P. Eng., M.ASCE

[1]Graduate Student, Civil & Environmental Engineering, University of North Carolina at Charlotte, Charlotte, NC 28223-0001; dpappus1@uncc.edu
[2]Associate Professor, Civil & Environmental Engineering, University of North Carolina at Charlotte, Charlotte, NC 28223-0001; mpando@uncc.edu

ABSTRACT: A wide variety of ocean energy converters (OEC) are being considered worldwide including tidal, wave and current ocean energy. Regardless of the type of ocean energy device a major technical and cost challenge for these systems is their foundation systems. The specific focus of this paper is for ocean stream energy converters supported on monopile foundations. One of the key design considerations for these OEC devices is related to the long term performance of the foundation system which during its design life can be subjected to more than 10^7 cycles of low magnitude lateral loading. This very high number of lateral load cycles can result in a gradual long-term accumulation of pile head deformations and rotations. In some cases these pile head deformations can exceed the OEC serviceability limits. Analyses of laterally loaded piles, based on conventional p-y methodologies, typically cannot predict these accumulations of deformations and rotations. This manuscript presents results of 3D dynamic finite element analyses for this soil-structure interaction (SSI) problem. Results from these analyses reveal that this gradual accumulation of pile head deformation could be an important design consideration for certain monopile conditions. These numerical results correspond well with published experimental data. Finite Element Analysis (FEA) parametric studies, which investigate the influence of different pile installation variables, are presented to shed light on possible design options that mitigate or minimize the accumulation of pile head deformations.

INTRODUCTION

There are a number of ongoing efforts in offshore wind and marine energy converters. Offshore renewable energy projects, for the most part, are in the research stages, although some wind turbines have been installed in Europe. A key challenge for offshore energy sources is to make them cost competitive with traditional energy production methods. This cost challenge is applicable to both initial OEC installation costs as well as for costs associated with operation and maintenance of the foundation system of the offshore energy device. Byrne and Houlsby (2003) have reported the foundation system of offshore wind turbines as costing up to about 35% of the initial installation cost of the facility. Therefore, an optimum design of the foundation system is paramount to help reduce the installation costs of these OEC devices.

The foundation system for offshore energy converters depends largely on the water depth and current conditions. Doherty et al. (2011) report various foundation systems commonly being considered for offshore wind installations. These foundation systems also apply to OEC installations. As shown in Figure 1a, gravity structures are commonly considered the most economical alternative in shallow waters of less than 15 m. Use of monopile foundations (Figure 1b) have been reported as the selected system in European projects with water depths of less than

30 m. In the U.S., monopiles are currently being considered for several offshore wind projects, including in North Carolina, where a recent study by UNC Chapel Hill (2009) suggests wind turbines could be supported on monopiles with embedment depths of 30 to 35 meters to be installed in state coastal waters not exceeding 30 m in depth. For locations with water depths above 50 m, conventional jacket structures (Figure 1c) or floating devices with mooring foundation systems (Figure 1d) are often considered (Doherty et al., 2011).

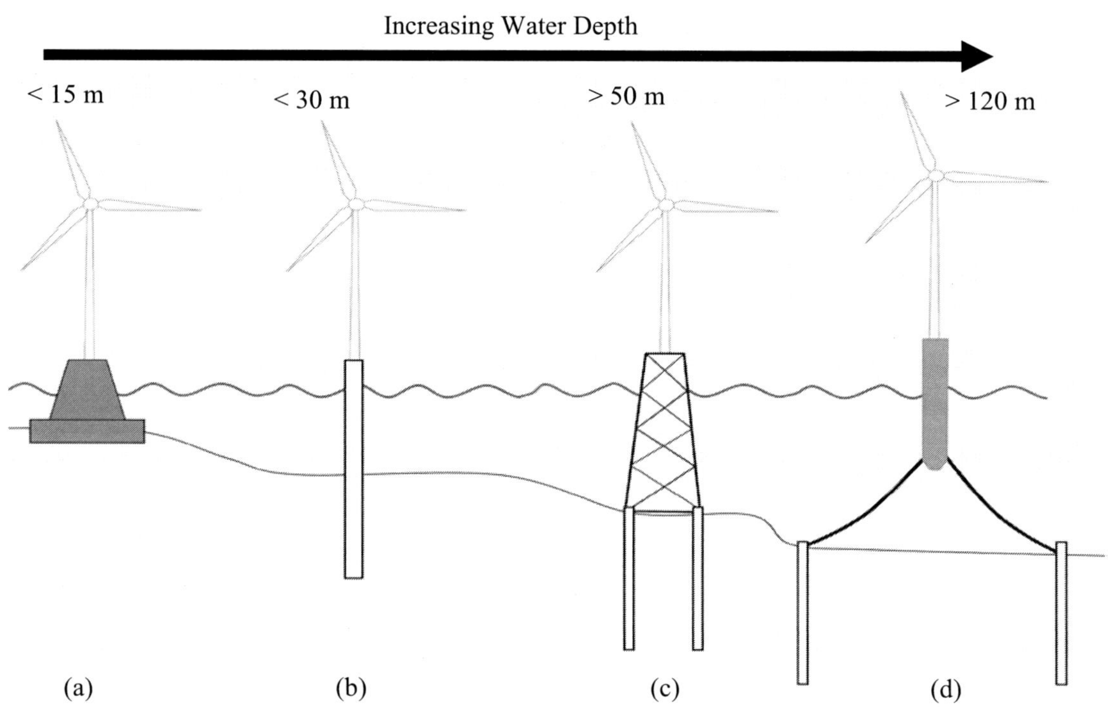

Figure 1. Common types of offshore foundation systems (Doherty et al. 2011).

The focus of this paper is on marine current energy converters with monopile foundation systems (Figure 1b). The main objective of this paper is to assess the key variables that affect the gradual accumulation of pile head displacement and rotations when the foundation is subjected to thousands of low amplitude load cycles during its design life. For an OEC with a 30 year design life a monopile foundation may experience in the order of 10^7 lateral load cycles which can result in a gradual accumulation of considerable pile head displacements beyond acceptable values governed by serviceability considerations. The following section provides a brief background on the loading conditions applicable to offshore monopile systems. This paper will also summarize experimental data reported in the literature that provide evidence of this important design consideration. Additionally, the paper also present results of an analytical parametric study involving dynamic 3D finite element analyses to help shed light on which monopile design parameters are most important in influencing the rate and magnitude of accumulation of pile head rotations and displacements.

OVERVIEW OF LOADING CONDITIONS ON OFFSHORE MONOPILE

The loading condition of marine energy converters is a complex and difficult phenomenon to predict. It depends on the actual marine environment of the installation, which in turn depends on geographic location, depth, time of the year or season, and several other factors. Some marine energy converters are tidal where the ocean current undergoes reversals of direction every day. Other marine energy converters involve oceanic currents with a predominant and persistent flow direction that have temporal fluctuations in magnitude and can have a complex frequency content. For the case of marine current, a typical current velocity time history is shown schematically in Figure 2a. As shown in this figure, an ocean current OEC has an average stream velocity above zero. There are no flow direction reversals, as in tidal OEC's, and the velocity time history typically has a complex frequency content and varying amplitudes. Furthermore, the stream velocity time history has a seasonal variation throughout the year. For the sake of simplicity, and to mimic conditions of published experimental data presented later in the paper, we assume an offset sinusoidal velocity time history as shown in Figure 2b. Based on information from ongoing marine current research projects, the mean current velocity that typically makes these installations feasible in terms of energy output is between 2 and 2.5 m/s (Fraenkel, 2002). The predominant frequency (Figure 2b) has been reported as being between 0.05 to 0.3 Hz and the average velocity amplitude ranges between 10 and 40 % of the mean current velocity.

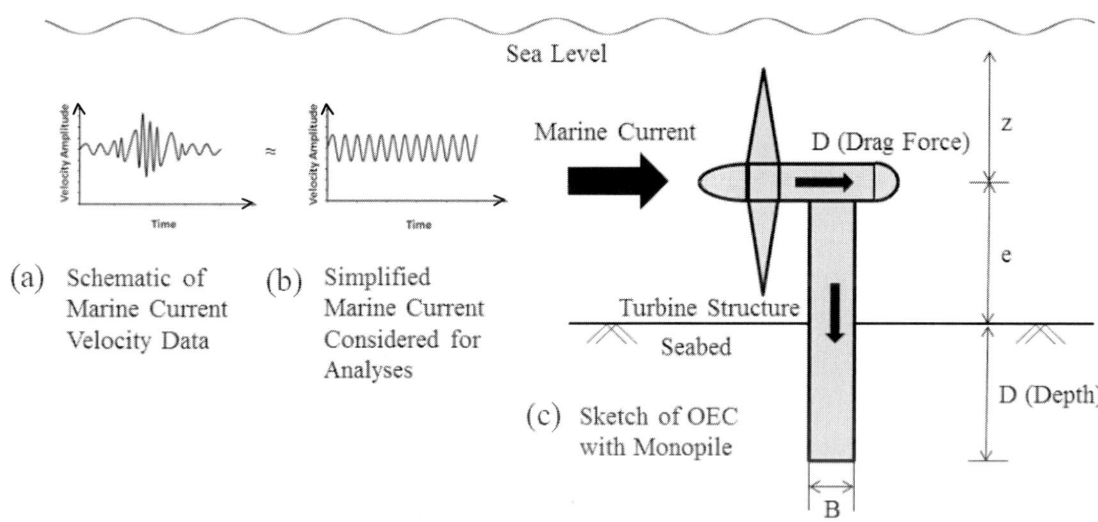

Figure 2. Schematic of marine current velocity data and the associated loading on Monopile.

The net horizontal force acting on the monopile foundation of the OEC can be estimated on the basis of computational fluid dynamics and depends on converter geometry (Figure 2c) and the current velocity (Figures 2a and b). Fraenkel (2002) reports that depending on maximum stream velocity and rotor size, a turbine may be subjected to a horizontal force ranging between 1,000 to 3,000 kN. This horizontal or drag force acting on turbine can be estimated as follows:

$$D = \frac{1}{2} C_d \rho U^2 A \tag{1}$$

where D is the drag force, C_d is the drag coefficient which depends on turbine geometry, ρ is the density of the seawater, U is the ocean current speed, and A is the rotor swept area calculated by $A = \pi r^2$ where r is rotor radius. With the aid of this equation, the average horizontal drag force acting on the turbine that will be transferred to the monopile can be computed for a mean current velocity and rotor diameter. An additional lateral load is related to drag forces acting on the support column. Lateral forces can also be estimated using more advanced approaches such as computational fluid dynamics (CFD) which are outside the scope of this paper.

SUMMARY OF KEY PUBLISHED EXPERIMENTAL STUDIES

Studying the effect of sustained cyclic loading on monopiles is important to ensure adequate performance of the OEC. This section summarizes experimental studies which provide evidence of gradual accumulation of pile head deformations. The main mechanism for this gradual accumulation of lateral pile head deformation is related to induced plastic strains in certain regions of the soil surrounding the monopile when subjected to sustained cyclic lateral loading. Several researchers have shown evidence of plastic strains accumulation during sustained cyclic or repetitive loading (e.g., Andersen 2004; Peng et al. 2006; Leblanc et al. 2010). Peng et al. (2006) and Leblanc et al. (2010) have reported experimental data that show how cyclic loading lead to accumulation of pile head displacements and rotations and have highlighted the influence of important variables such as cyclic load amplitude and frequency.

The experimental study performed by Peng et al. (2006) involved tests on small scale monopiles installed in a dry sand tank. Their experimental setup was capable of applying numerous uniform load cycles to the monopile. Lateral loading was applied in a harmonic fashion. Other experimental factors studied by the authors included amplitude of loading and frequency. Model tests included piles subjected to one-way cyclic loading at three different frequencies with a constant force amplitude, one-way loading with three different force amplitudes at a constant frequency of loading, and two-way balanced and unbalanced loading. In general, this study showed that the magnitude and rate of pile head displacement accumulation increased with increasing lateral force amplitude and with increasing frequency of the loading. Figure 3 (Peng, 2005) shows the variation of the peak pile head lateral displacement as a function of applied load cycle. This figure corresponds to results for medium dense sand and a monopile of diameter B, equal to 4.45 cm and embedment depth D, equal to 40 cm. As shown, the monopile initially displaces laterally about 1.5 mm but after 10^4 lateral load cycles the pile head deforms laterally about 10 mm. This represents an increase of about 567%. Figure 3 also shows the predicted pile head displacement using conventional p-y based analyses. These predictions use p-y curves as per API (2000), which recommends using p-y curves with a 10% reduction for offshore piles under cyclic loads. The API (2000) predictions under-predict experimental observations even if the reduced p-y curves are used. Furthermore this approach has the limitation that the predicted pile head deflection is independent of the number of load cycles applied. Clearly there is room for improvements in p-y based analyses. Later in this paper we present some reported improvements in p-y curves for cyclic loading.

Leblanc et al. (2010) also performed an extensive experimental program involving small scale 1g laterally loaded monopiles designed specifically to gather insight on the behavior of monopiles for offshore wind turbines. The experimental setup had similar capabilities to the one by Peng et al. (2006) in terms of flexibility to apply large number of uniform lateral load cycles, modes of loading, and variation in amplitude of loading. However, their setup had a constant frequency of lateral loading of 0.106 Hz. This study included results for single monopiles subjected up to 10,000 load cycles tested under two different soil density conditions. Their experimental tests applied different amplitude of loadings over monopile under both one way and two way conditions. The authors developed the following equation to estimate pile head rotation of a stiff monopile as a function of lateral load cycle N:

$$\frac{\Delta\theta(N)}{\theta_s} = T_b . T_c . N^{0.31} \tag{2}$$

where $\Delta\theta(N)$ is the pile head rotation incurred at N^{th} load cycle, θ_s is the static pile rotation, and T_b and T_c are dimensional functions which depend on loading characteristics and the relative density of the sand. Leblanc et al. (2010) provided valuable evidence of dependency of accumulated lateral pile head displacement on loading mode and loading direction history, among other factors.

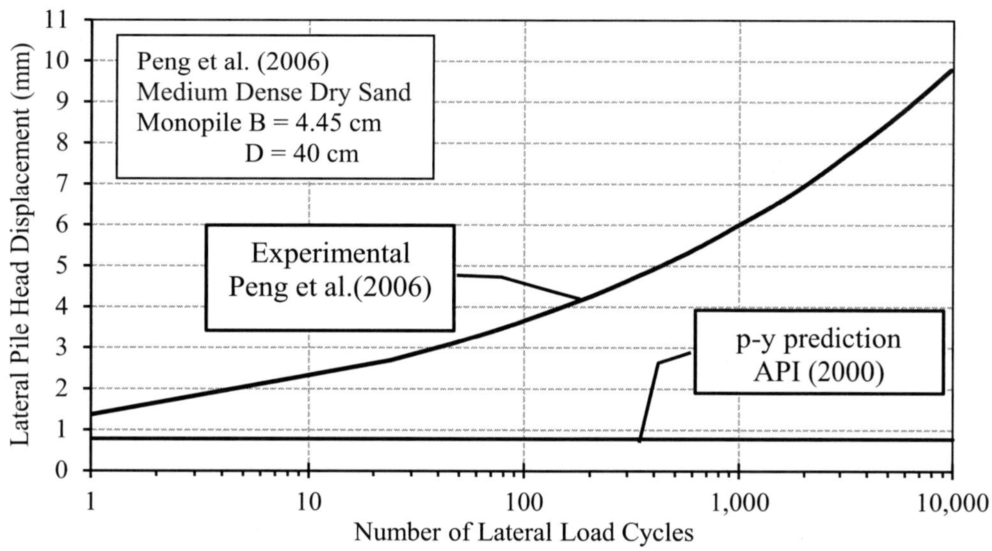

Figure 3. Experimental evidence of gradual pile head displacement accumulation for monopile under sustained cyclic loading.

ANALYSIS METHODS FOR MONOPILES UNDER CYCLIC LOADING

In general, assessment of accumulated pile head displacements of single monopiles under cyclic lateral loading is typically carried out by using p-y based analyses (e.g., Reese et al. 1974) or by means of finite element analyses (e.g., Achmus et al. 2010). A brief overview of how these two methods are typically implemented for this type of SSI problem is presented below.

Analyses using p-y curves

Analyses of laterally loaded piles based on p-y curves are routinely used in design practice, particularly for static loading conditions. For offshore platforms, piles are usually designed based on guidelines by API (2000) which for lateral loading recommends the use of p-y based procedures. For example, for piles installed in sandy soils, API (2000) recommends using p-y curves by Reese et al. (1974), as follows:

$$p = A \times p_u \times tanh\left[\frac{k \times H}{A \times p_u} \times y\right] \quad (3)$$

where p_u is ultimate soil resistance at depth H, k is the initial modulus of subgrade reaction, y is the lateral deflection of the pile at this same depth, and A is the factor to account for cyclic or static loading condition. For this p-y curve type, the effect of cyclic loading is primarily reflected in the value of the factor 'A', which goes from 1.0 for static to 0.9 for cyclic loading. In other words, the nonlinear p-y curve is degraded by 10% to account for p-y resistance degradation due to cyclic loading. Reese et al. (1974) indicate this p-y curve reduction was determined empirically from instrumented load tests involving piles subjected to less than 100 lateral load cycles. It is important to note that this degradation factor is applied only once so it does not allow computation of pile head displacements as a function of the applied load cycle N.

Little and Briaud (1988) developed cyclic p-y curves that consider the effect of number of cyclic loads. This study was based on experiments of single piles installed in sand and subjected to up to 20 cycles of lateral loading. The load cycles were of constant load amplitude where the pile load after reaching the peak amplitude was decreased to zero before applying the subsequent load cycle. Based on the experimental results the authors proposed p-y curves varying as a function of load cycle *(N)*. Using the p-y curve for the first cycle as the starting point, the authors gradually modified the p-y curves by increasing the abscissas *(y)* for a given soil reaction level (p_N) for the Nth cycle as follows:

$$y_N = y_1 . N^a \tag{4}$$

where y_N is the lateral displacement corresponding to the soil resistance (p_N) at the Nth cycle, y_1 is the static displacement for the monotonic loading condition (or cycle 1), and *'a'* is a degradation parameter developed from the results of cyclic pressuremeter tests. By incorporating Eq. 4 into non-linear p-y curves, cyclic p-y curves can be constructed for different load cycles and thus be used for analyzing laterally loaded piles subjected to cyclic loads.

Long and Vanneste (1994) also proposed a methodology for degradation of p-y curves for piles under cyclic lateral loads. The methodology by Long and Vanneste (1994) is similar to the one proposed by Little and Briaud (1988). It is based on experimental results from 34 cyclic lateral load tests carried out on piles installed in both sand and clay. The load cycles were similar to Little and Briaud (1988) in that the pile was loaded to a constant peak lateral load and then unloaded to zero before applying the subsequent load cycle. This study included results from tests with mostly less than 100 load cycles, but a few tests involved up to 500 cycles.

Long and Vanneste (1994) used the experimental results to develop expression for both *'p'* and *'y'* as a function of load cycle N, as follows:

$$p_N = p_1 . N^{-0.4t} \tag{5}$$

$$y_N = y_1 . N^{0.6t} \tag{6}$$

where p_N is the soil reaction for the Nth cycle, p_1 is as defined in Eq. (3), and y_N and y_1 are the lateral deflection at the p-y curve at the Nth and first cycle, respectively. The exponent t is a degradation parameter.

Based on the above, both of these proposed cyclic p-y curves were based on tests involving 500 lateral load cycles or less. This may be an important limitation when trying to apply them to offshore OEC monopile installations where lateral load cycles during the design life of the facility can reach $N=10^7$. Additionally, lateral load cycles for OEC current installations have lateral load cycles of varying frequency content and load cycles that do not load and unload to zero as the ones described in the above studies.

Finite element analyses

Finite element analyses (FEA) have often been the less preferred approach of analysis for this SSI problem primarily due to the required computational effort and challenges to select adequate constitutive soil models to capture the complex behavior for soils undergoing large numbers of load cycles. A few studies performed on monopiles using finite element analysis are summarized in this section.

Achmus et al. (2005) conducted three dimensional finite element analyses to predict behavior of large diameter monopiles for offshore wind installations requiring resisting high horizontal loads due to wind and waving action. The objective of their study was to identify the difference in behavior of monopile when analyzed using the conventional p-y method versus advanced FE analysis. A 7.5 m diameter monopile with an embedded length of 30 m was considered as the base case for their study. In order to model a sandy type of soil, the authors used an elasto-plastic material law with Mohr-Coulomb failure. The authors used ABAQUS and the soil model allowed the use of parameters to represent the stress dependency of the soil stiffness. Their numerical study showed that the results predicted using p-y analyses following the API (2000) lead to significant underestimation in lateral deflections when compared to the results from the more detailed 3D FE analyses.

Achmus et al. (2010) reported additional results from 3D FE analyses where they improved the soil model by calibrating with results from drained cyclic triaxial tests. From the results of cyclic triaxial tests, a relationship between plastic strains and stiffness of the soil was used to update soil properties as a function of load cycle. With the help of this approach the stiffness of the soil at a considered load cycle was updated as a degraded stiffness in the FE model. The numerical results were then integrated in order to estimate the complete behavior of monopile subjected to up to 10^4 lateral load cycles.

CALIBRATION OF FE MODELS USED IN THIS STUDY

A series of calibration analyses were performed for the FE model used herein. All analyses were carried out using the finite element software Plaxis 3D Dynamics (Plaxis 2011). Calibration studies included static analysis with both monotonic and cyclic loading and a case involving dynamic cyclic loading. Numerical studies were calibrated with the experimental results reported by Peng et al. (2004) and Peng et al. (2006). Details of the FEA models are presented in the next section.

The experiments performed by Peng et al. (2004) involved a soil test chamber with dimensions 1 m x 1 m x 1 m. The small monopile tested was a steel pipe tube with an outside diameter of 4.45 cm and a wall thickness of 2.15 mm. The monopile was embedded 40 cm and had a 3 cm stick up above ground. The lateral load was applied at the pile head elevation. The sand was uniform quartz sand placed dry at a relative density of 71.7%, which was reported as having a dry unit weight of 16.5 kN/m^3 and an effective internal friction angle of 35.2 degrees.

The first calibration analysis consisted of predicting a static pushover test reported by Peng et al. (2004). Figure 4 (a) shows the predicted pile head lateral displacement as a function of applied lateral load. This analysis used a Mohr-Coulomb (MC) model for the soil and from the figure it can be seen the FE model predicted fairly well the experimental data.

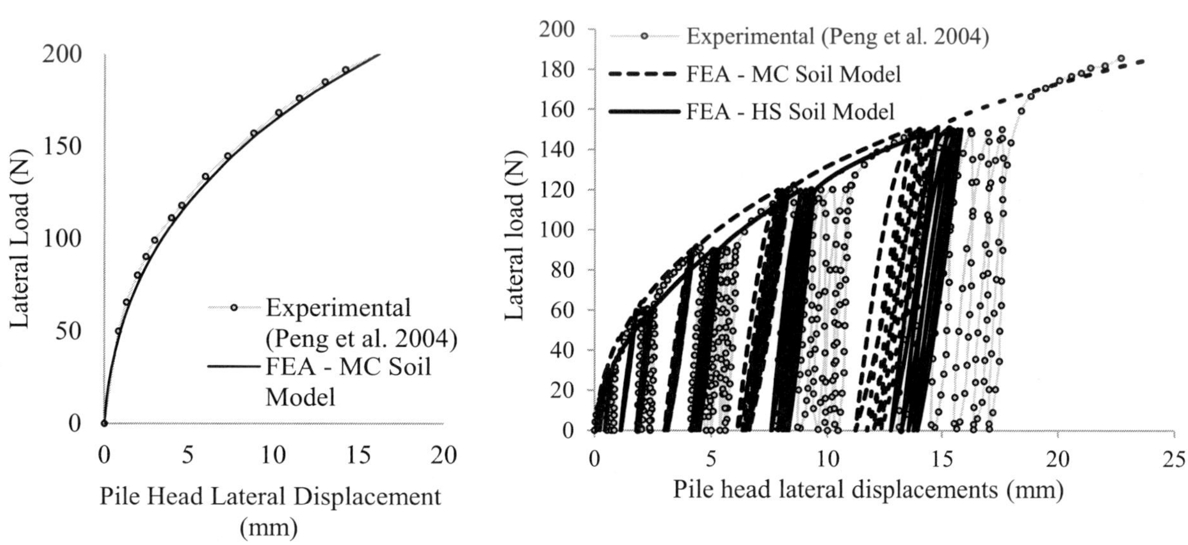

(a) Static Calibration (b) Pseudo Static Loading

Figure 4: Calibration of 3D FE models: Static cases using experiments by Peng et al. (2004).

The second static calibration analysis consisted on a pseudo-static cyclic lateral load test where the monopile was subjected to a monotonic lateral load that gradually increased from 0 N to 30 N. At this load level, the pile was subjected to five unloading and reloading cycles. After the fifth reloading cycle, the load was then increased to the next target load level where again 5 unload-reload cycles were applied. These 5 unload-reload cycles were applied when the lateral load reached 30, 60, 120 and 150 N. For this calibration test we modeled the soil with MC model and the Hardening Soil (HS) model. The predicted pile head displacements for both soil model types are compared with the experimental data in Figure 4b. Both soil models predict reasonably well the experimental data; however, the HS model better captured the slope of the unloading-reloading load cycles.

The calibration of the FE model using an experiment involving dynamic cyclic loading was the third stage of this process. For this case, the cyclic loads were sinusoidal dynamic loads with frequency and amplitude equal to 0.45 Hz and 50 N, respectively. The dynamic experiments were reported by Peng et al. (2006). The FE model for dynamic loading included viscous boundary conditions to absorb the waves approaching the boundaries and restrict them from reflecting back. Damping was introduced in the model by means of Rayleigh damping coefficients. The first dynamic FE calibration analysis used an MC model for the soil. The results obtained with the MC model are shown in Figure 5. The MC model did not capture the observed increase of pile head displacements with increasing number of applied lateral load cycles. The second dynamic FE calibration analysis used an HS model and the results obtained with this constitutive soil model are also shown in Figure 5. This figure shows a close match between experimental and numerical results using the HS model. However, some discrepancy was observed in the initial portion of the curve (1-200 cycles). The HS model analytical predictions show a better match with the experimental results for load cycles above 300.

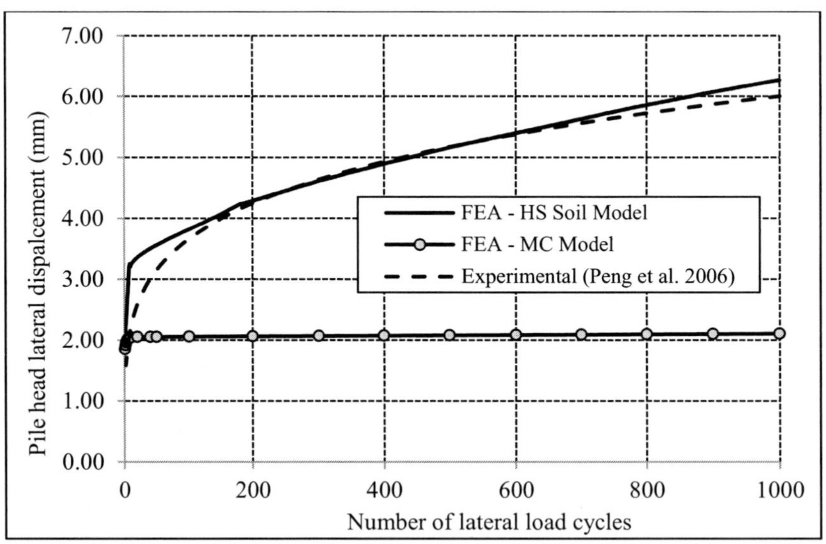

Figure 5. Calibration of dynamic 3D FE model using Peng et al. (2006).

PARAMETRIC FE ANALYSES

Geometry and Boundary Conditions of the Base Problem

A hypothetical case, shown in Figure 6, was considered as a base case. This geometry is considered representative of most ongoing OEC research projects. The OEC considered is an ocean current turbine designed for uni-directional marine current or stream flow. A concrete filled tubular (CFT) pile, as shown in Figure 6, with a 2 m diameter and a wall thickness of 25.4 mm, was assumed to represent the monopile. The pile head, representing the center of gravity of the turbine, was assumed to be located at 10.3 m above the seabed. The embedded length D, of the pile was changed as part of the parametric analyses to assess the effect of the D/B ratio (embedment length/diameter ratio). The mean lateral horizontal force acting on the turbine is evaluated as 600 kN for a rotor diameter of 16.5 m and a mean current speed of 2.4 m/s. The cyclic loading was assumed sinusoidal with a frequency, f = 0.1 or 0.2 Hz. The ratio between the minimum and maximum load amplitude ($R = F_{min}/F_{max}$) was equal to 0.6 or 0.67 for the FE parametric analyses. Figure 7 illustrates the loading time history used for all the FE simulations to define the load amplitude.

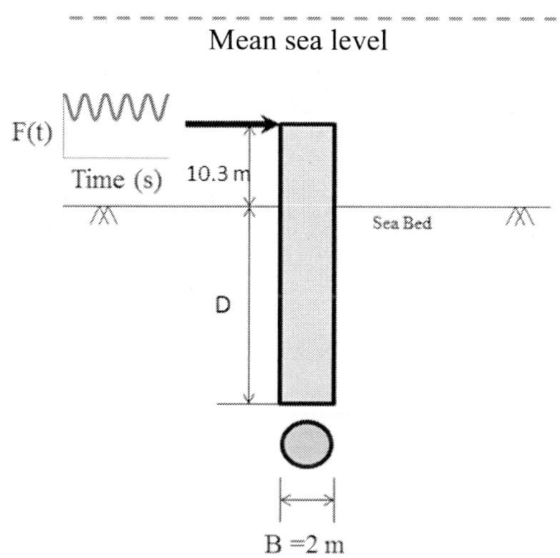

Figure 6. Geometry of the monopile used for the parametric analyses (base problem).

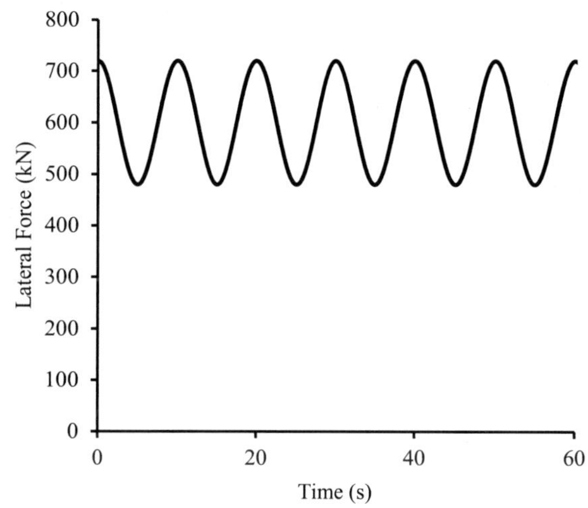

Figure 7. Load time history for mean lateral load of 600 kN, f = 0.1 Hz and R = 0.67.

FE Model and Constitutive Model

The FE model used for the parametric study is shown in Figure 8. The mesh had a total of 20,400 elements and 28,400 nodes. The finite element type used was ten node tetrahedral elements for both the pile and the soil. This is the preferred finite element type used by the automatic mesh generator of the program Plaxis 3D. Interface elements with a reduction factor, R_{inter} equal to 0.7, were used to consider the contact between the pile and the surrounding soil. R_{inter} is analogous to the ratio of the pile-soil interface friction angle to the soil effective friction angle. The FE mesh used had three zones of refinement. The zone closest to the pile which is subjected to maximum stress concentrations had the finest elements and the elements became gradually coarser as the zones move away from the pile. The pile was modeled as a volume pile with EI equivalent to the CFT pile. The FE mesh extended laterally at least 7.5 times the monopile diameter in both directions from the centerline of the monopile. The nodes located along the outer vertical boundaries of the mesh were restricted to only move along the vertical boundary plane. The nodes along the base plane of the model were restricted to move in all directions. Non-reflective boundaries were used along the outer boundaries to minimize dynamic boundary effects due to reflecting waves.

The pile was modeled as linear elastic with properties selected to match the flexural stiffness of the CFT pile. The soil model used was the HS model as described in Plaxis (2011). In this model the soil stiffness stress dependency is captured using three different input stiffness parameters, namely, the triaxial loading stiffness E_{50}^{ref}, the triaxial unloading stiffness E_{ur}^{ref}, and the oedometer loading stiffness E_{oed}^{ref}. All these parameters are selected for a reference stress P^{ref}. Table 1 presents a summary of the soil parameters used for the two relative densities considered.

Table 1. Soil Parameters Used.

Soil Condition		Soil Density	
		Dense	Medium Dense
γ'	(kN/m^3)	11	11
c'	(kN/m^2)	0	0
ϕ'	(Degrees)	37.5	35
ψ'	(Degrees)	7.5	5
E_{50}^{ref}	(MN/m^2)	60	40
E_{oed}^{ref}	(MN/m^2)	60	40
p^{ref}	(kN/m^2)	100	100
E_{ur}^{ref}	(MN/m^2)	180	120

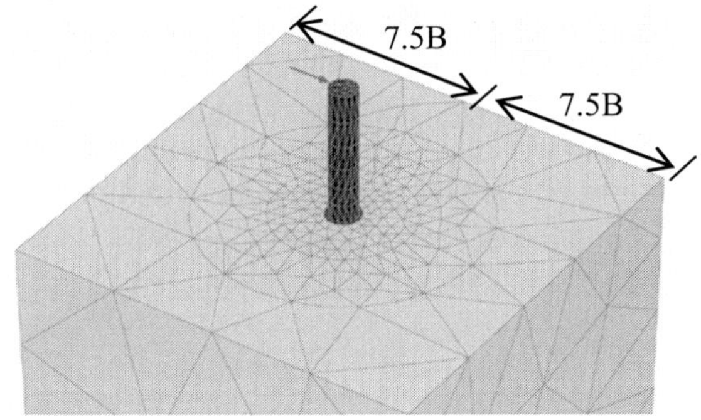

Figure 8. FE mesh used for parametric study

The different FE analyses considered for the parametric study are summarized in Figure 9. However for the sake of brevity, results for only the dense soil relative density condition with diameter, $D = 2$ m, $f = 0.1$ Hz, and $R = 0.67$ are presented in this paper. For the parametric study carried out, the values of the ratio of the predominant period of the applied lateral loading and the first mode period of the pile-soil system ranged from 10 to 50. These period ratios suggest the applied lateral loading is rapid (i.e., not pseudo-static) thus warranting the use of dynamic FEA.

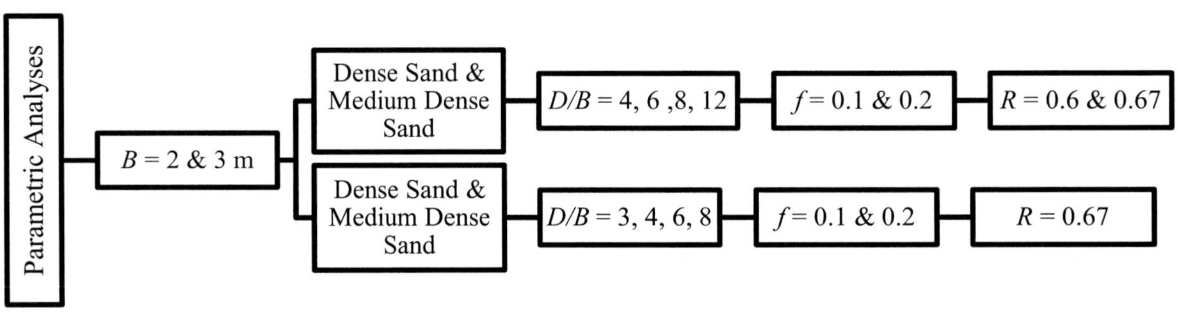

Figure 9. Summary of Cases Considered in Parametric study

Summary of Results

The results of the FE analyses for the dense relative density case are summarized in Figure 10. This figure shows the normalized accumulated lateral pile head displacement of the pile as a function of the number of load cycles. The normalized lateral displacement of the pile head corresponds to the dimensionless ratio $\Delta H_N/\Delta H_1$, where ΔH_N is the magnitude of the lateral displacement at the Nth load cycle, and ΔH_1 is the lateral displacement of the pile head at the first load cycle.

The lateral displacements, at the end of the first load cycle, for D/B ratios of 4, 6, 8 and 12 are equal to 23.1 mm, 6.3 mm, 3.8 mm, and 3.7 mm, respectively. Therefore, results show ΔH_1 values decreasing with increasing D/B ratio. Furthermore the rate of accumulation of pile head lateral displacement was found to significantly decrease with increasing D/B ratio. However, beyond D/B ratio of 8 there is not much difference in magnitude or rate of increase of accumulated displacements. It can be concluded from the above discussion that, for the given soil condition, properties of the pile, and loading conditions, an optimum D/B ratio of about 8 would be appropriate to minimize the cyclic loading effects of progressive accumulation of pile head deformations.

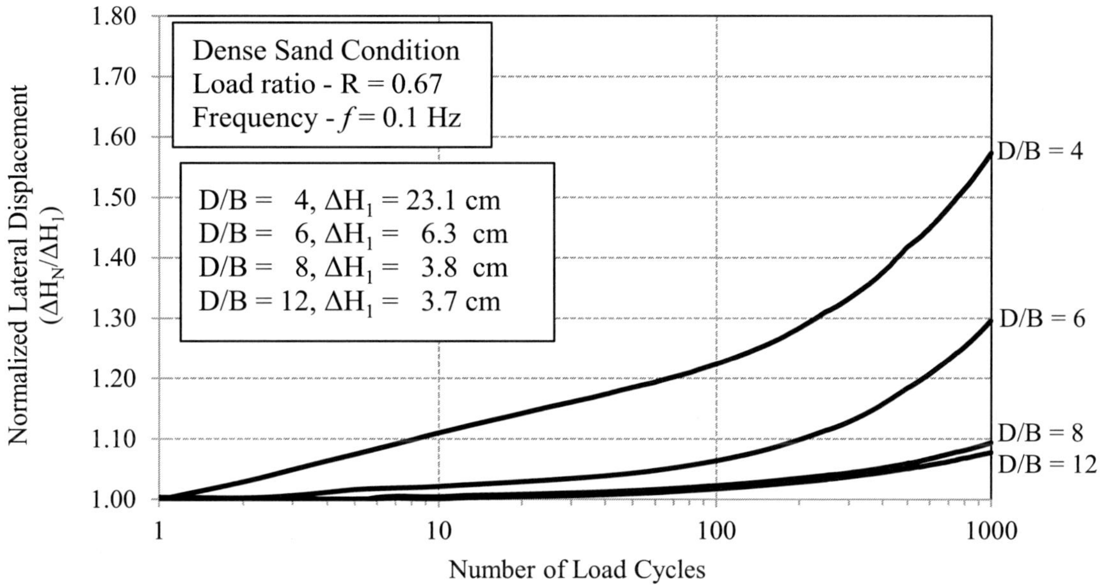

Figure 10. Normalized lateral head displacements with respect to number of lateral loads cycles.

SUMMARY AND CONCLUSIONS

A summary of published experimental data (Peng et al., 2004, 2006; Leblanc et al. 2010) was presented to help highlight the importance of considering the effect of sustained cyclic lateral loading on monopiles, which can result in considerable gradual accumulation of pile head displacements and rotations which for some cases can go beyond the allowable serviceability limits for these OEC installations, particularly since OEC projects with a design a life of about 30 years can experience 10^7 lateral load cycles or more.

The paper presents results of a numerical parametric study involving 3D dynamic FE analyses using the specialized geotechnical software Plaxis 3D dynamic (Plaxis, 2011). Specifically, the paper presents the result of FEA for a representative OEC monopile installed in dense sand for four embedment to diameter (D/B) ratios. However, for all the analyses the diameter of the pile was kept constant and only the embedment depth of the pile was varied. The pile was subjected to sinusoidal cyclic lateral load with a constant frequency. The results showed that for increasing D/B ratios the pile head displacements decreased. Similarly the rate of accumulation of lateral displacement of the pile head with lateral load cycles was found to sharply decrease with increasing D/B ratio. However, for the pile conditions considered, the pile head deflections (magnitude and rate of accumulation) did not seem to change much for D/B ratios over 8. This suggests that this monopile for the loading and soil conditions considered, would have an optimum D/B ratio of about 8 or higher.

For design of OEC monopiles it is recommended that the designer considers pile head displacement limits based on serviceability considerations of the turbine and selects a D/B ratio such that these serviceability limits are not exceeded for the maximum lateral load cycles anticipated for the design life of the project.

ACKNOWLEDGMENTS

The authors would like to acknowledge the financial support received from the NC Ocean Energy Program which provided financial support for the first year of graduate studies of the first author. However the opinions presented herein correspond solely to the authors and do not necessarily reflect the views of this program. The authors acknowledge the review comments of the two anonymous reviewers. The authors are honored to be part of this Special GSP which honors and acknowledges the many significant contributions made by Professor Emeritus Roy E. Olson, University of Texas at Austin, to the profession of geotechnical engineering.

REFERENCES

Abdel-Rahman, K. and Achmus, M. "Finite element modelling of horizontally loaded monopile foundations for offshore wind energy converters in Germany." Proc., 1st International Symposium on Frontiers in Offshore Geotechnics, ISFOG 2005, September 19 -21, 2005, Taylor and Francis, Balkema, 391-396.

Achmus, M., Kuo, Y.-S., and Abdel-Rahman, K. (2009). "Behavior of Monopile foundations under cyclic lateral load." *Computers and Geotechnics*, 36(5), 725-735.

Andersen, K. (2004). "Cyclic clay data for foundation design of structures subjected to wave loading." *Cyclic Behaviour of Soils and Liquefaction Phenomena*, 371-387.

API (2000) American Petroleum Institute. Recommended Practice for Planning, Design and Constructing Fixed Offshore Platform, *API Recommended Practice 2A-WSD (RP2A-WSD)* 21st edition, Dallas.

Byrne, B. W. and Houlsby, G. T. (2003). "Foundations for offshore wind turbines." Philosophical Transactions of the Royal Society London, Series A (Mathematical, Physical and Engineering Sciences), 361(1813), 2909-2930.

Doherty, P., Gavin, K., and Casey, B. (2011). "The Geotechnical Challenges Facing the Offshore Wind Sector." Geo-Frontiers 2011, American Society of Civil Engineers, 162-171.

Fraenkel, P. L. (2002). "Power from marine currents." *Proceedings of the Institution of Mechanical Engineers, Part A (Journal of Power and Energy)*, 216(A1), 1-14.

Leblanc, C., Houlsby, G. T., and Byrne, B. W. (2010). "Response of stiff piles in sand to long-term cyclic lateral loading." *Geotechnique*, 60(2), 11.

Little, R. L. and Briaud, J. L. (1988). "Full scale cyclic lateral load tests on six single piles in sand." *Miscellaneous Paper GL-88-27, Geotechnical Div.* Texas A&M Univ., College Station, Texas.

Long, J. and Vanneste, G. (1994). "Effects of Cyclic Lateral Loads on Piles in Sand." *Journal of Geotechnical Engineering*, 120(1), 225-244.

Peng, J. R., Clarke, B. G., and Rouinia, M. (2006). "A Device to Cyclic Lateral Loaded Model Piles." *Geotechnical Testing Journal (GTJ)*, 29(4), 1-7.

Peng, J. R., Rouainia, M., Clarke, B. G., Allan, P., and Irvine, J. (2004). "Lateral resistance of fin piles established from model tests." *Proceedings of the International Conference on Geotechnical Engineering,* Beirut, pp. 565-571.

PLAXIS (2011). Finite Element Code for Soil and Rock Analysis, PLAXIS b.v., Netherlands.

Randolph, M. and Gourvenec, S. (2011). *Offshore Geotechnical Engineering*, Spon Press, Abingdon.

Reese, L. C., Cox, W. R., and Koop, F. D. (1974). "Analysis of Laterally Loaded Piles in Sand." *Offshore Technology Conference*, Offshore Technology Conference, Paper OTC 2080, Houston, TX, 473-483.

UNC (2009). University of North Carolina at Chapel Hill. "Coastal Wind - Energy for North Carolina's Future." <http://offshorewindhub.org/sites/default/files/resources/unc_6-18-2009_coastalwindncoffshorewindfeasibility.pdf>, (Mar. 22, 2012)

Column Groups Analyses for Stone Column Reinforced Foundation

S.A Tan[1], K.S. Ng[2], and Jie Sun[3]

[1]Associate Professor, Dept. of Civil Engineering, National Univ. of Singapore, 1 Engineering Dr. 2, Singapore 117576; cvetansa@ nus.edu.sg
[2]Lecturer, Faculty of Civil Engineering, Universiti Teknologi MARA, 40450 Shah Alam, Selangor, Malaysia. E-mail: kokshien@yahoo.com
[3]Ph.D candidate, Dept. of Civil Engineering, National Univ. of Singapore, 1 Engineering Dr. 2, Singapore 117576; sunjie@ nus.edu.sg

ABSTRACT: Stone columns reinforced foundation has gained popularity as an effective ground improvement technique available in practice. The design of large group of columns is normally done assuming infinite columns grid under wide loaded area (i.e. unit cell concept). However, the design for small column groups is often being approached empirically due to the lack of research on this issue. This study is aimed at investigating the behavior of small column groups by establishing the relationship between the number of columns and the settlement improvement factors. Two-dimensional (2D) concentric ring approach is adopted. This 2D approach is validated by comparison with a three dimensional model before being used subsequently for detailed analyses. From the results, the optimum length for the columns group is determined where further lengthening of the column provides no further improvement for the system. Simple design method is suggested for rigid spread footings supported by small column groups and the comparison is made to the settlement performance obtained from the unit cell concept.

INTRODUCTION

Land suitable for development is dwindling due to the intensive land use as a result of growing population in growing economies. Geotechnical engineers would have to treat marginal sites before they can be used to construct any facilities. The use of stone columns is an effective ground improvement method in construction practice to meet the industry demand for transmitting larger loads through shallow foundations. Nevertheless, the ability to predict settlements of small foundations resting on small groups of stone column has been slow to follow.

Prediction models for stone column reinforced ground based on the unit cell concept (e.g. Balaam & Booker 1981, Barksdale & Bachus 1983, Priebe 1995, Chai et al. 2009) are not applicable for small column groups due to the breakdown of boundary assumptions. Empirical approaches (e.g. Lawton & Fox 1994, Sehn & Blackburn 2008) which adopt simplified spring analogies do not adequately capture the observed load-displacement response of footings due to the use of simple elastic theory. Based on large amount of field load tests, Stuedlein (2008) has proposed semi-empirical regression models to predict the displacement of spread footings on stone column improved clay. Due to nearly undrained loading, as in most field load tests, the model is therefore limited to short term loading condition of footing behavior.

Small group of stone columns either fully or partially penetrated (known as floating columns) received load applied at the surface, usually through a spread or mat foundation. The load is then transferred to the composite ground by a complex mechanism of load sharing and strain compatibility. The intricacy of this load transfer mechanism is less understood than the behavior of single stone column and columns under infinite loading condition where unit cell concept is applied. Unlike piles with very large contrast of stiffness, stone columns and native soil have typical stiffness contrast of 10 – 40 times. In addition, the nonlinearity of stress-strain response and the drainage ability of the soil and column's aggregate further complicate the design of foundation resting on small group of columns. Small model tests (Wood et al 1997, McKelvey et al. 2004, Shahu & Reddy 2011) and numerical models (Wehr 2006, Killeen & B. McCabe 2010) have, however, provided qualitative insight into the likely behavior of group stone columns.

For loading by a rigid plate or footing, the diagonal shear plane starts from the edge of footing and extends to the inner columns. Apart from this, columns near the outer edge of a footing will suffer some degree of buckling towards the unconfined sides. Bulging mode of failure are also observed for inner column similar to a loaded single column but to a lesser magnitude. Short columns or columns that are floating in the native soil (i.e., not founded on a hard layer) may create a potential situation where the toe may punch into the soil below. This is due to insufficient load shedding along the shaft of the column analogous to a skin friction pile with inadequate shaft resistance. The deformation characteristics of a small footing with columns are less sensitive to the stone column arrangement according to Killeen and B. McCabe (2010). The loading behaviors of flexible footing are not well studied.

The scope of this study is to investigate the settlement performance of a rigid shallow foundation supported by group columns and to propose design methods to improve the prediction of settlement improvement factor (n=settlement of untreated/settlement of treated ground). Numerical method with drained analyses was adopted in this quantitative research where a well-calibrated 2D finite element model was employed.

NUMERICAL MODEL

Finite element code PLAXIS 2D 2011 was adopted to analyze the spread footing supported by group of columns (i.e. 4, 9, 16, 25, 36, 49, 64, 81 and 100 columns). Investigations of the settlement behavior of isolated single column are common but are of little practical use (Kirsch & Kirsch 2010). Axisymmetrical concentric ring model proposed by Elshazly et al. (2008) was used to convert the off center columns to cylindrical equivalent rings. The properties in the ring element are kept the same as the stone column material while the radius and thickness of the rings are calculated so that the area ratio between column and the footings (footprint replacement ratio, $A_F = A_c/A$, where A_c = area of columns, and A = footing area) remains the same. Similar to area replacement ratio in infinite grid columns; this is a measure of the extent to which the soil area under the footing is replaced by column material.

Figure 1 shows the stone column model for 9 columns group. A uniform footing pressure of intensity q is applied over a footing of diameter (D) on a granular bed of 0.5 m thick. This layer of granular bed can also be treated as a layer of crust or a transfer layer. It was assumed that this layer exists as part of the original ground and there is no influence of the placement of this layer to the soil below. All columns studied are of floating type. The length of the columns, L is a key variable in settlement design of stone columns to support spread footings. The boundary and mesh sensitivity analyses were conducted for all the column configurations in order to reduce the influences of the mesh and boundary on the results of the simulations. Refinement of mesh especially around the footing was done until insignificant change in the results due to this refinement could be observed anymore. Generally, boundary effect becomes negligible when the model width and depth is greater than $4D$ (D = footing diameter). The standard boundary conditions in the model were assumed such that the vertical boundaries are free vertically and constrained horizontally ($u_x = 0$; u_y = free) while the bottom horizontal boundary is fully fixed ($u_{x\&y}=0$). This can be easily done by choosing the standard fixities option in PLAXIS 2D.

(a)

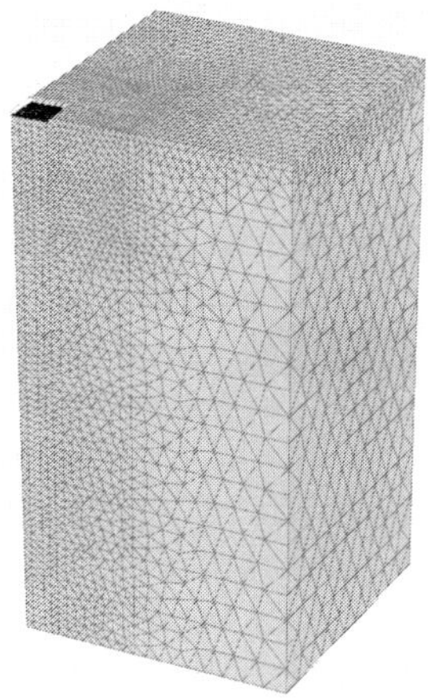

(b)

Figure 1. 9 column group for (a) 2D ring model (b) 3D

Material properties are presented in Table 1. The engineering properties of the materials fall between the typical ranges of actual problems encountered at site. Mohr-Coulomb yield criterion was adopted and a non-associated flow rule (dilatancy angle, $\psi = 0$ for stone and clay). This simple and robust elastic-plastic model is able to predict the bearing capacities and collapse loads of footings reasonably. Footing of stiffness EI = 2.1 x 10^5 kN/m^2 can be taken as relatively rigid material compared to soils below. In view of that, the loading can be regarded as vertically and uniformly distributed on the footing while the self-weight of footings are neglected.

The columns were simulated as "wish in place" whereby the installation process was not modeled. The stress changes induced (i.e. increase in total stress) during the construction process of stone column have been reported by Watts et al. (2000) and Kirsch (2008). However, it is the equilibrated effective stresses around columns (after dissipation of excess pore pressure) which influence column performance under load, and these have not been measured (McCabe et al. 2009). In this study, the at rest earth pressure coefficient is assumed to be 0.7 for all the materials which value is higher than the one estimated by the Jaky's equation ($K_0 = 1 - \sin\phi'$) for normally consolidated soils but lower than hydrostatic value of 1 adopted by Priebe (1995) and Goughnour & Bayuk (1979). In numerical simulations, same initial stress was imposed for column and soil to reduce the problems of unbalanced forces during the loading stage.

Table 1. Materials properties for group columns.

Name		Soft soil	Stone column	Granular bed
Type		Drained	Drained	Drained
γ_b/γ_{sat}	[kN/m^3]	18	18	18
v'		0.3	0.3	0.3
E'	[kN/m^2]	3000	30000	10000
c'	[kN/m^2]	0.1	0.1	0.1
ϕ'	[°]	25	40	30
K_o		0.7	0.7	0.7

Ground water table is located just below the granular bed to represent the worst-case scenario. Drained analysis was conducted for all simulations (long-term conditions). Actual construction of facilities founded on small footing is normally a drained process where the loading is applied slowly and the excess pore pressure is assumed to dissipate during the construction period. The results of drained analysis are somewhat different from the field load test as loading in load test is imposed rapidly. Owing to the installation process of stone column, the stone is tightly interlocked with the native soil and hence it can be regarded that near perfect bonding occurs along this interface, which makes interface elements unnecessary

The reliability of the ring model is validated through 3D model where three dimensional finite element analyses were carried out using PLAXIS 3D 2011. In 2D model, 4th order 15-node quartic triangular elements are used while it is quadratic tetrahedral 10-node elements in 3D model. To compensate for the lower order elements used in 3D model, finer meshes are required especially near the upper part of the footing. In three-dimensional analyses, the square footing with equivalent area is used instead of circular footing. Good agreements are obtained for both models, which proved the ability of ring model to reproduce the behavior of stone columns reinforced ground (Figure 2). The stone column radius, a_{rc} is kept at 500 mm. For the same number of columns, different footprint replacement ratios are achieved by changing the footing size.

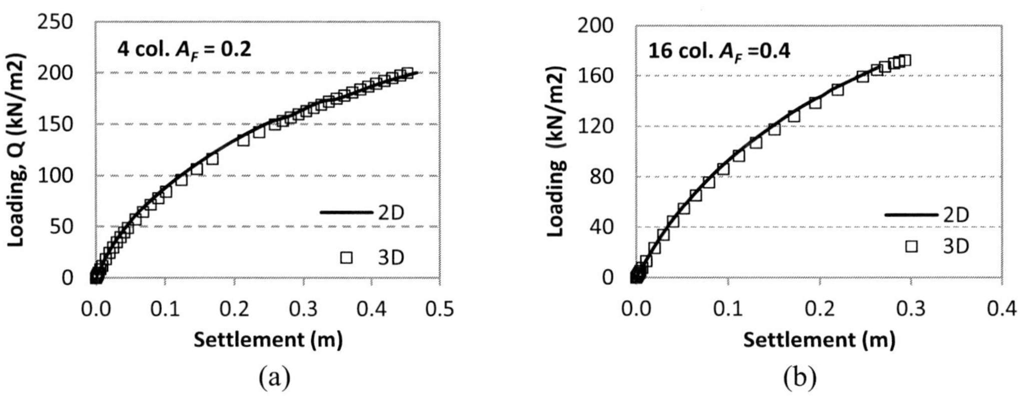

Figure 2. Comparison of 2D ring model and 3D model at column optimum length for groups of (a) 4 columns, and (b) 16 columns.

NUMERICAL SIMULATION AND RESULTS

Unlike most of infinite group columns, small group columns are normally constructed to "float" (toe does not reach the competent hard layer). The first part of the analysis looked into the influence of column length over the settlement performance. Wood et al. (2000) described that column length is relevant only up to a certain point, beyond that point, increasing the length of the columns, L confers no advantage. Contrarily, Killen & McCabe (2010) showed further improvement in settlement with increase of length obtained through numerical simulation. The classical Boussinesq's or Westergaad's solution for vertical stress distribution under a footing would suggest that the stress increment applied to the footing is small at L/D= 2. Therefore, in this study, more studies were conducted to search the existence of critical (or optimum) length and the relationship of optimum column length with size of footings. The analysis was done for every 2m increment and only the key results are reported herein.

The reinforced foundations have failure stress state, which is higher than the unreinforced foundation. Since the failure stress for unreinforced foundation is low, and also the focus of this study is on the settlement improvement factors, n (n = settlement without stone column/ settlement with stone column), the maximum pressure is set as 150 kPa which is lower than the ultimate bearing capacity of the footing system in this study. In practice, columns supported foundations are not loaded to failure, and working loads are considerably smaller than the ultimate bearing capacity.

Study was done for groups of 4, 9, 25, 36, 49, 64, 81 and 100 columns supporting a circular footing with a maximum radius of 11.2 m. Figure 3 illustrates the settlement improvement factor, n versus column length ratio, λ ($\lambda = L/2a_{rc}$) under different bearing pressure for 25 columns in a group. A few generalizations can be made here. First, for the same number of columns, higher footprint replacement ratio, A_F, also means smaller diameter of footing and closer spacing of columns, therefore results in better settlement improvement factors when the same amount of loading is applied. The improvement factor, n value ranges from 1.1 to 3.2 (for all cases under loading of 25 kPa to 150 kPa). As the footprint replacement ratio increases, the height of plastic zone is decreasing as shown in Figure 4a & 4b for same number of columns. This indicated that the deformation mode of stone column is controlled by the footing dimension itself rather than the column spacing. However, for the similar dimension of footing, higher replacement ratio has reduced the plastic zone height (Figure 4a & 4c); this implies that the closer spacing indeed gives deterrence effect on the formation of failure mechanism. The extensive zone of plastic points developed around the columns also suggested that the simple design method based purely on elastic approach e.g. using equivalent composite stiffness or equivalent volume compressibility obtained from weighted average ratio is not appropriate. In addition, the classification of elastic zone (point B in conical wedge shown in Figure 5) directly under the footing by Wood et al. (2000) is therefore not always correct as the current study shows the columns in that region are already undergoing plastic straining even under low stress level especially when low footprint replacement ratio is adopted.

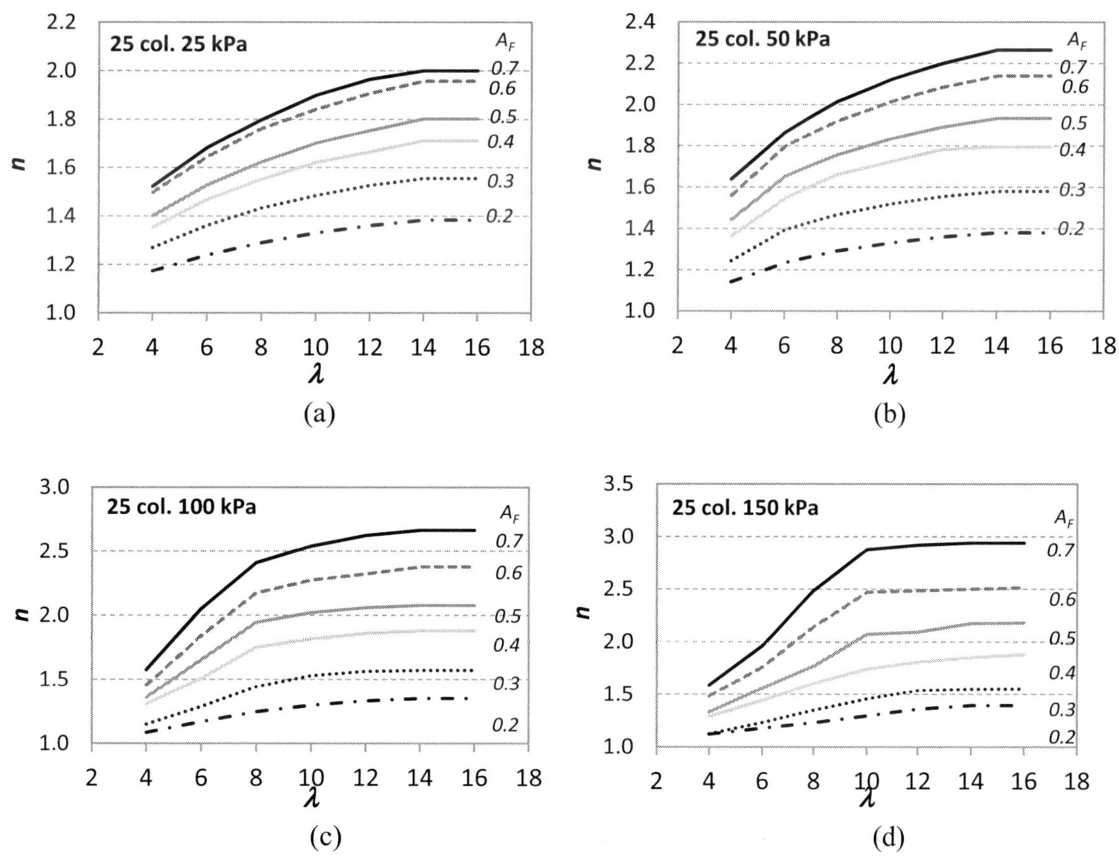

Figure 3. Settlement performance for 25 columns group under loading of (a) 25 kPa; (b) 50 kPa; (c) 100 kPa; (d) 150 kPa.

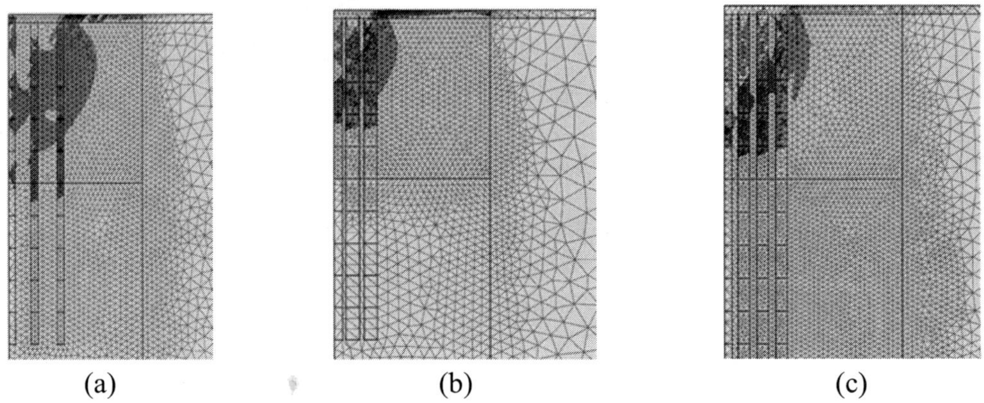

Figure 4. The extent of plastic points for loading of 150 kPa with (a) 25 columns $A_F = 0.3$; (b) 25 column $A_F = 0.7$ and (c) 49 columns $A_F = 0.6$.

Second, as the loading intensity increases, the n value also increases. Nonetheless, for footing with low footprint replacement ratio, the increase of settlement improvement factor is followed by relatively constant improvement factor as load increases. This is because the mobilized strength has achieved the maximum shear strength of the footing system even when lesser load applied. Failures of the stone column and surrounding soil at upper part occur early during loading, extending downward with increasing load. With the increase of load, the yielding would have exacerbated the settlement performance, both for treated and untreated ground.

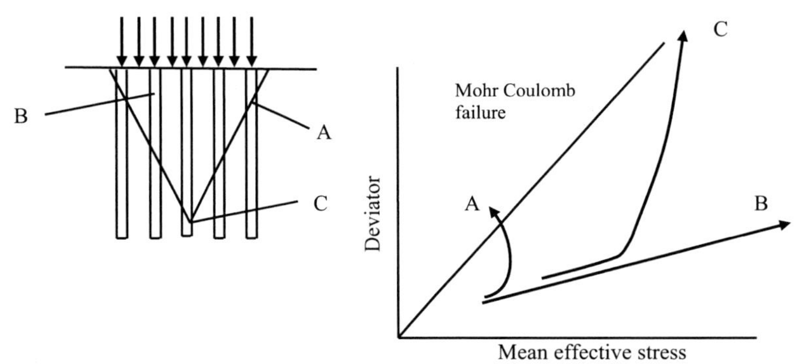

Figure 5. Schematic stress paths for elements in stone columns (Wood, et al, 2000)

Lastly, the optimum length ratio, λ_{opt} is slightly higher for low footprint replacement ratio than for high footprint replacement ratio but the difference is rather small. This relationship is not clearly shown in Figure 3 because the column length is plotted against settlement improvement ratio rather than the settlement value. For instance in 25 columns group, the optimum length ratio for $A_F = 0.2$ is $\lambda_{opt} \approx 14$ while for $A_F = 0.7$, it is $\lambda_{opt} \approx 12$. However, if optimum length, L_{opt} over footing diameter ratio, L_{opt}/D is used, then it was found that $L_{opt}/D \approx 1.25$ and $L_{opt}/D \approx 2.0$ for $A_F = 0.2$, and $A_F = 0.7$ respectively (where area of footings for $A_F = 0.2$ and $A_F = 0.7$ are 11.2 m^2 and 6 m^2 respectively for 25 columns group). The load-displacement curve for varying length is shown in Figure 6 for 25 columns group. Greater optimum length for higher A_F in fact explains the ability of longer columns to transfer the stress from the top to deeper depth and produce smaller settlements.

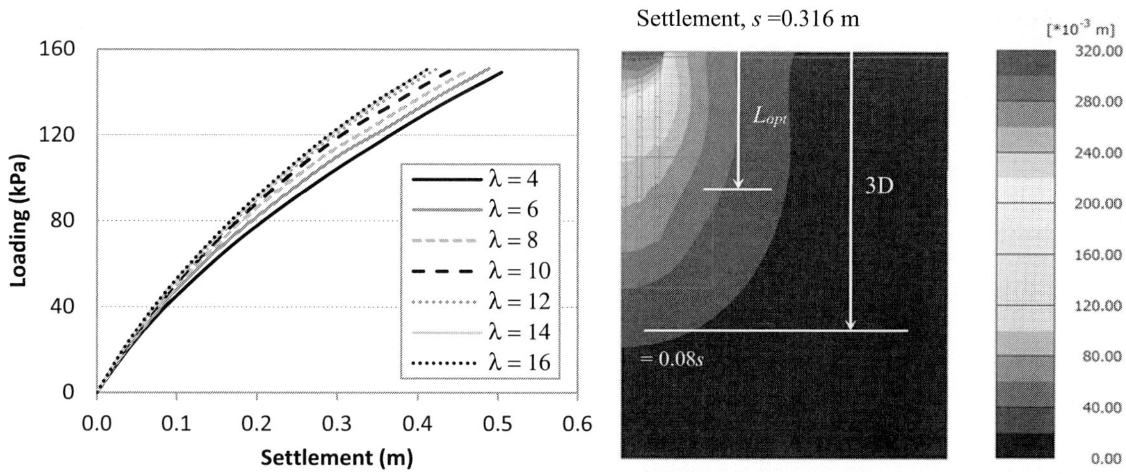

(a) (b)

Figure 6. (a) Load-displacement curve and (b) displacement profile for 25 columns group with $A_F = 0.3$.

This study has shown that for all the analyzed cases, the L_{opt}/D values are ranging from 1.2 to 2.2 for low to high footprint replacement ratio. Observations from the numerical study showed that the settlement performance of group columns is a combination result of few complex mechanisms (multiple failure modes: shear, bending, and punching) corresponding to the load applied. Key findings tell us it is the geometry of the footing that interacts with the individual columns in producing an ultimate mechanism of deformation in the columns system. The depth at which the prevalent strain is found is primarily controlled by the diameter of the footing itself, and somewhat influenced by the presence of the columns. Similar observation was obtained by Wood et al. (2000). Based on this study, it is recommended to build columns with length not more than 2.2 times the footing diameter (2.2D or approximately 2.3B) in order to achieve optimum performance for the columns group even though it is possible to build shorter columns if the footprint replacement ratios are low. It should be noted that the critical length in this study is different from the one suggested by Hughes et al (1976), where they discovered the critical length is approximately four column diameters, taken from the result of single column load test. Das (1987) suggested the minimum length of the stone column required to support the footing and to obtain the maximum increase in bearing capacity is 3B while Kirsch & Kirsch (2010) stated there would be no further settlement improvement for column length longer than 3B. The difference of optimum length for current study to Kirsch and Das is probably because they acknowledged the fact that stress distribution can be ignored when the depth is reaching 3B. It is correct that at larger depth than 3B, the stress influence is small (Figure 6b) but the study here also suggested that the contribution of columns is negligible even at shorter lengths. In order words, there is no need to build columns more than 3B but instead it is sufficient to use the suggested optimum lengths derived above.

SETTLEMENT IMPROVEMENT FACTOR

The results of settlement improvement factors for varied column groups with *optimum length* are shown in Figure 7. It can be concluded that the footprint replacement ratio, A_F, is a crucial parameter controlling the degree of improvement of the columns inclusion treatment in a soft ground. All the plots showed almost linear trend and they fall almost on the same line (a narrow region) especially for loading of 50 kPa and 100 kPa. In other words, same footprint replacement ratio produced similar settlement improvement factor regardless of number of columns in the group. This can be also interpreted as little interaction between adjacent columns for very small columns group.

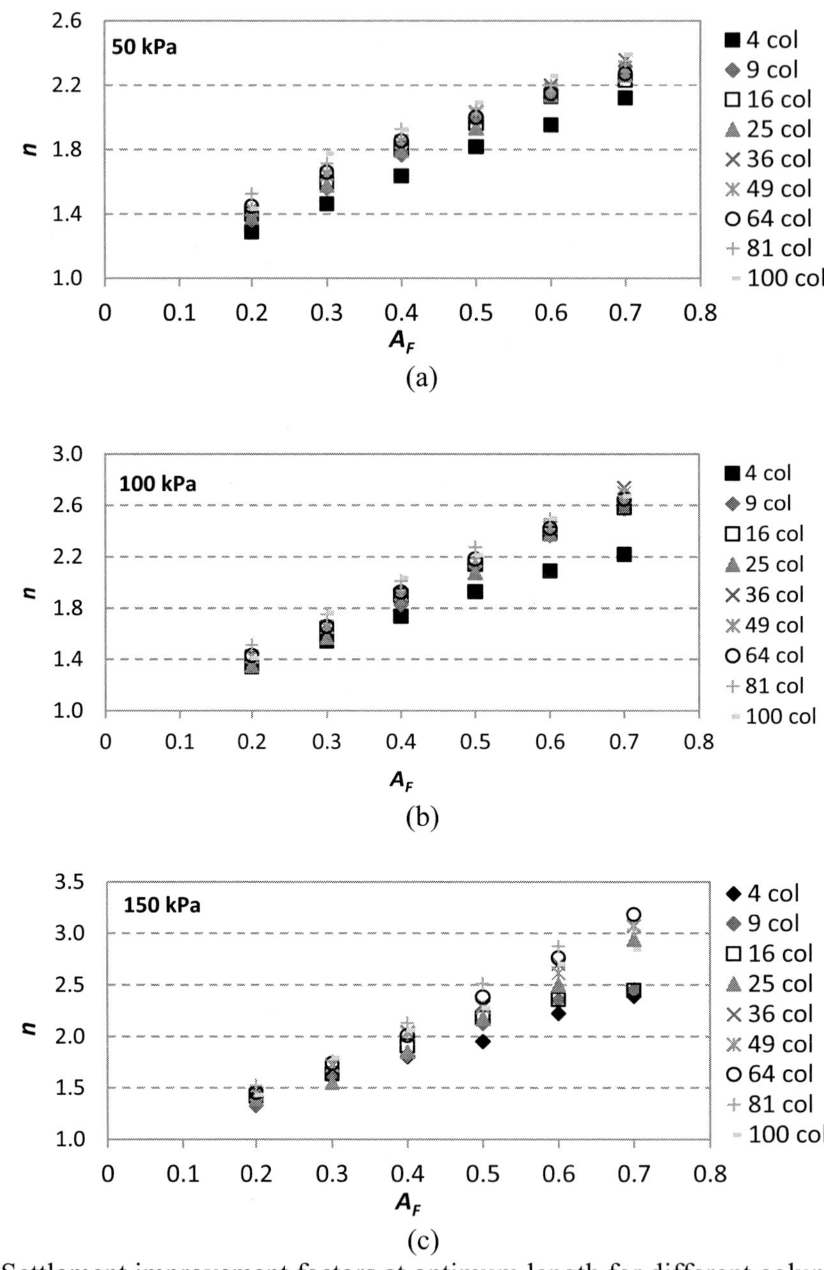

Figure 7. Settlement improvement factors at optimum length for different columns group at (a) 50 kPa; (b) 100 kPa; and (c) 150 kPa.

In contrast, larger groups of columns give stiffer response since more columns work as a group and relatively fewer columns at the outermost are subjected to lower confined stress; or in other words, the columns group achieved a lesser degree of deformation as the response of the array of columns within an overall mechanism that is driven by the footing. The enhanced rigidity of column group for larger columns groups due to the increased confining action has its limit, this explains the loading results for all number of column group which achieved the same degree of improvement.

The summary of above plots is presented in Figure 8a given by three linear lines for loading of 50 kPa, 100 kPa and 150 kPa. In this figure, the results of 4 columns group have been taken out so that the square of correlation coefficient, R^2 is beyond 0.9. Hence, the method proposed here excludes the 4 columns group. It is suggested the design of 4 columns group should be considered separately. The difference in settlement improvement factor for 50 kPa and 150 kPa with respect to 100 kPa is shown in Figure 8b. Based on these results, the settlement improvement factor, n can be obtained as

$$n = n_0\,(1 + C_Q) \tag{1}$$

where $n_0 = 2.575A_F + 0.931$ (line equation for 100 kPa) and C_Q = correction factor for loading of 50 kPa and 150 kPa, obtained from line equation in Figure 8b.

The above results are compared with numerical results obtained from unit cell model (Figure 9) assuming infinite column grids (width of footing, B = ∞) for fully penetrating columns. The unit cell model results produced design equation as follows (Ng, 2013):

$$n = 9.43\alpha^2 + 1.49\,\alpha + 1.06 \tag{2}$$

where α = area replacement ratio, defined as $\alpha = A_c/A$; A_c = area of column, A = total influence area in unit cell. Eq. 2 is derived from settlement improvement factors for end bearing columns for column's friction angle of 40^0 and 100 kPa. The area replacement ratio and footprint replacement ratio are plotted on the same axis.

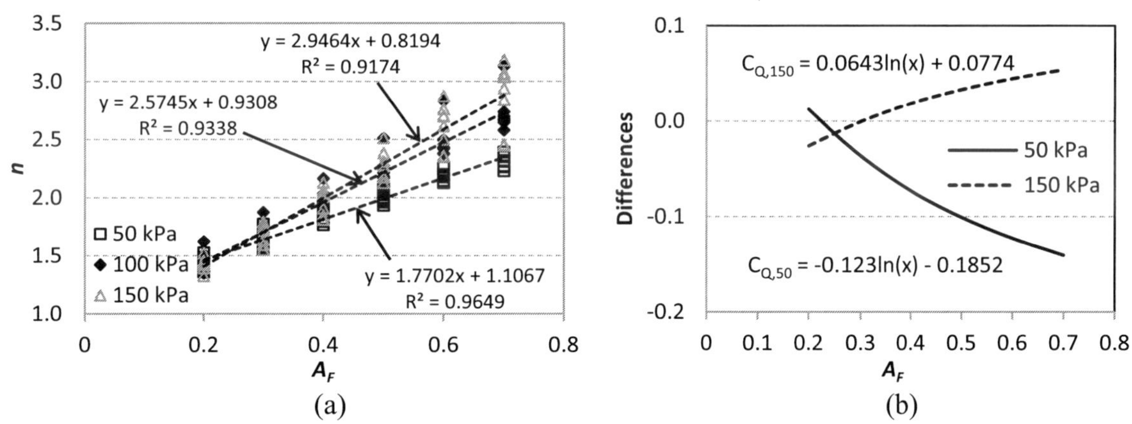

Figure 8. (a) The summary of settlement improvement plots for different loading intensity; and (b) Difference of n for 50 kPa and 150 kPa with respect to 100 kPa.

The distinct difference for this comparison is that when the replacement ratio increases, the improvement obtained from unit cell model increases as polynomial function while as a linear function for the spread footing. Hence, the degree of improvement is larger for unit cell model. This is correct because of the assumption of rigid boundary and equal strain in unit cell, unlike spread footings where the outer columns are free to bend outwards due to low confined stress together with the lateral displacement of soil underneath the footing. Furthermore, unit cell model involved mainly the volumetric strain unlike column groups where volumetric and shear strain are both significant. It should not be mistaken that stone column would produce less settlement under infinite column grids compare to small columns group albeit higher settlement improvement factor obtained in unit cell model, in fact the settlements is larger for infinite column grids than small column group when same loadings are applied. It is because the settlement for unimproved ground under wide spread loading is indeed much larger than unimproved ground with small footing. Therefore, it is the relative improvement that is higher for infinite column grid. Using the unit cell improvement factor for small group of column will then result in overestimation of its true settlement performance. Extrapolations of plots are made for Eq. 1 and Eq. 2 accordingly as the range of replacement ratios are outside of the studied values.

Based on a parametric study of small columns group, Kirsch (2004) has produced the results as shown in Figure 9 for end bearing type columns with $\eta = L/T = 1.0$ (where L = column length, and T = soft soil thickness. Even though the agreement is good, there are actually many differences in both studies especially in the model geometry considered. Most importantly, the relationship of column length, footing width and footprint replacement ratio is not clear in the Kirsh's study.

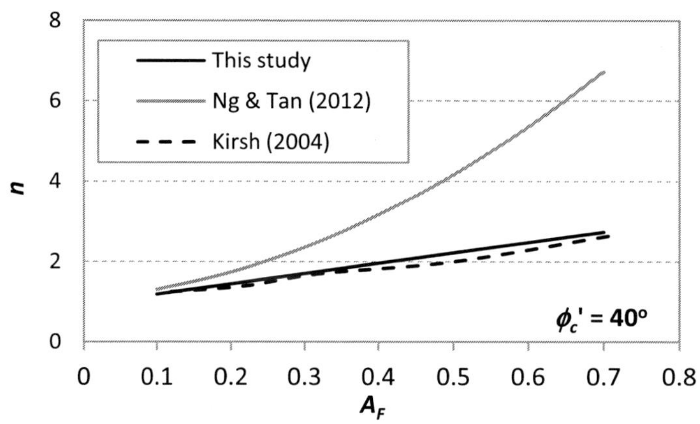

Figure 9. Comparison of results for spread footings and unit cell model.

A four column load test results from a clay fill site was adopted for comparison (Watts & Charles, 1991). In the southern area, the first meter was a fill layer consisting tarmac, hard core fill and lean mix concrete overlying 0.6 m thick of firm clay fill, 2.2m soft clay fill and 0.5m of sandy gravel. Bottom of excavation showed virgin clay deposit after 4.3m depth. The northern part of the site otherwise showed a firm clay fill in the top 1.6m, underlain by 2.0m of soft clay and silt but with organic materials. Stone columns were built to full depth of the fill. The foundation pad of 2.0 m square was casted over the four columns before the test load of 50kN/m^2 and 85kN/m^2 were applied for tests in northern area and southern area respectively. The settlements were measured for six-month period. The columns spacing was 1.0 m but the columns diameters are not provided in the literature. Based on the previous project information, the columns were assumed to be 0.6 m diameter, and the area ratio was calculated to be A_F = 0.28. Load tests were carried out on untreated ground for the references. The measured settlement for the northern area for treated ground and untreated ground were 13 mm and 18 mm respectively. In southern part, the untreated ground settled by 22 mm while treated ground was likely to have settled 13 mm obtained from the results of four load tests. The settlement improvement ratio for the northern part is therefore calculated to be n = 1.38 while n = 1.69 for the southern part. Simplified method above gives the settlements improvement factors of approximately 1.60 and 1.64 respectively for 50 kPa and 85 kPa loading. Slight over prediction is expected for 50 kPa test load since the simplified method is better suited for larger group of columns, that is, with more than 4 columns in the group.

In addition, a long-term settlement field observation result for treated and untreated strip footing was adopted for reference rather than direct comparison. The details of this well instrumented trial of stone column reinforced foundation can be found in the paper by Watts et al., (2000). The footprint replacement ratio, A_F adopted in their study is 0.44 and the footing is loaded to pressure of 123 kPa. The columns are of end bearing type and range from 3-5 m length. The foundation is placed above 1.5 m granular fill underlying a layer of drained silt. After 7 months' monitoring, the maximum settlements of the treated and untreated foundation were 16 mm and 26 mm respectively or equivalents to n = 1.63. Using Eq. (1), the settlement improvement factor is predicted to be 2.08, which higher than the field measurement. The difference in the value is mostly attributed to the thickness of granular fill where the simple methods here are obtained from analysis of 0.5 m thickness granular bed. As the thickness increases, the settlement improvement reduces.

This simple design method proposed above is used for footing under small working load (i.e. in this case $\leq$ 150 kPa) to obtain long-term settlement performance. Stuedlein (2008) reported field load tests on over-consolidated clay and obtained settlement improvement factor in the range of 0.91 to 2.47 when applied bearing pressure is less than 200 kPa. Comparison of this simple method to the footing load test may not be entirely appropriate due to the short testing time where excess pore pressure is allowed to build up resulting in less volumetric strains. The long-term settlement performance of footing is much dependent on the in-situ soil drainage capability. McCabe et al. (2009) presented a long-term load test on trial strip footing where the primary consolidation settlement was completed in 8 weeks.

This method involves calculating the treated settlement by dividing the untreated settlement by an improvement factor. The untreated settlement can be calculated using the conventional method and will not be discussed here. Generally, simple method conform qualitatively to field experience but need to be complemented by conducting analysis using advanced soil model to capture realistic non-linear soil behavior as well as stress dependent stiffness especially when unloading-reloading condition are required.

CONCLUSIONS

Numerical simulation has been conducted to study the drained performance of small foundation supported by floating stone columns. The results suggest many features of columns group behavior that will interest designers. In particular, this work offers insights into the relationship between replacement ratio, column length, and footing size. Several conclusions can be made here:

(i) The relationship between footprint replacement ratio and settlement improvement factor has been established. It was found that by maintaining the footprint replacement ratio regardless of number of columns, the settlement improvement factors obtained are about the same especially when the load levels are small. The exception is column groups of four.

(ii) Simple method is proposed in the form of linear equation with correction factors for various load levels (below failure state), that enable an expedient hand calculation to be made of settlement improvement factors best used during preliminary design.

(iii) For all the studied cases, the optimum length of column L_{opt}/D ranges from 1.3 to 2.2 for low to high footprint replacement ratios. The geometry of footing governs the depth of stress influence. However, higher footprint replacement ratio encourages load transfer to a greater depth and thus increases the optimum length.

(iv) Settlement improvement increases as the footprint replacement ratio increases but the magnitude of improvement is smaller than the improvement factors obtained through unit cell model.

REFERENCES

Balaam, N. P., & Booker, J. R. (1981). "Analysis of rigid rafts supported by granular piles". *International Journal for Numerical and Analytical Methods in Geomechanics*, *5*(4), 379–403.

Barksdale, R. D., & Bachus, R. C. (1983). "Design and construction of stone columns". *Federal Highway Administration Office of Engineering and Highway Operations.*

Chai, J., Miura, N., Kirekawa, T., & Hino, T. (2009). "Settlement prediction for soft ground improved by columns". *Proceedings of the ICE-Ground Improvement*, *163*(2), 109–119.

Das, B. M. (1987). "Bearing capacity of shallow foundation on granular column in weak clay". *foundation Engineering: Current Principles and Practices*, *Northwestern University*, ASCE, Vol 2, 1252-1263.

Elshazly, H. A., Hafez, D. H., & Mossaad, M. E. (2008). "Reliability of conventional settlement evaluation for circular foundations on stone columns". *Geotechnical and Geological Engineering*, *26*(3), 323–334.

Killeen, M. M., & McCabe, B. (2010). "A numerical study of factors affecting the performance of stone columns supporting rigid footings on soft clay". *Civil Research Session*.

Kirsch, F. (2004). "Experimentelle und numerische Untersuchungen zum Tragverhalten von Rüttelstopfsäulengruppen". *Disertation*. TU Braunschweig.

Kirsch F. (2008). "Evaluation of ground improvement by groups of vibro stone columns using field measurements and numerical analysis". *Geotechnics of Soft Soils: Focus on Ground Improvement*. Taylor & Francis.

Kirsch, K., & Kirsch, F. (2010). *Ground improvement by deep vibratory methods*. Taylor & Francis.

Lawton, E. C., & Fox, N. S. (1994). "Settlement of structures supported on marginal or inadequate soils stiffened with short aggregate piers". *Vertical and Horizontal Deformations of Foundations and Embankments (GSP 40)*, 962–974.

McCabe, B. A., Nimmons, G. J., & Egan, D. (2009). "A review of field performance of stone columns in soft soils". *Proceedings of the Institution of Civil Engineers-Geotechnical engineering*, *162*(6), 323–334.

McKelvey, D., Sivakumar, V., Bell, A., & Graham, J. (2004). "Modelling vibrated stone columns in soft clay". *Proceedings of the ICE - Geotechnical Engineering*, *157*(3), 137–149.

Ng, K.S. (2013). "Numerical study and design criteria of floating stone columns". *PhD thesis*. National University of Singapore.

Priebe, H. J. (1995). "The design of vibro replacement". *Ground engineering*, *28*, 31–37.

Sehn, A.L., and Blackburn, J.T. (2008) "Predicting performance of aggregate piers", *23rd Central Pennsylvania Geotechnical Conference,* Hershey, Pennsylvania, May 28-30. 2008.

Shahu, J., & Reddy, Y. (2011). "Clayey soil reinforced with stone column group: Model tests and analyses". *J. Geotech. Geoenviron. Eng.*, *137*(12), 1265–1274.

Stuedlein, A. W. (2008). *Bearing capacity and displacement of spread footings on aggregate pier reinforced clay*. ProQuest.

Watts, K. S., Johnson, D., Wood, L. A., & Saadi, A. (2000). "An instrumented trial of vibro ground treatment supporting strip foundations in a variable fill". *Ground and Soil Improvement*, 129.

Watts, K. S., & Charles, J. A. (1991). "The use, testing, and performance of vibrated stone columns in the United Kingdom". *ASTM Special Technical Publication*, (1089), 212–223.

Wehr W. (2006). "Stone columns - group behaviour and influence of footing flexibility". *Numerical Methods in Geotechnical Engineering*. Taylor & Francis.

Wood, D. M., Hu, W., & Nash, D. F. T. (2000). "Group effects in stone column foundations: model tests". *Geotechnique*, *50*(6), 689–698.

A Field Investigation of Pullout Resistance of Steel Anchors in Rock

Jiho Park[1], Tong Qiu[2], P.E., M.ASCE, and Younguk Kim[3]

[1]Post-doctoral Researcher, Department of Civil and Environmental Engineering, The Pennsylvania State University, University Park, PA 16802, USA; jup28@psu.edu
[2]Assistant Professor, Department of Civil and Environmental Engineering, The Pennsylvania State University, University Park, PA 16802, USA; tqiu@engr.psu.edu (corresponding author)
[3]Professor, Department of Civil and Environmental Engineering, Myongji University, Yongin-si, Gyeonggi-do, Korea; yukim@mju.ac.kr

ABSTRACT: This paper presents a field investigation of pullout resistance of steel anchors in rock. The field testing involves pullout tests of six steel anchors installed in biotite gneiss rock by gravity grouting and pressure grouting methods. The field pullout tests suggest a progressive failure mechanism, which was manifested through the progressive elongation of the bond zone. The pressure-grouted anchors yielded higher pullout resistance than did the gravity-grouted anchors. The increase in rock-grout bond strength from pressure grouting is likely due to a combination of several factors including the resultant higher grout strength, higher rock-grout contact stresses, and grout injected under pressure into the fractures/fissures that existed in the bedrock.

INTRODUCTION

The axial capacity of ground anchors, soil nails, and micropiles is developed primarily through the grout-ground bond that is influenced by a variety of factors including the strata type, groundwater, grouting method, and drilling method. Among these factors, the method of grouting is generally the most sensitive construction control over grout-ground bond capacity (FHWA 1999, 2000). Two grouting methods are widely used, namely the gravity grouting and pressure grouting. Gravity grouting may result in insufficient skin friction resistance due to the cavities and poor filling that frequently occur during installation. On the other hand, pressure grouting improves the grout-ground bond capacity and load/displacement performance, which can be attributed to a combination of recompaction or redensification of soil/rock mass around the borehole, an increase in the density and strength of the grout, and an increase in the effective diameter of the anchor or pile in the bond zone under pressure (Scott 1981; Yin et al. 2009; Seo et al. 2012).

The behaviors of ground anchors and micropiles in soils are generally well known. However, the performance of these systems in rock remains relatively unknown due to the limited case studies available (Gomez et al. 2003); consequently, current design practices of these systems are generally overconservative (Bruce 2005). Field tests of micropiles in rock have suggested that the maximum design load of these systems in rock is governed primarily by the allowable displacement rather than the application of factor of safety on the ultimate load (e.g., Finno et al. 2002; Gomez et al. 2003). Although pressure grouting is known to improve bond quality and load-carrying capacity over gravity grouting based on field observations, the mechanism for this improved capacity is not fully explored in geotechnical literature. This paper presents a field investigation of pullout resistance of steel anchors in rock. The field testing involves pullout tests of six steel anchors installed in rock by gravity grouting and pressure grouting methods.

FIELD TESTS

The field test site is located in the city of Yong-in, Gyeonggi province, South Korea. Geotechnical subsurface investigations consisted of soil borings, rock coring, the standard penetration tests, and pressuremeter tests (PMT). Figure 1 presents the geologic profile of the test site based on borings BH-2 and BH-3. The test site consists of fill soil, alluvial soil, residual soil, and biotite gneiss bedrock. For the bedrock, the upper weathered and fractured zone (UWFZ) was encountered at the depths between 8.8 m and 14.2 m in BH-2 and between 5.5 m and 6.2 m in BH-3 based on rock core samples. Two pressuremeter tests were conducted in borings BH-2 and BH-3 at the depth of 12 m. Based on the cavity strain between 0.1% and 0.2% as recommended by Clarke (1996), the pressuremeter tests yielded shear modulus values of 132 MPa and 364 MPa for the rock in borings BH-2 and BH-3, respectively. Figure 1 also presents the locations of the steel anchors in both cross-sectional view and plan view. As shown in Figure 1, the anchors were installed in the bedrock below the UWFZ.

For the anchor installation, a borehole with a diameter of 152 mm was first drilled to the required depth in the bedrock layer. The borehole was supported by a 150-mm diameter casing. A 65-mm diameter high-strength continuous-thread steel bar (ASTM A706 Grade 80) was then lowered through the casing to the rock socket. As shown in Figure 1, three underground displacement sensors were placed in the bond zone to monitor the displacements of these sensors during pullout tests for anchors #1, #2, #3, and #5. For pressure-grouted anchors, a packer system was installed right above the bond zone. Grout with a water/cement ratio of 0.45 was injected into the bond zone through a grouting hose until the air inside was bleed out and a grout pressure of 1.0 MPa was reached. The grout was then cured for 28 days. Gravity-grouted anchors were installed using a similar procedure except that the packer system was not needed and no grout pressure was applied.

The pullout loading system consisted of a loading frame, reaction beams, anchor piles, hydraulic jack, and data acquisition system. Two linear variable differential transducers (LVDT) were instrumented to monitor the displacement of anchors during pullout tests. A load cell with a capacity of 5000 kN was used to monitor pullout loads from the hydraulic jack. Care was exercised to ensure that the pullout load was aligned along the axial direction of the anchors. Six field pullout tests were conducted in accordance with ASTM D3689-07 (ASTM 2007). During the pullout tests, the anchors were subjected to a load increment of 100 kN for every 10 minutes until failure (i.e., excessive movement of anchor head without further load increase), except for Test #1, in which a sudden pullout occurred. LVDT readings were recorded at the one-minute, four-minute, seven-minute, and ten-minute (i.e., end of increment) marks of each load increment. Figure 2 presents the recorded pullout load versus anchor displacement. Various methods are available to interpret load tests on deep foundations, such as the Davisson's method (Davisson 1972) and Chin's method (Chin 1970). Comparisons of these methods are provided in detail by Fellenius (1980) and Fellenius (2001). The general shapes of the load-displacement plots for loading in both tension and compression are similar; therefore, most of these methods can be applied to either mode of loading (Kulhawy et al. 1983). Among these methods, the Davisson's method (Davisson 1972) is widely used in practice and is hence utilized to determine the ultimate pullout capacities, which can be presented as (Kulhawy et al. 1983)

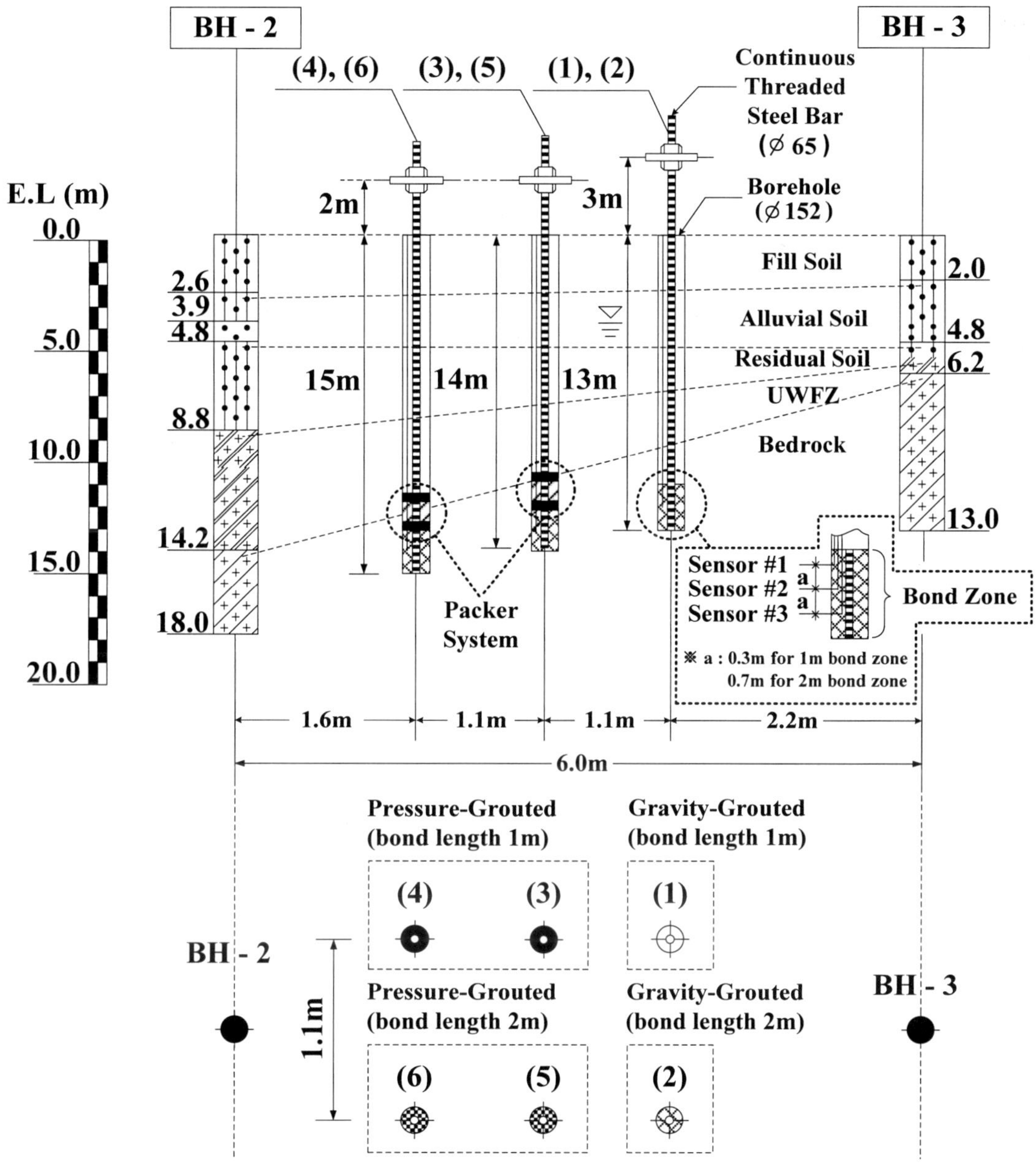

Figure 1. Geotechnical profile of test site and locations and details of anchors #1 through 6.

$$s_{ult}(\mathrm{mm}) = 3.8 + \frac{Q_u L}{AE} \tag{1}$$

where s_{ult} = anchor displacement at the ultimate capacity in mm; Q_u = ultimate pullout capacity in kN; L = anchor length in mm; A = area of anchor cross section in mm^2; and E = Yong's modulus of anchor material in kN/mm^2. For the anchors tested, A = 3318 mm^2 and E = 21 kN/mm^2. The method is illustrated in Figure 3 using anchor #1 as an example. The mobilized skin friction resistance corresponding to the ultimate pullout capacity can be calculated as

$$\tau_{ult} = \frac{Q_u}{\pi D_b L_b} \tag{2}$$

where D_b = effective grout column diameter (approximately the borehole diameter) and L_b = bond length. Values of Q_u and τ_{ult} from the pullout tests are presented in Table 1.

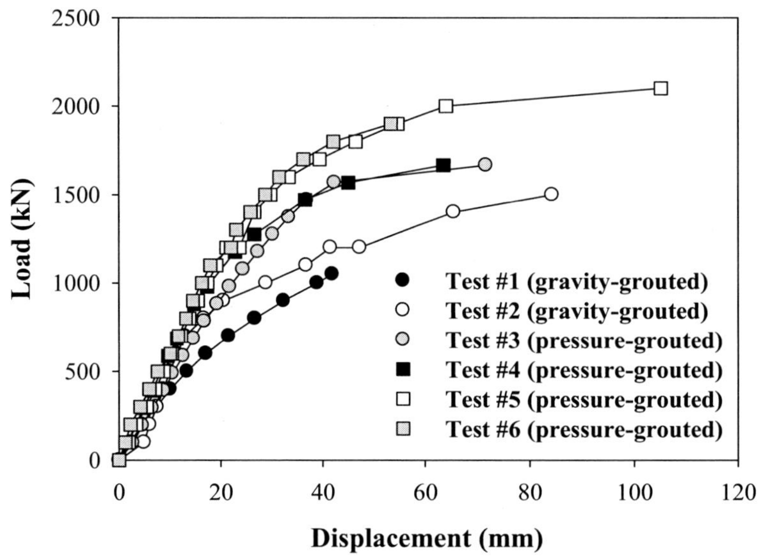

Figure 2. Recorded pullout load versus anchor displacement.

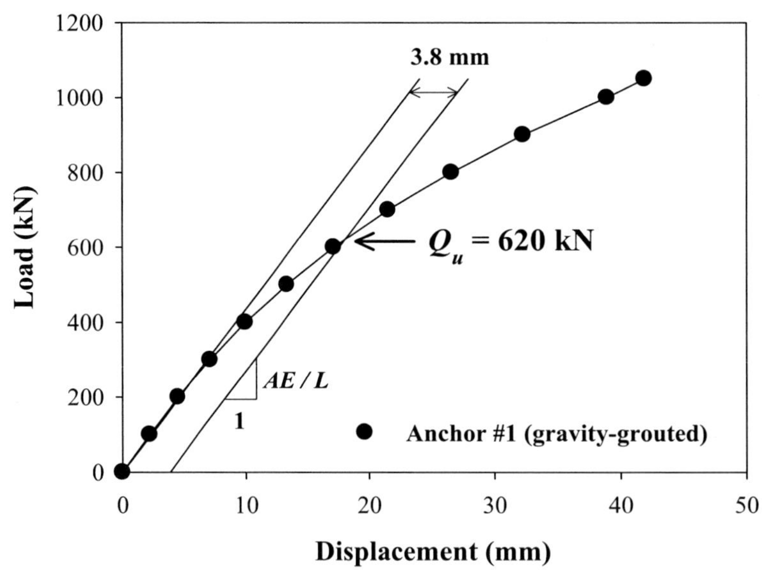

Figure 3. Ultimate pullout capacity according to Davisson's method for anchor #1.

Figures 2 and 3 demonstrate that the pullout load versus displacement curves initially follow a linear relationship, suggesting a non-yielding condition. When the load increases to the ultimate pullout load, yielding has occurred. The yielding and failure mechanisms of the pullout tests are discussed as follows. Four failure mechanisms are identified for ground anchor systems in FHWA (1999): failure of the reinforcement, failure of the ground mass, failure of the grout-reinforcement bond, and failure of the ground-grout bond. The 65-mm diameter steel rebar (ASTM A706 Grade 80) has a minimum yield load of 1825 kN, which is greater than the ultimate pullout loads presented in Table 1. Therefore, yielding of the steel reinforcement is not responsible for the yielding in the pullout tests. Failure of the rock mass may occur for shallow installations. Even in fractured rock, however, rock mass failure rarely occurs in anchors embedded more than 4.5 m below ground (FHWA 1999). As indicated in Figure 1, the anchors were embedded more than 10 meters below ground; therefore, this failure mechanism is also ruled out. For the anchors installed in this study, the mechanical interlock between the grout and the continuous thread along the rebars yields high bond strength (FHWA 1999); consequently, failure of the grout-reinforcement bond is unlikely to be the failure mechanism for the pullout tests conducted in this study.

Table 1. Summary of Field Test Results.

Test #	Grouting Method	Bond Length (m)	Ultimate Pullout Capacity (kN)	Skin Friction Resistance (MPa)
1	Gravity	1	620	1.30
2	Gravity	2	1120	1.17
3	Pressure	1	1570	3.29
4	Pressure	1	1350	2.83
5	Pressure	2	1770	1.85
6	Pressure	2	1820	1.91

The likely failure mechanism for the pullout tests is the failure of rock-grout bond. The pullout resistance is mobilized progressively along the bond zone through the skin friction at the rock-grout interface. The interface bond strength is dependent on the normal stress, adhesion, and friction mobilized. Figure 4 presents the displacement versus pullout load at three locations along the bond zone for anchors #1, 2, 3, and 5. The displacement values are based on the recordings of the underground displacement gauges/sensors embedded in the bond zone (see Figure 1). Sensor #2 is embedded in the middle elevation of the bond zone; sensors #1 and #3 are embedded at one third of the bond length above and below sensor #2, respectively. The ultimate pullout loads based on the Davisson's method are also shown in Figure 4. Figure 4 indicates that, to initiate a given displacement, higher load is required for pressure-grouted anchors (i.e., anchors # 3 and 5) than for gravity-grouted anchors (i.e., anchors # 1 and 2) with the same bond length. This is likely due to the higher rock-grout bond strength in pressure-grouted anchors.

To demonstrate the progressive load transfer along the bond length, Figure 5 shows the elongation of bond zone versus pullout load for anchors #1, 2, 3, and 5. As shown in Figure 4, the displacement sensors have different displacements throughout a pullout test. The difference in their displacements represents the elongation of the bond zone between them. Figure 5 indicates that when the pullout load increases from zero, the bond zone between sensors #1 and #2 is immediately elongated. As the load increases, the elongation and resistance in this portion of the bond zone continue to increase. At a certain load, as shown in Figure 5, load transfer to the lower portion of the bond zone (i.e., between sensors 2 and 3) is initiated. This load transfer is manifested through the elongation of the lower bond zone and doesn't exactly coincide with the peak elongation of the upper bond zone. After the load transfer, the elongation of the upper bond zone undergoes complex responses likely due to several factors including the potential tensile cracks in the grout and the load redistribution between the bond zones. The progressive elongation of bond zone is indicative of the progressive load transfer, which is consistent with the failure mechanism outlined in FHWA (1999).

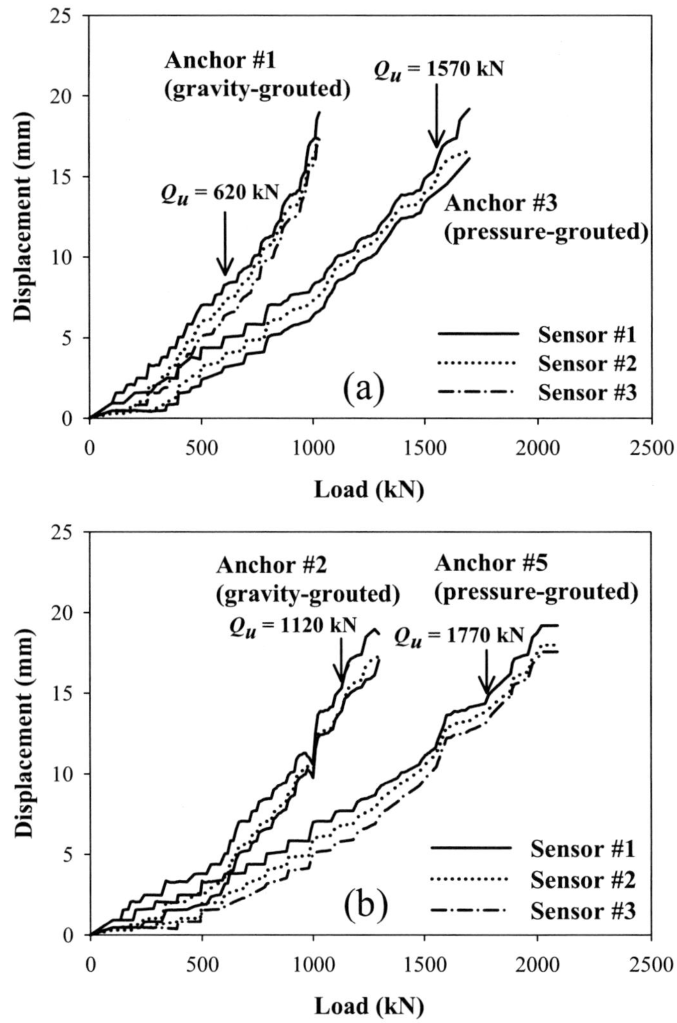

Figure 4. Displacement readings inside bond zone: (a) anchor #1 & 3, bond length = 1m; (b) anchor #2 & 5, bond length = 2m.

The field pullout tests indicate that the pressure-grouted anchors yielded higher pullout resistance than the gravity-grouted anchors. As shown in Table 1, the average skin friction resistance of the gravity-grouted anchors is 1.24 MPa and the corresponding value of the pressure-grouted anchors is 2.49 MPa, representing a 100% increase from pressure grouting. These values are within the range of typical rock-grout bond strength values listed in PTI (1996) and reported by Bruce (2005). The likely failure mechanism for the field pullout tests is the failure of rock-grout bond. In general, the strength of rock-grout bond is dependent on the borehole surface roughness, the strengths of the grout and rock, and rock fractures. The borehole surface roughness and rock strength are considered to be minor factors given that these boreholes were drilled using the same equipment and the proximity of the anchor locations (see Figure 1). Three primary factors may have contributed to the higher pullout resistance from pressure grouting. First, pressure grouting may have injected grout into the fractures/fissures preexisted in the bedrock. Second, pressure grouting may have resulted in higher grout strength and stiffness

(Park et al. 2013). Third, pressure grouting may have increased the rock-grout contact stresses. Although the grouting pressure may dissipate due to the seepage of grout into the rock fractures/fissures in the field tests, pressure grouting can increase the in-situ stresses between the grout and rock and hence improve the efficiency of anchoring (Seo et al. 2012).

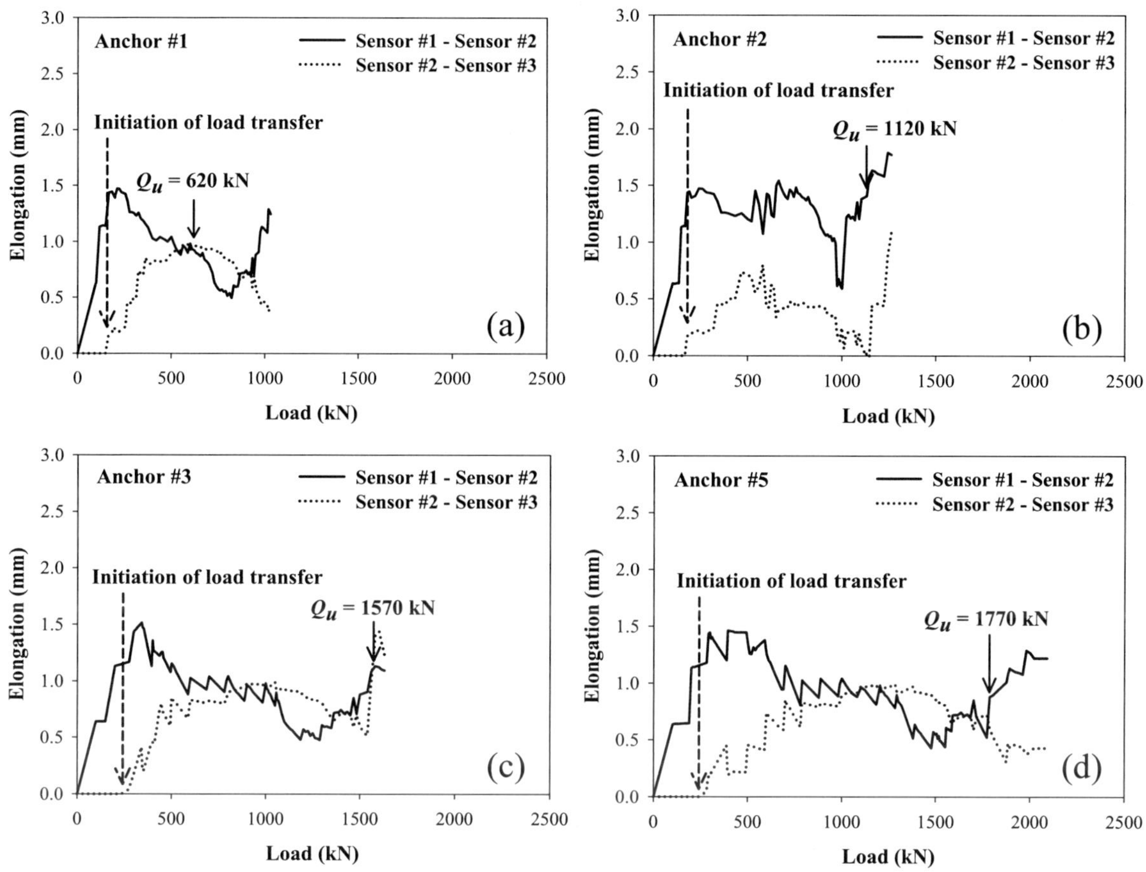

Figure 5. Elongation of bond zone: (a) anchor #1; (b) anchor #2; (c) anchor #3; (d) anchor #5.

CONCLUSION

A field investigation of pullout resistance of steel anchors in rock is presented in this paper. The field testing involves pullout tests of six steel anchors installed in rock by gravity grouting and pressure grouting methods. The field pullout tests suggest a progressive failure mechanism. The progressive failure along the bond zone was manifested through the progressive elongation of the bond zone. The failure of rock-grout bond is the likely failure mechanism of the steel anchors tested in this study. The pressure-grouted anchors yielded higher pullout resistance than the gravity-grouted anchors. Assuming other factors being equal (e.g., rock strength, bond length, and borehole surface roughness), the increase in pullout resistance from pressure grouting is due to the higher grout strength, higher rock-grout contact stresses, and the possibility of grout injected under pressure into the fractures/fissures preexisting in the bedrock.

ACKNOWLEDGMENTS

This research was supported by a grant (10TRPI-C057185-01) from the Construction Transportation RND Policy Infra Program, funded by Ministry of Land, Transport and Maritime Affairs of Korean government. This support is gratefully acknowledged.

REFERENCES

ASTM. (2007). "Standard Test Methods for Deep Foundations under static axial Tensile Load". D 3689-07, ASTM International, West Conshohocken, PA.

Bruce, D.A. (2005). "Evolution of rock anchor practice over three decades". *Proceedings of Annual Geotechnical Engineering Conference*, The Minnesota Geotechnical Society.

Chin, F.K. (1970). "Estimation of the ultimate load of piles not carried to failure". *Proc. 2nd Southeast Asian Conl. On Soil Engng.*, 81-90.

Clarke, B.G. (1996). "Pressuremeter testing in ground investigation. I: Site operations". *Geotechnical Engineering*, 119, 96-108.

Davisson, M.T (1972). "High capacity piles". *Proceedings, Lecture Series, Innovations in Foundation Construction*, ASCE, Illinois Section, 82-112.

Fellenius, B. H (1980). "The analysis of results from routine pile load tests". *Ground Engineering,* London, 13(6), 9-31.

Fellenius B. H (2001). "What Capacity Value to Choose from the Results of a Static Loading Test". *Fulcrum, Deep Foundation Institute*, New Jersey, 19-22.

FHWA (1999). *Ground Anchors and Anchored Systems, Geotechnical Engineering Circular No. 4,* Report No. FHWA-IF-99-015, Federal Highway Administration, U.S. Department of Transportation, Washington D.C.

FHWA (2000). *Micropile Design and Construction Guidelines: Implementation Manual,* Federal Highway Administration, U.S. Department of Transportation.

Finno, R.J., Scherer, S.D., Paineau, B., and Roboski, J. (2002). "Load Transfer Characteristics of Micropiles in Dolomite". *Deep Fndns*, ASCE GSP, 2(116), 1038-1053.

Gomez, J.E., Cadden, A.W., and Bruce, D.A. (2003). "Micropiles in Rock. Development and Evolution of Bond Stresses under Repeated Loading". *12th Pan-American Conference on Soil Mechanics and Geotechnical Engineering*, Cambridge, MA.

Kulhawy, F.H., Trautmann, C.H., Beech, J.F., O'Rourke, T.D., and McGuire, W. (1983). Transmission Line Structure Foundations for Uplift-Compression Loading, Electric Power Research Institute (EPRI) Report.

Park, J., Qiu, T., and Kim, Y. (2013). "A field and laboratory investigation of pullout resistance of steel anchors in rock", *J. Geotech. & Geoenv. Eng.,* in press (December issue).

Post Tensioning Institute (PTI). (1996). "Recommendations for prestressed rock and soil anchors". *Post Tensioning Manual*, 4th Edition, Phoenix, Arizona.

Seo, H.J., Jeong, K.H., Choi, H., and Lee, I-M. (2012). "Pullout resistance increase of soil nailing induced by pressurized grouting". *J. Geotech. & Geoenv. Eng.,* 138(5), 604-613.

Scott, R.F. (1981). *Foundation analysis*, Prentice-Hall, Englewood Cliffs, N.J.

Yin, J.H, Su, L.J., Cheung, R.W.M., Shiu, Y.K., and Tang C. (2009). "The influence of grouting on the pullout resistance of soil nails in compacted completely decomposed granite fill". *Geotechnique*, 59(2), 103-113.

Thermo-Mechanical Characterization of a Full-Scale Energy Foundation

Kyle D. Murphy[1], M.S., E.I.T., A.M.ASCE, John S. McCartney[2], Ph.D., P.E., M.ASCE, and Karen S. Henry[3], Ph.D., P.E. M.ASCE

[1]Graduate Research Assistant, University of Colorado Boulder, Dept. of Civil, Environmental and Architectural Eng., UCB 428, Boulder, CO 80309, kyle.murphy@colorado.edu
[2]Associate Professor and Lyall Faculty Fellow, University of Colorado Boulder, Dept. of Civil, Environmental and Architectural Eng., UCB 428, Boulder, CO 80309, john.mccartney@colorado.edu
[3]Associate Professor, United States Air Force Academy. Department of Civil Engineering, USAF Academy, CO 80840-6232, karen.henry@usafa.edu

ABSTRACT: This paper documents the thermo-mechanical characterization of an energy foundation constructed as part of a new single-story building at the U.S. Air Force Academy. During a 39 hour heating test, the foundation increased in temperature by 21.7 °C, while the sandstone subsurface only increased in temperature by less than 1 °C at a distance of 1.22 m from the center of the foundation. The distributions in thermal axial stress and displacement in the foundation were calculated from the measured thermal axial strain values, and a maximum compressive thermal axial stress of 5.2 MPa was observed near the middle of the foundation, and a maximum upward displacement of 1.55 mm was observed at the head. The thermal axial stresses and displacements measured were well within acceptable limits for structural elements.

INTRODUCTION

A ground source heat pump (GSHP) is used to transfer heat from a refrigerant to an anti-freeze solution circulating through the heat exchange elements, which serves to transfer heat to or from the subsurface to the building for the purpose of heating a building. A GSHP has higher efficiency than an air-source heat pump due to the relatively constant temperature of the ground compared to that of the outside air. If the subsurface heat exchanger is contained within a foundation, it is called an energy foundation. The most common configuration of energy foundations consists of drilled shafts with embedded closed-loop ground-source heat exchange elements (Brandl 2006; Laloui et al. 2006; McCartney 2011). Energy foundations help to reduce the installation costs of ground-source heat exchangers, which is one of the major barriers to implementation of GSHP systems (Hughes 2008). Further, their heat exchange response has been well established (Ooka et al. 2007; Adam and Markiewicz 2009; Wood et al. 2009; Ozudogru et al. 2012). However, the thermo-mechanical response of the foundations during heating and cooling due to the thermal expansion and contraction of the reinforced concrete is still a topic of importance. This paper presents measurements from strain gages and thermistors embedded within a full-scale energy foundation during a heating test to better understand the role of soil-structure interaction in the thermo-mechanical response of the foundation.

BACKGROUND

As an energy foundation is heated or cooled, it tends to expand or contract, respectively, depending on the end-restraint boundary conditions at the head and toe of the foundation. The free thermal axial strain in an unconstrained energy foundation is calculated as follows:

$$\varepsilon_{thermal,free} = \alpha_c \Delta T \tag{1}$$

where α_c is the coefficient of linear thermal expansion of reinforced concrete, and ΔT is the change in temperature. The value of $\varepsilon_{thermal,free}$ is the upper or lower limit of the thermal axial strains resulting from heating or cooling, respectively. To be consistent with geotechnical engineering conventions, the strain is defined as positive for compression. Accordingly, α_c is defined as negative as structural elements expand during heating (positive ΔT). If the energy foundation is vertically constrained, the actual thermal strains will be less than those predicted by Eq. 1. In this case, the thermal axial stresses induced in an energy foundation are calculated as follows:

$$\sigma_T = E(\varepsilon_{thermal} - \alpha_c \Delta T) \tag{2}$$

where E is the Young's modulus of reinforced concrete and $\varepsilon_{thermal}$ is the measured thermal axial strain. For full-scale energy foundations embedded in soil or rock, soil-structure interaction mechanisms will restrict the movement of the foundation during heating. Specifically, the side shear resistance, end bearing response, and the stiffness characteristics of the overlying building leads to a range of vertical distributions in axial stresses and strains in energy foundations. Soil-structure interaction mechanisms have been studied in centrifuge-scale tests for simplified soil profiles (McCartney and Rosenberg 2011; Stewart 2012; McCartney et al. 2013). However, evaluation of full-scale foundations in the field utilizes realistic boundary conditions and soil strata. Several full-scale energy foundations have been evaluated to study the thermo-mechanical stresses and strains during mechanical loading, heating, and cooling (Laloui et al. 2006; Bourne-Webb et al. 2009; Amatya et al. 2012; McCartney and Murphy 2012; Murphy 2013). A summary of results from these studies is presented in Table 1. The information from these studies has been implemented into thermo-mechanical load-transfer analyses (Knellwolf et al. 2011; Plaseied 2012) that can be used to predict the distributions in thermal axial displacement and stress.

Table 1. Results of previous studies on thermo-mechanical behavior of energy foundations.

Case	Laloui et al. (2006)	Bourne-Webb et al. (2009)	McCartney and Murphy (2012); Murphy (2013)
Load mechanism at foundation head	Building dead load	Load frame	Building dead load
Foundation diameter (m)	0.88	0.56	0.91
Foundation length (m)	25.8	23	14.8 (A), 13.4 (B)
Maximum mechanical load during heating test (kN)	0, 1300	1200	3840 (A), 3640 (B)
Range of ΔT (°C)	+20.9, +13.4	-19.0, +29.4	+14.0
Depth of minimum thermal axial strain after heating (m)	21.0	17.0	11.6
Maximum increase in thermal axial stress with temperature (kPa/°C)	104	192	260

ENERGY FOUNDATION SYSTEM

Building and Project Description

A one-story, shower-shave building was constructed at the Field Engineering and Readiness Laboratory (FERL) of the US Air Force Academy (USAFA) beginning March, 2012. At the time of this publication, the building is not yet in service (the interior plumbing and outside façade are being installed). The building will provide men's and women's restrooms, showers, and laundry facilities for USAFA cadets and other military personnel training at the FERL site. One purpose of the building is to evaluate the performance of energy efficient technologies to aid in the development of "net zero" energy consuming structures for the U.S. Department of Defense (DoD). The building includes energy foundations, and a radiant in-floor heating system; and, there are plans to install a solar water heating system in the future. Each building energy component will be monitored to evaluate the energy usage and related output. In addition to the ground-source heat pump coupled with the energy foundations, the building contains a conventional gas-powered boiler. Having both conventional and ground-source HVAC systems will permit comparison of their energy efficiencies under similar environmental conditions.

Subsurface Conditions

A geotechnical site investigation was performed in September 2011 which consisted of two 102 mm-diameter borings located within the building footprint, extending 12 and 7 m below the ground surface (Hernandez 2011). At selected intervals, disturbed samples were obtained by driving a split-spoon with a 622.75-N-hammer falling 762 mm. Exploration results from both boreholes were similar and showed three prominent strata (Table 2). The top layer is approximately 1 m thick and consists of sandy fill. Beneath the fill is a very dense 1 m-thick sandy gravelly layer. The bedrock is comprised of Dawson-Arkose sandstone that extends to the maximum depth explored. No groundwater was encountered during the site investigation or foundation installation, so it is assumed to be at a depth greater than 16 m. Laboratory tests indicate that all materials are non-plastic and non-expansive.

Table 2. Summary of stratigraphy encountered during subsurface exploration.

Layer	Depth to bottom of stratum (m)	Material encountered	SPT N-value (blows/300 mm)	Gravimetric water content (%)	Dry unit weight (kN/m^3)
1	1	sandy fill with silt and gravel	70	5	18.4
2	2	dense sands, silt, and gravel	85	7	19.2
3	12+	silty sandstone	50/25.4 mm	N/A	N/A

Energy Foundations and Heat Exchanger Configurations

Eight drilled shafts, each 15.2 m deep and 0.61 m in diameter, provide the foundation support for the structure. The one-story building could have been constructed with a slab-on-grade foundation, so the main purpose of incorporating the deep drilled shafts into the building was to evaluate the thermo-mechanical response of the energy foundations. Photos of construction of the energy foundations are shown in Figure 1. Each foundation contains a 0.46-m-diameter steel reinforcing cage that extends the full length of the shaft. The reinforcing cages are composed of six #7 longitudinal bars with #5 radial hoops spaced at 0.3 m on center throughout the length of the cage [Fig. 1(a)]. Each foundation contains a closed-loop heat exchanger that consists of 19 mm-diameter HDPE tubing [Fig. 1(b)]. The drilled shafts were constructed using the dry hole method as the subsurface material was competent enough to avoid sloughing and no groundwater was present [Fig. 1(e)]. The head of the shafts are spliced into a 0.91 m-deep by 0.61 m-wide grade beam that extends around the perimeter of the building [Fig. 1(h)]. At the head of each foundation (1 meter below grade), the heat exchanger loop is connected with tubing that is routed through the grade beam into a manifold within the mechanical room of the building.

Figure 1. Construction photos: (a) Reinforcing cages with heat exchangers; (b) Inside view of cage; (c) Vibrating wire strain gage; (d) Lifting with 3-point pick; (e) Lowering into uncased hole; (f) Concrete tremie; (g) Finished foundation; (h) Tubing in grade beam; (i) Completed grade beam and heat exchanger manifold.

The heat exchange tubing was attached to the inside of the reinforcing cages with zip-ties at a distance of at least 70 mm from each vertical reinforcing member. The inlet and outlet tubes were seperated diametrically by at least 90°, which minimizes the chances for thermal short-circuiting, where heat is lost by direct transmission from the inlet to outlet tubes. U-shaped couplings were used to connect the inlet and outlet tubes so that the tubing does not cross the bottom of the cage. Cables for the instrumentation that were embedded in the foundation [Fig. 1(c)] were bundled at the top of the cage until it was placed in the uncased hole. The reinforcing cages were lifted with a 3-point pick to minimize bending [Fig. 1(d)], and the cages were lowered into the hole with a crane [Fig. 1(e)] and were suspended on wooden beams to ensure that the top of the cage was at the base of the grade beam. Although efforts were made to clean the shafts, it is possible that some loose sandstone debris remained at the bottom of the holes. A concrete pump truck was used to place high-slump concrete with a compressive strength of 21 MPa in the holes following placement of the reinforcing cages. A tremie pipe was used to avoid excessive segragation of the concrete during free-fall [Fig. 1(f)]. The use of the tremie also minimized the risk of damage to the heat exchanger loops and embedded instrumentation. The grade beam was constructed around the foundations [Fig. 1(g)], and the heat exchangers were routed through the grade beam [Fig. 1(h)] into a manifold in the utility room [Fig. 1(i)].

Each shaft has either one, two, or three heat exhanger loops configured in different ways, (Fig. 2). Foundations 1 through 4 have identical heat exchanger configurations, with two continuous heat exchanger loops attached to the inside of the steel reinforcement cage. Foundation 5 has three individual loops; each having a supply and return line running to the mechanical room; this permits any combination of the loops to be operational in order to evaluate the efficiency of multiple loops in a single foundation. Foundation 6 has three continuous heat exchanger loops with only one supply and return line extending to the manifolds. Foundation 7 contains one loop connected to the interrior of the reinforcing cage. Foundation 8 has a single loop in the center of the foundation to simulate a retrofit where a heat exchanger would be inserted into a corehole bored into an existing foundation. This was constructed with a 100 mm-diameter plastic sleeve in the center of the foundation. After curing, a single heat exchanger loop was inserted into the plastic sleeve and the hole was backfilled with sand bentonite grout.

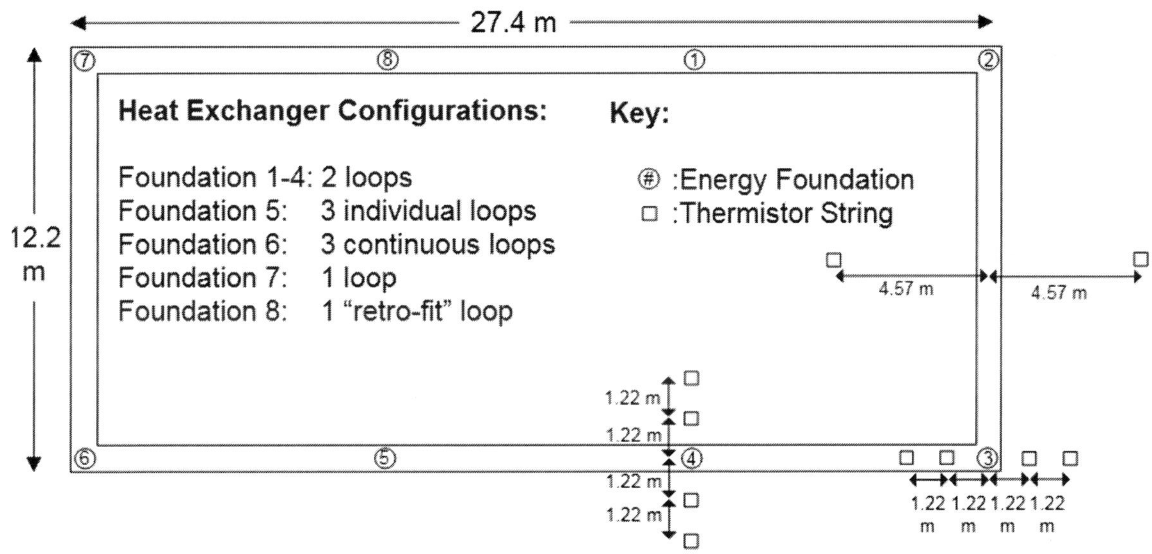

Figure 2. Heat exchanger loop configurations in the different energy foundations.

Instrumentation and Setup for the In-situ Heating Test

Instrumentation was incorporated into three of the eight energy foundations to capture the distribution of axial strain and temperature with depth [Fig. 3(a)]. Foundations 1 and 3 contain six Geokon Model 4200 vibrating wire strain gages (VWSGs), while Foundation 4 contains twelve. Foundation 4 has twice the number of gages as the other two with the intent of capturing detailed strain and temperature distribution along the length of the shaft. At three locations within Foundation 4, gages were located at the same depth on opposite sides of the reinforcing cage to gain redundancy in temperature and strain readings and to capture any differential strain measurements across the width of the shaft. All of the gages were oriented vertically and attached to brackets welded to longitudinal reinforcing bars, as shown in Figure 1(c). The sensor cables were routed to the mechanical room where they are connected to the data acquisition system.

A heating test was conducted to evaluate the thermo-mechanical response of Foundation 4. A schematic of the heating system is shown in Figure 3(b). A 125-liter-reservoir was first filled with a 20% propylene glycol-water mixture. This heat exchange fluid was circulated through a

12 kW tankless hot water heater using a submersible pump. The heated fluid passed into the inlet port of the heat exchanger tubing, circulated through the foundation, then passed back out of the outlet port. A flow rate of 6.25 L/min was measured from the outlet line at a fluid temperature of 30 °C. During this process, the inlet and outlet temperatures of the heat exchanger fluid were continuously monitored using pipe plug thermocouples installed in the manifold connected to the foundations. Temperature variations in the soil surrounding the energy foundations are monitored using a series of ten Geokon model 3810 thermistor strings which each have six sensors, installed in boreholes that were then backfilled with CETCO high thermal conductivity grout at the locations shown in Figure 2. The temperature of the subsurface around Foundations 3 and 4 was monitored using four thermistor strings around each foundation, while another thermistor string lies beneath the floor slab to observe any long-term changes in the ground temperature beneath the slab, and one other thermistor string is located outside the building footprint to observe seasonal variations in ground temperature.

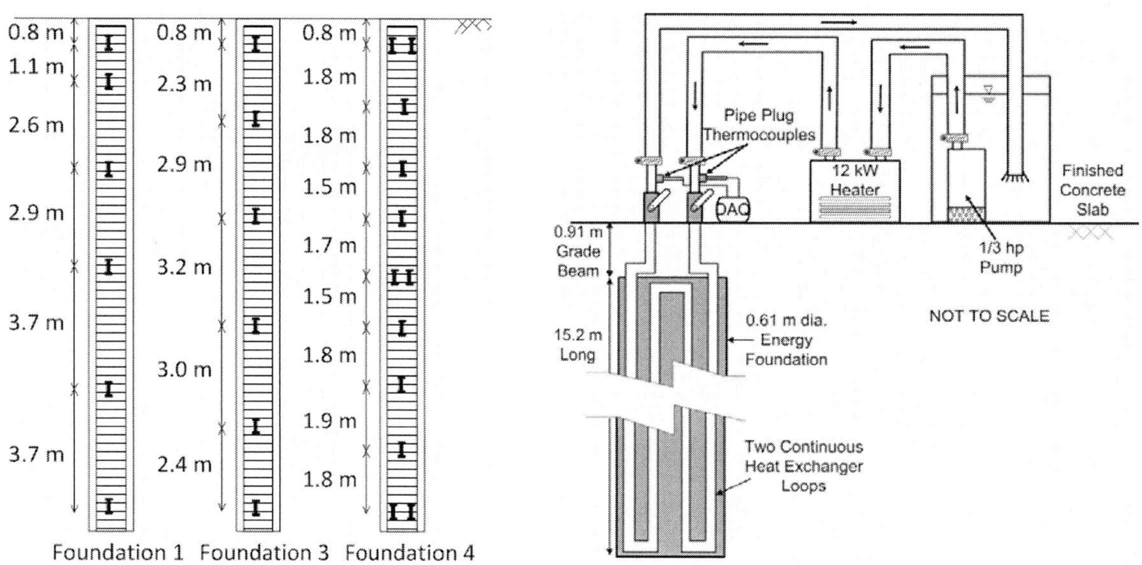

(a) (b)

Figure 3. Heating test setup: (a) Instrumentation locations; (b) Heating system.

RESULTS

Initial Conditions

Between the time of installation in April 2012 and the heating test (occurring just after the completion of the building roof) readings were obtained periodically from the embedded thermistors and VWSGs to characterize the ambient temperature profile and the strains induced by the curing process and mechanical loading. The ambient temperatures of the ground shown in Figure 4(a) indicate that the temperature of the ground is relatively stable at depths below 4 m. The depths in this figure (and other figures) are from the bottom of the grade beam, which is 0.91 m below the ground surface. The axial strains measured in July 2012 reflect the strains induced by curing are shown in Figure 4(b) in terms of micro-strain (με) because the remaining structure had not yet been built. The majority of the structure was constructed between July 2012 and February 2013, so the difference between these curves reflects the mechanical loading of the foundation. The axial strains due to the dead load of the building dissipate with depth due to side shear resistance in the foundation.

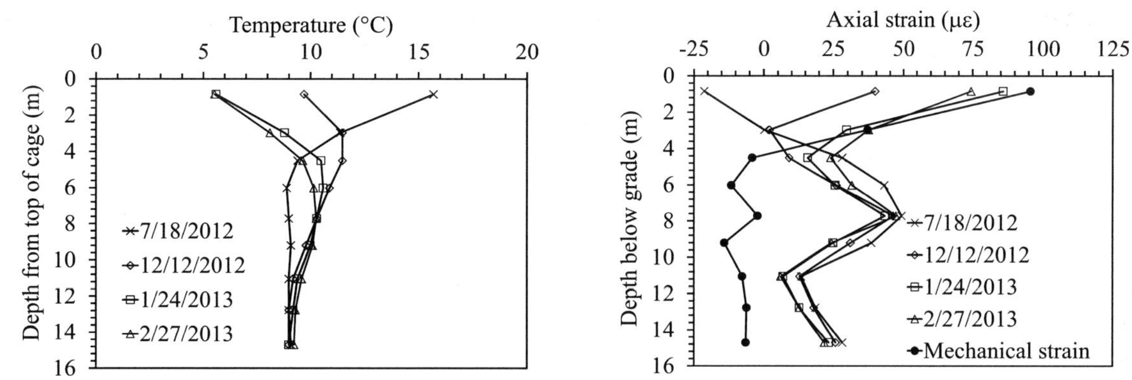

(a) (b)

Figure 4. (a) Ambient temperature fluctuations in Foundation 4 before heating test; (b) Initial axial strains induced by curing and the strain assumed to be due only to the building load.

In-situ Heating Test Results

During the heating test, the foundation was heated continuously for 39 hours. Since the ambient temperature at the site ranged from 0 to 10 °C during the test, the heat exchange fluid partially cooled in the reservoir before being recirculated. During the test, the temperature of the fluid reservoir gradually increased, causing the fluid entering the system to increase in temperature, as shown in Figure 5(a). The drop in temperature after 10 hours was due to the very cold overnight air temperature. The foundation temperature also increased steadily during circulation of the heat exchange fluid, as shown in Figure 5(b)., The foundation changed in temperature by 21.7 °C due to heating. The temperature of the foundation was relatively uniform with depth (within 2 °C) during heating (Fig. 5b).

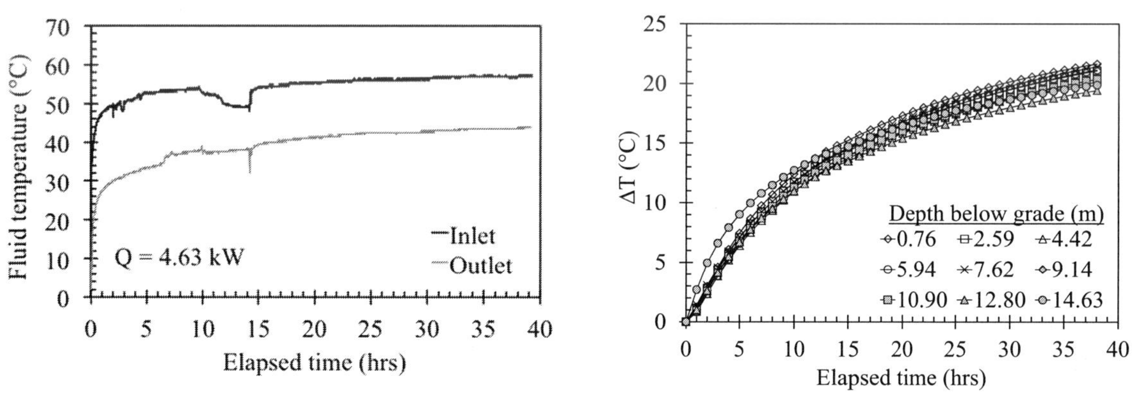

(a) (b)

Figure 5. Temperatures during heating test: (a) Inlet and outlet fluids; (b) Foundation.

The subsurface at a distance from 1.22 m from the center of the foundation increased in temperature less than 1◦C, as shown in Figure 6. The shape of the curves in Figure 6 also differs from that of the foundation, with a more convex shape during heating. Further, a lag of about 12 to 15 hours was observed before the soil at 1.22 m from the foundation started to change in temperature. The upper three sensors indicated the greatest rise in temperature, likely due to cooler initial ground temperatures in this zone (and greater thermal gradients), as this test was performed in March, 2013. The boreholes located at 2.44 m from the foundation showed negligible change in temperature during the testing period. Although a longer heating test may have affected a larger volume of soil around the foundation, potentially changing the impact of soil-structure interaction on the thermal axial strain distribution, the thermo-mechanical behavior observed in this test still provides useful information on the transient response of the foundation.

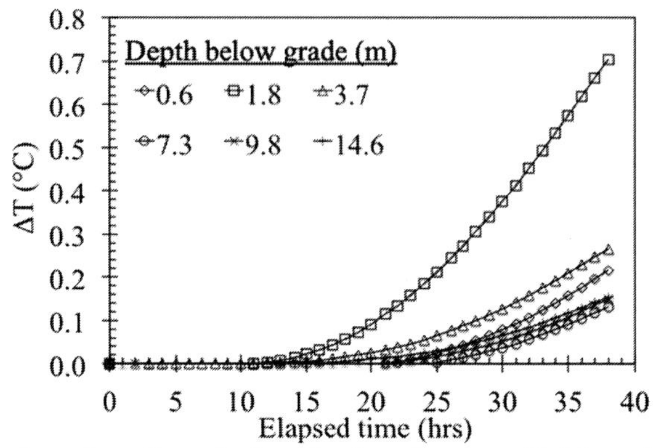

Figure 6. Temperatures in subsurface during the heating test (1.22 m from foundation center).

Profiles of temperatures with depth in the foundation are shown in Figure 7 for different times after the start of heating. It is clear that the temperature of the foundation is relatively uniform (within 2 °C). The reason for the slightly higher temperature at the toe of the foundation during the early stages of the heating process may be due to the proximity of the sensor to the U-bend of the pipe at the base, which may have heated this part of the foundation more than the rest of the foundation.

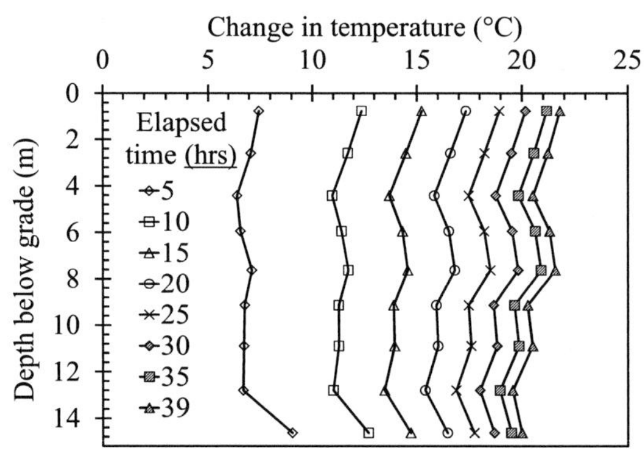

Figure 7. Profiles in foundation changes in temperature during the heating test.

To evaluate the thermo-mechanical response of the energy foundation, the resonant frequency values, *f*, from the VWSGs during the heating test were first converted into axial strain ε, as follows:

$$\varepsilon = -Gf^2 \tag{3}$$

where G is the gage factor of 3.304×10^{-3} and the units of ε are micro-strain. The negative sign follows the geotechnical sign convention for positive strains being compressive. The strain values calculated with Eq. 3 were then converted to thermal axial strains, as follows:

$$\varepsilon_{thermal} = [(\varepsilon_i - \varepsilon_0)B + (T_i - T_0)\alpha_s] \tag{4}$$

where B is the batch calibration factor of 0.975, ε_i and T_i are the axial strain and temperature at time i, ε_0 and T_0 are the initial axial strain and temperature, and α_s is the coefficient of thermal expansion of the steel wire of –12.2 με/°C. This equation accounts for the elongation of the steel wire in the gage during heating. The thermal axial strains calculated using Eq. 4 during the heating process are shown in Figure 8. The strain was greatest at the top and bottom of the foundation. The bottom of the foundation possibly experienced slightly greater strains due to the higher temperatures at the base of the foundation shown in Figure 6(a). The distribution in strain in this figure reflects the fact that soil-structure interaction due to mobilization of side shear resistance leads to a nonlinear distribution in thermal strain with depth, similar to the observations of Laloui et al. (2006) and Bourne-Webb et al. (2009) during the heating portions of their tests.

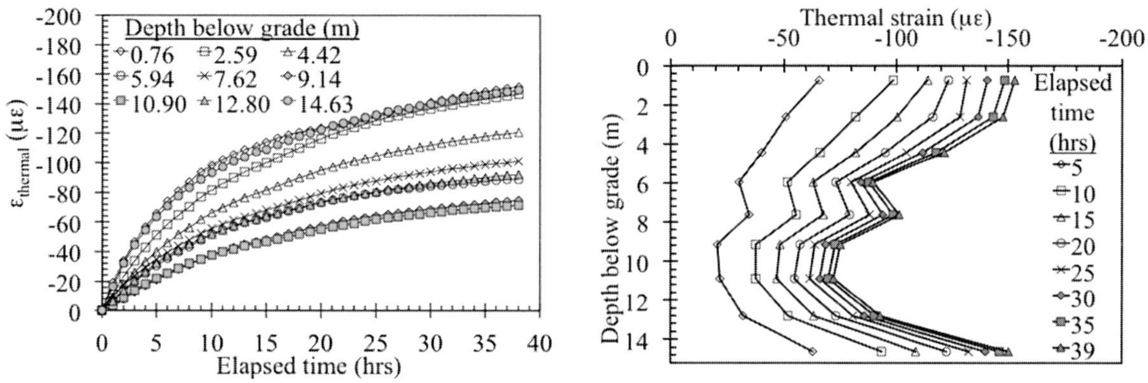

(a) (b)

Figure 8. Thermal axial strains induced during the heating test: (a) Time series; (b) Profiles.

ANALYSIS

Thermal Stresses

The thermal axial stresses in the energy foundation were calculated from the measured thermal axial strains shown in Figure 8 using Eq. 2, assuming a coefficient of thermal expansion of the reinforced concrete of –12 με/°C (Stewart 2012; Goode et al. 2014) and a Young's modulus of E = 30 GPa. The data shown in Figure 9 indicate that the greatest thermal axial stresses occur in the bottom half of the foundation with a maximum magnitude of 5.2 MPa, which is approximately 16% of the unconfined compressive strength of reinforced concrete ($0.16f'_c$). It is a general convention in reinforced concrete design to limit axial stresses to $0.33f'_c$. The thermal axial stresses are consistently compressive with depth during heating. This is expected as the foundation has the potential to expand during heating, but compressive stresses occur when this expansion is restrained. The formulation of the thermal stress in Eq. 2 indicates that the greatest stresses will occur at the locations of the lowest strain. It was expected that the greatest thermal axial stress would coincide with the point of zero displacement, which is referred to as the null point (Knellwolf et al. 2011), based on conceptual models of thermal soil-structure interaction (Amatya et al. 2012). If the energy foundation is resting on a rigid bedrock (i.e., an end-bearing boundary condition), then it would be expected that the null point should occur at the toe. McCartney et al. (2013) observed that the greatest thermal axial stress occurred near the toe for two different end-bearing foundations. However, the greatest thermal axial stress was observed in the lower half of the foundation. It is possible that the dry sandstone cuttings at the toe of the foundation were not thoroughly removed from the hole during construction, leading to a softer, non-rigid response at the toe. The maximum thermal axial stress at a depth of 10.9 m likely occurred due to a combination of end bearing from the toe and side shear resistance. The maximum increase in thermal axial stress with temperature of 252 kN/°C also occurred at a depth of 10.9 m, which is within the range of values from the literature presented in Table 1.

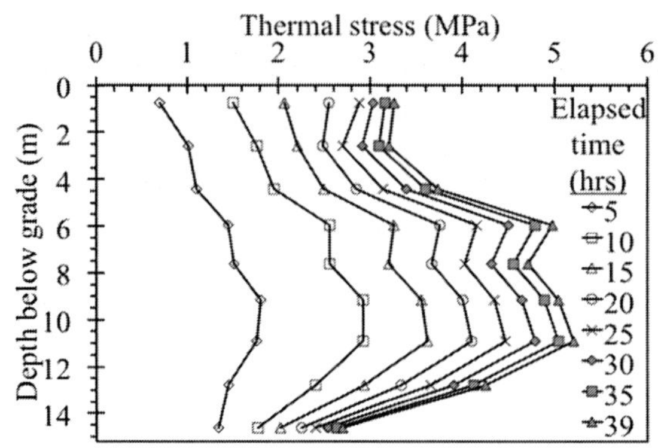

Figure 9. Thermally induced stresses in the foundation.

Another way to evaluate the thermal soil-structure interaction is through the relationships between thermal axial strains and the changes in temperature at the location of each gage, as shown in Figure 10(a). Specifically, the slopes of these relationships, which should be linear as reinforced concrete is a thermo-elastic material, reflect the mobilized coefficient of thermal expansion of the reinforced concrete. As mentioned, the maximum value of the mobilized coefficient of thermal expansion is that for free expansion conditions (α_c = –12 με/°C). The distribution in the mobilized coefficients of thermal expansion is shown in Figure 10(b). The mobilized coefficients of thermal expansion are consistently less than -12 με/°C, indicating that side shear resistance and the end restraint boundary conditions prevented the foundation from expanding as much as it possibly could in free-expansion conditions.

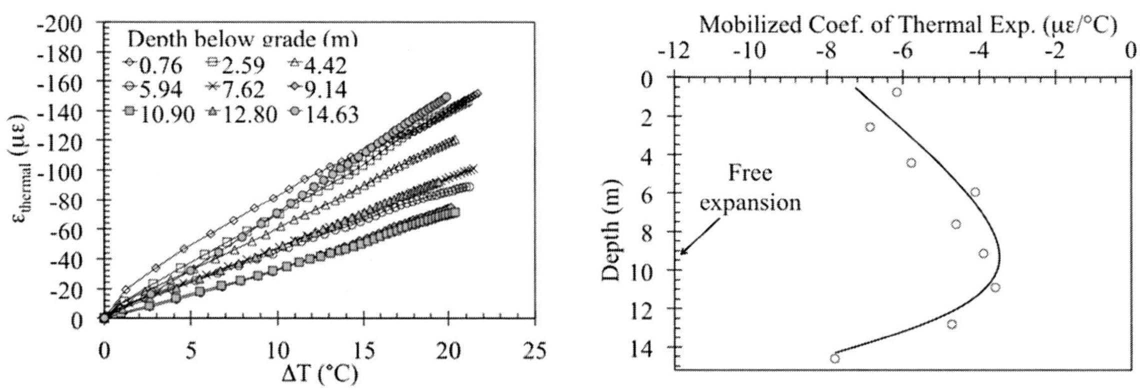

(a) (b)

Figure 10. (a) Thermal axial strain with change in foundation temperature at each depth; (b) Mobilized coefficient of thermal expansion with depth.

Relative Thermal Displacements

Thermal axial displacement at different depths in the energy foundation can be calculated from the thermal axial strain data as follows:

$$\delta_{thermal,i} = \delta_{thermal,i-1} + \frac{1}{2}(\varepsilon_{thermal,i-1} + \varepsilon_{thermal,i})\Delta l \tag{5}$$

where $\delta_{thermal,i}$ is the thermal axial displacement at the midpoint between gages, $\varepsilon_{thermal,i}$ is the thermal axial strain at the location of gage i, and Δl is the distance between gages. This analysis assumes that the bottom of the foundation is fixed from moving in the vertical direction. Since no actual displacement measurement was recorded in either case study, the assumption of a fixed base is the most practical approach to calculating relative foundation displacements. The assumption of a fixed base gives displacements during heating relative to the toe of the energy foundation. Thermal axial displacements were calculated using Eq. 5, and are shown in Figure 11. For a maximum increase in temperature of 21.7 °C, the displacement of the foundation head relative to the toe was determined to be approximately 1.55 mm. Since the foundation exhibited large strains near the toe during heating, it is possible that the toe of the foundation may have moved downward slightly during heating, thus reducing the head displacement trends shown in Figure 11. Nonetheless, if the displacement of the toe is zero, the maximum angular distortion over the 9.1 m span between foundations would be 1/5000, which is well within acceptable limits of a typical commercial structure (Skempton and MacDonald 1956, Bjerrum 1963).

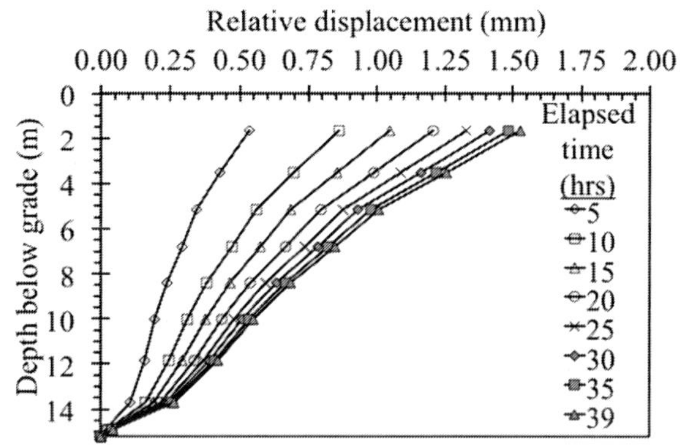

Figure 11. Displacement relative to the toe of the energy foundation during heating.

CONCLUSIONS

The thermo-mechanical behavior of an energy foundation during a 39-hour heating response test is presented in this paper. Temperature and thermal axial strain measurements were captured throughout the test. The thermal axial strains were used to calculate the distributions in thermal axial stress and relative displacement in the energy foundation. The following conclusions can be drawn from this study:

1. A nonlinear change in thermal axial stress with depth is attributed to side shear resistance, an issue that will be evaluated in future studies. The temperature of the foundation during heating was relatively constant during the heating process.
2. The maximum thermally induced stress of 5.2 MPa occurred at a depth of 10.9 m below the grade beam for an increase in foundation temperature of 21.7 °C.
3. The maximum displacement of the head of the energy foundation relative to the toe was 1.55 mm during heating for an increase in foundation temperature of 21.7 °C.
4. Both the thermal axial stress and the thermal axial displacement are consistent with values measured in the literature and are within the limits that the structure can safely withstand.
5. The fact that the maximum stress was not observed at the toe of the foundation indicates that the foundation does not behave strictly as an end-bearing foundation, possibly due to construction effects on the stiffness of the underlying material below the tip of the foundation, potentially preventing it from acting as a completely rigid boundary.

ACKNOWLEDGEMENTS

Support from DoD ESTCP project EW-201153 is gratefully acknowledged, as are the helpful contributions of the 819th Air Force RED HORSE Squadron, who constructed the building and provided support for the heating test. The views in the paper are those of the authors alone.

REFERENCES

Adam, D. and Markiewicz, R. (2009). "Energy from Earth-Coupled Structures, Foundations, Tunnels and Sewers." *Géotechnique*. 59(3), 229–236.

Amatya, B.L., Soga, K., Bourne-Webb, P.J., Amis, T. and Laloui, L. (2012). "Thermo-mechanical Behaviour of Energy Piles." *Géotechnique*. 62(6), 503–519.

Bjerrum, L. (1963). "Allowable Settlement of Structures." *Proc., European Conf. on Soil Mech. and Found. Engr.*, Wiesbaden, Germany, Vol. 3, pp. 135- 137.

Bourne-Webb, P., Amatya, B., Soga, K., Amis, T., Davidson, C. and Payne, P. (2009). "Energy Pile Test at Lambeth College, London: Geotechnical and Thermodynamic Aspects of Pile Response to Heat Cycles." *Géotechnique* 59(3), 237–248.

Brandl, H. (2006). "Energy Foundations and other Energy Ground Structures." *Géotechnique*. 56(2), 81-122.

Hernandez, R. (2011). *Geotechnical Investigation Report. Construct Shower/Shave Facility. United States Air Force Academy, Colorado*. Terracon Consultants Inc. November 10th, 2011.

Hughes, P.J. (2008). *Geothermal (Ground-Source) Heat Pumps: Market Status, Barriers to Adoption, and Actions to Overcome Barriers*. Oak Ridge Nat. Lab. Report ORNL-2008/232.

Knellwolf, C. Peron, H., and Laloui, L. (2011). "Geotechnical Analysis of Heat Exchanger Piles." *ASCE Journal of Geotechnical and Geoenvironmental Engineering*. 137(10). 890-902.

Laloui, L., Nuth, M., and Vulliet, L. (2006). "Experimental and Numerical Investigations of the Behaviour of a Heat Exchanger Pile." *International Journal of Numerical and Analytical Methods in Geomechanics*. 30, 763–781.

Goode, J.C., III, Zhang, M., and McCartney, J.S. (2014). "Centrifuge Modelling of Energy Foundations in Sand." *8th International Conference on Physical Modelling in Geotechnics*. Perth, Australia. January 14-17, 2014. 6 pg.

McCartney, J.S. and Murphy, K.D. (2012). "Strain Distributions in Full-Scale Energy Foundations." *DFI Journal*. 6(2). 28-36.

McCartney, J.S. and Rosenberg, J.E. (2011). "Impact of Heat Exchange on the Axial Capacity of Thermo-Active Foundations." *GeoFrontiers 2011*. Dallas, TX. 10 pg.

McCartney, J.S., Murphy, K.D., and Stewart, M.A. (2013). "Thermo-Mechanical Strain Distributions in Full-Scale Energy Foundations." *18th International Conference on Soil Mechanics and Foundation Engineering*. Paris, France. September 1-5, 2013. 4 pg.

McCartney, J.S. (2011). "Engineering Performance of Energy Foundations." *2011 PanAm CGS Geotechnical Conference*. Toronto, Canada. October 2-6, 2011. 14 pg.

Murphy, K.D. (2013). *Evaluation of Thermal and Thermo-mechanical Behavior of Full-scale Energy Foundations*. MS Thesis. University of Colorado Boulder.

Ooka, R., Sekine, K., Mutsumi, Y., Yoshiro, S. SuckHo, H. (2007). "Development of a Ground Source Heat Pump System with Ground Heat Exchanger Utilizing the Cast-in Place Concrete Pile Foundations of a Building." *EcoStock 2007*. 8 pp.

Ozudogru, T., Brettmann, T., Olgun, G., Martin, J., and Senol, A. (2012). "Thermal Conductivity Testing of Energy Piles: Field Testing and Numerical Modeling." *ASCE GeoCongress 2012*. Oakland, CA. March 25-29th, 2012. 10 pg. CD ROM.

Plaseied, N. (2011). *Load-Transfer Analysis of Energy Foundations*. M.S. Thesis. University of Colorado Boulder.

Skempton, A.W., and MacDonald, D.H. (1956), "Allowable Settlement of Buildings, *Proc. Institute of Civil Engineers*. London, Part 3, Vol. 5, pp. 727- 768.

Stewart, M.A. (2012). *Centrifuge Modeling of Strain Distributions in Energy Foundations*. MS Thesis. University of Colorado Boulder.

Wood, C.J., Liu, H. and Riffat, S.B. (2009). "Use of Energy Piles in a Residential Building, and Effects on Ground Temperature and Heat Pump Efficiency." *Géotechnique* 59(3), 287–290.

Building a Framework for Predicting Foundation Settlements on Granular Soil with Dynamically Measured Properties

Onur Kacar[1], S.M.ASCE, Kenneth H. Stokoe II[2] P.E, M.ASCE, Kwangsoo Park[3], P.E.

[1]Graduate Research Assistant, Civil, Architectural and Environmental Engineering, 301 E. Dean Keeton St. Stop C1700,University of Texas at Austin, Austin,TX 78712; onurkacar@gmail.com
[2]Professor, Civil, Architectural and Environmental Engineering, 301 E. Dean Keeton St. Stop C1792,University of Texas at Austin, Austin,TX 78712; k.stokoe@mail.utexas.edu
[3]Manager/Geotechnical Engineer, Offshore Wind Energy EPCI, Samsung Heavy Industries Co., Ltd. South Korea; kwangs.park@samsung.com

ABSTRACT: Settlements are usually the primary concern in designing shallow foundations on freely draining granular soil. Settlement predictions are traditionally based on empirical correlations with *in-situ* penetration tests due to difficulties in obtaining undisturbed samples. In the proposed framework, field measurements of shear wave velocity (V_s) are used to characterize the *in-situ* condition in terms of small-strain shear modulus (G_{max}) profiles. Disturbed field samples are recovered to determine gradation characteristics. Normalized shear modulus-shear strain relationships (G/G_{max}-log γ) that have been developed from dynamic resonant column testing of reconstituted sands and gravels are used to model stress state, strain amplitude and material dependencies. The G/G_{max}-log γ relationships are modified to extend to strains of several percent. A subroutine has been implemented in the finite element program PLAXIS to combine the field G_{max} profiles, modified G/G_{max}-log γ relationships and gradation characteristics to estimate stress-strain response under loading. The usefulness of the approach is demonstrated by comparing predicted and measured load-settlement curves from field tests with 0.46- and 0.91-m diameter footings. Nonlinearity and settlement values are predicted well at loads in the working stress range.

Key Words: shallow foundations, granular soil, settlements, field seismic tests, nonlinear shear moduli

INTRODUCTION

Permissible settlement and bearing capacity are the two main criteria to be considered when designing shallow foundations. Permissible settlement generally becomes the governing factor for foundations on granular soil. It is difficult and/or expensive to obtain undisturbed samples of granular soil. Hence, the stress-strain behavior of granular soil is not routinely measured in the laboratory. Instead, foundation settlements on such soil are predicted using empirical correlations that relate *in-situ* penetration test results with load-settlement tests or case histories. It is worth noting that most penetration test results incorporated in these correlations represent indirect measures of soil stiffness.

In this paper, a framework is proposed for predicting foundation settlements on granular soil using properties that are dynamically measured. The genesis of the framework comes from geotechnical earthquake engineering and the calculation of strains and displacements during shaking (Kramer, 1996). The framework is divided into three parts. The first part involves the field investigation. *In-situ* evaluation of soil stiffness is performed with field seismic tests in which shear wave velocities (V_s) are determined. Traditional disturbed sampling is used to characterize gradation characteristics of the soil and, if desired, penetration testing is conducted, mainly to identify thin softer layers. The shear modulus profile is calculated with the V_s profile and estimated total unit weight of the soil. The seismic V_s measurements are performed at the *in-situ* state and in the small-strain range; hence, shear moduli are strain independent and denoted as G_{max} or G_o. At small strains, values of G_{max} of nonplastic granular soil typically show little viscous effects, increasing by less than 10% as frequency increases from 1 to several hundred Hertz (Stokoe and Santamarina, 2000). In the second part, small-strain shear moduli are combined with normalized shear modulus-shear strain (G/G_{max}-log γ) relationships that are nonlinear. The G/G_{max}-log γ relationships depend on stress state and material gradation. The relationships are modified to extend to shear strains larger than 0.2% as discussed herein.

In the third part of the framework, foundation settlements under loading are calculated using the finite element program, PLAXIS with a subroutine, named MoDaMP, that combines the field G_{max} profiles and modified G/G_{max}-log γ relationships. Settlements are calculated based on shear moduli that are continually adjusted for the combined effects of increasing confining pressure and increasing shear strain. This framework has several important benefits including: (1) *in-situ* seismic testing which can readily be performed in all types of granular soil, including gravels and cobbles, (2) continuous load-settlement curves that are evaluated to stress states beyond the range of those expected under working loads, and (3) the potential to be adapted to all types of geotechnical materials, even those where effective stresses change with time such as consolidating clays.

METHODS USED TO PREDICT SETTLEMENT

One of the first methods to predict the settlements of foundations on granular soils was proposed by Terzaghi and Peck (1948). They used results from load tests with 300-mm square plates on sand. The sand had a wide range in standard penetration test (SPT) blow count (N_{60}). They related the observed settlements from the load tests to the expected settlements on actual foundations. Meyerhof (1965) proposed a method where the settlements were predicted based on SPT blow count, N_{60}. Schmertmann (1970) proposed a semi-empirical method which has become one of the most widely used methods today. He used elastic theory, model load tests, field cone penetration tests (CPT) and finite element analyses to develop the approach. He modified the method to assess strip footings and square footings separately by changing strain influence factors (Schmertmann et al., 1978). In this work, Schmertmann expressed soil stiffness as an equivalent elastic modulus which is based on CPT results. Burland and Burbidge (1985) developed a semi-empirical formulation by investigating a large set of case histories of foundation settlements. They used corrected SPT values ($N_{1,60}$) and correction factors for foundation shape, soil layer thickness and time dependency of the settlements. In all methods, the strain dependency of the soil stiffness is indirectly incorporated.

Berardi and Lancellotta (1991) proposed one of the earliest methods to take the strain dependency of the soil stiffness into account. In their method, using an iterative scheme, the soil stiffness was evaluated based on the corrected SPT blow count and varied according to the calculated relative strain level. Lee and Salgado (2001) proposed a model where soil stiffness is reduced based on the tolerable settlements and relative density of the soil. Another method in which strain level is considered in determining the soil stiffness was developed by Lehane and Fahey (2002). In their method, soil nonlinearity is taken into account by reducing the small-strain Young's modulus with increasing axial strain. Mayne (2000) used a modified hyperbolic model where the elastic modulus decreases as the ratio of footing pressure to bearing capacity increases.

None of these methods incorporate field seismic measurements of V_s to evaluate the G_{max} profile and then combine the profile with nonlinear G/G_{max}-log γ relationships to calculate linear and nonlinear strains. In addition, the following effects are not explicitly considered in the methods: (1) gradation effects on the nonlinear stress-strain behavior of granular soil, and (2) the combined effects on soil stiffness of changing levels of stress state and shear strain. The framework in which these factors are incorporated is presented below.

NONLINEAR BEHAVIOR OF GRANULAR SOIL

The shear stress-shear strain response of nonplastic granular soil consists of linear, moderately nonlinear and highly nonlinear parts. The linear part is usually referred as the "elastic" or "small-strain" range. In this part, shear modulus ($G=\tau/\gamma$) is essentially independent of shear strain (γ) and exhibits little effect of viscosity (Stokoe and Santamarina, 2000). An elastic threshold value of shear strain, γ_t^e, is defined to represent the shear strain beyond which the stress-strain response becomes strain dependent. The value of γ_t^e for granular soil varies with mean effective confining pressure, σ_0', and gradation characteristics, usually expressed by the uniformity coefficient, C_u (Menq, 2003). For working stresses associated with shallow foundations, γ_t^e likely ranges from 0.0001 to 0.003%. Shear modulus of soil in the small-strain range is denoted as G_{max} or G_o. (G_{max} is used herein.) G_{max} can be evaluated using various *in-situ* testing methods such as crosshole, downhole, Spectral-Analysis-of-Surface-Waves (SASW) and seismic cone penetration testing (SCPT) (Stokoe and Santamarina, 2000). The SASW and other surface-wave methods permit small-strain profiles to be evaluated noninvasively and near the ground surface (within 0.3 m) in all types of granular soils, even soils with large gravel and cobbles. On the other hand, the SCPT is well-suited for detecting thin soft layers.

Above γ_t^e, the value of shear modulus, G, decreases with increasing shear strain so that the stress-strain relationship is nonlinear. Dynamic laboratory testing methods, such as the torsional resonant column, have made it possible to investigate the linear and nonlinear behavior of granular soils over a wide range in shear strain (e.g., 0.0001 to 0.2 %). This range in shear strain encompasses the linear and moderately nonlinear ranges and extends somewhat into the highly nonlinear range. One of the earliest comprehensive studies of nonlinear shear modulus and the important parameters affecting it was conducted by Hardin and Drnevich (1972). They found that nonlinear shear moduli of nonplastic granular soil are significantly affected by shear strain, confining pressure (including increases due to overconsolidation), and void ratio. They proposed a hyperbolic model to define nonlinear stress-strain behavior as:

$$\tau = \frac{\gamma}{\frac{1}{G_{max}} + \frac{\gamma}{\tau_{max}}} \tag{1}$$

where γ is shear strain, τ is shear stress, G_{max} is the small-strain shear modulus and τ_{max} is shear strength. This hyperbolic model was modified by Darendeli (2001) based on a large dataset of combined resonant column and torsional shear tests (RCTS) as:

$$G/G_{max} = \frac{1}{1 + \left(\frac{\gamma}{\gamma_r}\right)^a} \tag{2}$$

where a = curvature coefficient; γ_r = reference shear strain at $G/G_{max} = 0.5$; and G is the shear modulus at shear strain = γ. The modified hyperbolic model was further refined for sands and gravels by Menq (2003). Using a large-scale, free-free resonant column device, Menq tested reconstituted specimens of sands and gravels. He found that gradation was an important factor affecting the nonlinear response and developed the following expressions for the curvature coefficient, a, and reference shear strain, γ_r:

$$\gamma_r = 0.12 C_u^{-0.6} \left(\frac{\sigma'_0}{P_a}\right)^{0.5 C_u^{-0.15}} \tag{3}$$

$$a = 0.86 + 0.1 \log\left(\frac{\sigma'_0}{P_a}\right) \tag{4}$$

where γ_r is in %; C_u = uniformity coefficient; σ'_0 = mean effective confining pressure in the same units as P_a; and P_a = reference mean effective confining pressure (equal to one atm). G_{max} values are calculated with the expression given by Menq (2003) as:

$$G_{max} = C_{G3} \cdot C_U^{b1} \cdot e^x \cdot \left(\frac{\sigma'_0}{P_a}\right)^{n_G} \tag{5}$$

where C_{G3} =67.1 MPa, b1 = -0.20, e = void ratio, x =$-1-(D_{50}/20)^{0.75}$ in which D_{50} is the median grain size in mm, and $n_G = 0.48 \cdot C_U^{0.09}$. Example log G_{max}-log σ'_0 relationships that were calculated using Equation (5) for different C_u, e and D_{50} values are presented in Figure 1. The G/G_{max}-log γ relationships using Darendeli's (2001) and Menq's (2003) formulations (Equations 2, 3 and 4) are shown in Figure 2 for different values of σ'_0 and C_u.

CHARACTERIZATION OF FIELD TEST SITE

Two, small-scale footings with diameters of 0.46 and 0.91 m were constructed at the National Geotechnical Experimentation Site (NGES) at Texas A&M University. The footings were constructed as a part of a long-term project to create a dataset of load-settlement curves of footings on granular soil with *in-situ* seismic measurements and either nonlinear dynamic G/G_{max}-log γ relationships measured in the laboratory or estimated from gradation characteristics. The

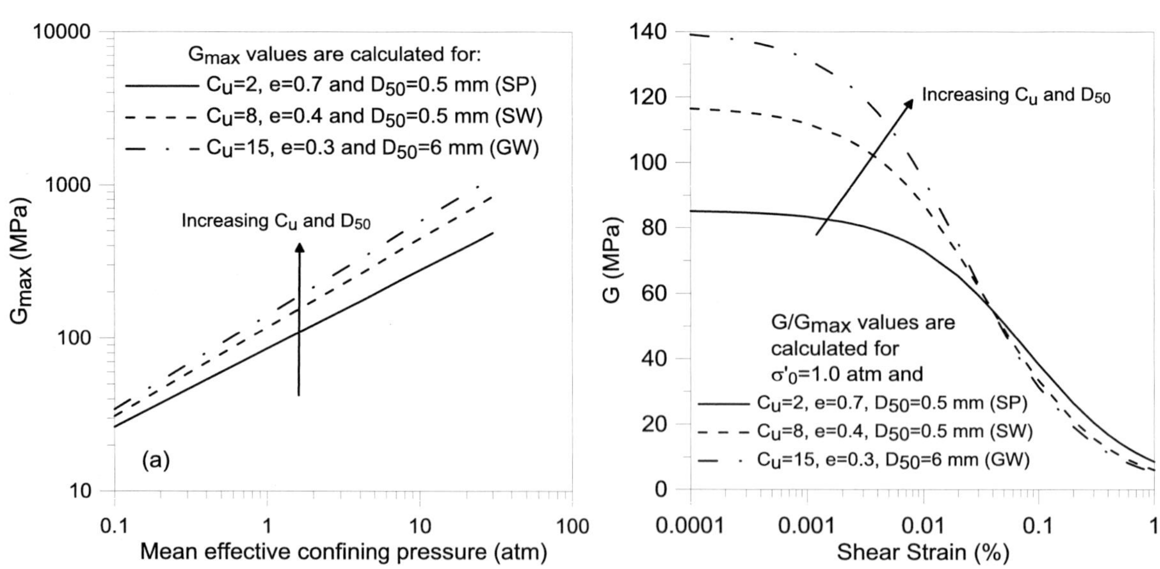

Figure 1. Effects of coefficient of uniformity (C_u) and median grain size (D_{50}) on the log G_{max} – log σ'_0 relationship (a) and G-log γ relationship (b) of granular soil based on Darendeli's (2001) and Menq's (2003) formulations (Note: Values of void ratio (e) represent medium-dense granular soils based on Kokusho et al., 1995.)

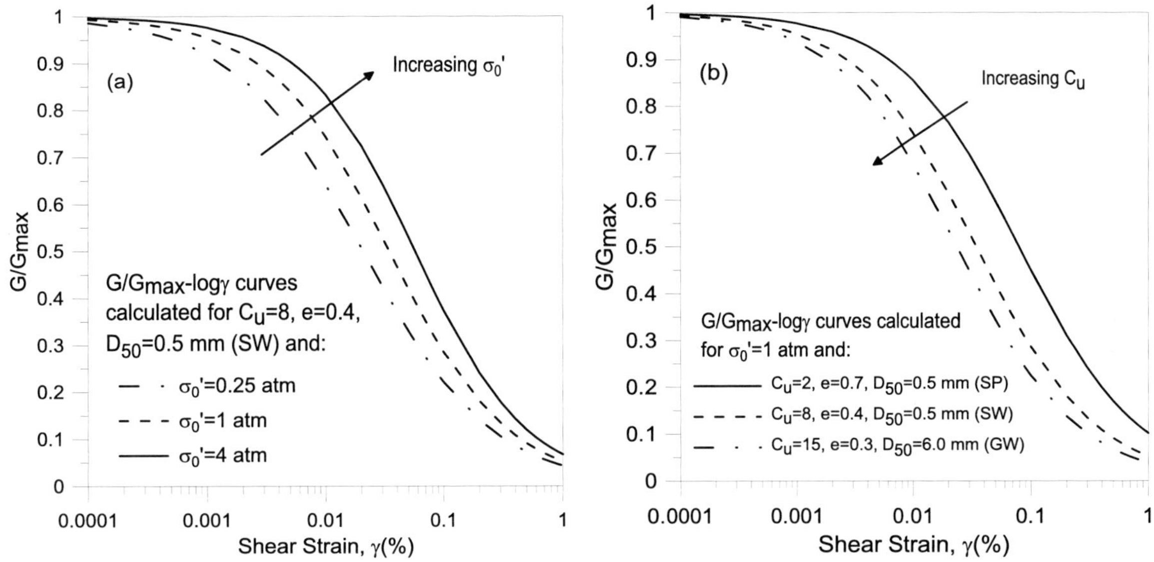

Figure 2. Effects of mean effective confining pressure (a) and coefficient of uniformity (b) on normalized shear modulus reduction curves of granular soil based on Darendeli's (2001) and Menq's (2003) formulations

small-scale footings were in the vicinity of the test site reported by Briaud and Gibbens (1994) where different sized footings were tested for a settlement prediction symposium. Several field tests, including SPT, CPT, pressuremeter, dilatometer and seismic crosshole tests, were conducted at the site (Briaud and Gibbens, 1994). Additional SASW tests near the small-scale footings were conducted by Park et al. (2009).

Soil Properties at NGES Test Site

The soil at the NGES site is predominantly sand extending to a depth of 11 m. The upper 3.5 m is medium dense, silty fine sand, classified as SP-SM in the USCS. The silty sand is underlain by a 3.5-m thick layer of medium dense sand with clay and gravel. From a depth of 7 m to 11 m, the soil is silty sand to sandy clay (Briaud and Gibbens 1994).The soil below this layer is hard clay which has a negligible effect on the field load-settlement tests because of its depth and stiffness. The water table is at a depth of 4.9 m. Briaud and Gibbens (1994) report that the 3.5-m thick layer of SP-SM material has a water content of about 12 to 14%, a total unit weight of about 19.8 kN/m^3, and a friction angle of 34 to 36 degrees. The results of SPT and CPT tests are shown in Figure 3. The results from the SASW tests conducted by Park et al. (2009) and the results from the crosshole tests reported by Briaud and Gibbens (1994) are presented in Figure 4. As seen in Figure 4, the results from the two types of seismic tests performed in different studies are in good agreement.

Representative Shear Wave Velocity Profile at Location of Footings

The small-strain shear modulus profile at the location of the small-scale footings was calculated from a representative V_s profile. The representative profile was based on the average of three SASW tests performed next to the footings and is shown in Figure 4 as the best-estimate V_s profile. The relationship between V_s and G_{max} is:

$$G_{max} = V_s^2 \cdot \rho \tag{6}$$

where ρ is the total mass density of the soil. The general V_s-σ'_0 relationship is expressed as:

$$V_s = V_{s\_1atm} \cdot \left(\frac{\sigma'_0}{P_a}\right)^{n_s} \tag{7}$$

where V_{s_1atm} = shear wave velocity at an effective confining pressure of one atm; σ'_0 = mean effective confining pressure, P_a is the reference mean effective confining pressure of one atm in

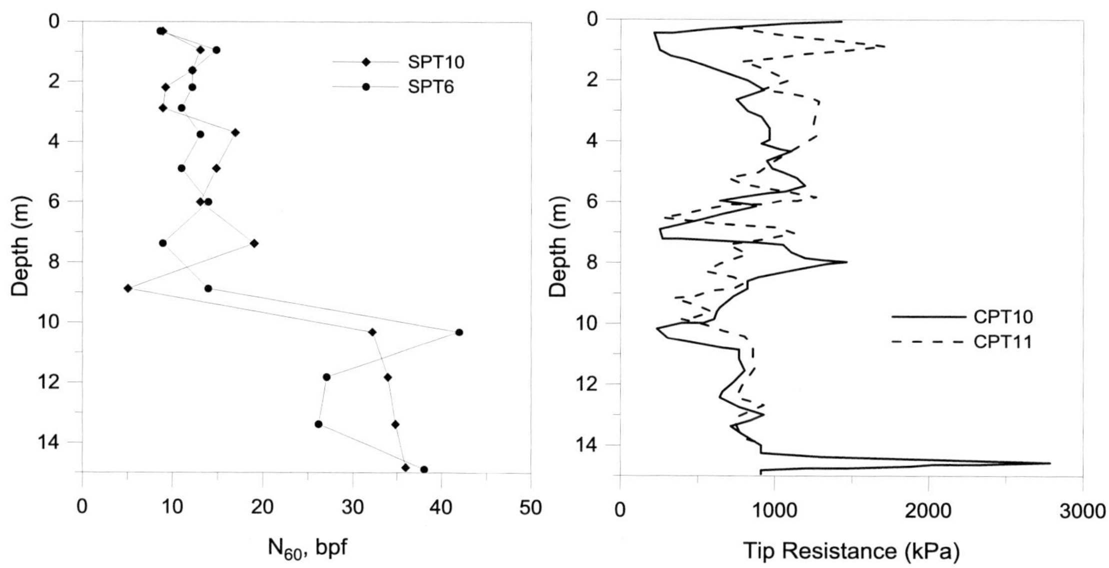

Figure 3. SPT and CPT profiles at the NGES sand site (from Briaud & Gibbens 1994)

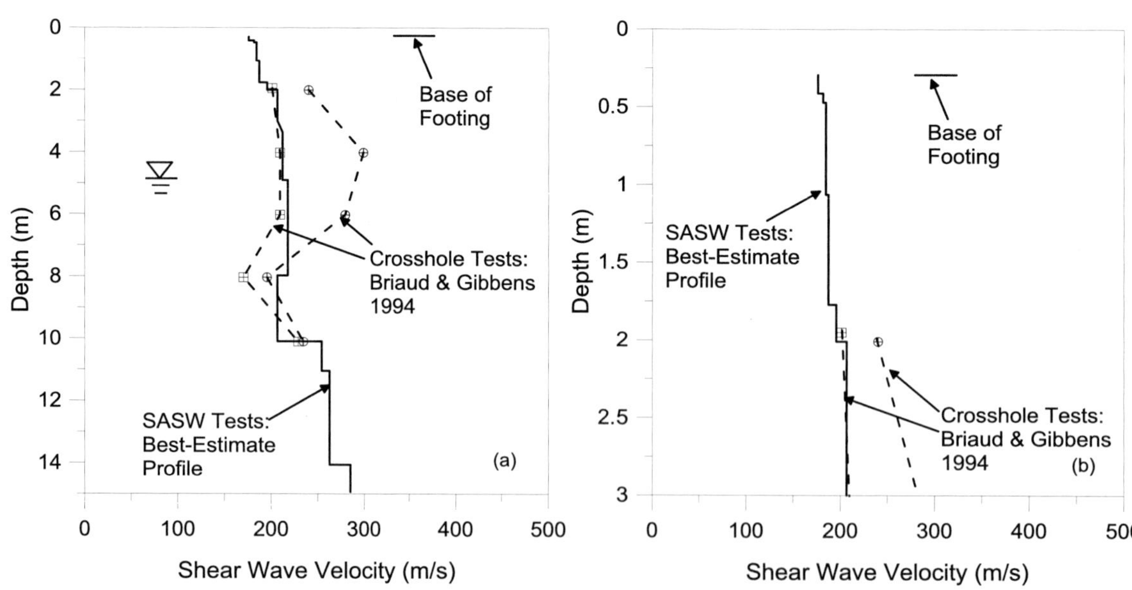

Figure 4. Shear wave velocity profiles at the NGES sand site: (a) to a depth of 15 m and (b) expanded to show the top 3 m (from Park et al. 2009)

the same units as σ'_0, and n_S is the slope of the log V_S-log σ'_0 relationship. To calculate V_{S_1atm} and n_S, points along the best-estimate V_s profile in Figure 4 were selected. A two-layer profile was developed. The resulting log V_S-log σ'_0 relationships are plotted as two lines in Figure 5. The mean effective confining pressures were calculated by:

$$\sigma'_0 = \frac{\sigma'_v + 2\sigma'_v K_0}{3} \tag{8}$$

where σ'_v is the vertical effective pressure and K_0 is the lateral earth pressure at rest. A value for K_0 of 0.75 was used in fitting the V_s results in Figure 5 because the shear wave velocities at shallow depths are considerably higher than those of a normally consolidated sand. Also, a constant value of 15 kPa was added to σ'_0 to approximate capillary stresses in the unsaturated silty sand.

The relationships in Figure 5 for the two-layer system compare well with the *in-situ* V_s profiles evaluated by SASW testing as seen in Figure 6. The two-layer profile has V_{s_1atm} values of 208 and 217 m/s for the upper and lower layers, respectively, and n_S values of 0.09 for both layers. These results are key input parameters for the subroutine, MoDaMP, that is implemented in PLAXIS. The small-strain shear modulus profile is calculated as follows:

$$G_{max} = \rho \cdot \left(V_{s\_1atm} \cdot \left(\frac{\sigma'_0}{P_a} \right)^{n_S} \right)^2 = G_{max\_1atm} \cdot \left(\frac{\sigma'_0}{P_a} \right)^{2n_S} \tag{9}$$

where G_{max_1atm} is the small-strain shear modulus at a mean effective confining pressure of one atm. The resulting G_{max_1atm} and n_G ($=2n_S$) parameters of the two layers are: Layer 1 - G_{max_1atm}= 87.3 MPa and n_G = 0.18; and Layer 2 - G_{max_1atm}= 95.0 MPa and n_G = 0.18.

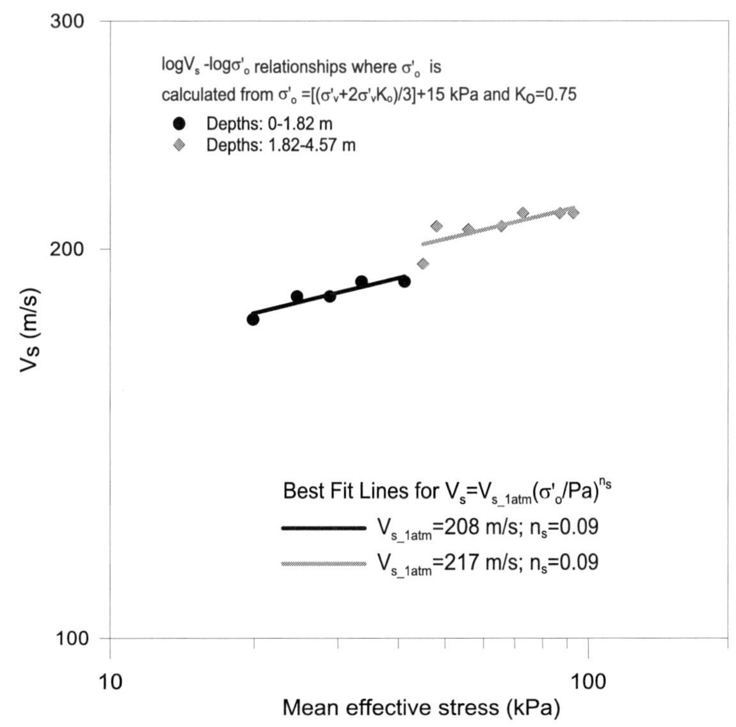

Figure 5. Log V_S-log σ'_o relationships for the two-layered system used to model the test site

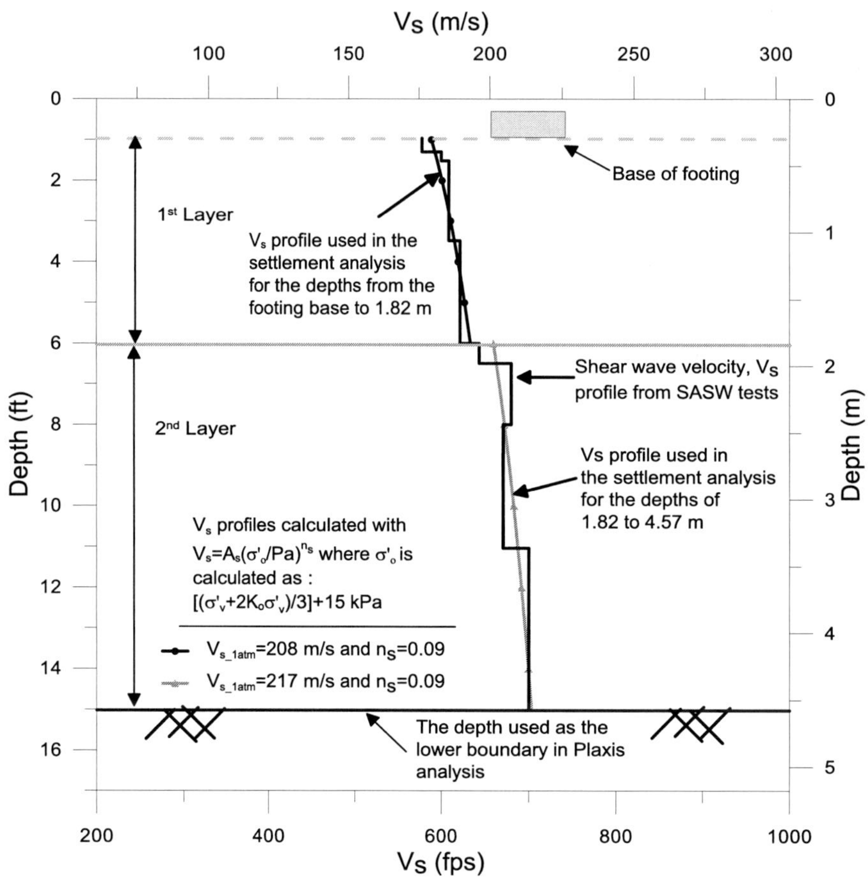

Figure 6. Idealized, two-layer shear wave velocity profile beneath the footing based on fitting the field SASW results with the typical log V_S-σ'_0 relationship used to model sands

IMPLEMENTATION OF G/G_{max}-log γ RELATIONSHIPS INTO MoDaMP

The G/G_{max}-log γ relationships described with Equations 2 through 4 have been primarily developed by torsional resonant column testing over small-to-moderate values of strains, typically shear strains less than 0.2 %. The relationships need to be modified to use them in MoDaMP at larger strains which occur in problems such as shallow foundation settlements. A two-step modification process was used. The first step involved adjusting the value of the curvature coefficient, "a", given in Equation 4 based on comparisons with results from laboratory triaxial tests. The adjustment was done by converting the G -log γ relationships into τ-γ (shear stress-shear strain) curves and comparing these curves at larger strains, generally above 2%, with τ-γ curves from triaxial tests. The triaxial tests were performed on undisturbed silty sand samples obtained at the NGES site from a depth of 0.2 m (Briaud and Gibbens, 1994). This step resulted in the following set of equations for "a" values. In the small-to-moderate strain range where the resonant column results are not modified (defined by values of $G/G_{max} > 0.4$), the "a" values are calculated from:

$$a = 0.86 + 0.1 \log\left(\frac{\sigma'_0}{P_a}\right) \quad \text{for } \frac{G}{G_{max}} > 0.4 \tag{10}$$

Values of "a" in the larger strain range, where the triaxial test results became a controlling factor, are calculated from:

$$a = -0.086 \ln\left(\frac{\sigma'_0}{P_a}\right) + 0.7879 \quad \text{for } \frac{G}{G_{max}} \leq 0.4 \tag{11}$$

The unmodified and modified G/G_{max}-log γ relationships for σ'_0= 345 kPa are presented in Figure 7a to show the effect of the modified curvature coefficient. As seen in Figure 7a, the modified "a" values result in increasingly larger G/G_{max} values as shear strain increases above γ = 0.16%. The shear stress-shear strain curves that are calculated with the modified G/G_{max}-log γ relationships are compared with the shear stress-shear strain curves from triaxial tests in Figure 7b. It can be seen that the stress-strain curves from the modified G/G_{max}-log γ relationships are stiffer (higher τ at given γ) at strains less than 1.5% than those from the triaxial tests. This difference results from higher calculated strains in the triaxial tests due to platen-to-platen displacement measurements instead of localized deformation measurements over the middle portion of the specimen (Tatsuoka et al., 1994). The sets of stress-strain curves match well at strains larger than about 2%.

Finite element analyses were performed to calculate the settlements of both small-scale footings. PLAXIS and the subroutine, now designated as MoDaMP1, were used. MoDaMP1 identified the subroutine as the first step in modifying the "a" values (Equation 11). The predicted settlements were unsatisfactory because they were 2 to 5 times larger than the measured settlements in the working stress range. It was concluded that the triaxial test results were exhibiting a softer stress-strain response than the actual soil beneath the footings. The stiffer stress-strain response of the soil beneath the footings was attributed to increases in horizontal stresses *in-situ* during loading which did not occur during loading in the triaxial test. Therefore, it was decided that a second step in modifying the "a" values was warranted so the *in-situ* stiffening response of the soil during loading would be taken into account.

The second modification to the G/G_{max}-log γ relationships was done using the measured load-settlement curves from the footing tests. Use of the load-settlement data did not change

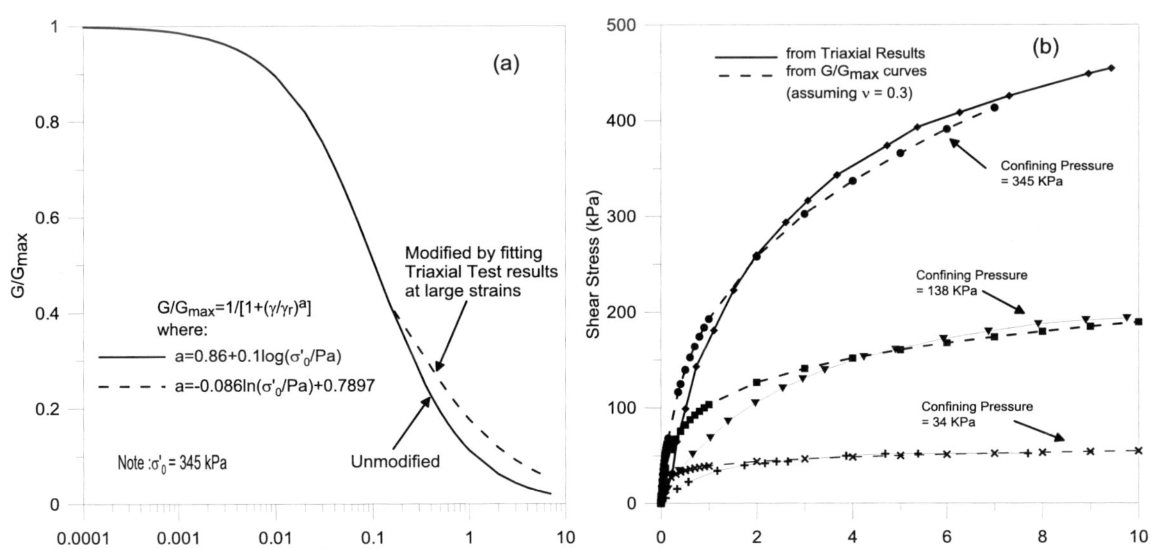

Figure 7. Effect of the modified curvature coefficient on (a) the G/G_{max}-log γ relationships with at σ'_0 = 345 kPa and (b) the shear stress-strain relationships from triaxial test results and from the modified G/G_{max}-log γ relationships

Equation 10; hence, no change occurred in the small-to-moderate strain range ($G/G_{max} > 0.4$). However, use of the settlement data did result in stiffening the G/G_{max}-log γ relationships at larger strains by decreasing the "a" values. The equation for "a" at larger strains is:

$$a = -0.109 \ln\left(\frac{\sigma'_0}{P_a}\right) + 0.6695 \quad \text{for } \frac{G}{G_{max}} \leq 0.4 \tag{12}$$

Equations 10 and 12 were used to calculate the G/G_{max}-log γ relationships in the subroutine, now designated as MoDaMP2 to identify the second step in the modification process. Calculation of shear moduli during loading as determined with MoDaMP2 is discussed below.

Example Modulus-Strain Calculations with MoDaMP2

Subroutine MoDaMP2 contains Equations 2, 3, 4, 9, 10 and 12 and is implemented in PLAXIS as a soil constitutive model. Since a comprehensive soil model must be created for three-dimensional stress and strain states, a substitution was made to Equation 2. A scalar measure of shear strain, octahedral shear strain, γ_{oct}, was used in place of γ. The model input parameters are G_{max_1atm}, Poisson's ratio, ν, and coefficient of uniformity for each layer. G_{max} values are calculated with Equation 9 based on the initial stress states. Values of G are calculated using the G_{max} values and Equations 2, 3, 4, 10 and 12. Next, values of G and ν are used to establish the nonlinear elastic stiffness matrix. New stresses are calculated based on the nonlinear elastic stiffness matrix and new G_{max} values are calculated with these stresses.

To illustrate this procedure, three reference points at different depths beneath the 0.91-m diameter footing are used (Figure 8a). The octahedral shear strain and shear modulus at each reference point are determined under five, increasing load levels applied to the footing (Figure 8b). Five G-log γ_{oct} relationships determined for the five load levels identified in Figure 8b are plotted as solid lines in Figures 8c, 8d and 8e for Reference Points X, Y, and Z, respectively. The sets of G, γ_{oct} coordinates that correspond to each of the five load levels on the load-settlement curve are identified on the appropriate G-log γ_{oct} relationship by the solid circles and associated load-level number. The increase in G_{max} and general upward shifting of the G-log γ_{oct} curves (solid lines) with increasing load level is seen in Figures 8c, 8d, and 8e. As loading increases, the net effect is G decreasing as γ_{oct} increases at each reference point. This effect diminishes with increasing depth because the incremental stress change decreases with depth. Additional details of the subroutine MoDaMP2 can be found in Kacar (2013).

Verification of MoDaMP2 in the Elastic Range

The settlement calculations with PLAXIS and MoDaMP2 begin in the elastic range of soil. Therefore, it is possible to verify calculations in this range by comparing with settlements predicted using the theory of elasticity. This verification is performed by calculating the settlement of a 0.91-m diameter, rigid circular footing on an elastic half-space. The theoretical solution is (Richart et al., 1970):

$$S = Q\frac{1 - \nu}{2GB} \tag{13}$$

where S is the settlement, Q is the load applied to the footing and B is the footing diameter. This settlement is compared with the settlement of the same footing calculated with PLAXIS and MoDaMP2. The soil is modeled by assuming γ_r=1000%, a=1.0 and n_G=0.0 so that the initial G value (G_{max}) stays constant during loading and thus MoDaMP2 becomes a linear elastic model. The depth of the elastic layer, H, was assumed equal to 100B. G and ν are assumed to be 65MPa and 0.3, respectively, the same values as in the theoretical elastic solution. The calculated

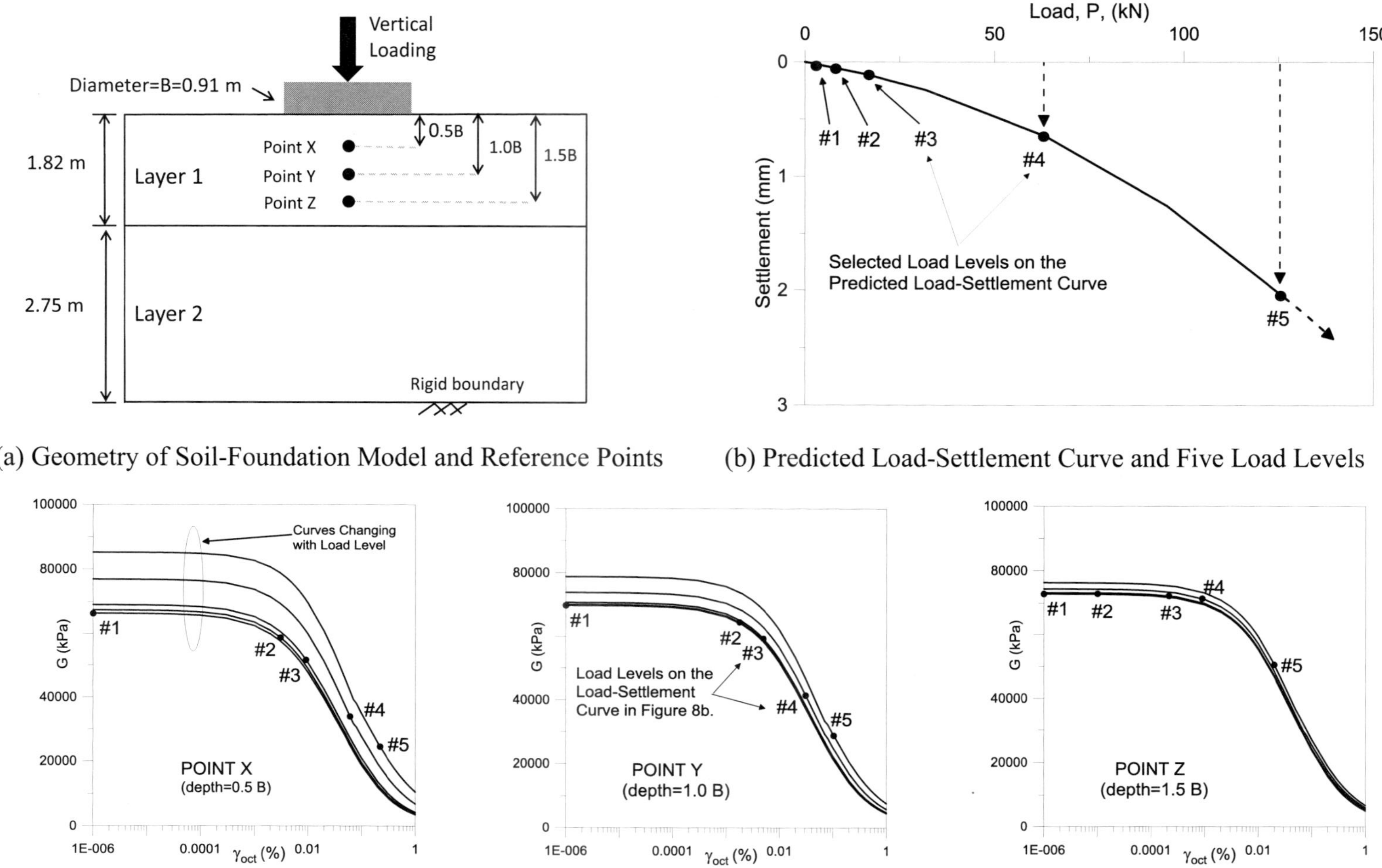

(a) Geometry of Soil-Foundation Model and Reference Points

(b) Predicted Load-Settlement Curve and Five Load Levels

(c) Secant Shear Modulus at Five Load Levels at Point X

(d) Secant Shear Modulus at Five Load Levels at Point Y

(e) Secant Shear Modulus at Five Load Levels at Point Z

Figure 8. Illustration showing how the modulus-strain relationships and secant shear moduli change in the numerical model during loading at three reference points beneath the 0.91-m diameter footing settlements are 1.16 mm and 1.17 mm with MoDaMP2 and the theory of elasticity, respectively. The difference in the predicted settlements between the two methods is about 1%, an excellent comparison. The difference is a result of $H/B = \infty$ in the elastic theory and 100 in the numerical model.

LOAD-SETTLEMENT TESTS

Each circular, concrete footing constructed at the NGES sand site was reinforced, 0.30 m thick, and cast in place. Prior to construction, about 0.3 m of soil was removed from the ground surface and 0.15 m of backfill was placed after footing construction. Two reference frames spanned each footing to support displacement potentiometers which were arranged in an equilateral triangle on the footing. The load was applied by jacking against the central loading point of a large vibroseis, named "T-Rex". T-Rex was positioned over the footing (Figure 9a) and the load was applied incrementally while measuring footing movement (Figure 9b).

Load-Settlement Test Results

The load-settlement results of the 0.46- and 0.91-m diameter footings are shown in Figures 10a and10b, respectively. As seen in the figures, each footing was loaded and unloaded in two or three stages, with increasing loads in each successive stage. A backbone curve was fit to the staged loading/unloading cycles to approximate a continuous load-settlement curve (dashed lines in Figures 10a and 10b). The backbone curves are compared with the predicted settlements below.

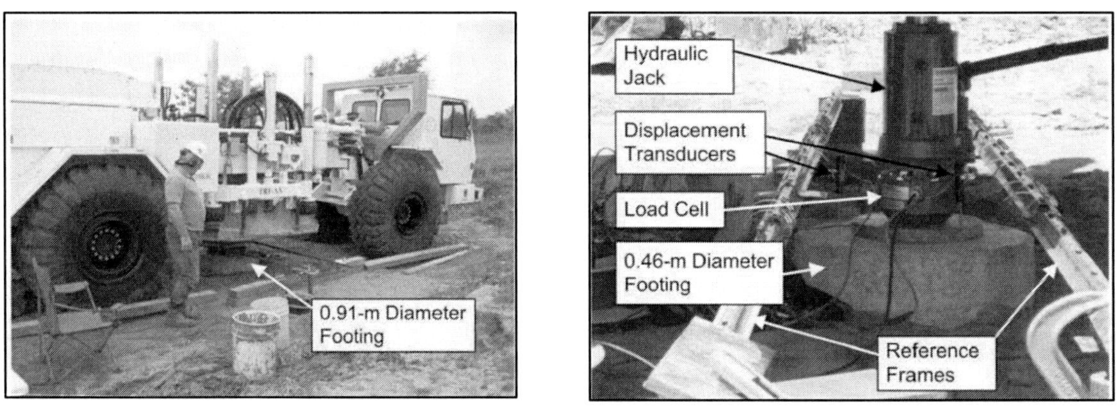

Figure 9. Load-settlement tests at the NGES sand site: (a) T-Rex loading the 0.91-m diameter footing, and (b) set-up of the instrumentation to load and measure the settlement of the 0.46-m diameter footing (Park et al. 2009)

LOAD-SETTLEMENT PREDICTIONS

Finite element analyses were carried out using MoDaMP2 implemented in PLAXIS to predict the settlements of the small-scale footings. A thickness of soil beneath the footing of 4.57 m was used in the settlement analysis to minimize boundary effects (Brinkgreve et al, 2011). The finite element mesh consisted of 1198, 15-node triangular elements.

The load-settlement predictions are compared with the measured load-settlement curves in Figure 11. The predicted settlements reasonably follow the measured settlements for both footings from the start of loading through the working stress range (200 to 400kPa). The working stress range is a typical range of design pressures set in building codes. The measured and predicted load-settlement curves exhibit nonlinearity throughout the loading.

The good comparison between measured and predicted settlements was expected because the curvature coefficient, "a", in MoDaMP2 was modified to increase stiffness of the soil beneath the footing at strains larger than 0.16%. In the modification process, the measured settlement curves were used as a guide to the effect of increasing horizontal stresses on G during loading. However, increasing horizontal pressures simply represent part of the physics of this deformation-based

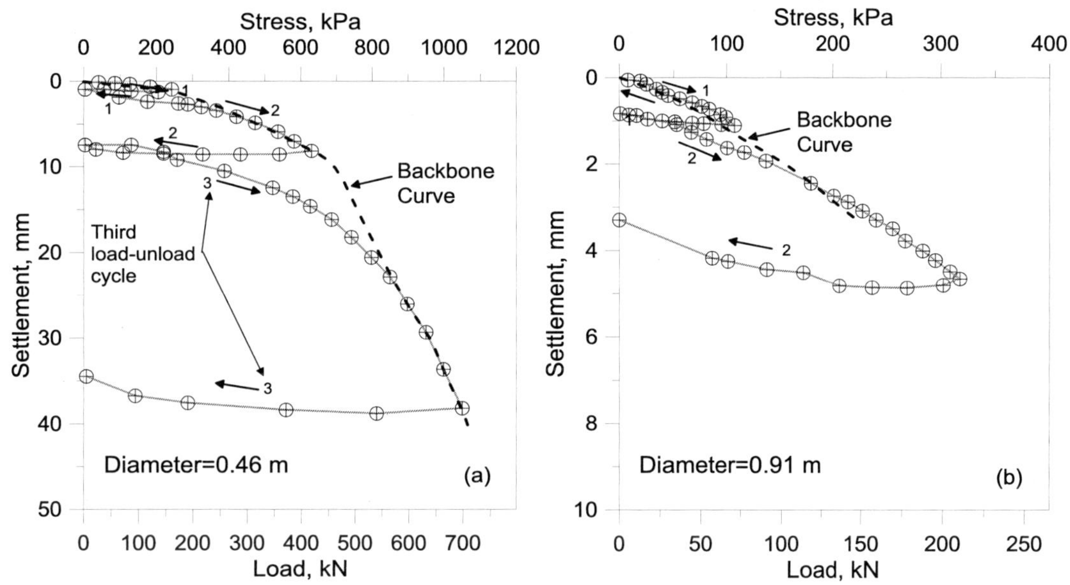

Figure 10. Load-unload-reload settlement results for the 0.46-m diameter footing (a) and for the 0.91-m diameter footing (b) with backbone curves fitted to approximate simple continuous loading (Note: Working stress range is approx. 200-400 kPa.) (from Park et. al, 2009)

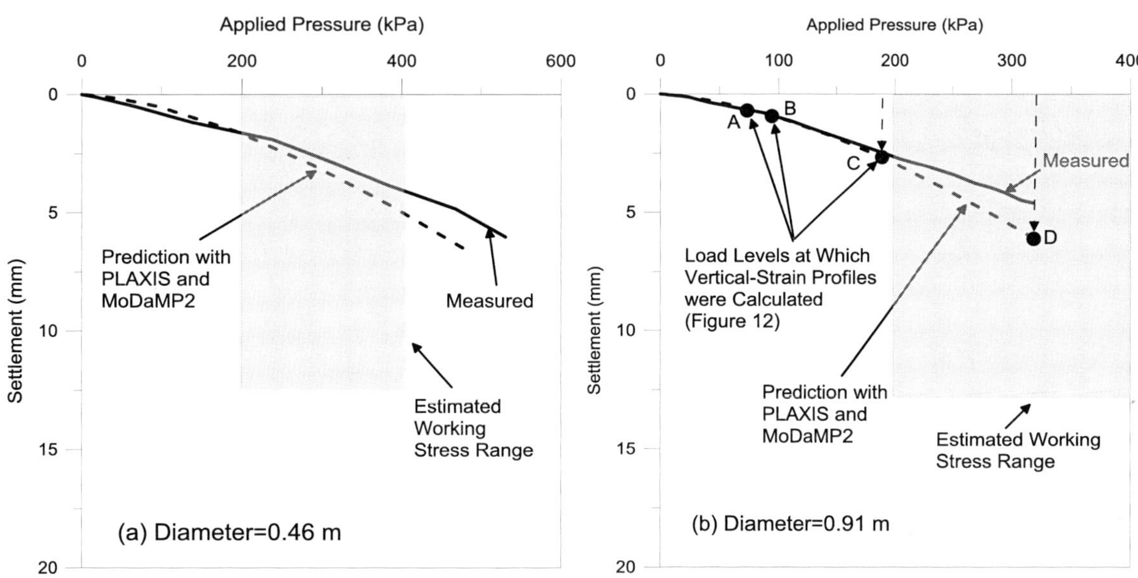

Figure 11. Comparison of predicted and measured load-settlement curves for the small-scale footings test

problem and the proposed framework includes incorporation of this effect. Clearly, more field load-settlement tests which incorporate dynamically measured properties are required to investigate, and likely improve, the soil model in MoDaMP2. Well-documented case histories are also needed in which dynamic soil properties are measured, such as settlement case histories at existing nuclear power plants. Another factor that will likely need to be modeled and included in comparisons with case histories is creep of granular soil under long-term loads as included in Schmertmann (1970).

Effects of Interactions between T-Rex and Footings during Load Tests

T-Rex had to be positioned over each footing during load testing. This process required the front tires to move past each footing. An analysis with PLAXIS and MoDaMP2 was carried out to investigate possible pre-loading of the soil beneath the footing. The tire was modeled as a circular loaded area, equal to the contact are of the tire which is about 0.34 m^2. The maximum increase in the vertical stress at the edge of the 0.91-m diameter footing due to the pressure applied by the moving tire was calculated to be 33 kPa. The maximum stress increase occurred at a depth of 0.3 m and the initial settlement due to the pre-loading was calculated to be 0.34 mm. This movement was neglected in analyzing the load tests because it was: (1) only beneath a small portion of opposite sides of the footing, and (2) small (about 6%) compared to the total settlement of 6 mm.

During the process of load testing, the soil beneath the tires of T-Rex was being unloaded. The unloading of the soil was modeled and was found to be a maximum stress decrease of about 5 kPa, a negligible change.

Vertical Strains beneath the 0.91-m Diameter Footing

One interesting deformation parameter that can be studied with the proposed framework is the vertical strain profile beneath the footing at various load levels. As an example, consider axial strains beneath the center of the 0.91-m diameter footing. Strain profiles were calculated at four load levels (A, B, C and D) on the predicted load-settlement curve (Figure 11b). The variation predicted in vertical strain with depth at each load level is shown in Figure 12. The general shape

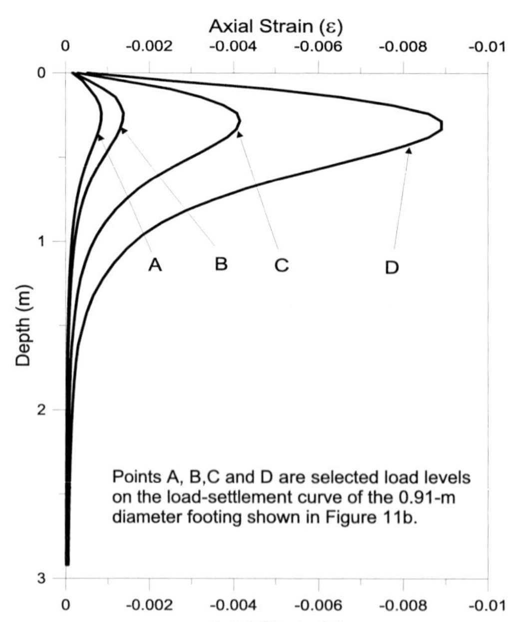

Figure 12. Predicted vertical-strain profiles at four load levels along the centerline beneath the 0.91-m diameter footing

of the strain profiles agree with Schmertmann's (1970) proposed profile. The depths where the maximum vertical strains occur are approximately 0.3B. The reasonableness of strain profiles support the validity of the proposed framework.

SUMMARY AND CONCLUSIONS

A deformation-based framework for predicting settlements of shallow foundations on granular soil is presented. The framework involves utilizing *in-situ* seismic measurements to characterize the granular soil in terms of small-strain shear moduli (G_{max}). The small-strain shear moduli are combined with nonlinear normalized shear modulus-shear strain (G/G_{max}-log γ) relationships that are stress dependent and material dependent. The G/G_{max}-log γ relationships have been modified following a two-step process to extend them to strains in the range of several percent. The first step involved comparison with laboratory triaxial τ-γ curves and the second step involved comparison with the field load-settlement curves. Each step resulted in values of G/G_{max} in the relationships increasing for $G/G_{max} \leq 0.4$. The field G_{max} profiles, modified G/G_{max}-log γ relationships and stress and material dependencies are used in the finite element program, PLAXIS, via a subroutine named MoDaMP2, with the "2" representing the final step in the two-step process.

The new framework was investigated by comparing measured and predicted load-settlement curves of two, small-scale footings with diameters of 0.46 and 0.91 m. Nonlinearity in the load-settlement curves which was measured in the field tests was captured in the predicted settlements. However, more full-scale load tests combined with well-documented case histories are required to understand, and likely improve, the soil model in MoDaMP2. It is also important to note that creep under sustained loads has not been incorporated into the soil model. Hence, the "framework" will need to continue to be improved.

The predictive method has several advantages over traditional CPT and SPT methods. First, field seismic measurements are used to characterize the soil *in-situ*. Field seismic measurements, especially those done with surface-wave tests, are readily applied to all granular soils, including soils containing large gravel and cobbles. Second, the nonlinear characterization of granular soil modeled with modified G/G_{max}-log γ relationships captures the nonlinear stress-strain curve of granular soil during loading. The modification at large strains, which includes the effect of increasing horizontal stresses during loading, needs further study. Additionally, limitations of the proposed framework due to stress and strain concentrations at large movements leading to bearing capacity failure need to be studied.

ACKNOWLEDGEMENTS

It is with great pleasure that the writers contribute this article to the symposium honoring Professor Roy E. Olson. He is an outstanding teacher, researcher, mentor and friend and our interactions with him have enriched our lives.

Financial support for the field portion of this study was provided through the National Science Foundation under grant CMS-0421275. Financial support for the development of the nees@utexas Equipment Site was provided by the George E. Brown, Jr. Network for Earthquake Engineering Simulation (NEES) under grant CMS-0086605. The writers wish to thank Dr. Thomas Benz and Dr. Lukas Kallivokas for their comments and guidance in implementation of the subroutine. The writers also wish to thank to Dr. Giovanna Biscontin, Dr. Jose Roesset, Dr. Luciana Barroso, Dr. Farn-Yuh Menq, Mr. Cecil Hoffpauir, Mr. Andrew Valentine, and Mr. Christopher Stanton for their help during the field effort.

The writers also appreciate the comments of the reviewers which helped significantly with content and balance of the article.

$168.00

REFERENCES

Berardi, R. and Lancellotta, R. (1991). "Stiffness of granular soil from field performance". *Geotechnique* 41 (1), 149-157

Briaud, J. and Gibbens. R. M., (1994). " Predicted and measured behavior of five spread footings on sand". *Proceedings of a Prediction Symposium at the Settlement'94 ASCE Conference* Geotechnical Special Publication No.41

Brinkgreve, R.B.J., Swolfs W.M. and Engin E. (2011). Plaxis 2D 2011. *Tutorial Manual*

Burland, J.B. and Burbidge M.C. (1985). "Settlement of foundations on sand and gravel". *Proceedings, Institution of Civil Engineers* 78 (1), 1325-1381

Darendeli, M.B. (2001). "Development of a new family of normalized modulus reduction and material damping curves". *Ph.D. Dissertation, University of Texas at Austin*, 362 p.

Hardin, B.O. and Drnevich, V.P. (1972). "Shear modulus and damping in soils; Design equations and curves". *Journal of the Soil Mechanics and Foundation engineering Division,* ASCE 98 (7), 667-692

Kacar, O. (2013) "Predicting settlement of shallow footings on granular soil using field seismic tests and laboratory dynamic nonlinear soil properties". *Ph.D. Dissertation, University of Texas at Austin,* expected December.

Kokusho, T., Yoshida Y and Tanaka Y. (1995) "Shear wave velocity in gravelly soils with different particle grading". *Static and Dynamic Properties of Gravelly Soils, Geotech. Spec.Pub.* (56), ASCE 92-105

Kramer, S.L. (1996) *Geotechnical Earthquake Engineering*, Prentice Hall, Upper Saddle River, New Jersey

Mayne, P.W. (2000) "Enhanced geotechnical site characterization by seismic piezocone penetration tests". *Invited lecture, 4th International Geotechnical Conference*, Cairo University, 95-120

Menq, F. (2003). "Dynamic properties of sandy and gravelly soils". *Ph.D. Dissertation, University of Texas at Austin,* 364 p.

Meyerhof, G.G. (1965). "Shallow foundations". *Journal of the Soil Mechanics and Foundations Division,* ASCE 91(SM2), 21-31

Lehane, B. and Fahey, M. (2002). "A simplified nonlinear settlement prediction model for foundations on sand". *Canadian Geotechnical Journal* 39 (2) 293-303

Lee, J. and Salgado, R. (2001). "Estimation of footing settlement in sand". *International Journal of Geomechanics* 2 (1) 1-28

Park, K., Stokoe, K.H. II, Olson, R.E. and Seo, W. (2009) "Settlement of footings in sand using dynamic soil properties". *Proceedings, International workshop on soil-foundation-structure interaction (SFSI 09)*, Auckland, New Zealand

Richart, F.E., Jr., Hall, J.R., and Woods, R.D. (1970) *Vibration of Soils and Foundations*, Prentice-Hall, Inc., Englewood Cliffs, New Jersey

Schmertmann, J.H. (1970). "Static cone to compute static settlement over sand". *Journal of the Soil Mechanics and Foundations Division,* ASCE 96(3), 1011-1043

Schmertmann, J.H., Hartman, J.P. and Brown P.R. (1978). "Improved strain influence factor diagrams". *Journal of the Geotechnical Engineering Division,* ASCE 104 (8), 1131-1135

Stokoe, K. H. II, and Santamarina,C. (2000). "The Increasing Role of Geophysically-Based Tests in Geotechnical Engineering," *International Conference on Geotechnical and Geological Engineering,* GeoEng, Melbourne, Australia. Vol.1, 1490-1536

Tatsuoka, F., Teachavorasinskun, S., Dong, J., Kohata, Y. and Sato, T. (1994) " Importance of measuring local strains in cyclic triaxial tests on granular materials". *In dynamic geotechnical testing, II,* ASTM, STP 1213, pp. 288-302

Terzaghi, K. and Peck, R.B. (1948). "*Soil Mechanics in Engineering Practice*". 1st edition, John Wiley & Sons, New York